INTERNATIONAL CODE COUNCIL®

International Residential Code®

for One- and Two-Family Dwellings

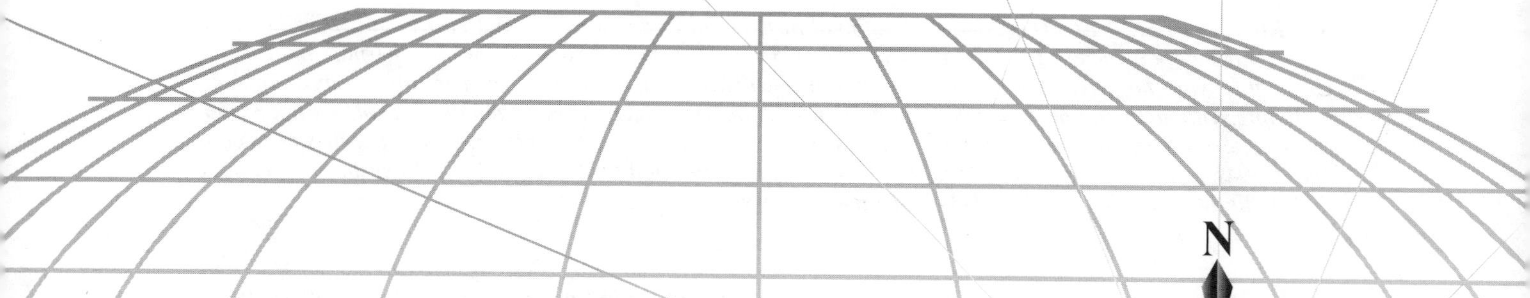

2000

BOCA BUILDING OFFICIALS CODE ADMINISTRATORS INTERNATIONAL®

INTERNATIONAL CONFERENCE OF BUILDING OFFICIALS ICBO 1922™

SBCCI®

Fourth Printing

Publication Date: January 2000

ISBN #1-892395-17-7 (soft-cover edition)
ISBN #1-892395-18-5 (loose-leaf edition)

Published in cooperation with:

BUILDING OFFICIALS AND CODE ADMINISTRATORS INTERNATIONAL, INC.
4051 West Flossmoor Road • Country Club Hills, Illinois 60478-5795
(800) 214-4231

INTERNATIONAL CONFERENCE OF BUILDING OFFICIALS
5360 Workman Mill Road • Whittier, California 90601-2298
(800) 284-4406

SOUTHERN BUILDING CODE CONGRESS INTERNATIONAL, INC.
900 Montclair Road • Birmingham, Alabama 35213-1206
(205) 591-1853

PRINTED IN THE U.S.A.

PREFACE

Introduction

Internationally, code officials recognize the need for a modern, up-to-date residential construction code addressing the design and construction of one- and two-family dwellings and townhouses. The *International Residential Code for One- and Two-Family Dwellings* is designed to meet these needs through model code regulations that safeguard the public health and safety and consider dwelling affordability in all communities, large and small.

This comprehensive, stand-alone residential construction code, which incorporates approved changes from the 1999 code development cycle, establishes minimum regulations for one- and two-family dwellings and townhouses using prescriptive provisions. Additionally, the *International Residential Code* is designed to be compatible with the BOCA National Codes published by Building Officials and Code Administrators International (BOCA), the Standard Codes published by the Southern Building Code Congress International (SBCCI), the Uniform Codes published by the International Conference of Building Officials (ICBO), and all the International Codes published by the International Code Council (ICC®).

The *International Residential Code* provides many benefits, among which is the model code development process that offers an international forum for construction professionals to discuss these prescriptive code requirements. This forum provides an excellent arena to debate proposed revisions. This model code also encourages international consistency in the application of provisions.

Development

On April 22, 1996, the ICC Board of Directors established the ICC/NAHB Task Force, charged with the examination of performance options of the CABO *One and Two Family Dwelling Code*. The Board further stated that the "purpose is to identify the issues (including the process of code development after 1998) and develop an action plan to resolve the issues."

Subsequently, the ICC/NAHB Task Force recommended to the ICC Board that the "ICC develop and maintain a stand-alone residential code that will cover the construction of detached one- and two-family dwellings and multiple single-family dwellings (i.e., townhouses) not more than three stories in height with separate means of egress. This code is to be called the *International Residential Code for One- and Two-Family Dwellings*."

A code drafting committee was formed by the International Code Council. The intent of the code drafting committee was to develop a draft of a comprehensive residential code that is consistent with and inclusive of the scope and content of the existing model building codes and the *International One- and Two-Family Dwelling Code*.

The technical content of the 1998 *International One- and Two-Family Dwelling Code*, the other *International Codes* (including the *International Building Code/First Draft*) and the latest building codes promulgated by BOCA, ICBO and SBCCI was the nucleus for the development of this document. Other sources of relevant technical information were also considered. While there were a great many similarities among the codes, careful consideration was given to identified differences. The code drafting committee used certain principles as guidance in the resolution of technical differences. The principles were based on the intent to establish provisions consistent with the scope of a residential code that adequately protects public health, safety and welfare; provisions that do not unnecessarily increase construction costs; provisions that do not restrict the use of new materials, products or methods of construction; and provisions that do not give preferential treatment to particular types or classes of materials, products or methods of construction.

Fuel gas provisions have been included in this edition through an agreement with the American Gas Association. The maintenance process for the fuel gas provisions of the *International Residential Code* will be undertaken by the International Fuel Gas Code Development Committee.

Electrical provisions have been included in this edition through an agreement with the National Fire Protection Association (NFPA). The maintenance process for the electrical provisions of future *International Residential Code* editions will be undertaken by NFPA.

Adoption

The *International Residential Code* is available for adoption and use by jurisdictions internationally. Its use within a governmental jurisdiction is intended to be accomplished through adoption by reference in accordance with proceedings establishing the jurisdiction's laws. At the time of adoption, jurisdictions should insert the appropriate information in provisions requiring specific local information, such as the name of the adopting jurisdiction. These locations are shown in bracketed words in small capital letters in the code and in the sample adoption ordinance. The sample adoption ordinance on page v addresses several key elements of a code adoption ordinance, including the information required for insertion into the code text.

Maintenance

The *International Residential Code* is kept up to date through the review of proposed changes submitted by code enforcement officials, industry representatives, design professionals and other interested parties. Proposed changes are carefully considered through an open code development process in which all interested and affected parties may participate.

For more information regarding the code development process, contact BOCA, 4051 West Flossmoor Road, Country Club Hills, Illinois 60478-5795; ICBO, 5360 Workman Mill Road, Whittier, California 90601-2298; or SBCCI, 900 Montclair Road, Birmingham, Alabama 35213-1206.

While the development procedure of the *International Residential Code* ensures the highest degree of care, BOCA, ICBO, SBCCI, their members and those participating in the development of this code do not accept any liability resulting from compliance or noncompliance with the provisions given herein, for any restrictions imposed on materials or processes, or for the completeness of the text. BOCA, ICBO and SBCCI do not have power or authority to police or enforce compliance with the contents of this code. Only the governmental body that enacts the code into law has such authority.

SAMPLE ORDINANCE FOR ADOPTION OF THE *INTERNATIONAL RESIDENTIAL CODE FOR ONE- AND TWO-FAMILY DWELLINGS*

ORDINANCE NO. ______

An ordinance of the _[JURISDICTION]_ adopting the 2000 edition of the *International Residential Code,* regulating and controlling the design, construction, quality of materials, erection, installation, alteration, repair, location, relocation, replacement, addition to, use or maintenance of one- and two-family dwellings and townhouses in the _[JURISDICTION]_; providing for the issuance of permits and collection of fees therefor when used with money; repealing Ordinance No. ______ of the _[JURISDICTION]_ and all other ordinances and parts of the ordinances in conflict therewith.

The _[GOVERNING BODY]_ of the _[JURISDICTION]_ does ordain as follows:

Section 1. That certain documents, three (3) copies of which are on file in the office of the _[JURISDICTION'S KEEPER OF RECORDS]_ and the _[JURISDICTION]_, being marked and designated as *International Residential Code,* including Appendix Chapters (fill in the applicable Appendix Chapters. [See *International Residential Code* Section R102.5, 2000 edition]), as published by the International Code Council and is hereby adopted as the code of the _[JURISDICTION]_ for regulating the design, construction, quality of materials, erection, installation, alteration, repair, location, relocation, replacement, addition to, use or maintenance of one- and two-family dwellings and townhouses not more than three stories in height in the _[JURISDICTION]_, and providing for the issuance of permits and collection of fees therefor; and each and all of the regulations, provisions, conditions and terms of such *International Residential Code,* 2000 edition, published by the International Code Council on file in the office of the _[JURISDICTION]_ are hereby referred to, adopted and made a part hereof as if fully set out in this ordinance.

Section 2. The following sections are hereby revised:

Section R101.1 Insert: [NAME OF JURISDICTION]

Table R301.2(1) Insert: [APPROPRIATE DESIGN CRITERIA]

Section 3. That Ordinance No. ______ of _[JURISDICTION]_ entitled *(fill in here the complete title of the present ordinance or ordinances in effect at the present time so that they will be repealed by definite mention)* and all other ordinances or parts of ordinances in conflict herewith are hereby repealed.

Section 4. That if any section, subsection, sentence, clause or phrase of this ordinance is, for any reason, held to be unconstitutional, such decision shall not affect the validity of the remaining portions of this ordinance. The _[GOVERNING BODY]_ hereby declares that it would have passed this ordinance, and each section, subsection, clause or phrase thereof, irrespective of the fact that any one or more sections, subsections, sentences, clauses and phrases be declared unconstitutional.

Section 5. That the _[JURISDICTION'S KEEPER OF RECORDS]_ is hereby ordered and directed to cause this ordinance to be published. *(An additional provision may be required to direct the number of times the ordinance is to be published and to specify that it is to be in a newspaper in general circulation. Posting may also be required.)*

Section 6. That this ordinance and the rules, regulations, provisions, requirements, orders and matters established and adopted hereby shall take effect and be in full force and effect _[TIME PERIOD]_ from and after the date of its final passage and adoption.

TABLE OF CONTENTS

CHAPTER 27 PLUMBING FIXTURES 387

CHAPTER 28 WATER HEATERS 393

CHAPTER 29 WATER SUPPLY AND DISTRIBUTION 395

CHAPTER 30 SANITARY DRAINAGE 405

CHAPTER 31 VENTS 413

CHAPTER 32 TRAPS 425

Part VIII—Electrical *427*

CHAPTER 33 GENERAL REQUIREMENTS 427

CHAPTER 34 ELECTRICAL DEFINITIONS 433

CHAPTER 35 SERVICES 437

Part I — Administrative

CHAPTER 1
ADMINISTRATION

SECTION R101
TITLE, SCOPE AND PURPOSE

R101.1 Title. These provisions shall be known as the *Residential Code for One- and Two-Family Dwellings* of [NAME OF JURISDICTION], and shall be cited as such and will be referred to herein as "this code."

R101.2 Scope. The provisions of the *International Residential Code for One- and Two-Family Dwellings* shall apply to the construction, alteration, movement, enlargement, replacement, repair, equipment, use and occupancy, location, removal and demolition of detached one- and two-family dwellings and multiple single-family dwellings (townhouses) not more than three stories in height with a separate means of egress and their accessory structures.

R101.3 Purpose. The purpose of this code is to provide minimum requirements to safeguard life or limb, health and public welfare.

SECTION R102
APPLICABILITY

R102.1 General. Where, in any specific case, different sections of this code specify different materials, methods of construction or other requirements, the most restrictive shall govern. Where there is a conflict between a general requirement and a specific requirement, the specific requirement shall be applicable.

R102.2 Other laws. The provisions of this code shall not be deemed to nullify any provisions of local, state or federal law.

R102.3 Application of references. References to chapter or section numbers, or to provisions not specifically identified by number, shall be construed to refer to such chapter, section or provision of this code.

R102.4 Referenced codes and standards. The codes and standards referenced in this code shall be considered part of the requirements of this code to the prescribed extent of each such reference. Where differences occur between provisions of this code and referenced codes and standards, the provisions of this code shall apply.

> **Exception:** Where enforcement of a code provision would violate the conditions of the listing of the equipment or appliance, the conditions of the listing and manufacturer's instructions shall apply.

R102.5 Appendices. Provisions in the appendices shall not apply unless specifically referenced in the adopting ordinance.

R102.6 Partial invalidity. In the event any part or provision of this code is held to be illegal or void, this shall not have the effect of making void or illegal any of the other parts or provisions.

R102.7 Existing structures. The legal occupancy of any structure existing on the date of adoption of this code shall be permitted to continue without change, except as is specifically covered in this code, the *International Property Maintenance Code* or the *International Fire Code*, or as is deemed necessary by the building official for the general safety and welfare of the occupants and the public.

R102.7.1 Additions, alterations or repairs. Additions, alterations or repairs to any structure shall conform to that required for a new structure without requiring the existing structure to comply with all of the requirements of this code, unless otherwise stated. Additions, alterations or repairs shall not cause an existing structure to become unsafe or adversely affect the performance of the building.

SECTION R103
DEPARTMENT OF BUILDING SAFETY

R103.1 Creation of enforcement agency. The department of building safety is hereby created and the official in charge thereof shall be known as the building official.

R103.2 Appointment. The building official shall be appointed by the chief appointing authority of the jurisdiction.

R103.3 Deputies. In accordance with the prescribed procedures of this jurisdiction and with the concurrence of the appointing authority, the building official shall have the authority to appoint a deputy building official, the related technical officers, inspectors, plan examiners and other employees. Such employees shall have powers as delegated by the building official.

SECTION R104
DUTIES AND POWERS OF THE BUILDING OFFICIAL

R104.1 General. The building official is hereby authorized and directed to enforce the provisions of this code. The building official shall have the authority to render interpretations of this code and to adopt policies and procedures in order to clarify the application of its provisions. Such interpretations, policies and procedures shall be in conformance with the intent and purpose of this code. Such policies and procedures shall not have the effect of waiving requirements specifically provided for in this code.

R104.2 Applications and permits. The building official shall receive applications, review construction documents and issue permits for the erection and alteration of buildings and structures, inspect the premises for which such permits have been issued and enforce compliance with the provisions of this code.

R104.3 Notices and orders. The building official shall issue all necessary notices or orders to ensure compliance with this code.

R104.4 Inspections. The building official is authorized to make all of the required inspections, or the building official shall have the authority to accept reports of inspection by approved agencies or individuals. Reports of such inspections shall be in writing and be certified by a responsible officer of such approved agency or by the responsible individual. The building official is authorized to engage such expert opinion as deemed necessary to report upon unusual technical issues that arise, subject to the approval of the appointing authority.

R104.5 Identification. The building official shall carry proper identification when inspecting structures or premises in the performance of duties under this code.

R104.6 Right of entry. Where it is necessary to make an inspection to enforce the provisions of this code, or where the building official has reasonable cause to believe that there exists in a structure or upon a premises a condition which is contrary to or in violation of this code which makes the structure or premises unsafe, dangerous or hazardous, the building official or designee is authorized to enter the structure or premises at reasonable times to inspect or to perform the duties imposed by this code, provided that if such structure or premises be occupied that credentials be presented to the occupant and entry requested. If such structure or premises be unoccupied, the building official shall first make a reasonable effort to locate the owner or other person having charge or control of the structure or premises and request entry. If entry is refused, the building official shall have recourse to the remedies provided by law to secure entry.

R104.7 Department records. The building official shall keep official records of applications received, permits and certificates issued, fees collected, reports of inspections, and notices and orders issued. Such records shall be retained in the official records for the period required for the retention of public records.

R104.8 Liability. The building official, member of the board of appeals or employee charged with the enforcement of this code, while acting for the jurisdiction in good faith and without malice in the discharge of the duties required by this code or other pertinent law or ordinance, shall not thereby be rendered liable personally and is hereby relieved from personal liability for any damage accruing to persons or property as a result of any act or by reason of an act or omission in the discharge of official duties. Any suit instituted against an officer or employee because of an act performed by that officer or employee in the lawful discharge of duties and under the provisions of this code shall be defended by legal representative of the jurisdiction until the final termination of the proceedings. The building official or any subordinate shall not be liable for cost in any action, suit or proceeding that is instituted in pursuance of the provisions of this code.

R104.9 Approved materials and equipment. Materials, equipment and devices approved by the building official shall be constructed and installed in accordance with such approval.

R104.9.1 Used materials and equipment. Used materials, equipment and devices shall not be reused unless approved by the building official.

R104.10 Modifications. Wherever there are practical difficulties involved in carrying out the provisions of this code, the building official shall have the authority to grant modifications for individual cases, provided the building official shall first find that special individual reason makes the strict letter of this code impractical and the modification is in compliance with the intent and purpose of this code and that such modification does not lessen health, life and fire safety requirements or structural. The details of action granting modifications shall be recorded and entered in the files of the department of building safety.

R104.10.1 Areas prone to flooding. The building official shall not grant modifications to any provision related to areas prone to flooding as established by Table R301.2(1) without the granting of a variance to such provisions by the board of appeals.

R104.11 Alternative materials, design and methods of construction and equipment. The provisions of this code are not intended to prevent the installation of any material or to prohibit any design or method of construction not specifically prescribed by this code, provided that any such alternative has been approved. An alternative material, design or method of construction shall be approved where the building official finds that the proposed design is satisfactory and complies with the intent of the provisions of this code, and that the material, method or work offered is, for the purpose intended, at least the equivalent of that prescribed in this code. Compliance with the specific performance-based provisions of the International Code Council (ICC) codes in lieu of specific requirements of this code shall also be permitted as an alternate.

R104.11.1 Tests. Whenever there is insufficient evidence of compliance with the provisions of this code, or evidence that a material or method does not conform to the requirements of this code, or in order to substantiate claims for alternative materials or methods, the building official shall have the authority to require tests as evidence of compliance to be made at no expense to the jurisdiction. Test methods shall be as specified in this code or by other recognized test standards. In the absence of recognized and accepted test methods, the building official shall approve the testing procedures. Tests shall be performed by an approved agency. Reports of such tests shall be retained by the building official for the period required for retention of public records.

SECTION R105
PERMITS

R105.1 Required. Any owner or authorized agent who intends to construct, enlarge, alter, repair, move, demolish, or change the occupancy of a building or structure, or to erect, install, enlarge, alter, repair, remove, convert or replace any electrical, gas, mechanical or plumbing system, the installation of which is regulated by this code, or to cause any such work to be done, shall first make application to the building official and obtain the required permit.

R105.2 Work exempt from permit. Permits shall not be required for the following. Exemption from the permit requirements of this code shall not be deemed to grant authorization for any work to be done in any manner in violation of the provisions of this code or any other laws or ordinances of this jurisdiction.

Building:

1. One-story detached accessory structures, provided the floor area does not exceed 200 square feet (18.58 m^2).
2. Fences not over 6 feet (1829 mm) high.
3. Retaining walls that are not over 4 feet (1219 mm) in height measured from the bottom of the footing to the top of the wall, unless supporting a surcharge.
4. Water tanks supported directly upon grade if the capacity does not exceed 5,000 gallons (18 927 L) and the ratio of height to diameter or width does not exceed 2 to 1.
5. Sidewalks and driveways not more than 30 inches (762 mm) above adjacent grade and not over any basement or story below.
6. Painting, papering, tiling, carpeting, cabinets, counter tops and similar finish work.
7. Prefabricated swimming pools that are less than 24 inches (610 mm) deep.
8. Swings and other playground equipment accessory to a one- or two-family dwelling.
9. Window awnings supported by an exterior wall.

Electrical:

Repairs and maintenance: A permit shall not be required for minor repair work, including the replacement of lamps or the connection of approved portable electrical equipment to approved permanently installed receptacles.

Gas:

1. Portable heating, cooking or clothes drying appliances.
2. Replacement of any minor part that does not alter approval of equipment or make such equipment unsafe.

Mechanical:

1. Portable heating appliance.
2. Portable ventilation appliances.
3. Portable cooling unit.
4. Steam, hot or chilled water piping within any heating or cooling equipment regulated by this code.
5. Replacement of any minor part that does not alter approval of equipment or make such equipment unsafe.
6. Portable evaporative cooler.
7. Self-contained refrigeration systems containing 10 pounds (4.54 kg) or less of refrigerant or that are actuated by motors of 1 horsepower (746 W) or less.

Plumbing:

The stopping of leaks in drains, water, soil, waste or vent pipe; provided, however, that if any concealed trap, drainpipe, water, soil, waste or vent pipe becomes defective and it becomes necessary to remove and replace the same with new material, such work shall be considered as new work and a permit shall be obtained and inspection made as provided in this code.

The clearing of stoppages or the repairing of leaks in pipes, valves or fixtures, and the removal and reinstallation of water closets, provided such repairs do not involve or require the replacement or rearrangement of valves, pipes or fixtures.

R105.2.1 Emergency repairs. Where equipment replacements and repairs must be performed in an emergency situation, the permit application shall be submitted within the next working business day to the building official.

R105.2.2 Repairs. Application or notice to the building official is not required for ordinary repairs to structures, replacement of lamps or the connection of approved portable electrical equipment to approved permanently installed receptacles. Such repairs shall not include the cutting away of any wall, partition or portion thereof, the removal or cutting of any structural beam or load bearing support, or the removal or change of any required means of egress, or rearrangement of parts of a structure affecting the egress requirements; nor shall ordinary repairs include addition to, alteration of, replacement or relocation of any, water supply, sewer, drainage, drain leader, gas, soil, waste, vent or similar piping, electric wiring or mechanical or other work affecting public health or general safety.

R105.2.3 Public service agencies. A permit shall not be required for the installation, alteration or repair of generation, transmission, distribution, metering or other related equipment that is under the ownership and control of public service agencies by established right.

R105.3 Application for permit. To obtain a permit, the applicant shall first file an application therefor in writing on a form furnished by the department of building safety for that purpose. Such application shall:

1. Identify and describe the work to be covered by the permit for which application is made.
2. Describe the land on which the proposed work is to be done by legal description, street address or similar description that will readily identify and definitely locate the proposed building or work.
3. Indicate the use and occupancy for which the proposed work is intended.
4. Be accompanied by construction documents and other information as required in Section R106.1.
5. State the valuation of the proposed work.
6. Be signed by the applicant, or the applicant's authorized agent.
7. Give such other data and information as required by the building official.

R105.3.1 Action on application. The building official shall examine or cause to be examined applications for permits and amendments thereto within a reasonable time after filing. If the application or the construction documents do not conform to the requirements of pertinent laws, the building official shall reject such application in writing, stating the reasons therefor. If the building official is satisfied that the proposed work conforms to the requirements of this code and laws and ordinances applicable thereto, the building official shall issue a permit therefor as soon as practicable.

R105.3.1.1 Substantially improved or substantially damaged existing buildings and structures. For applications for reconstruction, rehabilitation, addition, or other improvement of existing buildings or structures located in an area prone to flooding as established by Table R301.2(1), the building official shall examine or cause to be examined the construction documents and shall prepare a finding with regard to the value of the proposed work. For buildings that have sustained damage of any origin, the value of the proposed work shall include the cost to repair the building or structure to its predamage condition. If the building official finds that the value of proposed work equals or exceeds 50 percent of the market value of the building or structure, the finding shall be provided to the board of appeals for a determination of substantial improvement or substantial damage. Applications determined by the board of appeals to constitute substantial improvement or substantial damage shall meet the requirements of Section R327.

R105.3.2 Time limitation of application. An application for a permit for any proposed work shall be deemed to have been abandoned 180 days after the date of filing, unless such application has been pursued in good faith or a permit has been issued; except that the building official is authorized to grant one or more extensions of time for additional periods not exceeding 180 days each. The extension shall be requested in writing and justifiable cause demonstrated.

R105.4 Validity of permit. The issuance or granting of a permit shall not be construed to be a permit for, or an approval of, any violation of any of the provisions of this code or of any other ordinance of the jurisdiction. Permits presuming to give authority to violate or cancel the provisions of this code or other ordinances of the jurisdiction shall not be valid. The issuance of a permit based on construction documents and other data shall not prevent the building official from requiring the correction of errors in the construction documents and other data. The building official is also authorized to prevent occupancy or use of a structure where in violation of this code or of any other ordinances of this jurisdiction.

R105.5 Expiration. Every permit issued shall become invalid unless the work authorized by such permit is commenced within 180 days after its issuance, or if the work authorized by such permit is suspended or abandoned for a period of 180 days after the time the work is commenced. The building official is authorized to grant, in writing, one or more extensions of time, for periods not more than 180 days each. The extension shall be requested in writing and justifiable cause demonstrated.

R105.6 Suspension or revocation. The building official is authorized to suspend or revoke a permit issued under the provisions of this code wherever the permit is issued in error or on the basis of incorrect, inaccurate or incomplete information, or in violation of any ordinance or regulation or any of the provisions of this code.

R105.7 Placement of permit. The building permit or copy thereof shall be kept on the site of the work until the completion of the project.

R105.8 Responsibility. It shall be the duty of every person who performs work for the installation or repair of building, structure, electrical, gas, mechanical or plumbing systems, for which this code is applicable, to comply with this code.

SECTION R106
CONSTRUCTION DOCUMENTS

R106.1 Submittal documents. Construction documents, special inspection and structural observation programs, and other data shall be submitted in one or more sets with each application for a permit. The construction documents shall be prepared by a registered design professional where required by the statutes of the jurisdiction in which the project is to be constructed. Where special conditions exist, the building official is authorized to require additional construction documents to be prepared by a registered design professional.

Exception: The building official is authorized to waive the submission of construction documents and other data not required to be prepared by a registered design professional if it is found that the nature of the work applied for is such that reviewing of construction documents is not necessary to obtain compliance with this code.

R106.1.1 Information on construction documents. Construction documents shall be drawn upon suitable material. Electronic media documents are permitted to be submitted when approved by the building official. Construction documents shall be of sufficient clarity to indicate the location, nature and extent of the work proposed and show in detail that it will conform to the provisions of this code and relevant laws, ordinances, rules and regulations, as determined by the building official.

R106.1.2 Manufacturer's installation instructions. Manufacturer's installation instructions, as required by this code, shall be available on the job site at the time of inspection.

R106.1.3 Information for construction in areas prone to flooding. For buildings and structures in flood hazard areas as established by Table R301.2(1), construction documents shall include:

1. Delineation of flood hazard areas, floodway boundaries, and flood zones, and the design flood elevation, as appropriate;
2. The elevation of the proposed lowest floor, including basement; in areas of shallow flooding (AO Zones), the height of the proposed lowest floor, including basement, above the highest adjacent grade; and
3. If design flood elevations are not included on the community's Flood Insurance Rate Map (FIRM), the building official and the applicant shall obtain and reasonably utilize any design flood elevation and floodway data available from other sources.

R106.2 Site plan. The construction documents submitted with the application for permit shall be accompanied by a site plan showing the size and location of new construction and existing structures on the site and distances from lot lines. In the case of demolition, the site plan shall show construction to be demolished and the location and size of existing structures and construction that are to remain on the site or plot.

R106.3 Examination of documents. The building official shall examine or cause to be examined the accompanying construction documents and shall ascertain by such examinations whether the construction indicated and described is in accordance with the requirements of this code and other pertinent laws or ordinances.

R106.3.1 Approval of construction documents. When the building official issues a permit, the construction documents shall be approved in writing or by stamp. One set of construction documents so reviewed shall be retained by the building official. The other set shall be returned to the applicant, shall be kept at the site of work and shall be open to inspection by the building official or his or her authorized representative.

R106.3.2 Previous approvals. This code shall not require changes in the construction documents, construction or designated occupancy of a structure for which a lawful permit has been heretofore issued or otherwise lawfully authorized, and the construction of which has been pursued in good faith within 180 days after the effective date of this code and has not been abandoned.

R106.3.3 Phased approval. The building official is authorized to issue a permit for the construction of foundations or any other part of a building or structure before the construction documents for the whole building or structure have been submitted, provided that adequate information and detailed statements have been filed complying with pertinent requirements of this code. The holder of such permit for the foundation or other parts of a building or structure shall proceed at the holder's own risk with the building operation and without assurance that a permit for the entire structure will be granted.

R106.4 Amended construction documents. Work shall be installed in accordance with the reviewed construction documents, and any changes made during construction that are not in compliance with the approved construction documents shall be resubmitted for approval as an amended set of construction documents.

R106.5 Retention of construction documents. One set of approved construction documents shall be retained by the building official for a period of not less than 180 days from date of completion of the permitted work, or as required by state or local laws.

SECTION R107
TEMPORARY STRUCTURES AND USES

R107.1 General. The building official is authorized to issue a permit for temporary structures and temporary uses. Such permits shall be limited as to time of service, but shall not be permitted for more than 180 days. The building official is authorized to grant extensions for demonstrated cause.

R107.2 Conformance. Temporary structures and uses shall conform to the structural strength, fire safety, means of egress, light, ventilation and sanitary requirements of this code as necessary to ensure the public health, safety and general welfare.

R107.3 Temporary power. The building official is authorized to give permission to temporarily supply and use power in part of an electric installation before such installation has been fully completed and the final certificate of completion has been issued. The part covered by the temporary certificate shall comply with the requirements specified for temporary lighting, heat or power in the ICC *Electrical Code*.

R107.4 Termination of approval. The building official is authorized to terminate such permit for a temporary structure or use and to order the temporary structure or use to be discontinued.

SECTION R108
FEES

R108.1 Payment of fees. A permit shall not be valid until the fees prescribed by law have been paid. Nor shall an amendment to a permit be released until the additional fee, if any, has been paid.

R108.2 Schedule of permit fees. On buildings, structures, electrical, gas, mechanical, and plumbing systems or alterations requiring a permit, a fee for each permit shall be paid as required, in accordance with the schedule as established by the applicable governing authority.

R108.3 Building permit valuations. Building permit valuation shall include total value of the work for which a permit is being issued, such as electrical, gas, mechanical, plumbing equipment and other permanent systems, including materials and labor.

R108.4 Related fees. The payment of the fee for the construction, alteration, removal or demolition for work done in connection with or concurrently with the work authorized by a building permit shall not relieve the applicant or holder of the permit from the payment of other fees that are prescribed by law.

R108.5 Refunds. The building official is authorized to establish a refund policy.

SECTION R109
INSPECTIONS

R109.1 Types of inspections. For on-site construction, from time to time the building official, upon notification from the permit holder or his agent, shall make or cause to be made any necessary inspections and shall either approve that portion of the construction as completed or shall notify the permit holder or his or her agent wherein the same fails to comply with this code.

R109.1.1 Foundation inspection. Inspection of the foundation shall be made after poles or piers are set or trenches or basement areas are excavated and any required forms erected and any required reinforcing steel is in place and prior to the placing of concrete. The foundation inspection shall include excavations for thickened slabs intended for the support of bearing walls, partitions, structural supports, or equipment and special requirements for wood foundations.

R109.1.2 Plumbing, mechanical, gas and electrical systems inspection. Rough inspection of plumbing, mechanical, gas and electrical systems shall be made prior to covering or concealment, before fixtures or appliances are set or installed, and prior to framing inspection.

Exception: Ground-source heat pump loop systems tested in accordance with Section M2105.1 shall be permitted to be backfilled prior to inspection.

R109.1.3 Floodplain inspections. For construction permitted in areas prone to flooding as established by Table R301.2(1), upon placement of the lowest floor, including basement, and prior to further vertical construction, the building official shall require submission of a certification, prepared by a registered professional engineer or land surveyor, of the elevation of the lowest floor, including basement, required in Section R327.

R109.1.4 Frame and masonry inspection. Inspection of framing and masonry construction shall be made after the roof, masonry, all framing, firestopping, draftstopping and bracing are in place and after the plumbing, mechanical and electrical rough inspections are approved.

R109.1.5 Other inspections. In addition to the called inspections above, the building department may make or require any other inspections to ascertain compliance with this code and other laws enforced by the building department.

R109.1.5.1 Fire-resistance-rated construction inspection. Where fire-resistance-rated construction is required between dwelling units or due to location on property, the building official shall require an inspection of such construction after all lathing and/or wallboard is in place, but before any plaster is applied, or before wallboard joints and fasteners are taped and finished.

R109.1.6 Final inspection. Final inspection shall be made after the permitted work is complete and prior to occupancy.

R109.2 Inspection agencies. The building official is authorized to accept reports of approved agencies, provided such agencies satisfy the requirements as to qualifications and reliability.

R109.3 Inspection requests. It shall be the duty of the person doing the work authorized by a permit to notify the building official that such work is ready for inspection. It shall be the duty of the person requesting any inspections required by this code to provide access to and means for inspection of such work

R109.4 Approval required. Work shall not be done beyond the point indicated in each successive inspection without first obtaining the approval of the building official. The building official upon notification shall make the requested inspections and shall either indicate the portion of the construction that is satisfactory as completed, or shall notify the permit holder or an agent of the permit holder wherein the same fails to comply with this code. Any portions that do not comply shall be corrected and such portion shall not be covered or concealed until authorized by the building official.

SECTION R110
CERTIFICATE OF OCCUPANCY

R110.1 Use and occupancy. No building or structure shall be used or occupied, and no change in the existing occupancy classification of a building or structure or portion thereof shall be made until the building official has issued a certificate of occupancy therefor as provided herein. Issuance of a certificate of occupancy shall not be construed as an approval of a violation of the provisions of this code or of other ordinances of the jurisdiction. Certificates presuming to give authority to violate or cancel the provisions of this code or other ordinances of the jurisdiction shall not be valid.

R110.2 Change in use. Changes in the character or use of an existing structure shall not be made except as specified in Sections 3405 and 3406 of the *International Building Code.*

R110.3 Certificate issued. After the building official inspects the building or structure and finds no violations of the provisions of this code or other laws that are enforced by the department of building safety, the building official shall issue a certificate of occupancy which shall contain the following:

1. The building permit number.
2. The address of the structure.
3. The name and address of the owner.
4. A description of that portion of the structure for which the certificate is issued.
5. A statement that the described portion of the structure has been inspected for compliance with the requirements of this code.
6. The name of the building official.
7. The edition of the code under which the permit was issued.
8. If an automatic sprinkler system is provided.
9. Any special stipulations and conditions of the building permit.

R110.4 Temporary occupancy. The building official is authorized to issue a temporary certificate of occupancy before the completion of the entire work covered by the permit, provided that such portion or portions shall be occupied safely. The building official shall set a time period during which the temporary certificate of occupancy is valid.

R110.5 Revocation. The building official shall, in writing, suspend or revoke a certificate of occupancy issued under the provisions of this code wherever the certificate is issued in error, or on the basis of incorrect information supplied, or where it is determined that the building or structure or portion thereof is in violation of any ordinance or regulation or any of the provisions of this code.

SECTION R111
SERVICE UTILITIES

R111.1 Connection of service utilities. No person shall make connections from a utility, source of energy, fuel or power to any building or system that is regulated by this code for which a permit is required, until approved by the building official.

R111.2 Temporary connection. The building official shall have the authority to authorize and approve the temporary connection of the building or system to the utility source of energy, fuel or power.

R111.3 Authority to disconnect service utilities. The building official shall have the authority to authorize disconnection of utility service to the building, structure or system regulated by this code and the referenced codes and standards set forth in Section R102.4 in case of emergency where necessary to eliminate an immediate hazard to life or property or when such utility connection has been made without the approval required by Section R111.1 or R111.2. The building official shall notify the serving utility and whenever possible the owner and occupant of the building, structure or service system of the decision to disconnect prior to taking such action if not notified prior to disconnection. The owner or occupant of the building, structure or service system shall be notified in writing as soon as practical thereafter.

SECTION R112
BOARD OF APPEALS

R112.1 General. In order to hear and decide appeals of orders, decisions or determinations made by the building official relative to the application and interpretation of this code, there shall be and is hereby created a board of appeals. The building official shall be an ex officio member of said board but shall have no vote on any matter before the board. The board of appeals shall be appointed by the governing body and shall hold office at its pleasure. The board shall adopt rules of procedure for conducting its business, and shall render all decisions and findings in writing to the appellant with a duplicate copy to the building official.

R112.2 Limitations on authority. An application for appeal shall be based on a claim that the true intent of this code or the rules legally adopted thereunder have been incorrectly interpreted, the provisions of this code do not fully apply, or an equally good or better form of construction is proposed. The board shall have no authority to waive requirements of this code.

R112.2.1 Determination of substantial improvement in areas prone to flooding. When the building official provides a finding required in Section R105.3.1.1, the board of appeals shall determine whether the value of the proposed work constitutes a substantial improvement. A substantial improvement means any repair, reconstruction, rehabilitation, addition, or improvement of a building or structure, the cost of which equals or exceeds 50 percent of the market value of the building or structure before the improvement or repair is started. If the building or structure has sustained substantial damage, all repairs are considered substantial improvement regardless of the actual repair work performed. The term does not include:

1. Improvements of a building or structure required to correct existing health, sanitary or safety code violations identified by the building official and which are the minimum necessary to assure safe living conditions; or
2. Any alteration of an historic building or structure provided that the alteration will not preclude the continued designation as an historic building or structure.

R112.2.2 Criteria for issuance of a variance for areas prone to flooding. A variance shall only be issued upon:

1. A showing of good and sufficient cause that the unique characteristics of the size, configuration or topography of the site render the elevation standards in Section R327 inappropriate.
2. A determination that failure to grant the variance would result in exceptional hardship by rendering the lot undevelopable.
3. A determination that the granting of a variance will not result in increased flood heights, additional threats to public safety, extraordinary public expense, nor create nuisances, cause fraud on or victimization of the public, or conflict with existing local laws or ordinances.
4. A determination that the variance is the minimum necessary to afford relief, considering the flood hazard.
5. Submission to the applicant of written notice specifying the difference between the design flood elevation and the elevation to which the building is to be built, stating that the cost of flood insurance will be commensurate with the increased risk resulting from the reduced floor elevation, and stating that construction below the design flood elevation increases risks to life and property.

R112.3 Qualifications. The board of appeals shall consist of members who are qualified by experience and training to pass on matters pertaining to building construction and are not employees of the jurisdiction.

R112.4 Administration. The building official shall take immediate action in accordance with the decision of the board.

SECTION R113
VIOLATIONS

R113.1 Unlawful acts. It shall be unlawful for any person, firm or corporation to erect, construct, alter, extend, repair, move, remove, demolish or occupy any building, structure or equipment regulated by this code, or cause same to be done, in conflict with or in violation of any of the provisions of this code.

R113.2 Notice of violation. The building official is authorized to serve a notice of violation or order on the person responsible for the erection, construction, alteration, extension, repair, moving, removal, demolition or occupancy of a building or structure in violation of the provisions of this code, or in violation of a detail statement or a plan approved thereunder, or in violation of a permit or certificate issued under the provisions of this code. Such order shall direct the discontinuance of the illegal action or condition and the abatement of the violation.

R113.3 Prosecution of violation. If the notice of violation is not complied with promptly, the building official is authorized to request the legal counsel of the jurisdiction to institute the appropriate proceeding at law or in equity to restrain, correct or abate such violation, or to require the removal or termination of the unlawful occupancy of the building or structure in violation of the provisions of this code or of the order or direction made pursuant thereto.

R113.4 Violation penalties. Any person who violates a provision of this code or fails to comply with any of the requirements thereof or who erects, constructs, alters or repairs a building or structure in violation of the approved construction documents or directive of the building official, or of a permit or certificate issued under the provisions of this code, shall be subject to penalties as prescribed by law.

SECTION R114
STOP WORK ORDER

R114.1 Notice to owner. Upon notice from the building official that work on any building or structure is being prosecuted contrary to the provisions of this code or in an unsafe and dangerous manner, such work shall be immediately stopped. The stop work order shall be in writing and shall be given to the owner of the property involved, or to the owner's agent, or to the person doing the work; and shall state the conditions under which work will be permitted to resume.

R114.2 Unlawful continuance. Any person who shall continue any work in or about the structure after having been served with a stop work order, except such work as that person is directed to perform to remove a violation or unsafe condition, shall be subject to penalties as prescribed by law.

Part II — Definitions

CHAPTER 2
DEFINITIONS

SECTION R201
GENERAL

R201.1 Scope. Unless otherwise expressly stated, the following words and terms shall, for the purposes of this code, have the meanings indicated in this chapter.

R201.2 Interchangeability. Words used in the present tense include the future; words in the masculine gender include the feminine and neuter; the singular number includes the plural and the plural, the singular.

R201.3 Terms defined in other codes. Where terms are not defined in this code such terms shall have meanings ascribed to them as in other code publications of the International Code Council.

R201.4 Terms not defined. Where terms are not defined through the methods authorized by this section, such terms shall have ordinarily accepted meanings such as the context implies.

SECTION R202
DEFINITIONS

ACCESSIBLE. Signifies access that requires the removal of an access panel or similar removable obstruction.

ACCESSIBLE, READILY. Signifies access without the necessity for removing a panel or similar obstruction.

ACCESSORY STRUCTURE. In one- and two-family dwellings not more than three stories high with separate means of egress, a building, the use of which is incidental to that of the main building and which is located on the same lot.

ADDITION. An extension or increase in floor area or height of a building or structure.

AIR ADMITTANCE VALVE. A one-way valve designed to allow air into the plumbing drainage system when a negative pressure develops in the piping. This device shall close by gravity and seal the terminal under conditions of zero differential pressure (no flow conditions) and under positive internal pressure.

AIR BREAK, DRAINAGE SYSTEM. An arrangement in which a discharge pipe from a fixture, appliance or device drains indirectly into a receptor below the flood-level rim of the receptor.

AIR CIRCULATION, FORCED. A means of providing space conditioning utilizing movement of air through ducts or plenums by mechanical means.

AIR-CONDITIONING SYSTEM. A system that consists of heat exchangers, blowers, filters, supply, exhaust and return-air systems, and shall include any apparatus installed in connection therewith.

AIR GAP, DRAINAGE SYSTEM. The unobstructed vertical distance through free atmosphere between the outlet of a waste pipe and the flood-level rim of the fixture or receptor into which it is discharging.

AIR GAP, WATER-DISTRIBUTION SYSTEM. The unobstructed vertical distance through free atmosphere between the lowest opening from a water supply discharge to the flood-level rim of a plumbing fixture.

ALTERATION. Any construction or renovation to an existing structure other than repair or addition that requires a permit. Also, a change in a mechanical system that involves an extension, addition or change to the arrangement, type or purpose of the original installation that requires a permit.

ANCHORS. See "Supports."

ANTISIPHON. A term applied to valves or mechanical devices that eliminate siphonage.

APPLIANCE. A device or apparatus that is manufactured and designed to utilize energy and for which this code provides specific requirements.

APPROVED. Approved refers to approval by the building official as the result of investigation and tests conducted by him or her, or by reason of accepted principles or tests by nationally recognized organizations.

APPROVED AGENCY. An established and recognized agency regularly engaged in conducting tests or furnishing inspection services, when such agency has been approved by the building official.

ATTIC. The unfinished space between the ceiling joists of the top story and the roof rafters.

BACKFLOW, DRAINAGE. A reversal of flow in the drainage system.

BACKFLOW PREVENTER. A device or means to prevent backflow.

BACKFLOW PREVENTER, REDUCED-PRESSURE-ZONE TYPE. A backflow-prevention device consisting of two independently acting check valves, internally force loaded to a normally closed position and separated by an intermediate chamber (or zone) in which there is an automatic relief means of venting to atmosphere internally loaded to a normally open position between two tightly closing shutoff valves and with means for testing for tightness of the checks and opening of relief means.

BACKFLOW, WATER DISTRIBUTION. The flow of water or other liquids into the potable water-supply piping from any sources other than its intended source. Backsiphonage is one type of backflow.

BACKSIPHONAGE. The flowing back of used or contaminated water from piping into a potable water-supply pipe due to a negative pressure in such pipe.

BACKWATER VALVE. A device installed in a drain or pipe to prevent backflow of sewage.

BALCONY, EXTERIOR. An exterior floor projecting from and supported by a structure without additional independent supports.

BALL COCK. A valve that is used inside a gravity-type water closet flush tank to control the supply of water into the tank. It may also be called a flush-tank fill valve or water control.

BASEMENT. That portion of a building that is partly or completely below grade (see "Story above grade").

BASEMENT WALL. The opaque portion of a wall that encloses one side of a basement and has an average below grade wall area that is 50 percent or more of the total opaque and non-opaque area of that enclosing side.

BASIC WIND SPEED. Three-second gust speed at 33 feet (10 058 mm) above the ground in Exposure C (see Section R301.2.1) as given in Figure R301.2(4).

BATHROOM GROUP. A group of fixtures, including or excluding a bidet, consisting of a water closet, lavatory, and bathtub or shower. Such fixtures are located together on the same floor level.

BEND. A drainage fitting, designed to provide a change in direction of a drain pipe of less than the angle specified by the amount necessary to establish the desired slope of the line (see "Elbow" and "Sweep").

BOILER. A self-contained appliance from which hot water is circulated for heating purposes and then returned to the boiler, and which operates at water pressures not exceeding 160 pounds per square inch gage (psig) (1102 kPa gage) and at water temperatures not exceeding 250°F (121°C).

BOND BEAM. A horizontal grouted element within masonry in which reinforcement is embedded.

BRACED WALL LINE. A series of braced wall panels constructed in accordance with Section R602.10 for wood framing or Section R603.7 or R603.8.1.2 for cold-formed steel framing to resist racking from seismic and wind forces.

BRACED WALL PANEL. A section of a braced wall line constructed in accordance with Section R602.10 for wood framing or Section R603.7 or R603.8.1.2 for cold-formed steel framing, which extend the full height of the wall.

BRANCH. Any part of the piping system other than a riser, main or stack.

BRANCH, FIXTURE. See "Fixture branch, drainage."

BRANCH, HORIZONTAL. See "Horizontal branch, drainage."

BRANCH, INTERVAL. A distance along a soil or waste stack corresponding to a story height, but not less than 8 feet (2438 mm), within which the horizontal branches from one floor or story of a building are connected to the stack.

BRANCH, MAIN. A water-distribution pipe that extends horizontally off a main or riser to convey water to branches or fixture groups.

BRANCH, VENT. A vent connecting two or more individual vents with a vent stack or stack vent.

BTU/H. The listed maximum capacity of an appliance, absorption unit or burner expressed in British thermal units input per hour.

BUILDING. Building shall mean any one- and two-family dwelling or portion thereof, including townhouses, that is used, or designed or intended to be used for human habitation, for living, sleeping, cooking or eating purposes, or any combination thereof, and shall include accessory structures thereto.

BUILDING DRAIN. The lowest piping that collects the discharge from all other drainage piping inside the house and conveys it to the building sewer 30 inches (762 mm) outside the building wall.

BUILDING, EXISTING. Existing building is a building erected prior to the adoption of this code, or one for which a legal building permit has been issued.

BUILDING LINE. The line established by law, beyond which a building shall not extend, except as specifically provided by law.

BUILDING OFFICIAL. The officer or other designated authority charged with the administration and enforcement of this code.

BUILDING SEWER. That part of the drainage system that extends from the end of the building drain and conveys its discharge to a public sewer, private sewer, individual sewage-disposal system or other point of disposal.

BUILDING THERMAL ENVELOPE. The basement walls, exterior walls, floor, roof and any other building element that enclose conditioned spaces.

BUILT-UP ROOF COVERING. Two or more layers of felt cemented together and surfaced with a cap sheet, mineral aggregate, smooth coating or similar surfacing material.

CEILING HEIGHT. The clear vertical distance from the finished floor to the finished ceiling.

CHIMNEY. A primary vertical structure containing one or more flues, for the purpose of carrying gaseous products of combustion and air from a fuel-burning appliance to the outside atmosphere.

CHIMNEY CONNECTOR. A pipe that connects a fuel-burning appliance to a chimney.

CHIMNEY TYPES

Residential-type appliance. An approved chimney for removing the products of combustion from fuel-burning, residential-type appliances producing combustion gases not in excess of 1,000°F (538°C) under normal operating conditions, but capable of producing combustion gases of 1,400°F (760°C) during intermittent forces firing for periods up to 1 hour. All temperatures shall be measured at the appliance flue outlet. Residential-type appliance chimneys include masonry and factory-built types.

CIRCUIT VENT. A vent that connects to a horizontal drainage branch and vents two traps to a maximum of eight traps or trapped fixtures connected into a battery.

CLADDING. The exterior materials that cover the surface of the building envelope that is directly loaded by the wind.

CLEANOUT. An accessible opening in the drainage system used for the removal of possible obstruction.

CLOSET. A small room or chamber used for storage.

COMBINATION WASTE AND VENT SYSTEM. A specially designed system of waste piping embodying the horizontal wet venting of one or more sinks or floor drains by means of a common waste and vent pipe adequately sized to provide free movement of air above the flow line of the drain.

COMBUSTIBLE MATERIAL. Any material not defined as noncombustible.

COMBUSTION AIR. The air provided to fuel-burning equipment including air for fuel combustion, draft hood dilution and ventilation of the equipment enclosure.

COMMON VENT. A single pipe venting two trap arms within the same branch interval, either back-to-back or one above the other.

CONDENSATE. The liquid that separates from a gas due to a reduction in temperature, e.g., water that condenses from flue gases and water that condenses from air circulating through the cooling coil in air conditioning equipment.

CONDENSING APPLIANCE. An appliance that condenses water generated by the burning of fuels.

CONDITIONED AIR. Air treated to control its temperature, relative humidity or quality.

CONDITIONED AREA. That area within a building provided with heating and/or cooling systems or appliances capable of maintaining, through design or heat loss/gain, 68°F (20°C) during the heating season and/or 80°F (27°C) during the cooling season, or has a fixed opening directly adjacent to a conditioned area.

CONDITIONED FLOOR AREA. The horizontal projection of the floors associated with the conditioned space.

CONDITIONED SPACE. For energy purposes, space within a building that is provided with heating and/or cooling equipment or systems capable of maintaining, through design or heat loss/gain, 50°F (10°C) during the heating season and 85°F (29°C) during the cooling season, or communicates directly with a conditioned space. For mechanical purposes, an area, room or space being heated or cooled by any equipment or appliance.

CONFINED SPACE. A room or space having a volume less than 50 cubic feet per 1,000 Btu/h (4.83 L/W) of the aggregate input rating of all fuel-burning appliances installed in that space.

CONSTRUCTION DOCUMENTS. Written, graphic and pictorial documents prepared or assembled for describing the design, location and physical characteristics of the elements of a project necessary for obtaining a building permit. Construction drawings shall be drawn to an appropriate scale.

CONTINUOUS WASTE. A drain from two or more similar adjacent fixtures connected to a single trap.

CONTROL, LIMIT. An automatic control responsive to changes in liquid flow or level, pressure, or temperature for limiting the operation of an appliance.

CONTROL, PRIMARY SAFETY. A safety control responsive directly to flame properties that senses the presence or absence of flame and, in event of ignition failure or unintentional flame extinguishment, automatically causes shutdown of mechanical equipment.

CONVECTOR. A system-incorporating heating element in an enclosure in which air enters an opening below the heating element, is heated and leaves the enclosure through an opening located above the heating element.

CORROSION RESISTANT. Any nonferrous metal or any metal having an unbroken surfacing of nonferrous metal, or steel with not less than 10-percent chromium or with not less than 0.20-percent copper.

COURT. A space, open and unobstructed to the sky, located at or above grade level on a lot and bounded on three or more sides by walls or a building.

CROSS CONNECTION. Any connection between two otherwise separate piping systems whereby there may be a flow from one system to the other.

DALLE GLASS. A decorative composite glazing material made of individual pieces of glass that are embedded in a cast matrix of concrete or epoxy.

DAMPER, VOLUME. A device that will restrict, retard or direct the flow of air in any duct, or the products of combustion of heat-producing equipment, vent connector, vent or chimney.

DEAD END. A branch leading from a DWV system terminating at a developed length of 2 feet (610 mm) or more. Dead ends shall be prohibited except as an approved part of a rough-in for future connection.

DEAD LOADS. The weight of all materials of construction incorporated into the building, including but not limited to walls, floors, roofs, ceilings, stairways, built-in partitions, finishes, cladding, and other similarly incorporated architectural and structural items, and fixed service equipment.

DECK. An exterior floor system supported on at least two opposing sides by an adjoining structure and/or posts, piers, or other independent supports.

DECORATIVE GLASS. A carved, leaded or Dalle glass or glazing material whose purpose is decorative or artistic, not functional; whose coloring, texture or other design qualities or components cannot be removed without destroying the glazing material; and whose surface, or assembly into which it is incorporated, is divided into segments.

DESIGN PROFESSIONAL. See definition of "Registered design professional."

DEVELOPED LENGTH. The length of a pipeline measured along the center line of the pipe and fittings.

DIAMETER. Unless specifically stated, the term "diameter" is the nominal diameter as designated by the approved material standard.

DIAPHRAGM. A horizontal or nearly horizontal system acting to transmit lateral forces to the vertical resisting elements. When the term "diaphragm" is used, it includes horizontal bracing systems.

DILUTION AIR. Air that enters a draft hood or draft regulator and mixes with flue gases.

DIRECT-VENT APPLIANCE. A fuel-burning appliance with a sealed combustion system that draws all air for combustion from the outside atmosphere and discharges all flue gases to the outside atmosphere.

DRAFT. The pressure difference existing between the appliance or any component part and the atmosphere, that causes a continuous flow of air and products of combustion through the gas passages of the appliance to the atmosphere.

Induced draft. The pressure difference created by the action of a fan, blower or ejector, that is located between the appliance and the chimney or vent termination.

Natural draft. The pressure difference created by a vent or chimney because of its height, and the temperature difference between the flue gases and the atmosphere.

DRAFT HOOD. A device built into an appliance, or a part of the vent connector from an appliance, which is designed to provide for the ready escape of the flue gases from the appliance in the event of no draft, backdraft or stoppage beyond the draft hood; prevent a backdraft from entering the appliance; and neutralize the effect of stack action of the chimney or gas vent on the operation of the appliance.

DRAFT REGULATOR. A device that functions to maintain a desired draft in the appliance by automatically reducing the draft to the desired value.

DRAFT STOP. A material, device or construction installed to restrict the movement of air within open spaces of concealed areas of building components such as crawl spaces, floor-ceiling assemblies, roof-ceiling assemblies and attics.

DRAIN. Any pipe that carries soil and water-borne wastes in a building drainage system.

DRAINAGE FITTING. A pipe fitting designed to provide connections in the drainage system that have provisions for establishing the desired slope in the system. These fittings are made from a variety of both metals and plastics. The methods of coupling provide for required slope in the system (see "Durham fitting").

DUCT SYSTEM. A continuous passageway for the transmission of air which, in addition to ducts, includes duct fittings, dampers, plenums, fans and accessory air-handling equipment and appliances.

DURHAM FITTING. A special type of drainage fitting for use in the durham systems installations in which the joints are made with recessed and tapered threaded fittings, as opposed to bell and spigot lead/oakum or solvent/cemented or soldered joints. The tapping is at an angle (not 90 degrees) to provide for proper slope in otherwise rigid connections.

DURHAM SYSTEM. A term used to describe soil or waste systems where all piping is of threaded pipe, tube or other such rigid construction using recessed drainage fittings to correspond to the types of piping.

DWELLING. Any building that contains one or two dwelling units used, intended, or designed to be built, used, rented, leased, let or hired out to be occupied, or that are occupied for living purposes.

DWELLING UNIT. A single unit providing complete independent living facilities for one or more persons, including permanent provisions for living, sleeping, eating, cooking and sanitation.

DWV. Abbreviated term for drain, waste and vent piping as used in common plumbing practice.

EFFECTIVE OPENING. The minimum cross-sectional area at the point of water-supply discharge, measured or expressed in terms of diameter of a circle and if the opening is not circular, the diameter of a circle of equivalent cross-sectional area. (This is applicable to air gap.)

ELBOW. A pressure pipe fitting designed to provide an exact change in direction of a pipe run. An elbow provides a sharp turn in the flow path (see "Bend" and "Sweep").

EMERGENCY ESCAPE AND RESCUE OPENING. An operable window, door or similar device that provides for a means of escape and access for rescue in the event of an emergency.

EQUIPMENT. All piping, ducts, vents, control devices and other components of systems other than appliances that are permanently installed and integrated to provide control of environmental conditions for buildings. This definition shall also include other systems specifically regulated in this code.

EQUIVALENT LENGTH. For determining friction losses in a piping system, the effect of a particular fitting equal to the friction loss through a straight piping length of the same nominal diameter.

ESSENTIALLY NONTOXIC TRANSFER FLUIDS. Fluids having a Gosselin rating of 1, including propylene glycol; mineral oil; polydimenthyoil oxane; hydrochlorofluorocarbon, chlorofluorocarbon and hydrofluorocarbon refrigerants; and FDA-approved boiler water additives for steam boilers.

ESSENTIALLY TOXIC TRANSFER FLUIDS. Soil, water or gray water and fluids having a Gosselin rating of 2 or more including ethylene glycol, hydrocarbon oils, ammonia refrigerants and hydrazine.

EVAPORATIVE COOLER. A device used for reducing air temperature by the process of evaporating water into an airstream.

EXCESS AIR. Air that passes through the combustion chamber and the appliance flue in excess of that which is theoretically required for complete combustion.

EXHAUST HOOD, FULL OPENING. An exhaust hood with an opening at least equal to the diameter of the connecting vent.

EXISTING INSTALLATIONS. Any plumbing system regulated by this code that was legally installed prior to the effective date of this code, or for which a permit to install has been issued.

EXTERIOR INSULATION FINISH SYSTEMS (EIFS). Synthetic stucco cladding systems typically consisting of five layers: adhesive, insulation board, base coat into which fiberglass reinforcing mesh is embedded, and a finish coat in the desired color.

EXTERIOR WALL. An above-grade wall enclosing conditioned space. Includes between floor spandrels, peripheral edges of floors, roof and basement knee walls, dormer walls, gable end walls, walls enclosing a mansard roof, and basement walls with an average below grade wall area that is less than 50 percent of the total opaque and nonopaque area of that enclosing side.

FACTORY-BUILT CHIMNEY. A listed and labeled chimney composed of factory-made components assembled in the field in accordance with the manufacturer's instructions and the conditions of the listing.

FENESTRATION. Skylights, roof windows, vertical windows (whether fixed or moveable); opaque doors; glazed doors; glass block; and combination opaque/glazed doors.

FIREBLOCKING. Building materials installed to resist the free passage of flame to other areas of the building through concealed spaces.

FIREPLACE. An assembly consisting of a hearth and fire chamber of noncombustible material and provided with a chimney, for use with solid fuels.

Factory-built fireplace. A listed and labeled fireplace and chimney system composed of factory-made components, and assembled in the field in accordance with manufacturer's instructions and the conditions of the listing.

Masonry chimney. A field-constructed chimney composed of solid masonry units, bricks, stones or concrete.

Masonry fireplace. A field-constructed fireplace composed of solid masonry units, bricks, stones or concrete.

FIREPLACE STOVE. A free-standing, chimney-connected solid-fuel-burning heater designed to be operated with the fire chamber doors in either the open or closed position.

FIREPLACE THROAT. The opening between the top of the firebox and the smoke chamber.

FIRE SEPARATION DISTANCE. The distance measured from the building face to the closest interior lot line, to the centerline of a street, alley or public way, or to an imaginary line between two buildings on the property. The distance shall be measured at right angles from the lot line.

FIXTURE. See "Plumbing fixture."

FIXTURE BRANCH, DRAINAGE. A drain serving one or more fixtures that discharges into another portion of the drainage system.

FIXTURE BRANCH, WATER-SUPPLY. A water-supply pipe between the fixture supply and a main water-distribution pipe or fixture group main.

FIXTURE DRAIN. The drain from the trap of a fixture to the junction of that drain with any other drain pipe.

FIXTURE FITTING

Supply fitting. A fitting that controls the volume and/or directional flow of water and is either attached to or accessible from a fixture or is used with an open or atmospheric discharge.

Waste fitting. A combination of components that conveys the sanitary waste from the outlet of a fixture to the connection of the sanitary drainage system.

FIXTURE GROUP, MAIN. The main water-distribution pipe (or secondary branch) serving a plumbing fixture grouping such as a bath, kitchen or laundry area to which two or more individual fixture branch pipes are connected.

FIXTURE SUPPLY. The water-supply pipe connecting a fixture or fixture fitting to a fixture branch.

FIXTURE UNIT, DRAINAGE (d.f.u.). A measure of probable discharge into the drainage system by various types of plumbing fixtures, used to size DWV piping systems. The drainage fixture-unit value for a particular fixture depends on its volume rate of drainage discharge, on the time duration of a single drainage operation and on the average time between successive operations.

FIXTURE UNIT, WATER-SUPPLY (w.s.f.u.). A measure of the probable hydraulic demand on the water supply by various types of plumbing fixtures used to size water-piping systems. The water-supply fixture-unit value for a particular fixture depends on its volume rate of supply, on the time duration of a single supply operation and on the average time between successive operations.

FLAME SPREAD. The propagation of flame over a surface.

FLAME SPREAD INDEX. The numeric value assigned to a material tested in accordance with ASTM E 84.

FLOOD-LEVEL RIM. The edge of the receptor or fixture from which water overflows.

FLOOR DRAIN. A plumbing fixture for recess in the floor having a floor-level strainer intended for the purpose of the collection and disposal of waste water used in cleaning the floor and for the collection and disposal of accidental spillage to the floor.

FLOOR FURNACE. A self-contained furnace suspended from the floor of the space being heated, taking air for combustion from outside such space, and with means for lighting the appliance from such space.

FLOW PRESSURE. The static pressure reading in the water-supply pipe near the faucet or water outlet while the faucet or water outlet is open and flowing at capacity.

FLUE. See "Vent."

FLUE, APPLIANCE. The passages within an appliance through which combustion products pass from the combustion chamber to the flue collar.

FLUE COLLAR. The portion of a fuel-burning appliance designed for the attachment of a draft hood, vent connector or venting system.

FLUE GASES. Products of combustion plus excess air in appliance flues or heat exchangers.

FLUSH VALVE. A device located at the bottom of a flush tank that is operated to flush water closets.

FLUSHOMETER TANK. A device integrated within an air accumulator vessel that is designed to discharge a predetermined quantity of water to fixtures for flushing purposes.

FLUSHOMETER VALVE. A flushometer valve is a device that discharges a predetermined quantity of water to fixtures for flushing purposes and is actuated by direct water pressure.

FOAM PLASTIC INSULATION. A plastic that is intentionally expanded by the use of a foaming agent to produce a reduced-density plastic consisting open or closed cells distributed throughout the plastic and that has a density less than 20 pounds per cubic foot (320 kg/m^3).

FUEL-PIPING SYSTEM. All piping, tubing, valves and fittings used to connect fuel utilization equipment to the point of fuel delivery.

FULLWAY VALVE. A valve that in the full open position has an opening cross-sectional area equal to a minimum of 85 percent of the cross-sectional area of the connecting pipe.

FURNACE. A vented heating appliance designed or arranged to discharge heated air into a conditioned space or through a duct or ducts.

GLAZING AREA. The interior surface area of all glazed fenestration, including the area of sash, curbing or other framing elements, that enclose conditioned space. Includes the area of glazed fenestration assemblies in walls bounding conditioned basements.

GRADE. The finished ground level adjoining the building at all exterior walls.

GRADE FLOOR OPENING. A window or other opening located such that the sill height of the opening is not more than 44 inches (1118 mm) above or below the finished ground level adjacent to the opening.

GRADE, PIPING. See "Slope."

GRADE PLANE. A reference plane representing the average of the finished ground level adjoining the building at all exterior walls.

GROSS AREA OF EXTERIOR WALLS. The normal projection of all exterior walls, including the area of all windows and doors installed therein.

GROUND-SOURCE HEAT PUMP LOOP SYSTEM. Piping buried in horizontal or vertical excavations or placed in a body of water for the purpose of transporting heat transfer liquid to and from a heat pump. Included in this definition are closed loop systems in which the liquid is recirculated and open loop systems in which the liquid is drawn from a well or other source.

GUARD. A building component or a system of building components located near the open sides of elevated walking surfaces that minimizes the possibility of a fall from the walking surface to the lower level.

HABITABLE SPACE. A space in a building for living, sleeping, eating or cooking. Bathrooms, toilet rooms, closets, halls, storage or utility spaces and similar areas are not considered habitable spaces.

HANDRAIL. A horizontal or sloping rail intended for grasping by the hand for guidance or support.

HANGERS. See "Supports."

HAZARDOUS LOCATION. Any location considered to be a fire hazard for flammable vapors, dust, combustible fibers or other highly combustible substances.

HEATING DEGREE DAYS (HDD). The sum, on an annual basis, of the difference between 65°F (18°C) and the mean temperature for each day as determined from "NOAA Annual Degree Days to Selected Bases Derived from the 1960-1990 Normals" or other weather data sources acceptable to the code official.

HEAT PUMP. An appliance having heating or heating/cooling capability and that uses refrigerants to extract heat from air, liquid or other sources.

HEIGHT, BUILDING. The vertical distance from grade plane to the average height of the highest roof surface.

HEIGHT, STORY. The vertical distance from top to top of two successive tiers of beams or finished floor surfaces; and, for the topmost story, from the top of the floor finish to the top of the ceiling joists or, where there is not a ceiling, to the top of the roof rafters.

HIGH-TEMPERATURE (H.T.) CHIMNEY. A high temperature chimney complying with the requirements of UL 103. A Type H.T. chimney is identifiable by the markings "Type H.T." on each chimney pipe section.

HORIZONTAL BRANCH, DRAINAGE. A drain pipe extending laterally from a soil or waste stack or building drain, that receives the discharge from one or more fixture drains.

HORIZONTAL PIPE. Any pipe or fitting that makes an angle of less than 45 degrees (0.79 rad) with the horizontal.

HOT WATER. Water at a temperature greater than or equal to 110°F (43°C).

HURRICANE-PRONE REGIONS. Areas vulnerable to hurricanes, defined as the U.S. Atlantic Ocean and Gulf of Mexico coasts where the basic wind speed is greater than 110 miles per hour (177 km/h), and Hawaii, Puerto Rico, Guam, Virgin Islands, and America Samoa.

IGNITION SOURCE. A flame spark or hot surface capable of igniting flammable vapors or fumes. Such sources include appliance burners, burner ignitions and electrical switching devices.

INDIRECT WASTE PIPE. A waste pipe that discharges into the drainage system through an air gap into a trap, fixture or receptor.

INDIVIDUAL SEWAGE DISPOSAL SYSTEM. A system for disposal of sewage by means of a septic tank or mechanical treatment, designed for use apart from a public sewer to serve a single establishment or building.

INDIVIDUAL VENT. A pipe installed to vent a single-fixture drain that connects with the vent system above or terminates independently outside the building.

INDIVIDUAL WATER SUPPLY. A supply other than an approved public water supply that serves one or more families.

INSULATING CONCRETE FORM (ICF). A concrete forming system using stay-in-place forms of rigid foam plastic insulation, a hybrid of cement and foam insulation, a hybrid of cement and wood chips, or other insulating material for constructing cast-in-place concrete walls.

INSULATING SHEATHING. An insulating board having a minimum thermal resistance of R-2 of the core material.

JURISDICTION. The governmental unit that has adopted this code under due legislative authority.

KITCHEN. Kitchen shall mean an area used, or designated to be used, for the preparation of food.

LABEL. An identification applied on a product by the manufacturer which contains the name of the manufacturer, the function and performance characteristics of the product or material, and the name and identification of an approved agency and that indicates that the representative sample of the product or material has been tested and evaluated by an approved agency. (See also "Manufacturer's designation" and "Mark.")

LABELED. Devices, equipment or materials to which have been affixed a label, seal, symbol or other identifying mark of a testing laboratory, inspection agency or other organization concerned with product evaluation that maintains periodic inspection of the production of the above labeled items that attests to compliance with a specific standard.

LIGHT-FRAMED CONSTRUCTION. A type of construction whose vertical and horizontal structural elements are primarily formed by a system of repetitive wood or light gage steel framing members.

LISTED AND LISTING. Terms referring to equipment that is shown in a list published by an approved testing agency qualified and equipped for experimental testing and maintaining an adequate periodic inspection of current productions and whose listing states that the equipment complies with nationally recognized standards when installed in accordance with the manufacturer's installation instructions.

LIVE LOADS. Those loads produced by the use and occupancy of the building or other structure and do not include construction or environmental loads such as wind load, snow load, rain load, earthquake load, flood load or dead load.

LIVING SPACE. Space within a dwelling unit utilized for living, sleeping, eating, cooking, bathing, washing and sanitation purposes.

LOT. A portion or parcel of land considered as a unit.

LOT LINE. A line dividing one lot from another, or from a street or any public place.

MACERATING TOILET SYSTEMS. A system comprised of a sump with macerating pump and with connections for a water closet and other plumbing fixtures, that is designed to accept, grind and pump wastes to an approved point of discharge.

MAIN. The principal pipe artery to which branches may be connected.

MAIN SEWER. See "Public sewer."

MANIFOLD WATER DISTRIBUTION SYSTEMS. A fabricated piping arrangement in which a large supply main is fitted with multiple branches in close proximity in which water is distributed separately to fixtures from each branch.

MANUFACTURED HOME. Manufactured home means a structure, transportable in one or more sections, which in the traveling mode is 8 body feet (2438 body mm) or more in width or 40 body feet (12 192 body mm) or more in length, or, when erected on site, is 320 square feet (30 m^2) or more, and which is built on a permanent chassis and designed to be used as a dwelling with or without a permanent foundation when connected to the required utilities, and includes the plumbing, heating, air-conditioning and electrical systems contained therein; except that such term shall include any structure that meets all the requirements of this paragraph except the size requirements and with respect to which the manufacturer voluntarily files a certification required by the secretary (HUD) and complies with the standards established under this title. For mobile homes built prior to June 15, 1976, a label certifying compliance to the Standard for Mobile Homes, NFPA 501, in effect at the time of manufacture is required. For the purpose of these provisions, a mobile home shall be considered a manufactured home.

MANUFACTURER'S DESIGNATION. An identification applied on a product by the manufacturer indicating that a product or material complies with a specified standard or set of rules. (See also "Mark" and "Label.")

MANUFACTURER'S INSTALLATION INSTRUCTIONS. Printed instructions included with equipment as part of the conditions of listing and labeling.

MARK. An identification applied on a product by the manufacturer indicating the name of the manufacturer and the function of a product or material. (See also "Manufacturer's designation" and "Label.")

MASONRY CHIMNEY. A field-constructed chimney composed of solid masonry units, bricks, stones or concrete.

MASONRY, SOLID. Masonry consisting of solid masonry units laid contiguously with the joints between the units filled with mortar.

MASONRY UNIT. Brick, tile, stone, glass block or concrete block conforming to the requirements specified in Section 2103 of the *International Building Code*.

Clay. A building unit larger in size than a brick, composed of burned clay, shale, fire clay or mixtures thereof.

Concrete. A building unit or block larger in size than 12 inches by 4 inches by 4 inches (305 mm by 102 mm by 102 mm) made of cement and suitable aggregates.

Glass. Nonload-bearing masonry composed of glass units bonded by mortar.

Hollow. A masonry unit whose net cross-sectional area in any plane parallel to the loadbearing surface is less than 75 percent of its gross cross-sectional area measured in the same plane.

Solid. A masonry unit whose net cross-sectional area in every plane parallel to the loadbearing surface is 75 percent or more of its cross-sectional area measured in the same plane.

MASS WALL. Masonry or concrete walls having a mass greater than or equal to 30 pounds per square foot (146 kg/m^2), solid wood walls having a mass greater than or equal to 20 pounds per square foot (98 kg/m^2), and any other walls having a heat capacity greater than or equal to 6 Btu/ft^2°F [266 J/(m^2·k)].

MEAN ROOF HEIGHT. The average of the roof eave height and the height to the highest point on the roof surface, except that eave height shall be used for roof angle of less than or equal to 10 degrees (0.18 rad).

MECHANICAL DRAFT SYSTEM. A venting system designed to remove flue or vent gases by mechanical means, that consists of an induced draft portion under nonpositive static pressure or a forced draft portion under positive static pressure.

Forced-draft venting system. A portion of a venting system using a fan or other mechanical means to cause the removal of flue or vent gases under positive static pressure.

Induced draft venting system. A portion of a venting system using a fan or other mechanical means to cause the removal of flue or vent gases under nonpositive static vent pressure.

Power venting system. A portion of a venting system using a fan or other mechanical means to cause the removal of flue or vent gases under positive static vent pressure.

MECHANICAL EXHAUST SYSTEM. A system for removing air from a room or space by mechanical means.

MECHANICAL SYSTEM. A system specifically addressed and regulated in this code and composed of components, devices, appliances and equipment.

METAL ROOF PANEL. An interlocking metal sheet having a minimum installed weather exposure of at least 3 square feet (0.28 m^2) per sheet.

METAL ROOF SHINGLE. An interlocking metal sheet having an installed weather exposure less than 3 square feet (0.28 m^2) per sheet.

MEZZANINE, LOFT. An intermediate level or levels between the floor and ceiling of any story with an aggregate floor area of not more than one-third of the area of the room or space in which the level or levels are located.

MODIFIED BITUMEN ROOF COVERING. One or more layers of polymer modified asphalt sheets. The sheet materials shall be fully adhered or mechanically attached to the substrate or held in place with an approved ballast layer.

MULTIPLE STATION SMOKE ALARM. Two or more single station alarm devices that are capable of interconnection such that actuation of one causes all integral or separate audible alarms to operate.

NATURAL DRAFT SYSTEM. A venting system designed to remove flue or vent gases under nonpositive static vent pressure entirely by natural draft.

NONCOMBUSTIBLE MATERIAL. Materials that pass the test procedure for defining noncombustibility of elementary materials set forth in ASTM E 136.

NONCONDITIONED SPACE. A space that is not a conditioned space by insulated walls, floors or ceilings.

OCCUPIED SPACE. The total area of all buildings or structures on any lot or parcel of ground projected on a horizontal plane, excluding permitted projections as allowed by this code.

OFFSET. A combination of fittings that makes two changes in direction bringing one section of the pipe out of line but into a line parallel with the other section.

OWNER. Any person, agent, firm or corporation having a legal or equitable interest in the property.

PELLET FUEL-BURNING APPLIANCE. A closed combustion, vented appliance equipped with a fuel feed mechanism for burning processed pellets of solid fuel of a specified size and composition.

PELLET VENT. A vent listed and labeled for use with a listed pellet fuel-burning appliance.

PERMIT. An official document or certificate issued by the authority having jurisdiction that authorizes performance of a specified activity.

PERSON. An individual, heirs, executors, administrators or assigns, and also includes a firm, partnership or corporation, its or their successors or assigns, or the agent of any of the aforesaid.

PITCH. See "Slope."

PLATFORM CONSTRUCTION. A method of construction by which floor framing bears on load bearing walls that are not continuous through the story levels or floor framing.

PLENUM. A chamber that forms part of an air-circulation system other than the occupied space being conditioned.

PLUMBING. For the purpose of this code, plumbing refers to those installations, repairs, maintenance and alterations regulated by Chapters 25 through 32.

PLUMBING APPLIANCE. An energized household appliance with plumbing connections, such as a dishwasher, food-waste grinder, clothes washer or water heater.

PLUMBING APPURTENANCE. A device or assembly that is an adjunct to the basic plumbing system and demands no additional water supply nor adds any discharge load to the system. It is presumed that it performs some useful function in the operation, maintenance, servicing, economy or safety of the plumbing system. Examples include filters, relief valves and aerators.

PLUMBING FIXTURE. A receptor or device that requires both a water-supply connection and a discharge to the drainage system, such as water closets, lavatories, bathtubs and sinks. Plumbing appliances as a special class of fixture are further defined.

PLUMBING SYSTEM. Includes the water supply and distribution pipes, plumbing fixtures, supports and appurtenances; soil, waste and vent pipes; sanitary drains and building sewers to an approved point of disposal.

POSITIVE ROOF DRAINAGE. The drainage condition in which consideration has been made for all loading deflections of the roof deck, and additional slope has been provided to ensure drainage of the roof within 48 hours of precipitation.

POTABLE WATER. Water free from impurities present in amounts sufficient to cause disease or harmful physiological effects and conforming in bacteriological and chemical quality to the requirements of the public health authority having jurisdiction.

PRESSURE-RELIEF VALVE. A pressure-actuated valve held closed by a spring or other means and designed to automatically relieve pressure at the pressure at which it is set.

PUBLIC SEWER. A common sewer directly controlled by public authority.

PUBLIC WATER MAIN. A water-supply pipe for public use controlled by public authority.

PUBLIC WAY. Any street, alley or other parcel of land open to the outside air leading to a public street, which has been deeded, dedicated or otherwise permanently appropriated to the public for public use and that has a clear width and height of not less than 10 feet (3048 mm).

PURGE. To clear of air, gas or other foreign substances.

QUICK-CLOSING VALVE. A valve or faucet that closes automatically when released manually or controlled by mechanical means for fast-action closing.

***R*-VALUE, THERMAL RESISTANCE.** The inverse of the time rate of heat flow through a building thermal envelope element from one of its bounding surfaces to the other for a unit temperature difference between the two surfaces, under steady state conditions, per unit area ($h \cdot ft^2 {}^\circ F/Btu$).

RAMP. A walking surface that has a running slope steeper than 1 unit vertical in 20 units horizontal (5-percent slope).

RECEPTOR. A fixture or device that receives the discharge from indirect waste pipes.

REFRIGERANT. A substance used to produce refrigeration by its expansion or evaporation.

REFRIGERANT COMPRESSOR. A specific machine, with or without accessories, for compressing a given refrigerant vapor.

REFRIGERATING SYSTEM. A combination of interconnected parts forming a closed circuit in which refrigerant is circulated for the purpose of extracting, then rejecting, heat. A direct refrigerating system is one in which the evaporator or condenser of the refrigerating system is in direct contact with the air or other substances to be cooled or heated. An indirect refrigerating system is one in which a secondary coolant cooled or heated by the refrigerating system is circulated to the air or other substance to be cooled or heated.

REGISTERED DESIGN PROFESSIONAL. An individual who is registered or licensed to practice their respective design profession as defined by the statutory requirements of the professional registration laws of the state or jurisdiction in which the project is to be constructed.

RELIEF VALVE, VACUUM. A device to prevent excessive buildup of vacuum in a pressure vessel.

REPAIR. The reconstruction or renewal of any part of an existing building for the purpose of its maintenance.

REROOFING. The process of recovering or replacing an existing roof covering. See "Roof recover."

RESIDENTIAL BUILDING TYPE. The type of residential building for determining building thermal envelope criteria. Detached one- and two-family dwellings are Type A-1. Townhouses are Type A-2.

RETURN AIR. Air removed from an approved conditioned space or location and recirculated or exhausted.

RISER. A water pipe that extends vertically one full story or more to convey water to branches or to a group of fixtures.

ROOF ASSEMBLY. A system designed to provide weather protection and resistance to design loads. The system consists of a roof covering and roof deck or a single component serving as both the roof covering and the roof deck. A roof assembly includes the roof deck, vapor retarder, substrate or thermal barrier, insulation, vapor retarder, and roof covering.

ROOF COVERING. The covering applied to the roof deck for weather resistance, fire classification or appearance.

ROOF COVERING SYSTEM. See "Roof assembly."

ROOF DECK. The flat or sloped surface not including its supporting members or vertical supports.

ROOF RECOVER. The process of installing an additional roof covering over a prepared existing roof covering without removing the existing roof covering.

ROOF REPAIR. Reconstruction or renewal of any part of an existing roof for the purposes of its maintenance.

ROOFTOP STRUCTURE. An enclosed structure on or above the roof of any part of a building.

ROOM HEATER. A freestanding heating appliance installed in the space being heated and not connected to ducts.

ROUGH-IN. The installation of all parts of the plumbing system that must be completed prior to the installation of fixtures. This includes DWV, water supply and built-in fixture supports.

RUNNING BOND. The placement of masonry units such that head joints in successive courses are horizontally offset at least one-quarter the unit length.

SANITARY SEWER. A sewer that carries sewage and excludes storm, surface and groundwater.

SCUPPER. An opening in a wall or parapet that allows water to drain from a roof.

SEISMIC DESIGN CATEGORY. A classification assigned to a structure based on its Seismic Group and the severity of the design earthquake ground motion at the site.

SEPTIC TANK. A water-tight receptor that receives the discharge of a building sanitary drainage system and is constructed so as to separate solids from the liquid, digest organic matter through a period of detention, and allow the liquids to discharge into the soil outside of the tank through a system of open joint or perforated piping or a seepage pit.

SEWAGE. Any liquid waste containing animal matter, vegetable matter or other impurity in suspension or solution.

SEWAGE PUMP. A permanently installed mechanical device for removing sewage or liquid waste from a sump.

SHALL. The term, when used in the code, is construed as mandatory.

SHEAR WALL. A general term for walls that are designed and constructed to resist racking from seismic and wind by use of masonry, concrete, cold-formed steel or wood framing in accordance with Chapter 6 of this code and the associated limitations in Section R301.2 of this code.

SIDE VENT. A vent connecting to the drain pipe through a fitting at an angle less than 45 degrees (0.79 rad) to the horizontal.

SINGLE PLY MEMBRANE. A roofing membrane that is field applied using one layer of membrane material (either homogeneous or composite) rather than multiple layers.

SINGLE STATION SMOKE ALARM. An assembly incorporating the detector, control equipment and alarm sounding device in one unit that is operated from a power supply either in the unit or obtained at the point of installation.

SKYLIGHT AND SLOPED GLAZING. See Section R308.6.1.

SLIP JOINT. A mechanical-type joint used primarily on fixture traps. The joint tightness is obtained by compressing a friction-type washer such as rubber, nylon, neoprene, lead or special packing material against the pipe by the tightening of a (slip) nut.

SLOPE. The fall (pitch) of a line of pipe in reference to a horizontal plane. In drainage, the slope is expressed as the fall in units vertical per units horizontal (percent) for a length of pipe.

SMOKE-DEVELOPED RATING. A numerical index indicating the relative density of smoke produced by burning assigned to a material tested in accordance with ASTM E 84.

SOIL STACK OR PIPE. A pipe that conveys sewage containing fecal material.

SOLAR HEAT GAIN COEFFICIENT (SHGC). The solar heat gain through a fenestration or glazing assembly relative to the incident solar radiation (Btu/h · ft^2 · °F).

SOLID MASONRY. Load-bearing or nonload-bearing construction using masonry units where the net cross-sectional area of each unit in any plane parallel to the bearing surface is not less than 75 percent of its gross cross-sectional area. Solid masonry units shall conform to ASTM C 55, C 62, C 73, C 145 or C 216.

STACK. Any main vertical DWV line, including offsets, that extends one or more stories as directly as possible to its vent terminal.

STACK BOND. The placement of masonry units in a bond pattern is such that head joints in successive courses are vertically aligned. For the purpose of this code, requirements for stack bond shall apply to all masonry laid in other than running bond.

STACK VENT. The extension of soil or waste stack above the highest horizontal drain connected.

STACK VENTING. A method of venting a fixture or fixtures through the soil or waste stack without individual fixture vents.

STANDARD TRUSS. Any construction that does not permit the roof/ceiling insulation to achieve the required *R*-value over the exterior walls.

STORM SEWER, DRAIN. A pipe used for conveying rainwater, surface water, condensate, cooling water or similar liquid wastes.

STORY. That portion of a building included between the upper surface of a floor and the upper surface of the floor or roof next above.

STORY ABOVE GRADE. Any story having its finished floor surface entirely above grade, except that a basement shall be considered as a story above grade where the finished surface of the floor above the basement is:

1. More than 6 feet (1829 mm) above grade plane.
2. More than 6 feet (1829 mm) above the finished ground level for more than 50 percent of the total building perimeter.
3. More than 12 feet (3658 mm) above the finished ground level at any point.

STRUCTURAL INSULATED PANELS (SIPS). Factory fabricated panels of solid core insulation with structural skins of oriented strand board (OSB) or plywood.

SUMP. A tank or pit that receives sewage or waste, located below the normal grade of the gravity system and that must be emptied by mechanical means.

SUMP PUMP. A pump installed to empty a sump. The pump is chosen to handle the type of material to be pumped—either clear water waste or soil-type sewage. The pump is selected for the specific head and volume of the load and is usually operated by level controllers.

SUPPLY AIR. Air delivered to a conditioned space through ducts or plenums from the heat exchanger of a heating, cooling or ventilating system.

SUPPORTS. Devices for supporting, hanging and securing pipes, fixtures and equipment.

SWEEP. A drainage fitting designed to provide a change in direction of a drain pipe of less than the angle specified by the amount necessary to establish the desired slope of the line. Sweeps provide a longer turning radius than bends and a less turbulent flow pattern (see "Bend" and "Elbow").

TEMPERATURE- AND PRESSURE-RELIEF (T AND P) VALVE. A combination relief valve designed to function as both a temperature-relief and pressure-relief valve.

TEMPERATURE-RELIEF VALVE. A temperature-actuated valve designed to discharge automatically at the temperature at which it is set.

THERMAL RESISTANCE, *R*-VALUE. The inverse of the time rate of heat flow through a body from one of its bounding surfaces to the other for a unit temperature difference between the two surfaces, under steady state conditions, per unit area ($h \cdot ft^2 \cdot °F/Btu$).

THERMAL TRANSMITTANCE, *U*-FACTOR. The coefficient of heat transmission (air to air) through a building envelope component or assembly, equal to the time rate of heat flow per unit area and unit temperature difference between the warm side and cold side air films ($Btu/h \cdot ft^2 \cdot °F$).

TOWNHOUSE. A single-family dwelling unit constructed in a group of three or more attached units in which each unit extends from foundation to roof and with open space on at least two sides.

TRAP. A fitting, either separate or built into a fixture, that provides a liquid seal to prevent the emission of sewer gases without materially affecting the flow of sewage or waste water through it.

TRAP ARM. That portion of a fixture drain between a trap weir and the vent fitting.

TRAP PRIMER. A device or system of piping to maintain a water seal in a trap, typically installed where infrequent use of the trap would result in evaporation of the trap seal, such as floor drains.

TRAP SEAL. The trap seal is the maximum vertical depth of liquid that a trap will retain, measured between the crown weir and the top of the dip of the trap.

TRIM. Picture molds, chair rails, baseboards, handrails, door and window frames, and similar decorative or protective materials used in fixed applications.

TRUSS DESIGN DRAWING. The graphic depiction of an individual truss, which describes the design and physical characteristics of the truss.

TYPE L VENT. A listed and labeled vent conforming to UL 641 for venting oil-burning appliances listed for use with Type L vents or with listed gas appliances.

***U*-FACTOR, THERMAL TRANSMITTANCE.** The coefficient of heat transmission (air to air) through a building envelope component or assembly, equal to the time rate of heat flow per unit area and unit temperature difference between the warm side and cold side air films ($Btu/h \cdot ft^2$ °F).

UNCONFINED SPACE. A space having a volume not less than 50 cubic feet per 1,000 Btu/h (4.8 m^3/kW) of the aggregate input rating of all appliances installed in that space. Rooms communicating directly with the space in which the appliances are installed, through openings not furnished with doors, are considered a part of the unconfined space.

UNDERLAYMENT. One or more layers of felt, sheathing paper, nonbituminous saturated felt, or other approved material over which a s roof covering, with a slope of 2 to 12 (17-percent slope) or greater, is applied.

UNUSUALLY TIGHT CONSTRUCTION. Construction in which:

1. Walls and ceilings comprising the building thermal envelope have a continuous water vapor retarder with a rating of 1 perm [57.4 $ng/(s \cdot m^2 \cdot Pa)$] or less with openings therein gasketed or sealed.
2. Storm windows or weatherstripping is applied around the threshold and jambs of opaque doors and openable windows.
3. Caulking or sealants are applied to areas such as joints around window and door frames between sole plates and floors, between wall-ceiling joints, between wall panels, at penetrations for plumbing, electrical and gas lines, and at other openings.

VACUUM BREAKERS. A device which prevents backsiphonage of water by admitting atmospheric pressure through ports to the discharge side of the device.

VAPOR RETARDER. A material having a permeance rating of 1.0 or less when tested in accordance with ASTM E 96.

VENT. A passageway for conveying flue gases from fuel-fired appliances, or their vent connectors, to the outside atmosphere.

VENT COLLAR. See "Flue collar."

VENT CONNECTOR. That portion of a venting system which connects the flue collar or draft hood of an appliance to a vent.

VENT DAMPER DEVICE, AUTOMATIC. A device intended for installation in the venting system, in the outlet of an individual, automatically operated fuel burning appliance and that is designed to open the venting system automatically when the appliance is in operation and to close off the venting system automatically when the appliance is in a standby or shutdown condition.

VENT GASES. Products of combustion from fuel-burning appliances, plus excess air and dilution air, in the venting system above the draft hood or draft regulator.

VENT STACK. A vertical vent pipe installed to provide circulation of air to and from the drainage system and which extends through one or more stories.

VENT SYSTEM. Piping installed to equalize pneumatic pressure in a drainage system to prevent trap seal loss or blow-back due to siphonage or back pressure.

VENTILATION. The natural or mechanical process of supplying conditioned or unconditioned air to, or removing such air from, any space.

VENTING. Removal of combustion products to the outdoors.

VENTING SYSTEM. A continuous open passageway from the flue collar of an appliance to the outside atmosphere for the purpose of removing flue or vent gases. A venting system is usually composed of a vent or a chimney and vent connector, if used, assembled to form the open passageway.

VERTICAL PIPE. Any pipe or fitting that makes an angle of 45 degrees (0.79 rad) or more with the horizontal.

WALLS. Walls shall be defined as follows:

Load-bearing wall is a wall supporting any vertical load in addition to its own weight.

Nonbearing wall is a wall which does not support vertical loads other than its own weight.

WASTE. Liquid-borne waste that is free of fecal matter.

WASTE PIPE OR STACK. Piping that conveys only liquid sewage not containing fecal material.

WATER-DISTRIBUTION SYSTEM. Piping which conveys water from the service to the plumbing fixtures, appliances, appurtenances, equipment, devices or other systems served, including fittings and control valves.

WATER HEATER. Any heating appliance or equipment that heats potable water and supplies such water to the potable hot water distribution system.

WATER MAIN. A water-supply pipe for public use.

WATER OUTLET. A valved discharge opening, including a hose bibb, through which water is removed from the potable water system supplying water to a plumbing fixture or plumbing appliance that requires either an air gap or backflow prevention device for protection of the supply system.

WATER-SERVICE PIPE. The outside pipe from the water main or other source of potable water supply to the water-distribution system inside the building, terminating at the service valve.

WATER-SUPPLY SYSTEM. The water-service pipe, the water-distributing pipes and the necessary connecting pipes, fittings, control valves and all appurtenances in or adjacent to the building or premises.

WET VENT. A vent that also receives the discharge of wastes from other fixtures.

WIND BORNE DEBRIS REGION. Areas within hurricane-prone regions within one mile of the coastal mean high water line where the basic wind speed is 110 miles per hour (177 km/h) or greater; or where the basic wind speed is equal to or greater than 120 miles per hour (193 km/h); or Hawaii.

WOOD STRUCTURAL PANEL. A panel manufactured from veneers; or wood strands or wafers; bonded together with waterproof synthetic resins or other suitable bonding systems. Examples of wood structural panels are plywood, OSB or composite panels.

YARD. An open space, other than a court, unobstructed from the ground to the sky, except where specifically provided by this code, on the lot on which a building is situated.

Part III — Building Planning and Construction

CHAPTER 3
BUILDING PLANNING

SECTION R301
DESIGN CRITERIA

R301.1 Design. Buildings and structures, and all parts thereof, shall be constructed to safely support all loads, including dead loads, live loads, roof loads, flood loads, snow loads, wind loads and seismic loads as prescribed by this code. The construction of buildings and structures shall result in a system that provides a complete load path capable of transferring all loads from their point of origin through the load-resisting elements to the foundation. When a building of otherwise conventional construction contains structural elements that exceed the limits of Section R301, those elements shall be designed in accordance with accepted engineering practice.

R301.1.1 Construction systems. The requirements of this code are based on platform and balloon-frame construction for light-frame buildings. The requirements for concrete and masonry buildings are based on a balloon framing system. Other framing systems must have equivalent detailing to ensure force transfer, continuity and compatible deformations.

R301.1.2 Engineered design. When a building of otherwise conventional light-frame construction contains structural elements not conforming to this code, these elements shall be designed in accordance with accepted engineering practice. The extent of such design need only demonstrate compliance of nonconventional elements with other applicable provisions and shall be compatible with the performance of the conventional framed system.

R301.2 Climatic and geographic design criteria. Buildings shall be constructed in accordance with the provisions of this code as limited by the provisions of this section. Additional criteria shall be established by the local jurisdiction and set forth in Table R301.2(1).

R301.2.1 Wind limitations. Buildings and portions thereof shall be limited by wind speed, as defined in Table R301.2(1), and construction methods in accordance with this code. Basic wind speeds shall be determined from Figure R301.2(4). Where different construction methods and structural materials are used for various portions of a building, the applicable requirements of this section for each portion shall apply. Where loads for windows, skylights and exterior doors are not otherwise specified, the loads listed in Table R301.2(2) adjusted for height and exposure per Table R301.2(3) shall be used to determine design load performance requirements for windows and doors.

R301.2.1.1 Design criteria. Construction in regions where the basic wind speeds from Figure R301.2(4) equal or exceed 110 miles per hour (177.1 km/h) shall be designed in accordance with one of the following:

1. American Forest and Paper Association (AF&PA) Wood Frame Construction Manual for One- and Two-Family Dwellings (WFCM); or
2. Southern Building Code Congress International Standard for Hurricane Resistant Residential Construction (SSTD 10); or
3. Minimum Design Loads for Buildings and Other Structures (ASCE-7); or
4. Cold-formed steel construction shall be designed in accordance with the provisions of this code.

TABLE R301.2(1)
CLIMATIC AND GEOGRAPHIC DESIGN CRITERIA

GROUND SNOW LOAD	WIND	SEISMIC DESIGN CATEGORY[f,g]	SUBJECT TO DAMAGE FROM				WINTER DESIGN TEMP[f]	FLOOD HAZARDS[h]
	Speed[e] (mph)		Weathering[a]	Frost line depth[b]	Termite[c]	Decay[d]		

For SI: 1 pound per square foot = 0.0479 kN/m^2, 1 mile per hour = 1.609 km/h.

a. Weathering may require a higher strength concrete or grade of masonry than necessary to satisfy the structural requirements of this code. The weathering column shall be filled in with the weathering index (i.e., "negligible," "moderate" or "severe") for concrete as determined from the Weathering Probability Map [Figure R301.2(3)]. The grade of masonry units shall be determined from ASTM C 34, C 55, C 62, C 73, C 90, C 129, C 145, C 216 or C 652.

b. The frost line depth may require deeper footings than indicated in Figure R403.1(1). The jurisdiction shall fill in the frost line depth column with the minimum depth of footing below finish grade.

c. The jurisdiction shall fill in this part of the table with "very heavy," "moderate to heavy," "slight to moderate," or "none to slight" in accordance with Figure R301.2(6) depending on whether there has been a history of local damage.

d. The jurisdiction shall fill in this part of the table with "moderate to severe," "slight to moderate," or "none to slight" in accordance with Figure R301.2(7) depending on whether there has been a history of local damage.

e. The jurisdiction shall fill in this part of the table with the wind speed from the basic wind speed map [Figure R301.2(4)]. Wind exposure category shall be determined on a site-specific basis in accordance with Section R301.2.1.4.

f. The outdoor design dry-bulb temperature shall be selected from the columns of 97$^1/_2$-percent values for winter from Appendix D of the *International Plumbing Code*. Deviations from the Appendix D temperatures shall be permitted to reflect local climates or local weather experience as determined by the building official.

g. The jurisdiction shall fill in this part of the table with the Seismic Design Category determined from Section R301.2.2.1.

h. The jurisdiction shall fill in this part of the table with (a) the date of the jurisdiction's entry into the National Flood Insurance Program (date of adoption of the first code or ordinance for management of flood hazard areas), (b) the date(s) of the currently effective FIRM and FBFM, or other flood hazard map adopted by the community, as may be amended.

TABLE R301.2(2) COMPONENT AND CLADDING LOADS FOR A BUILDING WITH A MEAN ROOF HEIGHT OF 30 FEET LOCATED IN EXPOSURE B

	ZONE	EFFECTIVE WIND AREA (feet²)	BASIC WIND SPEED (mph—3-second gust) 85		90		100		105		110		120		125		130		140		145		150		170	
Roof > 0 to 10 degrees	1	10	10.0	-13.0	10.0	-14.6	10.0	-18.0	10.0	-19.8	10.0	-21.8	10.5	-25.9	11.4	-28.1	12.4	-30.4	14.3	-35.3	15.4	-37.8	16.5	-40.5	21.1	-52.0
	1	20	10.0	-12.7	10.0	-14.2	10.0	-17.5	10.0	-19.3	10.0	-21.2	10.0	-25.2	10.7	-27.4	11.6	-29.6	13.4	-34.4	14.4	-36.9	15.4	-39.4	19.8	-50.7
	1	50	10.0	-12.2	10.0	-13.7	10.0	-16.9	10.0	-18.7	10.0	-20.5	10.0	-24.4	10.0	-26.4	10.6	-28.6	12.3	-33.2	13.1	-35.6	14.1	-38.1	18.1	-48.9
	1	100	10.0	-11.9	10.0	-13.3	10.0	-18.5	10.0	-18.2	10.0	-19.9	10.0	-23.7	10.0	-25.7	10.0	-27.8	11.4	-32.3	12.2	-34.6	13.0	-37.0	16.7	-47.6
	2	10	10.0	-21.8	10.0	-24.4	10.0	-30.2	10.0	-33.3	10.0	-36.5	10.5	-43.5	11.4	-47.2	12.4	-51.0	14.3	-59.2	15.4	-63.5	16.5	-67.9	21.1	-87.2
	2	20	10.0	-19.5	10.0	-21.8	10.0	-27.0	10.0	-29.7	10.0	-32.6	10.0	-38.8	10.7	-42.1	11.6	-45.6	13.4	-52.9	14.4	-56.7	15.4	-60.7	19.8	-78.0
	2	50	10.0	-16.4	10.0	-18.4	10.0	-22.7	10.0	-25.1	10.0	-27.5	10.0	-32.7	10.0	-35.5	10.6	-38.4	12.3	-44.5	13.1	-47.8	14.1	-51.1	18.1	-65.7
	2	100	10.0	-14.1	10.0	-15.8	10.0	-19.5	10.0	-21.5	10.0	-23.6	10.0	-28.1	10.0	-30.5	10.0	-33.0	11.4	-38.2	12.2	-41.0	13.0	-43.9	16.7	-56.4
	3	10	10.0	-32.8	10.0	-36.8	10.0	-45.4	10.0	-50.1	10.0	-55.0	10.5	-65.4	11.4	-71.0	12.4	-76.8	14.3	-89.0	15.4	-95.5	16.5	-102.2	21.1	-131.3
	3	20	10.0	-27.2	10.0	-30.5	10.0	-37.6	10.0	-41.5	10.0	-45.5	10.0	-54.2	10.7	-58.8	11.6	-63.6	13.4	-73.8	14.4	-79.1	15.4	-84.7	19.8	-108.7
	3	50	10.0	-19.7	10.0	-22.1	10.0	-27.3	10.0	-30.1	10.0	-33.1	10.0	-39.3	10.0	-42.7	10.6	-46.2	12.3	-53.5	13.1	-57.4	14.1	-61.5	18.1	-78.9
	3	100	10.0	-14.1	10.0	-15.8	10.0	-19.5	10.0	-21.5	10.0	-23.6	10.0	-28.1	10.0	-30.5	10.0	-33.0	11.4	-38.2	12.2	-41.0	13.0	-43.9	16.7	-56.4
Roof > 10 to 30 degrees	1	10	10.0	-11.9	10.0	-13.3	10.4	-16.5	11.4	-18.2	12.5	-19.9	14.9	-23.7	16.2	-25.7	17.5	-27.8	20.3	-32.3	21.8	-34.6	23.3	-37.0	30.0	-47.6
	1	20	10.0	-11.6	10.0	-13.0	10.0	-16.0	10.4	-17.6	11.4	-19.4	13.6	-23.0	14.8	-25.0	16.0	-27.0	18.5	-31.4	19.9	-33.7	21.3	-36.0	27.3	-46.3
	1	50	10.0	-11.1	10.0	-12.5	10.0	-15.4	10.0	-17.0	10.0	-18.6	11.9	-22.2	12.9	-24.1	13.9	-26.0	16.1	-30.2	17.3	-32.4	18.5	-34.6	23.8	-44.5
	1	100	10.0	-10.8	10.0	-12.1	10.0	-14.9	10.0	-16.5	10.0	-18.1	10.5	-21.5	11.4	-23.3	12.4	-25.2	14.3	-29.3	15.4	-31.4	16.5	-33.6	21.1	-43.2
	2	10	10.0	-25.1	10.0	-28.2	10.4	-34.8	11.4	-38.3	12.5	-42.1	14.9	-50.1	16.2	-54.3	17.5	-58.7	20.3	-68.1	21.8	-73.1	23.3	-78.2	30.0	-100.5
	2	20	10.0	-22.8	10.0	-25.6	10.0	-31.5	10.4	-34.8	11.4	-38.2	13.6	-45.4	14.8	-49.3	16.0	-53.3	18.5	-61.8	19.9	-66.3	21.3	-71.0	27.3	-91.2
	2	50	10.0	-19.7	10.0	-22.1	10.0	-27.3	10.0	-30.1	10.0	-33.0	11.9	-39.3	12.9	-42.7	13.9	-46.1	16.1	-53.5	17.3	-57.4	18.5	-61.4	23.8	-78.9
	2	100	10.0	-17.4	10.0	-19.5	10.0	-24.1	10.0	-26.6	10.0	-29.1	10.5	-34.7	11.4	-37.6	12.4	-40.7	14.3	-47.2	15.4	-50.6	16.5	-54.2	21.1	-69.6
	3	10	10.0	-25.1	10.0	-28.2	10.4	-34.8	11.4	-38.3	12.5	-42.1	14.9	-50.1	16.2	-54.3	17.5	-58.7	20.3	-68.1	21.8	-73.1	23.3	-78.2	30.0	-100.5
	3	20	10.0	-22.8	10.0	-25.6	10.0	-31.5	10.4	-34.8	11.4	-38.2	13.6	-45.4	14.8	-49.3	16.0	-53.3	18.5	-61.8	19.9	-66.3	21.3	-71.0	27.3	-91.2
	3	50	10.0	-19.7	10.0	-22.1	10.0	-27.3	10.0	-30.1	10.0	-33.0	11.9	-39.3	12.9	-42.7	13.9	-46.1	16.1	-53.5	17.3	-57.4	18.5	-61.4	23.8	-78.9
	3	100	10.0	-17.4	10.0	-19.5	10.0	-24.1	10.0	-26.6	10.0	-29.1	10.5	-34.7	11.4	-37.6	12.4	-40.7	14.3	-47.2	15.4	-50.6	16.5	-54.2	21.1	-69.6
Roof > 30 to 45 degrees	1	10	11.9	-13.0	13.3	-14.6	16.5	-18.0	18.2	-19.8	19.9	-21.8	23.7	-25.9	25.7	-28.1	27.8	-30.4	32.3	-35.3	34.6	-37.8	37.0	-40.5	47.6	-52.0
	1	20	11.6	-12.3	13.0	-13.8	16.0	-17.1	17.6	-18.8	19.4	-20.7	23.0	-24.6	25.0	-26.7	27.0	-28.9	31.4	-33.5	33.7	-35.9	36.0	-38.4	46.3	-49.3
	1	50	11.1	-11.5	12.5	-12.8	15.4	-15.9	17.0	-17.5	18.6	-19.2	22.2	-22.8	24.1	-24.8	26.0	-25.8	30.2	-31.1	32.4	-33.3	34.6	-35.7	44.5	-45.8
	1	100	10.8	-10.8	12.1	-12.1	14.9	-14.9	16.5	-16.5	18.1	-18.1	21.5	-21.5	23.3	-23.3	25.2	-25.2	29.3	-29.3	31.4	-31.4	33.6	-33.6	43.2	-43.2
	2	10	11.9	-15.2	13.3	-17.0	16.5	-21.0	18.2	-23.2	19.9	-25.5	23.7	-30.3	25.7	-32.9	27.8	-35.6	32.3	-41.2	34.6	-44.2	37.0	-47.3	47.6	-60.8
	2	20	11.6	-14.5	13.0	-16.3	16.0	-20.1	17.6	-22.2	19.4	-24.3	23.0	-29.0	25.0	-31.4	27.0	-34.0	31.4	-39.4	33.7	-42.3	36.0	-45.3	46.3	-58.1
	2	50	11.1	-13.7	12.5	-15.3	15.4	-18.9	17.0	-20.8	18.6	-22.9	22.2	-27.2	24.1	-29.5	26.0	-32.0	30.2	-37.1	32.4	-39.8	34.6	-42.5	44.5	-54.6
	2	100	10.8	-13.0	12.1	-14.6	14.9	-18.0	16.5	-19.8	18.1	-21.8	21.5	-25.9	23.3	-28.1	25.2	-30.4	29.3	-35.3	31.4	-37.8	33.6	-40.5	43.2	-52.0
	3	10	11.9	-15.2	13.3	-17.0	16.5	-21.0	18.2	-23.2	19.9	-25.5	23.7	-30.3	25.7	-32.9	27.8	-35.6	32.3	-41.2	34.6	-44.2	37.0	-47.3	47.6	-60.8
	3	20	11.6	-14.5	13.0	-16.3	16.0	-20.1	17.6	-22.2	19.4	-24.3	23.0	-29.0	25.0	-31.4	27.0	-34.0	31.4	-39.4	33.7	-42.3	36.0	-45.3	46.3	-58.1
	3	50	11.1	-13.7	12.5	-15.3	15.4	-18.9	17.0	-20.8	18.6	-22.9	22.2	-27.2	24.1	-29.5	26.0	-32.0	30.2	-37.1	32.4	-39.8	34.6	-42.5	44.5	-54.5
	3	100	10.8	-13.0	12.1	-14.6	14.9	-18.0	16.5	-19.8	18.1	-21.8	21.5	-25.9	23.3	-28.1	25.2	-30.4	29.3	-35.3	31.4	-37.8	33.6	-40.5	43.2	-52.0
Wall	4	10	13.0	-14.1	14.6	-15.8	18.0	-19.5	19.8	-21.5	21.8	-23.6	25.9	-28.1	28.1	-30.5	30.4	-33.0	35.3	-38.2	37.8	-41.0	40.5	-43.9	52.0	-56.4
	4	20	12.4	-13.5	13.9	-15.1	17.2	-18.7	18.9	-20.6	20.8	-22.6	24.7	-26.9	26.8	-29.2	29.0	-31.6	33.7	-36.7	36.1	-39.3	38.7	-42.1	49.6	-54.1
	4	50	11.6	-12.7	13.0	-14.3	16.1	-17.6	17.8	-19.4	19.5	-21.3	23.2	-25.4	25.2	-27.5	27.2	-29.8	31.6	-34.6	33.9	-37.1	36.2	-39.7	46.6	-51.0
	4	100	11.1	-12.2	12.4	-13.6	15.3	-16.8	16.9	-18.5	18.5	-20.4	22.0	-24.2	23.9	-26.3	25.9	-28.4	30.0	-33.0	32.2	-35.4	34.4	-37.8	44.2	-48.6
	5	10	13.0	-17.4	14.6	-19.5	18.0	-24.1	19.8	-26.6	21.8	-29.1	25.9	-34.7	28.1	-37.6	30.4	-40.7	35.3	-47.2	37.8	-50.6	40.5	-54.2	52.0	-69.6
	5	20	12.4	-16.2	13.9	-18.2	17.2	-22.5	18.9	-24.8	20.8	-27.2	24.7	-32.4	26.8	-35.1	29.0	-38.0	33.7	-44.0	36.1	-47.2	38.7	-50.5	49.6	-64.9
	5	50	11.6	-14.7	13.0	-16.5	16.1	-20.3	17.8	-22.4	19.5	-24.6	23.2	-29.3	25.2	-31.8	27.2	-34.3	31.6	-39.8	33.9	-42.7	36.2	-45.7	46.6	-58.7
	5	100	11.1	-13.5	12.4	-15.1	15.3	-18.7	16.9	-20.6	18.5	-22.6	22.0	-26.9	23.9	-29.2	25.9	-31.6	30.0	-36.7	32.2	-39.3	34.4	-42.1	44.2	-54.1

For SI: 1 foot = 304.8 mm, 1 square foot = 0.0929 m^2, 1 mile per hour = 1.609 km/h.

NOTES: For effective areas between those given above the load may be interpolated, otherwise use the load associated with the lower effective area.

Table values shall be adjusted for height and exposure by multiplying by the adjustment coefficient in Table R301.2(3).

See Figure R301.2(8) for location of zones.

Plus and minus signs signify pressures acting toward and away from the building surfaces.

TABLE R301.2(3)
HEIGHT AND EXPOSURE ADJUSTMENT COEFFICIENTS FOR TABLE R301.2(2)

MEAN ROOF HEIGHT	EXPOSURE		
	B	C	D
15	1.00	1.21	1.47
20	1.00	1.29	1.55
25	1.00	1.35	1.61
30	1.00	1.40	1.66
35	1.05	1.45	1.70
40	1.09	1.49	1.74
45	1.12	1.53	1.78
50	1.16	1.56	1.81
55	1.19	1.59	1.84
60	1.22	1.62	1.87

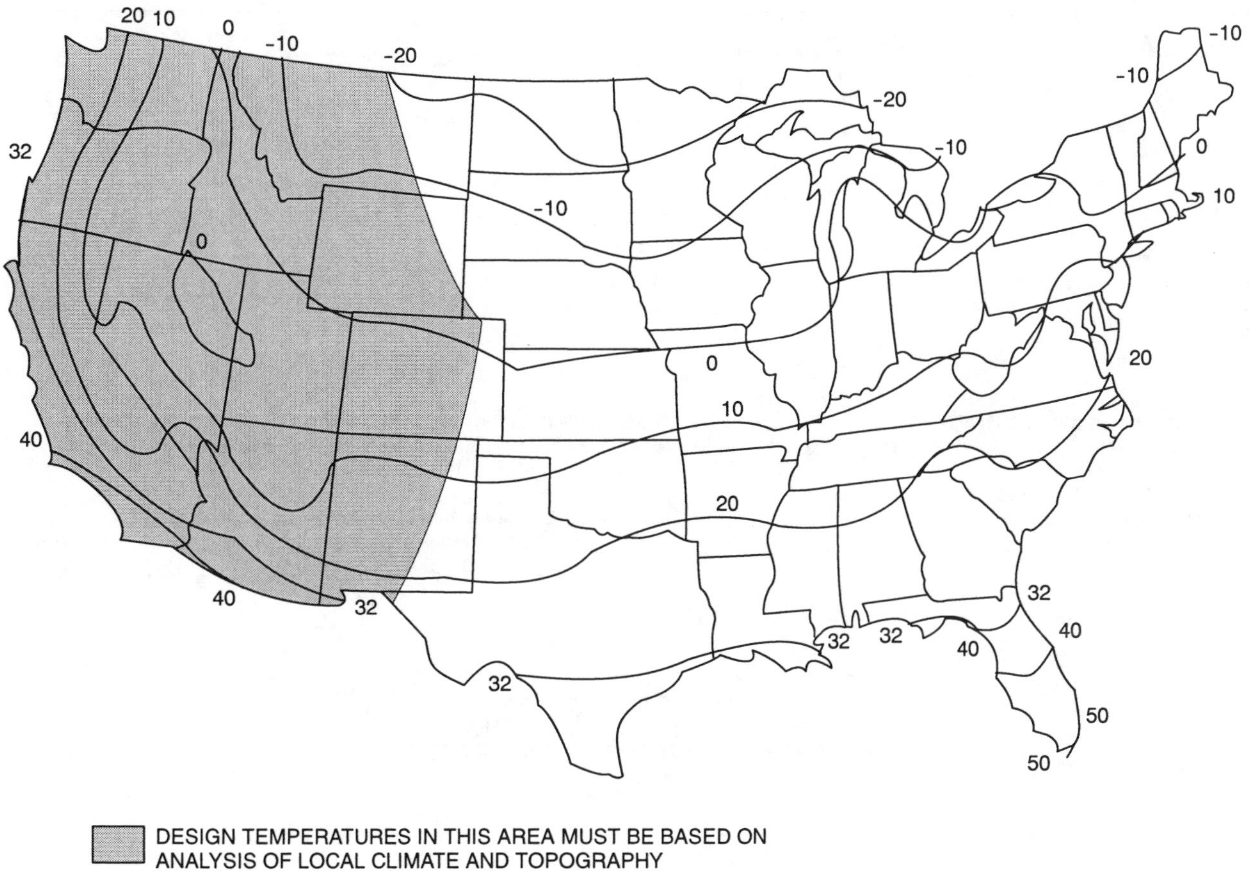

For SI: °C = [(°F)-32]/1.8.

FIGURE R301.2(1)
ISOLINES OF THE 97$^1/_2$ PERCENT WINTER (DECEMBER, JANUARY AND FEBRUARY) DESIGN TEMPERATURES (°F)

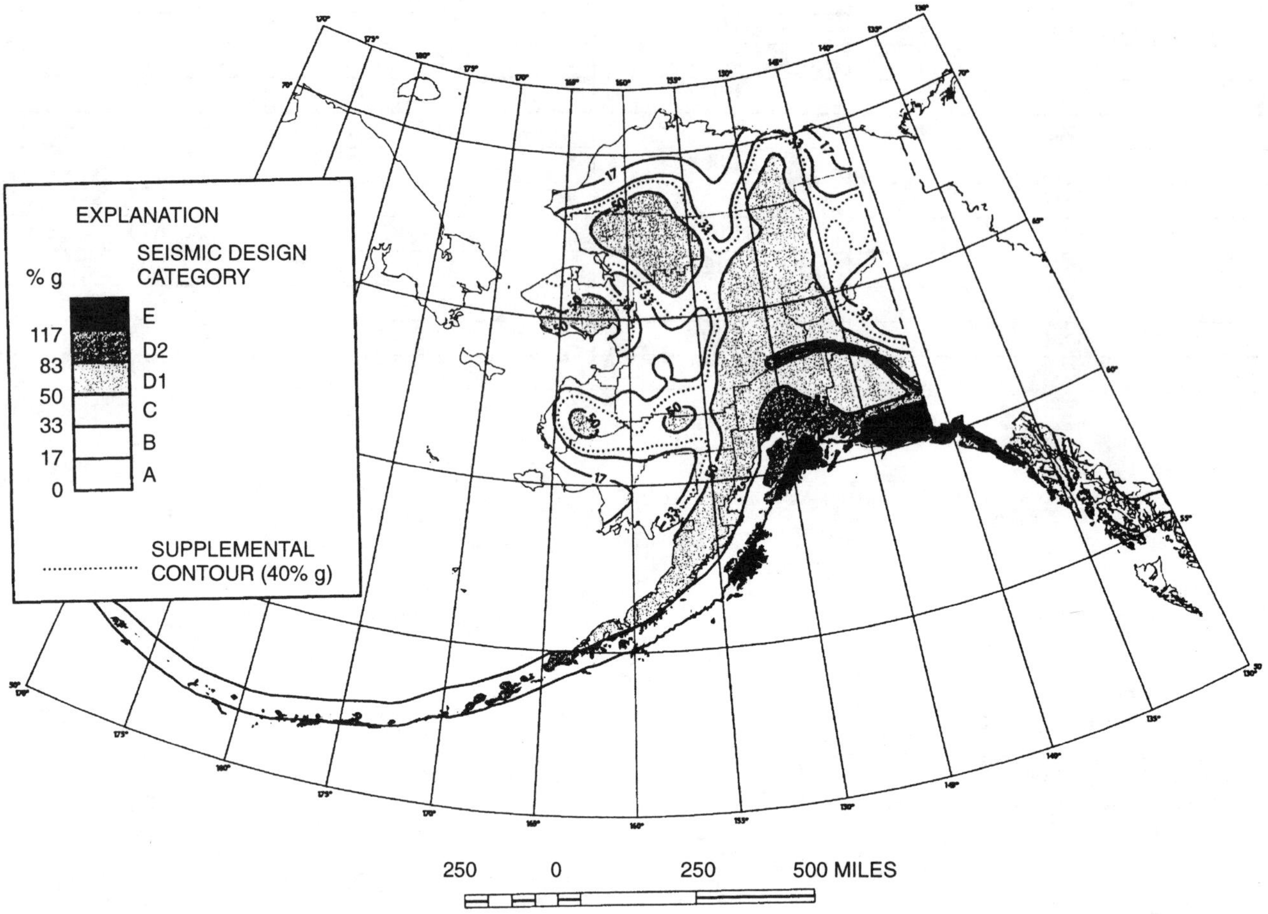

Prepared by U.S. Geological Survey

Scale 1:17,000,000

Digital data prepared with ARC/INFO 7.1.1
running under Solaris 2.5 on a UNIX workstation

Albers Equal-Area Conic Projection
Standard parallels 55°N and 65°N
Central Meridian 160°W

REFERENCES

U.S. Geological Survey National Seismic-Hazard Mapping Project, 1998 Alaska Seismic-Hazard Maps: Documentation: U.S. Geological Survey Open-File Report, in progress.

U.S. Geological Survey National Seismic-Hazard Mapping Project, 1998, Seismic-Hazard Maps of Alaska: U.S. Geological Survey Open-File Report, 6 sheet, scale 1:5,000,000, in progress.

For SI: 1 mile = 1.61 km.

FIGURE R301.2(2)
INTERNATIONAL RESIDENTIAL CODE
SEISMIC DESIGN CATEGORIES—SITE CLASS D
(continued)

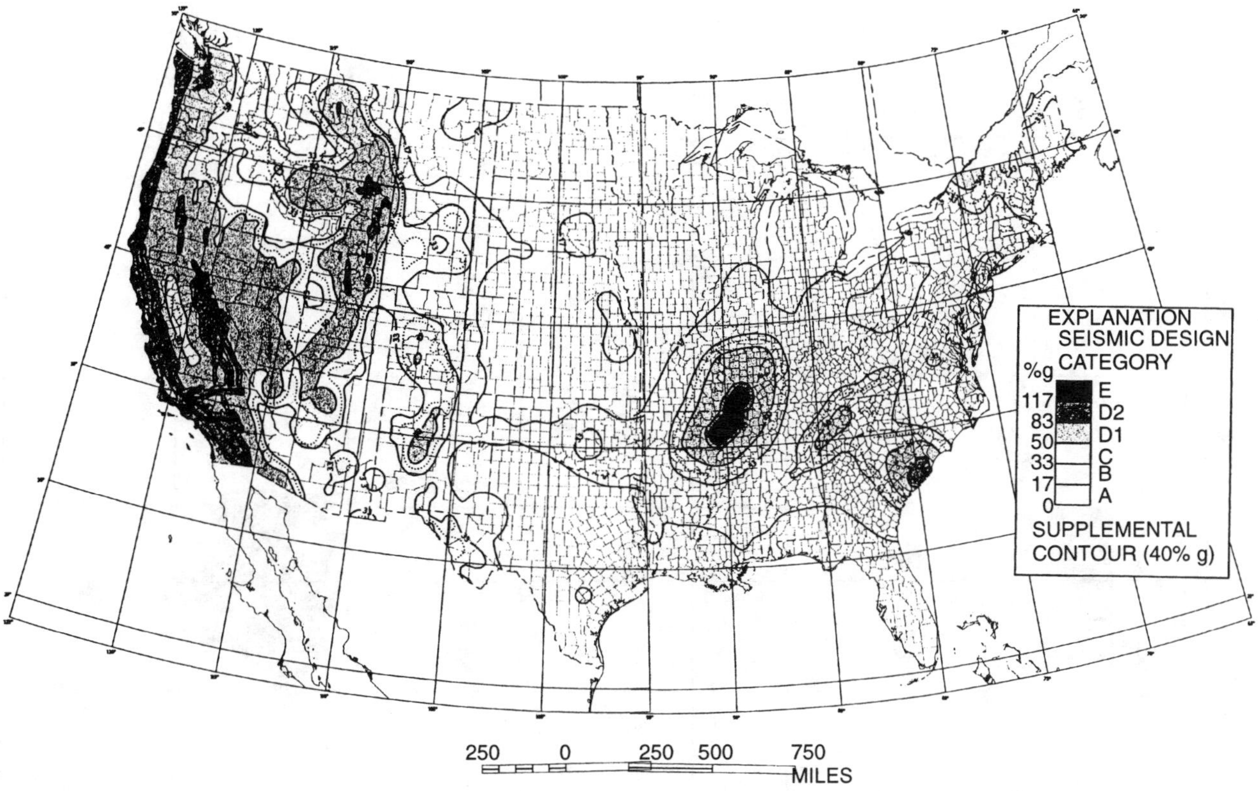

SCALE 1:15,000,000

Digital data prepared with ARC/INFO 7.1.1 running under Solaris 2.5 on a UNIX workstation

Albers Equal-Area Conic Projection
Standard Parallels 29.5°N and 45.5°N
Central Meridian 95°W

Prepared by U.S. Geological Survey

For SI: 1 mile = 1.61 km.

REFERENCES

Frankel, A. Mueller, C., Barnhard, T., Perkins, D., Leyendecker, E. V., Dickman, N., Hanson, S., and Hopper, M., 1996, National Seismic-Hazard Maps: Documentation June 1996: U.S. Geological Survey Open-File Report 96-532, 110 p.

Frankel, A., Muller, C., Barnhard, T., Perkins, D., Leyendecker, E. V., Dickman, N., Hanson, S., and Spectral Response Acceleration for 0.2 Second Period with 2% Probability of Exceedance in 50 Years: U. S. Geological Survey Open-File Report 97-131-F, scale 1:7,000,000.

Petersen, M., Bryant, W., Cramer, C., Cao, T., Reichle, M., Frankel, A., Lienkaemper, J., McCrory, P., and Schwartz, D., 1996, Probabilitic Seismic Hazard Assessment for the State of California: California Division of Mines and Geology Open-File Report 96-08, 66 p., and U.S. Geological Survey Open-File Report 96-706, 66 p.

FIGURE R301.2(2)—continued
INTERNATIONAL RESIDENTIAL CODE
SEISMIC DESIGN CATEGORIES—SITE CLASS D
(continued)

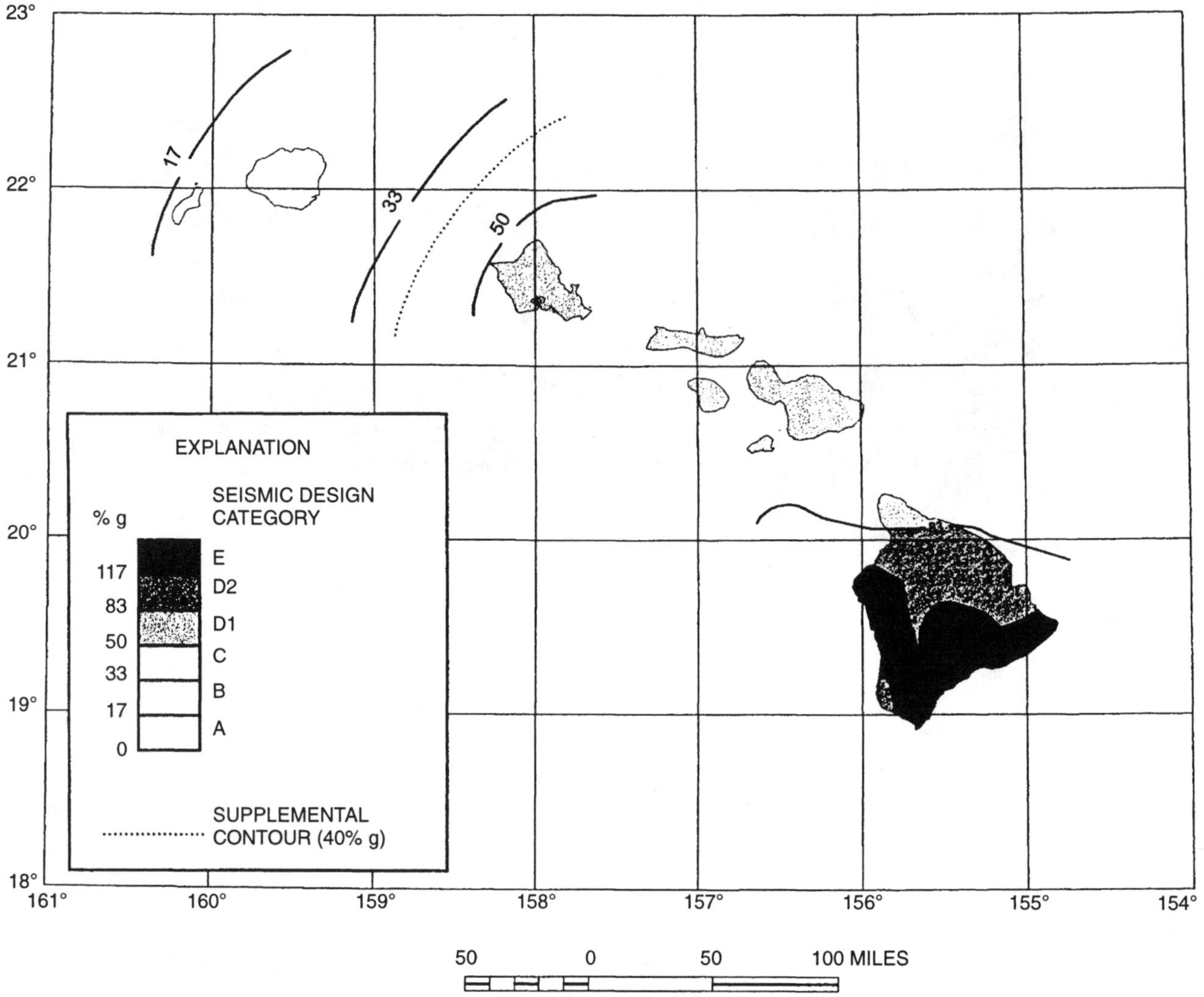

Prepared by U.S. Geological Survey

SCALE 1:3,500,000

Digital data prepared with ARC/INFO 7.1.1 running under Solaris 2.5 on a UNIX workstation

Albers Equal-Area Conic Projection
Standard parallels 8°N and 18°N
Central Meridian 157.5°W

For SI: 1 mile = 1.61 km.

REFERENCES

U.S. Geological Survey National Seismic-Hazard Mapping Project, 1998 Hawaii Seismic-Hazard Maps:
Documentation: U.S. Geological Survey Open-File Report, in progress.

U.S. Geological Survey National Seismic-Hazard Mapping Project, 1998, Seismic-Hazard Maps of Hawaii: U.S. Geological Survey Open-File Report, 6 sheet, scale 1:2,000,000, in progress.

FIGURE R301.2(2)—continued
INTERNATIONAL RESIDENTIAL CODE
SEISMIC DESIGN CATEGORIES—SITE CLASS D

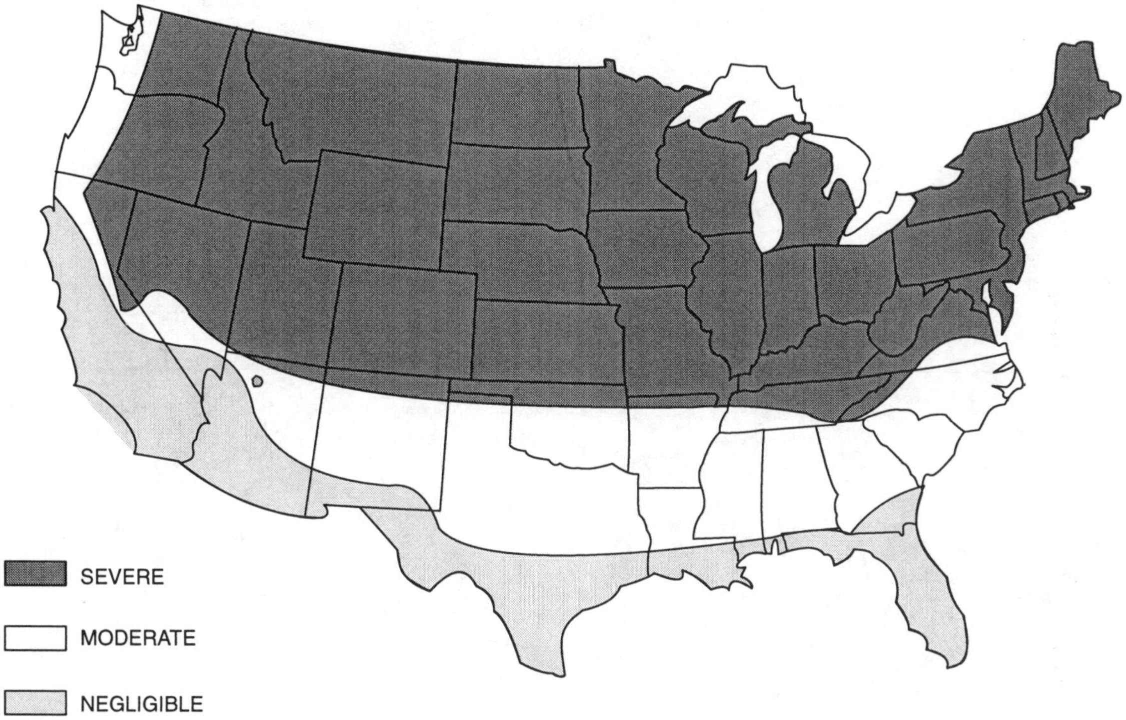

a. Alaska and Hawaii are classified as severe and negligible, respectively.
b. Lines defining areas are approximate only. Local conditions may be more or less severe than indicated by region classification. A severe classification is where weather conditions result in significant snowfall combined with extended periods during which there is little or no natural thawing causing deicing salts to be used extensively.

FIGURE R301.2(3)
WEATHERING PROBABILITY MAP FOR CONCRETE

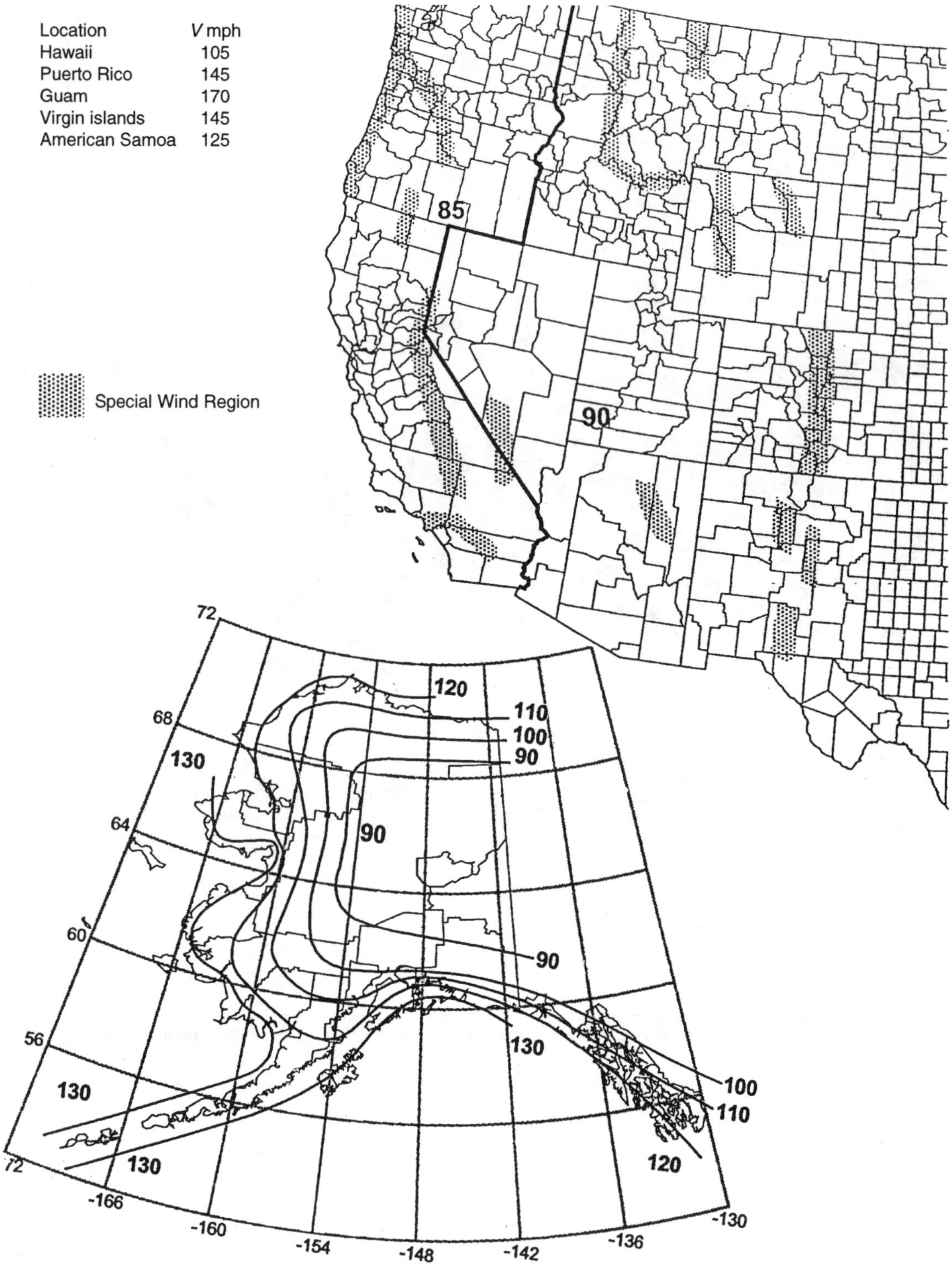

For SI: 1 foot = 304.8 mm, 1 mile per hour = 0.447 m/s.

a. Values are nominal design 3-second gust wind speeds in miles per hour at 33 feet above ground for Exposure C category.
b. Linear interpolation between wind contours is permitted.
c. Islands and coastal areas outside the last contour shall use the last wind speed contour of the coastal area.
d. Mountainous terrain, gorges, ocean promontories, and special wind regions shall be examined for unusual wind conditions.

FIGURE R301.2(4)
BASIC WIND SPEEDS FOR 50-YEAR MEAN RECURRENCE INTERVAL
(continued)

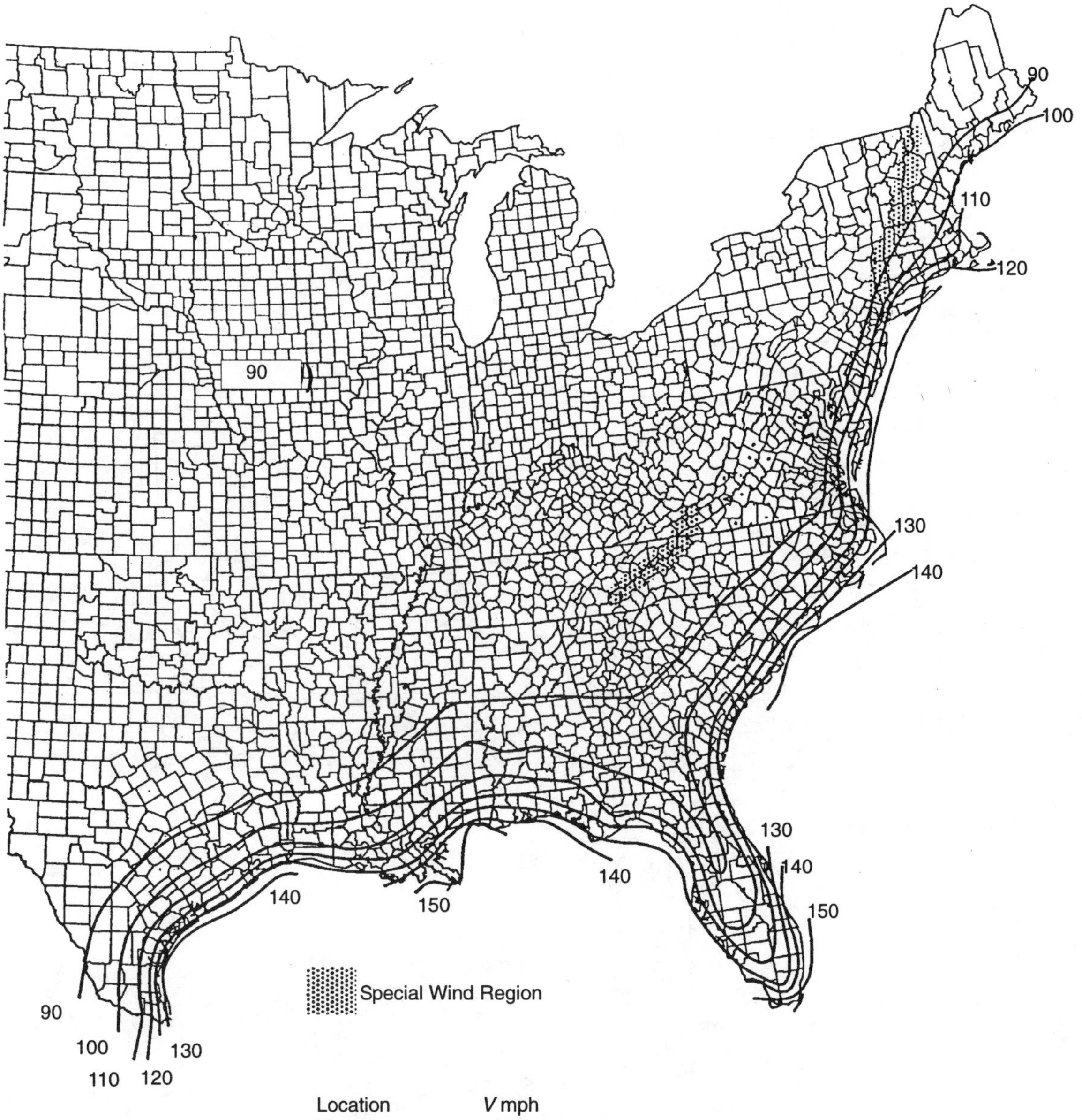

Location	*V* mph
Hawaii	105
Puerto Rico	145
Guam	170
Virgin islands	145
American Samoa	125

For SI: 1 foot = 304.8 mm, 1 mile per hour = 0.447 m/s.

a. Values are nominal design 3-second gust wind speeds in miles per hour at 33 feet above ground for Exposure C category.
b. Linear interpolation between wind contours is permitted.
c. Islands and coastal areas outside the last contour shall use the last wind speed contour of the coastal area.
d. Mountainous terrain, gorges, ocean promontories, and special wind regions shall be examined for unusual wind conditions.

FIGURE R301.2(4)—continued
BASIC WIND SPEEDS FOR 50-YEAR MEAN RECURRENCE INTERVAL
(continued)

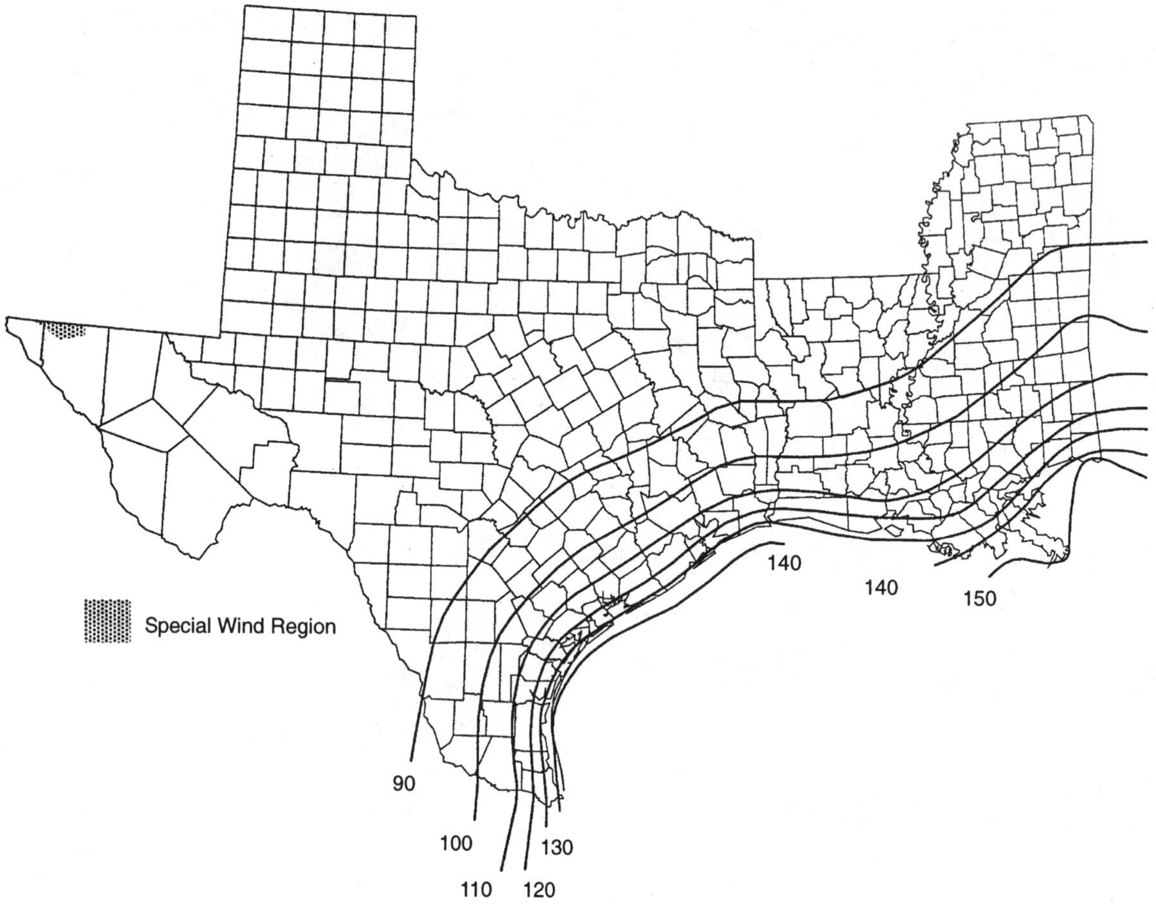

For SI: 1 foot = 304.8 mm, 1 mile per hour = 0.447 m/s.

a. Values are nominal design 3-second gust wind speeds in miles per hour at 33 feet above ground for Exposure C category.
b. Linear interpolation between wind contours is permitted.
c. Islands and coastal areas outside the last contour shall use the last wind speed contour of the coastal area.
d. Mountainous terrain, gorges, ocean promontories, and special wind regions shall be examined for unusual wind conditions.

FIGURE R301.2(4)—continued
BASIC WIND SPEEDS FOR 50-YEAR MEAN RECURRENCE INTERVAL
(continued)

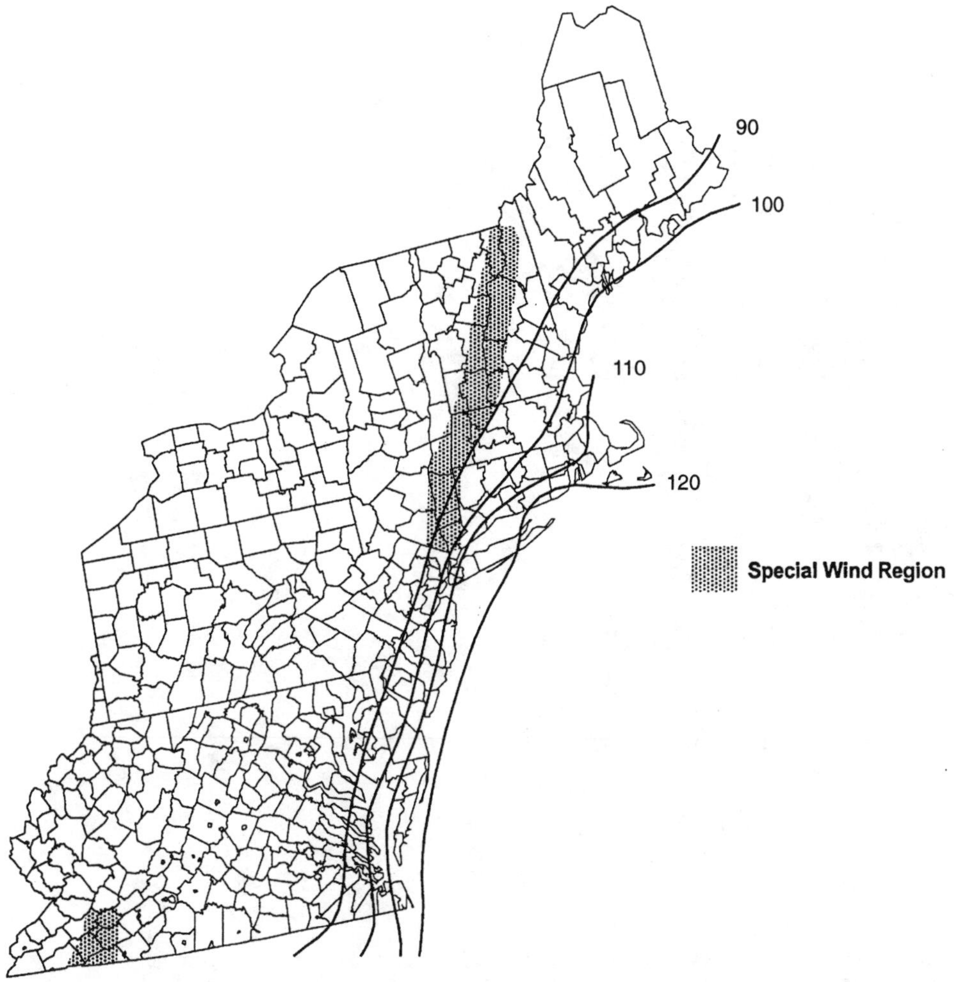

For SI: 1 foot = 304.8 mm, 1 mile per hour = 0.447 m/s.

a. Values are nominal design 3-second gust wind speeds in miles per hour at 33 feet above ground for Exposure C category.
b. Linear interpolation between wind contours is permitted.
c. Islands and coastal areas outside the last contour shall use the last wind speed contour of the coastal area.
d. Mountainous terrain, gorges, ocean promontories, and special wind regions shall be examined for unusual wind conditions.

FIGURE R301.2(4)—continued
BASIC WIND SPEEDS FOR 50-YEAR MEAN RECURRENCE INTERVAL
(continued)

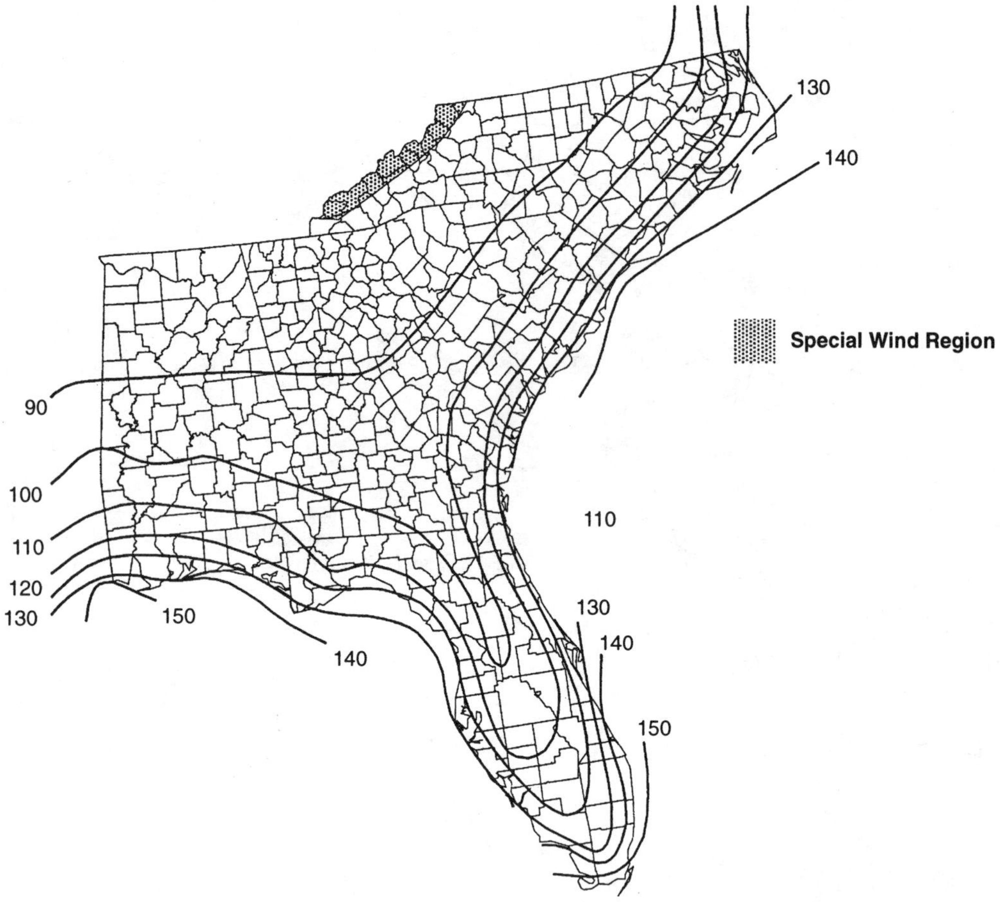

For SI: 1 foot = 304.8 mm, 1 mile per hour = 0.447 m/s.

a. Values are nominal design 3-second gust wind speeds in miles per hour at 33 feet above ground for Exposure C category.
b. Linear interpolation between wind contours is permitted.
c. Islands and coastal areas outside the last contour shall use the last wind speed contour of the coastal area.
d. Mountainous terrain, gorges, ocean promontories, and special wind regions shall be examined for unusual wind conditions.

FIGURE R301.2(4)—continued
BASIC WIND SPEEDS FOR 50-YEAR MEAN RECURRENCE INTERVAL

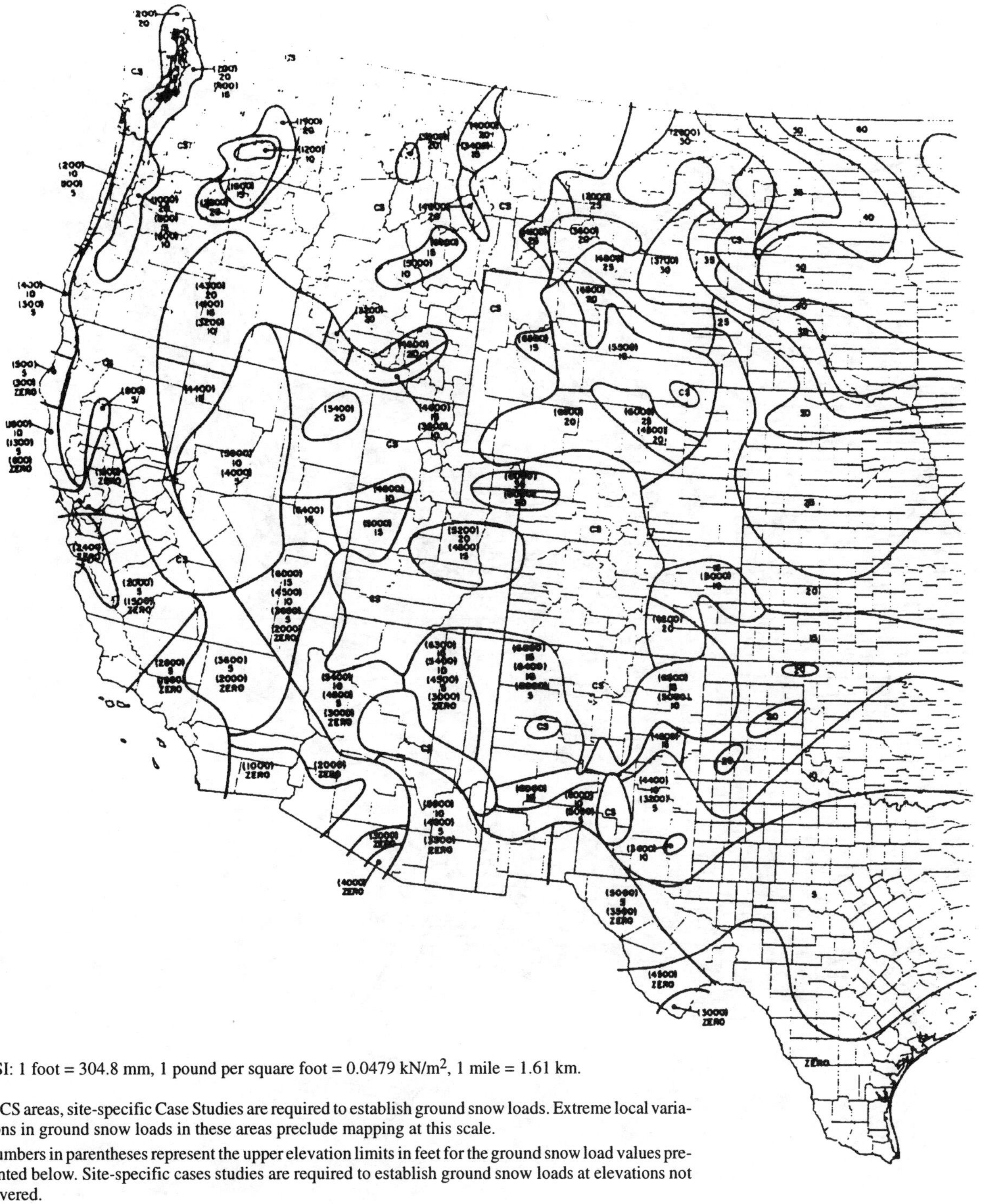

For SI: 1 foot = 304.8 mm, 1 pound per square foot = 0.0479 kN/m^2, 1 mile = 1.61 km.

a. In CS areas, site-specific Case Studies are required to establish ground snow loads. Extreme local variations in ground snow loads in these areas preclude mapping at this scale.

b. Numbers in parentheses represent the upper elevation limits in feet for the ground snow load values presented below. Site-specific cases studies are required to establish ground snow loads at elevations not covered.

0 100 200 300 MILES

FIGURE R301.2(5)
GROUND SNOW LOADS, P_g, FOR THE UNITED STATES (lb/ft^2)
(continued)

For SI: 1 foot = 304.8 mm, 1 pound per square foot = 0.0479 kN/m^2, 1 mile = 1.61 km.

FIGURE R301.2(5)—continued
GROUND SNOW LOADS, P_g, FOR THE UNITED STATES (lb/ft^2)

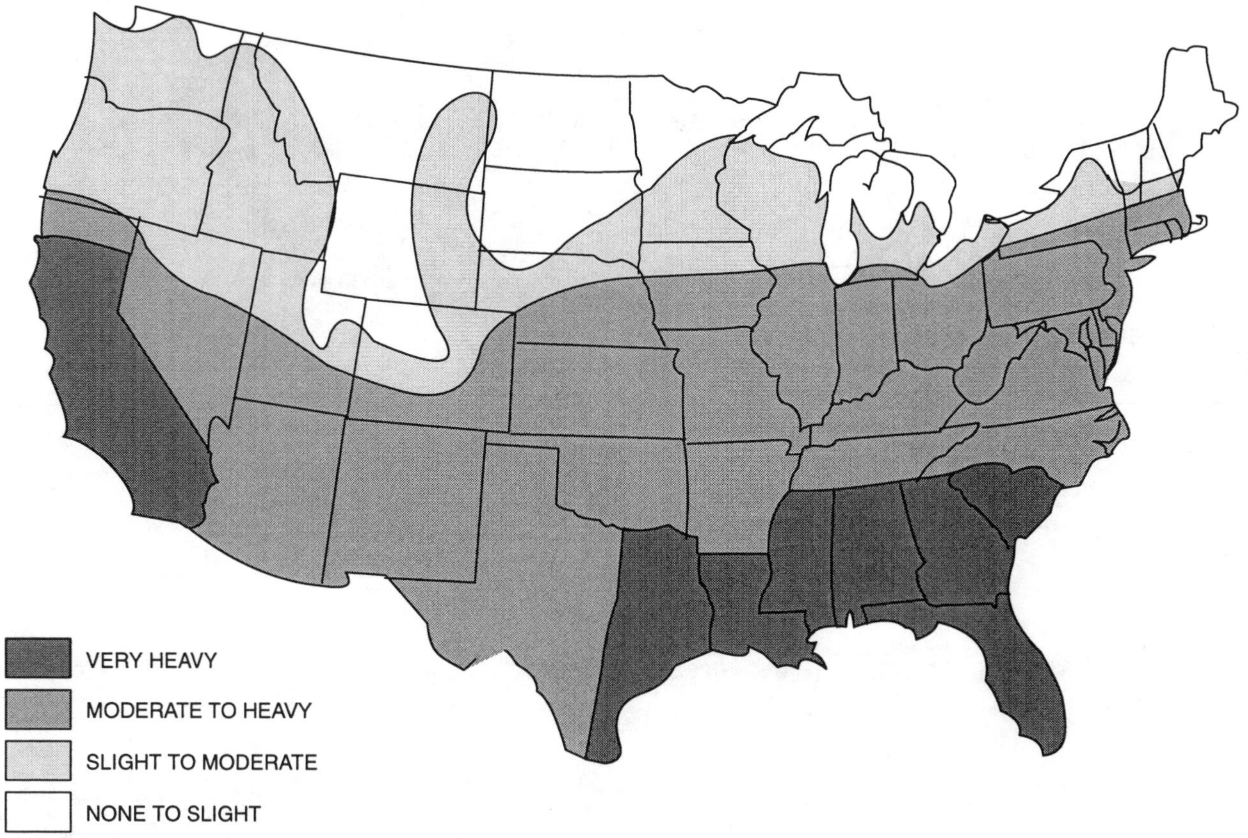

Note: Lines defining areas are approximate only. Local conditions may be more or less severe than indicated by the region classification.

FIGURE R301.2(6)
TERMITE INFESTATION PROBABILITY MAP

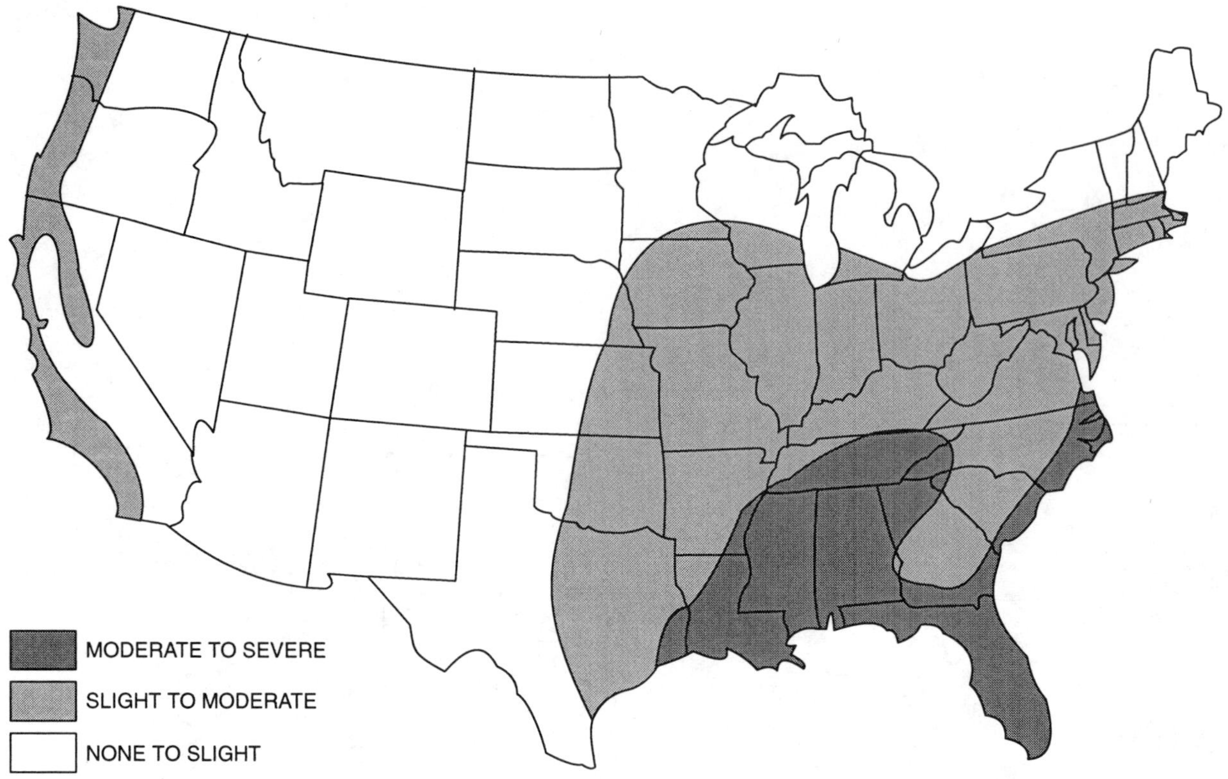

Notes: Lines defining areas are approximate only. Local conditions may be more or less severe than indicated by the region classification.

FIGURE R301.2(7)
DECAY PROBABILITY MAP

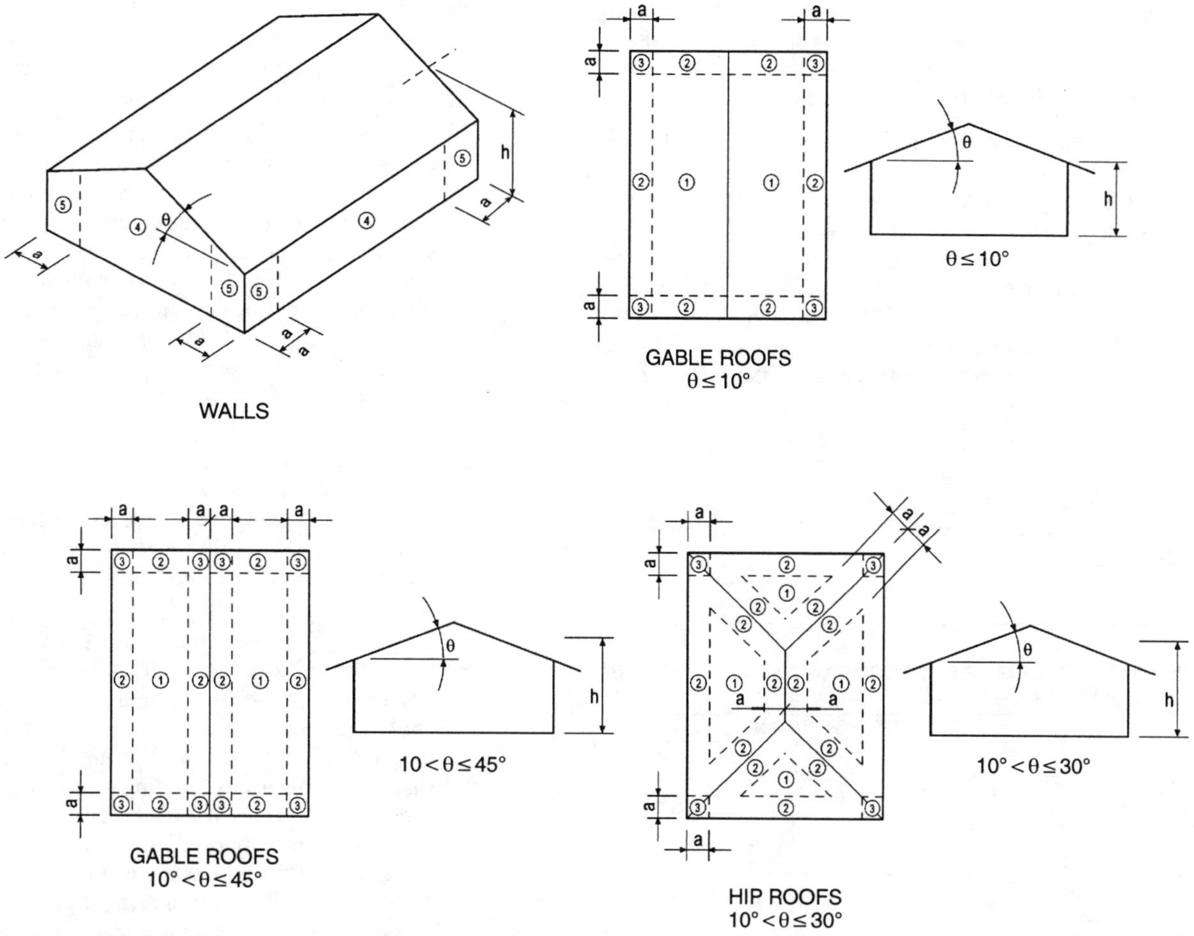

For SI: 1 foot = 304.8 mm, 1 degree = 0.009 rad.

Note: a = 4 feet in all cases

FIGURE R301.2(8)
COMPONENT AND CLADDING PRESSURE ZONES

R301.2.1.2 Internal pressure. Windows in buildings located in wind borne debris regions shall have glazed openings protected from windborne debris or the building shall be designed as a partially enclosed building in accordance with the *International Building Code.* Glazed opening protection for windborne debris shall meet the requirements of the Large Missile Test of ASTM E 1996 and of ASTM E 1886 referenced therein.

Exception: Wood structural panels with a minimum thickness of $^7/_{16}$ inch (11.1 mm) and a maximum span of 8 feet (2438 mm) shall be permitted for opening protection in one- and two-story buildings. Panels shall be precut to cover the glazed openings with attachment hardware provided. Attachments shall be provided in accordance with Table R301.2.1.2 or shall be designed to resist the components and cladding loads determined in accordance with the provisions of the *International Building Code.*

R301.2.1.3 Wind speed conversion. When referenced documents are based on fastest mile wind speeds, the three second gust wind velocities of Figure R301.2(4) shall be converted to fastest mile wind velocities using Table R301.2.1.3.

TABLE R301.2.1.2
WINDBORNE DEBRIS PROTECTION FASTENING SCHEDULE FOR WOOD STRUCTURAL PANELS[a,b,c,d]

FASTENER TYPE	FASTENER SPACING		
	Panel span ≤ 4 foot	4 foot < panel span ≤ 6 foot	6 foot < panel span ≤ 8 foot
$2^1/_2$″ #6 Wood screws	16″	12″	9″
$2^1/_2$″ #8 Wood screws	16″	16″	12″

For SI: 1 inch = 25.4 mm, 1 foot = 304.8 mm, 1 pound = 0.454 kg, 1 mile per hour = 1.609 km/h.

a. This table is based on 110 mph wind speeds and a 33-foot mean roof height.

b. Fasteners shall be installed at opposing ends of the wood structural panel.

c. Nails shall be 10d common or 12d box nails.

d. Where screws are attached to masonry or masonry/stucco, they shall be attached utilizing vibration-resistant anchors having a minimum ultimate withdrawal capacity of 490 pounds.

R301.2.1.4 Exposure category. For each wind direction considered, an exposure category that adequately reflects the characteristics of ground surface irregularities shall be determined for the site at which the building or structure is to be constructed. For a site located in the transition zone between categories, the category resulting in the largest wind forces shall apply. Account shall be taken of variations in ground surface roughness that arise from natural topography and vegetation as well as from constructed features. For any given wind direction, the exposure in which a specific building or other structure is sited shall be assessed as being one of the following categories:

1. Exposure A. Large city centers with at least 50 percent of the buildings having a height in excess of 70 feet (21 336 mm). Use of this exposure category shall be limited to those areas for which terrain representative of Exposure A prevails in the upwind direction for a distance of at least 0.5 mile (0.8 km) or 10 times the height of the building or other structure, whichever is greater. Possible channeling effects or increased velocity pressures due to the building or structure being located in the wake of adjacent buildings shall be taken into account.
2. Exposure B. Urban and suburban areas, wooded areas, or other terrain with numerous closely spaced obstructions having the size of single-family dwellings or larger. Exposure B shall be assumed unless the site meets the definition of another type exposure.
3. Exposure C. Open terrain with scattered obstructions, including surface undulations or other irregularities, having heights generally less than 30 feet (9144 mm) extending more than 1,500 feet (457 m) from the building site in any quadrant. This exposure shall also apply to any building located within Exposure B type terrain where the building is directly adjacent to open areas of Exposure C type terrain in any quadrant for a distance of more than 600 feet (183 m). This category includes flat open country, grasslands and shorelines in hurricane prone regions.
4. Exposure D. Flat, unobstructed areas exposed to wind flowing over open water (excluding shorelines in hurricane prone regions) for a distance of at least 1 mile (1.61 km). Shorelines in Exposure D include inland waterways, the Great Lakes and coastal areas of California, Oregon, Washington and Alaska. This exposure shall apply only to those buildings and other structures exposed to the wind coming from over the water. Exposure D extends inland from the shoreline a distance of 1,500 feet (457 m) or 10 times the height of the building or structure, whichever is greater.

TABLE R301.2.1.3
EQUIVALENT BASIC WIND SPEEDS[a]

3-second gust	85	90	100	105	110	120	125	130	140	145	150	160	170
Fastest mile	70	75	80	85	90	100	105	110	120	125	130	140	150

For SI: 1 mile per hour = 1.609 km/h.

a. Linear interpolation is permitted.

R301.2.2 Seismic provisions. The seismic provisions of this code shall apply to buildings constructed in Seismic Design Categories C, D_1, and D_2, as determined in accordance with this section.

Exception: Detached one- and two-family dwellings located in Seismic Design Category C are exempt from the seismic requirements of this code.

The weight limitations of Section R301.2.2.4 shall apply to buildings in all Seismic Design Categories regulated by this code. Buildings in Seismic Design Category C shall be constructed in accordance with the additional requirements of Sections R301.2.2.3 and R301.2.2.4. Buildings in Seismic Design Categories D_1 and D_2 shall be constructed in accordance with the additional requirements of Sections R301.2.2.5 through R301.2.2.9. Buildings in Seismic Design Category E shall be designed in accordance with the *International Building Code*.

R301.2.2.1 Determination of seismic design category. Buildings shall be assigned a Seismic Design Category in accordance with Figure 301.2(2).

R301.2.2.1.1 Alternate determination of seismic design category. The Seismic Design Categories and corresponding Short Period Design Spectral Response Accelerations, S_{DS} shown in Figure R301.2(2) are based on soil Site Class D, as defined in Section 1615.1 of the *International Building Code*. If soil conditions are other than Site Class D, the Short Period Design Spectral Response Acceleration, S_{DS}, for a site can be determined according to Section 1615.1 of the *International Building Code*. The value of S_{DS} determined according to Section 1615.1 of the *International Building Code* is permitted to be used to set the Seismic Design Category according to Table R301.2.2.1.1, and to interpolate between values in Tables R602.10.1, R603.7, and other seismic design requirements of this code.

TABLE R301.2.2.1.1
SEISMIC DESIGN CATEGORY DETERMINATION

CALCULATED S_{DS}	SEISMIC DESIGN CATEGORY
$S_{DS} \leq 0.17g$	A
$0.17g < S_{DS} \leq 0.33g$	B
$0.33g < S_{DS} \leq 0.50g$	C
$0.50g < S_{DS} \leq 0.83g$	D_1
$0.83g < S_{DS} \leq 1.17g$	D_2
$1.17g < S_{DS}$	E

R301.2.2.1.2 Alternative determination of Seismic Design Category E. Buildings located in Seismic Design Category E in accordance with Figure R301.2(2) are permitted to be reclassified as being in Seismic Design Category D_2 provided one of the following is done:

1. A more detailed evaluation of the Seismic Design Category is made in accordance with the provisions and maps of the *International Building Code.* Buildings located in Seismic Design Category E per Table R301.2.2.1.1, but located in Seismic Design Category D per the *International Building Code*, may be designed using the Seismic Design Category D_2 requirements of this code.
2. Buildings located in Seismic Design Category E that conform to the following additional restrictions are permitted to be constructed in accordance with the provisions for Seismic Design Category D_2 of this code:
   - 2.1. All exterior shear wall lines or braced wall panels are in one plane vertically from the foundation to the uppermost story.
   - 2.2. Floors shall not cantilever past the exterior walls.
   - 2.3. The building is within all of the requirements of Section R301.2.2.6 for being considered as regular.

R301.2.2.2 Determination of seismic design category. It is permitted to determine the Seismic Design Category of a building in accordance with Figure R301.2(2), based on a default Site Class D soil. If soil conditions are other than Site Class D, as defined in the *International Building Code*, the Short Period Design Spectral Response Acceleration, S_{DS}, for a site shall be determined according to the *International Building Code.* The value of S_{DS} determined according to this section is permitted to be used to set the Seismic Design Category according to the legend of Figure R301.2(2) and to interpolate between values in Tables R602.10.3 and R603.7, and other seismic design requirements of this code.

R301.2.2.3 Anchored stone and masonry veneer in Seismic Design Category C. Anchored stone and masonry veneer in Seismic Design Category C shall be limited to the first story above grade and shall not exceed 5 inches (127 mm) in thickness.

Exception: In Seismic Design Category C, anchored stone and masonry veneer not exceeding 5 inches (127 mm) in thickness shall be permitted to the height allowed in Section R703.7. In other than the topmost story, the length of wall bracing shall be 1.5 times the length otherwise required in Table R602.10.1.

R301.2.2.4 Weights of materials. Average dead loads shall not exceed 15 psf (0.72 kN/m^2) for roofs/ceiling assemblies or 10 psf (0.48 kN/m^2) for floor assemblies, except as further limited by Section R301.2.2. Dead loads for walls above grade shall not exceed:

1. Fifteen psf (0.72 kN/m^2) for exterior light-frame wood walls.
2. Fourteen psf (0.67 kN/m^2) for exterior light-frame cold-formed steel walls.
3. Ten psf (0.48 kN/m^2) for interior light-frame wood walls.
4. Five psf (0.24 kN/m^2) for interior light-frame cold-formed steel walls.
5. Eighty psf (3.83 kN/m^2) for 8-inch-thick (203 mm) masonry walls.
6. Eighty-five psf (4.07 kN/m^2) for 6-inch-thick (152 mm) concrete walls.

Exception: Roof/ceiling dead loads not exceeding 25 psf (1.19 kN/m^2) shall be permitted in Seismic Design Category D_2 provided the wall bracing amounts in Chapter 6 are increased in accordance with Table R301.2.2.4.

TABLE R301.2.2.4
WALL BRACING ADJUSTMENT FACTORS BY ROOF COVERING DEAD LOAD[a]

WALL SUPPORTING	ROOF/CEILING DEAD LOAD 15 psf or less	ROOF/CEILING DEAD LOAD 25 psf
Roof only	1.0	1.2
Roof plus one story	1.0	1.1

For SI: 1 pound per square foot = 0.0479 kN/m^2.

a. Linear interpolation shall be permitted.

R301.2.2.5 Masonry construction in Seismic Design Category C. Masonry construction in Seismic Design Category C shall comply with the requirements of Section R606.11.2.

R301.2.2.6 Height limitations in Seismic Design Categories D_1 and D_2. Wood frame buildings shall be limited to three stories above grade or the limits given in Table R602.10.3. Cold-formed steel frame buildings shall be limited to two stories above grade in accordance with Sections R505.1.1, R603.1.1 and R804.1.1. Masonry construction in Seismic Design Category D_1 shall be limited in accordance with Section R606.11.3. Masonry construction in Seismic Design Category D_2 shall be limited in accordance with Section R606.11.4. Mezzanines as defined in Section R202 shall not be considered as stories.

R301.2.2.7 Irregular buildings. Conventional light-frame construction shall not be used in irregular portions of structures in Seismic Design Categories C, D_1 and D_2. Only such irregular portions of structures shall be designed in accordance with accepted engineering practice to the extent such irregular features affect the performance of the conventional framing system. A portion of a building shall be considered to be irregular when one or more of the following conditions occur:

1. Exterior shear wall lines or braced wall panels are not in one plane vertically from the foundation to the uppermost story in which they are required.

   Exception: For wood light-frame construction, floors with cantilevers or setbacks not exceeding four times the nominal depth of the wood floor joists are permitted to support braced wall panels that are out of plane with braced wall panels below provided that:

   1.1. Floor joists are nominal 2 inches by 10 inches (51 mm by 254 mm) or larger and spaced not more than 16 inches (406 mm) on center.

   1.2. The ratio of the back span to the cantilever is at least 2 to 1.

   1.3. Floor joists at ends of braced wall panels are doubled.

   1.4. For wood frame construction, a continuous rim joist is connected to ends of all cantilever joists. When spliced, the rim joists shall be spliced using a galvanized metal tie not less than 0.058 inch (1.47 mm) (16 gage) and $1^1/_2$ inches (38 mm) wide fastened with six 16d nails on each side of the splice or a block of the same size as the rim joist of sufficient length to fit securely between the joist space at which the splice occurs fastened with eight 16d nails on each side of the splice; and

   1.5. Gravity loads carried at the end of cantilevered joists are limited to uniform wall and roof load and the reactions from headers having span of 8 feet (2438 mm) or less.

2. A section of floor or roof is not laterally supported by shear walls or braced wall lines on all edges.

   Exception: Portions of floors that do not support shear walls or braced wall panels above, or roofs, shall be permitted to extend no more than 6 feet (1829 mm) beyond a shear wall or braced wall line.

3. The end of a braced wall panel occurs over an opening in the wall below and ends at a horizontal distance greater than 1 foot (305 mm) from the edge of the opening. This provision is applicable to shear walls and braced wall panels offset in plane and to braced wall panels offset out of plane as permitted by the exception to Item 1 above.

   Exception: For light-frame construction, a braced wall panel shall be permitted to extend more than 1 foot (305 mm) over an opening in the wall below provided that the opening includes a header in accordance with Chapter 6 of this code. The entire length of the braced wall panel shall not occur over an opening in the wall below.

4. An opening in a floor or roof exceeds the lesser of 12 feet (3657 mm) or 50 percent of the least floor or roof dimension.

5. Portions of a floor level are vertically offset.

   Exceptions:

   1. Framing supported directly by continuous foundations at the perimeter of the building.
   2. For wood light-frame construction, floors shall be permitted to be vertically offset when the floor framing is lapped or tied together as required by Section R502.6.1.

6. Shear walls and braced wall lines do not occur in two perpendicular directions.

7. Shear walls or braced wall lines are constructed of dissimilar bracing systems on any one story level above grade.

R301.2.2.8 Concrete construction in Seismic Design Categories D_1 and D_2. Buildings with above-grade concrete walls in Seismic Design Categories D_1 and D_2 shall be designed in accordance with accepted engineering practice.

R301.2.2.9 Irregular buildings in Seismic Design Categories D_1 and D_2. In Seismic Design Categories C, D_1

and D_2, irregular buildings shall have an engineered lateral-force-resisting system designed in accordance with accepted engineering practice. A building shall be considered to be irregular when one or more of the following conditions occur:

1. Exterior shear wall lines or braced wall panels are not in one plane vertically from the foundation to the uppermost story in which they are required.

   Exception: For light-frame construction, floors with cantilevers or setbacks not exceeding four times the nominal depth of the wood floor joists or 24 inches (610 mm) for cold-formed steel joists are permitted to support braced wall panels that are out of plane with braced wall panels below provided that:

   1. Floor joists are nominal 2 inches by 10 inches (51 mm by 254 mm) or larger and spaced not more than 16 inches (406 mm) on center;
   2. The ratio of the back span to the cantilever is at least 2 to 1;
   3. Floor joists at ends of braced wall panels are doubled;
   4. For wood-frame construction, a continuous rim joist is connected to ends of all cantilever joists. When spliced, the rim joists shall be spliced using a galvanized metal tie not less than 0.058 inch (1.47 mm) (16 gage) and 1.5 inches (38 mm) wide fastened with six 16d nails on each side; and
   5. Gravity loads carried at the end of cantilevered joists are limited touniform wall and roof load and the reactions from headers having span of 8 feet (2438 mm) or less.

2. A section of floor or roof is not laterally supported by shear walls or braced wall lines on all edges.

   Exception: Portions of floors which do not support shear walls or braced wall panels above, or roofs, shall be permitted to extend no more than 6 feet (1829 mm) beyond a shear wall or braced wall line.

3. The end of a braced wall panel occurs over an opening in the wall below and ends at a horizontal distance greater than 1 foot (305 mm) from the edge of the opening. This provision is applicable to shear walls and braced wall panels offset in plane and to braced wall panels offset out of plane as permitted by the exception to Item 1 above.

   Exception: For light-frame construction, a braced wall panel shall be permitted to extend more than 1 foot (305 mm) over an opening in the wall below provided that the opening includes a header in accordance with Chapter 6 of this code. The entire length of the braced wall panel shall not occur over an opening in the wall below.

4. An opening in a floor or roof exceeds the lesser of 12 feet (3657 mm) or 50 percent of the least floor or roof dimension.

5. Portions of a floor level are vertically offset.

   Exceptions:

   1. Framing supported directly by continuous foundations at the perimeter of the building.
   2. For light-frame construction, floors shall be permitted to be vertically offset when the floor framing is lapped or tied together as required by Section R502.6.1 for wood framing and Section R505.3.1 for cold-formed steel framing.

6. Shear walls and braced wall lines do not occur in two perpendicular directions.

7. Shear walls or braced wall lines are constructed of dissimilar bracing systems on any one-story level above grade.

R301.2.3 Snow loads. Wood frame construction, cold-formed steel frame construction and masonry and concrete construction in regions with ground snow loads 70 psf (3.35 kN/m^2) or less shall be in accordance with Chapters 5, 6 and 8. Buildings in regions with ground snow loads greater than 70 psf (3.35 kN/m^2) shall be designed in accordance with accepted engineering practice.

R301.2.4 Floodplain construction. Buildings and structures constructed in flood hazard areas (including A or V Zones) as established in Table R301.2(1) shall be designed and constructed in accordance with Section R327.

Exception: Buildings in floodways that are designated on the Flood Insurance Rate Maps (FIRM) or the Flood Boundary and Floodway Maps (FBFM) that are provided by the National Flood Insurance Program shall not be approved under this section; the provisions of the *International Building Code* shall apply.

R301.3 Dead load. The actual weights of materials and construction shall be used for determining dead load with consideration for the dead load of fixed service equipment.

R301.4 Live load. The minimum uniformly distributed live load shall be as provided in Table R301.4.

TABLE R301.4
MINIMUM UNIFORMLY DISTRIBUTED LIVE LOADS
(in pounds per square foot)

USE	LIVE LOAD
Exterior balconies	60
Decks[f]	40
Fire escapes	40
Passenger vehicle garages[a]	50[a]
Attics without storage[b,e]	10
Attics with storage[b,e]	20
Rooms other than sleeping rooms	40
Sleeping rooms	30
Stairs	40[c]
Guardrails and handrails[d]	200

For SI: 1 pound per square foot = 0.0479 kN/m^2, 1 square inch = 645 mm^2, 1 pound = 4.45 N.

a. Elevated garage floors shall be capable of supporting a 2,000-pound load applied over a 20-square-inch area.

b. No storage with roof slope not over 3 units in 12 units.

c. Individual stair treads shall be designed for the uniformly distributed live load or a 300-pound concentrated load acting over an area of 4 square inches, whichever produces the greater stresses.

(continued)

Notes to Table R301.4—continued

d. A single concentrated load applied in any direction at any point along the top.

e. Attics constructed with wood trusses shall be designed in accordance with Section R802.10.1.

f. See Section R502.2.1 for decks attached to exterior walls.

R301.5 Roof load. Roof shall be designed for the live load indicated in Table R301.5 or the snow load indicated in Table R301.2(1), whichever is greater.

TABLE R301.5
MINIMUM ROOF LIVE LOADS IN POUNDS-FORCE PER SQUARE FOOT OF HORIZONTAL PROJECTION

ROOF SLOPE	TRIBUTARY LOADED AREA IN SQUARE FEET FOR ANY STRUCTURAL MEMBER		
	0 to 200	201 to 600	Over 600
Flat or rise less than 4 inches per foot (1:3)	20	16	12
Rise 4 inches per foot (1:3) to less than 12 inches per foot (1:1)	16	14	12
Rise 12 inches per foot (1:1) and greater	12	12	12

For SI: 1 square foot = 0.0929 m^2, 1 pound per square foot = 0.0479 kN/m^2, 1 inch per foot = 0.0833 mm/m.

R301.6 Deflection. The allowable deflection of any structural member under the live load listed in Sections R301.4 and R301.5 shall not exceed the values in Table R301.6.

TABLE R301.6
ALLOWABLE DEFLECTION OF STRUCTURAL MEMBERS

STRUCTURAL MEMBER	ALLOWABLE DEFLECTION
Rafters having slopes greater than 3/12 with no finished ceiling attached to rafters	L/180
Interior walls and partitions	H/180
Floors and plastered ceilings	L/360
All other structural members	L/240
Exterior walls with plaster or stucco finish	H/360
Exterior walls—wind loads[a] with brittle finishes	L/240
Exterior walls—wind loads[a] with flexible finishes	L/120

Note: L = span length, H = span height.

a. The wind load shall be permitted to be taken as 0.7 times the Component and Cladding loads for the purpose of the determining deflection limits herein.

R301.7 Nominal sizes. For the purposes of this code, where dimensions of lumber are specified, they shall be deemed to be nominal dimensions unless specifically designated as actual dimensions.

SECTION R302
LOCATION ON LOT

R302.1 Exterior walls. Exterior walls with a fire separation distance less than 3 feet (914 mm) shall have not less than a one-hour fire-resistive rating with exposure from both sides. Projections shall not extend beyond the distance determined by the following two methods, whichever results in the lesser projections:

1. A point one-third the distance to the property line from an assumed vertical plane located where protected openings are required.
2. More than 12 inches (305 mm) into areas where openings are prohibited.

Projections extending into the fire separation distance shall have not less than one-hour fire-resistive construction on the underside. The above provisions shall not apply to walls which are perpendicular to the line used to determine the fire separation distance.

Exception: Tool and storage sheds, playhouses and similar structures exempted from permits by Section R105.2 are not required to provide wall protection based on location on the lot. Projections beyond the exterior wall shall not extend over the lot line.

R302.2 Openings. Openings shall not be permitted in the exterior wall of a dwelling or accessory building with a fire separation distance less than 3 feet (914 mm). This distance shall be measured perpendicular to the line used to determine the fire separation distance.

Exceptions:

1. Openings shall be permitted in walls that are perpendicular to the line used to determine the fire separation distance.
2. Foundation vents installed in compliance with this code are permitted.

R302.3 Penetrations. Penetrations located in the exterior wall of a dwelling with a fire separation distance less than 3 feet (914 mm) shall be protected in accordance with Section R321.3.

Exception: Penetrations shall be permitted in walls that are perpendicular to the line used to determine the fire separation distance.

SECTION R303
LIGHT, VENTILATION AND HEATING

R303.1 Habitable rooms. All habitable rooms shall be provided with aggregate glazing area of not less than 8 percent of the floor area of such rooms. Natural ventilation shall be through windows, doors, louvers or other approved openings to the outdoor air. Such openings shall be provided with ready access or shall otherwise be readily controllable by the building occupants. The minimum openable area to the outdoors shall be 4 percent of the floor area being ventilated.

Exceptions:

1. The glazed areas need not be openable where the opening is not required by Section R310 and an approved

mechanical ventilation system is provided capable of producing 0.35 air change per hour in the room or a whole-house mechanical ventilation system is installed capable of supplying outdoor ventilation air of 15 cubic feet per minute (cfm) (7.08 L/s) per occupant computed on the basis of two occupants for the first bedroom and one occupant for each additional bedroom.

2. The glazed areas need not be provided in rooms where Exception 1 above is satisfied and artificial light is provided capable of producing an average illumination of 6 footcandles (6.46 lux) over the area of the room at a height of 30 inches (762 mm) above the floor level.

R303.2 Adjoining rooms. For the purpose of determining light and ventilation requirements, any room shall be considered as a portion of an adjoining room when at least one-half of the area of the common wall is open and unobstructed and provides an opening of not less than one-tenth of the floor area of the interior room but not less than 25 square feet (2.32 m^2).

R303.3 Bathrooms. Bathrooms, water closet compartments and other similar rooms shall be provided with aggregate glazing area in windows of not less than 3 square feet (0.279 m^2), one-half of which must be openable.

Exception: The glazed areas shall not be required where artificial light and a mechanical ventilation system are provided. The minimum ventilation rates shall be 50 cfm (23.6 L/s) for intermittent ventilation or 20 cfm (9.4 L/s) for continuous ventilation. Ventilation air from the space shall be exhausted directly to the outside.

R303.4 Stairway illumination. All interior and exterior stairways shall be provided with a means to illuminate the stairs, including the landings and treads. Interior stairways shall be provided with an artificial light source located in the immediate vicinity of each landing of the stairway. Exterior stairways shall be provided with an artificial light source located in the immediate vicinity of the top landing of the stairway. Exterior stairways providing access to a basement from the outside grade level shall be provided with an artificial light source located in the immediate vicinity of the bottom landing of the stairway.

Exception: An artificial light source is not required at the top and bottom landing, provided an artificial light source is located directly over each stairway section.

R303.4.1 Light activation. The control for activation of the required interior stairway lighting shall be accessible at the top and bottom of each stair without traversing any step of the stair. The illumination of exterior stairs shall be controlled from inside the dwelling unit.

Exception: Lights that are continuously illuminated or automatically activated.

R303.5 Required glazed openings. Required glazed openings shall open directly onto a street or public alley, or a yard or court located on the same lot as the building.

R303.5.1 Roofed porches. Required glazed openings may face into a roofed porch where the porch abuts a street, yard or court and the longer side of the porch is at least 65 percent open and unobstructed and the ceiling height is not less than 7 feet (2134 mm).

R303.6 Required heating. When the winter design temperature in Table R301.2(1) is below 60°F (16°C), every dwelling unit shall be provided with heating facilities capable of maintaining a minimum room temperature of 68°F (20°C) at a point 3 feet (914 mm) above the floor and 2 feet (610 mm) from exterior walls in all habitable rooms at the design temperature.

SECTION R304
MINIMUM ROOM AREAS

R304.1 Minimum area. Every dwelling unit shall have at least one habitable room that shall have not less than 120 square feet (11.2 m^2) of gross floor area.

R304.2 Other rooms. Other habitable rooms shall have a floor area of not less than 70 square feet (6.5 m^2).

Exception: Every kitchen shall have not less than 50 square feet (4.64 m^2) of gross floor area.

R304.3 Minimum dimensions. Habitable rooms shall not be less than 7 feet (2134 mm) in any horizontal dimension.

Exception: Kitchens.

R304.4 Height effect on room area. Portions of a room with a sloping ceiling measuring less than 5 feet (1524 mm) or a furred ceiling measuring less than 7 feet (2134 mm) from the finished floor to the finished ceiling shall not be considered as contributing to the minimum required habitable area for that room.

SECTION R305
CEILING HEIGHT

R305.1 Minimum height. Habitable rooms, hallways, corridors, bathrooms, toilet rooms, laundry rooms and basements shall have a ceiling height of not less than 7 feet (2134 mm). The required height shall be measured from the finished floor to the lowest projection from the ceiling.

Exceptions:

1. Beams and girders spaced not less than 4 feet (1219 mm) on center may project not more than 6 inches (152 mm) below the required ceiling height.
2. Ceilings in basements without habitable spaces may project to within 6 feet, 8 inches (2032 mm) of the finished floor; and beams, girders, ducts or other obstructions may project to within 6 feet, 4 inches (1931 mm) of the finished floor.
3. Not more than 50 percent of the required floor area of a room or space is permitted to have a sloped ceiling less than 7 feet (2134 mm) in height with no portion of the required floor area less than 5 feet (1524 mm) in height.

SECTION R306
SANITATION

R306.1 Toilet facilities. Every dwelling unit shall be provided with a water closet, lavatory, and a bathtub or shower.

R306.2 Kitchen. Each dwelling unit shall be provided with a kitchen area and every kitchen area shall be provided with a sink.

R306.3 Sewage disposal. All plumbing fixtures shall be connected to a sanitary sewer or to an approved private sewage disposal system.

R306.4 Water supply to fixtures. All plumbing fixtures shall be connected to an approved water supply. Kitchen sinks, lavatories, bathtubs, showers, bidets, laundry tubs and washing machine outlets shall be provided with hot and cold water.

SECTION R307
TOILET, BATH AND SHOWER SPACES

R307.1 Space required. Fixtures shall be spaced as per Figure R307.2.

R307.2 Bathtub and shower spaces. Bathtub and shower floors and walls above bathtubs with installed shower heads and in shower compartments shall be finished with a nonabsorbent surface. Such wall surfaces shall extend to a height of not less than 6 feet (1829 mm) above the floor.

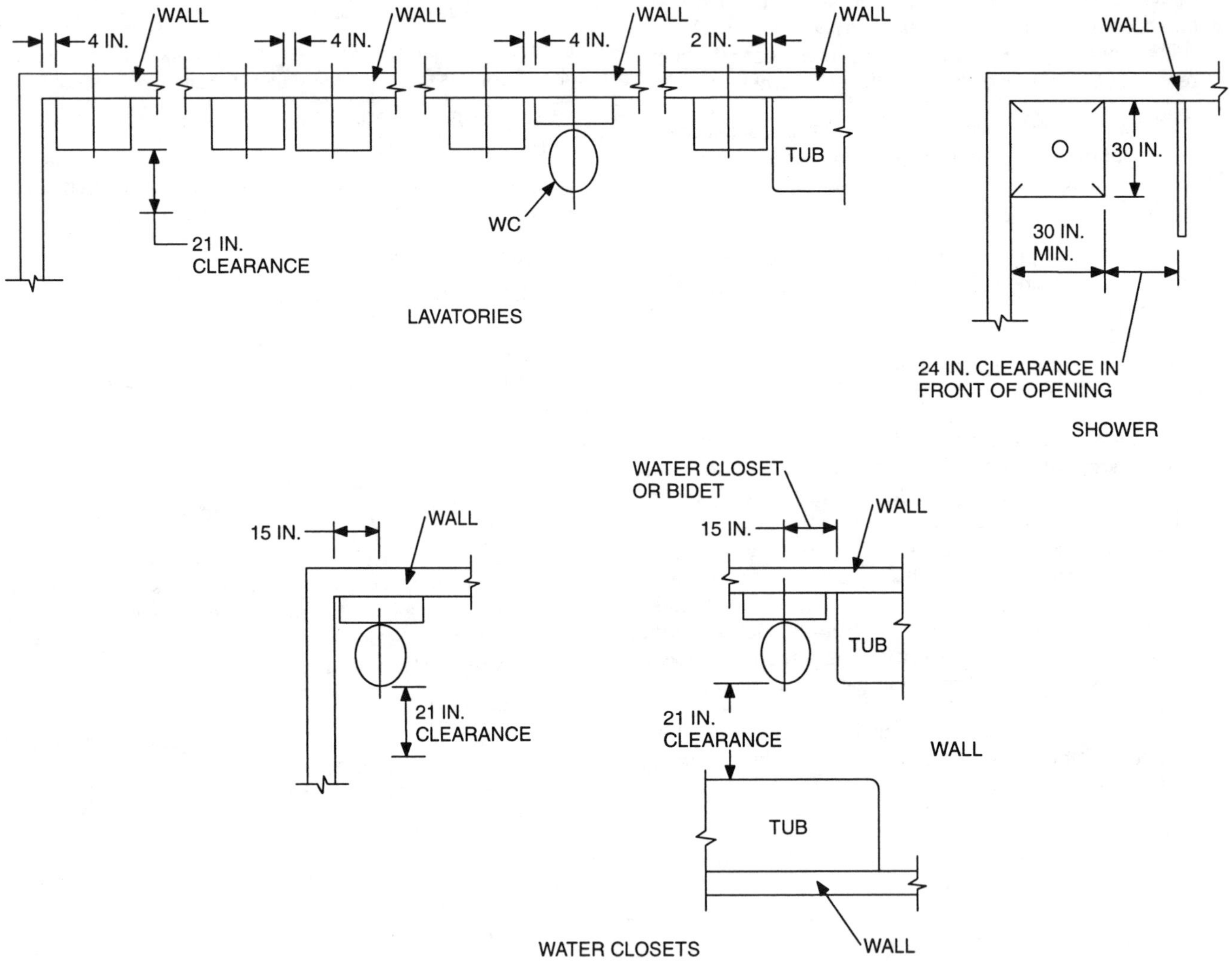

For SI: 1 inch = 25.4 mm.

FIGURE R307.2
MINIMUM FIXTURE CLEARANCES

SECTION R308
GLAZING

R308.1 Identification. Except as indicated in Section R308.1.1, each pane of glazing installed in hazardous locations as defined in Section R308.4 shall be provided with a manufacturer's or installer's label, designating the type and thickness of glass and the safety glazing standard with which it complies, which is visible in the final installation. The label shall be acid etched, sandblasted, ceramic-fired, embossed mark, or shall be of a type which once applied cannot be removed without being destroyed.

Exceptions:

1. For other than tempered glass, labels may be omitted provided the building official approves the use of a certificate, affidavit or other evidence confirming compliance with this code.
2. Tempered spandrel glass may be identified by the manufacturer with a removable paper label.

R308.1.1 Identification of multipane assemblies. Multipane assemblies having individual panes not exceeding 1 square foot ($0.09 m^2$) in exposed area shall have at least one pane in the assembly identified in accordance with Section R308.1. All other panes in the assembly shall be labeled "16 CFR 1201."

R308.2 Louvered windows or jalousies. Regular, float, wired or patterned glass in jalousies and louvered windows shall be no thinner than nominal $^3/_{16}$ inch (4.76 mm) and no longer than 48 inches (1219 mm). Exposed glass edges shall be smooth.

R308.2.1 Wired glass prohibited. Wired glass with wire exposed on longitudinal edges shall not be used in jalousies or louvered windows.

R308.3 Human impact loads. Individual glazed areas including glass mirrors in hazardous locations such as those indicated as defined in Section R308.4 shall pass the test requirements of CPSC 16 CFR, Part 1201. Glazing shall comply with the CPSC 16 CFR, Part 1201 criteria for Category I or II as indicated in Table R308.3.

Exceptions:

1. Polished wired glass for use in fire doors and other fire resistant locations shall comply with ANSI Z97.1.
2. Louvered windows and jalousies shall comply with Section R308.2.

TABLE R308.3
MINIMUM CATEGORY CLASSIFICATION OF GLAZING

EXPOSED SURFACE AREA OF ONE SIDE OF ONE LITE	GLAZING IN STORM OR COMBINATION DOORS (Category Class)	GLAZING IN DOORS (Category Class)	GLAZED PANELS REGULATED BY ITEM 7 OF SECTION R308.4 (Category Class)	GLAZED PANELS REGULATED BY ITEM 6 OF SECTION R308.4 (Category Class)	GLAZING IN DOORS AND ENCLOSURES REGULATED BY ITEM 5 OF SECTION R308.4 (Category Class)	SLIDING GLASS DOORS PATIO TYPE (Category Class)
9 sq. ft. or less	I	I	NR[a]	I	II	II
More than 9 sq. ft	II	II	II	II	II	II

For SI: 1 square foot = 0.0929 m^2

[a]NR means "No Requirement."

R308.4 Hazardous locations. The following shall be considered specific hazardous locations for the purposes of glazing:

1. Glazing in side-hinged doors except jalousies.
2. Glazing in fixed and sliding panels of sliding door assemblies and panels in sliding and bifold closet door assemblies.
3. Glazing in storm doors.
4. Glazing in all unframed swinging doors
5. Glazing in doors and enclosures for hot tubs, whirlpools, saunas, steam rooms, bathtubs and showers. Glazing in any part of a building wall enclosing these compartments where the bottom exposed edge of the glazing is less than 60 inches (1524 mm) measured vertically above any standing or walking surface.
6. Glazing, in an individual fixed or operable panel adjacent to a door where the nearest vertical edge is within a 24-inch (610 mm) arc of the door in a closed position and whose bottom edge is less than 60 inches (1524 mm) above the floor or walking surface.
7. Glazing in an individual fixed or operable panel, other than those locations described in Items 5 and 6 above, that meets all of the following conditions:
   - 7.1. Exposed area of an individual pane greater than 9 square feet ($0.836 m^2$).
   - 7.2. Bottom edge less than 18 inches (457 mm) above the floor.
   - 7.3. Top edge greater than 36 inches (914 mm) above the floor.
   - 7.4. One or more walking surfaces within 36 inches (914 mm) horizontally of the glazing.
8. All glazing in railings regardless of an area or height above a walking surface. Included are structural baluster panels and nonstructural in-fill panels.
9. Glazing in walls and fences enclosing indoor and outdoor swimming pools, hot tubs and spas where the bottom edge of the pool or spa side is less than 60 inches (1524 mm) above a walking surface and within 60 inches (1524 mm) horizontally of the water's edge. This shall apply to single glazing and all panes in multiple glazing.
10. Glazing in walls enclosing stairway landings or within 60 inches (1524 mm) of the top and bottom of stairways where the bottom edge of the glass is less than 60 inches (1524 mm) above the walking surface.

Exception: The following products, materials and uses are exempt from the above hazardous locations:

1. Openings in doors through which a 3-inch (76 mm) sphere is unable to pass.
2. Decorative glass in Item 1, 6 or 7.
3. Glazing in Section R308.4, Item 6, when there is an intervening wall or other permanent barrier between the door and the glazing.
4. Glazing in all unframed swinging doors.
5. Glazing in Section R308.4, Item 6, where access through the door is to a closet or storage area 3 feet (914 mm) or less in depth.
6. Glazing in Section R308.4, Item 7, when a protective bar is installed on the accessible side(s) of the glazing 36 inches ± 2 inches (914 mm ± 51 mm) above the floor. The bar shall be capable of withstanding a horizontal load of 50 pounds per linear foot (74.5 kg/m) without contacting the glass and be a minimum of 1.5 inches (38 mm) in height.
7. Outboard panes in insulating glass units and other multiple glazed panels in Section R308.4, Item 7, when the bottom edge of the glass is 25 feet (7620 mm) or more above grade, a roof, walking surface, or other horizontal [within 45 degrees (0.79 rad) of horizontal] surface adjacent to the glass exterior.
8. Louvered windows and jalousies complying with the requirements of Section R308.2.
9. Mirrors and other glass panels mounted or hung on a surface that provides a continuous backing support.

R308.5 Site built windows. Site built windows shall comply with Section 2404 of the *International Building Code*.

R308.6 Skylights and sloped glazing. Skylights and sloped glazing shall comply with the following sections.

R308.6.1 Definition. Sloped glazing means any installation of glass or other transparent or translucent glazing material installed at a slope of more than 15 degrees (0.26 rad) from vertical. Glazing materials in skylights, solariums, sun spaces, roofs and sloped walls are included in this definition.

R308.6.2 Permitted materials. The following types of glazing may be used:

1. Laminated glass with a minimum 0.015-inch (0.38 mm) polyvinyl butyral interlayer for glass panes 16 square feet (1.5 m^2) or less in area located such that the highest point of the glass is not more than 12 feet (3658 mm) above a walking surface or other accessible area; for higher or larger sizes, the minimum interlayer thickness shall be 0.030 inch (0.76 mm).
2. Fully tempered glass.
3. Heat-strengthened glass.
4. Wired glass.
5. Approved rigid plastics.

R308.6.3 Screens, general. For fully tempered or heat-strengthened glass, a retaining screen meeting the requirements of Section R308.6.7 shall be installed below the glass, except for fully tempered glass that meets either condition listed in Section R308.6.5.

R308.6.4 Screens with multiple glazing. When the inboard pane is fully tempered, heat-strengthened, or wired glass, a retaining screen meeting the requirements of Section R308.6.7 shall be installed below the glass, except for either condition listed in Section R308.6.5. All other panes in the multiple glazing may be of any type listed in Section R308.6.2.

R308.6.5 Screens not required. Screens shall not be required when fully tempered glass is used as single glazing or the inboard pane in multiple glazing and either of the following conditions are met:

1. Glass area 16 square feet (1.49 m^2) or less. Highest point of glass not more than 12 feet (3658 mm) above a walking surface or other accessible area, nominal glass thickness not more than $^3/_{16}$ inch (4.76 mm), and (for multiple glazing only) the other pane or panes fully tempered, laminated or wired glass.
2. Glass area greater than 16 square feet (1.49 m^2). Glass sloped 30 degrees (0.52 rad) or less from vertical, and highest point of glass not more than 10 feet (3048 mm) above a walking surface or other accessible area.

R308.6.6 Glass in greenhouses. Any glazing material is permitted to be installed without screening in the sloped areas of greenhouses, provided the greenhouse height at the ridge does not exceed 20 feet (6096 mm) above grade.

R308.6.7 Screen characteristics. The screen and its fastenings shall be capable of supporting twice the weight of the glazing, be firmly and substantially fastened to the framing members, and have a mesh opening of no more than 1 inch by 1 inch (25.4 mm by 25.4 mm).

R308.6.8 Curbs for skylights. All unit skylights installed in a roof with a pitch flatter than three units vertical in 12 units horizontal (25-percent slope) shall be mounted on a curb extending at least 4 inches (102 mm) above the plane of the roof unless otherwise specified in the manufacturer's installation instructions.

SECTION R309
GARAGES AND CARPORTS

R309.1 Opening protection. Openings from a private garage directly into a room used for sleeping purposes shall not be permitted. Other openings between the garage and residence shall be equipped with solid wood doors not less than $1^3/_8$ inch (35 mm) in thickness, solid or honeycomb core steel doors not less than $1^3/_8$ inches (35 mm) thick, or 20-minute fire-rated doors.

R309.1.1 Duct penetration. Ducts in the garage and ducts penetrating the walls or ceilings separating the dwelling from the garage shall be constructed of a minimum No. 26 gage (0.48 mm) sheet steel or other approved material and shall have no openings into the garage.

R309.2 Separation required. The garage shall be separated from the residence and its attic area by not less than $^1/_2$-inch

(12.7 mm) gypsum board applied to the garage side. Where the separation is a floor-ceiling assembly, the structure supporting the separation shall also be protected by not less than $^1/_2$-inch (12.7 mm) gypsum board or equivalent.

R309.3 Floor surface. Garage floor surfaces shall be of approved noncombustible material.

The area of floor used for parking of automobiles or other vehicles shall be sloped to facilitate the movement of liquids to a drain or toward the main vehicle entry doorway.

R309.4 Carports. Carports shall be open on at least two sides. Carport floor surfaces shall be of approved noncombustible material. Carports not open on at least two sides shall be considered a garage and shall comply with the provisions of this section for garages.

Exception: Asphalt surfaces shall be permitted at ground level in carports.

The area of floor used for parking of automobiles or other vehicles shall be sloped to facilitate the movement of liquids to a drain or toward the main vehicle entry doorway.

R309.5 Flood hazard areas. For buildings located in flood hazard areas as established by Table 301.2(1), garage floors shall be:

1. Elevated to or above the design flood elevation as determined in Section R327; or
2. Located below the design flood elevation provided they are at or above grade on all sides, are used solely for parking, building access, or storage, meet the requirements of Section R327, and are otherwise constructed in accordance with this code.

SECTION R310 EMERGENCY ESCAPE AND RESCUE OPENINGS

R310.1 Emergency escape and rescue required. Basements with habitable space and every sleeping room shall have at least one openable emergency escape and rescue window or exterior door opening for emergency escape and rescue. Where openings are provided as a means of escape and rescue they shall have a sill height of not more than 44 inches (1118 mm) above the floor. Where a door opening having a threshold below the adjacent ground elevation serves as an emergency escape and rescue opening and is provided with a bulkhead enclosure, the bulkhead enclosure shall comply with Section R310.3. The net clear opening dimensions required by this section shall be obtained by the normal operation of the window or door opening from the inside. Escape and rescue window openings with a finished sill height below the adjacent ground elevation shall be provided with a window well in accordance with Section R310.2.

R310.1.1 Minimum opening area. All emergency escape and rescue openings shall have a minimum net clear opening of 5.7 square feet (0.530 m^2).

Exception: Grade floor openings shall have a minimum net clear opening of 5 square feet (0.465 m^2).

R310.1.2 Minimum opening height. The minimum net clear opening height shall be 24 inches (610 mm).

R310.1.3 Minimum opening width. The minimum net clear opening width shall be 20 inches (508 mm).

R310.1.4 Operational constraints. Emergency escape and rescue openings shall be operational from the inside of the room without the use of keys or tools.

R310.2 Window wells. Window wells required for emergency escape and rescue shall have horizontal dimensions that allow the door or window of the emergency escape and rescue opening to be fully opened. The horizontal dimensions of the window well shall provide a minimum net clear area of 9 square feet (0.84 m^2) with a minimum horizontal projection and width of 36 inches (914 mm).

Exception: The ladder or steps required by Section R310.2.1 shall be permitted to encroach a maximum of 6 inches (152 mm) into the required dimensions of the window well.

R310.2.1 Ladder and steps. Window wells with a vertical depth greater than 44 inches (1118 mm) below the adjacent ground level shall be equipped with a permanently affixed ladder or steps usable with the window in the fully open position. Ladders or steps required by this section shall not be required to comply with Sections R314 and R315. Ladders or rungs shall have an inside width of at least 12 inches (305 mm), shall project at least 3 inches (76 mm) from the wall and shall be spaced not more than 18 inches (457 mm) on center vertically for the full height of the window well.

R310.3 Bulkhead enclosures. Bulkhead enclosures shall provide direct access to the basement. The bulkhead enclosure with the door panels in the fully open position shall provide the minimum net clear opening required by Section R310.1.1. Bulkhead enclosures shall also comply with Section R314.9.

R310.4 Bars, grills, covers and screens. Bars, grills, covers, screens or similar devices are permitted to be placed over emergency escape and rescue openings, bulkhead enclosures, or window wells that serve such openings, provided the minimum net clear opening size complies with Sections R310.1.1 to R310.1.3, and such devices shall be releasable or removable from the inside without the use of a key, tool or force greater than that which is required for normal operation of the escape and rescue opening.

SECTION R311 EXITS

R311.1 Exit door required. Not less than one exit door conforming to this chapter shall be provided from each dwelling unit. The required exit door shall provide for direct access from the habitable portions of the dwelling to the exterior without requiring travel through a garage.

R311.2 Type of lock or latch. All egress doors shall be readily openable from the side from which egress is to be made without the use of a key or special knowledge or effort.

R311.3 Type and size. The required exit door shall be a side-hinged door not less than 3 feet (914 mm) in width and 6 feet,

8 inches (2032 mm) in height. Other exterior hinged or sliding doors shall not be required to comply with these minimum dimensions.

R311.4 Hallways. The minimum width of a hallway shall be not less than 3 feet (914 mm).

R311.5 Exit facilities. Exterior exit balconies, stairs and similar exit facilities shall be positively anchored to the primary structure to resist both vertical and lateral forces. Such attachment shall not be accomplished by use of toenails or nails subject to withdrawal.

SECTION R312 LANDINGS

R312.1 General. Landings for stairways shall comply with this section.

R312.1.1 Landings for stairways. There shall be a floor or landing at the top and bottom of each stairway.

Exception: At the top of an interior flight of stairs, provided a door does not swing over the stairs.

R312.1.2 Landings at doors. There shall be a floor or landing on each side of each exterior door.

Exception: At the exterior side of sliding doors.

The floor or landing at a door shall not be more than 1.5 inches (38 mm) lower than the top of the threshold.

Exception: The landing at an exterior doorway shall not be more than 8 inches (197 mm) below the top of the threshold, provided that the door, other than an exterior storm or screen door, does not swing over the landing.

R312.2 Size. The width of each landing shall not be less than the stairway or door served. Every landing shall have a minimum dimension of 36 inches (914 mm) measured in the direction of travel.

SECTION R313 RAMPS

R313.1 Maximum slope. Ramps shall have a maximum slope of one unit vertical in eight units horizontal (12.5-percent slope).

R313.2 Handrails required. Handrails shall be provided on at least one side of all ramps exceeding a slope of one unit vertical in 12 units horizontal (8.33-percent slope).

R313.3 Landing required. A minimum 3-foot-by-3-foot (914 mm by 914 mm) landing shall be provided:

1. At the top and bottom of ramps.
2. Where doors open onto ramps.
3. Where ramps change direction.

SECTION R314 STAIRWAYS

R314.1 Width. Stairways shall not be less than 36 inches (914 mm) in clear width at all points above the permitted handrail height and below the required headroom height. Handrails shall not project more than 4.5 inches (114 mm) on either side of the stairway and the minimum clear width of the stairway at and below the handrail height, including treads and landings, shall not be less than 31.5 inches (787 mm) where a handrail is installed on one side and 27 inches (698 mm) where handrails are provided on both sides.

Exception: The width of spiral stairways shall be in accordance with Section R314.5.

R314.2 Treads and risers. The maximum riser height shall be $7^3/_4$ inches (196 mm) and the minimum tread depth shall be 10 inches (254 mm). The riser height shall be measured vertically between leading edges of the adjacent treads. The tread depth shall be measured horizontally between the vertical planes of the foremost projection of adjacent treads and at a right angle to the tread's leading edge. The walking surface of treads and landings of a stairway shall be sloped no steeper than one unit vertical in 48 units horizontal (2-percent slope). The greatest riser height within any flight of stairs shall not exceed the smallest by more than $^3/_8$ inch (9.5 mm). The greatest tread depth within any flight of stairs shall not exceed the smallest by more than $^3/_8$ inch (9.5 mm).

R314.2.1 Profile. The radius of curvature at the leading edge of the tread shall be no greater than $^9/_{16}$ inch (14.3 mm). A nosing not less than $^3/_4$ inch (19.1 mm) but not more than $1^1/_4$ inches (32 mm) shall be provided on stairways with solid risers. The greatest nosing projection shall not exceed the smallest nosing projection by more than $^3/_8$ inch (9.5 mm) between two stories, including the nosing at the level of floors and landings. Beveling of nosing shall not exceed $^1/_2$ inch (12.7 mm). Risers shall be vertical or sloped from the underside of the leading edge of the tread above at an angle not more than 30 degrees from the vertical. Open risers are permitted, provided that the opening between treads does not permit the passage of a 4-inch-diameter (102 mm) sphere.

Exceptions:

1. A nosing is not required where the tread depth is a minimum of 11 inches (279 mm).
2. The opening between adjacent treads is not limited on stairs with a total rise of 30 inches (762 mm) or less.

R314.3 Headroom. The minimum headroom in all parts of the stairway shall not be less than 6 feet, 8 inches (2032 mm) measured vertically from the sloped plane adjoining the tread nosing or from the floor surface of the landing or platform.

R314.4 Winders. Winders are permitted, provided that the width of the tread at a point not more than 12 inches (305 mm) from the side where the treads are narrower is not less than 10 inches (254 mm) and the minimum width of any tread is not less than 6 inches (152 mm). The continuous handrail required by Section R315.1 shall be located on the side where the tread is narrower.

R314.5 Spiral stairs. Spiral stairways are permitted, provided the minimum width shall be 26 inches (660 mm) with each tread having a $7^1/_2$-inch (190 mm) minimum tread width at 12 inches (305 mm) from the narrow edge. All treads shall be identical, and the rise shall be no more than $9^1/_2$ inches (241 mm). A minimum headroom of 6 feet, 6 inches (1982 mm) shall be provided.

R314.6 Circular stairways. Circular stairways shall have a tread depth at a point not more than 12 inches (305 mm) from the side where the treads are narrower of not less than 11 inches (279 mm) and the minimum depth of any tread shall not be less than 6 inches (152 mm). Tread depth at any walking line, measured a consistent distance from a side of the stairway, shall be uniform as specified in Section R314.2.

R314.7 Illumination. All stairs shall be provided with illumination in accordance with Section R303.4.

R314.8 Under stair protection. Enclosed accessible space under stairs shall have walls, under stair surface and any soffits protected on the enclosed side with $^1/_2$-inch (12.7 mm) gypsum board.

R314.9 Bulkhead enclosure stairways. Stairways serving bulkhead enclosures not part of the required building egress and providing access from the outside grade level to the basement shall be exempt from the requirements of Sections R312, R314 and R315 when the maximum height from the basement finished floor level to grade adjacent to the stairway does not exceed 8 feet (2438 mm), and the grade level opening to stairway is covered by a bulkhead enclosure with hinged doors or other approved means.

SECTION R315 HANDRAILS

R315.1 Handrails. Handrails having minimum and maximum heights of 34 inches and 38 inches (864 mm and 965 mm), respectively, measured vertically from the nosing of the treads, shall be provided on at least one side of stairways. All required handrails shall be continuous the full length of the stairs with two or more risers from a point directly above the top riser of a flight to a point directly above the lowest riser of the flight. Ends shall be returned or shall terminate in newel posts or safety terminals. Handrails adjacent to a wall shall have a space of not less than 1.5 inches (38 mm) between the wall and the handrail.

Exceptions:

1. Handrails shall be permitted to be interrupted by a newel post at a turn.
2. The use of a volute, turnout or starting easing shall be allowed over the lowest tread.

R315.2 Handrail grip size. The handgrip portion of handrails shall have a circular cross section of $1^1/_4$ inches (32 mm) minimum to $2^5/_8$ inches (67 mm) maximum. Other handrail shapes that provide an equivalent grasping surface are permissible. Edges shall have a minimum radius of $^1/_8$ inch (3.2 mm).

SECTION R316 GUARDS

R316.1 Guards required. Porches, balconies or raised floor surfaces located more than 30 inches (762 mm) above the floor or grade below shall have guards not less than 36 inches (914 mm) in height. Open sides of stairs with a total rise of more than 30 inches (762 mm) above the floor or grade below shall have guards not less than 34 inches (864 mm) in height measured vertically from the nosing of the treads.

R316.2 Guard opening limitations. Required guards on open sides of stairways, raised floor areas, balconies and porches shall have intermediate rails or ornamental closures that do not allow passage of a sphere 4 inches (102 mm) in diameter. Required guards shall not be constructed with horizontal rails or other ornamental pattern that results in a ladder effect.

Exception: The triangular openings formed by the riser, tread and bottom rail of a guard at the open side of a stairway are permitted to be of such a size that a sphere 6 inches (152 mm) cannot pass through.

SECTION R317 SMOKE ALARMS

R317.1 Single- and multiple-station smoke alarms. Single- and multiple-station smoke alarms shall be installed in the following locations:

1. In each sleeping room.
2. Outside of each separate sleeping area in the immediate vicinity of the bedrooms.
3. On each additional story of the dwelling, including basements and cellars but not including crawl spaces and uninhabitable attics. In dwellings or dwelling units with split levels and without an intervening door between the adjacent levels, a smoke alarm installed on the upper level shall suffice for the adjacent lower level provided that the lower level is less than one full story below the upper level.

When more than one smoke alarm is required to be installed within an individual dwelling unit the alarm devices shall be interconnected in such a manner that the actuation of one alarm will activate all of the alarms in the individual unit. The alarm shall be clearly audible in all bedrooms over background noise levels with all intervening doors closed.

All smoke alarms shall be listed and installed in accordance with the provisions of this code and the household fire warning equipment provisions of NFPA 72.

R317.1.1 Alterations, repairs and additions. When interior alterations, repairs or additions requiring a permit occur, or when one or more sleeping rooms are added or created in existing dwellings, the individual dwelling unit shall be provided with smoke alarms located as required for new dwellings; the smoke alarms shall be interconnected and hard wired.

Exceptions:

1. Smoke alarms in existing areas shall not be required to be interconnected and hard wired where the alterations or repairs do not result in the removal of

interior wall or ceiling finishes exposing the structure, unless there is an attic, crawl space, or basement available which could provide access for hard wiring and interconnection without the removal of interior finishes.

2. Repairs to the exterior surfaces of dwellings are exempt from the requirements of this section.

R317.2 Power source. In new construction, the required smoke alarms shall receive their primary power from the building wiring when such wiring is served from a commercial source, and when primary power is interrupted, shall receive power from a battery. Wiring shall be permanent and without a disconnecting switch other than those required for overcurrent protection. Smoke alarms shall be permitted to be battery operated when installed in buildings without commercial power or in buildings that undergo alterations, repairs or additions regulated by Section R317.1.1.

SECTION R318
FOAM PLASTIC

R318.1 General. The provisions of this section shall govern the requirements and uses of foam plastic insulation.

R318.1.1 Surface burning characteristics. Except where otherwise noted in Section R318.2, all foam plastic or foam plastic cores in manufactured assemblies used in building construction shall have a flame-spread rating of not more than 75 and shall have a smoke-developed rating of not more than 450 when tested in the maximum thickness intended for use in accordance with ASTM E 84.

R318.1.2 Thermal barrier. Foam plastic, except where otherwise noted, shall be separated from the interior of a building by minimum $^1/_2$-inch (12.7 mm) gypsum board or an approved finish material equivalent to a thermal barrier to limit the average temperature rise of the unexposed surface to no more than 250°F (121°C) after 15 minutes of fire exposure to the ASTM E 119 standard time temperature curve. The gypsum board shall be installed using a mechanical fastening system in accordance with Section R702.3.5. Reliance on adhesives to ensure that the gypsum board will remain in place when exposed to fire shall be prohibited.

R318.2 Specific requirements. The following requirements shall apply to all uses of foam plastic unless specifically approved in accordance with Section R318.3 or by other sections of the code.

R318.2.1 Masonry or concrete construction. Foam plastics may be used without the thermal barrier described in Section R318.1 when the foam plastic is protected by a minimum 1-inch (25.4 mm) thickness of masonry or concrete.

R318.2.2 Roofing. Foam plastic may be used in a roof-covering assembly without the thermal barrier when the foam is separated from the interior of the building by wood structural panel sheathing in accordance with Section R803, not less than $^{15}/_{32}$ inch (11.9 mm) in thickness bonded with exterior glue and identified as Exposure 1, with edge supported by blocking or tongue-and-groove joints. The smoke-developed rating shall not be limited.

R318.2.3 Attics and crawlspaces. Within attics and crawlspaces where entry is made only for service of utilities, foam plastics shall be protected against ignition by $1^1/_2$-inch-thick (38 mm) mineral fiber insulation, $^1/_4$-inch-thick (6.4 mm) wood structural panels, $^3/_8$-inch (9.5 mm) particleboard, $^1/_4$-inch (6.4 mm) hardboard, $^3/_8$-inch (9.5 mm) gypsum board, or corrosion-resistant steel having a base metal thickness of 0.016 inch (0.406 mm).

R318.2.4 Foam-filled doors. Foam-filled doors are exempt from the requirements of Section R318.1.

R318.2.5 Siding backer board. Foam plastic board of not more than $^1/_2$-inch (12.7 mm) thickness may be used as siding backer board when separated from interior spaces by not less than 2 inches (51 mm) of mineral fiber insulation or $^1/_2$-inch (12.7 mm) gypsum wallboard or installed over existing exterior wall finish in conjunction with re-siding, providing the plastic board does not have a potential heat of more than 2,000 Btu per square foot (22 720 kJ/m^2) when tested in accordance with NFPA 259.

R318.2.6 Interior trim. Foam plastic trim defined as picture molds, chair rails, baseboards, handrails, ceiling beams, door trim and window trim may be installed, provided:

1. The minimum density is 20 pounds per cubic foot (3.14 kg/m^3).
2. The maximum thickness of the trim is 0.5 inch (12.7 mm) and the maximum width is 4 inches (102 mm).
3. The trim constitutes no more than 10 percent of the area of any wall or ceiling.
4. The flame-spread rating does not exceed 75 when tested per ASTM E 84. The smoke-developed rating is not limited.

R318.3 Specific approval. Plastic foam not meeting the requirements of Sections R318.1 and R318.2 may be specifically approved on the basis of one of the following approved tests: ASTM E 84, FM 4880, UL 1040, ASTM E 152, or UL 1715, or fire tests related to actual end-use configurations. The specific approval may be based on the end use, quantity, location and similar considerations where such tests would not be applicable or practical.

R318.4 Interior finish. Foam plastics that are used as interior finish shall also meet the flame-spread requirements for interior finish.

R318.5 Termite damage. The use of foam plastics in areas of "very heavy" termite infestation probability shall be in accordance with Section R324.4.

SECTION R319
FLAME SPREAD AND SMOKE DENSITY

R319.1 Wall and ceiling. Wall and ceiling finishes shall have a flame-spread classification of not greater than 200.

Exception: Flame-spread requirements for finishes shall not apply to trim defined as picture molds, chair rails, baseboards and handrails; to doors and windows or their frames; or to materials that are less than $^{1}/_{28}$ inch (0.907 mm) in thickness cemented to the surface of walls or ceilings if these materials have a flame-spread characteristic no greater than paper of this thickness cemented to a noncombustible backing.

R319.2 Smoke-developed index. Wall and ceiling finishes shall have a smoke-developed index of not greater than 450.

R319.3 Testing. Tests shall be made in accordance with ASTM E 84.

SECTION R320 INSULATION

R320.1 Insulation. Insulation materials, including facings, such as vapor barriers or breather papers installed within floor-ceiling assemblies, roof-ceiling assemblies, wall assemblies, crawl spaces and attics shall have a flame-spread index not to exceed 25 with an accompanying smoke-developed index not to exceed 450 when tested in accordance with ASTM E 84.

Exceptions:

1. When such materials are installed in concealed spaces, the flame-spread and smoke-developed limitations do not apply to the facings, provided that the facing is installed in substantial contact with the unexposed surface of the ceiling, floor or wall finish.
2. Cellulose loose-fill insulation, which is not spray applied, complying with the requirements of Section R320.3, shall only be required to meet the smoke-developed index of not more than 450.

R320.2 Loose-fill insulation. Loose-fill insulation materials that cannot be mounted in the ASTM E 84 apparatus without a screen or artificial supports shall have a flame-spread rating not to exceed 25 with an accompanying smoke-developed factor not to exceed 450 when tested in accordance with CAN/ULC-S102.2.

Exception: Cellulose loose-fill insulation shall not be required to comply with this test method provided that such insulation complies with the requirements of Section R320.3.

R320.3 Cellulose loose-fill insulation. Cellulose loose-fill insulation shall comply with CPSC 16 CFR, Parts 1209 and 1404. Each package of such insulating material shall be clearly labeled in accordance with CPSC 16 CFR, Parts 1209 and 1404.

R320.4 Exposed attic insulation. All exposed insulation materials installed on attic floors shall have a critical radiant flux not less than 0.12 watt per square centimeter.

R320.5 Testing. Tests for critical radiant flux shall be made in accordance with ASTM E 970.

SECTION R321 DWELLING UNIT SEPARATION

R321.1 Two-family dwellings. Dwelling units in two-family dwellings shall be separated from each other by wall and/or floor assemblies of not less than 1-hour fire-resistive rating when tested in accordance with ASTM E 119. Fire-resistance-rated floor-ceiling and wall assemblies shall extend to and be tight against the exterior wall, and wall assemblies shall extend to the underside of the roof sheathing.

Exception: A fire resistance rating of $^{1}/_{2}$ hour shall be permitted in buildings equipped throughout with an automatic sprinkler system installed in accordance with NFPA 13.

R321.1.1 Supporting construction. When floor assemblies are required to be fire-resistance-rated by Section R321.1, the supporting construction of such assemblies shall have an equal or greater fire-resistive rating.

R321.2 Townhouses. Each townhouse shall be considered a separate building and shall be separated by fire-resistance-rated wall assemblies meeting the requirements of Section R302 for exterior walls.

Exception: A common 2-hour fire-resistance-rated wall is permitted for townhouses if such walls do not contain plumbing or mechanical equipment, ducts or vents in the cavity of the common wall. Electrical installations shall be installed in accordance with Chapters 33 through 42. Penetrations of electrical outlet boxes shall be in accordance with Section R321.3.

R321.2.1 Continuity. The common wall for townhouses shall be continuous from the foundation to the underside of the roof sheathing, deck or slab and shall extend the full length of the common wall.

R321.2.2 Parapets. Parapets constructed in accordance with Section R321.2.3 shall be provided for townhouses as an extension of common exterior or walls in accordance with the following:

1. Where roof surfaces adjacent to the wall or walls are at the same elevation, the parapet shall extend not less than 30 inches (762 mm) above the roof surfaces.
2. Where roof surfaces adjacent to the wall or walls are at different elevations and the higher roof is not more than 30 inches (762 mm) above the lower roof, the parapet shall extend not less than 30 inches (762 mm) above the lower roof surface.

   Exception: A parapet is not required in the two cases above when the roof is covered with a minimum class C roof covering, and the roof decking or sheathing is of noncombustible materials or approved fire-retardant-treated wood for a distance of 4 feet (1219 mm) on each side of the wall or walls, or one layer of $^{5}/_{8}$-inch (15.9 mm) Type X gypsum board is installed directly beneath the roof decking or sheathing for a distance of 4 feet (1219 mm) on each side of the wall or walls.
3. A parapet is not required where roof surfaces adjacent to the wall or walls are at different elevations and the higher roof is more than 30 inches (762 mm) above the lower roof. The common wall construction from the

lower roof to the underside of the higher roof deck shall not have less than a 1-hour fire-resistive rating. The wall shall be rated for exposure from both sides.

R321.2.3 Parapet construction. Parapets shall have the same fire-resistance rating as that required for the supporting wall or walls. On any side adjacent to a roof surface, the parapet shall have noncombustible faces for the uppermost 18 inches (457 mm), to include counterflashing and coping materials. Where the roof slopes toward a parapet at slopes greater than two units vertical in 12 units horizontal (16.7-percent slope), the parapet shall extend to the same height as any portion of the roof within a distance of 3 feet (914 mm), but in no case shall the height be less than 30 inches (762 mm).

R321.2.4 Structural independence. Each individual townhouse shall be structurally independent.

Exceptions:

1. Foundations supporting exterior walls or common walls.
2. Structural roof and wall sheathing from each unit may fasten to the common wall framing.
3. Nonstructural wall coverings.
4. Flashing at termination of roof covering over common wall.
5. Townhouses separated by a common 2-hour fire-resistance-rated wall as provided in Section R321.2.

R321.3 Rated penetrations. Penetrations of wall or floor/ceiling assemblies required to be fire-resistance-rated in accordance with Section R321.1 or R321.2 shall be protected in accordance with this section.

R321.3.1 Through penetrations. Through penetrations of fire-resistance-rated wall or floor assemblies shall comply with Section R321.3.1.1 or R321.3.1.2.

Exception: Where the penetrating items are steel, ferrous or copper pipes or steel conduits, the annular space shall be permitted to be protected as follows:

1. In concrete or masonry wall or floor assemblies where the penetrating item is a maximum 6 inches (152 mm) nominal diameter and the opening is a maximum 144 square inches (92 900 mm^2), concrete, grout or mortar shall be permitted where installed to the full thickness of the wall or floor assembly or the thickness required to maintain the fire-resistance rating.
2. The material used to fill the annular space shall prevent the passage of flame and hot gases sufficient to ignite cotton waste where subjected to ASTM E 119 time temperature fire conditions under a minimum positive pressure differential of 0.01 inch of water (3 Pa) at the location of the penetration for the time period equivalent to the fire-resistance rating of the construction penetrated.

R321.3.1.1 Fire-resistance-rated assembly. Penetrations shall be installed as tested in the approved fire-resistance-rated assembly.

R321.3.1.2 Penetration firestop system. Penetrations shall be protected by an approved penetration firestop system installed as tested in accordance with ASTM E 814, with a minimum positive pressure differential of 0.01 inch of water (3 Pa) and shall have an F rating of not less than the required fire-resistance rating of the wall or floor/ceiling assembly penetrated.

R321.3.2 Membrane penetrations. Membrane penetrations shall comply with Section R321.3.1. Where walls are required to have a minimum 1-hour fire-resistance rating, recessed light fixtures shall be so installed such that the required fire resistance will not be reduced.

Exceptions:

1. Steel electrical boxes that do not exceed 16 square inches (0.0103 m^2) in area provided the total area of such openings does not exceed 100 square inches (0.0645 m^2) for any 100 square feet (9.29 m^2) of wall area. Outlet boxes on opposite sides of the wall shall be separated as follows:
   - 1.1. By a horizontal distance of not less than 24 inches, or
   - 1.2. By a horizontal distance of not less than the depth of the wall cavity when the wall cavity is filled with cellulose loose-fill or mineral fiber insulation, or
   - 1.3. By molded fire blocking in accordance with Section R602.8, or
   - 1.4. By other listed materials and methods.
2. Two-hour fire-resistance-rated nonmetallic electrical outlet boxes shall be installed in accordance with their listings.
3. The annular space created by the penetration of a fire sprinkler provided it is covered by a metal escutcheon plate.

SECTION R322 MOISTURE VAPOR RETARDERS

R322.1 Moisture control. In all framed walls, floors and roof/ceilings comprising elements of the building thermal envelope, a vapor retarder shall be installed on the warm-in-winter side of the insulation.

Exceptions:

1. In construction where moisture or freezing will not damage the materials.
2. Where the framed cavity or space is ventilated to allow moisture to escape.
3. In counties identified with footnote a in Table N1101.2.

SECTION R323 PROTECTION AGAINST DECAY

R323.1 Location required. In areas subject to decay damage as established by Figure R301.2(7), the following locations shall require the use of an approved species and grade of lumber, pressure preservatively treated in accordance with AWPA C1, C2, C3, C4, C9, C15, C18, C22, C23, C24, C28, P1, P2 and

P3, or decay-resistant heartwood of redwood, black locust, or cedars.

1. Wood joists or the bottom of a wood structural floor when closer than 18 inches (457 mm) or wood girders when closer than 12 inches (305 mm) to exposed ground in crawl spaces or unexcavated area located within the periphery of the building foundation.
2. All sills or plates that rest on concrete or masonry exterior walls and are less than 8 inches (203 mm) from exposed ground.
3. Sills and sleepers on a concrete or masonry slab that is in direct contact with the ground unless separated from such slab by an impervious moisture barrier.
4. The ends of wood girders entering exterior masonry or concrete walls having clearances of less than 0.5 inch (12.7 mm) on tops, sides and ends.
5. Wood siding, sheathing and wall framing on the exterior of a building having a clearance of less than 6 inches (152 mm) from the ground.
6. Wood structural members supporting moisture-permeable floors or roofs that are exposed to the weather, such as concrete or masonry slabs, unless separated from such floors or roofs by animpervious moisture barrier.
7. Wood furring strips or other wood framing members attached directly to the interior of exterior masonry walls or concrete walls below grade except where an approved vapor retarder is applied between the wall and the furring strips or framing members.

R323.1.1 Ground contact. All wood in contact with the ground and that supports permanent structures intended for human occupancy shall be approved pressure preservatively treated wood suitable for ground contact use, except untreated wood may be used where entirely below groundwater level or continuously submerged in fresh water.

R323.1.2 Geographical areas. In geographical areas where experience has demonstrated a specific need, approved naturally durable or pressure preservatively treated wood shall be used for those portions of wood members that form the structural supports of buildings, balconies, porches or similar permanent building appurtenances when such members are exposed to the weather without adequate protection from a roof, eave, overhang or other covering that would prevent moisture or water accumulation on the surface or at joints between members. Depending on local experience, such members may include:

1. Horizontal members such as girders, joists and decking.
2. Vertical members such as posts, poles and columns.
3. Both horizontal and vertical members.

R323.1.3 Posts, poles and columns. Posts, poles and columns supporting permanent structures that are embedded in concrete in direct contact with the ground or embedded in concrete exposed to the weather shall be approved pressure preservatively treated wood suitable for ground contact use.

R323.1.4 Wood columns. Wood columns shall be approved wood of natural decay resistance or approved pressure preservatively treated wood.

Exception: Posts or columns supported by piers or metal pedestals projecting 1 inch (25.4 mm) above the floor or finish grade and are separated therefrom by an approved impervious moisture barrier.

R323.2 Quality mark. Lumber and plywood required to be pressure preservatively treated in accordance with Section R323.1 shall bear the quality mark of an approved inspection agency that maintains continuing supervision, testing and inspection over the quality of the product and that has been approved by an accreditation body that complies with the requirements of the American Lumber Standard Committee treated wood program.

R323.2.1 Required information. The required quality mark on each piece of pressure preservatively treated lumber or plywood shall contain the following information:

1. Identification of the treating plant.
2. Type of preservative.
3. The minimum preservative retention.
4. End use for which the product was treated.
5. Standard to which the product was treated.
6. Identity of the approved inspection agency.
7. The designation "Dry," if applicable.

Exception: Quality marks on lumber less than 1 inch (25.4 mm) nominal thickness, or lumber less than nominal 1 inch by 5 inches (25.4 mm by 127 mm) or 2 inches by 4 inches (51 mm by 102 mm) or lumber 36 inches (914 mm) or less in length shall be applied by stamping the faces of exterior pieces or by end labeling not less than 25 percent of the pieces of a bundled unit.

R323.3 Fasteners. Fasteners for pressure preservative and fire-retardant-treated wood shall be of hot-dipped galvanized steel, stainless steel, silicon bronze or copper.

Exception: One-half-inch (12.7 mm) diameter or greater steel bolts.

SECTION R324
PROTECTION AGAINST TERMITES

R324.1 Subterranean termite control. In areas favorable to termite damage as established by Table R301.2(1), methods of protection shall be by chemical soil treatment, pressure preservatively treated wood in accordance with the AWPA standards listed in Section R323.1, naturally termite-resistant wood or physical barriers (such as metal or plastic termite shields), or any combination of these methods.

R324.2 Chemical soil treatment. The concentration, rate of application and treatment method of the termiticide shall be consistent with and never less than the termiticide label.

R324.3 Pressure preservatively treated and naturally resistant wood. Heartwood of redwood and eastern red cedar shall be considered termite resistant. Pressure preservatively treated wood and naturally termite-resistant wood shall not be used as a physical barrier unless a barrier can be inspected for any termite shelter tubes around the inside and outside edges and joints of a barrier.

R324.3.1 Field treatment. Field cut ends, notches and drilled holes of pressure preservatively treated wood shall be retreated in the field in accordance with AWPA M4.

R324.4 Foam plastic protection. In areas where the probability of termite infestation is "very heavy" as indicated in Figure R301.2(6), extruded and expanded polystyrene, polyisocyanurate and other foam plastics shall not be installed on the exterior face or under interior or exterior foundation walls or slab foundations located below grade. The clearance between foam plastics installed above grade and exposed earth shall be at least 6 inches (152 mm).

Exceptions:

1. Buildings where the structural members of walls, floors, ceilings and roofs are entirely of noncombustible materials or pressure preservatively treated wood.
2. When in addition to the requirements of R324.1, an approved method of protecting the foam plastic and structure from subterranean termite damage is provided.
3. On the interior side of basement walls.

SECTION R325 SITE ADDRESS

R325.1 Premises identification. Approved numbers or addresses shall be provided for all new buildings in such a position as to be plainly visible and legible from the street or road fronting the property.

SECTION R326 ACCESSIBILITY

R326.1 Scope. Accessible dwelling units shall comply with Chapter 11 of the *International Building Code* as applicable.

SECTION R327 FLOOD-RESISTANT CONSTRUCTION

R327.1 General. All buildings and structures erected in areas prone to flooding as identified in Table R301.2(1) and classified as either flood hazard areas (including A Zones) or coastal high hazard areas (including V-Zones) shall be constructed and elevated as required by the provisions contained in this section.

Exception: All buildings and structures erected in identified floodways as established in Table R301.2(1) shall be designed and constructed as stipulated in the *International Building Code*.

R327.1.1 Structural systems. All structural systems of all buildings and structures shall be designed, connected and anchored to resist flotation, collapse or permanent lateral movement due to structural loads and stresses from flooding equal to the design flood elevation.

R327.1.2 Flood-resistant construction. All buildings and structures erected in areas prone to flooding shall be constructed by methods and practices that minimize flood damage.

R327.1.3 Establishing the design flood elevation. The design flood elevation shall be used to define areas prone to flooding, and shall describe, at a minimum, the base flood elevation at the depth of peak elevation of flooding (including wave height) which has a 1 percent (100-year flood) or greater chance of being equaled or exceeded in any given year.

R327.1.4 Lowest floor. The lowest floor shall be the floor of the lowest enclosed area, including basement, but excluding any unfinished flood-resistant enclosure that is useable solely for vehicle parking, building access or limited storage provided that such enclosure is not built so as to render the building or structure in violation of this section.

R327.1.5 Protection of mechanical and electrical systems. New and replacement electrical equipment, heating, ventilating, air conditioning, plumbing connections, and other service equipment shall be located at or above the design flood elevation. Electrical wiring and outlets, switches, junction boxes and panels shall be elevated to or above the design flood elevation unless they conform to the provisions of the electrical part of this code for location of such items in wet locations. Duct systems shall not be installed below the design flood elevation.

R327.1.6 Protection of water supply and sanitary sewage systems. New and replacement water supply systems shall be designed to minimize infiltration of flood waters into the systems in accordance with the plumbing provisions of this code. New and replacement sanitary sewage systems shall be designed to minimize infiltration of floodwaters into systems and discharges from systems into floodwaters in accordance with the plumbing provisions of this code and Chapter 3 of the *International Private Sewage Disposal Code*.

R327.1.7 Flood-resistant materials. Building materials used below the design flood elevation shall comply with the following:

1. All wood, including floor sheathing, shall be pressure preservatively treated in accordance with AWPA C1, C2, C3, C4, C9, C15, C18, C22, C23, C24, C28, P1, P2 and P3 or decay-resistant heartwood or redwood, black locust, or cedars.
2. Materials and installation methods used for flooring and interior and exterior walls shall conform to the provisions of FEMA/FIA-TB-2.

R327.1.8 Manufactured housing. New or replacement manufactured housing shall be elevated in accordance with Section R327.2 and the anchor and tie-down requirements of Sections AE604 and AE605 of Appendix E shall apply. The foundation and anchorage of manufactured housing to be located in identified flood ways as established in Table R301.2(1) shall be designed and constructed in accordance with the applicable provisions in the *International Building Code.*

R327.1.9 As-built elevation certifications. A licensed land surveyor or registered design professional shall certify that the building or structure is in compliance with the elevation requirements of Section R327.2 or R327.3.

R327.2 Flood hazard areas (including A Zones). All areas that have been determined to be prone to flooding but not subject to high velocity wave action shall be designated as flood hazard areas. All buildings and structures erected in flood hazard areas shall be designed and constructed in accordance with Sections R327.2.1 through R327.2.3.

R327.2.1 Elevation requirements.

1. Buildings and structures shall have the lowest floors elevated to or above the design flood elevation.
2. In areas of shallow flooding (AO Zones), buildings and structures shall have the lowest floor (including basement) elevated at least as high above the highest adjacent grade as the depth number specified in feet (mm) on the FIRM, or at least 2 feet (51 mm) if a depth number is not specified.
3. Basement floors that are below grade on all sides shall be elevated to or above the design flood elevation.

Exception: Enclosed areas below the design flood elevation, including basements whose floors are not below grade on all sides, shall meet the requirements of Section R327.2.2.

R327.2.2 Enclosed area below design flood elevation. Enclosed areas, including crawl spaces, that are below the design flood elevation shall:

1. Be used solely for parking of vehicles, building access or storage.
2. Be provided with flood openings which shall meet the following criteria:
   - 2.1. There shall be a minimum of two openings on different sides of each enclosed area; if a building has more than one enclosed area below the design flood elevation, each area shall have openings on exterior walls.
   - 2.2. The total net area of all openings shall be at least 1 square inch for each square foot (275 mm for each square meter) of enclosed area.
   - 2.3. The bottom of each opening shall be 1 foot (305 mm) or less above the adjacent ground level.
   - 2.4. Openings shall be at least 3 inches (76 mm) in diameter.
   - 2.5. Any louvers, screens or other opening covers shall allow the automatic flow of floodwaters into and out of the enclosed area.
   - 2.6. Openings installed in doors and windows, that meet requirements 2.1 through 2.5, are acceptable; however, doors and windows without installed openings do not meet the requirements of this section.

R327.2.3 Foundation design and construction. Foundation walls for all buildings and structures erected in flood hazard areas shall meet the requirements of Chapter 4.

Exception: Unless designed in accordance with Section 404:

1. The unsupported height of 6 inches (152 mm) plain masonry walls shall be no greater than 3 feet (914 mm).
2. The unsupported height of 8 inches (203 mm) plain masonry walls shall be no greater than 4 feet (1219 mm).
3. The unsupported height of 8 inches (203 mm) reinforced masonry walls shall be no greater than 8 feet (2438 mm).

For the purpose of this exception, unsupported height is the distance from the finished grade of the under-floor space and the top of the wall.

R327.3 Coastal high hazard areas (including V Zones). Areas that have been determined to be subject to wave heights in excess of 3 feet (914 mm) or subject to high velocity wave action or wave-induced erosion shall be designated as coastal high hazard areas. All buildings and structures erected in coastal high hazard areas shall be designed and constructed in accordance with Sections R327.3.1 through R327.3.5.

R327.3.1 Elevation requirements.

1. All buildings and structures erected within coastal high hazard areas shall be elevated so that the lowest portion of all structural members supporting the lowest floor, with the exception of mat or raft foundations, piling, pile caps, columns, grade beams and bracing, is located at or above the design flood elevation.
2. Basement floors that are below grade on all sides are prohibited.
3. The use of fill for structural support is prohibited.
4. The placement of fill beneath buildings and structures is prohibited.

Exception: Walls and partitions enclosing areas below the design flood elevation shall meet the requirements of Sections R327.3.3 and R327.3.4.

R327.3.2 Foundations. All buildings and structures erected in coastal high hazard areas shall be supported on pilings or columns and shall be adequately anchored to such pilings or columns. Piling shall have adequate soil penetrations to resist the combined wave and wind loads (lateral and uplift). Water loading values used shall be those associated with the design flood. Wind loading values shall be those required by this code. Pile embedment shall include consideration of decreased resistance capacity caused by scour of soil strata surrounding the piling. Pile systems design and installation shall be certified in accordance with Section R327.3.5. Mat, raft or other foundations that support columns shall not be permitted where soil investigations that are required in accordance with Section R401.4 indicate that soil material under the mat, raft or other foundation is subject to scour or erosion from wave-velocity flow conditions.

R327.3.3 Walls below design flood elevation. Walls and partitions are permitted below the elevated floor, provided that such walls and partitions are not part of the structural support of the building or structure and:

1. Are constructed with insect screening or open lattice.
2. Designed to break away or collapse without causing collapse, displacement or other structural damage to the elevated portion of the building or supporting foundation system. Such walls, framing and connections shall have a design safe loading resistance of not less than 10 pounds per square foot (0.48 kN/m^2) and no more than 20 pounds per square foot (0.96 kN/m^2); or
3. Where wind loading values of this code exceed 20 pounds per square foot (0.96 kN/m^2), a registered design professional shall certify the following:
   3.1. Collapse of walls and partitions below the design flood elevation shall result from a water load less than that which would occur during the design flood.
   3.2. The elevated portion of the building and supporting foundation system shall not be subject to collapse, displacement, or other structural damage due to the effects of wind and flood loads acting simultaneously on all building components (structural and nonstructural). Water loading values used shall be those associated with the design flood. Wind loading values used shall be those required by this code.

R327.3.4 Enclosed areas below design flood elevation. Enclosed areas below the design flood elevation shall be used solely for parking of vehicles, building access or storage.

R327.3.5 Design certificate. A registered design professional shall certify that the design and methods of construction to be used meet the applicable criteria of this section.

CHAPTER 4
FOUNDATIONS

SECTION R401
GENERAL

R401.1 Application. The provisions of this chapter shall control the design and construction of the foundation and foundation spaces for all buildings. Wood foundations shall be designed and installed in accordance with AF&PA Report No. 7.

Exceptions:

1. The provisions of this chapter shall be permitted to be used for wood foundations only in the following situations:
   1.1. In buildings that have no more than two floors and a roof.
   1.2. When no dimension in a basement room or crawl space area exceeds the smaller of either the building width or length.
2. In addition to the provisions of this chapter, the design and construction of foundations in areas prone to flooding as established by Table R301.2(1) shall meet the provisions of Section R327.

Wood foundations in Seismic Design Categories D_1 and D_2 shall be designed in accordance with accepted engineering practice.

R401.2 Requirements. Foundation construction shall be capable of accommodating all loads according to Section R301 and of transmitting the resulting loads to the supporting soil. Fill soils that support footings and foundations shall be designed, installed and tested in accordance with accepted engineering practice. Gravel fill used as footings for wood foundations shall comply with Section R403.

R401.3 Drainage. Surface drainage shall be diverted to a storm sewer conveyance or other approved point of collection so as to not create a hazard. Lots shall be graded so as to drain surface water away from foundation walls. The grade away from foundation walls shall fall a minimum of 6 inches (152 mm) within the first 10 feet (3048 mm).

Exception: Where lot lines, walls, slopes or other physical barriers prohibit 6 inches (152 mm) of fall within 10 feet (3048 mm), drains or swales shall be provided to ensure drainage away from the structure.

R401.4 Soil tests. In areas likely to have expansive, compressible, shifting or other unknown soil characteristics, the building official shall determine whether to require a soil test to determine the soil's characteristics at a particular location. This test shall be made by an approved agency using an approved method.

R401.4.1 Geotechnical evaluation. In lieu of a complete geotechnical evaluation, the load-bearing values in Table R401.4.1 shall be assumed.

TABLE R401.4.1
PRESUMPTIVE LOAD-BEARING VALUES OF FOUNDATION MATERIALS[a]

CLASS OF MATERIAL	LOAD-BEARING PRESSURE (pounds per square foot)
Crystalline bedrock	12,000
Sedimentary and foliated rock	4,000
Sandy gravel and/or gravel (GW and GP)	3,000
Sand, silty sand, clayey sand, silty gravel and clayey gravel (SW, SP, SM, SC, GM and GC)	2,000
Clay, sandy clay, silty clay, clayey silt, silt and sandy silt (CI, ML, MH and CH)	1,500[b]

For SI: 1 pound per square foot = 0.0479 kN/m^2.

a. When soil tests are required by Section R401.4, the allowable bearing capacities of the soil shall be part of the recommendations.

b. Where the building official determines that in-place soils with an allowable bearing capacity of less than 1,500 psf are likely to be present at the site, the allowable bearing capacity shall be determined by a soils investigation.

R401.5 Compressible or shifting soil. When top or subsoils are compressible or shifting, such soils shall be removed to a depth and width sufficient to assure stable moisture content in each active zone and shall not be used as fill or stabilized within each active zone by chemical, dewatering, or presaturation.

SECTION R402
MATERIALS

R402.1 Wood foundations. Wood foundation systems shall be designed and installed in accordance with the provisions of this code.

R402.1.1 Fasteners. Fasteners used below grade to attach plywood to the exterior side of exterior basement or crawlspace wall studs, or fasteners used in knee wall construction, shall be of Type 304 or 316 stainless steel. Fasteners used above grade to attach plywood and all lumber-to-lumber fasteners except those used in knee wall construction shall be of Type 304 or 316 stainless steel, silicon bronze, copper, hot-dipped galvanized (zinc coated) steel nails, or hot-tumbled galvanized (zinc coated) steel nails. Electrogalvanized steel nails and galvanized (zinc coated) steel staples shall not be permitted.

R402.1.2 Wood treatment. All lumber and plywood shall be treated in accordance with AWPA C22, and shall bear the label of an accredited agency showing 0.60 retention. Where lumber and/or plywood is cut or drilled after treatment, the treated surface shall be field treated with Copper Napthenate, the concentration of which shall contain a minimum of 2 percent copper metal, by repeated brushing, dipping or soaking until the wood absorbs no more preservative.

R402.2 Concrete. Concrete shall have a minimum specified compressive strength as shown in Table R402.2. Concrete subject to weathering as indicated in Table R301.2(1) shall be air entrained as specified in Table R402.2. The maximum weight of fly ash, other pozzolans, silica fume, or slag that is included in concrete mixtures for garage floor slabs and for exterior porches, carport slabs, and steps that will be exposed to deicing chemicals shall not exceed the percentages of the total weight of cementitious materials. Materials used to produce concrete and testing thereof shall comply with the applicable standards listed in ACI 318. In addition to the cements permitted by ACI 318, cement complying with ASTM C 1157 is permitted.

SECTION R403
FOOTINGS

R403.1 General. All exterior walls shall be supported on continuous solid or fully grouted masonry or concrete footings, wood foundations, or other approved structural systems which shall be of sufficient design to accommodate all loads according to Section R301 and to transmit the resulting loads to the soil within the limitations as determined from the character of the soil. Footings shall be supported on undisturbed natural soils or engineered fill.

R403.1.1 Minimum size. Minimum sizes for concrete and masonry footings shall be as set forth in Table R403.1 and Figure R403.1(1). The footing width, *W*, shall be based on the load-bearing value of the soil in accordance with Table R401.4.1. Spread footings shall be at least 6 inches (152 mm) in thickness. Footing projections, *P*, shall be at least 2 inches (51 mm) and shall not exceed the thickness of the footing. The size of footings supporting piers and columns shall be based on the tributary load and allowable soil pressure in accordance with Table R401.4.1. Footings for wood foundations shall be in accordance with the details set forth in Section R403.2, and Figures R403.1(2) and R403.1(3).

R403.1.2 Continuous footings in Seismic Design Categories D_1 and D_2. The braced wall panels at exterior walls of all buildings located in Seismic Design Categories D_1 and D_2 shall be supported by continuous footings. All required interior braced wall panels in buildings with plan dimensions greater than 50 feet (15 240 mm) shall also be supported by continuous footings.

R403.1.3 Seismic reinforcing. Concrete footings located in Seismic Design Categories D_1 and D_2, as established in Table R301.2(1), shall have minimum reinforcement. Bottom reinforcement shall be located a minimum of 3 inches (76 mm) clear from the bottom of the footing.

In Seismic Design Categories D_1 and D_2 where a construction joint is created between a concrete footing and stem wall, a minimum of one No. 4 bar shall be provided at not more than 4 feet (1219 mm) on center. The vertical bar shall extend to 3 inches (76 mm) clear of the bottom of the footing, have a standard hook and extend a minimum of 14 inches (357 mm) into the stem wall.

In Seismic Design Categories D_1 and D_2 where a grouted masonry stem wall is supported on a concrete footing and stem wall, a minimum of one No. 4 bar shall be provided at not more than four feet on center. The vertical bar shall extend to 3 inches (76 mm) clear of the bottom of the footing and have a standard hook.

In Seismic Design Categories D_1 and D_2 masonry stem walls without solid grout and vertical reinforcing shall not be permitted.

Exception: In detached one and two family dwellings which are three stories or less in height and constructed with stud bearing walls, plain concrete footings without longitudinal reinforcement supporting walls and isolated plain concrete footings supporting columns or pedestals are permitted.

R403.1.3.1 Foundations with stemwalls. Foundations with stemwalls shall be provided with a minimum of one No. 4 bar at the top of the wall and one No. 4 bar at the bottom of the footing.

R403.1.3.2 Slabs-on-ground with turned-down footings. Slabs-on-ground with turned down footings shall have a minimum of one No. 4 bar at the top and bottom of the footing.

Exception: For slabs-on-ground cast monolithically with a footing, one No. 5 bar or two No. 4 bars shall be located in the middle third of the footing depth.

R403.1.4 Minimum depth. All exterior footings and foundation systems shall extend below the frost line specified in Table R301.2(1). All exterior footings shall be placed at least 12 inches (305 mm) below the undisturbed ground.

Exception: Frost-protected footings constructed in accordance with Section R403.3 and footings and foundations erected on solid rock shall not be required to extend below the frost line.

In Seismic Design Categories D_1 and D_2, interior footings supporting bearing or bracing walls and cast monolithically with a slab on grade shall extend to a depth of not less than 12 inches (325 mm) below the top of slab.

R403.1.5 Slope. The top surface of footings shall be level. The bottom surface of footings shall not have a slope exceeding one unit vertical in 10 units horizontal (10-percent slope). Footings shall be stepped where it is necessary to change the elevation of the top surface of the footings or where the slope of the bottom surface of the footings will exceed one unit vertical in ten units horizontal (10-percent slope).

TABLE R402.2
MINIMUM SPECIFIED COMPRESSIVE STRENGTH OF CONCRETE

TYPE OR LOCATIONS OF CONCRETE CONSTRUCTION	MINIMUM SPECIFIED COMPRESSIVE STRENGTH[a] (f'_c)		
	Weathering potential[b]		
	Negligible	Moderate	Severe
Basement walls, foundations and other concrete not exposed to the weather	2,500	2,500	2,500[c]
Basement slabs and interior slabs on grade, except garage floor slabs	2,500	2,500	2,500[c]
Basement walls, foundation walls, exterior walls and other vertical concrete work exposed to the weather	2,500	3,000[d]	3,000[d]
Porches, carport slabs and steps exposed to the weather, and garage floor slabs	2,500	3,000[d,e]	3,500[d,e]

For SI: 1 pound per square inch = 6.895 kPa.

a. At 28 days psi.

b. See Table R301.2(1) for weathering potential.

c. Concrete in these locations that may be subject to freezing and thawing during construction shall be air-entrained concrete in accordance with Footnote d.

d. Concrete shall be air entrained. Total air content (percent by volume of concrete) shall not be less than 5 percent or more than 7 percent.

e. See Section R402.2 for minimum cement content.

TABLE R403.1
MINIMUM WIDTH OF CONCRETE OR MASONRY FOOTINGS (inches)

	LOAD-BEARING VALUE OF SOIL (psf)					
	1,500	2,000	2,500	3,000	3,500	4,000
Conventional light-frame construction						
1-story	16	12	10	8	7	6
2-story	19	15	12	10	8	7
3-story	22	17	14	11	10	9
4-inch brick veneer over light frame or 8-inch hollow concrete masonry						
1-story	19	15	12	10	8	7
2-story	25	19	15	13	11	10
3-story	31	23	19	16	13	12
8-inch solid or fully grouted masonry						
1-story	22	17	13	11	10	9
2-story	31	23	19	16	13	12
3-story	40	30	24	20	17	15

For SI: 1 inch = 25.4 mm, 1 pound per square foot = 0.0479 kN/m^2.

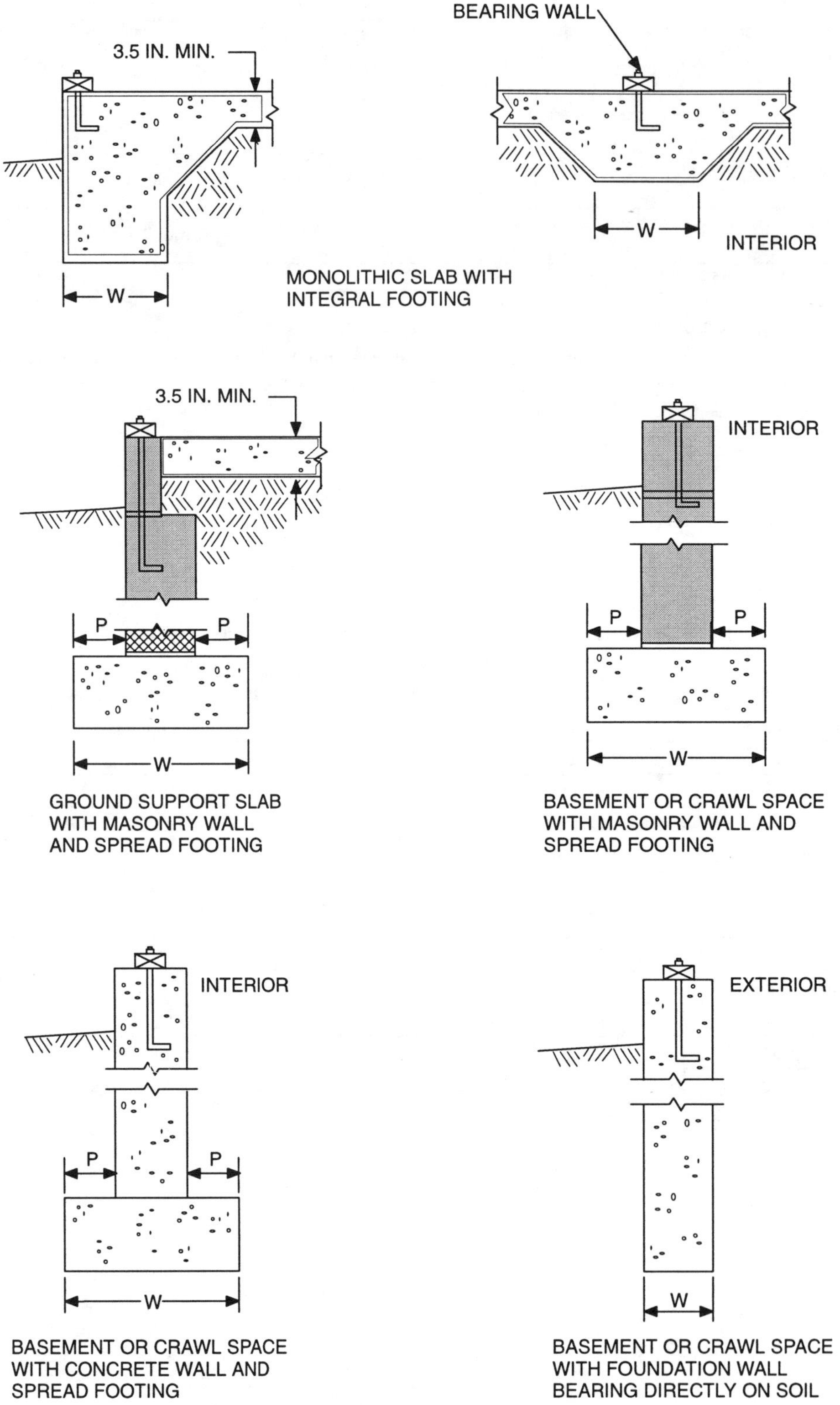

For SI: 1 inch = 25.4 mm.

FIGURE R403.1(1)
CONCRETE AND MASONRY FOUNDATION DETAILS

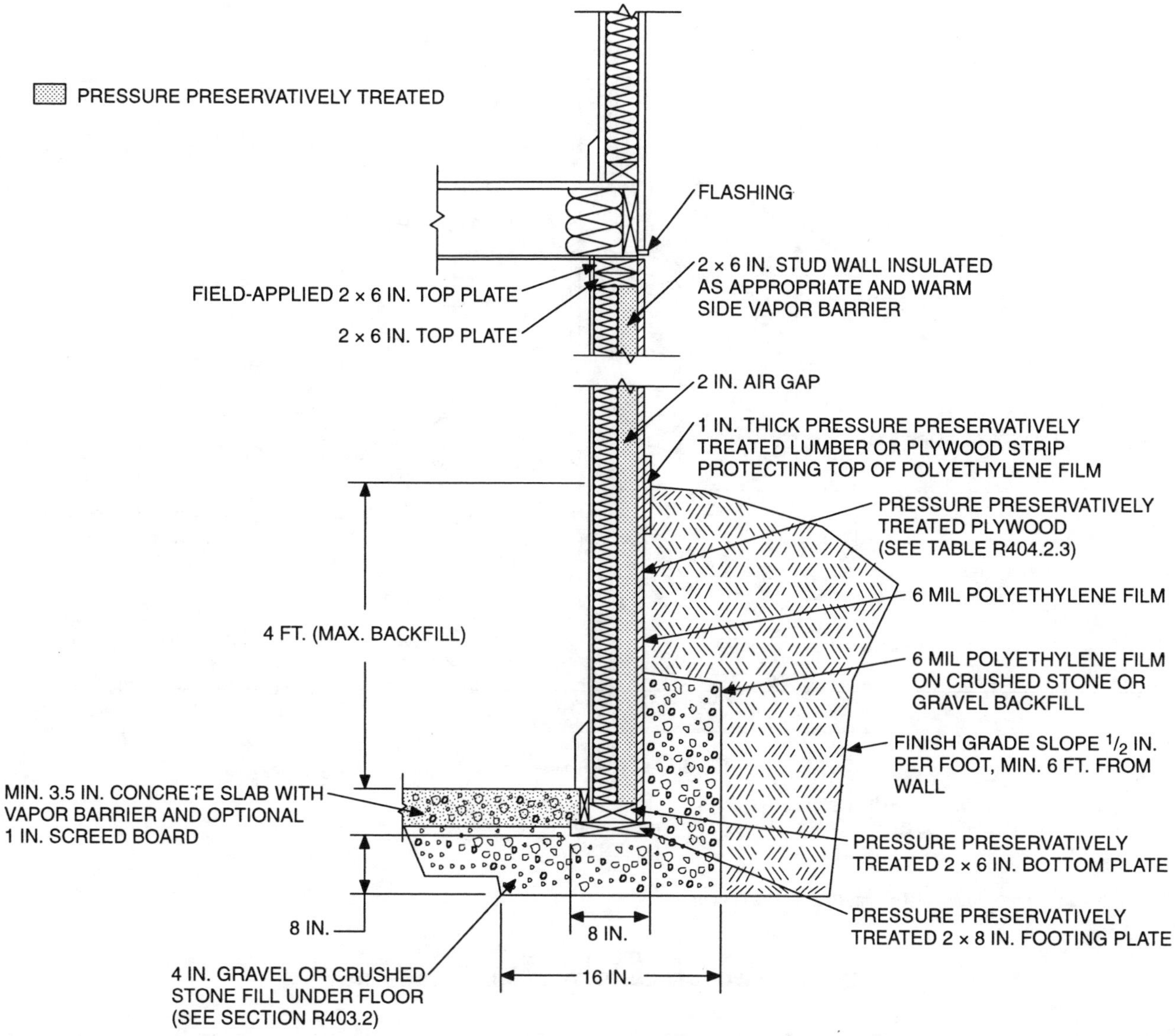

For SI: 1 inch = 25.4 mm, 1 foot = 304.8 mm, 1 mil = 0.0254 mm.

FIGURE R403.1(2)
PERMANENT WOOD FOUNDATION BASEMENT WALL SECTION

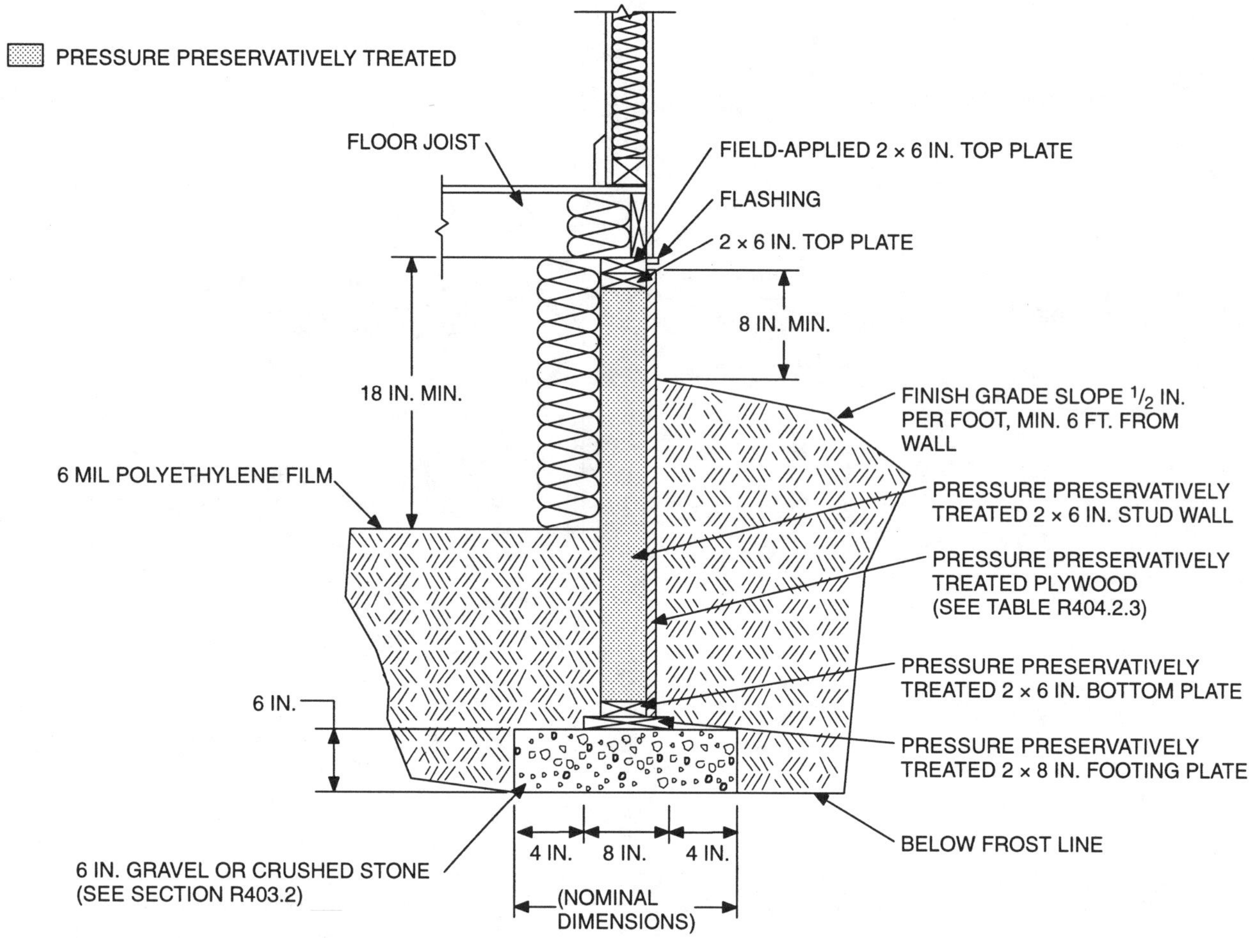

For SI: 1 inch = 25.4 mm, 1 foot = 304.8 mm, 1 mil = 0.0254 mm.

FIGURE R403.1(3)
PERMANENT WOOD FOUNDATION CRAWL SPACE SECTION

R403.1.6 Foundation anchorage. When braced wall panels are supported directly on continuous foundations, the wall wood sill plate or cold-formed steel bottom track shall be anchored to the foundation in accordance with this section. In Seismic Design Categories D_1 and D_2, the additional anchorage requirements of Section R602.11.1 shall apply for wood framing. In Seismic Design Categories D_1 and D_2 where continuous wood foundations in accordance with Section R404.2 are used, the force transfer shall have a capacity equal to or greater than the connections required by R602.11.1 or the braced wall panel shall be connected to the wood foundations in accordance with the braced wall panel-to-floor fastening requirements of Table R602.3(1).

The wood sole plate at exterior walls on monolithic slabs and wood sill plate shall be anchored to the foundation with anchor bolts spaced a maximum of 6 feet (1829 mm) on center. Anchor bolts shall also be located within 12 inches (305 mm) from the ends of each plate section. In Seismic Design Categories D_1 and D_2, anchor bolts shall also be spaced at 6 feet (1829 mm) on center and located within 12 inches (305 mm) from the ends of each plate section at interior braced wall lines when required by Section R602.10.9 to be supported on a continuous foundation. Bolts shall be at least $^1/_2$ inch (12.7 mm) in diameter and shall extend a minimum of 7 inches (178 mm) into masonry or concrete. Interior bearing wall sole plates on monolithic slab foundations shall be positively anchored with approved fasteners. A nut and washer shall be tightened on each bolt to the plate. Sills and sole plates shall be protected against decay and termites where required by Sections R322 and R323. Cold-formed steel framing systems shall be fastened to the wood sill plates or anchored directly to the foundation as required in Section R505.3.1 or R603.1.1.

Exception: Foundation anchor straps, spaced as required to provide equivalent anchorage to $^1/_2$-inch-diameter (12.7 mm) anchor bolts.

R403.1.6.1 Foundation anchorage in Seismic Design Categories D_1 and D_2. In addition to the requirements of Section R403.1.6, the following requirements shall apply to light-wood frame structures in Seismic Design Categories D_1 and D_2. Anchor bolts shall be located within 12 inches (305 mm) from the ends of each plate section at interior bearing walls, interior braced wall lines and at all exterior walls. Plate washers a minimum of 2 inches by

2 inches by $^{3}/_{16}$ inch (51 mm by 51 mm by 4.8 mm) thick shall be used on each bolt. The maximum anchor bolt spacing shall be 4 feet (1219 mm) for two-story structures.

R403.1.7 Footings on or adjacent to slopes. The placement of buildings and structures on or adjacent to slopes steeper than 1 unit vertical in 3 units horizontal (33.3-percent slope) shall conform to Sections R403.1.7.1 through R403.1.7.4.

R403.1.7.1 Building clearances from ascending slopes. In general, buildings below slopes shall be set a sufficient distance from the slope to provide protection from slope drainage, erosion and shallow failures. Except as provided in Section R403.1.7.4 and Figure R403.1.7.1, the following criteria will be assumed to provide this protection. Where the existing slope is steeper than one unit vertical in one unit horizontal (100-percent slope), the toe of the slope shall be assumed to be at the intersection of a horizontal plane drawn from the top of the foundation and a plane drawn tangent to the slope at an angle of 45 degrees (0.79 rad) to the horizontal. Where a retaining wall is constructed at the toe of the slope, the height of the slope shall be measured from the top of the wall to the top of the slope.

R403.1.7.2 Footing setback from descending slope surfaces. Footings on or adjacent to slope surfaces shall be founded in material with an embedment and setback from the slope surface sufficient to provide vertical and lateral support for the footing without detrimental settlement. Except as provided for in Section R403.1.7.4 and Figure R403.1.7.1, the following setback is deemed adequate to meet the criteria. Where the slope is steeper than one unit vertical in one unit horizontal (100-percent slope), the required setback shall be measured from an imaginary plane 45 degrees (0.79 rad) to the horizontal, projected upward from the toe of the slope.

R403.1.7.3 Foundation elevation. On graded sites, the top of any exterior foundation shall extend above the elevation of the street gutter at point of discharge or the inlet of an approved drainage device a minimum of 12 inches (305 mm) plus 2 percent. Alternate elevations are permitted subject to the approval of the building official, provided it can be demonstrated that required drainage to the point of discharge and away from the structure is provided at all locations on the site.

R403.1.7.4 Alternate setback and clearances. Alternate setbacks and clearances are permitted, subject to the approval of the building official. The building official is permitted to require an investigation and recommendation of a qualified engineer to demonstrate that the intent of this section has been satisfied. Such an investigation shall include consideration of material, height of slope, slope gradient, load intensity and erosion characteristics of slope material.

R403.1.8 Foundations on expansive soils. Foundation and floor slabs for buildings located on expansive soils shall be designed in accordance with Section 1805.8 of the *International Building Code*.

Exception: Slab-on-ground and other foundation systems which have performed adequately in soil conditions similar to those encountered at the building site are permitted subject to the approval of the building official.

R403.1.8.1 Expansive soils classifications. Soils meeting all four of the following provisions shall be considered expansive, except that tests to show compliance with Items 1, 2 and 3 shall not be required if the test prescribed in Item 4 is conducted:

1. Plasticity Index (PI) of 15 or greater, determined in accordance with ASTM D 4318.
2. More than 10 percent of the soil particles pass a No. 200 sieve (75 μm), determined in accordance with ASTM D 422.
3. More than 10 percent of the soil particles are less than 5 micrometers in size, determined in accordance with ASTM D 422.
4. Expansion Index greater than 20, determined in accordance with UBC Standard 18-1.

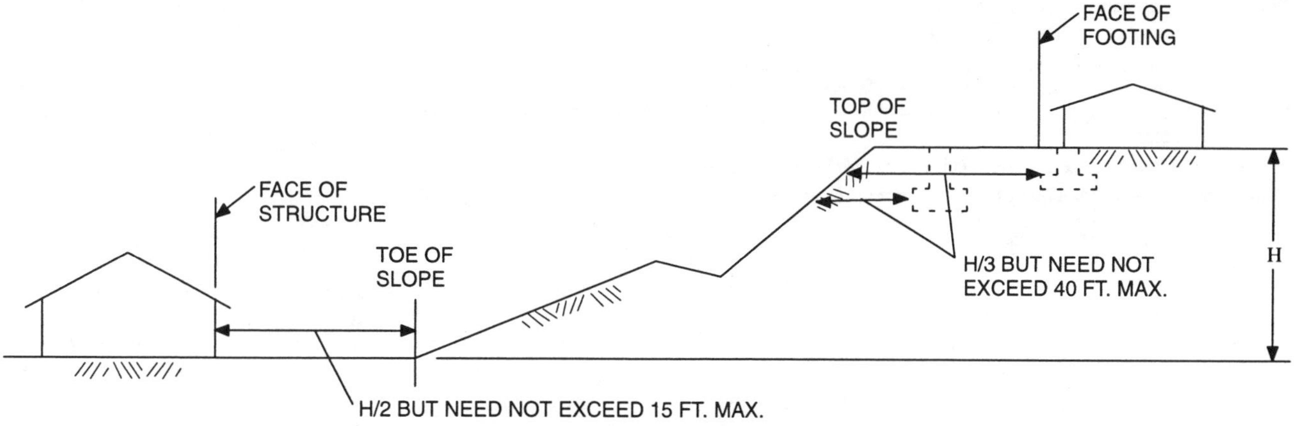

For SI: 1 foot = 304.8 mm.

FIGURE R403.1.7.1
FOUNDATION CLEARANCE FROM SLOPES

R403.2 Footings for wood foundations. Footings for wood foundations shall be in accordance with Figures R403.1(2) and R403.1(3). Gravel shall be washed and well graded. The maximum size stone shall not exceed $^3/_4$ inch (19.1 mm). Gravel shall be free from organic, clayey or silty soils. Sand shall be coarse, not smaller than $^1/_{16}$-inch (1.6 mm) grains and shall be free from organic, clayey or silty soils. Crushed stone shall have a maximum size of $^1/_2$ inch (12.7 mm).

R403.3 Frost protected shallow foundations. For buildings where the monthly mean temperature of the building is maintained at a minimum of 64°F (18°C), footings are not required to extend below the frost line when protected from frost by insulation in accordance with Figure R403.3(1) and Table R403.3.

Exceptions:

1. No foundation not so protected may be attached to frost-protected shallow foundations.
2. Unheated garages, porches, utility rooms and carports shall not be permitted to be attached to dwelling units with frost-protected shallow foundation.

Materials used below grade for the purpose of insulating footings against frost shall be labeled as complying with ASTM C 578.

R403.3.1 Protection of horizontal insulation below gound. Horizontal insulation placed less than 12 inches (305 mm) below the ground surface or that portion of horizontal insulation extending outward more than 24 inches (610 mm) from the foundation edge shall be protected against damage by use of a concrete slab or asphalt paving on the ground surface directly above the insulation or by cementitious board, plywood rated for below-ground use, or other approved materials placed below ground, directly above the top surface of the insulation.

R403.3.2 Drainage. Final grade shall be sloped in accordance with Section R401.3. In other than Group I Soils, as detailed in Table R405.1, gravel or crushed stone beneath horizontal insulation below ground shall drain to daylight or into an approved sewer system.

R403.3.3 Termite damage. The use of foam plastic in areas of "very heavy" termite infestation probability shall be in accordance with Section R324.4.

TABLE R403.3
MINIMUM INSULATION REQUIREMENTS FOR FROST-PROTECTED FOOTINGS IN HEATED BUILDINGS[a]

AIR FREEZING INDEX (°F-days)[b]	VERTICAL INSULATION *R*-VALUE[c,d]	HORIZONTAL INSULATION *R*-VALUE[c,e]		HORIZONTAL INSULATION DIMENSIONS PER FIGURE R403.3(1) (inches)		
		Along walls	At corners	A	B	C
1,500 or less	4.5	NR	NR	NR	NR	NR
2,000	5.6	NR	NR	NR	NR	NR
2,500	6.7	1.7	4.9	12	24	40
3,000	7.8	6.5	8.6	12	24	40
3,500	9.0	8.0	11.2	24	30	60
4,000	10.1	10.5	13.1	24	36	60

For SI: 1 inch = 25.4 mm, °C = [(°F)-32]/1.8.

a. Insulation requirements are for protection against frost damage in heated buildings. Greater values may be required to meet energy conservation standards. Interpolation between values is permissible.

b. See Figure R403.3(2) for Air Freezing Index values.

c. Insulation materials shall provide the stated minimum *R*-values under long-term exposure to moist, below-ground conditions in freezing climates. The following *R*-values shall be used to determine insulation thicknesses required for this application: Type II expanded polystyrene—2.4R per inch; Type IV extruded polystyrene—4.5R per inch; Type VI extruded polystyrene—4.5R per inch; Type IX expanded polystyrene—3.2R per inch; Type X extruded polystyrene—4.5R per inch. NR denotes "not required."

d. Vertical insulation shall be expanded polystyrene insulation or extruded polystyrene insulation.

e. Horizontal insulation shall be extruded polystyrene insulation.

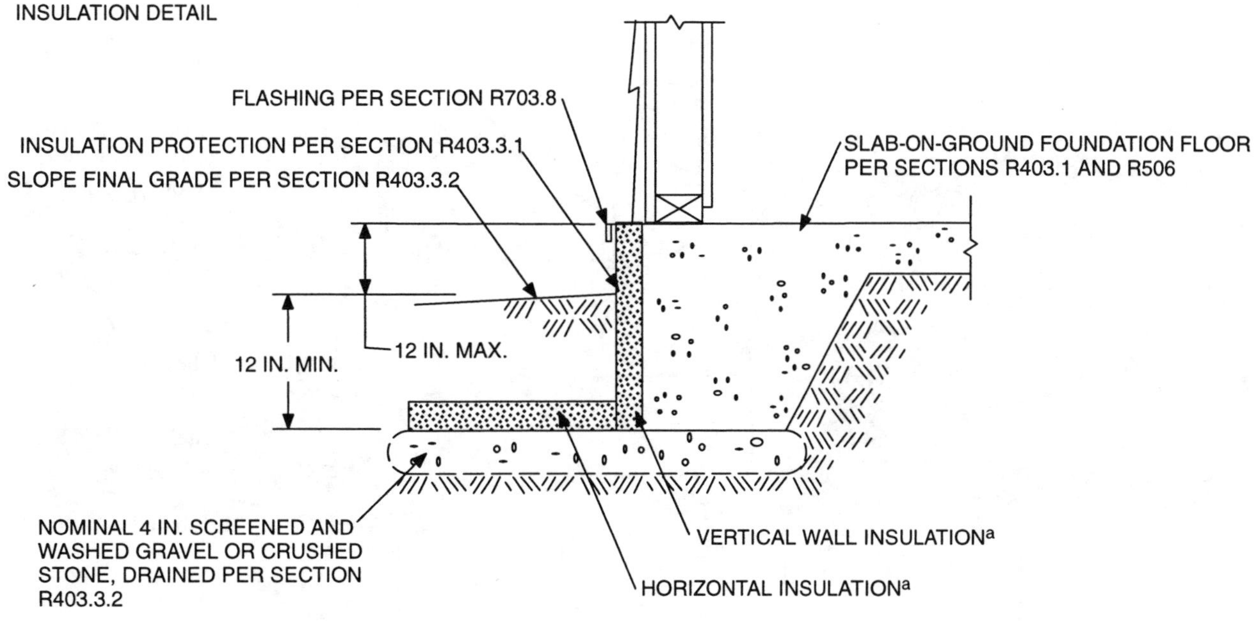

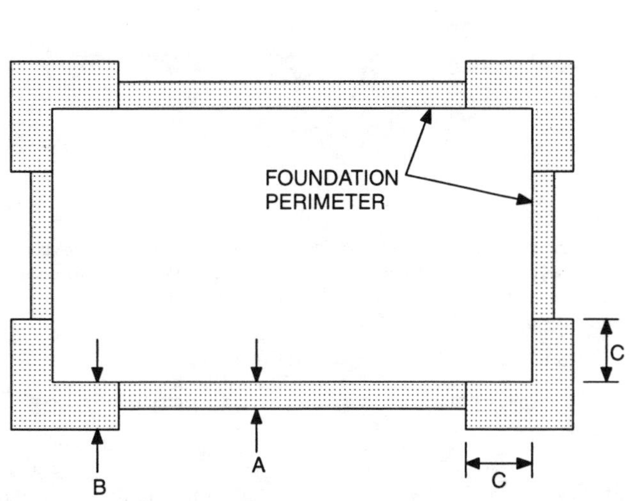

For SI: 1 inch = 25.4 mm.

a. See Table R403.3 for required dimensions and *R*-values for vertical and horizontal insulation.

FIGURE R403.3(1)
INSULATION PLACEMENT FOR FROST-PROTECTED FOOTINGS IN HEATED BUILDINGS

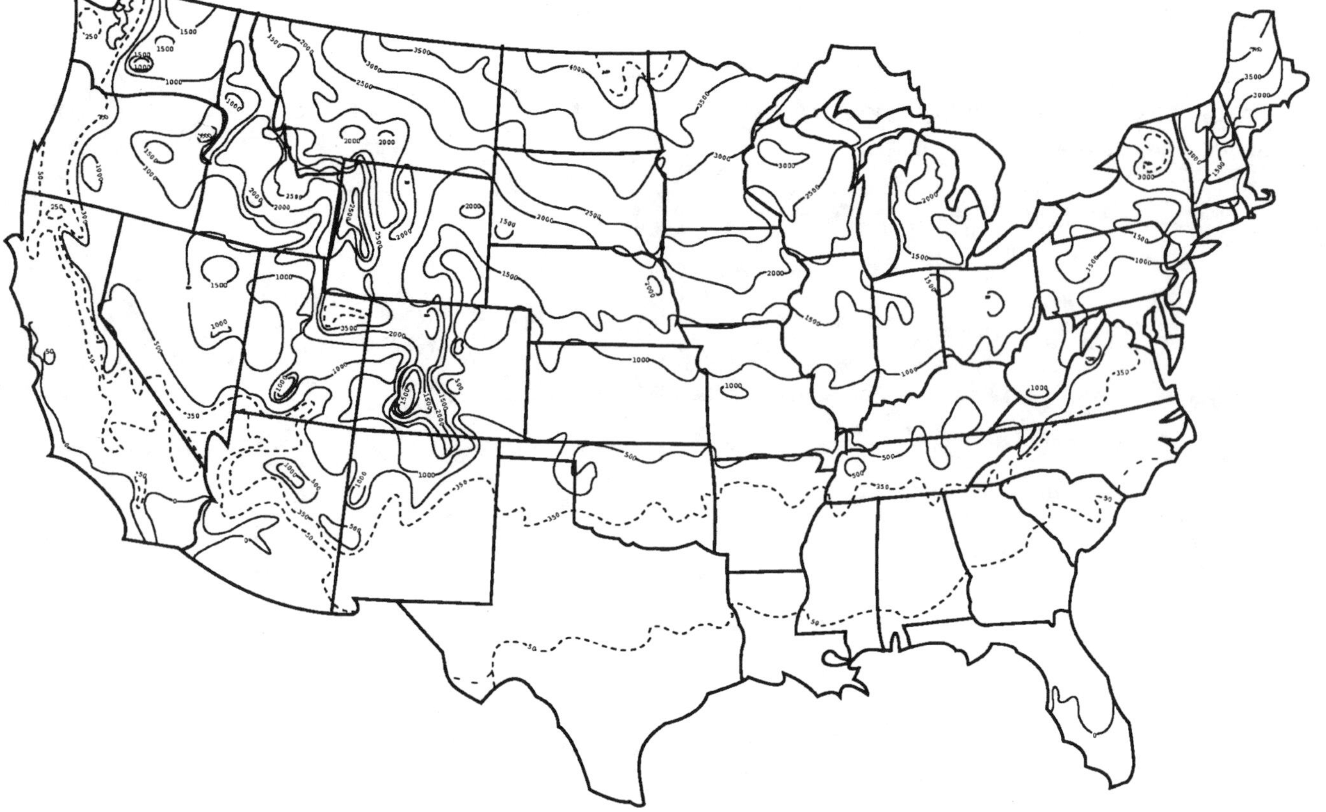

For SI: °C = [(°F)-32]/1.8.

Note: The air-freezing index is defined as cumulative degree days below 325°F. It is used as a measure of the combined magnitude and duration of air temperature below freezing. The index was computed over a 12-month period (July-June) for each of the 3,044 stations used in the above analysis. Data from the 1951-80 period were fitted to a Weibull probability distribution to produce an estimate of the 100-year return period.

FIGURE R403.3(2)
AIR-FREEZING INDEX
AN ESTIMATE OF THE 100-YEAR RETURN PERIOD

SECTION R404
FOUNDATION WALLS

R404.1 Concrete and masonry foundation walls. Concrete and masonry foundation walls shall be selected and constructed in accordance with the provisions of this section or in accordance with ACI 318, NCMA TR68-A or ACI 530/ASCE 5/TMS 402 or other approved structural standards. When ACI 318 or ACI 530/ASCE 5/TMS 402 or the provisions of this section are used to design concrete or masonry foundation walls, project drawings, typical details and specifications are not required to bear the seal of the architect or engineer responsible for design, unless otherwise required by the state law of the jurisdiction having authority.

R404.1.1 Masonry foundation walls. Concrete masonry and clay masonry foundation walls shall be constructed as set forth in Tables R404.1.1(1), R404.1.1(2), R404.1.1(3) and R404.1.1(4) and shall also comply with the provisions of this section and the applicable provisions of Sections R606, R607 and R608. In Seismic Design Categories D_1 and D_2, concrete masonry and clay masonry foundation walls shall comply with Section R404.1.4. Rubble stone masonry foundation walls shall be constructed in accordance with Sections R404.1.8 and R606.2.2. Rubble stone masonry walls shall not be used in Seismic Design Categories D_1 and D_2.

R404.1.2 Concrete foundation walls. Concrete foundation walls shall be constructed as set forth in Tables R404.1.1(1), R404.1.1(2), R404.1.1(3) and R404.1.1(4), and shall also comply with the provisions of this section and the applicable provisions of Section R402.2. In Seismic Design Categories D_1 and D_2, concrete foundation walls shall comply with Section R404.1.4.

R404.1.3 Design required. A design in accordance with accepted engineering practice shall be provided for concrete or masonry foundation walls when any of the following conditions exist:

1. Walls are subject to hydrostatic pressure from groundwater.

2. Walls supporting more than 48 inches (1219 mm) of unbalanced backfill that do not have permanent lateral support at the top and bottom.

R404.1.4 Seismic Design Categories D_1 and D_2. In addition to the requirements of Table R404.1.1(1), plain concrete and plain masonry foundation walls located in Seismic Design Categories D_1 and D_2, as established in Table R301.2(1), shall comply with the following:

1. Minimum reinforcement shall consist of one No. 4 (No. 13) horizontal bar located in the upper 12 inches (305 mm) of the wall,
2. Wall height shall not exceed 8 feet (2438 mm),
3. Height of unbalanced backfill shall not exceed 4 feet (1219 mm), and
4. A minimum thickness of 7.5 inches (191 mm) is required for plain concrete foundation walls except that a minimum thickness of 6 inches (152 mm) shall be permitted for plain concrete foundation walls with a maximum height of 4 feet, 6 inches (1372 mm).
5. Plain masonry foundation walls shall be a minimum of 8 inches (203 mm) thick.

Vertical reinforcement for masonry stem walls shall be tied to the horizontal reinforcement in the footings. Masonry stem walls located in Seismic Design Categories D_1 and D_2 shall have a minimum vertical reinforcement of one No. 3 bar located a maximum of 4 feet (1220 mm) on center in grouted cells.

Foundation walls located in Seismic Design Categories D_1 and D_2, as established in Table R301.2(1), supporting more than 4 feet (1219 mm) of unbalanced backfill or exceeding 8 feet (2438 mm) in height shall be constructed in accordance with Table R404.1.1(2), R404.1.1(3) or R404.1.1(4) and shall have two No. 4 (No. 13) horizontal bars located in the upper 12 inches (305 mm) of the wall.

TABLE R404.1.1(1)
PLAIN CONCRETE AND PLAIN MASONRY FOUNDATION WALLS

MAXIMUM WALL HEIGHT (feet)	MAXIMUM UNBALANCED BACKFILL HEIGHT[c] (feet)	PLAIN CONCRETE MINIMUM NOMINAL WALL THICKNESS (inches)			PLAIN MASONRY[a] MINIMUM NOMINAL WALL THICKNESS (inches)		
		Soil classes[b]					
		GW, GP, SW and SP	GM, GC, SM, SM-SC and ML	SC, MH, ML-CL and inorganic CL	GW, GP, SW and SP	GM, GC, SM, SM-SC and ML	SC, MH, ML-CL and inorganic CL
5	4	6	6	6	6 solid[d] or 8	6 solid[d] or 8	6 solid[d] or 8
	5	6	6	6	6 solid[d] or 8	8	10
6	4	6	6	6	6 solid[d] or 8	6 solid[d] or 8	6 solid[d] or 8
	5	6	6	6	6 solid[d] or 8	8	10
	6	6	8[g]	8[g]	8	10	12
7	4	6	6	6	6 solid[d] or 8	8	8
	5	6	6	8[g]	6 solid[d] or 8	10	10
	6	6	8	8	10	12	10 solid[d]
	7	8	8	10	12	10 solid[d]	12 solid[d]
8	4	6	6	6	6 solid[d] or 8	6 solid[d] or 8	8
	5	6	6	8	6 solid[d] or 8	10	12
	6	8[h]	8	10	10	12	12 solid[d]
	7	8	10	10	12	12 solid[d]	Footnote e
	8	10	10	12	10 solid[d]	12 solid[d]	Footnote e
9	4	6	6	6	6 solid[d] or 8	6 solid[d] or 8	8
	5	6	8[g]	8	8	10	12
	6	8	8	10	10	12	12 solid[d]
	7	8	10	10	12	12 solid[d]	Footnote e
	8	10	10	12	12 solid[d]	Footnote e	Footnote e
	9	10	12	Footnote f	Footnote e	Footnote e	Footnote e

For SI: 1 inch = 25.4 mm, 1 foot = 304.8 mm, 1 pound per square inch = 6.895 Pa.

a. Mortar shall be Type M or S and masonry shall be laid in running bond. Ungrouted hollow masonry units are permitted except where otherwise indicated.

b. Soil classes are in accordance with the Unified Soil Classification System. Refer to Table R405.1.

c. Unbalanced backfill height is the difference in height of the exterior and interior finish ground levels. Where an interior concrete slab is provided, the unbalanced backfill height shall be measured from the exterior finish ground level to the top of the interior concrete slab.

d. Solid grouted hollow units or solid masonry units.

e. Wall construction shall be in accordance with Table R404.1.1(2) or a design shall be provided.

f. A design is required.

g. Thickness may be 6 inches, provided minimum specified compressive strength of concrete, f_c, is 4,000 psi.

TABLE R404.1.1(2) REINFORCED CONCRETE AND MASONRY[a] FOUNDATION WALLS

MAXIMUM WALL HEIGHT (feet)	MAXIMUM UNBALANCED BACKFILL HEIGHT[e] (feet)	MINIMUM VERTICAL REINFORCEMENT SIZE AND SPACING[b, c] FOR 8-INCH NOMINAL WALL THICKNESS		
		Soil classes[d]		
		GW, GP, SW and SP soils	GM, GC, SM, SM-SC and ML soils	SC, MH, ML-CL and inorganic CL soils
6	5	#4 at 48″ o.c.	#4 at 48″ o.c.	#4 at 48″ o.c.
	6	#4 at 48″ o.c.	#4 at 40″ o.c.	#5 at 48″ o.c.
7	4	#4 at 48″ o.c.	#4 at 48″ o.c.	#4 at 48″ o.c.
	5	#4 at 48″ o.c.	#4 at 48″ o.c.	#4 at 40″ o.c.
	6	#4 at 48″ o.c.	#5 at 48″ o.c.	#5 at 40″ o.c.
	7	#4 at 40″ o.c.	#5 at 40″ o.c.	#6 at 48″ o.c.
8	5	#4 at 48″ o.c.	#4 at 48″ o.c.	#4 at 40″ o.c.
	6	#4 at 48″ o.c.	#5 at 48″ o.c.	#5 at 40″ o.c.
	7	#5 at 48″ o.c.	#6 at 48″ o.c.	#6 at 40″ o.c.
	8	#5 at 40″ o.c.	#6 at 40″ o.c.	#6 at 24″ o.c.
9	5	#4 at 48″ o.c.	#4 at 48″ o.c.	#5 at 48″ o.c.
	6	#4 at 48″ o.c.	#5 at 48″ o.c.	#6 at 48″ o.c.
	7	#5 at 48″ o.c.	#6 at 48″ o.c.	#6 at 32″ o.c.
	8	#5 at 40″ o.c.	#6 at 32″ o.c.	#6 at 24″ o.c.
	9	#6 at 40″ o.c.	#6 at 24″ o.c.	#6 at 16″ o.c.

For SI: 1 inch = 25.4 mm, 1 foot = 304.8 mm.

a. Mortar shall be Type M or S and masonry shall be laid in running bond.

b. Alternative reinforcing bar sizes and spacings having an equivalent cross-sectional area of reinforcement per lineal foot of wall shall be permitted provided the spacing of the reinforcement does not exceed 72 inches.

c. Vertical reinforcement shall be Grade 60 minimum. The distance from the face of the soil side of the wall to the center of vertical reinforcement shall be at least 5 inches.

d. Soil classes are in accordance with the Unified Soil Classification System. Refer to Table R405.1.

e. Unbalanced backfill height is the difference in height of the exterior and interior finish ground levels. Where an interior concrete slab is provided, the unbalanced backfill height shall be measured from the exterior finish ground level to the top of the interior concrete slab.

TABLE R404.1.1(3) REINFORCED CONCRETE AND MASONRY[a] FOUNDATION WALLS

MAXIMUM WALL HEIGHT (feet)	MAXIMUM UNBALANCED BACKFILL HEIGHT[e] (feet)	VERTICAL REINFORCEMENT SIZE AND SPACING[b,c] FOR 12-INCH NOMINAL WALL THICKNESS		
		Soil classes[d]		
		GW, GP, SW and SP soils	GM, GC, SM, SM-SC and ML soils	SC, MH, ML-CL and inorganic CL soils
7	4	#4 at 72″ o.c.	#4 at 72″ o.c.	#4 at 72″ o.c.
	5	#4 at 72″ o.c.	#4 at 72″ o.c.	#4 at 72″ o.c.
	6	#4 at 72″ o.c.	#4 at 64″ o.c.	#4 at 48″ o.c.
	7	#4 at 72″ o.c.	#4 at 48″ o.c.	#5 at 56″ o.c.
8	5	#4 at 72″ o.c.	#4 at 72″ o.c.	#4 at 72″ o.c.
	6	#4 at 72″ o.c.	#4 at 56″ o.c.	#5 at 72″ o.c.
	7	#4 at 64″ o.c.	#5 at 64″ o.c.	#4 at 32″ o.c.
	8	#4 at 48″ o.c.	#4 at 32″ o.c.	#5 at 40″ o.c.
9	5	#4 at 72″ o.c.	#4 at 72″ o.c.	#4 at 72″ o.c.
	6	#4 at 72″ o.c.	#4 at 56″ o.c.	#5 at 64″ o.c.
	7	#4 at 56″ o.c.	#4 at 40″ o.c.	#6 at 64″ o.c.
	8	#4 at 64″ o.c.	#6 at 64″ o.c.	#6 at 48″ o.c.
	9	#5 at 56″ o.c.	#7 at 72″ o.c.	#6 at 40″ o.c.

For SI: 1 inch = 25.4 mm, 1 foot = 304.8 mm.

a. Mortar shall be Type M or S and masonry shall be laid in running bond.

b. Alternative reinforcing bar sizes and spacing having an equivalent cross-sectional area of reinforcement per lineal foot of wall shall be permitted provided the spacing of the reinforcement does not exceed 72 inches.

c. Vertical reinforcement shall be Grade 60 minimum. The distance from the face of the soil side of the wall to the center of vertical reinforcement shall be at least 8.75 inches.

d. Soil classes are in accordance with the Unified Soil Classification System. Refer to Table R405.1.

e. Unbalanced backfill height is the difference in height of the exterior and interior finish ground levels. Where an interior concrete slab is provided, the unbalanced backfill height shall be measured from the exterior finish ground level to the top of the interior concrete slab.

TABLE R404.1.1(4)
REINFORCED CONCRETE AND MASONRY[a] FOUNDATION WALLS

MAXIMUM WALL HEIGHT (feet)	MAXIMUM UNBALANCED BACKFILL HEIGHT[e] (feet)	MINIMUM VERTICAL REINFORCEMENT SIZE AND SPACING[b,c] FOR 10-INCH NOMINAL WALL THICKNESS		
		Soil Classes[d]		
		GW, GP, SW and SP soils	GM, GC, SM, SM-SC and ML soils	SC, MH, ML-CL and inorganic CL soils
7	4	#4 at 56″ o.c.	#4 at 56″ o.c.	#4 at 56″ o.c
	5	#4 at 56″ o.c	#4 at 56″ o.c.	#4 at 56″ o.c.
	6	#4 at 56″ o.c.	#4 at 48″ o.c	#4 at 40″ o.c.
	7	#4 at 56″ o.c.	#5 at 56″ o.c.	#5 at 40″ o.c.
8	5	#4 at 56″ o.c.	#4 at 56″ o.c.	#4 at 48″ o.c.
	6	#4 at 56″ o.c	#4 at 48″ o.c	#5 at 56″ o.c
	7	#4 at 48″ o.c.	#4 at 32″ o.c.	#6 at 56″ o.c.
	8	#5 at 56″ o.c.	#5 at 40″ o.c.	#7 at 56″ o.c.
9	5	#4 at 56″ o.c.	#4 at 56″ o.c.	#4 at 48″ o.c.
	6	#4 at 56″ o.c	#4 at 40″ o.c	#4 at 32″ o.c
	7	#4 at 56″ o.c.	#5 at 48″ o.c.	#6 at 48″ o.c.
	8	#4 at 32″ o.c.	#6 at 48″ o.c.	#4 at 16″ o.c.
	9	#5 at 40″ o.c.	#6 at 40″ o.c.	#7 at 40″ o.c.

For SI: 1 inch = 25.4 mm, 1 foot = 304.8 mm.

a. Mortar shall be Type M or S and masonry shall be laid in running bond.

b. Alternative reinforcing bar sizes and spacings having an equivalent cross-sectional area of reinforcement per lineal foot of wall shall be permitted provided the spacing of the reinforcement does not exceed 72 inches.

c. Vertical reinforcement shall be Grade 60 minimum. The distance from the face of the soil side of the wall to the center of vertical reinforcement shall be at least 6.75 inches.

d. Soil classes are in accordance with the Unified Soil Classification System. Refer to Table R405.1.

e. Unbalanced backfill height is the difference in height of the exterior and interior finish ground levels. Where an interior concrete slab is provided, the unbalanced backfill height shall be measured from the exterior finish ground level to the top of the interior concrete slab.

R404.1.5 Foundation wall thickness based on walls supported. The thickness of concrete and masonry foundation walls shall not be less than the thickness of the wall supported, except that foundation walls of at least 8-inch (203 mm) nominal thickness shall be permitted under brick-veneered frame walls and under 10-inch-wide (254 mm) cavity walls where the total height of the wall supported, including gables, is not more than 20 feet (6096 mm), provided the requirements of Sections R404.1.1 and R404.1.2 are met.

R404.1.5.1 Pier and curtain wall foundations. Except in Seismic Design Categories D and E, pier and curtain wall foundations shall be permitted to be used to support light-frame construction not more than two stories in height, provided the following requirements are met:

1. All load-bearing walls shall be placed on continuous concrete footings placed integrally with the exterior wall footings.
2. The minimum actual thickness of a load-bearing masonry wall shall be not less than 4 inches (102 mm) nominal or $3^3/_8$ inches (92 mm) actual thickness, and shall be bonded integrally with piers spaced in accordance with R606.8.
3. Piers shall be constructed in accordance with Section R606.5 and Section R606.5.1, and shall be bonded into the load-bearing masonry wall in accordance with Section R608.1.1 or Section R608.1.1.2.
4. The maximum height of a 4-inch (102 mm) load-bearing masonry foundation wall supporting wood framed walls and floors shall not be more than 4 feet (1219 mm) in height.
5. Anchorage shall be in accordance with Section R403.1.6 or as specified by engineered design accepted by the building official.
6. The unbalanced fill for 4-inch (102 mm) foundation walls shall not exceed 24 inches (610 mm) for solid masonry or 12 inches (305 mm) for hollow masonry.

R404.1.6 Height above finished grade. Concrete and masonry foundation walls shall extend above the finished grade adjacent to the foundation at all points a minimum of 4 inches (102 mm) where masonry veneer is used and a minimum of 6 inches (152 mm) elsewhere.

R404.1.7 Backfill placement. Backfill shall not be placed against the wall until the wall has sufficient strength and has been anchored to the floor above, or has been sufficiently braced to prevent damage by the backfill.

Exception: Such bracing is not required for walls supporting less than 4 feet (1219 mm) of unbalanced backfill.

R404.1.8 Rubble stone masonry. Rubble stone masonry foundation walls shall have a minimum thickness of 16 inches (406 mm), shall not support an unbalanced backfill exceeding 8 feet (2438 mm) in height, shall not support a soil pressure greater than 30 psf (481 kg/m^2), and shall not be constructed in Seismic Design Categories D_1 or D_2 as established in Figure R301.2(2).

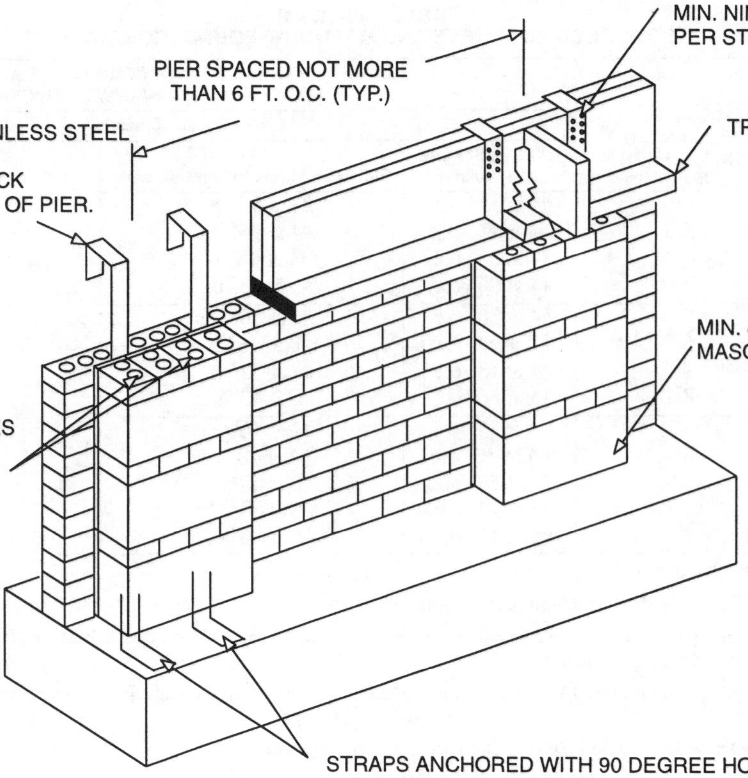

For SI: 1 inch = 25.4 mm, 1 foot = 304.8 mm, 1 degree = 0.79/45 rad.

FIGURE R404.1.5(1)
FOUNDATION WALL CLAY MASONRY CURTAIN WALL WITH CONCRETE MASONRY PIERS

R404.2 Wood foundation walls. Wood foundation walls shall be constructed in accordance with the provisions of Sections R404.2.1 through R404.2.6 and with the details shown in Figures R403.1(2) and R403.1(3).

R404.2.1 Wood grade. All load-bearing lumber shall be identified by the grade mark of a lumber grading or inspection agency which has been approved by an accreditation body that complies with DOC PS 20. In lieu of a grade mark, a certificate of inspection issued by a lumber grading or inspection agency meeting the requirements of this section shall be accepted. Wood structural panels shall conform to DOC PS 1 or DOC PS 2 and shall be identified by a grade mark or certificate of inspection issued by an approved agency.

R404.2.2 Stud size. The studs used in foundation walls shall be 2-inch-by-6-inch (51 mm by 152 mm) members. When spaced 16 inches (406 mm) on center, a wood species with an F_b value of not less than 1,250 (8612 kPa) as listed in AF&PA/NDS shall be used. When spaced 12 inches (305 mm) on center, an F_b of not less than 875 (6029 kPa) shall be required.

R404.2.3 Height of backfill. For wood foundations built in accordance with the provisions of this code, the height of backfill against a foundation wall shall not exceed 4 feet (1219 mm). When the height of fill is more than 12 inches (305 mm) above the interior grade of a crawl space or floor of a basement, the thickness of the plywood sheathing shall meet the requirements of Table R404.2.3.

TABLE R404.2.3[d]
PLYWOOD GRADE AND THICKNESS FOR WOOD FOUNDATION CONSTRUCTION
(30 pcf equivalent-fluid weight soil pressure)

HEIGHT OF FILL (inches)	STUD SPACING (inches)	FACE GRAIN ACROSS STUDS			FACE GRAIN PARALLEL TO STUDS		
		Grade[a]	Minimum thickness (inches)	Span rating	Grade[a]	Minimum thickness (inches)[b,c]	Span rating
24	12	B	15/32	32/16	A	15/32	32/16
					B	15/32[c]	32/16
	16	B	15/32	32/16	A	15/32[c]	32/16
					B	19/32[c] (4, 5 ply)	40/20
36	12	B	15/32	32/16	A	15/32	32/16
					B	15/32[c] (4, 5 ply)	32/16
					B	19/32 (4, 5 ply)	40/20
	16	B	15/32[c]	32/16	A	19/32	40/20
					B	23/32	48/24
48	12	B	15/32	32/16	A	15/32[c]	32/16
					B	19/32[c] (4, 5 ply)	40/20
	16	B	19/32	40/20	A	19/32[c]	40/20
					A	23/32	48/24

For SI: 1 inch = 25.4 mm, 1 foot = 304.8 mm, 1 pound per cubic foot = 0.1572 kN/m^3.

a. Plywood shall be of the following minimum grades in accordance with DOC PS 1 or DOC PS 2:
1. DOC PS 1 Plywood grades marked:
 1.1. Structural I C-D (Exposure 1)
 1.2. C-D (Exposure 1)
2. DOC PS 2 Plywood grades marked:
 2.1. Structural I Sheathing (Exposure 1)
 2.2. Sheathing (Exposure 1)
3. Where a major portion of the wall is exposed above ground and a better appearance is desired, the following plywood grades marked exterior are suitable:
 3.1. Structural I A-C, Structural I B-C or Structural I C-C (Plugged) in accordance with DOC PS 1
 3.2. A-C Group 1, B-C Group 1, C-C (Plugged) Group 1 or MDO Group 1 in accordance with DOC PS 1
 3.3. Single Floor in accordance with DOC PS 1 or DOC PS 2

b. Minimum thickness 15/32 inch, except crawl space sheathing may be 3/8 inch for face grain across studs 16 inches on center and maximum 2-foot depth of unequal fill.

c. For this fill height, thickness and grade combination, panels that are continuous over less than three spans (across less than three stud spacings) require blocking 16 inches above the bottom plate. Offset adjacent blocks and fasten through studs with two 16d corrosion-resistant nails at each end.

d. Fastening shall be in accordance with Section R323.3.

R404.2.4 Backfilling. Wood foundation walls shall not be backfilled until the basement floor and first floor have been constructed or the walls have been braced. For crawl space construction, backfill or bracing shall be installed on the interior of the walls prior to placing backfill on the exterior.

R404.2.5 Drainage and dampproofing. Wood foundation basements shall be drained and dampproofed in accordance with Sections R405 and R406, respectively.

R404.2.6 Fastening. Wood structural panel foundation wall sheathing shall be attached to framing in accordance with Table R602.3(1) and Section R402.1.1.

R404.3 Wood sill plates. Wood sill plates shall be a minimum of 2-inch by 4-inch (51 mm by 102 mm) nominal lumber. Sill plate anchorage shall be in accordance with Sections R403.1.6 and R602.11.

R404.4 Insulating concrete form foundation walls. Insulating concrete form (ICF) foundation walls shall be designed and constructed in accordance with the provisions of this section or in accordance with the provisions of ACI 318. When ACI 318 or the provisions of this section are used to design insulating concrete form foundation walls, project drawings, typical details and specifications are not required to bear the seal of the architect or engineer responsible for design unless otherwise required by the state law of the jurisdiction having authority.

TABLE R404.4(1)
5.5-INCH THICK FLAT ICF FOUNDATION WALLS[a, b,c]

HEIGHT OF BASEMENT WALL (feet)	MAXIMUM UNBALANCED BACKFILL HEIGHT[d] (feet)	MINIMUM VERTICAL REINFORCEMENT SIZE AND SPACING[e]		
		Soil group I[e]	Soil group II[e]	Soil group III[e]
8	4	#4@48″	#4@48″	#4@48″
	5	#4@48″	#3@12″; #4@22″; #5@32″	#3@8″; #4@14″; #5@20″; #6@26″
	6	#3@12″; #4@22″; #5@30″;	#3@8″; #4@14″; #5@20″; #6@24″	#3@6″; #4@10″: #5@14″; #6@20″
	7	#3@8″; #4@14″; #5@22″; #6@26″	#3@5″; #4@10″; #5@14″; #6@18″	#3@4″; #4@6″; #5@10″; #6@14″
9	4	#4@48″	#4@48″	#4@48″
	5	#4@48″	#3@12″; #4@20″; #5@28″; #6@36″	#3@8″; #4@14″; #5@20″; #6@22″
	6	#3@10″; #4@20″; #5@28″; #6@34″	#3@6″; #4@12″; #5@18″; #6@20″	#4@8″; #5@14″; #6@16″
	7	#3@8″; #4@14″; #5@20″; #6@22″	#4@8″; #5@12″; #6@16″	#4@6″; #5@10″; #6@12″
	8	#3@6″; #4@10″; #5@14″; #6@16″	#4@6″; #5@10″; #6@12″	#4@4″; #5@6″; #6@8″
10	4	#4@48″	#4@48″	#4@48″
	5	#4@48″	#3@10″; #4@18″; #5@26″; #6@30″	#3@6″; #4@14″; #5@18″; #6@20″
	6	#3@10″; #4@18″; #5@24″; #6@30″	#3@6″; #4@12″; #5@16″; #6@18″	#3@4″; #4@8″; #5@12″; #6@14″
	7	#3@6″; #4@12″; #5@16″; #6@18″	#3@4″; #4@8″; #5@12″	#4@6″; #5@8″; #6@10″
	8	#4@8″; #5@12″; #6@14″	#4@6″; #5@8″; #6@12″	#4@4″; #5@6″; #6@8″
	9	#4@6″; #5@10″; #6@12″	#4@4″; #5@6″; #6@8″	#5@4″; #6@6″

For SI: 1 inch = 25.4 mm, 1 foot = 304.8 mm, 1 pound per square inch = 6.895 kPa.

a. This table is based on concrete with a minimum specified compressive strength of 2,500 psi, reinforcing steel with a minimum yield strength of 40,000 psi and an assumed equivalent rectangular cross-section. This table is not intended to prohibit the use of an ICF manufacturer's tables based on engineering analysis in accordance with ACI 318.

b. Deflection criteria: *L*/240.

c. Interpolation between rebar sizes and spacing is not permitted.

d. Unbalanced backfill height is the difference in height of the exterior and interior finished ground. Where an interior concrete slab is provided, the unbalanced backfill height shall be measured from the exterior finished ground level to the top of the interior concrete slab.

e. Soil classes are in accordance with the Unified Soil Classification System. Refer to Table R405.1.

TABLE R404.4(2)
7.5-INCH THICK FLAT ICF FOUNDATION WALLS[a, b, c, d]

HEIGHT OF BASEMENT WALL (feet)	MAXIMUM UNBALANCED BACKFILL HEIGHT[e] (feet)	MINIMUM VERTICAL REINFORCEMENT SIZE AND SPACING[f]		
		Soil group I[f]	Soil group II[f]	Soil group III[f]
8	6	N/R	N/R	N/R
	7	N/R	#3@8″; #4@14″; #5@20″; #6@28″	#3@6″; #4@10″; #5@16″; #6@20″
9	6	N/R	N/R	#3@8″; #4@14″; #5@20″; #6@28″
	7	N/R	#3@6″; #4@12″; #5@18″; #6@26″	#3@4″; #4@8″; #5@14″; #6@18″
	8	#3@8″; #4@14″; #5@22″; #6@28″	#3@4″; #4@8″; #5@14″; #6@18″	#3@4″; #4@6″; #5@10″; #6@14″
10	6	N/R	N/R	#3@6″; #4@12″; #5@18″; #6@26″
	7	N/R	#3@6″; #4@12″; #5@18″; #6@24″	#3@4″; #4@8″; #5@12″; #6@18″
	8	#3@6″; #4@12″; #5@20″; #6@26″	#3@4″; #4@8″; #5@12″; #6@16″	#3@4″; #4@6″; #5@8″; #6@12″
	9	#3@6″; #4@10″; #5@14″; #6@20″	#3@4″; #4@6″; #5@10″; #6@12″	#4@4″; #5@6″; #6@10″

For SI: 1 inch = 25.4 mm, 1 foot = 304.8 mm, 1 pound per square inch = 6.895 kPa.

a. This table is based on concrete with a minimum specified compressive strength of 2,500 psi, reinforcing steel with a minimum yield strength of 40,000 psi and an assumed equivalent rectangular cross-section. This table is not intended to prohibit the use of an ICF manufacturer's tables based on engineering analysis in accordance with ACI 318.

b. N/R denotes "not required."

c. Deflection criteria: *L*/240.

d. Interpolation between rebar sizes and spacing is not permitted.

e. Unbalanced backfill height is the difference in height of the exterior and interior finished ground. Where an interior concrete slab is provided, the unbalanced backfill height shall be measured from the exterior finished ground level to the top of the interior concrete slab.

f. Soil classes are in accordance with the Unified Soil Classification System. Refer to Table R405.1.

TABLE R404.4(3)
9.5-INCH THICK FLAT ICF FOUNDATION WALLS[a, b, c, d]

HEIGHT OF BASEMENT WALL (feet)	MAXIMUM UNBALANCED BACKFILL HEIGHT[e] (feet)	MINIMUM VERTICAL REINFORCEMENT SIZE AND SPACING[f]		
		Soil I[f]	Soil II[f]	Soil III[f]
8	7	N/R	N/R	N/R
9	6	N/R	N/R	N/R
	7	N/R	N/R	#3@6″; #4@12″; #5@18″; #6@26″
	8	N/R	#3@6″; #4@12″; #5@18″; #6@26″	#3@4″; #4@8″; #5@14″; #6@18″
10	5	N/R	N/R	N/R
	6	N/R	N/R	#3@10″; #4@18″; #5@26″; #6@36″
	7	N/R	N/R	#3@6″; #4@10″; #5@18″; #6@24″
	8	N/R	#3@6″; #4@12″; #5@16″; #6@24″	#3@4″; #4@8″; #5@12″; #6@16″
	9	N/R	#3@4″; #4@8″; #5@12″; #6@18″	#3@4″; #4@6″; #5@10″; #6@12″

For SI: 1 inch = 25.4 mm, 1 foot = 304.8 mm, 1 pound per square inch = 6.895 kPa.

a. This table is based on concrete with a minimum specified compressive strength of 2,500 psi, reinforcing steel with a minimum yield strength of 40,000 psi and an assumed equivalent rectangular cross-section. This table is not intended to prohibit the use of an ICF manufacturer's tables based on engineering analysis in accordance with ACI 318.

b. N/R denotes "not required."

c. Deflection criteria: *L*/240.

d. Interpolation between rebar sizes and spacing is not permitted.

e. Unbalanced backfill height is the difference in height of the exterior and interior finished ground. Where an interior concrete slab is provided, the unbalanced backfill height shall be measured from the exterior finished ground level to the top of the interior concrete slab.

f. Soil classes are in accordance with the Unified Soil Classification System. Refer to Table R405.1.

TABLE R404.4(4)
WAFFLE GRID ICF FOUNDATION WALLS[a,b,c,d]

MINIMUM NOMINAL WALL THICKNESS[e] (inches)	HEIGHT OF BASEMENT WALL (feet)	MAXIMUM UNBALANCED BACKFILL HEIGHT[e] (feet)	MINIMUM VERTICAL REINFORCEMENT SIZE AND SPACING		
			Soil group I[f]	Soil group II[f]	Soil group III[f]
6	8	4	#4@48″	#3@12″; #4@24″	#3@12″
		5	#3@12″; #5@24″	#4@12″	#7@12″
		6	#4@12″	Design required	Design required
		7	#7@12″	Design required	Design required
	9	4	#4@48″	#3@12″; #5@24″	#3@12″
		5	#3@12″	#4@12″	Design required
		6	#5@12″	Design required	Design required
		7	Design required	Design required	Design required
	10	4	#4@48″	#4@12″	#5@12″
		5	#3@12″	Design required	Design required
		6	Design required	Design required	Design required
		7	Design required	Design required	Design required
8	8	4	N/R	N/R	N/R
		5	N/R	#3@12″; #4@24″; #5@36″	#3@12″; #5@24″
		6	#3@12″; #4@24″; #5@36″	#4@12″; #5@24″	#4@12″
		7	#3@12″; #6@24″	#4@12″	#5@12″
	9	4	N/R	N/R	N/R
		5	N/R	#3@12″; #5@24″	#3@12″; #5@24″
		6	#3@12″; #4@24″	#4@12″	#4@12″
		7	#4@12″; #5@24″	#5@12″	#5@12″
		8	#4@12″	#5@12″	#8@12″
	10	4	N/R	#3@12″; #4@24″; #6@36″	#3@12″; #5@24″
		5	N/R	#3@12″; #4@24″; #6@36″	#4@12″; #5@24″
		6	#3@12″; #5@24″	#4@12″	#5@12″
		7	#4@12″	#5@12″	#6@12″
		8	#4@12″	#6@12″	Design required
		9	#5@12″	Design required	Design required

For SI: 1 inch = 25.4 mm, 1 foot = 304.8 mm, 1 pound per square foot = 0.0479 kN/m^2.

a. N/R denotes "not required."

b. Deflection criteria: *L*/240.

c. Interpolation between rebar sizes and spacing is not permitted.

d. Refer to Table R611.4(2) for wall dimensions.

e. Unbalanced backfill height is the difference in height of the exterior and interior finished ground. Where an interior concrete slab is provided, the unbalanced backfill height shall be measured from the exterior finished ground level to the top of the interior concrete slab.

f. Soil classes are in accordance with the Unified Soil Classification System. Refer to Table R405.1.

TABLE R404.4(5)
SCREEN-GRID ICF FOUNDATION WALLS[a,b,c]

MINIMUM NOMINAL WALL THICKNESS[d] (inches)	MAXIMUM WALL HEIGHT (feet)	MAXIMUM UNBALANCED BACKFILL HEIGHT[e] (feet)	MINIMUM VERTICAL REINFORCEMENT SIZE AND SPACING		
			Soil classes		
			Soil group I[f]	Soil group II[f]	Soil group III[f]
6	8	4	#4@48″	#3@12″; #4@24″; #5@36″	#3@12″; #5@24″
		5	#3@12″; #4@24″	#3@12″	#4@12″
		6	#4@12″	#5@12″	Design required
		7	#4@12″	Design required	Design required
	9	4	#4@48″	#3@12″; #4@24″	#3@12″; #6@24″
		5	#3@12″; #5@24″	#4@12″	#7@12″
		6	#4@12″	Design required	Design required
		7	Design required	Design required	Design required
		8	Design required	Design required	Design required
	10	4	#4@48″	#3@12″; #5@24″	#3@12″
		5	#3@12″	#4@12″	#7@12″
		6	#4@12″	Design required	Design required
		7	Design required	Design required	Design required
		8	Design required	Design required	Design required

For SI: 1 inch = 25.4 mm, 1 foot = 304.8 mm, 1 pound per square foot = 0.0479 kN/m^2.

a. N/R denotes "not required."

b. Deflection criteria: *L*/240.

c. Interpolation between rebar sizes and spacing is not permitted.

d. Refer to Table R611.4(2) for wall dimensions.

e. Unbalanced backfill height is the difference in height of the exterior and interior finished ground. Where an interior concrete slab is provided, the unbalanced backfill height shall be measured from the exterior finished ground level to the top of the interior concrete slab.

f. Soil classes are in accordance with the Unified Soil Classification System. Refer to Table R405.1.

R404.4.1 Applicability limits. The provisions of this section shall apply to the construction of insulating concrete form foundation walls for buildings not greater than 60 feet (18 288 mm) in plan dimensions, and floors not greater than 32 feet (9754 mm) or roofs not greater than 40 feet (12 192 mm) in clear span. Buildings shall not exceed two stories in height above-grade with each story not greater than 10 feet (3048 mm) high. Foundation walls constructed in accordance with the provisions of this section shall be limited to buildings subjected to a maximum ground snow load of 70 psf (3.35 kN/m^2) and located in Seismic Design Category A, B or C.

R404.4.2 Flat insulating concrete form wall systems. Flat ICF wall systems shall comply with Figure R611.3, shall have a minimum concrete thickness of 5.5 inches (140 mm), and shall have reinforcement in accordance with Table R404.4(1), R404.4(2) or R404.4(3).

R404.4.3 Waffle grid insulating concrete form wall systems. Waffle-grid wall systems shall have a minimum nominal concrete thickness of 6 inches (152 mm) for the horizontal and vertical concrete members (cores) and shall be reinforced in accordance with Table R404.4(4). The minimum core dimension shall comply with Table R611.4(2) and Figure R611.4.

R404.4.4 Screen grid insulating concrete form wall systems. Screen-grid ICF wall systems shall have a minimum nominal concrete thickness of 6 inches (152 mm) for the horizontal and vertical concrete members (cores). The minimum core dimensions shall comply with Table R611.4(2) and Figure R611.5. Walls shall have reinforcement in accordance with Table R404.4(5).

R404.4.5 Concrete material. Ready-mixed concrete for insulating concrete form walls shall be in accordance with Section R402.2. Maximum slump shall not be greater than 6 inches (152 mm) as determined in accordance with ASTM C 143. Maximum aggregate size shall not be larger than $^3/_4$ inch (19.1 mm).

> **Exception:** Concrete mixes conforming to the ICF manufacturer's recommendations.

R404.4.6 Reinforcing steel.

R404.4.6.1 General. Reinforcing steel shall meet the requirements of ASTM A 615, A 616, A 617 or A 706. The minimum yield strength of reinforcing steel shall be 40,000 psi (Grade 40) (276 MPa). Vertical and horizontal wall reinforcements shall be placed no closer to the outside face of the wall than one-half the wall thickness. Steel reinforcement for foundation walls shall have concrete cover in accordance with ACI 318.

> **Exception:** Where insulated concrete forms are used and the form remains in place as cover for the concrete, the minimum concrete cover for the reinforcing steel is permitted to be reduced to $^3/_4$ inch (19.1 mm).

R404.4.6.2 Horizontal reinforcement. When vertical reinforcement is required, ICF foundation walls shall have horizontal reinforcement in accordance with this section. ICF foundation walls up to 8 feet (2438 mm) in height shall have a minimum of one continuous No. 4 horizontal reinforcing bar placed at 48 inches (1219 mm) on center with one bar located within 12 inches (305 mm) of the top of the wall story. ICF Foundation walls greater than 8 feet (2438 mm) in height shall have a minimum of one continuous No. 4 horizontal reinforcing bar placed at 36 inches (914 mm) on center with one bar located within 12 inches (305 mm) of the top of the wall story.

R404.4.6.3 Wall openings. Vertical wall reinforcement required by Section R404.4.2, R404.4.3 or R404.4.4 that is interrupted by wall openings shall have additional vertical reinforcement of the same size placed within 12 inches (305 mm) of each side of the opening.

R404.4.7 Foam plastic insulation. Foam plastic insulation in insulating concrete foam construction shall comply with this section.

R404.4.7.1 Material. Insulating concrete form material shall meet the surface burning characteristics of Section R318.1.1. A thermal barrier shall be provided on the building interior in accordance with Section R318.1.2.

R404.4.7.2 Termite hazards. In areas where hazard of termite damage is very heavy in accordance with Figure R301.2(6), foam plastic insulation shall be permitted below grade on foundation walls in accordance with one of the following conditions:

1. When in addition to the requirements in Section R324.1, an approved method of protecting the foam plastic and structure from subterranean termite damage is provided.
2. The structural members of walls, floors, ceilings and roofs are entirely of noncombustible materials or pressure preservatively treated wood.
3. On the interior side of basement walls.

R404.4.8 Foundation wall thickness based on walls supported. The thickness of ICF foundation walls shall not be less than the thickness of the wall supported above.

R404.4.9 Height above finished ground. ICF foundation walls shall extend above the finished ground adjacent to the foundation at all points a minimum of 4 inches (102 mm) where masonry veneer is used and a minimum of 6 inches (152 mm) elsewhere.

R404.4.10 Backfill placement. Backfill shall be placed in accordance with Section R404.1.7.

R404.4.11 Drainage and dampproofing/waterproofing. ICF foundation basements shall be drained and dampproofed/waterproofed in accordance with Sections R405 and R406.

SECTION R405
FOUNDATION DRAINAGE

R405.1 Concrete or masonry foundations. Drains shall be provided around all concrete or masonry foundations that retain earth and enclose habitable or usable spaces located below grade. Drainage tiles, gravel or crushed stone drains, perforated pipe or other approved systems or materials shall be installed at or below the area to be protected and shall discharge by gravity or mechanical means into an approved drainage system. Gravel or crushed stone drains shall extend at least 1 foot (305 mm) beyond the outside edge of the footing and 6 inches (153 mm) above the top of the footing and be covered with an approved filter membrane material. The top of open joints of drain tiles shall be protected with strips of building paper, and the drainage tiles or perforated pipe shall be placed on a minimum of 2 inches (51 mm) of washed gravel or crushed rock at least one sieve size larger than the tile joint opening or perforation and covered with not less than 6 inches (153 mm) of the same material.

Exception: A drainage system is not required when the foundation is installed on well-drained ground or sand-gravel mixture soils according to the Unified Soil Classification System, Group I Soils, as detailed in Table R405.1.

R405.2 Wood foundations. Wood foundations enclosing habitable or usable spaces located below grade shall be adequately drained in accordance with Sections R405.2.1 through R405.2.3.

R405.2.1 Base. A porous layer of gravel, crushed stone or coarse sand shall be placed to a minimum thickness of 4 inches (102 mm) under the basement floor. Provision shall be made for automatic draining of this layer and the gravel or crushed stone wall footings.

R405.2.2 Moisture barrier. A 6-mil-thick (0.15 mm) polyethylene moisture barrier shall be applied over the porous layer with the basement floor constructed over the polyethylene.

R405.2.3 Drainage system. In other than Group I soils, a sump shall be provided to drain the porous layer and footings. The sump shall be at least 24 inches (610 mm) in diameter or 20 inches square (0.0129 m^2), shall extend at least 24 inches (610 mm) below the bottom of the basement floor and shall be capable of positive gravity or mechanical drainage to remove any accumulated water. The drainage system shall discharge into an approved sewer system or to daylight.

SECTION R406
FOUNDATION WATERPROOFING AND DAMPPROOFING

R406.1 Concrete and masonry foundation dampproofing. Except where required to be waterproofed by Section R406.2, foundation walls that retain earth and enclose habitable or usable spaces located below grade shall be dampproofed from the top of the footing to the finished grade. Masonry walls shall have not less than $^3/_8$ inch (9.5 mm) portland cement parging applied to the exterior of the wall. The parging shall be dampproofed with a bituminous coating, 3 pounds per square yard (1.63 kg/m^2) of acrylic modified cement, $^1/_8$-inch (3.2 mm) coat of surface-bonding mortar complying with ASTM C 887 or any material permitted for waterproofing in Section R406.2. Concrete walls shall be dampproofed by applying any one of the above listed dampproofing materials or any one of the waterproofing materials listed in Section R406.2 to the exterior of the wall.

TABLE R405.1
PROPERTIES OF SOILS CLASSIFIED ACCORDING TO THE UNIFIED SOIL CLASSIFICATION SYSTEM

SOIL GROUP	UNIFIED SOIL CLASSIFICATION SYSTEM SYMBOL	SOIL DESCRIPTION	DRAINAGE CHARACTERISTICS[a]	FROST HEAVE POTENTIAL	VOLUME CHANGE POTENTIAL EXPANSION[b]
Group I	GW	Well-graded gravels, gravel sand mixtures, little or no fines.	Good	Low	Low
	GP	Poorly graded gravels or gravel sand mixtures, little or no fines.	Good	Low	Low
	SW	Well-graded sands, gravelly sands, little or no fines.	Good	Low	Low
	SP	Poorly graded sands or gravelly sands, little or no fines.	Good	Low	Low
	GM	Silty gravels, gravel-sand-silt mixtures.	Good	Medium	Low
	SM	Silty sand, sand-silt mixtures.	Good	Medium	Low
Group II	GC	Clayey gravels, gravel-sand-clay mixtures.	Medium	Medium	Low
	SC	Clayey sands, sand-clay mixture.	Medium	Medium	Low
	ML	Inorganic silts and very fine sands, rock flour, silty or clayey fine sands or clayey silts with slight plasticity.	Medium	High	Low
	CL	Inorganic clays of low to medium plasticity, gravelly clays, sandy clays, silty clays, lean clays.	Medium	Medium	Medium to Low
Group III	CH	Inorganic clays of high plasticity, fat clays.	Poor	Medium	High
	MH	Inorganic silts, micaceous or diatomaceous fine sandy or silty soils, elastic silts.	Poor	High	High
Group IV	OL	Organic silts and organic silty clays of low plasticity.	Poor	Medium	Medium
	OH	Organic clays of medium to high plasticity, organic silts.	Unsatisfactory	Medium	High
	Pt	Peat and other highly organic soils.	Unsatisfactory	Medium	High

For SI: 1 inch = 25.4 mm.

a. The percolation rate for good drainage is over 4 inches per hour, medium drainage is 2 inches to 4 inches per hour, and poor is less than 2 inches per hour.

b. Soils with a low potential expansion typically have a plasticity index (PI) of 0 to 15, soils with a medium potential expansion have a PI of 10 to 35 and soils with a high potential expansion have a PI greater than 20.

R406.2 Concrete and masonry foundation waterproofing. In areas where a high water table or other severe soil-water conditions are known to exist, exterior foundation walls that retain earth and enclose habitable or usable spaces located below grade shall be waterproofed with a membrane extending from the top of the footing to the finished grade. The membrane shall consist of 2-ply hot-mopped felts, 55 pound (25 kg) roll roofing, 6-mil (0.15 mm) polyvinyl chloride, 6-mil (0.15 mm) polyethylene or 40-mil (1 mm) polymer-modified asphalt. The joints in the membrane shall be lapped and sealed with an adhesive compatible with the waterproofing membrane.

> **Exception:** Organic solvent based products such as hydrocarbons, chlorinated hydrocarbons, ketons and esters shall not be used for ICF walls with expanded polystyrene form material. Plastic roofing cements, acrylic coatings, latex coatings, mortars and pargings are permitted to be used to seal ICF walls. Cold setting asphalt or hot asphalt shall conform to type C of ASTM D 449. Hot asphalt shall be applied at a temperature of less than 200 degrees.

R406.3 Dampproofing for wood foundations. Wood foundations enclosing habitable or usable spaces located below grade shall be dampproofed in accordance with Sections R406.3.1 through R406.3.4.

R406.3.1 Panel joint sealed. Plywood panel joints in the foundation walls shall be sealed full length with a caulking compound capable of producing a moisture-proof seal under the conditions of temperature and moisture content at which it will be applied and used.

R406.3.2 Below grade moisture barrier. A 6-mil-thick (0.15 mm) polyethylene film shall be applied over the below-grade portion of exterior foundation walls prior to backfilling. Joints in the polyethylene film shall be lapped 6 inches (152 mm) and sealed with adhesive. The top edge of the polyethylene film shall be bonded to the sheathing to form a seal. Film areas at grade level shall be protected from mechanical damage and exposure by a pressure preservatively treated lumber or plywood strip attached to the wall several inches above finish grade level and extending approximately 9 inches (229 mm) below grade. The joint between the strip and the wall shall be caulked full length prior to fastening the strip to the wall. Other coverings appropriate to the architectural treatment may also be used. The polyethylene film shall extend down to the bottom of the wood footing plate but shall not overlap or extend into the gravel or crushed stone footing.

R406.3.3 Porous fill. The space between the excavation and the foundation wall shall be backfilled with the same material used for footings, up to a height of 1 foot (305 mm) above the footing for well-drained sites, or one-half the total backfill height for poorly drained sites. The porous fill shall be covered with strips of 30-pound (13.6 kg) asphalt paper or 6-mil (0.15 mm) polyethylene to permit water seepage while avoiding infiltration of fine soils.

R406.3.4 Backfill. The remainder of the excavated area shall be backfilled with the same type of soil as was removed during the excavation.

SECTION R407 COLUMNS

R407.1 Wood column protection. Wood columns shall be protected against decay as set forth in Section R323.

R407.2 Steel column protection. All surfaces (inside and outside) of steel columns shall be given a shop coat of rust-inhibitive paint, except for corrosion-resistant steel and steel treated with coatings to provide corrosion resistance.

R407.3 Structural requirements. The columns shall be restrained to prevent lateral displacement at the bottom end. Wood columns shall not be less in nominal size than 4 inches by 4 inches (102 mm by 102 mm) and steel columns shall not be less than 3-inch-diameter (76 mm) standard pipe or approved equivalent.

Exception: In Seismic Design Categories A, B and C columns no more than 48 inches (1219 mm) in height on a pier or footing are exempt from the bottom end lateral displacement requirement within underfloor areas enclosed by a continuous foundation.

SECTION R408 UNDER-FLOOR SPACE

R408.1 Ventilation. The under-floor space between the bottom of the floor joists and the earth under any building (except space occupied by a basement or cellar) shall be provided with ventilation openings through foundation walls or exterior walls. The minimum net area of ventilation openings shall not be less than 1 square foot for each 150 square feet (0.67 m^2 for each 100 m^2) of under-floor space area. One such ventilating opening shall be within 3 feet (914 mm) of each corner of said building.

R408.2 Openings for under-floor ventilation. The minimum net area of ventilation openings shall not be less than 1 square foot (0.0929 m^2) for each 150 square feet (100 m^2) of underfloor space area. One such ventilating opening shall be within 3 feet (914 mm) of each corner of the building. Ventilation openings shall be covered for their height and width with any of the following materials provided that the least dimension of the covering shall not exceed $^1/_4$ inch (6.4 mm):

1. Perforated sheet metal plates not less than 0.070 inch (1.8 mm) thick.
2. Expanded sheet metal plates not less than 0.047 inch (1.2 mm) thick.
3. Cast iron grills or grating.
4. Extruded load-bearing brick vents.
5. Hardware cloth of 0.035 inch (0.89 mm) wire or heavier.
6. Corrosion-resistant wire mesh, with the least dimension being $^1/_8$ inch (3.2 mm).

Exceptions:

1. Where warranted by climatic conditions, ventilation openings to the outdoors are not required if ventilation openings to the interior are provided.
2. The total area of ventilation openings may be reduced to 1/1,500 of the under-floor area where the ground surface is treated with an approved vapor retarder material and the required openings are placed so as to provide cross-ventilation of the space. The installation of operable louvers shall not be prohibited.
3. Under-floor spaces used as supply plenums for distribution of heated and cooled air shall comply with the requirements of Section M1601.4.
4. Ventilation openings are not required where continuously operated mechanical ventilation is provided at a rate of 1.0 cfm (10 m^2) for each 50 square feet (1.02 L/s) of underfloor space floor area and ground surface is covered with an approved vapor retarder material.
5. Ventilation openings are not required when the ground surface is covered with an approved vapor retarder material, the space is supplied with conditioned air and the perimeter walls are insulated in accordance with Section N1102.1.7.

R408.3 Access. An access opening 18 inches by 24 inches (457 mm by 610 mm) shall be provided to the under-floor space. See Section M1305.1.4 for access requirements where mechanical equipment is located under floors.

R408.4 Removal of debris. The under-floor grade shall be cleaned of all vegetation and organic material. All wood forms used for placing concrete shall be removed before a building is occupied or used for any purpose. All construction materials shall be removed before a building is occupied or used for any purpose.

R408.5 Finished grade. The finished grade of under-floor surface may be located at the bottom of the footings; however, where there is evidence that the groundwater table can rise to within 6 inches (152 mm) of the finished floor at the building perimeter or where there is evidence that the surface water does not readily drain from the building site, the grade in the under-floor space shall be as high as the outside finished grade, unless an approved drainage system is provided.

R408.6 Flood resistance. For buildings located in areas prone to flooding as established in Table R301.2(1), the walls enclosing the underfloor space shall be provided with flood openings in accordance with Section R327.2.2.

CHAPTER 5
FLOORS

SECTION R501
GENERAL

R501.1 Application. The provisions of this chapter shall control the design and construction of the floors for all buildings including the floors of attic spaces used to house mechanical and/or plumbing fixtures and equipment.

R501.2 Requirements. Floor construction shall be capable of accommodating all loads according to Section R301 and of transmitting the resulting loads to the supporting structural elements.

SECTION R502
WOOD FLOOR FRAMING

R502.1 General. Load-bearing dimension lumber for joists, beams and girders shall be identified by a grade mark of a lumber grading or inspection agency that has been approved by an accreditation body that complies with DOC PS 20. In lieu of a grade mark, a certificate of inspection issued by a lumber grading or inspection agency meeting the requirements of this section shall be accepted.

R502.1.1 Preservatively treated lumber. Preservatively treated dimension lumber shall also be identified as required by Section R323.1.

R502.1.2 Blocking and subflooring. Blocking shall be a minimum of utility grade lumber. Subflooring may be a minimum of utility grade lumber or No. 4 common grade boards.

R502.1.3 End-jointed lumber. Approved end-jointed lumber identified by a grade mark conforming to Section R501.2 may be used interchangeably with solid-sawn members of the same species and grade.

R502.1.4 Prefabricated wood I-joists. Structural capacities and design provisions for prefabricated wood I-joists shall be established and monitored in accordance with ASTM D 5055.

R502.2 Design and construction. Floors shall be designed and constructed in accordance with the provisions of this chapter, Figure R502.2 and Sections R323 and R324 or in accordance with AF&PA/NDS.

R502.2.1 Decks. Where supported by attachment to an exterior wall, decks shall be positively anchored to the primary structure and designed for both vertical and lateral loads as applicable. Such attachment shall not be accomplished by the use of toenails or nails subject to withdrawal. Where positive connection to the primary building structure cannot be verified during inspection, decks shall be self-supporting. For decks with cantilevered framing members, connections to exterior walls or other framing members, shall be designed and constructed to resist uplift resulting from the full live load specified in Table R301.4 acting on the cantilevered portion of the deck.

R502.3 Allowable joist spans. Spans for floor joists shall be in accordance with Tables R502.3.1(1) and R502.3.1(2). For other grades and species and for other loading conditions, refer to the AF&PA Span Tables for Joists and Rafters.

R502.3.1 Sleeping areas and attic joists. Table R502.3.1(1) shall be utilized to determine the maximum allowable span of floor joists that support sleeping areas and attics that are accessed by means of a fixed stairway provided that the design live load does not exceed 30 psf (1.44 kN/m^2) and the design dead load does not exceed 10 psf (0.48 kN/m^2). The allowable span of ceiling joists that support attics utilized for limited storage or no storage shall be determined in accordance with Section R802.4.

R502.3.2 Other floor joists. Table R502.3.1(2) shall be utilized to determine the maximum allowable span of floor joists that support all areas of the building, other than sleeping and attics, provided that the design live load does not exceed 40 psf (1.92 kN/m^2) and the design dead load does not exceed 10 psf (0.48 kN/m^2).

R502.4 Joists under bearing partitions. Joists under parallel bearing partitions shall be doubled or a beam of adequate size to support the load shall be provided. Double joists that are separated to permit the installation of piping or vents shall be full depth solid blocked with lumber not less than 2 inches (51 mm) in nominal thickness spaced not more than 4 feet (1219 mm) on center.

R502.5 Allowable girder spans. The allowable spans of girders fabricated of dimension lumber shall not exceed the values set forth in Tables R502.5(1) and R502.5(2).

R502.6 Bearing. The ends of each joist, beam or girder shall have not less than 1.5 inches (38 mm) of bearing on wood or metal and not less than 3 inches (76 mm) on masonry or concrete except where supported on a 1-inch-by-4-inch (25.4 mm by 102 mm) ribbon strip and nailed to the adjacent stud or by the use of approved joist hangers.

R502.6.1 Floor systems. Joists framing from opposite sides over a bearing support shall lap a minimum of 3 inches (76 mm) and shall be nailed together with a minimum three 10d face nails. A wood or metal splice with strength equal to or greater than that provided by the nailed lap is permitted.

R502.6.2 Joist framing. Joists framing into the side of a wood girder shall be supported by approved framing anchors or on ledger strips not less than nominal 2 inches by 2 inches (51 mm by 51 mm).

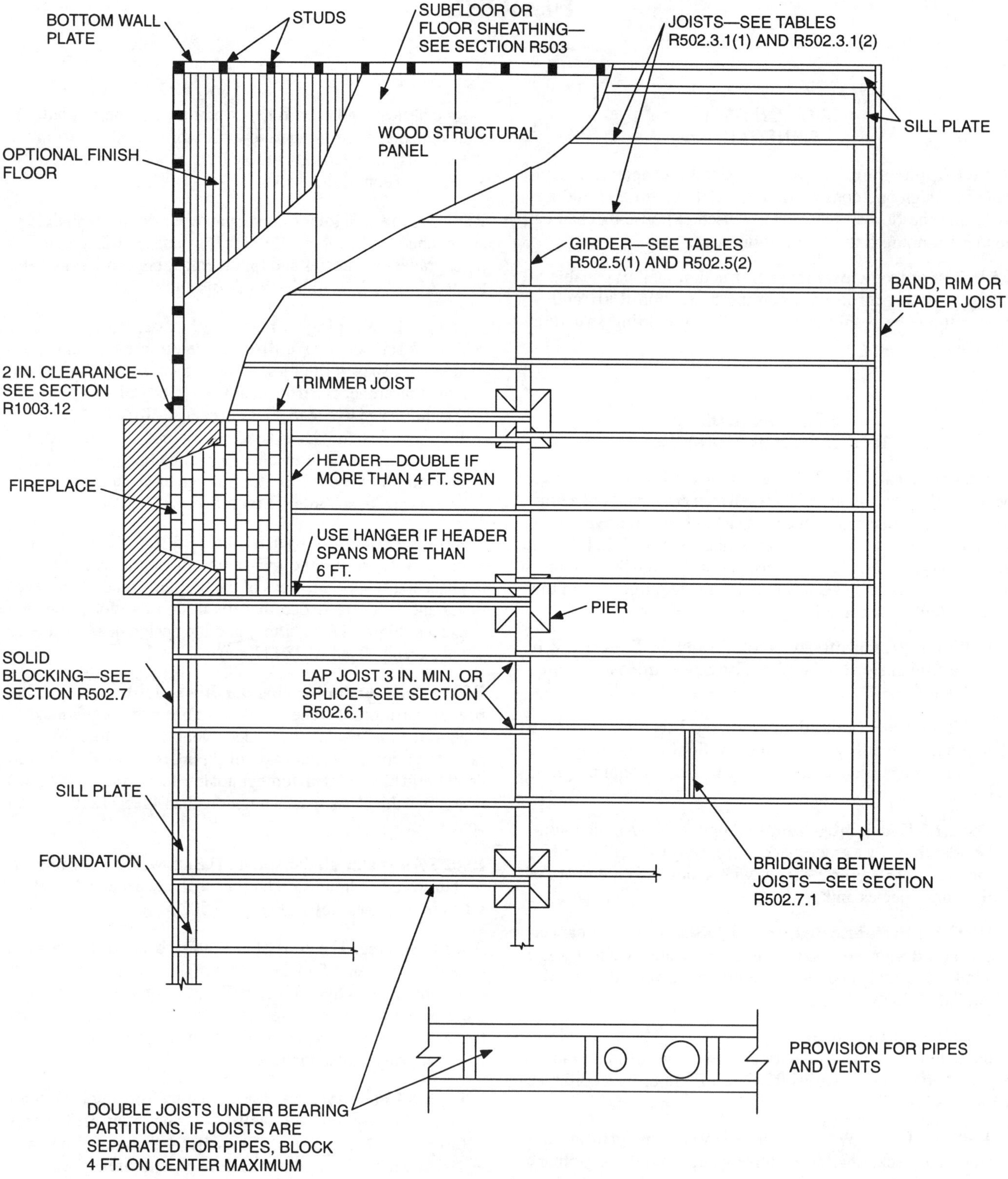

For SI: 1 inch = 25.4 mm, 1 foot = 304.8 mm.

FIGURE R502.2
FLOOR CONSTRUCTION

TABLE R502.3.1(1) FLOOR JOIST SPANS FOR COMMON LUMBER SPECIES (Residential sleeping areas, live load=30 psf, L/Δ=360)

		DEAD LOAD = 10 psf				DEAD LOAD = 20 psf			
		2x6	2x8	2x10	2x12	2x6	2x8	2x10	2x12
		Maximum floor joist spans							
JOIST SPACING (inches)	SPECIE AND GRADE	(ft.- in.)	(ft.- in.)	(ft.- in.)	(ft.- in.)	(ft.- in.)	(ft.- in.)	(ft.- in.)	(ft.- in.)
12	Douglas fir-larch SS	12-6	16-6	21-0	25-7	12-6	16-6	21-0	25-7
	Douglas fir-larch #1	12-0	15-10	20-3	24-8	12-0	15-7	19-0	22-0
	Douglas fir-larch #2	11-10	15-7	19-10	23-0	11-6	14-7	17-9	20-7
	Douglas fir-larch #3	9-8	12-4	15-0	17-5	8-8	11-0	13-5	15-7
	Hem-fir SS	11-10	15-7	19-10	24-2	11-10	15-7	19-10	24-2
	Hem-fir #1	11-7	15-3	19-5	23-7	11-7	15-2	18-6	21-6
	Hem-fir #2	11-0	14-6	18-6	22-6	11-0	14-4	17-6	20-4
	Hem-fir #3	9-8	12-4	15-0	17-5	8-8	11-0	13-5	15-7
	Southern pine SS	12-3	16-2	20-8	25-1	12-3	16-2	20-8	25-1
	Southern pine #1	12-0	15-10	20-3	24-8	12-0	15-10	20-3	24-8
	Southern pine #2	11-10	15-7	19-10	18-8	11-10	15-7	18-7	21-9
	Southern pine #3	10-5	13-3	15-8	18-8	9-4	11-11	14-0	16-8
	Spruce-pine-fir SS	11-7	15-3	19-5	23-7	11-7	15-3	19-5	23-7
	Spruce-pine-fir #1	11-3	14-11	19-0	23-0	11-3	14-7	17-9	20-7
	Spruce-pine-fir #2	11-3	14-11	19-0	23-0	11-3	14-7	17-9	20-7
	Spruce-pine-fir #3	9-8	12-4	15-0	17-5	8-8	11-0	13-5	15-7
16	Douglas fir-larch SS	11-4	15-0	19-1	23-3	11-4	15-0	19-1	23-0
	Douglas fir-larch #1	10-11	14-5	18-5	21-4	10-8	13-6	16-5	19-1
	Douglas fir-larch #2	10-9	14-1	17-2	19-11	9-11	12-7	15-5	17-10
	Douglas fir-larch #3	8-5	10-8	13-0	15-1	7-6	9-6	11-8	13-6
	Hem-fir SS	10-9	14-2	18-0	21-11	10-9	14-2	18-0	21-11
	Hem-fir #1	10-6	13-10	17-8	20-9	10-4	13-1	16-0	18-7
	Hem-fir #2	10-0	13-2	16-10	19-8	9-10	12-5	15-2	17-7
	Hem-fir #3	8-5	10-8	13-0	15-1	7-6	9-6	11-8	13-6
	Southern pine SS	11-2	14-8	18-9	22-10	11-2	14-8	18-9	22-10
	Southern pine #1	10-11	14-5	18-5	22-5	10-11	14-5	17-11	21-4
	Southern pine #2	10-9	14-2	18-0	21-1	10-5	13-6	16-1	18-10
	Southern pine #3	9-0	11-6	13-7	16-2	8-1	10-3	12-2	14-6
	Spruce-pine-fir SS	10-6	13-10	17-8	21-6	10-6	13-10	17-8	21-4
	Spruce-pine-fir #1	10-3	13-6	17-2	19-11	9-11	12-7	15-5	17-10
	Spruce-pine-fir #2	10-3	13-6	17-2	19-11	9-11	12-7	15-5	17-10
	Spruce-pine-fir #3	8-5	10-8	13-0	15-1	7-6	9-6	11-8	13-6
19.2	Douglas fir-larch SS	10-8	14-1	18-0	21-10	10-8	14-1	18-0	21-0
	Douglas fir-larch #1	10-4	13-7	16-9	19-6	9-8	12-4	15-0	17-5
	Douglas fir-larch #2	10-1	12-10	15-8	18-3	9-1	11-6	14-1	16-3
	Douglas fir-larch #3	7-8	9-9	11-10	13-9	6-10	8-8	10-7	12-4
	Hem-fir SS	10-1	13-4	17-0	20-8	10-1	13-4	17-0	20-7
	Hem-fir #1	9-10	13-0	16-4	19-0	9-6	12-0	14-8	17-0
	Hem-fir #2	9-5	12-5	15-6	17-1	8-11	11-4	13-10	16-1
	Hem-fir #3	7-8	9-9	11-10	13-9	6-10	8-8	10-7	12-4
	Southern pine SS	10-6	13-10	17-8	21-6	10-6	13-10	17-8	21-6
	Southern pine #1	10-4	13-7	17-4	21-1	10-4	13-7	16-4	19-6
	Southern pine #2	10-1	13-4	16-5	19-3	9-6	12-4	14-8	17-2
	Southern pine #3	8-3	10-6	12-5	14-9	7-4	9-5	11-1	13-2
	Spruce-pine-fir SS	9-10	13-0	16-7	20-2	9-10	13-0	16-7	19-6
	Spruce-pine-fir #1	9-8	12-9	15-8	18-3	9-1	11-6	14-1	16-3
	Spruce-pine-fir #2	9-8	12-9	15-8	18-3	9-1	11-6	14-1	16-3
	Spruce-pine-fir #3	7-8	9-9	11-10	13-9	6-10	8-8	10-7	12-4
24	Douglas fir-larch SS	9-11	13-1	16-8	20-3	9-11	13-1	16-2	18-9
	Douglas fir-larch #1	9-7	12-4	15-0	17-5	8-8	11-0	13-5	15-7
	Douglas fir-larch #2	9-1	11-6	14-1	16-3	8-1	10-3	12-7	14-7
	Douglas fir-larch #3	6-10	8-8	10-7	12-4	6-2	7-9	9-6	11-0
	Hem-fir SS	9-4	12-4	15-9	19-2	9-4	12-4	15-9	18-5
	Hem-fir #1	9-2	12-0	14-8	17-0	8-6	10-9	13-1	15-2
	Hem-fir #2	8-9	11-4	13-10	16-1	8-0	10-2	12-5	14-4
	Hem-fir #3	6-10	8-8	10-7	12-4	6-2	7-9	9-6	11-0
	Southern pine SS	9-9	12-10	16-5	19-11	9-9	12-10	16-5	19-11
	Southern pine #1	9-7	12-7	16-1	19-6	9-7	12-4	14-7	17-5
	Southern pine #2	9-4	12-4	14-8	17-2	8-6	11-0	13-1	15-5
	Southern pine #3	7-4	9-5	11-1	13-2	6-7	8-5	9-11	11-10
	Spruce-pine-fir SS	9-2	12-1	15-5	18-9	9-2	12-1	15-0	17-5
	Spruce-pine-fir #1	8-11	11-6	14-1	16-3	8-1	10-3	12-7	14-7
	Spruce-pine-fir #2	8-11	11-6	14-1	16-3	8-1	10-3	12-7	14-7
	Spruce-pine-fir #3	6-10	8-8	10-7	12-4	6-2	7-9	9-6	11-0

For SI: 1 inch = 25.4 mm, 1 foot = 304.8 mm, 1 pound per square foot = 0.0479 kN/m^2.

NOTE: Check sources for availability of lumber in lengths greater than 20 feet.

TABLE R502.3.1(2)
FLOOR JOIST SPANS FOR COMMON LUMBER SPECIES (Residential living areas, live load=40 psf, L/Δ=360)

			DEAD LOAD = 10 psf				DEAD LOAD = 20 psf			
			2x6	2x8	2x10	2x12	2x6	2x8	2x10	2x12
			Maximum floor joist spans							
JOIST SPACING (inches)	SPECIE AND GRADE		(ft.- in.)	(ft.- in.)	(ft.- in.)	(ft.- in.)	(ft.- in.)	(ft.- in.)	(ft.- in.)	(ft.- in.)
12	Douglas fir-larch	SS	11- 4	15- 0	19- 1	23- 3	11- 4	15- 0	19- 1	23- 3
	Douglas fir-larch	#1	10-11	14- 5	18- 5	22- 0	10-11	14- 2	17- 4	20- 1
	Douglas fir-larch	#2	10- 9	14- 2	17- 9	20- 7	10- 6	13- 3	16- 3	18-10
	Douglas fir-larch	#3	8- 8	11- 0	13- 5	15- 7	7-11	10- 0	12- 3	14- 3
	Hem-fir	SS	10- 9	14- 2	18- 0	21-11	10- 9	14- 2	18- 0	21-11
	Hem-fir	#1	10- 6	13-10	17- 8	21- 6	10- 6	13-10	16-11	19- 7
	Hem-fir	#2	10- 0	13- 2	16-10	20- 4	10- 0	13- 1	16- 0	18- 6
	Hem-fir	#3	8- 8	11- 0	13- 5	15- 7	7-11	10- 0	12- 3	14- 3
	Southern pine	SS	11- 2	14- 8	18- 9	22-10	11- 2	14- 8	18- 9	22-10
	Southern pine	#1	10-11	14- 5	18- 5	22- 5	10-11	14- 5	18- 5	22- 5
	Southern pine	#2	10- 9	14- 2	18- 0	21- 9	10- 9	14- 2	16-11	19-10
	Southern pine	#3	9- 4	11-11	14- 0	16- 8	8- 6	10-10	12-10	15- 3
	Spruce-pine-fir	SS	10- 6	13-10	17- 8	21- 6	10- 6	13-10	17- 8	21- 6
	Spruce-pine-fir	#1	10- 3	13- 6	17- 3	20- 7	10- 3	13- 3	16- 3	18-10
	Spruce-pine-fir	#2	10- 3	13- 6	17- 3	20- 7	10- 3	13- 3	16- 3	18-10
	Spruce-pine-fir	#3	8- 8	11- 0	13- 5	15- 7	7-11	10- 0	12- 3	14- 3
16	Douglas fir-larch	SS	10- 4	13- 7	17- 4	21- 1	10- 4	13- 7	17- 4	21- 0
	Douglas fir-larch	#1	9-11	13- 1	16- 5	19- 1	9- 8	12- 4	15- 0	17- 5
	Douglas fir-larch	#2	9- 9	12- 7	15- 5	17-10	9- 1	11- 6	14- 1	16- 3
	Douglas fir-larch	#3	7- 6	9- 6	11- 8	13- 6	6-10	8- 8	10- 7	12- 4
	Hem-fir	SS	9- 9	12-10	16- 5	19-11	9- 9	12-10	16- 5	19-11
	Hem-fir	#1	9- 6	12- 7	16- 0	18- 7	9- 6	12- 0	14- 8	17- 0
	Hem-fir	#2	9- 1	12- 0	15- 2	17- 7	8-11	11- 4	13-10	16- 1
	Hem-fir	#3	7- 6	9- 6	11- 8	13- 6	6-10	8- 8	10- 7	12- 4
	Southern pine	SS	10- 2	13- 4	17- 0	20- 9	10- 2	13- 4	17- 0	20- 9
	Southern pine	#1	9-11	13- 1	16- 9	20- 4	9-11	13- 1	16- 4	19- 6
	Southern pine	#2	9- 9	12-10	16- 1	18-10	9- 6	12- 4	14- 8	17- 2
	Southern pine	#3	8- 1	10- 3	12- 2	14- 6	7- 4	9- 5	11- 1	13- 2
	Spruce-pine-fir	SS	9- 6	12- 7	16- 0	19- 6	9- 6	12- 7	16- 0	19- 6
	Spruce-pine-fir	#1	9- 4	12- 3	15- 5	17-10	9- 1	11- 6	14- 1	16- 3
	Spruce-pine-fir	#2	9- 4	12- 3	15- 5	17-10	9- 1	11- 6	14- 1	16- 3
	Spruce-pine-fir	#3	7- 6	9- 6	11- 8	13- 6	6-10	8- 8	10- 7	12- 4
19.2	Douglas fir-larch	SS	9- 8	12-10	16- 4	19-10	9- 8	12-10	16- 4	19- 2
	Douglas fir-larch	#1	9- 4	12- 4	15- 0	17- 5	8-10	11- 3	13- 8	15-11
	Douglas fir-larch	#2	9- 1	11- 6	14- 1	16- 3	8- 3	10- 6	12-10	14-10
	Douglas fir-larch	#3	6-10	8- 8	10- 7	12- 4	6- 3	7-11	9- 8	11- 3
	Hem-fir	SS	9- 2	12- 1	15- 5	18- 9	9- 2	12- 1	15- 5	18- 9
	Hem-fir	#1	9- 0	11-10	14- 8	17- 0	8- 8	10-11	13- 4	15- 6
	Hem-fir	#2	8- 7	11- 3	13-10	16- 1	8- 2	10- 4	12- 8	14- 8
	Hem-fir	#3	6-10	8- 8	10- 7	12- 4	6- 3	7-11	9- 8	11- 3
	Southern pine	SS	9- 6	12- 7	16- 0	19- 6	9- 6	12- 7	16- 0	19- 6
	Southern pine	#1	9- 4	12- 4	15- 9	19- 2	9- 4	12- 4	14-11	17- 9
	Southern pine	#2	9- 2	12- 1	14- 8	17- 2	8- 8	11- 3	13- 5	15- 8
	Southern pine	#3	7- 4	9- 5	11- 1	13- 2	6- 9	8- 7	10- 1	12- 1
	Spruce-pine-fir	SS	9- 0	11-10	15- 1	18- 4	9- 0	11-10	15- 1	17- 9
	Spruce-pine-fir	#1	8- 9	11- 6	14- 1	16- 3	8- 3	10- 6	12-10	14-10
	Spruce-pine-fir	#2	8- 9	11- 6	14- 1	16- 3	8- 3	10- 6	12-10	14-10
	Spruce-pine-fir	#3	6-10	8- 8	10- 7	12- 4	6- 3	7-11	9- 8	11- 3
24	Douglas fir-larch	SS	9- 0	11-11	15- 2	18- 5	9- 0	11-11	14- 9	17- 1
	Douglas fir-larch	#1	8- 8	11- 0	13- 5	15- 7	7-11	10- 0	12- 3	14- 3
	Douglas fir-larch	#2	8- 1	10- 3	12- 7	14- 7	7- 5	9- 5	11- 6	13- 4
	Douglas fir-larch	#3	6- 2	7- 9	9- 6	11- 0	5- 7	7- 1	8- 8	10- 1
	Hem-fir	SS	8- 6	11- 3	14- 4	17- 5	8- 6	11- 3	14- 4	16-10[a]
	Hem-fir	#1	8- 4	10- 9	13- 1	15- 2	7- 9	9- 9	11-11	13-10
	Hem-fir	#2	7-11	10- 2	12- 5	14- 4	7- 4	9- 3	11- 4	13- 1
	Hem-fir	#3	6- 2	7- 9	9- 6	11- 0	5- 7	7- 1	8- 8	10- 1
	Southern pine	SS	8-10	11- 8	14-11	18- 1	8-10	11- 8	14-11	18- 1
	Southern pine	#1	8- 8	11- 5	14- 7	17- 5	8- 8	11- 3	13- 4	15-11
	Southern pine	#2	8- 6	11- 0	13- 1	15- 5	7- 9	10- 0	12- 0	14- 0
	Southern pine	#3	6- 7	8- 5	9-11	11-10	6- 0	7- 8	9- 1	10- 9
	Spruce-pine-fir	SS	8- 4	11- 0	14- 0	17- 0	8- 4	11- 0	13- 8	15-11
	Spruce-pine-fir	#1	8- 1	10- 3	12- 7	14- 7	7- 5	9- 5	11- 6	13- 4
	Spruce-pine-fir	#2	8- 1	10- 3	12- 7	14- 7	7- 5	9- 5	11- 6	13- 4
	Spruce-pine-fir	#3	6- 2	7- 9	9- 6	11- 0	5- 7	7- 1	8- 8	10- 1

Check sources for availability of lumber in lengths greater than 20 feet.

For SI: 1 inch = 25.4 mm, 1 foot = 308.4 mm, 1 pound per square foot = 0.0479 kN/m.

a. End bearing length shall be increased to 2 inches.

TABLE R502.5(1)
GIRDER SPANS[a] AND HEADER SPANS[a] FOR EXTERIOR BEARING WALLS
(Maximum header spans for douglas fir-larch, hem-fir, southern pine and spruce-pine-fir[b] and required number of jack studs)

HEADERS SUPPORTING	SIZE	GROUND SNOW LOAD (psf)[e]											
		30						50					
		Building width[c] (feet)											
		20		28		36		20		28		36	
		Span	NJ[d]	Span	NJ[d]	Span	NJ[d]	Span	NJ[d]	Span	NJ[d]	Span	NJ[d]
Roof and ceiling	2-2×4	3-6	1	3-2	1	2-10	1	3-2	1	2-9	1	2-6	1
	2-2×6	5-5	1	4-8	1	4-2	1	4-8	1	4-1	1	3-8	2
	2-2×8	6-10	1	5-11	2	5-4	2	5-11	2	5-2	2	4-7	2
	2-2×10	8-5	2	7-3	2	6-6	2	7-3	2	6-3	2	5-7	2
	2-2×12	9-9	2	8-5	2	7-6	2	8-5	2	7-3	2	6-6	2
	3-2×8	8-4	1	7-5	1	6-8	1	7-5	1	6-5	2	5-9	2
	3-2×10	10-6	1	9-1	2	8-2	2	9-1	2	7-10	2	7-0	2
	3-2×12	12-2	2	10-7	2	9-5	2	10-7	2	9-2	2	8-2	2
	4-2×8	7-0	1	6-1	2	5-5	2	6-1	2	5-3	2	4-8	2
	4-2×10	11-8	1	10-6	1	9-5	2	10-6	1	9-1	2	8-2	2
	4-2×12	14-1	1	12-2	2	10-11	2	12-2	2	10-7	2	9-5	2
Roof, ceiling and one center-bearing floor	2-2×4	3-1	1	2-9	1	2-5	1	2-9	1	2-5	1	2-2	1
	2-2×6	4-6	1	4-0	1	3-7	2	4-1	1	3-7	2	3-3	2
	2-2×8	5-9	2	5-0	2	4-6	2	5-2	2	4-6	2	4-1	2
	2-2×10	7-0	2	6-2	2	5-6	2	6-4	2	5-6	2	5-0	2
	2-2×12	8-1	2	7-1	2	6-5	2	7-4	2	6-5	2	5-9	3
	3-2×8	7-2	1	6-3	2	5-8	2	6-5	2	5-8	2	5-1	2
	3-2×10	8-9	2	7-8	2	6-11	2	7-11	2	6-11	2	6-3	2
	3-2×12	10-2	2	8-11	2	8-0	2	9-2	2	8-0	2	7-3	2
	4-2×8	5-10	2	5-2	2	4-8	2	5-3	2	4-7	2	4-2	2
	4-2×10	10-1	1	8-10	2	8-0	2	9-1	2	8-0	2	7-2	2
	4-2×12	11-9	2	10-3	2	9-3	2	10-7	2	9-3	2	8-4	2
Roof, ceiling and one clear span floor	2-2×4	2-8	1	2-4	1	2-1	1	2-7	1	2-3	1	2-0	1
	2-2×6	3-11	1	3-5	2	3-0	2	3-10	2	3-4	2	3-0	2
	2-2×8	5-0	2	4-4	2	3-10	2	4-10	2	4-2	2	3-9	2
	2-2×10	6-1	2	5-3	2	4-8	2	5-11	2	5-1	2	4-7	3
	2-2×12	7-1	2	6-1	3	5-5	3	6-10	2	5-11	3	5-4	3
	3-2×8	6-3	2	5-5	2	4-10	2	6-1	2	5-3	2	4-8	2
	3-2×10	7-7	2	6-7	2	5-11	2	7-5	2	6-5	2	5-9	2
	3-2×12	8-10	2	7-8	2	6-10	2	8-7	2	7-5	2	6-8	2
	4-2×8	5-1	2	4-5	2	3-11	2	4-11	2	4-3	2	3-10	2
	4-2×10	8-9	2	7-7	2	6-10	2	8-7	2	7-5	2	6-7	2
	4-2×12	10-2	2	8-10	2	7-11	2	9-11	2	8-7	2	7-8	2
Roof, ceiling and two center-bearing floors	2-2×4	2-7	1	2-3	1	2-0	1	2-6	1	2-2	1	1-11	1
	2-2×6	3-9	2	3-3	2	2-11	2	3-8	2	3-2	2	2-10	2
	2-2×8	4-9	2	4-2	2	3-9	2	4-7	2	4-0	2	3-8	2
	2-2×10	5-9	2	5-1	2	4-7	3	5-8	2	4-11	2	4-5	3
	2-2×12	6-8	2	5-10	3	5-3	3	6-6	2	5-9	3	5-2	3
	3-2×8	5-11	2	5-2	2	4-8	2	5-9	2	5-1	2	4-7	2
	3-2×10	7-3	2	6-4	2	5-8	2	7-1	2	6-2	2	5-7	2
	3-2×12	8-5	2	7-4	2	6-7	2	8-2	2	7-2	2	6-5	3
	4-2×8	4-10	2	4-3	2	3-10	2	4-9	2	4-2	2	3-9	2
	4-2×10	8-4	2	7-4	2	6-7	2	8-2	2	7-2	2	6-5	2
	4-2×12	9-8	2	8-6	2	7-8	2	9-5	2	8-3	2	7-5	2

For SI: 1 inch = 25.4 mm, 1 pound per square foot = 0.0479 kN/m^2.

a. Spans are given in feet and inches.

b. Tabulated values assume #2 grade lumber.

c. Building width is measured perpendicular to the ridge. For widths between those shown, spans are permitted to be interpolated.

d. NJ - Number of jack studs required to support each end. Where the number of required jack studs equals one, the header is permitted to be supported by an approved framing anchor attached to the full-height wall stud and to the header.

e. Use 30 psf ground snow load for cases in which ground snow load is less than 30 psf and the roof live load is equal to or less than 20 psf.

TABLE R502.5(2)
GIRDER SPANS[a] AND HEADER SPANS[a] FOR INTERIOR BEARING WALLS
(Maximum header spans for douglas fir-larch, hem-fir, southern pine and spruce-pine-fir[b] and required number of jack studs)

HEADERS AND GIRDERS SUPPORTING	SIZE	BUILDING WIDTH[c] (feet) 20 Span	20 NJ[d]	28 Span	28 NJ[d]	36 Span	36 NJ[d]
One floor only	2-2×4	3-1	1	2-8	1	2-5	1
	2-2×6	4-6	1	3-11	1	3-6	1
	2-2×8	5-9	1	5-0	2	4-5	2
	2-2×10	7-0	2	6-1	2	5-5	2
	2-2×12	8-1	2	7-0	2	6-3	2
	3-2×8	7-2	1	6-3	1	5-7	2
	3-2×10	8-9	1	7-7	2	6-9	2
	3-2×12	10-2	2	8-10	2	7-10	2
	4-2×8	5-10	1	5-1	2	4-6	2
	4-2×10	10-1	1	8-9	1	7-10	2
	4-2×12	11-9	1	10-2	2	9-1	2
Two floors	2-2×4	2-2	1	1-10	1	1-7	1
	2-2×6	3-2	2	2-9	2	2-5	2
	2-2×8	4-1	2	3-6	2	3-2	2
	2-2×10	4-11	2	4-3	2	3-10	3
	2-2×12	5-9	2	5-0	3	4-5	3
	3-2×8	5-1	2	4-5	2	3-11	2
	3-2×10	6-2	2	5-4	2	4-10	2
	3-2×12	7-2	2	6-3	2	5-7	3
	4-2×8	4-2	2	3-7	2	3-2	2
	4-2×10	7-2	2	6-2	2	5-6	2
	4-2×12	8-4	2	7-2	2	6-5	2

For SI: 1 inch = 25.4 mm, 1 foot = 304.8 mm.

a. Spans are given in feet and inches.

b. Tabulated values assume #2 grade lumber.

c. Building width is measured perpendicular to the ridge. For widths between those shown, spans are permitted to be interpolated.

d. NJ - Number of jack studs required to support each end. Where the number of required jack studs equals one, the headers are permitted to be supported by an approved framing anchor attached to the full-height wall stud and to the header.

R502.7 Lateral restraint at supports. Joists shall be supported laterally at the ends by full-depth solid blocking not less than 2 inches (51 mm) nominal in thickness; or by attachment to a header, band, or rim joist, or to an adjoining stud; or shall be otherwise provided with lateral support to prevent rotation.

Exception: In Seismic Design Categories D_1 and D_2, lateral restraint shall also be provided at each intermediate support.

R502.7.1 Bridging. Joists exceeding a nominal 2 by 12 shall be supported laterally by solid blocking, diagonal bridging (wood or metal), or a continuous 1-inch-by-3-inch (25 mm by 76 mm) strip nailed across the bottom of joists perpendicular to joists at intervals not exceeding 8 feet (2438 mm).

R502.8 Drilling and notching. Structural floor members shall not be cut, bored or notched in excess of the limitations specified in this section. See Figure R502.8.

R502.8.1 Sawn lumber. Notches in solid lumber joists, rafters and beams shall not exceed one-sixth of the depth of the member, shall not be longer than one-third of the depth of the member and shall not be located in the middle one-third of the span. Notches at the ends of the member shall not exceed one-fourth the depth of the member. The tension side of members 4 inches (102 mm) or greater in nominal thickness shall not be notched except at the ends of the members. The diameter of holes bored or cut into members shall not exceed one-third the depth of the member. Holes shall not be closer than 2 inches (51 mm) to the top or bottom of the member, or to any other hole located in the member. Where the member is also notched, the hole shall not be closer than 2 inches (51 mm) to the notch.

R502.8.2 Engineered wood products. Cuts, notches and holes bored in trusses, laminated veneer lumber, glue-laminated members or I-joists are not permitted unless the effects of such penetrations are specifically considered in the design of the member.

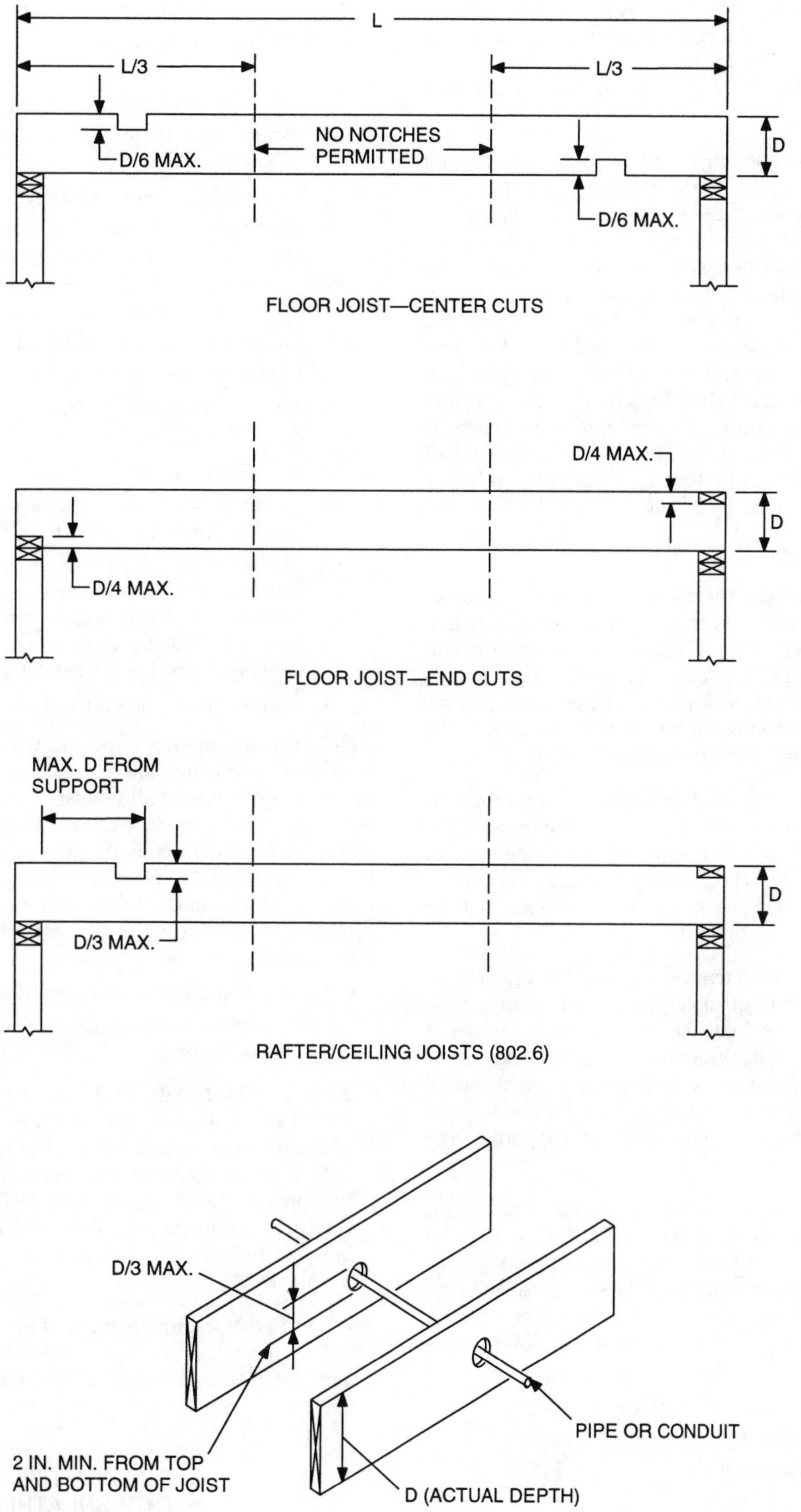

For SI: 1 inch = 25.4 mm.

FIGURE R502.8
CUTTING, NOTCHING AND DRILLING

R502.9 Fastening. Floor framing shall be nailed in accordance with Table R602.3(1). Where posts and beam or girder construction is used to support floor framing, positive connections shall be provided to ensure against uplift and lateral displacement.

R502.10 Framing of openings. Openings in floor framing shall be framed with a header and trimmer joists. When the header joist span does not exceed 4 feet (1219 mm), the header joist may be a single member the same size as the floor joist. Single trimmer joists may be used to carry a single header joist that is located within 3 feet (914 mm) of the trimmer joist bearing. When the header joist span exceeds 4 feet (1219 mm), the trimmer joists and the header joist shall be doubled and of sufficient cross section to support the floor joists framing into the header. Approved hangers shall be used for the header joist to trimmer joist connections when the header joist span exceeds 6 feet (1829 mm). Tail joists over 12 feet (3658 mm) long shall be supported at the header by framing anchors or on ledger strips not less than 2 inches by 2 inches (51 mm by 51 mm).

R502.11 Wood trusses.

R502.11.1 Design. Wood trusses shall be designed in accordance with approved engineering practice. The design and manufacture of metal plate connected wood trusses shall comply with ANSI/TPI 1. The truss design drawings shall be prepared by a registered professional where required by the statutes of the jurisdiction in which the project is to be constructed in accordance with Section R106.1.

R502.11.2 Bracing. Trusses shall be braced to prevent rotation and provide lateral stability in accordance with the requirements specified in the construction documents for the building and on the individual truss design drawings. In the absence of specific bracing requirements, trusses shall be braced in accordance with the TPI, HIB.

R502.11.3 Alterations to trusses. Truss members and components shall not be cut, notched, spliced or otherwise altered in any way without the approval of a registered design professional. Alterations resulting in the addition of load (e.g. HVAC equipment, water heater, etc.), that exceed the design load for the truss, shall not be permitted without verification that the truss is capable of supporting the additional loading.

R502.11.4 Truss design drawings. Truss design drawings, prepared in compliance with Section R502.11.1, shall be provided to the building official and approved prior to installation. Truss design drawing shall be provided with the shipment of trusses delivered to the jobsite. Truss design drawings shall include, at a minimum, the information specified below:

1. Slope or depth, span, and spacing.
2. Location of all joints.
3. Required bearing widths.
4. Design loads as applicable.
   - 4.1. Top chord live load (including snow loads).
   - 4.2. Top chord dead load.
   - 4.3. Bottom chord live load.
   - 4.4. Bottom chord dead load.
   - 4.5. Concentrated loads and their points of application.
   - 4.6. Controlling wind and earthquake loads.
5. Adjustments to lumber and joint connector design values for conditions of use.
6. Each reaction force and direction.
7. Joint connector type and description (e.g., size, thickness or gauge); and the dimensioned location of each joint connector except where symmetrically located relative to the joint interface.
8. Lumber size, species and grade for each member.
9. Connection requirements for:
   - 9.1. Truss-to-truss girder.
   - 9.2. Truss ply-to-ply.
   - 9.3. Field splices.
10. Calculated deflection ratio and/or maximum description for live and total load.
11. Maximum axial compression forces in the truss members to enable the building designer to design the size, connections and anchorage of the permanent continuous lateral bracing. Forces shall be shown on the truss drawing or on supplemental documents.
12. Required permanent truss member bracing location.

R502.12 Draftstopping required. When there is usable space both above and below the concealed space of a floor/ceiling assembly, draftstops shall be installed so that the area of the concealed space does not exceed 1,000 square feet (92.9 m^2). Draftstopping shall divide the concealed space into approximately equal areas. Where the assembly is enclosed by a floor membrane above and a ceiling membrane below draftstopping shall be provided in floor/ceiling assemblies under the following circumstances:

1. Ceiling is suspended under the floor framing.
2. Floor framing is constructed of truss-type open-web or perforated members.

R502.12.1 Materials. Draftstopping materials shall not be less than $^1/_2$-inch (12.7 mm) gypsum board, $^3/_8$-inch (9.5 mm) wood structural panels, $^3/_8$-inch (9.5 mm) Type 2-M-W particleboard or other approved materials adequately supported. Draftstopping shall be installed parallel to the floor framing members unless otherwise approved by the building official. The integrity of all draftstops shall be maintained.

R502.13 Fireblocking required. Fireblocking shall be provided in wood-frame floor construction and floor-ceiling assemblies in accordance with Section R602.8.

SECTION R503
FLOOR SHEATHING

R503.1 Lumber sheathing. Maximum allowable spans for lumber used as floor sheathing shall conform to Tables R503.1, R503.2.1.1(1) and R503.2.1.1(2).

TABLE R503.1
MINIMUM THICKNESS OF LUMBER FLOOR SHEATHING

JOIST OR BEAM SPACING (inches)	MINIMUM NET THICKNESS	
	Perpendicular to joist	Diagonal to joist
24	$^{11}/_{16}$	$^{3}/_{4}$
16	$^{5}/_{8}$	$^{5}/_{8}$
48[a]	$1^{1}/_{2}$ T & G	N/A
54[b]		
60[c]		

For SI: 1 inch = 25.4 mm, 1 pound per square inch = 6.895 kPa.

a. For this support spacing, lumber sheathing shall have a minimum F_b of 675 and minimum E of 1,100,000 (see AF&PA/NDS).

b. For this support spacing, lumber sheathing shall have a minimum F_b of 765 and minimum E of 1,400,000 (see AF&PA/NDS).

c. For this support spacing, lumber sheathing shall have a minimum F_b of 855 and minimum E of 1,700,000 (see AF&PA/NDS).

R503.1.1 End joints. End joints in lumber used as subflooring shall occur over supports unless end-matched lumber is used, in which case each piece shall bear on at least two joists. Subflooring may be omitted when joist spacing does not exceed 16 inches (406 mm) and a 1-inch (25.4 mm) nominal tongue-and-groove wood strip flooring is applied perpendicular to the joists.

R503.2 Wood structural panel sheathing.

R503.2.1 Identification and grade. Wood structural panel sheathing used for structural purposes shall conform to DOC PS 1, DOC PS 2 or, when manufactured in Canada, CSA 0437 or CSA 0325. All panels shall be identified by a grade mark of certificate of inspection issued by an approved agency.

R503.2.1.1 Subfloor and combined subfloor underlayment. Where used as subflooring or combination subfloor underlayment, wood structural panels shall be of one of the grades specified in Table R503.2.1.1(1). When sanded plywood is used as combination subfloor underlayment, the grade shall be as specified in Table R503.2.1.1(2).

TABLE R503.2.1.1(1)
ALLOWABLE SPANS AND LOADS FOR WOOD STRUCTURAL PANELS FOR ROOF AND SUBFLOOR SHEATHING AND COMBINATION SUBFLOOR UNDERLAYMENT[a,b,c]

SPAN RATING	MINIMUM NOMINAL PANEL THICKNESS (inch)	MAXIMUM SPAN (inches)[d]		LOAD (pounds per square foot, at maximum span)		MAXIMUM SPAN (inches)
		With edge support	Without edge support	Total load	Live load	
Sheathing[e]		Roof[f]				Subfloor[j]
12/0	$^{5}/_{16}$	12	12	40	30	0
16/0	$^{5}/_{16}$	16	16	40	30	0
20/0	$^{5}/_{16}$	20	20	40	30	0
24/0	$^{3}/_{8}$	24	20[g]	40	30	0
24/16	$^{7}/_{16}$	24	24	50	40	16
32/16	$^{15}/_{32}$, $^{1}/_{2}$	32	28	40	30	16[h]
40/20	$^{19}/_{32}$, $^{5}/_{8}$	40	32	40	30	20[h,i]
48/24	$^{23}/_{32}$, $^{3}/_{48}$	48	36	45	35	24
60/32	$^{7}/_{8}$	60	48	45	35	32
Underlayment, C-C plugged, single floor[e]		Roof[f]				Combination subfloor underlayment[k]
16 o.c.	$^{19}/_{32}$, $^{5}/_{8}$	24	24	50	40	16[i]
20 o.c.	$^{19}/_{32}$, $^{5}/_{8}$	32	32	40	30	20[i,j]
24 o.c.	$^{23}/_{32}$, $^{3}/_{4}$	48	36	35	25	24
32 o.c.	$^{7}/_{8}$	48	40	50	40	32
48 o.c.	$1^{3}/_{32}$, $1^{1}/_{8}$	60	48	50	40	48

For SI: 1 inch = 25.4 mm, 1 pound per square foot = 0.0479 kN/m^2.

a. The allowable total loads were determined using a dead load of 10 psf. If the dead load exceeds 10 psf, then the live load shall be reduced accordingly.

b. Panels continuous over two or more spans with long dimension perpendicular to supports. Spans shall be limited to values shown because of possible effect of concentrated loads.

c. Applies to panels 24 inches or wider.

d. Lumber blocking, panel edge clips (one midway between each support, except two equally spaced between supports when span is 48 inches), tongue-and-groove panel edges, or other approved type of edge support.

e. Includes Structural 1 panels in these grades.

f. Uniform load deflection limitation: $^{1}/_{180}$ of span under live load plus dead load, $^{1}/_{240}$ of span under live load only.

g. Maximum span 24 inches for $^{15}/_{32}$- and $^{1}/_{2}$-inch panels.

h. Maximum span 24 inches where $^{3}/_{4}$-inch wood finish flooring is installed at right angles to joists.

i. Maximum span 24 inches where 1.5 inches of lightweight concrete or approved cellular concrete is placed over the subfloor.

j. Unsupported edges shall have tongue-and-groove joints or shall be supported with blocking unless minimum nominal $^{1}/_{4}$-inch thick underlayment with end and edge joints offset at least 2 inches or 1.5 inches of lightweight concrete or approved cellular concrete is placed over the subfloor, or $^{3}/_{4}$-inch wood finish flooring is installed at right angles to the supports. Allowable uniform live load at maximum span, based on deflection of $^{1}/_{360}$ of span, is 100 psf.

k. Unsupported edges shall have tongue-and-groove joints or shall be supported by blocking unless nominal $^{1}/_{4}$-inch-thick underlayment with end and edge joints offset at least 2 inches or $^{3}/_{4}$-inch wood finish flooring is installed at right angles to the supports. Allowable uniform live load at maximum span, based on deflection of $^{1}/_{360}$ of span, is 100 psf, except panels with a span rating of 48 on center are limited to 65 psf total uniform load at maximum span.

TABLE R503.2.1.1(2)
ALLOWABLE SPANS FOR PLYWOOD COMBINATION SUBFLOOR UNDERLAYMENT[a]

IDENTIFICATION	SPACING OF JOISTS (inches)		
	16	20	24
Species group[b]			
1	$^1/_2$	$^5/_8$	$^3/_4$
2, 3	$^5/_8$	$^3/_4$	$^7/_8$
4	$^3/_4$	$^7/_8$	1

For SI: 1 inch = 25.4 mm, 1 pound per square foot = 0.0479 kN/m^2.

a. Plywood continuous over two or more spans and face grain perpendicular to supports. Unsupported edges shall be tongue-and-groove or blocked except where nominal $^1/_4$-inch-thick underlayment or $^3/_4$-inch wood finish floor is used. Allowable uniform live load at maximum span based on deflection of $^1/_{360}$ of span is 100 psf.

b. Applicable to all grades of sanded exterior-type plywood.

R503.2.2 Allowable spans. The maximum allowable span for wood structural panels used as subfloor or combination subfloor underlayment shall be as set forth in Table R503.2.1.1(1). The maximum span for sanded plywood combination subfloor underlayment shall be as set forth in Table R503.2.1.1(2).

R503.2.3 Installation. Wood structural panels used as subfloor or combination subfloor underlayment shall be attached to wood framing in accordance with Table R602.3(1) and shall be attached to cold-formed steel framing in accordance with Table R505.3.1(2).

R503.3 Particleboard.

R503.3.1 Identification and grade. Particleboard shall conform to ANSI A208.1 and shall be so identified by a grade mark or certificate of inspection issued by an approved agency.

R503.3.2 Floor underlayment. Particleboard floor underlayment shall conform to Type PBU and shall not be less than $^1/_4$ inch (6.4 mm) in thickness.

R503.3.3 Installation. Particleboard underlayment shall be installed in accordance with the recommendations of the manufacturer and attached to framing in accordance with Table R602.3(1).

SECTION R504
PRESSURE PRESERVATIVELY TREATED-WOOD FLOORS (ON GROUND)

R504.1 General. Pressure preservatively treated-wood basement floors and floors on ground shall be designed to withstand axial forces and bending moments resulting from lateral soil pressures at the base of the exterior walls and floor live and dead loads. Floor framing shall be designed to meet joist deflection requirements in accordance with Section R301.

R504.1.1 Unbalanced soil loads. Unless special provision is made to resist sliding caused by unbalanced lateral soil loads, wood basement floors shall be limited to applications where the differential depth of fill on opposite exterior foundation walls is 2 feet (610 mm) or less.

R504.1.2 Construction. Joists in wood basement floors shall bear tightly against the narrow face of studs in the foundation wall or directly against a band joist that bears on the studs. Plywood subfloor shall be continuous over lapped joists or over butt joints between in-line joists. Sufficient blocking shall be provided between joists to transfer lateral forces at the base of the end walls into the floor system.

R504.1.3 Uplift and buckling. Where required, resistance to uplift or restraint against buckling shall be provided by interior bearing walls or properly designed stub walls anchored in the supporting soil below.

R504.2 Site preparation. The area within the foundation walls shall have all vegetation, topsoil and foreign material removed, and any fill material that is added shall be free of vegetation and foreign material. The fill shall be compacted to assure uniform support of the pressure preservatively treated-wood floor sleepers.

R504.2.1 Base. A minimum 4-inch-thick (102 mm) granular base of gravel having a maximum size of $^3/_4$ inch (19.1 mm) or crushed stone having a maximum size of $^1/_2$ inch (12.7 mm) shall be placed over the compacted earth.

R504.2.2 Moisture barrier. Polyethylene sheeting of minimum 6-mil (0.15 mm) thickness shall be placed over the granular base. Joints shall be lapped 6 inches (152 mm) and left unsealed. The polyethylene membrane shall be placed over the pressure preservatively treated-wood sleepers and shall not extend beneath the footing plates of the exterior walls.

R504.3 Materials. All framing materials, including sleepers, joists, blocking and plywood subflooring, shall be pressure preservatively treated and dried after treatment in accordance with AWPA C22.

SECTION R505
STEEL FLOOR FRAMING

R505.1 Cold-formed steel floor framing. Elements shall be straight and free of any defects that would significantly affect structural performance. Cold-formed steel floor framing members shall comply with the requirements of this section.

R505.1.1 Applicability limits. The provisions of this section shall control the construction of steel floor framing for buildings not greater than 60 feet (18 288 mm) in length perpendicular to the joist span, not greater than 36 feet (10 973 mm) in width parallel to the joist span, and not greater than two stories in height with each story not greater than 10 feet (3048 mm) high. Steel floor framing constructed in accordance with the provisions of this section shall be limited to sites subjected to a maximum design wind speed of 130 miles per hour (209 km/hr) Exposure A, B or C and a maximum ground snow load of 70 pounds per square foot (3.35 kN/m^2).

R505.1.2 In-line framing. When supported by steel-framed walls in accordance with Section R603, steel floor framing shall be constructed with floor joists located directly in-line with load-bearing studs located below the joists with a maximum tolerance of $^3/_4$ inch (19.1 mm) between the center lines of the joist and the stud.

R505.2 Structural framing. Load-bearing floor framing members shall comply with Figure R505.2(1) and with the dimensional and minimum thickness requirements specified in Tables R505.2(1) and R505.2(2). Tracks shall comply with Figure R505.2(2) and shall have a minimum flange width of $1^1/_4$ inches (32 mm). The maximum inside bend radius for members shall be the greater of $^3/_{32}$ inch (2.4 mm) or twice the uncoated steel thickness. Holes in joist webs shall conform to Figure R505.2(3) and to the dimensional requirements specified in Table R505.2(3). Holes shall be permitted only along the centerline of the web of the framing member. Holes for 800S162-33, 1000S162-43, 1200S162-43 and 1200S162-54 nominal joist sizes located less than 10 inches (254 mm) from the edge of load-bearing surface shall be patched in accordance with Section R505.3.6.

TABLE R505.2(1)
COLD-FORMED STEEL JOIST SIZES

MEMBER DESIGNATION[a]	WEB DEPTH (inches)	MINIMUM FLANGE WIDTH (inches)	MAXIMUM FLANGE WIDTH (inches)	MINIMUM LIP SIZE (inches)
550S162-t	5.5	1.625	2	0.5
800S162-t	8	1.625	2	0.5
1000S162-t	10	1.625	2	0.5
1200S162-t	12	1.625	2	0.5

For SI: 1 inch = 25.4 mm.

a. The member designation is defined by the first number representing the member depth in $^1/_{100}$ inches, the letter "S" representing a stud or joist member, the second number representing the flange width in $^1/_{100}$ inches, and the letter "t" shall be a number representing the minimum base metal thickness in mils [See Table R505.2(2)].

TABLE R505.2(2)
MINIMUM THICKNESS OF COLD-FORMED STEEL MEMBERS

DESIGNATION (mils)	MINIMUM UNCOATED THICKNESS (inches)	REFERENCE GAGE NUMBER
33	0.033	20
43	0.043	18
54	0.054	16
68	0.068	14

For SI: 1 inch = 25.4 mm, 1 mil = 0.0254 mm.

TABLE R505.2(3)
MAXIMUM HOLE DIMENSIONS AND SPACING IN JOIST WEBS

NOMINAL MEMBER SIZE	MAXIMUM HOLE DEPTH[a] (inches)	MAXIMUM HOLE LENGTH[b] (inches)	MINIMUM HOLE SPACING (inches)	MINIMUM HOLE EDGE DISTANCE[c] (inches)
550S162-33	2	5.25	16.5	10
550S162-43	2	5.25	16.5	10
550S162-54	2	5.25	16.5	10
550S162-68	2	5.25	16.5	10
800S162-33	1.5	4	24	10
800S162-43	3	6	24	10
800S162-54	3	6	24	10
800S162-68	3	6	24	10
1000S162-43	1.5	4	24	10
1000S162-54	4	6	24	10
1000S162-68	4	6	24	10
1200S162-43	1.5	4	24	10
1200S162-54	1.5	4	24	10
1200S162-68	4.75	6	24	10

For SI: 1 inch = 25.4 mm.

a. The dimension of the hole measured across the depth of the joist web.

b. The dimension of the hole measured along the length of the joist.

c. Edge distance is measured from the edge of the hole to the edge of bearing support.

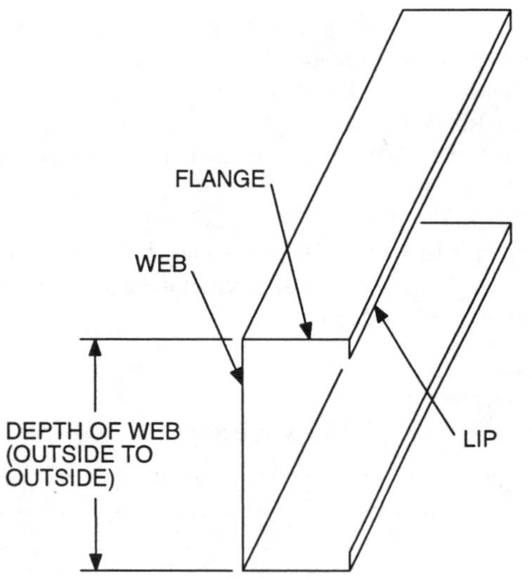

FIGURE R505.2(1)
C-SECTION

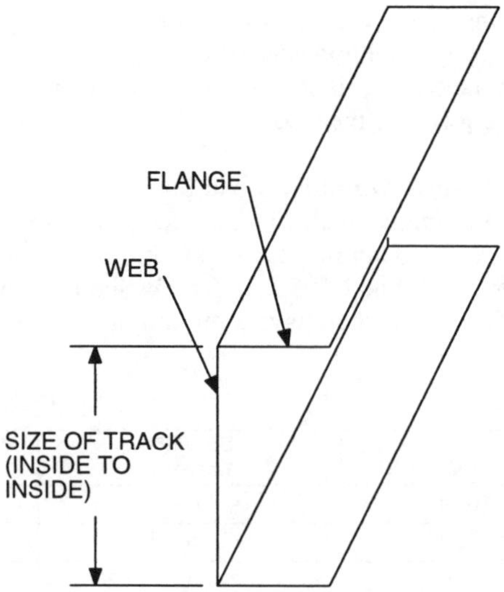

FIGURE R505.2(2)
TRACK SECTION

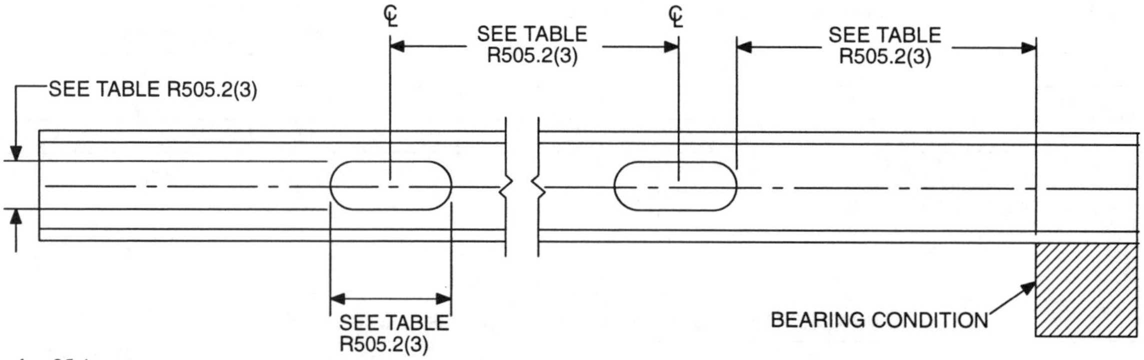

For SI: 1 inch = 25.4 mm.

FIGURE R505.2(3)
FLOOR JOIST WEB HOLES

R505.2.1 Material. Load-bearing members utilized in steel floor construction shall be cold formed to shape from structural quality sheet steel complying with the requirements of one of the following:

1. ASTM A 653: Grades 33, 37, 40 and 50 (Class 1 and 3).
2. ASTM A 792: Grades 33, 37, 40 and 50A.
3. ASTM A 875: Grades 33, 37, 40 and 50 (Class 1 and 3).
4. Steels that comply with ASTM A 653, except for tensile and elongation, shall be permitted provided the ratio of tensile strength to yield point is at least 1.08 and the total elongation is at least 10 percent for a 2-inch (51 mm) gage length or 7 percent for an 8-inch (203 mm) gage length.

R505.2.2 Identification. Load-bearing steel framing members shall have a legible label, stencil, stamp or embossment with the following information as a minimum:

1. Manufacturer's identification.
2. Minimum uncoated steel thickness in inches (mm).
3. Minimum coating designation.
4. Minimum yield strength, in kips per square inch (ksi) (kPa).

R505.2.3 Corrosion protection. Load-bearing steel framing shall have a metallic coating complying with one of the following:

1. A minimum of G 60 in accordance with ASTM A 653.
2. A minimum of AZ 50 in accordance with ASTM A 792.
3. A minimum of GF 60 in accordance with ASTM A 875.

R505.2.4 Fastening requirements. Screws for steel-to-steel connections shall be installed with a minimum edge distance and center-to-center spacing of 0.5 inch (12.7 mm), shall be self-drilling tapping, and shall conform to SAE J78. Floor sheathing shall be attached to steel joists with minimum No. 8 self-drilling tapping screws that conform to SAE J78. Screws attaching floor-sheathing-to-steel joists shall have a minimum head diameter of 0.292 inch (7.4 mm) with countersunk heads and shall be installed with a minimum edge distance of 0.375 inch (9.5 mm). Gypsum board ceilings shall be attached to steel joists with minimum No. 6 screws conforming to ASTM C 954 and shall be installed in accordance with Section R702. For all connections, screws shall extend through the steel a minimum of three exposed threads. All self-drilling tapping screws conforming to SAE J78 shall have a Type II coating in accordance with ASTM B 633.

Where No. 8 screws are specified in a steel to steel connection the required number of screws in the connection is permitted to be reduced in accordance with the reduction factors in Table R505.2.4 when larger screws are used or when one of the sheets of steel being connected is thicker than 33 mils (0.84 mm). When applying the reduction factor the resulting number of screws shall be rounded up.

TABLE R505.2.4
SCREW SUBSTITUTION FACTOR

SCREW SIZE	THINNEST CONNECTED STEEL SHEET (mils)	
	33	43
#8	1.0	0.67
#10	0.93	0.62
#12	0.86	0.56

For SI: 1 mil = 0.0254 mm.

R505.3 Floor construction. Cold-formed steel floors shall be constructed in accordance with this section and Figure R505.3.

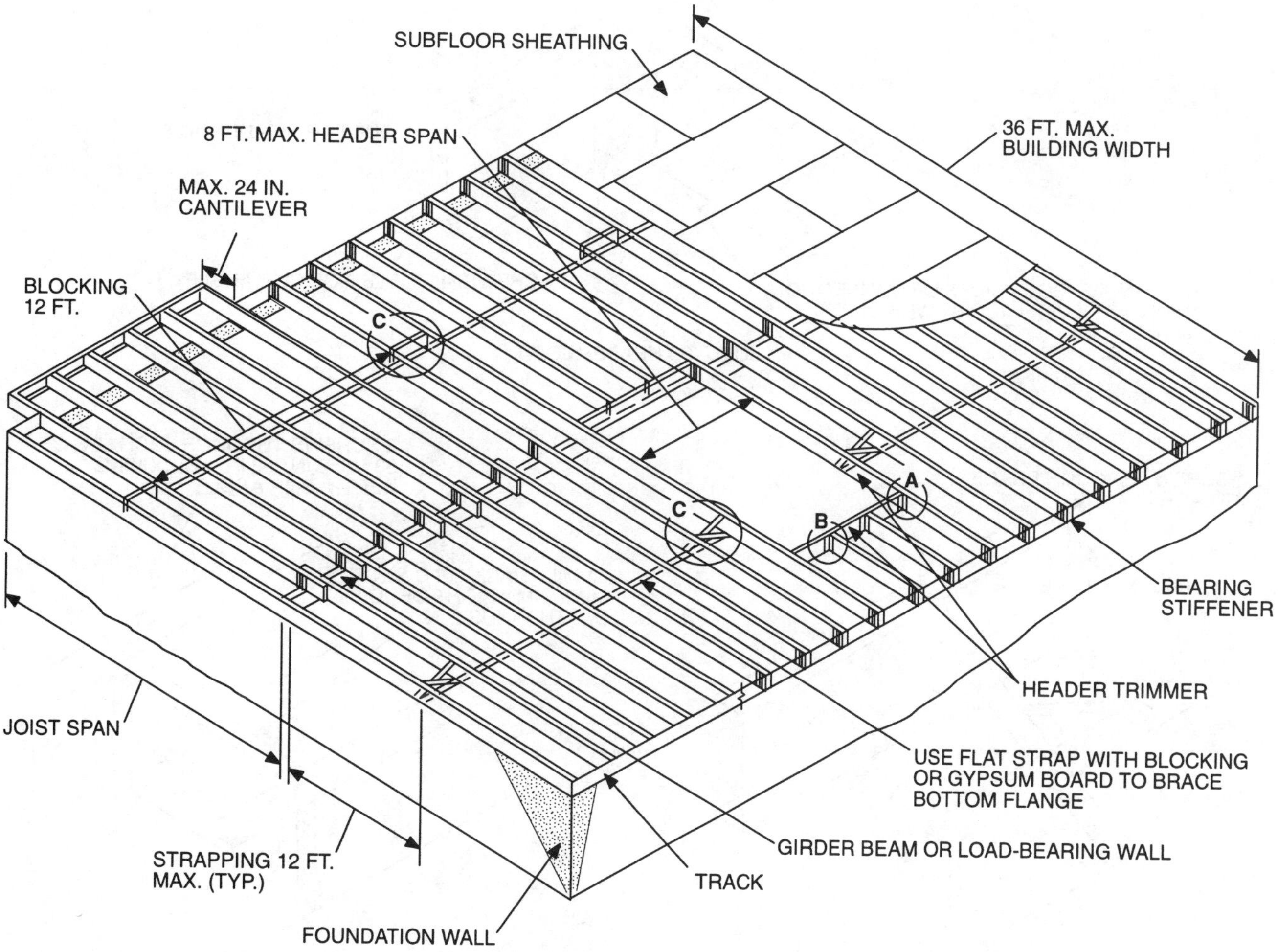

For SI: 1 inch = 25.4 mm, 1 foot = 304.8 mm.

FIGURE R505.3
STEEL FLOOR CONSTRUCTION
(continued)

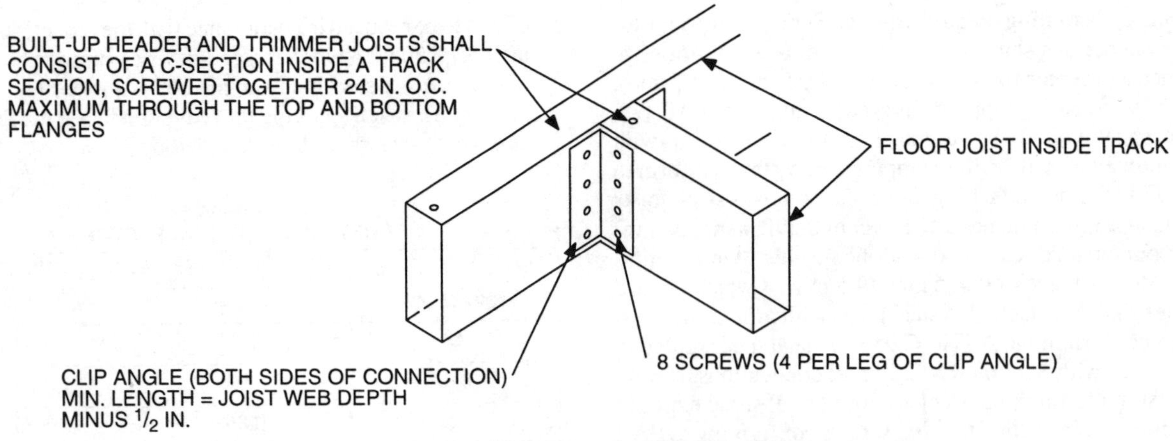

BUILT-UP HEADER OR TRIMMER JOIST

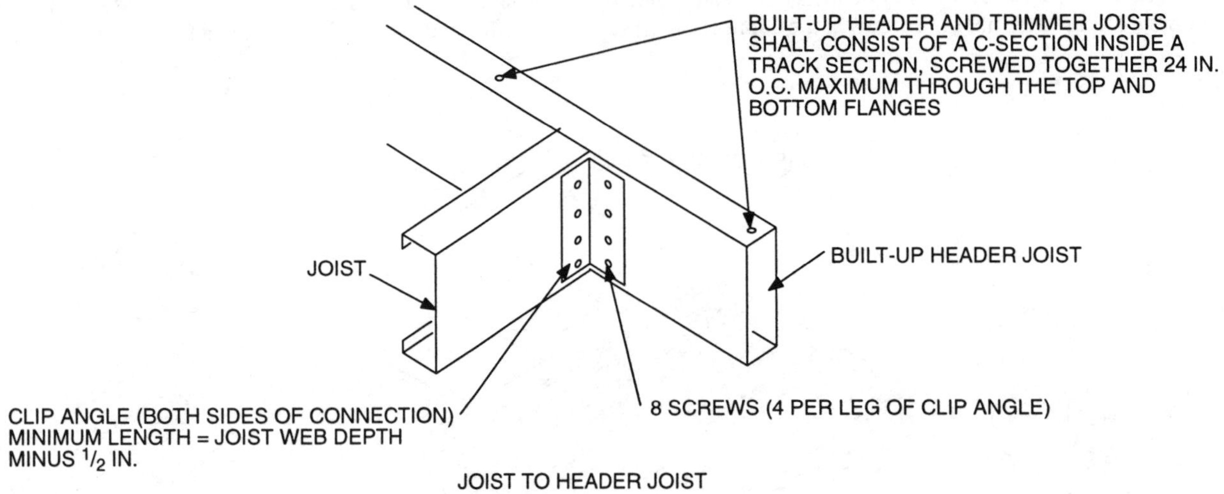

JOIST TO HEADER JOIST

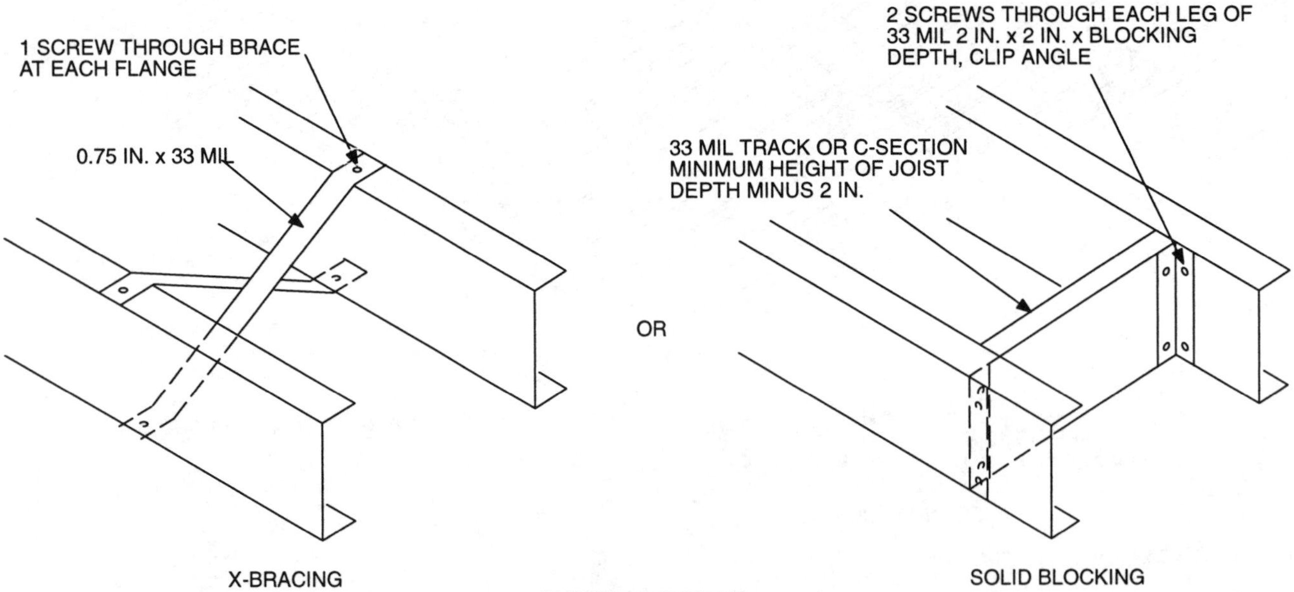

BLOCKING OPTIONS

For SI: 1 inch = 25.4 mm, 1 mil = 0.0254 mm.

FIGURE R505.3—continued
STEEL FLOOR CONSTRUCTION

R505.3.1 Floor to foundation or bearing wall connections. Cold-formed steel floors shall be anchored to foundations, wood sills or load-bearing walls in accordance with Table R505.3.1(1) and Figure R505.3.1(1), R505.3.1(2), R505.3.1(3), R505.3.1(4), R505.3.1(5) or R505.3.1(6). Continuous steel joists supported by interior load-bearing walls shall be constructed in accordance with Figure R505.3.1(7). Lapped steel joists shall be constructed in accordance with Figure R505.3.1(8). Fastening of steel joists to other framing members shall be in accordance with Table R505.3.1(2).

TABLE R505.3.1(1)
FLOOR TO FOUNDATION OR BEARING WALL CONNECTION REQUIREMENTS [a,b,c]

FRAMING CONDITION	WIND SPEED (mph) AND EXPOSURE		
	Up to 110 A/B or 85 C or Seismic Design Categories A, B, C and D_1	Up to 130 B or 110 C or Seismic Design Category D_2	Up to 130C
Floor joist to wall track of exterior steel load-bearing wall per Figure R505.3.1(1)	2-No. 8 screws	3-No. 8 screws	4-No. 8 screws
Floor joist track to wood sill per Figure R505.3.1(2)	Steel plate spaced at 3′ o.c., with 4-No. 8 screws and 4-10d or 6-8d common nails	Steel plate, spaced at 2′ o.c., with 4-No. 8 screws and 4-10d or 6-8d common nails	Steel plate spaced at 1′ o.c. with 4-No. 8 screws and 4-10d or 6-8d common nails
Floor joist track to foundation per Figure R505.3.1(3)	$^1/_2$″ minimum diameter anchor bolt and clip angle spaced at 6′ o.c. with 8-No. 8 screws	$^1/_2$″ minimum diameter anchor bolt and clip angle spaced at 4′ o.c. with 8-No. 8 screws	$^1/_2$″ minimum diameter anchor bolt and clip angle spaced at 2′ o.c. with 8-No. 8 screws
Joist cantilever to wall track per Figure R505.3.1(4)	2-No. 8 screws per stiffener or bent plate	3-No. 8 screws per stiffener or bent plate	$^1/_2$″ minimum diameter anchor bolt and clip angle spaced at 2′ o.c. with 8-No. 8 screws
Joist cantilever to wood sill per Figure R505.3.1(5)	Steel plate spaced at 3′ o.c., with 4-No. 8 screws and 4-10d or 6-8d common nails	Steel plate spaced at 2′ o.c., with 4-No. 8 screws and 4-10d or 6-8d common nails	Steel plate spaced at 1′ o.c. with 4-No. 8 screws and 4-10d or 6-8d common nails
Joist cantilever to foundation per Figure R505.3.1(6)	$^1/_2$″ minimum diameter anchor bolt and clip angle spaced at 6′ o.c. with 8-No. 8 screws	$^1/_2$″ minimum diameter anchor bolt and clip angle spaced at 4′ o.c. with 8-No. 8 screws	$^1/_2$″ minimum diameter anchor bolt and clip angle spaced at 2′ o.c. with 8-No. 8 screws

For SI: 1 inch = 25.4 mm, 1 foot = 304.8 mm, 1 mile per hour = 1.609 km/h.

a. Anchor bolts shall be located not more than 12 inches from corners or the termination of bottom tracks (e.g., at door openings). Bolts shall extend a minimum of 15 inches into masonry or 7 inches into concrete.

b. All screw sizes shown are minimum.

c. In Seismic Design Category D_1 and D_2 or where the basic wind speed equals or exceeds 110 mph, connection shall comply with requirements in Section R603.8, but shall be no less than the minimum required herein.

TABLE R505.3.1(2)
FLOOR FASTENING SCHEDULE[a]

DESCRIPTION OF BUILDING ELEMENTS	NUMBER AND SIZE OF FASTENERS	SPACING OF FASTENERS
Floor joist to track of an interior load-bearing wall per Figures R505.3.1(7) and R505.3.1(8)	2 No. 8 screws	Each joist
Floor joist to track at end of joist	2 No. 8 screws	One per flange or two per bearing stiffener
Subfloor to floor joists	No. 8 screws	6″ o.c. on edges and 10″ o.c. at intermediate supports

For SI: 1 inch = 25.4 mm.

a. All screw sizes shown are minimum.

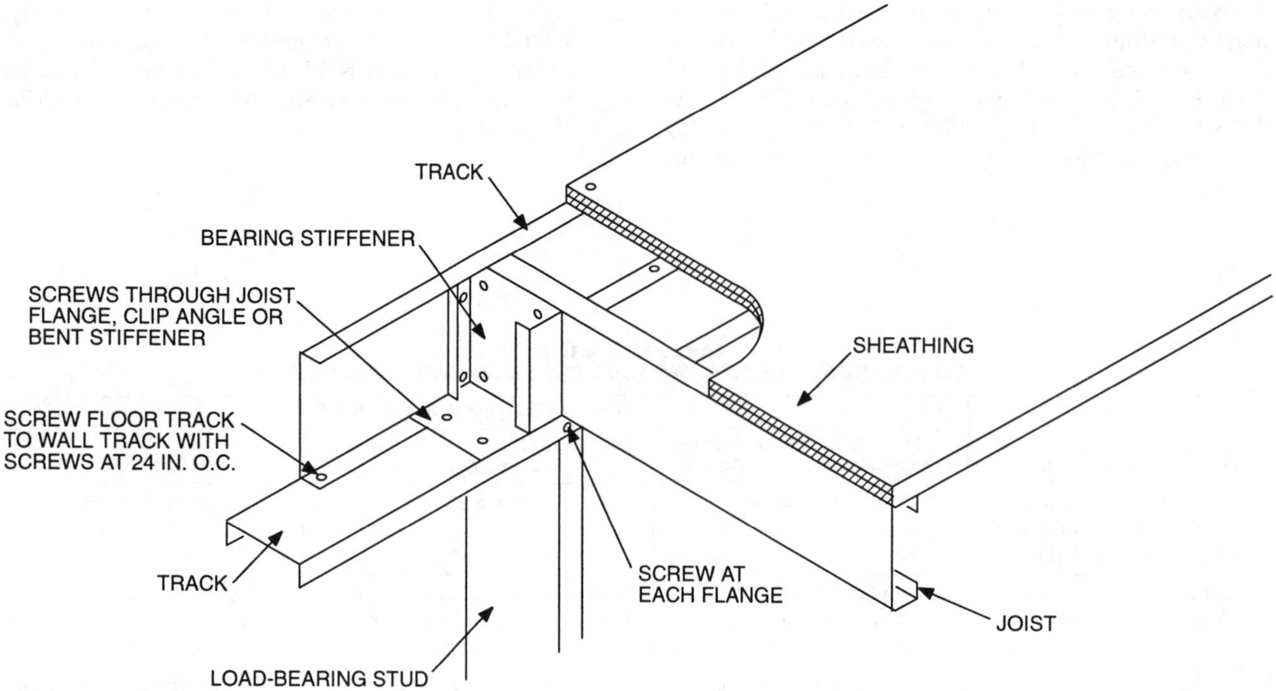

For SI: 1 inch = 25.4 mm, 1 mil = 0.0254 mm.

FIGURE R505.3.1(1)
FLOOR TO LOAD-BEARING WALL STUD CONNECTION

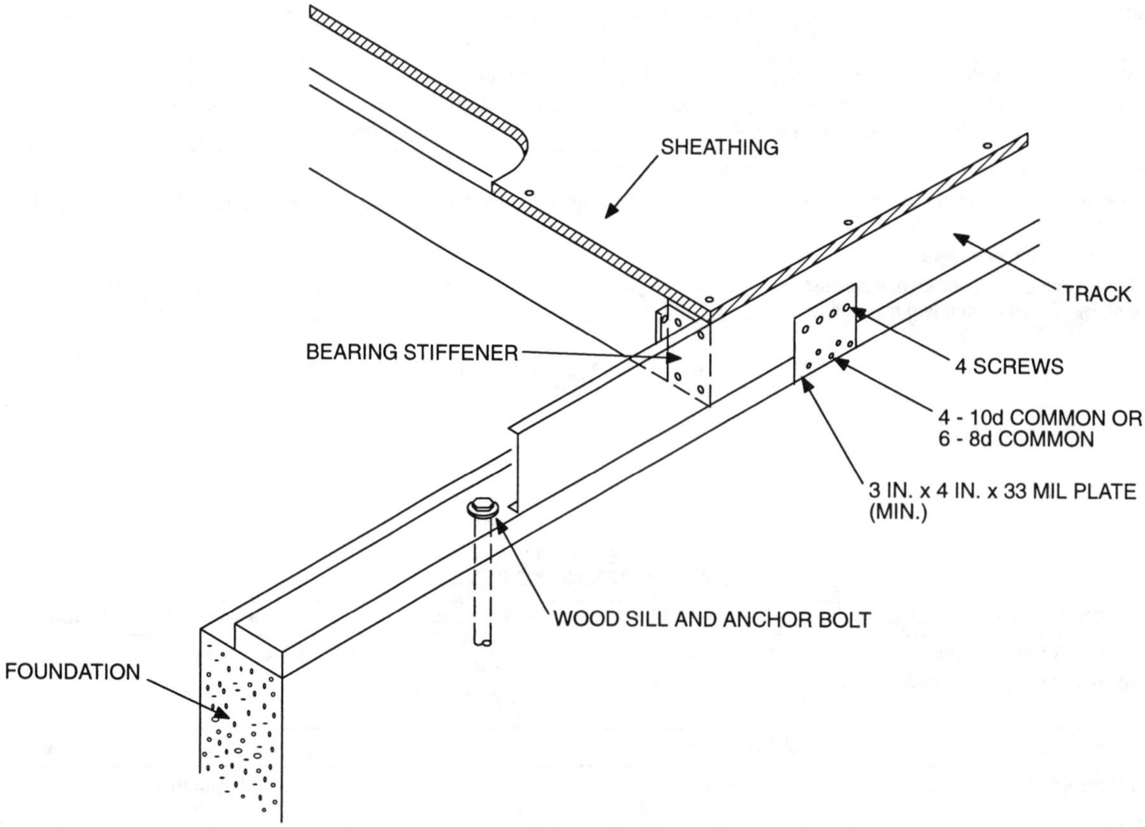

For SI: 1 inch = 25.4 mm, 1 mil = 0.0254 mm.

FIGURE R505.3.1(2)
FLOOR TO WOOD SILL CONNECTION

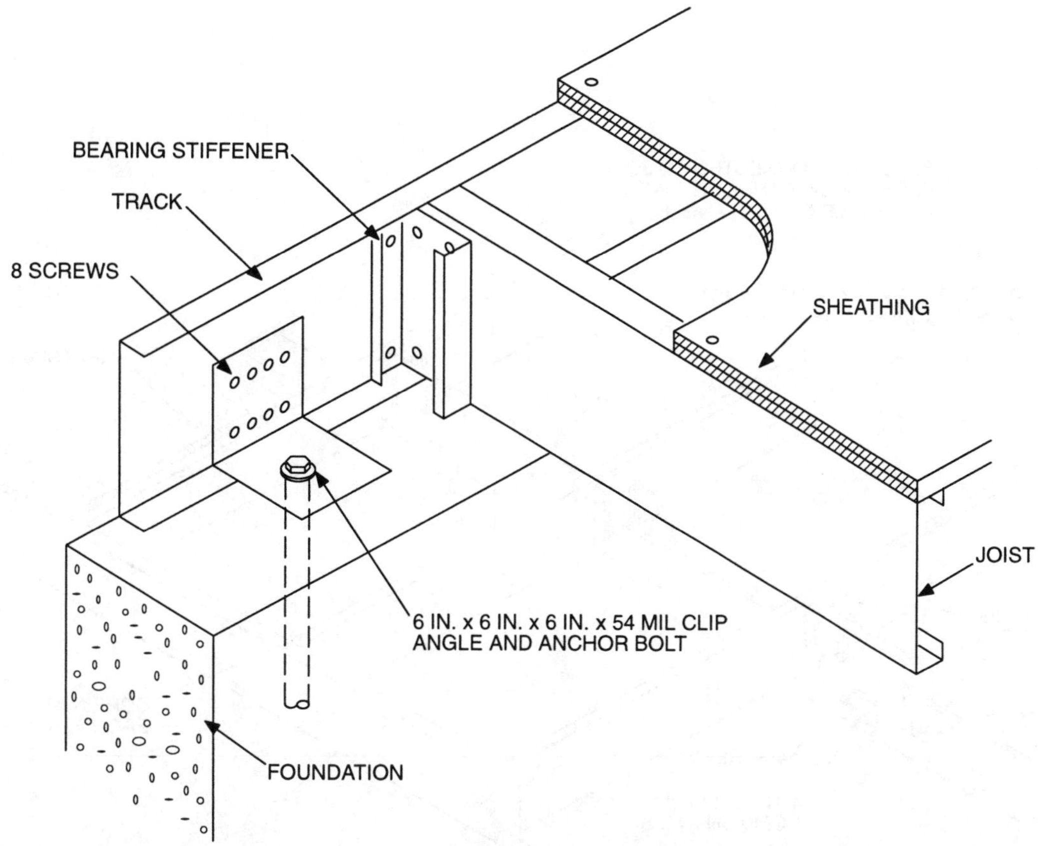

For SI: 1 inch = 25.4 mm, 1 mil = 0.0254 mm.

FIGURE R505.3.1(3)
FLOOR TO FOUNDATION CONNECTION

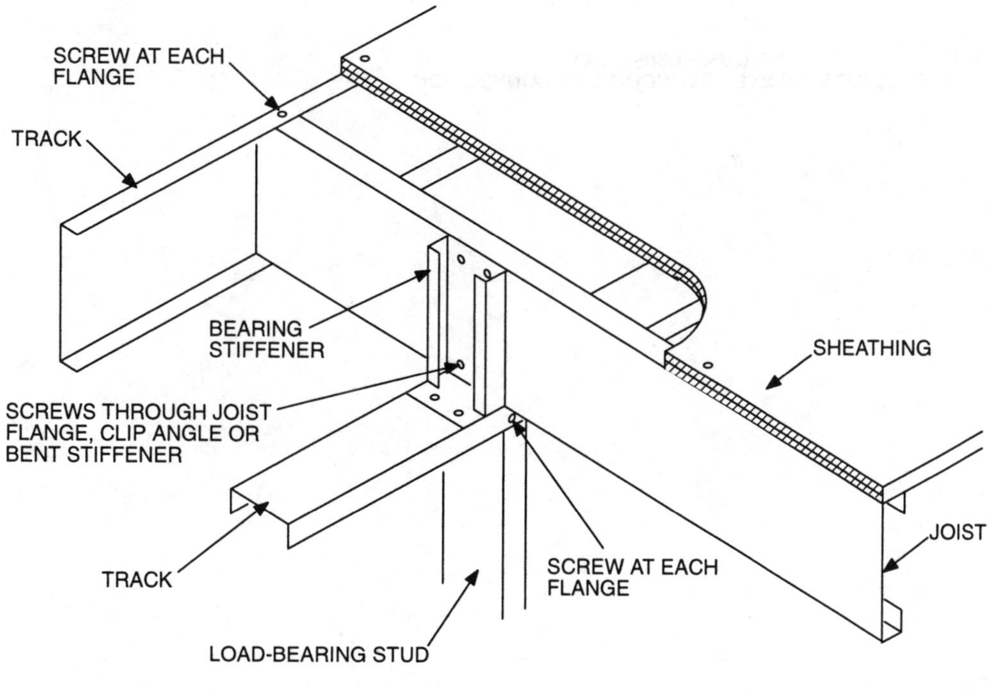

FIGURE R505.3.1(4)
FLOOR CANTILEVER TO LOAD-BEARING WALL CONNECTION

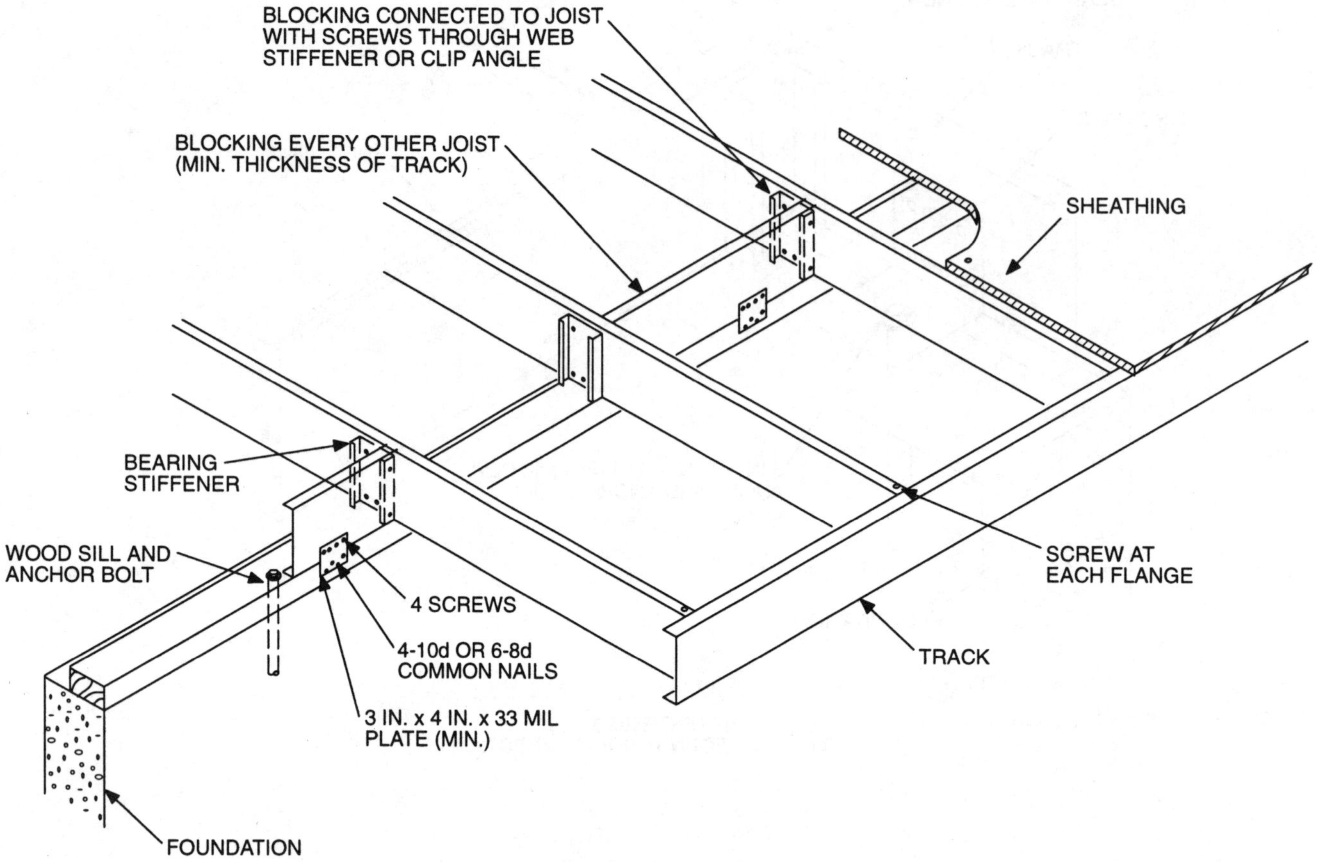

For SI: 1 inch = 25.4 mm, 1 mil = 0.0254 mm.

FIGURE R505.3.1(5)
FLOOR CANTILEVER TO WOOD SILL CONNECTION

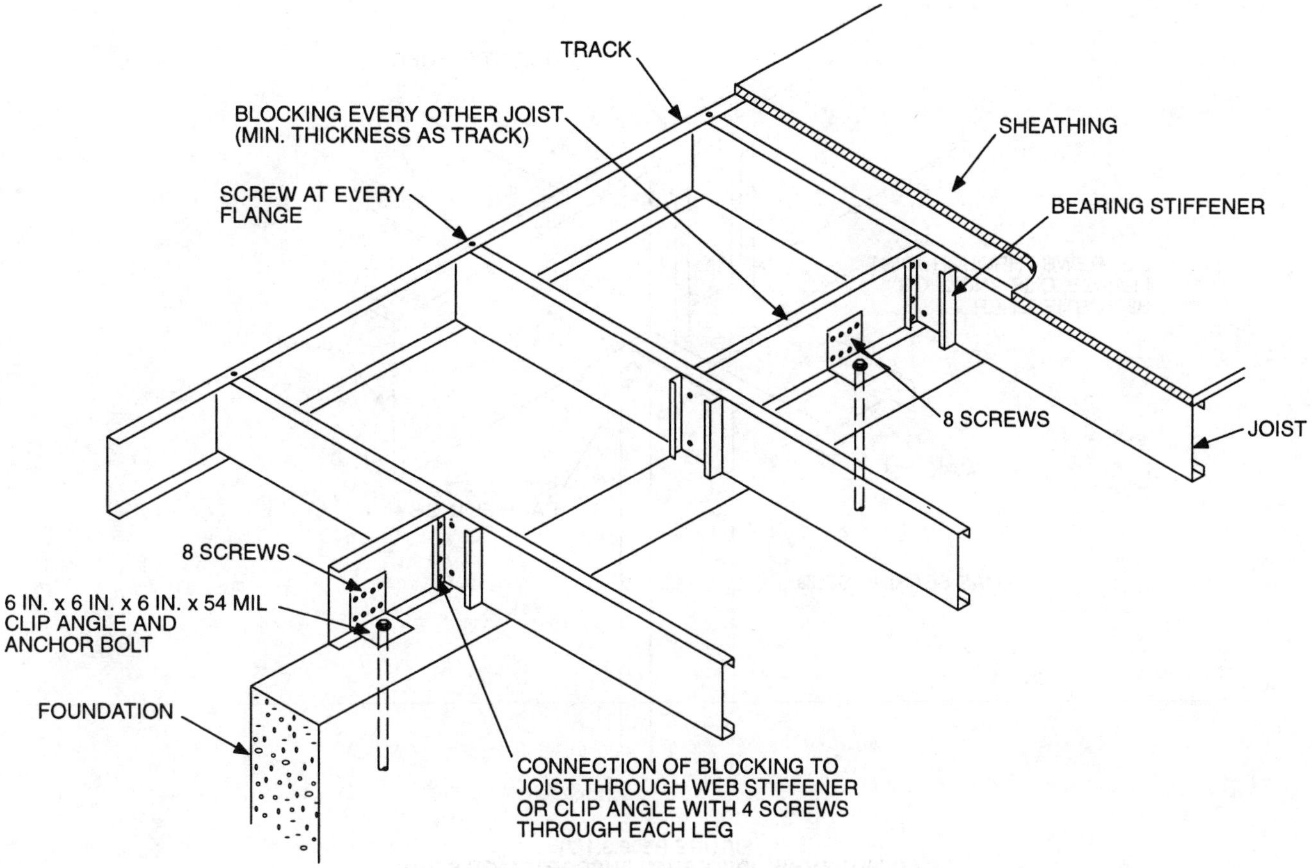

For SI: 1 inch = 25.4 mm, 1 mil = 0.0254 mm.

FIGURE R505.3.1(6)
FLOOR CANTILEVER TO FOUNDATION CONNECTION

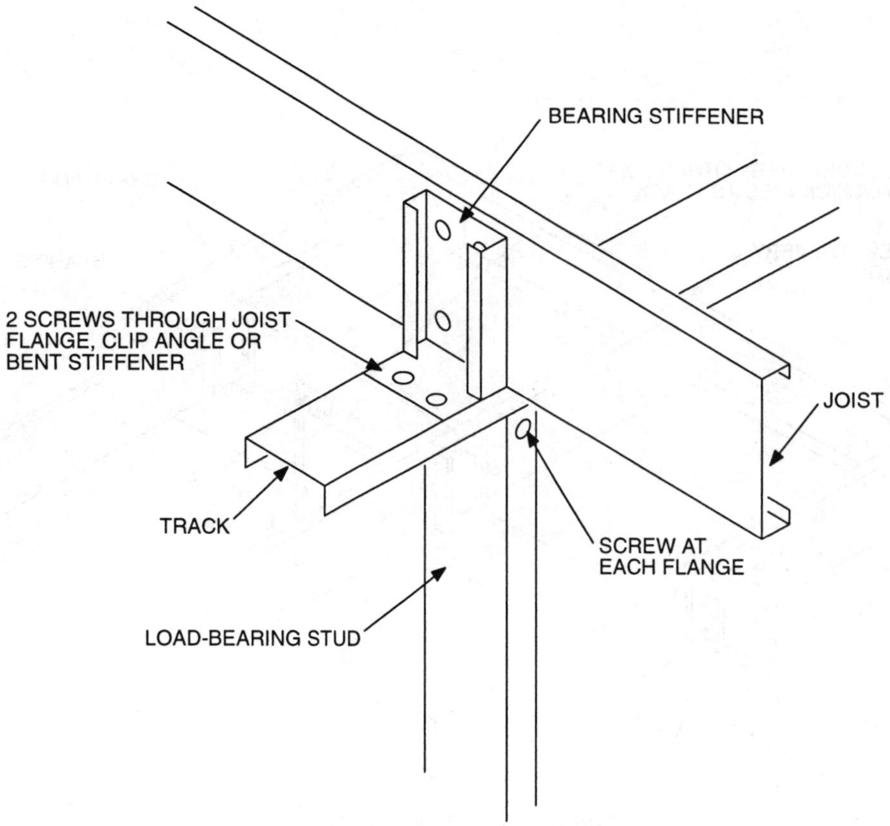

**FIGURE R505.3.1(7)
CONTINUOUS JOIST SPAN SUPPORTED ON STUD**

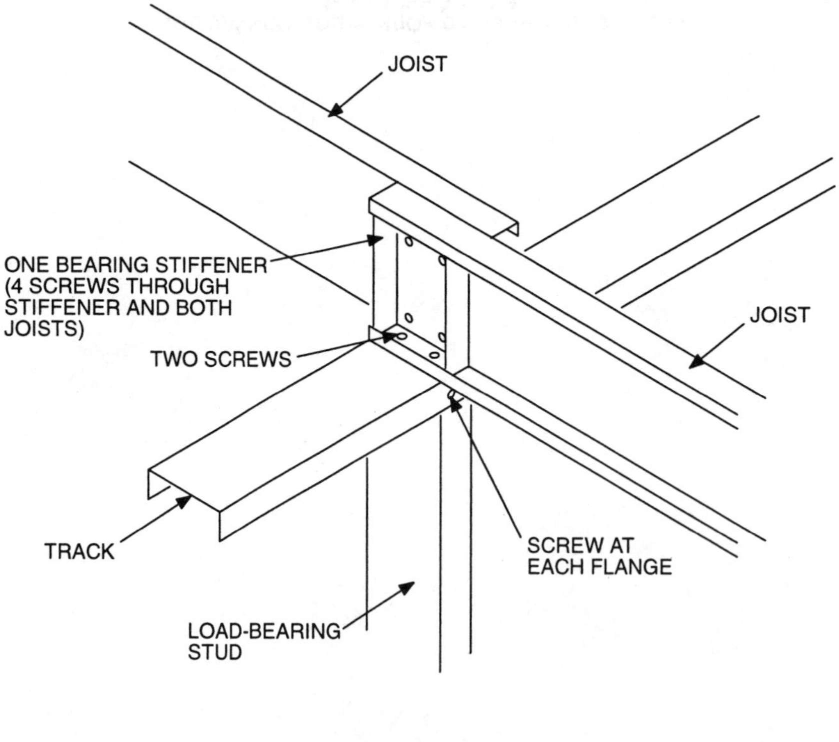

**FIGURE R505.3.1(8)
LAPPED JOISTS SUPPORTED ON STUD**

R505.3.2 Allowable joist spans. The clear span of cold-formed steel floor joists shall not exceed the limits set forth in Table R505.3.2. Floor joists shall have a minimum bearing length of 1.5 inches (38 mm). When continuous joists are used the interior bearing supports shall be located within 2 feet (610 mm) of mid span of the steel joists, and the individual spans shall not exceed the spans in Table R505.3.2. Bearing stiffeners shall be installed at each bearing location in accordance with Section R505.3.4 and as shown in Figure R505.3.

R505.3.3 Joist bracing. The top flanges of steel joists shall be laterally braced by the application of floor sheathing fastened to the joists in accordance with Table R505.3.1(2). Floor joists with spans that exceed 12 feet (3658 mm) shall have the bottom flanges laterally braced in accordance with one of the following:

1. Gypsum board installed with minimum No. 6 screws in accordance with Section R702.
2. Continuous steel strapping installed in accordance with Figure R505.3. Steel straps shall be at least 1.5 inches (38 mm) in width and 33 mils (0.84 mm) in thickness. Straps shall be fastened to the bottom flange at each joist with at least one No. 8 screw and shall be fastened to blocking with at least two No. 8 screws. Blocking or bridging (X-bracing) shall be installed between joists in-line with straps at a maximum spacing of 12 feet (3658 mm) measured perpendicular to the joist run and at the termination of all straps.

R505.3.4 Bearing stiffeners. Bearing stiffeners shall be installed at all bearing locations for steel floor joists. A bearing stiffener shall be fabricated from a minimum 33 mil (0.84 mm) C-section or 43 mil (1.09 mm) track section. Each stiffener shall be fastened to the web of the joist with a minimum of four No. 8 screws equally spaced as shown in Figure R505.3.4. Stiffeners shall extend across the full depth of the web and shall be installed on either side of the web.

R505.3.5 Cutting and notching. Flanges and lips of load-bearing steel floor framing members shall not be cut or notched.

R505.3.6 Hole patching. Web holes for 800S162-33, 1000S162-43, 1200S162-43 and 1200S162-54 nominal joist sizes with dimensions conforming to Section R505.2 that are closer than 10 inches (305 mm) from the edge of the hole to the edge of the bearing surface shall be patched with a solid steel plate, C-section or track section in accordance with Figure R505.3.6. The steel patch shall be of a minimum thickness as the receiving member and shall extend at least 1 inch (25.4 mm) beyond all edges of the hole. The steel patch shall be fastened to the web with No. 8 screws (minimum) spaced no greater than 1 inch (25.4 mm) center-to-center along the edges of the patch, with a minimum edge distance of 0.5 inch (12.7 mm).

R505.3.7 Floor cantilevers. Floor cantilevers shall not exceed 24 inches (610 mm) as illustrated in Figure R505.3. The cantilever back-span shall extend a minimum of 6 feet (1830 mm) within the building, and shall be fastened to a bearing condition in accordance with Section R505.3.1. Floor cantilevers shall be permitted only on the second floor of a two-story building or the first floor of a one-story building. Floor framing that is cantilevered and supports the cantilevered floor only shall consist of single joist members in accordance with Section R505.3.2. Floor framing that is cantilevered and supports the cantilevered floor and the roof framing load above shall consist of double joist members of the same size and material thickness as that for single joist members in accordance with Section R505.3.2, and shall be fastened web-to-web with minimum No. 8 screws at 24 inches (610 mm) maximum on-center spacing top and bottom. Built-up floor framing consisting of a C-section inside a track section, fastened at the top and bottom flanges by minimum No. 8 screws at 24 inches (610 mm) maximum on center spacing, is permitted in lieu of the web-to-web double joist method.

TABLE R505.3.2
ALLOWABLE SPANS FOR COLD-FORMED STEEL JOISTS[a,b]

NOMINAL JOIST SIZE	30 PSF LIVE LOAD		40 PSF LIVE LOAD	
	Spacing (inches)		Spacing (inches)	
	16	24	16	24
550S162-33	10′-7″	9′-1″	9′-7″	8′-1″
550S162-43	11′-6″	10′-0″	10′-5″	9′-1″
550S162-54	12′-4″	10′-9″	11′-2″	9′-9″
550S162-68	13′-2″	11′-6″	12′-0″	10′-6″
800S162-33	13′-3″	8′-10″	10′-7″	7′-1″
800S162-43	15′-6″	13′-7″	14′-1″	12′-3″
800S162-54	16′-8″	14′-7″	15′-2″	13′-3″
800S162-68	17′-11″	15′-7″	16′-3″	14′-2″
1000S162-43	18′-8″	15′-3″	16′-8″	13′-1″
1000S162-54	20′-1″	17′-6″	18′-3″	15′-11″
1000S162-68	21′-6″	18′-10″	19′-7″	17′-1″
1200S162-43	20′-3″	14′-1″	16′-10″	11′-3″
1200S162-54	23′-4″	19′-7″	21′-3″	17′-6″
1200S162-68	25′-1″	21′-11″	22′-10″	19′-11″

For SI: 1 inch = 25.4 mm, 1 foot = 304.8 mm, 1 pound per square foot = 0.0479 kN/m^2.

a. Deflection criteria: *L*/480 for live loads, *L*/360 for total loads.

b. Floor dead load = 10 psf.

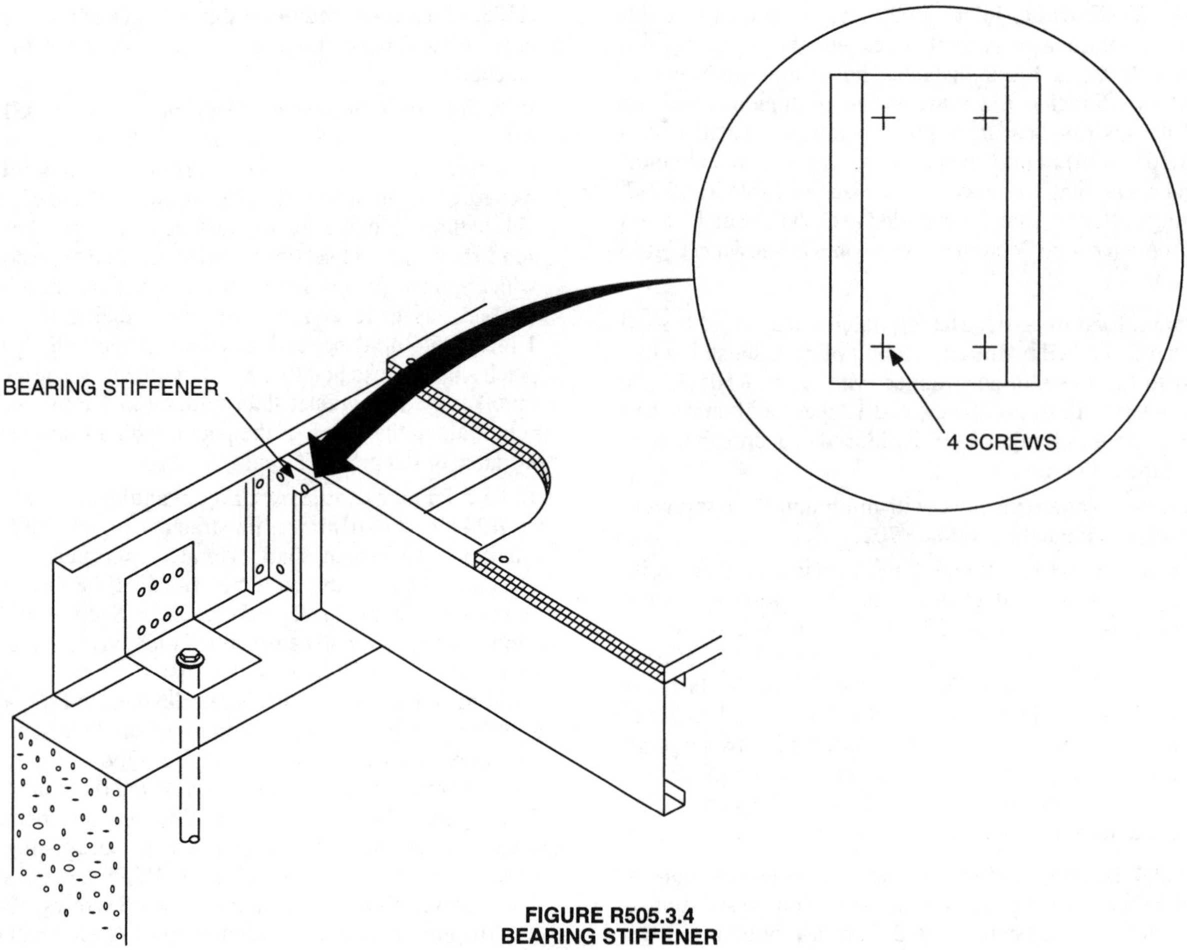

FIGURE R505.3.4
BEARING STIFFENER

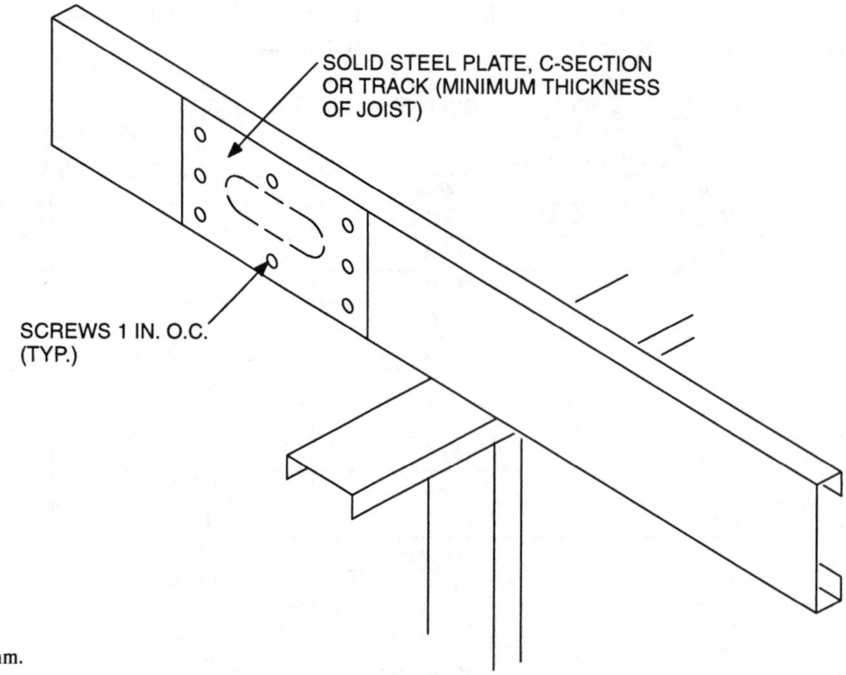

For SI: 1 inch = 25.4 mm.

FIGURE R505.3.6
HOLE PATCH

R505.3.8 Splicing. Joists and other structural members shall not be spliced. Splicing of tracks shall conform with Figure R505.3.8.

R505.3.9 Framing of openings. Openings in floor framing shall be framed with header and trimmer joists. Header joist spans shall not exceed 8 feet (2438 mm). Header and trimmer joists shall be fabricated from joist and track sections, which shall be of a minimum size and thickness as the adjacent floor joists and shall be installed in accordance with Figure R505.3. Each header joist shall be connected to trimmer joists with a minimum of four 2-inch-by-2-inch (51 mm by 51 mm) clip angles. Each clip angle shall be fastened to both the header and trimmer joists with four No. 8 screws, evenly spaced, through each leg of the clip angle. The clip angles shall have a steel thickness not less than that of the floor joist.

SECTION R506
CONCRETE FLOORS (ON GROUND)

R506.1 General. Concrete slab-on-ground floors shall be a minimum 3.5 inches (89 mm) thick (for expansive soils, see Section R403.1.8). The specified compressive strength of concrete shall be as set forth in Section R402.2.

R506.2 Site preparation. The area within the foundation walls shall have all vegetation, top soil and foreign material removed.

R506.2.1 Fill. Fill material shall be free of vegetation and foreign material. The fill shall be compacted to assure uniform support of the slab, and except where approved, the fill depths shall not exceed 24 inches (610 mm) for clean sand or gravel and 8 inches (203 mm) for earth.

R506.2.2 Base. A 4-inch-thick (102 mm) base course consisting of clean graded sand, gravel, crushed stone or crushed blast-furnace slag passing a 2-inch (51 mm) sieve shall be placed on the prepared subgrade when the slab is below grade.

Exception: A base course is not required when the concrete slab is installed on well-drained or sand-gravel mixture soils classified as Group I according to the United Soil Classification System in accordance with Table R405.1.

R506.2.3 Vapor retarder. An approved vapor retarder with joints lapped not less than 6 inches (153 mm) shall be placed between the concrete floor slab and the base course or the prepared subgrade where no base course exists.

Exception: The vapor retarder may be omitted:

1. From detached garages, utility buildings and other unheated accessory structures.
2. From driveways, walks, patios and other flatwork not likely to be enclosed and heated at a later date.
3. Where approved by the building official, based on local site conditions.

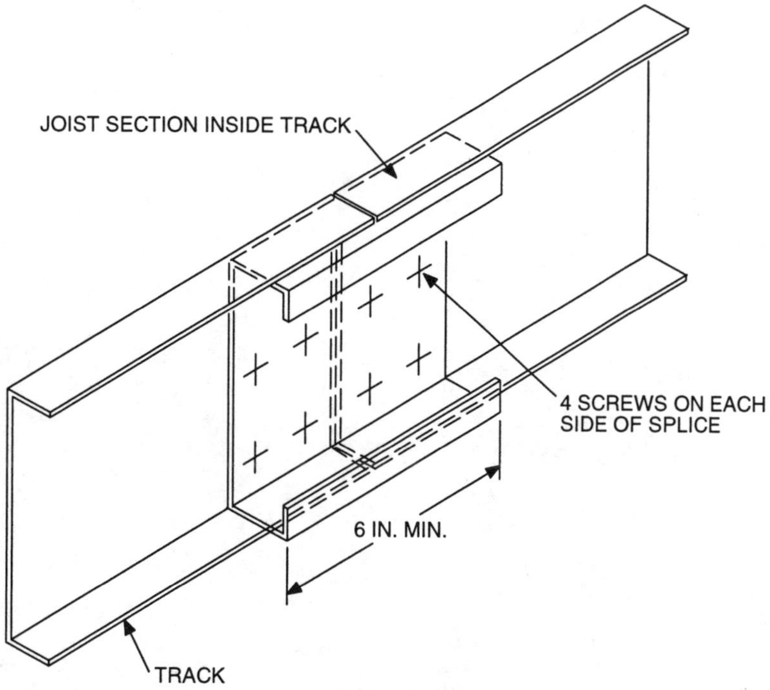

For SI: 1 inch = 25.4 mm.

FIGURE R505.3.8
TRACK SPLICE

CHAPTER 6
WALL CONSTRUCTION

SECTION R601
GENERAL

R601.1 Application. The provisions of this chapter shall control the design and construction of all walls and partitions for all buildings.

R601.2 Requirements. Wall construction shall be capable of accommodating all loads imposed according to Section R301 and of transmitting the resulting loads to the supporting structural elements.

R601.2.1 Compressible floor-covering materials. Compressible floor-covering materials that compress more than $^1/_{32}$ inch (0.794 mm) when subjected to 50 pounds (23 kg) applied over 1 inch square (645 mm) of material and are greater than $^1/_8$ inch (3.2 mm) in thickness in the uncompressed state shall not extend beneath walls, partitions or columns, which are fastened to the floor.

SECTION R602
WOOD WALL FRAMING

R602.1 Identification. Load-bearing dimension lumber for studs, plates and headers shall be identified by a grade mark of a lumber grading or inspection agency that has been approved by an accreditation body that complies with DOC PS 20. In lieu of a grade mark, a certification of inspection issued by a lumber grading or inspection agency meeting the requirements of this section shall be accepted.

R602.1.1 End-jointed lumber. Approved end-jointed lumber identified by a grade mark conforming to Section R602.1 may be used interchangeably with solid-sawn members of the same species and grade.

R602.2 Grade. Studs shall be a minimum No. 3, standard or stud grade lumber.

Exception: Bearing studs not supporting floors and nonbearing studs may be utility grade lumber, provided the studs are spaced in accordance with Table R602.3(5).

R602.3 Design and construction. Exterior walls of wood-frame construction shall be designed and constructed in accordance with the provisions of this chapter and Figures R602.3(1) and R602.3(2) or in accordance with AF&PA's NDS. Components of exterior walls shall be fastened in accordance with Tables R602.3(1) through R602.3(4).

R602.3.1 Stud spacing. In bearing walls, studs that are not more than 10 feet (3048 mm) in length shall be spaced not more than is specified in Table R602.3(5). In bearing walls, studs that are more than 10 feet (3048 mm) in height shall be spaced not more than specified in Table R602.3.1.

R602.3.2 Top plate. Wood stud walls shall be capped with a double top plate installed to provide overlapping at corners and intersections with bearing partitions. End joints in top plates shall be offset at least 24 inches (610 mm).

Exception: A single top plate may be installed in stud walls, provided the plate is adequately tied at joints, corners and intersecting walls by a minimum 3-inch-by-6-inch by a 0.036-inch-thick (76 mm by 152 mm by 0.914 mm) galvanized steel plate that is nailed to each wall or segment of wall by six 8d nails on each side, provided the rafters or joists are centered over the studs with a tolerance of no more than 1 inch (25.4 mm). The top plate may be omitted over lintels that are adequately tied to adjacent wall sections with steel plates or equivalent as previously described.

R602.3.3 Bearing studs. Where joists, trusses or rafters are spaced more than 16 inches (406 mm) on center and the bearing studs below are spaced 24 inches (610 mm) on center, such members shall bear within 5 inches (127 mm) of the studs beneath.

Exceptions:

1. The top plates are two 2-inch by 6-inch (38 mm by 140 mm) or two 3-inch by 4-inch (64 mm by 89 mm) members.
2. A third top plate is installed.
3. Solid blocking equal in size to the studs is installed to reinforce the double top plate.

R602.3.4 Bottom (sole) plate. Studs shall have full bearing on a nominal 2 by (38 mm) or larger plate or sill having a width at least equal to the width of the studs.

R602.4 Interior load-bearing walls. Interior load-bearing walls shall be constructed, framed and fireblocked as specified for exterior walls.

R602.5 Interior nonbearing walls. Interior nonbearing walls shall be permitted to be constructed with 2-inch-by-3-inch (51 mm by 76 mm) studs spaced 24 inches (610 mm) on center or, when not part of a braced wall line, 2-inch-by-4-inch (51 mm by 102 mm) flat studs spaced at 16 inches (406 mm) on center. Interior nonbearing walls shall be capped with at least a single top plate. Interior nonbearing walls shall be fireblocked in accordance with Section R602.8.

TABLE R602.3(1)
FASTENER SCHEDULE FOR STRUCTURAL MEMBERS

DESCRIPTION OF BUILDING ELEMENTS		NUMBER AND TYPE OF FASTENER[a,b,c,d]	SPACING OF FASTENERS
Joist to sill or girder, toe nail		3-8d	—
1″ × 6″ subfloor or less to each joist, face nail		2-8d 2 staples, $1^{3}/_{4}$	— —
2″ subfloor to joist or girder, blind and face nail		2-16d	—
Sole plate to joist or blocking, face nail		16d	16″ o.c.
Top or sole plate to stud, end nail		2-16d	—
Stud to sole plate, toe nail		3-8d or 2-16d	—
Double studs, face nail		10d	24″ o.c.
Double top plates, face nail		10d	24″ o.c.
Sole plate to joist or blocking at braced wall panels		3-16d	16″ o.c.
Double top plates, minimum 48-inch offset of end joints, face nail in lapped area		8-16d	—
Blocking between joists or rafters to top plate, toe nail		3-8d	—
Rim joist to top plate, toe nail		8d	6″ o.c.
Top plates, laps at corners and intersections, face nail		2-10d	—
Built-up header, two pieces with $^{1}/_{2}$″ spacer		16d	16″ o.c. along each edge
Continued header, two pieces		16d	16″ o.c. along each edge
Ceiling joists to plate, toe nail		3-8d	—
Continuous header to stud, toe nail		4-8d	—
Ceiling joist, laps over partitions, face nail		3-10d	—
Ceiling joist to parallel rafters, face nail		3-10d	—
Rafter to plate, toe nail		2-16d	—
1″ brace to each stud and plate, face nail		2-8d 2 staples, $1^{3}/_{4}$	— —
1″ x 6″ sheathing to each bearing, face nail		2-8d 2 staples, $1^{3}/_{4}$	— —
1″ x 8″ sheathing to each bearing, face nail		2-8d 3 staples, $1^{3}/_{4}$	— —
Wider than 1″ x 8″ sheathing to each bearing, face nail		3-8d 4 staples, $1^{3}/_{4}$	— —
Built-up corner studs		10d	24″ o.c.
Built-up girders and beams, 2-inch lumber layers		10d	Nail each layer as follows: 32″ o.c. at top and bottom and staggered. Two nails at ends and at each splice.
2″ planks		2-16d	At each bearing
Roof rafters to ridge, valley or hip rafters: toe nail face nail		 4-16d 3-16d	 — —
Rafter ties to rafters, face		3-8d	—
Wood structural panels, subfloor, roof and wall sheathing to framing, and particleboard wall sheathing to framing			
$^{5}/_{16}$-$^{1}/_{2}$	6d common nail (subfloor, wall) 8d common nail (roof)[f]	6	12[g]
$^{19}/_{32}$ -1	8d common nail	6	12[g]
$1^{1}/_{8}$-$1^{1}/_{4}$	10d common nail or 8d deformed nail	6	12

(continued)

TABLE R602.3(1)—continued
FASTENER SCHEDULE FOR STRUCTURAL MEMBERS

DESCRIPTION OF BUILDING MATERIALS	DESCRIPTION OF FASTENER[b,c,d,e]	SPACING OF FASTENERS	
		Edges (inches)[i]	Intermediate supports[c,e] (inches)
Other wall sheathing[h]			
1/2″ regular cellulosic fiberboardsheathing	1 1/2 galvanized roofing nail 6d common nail staple 16 ga., 1 1/2 long	3	6
1/2 structural cellulosic fiberboard sheathing	1 1/2 galvanized roofing nail 8d common nail staple 16 ga., 1 1/2 long	3	6
25/32 structural cellulosic fiberboard sheathing	1 3/4 galvanized roofing nail 8d common nail staple 16 ga., 1 3/4 long	3	6
1/2 gypsum sheathing	1 1/2 galvanized roofing nail; 6d common nail; staple galvanized, 1 1/2 long; 1 1/4 screws, Type W or S	4	8
5/8 gypsum sheathing	1 3/4 galvanized roofing nail; 8d common nail; staple galvanized, 1 5/8 long; 1 5/8 screws, Type W or S	4	8
Wood structural panels, combination subfloor underlayment to framing			
3/4 and less	6d deformed nail or 8d common nail	6	12
7/8-1	8d common nail or 8d deformed nail	6	12
1 1/8-1 1/4	10d common nail or 8d deformed nail	6	12

For SI: 1 inch = 25.4 mm, 1 foot = 304.8 mm, 1 mile per hour = 1.609 km/h.

a. All nails are smooth-common, box or deformed shanks except where otherwise stated.

b. Staples are 16 gage wire and have a minimum 7/16-inch on diameter crown width.

c. Nails shall be spaced at not more than 6 inches on center at all supports where spans are 48 inches or greater.

d. Four-foot-by-8-foot or 4-foot-by-9-foot panels shall be applied vertically.

e. Spacing of fasteners not included in this table shall be based on Table R602.3(1).

f. For regions having basic wind speed of 110 mph or greater, 8d deformed nails shall be used for attaching plywood and wood structural panel roof sheathing to framing within minimum 48-inch distance from gable end walls, if mean roof height is more than 25 feet, up to 35 feet maximum.

g. For regions having basic wind speed of 100 mph or less, nails for attaching wood structural panel roof sheathing to gable end wall framing shall be spaced 6 inches on center When basic wind speed is greater than 100 mph, nails for attaching panel roof sheathing to intermediate supports shall be spaced 6 inches on center for minimum 48-inch distance from ridges, eaves and gable end walls; and 4 inches on center to gable end wall framing.

h. Gypsum sheathing shall conform to ASTM C 79 and shall be installed in accordance with GA 253. Fiberboard sheathing shall conform to either AHA 194.1 or ASTM C 208.

i. Spacing of fasteners on floor sheathing panel edges applies to panel edges supported by framing members and at all floor perimeters only. Spacing of fasteners on roof sheathing panel edges applies to panel edges supported by framing members and at all roof plane perimeters. Blocking of roof or floor sheathing panel edges perpendicular to the framing members shall not be required except at intersection of adjacent roof planes. Floor and roof perimeter shall be supported by framing members or solid blocking.

TABLE R602.3(2) ALTERNATE ATTACHMENTS

NOMINAL MATERIAL THICKNESS (inches)	DESCRIPTION[a, b] OF FASTENER AND LENGTH (inches)	SPACING[c] OF FASTENERS	
		Edges (inches)	Intermediate supports (inches)
Wood structural panels subfloor, roof and wall sheathing to framing and particleboard wall sheathing to framing[f]			
$^{5}/_{16}$	0.097 - 0.099 Nail $1^{1}/_{2}$ Staple 15 ga. $1^{3}/_{8}$ Staple 16 ga. $1^{3}/_{4}$	6	12
$^{3}/_{8}$	Staple 15 ga. $1^{3}/_{8}$	6	12
	0.097 - 0.099 Nail $1^{1}/_{2}$	4	10
	Staple 16 ga. $1^{3}/_{4}$	6	12
$^{15}/_{32}$ and $^{1}/_{2}$	Staple 15 ga. $1^{1}/_{2}$	6	12
	0.097 - 0.099 Nail $1^{5}/_{8}$	3	6
	Staple 16 ga. $1^{3}/_{4}$	6	12
$^{19}/_{32}$ and $^{5}/_{8}$	0.113 Nail $1^{7}/_{8}$ Staple 15 and 16 ga. $1^{5}/_{8}$	6	12
	0.097 - 0.099 Nail $1^{3}/_{4}$	3	6
$^{23}/_{32}$ and $^{3}/_{4}$	Staple 14 ga. $1^{3}/_{4}$	6	12
	Staple 15 ga. $1^{3}/_{4}$	5	10
	0.097 - 0.099 Nail $1^{7}/_{8}$	3	6
	Staple 16 ga. 2	4	8
1	Staple 14 ga. 2	5	10
	0.113 Nail $2^{1}/_{4}$, Staple 15 ga. 2	4	8
	0.097 - 0.099 Nail $2^{1}/_{8}$	3	6
NOMINAL MATERIAL THICKNESS (inches)	**DESCRIPTION[a,b] OF FASTENER AND LENGTH**	**SPACING[c] OF FASTENERS**	
		Edges (inches)	**Body of panel (inches)**
Floor underlayment; plywood-hardboard-particleboard[f]			
Plywood			
$^{1}/_{4}$ and $^{5}/_{16}$	$1^{1}/_{4}$ ring or screw shank nail—minimum $12^{1}/_{2}$ ga. (0.099″) shank diameter	3	6
	Staple 18 ga., $^{7}/_{8}$, $^{3}/_{16}$ crown width	2	5
$^{11}/_{32}$, $^{3}/_{8}$, $^{15}/_{32}$ and $^{1}/_{2}$	$1^{1}/_{4}$ ring or screw shank nail—minimum $12^{1}/_{2}$ ga. (0.099) shank diameter	6	8[e]
$^{19}/_{32}$, $^{5}/_{8}$, $^{23}/_{32}$ and $^{3}/_{4}$	$1^{1}/_{2}$ ring or screw shank nail—minimum $12^{1}/_{2}$ ga. (0.099) shank diameter	6	12
	Staple 16 ga. $1^{1}/_{4}$	6	8
Hardboard[f]			
0.200	$1^{1}/_{2}$ long ring-grooved underlayment nail	6	6
	4d cement-coated sinker nail	6	6
	Staple 18 ga., $^{7}/_{8}$ long (plastic coated)	3	6
Particleboard			
$^{1}/_{4}$	4d ring-grooved underlayment nail	3	6
	Staple 18 ga., $^{7}/_{8}$ long, $^{3}/_{16}$ crown	3	6
$^{3}/_{8}$	6d ring-grooved underlayment nail	6	10
	Staple 16 ga., $1^{1}/_{8}$ long, $^{3}/_{8}$ crown	3	6
$^{1}/_{2}$, $^{5}/_{8}$	6d ring-grooved underlayment nail	6	10
	Staple 16 ga., $1^{5}/_{8}$ long, $^{3}/_{8}$ crown	3	6

For SI: 1 inch = 25.4 mm.

a. Nail is a general description and may be T-head, modified round head or round head.

b. Staples shall have a minimum crown width of $^{7}/_{16}$-inch on diameter except as noted.

c. Nails or staples shall be spaced at not more than 6 inches on center at all supports where spans are 48 inches or greater. Nails or staples shall be spaced at not more than 12 inches on center at intermediate supports for floors.

d. Fasteners shall be placed in a grid pattern throughout the body of the panel.

e. For 5-ply panels, intermediate nails shall be spaced not more than 12 inches on center each way.

f. Hardboard underlayment shall conform to ANSI/AHA A135.4.

TABLE R602.3(3)
ALLOWABLE STUD SPACING FOR WOOD STRUCTURAL PANEL WALL SHEATHING

PANEL SPAN RATING	PANEL NOMINAL THICKNESS (inch)	MAXIMUM STUD SPACING (inches) Siding nailed to:[a]	
		Stud	Sheathing
12/0, 16/0, 20/0, or wall —16 o.c.	$^5/_{16}$, $^3/_8$	16	16[b]
24/0, 24/16, 32/16 or wall—24 o.c.	$^3/_8$, $^7/_{16}$, $^{15}/_{32}$, $^1/_2$	24	24[c]

For SI: 1 inch = 25.4 mm.

a. Blocking of horizontal joints shall not be required.

b. Plywood sheathing $^3/_8$-inch thick or less shall be applied with long dimension across studs.

c. Three-ply plywood panels shall be applied with long dimension across studs.

TABLE R602.3(4)
ALLOWABLE SPANS FOR PARTICLEBOARD WALL SHEATHING[a]

THICKNESS (inch)	GRADE	STUD SPACING (inches)	
		When siding is nailed to studs	When siding is nailed to sheathing
$^3/_8$	M-1 Exterior glue	16	—
$^1/_2$	M-2 Exterior glue	16	16

For SI: 1 inch = 25.4 mm.

a. Wall sheathing not exposed to the weather. If the panels are applied horizontally, the end joints of the panel shall be offset so that four panels corners will not meet. All panel edges must be supported. Leave a $^1/_{16}$-inch gap between panels and nail no closer than $^3/_8$ inch from panel edges.

TABLE R602.3(5)
MAXIMUM STUD SPACING

STUD SIZE (inches)	SUPPORTING ROOF AND CEILING ONLY (inches)	SUPPORTING ONE FLOOR ROOF AND CEILING (inches)	SUPPORTING TWO FLOORS ROOF AND CEILING (inches)	SUPPORTING ONE FLOOR ONLY (inches)
2 × 4	24[a]	16	—	24[a]
3 × 4	24[a]	24	16	24
2 × 5	24	24	—	24
2 × 6	24	24	16	24

For SI: 1 inch = 25.4 mm.

a. Shall be reduced to 16 inches if utility grade studs are used.

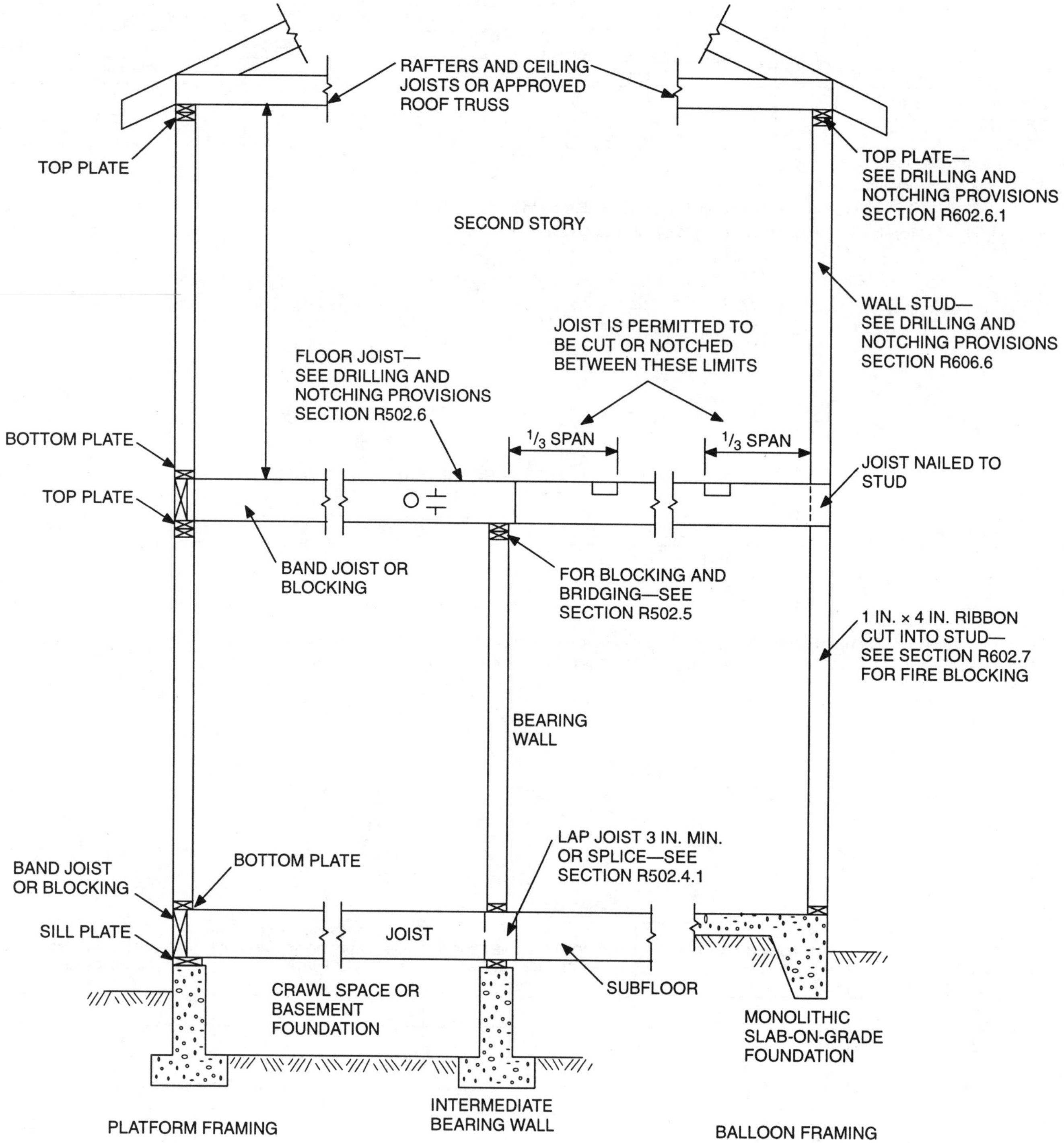

For SI: 1 inch = 25.4 mm.

FIGURE R602.3(1)
TYPICAL WALL, FLOOR AND ROOF FRAMING

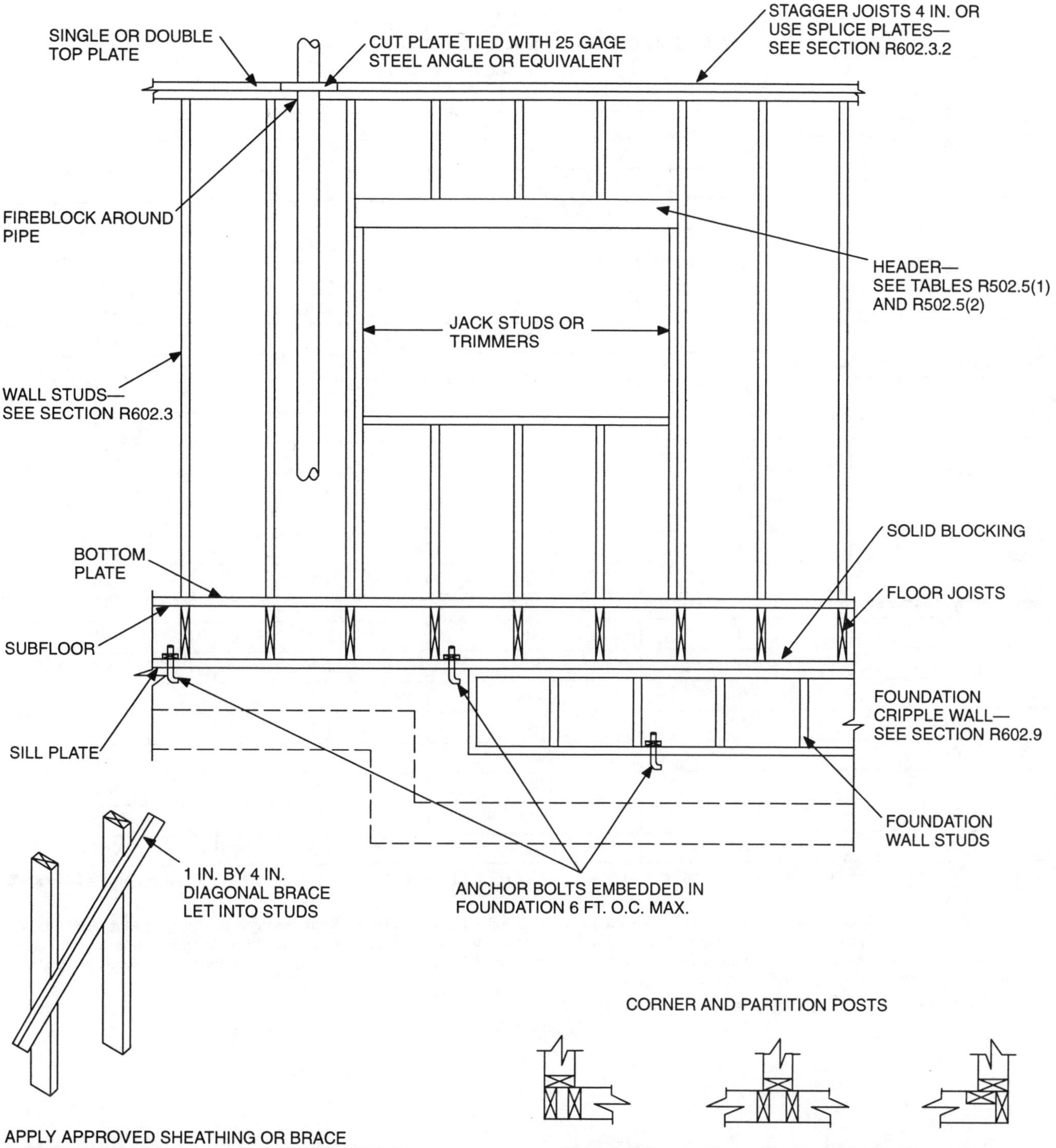

APPLY APPROVED SHEATHING OR BRACE EXTERIOR WALLS WITH 1 IN. BY 4 IN. BRACES LET INTO STUDS AND PLATES AND EXTENDING FROM BOTTOM PLATE TO TOP PLATE, OR OTHER APPROVED METAL STRAP DEVICES INSTALLED IN ACCORDANCE WITH THE MANUFACTURER'S SPECIFICATIONS. SEE SECTION R602.10.

Note: A third stud and/or partition intersection backing studs shall be permitted to be omitted through the use of wood backup cleats, metal drywall clips or other approved devices that will serve as adequate backing for the facing materials.

For SI: 1 inch = 25.4 mm, 1 foot = 304.8 mm.

FIGURE R602.3(2)
FRAMING DETAILS

TABLE R602.3.1
MAXIMUM ALLOWABLE LENGTH OF WOOD WALL STUDS EXPOSED TO WIND SPEEDS OF 100 MPH OR LESS IN SEISMIC DESIGN CATEGORIES A, B, C and D[a,b,c]

HEIGHT (feet)	ON-CENTER SPACING (inches)			
	24	16	12	8
Supporting a roof only				
>10	2×4	2×4	2×4	2×4
12	2×6	2×4	2×4	2×4
14	2×6	2×6	2×6	2×4
16	2×6	2×6	2×6	2×4
18	NA[a]	2×6	2×6	2×6
20	NA[a]	NA[a]	2×6	2×6
24	NA[a]	NA[a]	NA[a]	2×6
Supporting one floor and a roof				
>10	2×6	2×4	2×4	2×4
12	2×6	2×6	2×6	2×4
14	2×6	2×6	2×6	2×6
16	NA[a]	2×6	2×6	2×6
18	NA[a]	2×6	2×6	2×6
20	NA[a]	NA[a]	2×6	2×6
24	NA[a]	NA[a]	NA[a]	2×6
Supporting two floors and a roof				
>10	2×6	2×6	2×4	2×4
12	2×6	2×6	2×6	2×6
14	2×6	2×6	2×6	2×6
16	NA[a]	NA[a]	2×6	2×6
18	NA[a]	NA[a]	2×6	2×6
20	NA[a]	NA[a]	NA[a]	2×6
22	NA[a]	NA[a]	NA[a]	NA[a]
24	NA[a]	NA[a]	NA[a]	NA[a]

For SI: 1 inch = 25.4 mm, 1 foot = 304.8 mm, 1 pound per square foot = 0.0479 kN/m^2, 1 pound per square inch = 6.895 kPa, 1 mile per hour = 1.609 km/h.

a. Design required.

b. Applicability of this table assumes the following: Snow load not exceeding 25 psf, but not less than 1310 psi determined by multiplying the AF&PA NDS tabular base design value by the repetitive use factor, and by the size factor for all species except southern pine, E not less than 1.6 by 10^6 psi, tributary dimensions for floors and roofs not exceeding 6 feet, maximum span for floors and roof not exceeding 12 feet, eaves not greater than 2 feet in dimension and exterior sheathing. Where the conditions are not within these parameters, design is required.

c. Utility, standard, stud and No. 3 grade lumber of any species are not permitted.

(continued)

TABLE R602.3.1—continued
MAXIMUM ALLOWABLE LENGTH OF WOOD WALL STUDS EXPOSED TO WIND SPEEDS OF 100 MPH OR LESS IN SEISMIC DESIGN CATEGORIES A, B, C and D_1

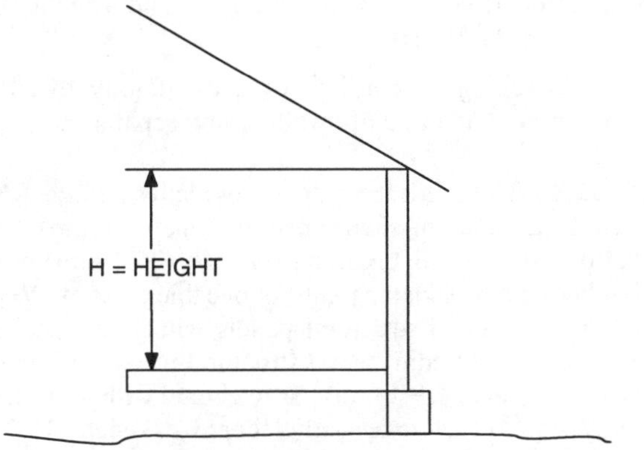

1 ROOF LOAD

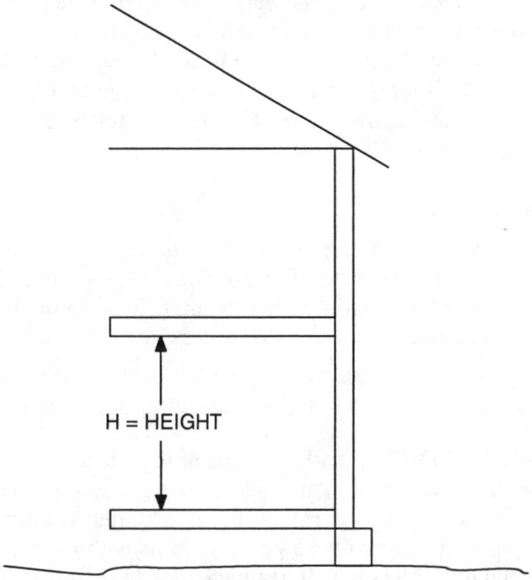

1 ROOF LOAD
1 FLOOR LOAD

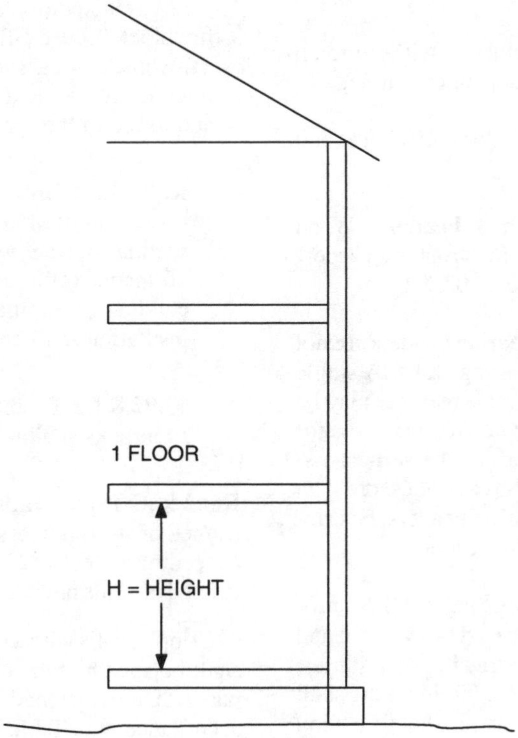

1 ROOF LOAD
2 FLOOR LOADS

R602.6 Drilling and notching—studs. Any stud in an exterior wall or bearing partition may be cut or notched to a depth not exceeding 25 percent of its width. Studs in nonbearing partitions may be notched to a depth not to exceed 40 percent of a single stud width. Any stud may be bored or drilled, provided that the diameter of the resulting hole is no greater than 40 percent of the stud width, the edge of the hole is no closer than $^5/_8$ inch (15.9 mm) to the edge of the stud, and the hole is not located in the same section as a cut or notch. See Figures R602.6(1) and R602.6(2).

Exceptions:

1. A stud may be bored to a diameter not exceeding 60 percent of its width, provided that such studs located in exterior walls or bearing partitions are doubled and that not more than two successive studs are bored.
2. Approved stud shoes may be used when installed in accordance with the manufacturer's recommendation.

R602.6.1 Drilling and notching of top plate. When piping or ductwork is placed in or partly in an exterior wall or interior, braced or load-bearing wall, necessitating a cutting of the top plate by more than 50 percent of its width, a galvanized metal tie not less than 0.054 inch thick (1.37 mm) (16 gage) and 1.5 inches (38 mm) wide shall be fastened to each plate across and to each side of the opening with not less than six 16d nails. (See Figure R602.6.1.)

Exception: When the entire side of the wall with the notch or cut is covered by wood structural panel sheathing.

R602.7 Headers. For header spans see Tables R502.5(1) and R502.5(2).

R602.7.1 Wood structural panel box headers. Wood structural panel box headers shall be constructed in accordance with Figure R602.7.2 and Table R602.7.2.

R602.7.2 Nonbearing walls. Load-bearing headers are not required in interior or exterior nonbearing walls. A single flat 2-inch-by-4-inch (51 mm by 102 mm) member may be used as a header in interior or exterior nonbearing walls for openings up to 8 feet (2438 mm) in width if the vertical distance to the parallel nailing surface above is not more than 24 inches (610 mm). For such nonbearing headers, no cripples or blocking are required above the header.

R602.8 Fireblocking required. Fireblocking shall be provided to cut off all concealed draft openings (both vertical and horizontal) and to form an effective fire barrier between stories, and between a top story and the roof space. Fireblocking shall be provided in wood-frame construction in the following locations:

1. In concealed spaces of stud walls and partitions, including furred spaces, at the ceiling and floor level and at 10 foot (3048 mm) intervals both vertical and horizontal. Batts or blankets of mineral or glass fiber or other approved non-rigid materials shall be allowed as fireblocking in walls constructed using parallel rows of studs or staggered studs.
2. At all interconnections between concealed vertical and horizontal spaces such as occur at soffits, drop ceilings and cove ceilings.
3. In concealed spaces between stair stringers at the top and bottom of the run. Enclosed spaces under stairs shall comply with Section R314.8.
4. At openings around vents, pipes, and ducts at ceiling and floor level, with an approved material to resist the free passage of flame and products of combustion.
5. For the fireblocking of chimneys and fireplaces, see Section R1001.16.
6. Fireblocking of cornices of a two-family dwelling is required at the line of dwelling unit separation.

R602.8.1 Materials. Except as provided in Section R602.8, Item 4, fireblocking shall consist of 2-inch (51 mm) nominal lumber, or two thicknesses of 1-inch (25.4 mm) nominal lumber with broken lap joints, or one thickness of $^{23}/_{32}$-inch (19.8 mm) wood structural panels with joints backed by $^{23}/_{32}$-inch (19.8 mm) wood structural panels or one thickness of $^3/_4$-inch (19.1 mm) particleboard with joints backed by $^3/_4$-inch (19.1 mm) particleboard, $^1/_2$-inch (12.7 mm) gypsum board, or $^1/_4$-inch (6.4 mm) cement-based millboard. Batts or blankets of mineral wool or glass fiber or other approved materials installed in such a manner as to be securely retained in place shall be permitted as an acceptable fire block. Loose-fill insulation material shall not be used as a fire block unless specifically tested in the form and manner intended for use to demonstrate its ability to remain in place and to retard the spread of fire and hot gases.

R602.8.1.1 Unfaced fiberglass. Unfaced fiberglass batt insulation used as fireblocking shall fill the entire cross section of the wall cavity to a minimum height of 16 inches (406 mm) measured vertically. When piping, conduit or similar obstructions are encountered, the insulation shall be packed tightly around the obstruction.

R602.8.1.2 Fireblocking integrity. The integrity of all fireblocks shall be maintained.

R602.9 Cripple walls. Foundation cripple walls shall be framed of studs not less in size than the studding above. When exceeding 4 feet (1219 mm) in height, such walls shall be framed of studs having the size required for an additional story.

Cripple walls with a stud height less than 14 inches (356 mm) shall be sheathed on at least one side with a wood structural panel that is fastened to both the top and bottom plates in accordance with Table R602.3(1), or the cripple walls shall be constructed of solid blocking. Cripple walls shall be supported on continuous foundations.

R602.10 Wall bracing. Walls shall be braced in accordance with this section. For buildings in Seismic Design Categories D_1 and D_2, walls shall be constructed in accordance with the additional requirements of Sections R602.10.9, R602.10.11 and R602.11.

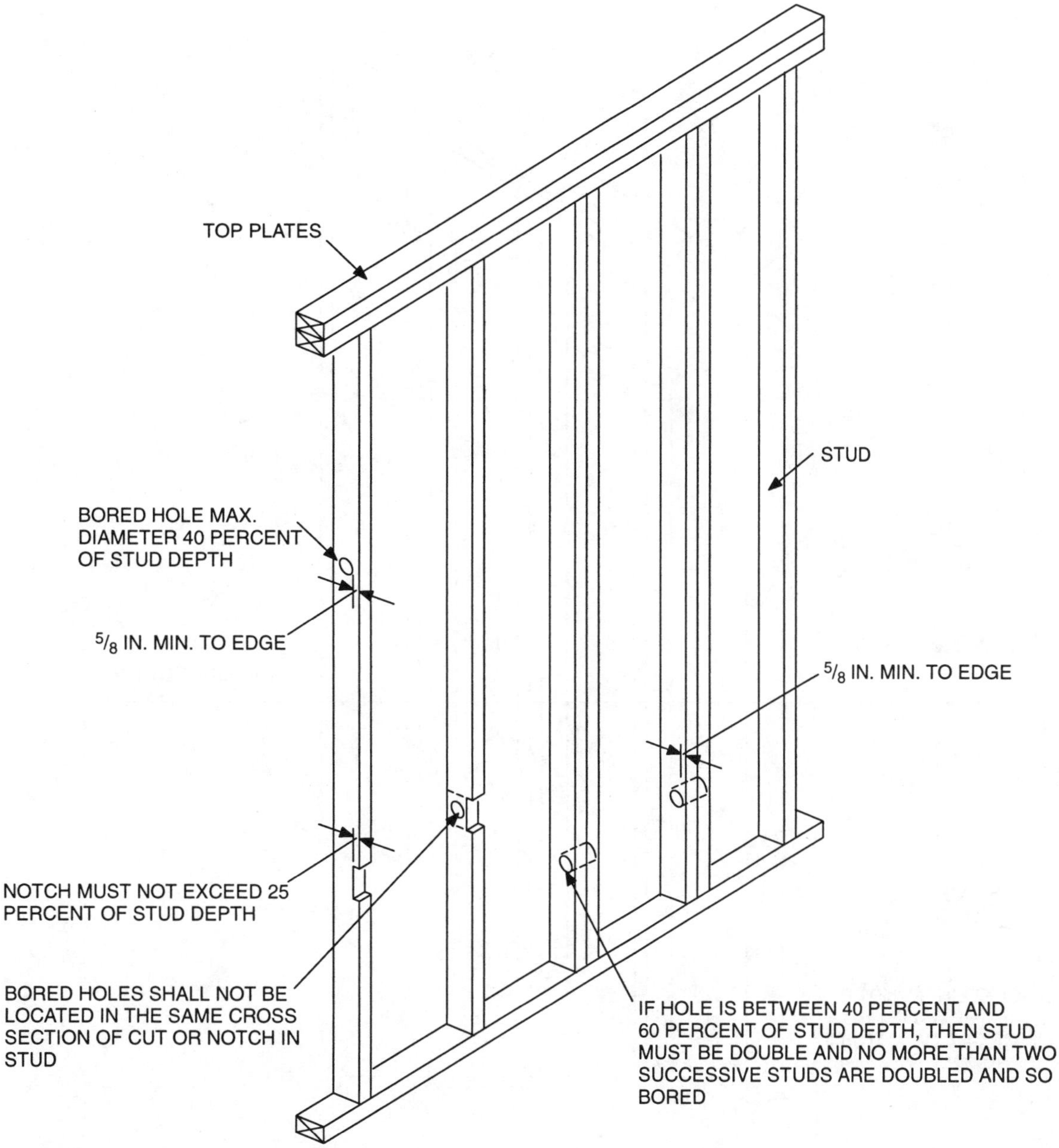

For SI: 1 inch = 25.4 mm.

Note: Condition for exterior and bearing walls.

FIGURE R602.6(1)
NOTCHING AND BORED HOLE LIMITATIONS FOR EXTERIOR WALLS AND BEARING WALLS

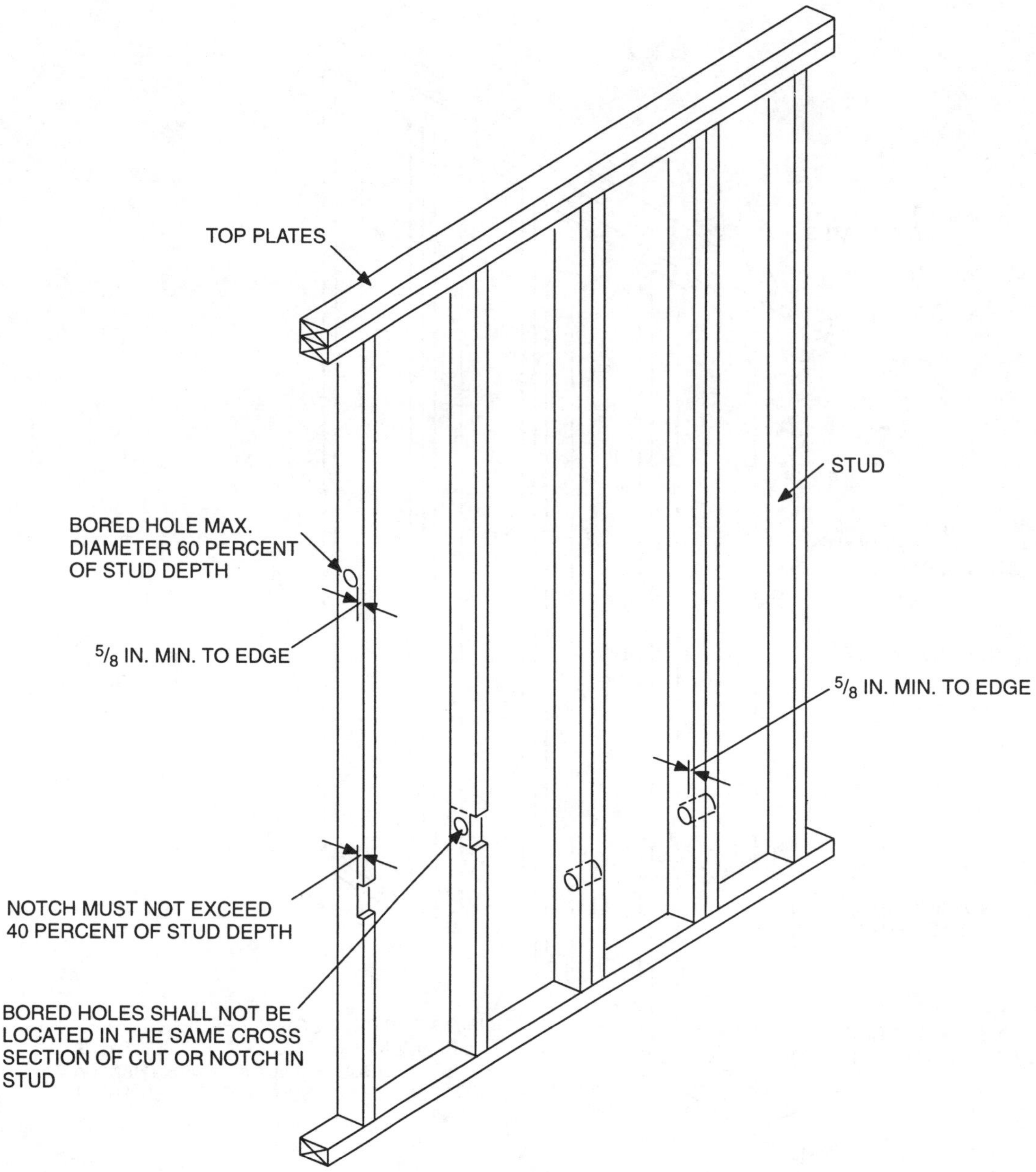

For SI: 1 inch = 25.4 mm.

FIGURE R602.6(2)
NOTCHING AND BORED HOLE LIMITATIONS FOR INTERIOR NONBEARING WALLS

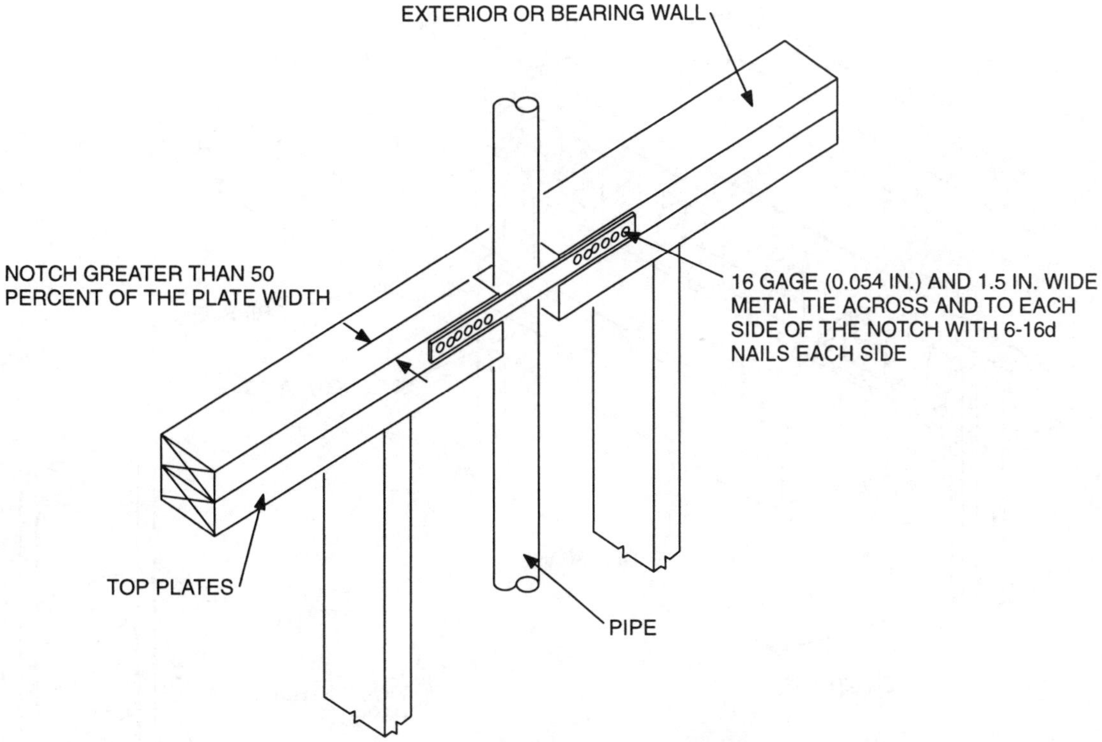

For SI: 1 inch = 25.4 mm.

FIGURE R602.6.1
TOP PLATE FRAMING TO ACCOMMODATE PIPING

TABLE R602.7.2
MAXIMUM SPANS FOR WOOD STRUCTURAL PANEL BOX HEADERS[a]

HEADER CONSTRUCTION[b]	HEADER DEPTH (inches)	HOUSE DEPTH (feet)				
		24	26	28	30	32
Wood structural panel—one side	9 15	4 5	4 5	3 4	3 3	— 3
Wood structural panel—both sides	9 15	7 8	5 8	5 7	4 7	3 6

For SI: 1 inch = 25.4 mm, 1 foot = 304.8 mm.

a. Spans are based on single story with clear-span trussed roof or two-story with floor and roof supported by interior-bearing walls.

b. See Figure R602.7.2 for construction details.

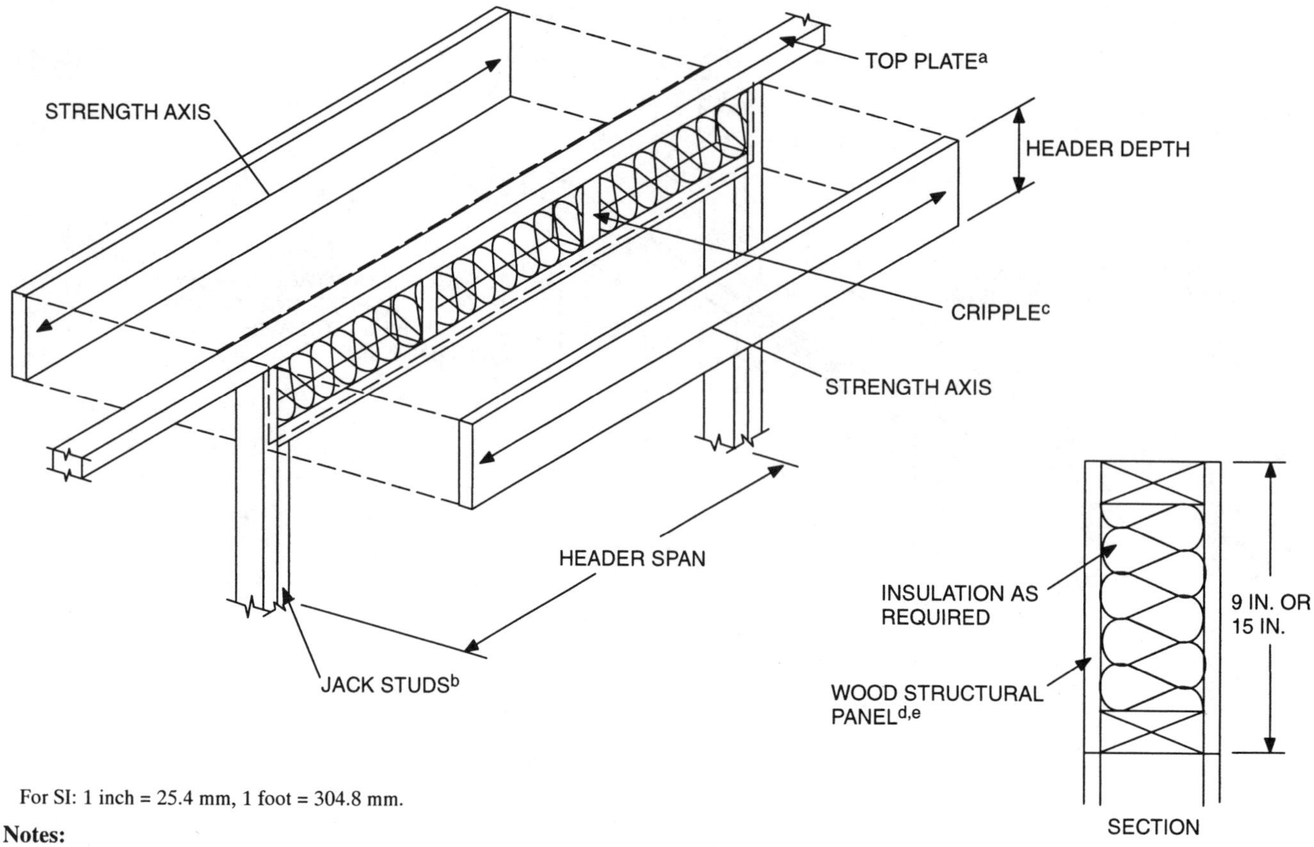

For SI: 1 inch = 25.4 mm, 1 foot = 304.8 mm.

Notes:

a. The top plate shall be continuous over header.
b. Jack studs shall be used for spans over 4 feet.
c. Cripple spacing shall be the same as for studs.
d. Wood structural panel faces shall be single pieces of $^{15}/_{32}$-inch-thick Exposure 1 (exterior glue) or thicker, installed on the interior or exterior or both sides of the header.
e. Wood structural panel faces shall be nailed to framing and cripples with 8d common or galvanized box nails spaced 3 inches on center, staggering alternate nails $^{1}/_{2}$ inch.
f. Galvanized nails shall be hot-dipped or tumbled.

FIGURE R602.7.2
TYPICAL WOOD STRUCTURAL PANEL BOX HEADER CONSTRUCTION

R602.10.1 Braced wall lines. Braced wall lines shall consist of braced wall panel construction methods in accordance with Section R602.10.3. The amount and location of bracing shall be in accordance with Table R602.10.1 and the amount of bracing shall be the greater of that required by the Seismic Design Category or the design wind speed. Braced wall panels shall begin no more than 12.5 feet (3810 mm) from each end of a braced wall line. Braced wall panels that are counted as part of a braced wall line shall be in line, except that offsets out-of-plane of up to 4 feet (1219 mm) shall be permitted provided that the total out-to-out offset dimension in any braced wall line is not more than 8 feet (2438 mm).

A designed collector shall be provided if the bracing begins more than 12 feet (3658 mm) from each end of a braced wall line.

R602.10.2 Cripple wall bracing. Cripple walls shall be braced with an amount and type of bracing as required for the wall above in accordance with Table R602.10.1 with the following modifications for the cripple wall bracing:

1. The percent bracing amount as determined from Table R602.10.3 shall be increased by 15 percent.
2. The wall panel spacing shall be decreased to 18 feet (5486 mm) instead of 25 feet (7620 mm).

R602.10.3 Braced wall panel construction methods. The construction of braced wall panels shall be in accordance with one of the following methods:

1. Nominal 1-inch-by-4-inch (25.4 mm by 102 mm) continuous diagonal braces let in to the top and bottom plates and the intervening studs or approved metal strap devices installed in accordance with the manufacturer's specifications. The let-in bracing shall be placed at an angle not more than 60 degrees (1.06 rad) or less than 45 degrees (0.79 rad) from the horizontal.
2. Wood boards of $^{5}/_{8}$ inch (15.9 mm) net minimum thickness applied diagonally on studs spaced a maximum of 24 inches (610 mm). Diagonal boards shall be attached to studs in accordance with Table R602.3(1).

3. Wood structural panel sheathing with a thickness not less than $^5/_{16}$ inch (7.9 mm) for 16-inch (406 mm) stud spacing and not less than $^3/_8$ inch (9.5 mm) for 24-inch (610 mm) stud spacing. Wood structural panels shall be installed in accordance with Table R602.3(3).
4. One-half-inch (12.7 mm) or $^{25}/_{32}$-inch (19.8 mm) thick structural fiberboard sheathing applied vertically on studs spaced a maximum of 16 inches (406 mm) on center. Structural fiberboard sheathing shall be installed in accordance with Table R602.3(1).
5. Gypsum board with minimum $^1/_2$-inch (12.7 mm) thickness placed on studs spaced a maximum of 24 inches (610 mm) on center and fastened at 7 inches (178 mm) on center with nails in accordance with Table R602.3(1).
6. Particleboard wall sheathing panels installed in accordance with Table R602.3(4).
7. Portland cement plaster on studs spaced a maximum of 16 inches (406 mm) on center and installed in accordance with Section R703.6.
8. Hardboard panel siding when installed in accordance with Table R703.4.

Exception: Alternate braced wall panels constructed in accordance with Section R602.10.6 shall be permitted to replace any of the above methods of braced wall panels.

R602.10.4 Length of braced panels. For Methods 2, 3, 4, 6 and 7 above, each braced wall panel shall be at least 48 inches (1219 mm) in length, covering a minimum of three stud spaces where studs are spaced 16 inches (406 mm) on center and covering a minimum of two stud spaces where studs are spaced 24 inches (610 mm) on center. For Method 5 above, each braced wall panel shall be at least 96 inches (2438 mm) in length where applied to one face of a braced wall panel and at least 48 inches (1219 mm) where applied to both faces.

Exceptions:

1. Lengths of braced wall panels for continuous wood structural panel sheathing shall be in accordance with Section R602.10.5.
2. Lengths of alternate braced wall panels shall be in accordance with Section R602.10.6.

R602.10.5 Continuous structural panel sheathing. When continuous wood structural panel sheathing is provided in accordance with Method 3 of Section R602.10.3, including areas above and below openings, braced wall panel lengths shall be in accordance with Table R602.10.5. Wood structural panel sheathing at corners shall be installed in accordance with Figure R602.10.5. The bracing amounts in Table R602.10.3 for Method 3 shall be permitted to be multiplied by a factor of 0.9 for walls with a maximum opening height that does not exceed 85 percent of the wall height or a factor of 0.8 for walls with a maximum opening height that does not exceed 67 percent of the wall height.

R602.10.6 Alternate braced wall panels. Alternate braced wall lines constructed in accordance with one of the following provisions shall be permitted to replace each 4 feet (1219 mm) of braced wall panel as required by Section R602.10.4:

1. In one-story buildings, each panel shall have a length of not less than 2 feet, 8 inches (813 mm) and a height of not more than 10 feet (3048 mm). Each panel shall be sheathed on one face with $^3/_8$-inch-minimum-thickness (9.5 mm) wood structural panel sheathing nailed with 8d common or galvanized box nails in accordance with Table R602.3(1) and blocked at all wood structural panel sheathing edges. Two anchor bolts installed in accordance with Figure R403.1(1) shall be provided in each panel. Anchor bolts shall be placed at panel quarter points. Each panel end stud shall have a tie-down device fastened to the foundation, capable of providing an uplift capacity of at least 1,800 pounds (816.5 kg). The tie-down device shall be installed in accordance with the manufacturer's recommendations. The panels shall be supported directly on a foundation or on floor framing supported directly on a foundation which is continuous across the entire length of the braced wall line. This foundation shall be reinforced with not less than one No. 4 bar top and bottom. When the continuous foundation is required to have a depth greater than 12 inches (305 mm), a minimum 12-inch-by-12-inch (305 mm by 305 mm) continuous footing or turned down slab edge is permitted at door openings in the braced wall line. This continuous footing or turned down slab edge shall be reinforced with not less than one No. 4 bar top and bottom. This reinforcement shall be lapped 15 inches (381 mm) with the reinforcement required in the continuous foundation located directly under the braced wall line.
2. In the first story of two-story buildings, each braced wall panel shall be in accordance with Item 1 above, except that the wood structural panel sheathing shall be provided on both faces, at least three anchor bolts shall be placed at one-fifth points, and tie-down device uplift capacity shall not be less than 3,000 pounds (1360.8 kg).

R602.10.7 Panel joints. All vertical joints of panel sheathing shall occur over studs. Horizontal joints in braced wall panels shall occur over blocking of a minimum of $1^1/_2$ inch (38 mm) thickness.

Exception: Blocking is not required behind horizontal joints where permitted by the manufacturer's installation requirements for the specific sheathing material.

R602.10.8 Connections. Braced wall panel sole plates shall be fastened to the floor framing and top plates shall be connected to the framing above in accordance with Table R602.3(1). Sills shall be fastened to the foundation or slab in accordance with Sections R403.1.6 and R602.11. Where joists are perpendicular to the braced wall lines above, blocking shall be provided under and in line with the braced wall panels.

TABLE R602.10.3
WALL BRACING

SEISMIC DESIGN CATEGORY OR WIND SPEED	CONDITION	TYPE OF BRACE[b,c]	AMOUNT OF BRACING[a,d,e]
Category A and B ($S_s \leq 0.35g$ and $S_{ds} \leq 0.33g$) or 100 mph and less	One story Top of two or three story	Methods 1, 2, 3, 4, 5, 6, 7 or 8	Located at each end and at least every 25 feet on center but not less than 16% of braced wall line.
	First story of two story Second story of three story	Methods 1, 2, 3, 4, 5, 6, 7 or 8	Located at each end and at least every 25 feet on center but not less than 16% of braced wall line for Method 3 and 25% of braced wall line for Methods 2, 4, 5, 6, 7 or 8.
	First story of three story	Methods 2, 3, 4, 5, 6, 7 or 8	Minimum 48-inch-wide panels located at each end and at least every 25 feet on center but not less than 25% of braced wall line for method 3 and 35% of braced wall line for Methods 2, 4, 5, 6, 7 or 8.
Category C ($S_s \leq 0.6g$ and $S_{ds} \leq 0.53g$) or less than 110 mph	One story Top of two or three story	Methods 1, 2, 3, 4, 5, 6, 7 or 8	Located at each end and at least every 25 feet on center but not less than 16% of braced wall line for Method 3 and 25% of braced wall line for Methods 2, 4, 5, 6, 7 or 8.
	First story of two story Second story of three story	Methods 2, 3, 4, 5, 6, 7 or 8	Located at each end and at least every 25 feet on center but not less than 30% of braced wall line for Method 3 and 45% of braced wall line for Methods 2, 4, 5, 6, 7 or 8.
	First story of three story	Methods 2, 3, 4, 5, 6, 7 or 8	Located at each end and at least every 25 feet on center but not less than 45% of braced wall line for Method 3 and 60% of braced wall line for Methods 2, 4, 5, 6, 7 or 8.
Category D_1 ($S_s \leq 1.25g$ and $S_{ds} \leq 0.83g$) or less than 110 mph	One story Top of two or three story	Methods 2, 3, 4, 5, 6, 7 or 8	Located at each end and at least every 25 feet on center but not less than 20% of braced wall line for Method 3 and 30% of braced wall line for Methods 2, 4, 5, 6, 7 or 8.
	First story of two story Second story of three story	Methods 2, 3, 4, 5, 6, 7 or 8	Located at each end and not more than 25 feet on center but not less than 45% of braced wall line for Method 3 and 60% of braced wall line for Methods 2, 4, 5, 6, 7 or 8.
	First story of three story	Methods 2, 3, 4, 5, 6, 7 or 8	Located at each end and not more than 25 feet on center but not less than 60% of braced wall line for Method 3 and 85% of braced wall line for Method 2, 4, 5, 6, 7 or 8.
Category D_2 or less than 110 mph	One story Top of two story	Methods 2, 3, 4, 5, 6, 7 or 8	Located at each end and at least every 25 feet on center but not less than 25% of braced wall line for Method 3 and 40% of braced wall line for Methods 2, 4, 5, 6, 7 or 8.
	First story of two story	Methods 2, 3, 4, 5, 6, 7 or 8	Located at each end and not more than 25 feet on center but not less than 55% of braced wall line for Method 3 and 75% of braced wall line for Methods 2, 4, 5, 6, 7 or 8.
	Cripple walls	Method 3	Located at each end and not more than 25 feet on center but not less than 75% of braced wall line.

For SI: 1 inch = 25.4 mm, 1 foot = 304.8 mm, 1 pound per square foot = 0.0479 kN/m^2, 1 mile per hour = 1.609 km/h.

a. Wall bracing amounts are based on a soil site class "D." Interpolation of bracing amounts between the S_{ds} values associated with the Seismic Design Categories shall be permitted when a site specific S_{ds} value is determined in accordance with Section 1615 of the *International Building Code.*

b. Foundation cripple wall panels shall be braced in accordance with Section R602.10.2.

c. Methods of bracing shall be as described in Section R602.10.3. The alternate braced wall panels described in Section R602.10.6 shall also be permitted.

d. The bracing amounts for Seismic Design Categories are based on a 15 psf wall dead load. For walls with a dead load of 8 psf or less, the bracing amounts shall be permitted to be multiplied by 0.85 provided that the adjusted bracing amount is not less than that required for the site's wind speed. The minimum length of braced panel shall not be less than required by Section R602.10.3.

e. When the dead load of the roof/ceiling exceeds 15 psf, the bracing amounts shall be increased in accordance with Section R301.2.2.4. Bracing required for a site's wind speed shall not be adjusted.

TABLE R602.10.5
LENGTH REQUIREMENTS FOR BRACED WALL PANELS IN A CONTINUOUSLY SHEATHED WALL[a,b]

LENGTH OF BRACED WALL PANEL (inches)			MAXIMUM OPENING HEIGHT NEXT TO THE BRACED WALL PANEL (% of wall height)
8-foot wall	9-foot wall	10-foot wall	
48	54	60	100%
32	36	40	85%
24	27	30	65%

For SI: 1 inch = 25.4 mm, 1 foot = 305 mm, 1 pound per square foot = 0.0479 kN/m^2.

a. Linear interpolation shall be permitted.

b. Full-height sheathed wall segments to either side of garage openings that support light frame roofs with roof covering dead loads of 3 psf or less shall be permitted to have a 4:1 aspect ratio.

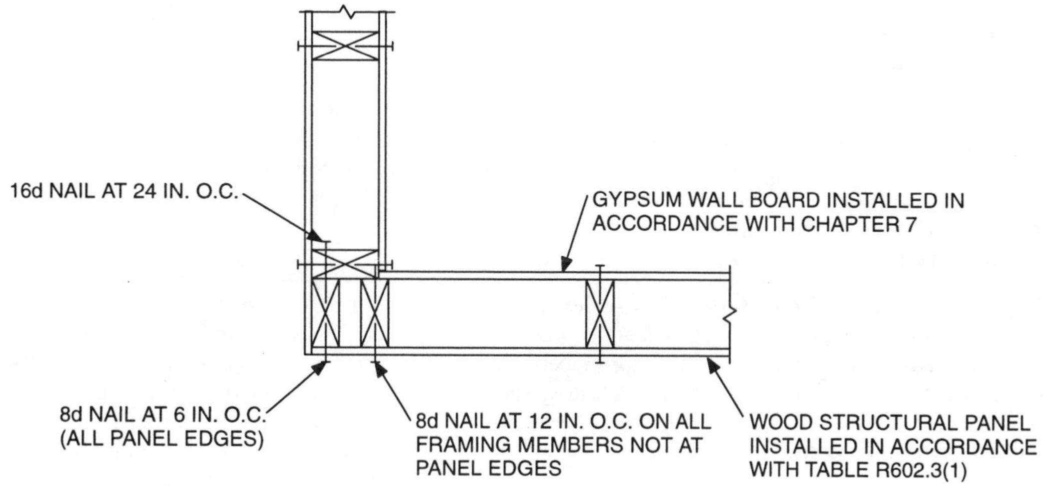

(a) OUTSIDE CORNER DETAIL

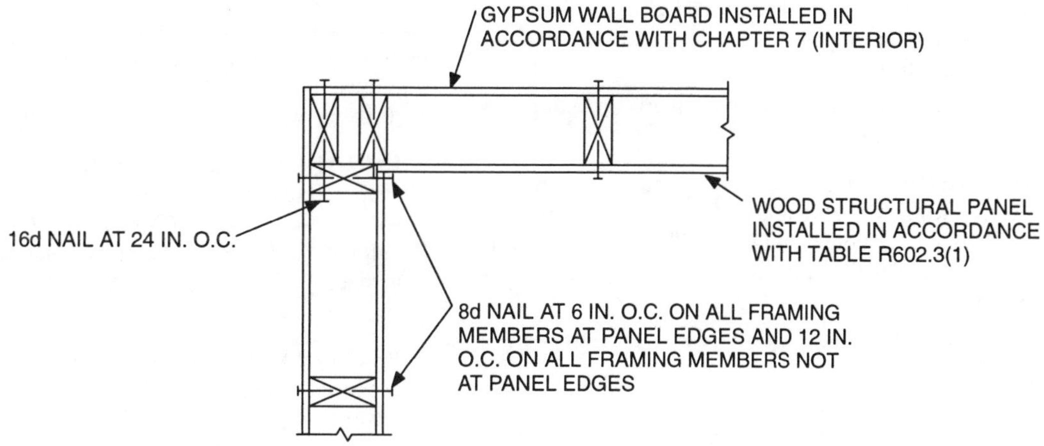

(b) INSIDE CORNER DETAIL

For SI: 1 inch = 25.4 mm.

FIGURE R602.10.5
EXTERIOR CORNER FRAMING

R602.10.9 Interior braced wall support. In one-story buildings located in Seismic Design Category D_2, interior braced wall lines shall be supported on continuous foundations at intervals not exceeding 50 feet (15 240 mm). In two story buildings located in Seismic Design Category D_2, all interior braced wall panels shall be supported on continuous foundations.

Exception: Two-story buildings shall be permitted to have interior braced wall lines supported on continuous foundations at intervals not exceeding 50 feet (15 240 mm) provided that:

1. The height of cripple walls does not exceed 4 feet (1219 mm).
2. First-floor braced wall panels are supported on doubled floor joists, continuous blocking or floor beams.
3. The distance between bracing lines does not exceed twice the building width measured parallel to the braced wall line.

R602.10.10 Design of structural elements. Where a building, or portion thereof, does not comply with one or more of the bracing requirements in this section, those portions shall be designed and constructed in accordance with accepted engineering practice.

R602.10.11 Bracing in Seismic Design Categories D_1 and D_2. Structures located in Seismic Design Categories D_1 and D_2 shall be provided with exterior and interior braced wall lines. Spacing between braced wall lines in each story shall not exceed 25 feet (7620 mm) on center in both the longitudinal and transverse directions.

Exception: In one- and two-story buildings, spacing between braced wall lines shall not exceed 35 feet (10 363 mm) on center in order to accommodate one single room not exceeding 900 square feet (83.61 m^2) in each dwelling unit. The length of wall bracing in braced wall lines spaced greater or less other than 25 feet (7620 mm) apart shall be the length required by Table R602.10.1 multiplied by the appropriate adjustment factor from Table R602.10.11.

Exterior braced wall lines shall have a braced wall panel located at each end of the braced wall line.

Exception: For braced wall panel construction Method 3 of Section R602.10.3, the braced wall panel shall be permitted to begin no more than 12 feet (3810 mm) from each end of the braced wall line provided one of the following is satisfied:

1. A minimum 24-inch-wide (610 mm) panel is applied to each side of the building corner and the two 24-inch-wide (610 mm) panels at the corner shall be attached to framing in accordance with Figure R602.10.5 or,
2. The end of each braced wall panel closest to the corner shall have a tie-down device fastened to the stud at the edge of the braced wall panel closest to the corner and to the foundation or framing below. The tie-down device shall be capable of providing an uplift allowable design value of at least 1,800 pounds (816.5 kg). The tie-down device shall be installed in accordance with the manufacturer's recommendations.

A designed collector shall be provided if the bracing is not located at each end of a braced wall line as indicated above or more than 8 feet (2438 mm) from each end of a braced wall line as indicated in the exception.

TABLE R602.10.11
ADJUSTMENT OF BRACING AMOUNTS FOR INTERIOR BRACED WALL LINES ACCORDING TO BRACED WALL LINE SPACING[a,b]

BRACED WALL LINE SPACING (feet)	MULTIPLY BRACING AMOUNT IN TABLE R602.10.3 BY:
15 or less	0.6
20	0.8
25	1.0
30	1.2
35	1.4

For SI: 1 foot = 304.8 mm.

a. Linear interpolation is permissible.

b. The adjustment is limited to the larger spacing between braced wall lines to either side of an interior braced wall line.

R602.10.11.1 Cripple wall bracing. In addition to the requirements of Section R602.10.2, where interior braced wall lines occur without a continuous foundation below, the length of parallel exterior cripple wall bracing shall be one and one-half times the length required by Table R602.10.3. Where cripple walls braced using Method 3 of Section R602.10.3 cannot provide this additional length, the capacity of the sheathing shall be increased by reducing the spacing of fasteners along the perimeter of each piece of sheathing to 4 inches (102 mm) on center.

R602.10.11.2 Sheathing attachment. Adhesive attachment of wall sheathing shall not be permitted in Seismic Design Categories C, D_1 and D_2.

R602.11 Framing and connections for Seismic Design Categories D_1 and D_2. The framing and connection details of buildings located in Seismic Design Categories D_1 and D_2 shall be in accordance with Sections R602.11.1 through R602.11.3.

R602.11.1 Wall anchorage. Braced wall line sills shall be anchored to concrete or masonry foundations in accordance with Sections R403.1.6 and R602.11. Plate washers, a minimum of $^3/_{16}$ inch by 2 inches by 2 inches (6.4 mm by 51 mm) in size, shall be provided between the foundation sill plate and the nut.

R602.11.2 Interior braced wall panel connections. Interior braced wall lines shall be fastened to floor and roof framing in accordance with Table R602.3(1), to required foundations in accordance with Section R602.11.1, and in accordance with the following requirements:

1. Floor joists parallel to the top plate shall be toe-nailed to the top plate with at least 8d nails spaced a maximum of 6 inches (150 mm) on center.
2. Top plate laps shall be face-nailed with at least eight 16d nails on each side of the splice.

R602.11.3 Stepped foundations. Where stepped foundations occur, the following requirements apply:

1. Where the height of a required braced wall panel that extends from foundation to floor above varies more than 4 feet (1220 mm), the braced wall panel shall be constructed in accordance with Figure R602.11.3.
2. Where the lowest floor framing rests directly on a sill bolted to a foundation not less than 8 feet (2440 mm) in length along a line of bracing, the line shall be considered as braced. The double plate of the cripple stud wall beyond the segment of footing that extends to the lowest framed floor shall be spliced by extending the upper top plate a minimum of 4 feet (1219 mm) along the foundation. Anchor bolts shall be located a maximum of 1 foot and 3 feet (305 and 914 mm) from the step in the foundation.
3. Where cripple walls occur between the top of the foundation and the lowest floor framing, the bracing requirements for a story shall apply.
4. Where only the bottom of the foundation is stepped and the lowest floor framing rests directly on a sill bolted to the foundations, the requirements of Section R602.11.1 shall apply.

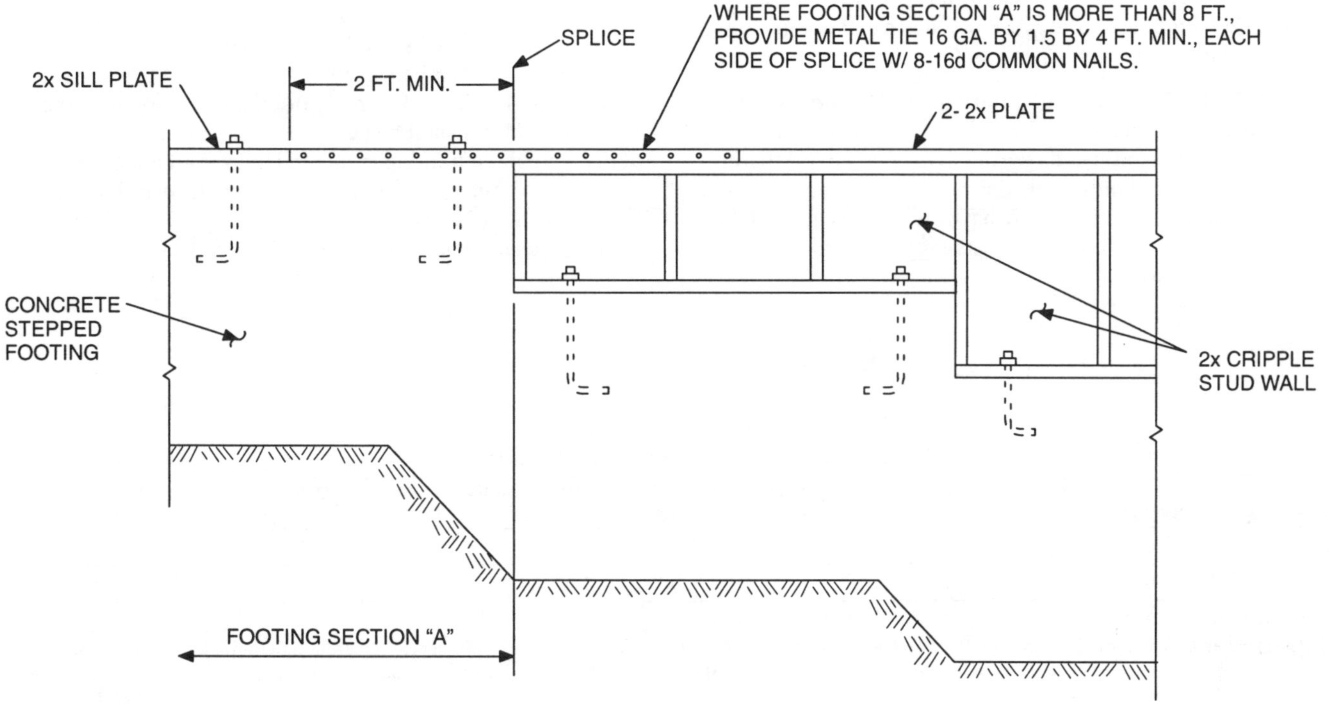

For SI: 1 inch = 25.4 mm, 1 foot = 304.8 mm.

Note: Where footing Section "A" is less than 8 feet long in a 25 feet total length wall, provide bracing at cripple stud wall.

FIGURE R602.11.3
STEPPED FOUNDATION CONSTRUCTION

SECTION R603
STEEL WALL FRAMING

R603.1 General. Elements shall be straight and free of any defects that would significantly affect structural performance. Cold-formed steel wall framing members shall comply with the requirements of this section.

R603.1.1 Applicability limits. The provisions of this section shall control the construction of exterior steel wall framing and interior load-bearing steel wall framing for buildings not greater than 60 feet (18 288 mm) in length perpendicular to the joist or truss span, not greater than 36 feet (10 973 mm) in width parallel to the joist span or truss, and not greater than two stories in height with each story not greater than 10 feet (3048 mm) high. All exterior walls installed in accordance with the provisions of this section shall be considered as load-bearing walls. Steel walls constructed in accordance with the provisions of this section shall be limited to sites subjected to a maximum design wind speed of 130 miles per hour (209 km/h) Exposure A, B or C and a maximum ground snow load of 70 pounds per foot (3.35 kN/m^2).

R603.1.1.1 Additional limits in high wind and high seismic regions. Braced wall lines and diaphragms in regions with wind speeds greater than or equal to 110 miles per hour (177 km/hr) or in Seismic Design Category D_1 or greater shall be permitted to have offsets of no greater than 4 feet (1219 mm). When offsets exceed 4 feet (1219 mm) the wall to either side shall be considered as a separate braced wall line with bracing amounts in accordance with Table R603.7.

R603.1.2 In-line framing. Load-bearing steel studs constructed in accordance with Section R603 shall be located directly in-line with joists, trusses and rafters with a maximum tolerance of $^3/_4$ inch (19.1 mm) between their center lines. Interior load-bearing steel stud walls shall be supported on foundations or shall be located directly above load-bearing walls with a maximum tolerance of $^3/_4$ inch (19.1 mm) between the centerline of the studs.

R603.2 Structural framing. Load-bearing steel wall framing members shall comply with Figure R603.2(1) and the dimensional and minimum thickness requirements specified in Tables R603.2(1) and R603.2(2). Tracks shall comply with Figure R603.2(2) and shall have a minimum flange width of $1^1/_4$ inches (32 mm). The maximum inside bend radius for load-bearing members shall be the greater of $^3/_{32}$ inch (2.4 mm) or twice the uncoated steel thickness. Holes in wall studs and other structural members shall not exceed 1.5 inches (38 mm) in width or 4 inches (102 mm) in length as shown in Figure R603.2(3). Holes shall be permitted only along the centerline of the web of the framing member. Holes shall not be less than 24 inches (610 mm) center to center and shall not be located less than 10 inches (254 mm) from edge of hole to end of member unless patched in accordance with Section R603.3.5.

TABLE R603.2(1)
LOAD-BEARING COLD-FORMED STEEL STUD SIZES

MEMBER DESIGNATION[a]	WEB DEPTH (inches)	MINIMUM FLANGE WIDTH (inches)	MAXIMUM FLANGE WIDTH (inches)	MINIMUM LIP SIZE (inches)
350S162-t	3.5	1.625	2	0.5
550S162-t	5.5	1.625	2	0.5

For SI: 1 inch = 25.4 mm.

a. The member designation is defined by the first number representing the member depth in $^1/_{100}$ inches, the letter "S" representing a stud or joist member, the second number representing the flange width in $^1/_{100}$ inches, and the letter "t" shall be a number representing the minimum base metal thickness in mils [See Table R603.2(2)].

TABLE R603.2(2)
MINIMUM THICKNESS OF COLD-FORMED STEEL STUDS

DESIGNATION (mils)	MINIMUM UNCOATED THICKNESS (inches)	REFERENCE GAGE NUMBER
33	0.033	20
43	0.043	18
54	0.054	16
68	0.068	14

For SI: 1 inch = 25.4 mm, 1 mil = 0.0254 mm.

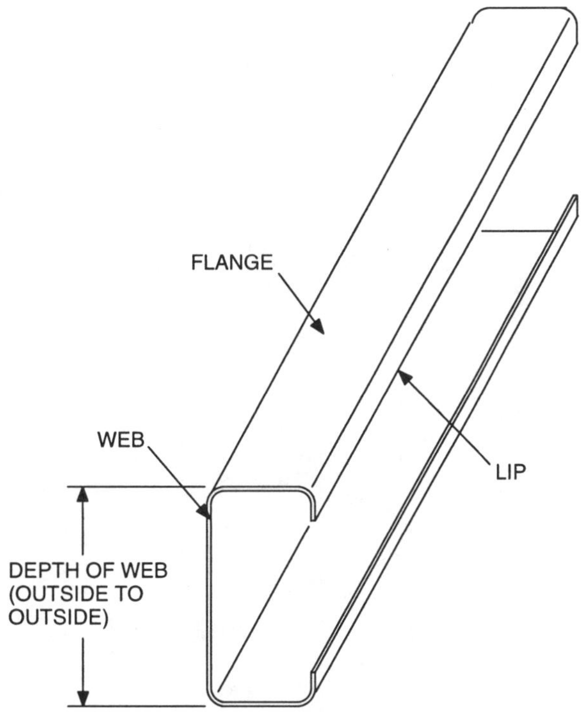

FIGURE R603.2(1)
C-SECTION

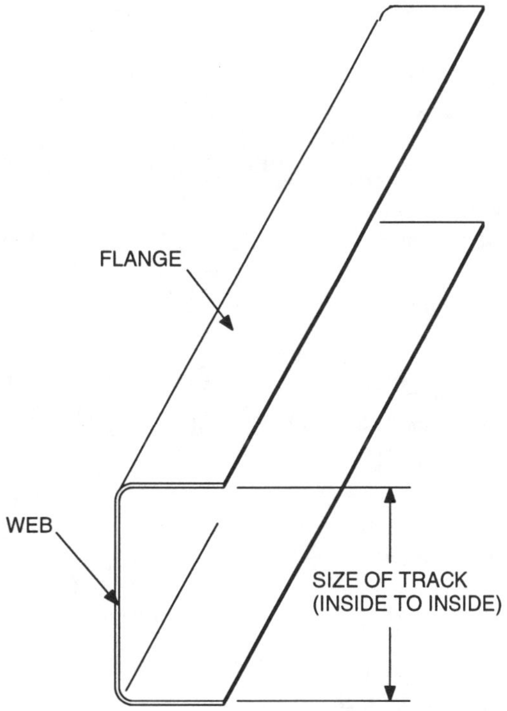

FIGURE R603.2(2)
TRACK SECTION

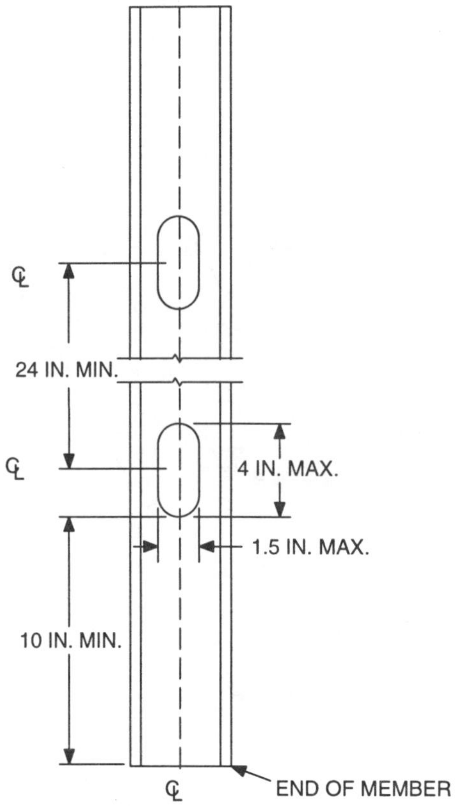

For SI: 1 inch = 25.4 mm.

FIGURE R603.2(3)
WEB HOLES

R603.2.1 Material. Load-bearing steel framing members shall be cold-formed to shape from structural quality sheet steel complying with the requirements of one of the following:

1. ASTM A 653: Grades 33, 37, 40 and 50 (Classes 1 and 3).
2. ASTM A 792: Grades 33, 37, 40 and 50A.
3. ASTM A 875: Grades 33, 37, 40 and 50 (Classes 1 and 3).
4. Steels that comply with ASTM A 653, except for tensile and elongation, shall be permitted, provided the ratio of tensile strength to yield point is at least 1.08 and the total elongation is at least 10 percent for a 2-inch (51 mm) gage length or 7 percent for an 8-inch (203 mm) gage length.

R603.2.2 Identification. Load-bearing steel framing members shall have a legible label, stencil, stamp or embossment with the following information as a minimum:

1. Manufacturer's identification.
2. Minimum uncoated steel thickness in inches (mm).
3. Minimum coating designation.
4. Minimum yield strength, in kips per square inch (ksi) (kPa).

R603.2.3 Corrosion protection. Load-bearing steel framing shall have a metallic coating complying with one of the following:

1. A minimum of G 60 in accordance with ASTM A 653.
2. A minimum of AZ 50 in accordance with ASTM A 792.
3. A minimum of GF 60 in accordance with ASTM A 875.

R603.2.4 Fastening requirements. Screws for steel-to-steel connections shall be installed with a minimum edge distance and center-to-center spacing of $^1/_2$ inch (12.7 mm), shall be self-drilling tapping and shall conform to SAE J 78. Structural sheathing shall be attached to steel studs with minimum No. 8 self-drilling tapping screws that conform to SAE J78. Screws for attaching structural sheathing to steel wall framing shall have a minimum head diameter of 0.292 inch (7.4 mm) with countersunk heads and shall be installed with a minimum edge distance of $^3/_8$ inch (9.5 mm). Gypsum board shall be attached to steel wall framing with minimum No. 6 screws conforming to ASTM C 954 and shall be installed in accordance with Section R702. For all connections, screws shall extend through the steel a minimum of three exposed threads. All self-drilling tapping screws conforming to SAE J 78 shall have a Type II coating in accordance with ASTM B 633.

Where No. 8 screws are specified in a steel to steel connection the required number of screws in the connection is permitted to be reduced in accordance with the reduction factors in Table R505.2.4, when larger screws are used or when one of the sheets of steel being connected is thicker than 33 mils (0.84 mm). When applying the reduction factor the resulting number of screws shall be rounded up.

TABLE R603.2.4
SCREW SUBSTITUTION FACTOR

SCREW SIZE	THINNEST CONNECTED STEEL SHEET (mils)	
	33	43
#8	1.0	0.67
#10	0.93	0.62
#12	0.86	0.56

For SI: 1 mil = 0.0254 mm.

R603.3 Wall construction. All exterior steel framed walls and interior load-bearing steel framed walls shall be constructed in accordance with the provisions of this section and Figure R603.3.

R603.3.1 Wall to foundation or floor connections. Steel framed walls shall be anchored to foundations or floors in accordance with Table R603.3.1 and Figure R603.3.1(1) or R603.3.1(2).

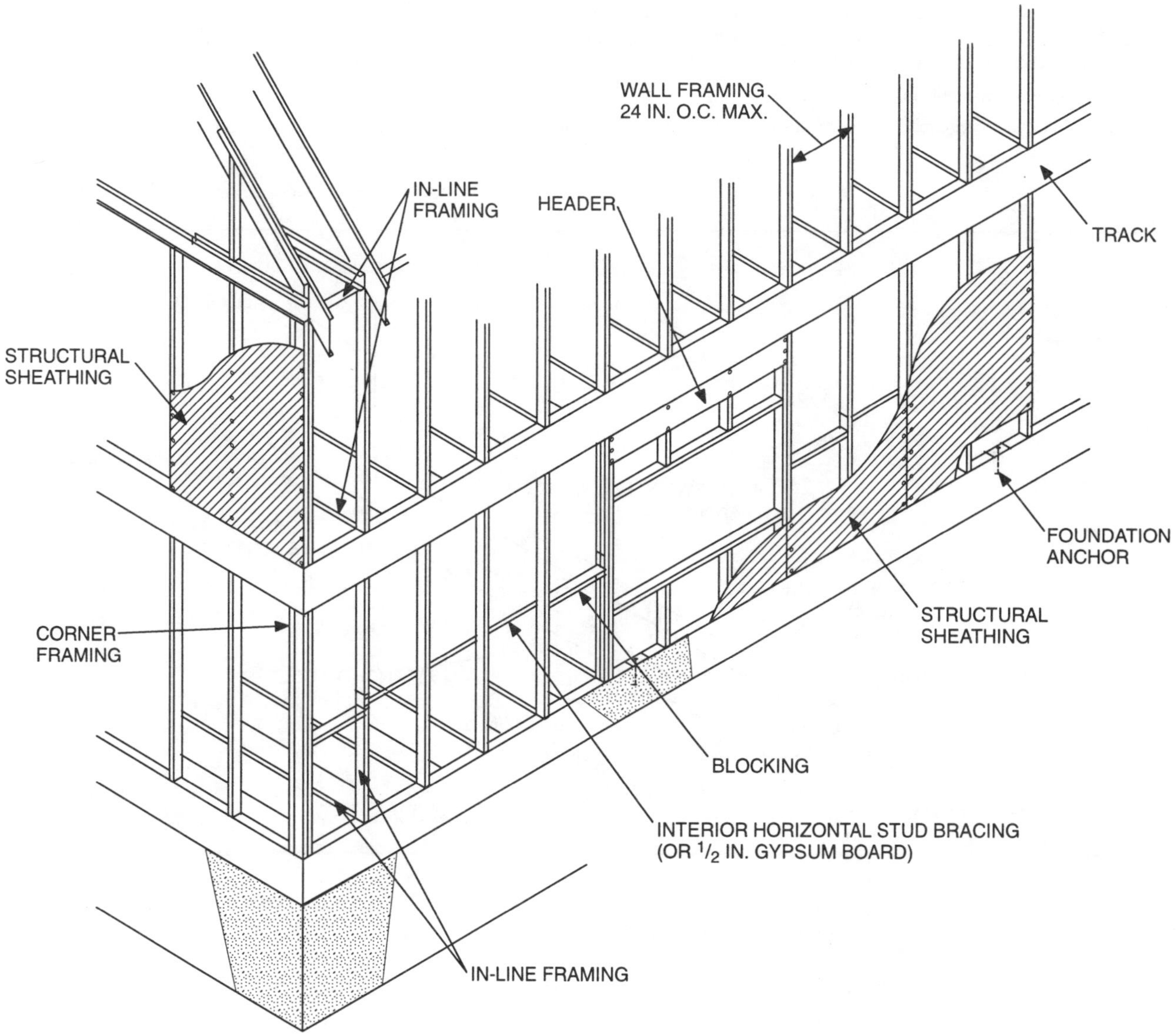

For SI: 1 inch = 25.4 mm.

FIGURE R603.3
STEEL WALL CONSTRUCTION

TABLE R603.3.1
WALL TO FOUNDATION OR FLOOR CONNECTION REQUIREMENTS[a,b,c]

FRAMING CONDITION	BASIC WIND SPEED (mph) AND EXPOSURE		
	85 A/B or Seismic Design Categories A, B and C	85 C or less than 110 A/B	Less than 110 C
Wall bottom track to floor joist or track	1-No. 8 screw at 12″ o.c.	1-No. 8 screw at 12″ o.c.	2-No. 8 screw at 12″ o.c.
Wall bottom track to wood sill per Figure R603.3.1(2)	Steel plate spaced at 4′ o.c., with 4-No. 8 screws and 4-10d or 6-8d common nails	Steel plate spaced at 3′ o.c., with 4-No. 8 screws and 4-10d or 6-8d common nails	Steel plate spaced at 2′ o.c., with 4-No. 8 screws and 4-10d or 6-8d common nails
Wall bottom track to foundation per Figure R603.3.1(1)	$^1/_2$″ minimum diameter anchor bolt at 6′ o.c.	$^1/_2$″ minimum diameter anchor bolt at 6′ o.c.	$^1/_2$″ minimum diameter anchor bolt at 4′ o.c.
Wind uplift connector capacity for 16-inch stud spacing[c]	N/R	N/R	65 lbs.
Wind uplift connector capacity for 24-inch stud spacing[c]	N/R	N/R	100 lbs.

For SI: 1 inch = 25.4 mm, 1 foot = 304.8 mm, 1 mile per hour = 1.609 km/hr, 1 pound = 4.4 N.

a. Anchor bolts shall be located not more than 12 inches from corners or the termination of bottom tracks (e.g. at door openings or corners). Bolts shall extend a minimum of 7 inches into concrete or masonry.

b. All screw sizes shown are minimum.

c. N/R = uplift connector not required. Uplift connectors are in addition to other connection requirements and shall be applied in accordance with Section R603.8.

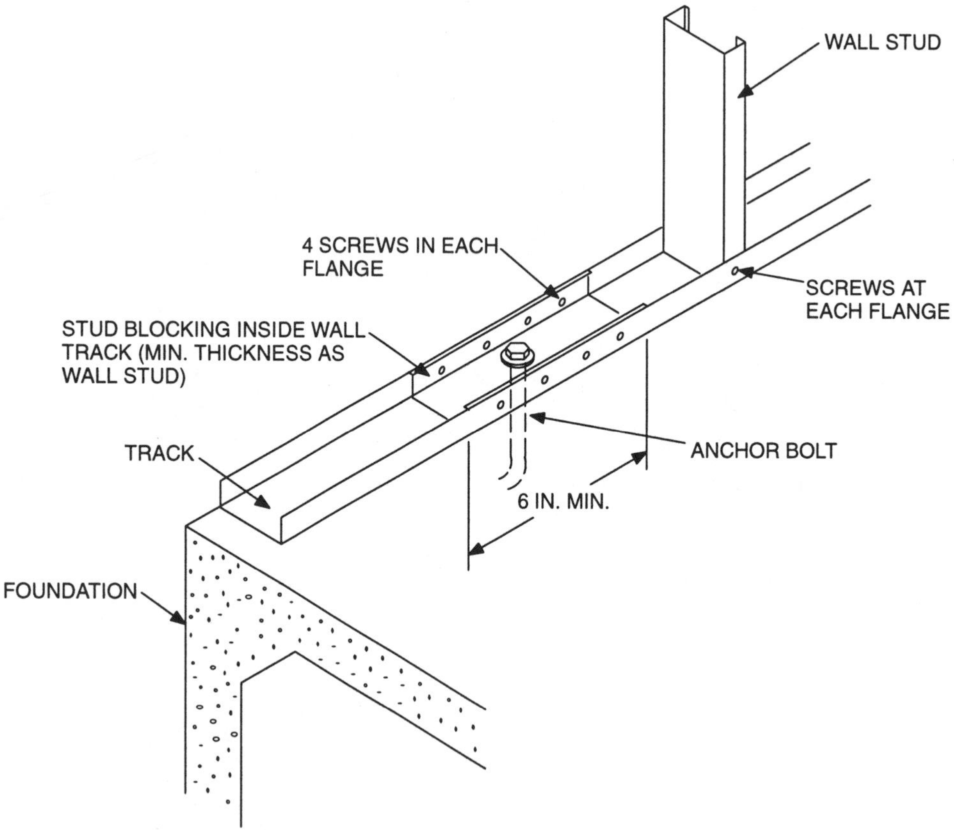

For SI: 1 inch = 25.4 mm.

FIGURE R603.3.1(1)
WALL TO FOUNDATION CONNECTION

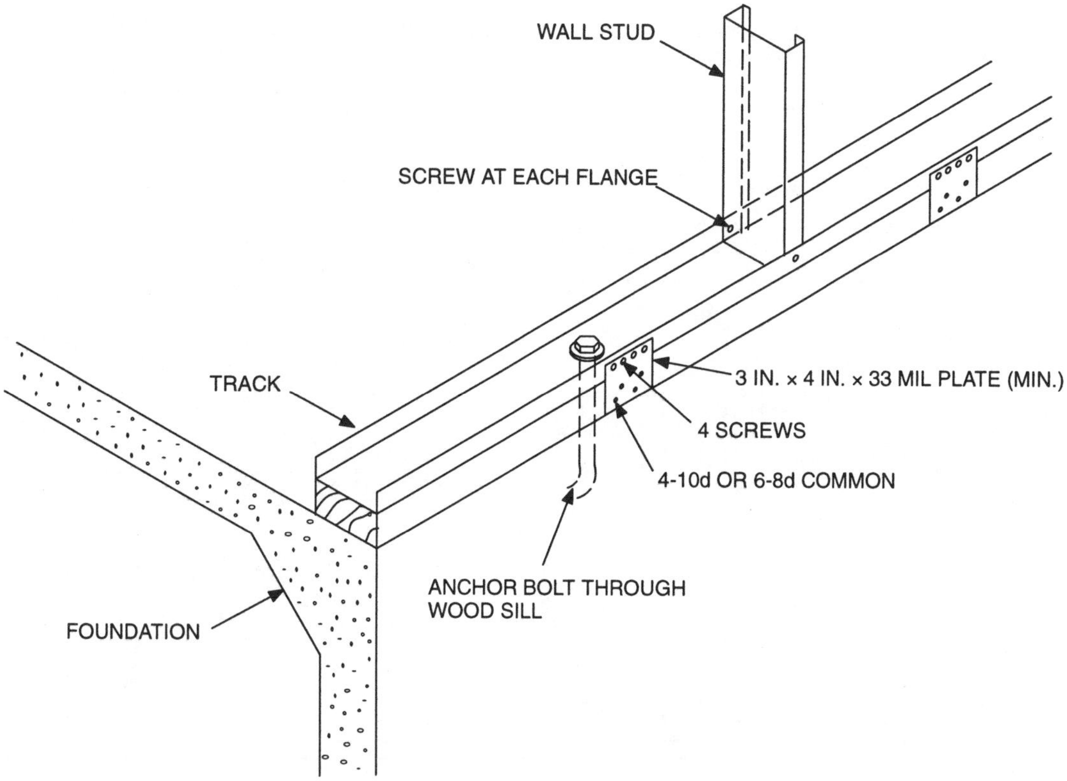

For SI: 1 inch = 25.4 mm, 1 mil = 0.0254 mm.

FIGURE R603.3.1(2)
WALL TO WOOD SILL CONNECTION

R603.3.2 Load-bearing walls. Steel studs shall comply with Tables R603.3.2(2) through R603.3.2(7) for steels with minimum yield strength of 33 ksi (227.7 MPa) and Tables R603.3.2(8) through R603.3.2(13) for steels with minimum yield strength of 50 ksi (345 MPa). Fastening requirements shall be in accordance with Section R603.2.4 and Table R603.3.2(1). Tracks shall have the same minimum steel thickness as the wall studs. Exterior walls with a minimum of $^1/_2$-inch (12.7 mm) gypsum board installed in accordance with Section R702 on the interior surface and wood structural panels of minimum $^7/_{16}$-inch-thick (11.1 mm) oriented strand board or $^{15}/_{32}$-inch-thick (11.9 mm) plywood installed in accordance with Table R603.3.2(1) on the outside surface shall be permitted to use the next thinner stud, from Tables R603.3.2(2) through R603.3.2(13) but not less than 33 mils (0.84 mm). Interior load-bearing walls with a minimum of $^1/_2$-inch (12.7 mm) gypsum board installed in accordance with Section R702 on both sides of the wall shall be permitted to use the next thinner stud, from Tables R603.3.2(2) through R603.3.2(13) but not less than 33 mils (0.84 mm).

TABLE R603.3.2(1)
WALL FASTENING SCHEDULE[a]

DESCRIPTION OF BUILDING ELEMENT	NUMBER AND SIZE OF FASTENERS[a]	SPACING OF FASTENERS
Floor joist to track of load-bearing wall	2-No. 8 screws	Each joist
Wall stud to top or bottom track	2-No. 8 screws	Each end of stud, one per flange
Structural sheathing to wall studs	No. 8 screws	6″ o.c. on edges and 12″ o.c. at intermediate supports
Roof framing to wall	Approved design or tie down in accordance with Section R802.11	

For SI: 1 inch = 25.4 mm.

a. All screw sizes shown are minimum.

TABLE R603.3.2(2)
COLD-FORMED STEEL STUD THICKNESS FOR 8-FOOT WALLS
Studs supporting roof and ceiling only (one-story building or second floor of a two-story building) 33 ksi steel

WIND SPEED				STUD THICKNESS (mils)[a,b]															
				Building width (feet)[d]															
				24				28				32				36			
			MEMBER SPACING	Ground snow load (psf)				Ground snow load (psf)				Ground snow load (psf)				Ground snow load (psf)			
Exp. A/B	Exp. C	MEMBER SIZE[c]	(inches)	20	30	50	70	20	30	50	70	20	30	50	70	20	30	50	70
85 mph	—	350S162	16	33	33	33	33	33	33	33	33	33	33	33	33	33	33	33	33
			24	33	33	33	33	33	33	33	33	33	33	33	33	33	33	33	43
		550S162	16	33	33	33	33	33	33	33	33	33	33	33	33	33	33	33	33
			24	33	33	33	33	33	33	33	33	33	33	33	33	33	33	33	33
100 mph	85 mph	350S162	16	33	33	33	33	33	33	33	33	33	33	33	33	33	33	33	33
			24	33	33	33	33	33	33	33	33	33	33	33	43	33	33	43	43
		550S162	16	33	33	33	33	33	33	33	33	33	33	33	33	33	33	33	33
			24	33	33	33	33	33	33	33	33	33	33	33	33	33	33	33	33
110 mph	100 mph	350S162	16	33	33	33	33	33	33	33	33	33	33	33	33	33	33	33	33
			24	33	33	43	43	33	33	43	43	33	43	43	43	43	43	43	43
		550S162	16	33	33	33	33	33	33	33	33	33	33	33	33	33	33	33	33
			24	33	33	33	33	33	33	33	33	33	33	33	33	33	33	33	33
120 mph	110 mph	350S162	16	33	33	33	33	33	33	33	33	33	33	33	33	33	33	33	33
			24	43	43	43	43	43	43	43	54	43	43	54	54	43	43	54	54
		550S162	16	33	33	33	33	33	33	33	33	33	33	33	33	33	33	33	33
			24	33	33	33	33	33	33	33	33	33	33	33	33	33	33	33	33
130 mph	120 mph	350S162	16	33	33	43	43	33	43	43	43	43	43	43	43	43	43	43	43
			24	54	54	54	54	54	54	54	68	54	54	68	68	54	54	68	68
		550S162	16	33	33	33	33	33	33	33	33	33	33	33	33	33	33	33	33
			24	33	33	33	33	33	33	33	43	33	33	33	43	33	33	43	43
—	130 mph	350S162	16	43	43	43	43	43	43	43	43	43	43	43	54	43	43	43	54
			24	68	68	68	68	68	68	68	68	68	68	68	68	68	68	68	(d)
		550S162	16	33	33	33	33	33	33	33	33	33	33	33	33	33	33	33	33
			24	33	43	43	43	43	43	43	43	43	43	43	43	43	43	43	43

For SI: 1 inch = 25.4 mm, 1 foot = 304.8 mm, 1 mil = 0.0254 mm, 1 mile per hour = 1.609 km/h, 1 pound per square foot = 0.0479 kN/m^2, 1 kilogram per square inch = 6.895 MPa.

a. Deflection criteria: *L*/240.

b. Building width is in the direction of horizontal framing members supported by the wall studs.

c. Design load assumptions:

Roof dead load is 12 psf.

Attic live load is 10 psf.

d. 68-mil-thick stud is allowed if wall is fully sheathed per Section R603.3.2.

TABLE R603.3.2(3) COLD-FORMED STEEL STUD THICKNESS FOR 8-FOOT WALLS
Studs supporting one floor, roof and ceiling (first story of a two-story building) 33 ksi steel

WIND SPEED				STUD THICKNESS (mils)[a,b]															
				Building width (feet)[d]															
				24				28				32				36			
			MEMBER SPACING (inches)	Ground snow load (psf)				Ground snow load (psf)				Ground snow load (psf)				Ground snow load (psf)			
Exp. A/B	Exp. C	MEMBER SIZE[c]		20	30	50	70	20	30	50	70	20	30	50	70	20	30	50	70
85 mph	—	350S162	16	33	33	33	33	33	33	33	33	33	33	33	43	33	33	33	43
			24	43	43	43	43	43	43	43	43	43	43	43	54	43	43	54	54
		550S162	16	33	33	33	33	33	33	33	33	33	33	33	33	33	33	33	33
			24	33	33	33	33	33	33	33	43	33	33	43	43	33	43	43	54
100 mph	85 mph	350S162	16	33	33	33	33	33	33	33	33	33	33	33	43	33	33	43	43
			24	43	43	43	54	43	43	54	54	54	54	54	54	54	54	54	54
		550S162	16	33	33	33	33	33	33	33	33	33	33	33	33	33	33	33	43
			24	33	33	33	33	33	33	33	43	33	33	43	43	43	43	43	54
110 mph	100 mph	350S162	16	33	33	33	33	33	33	43	43	43	43	43	43	43	43	43	43
			24	54	54	54	54	54	54	54	54	54	54	54	68	54	54	68	68
		550S162	16	33	33	33	33	33	33	33	33	33	33	33	33	33	33	33	33
			24	33	33	33	33	33	33	43	43	43	43	43	43	43	43	43	54
120 mph	110 mph	350S162	16	43	43	43	43	43	43	43	43	43	43	43	43	43	43	43	54
			24	54	54	54	68	54	68	68	68	68	68	68	68	68	68	68	68
		550S162	16	33	33	33	33	33	33	33	33	33	33	33	33	33	33	33	33
			24	33	43	43	43	43	43	43	43	43	43	43	43	43	43	43	54
130 mph	120 mph	350S162	16	43	43	43	54	43	54	54	54	54	54	54	54	54	54	54	54
			24	68	68	68	68	68	68	(d)	(d)	(d)	(d)	(d)	(d)	(d)	(d)	(d)	(d)
		550S162	16	33	33	33	33	33	33	33	33	33	33	33	33	33	33	33	43
			24	43	43	43	43	43	43	43	54	43	43	54	54	54	54	54	54
—	130 mph	350S162	16	54	54	54	54	54	54	54	54	54	54	54	68	54	54	68	68
			24	(d)	(d)	(d)	(d)	(d)	(d)	(d)	(d)	(d)	(d)	(d)	(d)	(d)	(d)	(d)	(d)
		550S162	16	33	33	33	33	33	33	33	43	33	33	43	43	43	43	43	43
			24	43	43	54	54	54	54	54	54	54	54	54	54	54	54	54	54

For SI: 1 inch = 25.4 mm, 1 foot = 304.8 mm, 1 mil = 0.0254 mm, 1 mile per hour = 1.609 km/h, 1 pound per square foot = 0.0479 kN/m^2, 1 kilogram per square inch = 6.895 MPa.

a. Deflection criteria: *L*/240.

b. Building width is in the direction of horizontal framing members supported by the wall studs.

c. Design load assumptions:

Roof dead load is 12 psf.

Attic live load is 10 psf.

d. 68-mil-thick stud is allowed used if wall is fully sheathed per Section R603.3.2.

TABLE R603.3.2(4)
COLD-FORMED STEEL STUD THICKNESS FOR 9-FOOT WALLS
Studs supporting roof and ceiling only (one-story building or second floor of a two-story building) 33 ksi steel

WIND SPEED				STUD THICKNESS (mils)[a,b]															
				Building width (feet)[d]															
				24				28				32				36			
			MEMBER SPACING	Ground snow load (psf)				Ground snow load (psf)				Ground snow load (psf)				Ground snow load (psf)			
Exp. A/B	Exp. C	MEMBER SIZE[c]	(inches)	20	30	50	70	20	30	50	70	20	30	50	70	20	30	50	70
85 mph	—	350S162	16	33	33	33	33	33	33	33	33	33	33	33	33	33	33	33	33
			24	33	33	33	33	33	33	33	33	33	33	33	33	33	33	33	43
		550S162	16	33	33	33	33	33	33	33	33	33	33	33	33	33	33	33	33
			24	33	33	33	33	33	33	33	33	33	33	33	33	33	33	33	33
100 mph	85 mph	350S162	16	33	33	33	33	33	33	33	33	33	33	33	33	33	33	33	33
			24	33	33	43	43	33	33	43	43	33	43	43	43	43	43	43	43
		550S162	16	33	33	33	33	33	33	33	33	33	33	33	33	33	33	33	33
			24	33	33	33	33	33	33	33	33	33	33	33	33	33	33	33	33
110 mph	100 mph	350S162	16	33	33	33	33	33	33	33	33	33	33	33	33	33	33	33	33
			24	43	43	43	43	43	43	43	43	43	43	43	54	43	43	43	54
		550S162	16	33	33	33	33	33	33	33	33	33	33	33	33	33	33	33	33
			24	33	33	33	33	33	33	33	33	33	33	33	33	33	33	33	33
120 mph	110 mph	350S162	16	33	33	33	33	33	33	33	43	33	33	43	43	33	33	43	43
			24	54	54	54	54	54	54	54	54	54	54	54	54	54	54	54	68
		550S162	16	33	33	33	33	33	33	33	33	33	33	33	33	33	33	33	33
			24	33	33	33	33	33	33	33	33	33	33	33	33	33	33	33	43
130 mph	120 mph	350S162	16	43	43	43	43	43	43	43	43	43	43	43	54	43	43	43	54
			24	68	68	68	68	68	68	68	68	68	68	68	68	68	68	68	(d)
		550S162	16	33	33	33	33	33	33	33	33	33	33	33	33	33	33	33	33
			24	33	43	43	43	43	43	43	43	43	43	43	43	43	43	43	43
—	130 mph	350S162	16	54	54	54	54	54	54	54	54	54	54	54	54	54	54	54	54
			24	(d)	(d)	(d)	(d)	(d)	(d)	(d)	(d)	(d)	(d)	(d)	(d)	(d)	(d)	(d)	(d)
		550S162	16	33	33	33	33	33	33	33	33	33	33	33	33	33	33	33	33
			24	43	43	43	43	43	43	43	43	43	43	43	54	43	43	43	54

For SI: 1 inch = 25.4 mm, 1 foot = 304.8 mm, 1 mil = 0.0254 mm, 1 mile per hour = 1.609 km/h, 1 pound per square foot = 0.0479 kN/m^2, 1 kilogram per square inch = 6.895 MPa.

a. Deflection criteria: *L*/240.

b. Building width is in the direction of horizontal framing members supported by the wall studs.

c. Design load assumptions:
Roof dead load is 12 psf.
Attic live load is 10 psf.

d. 68-mil-thick stud is allowed if wall is fully sheathed per Section R603.3.2.

TABLE R603.3.2(5)
COLD-FORMED STEEL STUD THICKNESS FOR 9-FOOT WALLS
Studs supporting one floor, roof and ceiling (first story of a two-story building) 33 ksi steel

WIND SPEED		MEMBER SIZE[c]	MEMBER SPACING (inches)	STUD THICKNESS (mils)[a,b] Building width (feet)[d]															
				24				28				32				36			
				Ground snow load (psf)				Ground snow load (psf)				Ground snow load (psf)				Ground snow load (psf)			
Exp. A/B	Exp. C			20	30	50	70	20	30	50	70	20	30	50	70	20	30	50	70
85 mph	—	350S162	16	33	33	33	33	33	33	33	33	33	33	33	33	33	33	33	43
			24	43	43	43	43	43	43	43	54	43	43	54	54	54	54	54	54
		550S162	16	33	33	33	33	33	33	33	33	33	33	33	33	33	33	33	33
			24	33	33	33	33	33	33	33	43	33	33	43	43	33	33	43	43
100 mph	85 mph	350S162	16	33	33	33	33	33	33	33	43	33	43	43	43	43	43	43	43
			24	43	54	54	54	54	54	54	54	54	54	54	54	54	54	68	68
		550S162	16	33	33	33	33	33	33	33	33	33	33	33	33	33	33	33	33
			24	33	33	33	33	33	33	33	43	33	33	43	43	43	43	43	43
110 mph	100 mph	350S162	16	43	43	43	43	43	43	43	43	43	43	43	43	43	43	43	43
			24	54	54	54	54	54	54	54	54	54	54	54	54	54	54	68	68
		550S162	16	33	33	33	33	33	33	33	33	33	33	33	33	33	33	33	33
			24	33	33	43	43	43	43	43	43	43	43	43	43	43	43	43	54
120 mph	110 mph	350S162	16	43	43	43	43	43	43	43	54	43	43	54	54	54	54	54	54
			24	68	68	68	68	68	68	68	68	68	68	68	(d)	68	68	(d)	(d)
		550S162	16	33	33	33	33	33	33	33	33	33	33	33	33	33	33	33	33
			24	43	43	43	43	43	43	43	43	43	43	43	54	43	43	54	54
130 mph	120 mph	350S162	16	54	54	54	54	54	54	54	54	54	54	54	68	54	54	68	68
			24	(d)	(d)	(d)	(d)	(d)	(d)	(d)	(d)	(d)	(d)	(d)	(d)	(d)	(d)	(d)	(d)
		550S162	16	33	33	33	33	33	33	33	43	33	33	43	43	33	43	43	43
			24	43	43	54	54	54	54	54	54	54	54	54	54	54	54	54	54
—	130 mph	350S162	16	68	68	68	68	68	68	68	68	68	68	68	68	68	68	68	68
			24	(d)	(d)	(d)	(d)	(d)	(d)	(d)	—	(d)	(d)	—	—	—	—	—	—
		550S162	16	33	33	43	43	43	43	43	43	43	43	43	43	43	43	43	43
			24	54	54	54	54	54	54	54	54	54	54	54	68	54	54	68	68

For SI: 1 inch = 25.4 mm, 1 foot = 304.8 mm, 1 mil = 0.0254 mm, 1 mile per hour = 1.609 km/h, 1 pound per square foot = 0.0479 kN/m^2, 1 kilogram per square inch = 6.895 MPa.

a. Deflection criteria: *L*/240.

b. Building width is in the direction of horizontal framing members supported by the wall studs.

c. Design load assumptions:
Roof dead load is 12 psf.
Attic live load is 10 psf.

d. 68-mil-thick stud is allowed if wall is fully sheathed per Section R603.3.2.

TABLE R603.3.2(6)
COLD-FORMED STEEL STUD THICKNESS FOR 10-FOOT WALLS
Studs supporting roof and ceiling only (one-story building or second floor of a two-story building) 33 ksi steel

WIND SPEED		MEMBER SIZE[c]	MEMBER SPACING (inches)	STUD THICKNESS (mils)[a,b]															
				Building width (feet)[d]															
				24				28				32				36			
				Ground snow load (psf)				Ground snow load (psf)				Ground snow load (psf)				Ground snow load (psf)			
Exp. A/B	Exp. C			20	30	50	70	20	30	50	70	20	30	50	70	20	30	50	70
85 mph	—	350S162	16	33	33	33	33	33	33	33	33	33	33	33	33	33	33	33	33
			24	33	33	33	43	33	33	43	43	33	33	43	43	33	43	43	43
		550S162	16	33	33	33	33	33	33	33	33	33	33	33	33	33	33	33	33
			24	33	33	33	33	33	33	33	33	33	33	33	33	33	33	33	33
100 mph	85 mph	350S162	16	33	33	33	33	33	33	33	33	33	33	33	33	33	33	33	33
			24	43	43	43	43	43	43	43	54	43	43	43	54	43	43	54	54
		550S162	16	33	33	33	33	33	33	33	33	33	33	33	33	33	33	33	33
			24	33	33	33	33	33	33	33	33	33	33	33	33	33	33	33	33
110 mph	100 mph	350S162	16	33	33	33	43	33	33	43	43	33	33	43	43	33	33	43	43
			24	54	54	54	54	54	54	54	54	54	54	54	54	54	54	54	68
		550S162	16	33	33	33	33	33	33	33	33	33	33	33	33	33	33	33	33
			24	33	33	33	33	33	33	33	33	33	33	33	33	33	33	33	43
120 mph	110 mph	350S162	16	43	43	43	43	43	43	43	43	43	43	43	43	43	43	43	54
			24	68	68	68	68	68	68	68	68	68	68	68	68	68	68	68	68
		550S162	16	33	33	33	33	33	33	33	33	33	33	33	33	33	33	33	33
			24	33	33	43	43	33	33	43	43	33	43	43	43	43	43	43	43
130 mph	120 mph	350S162	16	54	54	54	54	54	54	54	54	54	54	54	54	54	54	54	68
			24	(d)	(d)	(d)	(d)	(d)	(d)	(d)	(d)	(d)	(d)	(d)	(d)	(d)	(d)	(d)	(d)
		550S162	16	33	33	33	33	33	33	33	33	33	33	33	33	33	33	33	33
			24	43	43	43	43	43	43	43	54	43	43	54	54	43	43	54	54
—	130 mph	350S162	16	68	68	68	68	68	68	68	68	68	68	68	68	68	68	68	68
			24	(d)	(d)	(d)	—	(d)	(d)	—	—	(d)	(d)	—	—	(d)	—	—	—
		550S162	16	33	33	33	43	33	33	43	43	33	33	43	43	33	43	43	43
			24	54	54	54	54	54	54	54	54	54	54	54	54	54	54	54	68

For SI: 1 inch = 25.4 mm, 1 foot = 304.8 mm, 1 mil = 0.0254 mm, 1 mile per hour = 1.609 km/h, 1 pound per square foot = 0.0479 kN/m^2, 1 kilogram per square inch = 6.895 MPa.

a. Deflection criteria: *L*/240.

b. Building width is in the direction of horizontal framing members supported by the wall studs.

c. Design load assumptions:

Roof dead load is 12 psf.

Attic live load is 10 psf.

d. 68-mil-thick stud is allowed if wall is fully sheathed per Section R603.3.2.

TABLE R603.3.2(7)
COLD-FORMED STEEL STUD THICKNESS FOR 10-FOOT WALLS
Studs supporting one floor, roof and ceiling (first story of a two-story building) 33 ksi steel

WIND SPEED		MEMBER SIZE[c]	MEMBER SPACING (inches)	STUD THICKNESS (mils)[a,b]															
				Building width (feet)[d]															
				24				28				32				36			
				Ground snow load (psf)				Ground snow load (psf)				Ground snow load (psf)				Ground snow load (psf)			
Exp. A/B	Exp. C			20	30	50	70	20	30	50	70	20	30	50	70	20	30	50	70
85 mph	—	350S162	16	33	33	33	43	33	33	43	43	43	43	43	43	43	43	43	43
			24	54	54	54	54	54	54	54	54	54	54	54	68	54	54	68	68
		550S162	16	33	33	33	33	33	33	33	33	33	33	33	33	33	33	33	33
			24	33	33	33	33	33	33	33	43	33	33	43	43	43	43	43	54
100 mph	85 mph	350S162	16	43	43	43	43	43	43	43	43	43	43	43	43	43	43	43	54
			24	54	54	68	68	68	68	68	68	68	68	68	68	68	68	68	(d)
		550S162	16	33	33	33	33	33	33	33	33	33	33	33	33	33	33	33	33
			24	33	33	43	43	43	43	43	43	43	43	43	43	43	43	43	54
110 mph	100 mph	350S162	16	43	43	43	43	43	43	54	54	43	54	54	54	54	54	54	54
			24	68	68	68	68	68	68	68	(d)	68	68	(d)	(d)	(d)	(d)	(d)	(d)
		550S162	16	33	33	33	33	33	33	33	33	33	33	33	33	33	33	33	33
			24	43	43	43	43	43	43	43	43	43	43	43	54	43	43	54	54
120 mph	110 mph	350S162	16	54	54	54	54	54	54	54	54	54	54	54	54	54	54	54	54
			24	(d)	(d)	(d)	(d)	(d)	(d)	(d)	(d)	(d)	(d)	(d)	(d)	(d)	(d)	(d)	(d)
		550S162	16	33	33	33	33	33	33	33	33	33	33	33	43	33	33	43	43
			24	43	43	43	54	43	54	54	54	54	54	54	54	54	54	54	54
130 mph	120 mph	350S162	16	68	68	68	68	68	68	68	68	68	68	68	68	68	68	68	(4)
			24	(d)	(d)	—	—	—	—	—	—	—	—	—	—	—	—	—	—
		550S162	16	43	43	43	43	43	43	43	43	43	43	43	43	43	43	43	43
			24	54	54	54	54	54	54	54	68	54	68	68	68	68	68	68	68
—	130 mph	350S162	16	68	(d)	(d)	(d)	(d)	(d)	(d)	(d)	(d)	(d)	(d)	(d)	(d)	(d)	(d)	(d)
			24	—	—	—	—	—	—	—	—	—	—	—	—	—	—	—	—
		550S162	16	43	43	43	43	43	43	43	43	43	43	43	54	43	43	54	54
			24	68	68	68	68	68	68	68	68	68	68	68	68	68	68	68	(d)

For SI: 1 inch = 25.4 mm, 1 foot = 304.8 mm, 1 mil = 0.0254 mm, 1 mile per hour = 1.609 km/h, 1 pound per square foot = 0.0479 kN/m^2, 1 kilogram per square inch = 6.895 MPa.

a. Deflection criteria: *L*/240.

b. Building width is in the direction of horizontal framing members supported by the wall studs.

c. Design load assumptions:

Roof dead load is 12 psf.

Attic live load is 10 psf.

d. 68-mil-thick stud is allowed if wall is fully sheathed per Section R603.3.2.

TABLE R603.3.2(8)
COLD-FORMED STEEL STUD THICKNESS FOR 8-FOOT WALLS
Studs supporting roof and ceiling only (one-story building or second floor of a two-story building) 50 ksi steel

WIND SPEED		MEMBER SIZE[c]	MEMBER SPACING (inches)	STUD THICKNESS (mils)[a,b]															
				Building width (feet)[d]															
				24				28				32				36			
				Ground snow load (psf)				Ground snow load (psf)				Ground snow load (psf)				Ground snow load (psf)			
Exp. A/B	Exp. C			20	30	50	70	20	30	50	70	20	30	50	70	20	30	50	70
85 mph	—	350S162	16	33	33	33	33	33	33	33	33	33	33	33	33	33	33	33	33
			24	33	33	33	33	33	33	33	33	33	33	33	33	33	33	33	33
		550S162	16	33	33	33	33	33	33	33	33	33	33	33	33	33	33	33	33
			24	33	33	33	33	33	33	33	33	33	33	33	33	33	33	33	33
100 mph	85 mph	350S162	16	33	33	33	33	33	33	33	33	33	33	33	33	33	33	33	33
			24	33	33	33	33	33	33	33	33	33	33	33	33	33	33	33	33
		550S162	16	33	33	33	33	33	33	33	33	33	33	33	33	33	33	33	33
			24	33	33	33	33	33	33	33	33	33	33	33	33	33	33	33	33
110 mph	100 mph	350S162	16	33	33	33	33	33	33	33	33	33	33	33	33	33	33	33	33
			24	33	33	33	33	33	33	33	33	33	33	33	33	33	33	33	43
		550S162	16	33	33	33	33	33	33	33	33	33	33	33	33	33	33	33	33
			24	33	33	33	33	33	33	33	33	33	33	33	33	33	33	33	33
120 mph	110 mph	350S162	16	33	33	33	33	33	33	33	33	33	33	33	33	33	33	33	33
			24	33	33	33	43	33	33	43	43	33	33	43	43	33	43	43	43
		550S162	16	33	33	33	33	33	33	33	33	33	33	33	33	33	33	33	33
			24	33	33	33	33	33	33	33	33	33	33	33	33	33	33	33	33
130 mph	120 mph	350S162	16	33	33	33	33	33	33	33	33	33	33	33	33	33	33	33	33
			24	43	43	43	43	43	43	43	43	43	43	43	54	43	43	43	54
		550S162	16	33	33	33	33	33	33	33	33	33	33	33	33	33	33	33	33
			24	33	33	33	33	33	33	33	33	33	33	33	33	33	33	33	33
—	130 mph	350S162	16	33	33	33	33	33	33	33	43	33	33	43	43	33	33	43	43
			24	43	54	54	54	54	54	54	54	54	54	54	54	54	54	54	54
		550S162	16	33	33	33	33	33	33	33	33	33	33	33	33	33	33	33	33
			24	33	33	33	33	33	33	33	33	33	33	33	33	33	33	33	33

For SI: 1 inch = 25.4 mm, 1 foot = 304.8 mm, 1 mil = 0.0254 mm, 1 mile per hour = 1.609 km/h, 1 pound per square foot = 0.0479 kN/m^2, 1 kilogram per square inch = 6.895 MPa.

a. Deflection criteria: *L*/240.

b. Building width is in the direction of horizontal framing members supported by the wall studs.

c. Design load assumptions:

Roof dead load is 12 psf.

Attic live load is 10 psf.

d. 68-mil-thick stud is allowed if wall is fully sheathed per Section R603.3.2.

TABLE R603.3.2(9)
COLD-FORMED STEEL STUD THICKNESS FOR 8-FOOT WALLS
Studs supporting one floor, roof and ceiling (first story of a two-story building) 50 ksi steel

WIND SPEED		MEMBER SIZE[c]	MEMBER SPACING (inches)	STUD THICKNESS (mils)[a,b]															
				Building width (feet)[d]															
				24				28				32				36			
				Ground snow load (psf)				Ground snow load (psf)				Ground snow load (psf)				Ground snow load (psf)			
Exp. A/B	Exp. C			20	30	50	70	20	30	50	70	20	30	50	70	20	30	50	70
85 mph	—	350S162	16	33	33	33	33	33	33	33	33	33	33	33	33	33	33	33	43
			24	33	33	33	43	33	33	43	43	43	43	43	43	43	43	43	54
		550S162	16	33	33	33	33	33	33	33	33	33	33	33	33	33	33	33	33
			24	33	33	33	33	33	33	33	33	33	33	33	43	33	33	43	43
100 mph	85 mph	350S162	16	33	33	33	33	33	33	33	33	33	33	33	33	33	33	33	43
			24	33	43	43	43	43	43	43	43	43	43	43	43	43	33	33	54
		550S162	16	33	33	33	33	33	33	33	33	33	33	33	33	33	33	33	43
			24	33	33	33	33	33	33	33	33	33	33	33	43	33	33	43	43
110 mph	100 mph	350S162	16	33	33	33	33	33	33	33	33	33	33	33	33	33	33	33	43
			24	43	43	43	43	43	43	43	43	43	43	54	54	43	54	54	54
		550S162	16	33	33	33	33	33	33	33	33	33	33	33	33	33	33	33	33
			24	33	33	33	33	33	33	33	33	33	33	33	43	33	33	43	43
120 mph	110 mph	350S162	16	33	33	33	33	33	33	33	43	33	33	43	43	33	43	43	43
			24	43	43	43	54	43	54	54	54	54	54	54	54	54	54	54	54
		550S162	16	33	33	33	33	33	33	33	33	33	33	33	33	33	33	33	33
			24	33	33	33	33	33	33	33	33	33	33	33	43	33	33	43	43
130 mph	120 mph	350S162	16	43	43	43	43	43	43	43	43	43	43	43	43	43	43	43	43
			24	54	54	54	54	54	54	54	68	54	54	68	68	68	68	68	68
		550S162	16	33	33	33	33	33	33	33	33	33	33	33	33	33	33	33	43
			24	33	33	33	43	33	43	43	43	43	43	43	43	43	43	43	43
—	130 mph	350S162	16	43	43	43	43	43	43	43	43	43	43	43	54	43	43	54	54
			24	54	68	68	68	68	68	68	68	68	68	68	68	68	68	68	68
		550S162	16	33	33	33	33	33	33	33	33	33	33	33	33	33	33	33	33
			24	43	43	43	43	43	43	43	43	43	43	43	43	43	43	43	43

For SI: 1 inch = 25.4 mm, 1 foot = 304.8 mm, 1 mil = 0.0254 mm, 1 mile per hour = 1.609 km/h, 1 pound per square foot = 0.0479 kN/m^2, 1 kilogram per square inch = 6.895 MPa.

a. Deflection criteria: *L*/240.

b. Building width is in the direction of horizontal framing members supported by the wall studs.

c. Design load assumptions:

Roof dead load is 12 psf.

Attic live load is 10 psf.

d. 68-mil-thick stud is allowed if wall is fully sheathed per Section R603.3.2.

TABLE R603.3.2(10)
COLD-FORMED STEEL STUD THICKNESS FOR 9-FOOT WALLS
Studs supporting roof and ceiling only (one-story building or second floor of a two-story building) 50 ksi steel

WIND SPEED		MEMBER SIZE[c]	MEMBER SPACING (inches)	STUD THICKNESS (mils)[a,b]															
				Building width (feet)[d]															
				24				28				32				36			
				Ground snow load (psf)				Ground snow load (psf)				Ground snow load (psf)				Ground snow load (psf)			
Exp. A/B	Exp. C			20	30	50	70	20	30	50	70	20	30	50	70	20	30	50	70
85 mph	—	350S162	16	33	33	33	33	33	33	33	33	33	33	33	33	33	33	33	33
			24	33	33	33	33	33	33	33	33	33	33	33	33	33	33	33	33
		550S162	16	33	33	33	33	33	33	33	33	33	33	33	33	33	33	33	33
			24	33	33	33	33	33	33	33	33	33	33	33	33	33	33	33	33
100 mph	85 mph	350S162	16	33	33	33	33	33	33	33	33	33	33	33	33	33	33	33	33
			24	33	33	33	33	33	33	33	33	33	33	33	33	33	33	33	33
		550S162	16	33	33	33	33	33	33	33	33	33	33	33	33	33	33	33	33
			24	33	33	33	33	33	33	33	33	33	33	33	33	33	33	33	33
110 mph	100 mph	350S162	16	33	33	33	33	33	33	33	33	33	33	33	33	33	33	33	33
			24	33	33	33	33	33	33	33	43	33	33	33	43	33	33	43	43
		550S162	16	33	33	33	33	33	33	33	33	33	33	33	33	33	33	33	33
			24	33	33	33	33	33	33	33	33	33	33	33	33	33	33	33	33
120 mph	110 mph	350S162	16	33	33	33	33	33	33	33	33	33	33	33	33	33	33	33	33
			24	43	43	43	43	43	43	43	43	43	43	43	43	43	43	43	43
		550S162	16	33	33	33	33	33	33	33	33	33	33	33	33	33	33	33	33
			24	33	33	33	33	33	33	33	33	33	33	33	33	33	33	33	33
130 mph	120 mph	350S162	16	33	33	33	33	33	33	33	33	33	33	33	43	33	33	43	43
			24	43	43	54	54	43	54	54	54	54	54	54	54	54	54	54	54
		550S162	16	33	33	33	33	33	33	33	33	33	33	33	33	33	33	33	33
			24	33	33	33	33	33	33	33	33	33	33	33	33	33	33	33	33
—	130 mph	350S162	16	43	43	43	43	43	43	43	43	43	43	43	43	43	43	43	43
			24	54	54	54	54	54	54	54	54	54	54	54	68	54	54	54	68
		550S162	16	33	33	33	33	33	33	33	33	33	33	33	33	33	33	33	33
			24	33	33	33	33	33	33	33	43	33	33	43	43	33	33	43	43

For SI: 1 inch = 25.4 mm, 1 foot = 304.8 mm, 1 mil = 0.0254 mm, 1 mile per hour = 1.609 km/h, 1 pound per square foot = 0.0479 kN/m^2, 1 kilogram per square inch = 6.895 MPa.

a. Deflection criteria: *L*/240.

b. Building width is in the direction of horizontal framing members supported by the wall studs.

c. Design load assumptions:

Roof dead load is 12 psf.

Attic live load is 10 psf.

d. 68-mil-thick stud is allowed if wall is fully sheathed per Section R603.3.2.

TABLE R603.3.2(11)
COLD-FORMED STEEL STUD THICKNESS FOR 9-FOOT WALLS
Studs supporting one floor, roof and ceiling (first story of a two-story building) 50 ksi steel

WIND SPEED		MEMBER SIZE[c]	MEMBER SPACING (inches)	STUD THICKNESS (mils)[a,b]															
				Building width (feet)[d]															
				24				28				32				36			
				Ground snow load (psf)				Ground snow load (psf)				Ground snow load (psf)				Ground snow load (psf)			
Exp. A/B	Exp. C			20	30	50	70	20	30	50	70	20	30	50	70	20	30	50	70
85 mph	—	350S162	16	33	33	33	33	33	33	33	33	33	33	33	33	33	33	33	33
			24	33	33	33	43	33	43	43	43	43	43	43	43	43	43	43	43
		550S162	16	33	33	33	33	33	33	33	33	33	33	33	33	33	33	33	33
			24	33	33	33	33	33	33	33	33	33	33	33	33	33	33	33	43
100 mph	85 mph	350S162	16	33	33	33	33	33	33	33	33	33	33	33	33	33	33	33	33
			24	43	43	43	43	43	43	43	43	43	43	43	43	43	43	54	54
		550S162	16	33	33	33	33	33	33	33	33	33	33	33	33	33	33	33	33
			24	33	33	33	33	33	33	33	33	33	33	33	33	33	33	33	43
110 mph	100 mph	350S162	16	33	33	33	33	33	33	33	33	33	33	43	33	33	33	43	43
			24	43	43	43	43	43	43	43	54	43	54	54	54	54	54	54	54
		550S162	16	33	33	33	33	33	33	33	33	33	33	33	33	33	33	33	33
			24	33	33	33	33	33	33	33	33	33	33	33	33	33	33	33	43
120 mph	110 mph	350S162	16	33	33	33	43	33	33	43	43	43	43	43	43	43	43	43	43
			24	54	54	54	54	54	54	54	54	54	54	54	54	54	54	54	68
		550S162	16	33	33	33	33	33	33	33	33	33	33	33	33	33	33	33	33
			24	33	33	33	33	33	33	33	43	33	33	43	43	43	43	43	43
130 mph	120 mph	350S162	16	43	43	43	43	43	43	43	43	43	43	43	43	43	43	43	54
			24	54	54	68	68	68	68	68	68	68	68	68	68	68	68	68	68
		550S162	16	33	33	33	33	33	33	33	33	33	33	33	33	33	33	33	33
			24	43	43	43	43	43	43	43	43	43	43	43	43	43	43	43	43
—	130 mph	350S162	16	43	43	43	54	43	43	54	54	54	54	54	54	54	54	54	54
			24	68	68	68	68	68	68	68	68	68	68	(d)	(d)	68	(d)	(d)	(d)
		550S162	16	33	33	33	33	33	33	33	33	33	33	33	33	33	33	33	33
			24	43	43	43	43	43	43	43	43	43	43	43	43	43	43	43	54

For SI: 1 inch = 25.4 mm, 1 foot = 304.8 mm, 1 mil = 0.0254 mm, 1 mile per hour = 1.609 km/h, 1 pound per square foot = 0.0479 kN/m^2, 1 kilogram per square inch = 6.895 MPa.

a. Deflection criteria: *L*/240.

b. Building width is in the direction of horizontal framing members supported by the wall studs.

c. Design load assumptions:

Roof dead load is 12 psf.

Attic live load is 10 psf.

d. 68-mil-thick stud is allowed if wall is fully sheathed per Section R603.3.2.

TABLE R603.3.2(12)
COLD-FORMED STEEL STUD THICKNESS FOR 10-FOOT WALLS
Studs supporting roof and ceiling only (one-story building or second floor of a two-story building) 50 ksi steel

WIND SPEED		MEMBER SIZE[c]	MEMBER SPACING (inches)	STUD THICKNESS (mils)[a,b]															
				Building width (feet)[d]															
				24				28				32				36			
				Ground snow load (psf)				Ground snow load (psf)				Ground snow load (psf)				Ground snow load (psf)			
Exp. A/B	Exp. C			20	30	50	70	20	30	50	70	20	30	50	70	20	30	50	70
85 mph	—	350S162	16	33	33	33	33	33	33	33	33	33	33	33	33	33	33	33	33
			24	33	33	33	33	33	33	33	33	33	33	33	33	33	33	33	43
		550S162	16	33	33	33	33	33	33	33	33	33	33	33	33	33	33	33	33
			24	33	33	33	33	33	33	33	33	33	33	33	33	33	33	33	33
100 mph	85 mph	350S162	16	33	33	33	33	33	33	33	33	33	33	33	33	33	33	33	33
			24	33	33	33	43	33	33	33	43	33	33	43	43	33	33	43	43
		550S162	16	33	33	33	33	33	33	33	33	33	33	33	33	33	33	33	33
			24	33	33	33	33	33	33	33	33	33	33	33	33	33	33	33	33
110 mph	100 mph	350S162	16	33	33	33	33	33	33	33	33	33	33	33	33	33	33	33	33
			24	43	43	43	43	43	43	43	43	43	43	43	43	43	43	43	54
		550S162	16	33	33	33	33	33	33	33	33	33	33	33	33	33	33	33	33
			24	33	33	33	33	33	33	33	33	33	33	33	33	33	33	33	33
120 mph	110 mph	350S162	16	33	33	33	33	33	33	33	33	33	33	33	43	33	33	33	43
			24	43	43	54	54	43	43	54	54	43	54	54	54	54	54	54	54
		550S162	16	33	33	33	33	33	33	33	33	33	33	33	33	33	33	33	33
			24	33	33	33	33	33	33	33	33	33	33	33	33	33	33	33	33
130 mph	120 mph	350S162	16	43	43	43	43	43	43	43	43	43	43	43	43	43	43	43	43
			24	54	54	54	68	54	54	68	68	54	54	68	68	54	68	68	68
		550S162	16	33	33	33	33	33	33	33	33	33	33	33	33	33	33	33	33
			24	33	33	33	43	33	33	43	43	33	33	43	43	33	43	43	43
—	130 mph	350S162	16	43	43	54	54	43	43	54	54	43	54	54	54	54	54	54	54
			24	68	68	68	68	68	68	68	(d)	68	68	68	(d)	68	68	(d)	(d)
		550S162	16	33	33	33	33	33	33	33	33	33	33	33	33	33	33	33	33
			24	43	43	43	43	43	43	43	43	43	43	43	43	43	43	43	43

For SI: 1 inch = 25.4 mm, 1 foot = 304.8 mm, 1 mil = 0.0254 mm, 1 mile per hour = 1.609 km/h, 1 pound per square foot = 0.0479 kN/m^2, 1 kilogram per square inch = 6.895 MPa.

a. Deflection criteria: *L*/240.

b. Building width is in the direction of horizontal framing members supported by the wall studs.

c. Design load assumptions:

Roof dead load is 12 psf.

Attic live load is 10 psf.

d. 68-mil-thick stud is allowed if wall is fully sheathed per Section R603.3.2.

TABLE R603.3.2(13)
COLD-FORMED STEEL STUD THICKNESS FOR 10-FOOT WALLS
Studs supporting one floor, roof and ceiling (first story of a two-story building) 50 ksi steel

WIND SPEED		MEMBER SIZE[c]	MEMBER SPACING (inches)	STUD THICKNESS (mils)[a,b]															
				Building width (feet)[d]															
				24				28				32				36			
				Ground snow load (psf)				Ground snow load (psf)				Ground snow load (psf)				Ground snow load (psf)			
Exp. A/B	Exp. C			20	30	50	70	20	30	50	70	20	30	50	70	20	30	50	70
85 mph	—	350S162	16	33	33	33	33	33	33	33	33	33	33	33	33	33	33	33	43
			24	43	43	43	43	43	43	43	43	43	43	54	54	43	54	54	54
		550S162	16	33	33	33	33	33	33	33	33	33	33	33	33	33	33	33	33
			24	33	33	33	33	33	33	33	33	33	33	33	43	33	33	33	43
100 mph	85 mph	350S162	16	33	33	33	33	33	33	33	43	33	33	43	43	43	43	43	43
			24	43	43	54	54	54	54	54	54	54	54	54	54	54	54	54	54
		550S162	16	33	33	33	33	33	33	33	33	33	33	33	33	33	33	33	33
			24	33	33	33	33	33	33	33	33	33	33	33	43	33	33	43	43
110 mph	100 mph	350S162	16	33	33	43	43	43	43	43	43	43	43	43	43	43	43	43	43
			24	54	54	54	54	54	54	54	54	54	54	68	68	54	68	68	68
		550S162	16	33	33	33	33	33	33	33	33	33	33	33	33	33	33	33	33
			24	33	33	33	33	33	33	33	43	33	43	43	43	43	43	43	43
120 mph	110 mph	350S162	16	43	43	43	43	43	43	43	43	43	43	43	54	43	43	54	54
			24	54	54	68	68	68	68	68	68	68	68	68	68	68	68	68	(d)
		550S162	16	33	33	33	33	33	33	33	33	33	33	33	33	33	33	33	33
			24	43	43	43	43	43	43	43	43	43	43	43	43	43	43	43	43
130 mph	120 mph	350S162	16	54	54	54	54	54	54	54	54	54	54	54	54	54	54	54	54
			24	68	68	(d)	(d)	(d)	(d)	(d)	(d)	(d)	(d)	(d)	(d)	(d)	(d)	(d)	(d)
		550S162	16	33	33	33	33	33	33	33	33	33	33	33	33	33	33	33	43
			24	43	43	43	43	43	43	43	43	43	43	43	54	43	43	54	54
—	130 mph	350S162	16	54	54	54	54	54	54	54	54	54	54	68	68	54	68	68	68
			24	(d)	(d)	(d)	(d)	(d)	(d)	(d)	(d)	(d)	(d)	(d)	(d)	(d)	(d)	(d)	(d)
		550S162	16	33	33	33	33	33	33	43	43	33	43	43	43	43	43	43	43
			24	43	43	54	54	54	54	54	54	54	54	54	54	54	54	54	54

For SI: 1 inch = 25.4 mm, 1 foot = 304.8 mm, 1 mil = 0.0254 mm, 1 mile per hour = 1.609 km/h, 1 pound per square foot = 0.0479 kN/m², 1 kilogram per square inch = 6.895 MPa.

a. Deflection criteria: *L*/240.

b. Building width is in the direction of horizontal framing members supported by the wall studs.

c. Design load assumptions:
Roof dead load is 12 psf.
Attic live load is 10 psf.

d. 68-mil-thick stud is allowed if wall is fully sheathed per Section R603.3.2.

R603.3.3 Stud bracing. The flanges of steel studs shall be laterally braced in accordance with one of the following:

1. Gypsum board installed with minimum No. 6 screws in accordance with Section R702 or structural sheathing installed in accordance with Table R603.3.2.
2. Horizontal steel strapping installed in accordance with Figure R603.3 at mid-height for 8-foot (2438 mm) walls, and one-third points for 9-foot and 10-foot (2743 mm and 3048 mm) walls. Steel straps shall be at least 1.5 inches in width and 33 mils in thickness (38 mm by 0.836 mm). Straps shall be attached to the flanges of studs with at least one No. 8 screw. In-line blocking shall be installed between studs at the termination of all straps. Straps shall be fastened to the blocking with at least two No. 8 screws.

R603.3.4 Cutting and notching. Flanges and lips of steel studs and headers shall not be cut or notched.

R603.3.5 Hole patching. Stud web holes with dimensions conforming to Section R603.2 that are closer than 10 inches (305 mm) from the edge of the hole to the end of the member shall be patched with a solid steel plate, C-section or track section in accordance with Figure R603.3.5. The patch shall be of a minimum thickness as the stud member and shall extend at least 1 inch (25.4 mm) beyond all edges of the hole.

The patch shall be fastened to the web with No. 8 screws (minimum) spaced not greater than 1 inch (25.4 mm) center to center along the edges of the patch, with a minimum edge distance of $^1/_2$ inch (12.7 mm).

R603.3.6 Splicing. Steel studs and other structural members shall not be spliced. Tracks shall be spliced in accordance with Figure R603.3.6.

R603.4 Corner framing. Corner studs and the top tracks shall be installed in accordance with Figure R603.4.

R603.5 Exterior wall covering. The method of attachment of exterior wall covering materials to cold-formed steel stud wall framing shall conform to the manufacturer's installation instructions.

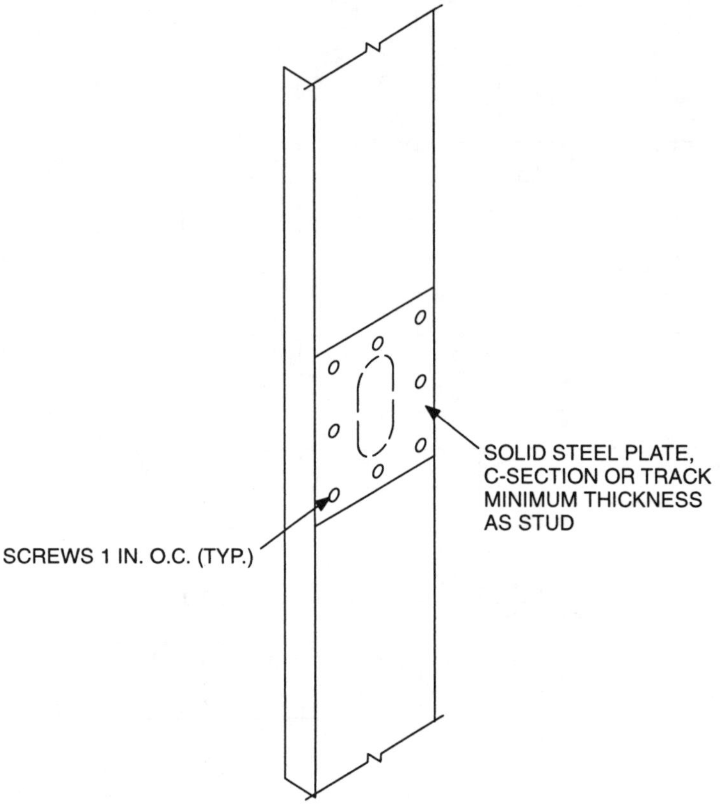

For SI: 1 inch = 25.4 mm.

FIGURE R603.3.5
HOLE PATCH

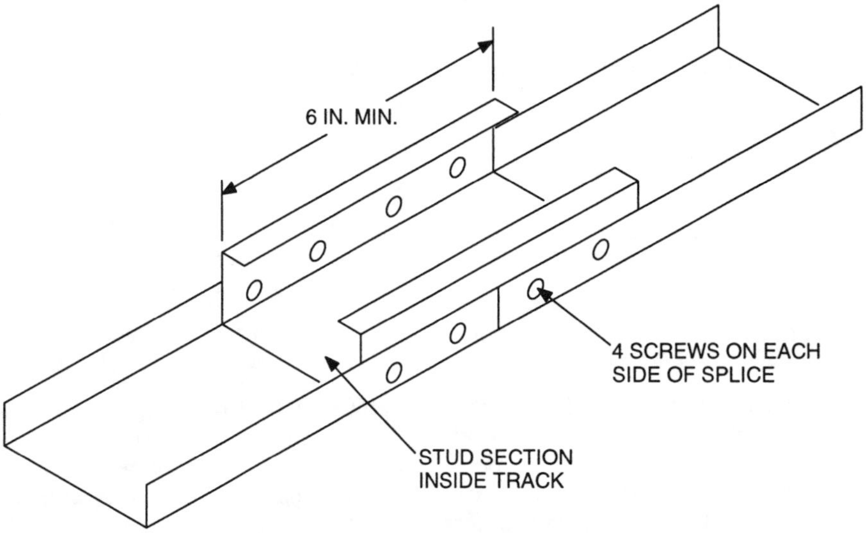

For SI: 1 inch = 25.4 mm.

FIGURE R603.3.6
TRACK SPLICE

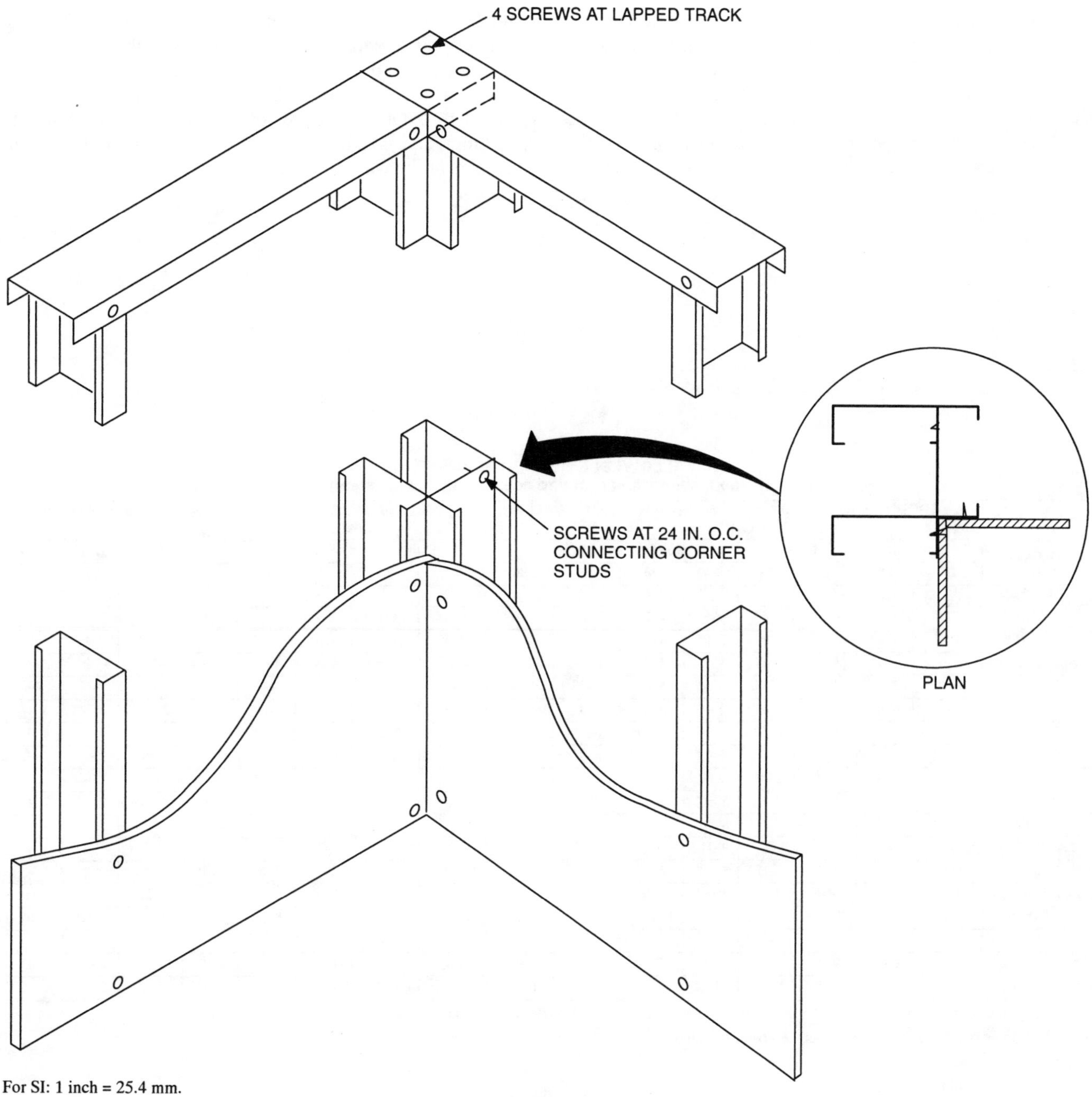

For SI: 1 inch = 25.4 mm.

FIGURE R603.4
CORNER FRAMING

R603.6 Headers. Headers shall be installed above wall openings in all exterior walls and interior load-bearing walls in accordance with Figure R603.6 and Tables R603.6(1), R603.6(2), and R603.6(3). The number of jack and king studs shall comply with Table R603.6(4). King and jack studs shall be of the same dimension and thickness as the adjacent wall studs. Headers shall be connected to king studs in accordance with Table R603.6(5). One-half of the total number of screws shall be applied to the header and one-half to the king stud by use of a minimum 2-inch by 2-inch (51 mm by 51 mm) clip angle or 4-inch (102 mm) wide steel plate. The clip angle or plate shall extend the depth of the header minus 1/2 inch (12.7 mm) and shall have a minimum thickness of the header members or the wall studs, whichever is thicker.

TABLE R603.6(1) ALLOWABLE HEADER SPANS[a,b] Headers supporting roof and ceiling only (33 ksi steel)

NOMINAL MEMBER	GROUND SNOW LOAD (20 psf)				GROUND SNOW LOAD (30 psf)				GROUND SNOW LOAD (50 psf)				GROUND SNOW LOAD (70 psf)			
	Building width[c] (feet)				Building width[c] (feet)				Building width[c] (feet)				Building width[c] (feet)			
	24	28	32	36	24	28	32	36	24	28	32	36	24	28	32	36
2-350S162-33	3'-11"	3'-8"	3'-5"	3'-3"	3'-8"	3'-5"	3'-2"	2'-10"	3'-0"	2'-7"	2'-4"	2'-1"	2'-4"	2'-1"	—	—
2-350S162-43	4'-9"	4'-5"	4'-2"	4'-0"	4'-5"	4'-2"	3'-11"	3'-9"	3'-10"	3'-7"	3'-4"	3'-2"	3'-5"	3'-2"	3'-0"	2'-9"
2-350S162-54	5'-4"	5'-0"	4'-9"	4'-6"	5'-0"	4'-8"	4'-5"	4'-2"	4'-3"	4'-0"	3'-9"	3'-7"	3'-10"	3'-7"	3'-4"	3'-2"
2-350S162-68	6'-0"	5'-7"	5'-3"	5'-0"	5'-7"	5'-3"	4'-11"	4'-8"	4'-10"	4'-6"	4'-3"	4'-0"	4'-3"	4'-0"	3'-9"	3'-7"
2-550S162-33	3'-11"	3'-5"	3'-0"	2'-9"	3'-5"	3'-0"	2'-8"	2'-5"	2'-6"	2'-2"	—	—	—	—	—	—
2-550S162-43	6'-5"	6'-0"	5'-8"	5'-5"	6'-0"	5'-8"	5'-4"	5'-0"	5'-2"	4'-10"	4'-4"	3'-11"	4'-5"	3'-10"	3'-5"	3'-1"
2-550S162-54	7'-3"	6'-10"	6'-5"	6'-1"	6'-9"	6'-4"	6'-0"	5'-8"	5'-10"	5'-5"	5'-1"	4'-10"	5'-2"	4'-10"	4'-7"	4'-4"
2-550S162-68	8'-2"	7'-8"	7'-2"	6'-11"	7'-7"	7'-2"	6'-9"	6'-4"	6'-6"	6'-1"	5'-9"	5'-6"	5'-10"	5'-5"	5'-1"	4'-10"
2-800S162-33	3'-0"	2'-8"	2'-4"	2'-1"	2'-7"	2'-3"	—	—	—	—	—	—	—	—	—	—
2-800S162-43	6'-8"	5'-10"	5'-2"	4'-8"	5'-10"	5'-1"	4'-6"	4'-1"	4'-3"	3'-9"	3'-4"	3'-0"	3'-4"	2'-11"	2'-7"	2'-4"
2-800S162-54	9'-6"	8'-10"	8'-4"	7'-11"	8'-10"	8'-3"	7'-9"	7'-5"	7'-7"	7'-1"	6'-7"	5'-11"	6'-9"	5'-10"	5'-3"	4'-8"
2-800S162-68	10'-8"	10'-0"	9'-5"	8'-11"	9'-11"	9'-4"	8'-9"	8'-4"	8'-6"	8'-0"	7'-6"	7'-2"	7'-7"	7'-1"	6'-8"	6'-4"
2-1000S162-43	5'-7"	4'-10"	4'-4"	3'-11"	4'-10"	4'-3"	3'-9"	3'-5"	3'-7"	3'-1"	2'-9"	2'-6"	2'-10"	2'-6"	2'-2"	—
2-1000S162-54	10'-6"	9'-8"	8'-7"	7'-9"	9'-8"	8'-5"	7'-6"	6'-9"	7'-1"	6'-2"	5'-6"	4'-11"	5'-7"	4'-11"	4'-4"	3'-11"
2-1000S162-68	12'-7"	11'-9"	11'-1"	10'-6"	11'-9"	10'-12"	10'-4"	9'-10"	10'-1"	9'-5"	8'-10"	8'-5"	8'-11"	8'-4"	7'-11"	7'-6"
2-1200S162-43	—	—	—	—	—	—	—	—	—	—	—	—	—	—	—	—
2-1200S162-54	9'-6"	8'-3"	7'-4"	6'-7"	8'-3"	7'-3"	6'-5"	5'-9"	6'-1"	5'-4"	4'-9"	4'-3"	4'-10"	4'-2"	3'-9"	3'-4"
2-1200S162-68	13'-5"	12'-7"	11'-10"	11'-3"	12'-6"	11'-9"	11'-1"	10'-6"	10'-9"	10'-1"	9'-6"	8'-6"	9'-7"	8'-5"	7'-6"	6'-9"

For SI: 1 inch = 25.4 mm, 1 foot = 304.8 mm, 1 pound per square foot = 0.0479 kN/m^2, 1 kilogram per square inch = 6.895 MPa.

a. Deflection criteria: *L*/360 for live loads, *L*/240 for total loads.

b. Design load assumptions:

Roof dead load is 7 psf.

Ceiling dead load is 5 psf.

Attic live load is 10 psf.

c. Building width is in the direction of horizontal framing members supported by the header.

TABLE R603.6(2)
ALLOWABLE HEADER SPANS[a,b]
Headers supporting one floor, roof and ceiling (33 ksi steel)

NOMINAL MEMBER	GROUND SNOW LOAD (20 psf)				GROUND SNOW LOAD (30 psf)				GROUND SNOW LOAD (50 psf)				GROUND SNOW LOAD (70 psf)			
	Building width[c] (feet)				Building width[c] (feet)				Building width[c] (feet)				Building width[c] (feet)			
	24	28	32	36	24	28	32	36	24	28	32	36	24	28	32	36
2-350S162-42	3′-3″	3′-1″	2′-11″	2′-8″	3′-2″	3′-0″	2′-9″	2′-6″	2′-11″	2′-8″	2′-4″	2′-2″	2′-10″	2′-6″	2′-3″	2′-1″
2-350S162-54	3′-8″	3′-6″	3′-3″	3′-2″	3′-6″	3′-4″	3′-2″	3′-0″	3′-4″	3′-1″	2′-11″	2′-9″	3′-3″	3′-0″	2′-10″	2′-9″
2-350S162-68	4′-2″	3′-11″	3′-8″	3′-6″	4′-0″	3′-9″	3′-6″	3′-4″	3′-8″	3′-6″	3′-3″	3′-1″	3′-7″	3′-5″	3′-2″	3′-1″
2-550S162-43	4′-1″	3′-8″	3′-3″	3′-0″	3′-10″	3′-4″	3′-0″	2′-9″	3′-3″	2′-11″	2′-7″	2′-4″	3′-2″	2′-9″	2′-6″	2′-3″
2-550S162-54	5′-0″	4′-9″	4′-6″	4′-3″	4′-10″	4′-6″	4′-3″	4′-1″	4′-6″	4′-2″	4′-0″	3′-9″	4′-4″	4′-1″	3′-11″	3′-8″
2-550S162-68	5′-7″	5′-4″	5′-0″	4′-9″	5′-5″	5′-1″	4′-10″	4′-7″	5′-0″	4′-9″	4′-6″	4′-3″	4′-11″	4′-7″	4′-4″	4′-2″
2-800S162-43	3′-2″	2′-10″	2′-6″	2′-3″	2′-11″	2′-7″	2′-4″	—	2′-6″	2′-3″	—	—	2′-5″	2′-2″	—	—
2-800S162-54	6′-3″	5′-7″	5′-0″	4′-6″	5′-10″	5′-2″	4′-7″	4′-2″	5′-0″	4′-5″	4′-0″	3′-7″	4′-10″	4′-3″	3′-10″	3′-5″
2-800S162-68	7′-4″	6′-11″	6′-7″	6′-3″	7′-1″	6′-8″	6′-3″	6′-0″	6′-7″	6′-2″	5′-10″	5′-7″	6′-5″	6′-0″	5′-8″	5′-5″
2-1000S162-43	2′-8″	2′-4″	—	—	2′-5″	—	—	—	—	—	—	—	—	—	—	—
2-1000S162-54	5′-3″	4′-8″	4′-2″	3′-9″	4′-10″	4′-3″	3′-10″	3′-6″	4′-2″	3′-8″	3′-4″	3′-0″	4′-0″	3′-6″	3′-2″	2′-10″
2-1000S162-68	8′-8″	8′-2″	7′-9″	7′-4″	8′-4″	7′-10″	7′-5″	7′-0″	7′-9″	7′-3″	6′-8″	6′-0″	7′-7″	7′-1″	6′-5″	5′-9″
2-1000S162-54	4′-6″	4′-0″	3′-7″	3′-3″	4′-2″	3′-8″	3′-3″	3′-0″	3′-7″	3′-2″	2′-10″	2′-7″	3′-5″	3′-0″	2′-9″	2′-5″
2-1000S162-68	9′-1″	8′-0″	7′-2″	6′-6″	8′-4″	7′-5″	6′-8″	6′-0″	7′-3″	6′-5″	5′-9″	5′-2″	6′-11″	6′-1″	5′-6″	4′-11″

For SI: 1 inch = 25.4 mm, 1 foot = 304.8 mm, 1 pound per square foot = 0.0479 kN/m^2, 1 kilogram per square inch = 6.895 MPa.

a. Deflection criteria: *L*/360 for live loads, *L*/240 for total loads.

b. Design load assumptions:
Roof dead load is 7 psf.
Ceiling dead load is 5 psf.
Attic live load is 10 psf.
Second floor live load is 30 psf.
Second floor dead load is 10 psf.
Second floor wall dead load is 10 psf.

c. Building width is in the direction of horizontal framing members supported by the header.

TABLE R603.6(3)
ALLOWABLE HEADER SPANS[a,b,c]
Headers supporting one floor, roof and ceiling first story of a two-story building with center load bearing beam (33 ksi steel)

NOMINAL MEMBER	GROUND SNOW LOAD (20 psf)				GROUND SNOW LOAD (30 psf)				GROUND SNOW LOAD (50 psf)				GROUND SNOW LOAD (70 psf)			
	Building width[c] (feet)				Building width[c] (feet)				Building width[c] (feet)				Building width[c] (feet)			
	24	28	32	36	24	28	32	36	24	28	32	36	24	28	32	36
2-350S162-33	2′-10″	2′-6″	2′-3″	—	2′-7″	2′-3″	—	—	2′-2″	—	—	—	—	—	—	—
2-350S162-43	3′-9″	3′-6″	3′-4″	3′-2″	3′-7″	3′-4″	3′-2″	3′-0″	3′-3″	3′-1″	2′-11″	2′-8″	3′-0″	2′-9″	2′-6″	2′-3″
2-350S162-54	4′-2″	4′-0″	3′-9″	3′-7″	4′-0″	3′-9″	3′-7″	3′-5″	3′-8″	3′-5″	3′-3″	3′-1″	3′-5″	3′-2″	3′-0″	2′-10″
2-350S162-68	4′-8″	4′-5″	4′-2″	4′-0″	4′-6″	4′-2″	4′-0″	3′-10″	4′-1″	3′-10″	3′-8″	3′-6″	3′-9″	3′-7″	3′-4″	3′-2″
2-550S162-33	2′-5″	2′-2″	—	—	2′-2″	—	—	—	—	—	—	—	—	—	—	—
2-550S162-43	5′-1″	4′-9″	4′-3″	3′-10″	4′-10″	4′-3″	3′-10″	3′-6″	4′-0″	3′-7″	3′-2″	2′-11″	3′-5″	3′-1″	2′-9″	2′-6″
2-550S162-54	5′-8″	5′-4″	5′-1″	4′-10″	5′-5″	5′-1″	4′-10″	4′-7″	4′-11″	4′-8″	4′-5″	4′-2″	4′-7″	4′-4″	4′-1″	3′-11″
2-550S162-68	6′-5″	6′-0″	5′-9″	5′-6″	6′-1″	5′-9″	5′-5″	5′-2″	5′-7″	5′-3″	4′-11″	4′-9″	5′-2″	4′-10″	4′-7″	4′-4″
2-800S162-33	4′-2″	3′-8″	3′-3″	3′-0″	3′-8″	3′-3″	2′-11″	2′-8″	3′-1″	2′-9″	2′-5″	2′-3″	2′-8″	2′-4″	—	—
2-800S162-43	7′-5″	7′-0″	6′-6″	6′-0″	7′-1″	6′-6″	5′-10″	5′-4″	6′-2″	5′-5″	4′-11″	4′-5″	5′-3″	4′-8″	4′-2″	3′-9″
2-800S162-54	8′-4″	7′-11″	7′-6″	7′-1″	7′-11″	7′-6″	7′-1″	6′-9″	7′-3″	6′-10″	6′-6″	6′-2″	6′-9″	6′-4″	6′-0″	5′-8″
2-800S162-68	10′-0″	9′-9″	9′-0″	8′-6″	9′-6″	9′-0″	8′-6″	8′-1″	8′-8″	8′-2″	7′-9″	7′-5″	8′-1″	7′-7″	7′-2″	6′-10″
2-1000S162-43	3′-5″	3′-0″	2′-9″	2′-6″	3′-1″	2′-9″	2′-6″	2′-3″	2′-7″	2′-3″	2′-1″	—	2′-3″	—	—	—
2-1000S162-54	6′-10″	6′-0″	5′-6″	5′-0″	6′-2″	5′-5″	4′-11″	4′-5″	5′-1″	4′-6″	4′-1″	3′-8″	4′-5″	3′-11″	3′-6″	3′-2″
2-1000S162-68	9′-10″	9′-4″	8′-10″	8′-5″	9′-4″	8′-10″	8′-4″	8′-0″	8′-7″	8′-1″	7′-8″	7′-3″	7′-	7′-6″	7′-0″	6′-4″
2-1200S162-43	2′-11″	2′-7″	2′-4″	2′-1″	2′-8″	2′-4″	—	—	2′-3″	—	—	—	—	—	—	—
2-1200S162-54	5′-10″	5′-2″	4′-8″	4′-3″	5′-3″	4′-8″	4′-2″	3′-10″	4′-5″	3′-11″	3′-6″	3′-2″	3′-9″	3′-4″	3′-0″	2′-8″
2-1200S162-68	10′-6″	10′-0″	9′-5″	8′-6″	9′-12″	9′-5″	8′-5″	7′-8″	8′-10″	7′-10″	7′-0″	6′-4″	7′-7″	6′-9″	6′-0″	5′-5″

For SI: 1 inch = 25.4 mm, 1 foot = 304.8 mm, 1 pound per square foot = 0.0479 kN/m^2, 1 kilogram per square inch = 6.895 MPa.

a. Deflection criteria: *L*/360 for live loads, *L*/240 for total loads.

b. Design load assumptions:
Roof dead load is 7 psf.
Ceiling dead load is 5 psf.
Attic live load is 10 psf.
Second floor live load is 30 psf.
Second floor dead load is 10 psf.
Second floor wall dead load is 10 psf.

c. Building width is in the direction of horizontal framing members supported by the header.

**TABLE R603.6(4)
TOTAL NUMBER OF JACK AND KING STUDS REQUIRED
AT EACH END OF AN OPENING**

SIZE OF OPENING (feet-inches)	24″ O.C. STUD SPACING		16″ O.C. STUD SPACING	
	No. of jack studs	No. of king studs	No. of jack studs	No. of king studs
Up to 3′-6″	1	1	1	1
>3′-6″ to 5′-0″	1	2	1	2
>5′-0″ to 5′-6″	1	2	2	2
>5′-6″ to 8′-0″	1	2	2	2
>8′-0″ to 10′-6″	2	2	2	3
>10′-6″ to 12′-0″	2	2	3	3
>12′-0″ to 13′-0″	2	3	3	3
>13′-0″ to 14′-0″	2	3	3	4

For SI: 1 inch = 25.4 mm, 1 foot = 304.8 mm.

**TABLE R603.6(5)
HEADER TO KING STUD CONNECTION REQUIREMENTS[a,b,c,d,e]**

HEADER SPAN (feet)	BASIC WIND SPEED (mph), EXPOSURE		
	85 A/B or Seismic Design Categories A, B, C and D	85 C or less than 110 A/B	Less than 110 C
≤ 4′	4-No. 8 screws	4-No. 8 screws	6-No. 8 screws
> 4′ to 8′	4-No. 8 screws	4-No. 8 screws	8-No. 8 screws
> 8′ to 12′	4-No. 8 screws	6-No. 8 screws	10-No. 8 screws
> 12′ to 16′	4-No. 8 screws	8-No. 8 screws	12-No. 8 screws

For SI: 1 inch = 25.4 mm, 1 foot = 304.8 mm, 1 mile per hour = 1.609 km/h, 1 pound = 0.454 kg.

a. All screw sizes shown are minimum.

b. For headers located on the first floor of a two-story building, the total number of screws may be reduced by two screws, but the total number of screws shall be no less than four.

c. For roof slopes of 6:12 or greater, the required number of screws may be reduced by half, but the total number of screws shall be no less than four.

d. Screws can be replaced by a up-lift connector which has a capacity of the number of screws multiplied by 164 pounds (e.g., 12-No. 8 screws can be replaced by an up-lift connector whose capacity exceeds 12 by 164 pounds = 1,968 pounds)

e. In Seismic Design Category D_1 and D_2 or where the basic wind speed equals or exceeds 110 miles per hour, connection shall comply with the requirements in Section R603.8, but shall be no less than the minimum required herein.

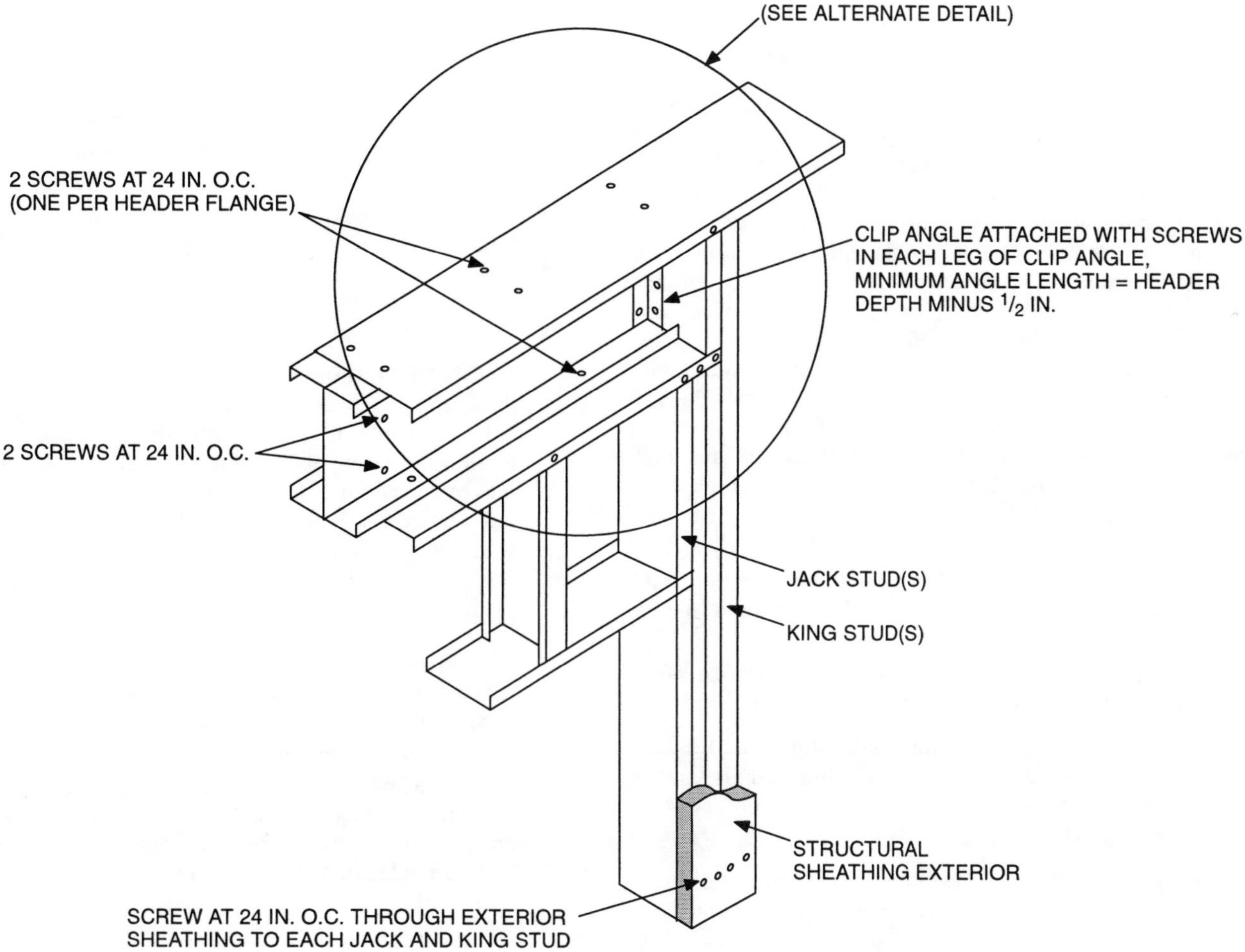

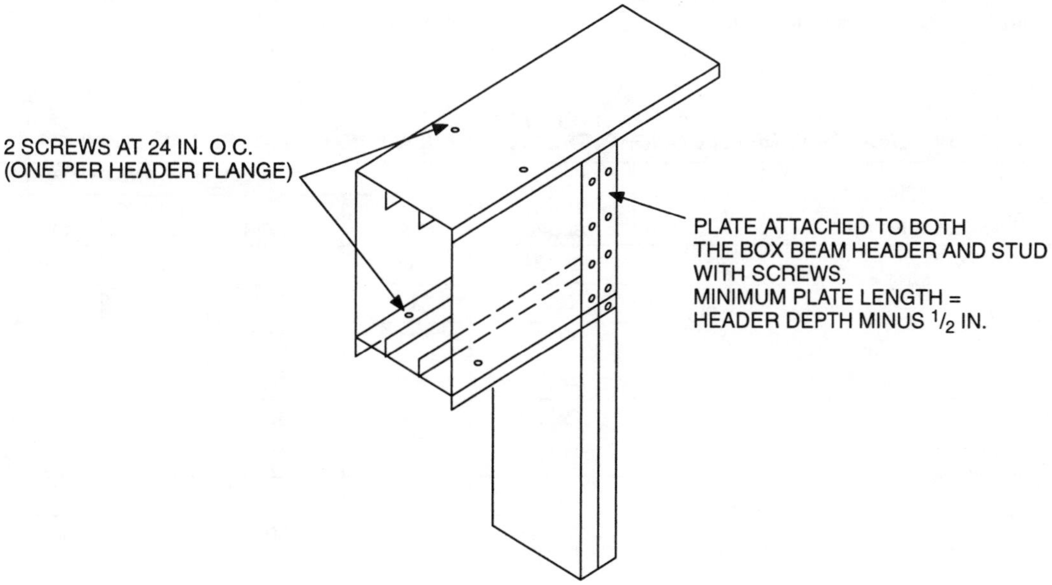

ALTERNATE DETAIL
(BOX BEAM HEADER)

For SI: 1 inch = 25.4 mm.

FIGURE R603.6
HEADER DETAIL

R603.7 Structural sheathing. In areas where the basic wind speed is less than 110 miles per hour (177 km/h), wood structural sheathing panels shall be installed on all exterior walls of buildings in accordance with this section. Wood structural sheathing panels shall consist of minimum $^{7}/_{16}$-inch (11.1 mm) thick oriented strand board or $^{15}/_{32}$-inch (11.9 mm) thick plywood and shall be installed on all exterior wall surfaces in accordance with Section R603.7.1 and Figure R603.3. The minimum length of full height sheathing on exterior walls shall be determined in accordance with Table R603.7, but shall not be less than 20 percent of the braced wall length in any case. The minimum percentage of full height sheathing in Table R603.7 shall include only those sheathed wall sections, uninterrupted by openings, which are a minimum of 48 inches (1120 mm) wide. The minimum percentage of full-height structural sheathing shall be multiplied by 1.10 for 9-foot (2743 mm) high walls and multiplied by 1.20 for 10-foot (3048 mm) high walls. In addition, structural sheathing shall:

1. Be installed with the long dimension parallel to the stud framing and shall cover the full vertical height of studs, from the bottom of the bottom track to the top of the top track of each story.
2. Be applied to each end (corners) of each of the exterior walls with a minimum 48-inch-wide (1220 mm) panel.

R603.7.1 Structural sheathing fastening. All edges and interior areas of wood, structural sheathing panels shall be fastened to a framing member and tracks in accordance with Table R603.3.2(1).

R603.7.2 Hold-down requirements. The percent of structural sheathing, required in Table R603.7, is permitted to be multiplied by 0.6 where a hold down anchor with a capacity of 4,300 pounds (1952.2 kg) is provided at each end of exterior walls. A single hold down anchors is permitted to be installed at wall corners per Figure R603.8.1.3.

R603.8 Braced walls and diaphragms in high seismic and high wind regions. Braced walls and diaphragms in high seismic and high wind regions shall comply with this section.

R603.8.1 General. Buildings in Seismic Design Categories D_1 and D_2 and buildings in areas where the basic wind speed equals or exceeds 110 miles per hour (177 km/h) shall have braced wall lines and diaphragms constructed in accordance with the additional provisions of Section R603.8. Where a building, or portion thereof, does not comply with the provisions of this section, those portions shall be designed and constructed in compliance with Section R301.

R603.8.1.1 Building configuration. Floor and roof diaphragm plan aspect ratios shall not exceed four to one (4:1). The diaphragm aspect ratio shall be determined by dividing the length of the diaphragm between braced wall lines by the width of the diaphragm between braced wall lines. Braced wall lines shall be placed on all exterior walls and on interior walls as required.

Floor and roof diaphragm plan offsets shall not exceed 4 feet (1219 mm).

Exception: Buildings where diaphragm plan offsets exceed 4 feet (1219 mm), buildings shall be analyzed as separate buildings, separated by a braced wall line or lines (see Figure R603.8.1).

Where a braced wall line separates different portions of a building, the required length of braced wall panels separating the two portions shall be determined by summing the required lengths of braced wall panels for each portion of the building.

Vertical offsets in floor and roof diaphragms shall be supported by braced wall lines. Braced wall lines shall be in a single vertical plane from the foundation to the uppermost story in which they are required.

TABLE R603.7
MINIMUM PERCENTAGE OF FULL HEIGHT STRUCTURAL SHEATHING ON EXTERIOR WALLS[a,b,c,d,e]

WALL SUPPORTING	ROOF SLOPE	WIND SPEED (mph) AND EXPOSURE				
		85 A/B	100 A/B	110 A/B or 85 C	100 C	110 C
Roof and ceiling only	3:12	8	9	12	16	20
	6:12	12	15	20	26	35
	9:12	21	25	30	50	58
	12:12	30	35	40	66	75
One story, roof and ceiling	3:12	24	30	35	50	66
	6:12	25	30	40	58	74
	9:12	35	40	55	74	91
	12:12	40	50	65	100	115

For SI: 1 mile per hour = 1.609 km/h.

a. Linear interpolation shall be permitted.

b. Bracing amount shall not be less than 20 percent of the wall length after all applicable adjustments are made.

c. Minimum percentages are based on a building aspect ratio of 1:1. Minimum percentages for the shorter walls of a building shall be multiplied by a factor of 1.5 and 2.0 for building aspect ratios of 1.5:1 and 2:1 respectively.

d. For hip roofed homes with continuous structural sheathing, the amount of bracing shall be permitted to be multiplied by a factor of 0.95 for roof slopes not exceeding 7:12 and a factor of 0.9 for roof slopes greater than 7:12.

e. Sheathing percentage are permitted to be reduced in accordance with Section R603.7.2.

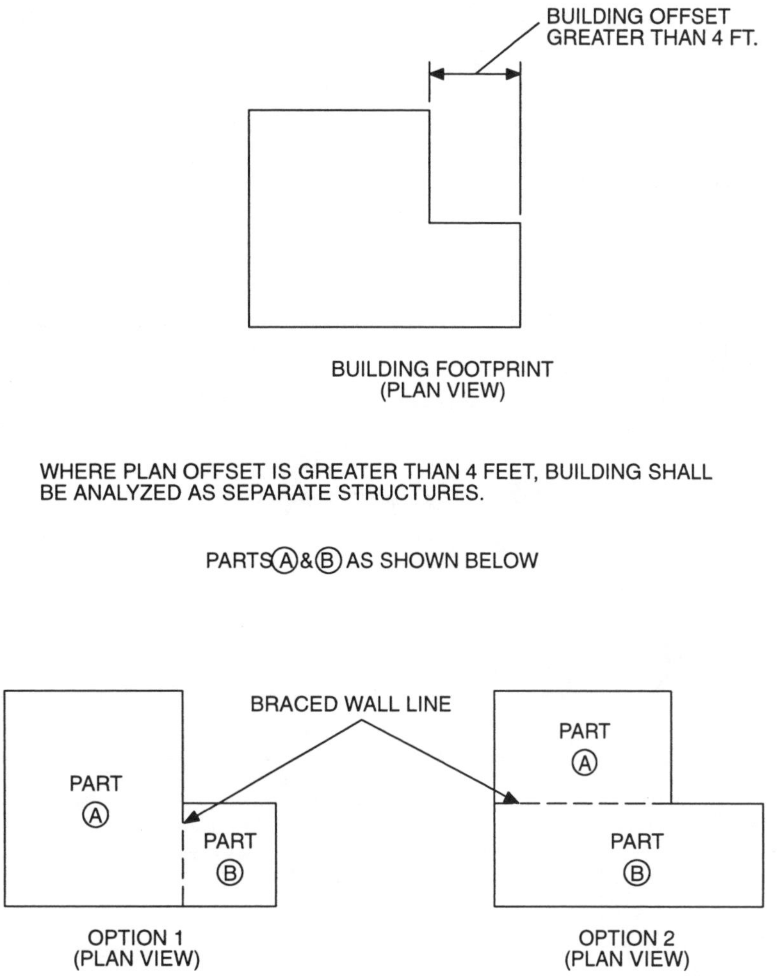

For SI: 1 foot = 304.8 mm.

FIGURE R603.8.1
BUILDING CONFIGURATION

R603.8.1.2 Braced wall lines. Braced wall lines shall be composed of either Type I braced wall panels or Type II (perforated) braced walls as shown in Figure R603.8.1.2. The required length of bracing shall be determined in accordance with the greater requirement for seismic or wind. Each braced wall line shall have not less than two full height braced wall panels, each meeting the minimum height to width aspect ratios in Sections R603.8.2 or R603.8.3. Braced wall panels shall begin not more than 8 feet (2438 mm) from each end.

Sheathing on Type I and Type II braced walls shall have wood structural sheathing panels one side and $^{1}/_{2}$-inch (12.7 mm) gypsum wallboard on the other. Wood structural sheathing panels shall be as specified in Section R603.7 and shall be attached to framing members with minimum No. 8 screws spaced as required by Table R603.8.1.2 on all edges and 12 inches (305 mm) in the field. Gypsum wallboard shall be attached as required by Table R702.3.4. In braced wall lines, all edges of wood structural sheathing panels shall be attached to framing members or 2-inch-by-33-mil (51 mm by 0.84 mm) horizontal strapping. Where horizontal strapping is used to attach adjacent sheets of sheathing material, solid blocking as shown in Figure R603.3 shall not be required.

TABLE R603.8.1.2
EDGE SCREW SPACING SHEARWALL LENGTH ADJUSTMENT FACTORS

EDGE SCREW SPACING (inches)			
6	4	3	2
Shearwall length adjustment factors			
1.00	0.72	0.60	0.56

For SI: 1 inch = 25.4 mm.

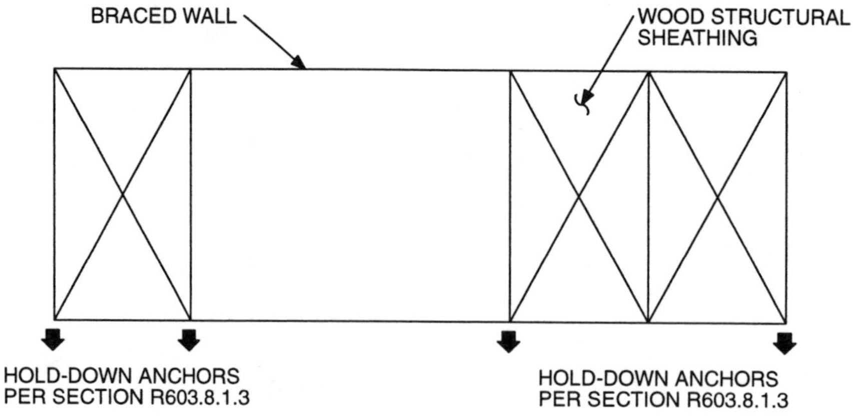

TYPE I BRACED WALL PANEL

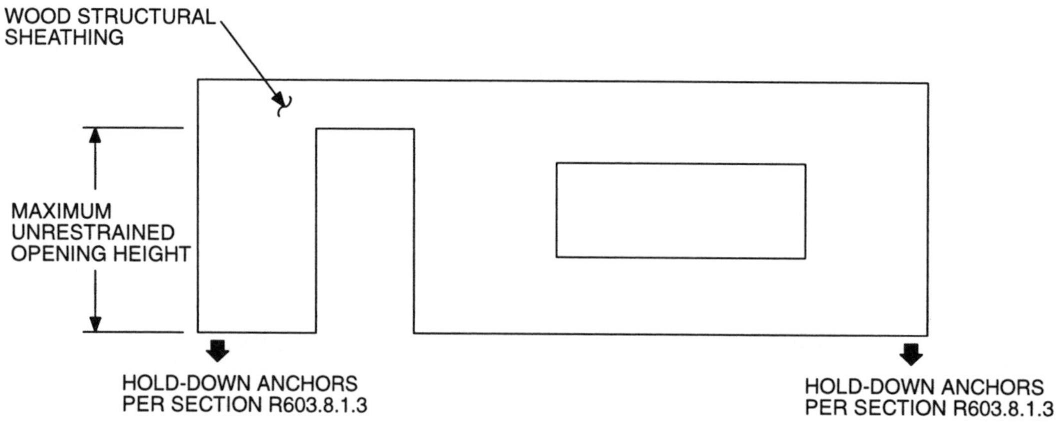

TYPE II BRACED WALL

FIGURE R603.8.1.2
TYPE I AND TYPE II BRACED WALLS

R603.8.1.2.1 Type I braced wall panels. Type I braced wall panels shall have no openings and shall be continuous between hold-down anchors. Hold-down anchors shall be as required by Section R603.8.1.3. The required length of Type I braced wall panels shall be determined by Section R603.8.2 for seismic applications and Section R603.8.3 for wind applications. The required length is permitted to be adjusted by the edge screw spacing adjustment factors in Table R603.8.1.2. The total length of Type I braced wall panels on a braced wall line shall be the sum of the lengths of panels conforming to the minimum height to width aspect ratio herein.

R603.8.1.2.2 Type II braced wall lines. Type II braced wall lines are fully sheathed and are permitted to have openings. Where sheathing does not extend above and below window openings and above door openings, the height of unrestrained opening shall be defined as the full height of the wall.

The required length of full height sheathing panels in Type II braced wall lines shall be determined by multiplying the required length of Type I braced wall panels, including adjustments permitted by Tables R603.8.1.2 and R603.8.2.2, by the length adjustment factors in Table R603.8.1.2.2. The length adjustment factors in Table R603.8.1.2.2 are permitted only in seismic applications. For a segment to count toward the required length, it shall have a height to width ratio of two to one (2:1). Type II braced wall lines shall have hold-down anchors as required by Section R603.8.1.3 at each end, attached to wall segments meeting the minimum height-to-width ratio.

TABLE R603.8.1.2.2
TYPE II BRACED WALL LINE LENGTH ADJUSTMENT FACTORS

	MAXIMUM UNRESTRAINED OPENING HEIGHT (feet)					
Wall height H (feet)	1/3H	1/2H	2/3H	3/4H	5/6H	H
8	2.67	4.00	5.33	5.93	6.67	8.00
9	3.00	4.50	6.00	6.67	7.50	9.00
10	3.33	5.00	6.67	7.41	8.33	10.00
Percent fully sheathed wall	Type II braced wall line length adjustment factors					
0	1.00	1.50	2.00	2.22	2.50	3.00
20	1.00	1.36	1.67	1.79	1.92	2.14
40	1.00	1.25	1.43	1.49	1.56	1.67
60	1.00	1.15	1.25	1.28	1.32	1.36
80	1.00	1.07	1.11	1.12	1.14	1.15
100	1.00	1.00	1.00	1.00	1.00	1.00

For SI: 1 foot = 304.8 mm.

R603.8.1.3 Braced wall hold-down anchorage and support. Hold-down anchors shall be installed at each end of Type I braced wall panels and at each end of Type II braced wall lines. The required hold-down force shall be determined by Table R603.8.1.3. Hold-down anchors shall attach to the foundation or to framing members below. Where hold-down anchors are attached to framing members below, the required anchor force shall be transferred to the foundation. Where hold-down anchors from an upper story align with those in the lower story, the required lower-story anchor force shall be determined by summing the upper-story anchor force and the lower-story anchor force. Hold-down anchors are permitted to be attached to blocking members provided the required hold-down forces are transferred to structural framing members. Hold-down anchors shall be attached to a minimum of two back-to-back, 33 mil (0.84 mm) studs in accordance with the anchor manufacturer's instructions.

A single hold-down anchor installed in accordance with Figure R603.8.1.3 is permitted at the corners of buildings.

Hold-down anchors shall be fastened to studs that have the required sheathing edge fastening. A minimum of two studs shall support end studs of Type I braced wall panel above. These supporting studs shall be in addition to load bearing studs required at the level below.

TABLE R603.8.1.3
REQUIRED HOLD-DOWN ANCHOR FORCE

Wall height (feet)	REQUIRED HOLD-DOWN ANCHOR FORCE (pounds)			
	Panel edge screw spacing (inches)			
	6	4	3	2
8	3440	4760	5720	6120
9	3870	5355	6435	6885
10	4300	5950	7150	7650

For SI: 1 inch = 25.4 mm, 1 foot = 304.8 mm, 1 pound = 4.4 N.

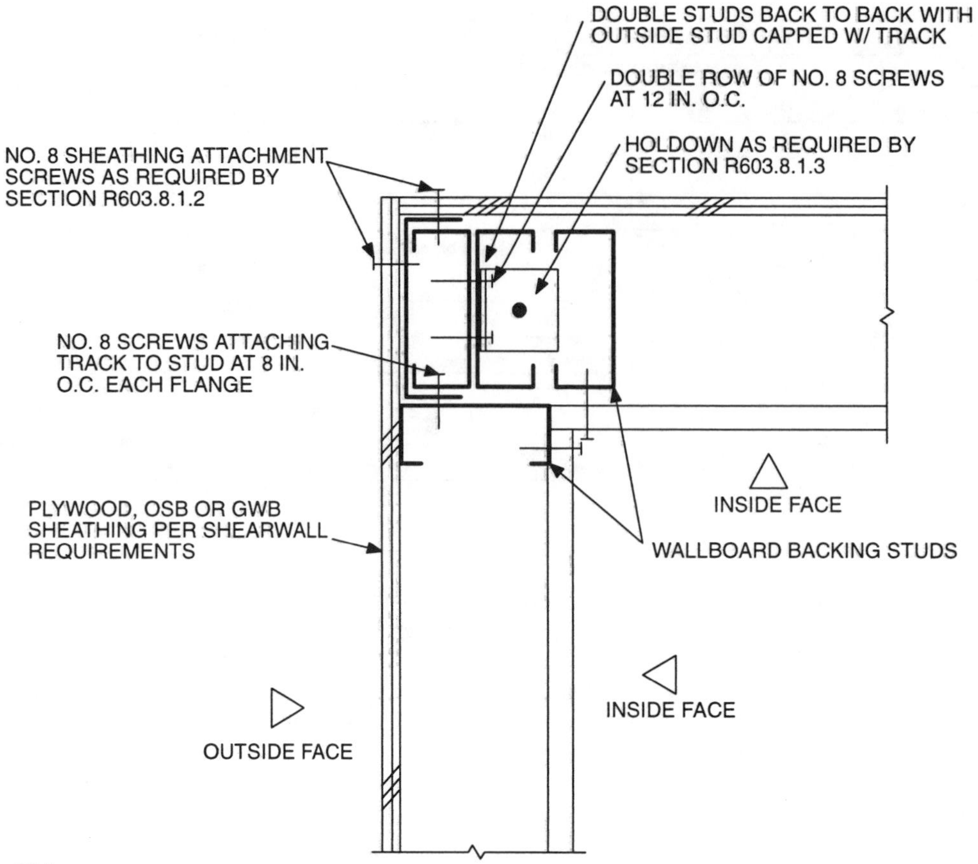

For SI: 1 inch = 25.4 mm.

FIGURE R603.8.1.3
CORNER STUD HOLD-DOWN

R603.8.1.4 Attachment of braced walls to foundations and floor and roof diaphragms. The top track of braced wall lines shall be attached directly to the roof sheathing in accordance with Figure R603.8.1.4(1) or shall have blocking, installed in accordance with Figure R603.8.1.4(2), connecting the top track to the roof sheathing at locations specified herein. Blocking shall be installed at each end of braced wall lines. Continuous blocking shall be installed above all Type I braced wall panels and above full height sheathing panels in Type II braced wall lines. Where the roof diaphragm spans more than 40 feet (12 192 mm) between braced wall lines, continuous blocking shall also be installed in the middle one-third of the braced wall lines.

Splices in top tracks in braced walls shall comply with Figure R603.3.6 except the number of screws in the splice shall be as shown in Table R603.8.1.4(1). Screws used to attach blocking to the top track are permitted to be counted toward the required number of track splice screws. Splices in the top track and the strap at the roof sheathing shall not occur in the same stud bay.

The top and bottom track of braced walls shall be attached to floor diaphragms in accordance with Figure R603.8.1.4.(3).

Splices in the floor track shall not occur in the same bay as splices in the wall track immediately above or below the floor track splice.

The bottom track of braced walls supported on foundations shall have anchor bolts installed in accordance with Figure R603.3.1(1). Floor track or rim joists supporting braced wall lines shall be attached to foundations in accordance with Figure R505.3.1(3). The maximum spacing between anchor bolts shall be as shown in Table R603.8.1.4(2). Anchor bolts shall extend 7 inches (178 mm) into concrete or masonry. An anchor bolt shall be located not more than 12 inches (305 mm) from wall corners or the termination of bottom track sections.

In regions where the basic wind speed equals or exceeds 110 miles per hour (177 km/h), the bottom track in exterior walls shall also comply with the provisions of Section R603.8.3.2.6 for uplift.

TABLE R603.8.1.4(1)
TOP TRACK SPLICE SCREW SCHEDULE

UP TO 120 MPH OR SDC D1	UP TO 130 MPH OR SDC D2
6 #8 screws each side of splice	11 #8 screws each side of splice

For SI: 1 mile per hour = 1.609 km/h.

NOTE: SDC = Seismic Design Category.

TABLE R603.8.1.4(2)
REQUIRED SHEAR ANCHORAGE FOR BRACED WALLS

ANCHOR BOLT DIAMETER (inch)	REQUIRED ANCHOR BOLT SPACING (feet on center)			
	Panel edge screw spacing (inches)			
	6	4	3	2
$^{1}/_{2}$	4.5	3	3	2.5
$^{5}/_{8}$	5.5	4	3	3

For SI: 1 inch = 25.4 mm.

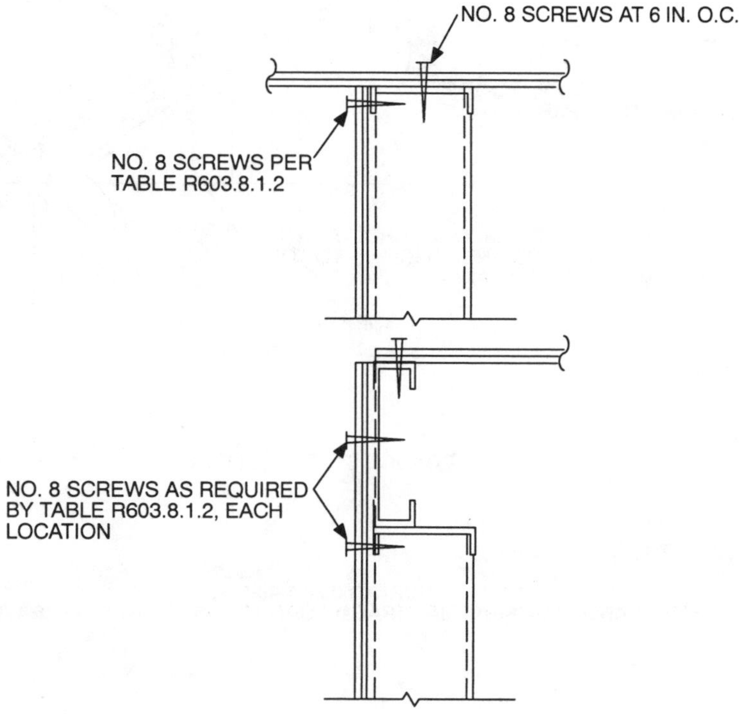

For SI: 1 inch = 25.4 mm.

FIGURE R603.8.1.4(1)
GABLE ROOF SHEATHING ATTACHMENT TO BRACE WALLS

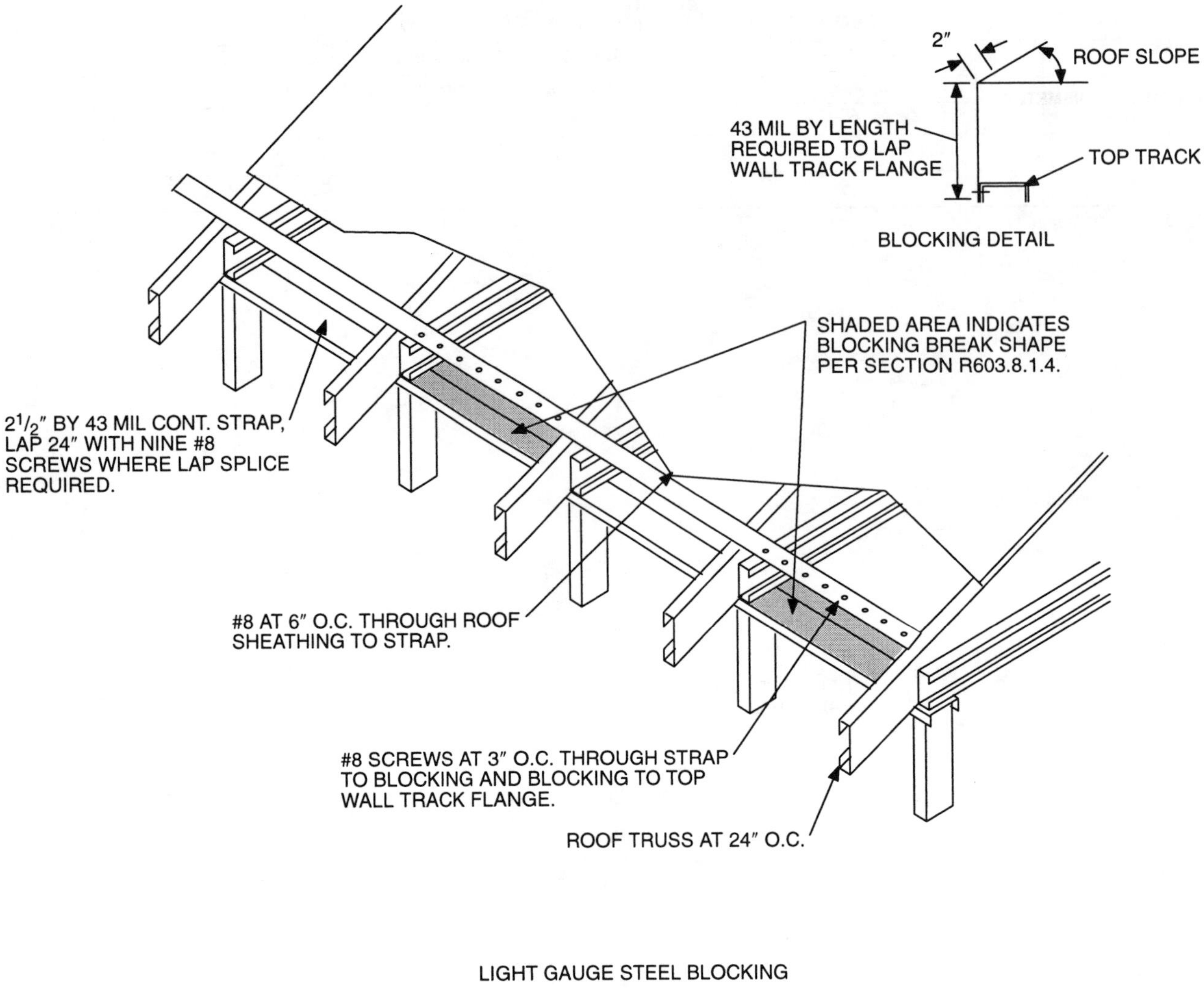

For SI: 1 inch = 25.4 mm, 1 mil = 0.0254 mm.

FIGURE R603.8.1.4(2)
STRAP AND BLOCKING DIAPHRAGM LOAD TRANSFER AT ROOF EAVE

SHEATHING SCREW SPACING IN WALL			
6 IN. O.C.	4 IN. O.C.	3 IN. O.C.	2 IN. O.C.
SCREW SPACING - WALL TRACK TO FRAMING BELOW			
4 NO. 8 AT 24 IN. O.C.	5 NO. 8 AT 24 IN. O.C.	7 NO. 8 AT 24 IN. O.C.	8 NO. 8 AT 24 IN. O.C.

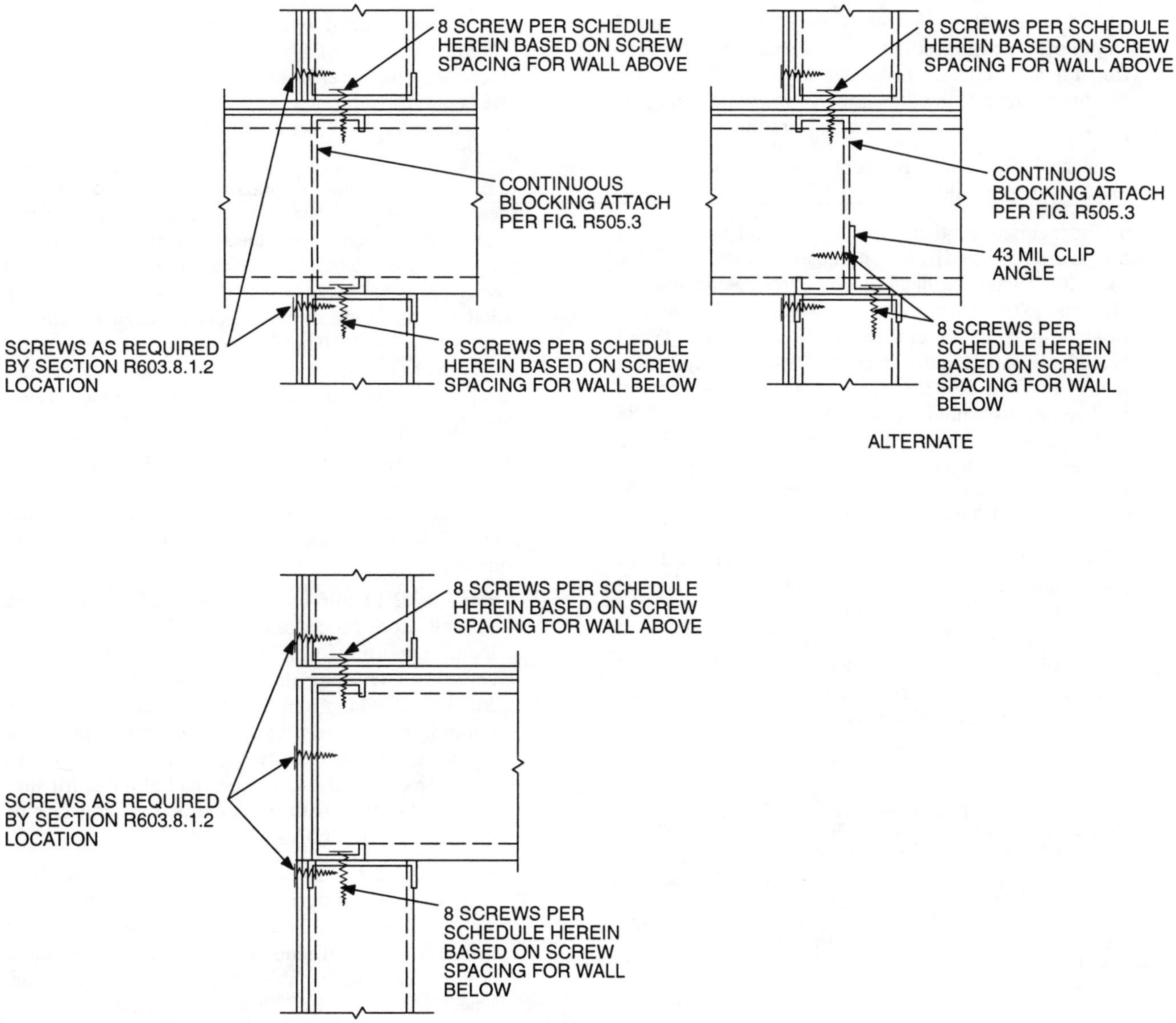

For SI: 1 inch = 25.4 mm, 1 mil = 0.0254 mm.

FIGURE R603.8.1.4(3)
FLOOR DIAPHRAGM—ATTACHMENT TO BRACED WALLS

R603.8.2 Seismic Design Categories D_1 and D_2.

R603.8.2.1 Limitations. The construction of buildings in Seismic Design Categories D_1 and D_2 shall comply with the provisions of this section. Buildings shall be limited to slab on grade or continuous concrete or masonry foundations and the weight of floors, roofs or walls shall not exceed the limits set forth in Section 301.2.2.4.

Exception: Buildings constructed in accordance with Section 301.1.

R603.8.2.2 Required length of Type I braced wall panels. For the purposes of this section, the diaphragm aspect ratio shall be taken as the dimension of the diaphragm perpendicular to the walls under consideration divided by the dimension of the diaphragm parallel to the walls under consideration, and shall be not less than one-quarter to one (0.25:1) nor more than four to one (4:1). The required length of Type I braced wall panels shall be determined from Figures R603.8.2(1) through R603.8.2(4) where the diaphragm span is the dimension of the diaphragm perpendicular to the walls under consideration.

Interpolation shall be permitted for determining diaphragm span values.

The required length of Type I braced wall panels is permitted to be adjusted by length adjustment factors in Table R603.8.2.2 where the dead weights of roof/ceiling assemblies and exterior walls are less than or equal to the unit weights specified therein. The required length of Type I braced wall panels shall be increased by length adjustment factors in Table R603.8.2.2 where the dead weights of roof/ceiling assemblies is greater than 15 psf (0.7185 kN/m^2) and less than or equal to 25 psf (1.1975 kN/m^2). The length adjustment factors in Tables R301.2.2.4 and R603.8.2.2 are permitted to be compounded with those in Table R603.8.1.2.

The height to width aspect ratio in Type I braced wall panels and full height sheathing segments in Type II braced walls shall be limited to two to one (2:1).

Exception: Type I braced wall panels on either side of garage openings, that support roofs only, are permitted to have an aspect ratio of four to one (4:1).

TABLE R603.8.2.2
LIGHT WEIGHT ROOF AND LIGHT WEIGHT EXTERIOR WALL TYPE I LENGTH ADJUSTMENT FACTORS

BRACED WALL SUPPORTING	LIGHT WEIGHT ROOF/CEILING —UNIT WEIGHT LESS THAN 11 PSF	LIGHT WEIGHT EXTERIOR WALLS—UNIT WEIGHT LESS THAN 7 PSF	BUILDINGS HAVING BOTH LIGHT WEIGHT WALLS AND ROOFS
Roof/ceiling only	0.80	0.95	0.70
One floor and roof/ceiling	0.90	0.90	0.75

For SI: 1 pound per square foot = 0.0479 kN/m^2.

R603.8.3 High wind requirements.

R603.8.3.1 Braced wall design. Buildings in areas where the basic wind speed is 110 miles per hour (177 km/h) or greater shall have braced wall lines complying with Section R603.8.1 as required by this section.

For the purpose of this section, the endwall shall be defined as the exterior wall of the building perpendicular to the roof ridge and the sidewall shall be defined as the exterior wall of the building parallel to the roof ridge. For the purposes of determining uplift and lateral bracing requirements, the attic shall be considered an additional story when the roof slope is 7 in 12 or greater.

The minimum and maximum allowable sidewall lengths shall be determined from Tables R603.8.3.1(1) and R603.8.3.1(2). The required length of Type I braced wall panels shall be determined from Tables R603.8.3.1(3) and R603.8.3.1(4).

Braced wall hold down anchors shall comply with Section R603.8.1.3. The height to width aspect ratio of Type I braced wall panels shall be limited to four to one (4:1). The height to width aspect ratio of full height sheathing segments in Type II braced walls shall be limited to two to one (2:1).

Exterior walls shall be sheathed with wood structural sheathing panels or other approved materials. Wood structural sheathing panels, and their attachments, shall comply with Section R603.8.1.2 except in regions where the basic wind speed exceeds 110 miles per hour (177 km/h) wood structural sheathing panels attached to framing spaced 24 inches (610 mm) on center shall be a minimum of $^{19}/_{32}$ inch (15.1 mm). Attachment of wall sheathing materials other than wood structural sheathing panels shall comply with the manufacturer's instructions.

R603.8.3.2 High wind connections. Connections, as required in this section, shall be provided to ensure a continuous load path capable of transferring shear and uplift loads from floors, studs and roof framing to the foundation.

R603.8.3.2.1 Uplift connection—roof rafter or truss to wall. Roof rafters and trusses shall be attached to their supporting wall assemblies by connections capable of resisting the uplift loads listed in Table R603.8.3.2.1(1). Alternatively, a 1.25-inch-by-minimum-33-mil (32 mm by 0.84 mm) steel uplift strap connecting the rafter or truss to the in-line framing stud below shall be permitted. Each end of the uplift strap shall be fastened with minimum No. 8 screws as required by Table R603.8.3.2.1(2).

R603.8.3.2.2 Uplift connection—wall assembly to wall assembly. Exterior wall studs in the upper story wall of a two-story building shall be attached to the in-line framing wall studs in the supporting wall below, with connections capable of resisting the uplift loads listed in Table R603.8.3.2.2(1). Alternatively, a 1.25-inch-by-33-mil (32 mm by 0.84 mm) steel uplift strap shall be permitted with minimum No. 8 screws attached to each stud, as required by Table R603.8.3.2.2(2).

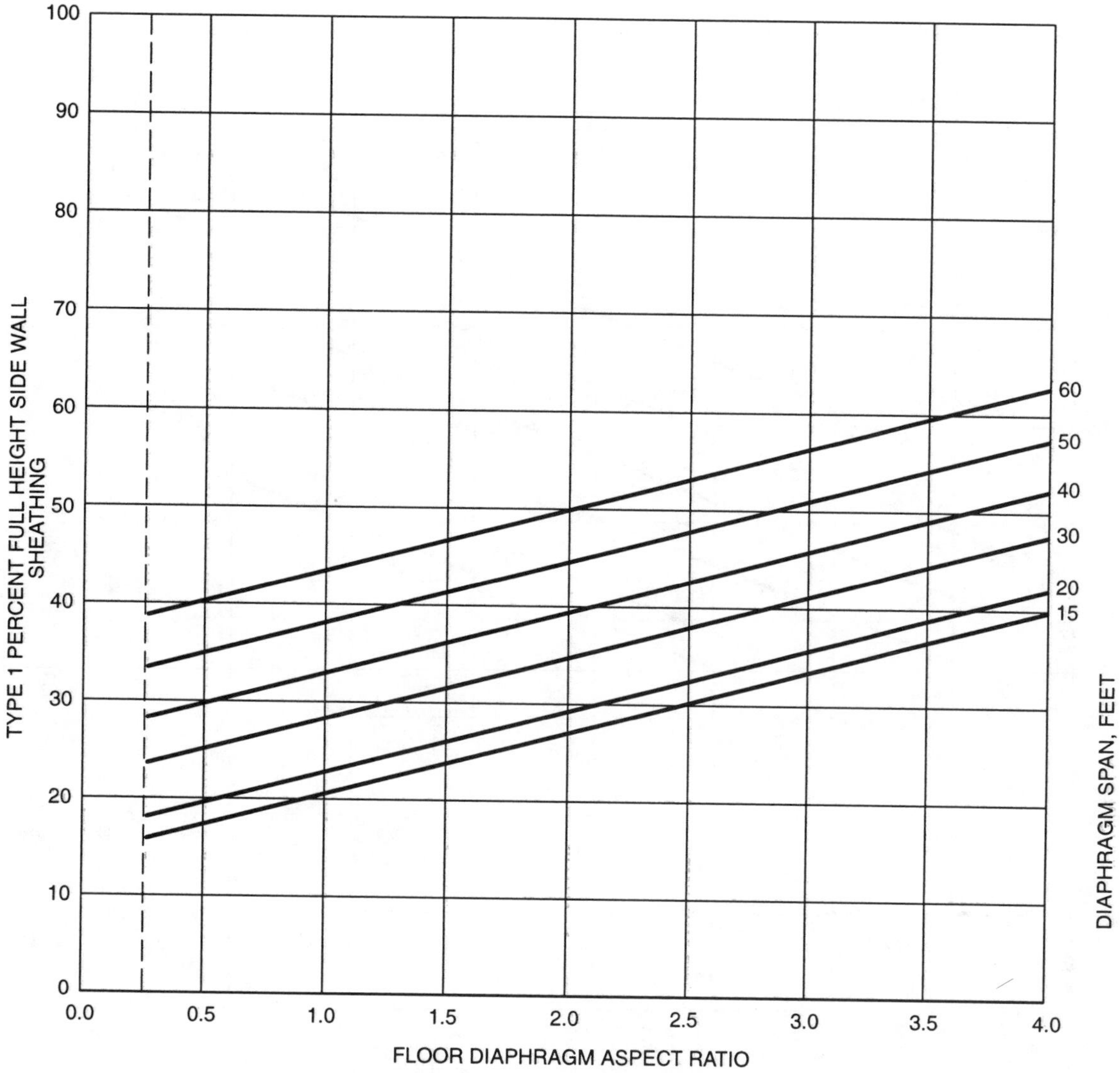

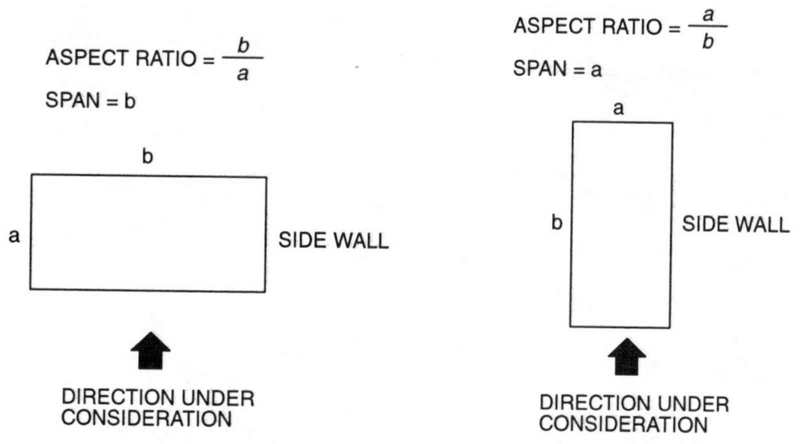

For SI: 1 foot = 304.8 mm.

FIGURE R603.8.2(1)
SDC D_1 SINGLE STORY OR TOP OF TWO STORY

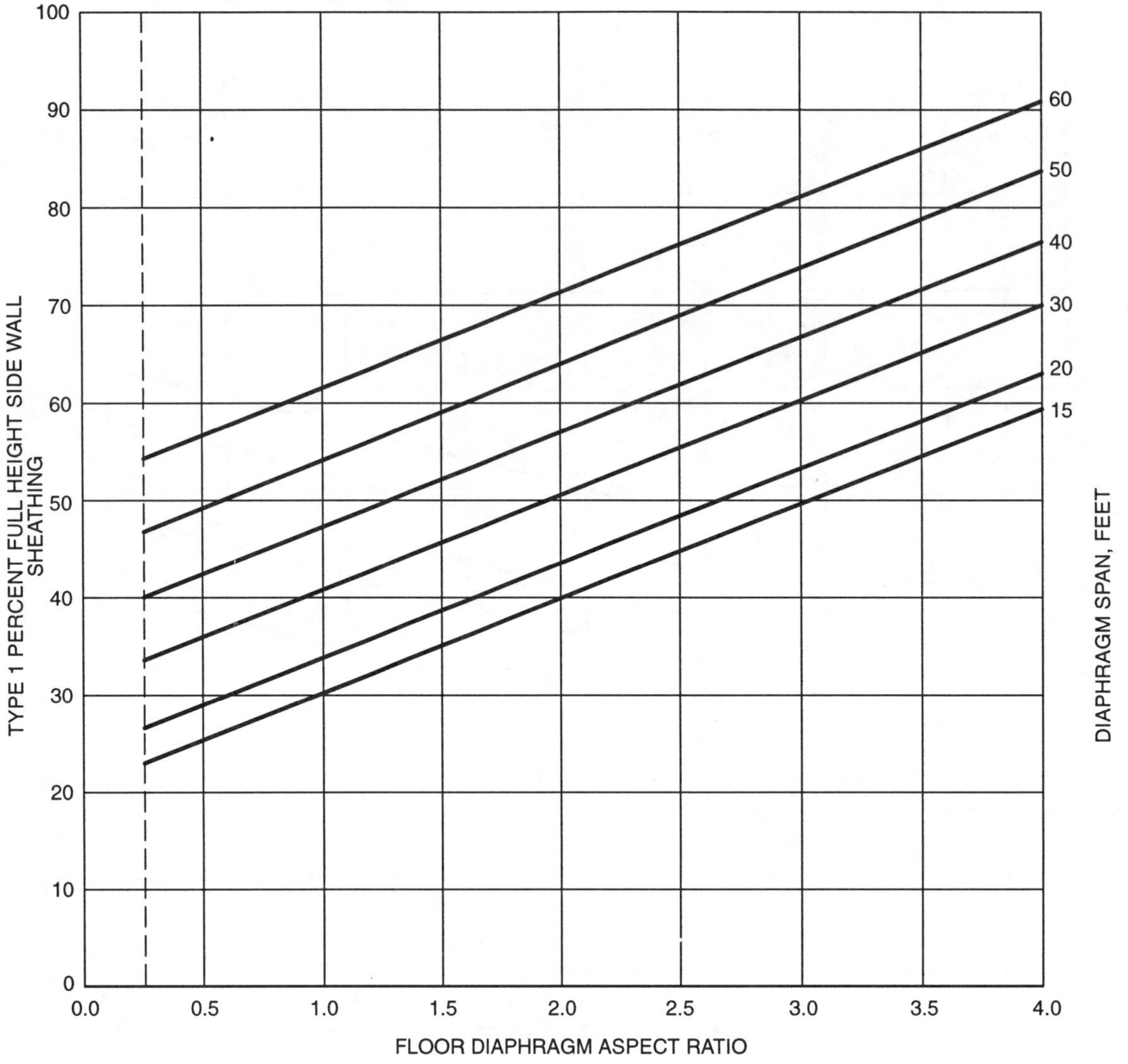

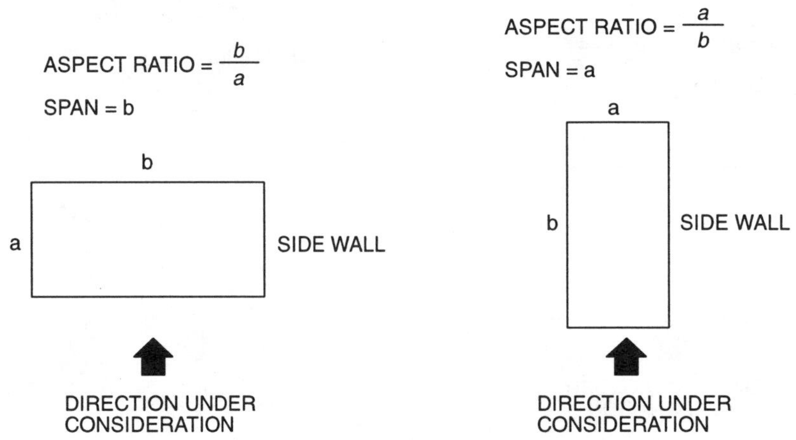

For SI: 1 foot = 304.8 mm.

FIGURE R603.8.2(2)
SDC D_1 BOTTOM OF TWO STORY

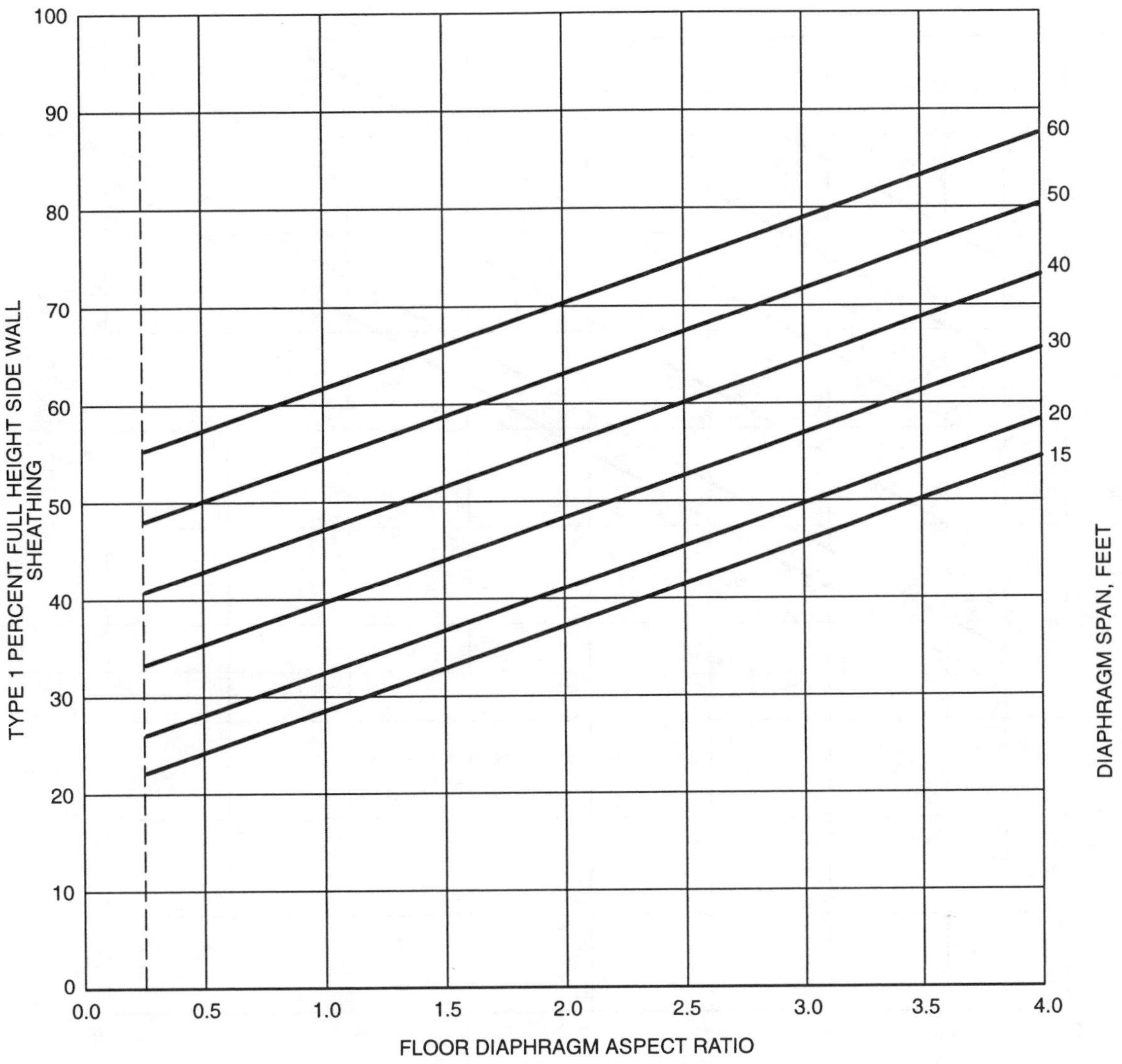

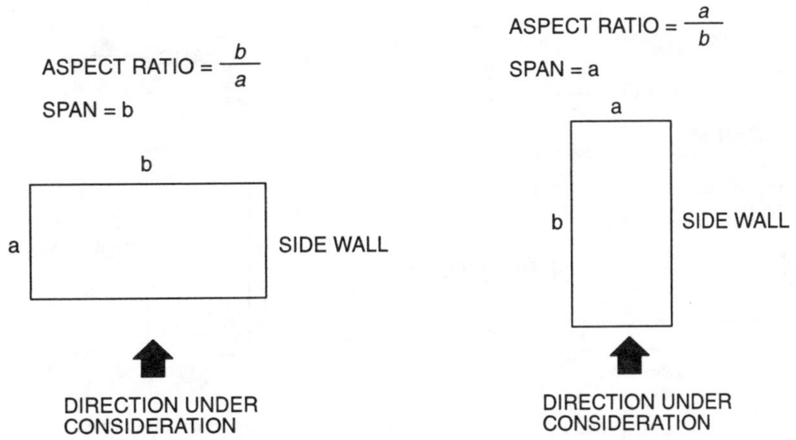

For SI: 1 foot = 304.8 mm.

FIGURE R603.8.2(3)
SDC D_2 SINGLE STORY OR TOP OF TWO STORY

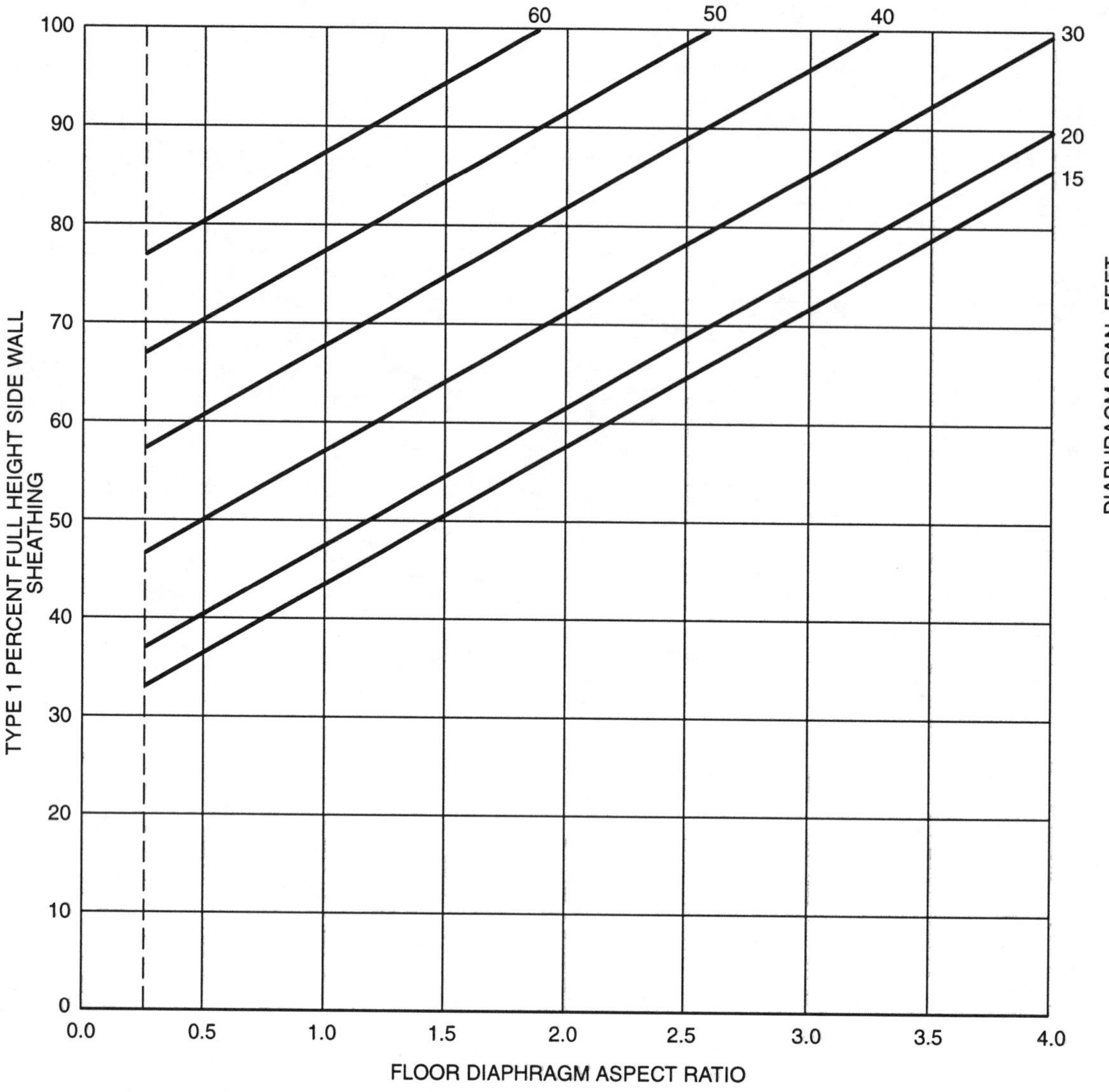

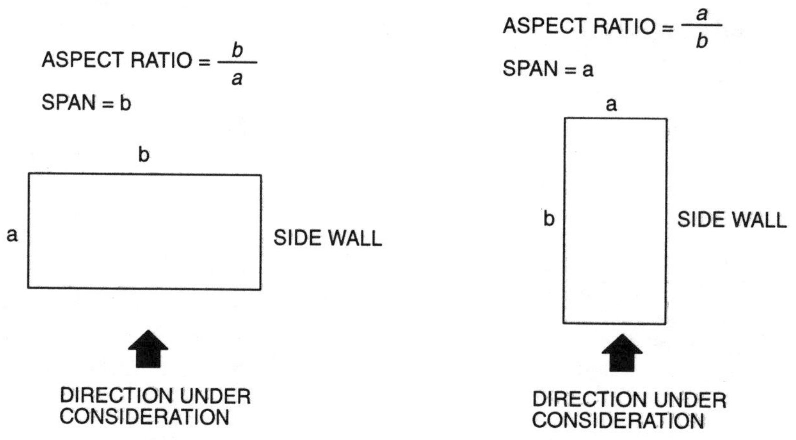

For SI: 1 foot = 304.8 mm.

FIGURE R603.8.2(4)
SDC D_2 BOTTOM OF TWO STORY

TABLE R603.8.3.1(1)
RANGE OF ALLOWABLE SIDEWALL LENGTHS ONE-STORY SLAB ON GRADE

FOUNDATION SUPPORTING	BUILDING ENDWALL WIDTH (FEET)	BASIC WIND SPEED (mph)					
		110		120		130	
		Allowable building sidewall length (feet)					
		Minimum	Maximum	Minimum	Maximum	Minimum	Maximum
One-story slab on grade	12	10	48	10	48	10	43
	16	10	60	10	60	10	58
	20	10	60	10	60	11	60
	24	10	60	11	60	13	60
	28	10	60	12	60	15	60
	32	11	60	14	60	17	60
	36	13	60	16	60	19	60

For SI: 1 foot = 304.8 mm, 1 mile per hour = 1.609 km/h.

TABLE R603.8.3.1(2)
RANGE OF ALLOWABLE SIDEWALL LENGTHS ALL OTHER CASES

FOUNDATION SUPPORTING	BUILDING ENDWALL WIDTH (feet)	BASIC WIND SPEED (mph)					
		110		120		130	
		Allowable building sidewall length (feet)					
		Minimum	Maximum	Minimum	Maximum	Minimum	Maximum
One to two stories	12	10	29	10	23	10	19
	16	10	38	10	31	10	26
	20	10	48	10	39	12	32
	24	10	58	12	47	14	39
	28	11	60	14	55	17	45
	32	13	60	16	60	19	51
	36	15	60	18	60	22	58

For SI: 1 foot = 304.8 mm, 1 mile per hour = 1.609 km/h.

TABLE R603.8.3.1(3)
TYPE I BRACED WALL PANEL SIDEWALL SHEATHING LENGTH REQUIREMENTS

BRACED WALL SUPPORTING	BUILDING ENDWALL LENGTH (W) (feet)	BASIC WIND SPEED (mph)		
		110	120	130
		Minimum length of full height sheathing on building sidewall (L)[a,b,c] (feet)		
Roof/ceiling only[d]	12	5	5	5
	16	5	5	5
	20	5	5	6
	24	5	6	7
	28	5	6	8
	32	6	7	9
	36	7	8	10
One floor and roof/ceiling[e]	20	8	10	11
	24	9	11	14
	28	11	13	16
	32	12	15	18
	36	14	17	20

For SI: 1 inch = 25.4 mm, 1 foot = 304.8 mm, 1 mile per hour = 1.609 km/h.

a. Tabulated sheathing lengths are based on 8-foot wall heights. For 9-foot wall heights, the tabulated values shall be multiplied by 1.13. For 10-foot wall heights, the tabulated values shall be multiplied by 1.25.

b. Tabulated sheathing lengths assume a mean roof height of 33 feet. For mean roof heights of 15 feet or less, the tabulated values shall be permitted to be multiplied by 0.8.

c. Tabulated sheathing lengths assume a 6-inch edge screw spacing. Required lengths shall be permitted to be multiplied by the adjustment factors in Table R603.8.1.2 for edge screw spacing other than 6 inches, but the resulting sheathing length shall not be less than 5 feet.

d. Applies to a one-story building or the top story of a two-story building.

e. Applies to the lower story of a two-story building.

TABLE R603.8.3.1(4)
TYPE I BRACED WALL PANEL ENDWALL SHEATHING LENGTH REQUIREMENTS

BRACED WALL SUPPORTING	BUILDING SIDEWALL LENGTH (W) (feet)	BASIC WIND SPEED (mph)		
		110	120	130
		Minimum length of full height sheathing on building endwall (L)[a,b,c] (feet)		
Roof/ceiling only[d]	12	5	5	5
	16	5	5	5
	20	5	5	5
	24	5	5	5
	28	5	5	5
	32	5	5	6
	36	5	6	7
	40	5	6	8
	50	6	8	9
	60	8	9	11
One floor and roof/ceiling[e]	20	8	10	12
	24	10	12	14
	28	11	14	17
	32	13	16	19
	36	14	18	21
	40	16	20	24
	50	20	24	29
	60	24	29	35

For SI: 1 inch = 25.4 mm, 1 foot = 304.8 mm, 1 mile per hour = 1.609 km/h.

a. Tabulated sheathing lengths are based on 8-foot wall heights. For 10-foot wall heights, the tabulated values shall be multiplied by 1.25.

b. Tabulated sheathing lengths are based on a mean roof height of 33 feet. For mean roof heights of 15 feet or less, the tabulated values shall be permitted to be multiplied by 0.8.

c. Tabulated sheathing lengths are based on a 6-inch edge screw spacing. Required lengths shall be permitted to be multiplied by the adjustment factors in Table R603.8.1.2 for edge screw spacing other than 6 inches, but the resulting sheathing length shall not be less than 5 feet.

d. Applies to a one-story building or the top story of a two-story building.

e. Applies to the lower story of a two-story building.

R603.8.3.2.3 Uplift connection—wall assembly to foundation or floor assembly. Exterior wall studs in bottom-story walls shall be attached to a wood sill plate or directly attached to the foundation by connections capable of resisting the uplift loads listed in Table R603.8.3.2.3(1). Alternatively, a continuous 1.25-inch-by-33-mil (32 mm by 0.84 mm) steel uplift strap is permitted when placed under the wood sill and attached to both flanges of the exterior stud as shown in Figure R603.8.3.2.3. The uplift strap shall be fastened to each flange with minimum No. 8 screws as required by Table R603.8.3.2.3(2).

R603.8.3.2.4 Ridge strap connection. Roof rafters shall be provided with a connection at the ridge line to transfer tension loads. The ridge connection shall be capable of resisting the unit loads listed in Table R804.3.3.1.1(1), multiplied by the appropriate spacing multiplier. Alternatively, steel ridge strap shall be provided with minimum No. 8 screws on each end of the strap as required in Table R804.3.3.1.1(1). The number of screws shall be increased to account for the spacing multipliers shown in the table. The width and thickness of the steel ridge strap shall be as shown in Table R804.3.3.1.1(2), based upon the required number of screws on one side of the strap.

R603.8.3.2.5 Header uplift connections.

R603.8.3.2.5.1 Single story or top story of a two-story building. Uplift connections shall be provided to fasten the rafter or roof trusses to the header by connectors capable of resisting the uplift loads listed in Table R603.8.3.2.1(1).

Uplift connections shall be provided to fasten the header to the jack studs by connectors capable of resisting the uplift loads listed in Table R603.8.3.2.1(1), multiplied by the number of framing members displaced, divided by two. An additional uplift strap shall be provided to fasten rafters or roof trusses to king studs that provide in-line framing support in accordance with Table R603.8.3.2.1(1).

Uplift connections shall be provided to fasten the jack studs to the foundation by connectors capable of resisting the uplift loads listed in Table R603.8.3.2.1(1), multiplied by the number of framing members displaced, divided by two.

R603.8.3.2.5.2 Bottom story of a two-story building. Uplift connections shall be provided to fasten the exterior wall studs in the upper story wall of a two-story building to the header below by connections capable of resisting the uplift loads listed in Table R603.8.3.2.2(1).

Uplift connections shall be provided to fasten the header to the jack studs by connectors capable of resisting the uplift loads listed in Table R603.8.3.2.2(1), multiplied by the number of framing members displaced, divided by two. An additional uplift strap shall be provided to fasten exterior wall studs in the upper story to king studs that provide in-line framing support in accordance with Table R603.8.3.2.2(1).

Uplift connections shall be provided to fasten the jack studs to the foundation by connectors capable of resisting the uplift loads listed in Table R603.8.3.2.2(1), multiplied by the number of framing members displaced, divided by two.

R603.8.3.2.6 Wall bottom track to foundation. The bottom track of exterior walls shall be connected to a wood sill plate as shown in Figure R603.3.1(2). The track shall be attached with steel plates spaced at 2 feet (610 mm) on center and fastened with four No. 8 screws with four 10d or six 8d common nails. The bottom track of interior braced walls shall be connected to supporting floors or foundations as required by Section R603.8.1.4.

The bottom track shall be connected to the foundation with $^1/_2$ inch (12.7 mm) anchor bolts extending 15 inches (381 mm) into masonry or 7 inches (178 mm) into concrete. An anchor bolt shall be located not more than 12 inches (305 mm) from wall corners or the termination of bottom track sections. Anchor bolts shall be spaced in accordance with Table R603.8.1.4(2), but shall be a maximum of 3 feet (915 mm) on center.

Exception: In regions with a wind speed of 120 miles per hour (193 km/h) or greater, anchor bolts located within the 8 feet (2438 mm) end zone shall be spaced a maximum of 2 feet (610 mm) on center.

TABLE R603.8.3.2.1(1) REQUIRED UPLIFT CAPACITY ROOF TRUSS OR RAFTER TO WALL

FRAMING SPACING[c] (inches)	ROOF SPAN (feet)	BASIC WIND SPEED (mph)		
		110	120	130
		Required connection capacity[a,b] (pounds)		
12	24	245	336	435
	28	279	382	495
	32	312	428	555
	36	346	474	615
16	24	327	447	580
	28	371	509	660
	32	416	570	740
	36	461	632	820
19.2	24	392	536	696
	28	446	610	792
	32	499	684	888
	36	553	758	984
24	24	490	671	868
	28	557	763	990
	32	624	855	1110
	36	691	947	1230

For SI: 1 inch = 25.4 mm, 1 foot = 304.8 mm, 1 pound = 4.4 N, 1 mile per hour = 1.609 km/h, 1 pound per square foot = 0.0479 kN/m^2.

a. Uplift requirements assume a roof/ceiling dead load of 10 psf ($^2/_3$ by 15 psf).

b. Uplift connection requirements shall be permitted to be multiplied by 0.70 for framing not located within 8 feet of building corners.

c. The 12-inch and 19.2-inch framing spacing provides options for design, but shall not negate the in-line framing requirement of Section R603.1.2.

TABLE R603.8.3.2.1(2) UPLIFT STRAP CONNECTION REQUIREMENTS ROOF RAFTER OR TRUSS TO WALL

FRAMING SPACING[a] (inches)	ROOF SPAN (feet)	BASIC WIND SPEED (mph)		
		110	120	130
		Number of #8 screws in each end of a $1^1/_4$ inch by 33 mil steel strap		
12	24	2	2	2
	28	2	2	3
	32	2	2	3
	36	2	3	3
16	24	2	3	3
	28	2	3	3
	32	2	3	4
	36	3	3	4
19.2	24	2	3	4
	28	3	3	4
	32	3	4	5
	36	3	4	5
24	24	3	4	4
	28	3	4	5
	32	3	4	6[b]
	36	4	5	6[b]

For SI: 1 inch = 25.4 mm, 1 foot = 304.8 mm, 1 mile per hour = 1.609 km/h, 1 mil = 0.0254 mm.

a. The 12-inch and 19.2-inch framing spacing provides options for design, but shall not negate the in-line framing requirement of Section R603.1.2.

b. Strap width shall be a minimum of 1.5 inches or thickness shall be a minimum of 0.043 inches.

TABLE R603.8.3.2.2(1)
REQUIRED UPLIFT CAPACITY WALL ASSEMBLY TO WALL ASSEMBLY

FRAMING SPACING[c] (inches)	ROOF SPAN (feet)	BASIC WIND SPEED (mph)		
		110	120	130
		Required connection capacity[a,b] (pounds)		
12	24	185	276	375
	28	219	322	435
	32	252	368	495
	36	286	414	555
16	24	242	362	495
	28	286	424	575
	32	331	485	655
	36	376	547	735
19.2	24	287	431	591
	28	341	505	687
	32	394	579	784
	36	448	653	879
24	24	355	536	733
	28	422	628	855
	32	489	720	975
	36	556	812	1095

For SI: 1 inch = 25.4 mm, 1 foot = 304.8 mm, 1 pound = 4.4 N, 1 mile per hour = 1.609 km/h, 1 pound per square foot = 0.0479 kN/m^2.

a. Uplift requirements assume a roof/ceiling dead load of 10 psf ($^2/_3$ by 15 psf).

b. Uplift connection requirements shall be permitted to be multiplied by 0.70 for framing not located within 8 feet of building corners.

c. The 12-inch and 19.2-inch framing spacing provides options for design, but shall not negate the in-line framing requirement of Section R603.1.2.

TABLE R603.8.3.2.2(2)
UPLIFT STRAP CONNECTION REQUIREMENTS WALL ASSEMBLY TO WALL ASSEMBLY

FRAMING SPACING[a] (inches)	ROOF SPAN (feet)	BASIC WIND SPEED (mph)		
		110	120	130
		Number of #8 screws in each of a $1^1/_4$ inch by 33 mil steel strap		
12	24	2	2	2
	28	2	2	2
	32	2	2	3
	36	2	2	3
16	24	2	2	3
	28	2	2	3
	32	2	3	3
	36	2	3	4
19.2	24	2	2	3
	28	2	3	4
	32	2	3	4
	36	3	3	4
24	24	2	3	4
	28	2	3	4
	32	3	4	5
	36	3	4	5

For SI: 1 inch = 25.4 mm, 1 foot = 304.8 mm, 1 mile per hour = 1.609 km/h, 1 mil = 0.0254 mm.

a. The 12-inch and 19.2-inch framing spacing provides options for design, but shall not negate the in-line framing requirement of Section R603.1.2.

TABLE R603.8.3.2.3(1)
REQUIRED UPLIFT CAPACITY WALL ASSEMBLY TO FOUNDATION OR FLOOR ASSEMBLY

FRAMING SPACING[c] (inches)	ROOF SPAN (feet)	BASIC WIND SPEED (mph)		
		110	120	130
		Required connection capacity[a,b] (pounds)		
12	24	170	261	360
	28	204	307	420
	32	237	353	480
	36	271	399	540
16	24	227	347	480
	28	271	409	560
	32	316	470	640
	36	361	532	720
19.2	24	272	416	576
	28	326	490	672
	32	379	564	768
	36	433	638	864
24	24	340	521	718
	28	407	613	840
	32	474	705	960
	36	541	797	1080

For SI: 1 inch = 25.4 mm, 1 foot = 304.8 mm, 1 pound = 4.4 N, 1 mile per hour = 1.609 km/h, 1 pound per square foot = 0.0479 kN/m^2.

a. Uplift requirements assume a roof/ceiling dead load of 10 psf ($^2/_3$ by15 psf).

b. Uplift connection requirements shall be permitted to be multiplied by 0.70 for framing not located within 8 feet of building corners.

c. The 12-inch and 19.2-inch framing spacing provides options for design, but shall not negate the in-line framing requirement of Section R603.1.2.

TABLE R603.8.3.2.3(2)
UPLIFT STRAP CONNECTION REQUIREMENTS WALL ASSEMBLY TO FOUNDATION OR FLOOR ASSEMBLY

FRAMING SPACING[a] (inches)	ROOF SPAN (feet)	BASIC WIND SPEED (mph)		
		110	120	130
		Number of #8 screw in a 1.25 inch by 33 mil strap in each flange of studs (see Figure R603.8.3.2.3)		
12	24	1	1	1
	28	1	1	1
	32	1	1	2
	36	1	1	2
16	24	1	1	2
	28	1	1	2
	32	1	2	2
	36	1	2	2
19.2	24	1	1	2
	28	1	2	2
	32	1	2	2
	36	1	2	2
24	24	1	2	2
	28	1	2	2
	32	2	2	3
	36	2	2	3

For SI: 1 inch = 25.4 mm, 1 foot = 304.8 mm, 1 mile per hour = 1.609 km/h, 1 mil = 0.0254 mm.

a. The 12-inch and 19.2-inch framing spacing provides options for design, but shall not negate the in-line framing requirement of Section R603.1.2.

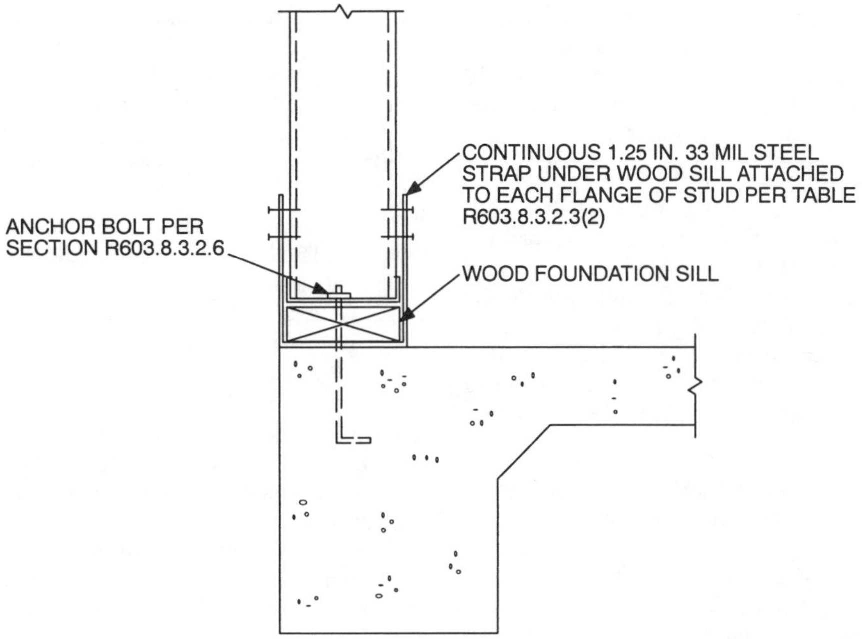

For SI: 1 inch = 25.4 mm, 1 mil = 0.0254 mm.

FIGURE R603.8.3.2.3
WIND UPLIFT CONNECTOR

SECTION R604
WOOD STRUCTURAL PANELS

R604.1 Identification and grade. Wood structural panels shall conform to DOC PS 1 or DOC PS 2. All panels shall be identified by a grade mark or certificate of inspection issued by an approved agency.

R604.2 Allowable spans. The maximum allowable spans for wood structural panel wall sheathing shall not exceed the values set forth in Table R602.3(3).

R604.3 Installation. Wood structural panel wall sheathing shall be attached to framing in accordance with Table R602.3(1). Wood structural panels marked Exposure 1 or Exterior are considered water-repellent sheathing under the code.

SECTION R605
PARTICLEBOARD

R605.1 Identification and grade. Particleboard shall conform to ANSI A208.1 and shall be so identified by a grade mark or certificate of inspection issued by an approved agency. Particleboard shall comply with the grades specified in Table R602.3(4).

SECTION R606
GENERAL MASONRY CONSTRUCTION

R606.1 General. Masonry construction shall be designed and constructed in accordance with the provisions of this section or in accordance with the provisions of ACI 530/ASCE 5/TMS 402.

R606.1.1 Professional registration not required. When the empirical design provisions of ACI 530/ASCE 5/TMS 402 Chapter 5 or the provisions of this section are used to design masonry, project drawings, typical details and specifications are not required to bear the seal of the architect or engineer responsible for design, unless otherwise required by the state law of the jurisdiction having authority.

R606.2 Thickness of masonry. The nominal thickness of masonry walls shall conform to the requirements of Sections R606.2.1 through R606.2.4.

R606.2.1 Minimum thickness. The minimum thickness of masonry bearing walls more than one story high shall be 8 inches (203 mm). Solid masonry walls of one-story dwellings and garages shall not be less than 6 inches (152 mm) in thickness when not greater than 9 feet (2743 mm) in height, provided that when gable construction is used, an additional 6 feet (1829 mm) is permitted to the peak of the gable. Masonry walls shall be laterally supported in either the horizontal or vertical direction at intervals as required by Section R606.8.

R606.2.2 Rubble stone masonry wall. The minimum thickness of rough, random or coursed rubble stone masonry walls shall be 16 inches (406 mm).

R606.2.3 Change in thickness. Where walls of masonry of hollow units or masonry bonded hollow walls are decreased in thickness, a course of solid masonry shall be constructed between the wall below and the thinner wall above, or special units or construction shall be used to transmit the loads from face shells or wythes above to those below.

R606.2.4 Parapet walls. Unreinforced solid masonry parapet walls shall not be less than 8 inches (203 mm) in thickness and their height shall not exceed four times their thickness. Unreinforced hollow unit masonry parapet walls shall be not less than 8 inches (203 mm) in thickness, and their height shall not exceed three times their thickness. Masonry parapet walls in areas subject to wind loads of 30 pounds per square foot (1.44 kN/m^2) or located in Seismic Design Category D_1 or D_2 shall be reinforced in accordance with Section R606.11.

R606.3 Corbeled masonry. Solid masonry units shall be used for corbeling. The maximum corbeled projection beyond the face of the wall shall not be more than one-half of the wall thickness or one-half the wythe thickness for hollow walls; the maximum projection of one unit shall not exceed one-half the height of the unit or one-third the thickness at right angles to the wall. The top course of corbels shall be a header course when the corbeled masonry is used to support floor or roof-framing members.

R606.3.1 Support conditions. Cavity wall or masonry veneer construction may be supported on an 8-inch (203 mm) foundation wall, provided the 8-inch (203 mm) wall is corbeled with solid masonry to the width of the wall system above. The total horizontal projection of the corbel shall not exceed 2 inches (51 mm) with individual corbels projecting not more than one-third the thickness of the unit or one-half the height of the unit. The top course of all corbels shall be a header course.

R606.4 Allowable stresses. Allowable compressive stresses in masonry shall not exceed the values prescribed in Table R606.4. In determining the stresses in masonry, the effects of all loads and conditions of loading and the influence of all forces affecting the design and strength of the several parts shall be taken into account.

R606.4.1 Combined units. In walls or other structural members composed of different kinds or grades of units, materials or mortars, the maximum stress shall not exceed the allowable stress for the weakest of the combination of units, materials and mortars of which the member is composed. The net thickness of any facing unit that is used to resist stress shall not be less than 1.5 inches (38 mm).

R606.5 Piers. The unsupported height of masonry piers shall not exceed ten times their least dimension. When structural clay tile or hollow concrete masonry units are used for isolated piers to support beams and girders, the cellular spaces shall be filled solidly with concrete or Type M or S mortar, except that unfilled hollow piers may be used if their unsupported height is not more than four times their least dimension. Where hollow masonry units are solidly filled with concrete or Type M, S or N mortar, the allowable compressive stress shall be permitted to be increased as provided in Table R606.4.

R606.5.1 Pier cap. Hollow piers shall be capped with 4 inches (102 mm) of solid masonry or concrete or shall have cavities of the top course filled with concrete or grout or other approved methods.

R606.6 Chases. Chases and recesses in masonry walls shall not be deeper than one-third the wall thickness, and the maximum length of a horizontal chase or horizontal projection shall not exceed 4 feet (1219 mm), and shall have at least 8 inches (203 mm) of masonry in back of the chases and recesses and between adjacent chases or recesses and the jambs of openings. Chases and recesses in masonry walls shall be designed and constructed so as not to reduce the required strength or required fire resistance of the wall and in no case shall a chase or recess be permitted within the required area of a pier. Masonry directly above chases or recesses wider than 12 inches (305 mm) shall be supported on noncombustible lintels

R606.7 Stack bond. In unreinforced masonry where masonry units are laid in stack bond, longitudinal reinforcement consisting of not less than two continuous wires each with a minimum aggregate cross-sectional area of 0.017 square inch (11 mm^2) shall be provided in horizontal bed joints spaced not more than 16 inches (406 mm) on center vertically.

R606.8 Lateral support. Masonry walls shall be laterally supported in either the horizontal or the vertical direction. The maximum spacing between lateral supports shall not exceed the distances in Table R606.8. Lateral support shall be provided by cross walls, pilasters, buttresses or structural frame members when the limiting distance is taken horizontally, or by floors or roofs when the limiting distance is taken vertically.

R606.8.1 Horizontal lateral support. Lateral support in the horizontal direction provided by intersecting masonry walls shall be provided by one of the methods in Section R606.8.1.1 or Section R606.8.1.2.

R606.8.1.1 Bonding pattern. Fifty percent of the units at the intersection shall be laid in an overlapping masonry bonding pattern, with alternate units having a bearing of not less than 3 inches (76 mm) on the unit below.

R606.8.1.2 Metal reinforcement. Interior nonloadbearing walls shall be anchored at their intersections, at vertical intervals of not more than 16 inches (406 mm) with joint reinforcement of at least 9 gage, or $^1/_4$ inch (6.4 mm) galvanized mesh hardware cloth. Intersecting masonry walls, other than interior nonloadbearing walls, shall be anchored at vertical intervals of not more than 8 inches (203 mm) with joint reinforcement of at least 9 gage and shall extend at least 30 inches (762 mm) in each direction at the intersection. Other metal ties, joint reinforcement or anchors, if used, shall be spaced to provide equivalent area of anchorage to that required by this section.

R606.8.2 Vertical lateral support. Vertical lateral support of masonry walls in Seismic Design Category A, B or C shall be provided in accordance with one of the methods in Section R606.8.2.1 or Section R606.8.2.2.

R606.8.2.1 Roof structures. Masonry walls shall be anchored to roof structures with metal strap anchors spaced in accordance with the manufacturer's instructions, $^1/_2$-inch (12.7 mm) bolts spaced not more than 6 feet (1829 mm) on center, or other approved anchors. Anchors shall be embedded at least 16 inches (406 mm) into the masonry, or be hooked or welded to bond beam reinforcement placed not less than 6 inches (152 mm) from the top of the wall.

TABLE R606.4
ALLOWABLE COMPRESSIVE STRESSES FOR EMPIRICAL DESIGN OF MASONRY

CONSTRUCTION; COMPRESSIVE STRENGTH OF UNIT, GROSS AREA	ALLOWABLE COMPRESSIVE STRESSES[a] GROSS CROSS-SECTIONAL AREA[b]	
	Type M or S mortar	Type N mortar
Solid masonry of brick and other solid units of clay or shale; sand-lime or concrete brick:		
8,000 + psi	350	300
4,500 psi	225	200
2,500 psi	160	140
1,500 psi	115	100
Grouted[c] masonry, of clay or shale; sand-lime or concrete:		
4,500+ psi	225	200
2,500 psi	160	140
1,500 psi	115	100
Solid masonry of solid concrete masonry units:		
3,000+ psi	225	200
2,000 psi	160	140
1,200 psi	115	100
Masonry of hollow load-bearing units:		
2,000+ psi	140	120
1,500 psi	115	100
1,000 psi	75	70
700 psi	60	55
Hollow walls (cavity or masonry bonded[d]) solid units:		
2,500+ psi	160	140
1,500 psi	115	100
Hollow units	75	70
Stone ashlar masonry:		
Granite	720	640
Limestone or marble	450	400
Sandstone or cast stone	360	320
Rubble stone masonry:		
Coarse, rough or random	120	100

For SI: 1 pound per square inch = 6.895 kPa.

a. Linear interpolation shall be used for determining allowable stresses for masonry units having compressive strengths that are intermediate between those given in the table.

b. Gross cross-sectional area shall be calculated on the actual rather than nominal dimensions.

c. See Section R607.

d. Where floor and roof loads are carried upon one wythe, the gross cross-sectional area is that of the wythe under load; if both wythes are loaded, the gross cross-sectional area is that of the wall minus the area of the cavity between the wythes. Walls bonded with metal ties shall be considered as cavity walls unless the collar joints are filled with mortar or grout.

TABLE R606.8
SPACING OF LATERAL SUPPORT FOR MASONRY WALLS

CONSTRUCTION	MAXIMUM WALL LENGTH TO THICKNESS OR WALL HEIGHT TO THICKNESS[a,b]
Bearing walls:	
Solid or solid grouted	20
All other	18
Nonbearing walls:	
Exterior	18
Interior	36

For SI: 1 foot = 304.8 mm.

a. Except for cavity walls and cantilevered walls, the thickness of a wall shall be its nominal thickness measured perpendicular to the face of the wall. For cavity walls, the thickness shall be determined as the sum of the nominal thicknesses of the individual wythes. For cantilever walls, except for parapets, the ratio of height to nominal thickness shall not exceed 6 for solid masonry, or 4 for hollow masonry. For parapets, see Section R606.2.4.

b. An additional unsupported height of 6 feet is permitted for gable end walls.

R606.8.2.2 Floor diaphragms. Masonry walls shall be anchored to floor diaphragm framing by metal strap anchors spaced in accordance with the manufacturer's instructions, $^1/_2$-inch-diameter (12.7 mm) bolts spaced at intervals not to exceed 6 feet (1829 mm) and installed as shown in Figure R606.10(1), or by other approved methods.

R606.9 Lintels. Masonry over openings shall be supported by steel lintels, reinforced concrete or masonry lintels or masonry arches, designed to support load imposed.

R606.10 Anchorage. Masonry walls shall be anchored to floor and roof systems in accordance with the details shown in Figure R606.10(1), R606.10(2) or R606.10(3). Footings may be considered as points of lateral support.

R606.11 Seismic requirements. The seismic requirements of this section shall apply to the design of masonry and the construction of masonry building elements located in Seismic Design Category C, D_1 or D_2. These requirements shall not apply to glass unit masonry conforming to Section R610 or masonry veneer conforming to Section R703.7.

R606.11.1 General. Masonry structures and masonry elements shall comply with the requirements of Sections R606.11.2 through R606.11.4 based on the seismic design category established in Table R301.2(1). Masonry structures and masonry elements shall comply with the requirements of Section R606.11 and Figures R606.10(1), R606.10(2) and R606.10(3) or shall be designed in accordance with ACI 530/ASCE 5/TMS 402.

R606.11.1.1 Floor and roof diaphragm construction. Floor and roof diaphragms shall be constructed of structural wood sheathing panels, attached to wood framing in accordance with Table R602.3(1) or to cold-formed steel floor framing in accordance with Table R505.3.1(2) or to cold-formed steel roof framing in accordance with Table R804.3. Additionally, sheathing panel edges perpendicular to framing members shall be backed by blocking, and sheathing shall be connected to the blocking with fasteners at the edge spacing. For Seismic Design Categories C, D_1 and D_2, where the width-to-thickness dimension of the diaphragm exceeds 2-to-1, edge spacing of fasteners shall be 4 inches (102 mm) on center.

R606.11.2 Seismic Design Category C. Structures located in Seismic Design Category C shall comply with the requirements of this section.

R606.11.2.1 Design of elements not part of the lateral force-resisting system.

R606.11.2.1.1 Load-bearing frames or columns. Elements not part of the lateral-force-resisting system shall be analyzed as to their effect on the response of the system. Such frames or columns shall be adequate for vertical load carrying capacity and induced moment due to the design story drift.

R606.11.2.1.2 Masonry partition walls. Masonry partition walls, masonry screen walls and other masonry elements that are not designed to resist vertical or lateral loads, other than those induced by their own weight, shall be isolated from the structure so that vertical and lateral forces are not imparted to these elements. Isolation joints and connectors between these elements and the structure shall be designed to accommodate the design story drift.

R606.11.2.1.3 Reinforcement requirements for masonry elements. Masonry elements listed in Section R606.11.2.1.2 shall be reinforced in either the horizontal or vertical direction as shown in Figure R606.10(2) and in accordance with the following:

1. Horizontal reinforcement. Horizontal joint reinforcement shall consist of at least two longitudinal W1.7 wires spaced not more than 16 inches (406 mm) for walls greater than 4 inches (102 mm) in width and at least one longitudinal W1.7 wire spaced not more than 16 inches (406 mm) for walls not exceeding 4 inches (102 mm) in width; or at least one No. 4 bar spaced not more than 48 inches (1219 mm). Where two longitudinal wires of joint reinforcement are used, the space between these wires shall be the widest that the mortar joint will accommodate. Horizontal reinforcement shall be provided within 16 inches (406 mm) of the top and bottom of these masonry elements.
2. Vertical reinforcement. Vertical reinforcement shall consist of at least one No. 4 bar spaced not more than 48 inches (1219 mm). Vertical reinforcement shall be located within 16 inches (406 mm) of the ends of masonry walls.

R606.11.2.2 Design of elements part of the lateral-force-resisting system.

R606.11.2.2.1 Connections to masonry shear walls. Connectors shall be provided to transfer forces between masonry walls and horizontal elements in accordance with the requirements of Section 2.1.6 of ACI 530/ASCE 5/TMS 402. Connectors shall be designed to transfer horizontal design forces acting either perpendicular or parallel to the wall, but not less than 200 pounds per linear foot (2919 N/m) of wall. The maximum spacing between connectors shall be 4 feet (1219 mm). Such anchorage mechanisms shall not induce tension stresses perpendicular to grain in ledgers or nailers.

R606.11.2.2.2 Connections to masonry columns. Connectors shall be provided to transfer forces between masonry columns and horizontal elements in accordance with the requirements of Section 2.1.6 of ACI 530/ASCE 5/TMS 402. Where anchor bolts are used to connect horizontal elements to the tops of columns, anchor bolts shall be placed within lateral ties. Lateral ties shall enclose both the vertical bars in the column and the anchor bolts. There shall be a minimum of two No. 4 lateral ties provided in the top 5 inches (127 mm) of the column.

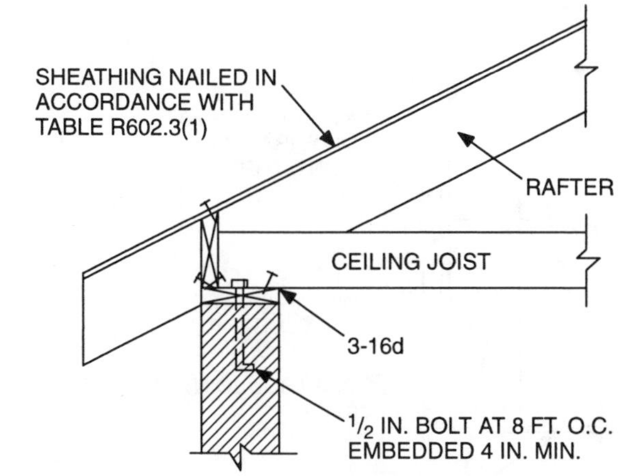

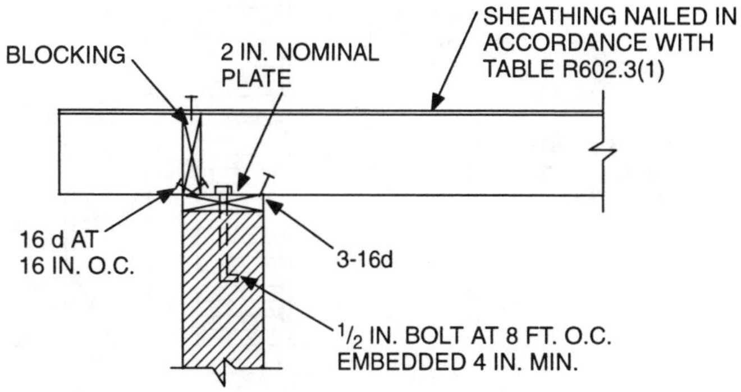

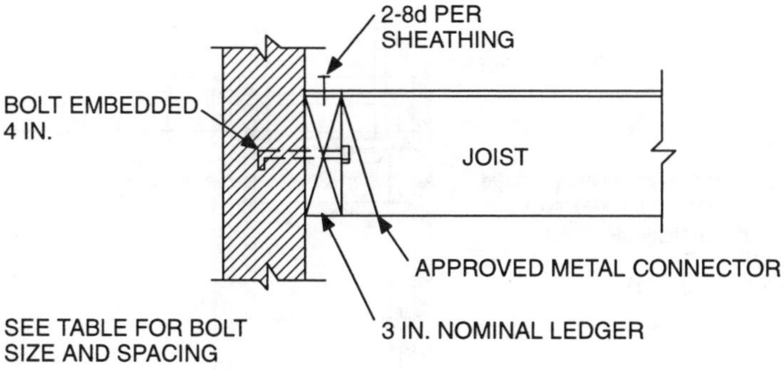

LEDGER BOLT SIZE AND SPACING

JOIST SPAN	BOLT SIZE AND SPACING	
	ROOF	FLOOR
10 FT.	1/2 AT 2 FT. 6 IN. 7/8 AT 3 FT. 6 IN.	1/2 AT 2 FT. 0 IN. 7/8 AT 2 FT. 9 IN.
10-15 FT.	1/2 AT 1 FT. 9 IN. 7/8 AT 2 FT. 6 IN.	1/2 AT 1 FT. 4 IN. 7/8 AT 2 FT. 0 IN.
15-20 FT.	1/2 AT 1 FT. 3 IN. 7/8 AT 2 FT. 0 IN.	1/2 AT 1 FT. 0 IN. 7/8 AT 1 FT. 6 IN.

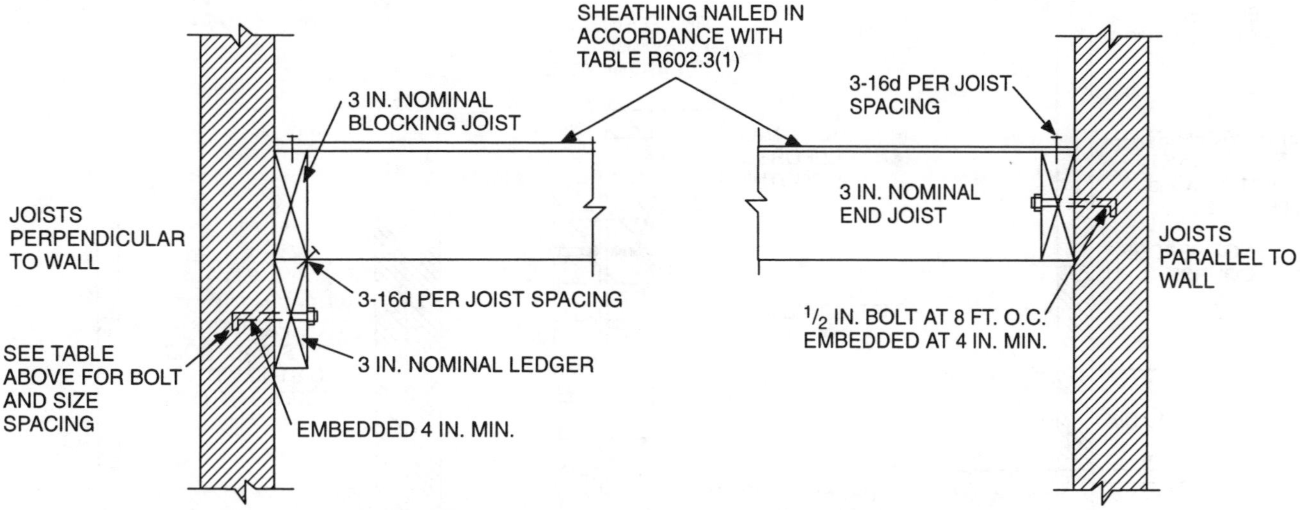

Note: Where bolts are located in hollow masonry, the cells in the courses receiving the bolt shall be grouted solid.

For SI: 1 inch = 25.4 mm, 1 foot = 304.8 mm, 1 pound per square foot = 0.0479 kN/m^2.

FIGURE R606.10(1)
ANCHORAGE REQUIREMENTS FOR MASONRY WALLS LOCATED IN SEISMIC DESIGN CATEGORY A, B OR C AND WHERE WIND LOADS ARE LESS THAN 30 PSF

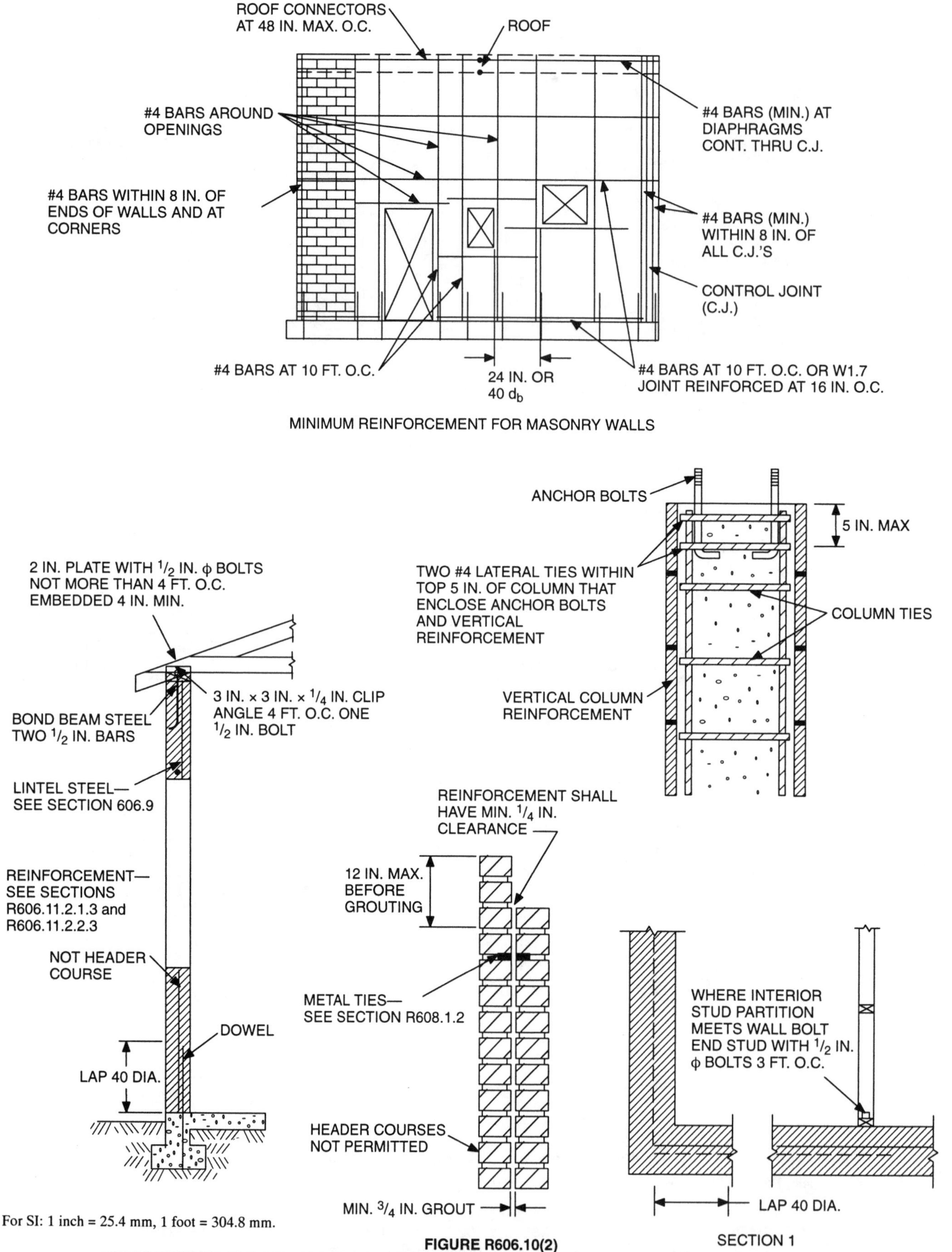

For SI: 1 inch = 25.4 mm, 1 foot = 304.8 mm.

FIGURE R606.10(2)
REQUIREMENTS FOR REINFORCED GROUTED MASONRY CONSTRUCTION IN SEISMIC DESIGN CATEGORY C

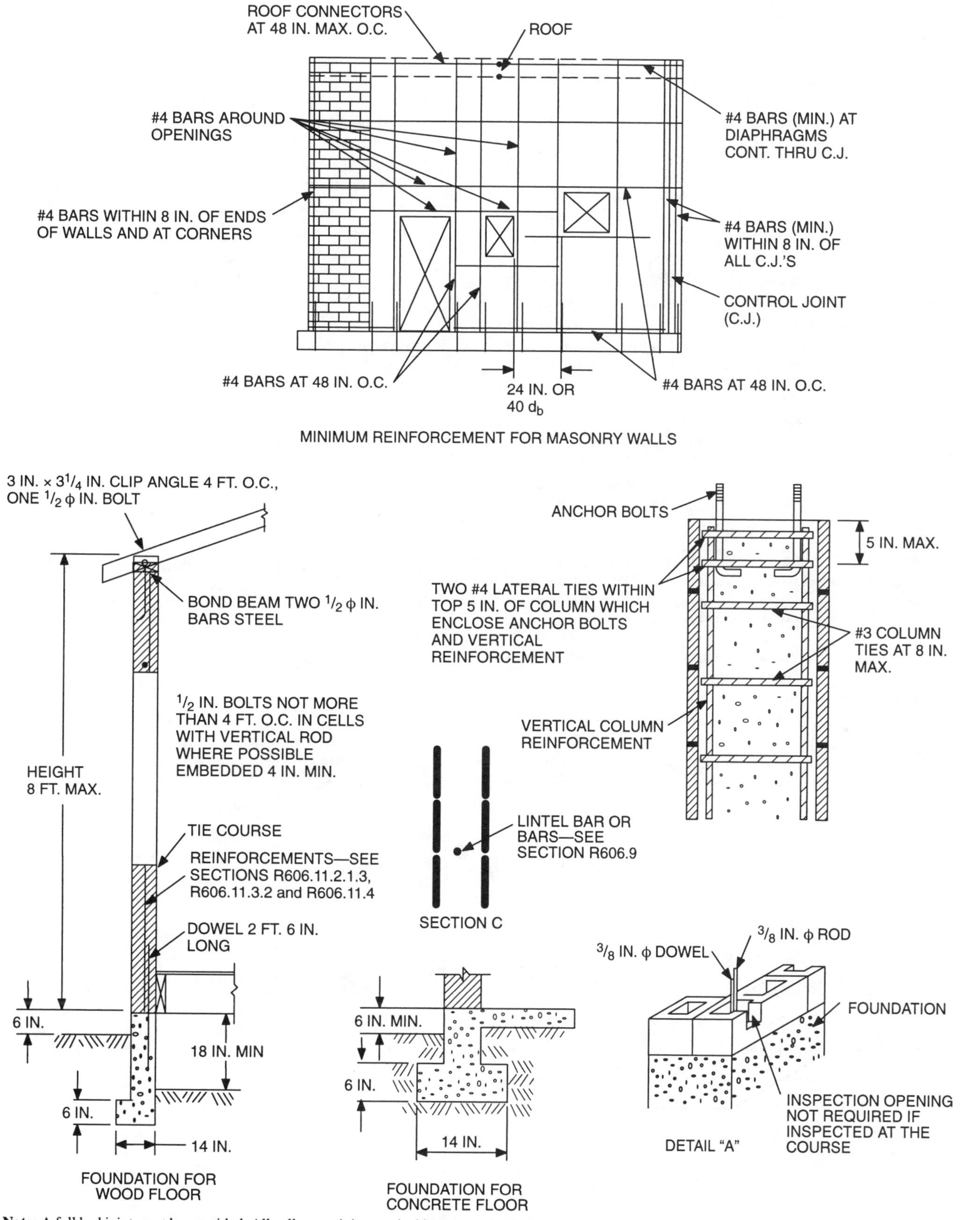

Note: A full bed joint must be provided. All cells containing vertical bars are to be filled to the top of wall and provide inspection opening as shown on detail "A." Horizontal bars are to be laid as shown on detail "B." Lintel bars are to be laid as shown on Section C.

For SI: 1 inch = 25.4 mm, 1 foot = 304.8 mm.

FIGURE R606.10(3)
REQUIREMENTS FOR REINFORCED MASONRY CONSTRUCTION IN SEISMIC DESIGN CATEGORY D_1 OR D_2

R606.11.2.2.3 Minimum reinforcement requirements for masonry shear walls. Vertical reinforcement of at least one No. 4 bar shall be provided at corners, within 16 inches (406 mm) of each side of openings, within 8 inches (203 mm) of each side of movement joints, within 8 inches (203 mm) of the ends of walls, and at a maximum spacing of 10 feet (3048 mm).

Horizontal joint reinforcement shall consist of at least two wires of W1.7 spaced not more than 16 inches (406 mm); or bond beam reinforcement shall be provided of at least one No. 4 bar spaced not more than 10 feet (3048 mm). Horizontal reinforcement shall also be provided at the bottom and top of wall openings and shall extend not less than 24 inches (610 mm) nor less than 40 bar diameters past the opening; continuously at structurally connected roof and floor levels; and within 16 inches (406 mm) of the top of walls.

R606.11.3 Seismic Design Category D_1. Structures in Seismic Design Category D_1 shall comply with the requirements of Seismic Design Category C and to the additional requirements of this section. Masonry structures in Seismic Design Category D_1 shall be limited in height to one story and 9 feet (2743 mm) between lateral supports.

R606.11.3.1 Design requirements. Masonry elements other than those covered by Section R606.11.2.1.2 shall be designed in accordance with the requirements of Chapter 2 of ACI 530/ASCE 5/TMS 402.

R606.11.3.2 Minimum reinforcement requirements for masonry walls. Masonry walls other than those covered by Section R606.11.2.1.3 shall be reinforced in both the vertical and horizontal direction. The sum of the cross-sectional area of horizontal and vertical reinforcement shall be at least 0.002 times the gross cross-sectional area of the wall, and the minimum cross-sectional area in each direction shall be not less than 0.0007 times the gross cross-sectional area of the wall. Reinforcement shall be uniformly distributed. Table R606.11.3.2 provides the minimum reinforcing bar sizes required for varying thicknesses of masonry walls. The maximum spacing of reinforcement shall be 48 inches (1219 mm) provided that the walls are solid grouted and constructed of hollow open-end units, hollow units laid with full head joints or two wythe of solid units. The maximum spacing of reinforcement shall be 24 inches (610 mm) for all other masonry.

R606.11.3.2.1 Shear wall reinforcement requirements. The maximum spacing of vertical and horizontal reinforcement shall be the smaller of one-third the length of the shear wall, one-third the height of the shear wall, or 48 inches (1219 mm). The minimum cross-sectional area of vertical reinforcement shall be one-third of the required shear reinforcement. Shear reinforcement shall be anchored around vertical reinforcing bars with a standard hook.

R606.11.3.3 Minimum reinforcement for masonry columns. Lateral ties in masonry columns shall be spaced not more than 8 inches (203 mm) on center and shall be at least $^3/_8$ inch (9.5 mm) diameter. Lateral ties shall be embedded in grout.

R606.11.3.4 Material restrictions. Type N mortar or masonry cement shall not be used as part of the lateral-force-resisting system.

R606.11.3.5 Lateral tie anchorage. Standard hooks for lateral tie anchorage shall be either a 135-degree (2.4 rad) standard hook or a 180-degree (3.2 rad) standard hook.

R606.11.4 Seismic Design Category D_2. Structures in Seismic Design Category D_2 shall comply with the requirements of Seismic Design Category D_1 and to the additional requirements of this section.

TABLE R606.11.3.2
MINIMUM DISTRIBUTED WALL REINFORCEMENT FOR BUILDINGS ASSIGNED TO SEISMIC DESIGN CATEGORY D_1

NOMINAL WALL THICKNESS (inches o.c.)	MINIMUM SUM OF THE VERTICAL AND HORIZONTAL REINFORCEMENT AREAS[a] (inches per square foot)	MINIMUM REINFORCEMENT AS DISTRIBUTED IN BOTH HORIZONTAL AND VERTICAL DIRECTIONS[b] (inches per square foot)	MINIMUM BAR SIZE FOR REINFORCEMENT SPACED AT 48 INCHES
6	0.135	0.047	#4
8	0.183	0.064	#5
10	0.231	0.081	#6
12	0.279	0.098	#6

For SI: 1 inch = 25.4 mm, 1 foot = 304.8 mm, 1 square inch per foot = 2.12 mm^2/mm.

a. Based on the minimum reinforcing ratio of 0.002 times the gross cross-sectional area of the wall.

b. Based on the minimum reinforcing ratio each direction of 0.0007 times the gross cross-sectional area of the wall.

R606.11.4.1 Design of elements not part of the lateral-force-resisting system. Stack bond masonry that is not part of the lateral-force-resisting system shall have a horizontal cross-sectional area of reinforcement of at least 0.0015 times the gross cross-sectional area of masonry. Table R606.11.4.1 provides minimum reinforcing bar sizes for masonry walls. The maximum spacing of horizontal reinforcement shall be 24 inches (610 mm). These elements shall be solidly grouted and shall be constructed of hollow open-end units or two wythes of solid units.

TABLE R606.11.4.1
MINIMUM REINFORCING FOR STACKED BONDED MASONRY WALLS IN SEISMIC DESIGN CATEGORY D_2

NOMINAL WALL THICKNESS (inches)	MINIMUM BAR SIZE SPACED AT 24 INCHES
6	#4
8	#5
10	#5
12	#6

For SI: 1 inch = 25.4 mm.

R606.11.4.2 Design of elements part of the lateral-force-resisting system. Stack bond masonry that is part of the lateral-force-resisting system shall have a horizontal cross-sectional area of reinforcement of at least 0.0025 times the gross cross-sectional area of masonry. Table R606.11.4.2 provides minimum reinforcing bar sizes for masonry walls. The maximum spacing of horizontal reinforcement shall be 16 inches (406 mm). These elements shall be solidly grouted and shall be constructed of hollow open-end units or two wythes of solid units.

TABLE R606.11.4.2
MINIMUM REINFORCING FOR STACKED BONDED MASONRY WALLS IN SEISMIC DESIGN CATEGORY D_2

NOMINAL WALL THICKNESS (inches)	MINIMUM BAR SIZE SPACED AT 16 INCHES
6	#4
8	#5
10	#5
12	#6

For SI: 1 inch = 25.4 mm.

R606.12 Protection for reinforcement. All bars shall be completely embedded in mortar or grout. Joint reinforcement embedded in horizontal mortar joints shall not have less than $^5/_8$-inch (15.9 mm) mortar coverage from the exposed face. All other reinforcement shall have a minimum coverage of one bar diameter over all bars, but not less than $^3/_4$ inch (19.1 mm), except where exposed to weather or soil, in which case the minimum coverage shall be 2 inches (51 mm).

R606.13 Beam supports. Beams, girders or other concentrated loads supported by a wall or column shall have a bearing of at least 3 inches (76 mm) in length measured parallel to the beam upon solid masonry not less than 4 inches (102 mm) in thickness, or upon a metal bearing plate of adequate design and dimensions to distribute the load safely, or upon a continuous reinforced masonry member projecting not less than 4 inches (102 mm) from the face of the wall.

R606.13.1 Joist bearing. Joists shall have a bearing of not less than $1^1/_2$ inches (38 mm), except as provided in Section R606.13, and shall be supported in accordance with Figure R606.10(1).

R606.14 Metal accessories. Joint reinforcement, anchors, ties and wire fabric shall conform to the following: ASTM A 82 for wire anchors and ties; ASTM A 36 for plate, headed and bent-bar anchors; ASTM A 510 for corrugated sheet metal anchors and ties; ASTM A 951 for joint reinforcement; ASTM B 227 for copper-clad steel wire ties; or ASTM A 167 for stainless steel hardware.

R606.14.1 Corrosion protection. Minimum corrosion protection of joint reinforcement, anchor ties and wire fabric for use in masonry wall construction shall conform to Table R606.14.1.

TABLE R606.14.1
MINIMUM CORROSION PROTECTION

MASONRY METAL ACCESSORY	STANDARD
Joint reinforcement, interior walls	ASTM A 641, Class 1
Wire ties or anchors in exterior walls completely embedded in mortar or grout	ASTM A 641, Class 3
Wire ties or anchors in exterior walls not completely embedded in mortar or grout	ASTM A 153, Class B-2
Joint reinforcement in exterior walls or interior walls exposed to moist environment	ASTM A 153, Class B-2
Sheet metal ties or anchors exposed to weather	ASTM A 153, Class B-2
Sheet metal ties or anchors completely embedded in mortar or grout	ASTM A 525, Class G-60
Stainless steel hardware for any exposure	ASTM A 167, Type 304

SECTION R607
UNIT MASONRY

R607.1 Mortar. Mortar for use in masonry construction shall comply with ASTM C 270. The type of mortar shall be in accordance with Sections R607.1.1, R607.1.2 and R607.1.3 and shall meet the proportion specifications of Table R607.1 or the property specifications of ASTM C 270.

R607.1.1 Foundation walls. Masonry foundation walls constructed as set forth in Tables R404.1.1(1) through R404.1.1(4) and mortar shall be Type M or S.

R607.1.2 Masonry in Seismic Design Categories A, B and C. Mortar for masonry serving as the lateral-force-resisting system in Seismic Design Categories A, B and C shall be Type M, S or N mortar.

R607.1.3 Masonry in Seismic Design Categories D_1 and D_2. Mortar for masonry serving as the lateral-force-resisting system in Seismic Design Categories D_1 and D_2 shall be Type M or S portland cement-lime or mortar cement mortar.

R607.2 Placing mortar and masonry units.

R607.2.1 Bed and head joints. Unless otherwise required or indicated on the project drawings, head and bed joints shall be $^3/_8$ inch (9.5 mm) thick, except that the thickness of the bed joint of the starting course placed over foundations shall not be less than $^1/_4$ inch (6.4 mm) and not more than $^3/_4$ inch (19.1 mm).

R607.2.1.1 Mortar joint thickness tolerance. Mortar joint thickness shall be within the following tolerances from the specified dimensions:

1. Bed joint: + $^1/_8$ inch (3.2 mm).
2. Head joint: $^1/_4$ inch (6.4 mm), + $^3/_8$ inch (9.5 mm).
3. Collar joints: $^1/_4$ inch (6.4 mm), + $^3/_8$ inch (9.5 mm).

Exception: Nonload-bearing masonry elements and masonry veneers designed and constructed in accordance with Section R703.7 are not required to meet these tolerances.

R607.2.2 Masonry unit placement. The mortar shall be sufficiently plastic and units shall be placed with sufficient pressure to extrude mortar from the joint and produce a tight joint. Deep furrowing of bed joints that produces voids shall not be permitted. Any units disturbed to the extent that initial bond is broken after initial placement shall be removed and relaid in fresh mortar. Surfaces to be in contact with mortar shall be clean and free of deleterious materials.

R607.2.2.1 Solid masonry. All solid masonry units shall be laid with full head and bed joints and all interior vertical joints that are designed to receive mortar shall be filled.

R607.2.2.2 Hollow masonry. For hollow masonry units, all head and bed joints shall be filled solidly with mortar for a distance in from the face of the unit not less than the thickness of the face shell.

R607.3 Installation of wall ties. The installation of wall ties shall be as follows:

1. The ends of wall ties shall be embedded in mortar joints. Wall tie ends shall engage outer face shells of hollow units by at least $^1/_2$ inch (12.7 mm). Wire wall ties shall be embedded at least $1^1/_2$ inches (38 mm) into the mortar bed of solid masonry units or solid grouted hollow units.
2. Wall ties shall not be bent after being embedded in grout or mortar.

TABLE R607.1
MORTAR PROPORTIONS[a,b]

		PROPORTIONS BY VOLUME (cementitious materials)								
			Mortar cement			Masonry cement				
Mortar	Type	Portland cement or blended cement	M	S	N	M	S	N	Hydrated lime or lime putty	Aggregate ratio (measured in damp, loose conditions)
Cement-lime	M	1	—	—	—	—	—	—	$^1/_4$	Not less than $2^1/_4$ and not more than 3 times the sum of separate volumes of lime, if used, and cement
	S	1	—	—	—	—	—	—	over $^1/_4$ to $^1/_2$	
	N	1	—	—	—	—	—	—	over $^1/_2$ to $1^1/_4$	
	O	1	—	—	—	—	—	—	over $1^1/_4$ to $2^1/_2$	
Mortar cement	M	1	—	—	1	—	—	—	—	
	M	—	1	—	—	—	—	—		
	S	$^1/_2$	—	—	1	—	—	—		
	S	—	—	1	—	—	—	—		
	N	—	—	—	1	—	—	—		
	O	—	—	—	1	—	—	—		
Masonry cement	M	1				—	—	1	—	
	M	—				1	—	—		
	S	$^1/_2$				—	—	1		
	S	—				—	1	—		
	N	—				—	—	1		
	O	—				—	—	1		

For SI: 1 cubic foot = 0.0283 m^3, 1 pound = 0.454 kg.

a. For the purpose of these specifications, the weight of 1 cubic foot of the respective materials shall be considered to be as follows:

Portland Cement	94 pounds	Masonry Cement	Weight printed on bag
Mortar Cement	Weight printed on bag	Hydrated Lime	40 pounds
Lime Putty (Quicklime)	80 pounds	Sand, damp and loose	80 pounds of dry sand

b. Two air-entraining materials shall not be combined in mortar.

SECTION R608
MULTIPLE WYTHE MASONRY

R608.1 General. The facing and backing of multiple wythe masonry walls shall be bonded in accordance with Section R608.1.1, R608.1.2 or R608.1.3. In cavity walls, neither the facing nor the backing shall be less than 3 inches (76 mm) nominal in thickness and the cavity shall not be more than 4 inches (102 mm) nominal in width. The backing shall be at least as thick as the facing.

Exception: Cavities shall be permitted to exceed the 4-inch (102 mm) nominal dimension provided tie size and tie spacing have been established by calculation.

R608.1.1 Bonding with masonry headers. Bonding with solid or hollow masonry headers shall comply with Sections R608.1.1.1 and R608.1.1.2.

R608.1.1.1 Solid units. Where the facing and backing (adjacent wythes) of solid masonry construction are bonded by means of masonry headers, no less than 4 percent of the wall surface of each face shall be composed of headers extending not less than 3 inches (76 mm) into the backing. The distance between adjacent full-length headers shall not exceed 24 inches (610 mm) either vertically or horizontally. In walls in which a single header does not extend through the wall, headers from the opposite sides shall overlap at least 3 inches (76 mm), or headers from opposite sides shall be covered with another header course overlapping the header below at least 3 inches (76 mm).

R608.1.1.2 Hollow units. Where two or more hollow units are used to make up the thickness of a wall, the stretcher courses shall be bonded at vertical intervals not exceeding 34 inches (864 mm) by lapping at least 3 inches (76 mm) over the unit below, or by lapping at vertical intervals not exceeding 17 inches (432 mm) with units that are at least 50 percent greater in thickness than the units below.

R608.1.2 Bonding with wall ties or joint reinforcement. Bonding with wall ties or joint reinforcement shall comply with Sections R608.1.2.1 through R608.1.2.3.

R608.1.2.1 Bonding with wall ties. Bonding with wall ties, except as required by Section R610, where the facing and backing (adjacent wythes) of masonry walls are bonded with $^3/_{16}$-inch-diameter (4.8 mm) wall ties embedded in the horizontal mortar joints, there shall be at least one metal tie for each 4.5 square feet (0.418 m^2) of wall area. Ties in alternate courses shall be staggered. The maximum vertical distance between ties shall not exceed 24 inches (610 mm), and the maximum horizontal distance shall not exceed 36 inches (914 mm). Rods or ties bent to rectangular shape shall be used with hollow masonry units laid with the cells vertically. In other walls, the ends of ties shall be bent to 90-degree (0.79 rad) angles to provide hooks no less than 2 inches (51 mm) long. Additional bonding ties shall be provided at all openings, spaced not more than 3 feet (914 mm) apart around the perimeter and within 12 inches (305 mm) of the opening.

R608.1.2.2 Bonding with adjustable wall ties. Where the facing and backing (adjacent wythes) of masonry are bonded with adjustable wall ties, there shall be at least one tie for each 2.67 square feet (0.248 m^2) of wall area. Neither the vertical nor horizontal spacing of the adjustable wall ties shall exceed 24 inches (610 mm). The maximum vertical offset of bed joints from one wythe to the other shall be 1.25 inches (32 mm). The maximum clearance between connecting parts of the ties shall be $^1/_{16}$ inch (1.6 mm). When pintle legs are used, ties shall have at least two $^3/_{16}$-inch-diameter (4.8 mm) legs.

R608.1.2.3 Bonding with prefabricated joint reinforcement. Where the facing and backing (adjacent wythes) of masonry are bonded with prefabricated joint reinforcement, there shall be at least one cross wire serving as a tie for each 2.67 square feet (0.248 m^2) of wall area. The vertical spacing of the joint reinforcement shall not exceed 16 inches (406 mm). Cross wires on prefabricated joint reinforcement shall not be smaller than No. 9 gage. The longitudinal wires shall be embedded in the mortar.

R608.1.3 Bonding with natural or cast stone. Bonding with natural and cast stone shall conform to Sections R608.1.3.1 and R608.1.3.2.

R608.1.3.1 Ashlar masonry. In ashlar masonry, bonder units, uniformly distributed, shall be provided to the extent of not less than 10 percent of the wall area. Such bonder units shall extend not less than 4 inches (102 mm) into the backing wall.

R608.1.3.2 Rubble stone masonry. Rubble stone masonry 24 inches (610 mm) or less in thickness shall have bonder units with a maximum spacing of 3 feet (914 mm) vertically and 3 feet (914 mm) horizontally, and if the masonry is of greater thickness than 24 inches (610 mm), shall have one bonder unit for each 6 square feet (0.557 m^2) of wall surface on both sides.

R608.2 Masonry bonding pattern. Masonry laid in running and stack bond shall conform to Sections R608.2.1 and R608.2.2.

R608.2.1 Masonry laid in running bond. In each wythe of masonry laid in running bond, head joints in successive courses shall be offset by not less than one-fourth the unit length, or the masonry walls shall be reinforced longitudinally as required in Section R608.2.2.

R608.2.2 Masonry laid in stack bond. Where unit masonry is laid with less head joint offset than in Section R607.2.1, the minimum area of horizontal reinforcement placed in mortar bed joints or in bond beams spaced not more than 48 inches (1219 mm) apart, shall be 0.0007 times the vertical cross-sectional area of the wall.

SECTION R609
GROUTED MASONRY

R609.1 General. Grouted multiple-wythe masonry is a form of construction in which the space between the wythes is solidly filled with grout. It is not necessary for the cores of masonry units to be filled with grout. Grouted hollow unit masonry is a form of construction in which certain cells of hollow units are continuously filled with grout.

R609.1.1 Grout. Grout shall consist of cementitious material and aggregate in accordance with ASTM C 476 and the proportion specifications of Table R609.1.1. Type M or Type S mortar to which sufficient water has been added to produce pouring consistency can be used as grout.

R609.1.2 Grouting requirements. Maximum pour heights and the minimum dimensions of spaces provided for grout placement shall conform to Table R609.1.2. If the work is stopped for one hour or longer, the horizontal construction joints shall be formed by stopping all tiers at the same elevation and with the grout 1 inch (25.4 mm) below the top.

R609.1.3 Grout space (cleaning). Provision shall be made for cleaning grout space. Mortar projections that project more than 0.5 inch (12.7 mm) into grout space and any other foreign matter shall be removed from grout space prior to inspection and grouting.

R609.1.4 Grout placement. Grout shall be a plastic mix suitable for pumping without segregation of the constituents and shall be mixed thoroughly. Grout shall be placed by pumping or by an approved alternate method and shall be placed before any initial set occurs and in no case more than $1^1/_2$ hours after water has been added. Grouting shall be done in a continuous pour, in lifts not exceeding 5 feet (1524 mm). It shall be consolidated by puddling or mechanical vibrating during placing and reconsolidated after excess moisture has been absorbed but before plasticity is lost.

R609.1.4.1 Grout pumped through aluminum pipes. Grout shall not be pumped through aluminum pipes.

R609.1.5 Cleanouts. Where required by the building official, cleanouts shall be provided as specified in this section. The cleanouts shall be sealed before grouting and after inspection.

R609.1.5.1 Grouted multiple-wythe masonry. Cleanouts shall be provided at the bottom course of the exterior wythe at each pour of grout where such pour exceeds 5 feet (1524 mm) in height.

R609.1.5.2 Grouted hollow unit masonry. Cleanouts shall be provided at the bottom course of each cell to be grouted at each pour of grout, where such pour exceeds 4 feet (1219 mm) in height.

TABLE R609.1.1
GROUT PROPORTIONS BY VOLUME FOR MASONRY CONSTRUCTION

TYPE	PORTLAND CEMENT OR BLENDED CEMENT SLAG CEMENT	HYDRATED LIME OR LIME PUTTY	AGGREGATE MEASURED IN A DAMP, LOOSE CONDITION	
			Fine	Coarse
Fine	1	0 to 1/10	$2^1/_4$ to 3 times the sum of the volume of the cementitious materials	—
Coarse	1	0 to 1/10	$2^1/_4$ to 3 times the sum of the volume of the cementitious materials	1 to 2 times the sum of the volumes of the cementitious materials

TABLE R609.1.2
GROUT SPACE DIMENSIONS AND POUR HEIGHTS

GROUT TYPE	GROUT POUR MAXIMUM HEIGHT (feet)	MINIMUM WIDTH OF GROUT SPACES[a,b] (inches)	MINIMUM GROUT[b,c] SPACE DIMENSIONS FOR GROUTING CELLS OF HOLLOW UNITS (inches x inches)
Fine	1	0.75	1.5 × 2
	5	2	2 × 3
	12	2.5	2.5 × 3
	24	3	3 × 3
Coarse	1	1.5	1.5 × 3
	5	2	2.5 × 3
	12	2.5	3 × 3
	24	3	3 × 4

For SI: 1 inch = 25.4 mm, 1 foot = 304.8 mm.

a. For grouting between masonry wythes.

b. Grout space dimension is the clear dimension between any masonry protrusion and shall be increased by the horizontal projection of the diameters of the horizontal bars within the cross section of the grout space.

c. Area of vertical reinforcement shall not exceed 6 percent of the area of the grout space.

R609.2 Grouted multiple-wythe masonry. Grouted multiple-wythe masonry shall conform to all the requirements specified in Section R609.1 and the requirements of this section.

R609.2.1 Bonding of backup wythe. Where all interior vertical spaces are filled with grout in multiple-wythe construction, masonry headers shall not be permitted. Metal wall ties shall be used in accordance with Section R608.1.2 to prevent spreading of the wythes and to maintain the vertical alignment of the wall. Wall ties shall be installed in accordance with Section R608.1.2 when the backup wythe in multiple-wythe construction is fully grouted.

R609.2.2 Grout spaces. Fine grout shall be used when interior vertical space to receive grout does not exceed 2 inches (51 mm) in thickness. Interior vertical spaces exceeding 2 inches (51 mm) in thickness shall use coarse or fine grout.

R609.2.3 Grout barriers. Vertical grout barriers or dams shall be built of solid masonry across the grout space the entire height of the wall to control the flow of the grout horizontally. Grout barriers shall not be more than 25 feet (7620 mm) apart. The grouting of any section of a wall between control barriers shall be completed in one day with no interruptions greater than one hour.

R609.3 Reinforced grouted multiple-wythe masonry. Reinforced grouted multiple-wythe masonry shall conform to all the requirements specified in Sections R609.1 and R609.2 and the requirements of this section.

R609.3.1 Construction. The thickness of grout or mortar between masonry units and reinforcement shall not be less than $^1/_4$ inch (6.4 mm), except that $^1/_4$-inch (6.4 mm) bars may be laid in horizontal mortar joints at least $^1/_2$ inch (12.7 mm) thick, and steel wire reinforcement may be laid in horizontal mortar joints at least twice the thickness of the wire diameter.

R609.4 Reinforced hollow unit masonry. Reinforced hollow unit masonry shall conform to all the requirements of Section R609.1 and the requirements of this section.

R609.4.1 Construction. Requirements for construction shall be as follows:

1. All reinforced hollow-unit masonry shall be built to preserve the unobstructed vertical continuity of the cells to be filled. Walls and cross webs forming such cells to be filled shall be full-bedded in mortar to prevent leakage of grout. All head and end joints shall be solidly filled with mortar for a distance in from the face of the wall or unit not less than the thickness of the longitudinal face shells. Bond shall be provided by lapping units in successive vertical courses.
2. Cells to be filled shall have vertical alignment sufficient to maintain a clear, unobstructed continuous vertical cell of dimensions prescribed in Table R609.1.2.
3. Vertical reinforcement shall be held in position at top and bottom and at intervals not exceeding 200 diameters of the reinforcement.
4. All cells containing reinforcement shall be filled solidly with grout. Grout shall be poured in lifts of 8-foot (2438 mm) maximum height. When total grout pour exceeds 8 feet (2438 mm) in height, the grout shall be placed in lifts not exceeding 5 feet (1524 mm) and special inspection during grouting shall be required.
5. Horizontal steel shall be fully embedded by grout in an uninterrupted pour.

SECTION R610 GLASS UNIT MASONRY

R610.1 General. Panels of glass unit masonry located in load-bearing and nonload-bearing exterior and interior walls shall be constructed in accordance with this section.

R610.2 Materials. Hollow glass units shall be partially evacuated and have a minimum average glass face thickness of $^3/_{16}$ inch (4.8 mm). The surface of units in contact with mortar shall be treated with a polyvinyl butyral coating or latex-based paint. The use of reclaimed units is prohibited.

R610.3 Units. Hollow or solid glass block units shall be standard or thin units.

R610.3.1 Standard units. The specified thickness of standard units shall be at least $3^7/_8$ inches (98 mm).

R610.3.2 Thin units. The specified thickness of thin units shall be at least $3^1/_8$ inches (79 mm) for hollow units and at least 3 inches (76 mm) for solid units.

R610.4 Isolated panels. Isolated panels of glass unit masonry shall conform to the requirements of this section.

R610.4.1 Exterior standard-unit panels. The maximum area of each individual standard-unit panel shall be 144 square feet (13.4 m^2) when the design wind pressure is 20 psf (958 Pa). The maximum area of such panels subjected to design wind pressures other than 20 psf (958 Pa) shall be in accordance with Figure R610.4.1. The maximum panel dimension between structural supports shall be 25 feet (7620 mm) in width or 20 feet (6096 mm) in height.

R610.4.2 Exterior thin-unit panels. The maximum area of each individual thin-unit panel shall be 85 square feet (7.9 m^2). The maximum dimension between structural supports shall be 15 feet (4572 mm) in width or 10 feet (3048 mm) in height. Thin units shall not be used in applications where the design wind pressure as stated in Table R301.2(1) exceeds 20 psf (958 Pa).

R610.4.3 Interior panels. The maximum area of each individual standard-unit panel shall be 250 square feet (23.2 m^2). The maximum area of each thin-unit panel shall be 150 square feet (13.9 m^2). The maximum dimension between structural supports shall be 25 feet (7620 mm) in width or 20 feet (6096 mm) in height.

R610.4.4 Curved panels. The width of curved panels shall conform to the requirements of Sections R610.4.1, R610.4.2 and R610.4.3, except additional structural supports shall be provided at locations where a curved section joins a straight section, and at inflection points in multicurved walls.

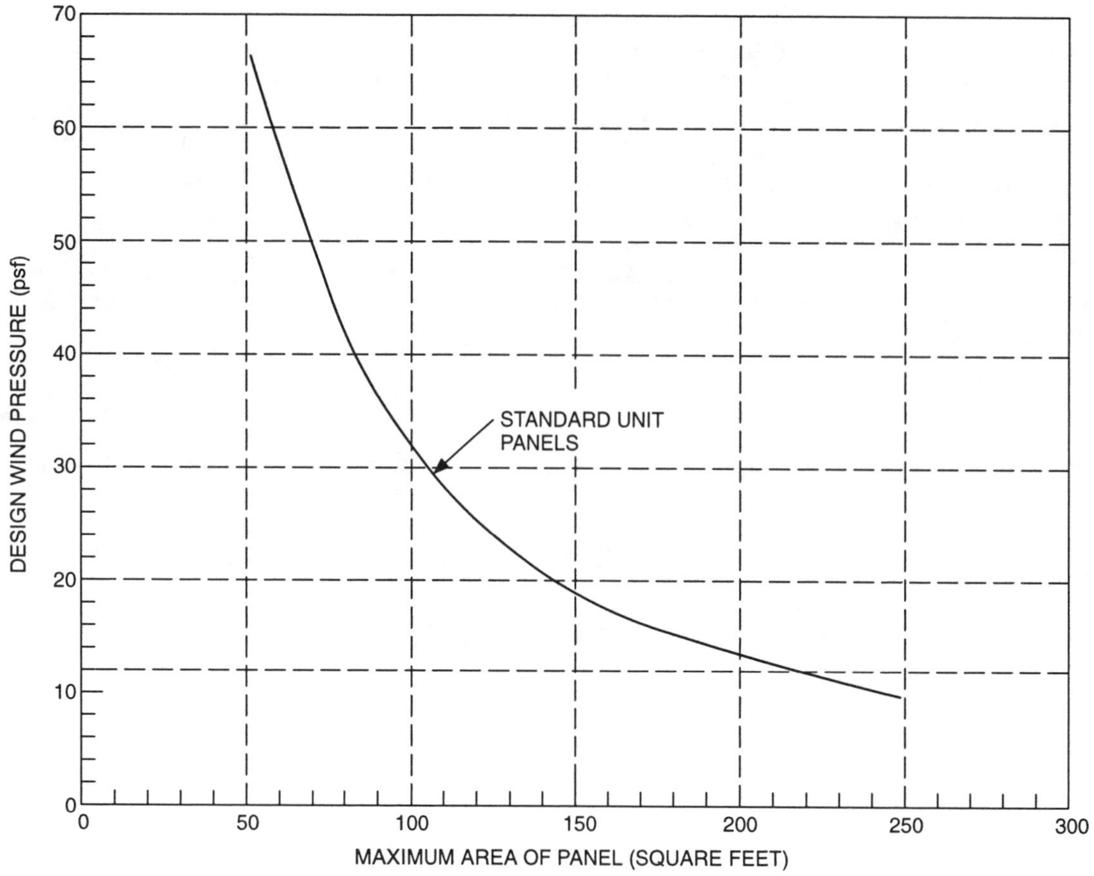

For SI: 1 square foot = 0.0929 m^2, 1 pound per square foot = 0.0479 kN/m^2.

FIGURE R610.4.1
GLASS UNIT MASONRY DESIGN WIND LOAD RESISTANCE

R610.5 Panel support. Glass unit masonry panels shall conform to the support requirements of this section.

R610.5.1 Deflection. The maximum total deflection of structural members that support glass unit masonry shall not exceed $^1/_{600}$.

R610.5.2 Lateral support. Glass unit masonry panels shall be laterally supported along the top and sides of the panel. Lateral supports for glass unit masonry panels shall be designed to resist a minimum of 200 pounds per lineal feet (2918 N/m) of panel, or the actual applied loads, whichever is greater. Except for single unit panels, lateral support shall be provided by panel anchors along the top and sides spaced a maximum of 16 inches (406 mm) on center or by channel-type restraints. Single unit panels shall be supported by channel-type restraints.

Exceptions:

1. Lateral support is not required at the top of panels that are one unit wide.
2. Lateral support is not required at the sides of panels that are one unit high.

R610.5.2.1 Panel anchor restraints. Panel anchors shall be spaced a maximum of 16 inches (406 mm) on center in both jambs and across the head. Panel anchors shall be embedded a minimum of 12 inches (305 mm) and shall be provided with two fasteners so as to resist the loads specified in Section R610.5.2.

R610.5.2.2 Channel-type restraints. Glass unit masonry panels shall be recessed at least 1 inch (25.4 mm) within channels and chases. Channel-type restraints shall be oversized to accommodate expansion material in the opening, packing and sealant between the framing restraints, and the glass unit masonry perimeter units.

R610.6 Sills. Before bedding of glass units, the sill area shall be covered with a water base asphaltic emulsion coating. The coating shall shall be a minimum of $^1/_8$ inch (3.2 mm) thick.

R610.7 Expansion joints. Glass unit masonry panels shall be provided with expansion joints along the top and sides at all structural supports. Expansion joints shall be a minimum of $^3/_8$ inch (9.5 mm) in thickness and shall have sufficient thickness to accommodate displacements of the supporting structure. Expansion joints shall be entirely free of mortar and other debris and shall be filled with resilient material.

R610.8 Mortar. Glass unit masonry shall be laid with Type S or N mortar. Mortar shall not be retempered after initial set.

Mortar unused within $1^1/_2$ hours after initial mixing shall be discarded.

R610.9 Reinforcement. Glass unit masonry panels shall have horizontal joint reinforcement spaced a maximum of 16 inches (406 mm) on center located in the mortar bed joint. Horizontal joint reinforcement shall extend the entire length of the panel but shall not extend across expansion joints. Longitudinal wires shall be lapped a minimum of 6 inches (152 mm) at splices. Joint reinforcement shall be placed in the bed joint immediately below and above openings in the panel. The reinforcement shall have not less than two parallel longitudinal wires of size W1.7 or greater, and have welded cross wires of size W1.7 or greater.

R610.10 Placement. Glass units shall be placed so head and bed joints are filled solidly. Mortar shall not be furrowed. Head and bed joints of glass unit masonry shall be $^1/_4$ inch (6.4 mm) thick, except that vertical joint thickness of radial panels shall not be less than $^1/_8$ inch (3.2 mm) or greater than $^5/_8$ inch (15.9 mm). The bed joint thickness tolerance shall be minus $^1/_{16}$ inch (1.6 mm) and plus $^1/_8$ inch (3.2 mm). The head joint thickness tolerance shall be plus or minus $^1/_8$ inch (3.2 mm).

SECTION R611
INSULATING CONCRETE FORM WALL CONSTRUCTION

R611.1 General. Insulating Concrete Form walls shall be designed and constructed in accordance with the provisions of this section or in accordance with the provisions of ACI 318. When ACI 318 or the provisions of this section are used to design insulating concrete form walls, project drawings, typical details and specifications are not required to bear the seal of the architect or engineer responsible for design, unless otherwise required by the state law of the jurisdiction having authority.

R611.2 Applicability limits. The provisions of this section shall apply to the construction of insulating concrete form walls for buildings not greater than 60 feet (18 288 mm) in plan dimensions, and floors not greater than 32 feet (9754 mm) or roofs not greater than 40 feet (12 192 mm) in clear span. Buildings shall not exceed two stories in height above-grade with each story not greater than 10 feet (3048 mm) high. Walls constructed in accordance with the provisions of this section shall be limited to buildings subjected to a maximum design wind speed of 130 miles per hour (209 km/h), a maximum ground snow load of 70 psf (3.35 kN/m^2), and Seismic Design Categories A, B and C.

R611.3 Flat insulating concrete form wall systems. Flat ICF wall systems shall comply with Figure R611.3 and shall have reinforcement in accordance with Table R611.3 and Section R611.7.

R611.4 Waffle-grid insulating concrete form wall systems. Waffle-grid wall systems shall comply with Figure R611.4 and shall have reinforcement in accordance with Table R611.4(1) and Section R611.7. The minimum core dimensions shall comply with Table R611.4(2).

R611.5 Screen-grid insulating concrete form wall systems. Screen-grid ICF wall systems shall comply with Figure R611.5 and shall have reinforcement in accordance with Table R611.5. The minimum core dimensions shall comply with Table R611.4(2).

R611.6 Material. Insulating concrete form wall materials shall comply with this section.

R611.6.1 Concrete material. Ready-mixed concrete for insulating concrete form walls shall be in accordance with Section R402.2. Maximum slump shall not be greater than 6 inches (152 mm) as determined in accordance with ASTM C 143. Maximum aggregate size shall not be larger than $^3/_4$ inch (19.1 mm).

Exception: Concrete mixes conforming to the ICF manufacturer's recommendations.

R611.6.2 Reinforcing steel. Reinforcing steel shall meet the requirements of ASTM A 615, A 616, A 617 or A 706. The minimum yield strength of reinforcing steel shall be 40,000 psi (Grade 40) (276 MPa).

R611.6.3 Insulation materials. Insulating concrete forms material shall meet the surface burning characteristics of Section R318.1.1. A thermal barrier shall be provided on the building interior in accordance with Section R318.1.2 or Section R702.3.4.

R611.7 Wall construction. Insulating concrete form walls shall be constructed in accordance with the provisions of this section and Figure R611.7(1).

R611.7.1 Reinforcement.

R611.7.1.1 Location. Vertical and horizontal wall reinforcements shall not be placed within the middle third of the wall. Steel reinforcement shall have a minimum concrete cover in accordance with ACI 318.

Exception: Where insulated concrete forms are used and the form remains in place as cover for the concrete, the minimum concrete cover for the reinforcing steel is permitted to be reduced to $^3/_4$ inch (19.1 mm).

R611.7.1.2 Vertical steel. Above-grade concrete walls shall have reinforcement in accordance with Sections R611.3, R611.4, or R611.5 and R611.7.2. All vertical reinforcement in the top-most ICF story shall terminate with a bend or a standard hook and be provided with a minimum lap splice of 24 inches (610 mm) with the top horizontal reinforcement.

R611.7.1.3 Horizontal steel. Concrete walls with minimum thickness of 4 inches (102 mm) shall have a minimum of one continuous No. 4 horizontal reinforcing bar placed at 32 inches (812 mm) on center with one bar within 12 inches (305 mm) of the top of the wall story. Concrete walls 5.5 inches (140 mm) thick or greater shall have a minimum of one continuous No. 4 horizontal reinforcing bar placed at 48 inches (1219 mm) on center with one bar located within 12 inches (305 mm) of the top of the wall story.

Horizontal reinforcement shall be continuous around building corners using corner bars or by bending the bars. In either case, the minimum lap splice shall be 24 inches (610 mm).

TABLE R611.3 MINIMUM VERTICAL WALL REINFORCEMENT FOR FLAT ICF ABOVE-GRADE WALLS[a,b,c,d]

MAXIMUM WIND SPEED (mph)	MAXIMUM WALL HEIGHT PER STORY (feet)	MINIMUM VERTICAL REINFORCEMENT					
		Supporting light-frame roof only		Supporting light-frame second story and roof		Supporting ICF second story and light-frame roof	
		Minimum wall thickness (inches)					
		3.5[e]	5.5	3.5[e]	5.5	3.5[e]	5.5
85	8	#4@32″	N/R	#4@32″	N/R	#4@32″	N/R
	9	#4@32″	N/R	#4@32″	N/R	#3@20″; #4@24″; #5@26″	N/R
	10	#4@32″	N/R	#4@32″	N/R	#3@14″; #4@16″; #5@18″	N/R
100	8	#4@32″	N/R	#4@32″	N/R	#4@32″	N/R
	9	#4@32″	N/R	#3@24″; #4@32″; #5@34″	N/R	#3@14″; #4@18″; #5@20″	N/R
	10	#3@16″; #4@26″; #5@34″	N/R	#3@16″; #4@20″; #5@22″	N/R	Design required	N/R
110	8	#4@32″	N/R	#4@32″	N/R	#4@32″	N/R
	9	#3@16″; #4@26″; #5@34″	N/R	#3@18″; #4@20″; #5@22″	N/R	Design required	N/R
	10	Design required	N/R	#3@10″; #4@12″; #5@14″	N/R	Design required	N/R
120	8	#3@18″; #4@30″; #5@40″	#4@96″	#3@18″; #4@30″; #5@30″	#4@96″	#4@32″	#4@96″
	9	#3@12″; #4@22″; #5@30″	#4@96″	#3@12″; #4@16″; #5@16″	#4@96″	Design required	#4@96″
	10	Design required	#4@96″	Design required	#4@96″	Design required	#4@96″
130	8	#3@14″; #4@20″; #5@24″	#4@96″	#3@14″; #4@18″; #5@20″	#4@96″	Design required	#4@96″
	9	Design required	#4@96″	Design required	#4@96″	Design required	#4@96″
	10	Design required	#4@96″	Design required	#4@96″	Design required	#4@96″

For SI: 1 inch = 25.4 mm, 1 foot = 304.8 mm, 1 mile per hour = 1.6093 km/h, 1 pound per square inch = 6.895 kPa.

NOTE: This table is based on concrete with a minimum specified compressive strength of 2,500 psi, reinforcing steel with a minimum yield strength of 40,000 psi and an assumed equivalent rectangular cross-section. This table is not intended to prohibit the use of ICF manufacturer's tables based on engineering analysis in accordance with ACI 318.

a. N/R indicates no vertical wall reinforcement is required.

b. Deflection criterion is *L*/240, where *L* is the height of the wall story in inches.

c. Interpolation shall not be permitted.

d. See Section R611.7.1.4 for additional reinforcement requirements for dwellings in Seismic Design Category C.

e. A 3.5-inch wall is not permitted if wood ledgers are to be used to support the second floor or roof loads. See Section R611.8.

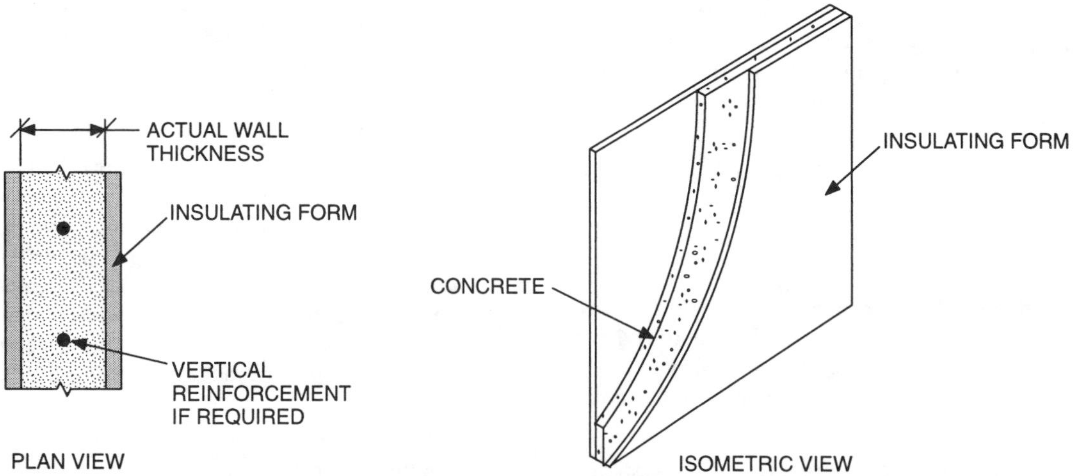

FIGURE R611.3 FLAT ICF WALL SYSTEM

TABLE R611.4(1)
MINIMUM VERTICAL WALL REINFORCEMENT FOR WAFFLE-GRID ICF ABOVE-GRADE WALLS[a, b, c, d, e]

MAXIMUM WIND SPEED (mph)	MAXIMUM WALL HEIGHT PER STORY (feet)	MINIMUM VERTICAL REINFORCEMENT					
		Supporting light-frame roof only		Supporting light-frame second story and roof		Supporting ICF second story and light-frame roof	
		Minimum wall thickness[d] (inches)					
		6	8	6	8	6	8
85	8	N/R	N/R	N/R	N/R	N/R	N/R
	9	N/R	N/R	N/R	N/R	N/R	N/R
	10	N/R	N/R	N/R	N/R	N/R	N/R
100	8	N/R	N/R	N/R	N/R	N/R	N/R
	9	N/R	N/R	N/R	N/R	N/R	N/R
	10	N/R	N/R	N/R	N/R	N/R	N/R
110	8	#4@96″	#4@96″	#4@96″	#4@96″	#4@96″	#4@96″
	9	#4@96″	#4@96″	#4@96″	#4@96″	#4@96″	#4@96″
	10	#3@12″; #4@24″; #5@36″	N/R	#3@12″; #4@24″; #5@24″	N/R	#3@12″; #4@24″; #5@24″	N/R
120	8	#4@96″	#4@96″	#4@96″	#4@96″	#4@96″	#4@96″
	9	#3@12″; #4@24″; #5@36″	#4@96″	#3@12″; #4@24″; #5@36″	#4@96″	#3@12″; #4@24″; #5@24″	#4@96″
	10	#3@12″; #4@24″; #5@24″	#4@96″	#3@12″; #4@24″; #5@24″	#4@96″	#3@12″; #4@12″	#4@96″
130	8	#3@12″; #4@24″; #5@36″	#4@96″	#3@12″; #4@24″; #5@36″	#4@96″	#3@12″; #4@24″; #5@24″	#4@96″
	9	#3@12″; #4@24″; #5@24″	#4@96″	#3@12″; #4@24″; #5@24″	#4@96″	#3@12″; #4@12″	#4@96″
	10	#3@12″; #4@12″; #5@24″	#4@96″	Design required	#4@96″	Design required	#4@96″

For SI: 1 inch = 25.4 mm, 1 foot = 304.8 mm, 1 pound per square inch = 6.895 kPa, 1 mile per hour = 1.6093 km/h.

NOTE: This table is based on concrete with a minimum specified compressive strength of 2,500 psi, reinforcing steel with a minimum yield strength of 40,000 psi and an assumed equivalent rectangular cross-section. This table is not intended to prohibit the use of ICF manufacturer's tables based on engineering analysis in accordance with ACI 318.

a. N/R indicates no vertical wall reinforcement is required.

b. Deflection criterion is $L/240$, where L is the height of the wall story in inches.

c. Interpolation shall not be permitted.

d. Nominal thickness is given; refer to Table R611.4(2) for actual concrete wall thickness.

e. See Section R611.7.1.4 for additional reinforcement requirements for dwellings in Seismic Design Category C.

TABLE R611.4(2)
DIMENSIONAL REQUIREMENTS FOR CORES AND WEBS IN WAFFLE-GRID AND SCREEN-GRID ICF WALLS[a,b]

WALL TYPE AND NOMINAL SIZE (inches)	MINIMUM WIDTH OF CORE (inches)	MINIMUM THICKNESS OF VERTICAL CORE (inches)	MAXIMUM SPACING OF VERTICAL CORES (inches)	MAXIMUM SPACING OF HORIZONTAL CORES (inches)	MINIMUM WEB THICKNESS (inches)
6″ Waffle-Grid	6.25	5	12	16	2
8″ Waffle-Grid	7	7	12	16	2
6″ Screen-Grid	5.5	5.5	12	12	N/A

For SI: 1 inch = 25.4 mm.

a. For width "W," thickness "T"; spacing, and web thickness, refer to Figures R611.4 and R611.5.

b. N/A indicates not applicable.

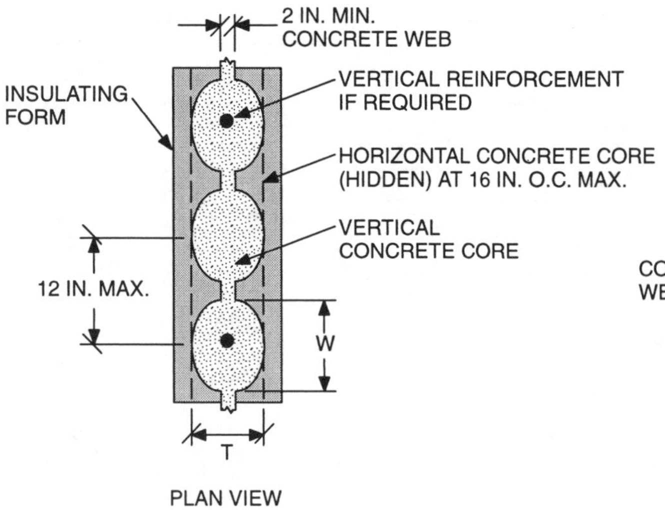

For SI: 1 inch = 25.4 mm.

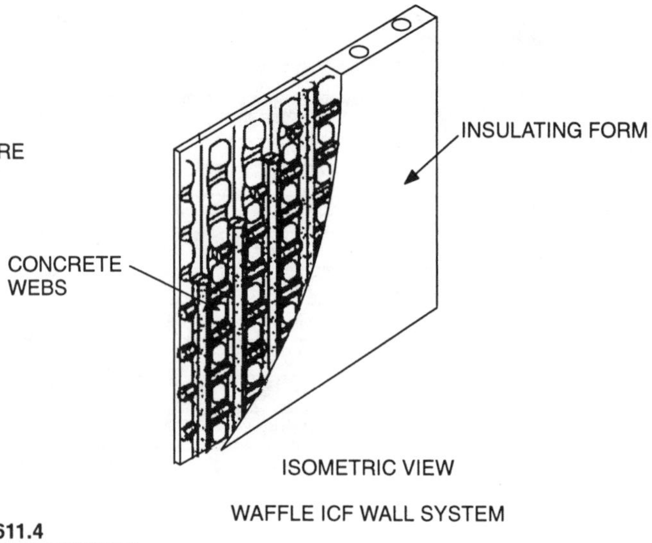

FIGURE R611.4
WAFFLE-GRID ICF WALL SYSTEM

TABLE R611.5
MINIMUM VERTICAL WALL REINFORCEMENT FOR SCREEN-GRID ICF ABOVE-GRADE WALLS[a,b,c,d,e]

MAXIMUM WIND SPEED (mph)	MAXIMUM WALL HEIGHT PER STORY (feet)	MINIMUM VERTICAL REINFORCEMENT		
		Supporting light-frame roof only	Supporting light-frame second story and roof	Supporting ICF second story and light-frame roof
		Minimum wall thickness[d] (inches)		
		6	6	6
85	8	N/R	N/R	N/R
	9	N/R	N/R	N/R
	10	N/R	N/R	N/R
100	8	N/R	N/R	N/R
	9	N/R	N/R	N/R
	10	N/R	N/R	N/R
110	8	#4@96″	#4@96″	#4@96″
	9	#4@96″	#4@96″	#4@96″
	10	#3@12″; #4@24″; #5@36″	#3@12″; #4@24″; #5@36″	N/R
120	8	#4@96″	#4@96″	#4@96″
	9	#3@24″; #4@36″	#3@24″; #4@36″; #5@36″	#4@96″
	10	#3@12″; #4@24″; #5@36″	#4@24″; #5@24″	#3@12″; #4@24″; #5@24″
130	8	#3@24″; #4@36″; #5@48″	#3@24″; #4@36″; #5@48″	#4@96″
	9	#3@12″; #4@24″; #5@36″	#3@12″; #4@24″; #5@24″	#3@12″; #4@24″; #5@24″
	10	#3@12″; #4@12″	#4@24″; #5@24″	#3@12″; #4@12″; #5@24″

For SI: 1 inch = 25.4 mm, 1 foot = 304.8 mm, 1 pound per square inch = 6.895 kPa, 1 mile per hour = 1.6093 km/h.

NOTE: This table is based on concrete with a minimum specified compressive strength of 2,500 psi, reinforcing steel with a minimum yield strength of 40,000 psi and an assumed equivalent rectangular cross-section. This table is not intended to prohibit the use of ICF manufacturer's tables based on engineering analysis in accordance with ACI 318.

a. N/R indicates no vertical wall reinforcement is required.

b. Deflection criterion is *L*/240, where *L* is the height of the wall story in inches.

c. Interpolation shall not be permitted.

d. Nominal thickness is given; refer to Table R611.4(2) for actual concrete wall thickness.

e. See Section R611.7.1.4 for additional requirements for dwellings in Seismic Design Category C.

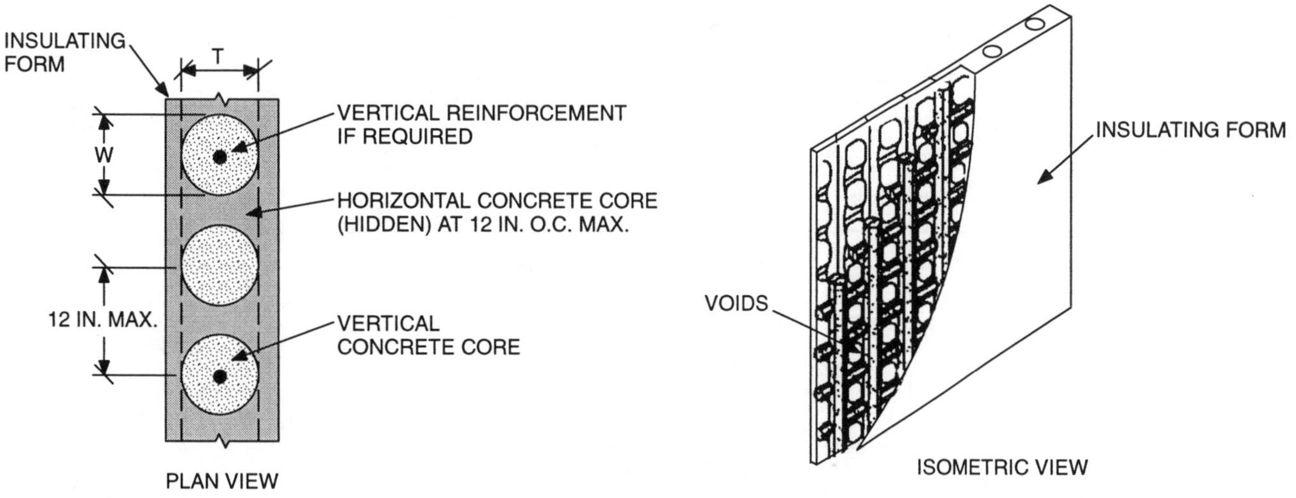

For SI: 1 inch = 25.4 mm.

FIGURE R611.5
SCREEN-GRID IFC WALL SYSTEM

TABLE R611.7(1)
MINIMUM WALL OPENING REINFORCEMENT REQUIREMENTS IN ICF WALLS

WALL TYPE AND OPENING WIDTH (L) (feet)	MINIMUM HORIZONTAL OPENING REINFORCEMENT	MINIMUM VERTICAL OPENING REINFORCEMENT
Flat, Waffle-, and Screen-Grid: L < 2 (0.61)	None required	None required
Screen-Grid: 2(0.61) ≤ L < 4 (1.2)	One No. 4 bar a minimum of 1.5 inches and a maximum of 2.5 inches above the top of the opening. One No. 4 bar within 12 inches below the bottom of the opening. Each No. 4 bar shall extend a minimum of 24 inches beyond the limits of the opening.	None required
Flat and Waffle-Grid: 2(0.61) ≤ L < 4 (1.2)	One No. 4 bar within 12 inches above the top of the opening. One No. 4 bar within 12 inches below the bottom of the opening. Each No. 4 bar shall extend a minimum of 24 inches beyond the limits of the opening.	None required
Flat, Waffle-, and Screen-Grid: L ≥ 4 (1.2)	Provide lintels in accordance with Section R611.7.3. Top and bottom lintel reinforcement shall extend a minimum of 24 inches beyond the limits of the opening.	110 mph or less - One No. 4 bar. Greater than 110 mph - Two No. 4 bars. Bars shall extend for the full height of each wall story and be within 12 inches of each side of the opening.

For SI: 1 inch = 25.4 mm, 1 foot = 304.8 mm, 1 pound per square inch = 6.895 kPa, 1 mile per hour = 1.609 km/h.

NOTE: This table is based on concrete with a minimum specified compressive strength of 2,500 psi, reinforcing steel with a minimum yield strength of 40,000 psi and an assumed equivalent rectangular cross section. This table is not intended to prohibit the use of ICF manufacturer's tables based on engineering analysis in accordance with ACI 318.

TABLE R611.7(2)
MAXIMUM ALLOWABLE CLEAR SPANS FOR ICF LINTELS FOR FLAT AND SCREEN-GRID LOAD-BEARING WALLS[a,b,c,d]
NO. 4 BOTTOM BAR SIZE

MINIMUM LINTEL THICKNESS (T) (inches)	LINTEL DEPTH (D) (inches)	MAXIMUM CLEAR SPAN (feet-inches)					
		Supporting light-framed roof		Supporting light framed second story and roof		Supporting ICF second story and light-frame roof	
		Maximum ground snow load (psf)					
		30	70	30	70	30	70
4	8	4-9	4-2	3-10	3-4	3-5	3-1
	12	6-8	5-5	5-0	4-5	4-6	4-0
	16	7-11	6-5	6-0	5-3	5-4	4-10
	20	8-11	7-4	6-9	6-0	6-1	5-6
	24	9-10	8-1	7-6	6-7	6-9	6-1
6	8	5-2	4-2	3-10	3-5	3-5	3-1
	12	6-8	5-5	5-0	4-5	4-6	4-1
	16	7-10	6-5	6-0	5-3	5-4	4-10
	20	8-10	7-3	6-9	6-0	6-1	5-6
	24	9-8	8-0	7-5	6-7	6-8	6-0
8	8	5-2	4-2	3-11	3-5	3-6	3-2
	12	6-7	5-5	5-0	4-5	4-6	4-1
	16	7-9	6-5	5-11	5-3	5-4	4-10
	20	8-8	7-2	6-8	5-11	6-0	5-5
	24	9-6	7-11	7-4	6-6	6-7	6-0
10	8	5-2	4-2	3-11	3-5	3-6	3-2
	12	6-7	5-5	5-0	4-5	4-6	4-1
	16	7-8	6-4	5-11	5-3	5-4	4-10

For SI: 1 inch = 25.4 mm, 1 foot = 304.8 mm, 1 pound per square inch = 6.895 kPa, 1 pound per square foot = 0.0479 kN/m^2.

NOTE: This table is based on concrete with a minimum specified compressive strength of 2,500 psi, reinforcing steel with a minimum yield strength of 40,000 psi and an assumed equivalent rectangular cross section. This table is not intended to prohibit the use of ICF manufacturer's tables based on engineering analysis in accordance with ACI 318.

a. Deflection criteria: *L*/240.

b. Design load assumptions:
Floor dead load is 10 psf — Attic live load is 20 psf
Floor live load is 30 psf — Roof dead load is 15 psf
Building width is 32 feet — ICF wall dead load is 69 psf
Light-framed wall dead load is 10 psf

c. No. 3 stirrups are required at *d*/2 spacing.

d. Interpolation is permitted between ground snow loads and between lintel depths.

TABLE R611.7(3)
MAXIMUM ALLOWABLE CLEAR SPANS FOR ICF LINTELS FOR FLAT AND SCREEN-GRID LOAD-BEARING WALLS[a,b,c,d]
NO. 5 BOTTOM BAR SIZE

MINIMUM LINTEL THICKNESS (T) (inches)	LINTEL DEPTH (D) (inches)	MAXIMUM CLEAR SPAN (feet-inches)					
		Supporting light-framed roof		Supporting light framed 2nd story and roof		Supporting ICF second story and light-frame roof	
		Maximum ground snow load (psf)					
		30	70	30	70	30	70
4	8	4-9	4-2	3-11	3-7	3-7	3-5
	12	7-2	6-3	5-11	5-5	5-5	5-0
	16	9-6	8-0	7-4	6-6	6-7	5-11
	20	11-1	9-1	8-4	7-5	7-6	6-9
	24	12-2	10-0	9-3	8-2	8-4	7-6
5.5	8	5-6	4-10	4-7	4-2	4-2	3-10
	12	8-3	6-9	6-3	5-6	5-7	5-0
	16	9-9	8-0	7-5	6-6	6-7	6-0
	20	10-11	9-0	8-4	7-5	7-6	6-9
	24	12-0	9-11	9-3	8-2	8-3	7-6
7.5	8	6-1	5-2	4-9	4-3	4-3	3-10
	12	8-2	6-9	6-3	5-6	5-7	5-0
	16	9-7	7-11	7-4	6-6	6-7	6-0
	20	10-10	8-11	8-4	7-4	7-6	6-9
	24	11-10	9-10	9-2	8-1	8-3	7-5
9.5	8	6-4	5-2	4-10	4-3	4-4	3-11
	12	8-2	6-8	6-2	5-6	5-7	5-0
	16	9-6	7-11	7-4	6-6	6-7	5-11
	20	10-8	8-10	8-3	7-4	7-5	6-9
	24	11-7	9-9	9-0	8-1	8-2	7-5

For SI: 1 inch = 25.4 mm, 1 foot = 304.8 mm, 1 pound per square inch = 6.895 kPa, 1 pound per square foot = 0.0479 kN/m^2.

NOTE: This table is based on concrete with a minimum specified compressive strength of 2,500 psi, reinforcing steel with a minimum yield strength of 40,000 psi and an assumed equivalent rectangular cross section. This table is not intended to prohibit the use of ICF manufacturer's tables based on engineering analysis in accordance with ACI 318.

a. Deflection criteria: *L*/240.

b. Design load assumptions:
Floor dead load is 10 psf — Attic live load is 20 psf
Floor live load is 30 psf — Roof dead load is 15 psf
Building width is 32 feet — ICF wall dead load is 69 psf
Light-framed wall dead load is 10 psf

c. No. 3 stirrups are required at *d*/2 spacing.

d. Interpolation is permitted between ground snow loads and between lintel depths.

TABLE R611.7(4) MAXIMUM ALLOWABLE CLEAR SPANS FOR WAFFLE-GRID ICF WALL LINTELS[a,b,c,d] NO. 4 BOTTOM BAR SIZE

MINIMUM LINTEL THICKNESS (T)[e,f] (inches)	LINTEL DEPTH (D) (inches)	MAXIMUM CLEAR SPAN (feet-inches)					
		Supporting light-framed roof		Supporting light framed 2nd story and roof		Supporting ICF second story and light-frame roof[g]	
		Maximum ground snow load (psf)					
		30	70	30	70	30	70
6	8	5-2	4-2	3-10	3-5	3-6	3-2
	12	6-8	5-5	5-0	4-5	4-7	4-2
	16	7-11	6-6	6-0	5-3	5-6	4-11
	20	8-11	7-4	6-9	6-0	6-3	5-7
	24	9-10	8-1	7-6	6-7	6-10	6-2
8	8	5-2	4-3	3-11	3-5	3-7	3-2
	12	6-8	5-5	5-1	4-5	4-8	4-2
	16	7-10	6-5	6-0	5-3	5-6	4-11
	20	8-10	7-3	6-9	6-0	6-2	5-7
	24	9-8	8-0	7-5	6-7	6-10	6-2

For SI: 1 inch = 25.4 mm, 1 foot = 304.8 mm, 1 pound per square inch = 6.895 kPa, 1 pound per square foot = 0.0479 kN/m^2.

NOTE: This table is based on concrete with a minimum specified compressive strength of 2,500 psi, reinforcing steel with a minimum yield strength of 40,000 psi and an assumed equivalent rectangular cross section. This table is not intended to prohibit the use of ICF manufacturer's tables based on engineering analysis in accordance with ACI 318.

a. Deflection criteria: *L*/240.

b. Design load assumptions:
Floor dead load is 10 psf — Attic live load is 20 psf
Floor live load is 30 psf — Roof dead load is 15 psf
Building width is 32 feet — ICF wall dead load is 69 psf
Light-framed wall dead load is 10 psf

c. No. 3 stirrups are required at *d*/2 spacing.

d. Interpolation is permitted between ground snow loads and between lintel depths.

e. For actual wall lintel width, refer to Table R611.4(2).

f. Lintel width corresponds to the nominal waffle-grid ICF wall thickness with a minimum web thickness of 2 inches.

g. ICF wall dead load used is 55 psf.

TABLE R611.7(5)
MAXIMUM ALLOWABLE CLEAR SPANS FOR WAFFLE-GRID ICF WALL LINTELS[a,b,c,d]
NO. 5 BOTTOM BAR SIZE

MINIMUM LINTEL THICKNESS (T)[e,f] (inches)	LINTEL DEPTH (D) (inches)	MAXIMUM CLEAR SPAN (feet-inches)					
		Supporting light-framed roof		Supporting light framed 2nd story and roof		Supporting ICF second story and light-frame roof[g]	
		Maximum ground snow load (psf)					
		30	70	30	70	30	70
6	8	5-4	4-8	4-5	4-1	4-5	3-10
	12	8-0	6-9	6-3	5-6	6-3	5-1
	16	9-9	8-0	7-5	6-6	7-5	6-1
	20	11-0	9-1	8-5	7-5	8-5	6-11
	24	12-2	10-0	9-3	8-2	9-3	7-8
8	8	6-0	5-2	4-9	4-3	4-9	3-11
	12	8-3	6-9	6-3	5-6	6-3	5-2
	16	9-9	8-0	7-5	6-6	7-5	6-1
	20	10-11	9-0	8-4	7-5	8-4	6-11
	24	12-0	9-11	9-2	8-2	9-2	7-8

For SI: 1 inch = 25.4 mm, 1 foot = 304.8 mm, 1 pound per square inch = 6.895 kPa, 1 pound per square foot = 0.0479 kN/m^2.

NOTE: This table is based on concrete with a minimum specified compressive strength of 2,500 psi, reinforcing steel with a minimum yield strength of 40,000 psi and an assumed equivalent rectangular cross section. This table is not intended to prohibit the use of ICF manufacturer's tables based on engineering analysis in accordance with ACI 318.

a. Deflection criteria: *L*/240.

b. Design load assumptions:
Floor dead load is 10 psf — Attic live load is 20 psf
Floor live load is 30 psf — Roof dead load is 15 psf
Building width is 32 feet — ICF wall dead load is 69 psf
Light-framed wall dead load is 10 psf

c. Two #3 stirrups are required.

d. Interpolation is permitted between ground snow loads and between lintel depths.

e. For actual wall lintel width, refer to Table R611.4(2).

f. Lintel width corresponds to the nominal waffle-grid ICF wall thickness with a minimum web thickness of 2 inches.

g. ICF wall dead load used is 55 psf.

TABLE R611.7(6)
MINIMUM BOTTOM BAR ICF LINTEL REINFORCEMENT FOR LARGE CLEAR SPANS IN LOAD-BEARING WALLS[a,b,c,d]

MINIMUM LINTEL THICKNESS (T[e]) (inches)	MINIMUM LINTEL DEPTH (D) (inches)	MINIMUM BOTTOM LINTEL REINFORCEMENT					
		Supporting light-frame roof only		Supporting light-frame second story and roof		Supporting ICF second story and light-frame roof[f]	
		Maximum ground snow load (psf)					
		30	70	30	70	30	70
Flat ICF lintel, 12 feet maximum clear span							
3.5	24	1-#5	1-#7	D/R	D/R	D/R	D/R
5.5	20	1-#6	1-#7	D/R	D/R	D/R	D/R
	24	1-#5	1-#7	1-#7	1-#8	1-#8	D/R
7.5	16	1-#7; 2-#5	D/R	D/R	D/R	D/R	D/R
	20	1-#6; 2-#4	1-#7; 2-#5	1-#8; 2-#6	D/R	D/R	D/R
	24	1-#6; 2-#4	1-#7; 2-#5	1-#7; 2-#5	1-#8; 2-#6	1-#8; 2-#6	1-#8; 2-#6
9.5	16	1-#7; 2-#5	D/R	D/R	D/R	D/R	D/R
	20	1-#6; 2-#4	1-#7; 2-#5	1-#8; 2-#6	1-#8; 2-#6	1-#8; 2-#6	1-#9; 2-#6
	24	1-#6; 2-#4	1-#7; 2-#5	1-#7; 2-#5	1-#8; 2-#6	1-#8; 2-#6	1-#9; 2-#6
Flat ICF lintel, 16 feet maximum clear span							
5.5	24	1-#7	D/R	D/R	D/R	D/R	D/R
7.5	24	1-#7; 2-#5	D/R	D/R	D/R	D/R	D/R
9.5	24	1-#7; 2-#5	1-#9; 2-#6	1-#9; 2-#6	D/R	D/R	D/R
Waffle-Grid ICF lintel, 12 feet maximum clear span							
6	20	1-#6	D/R	D/R	D/R	D/R	D/R
	24	1-#5	1-#7; 2-#5	1-#7; 2-#5	1-#8; 2-#6	1-#8; 2-#6	D/R
8	16	1-#7; 2-#5	D/R	D/R	D/R	D/R	D/R
	20	1-#6; 2-#4	1-#7; 2-#5	1-#8; 2-#6	D/R	D/R	D/R
	24	1-#5	1-#7; 2-#5	1-#7; 2-#5	1-#8; 2-#6	1-#8; 2-#6	1-#8; 2-#6

For SI: 1 inch = 25.4 mm, 1 foot = 304.8 mm, 1 pound per square inch = 6.895 kPa, 1 pound per square foot = 0.0479 kN/m^2.

NOTE: This table is based on concrete with a minimum specified compressive strength of 2,500 psi, reinforcing steel with a minimum yield strength of 40,000 psi and an assumed equivalent rectangular cross section. This table is not intended to prohibit the use of ICF manufacturer's tables based on engineering analysis in accordance with ACI 318.

a. D/R indicates design is required.
b. Deflection criterion is *L*/240, where *L* is the clear span of the lintel in inches.
c. Interpolation is permitted between ground snow loads and between lintel depths.
d. No. 3 stirrups are required at maximum *d*/2 spacing for spans greater than 4 feet.
e. Actual thickness is shown for flat lintels while nominal thickness is given for waffle-grid lintels. Lintel thickness corresponds to the nominal waffle-grid ICF wall thickness with a minimum web thickness of 2 inches. Refer to Section R611.4 for actual wall lintel width.
f. ICF wall dead load is maximum 55 psf.

TABLE R611.7(7)
MAXIMUM ALLOWABLE CLEAR SPANS FOR ICF LINTELS IN NONLOAD-BEARING WALLS[a,b,c]
NO. 4 BOTTOM BAR SIZE

MINIMUM LINTEL DEPTH (D) (inches)	MAXIMUM CLEAR SPAN	
	Supporting light-frame nonload-bearing gable end wall (feet)	Supporting ICF second story nonload-bearing wall[d] (feet)
8	12	6
12	16	8
16	16	10
20	16	12
24	16	16

For SI: 1 inch = 25.4 mm, 1 foot = 304.8 mm, 1 pound per square inch = 6.895 kPa, 1 pound per square foot = 0.0479 kN/m^2.

NOTE: This table is based on concrete with a minimum specified compressive strength of 2,500 psi, reinforcing steel with a minimum yield strength of 40,000 psi and an assumed equivalent rectangular cross-section. This table is not intended to prohibit the use of ICF manufacturer's tables based on engineering analysis in accordance with ACI 318.

a. Deflection criterion is *L*/240, where *L* is the clear span of the lintel in inches.
b. Linear interpolation is permitted between lintel depths.
c. No. 3 stirrups are required at maximum *d*/2 spacing for spans greater than 4 feet.
d. ICF wall dead load is maximum 69 psf.

TABLE R611.7(8) MINIMUM PERCENTAGE OF SOLID WALL LENGTH ALONG EXTERIOR WALL LINES[a,b,c]

ICF WALL TYPE AND MINIMUM WALL THICKNESS (inches)[d]	MAXIMUM ROOF SLOPE	MINIMUM SOLID WALL LENGTH (percent)[e]									
		Wall supporting light-frame roof and ceiling					Wall supporting one ICF story and light-frame roof and ceiling				
		Maximum wind speed (mph)									
		85	100	110	120	130	85	100	110	120	130
Flat 3.5	3:12	15	15	15	15	20	30	35	40	50	55
	6:12	15	15	20	20	25	30	40	50	55	60
	9:12	20	25	30	40	45	45	60	70	85	95
	12:12	25	35	40	50	60	50	65	80	95	100
Flat 5.5	3:12	15	15	15	15	15	20	25	30	40	40
	6:12	15	15	15	15	20	20	30	35	40	45
	9:12	15	15	20	25	30	35	45	50	60	70
	12:12	20	20	25	35	40	35	50	55	70	75
Flat 7.5	3:12	15	15	15	15	15	20	20	25	30	30
	6:12	15	15	15	15	15	20	20	25	30	35
	9:12	15	15	15	20	25	25	30	40	45	50
	12:12	15	20	20	25	30	30	35	40	50	55
Waffle-Grid 6	3:12	15	15	15	15	20	25	30	35	45	50
	6:12	15	15	20	20	25	25	35	45	50	55
	9:12	20	20	25	35	40	40	55	60	75	85
	12:12	25	30	35	45	50	45	60	70	85	90
Waffle-Grid 8	3:12	15	15	15	15	15	20	25	30	35	35
	6:12	15	15	15	15	20	20	25	30	35	40
	9:12	15	15	20	25	30	30	40	45	55	60
	12:12	20	20	25	30	35	35	40	50	60	65
Screen-Grid 6	3:12	15	15	20	20	25	30	40	45	55	60
	6:12	15	20	25	30	35	30	40	50	60	70
	9:12	25	30	45	50	65	50	65	80	90	100
	12:12	35	40	55	65	80	55	70	85	100	100

For SI: 1 inch = 25.4 mm, 1 foot = 304.8 mm, 1 pound per square inch = 6.895 kPa, 1 mile per hour = 1.6093 km/h.

NOTE: This table is based on concrete with a minimum specified compressive strength of 2,500 psi, reinforcing steel with a minimum yield strength of 40,000 psi and an assumed equivalent rectangular cross section. This table is not intended to prohibit the use of ICF manufacturer's tables based on engineering analysis in accordance with ACI 318.

a. Linear interpolation between roof slopes shall be permitted.

b. Minimum percentages are applicable for maximum 10-foot wall story heights.

c. N/A indicates not applicable.

d. Actual thickness is shown for flat walls while nominal thickness is given for waffle- and screen-grid walls. Refer to Table R611.4(2) for actual waffle- and screen-grid thickness and dimensions.

e. The minimum solid wall length requirements are based on a 2:1 building aspect ratio, L/W, where L is the length parallel to the ridge of the roof and W is the length perpendicular to the ridge of the roof. For a different ration the tabular values may be adjusted by the following multipliers provided the minimum wall lengths in Section R611.7.4 are met. Interpolation is permitted.

L/W	Endwall W	Sidewall, L
2.0	1.00	0.25
1.8	0.90	0.30
1.6	0.80	0.35
1.4	0.70	0.40
1.2	0.60	0.45
1.0	0.50	0.50

TABLE R611.7(9)
MINIMUM PERCENTAGE OF SOLID WALL LENGTH FOR SEISMIC DESIGN CATEGORY C[a]

ICF WALL TYPE AND MINIMUM WALL THICKNESS (inches)[b]	MINIMUM SOLID WALL LENGTH (percent)[c]	
	Wall supporting light-frame roof and ceiling	Wall supporting one ICF story and light-frame roof and ceiling
Flat, 3.5	20	35
Flat, 5.5	15	30
Flat 7.5	15	25
Waffle-grid, 6	20	35
Waffle-grid, 8	20	30
Screen-grid	25	45

For SI: 1 inch = 25.4 mm, 1 foot = 304.8 mm, 1 pound per square foot = 0.0479 kN/m^2.

a. Minimum percentages are for maximum 10-foot wall story height.

b. Actual thickness is shown for flat walls while nominal thickness is given for waffle-grid and screen-grid walls. Refer to Table R611.4(2) for actual waffle-grid and screen-grid thickness and dimension.

c. The minimum solid wall length requirements are based on a 2:1 building aspect ratio, L/W, where L is the length parallel to the ridge of the roof and W is the length perpendicular to the ridge of the roof. For a different ratio, the tabular values may be adjusted by the following multipliers provided the minimum wall lengths in Section R611.7.4 are met. Interpolation is permitted.

L/W	Endwall W	Sidewall, L
2.0	1.00	0.25
1.8	0.90	0.30
1.6	0.80	0.35
1.4	0.70	0.40
1.2	0.60	0.45
1.0	0.50	0.50

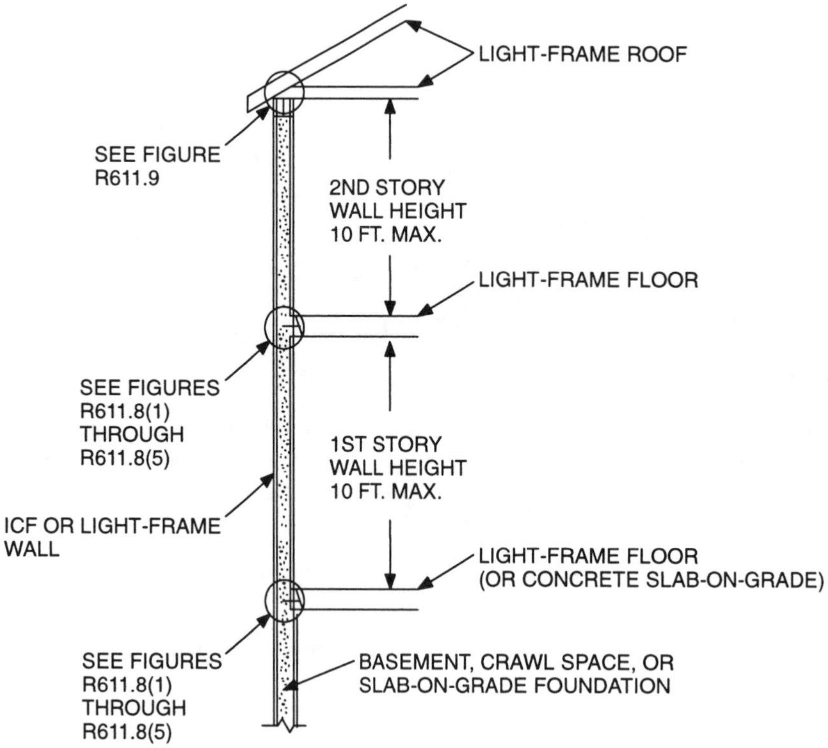

For SI: 1 foot = 304.8 mm.

Note: Section cut through flat wall or vertical core of waffle- or screen-grid wall.

FIGURE R611.7(1)
ICF WALL CONSTRUCTION

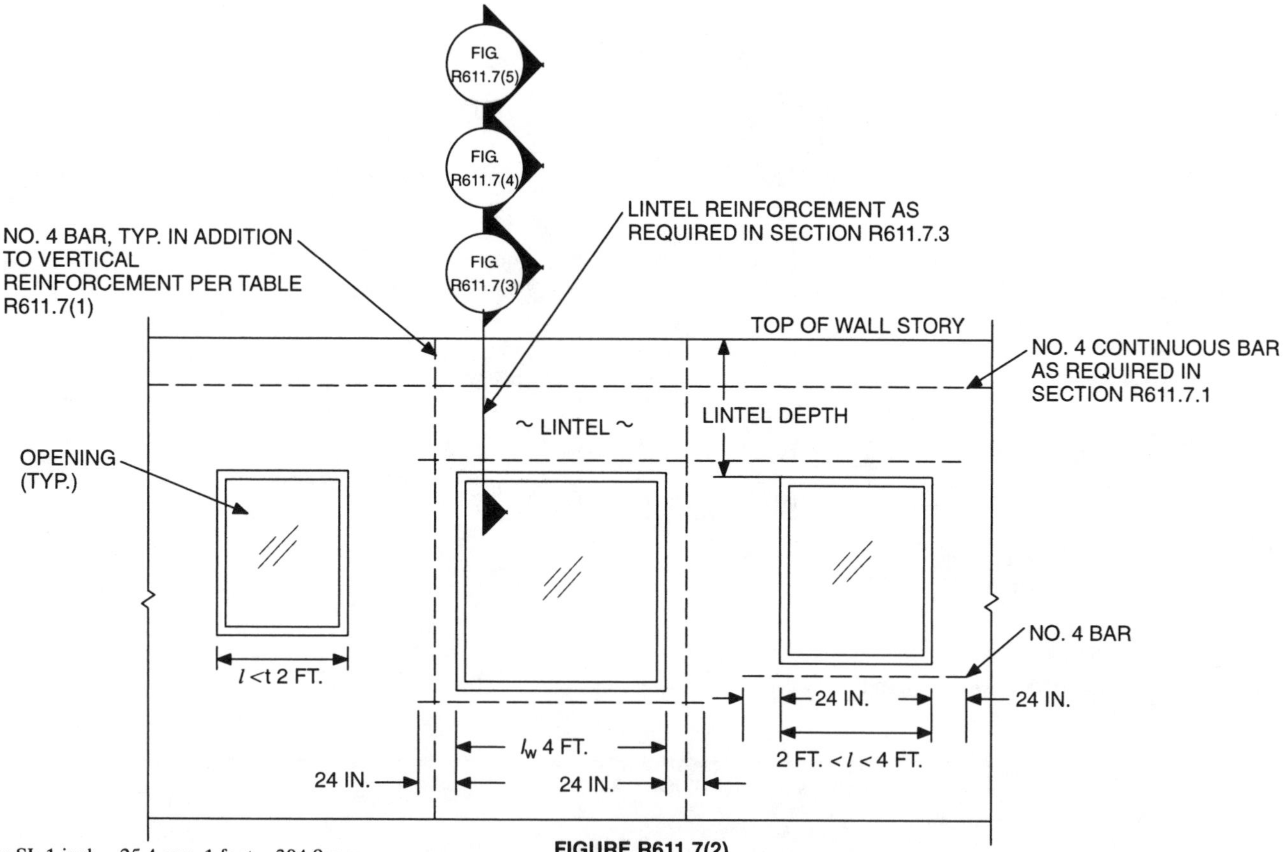

For SI: 1 inch = 25.4 mm, 1 foot = 304.8 mm.

FIGURE R611.7(2)
REINFORCEMENT OF OPENINGS

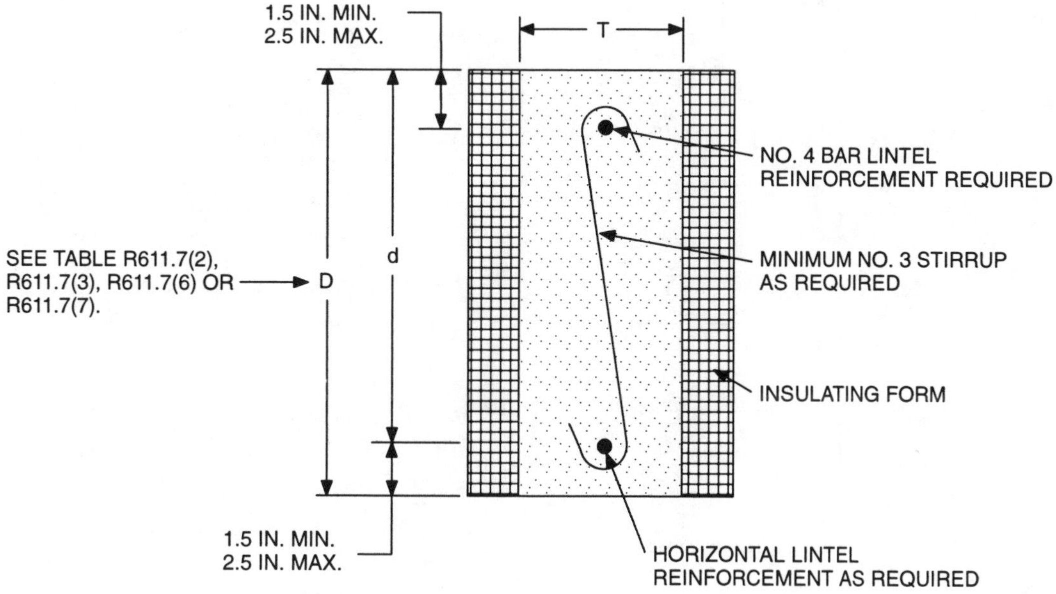

For SI: 1 inch = 25.4 mm.

Note: Section cut through flat wall.

FIGURE R611.7(3)
ICF LINTELS FOR FLAT AND SCREEN-GRID WALLS

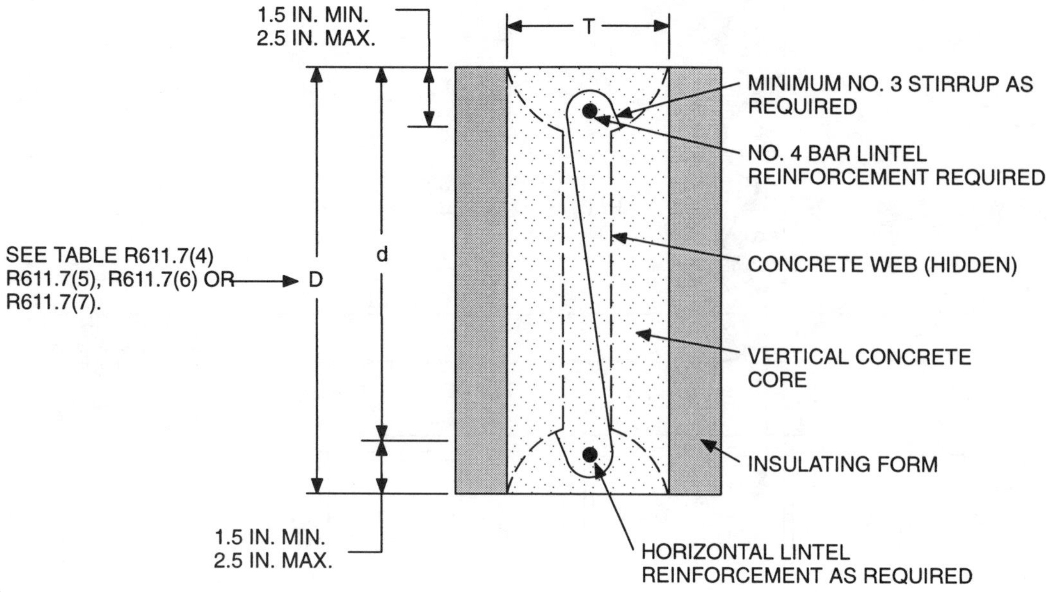

For SI: 1 inch = 25.4 mm.

Note: Section cut through vertical core of a waffle-grid lintel.

FIGURE R611.7(4)
SINGLE FORM HEIGHT WAFFLE-GRID LINTEL

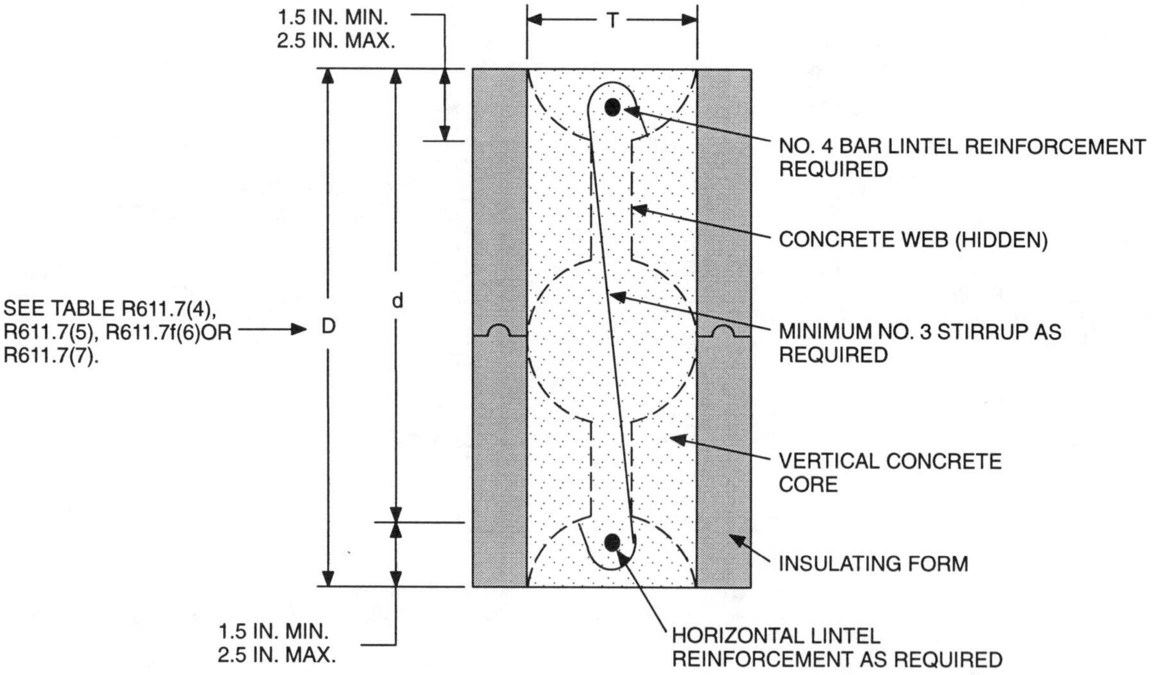

For SI: 1 inch = 25.4 mm.

Note: Section cut through vertical core of a waffle-grid lintel.

FIGURE R611.7(5)
DOUBLE FORM HEIGHT WAFFLE-GRID LINTEL

R611.7.1.4 Dwellings in Seismic Design Category C. Dwellings in Seismic Design Category C shall have horizontal and vertical reinforcement in accordance with the following:

1. Vertical reinforcement consisting of at least one No. 4 reinforcing bar, extending continuously from support to support, shall be provided at each corner, at each side of each opening, and at the ends of walls.
2. Horizontal reinforcement consisting of at least one No. 4 reinforcing bar, continuously at structurally connected roof and floor levels and at the top of the wall, at the bottom of load-bearing walls or in the top of foundations where doweled to the wall and at a maximum spacing of 10 feet (3048 mm).

Vertical reinforcement provided in accordance with Sections R611.7.1.2 and R611.7.2, and horizontal reinforcement provided in accordance with Sections R611.7.1.3 and R611.7.3, shall be permitted to be used to meet the requirements of this section.

R611.7.1.5 Lap splices. Where lap splicing of reinforcing steel is necessary, the lap splice shall be a minimum of $40d_b$, where d_b is the diameter of the smaller bar. The maximum distance between noncontact bars at a lap splice shall not exceed $8d_b$.

R611.7.2 Wall openings. Wall openings shall have a minimum of 8 inches (203 mm) of depth of concrete for flat and waffle-grid ICF walls and 12 inches (305 mm) for screen-grid walls over the length of the opening. Reinforcement around openings shall be provided in accordance with Table R611.7(1) and Figure R611.7(2). All reinforcement placed horizontally above or below an opening shall extend a minimum of 24 inches (610 mm) beyond the limits of the opening. Wall opening reinforcement shall be provided in addition to the reinforcement required by Sections R611.3, R611.4, R611.5 and R611.7.1. The perimeter of all wall openings shall be framed with a minimum 2-by-4 plate, anchored to the wall with $^1/_2$-inch (12.7 mm) diameter anchor bolts spaced a maximum of 24 inches (610 mm) on center. The bolts shall be embedded into the concrete a minimum of 4 inches (102 mm) and have a minimum of $1^1/_2$ (38 mm) inches of concrete cover to the face of the wall.

Exception: The 2-by-4 plate is not required where the wall is formed to provide solid concrete around the perimeter of the opening with a minimum depth of 4 inches (102 mm) for the full thickness of the wall.

R611.7.3 Lintels.

R611.7.3.1 General requirements. Lintels shall be provided over all openings greater than or equal to 4 feet (1219 mm) in width for flat or waffle-grid ICF walls, and greater than 2 feet (610 mm) for screen-grid ICF walls. Lintels for flat ICF walls and screen-grid ICF walls shall be constructed in accordance with Figure R611.7(3). Lintels for waffle-grid ICF walls shall be constructed in accordance with Figure R611.7(4) or Figure R611.7(5). Lintel depths are permitted to be increased by the height of ICF wall located directly above the lintels, provided that the lintel depth spans the entire length of the opening.

R611.7.3.2 Stirrups. A minimum of No. 3 stirrups shall be installed for all flat and screen-grid wall lintels at a maximum spacing of $d/2$ where d equals the depth of the lintel (D) minus the bottom cover of concrete as shown in Figure R611.7(3), R611.7(4) or R611.7(5). A minimum of two No. 3 stirrups shall be placed in each vertical core of waffle-grid lintels.

R611.7.3.3 Horizontal reinforcement. One No. 4 horizontal bar shall be provided in the top of the lintel. Horizontal reinforcement placed within 12 inches (305 mm) of the top of the wall in accordance with Section R611.7.1.3 shall be permitted to serve as the top or bottom reinforcement in the lintel provided the reinforcement meets the location requirements in Figure R611.7(2), R611.7(3), R611.7(4) or R611.7(5), and the size requirements in Table R611.7(2), R611.7(3), R611.7(4), R611.7(5), R611.7(6) or R611.7(7).

R611.7.3.4 Load-bearing walls. Lintels for flat and screen-grid ICF walls supporting roof or floor loads shall comply with Table R611.7(2) or Table R611.7(3). Lintels for waffle-grid ICF walls supporting roof or floor loads shall comply with Table R611.7(4) or Table R611.7(5).

Exception: Where spans larger than those permitted in Table R611.7(2), R611.7(3), R611.7(4) or R611.7(5) are required, the lintels shall comply with Table R611.7(6).

R611.7.3.5 Nonload-bearing walls. Lintels for nonload-bearing flat, waffle-grid and screen-grid ICF walls shall comply with Table R611.7(7).

R611.7.4 Minimum length of wall without openings. Exterior ICF walls shall have a minimum of solid wall length to total wall length in accordance with Table R611.7(8), but not less than 15 percent for ICF walls supporting a light framed roof or 20 percent for ICF walls supporting an ICF or light framed second story and light framed roof. For attached dwellings in Seismic Design Category C, the minimum percentage of solid wall length shall be greater than or equal to the requirements in Table R611.7(9). The minimum percentage of solid wall length shall include only those solid wall segments that are a minimum of 24 inches (610 mm) in length. The maximum distance between wall segments included in determining solid wall length shall not exceed 18 feet (5486 mm). A minimum length of 24 inches (610 mm) of solid wall segment, extending the full height of each wall story, shall occur at all corners of exterior walls.

R611.8 ICF wall-to-floor connections.

R611.8.1 Top bearing. Floors bearing on the top of ICF foundation walls in accordance with Figure R611.8(1) shall have the wood sill plate anchored to the ICF wall with minimum $^1/_2$-inch (12.7 mm) diameter bolts embedded a minimum of 7 inches (178 mm) and placed at a maximum spacing of 6 feet (1829 mm) on center and not more than 12 inches (305 mm) from corners. Anchor bolts for waffle-grid and screen-grid walls shall be located in the cores. In conditions where wind speeds are in excess of 90 miles per hour (144 km/h), the $^1/_2$-inch (12.7 mm) diameter anchor bolts shall be placed at a maximum spacing of 4 feet (1219 mm) on center. Bolts shall extend a minimum of 7 inches (178 mm) into concrete. Sill plates shall be protected against decay where required by Section R323. Cold-formed steel framing systems shall be anchored to the concrete in accordance with Section R505.3.1 or Section R603.3.1.

R611.8.2 Ledger bearing. Wood ledger boards supporting bearing ends of joists or trusses shall be anchored to flat ICF walls with minimum thickness of 5.5 inches (140 mm) and to waffle- or screen-grid ICF walls with minimum nominal thickness of 6 inches (152 mm) in accordance with Figure R611.8(2), R611.8(3), R611.8(4) or R611.8(5) and Table R611.8. The ledger shall be a minimum 2 by 8, No. 2 Southern Pine or No. 2 Douglas Fir. Ledgers anchored to nonload-bearing walls to support floor or roof sheathing shall be attached with $^1/_2$-inch (12.7 mm) diameter or headed anchor bolts spaced a maximum of 6 feet (1829 mm) on center. Anchor bolts shall be embedded a minimum of 4 inches (102 mm) into the concrete.

R611.8.2.1 Seismic Design Category C. In Seismic Design Category C, additional anchorage mechanisms shall be installed at a maximum spacing of 6 feet (1829 mm) on center. Such anchorage mechanisms shall not induce tension stresses perpendicular to grain in ledgers or nailers. The capacity of such anchors shall result in connections capable of resisting 200 pounds per linear foot (2.9 kN/m). Such anchorage mechanisms shall not induce tension stresses perpendicular to grain in ledgers or nailers.

R611.8.3 Floor and roof diaphragm construction. Floor and roof diaphragms shall be constructed of structural wood sheathing panels, attached to wood framing in accordance with Table R602.3(1) or to cold-formed steel floor framing in accordance with Table R505.3.1(2) or to cold-formed steel roof framing in accordance with Table R804.3. Additionally, sheathing panel edges perpendicular to framing members shall be backed by blocking and sheathing shall be connected to the blocking with fasteners at the edge spacing. For Seismic Design Category C, where the width-to-thickness dimension of the diaphragm exceeds 2-to-1, edge spacing of fasteners shall be 4 inches (102 mm) on center.

TABLE R611.8
FLOOR LEDGER-ICF WALL CONNECTION (SIDE-BEARING CONNECTION) REQUIREMENTS[a,b,c]

MAXIMUM FLOOR CLEAR SPAN[d] (feet)	MAXIMUM ANCHOR BOLT SPACING[e] (inches)			
	Staggered $^1/_2$-inch-diameter anchor bolts	Staggered $^5/_8$-inch-diameter anchor bolts	Two $^1/_2$-inch-diameter anchor bolts[f]	Two $^5/_8$-inch-diameter anchor bolts[f]
8	18	20	36	40
10	16	18	32	36
12	14	18	28	36
14	12	16	24	32
16	10	14	20	28
18	9	13	18	26
20	8	11	16	22
22	7	10	14	20
24	7	9	14	18
26	6	9	12	18
28	6	8	12	16
30	5	8	10	16
32	5	7	10	14

For SI: 1 inch = 25.4 mm, 1 foot = 304.8 mm.

a. Minimum ledger board nominal depth shall be 8 inches. The thickness of the ledger board shall be a minimum of 2 inches. Thickness of ledger board is in nominal lumber dimensions. Ledger board shall be minimum No. 2 Grade.

b. Minimum edge distance shall be 2 inches for $^1/_2$-inch-diameter anchor bolts and 2.5 inches for $^5/_8$-inch-diameter anchor bolts.

c. Interpolation is permitted between floor spans.

d. Floor span corresponds to the clear span of the floor structure (i.e., joists or trusses) spanning between load-bearing walls or beams.

e. Anchor bolts shall extend through the ledger to the center of the flat ICF wall thickness or the center of the horizontal or vertical core thickness of the waffle-grid or screen-grid ICF wall system.

f. Minimum vertical distance between bolts shall be 1.5 inches for $^1/_2$-inch-diameter anchor bolts and 2 inches for $^5/_8$-inch-diameter anchor bolts.

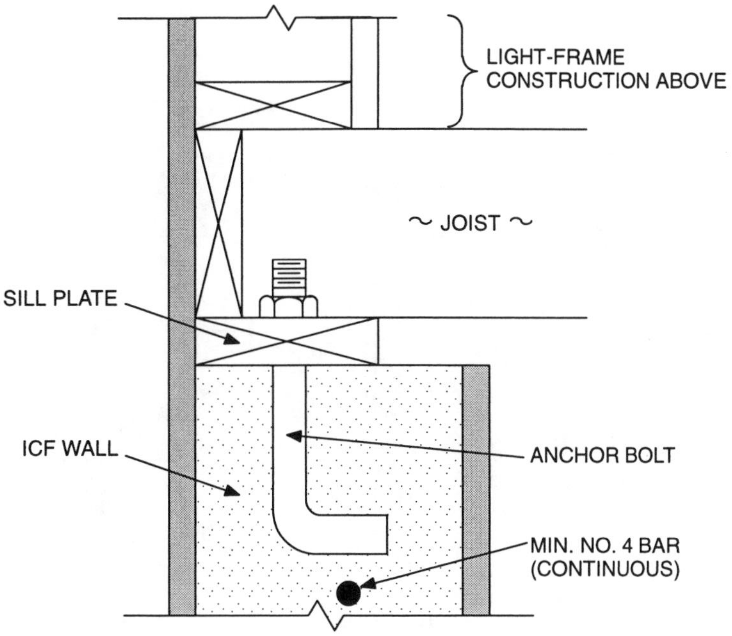

FIGURE R611.8(1)
SECTION CUT THROUGH FLAT WALL OR VERTICAL CORE OF WAFFLE- OR SCREEN-GRID WALL

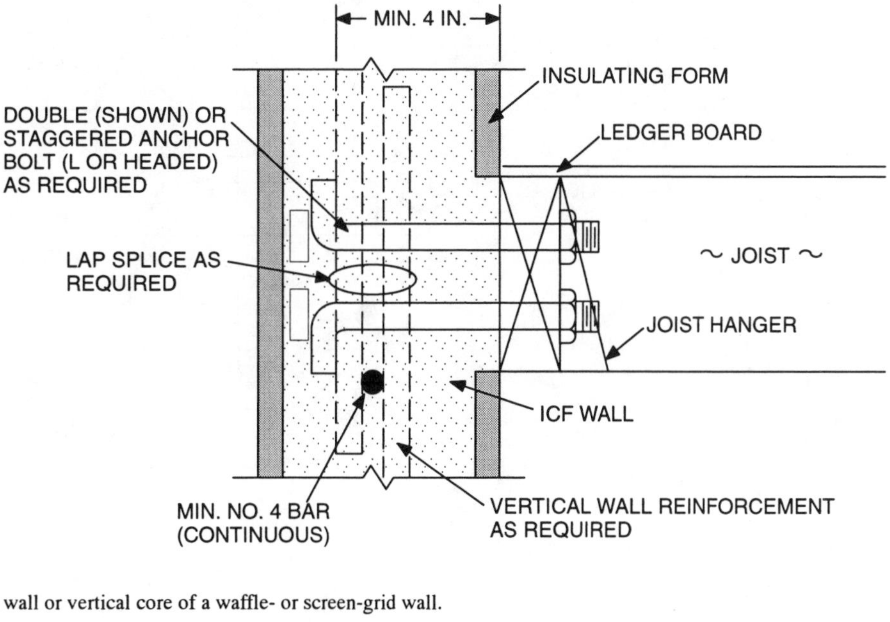

For SI: 1 inch = 25.4 mm.

Note: Section cut through flat wall or vertical core of a waffle- or screen-grid wall.

FIGURE R611.8(2)
FLOOR LEDGER—ICF WALL CONNECTION (SIDE-BEARING CONNECTION)

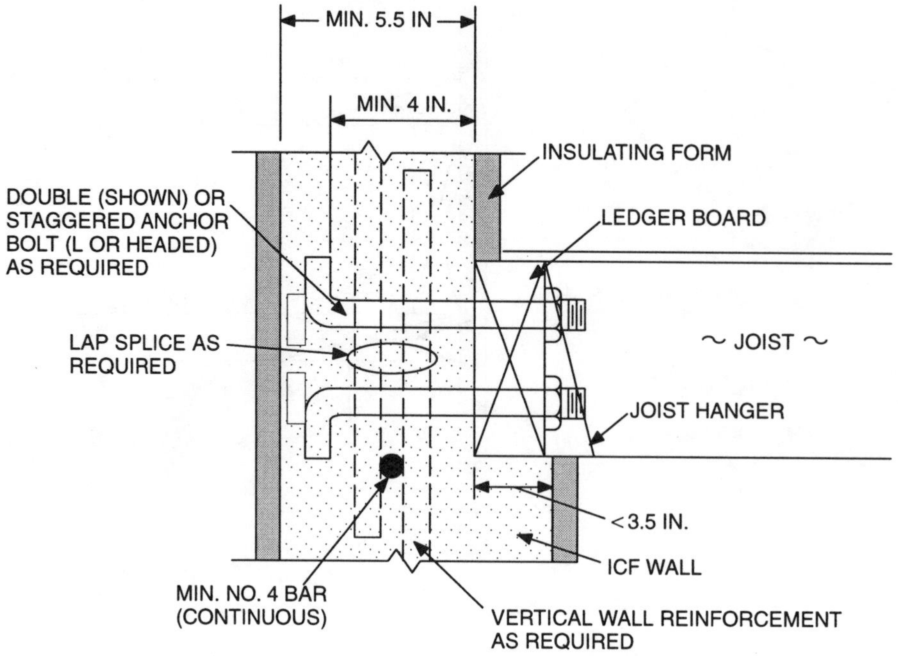

For SI: 1 inch = 25.4 mm.

Note: Section cut through flat wall or vertical core of a waffle- or screen-grid wall.

FIGURE R611.8(3)
FLOOR LEDGER—ICF WALL CONNECTION (SIDE-BEARING CONNECTION)

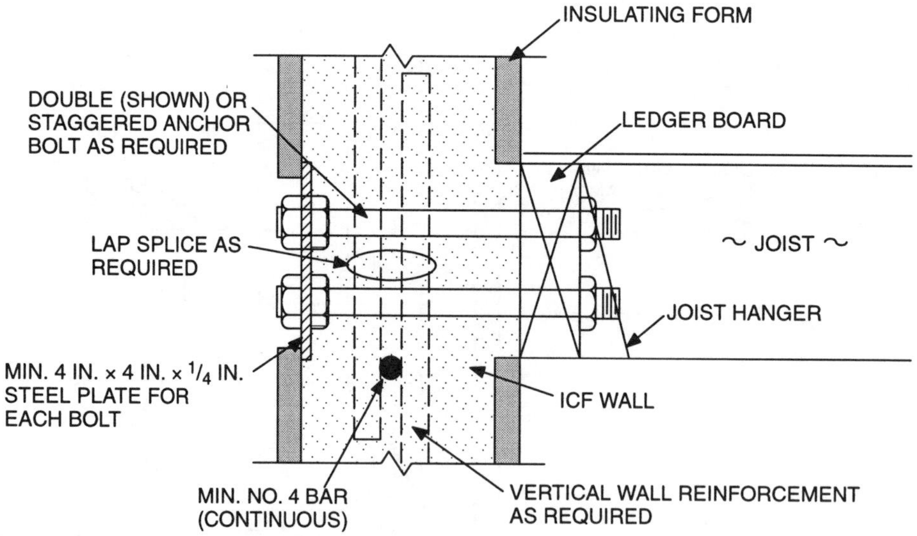

For SI: 1 inch = 25.4 mm.

Note: Section cut through flat wall.

FIGURE R611.8(4)
WOOD FLOOR LEDGER—ICF WALL SYSTEM CONNECTION (THROUGH-BOLT CONNECTION)

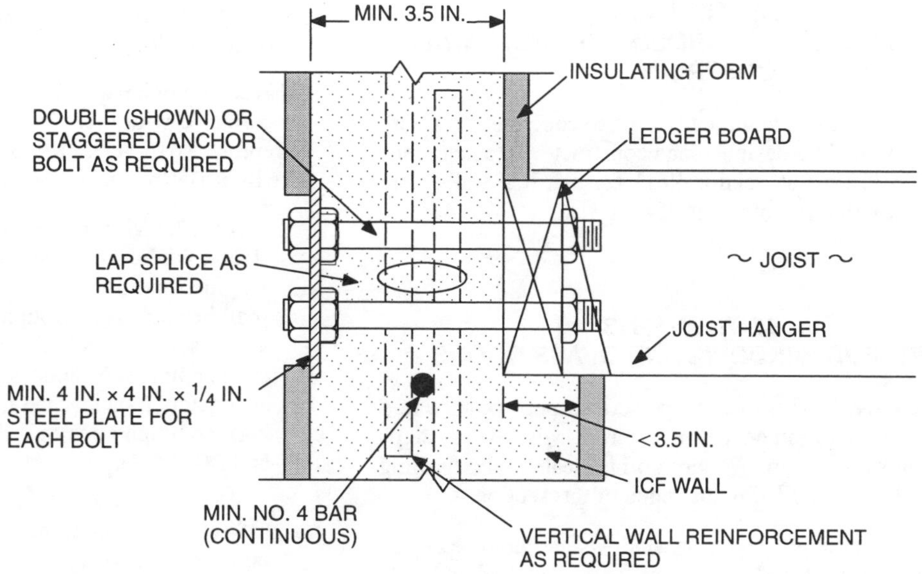

For SI: 1 inch = 25.4 mm.

Note: Section cut through flat wall.

FIGURE R611.8(5)
WOOD FLOOR LEDGER TO ICF WALL SYSTEM CONNECTION DETAIL

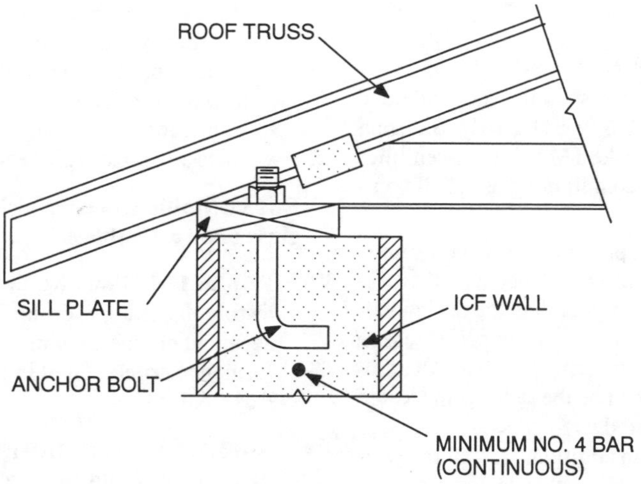

Note: Section cut through flat wall or vertical cove of waffle- or screen-grid wall.

FIGURE R611.9
ICF WALL TO TOP SILL PLATE (ROOF) CONNECTION

R611.9 ICF wall to top sill plate (roof) connections. Wood sill plates attaching roof framing to ICF walls shall be anchored with minimum $^1/_2$ inch (12.7 mm) diameter anchor bolt embedded a minimum of 7 inches (178 mm) and placed at 6 feet (1829 mm) on center in accordance with Figure R611.9. Anchor bolts shall be located in the cores of waffle-grid and screen-grid ICF walls. Roof assemblies subject to wind uplift pressure of 20 pounds per square foot (1.44 kN/m^2) or greater as established in Table R301.2(2) shall have rafter or truss ties provided in accordance with Table R802.11.

SECTION R612 CONVENTIONALLY FORMED CONCRETE WALL CONSTRUCTION

R612.1 General. Conventionally formed concrete walls with flat surfaces shall be designed and constructed in accordance with the provisions of Section R611 for Flat ICF walls or in accordance with the provisions of ACI 318.

SECTION R613 EXTERIOR WINDOWS AND GLASS DOORS

R613.1 General. This section prescribes performance and construction requirements for exterior window systems installed in wall systems. Waterproofing, sealing and flashing systems are not included in the scope of this section.

R613.2 Performance. Exterior windows and doors shall be designed to resist the design wind loads specified in Table R301.2(2) adjusted for height and exposure per Table R301.2(3).

R613.3 Testing and labeling. Exterior windows and glass doors shall be tested by an approved independent laboratory, and bear a label identifying manufacturer, performance characteristics and approved inspection agency to indicate compliance with the requirements of the following specification:

AAMA/NWWDA 101/ I.S.2

Exceptions:

1. Decorative glazed openings.
2. Exterior window and door assemblies not included within the scope of AAMA/NWWDA 101/I.S.2 shall be tested in accordance with ASTM E 330. Assemblies covered by this exception containing glass shall comply with Section R308.5.
3. Structural wind load design pressures for window units smaller than the size tested in accordance with this section shall be permitted to be higher than the design value of the tested unit provided such higher pressures are determined by accepted engineering analysis. All components of the small unit shall be the same as the tested unit. Where such calculated design pressures are used they shall be validated by an additional test of the window unit having the highest allowable design pressure.

R613.4 Windborne debris protection. Protection of exterior windows and glass doors in buildings located in hurricane-prone regions from windborne debris shall be in accordance with Section R301.2.1.2.

R613.5 Anchorage methods. The methods cited in this section apply only to anchorage of window and glass door assemblies to the main force-resisting system.

R613.5.1 Anchoring requirements. Window and glass door assemblies shall be anchored in accordance with the published manufacturer's recommendations to achieve the design pressure specified. Substitute anchoring systems used for substrates not specified by the fenestration manufacturer shall provide equal or greater anchoring performance as demonstrated by accepted engineering practice.

R613.5.2 Anchorage details. Products shall be anchored in accordance with the minimum requirements illustrated in Figures R613.5(1), R613.5(2), R613.5(3), R613.5(4), R613.5(5), R613.5(6), R613.5(7) and R613.5(8).

R613.5.2.1 Masonry, concrete or other structural substrate. Where the wood shim or buck thickness is less than $1^1/_2$ inches (38 mm), window and glass door assemblies shall be anchored through the jamb, or by jamb clip and anchors shall be embedded directly into the masonry, concrete or other substantial substrate material. Anchors shall adequately transfer load from the window or door frame into the rough opening substrate [see Figures R613.5(1) and R613.5(2).]

Where the wood shim or buck thickness is 1.5 inches (38 mm) or greater, the buck is securely fastened to the masonry, concrete or other substantial substrate and the buck extends beyond the interior face of the window or door frame, window and glass door assemblies shall be anchored through the jamb, or by jamb clip, or through the flange to the secured wood buck. Anchors shall be embedded into the secured wood buck to adequately transfer load from the window or door frame assembly [see Figures R613.5(3), R613.5(4) and R613.5(5)].

R613.5.2.2 Wood or other approved framing material. Where the framing material is wood or other approved framing material, window and glass door assemblies shall be anchored through the frame, or by frame clip, or through the flange. Anchors shall be embedded into the frame construction to adequately transfer load [Figures R613.5(6), R613.5(7) and R613.5(8)].

R613.6 Mullions occurring between individual window and glass door assemblies.

R613.6.1 Mullions. Mullions shall be tested by an approved testing laboratory or be engineered in accordance with accepted engineering practice. Both methods shall use performance criteria cited in Sections R613.6.2, R613.6.3 and R613.6.4.

R613.6.2 Load transfer. Mullions shall be designed to transfer the design pressure loads applied by the window and door assemblies to the rough opening substrate.

R613.6.3 Deflection. Mullions shall be capable of resisting the design pressure loads applied by the window and door assemblies to be supported without deflecting more than $L/175$, where L = the span of mullion in inches.

R613.6.4 Structural safety factor. Mullions shall be capable of resisting a load of 1.5 times the design pressure loads applied by the window and door assemblies to be supported without exceeding the appropriate material stress levels. If tested by an approved laboratory, the 1.5 times the design pressure load shall be sustained for 10 seconds, and the permanent deformation shall not exceed 0.4 percent of the mullion span after the 1.5 times design pressure load is removed.

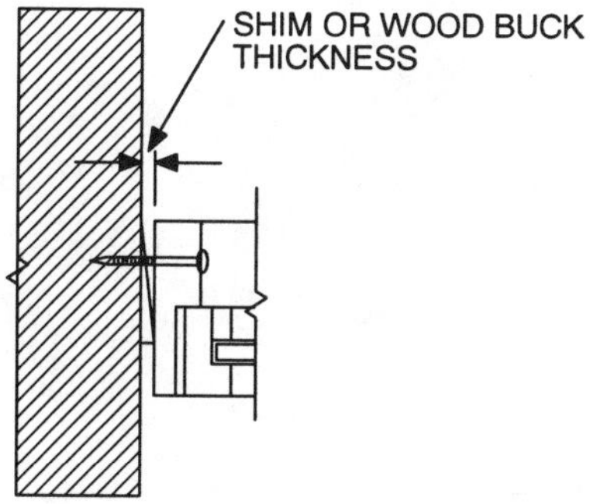

FIGURE R613.5(1)
THROUGH THE FRAME

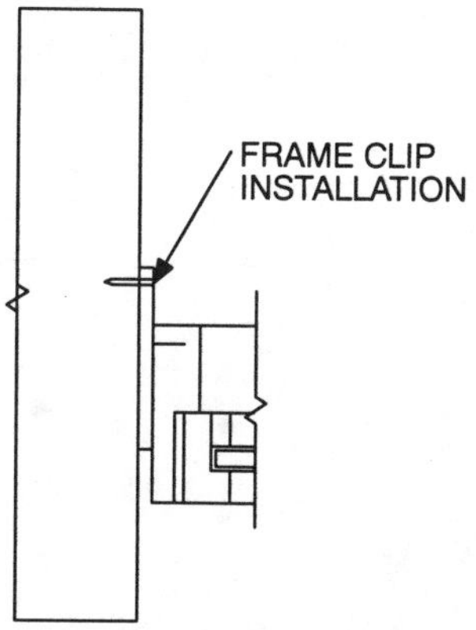

APPLY FRAME CLIP TO WINDOW OR DOOR IN ACCORDANCE WITH PUBLISHED MANUFACTURER'S RECOMMENDATIONS.

FIGURE R613.5(2)
FRAME CLIP

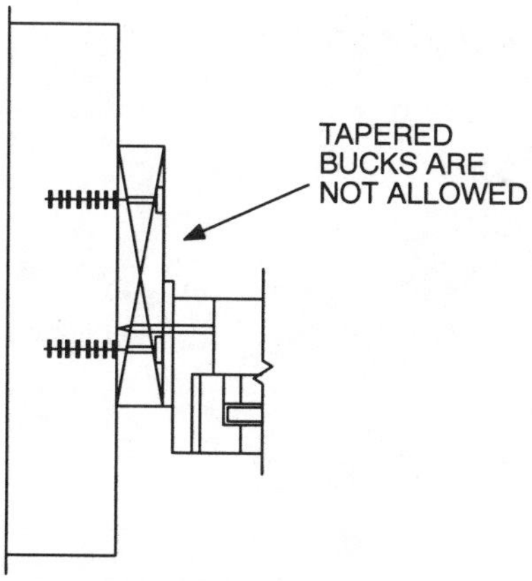

THROUGH THE FRAME ANCHORING METHOD. ANCHORS SHALL BE PROVIDED TO TRANSFER LOAD FROM THE WINDOW OR DOOR FRAME INTO THE ROUGH OPENING SUBSTRATE.

FIGURE R613.5(3)
THROUGH THE FRAME

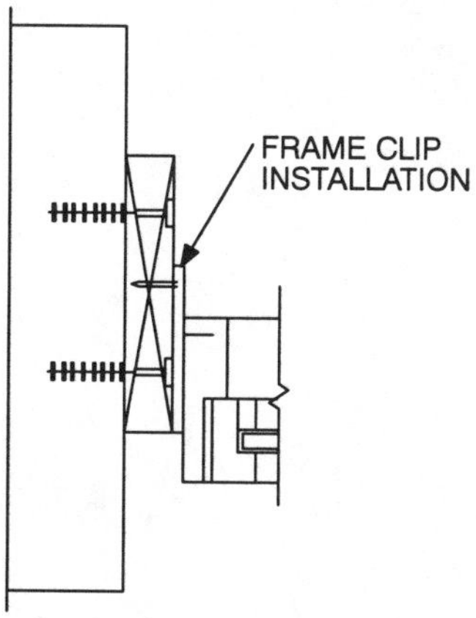

APPLY FRAME CLIP TO WINDOW OR DOOR FRAME IN ACCORDANCE WITH PUBLISHED MANUFACTURER'S RECOMMENDATIONS. ANCHORS SHALL BE PROVIDED TO TRANSFER LOAD FROM THE FRAME CLIP INTO THE ROUGH OPENING SUBSTRATE

FIGURE R613.5(4)
FRAME CLIP

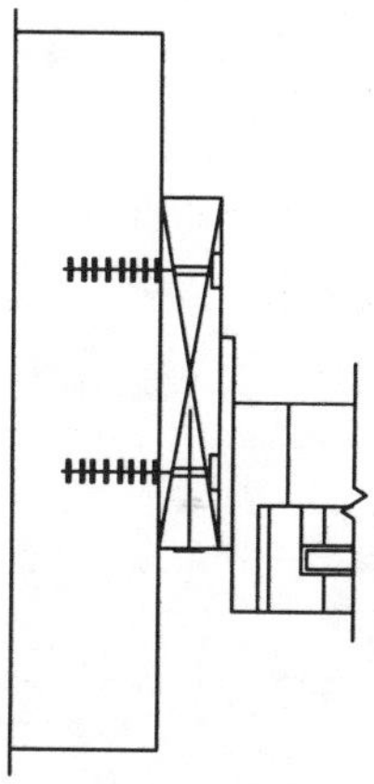

APPLY ANCHORS THROUGH FLANGE IN ACCORDANCE WITH PUBLISHED MANUFACTURER'S RECOMMENDATIONS.

FIGURE R613.5(5)
THROUGH THE FLANGE

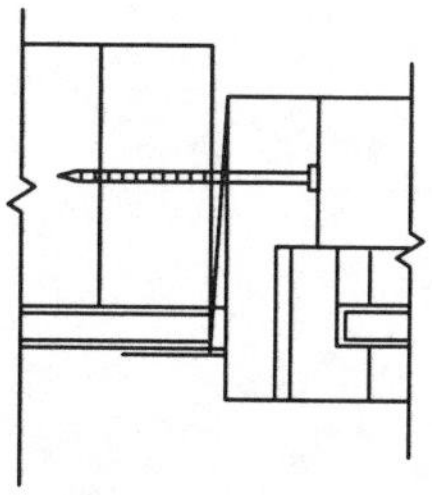

FIGURE R613.5(6)
THROUGH THE FRAME

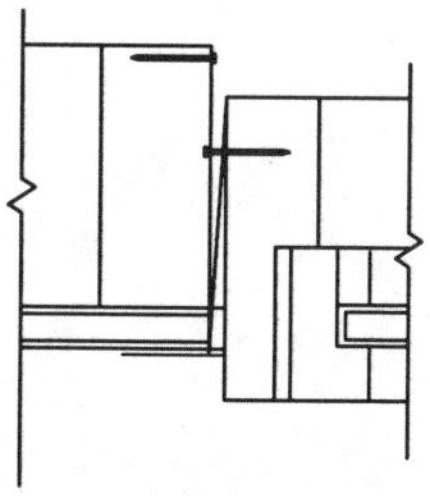

FIGURE R613.5(7)
FRAME CLIP

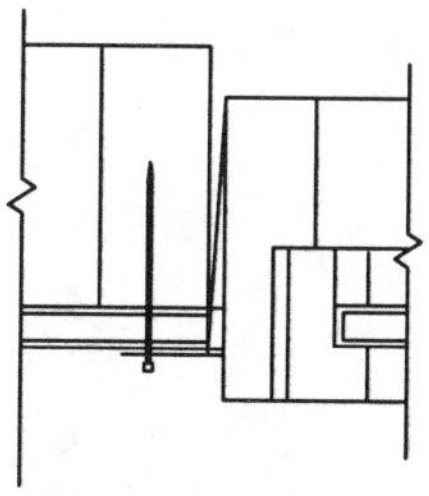

FIGURE R613.5(8)
THROUGH THE FLANGE

CHAPTER 7
WALL COVERING

SECTION R701
GENERAL

R701.1 Application. The provisions of this chapter shall control the design and construction of the interior and exterior wall covering for all buildings.

R701.2 Installation. Products sensitive to adverse weather shall not be installed until adequate weather protection for the installation is provided. Exterior sheathing shall be dry before applying exterior cover.

SECTION R702
INTERIOR COVERING

R702.1 General. Interior coverings shall be installed in accordance with this chapter and Tables R702.1(1), R702.1(2), R702.1(3) and R702.3.5. Interior finishes and materials shall conform to the flame spread and smoke-density requirements of Section R319.

TABLE R702.1(1)
THICKNESS OF PLASTER

PLASTER BASE	FINISHED THICKNESS OF PLASTER FROM FACE OF LATH, MASONRY, CONCRETE (inches)	
	Gypsum plaster	Portland cement mortar
Expanded metal lath	$^5/_8$, minimum[a]	$^5/_8$, minimum[a]
Wire lath	$^5/_8$, minimum[a]	$^3/_4$, minimum (interior)[b] $^7/_8$, minimum (exterior)[b]
Gypsum lath[g]	$^1/_2$, minimum	$^3/_4$, minimum (interior)[b]
Masonry walls[c]	$^1/_2$, minimum	$^1/_2$, minimum
Monolithic concrete walls[c,d]	$^5/_8$, maximum	$^7/_8$, maximum
Monolithic concrete ceilings[c,d]	$^3/_8$, maximum[e]	$^1/_2$, maximum
Gypsum veneer base[f,g]	$^1/_{16}$, minimum	$^3/_4$, minimum (interior)[b]
Gypsum sheathing[g]	—	$^3/_4$, minimum (interior)[b] $^7/_8$, minimum (exterior)[b]

For SI: 1 inch = 25.4 mm.

a. When measured from back plane of expanded metal lath, exclusive of ribs, or self-furring lath, plaster thickness shall be $^3/_4$ inch minimum.

b. When measured from face of support or backing.

c. Because masonry and concrete surfaces may vary in plane, thickness of plaster need not be uniform.

d. When applied over a liquid bonding agent, finish coat may be applied directly to concrete surface.

e. Approved acoustical plaster may be applied directly to concrete or over base coat plaster, beyond the maximum plaster thickness shown.

f. Attachment shall be in accordance with Table R702.3.5.

g. Where gypsum board is used as a base for portland cement plaster, weather-resistant sheathing paper complying with Section R703.2 shall be provided.

TABLE R702.1(2)
GYPSUM PLASTER PROPORTIONS[a]

NUMBER	COAT	PLASTER BASE OR LATH	MAXIMUM VOLUME AGGREGATE PER 100 POUNDS NEAT PLASTER[b] (cubic feet)	
			Damp loose sand[a]	Perlite or vermiculite[c]
Two-coat work	Base coat	Gypsum lath	2.5	2
	Base coat	Masonry	3	3
Three-coat work.	First coat	Lath	2[d]	2
	Second coat	Lath	3[d]	2[e]
	First and second coats	Masonry	3	3

For SI: 1 inch = 25.4 mm, 1 cubic foot = 0.0283 m^3, 1 pound = 0.454 kg.

a. Wood-fibered gypsum plaster may be mixed in the proportions of 100 pounds of gypsum to not more than 1 cubic foot of sand where applied on masonry or concrete.

b. When determining the amount of aggregate in set plaster, a tolerance of 10 percent shall be allowed.

c. Combinations of sand and lightweight aggregate may be used, provided the volume and weight relationship of the combined aggregate to gypsum plaster is maintained.

d. If used for both first and second coats, the volume of aggregate may be 2.5 cubic feet.

e. Where plaster is 1 inch or more in total thickness, the proportions for the second coat may be increased to 3 cubic feet.

TABLE R702.1(3)
PORTLAND CEMENT PLASTER

MAXIMUM VOLUME AGGREGATE PER VOLUME CEMENTITIOUS MATERIAL[a]					MINIMUM PERIOD MOIST COATS	MINIMUM INTERVAL BETWEEN
		Portland cement-lime plaster[c]				
Coat	Portland cement plaster[b] maximum volume aggregate per volume cement	Maximum volume lime per volume cement	Maximum volume sand per volume cement and lime	Approximate minimum thickness[d] curing (inches)		
First	4	$^3/_4$	4	$^3/_8$[e]	48[f] Hours	48[g] Hours
Second	5	$^3/_4$	5	First and second coats	48 Hours	7 Days[h]
Finished	3[i]	—	3[i]	$^1/_8$	—	Note[h]

For SI: 1 inch = 25.4 mm, 1 pound = 0.454 kg.

a. When determining the amount of aggregate in set plaster, a tolerance of 10 percent may be allowed.

b. From 10 to 20 pounds of dry hydrated lime (or an equivalent amount of lime putty) may be added as a plasticizing agent to each sack of Type I and Type II standard portland cement in base coat plaster.

c. No additions of plasticizing agents shall be made.

d. See Table R702.1(1).

e. Measured from face of support or backing to crest of scored plaster.

f. Twenty-four-hour minimum period for moist curing of interior portland cement plaster.

g. Twenty-four hour minimum interval between coats of interior portland cement plaster.

h. Finish coat plaster may be applied to interior portland cement base coats after a 48-hour period.

i. For finish coat, plaster up to an equal part of dry hydrated lime by weight (or an equivalent volume of lime putty) may be added to Type I, Type II and Type III standard portland cement.

R702.2 Interior plaster. Gypsum plaster or portland cement plastering materials shall conform to ASTM C 5, C 28, C 35, C 37, C 59, C 61, C 587, C 588, C 631, C 847, C 897, C 933, C 1032 and C 1047, and shall be installed or applied in conformance with ASTM C 843, C 844 and C 1063. Plaster shall not be less than three coats when applied over metal lath and not less than two coats when applied over other bases permitted by this section, except that veneer plaster may be applied in one coat not to exceed $^3/_{16}$ inch (4.76 mm) thickness, provided the total thickness is as set forth in Table R702.1(1).

R702.2.1 Support. Support spacing for gypsum or metal lath on walls or ceilings shall not exceed 16 inches (406 mm) for $^3/_8$ inch thick (9.5 mm) or 24 inches (610 mm) for $^1/_2$-inch-thick (12.7 mm) plain gypsum lath. Gypsum lath shall be installed at right angles to support framing with end joints in adjacent courses staggered by at least one framing space.

R702.3 Gypsum board.

R702.3.1 Materials. All gypsum board materials and accessories shall conform to ASTM C 36, C 79, C 475, C 514, C 630, C 960, C 1002, C 1047, C 1177, C1178, C1278 and C 1395 and shall be installed in accordance with the provisions of this section. Adhesives for the installation of gypsum board shall conform to ASTM C 557.

R702.3.2 Wood framing. Wood framing supporting gypsum board shall not be less than 2 inches (51 mm) nominal thickness in the least dimension except that wood furring strips not less than 1-inch-by-2 inch (25.4 mm by 51 mm) nominal dimension may be used over solid backing or framing spaced not more than 24 inches (610 mm) on center.

R702.3.3 Steel framing. Steel framing supporting gypsum board shall not be less than 1.25 inches (32 mm) wide in the least dimension. Light-gage nonload-bearing steel framing shall comply with ASTM C 645. Load-bearing steel framing and steel framing from 0.033 inch to 0.112 inch (0.838 mm to 2.84 mm) thick shall comply with ASTM C 955.

R702.3.4 Insulating concrete form walls. Foam plastics for insulating concrete form walls constructed in accordance with Sections R404.4 and R611 on the interior of habitable spaces shall be covered in accordance with Section R318.1.2. Adhesives are permitted to be used in conjunction with mechanical fasteners. Adhesives used for interior and exterior finishes shall be compatible with the insulating form materials.

R702.3.5 Application. Maximum spacing of supports and the size and spacing of fasteners used to attach gypsum board shall comply with Table R702.3.5. Gypsum sheathing shall be attached to exterior walls in accordance with Table R602.3(1). Gypsum board shall be applied at right angles or parallel to framing members. All edges and ends of gypsum board shall occur on the framing members, except those edges and ends that are perpendicular to the framing members. Interior gypsum board shall not be installed where it is exposed to the weather.

R702.3.6 Fastening. Screws for attaching gypsum board to wood framing shall be Type W or Type S in accordance with ASTM C 1002 and shall penetrate the wood not less than $^5/_8$ inch (15.9 mm). Screws for attaching gypsum board to light-gage steel framing shall be Type S in accordance with ASTM C 1002 and shall penetrate the steel not less than $^3/_8$ inch (9.5 mm). Screws for attaching gypsum board to steel framing 0.033 inch to 0.112 inch (0.838 mm to 2.84 mm) thick shall comply with ASTM C 954.

TABLE R702.3.5
MINIMUM THICKNESS AND APPLICATION OF GYPSUM BOARD

THICKNESS OF GYPSUM BOARD (inches)	APPLICATION	ORIENTATION OF GYPSUM BOARD TO FRAMING	MAXIMUM SPACING OF FRAMING MEMBERS (inches o.c.)	MAXIMUM SPACING OF FASTENERS (inches)		SIZE OF NAILS FOR APPLICATION TO WOOD FRAMING[c]
				Nails[a]	Screws[b]	
Application without adhesive						
3/8	Ceiling[d]	Perpendicular	16	7	12	13 gage, 1 1/4″ long, 19/64″ head; 0.098 diameter, 1 1/4″ long, annular-ringed; or 4d cooler nail, 0.080″ diameter, 1 3/8″ long, 7/32″ head.
	Wall	Either direction	16	8	16	
1/2	Ceiling	Either direction	16	7	12	13 gage, 1 3/8″ long, 19/64″ head; 0.098 diameter, 1 1/4″ long, annular-ringed; 5d cooler nail, 0.086 diameter, 1 5/8″ long, 15/64″ head; or gypsum board nail, 0.086 diameter, 1 5/8″ long, 9/32″ head.
	Ceiling[d]	Perpendicular	24	7	12	
	Wall	Either direction	24	8	12	
	Wall	Either direction	16	8	16	
5/8	Ceiling	Either direction	16	7	12	13 gage, 1 5/8″ long, 19/64″ head; 0.098 diameter, 1 3/8″ long, annular-ringed; 6d cooler nail, 0.092 diameter, 1 7/8″ long, 1/4″ head; or gypsum board nail, 0.0915 diameter, 1 7/8″ long, 19/64″ head.
	Ceiling	Perpendicular	24	7	12	
	Wall	Either direction	24	8	12	
	Wall	Either direction	16	8	16	
Application with adhesive						
3/8	Ceiling[d]	Perpendicular	16	16	16	Same as above for 3/8″ gypsum board
	Wall	Either direction	16	16	24	
1/2 or 5/8	Ceiling	Either direction	16	16	16	Same as above for 1/2″ and 5/8″ gypsum board, respectively
	Ceiling[d]	Perpendicular	24	12	16	
	Wall	Either direction	24	16	24	
two 3/8 layers	Ceiling	Perpendicular	16	16	16	Base ply nailed as above for 1/2″ gypsum board; face ply installed with adhesive
	Wall	Either direction	24	24	24	

For SI: 1 inch = 25.4 mm.

a. For application without adhesive, a pair of nails spaced not less than 2 inches apart or more than 2 1/2 inches apart may be used with the pair of nails spaced 12 inches on center.

b. Screws shall be Type S or W per ASTM C 1002 and shall be sufficiently long to penetrate wood framing not less than 5/8 inch and metal framing not less than 3/8 inch.

c. Where metal framing is used with a clinching design to receive nails by two edges of metal, the nails shall be not less than 5/8 inch longer than the gypsum board thickness and shall have ringed shanks. Where the metal framing has a nailing groove formed to receive the nails, the nails shall have barbed shanks or be 5d, 13 1/2 gage, 1 5/8 inches long, 15/64-inch head for 1/2-inch gypsum board; and 6d, 13 gage, 1 7/8 inches long, 15/64-inch head for 5/8-inch gypsum board.

d. Three-eighths-inch-thick single-ply gypsum board shall not be used on a ceiling where a water-based textured finish is to be applied, or where it will be required to support insulation above a ceiling. On ceiling applications to receive a water-based texture material, either hand or spray applied, the gypsum board shall be applied perpendicular to framing. When applying a water-based texture material, the minimum gypsum board thickness shall be increased from 3/8 inch to 1/2 inch for 16-inch on center framing, and from 1/2 inch to 5/8 inch for 24-inch on center framing or 1/2-inch sag-resistant gypsum ceiling board shall be used.

R702.4 Ceramic tile.

R702.4.1 General. Ceramic tile surfaces shall be installed in accordance with ANSI A108.1, A108.4, A108.5, A108.6, A108.11, A118.1, A118.3, A136.1 and A137.1.

R702.4.2 Gypsum backer. Gypsum board utilized as the base or backer board for adhesive application of ceramic tile or other nonabsorbent finish material shall conform with ASTM C 630 or C 1178. Water-resistant gypsum backing board shall be permitted to be used on ceilings where framing spacing does not exceed 12 inches (305 mm) on center for 1/2 inch thick (12.7 mm) or 16 inches (406 mm) for 5/8-inch-thick (15.9 mm) gypsum board. All cut or exposed edges, including those at wall intersections, shall be sealed as recommended by the manufacturer.

R702.5 Other finishes. Wood veneer paneling and hardboard paneling shall be placed on wood or cold-formed steel framing spaced not more than 16 inches (406 mm) on center. Wood veneer and hard board paneling less than 1/4 inch (6.4 mm) nominal thickness shall not have less than a 3/8-inch (9.5 mm) gypsum board backer. Wood veneer paneling not less than 1/4-inch (6.4 mm) nominal thickness shall conform to ANSI/HPVA HP-1. Hardboard paneling shall conform to ANSI/AHA A135.5.

R702.6 Wood shakes and shingles. Wood shakes and shingles shall conform to CSSB *Grading Rules for Wood Shakes and Shingles* and shall be permitted to be installed directly to the studs with maximum 24 inches (610 mm) on center spacing.

R702.6.1 Attachment. Nails, staples or glue are permitted for use in attaching shakes or shingles to the wall, and the shakes or shingles shall be permitted to be attached directly to the surface provided the fasteners are appropriate for the type of wall surface material. When nails or staples are used, two fasteners shall be provided and shall be placed so that they are covered by the course above.

R702.6.2 Furring strips. Where furring strips are used, they shall be 1 inch by 2 inches or 1 inch by 3 inches (25 mm by 51 mm or 25 mm by 76 mm), spaced a distance on center equal to the desired exposure, and shall be attached to the wall by nailing through other wall material into the studs.

SECTION R703 EXTERIOR COVERING

R703.1 General. Exterior walls shall provide the building with a weather-resistant exterior wall envelope. The exterior wall envelope shall include flashing as described in Section R703.8. The exterior wall envelope shall be designed and constructed in such a manner as to prevent the accumulation of water within the wall assembly by providing a water-resistive barrier behind the exterior veneer as required by Section R703.2.

R703.2 Weather-resistant sheathing paper. Asphalt-saturated felt free from holes and breaks, weighing not less tha 14 pounds per 100 square feet (0.683 kg/m^2) and complying with ASTM D 226 or other approved weather-resistant material shall be applied over studs or sheathing of all exterior walls as required by Table R703.4. Such felt or material shall be applied horizontally, with the upper layer lapped over the lower layer not less than 2 inches (51 mm). Where joints occur, felt shall be lapped not less than 6 inches (152 mm).

Exception: Such felt or material is permitted to be omitted in the following situations:

1. In detached accessory buildings.
2. Under panel siding with shiplap joints or battens.
3. Under exterior wall finish materials as permitted in Table R703.4.
4. Under paperbacked stucco lath.

R703.3 Wood, hardboard and wood structural panel siding.

R703.3.1 Panel siding. Joints in wood, hardboard or wood structural panel siding shall be made as follows unless otherwise approved. Vertical joints in panel siding shall occur over framing members, unless wood or wood structural panel sheathing is used, and shall be shiplapped or covered with a batten. Horizontal joints in panel siding shall be lapped a minimum of 1 inch (25.4 mm) or shall be shiplapped or shall be flashed with Z-flashing and occur over solid blocking, wood or wood structural panel sheathing.

R703.3.2 Horizontal siding. Horizontal lap siding shall be lapped a minimum of 1 inch (25.4 mm), or 0.5 inch (12.7 mm) if rabbeted, and shall have the ends caulked, covered with a batten, or sealed and installed over a strip of flashing.

R703.4 Attachments. Unless specified otherwise, all wall coverings shall be securely fastened in accordance with Table R703.4 or with other approved aluminum, stainless steel, zinc-coated or other approved corrosion-resistive fasteners.

R703.5 Wood shakes and shingles. Wood shakes and shingles shall conform to CSSB “Grading Rules for Wood Shakes and Shingles.”

R703.5.1 Application. Wood shakes or shingles shall be applied either single-course or double-course over nominal $^1/_2$-inch (12.7 mm) wood-based sheathing or to furring strips over $^1/_2$-inch (12.7 mm) nominal nonwood sheathing. A weather-resistant permeable membrane shall be provided over all sheathing, with horizontal overlaps in the membrane of not less than 2 inches (51 mm) and vertical overlaps of not less than 6 inches (152 mm). Where furring strips are used, they shall be 1 inch by 3 inches or 1 inch by 4 inches (25 mm by 76 mm or 25 mm by 102 mm) and shall be fastened horizontally to the studs with 7d or 8d box nails and shall be spaced a distance on center equal to the actual weather exposure of the shakes or shingles, not to exceed the maximum exposure specified in Table R703.5.2. The spacing between adjacent shingles to allow for expansion shall not exceed $^1/_4$ inch (6.4 mm), and between adjacent shakes, it shall not exceed $^1/_2$ inch (12.7 mm). The offset spacing between joints in adjacent courses shall be a minimum of $1^1/_2$ inches (38 mm).

R703.5.2 Weather exposure. The maximum weather exposure for shakes and shingles shall not exceed that specified in Table R703.5.2.

R703.5.3 Attachment. Each shake or shingle shall be held in place by two hot-dipped zinc-coated, stainless steel, or aluminum nails or staples. The fasteners shall be long enough to penetrate the sheathing or furring strips by a minimum of $^1/_2$ inch (12.7 mm) and shall not be overdriven.

R703.5.3.1 Staple attachment. Staples shall not be less than16 gage and shall have a crown width of not less than $^7/_{16}$ inch (11.1 mm), and the crown of the staples shall be parallel with the butt of the shake or shingle. In single-course application, the fasteners shall be concealed by the course above and shall be driven approximately 1 inch (25 mm) above the butt line of the succeeding course and $^3/_4$ inch (19.1 mm) from the edge. In double-course applications, the exposed shake or shingle shall be face-nailed with two casing nails, driven approximately 2 inches (51 mm) above the butt line and $^3/_4$ inch (19.1 mm) from each edge. In all applications, staples shall be concealed by the course above. With shingles wider than 8 inches (203 mm) two additional nails shall be required and shall be nailed approximately 1 inch (25.4 mm) apart near the center of the shingle.

R703.5.4 Bottom courses. The bottom courses shall be doubled.

TABLE R703.4
WEATHER-RESISTANT SIDING ATTACHMENT AND MINIMUM THICKNESS

SIDING MATERIAL		NOMINAL THICKNESS[a] (inches)	JOINT TREATMENT	SHEATHING PAPER REQUIRED	TYPE OF SUPPORTS FOR THE SIDING MATERIAL AND FASTENERS[b,c,d]				
					Wood or wood structural panel sheathing	Fiberboard sheathing into stud	Gypsum sheathing into stud	Direct to studs	Number or spacing of fasteners
Horizontal aluminum[e]	Without insulation	0.019[f]	Lap	No	0.120 nail 1 1/2″ long	0.120 nail 2″ long	0.120 nail 2″ long	Not allowed	Same as stud spacing
		0.024	Lap	No	0.120 nail 1 1/2″ long	0.120 nail 2″ long	0.120 nail 2″ long	Not allowed	
	With insulation	0.019	Lap	No	0.120 nail 1 1/2″ long	0.120 nail 2 1/2″ long	0.120 nail 2 1/2″ long	0.120 nail 1 1/2″ long	
Brick veneer Concrete masonry veneer		2 2	Section R703	Yes Note m	See Section R703 and Figure R703.7[h]				
Hardboard[l] Panel siding-vertical		7/16	Note r	Note r	Note o	Note o	Note o	Note o	6″ panel edges 12″ inter. sup.[p]
Hardboard[l] Siding vertical Lap-siding-horizontal		7/16 7/16	Note g Note r	Note g Note r	Note q	Note q	Note q	Note q	Same as stud spacing 2 per bearing
Steel[i]		29 ga.	Lap	No	0.113 nail 1 3/4″ Staple-1 3/4″	0.113 nail 2 3/4″ Staple-2 1/2″	0.113 nail 2 1/2″ Staple-2 1/4″	Not allowed	Same as stud spacing
Stone veneer		2	Section R703	Yes	See Section R703 and Figure R703.7				
Particleboard panels		3/8 - 1/2	Note g	Note g	6d box nail	6d box nail	6d box nail	6d box nail, 3/8 not allowed	6″ panel edges 12″ inter. sup.
		5/8	Note g	Note g	6d box nail	8d box nail	8d box nail	6d box nail	
Plywood panel[j] (exterior grade)		3/8	Note g	Note g	0.099 nail-2″	0.113 nail-2 1/2″	0.099 nail-2″	0.099 nail-2″	6″ on edges
Vinyl Siding[n]		0.035	Lap	No	0.120 nail 1 1/2″ Staple-1 3/4″	0.120 nail 2″ Staple-2 1/2″	0.120 nail 2″ Staple-2 1/2″	Not allowed	Same as stud spacing
Wood[k] Rustic, drop		3/8 Min	Lap	No	Fastener penetration into stud-1″			0.113 nail-2 1/2″ Staple-2″	Face nailing up to 6″ widths, 1 nail per bearing; 8″ widths and over, 2 nails per bearing
Shiplap		19/32 Average	Lap	No					
Bevel		7/16	Lap	No					
Butt tip		3/16	Lap	No					

For SI: 1 inch = 25.4 mm.

a. Based on stud spacing of 16 inches on center Where studs are spaced 24 inches, siding shall be applied to sheathing approved for that spacing.

b. Nail is a general description and shall be T-head, modified round head, or round head with smooth or deformed shanks.

c. Staples shall have a minimum crown width of 7/16-inch outside diameter and be manufactured of minimum No. 16 gage wire.

d. Nails or staples shall be aluminum, galvanized, or rust-preventive coated and shall be driven into the studs for fiberboard or gypsum backing.

e. Aluminum nails shall be used to attach aluminum siding.

f. Aluminum (0.019 inch) shall be unbacked only when the maximum panel width is 10 inches and the maximum flat area is 8 inches. The tolerance for aluminum siding shall be +0.002 inch of the nominal dimension.

g. If boards or panels are applied over sheathing or a weather-resistant membrane, joints need not be treated. Otherwise, vertical joints shall occur at studs and be covered with battens or be lapped.

h. All attachments shall be coated with a corrosion-resistive coating.

i. Shall be of approved type.

j. Three-eighths-inch plywood shall not be applied directly to studs spaced greater than 16 inches on center when long dimension is parallel to studs. One-half-inch plywood shall not may be applied directly to studs spaced greater than 24 inches on center. The stud spacing shall not exceed the panel span rating provided by the manufacturer unless the panels are installed with the face grain perpendicular to studs or over sheathing approved for that stud spacing.

(continued)

NOTES TO TABLE R703.4—continued

k. Woodboard sidings applied vertically shall be nailed to horizontal nailing strips or blocking set 24 inches on center Nails shall penetrate 1.5 inches into studs, studs and wood sheathing combined, or blocking. A weather-resistant membrane shall be installed weatherboard fashion under the vertical siding unless the siding boards are lapped or battens are used.

l. Hardboard siding shall comply with AHA A135.6.

m. For masonry veneer, a weather-resistant membrane or building paper is not required over water-repellent sheathing materials when a 1-inch air space is provided between the veneer and the sheathing. When the 1-inch space is filled with mortar, a weather-resistant membrane or building paper is required over studs or sheathing.

n. Vinyl siding shall comply with ASTM D 3679.

o. Minimum shank diameter of 0.092 inch, minimum head diameter of 0.225 inch, and nail length must accommodate sheathing and penetrate framing 1.5 inches.

p. When used to resist shear forces, the spacing must be 4 inches at panel edges and 8 inches on interior supports.

q. Minimum shank diameter of 0.099 inch, minimum head diameter of 0.240 inch, and nail length must accommodate sheathing and penetrate framing 1.5 inches.

r. Vertical end joints shall occur at studs and shall be covered with a joint cover or shall be caulked.

TABLE R703.5.2
MAXIMUM WEATHER EXPOSURE FOR WOOD SHAKES AND SHINGLES ON EXTERIOR WALLS[a,b,c]
(Dimensions are in inches)

LENGTH	EXPOSURE FOR SINGLE COURSE	EXPOSURE FOR DOUBLE COURSE
Shingles[a]		
16	$7^1/_2$	12[b]
18	$8^1/_2$	14[c]
24	$11^1/_2$	16
Shakes[a]		
18	$8^1/_2$	14
24	$11^1/_2$	18

For SI: 1 inch = 25.4 mm.

a. Dimensions given are for No. 1 grade.

b. A maximum 10-inch exposure is permitted for No. 2 grade.

c. A maximum 11-inch exposure is permitted for No. 2 grade.

R703.6 Exterior plaster.

R703.6.1 Lath. All lath and lath attachments shall be of corrosion-resistant materials. Expanded metal or woven wire lath shall be attached with $1^1/_2$-inch-long (38 mm), 11 gage nails having a $^7/_{16}$-inch (11.1 mm) head, or $^7/_8$-inch-long (22.2 mm), 16 gage staples, spaced at no more than 6 inches (152 mm), or as otherwise approved.

R703.6.2 Plaster. Plastering with portland cement plaster shall be not less than three coats when applied over metal lath or wire lath and shall be not less than two coats when applied over masonry, concrete or gypsum backing. If the plaster surface is completely covered by veneer or other facing material or is completely concealed, plaster application need be only two coats, provided the total thickness is as set forth in Table R702.1(1).

On wood-frame construction with an on-grade floor slab system, exterior plaster shall be applied in such a manner as to cover, but not extend below, lath, paper and screed.

The proportion of aggregate to cementitious materials shall be as set forth in Table R702.1(3).

R703.7 Stone and masonry veneer, general. All stone and masonry veneer shall be installed in accordance with this chapter, Table R703.4 and Figure R703.7. Such veneers installed over a backing of wood or cold-formed steel shall be limited to the first story above grade and shall not exceed 5 inches (127 mm) in thickness.

Exceptions:

1. In Seismic Design Categories A and B, exterior masonry veneer with a backing of wood or cold-formed steel framing shall not exceed 30 feet (9144 mm) in height above the noncombustible foundation, with an additional 8 feet (2348 mm) permitted for ends.
2. In Seismic Design Category C, exterior masonry veneer with a backing of wood or cold-formed steel framing shall not exceed 30 feet (9144 mm) in height above the noncombustible foundation, with an additional 8 feet (2348 mm) permitted for gabled ends. In other than the topmost story, the length of bracing shall be 1.5 times the length otherwise required in Chapter 6.

R703.7.1 Interior veneer support. Veneers used as interior wall finishes shall be permitted to be supported on wood or cold-formed steel floors that are designed to support the loads imposed.

TABLE R703.7.3
ALLOWABLE SPANS FOR LINTELS SUPPORTING MASONRY VENEER[a,b,c]

SIZE OF STEEL ANGLE[a,c] (inches)	NO STORY ABOVE	ONE STORY ABOVE	TWO STORIES ABOVE	NO. OF $^1/_2''$ OR EQUIVALENT REINFORCING BARS[c]
$3 \times 3 \times {}^1/_4$	6′-0″	3′-6″	3′-0″	1
$4 \times 3 \times {}^1/_4$	8′-0″	5′-0″	3′-0″	1
$6 \times 3^1/_2 \times {}^1/_4$	14′-0″	8′-0″	3′-6″	2
$2\text{-}6 \times 3^1/_2 \times {}^1/_4$	20′-0″	11′-0″	5′-0″	4

For SI: 1 inch = 25.4 mm, 1 foot =304.8 mm.

a. Long leg of the angle shall be placed in a vertical position.

b. Depth of reinforced lintels shall not be less than 8 inches and all cells of hollow masonry lintels shall be grouted solid. Reinforcing bars shall extend not less than 8 inches into the support.

c. Steel members indicated are adequate typical examples; other steel members meeting structural design requirements may be used.

R703.7.2 Exterior veneer support. Except in Seismic Design Categories D_1 and D_2, exterior masonry veneers having an installed weight of 40 pounds per square foot (195 kg/m^2) or less shall be permitted to be supported on wood or cold-formed steel construction. When masonry veneer supported by wood or cold-formed steel construction adjoins masonry veneer supported by the foundation, there shall be a movement joint between the veneer supported by the wood or cold-formed steel construction and the veneer supported by the foundation. The wood or cold-formed steel construction supporting the masonry veneer shall be designed to limit the deflection to $^1/_{600}$ of the span for the supporting members. The design of the wood or cold-formed steel construction shall consider the weight of the veneer and any other loads. The method of support for the masonry veneer on wood construction shall be constructed in accordance with Figure R703.7.1 or as approved by the building official.

R703.7.2.1 Support by steel angle. A minimum 6 inches by 4 inches by $^5/_{16}$ inch (152 mm by 102 mm by 8 mm) steel angle, with the long leg placed vertically, shall be anchored to double 2 inches by 4 inches (51 mm by 102 mm) wood studs at a maximum on center spacing of 16 inches (406 mm). Anchorage of the steel angle at every double stud spacing shall be a minimum of two $^7/_{16}$ inch (11.1 mm) diameter by 4 inches (102 mm) lag screws. The steel angle shall have a minimum clearance to underlying construction of $^1/_{16}$ inch (1.6 mm). A minimum of two-thirds the width of the masonry veneer thickness shall bear on the steel angle. Flashing and weep holes shall be located in the masonry veneer wythe in accordance with Figure R 703.7.1. The maximum height of masonry veneer above the steel angle support shall be 12 feet, 8 inches (3861 mm). The maximum slope of the roof construction shall be not more than 7:12. The air space separating the masonry veneer from the wood backing shall be in accordance with Sections R703.7.4 and R703.7.4.2.

R703.7.2.2 Support by roof construction. A steel angle shall be placed directly on top of the roof construction. The roof supporting construction for the steel angle shall consist of a minimum of three 2-inch by 6-inch (51 mm by 152 mm) wood members. The wood member abutting the vertical wall stud construction shall be anchored with a minimum of three $^5/_8$-inch (15.9 mm) diameter by 5-inch (127 mm) lag screws to every wood stud spacing. Each additional roof member shall be anchored by the use of two 10d nails at every wood stud spacing. A minimum of two-thirds the width of the masonry veneer thickness shall bear on the steel angle. Flashing and weep holes shall be located in the masonry veneer wythe in accordance with Figure R703.7.1. The maximum height of the masonry veneer above the steel angle support shall be 12 feet, 8 inches (3861 mm). The maximum slope of the roof construction shall be not more than 7:12. The air space separating the masonry veneer from the wood backing shall be in accordance with Sections R703.7.4 and R703.7.4.2.

R703.7.3 Lintels. Masonry veneer shall not support any vertical load other than the dead load of the veneer above. Veneer above openings shall be supported on lintels of noncombustible materials and the allowable span shall not exceed the values set forth in Table R703.7.3. The lintels shall have a length of bearing of not less than 4 inches (102 mm).

R703.7.4 Anchorage. Masonry veneer shall be anchored to the supporting wall with corrosion-resistant metal ties. Where veneer is anchored to wood backings through the use of corrugated sheet metal ties, the distance separating the veneer from the sheathing material shall be a maximum of 1 inch (25 mm). Where the veneer is anchored to wood backings through the use of metal strand wire ties, the distance separating the veneer from the sheathing material shall be a maximum of $4^1/_2$ inches (114 mm). Where the veneer is anchored to cold-formed steel backings, adjustable metal strand wire ties shall be used. Where veneer is anchored to cold-formed steel backings, the distance separating the veneer from the sheathing material shall be a maximum of 4.5 inches (114 mm).

R703.7.4.1 Size and spacing. Veneer ties, if strand wire, shall not be less in thickness than No. 9 U.S. gage wire and shall have a hood embedded in the mortar joint, or if sheet metal, shall be not less than No. 22 U.S. gage by $^7/_8$ inch (22.3 mm) corrugated. Each tie shall be spaced not more than 24 inches (610 mm) on center horizontally and shall support not more than $3^1/_4$ square feet (0.302 m^2) of wall area.

Exception: In Seismic Design Category D_1 or D_2 and in wind areas of more than 30 pounds per square foot

($1.44\ kN/m^2$), each tie shall support not more than 2 square feet ($0.186\ m^2$) of wall area.

R703.7.4.1.1 Veneer ties around wall openings. Additional metal ties shall be provided around all wall openings greater than 16 inches (406 mm) in either dimension. Metal ties around the perimeter of openings shall be spaced not more than 3 feet (9144 mm) on center and placed within 12 inches (305 mm) of the wall opening.

R703.7.4.1.2 Seismic Design Categories D_1 and D_2. In Seismic Design Categories D_1 and D_2, veneer ties shall be mechanically attached to horizontal joint reinforcement wire a minimum of No. 9 gage. The horizontal joint reinforcement shall be continuous in the veneer bed joint, with lap splices permitted between the veneer tie spacing.

R703.7.4.2 Air space. The veneer shall be separated from the sheathing by an air space of a minimum of 1 inch (25.4 mm) but not more than 4.5 inches (114 mm). The weather-resistant membrane or asphalt-saturated felt required by Section R703.2 is not required over water-repellent sheathing materials.

R703.7.4.3 Mortar or grout fill. As an alternate to the air space required by Section R703.7.4.2, mortar or grout shall be permitted to fill the air space. When the 1-inch (25.4 mm) space is filled with mortar, a weather-resistant membrane or building paper is required over studs or sheathing. When filling the air space, it is permitted to replace the sheathing and weather-resistant membrane or asphalt-saturated felt paper with a wire mesh and approved paper or an approved paper-backed reinforcement attached directly to the studs.

R703.7.5 Flashing. Flashing shall be located beneath the first course of masonry above finished ground level above the foundation wall or slab and at other points of support, including structural floors, shelf angles and lintels when masonry veneers are designed in accordance with Section R703.7. See Section R703.8 for additional requirements.

R703.7.6 Weepholes. Weepholes shall be provided in the outside wythe of masonry walls at a maximum spacing of 33 inches (838 mm) on center. Weepholes shall not be less than $^3/_{16}$ inch (4.8 mm) in diameter. Weepholes shall be located immediately above the flashing.

R703.8 Flashing. Approved corrosion-resistive flashing shall be provided in the exterior wall envelope in such a manner as to prevent entry of water into the wall cavity or penetration of water to the building structural framing components. The flashing shall extend to the surface of the exterior wall finish and shall be installed to prevent water from reentering the exterior wall envelope. Approved corrosion-resistant flashings shall be installed at all of the following locations:

1. At top of all exterior window and door openings in such a manner as to be leakproof, except that self-flashing windows having a continuous lap of not less than $1^1/_8$ inches (28 mm) over the sheathing material around the perimeter of the opening, including corners, do not require additional flashing; jamb flashing may also be omitted when specifically approved by the building official.
2. At the intersection of chimneys or other masonry construction with frame or stucco walls, with projecting lips on both sides under stucco copings.
3. Under and at the ends of masonry, wood or metal copings and sills.
4. Continuously above all projecting wood trim.
5. Where exterior porches, decks or stairs attach to a wall or floor assembly of wood-frame construction.
6. At wall and roof intersections.
7. At built-in gutters.

R703.9 Exterior insulation finish systems, general. All Exterior Insulation Finish Systems (EIFS) shall be installed in accordance with the manufacturer's installation instructions and the requirements of this section. Decorative trim shall not be face nailed through the EIFS. The EIFS shall terminate not less than 6 inches (152 mm) above the finished ground level.

R703.9.1 Weather-resistive barrier. All EIFS shall have a weather-resistive barrier applied between the underlying water-sensitive building components and the exterior insulation, and a means of draining water to the exterior of the veneer. A weather-resistive barrier shall be compliant with ASTM D 226 Type I asphalt saturated felt or equivalent, shall be applied horizontally with the upper layer lapped over the lower layer not less than 2 inches (51 mm), and shall have all vertical joints lapped not less than 6 inches (152 mm).

R703.9.2 Flashing, general. Flashing of EIFS shall be provided in accordance with the requirements of Section R703.8.

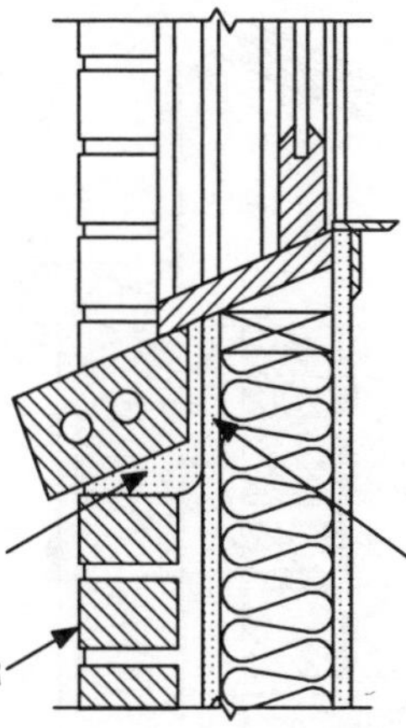

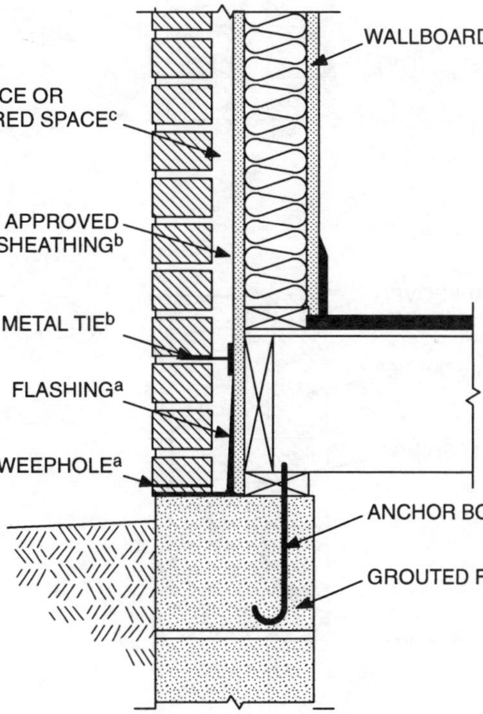

For SI: 1 inch = 25.4 mm.

FIGURE R703.7
MASONRY VENEER WALL DETAILS
(continued)

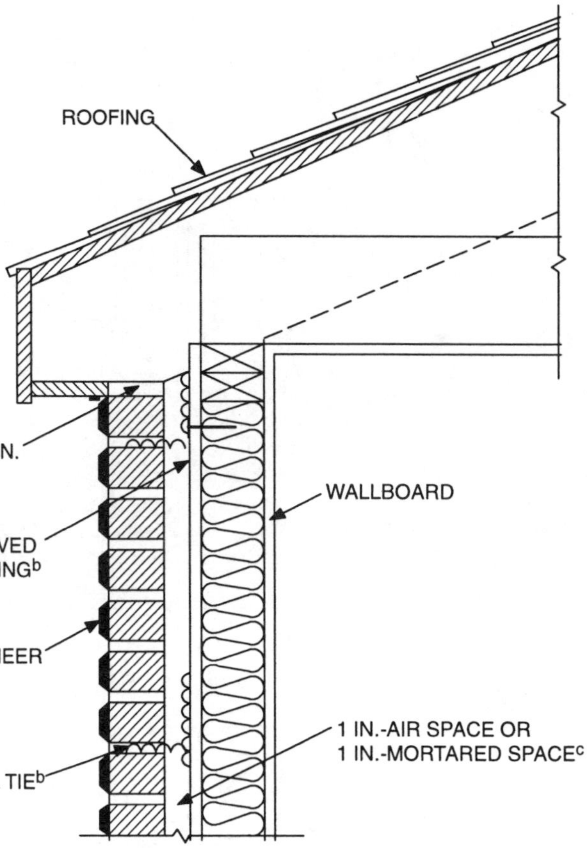

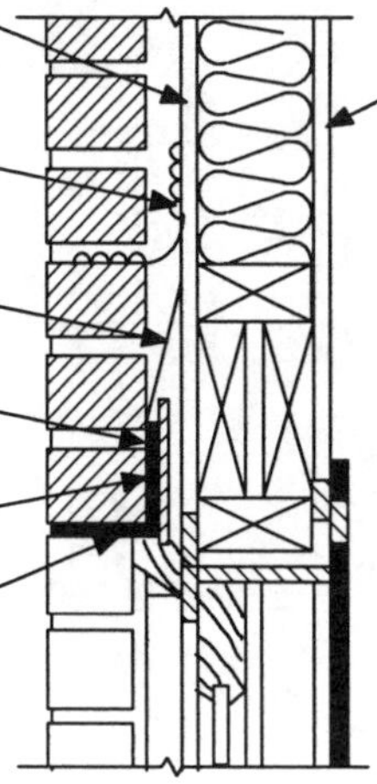

For SI: 1 inch = 25.4 mm.

a. See Sections R703.7.5 and R703.7.6.
b. See Sections R703.2 and R703.7.4.
c. See Section R703.7.4.2 and R703.7.4.3.
d. See Section R703.7.2.1.

FIGURE R703.7—continued
MASONRY VENEER WALL DETAILS

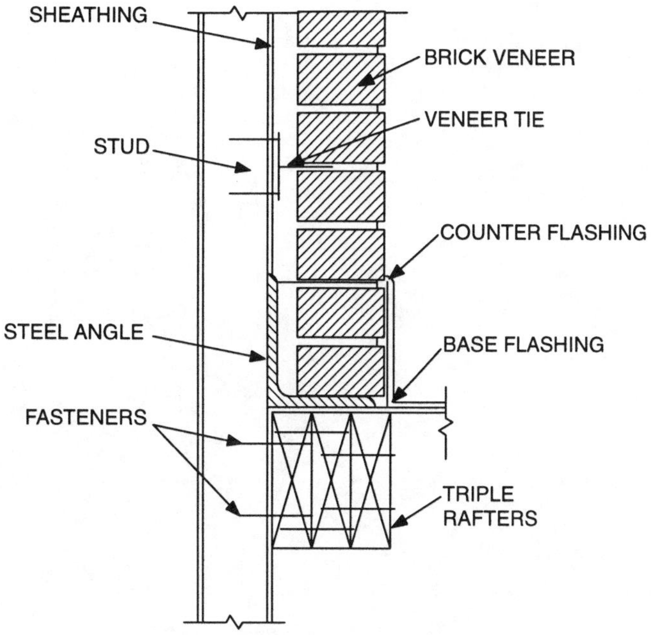

SUPPORT BY ROOF MEMBERS

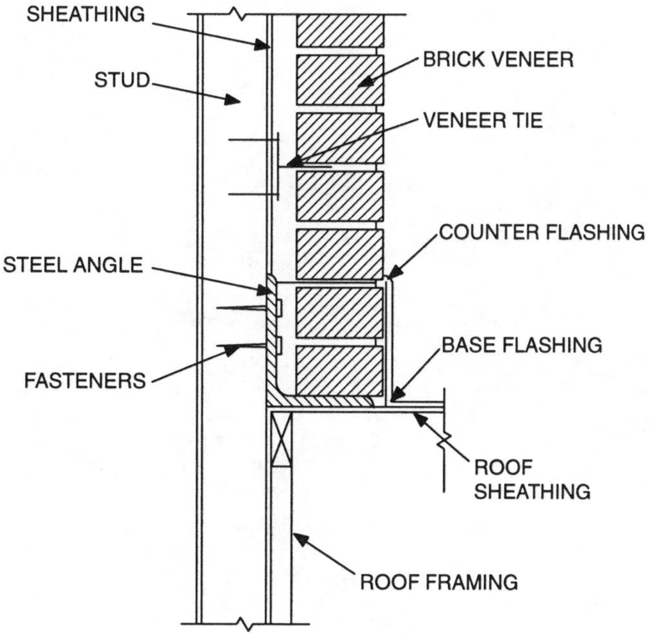

SUPPORT BY STEEL ANGLE

FIGURE R703.7.1
EXTERIOR MASONRY VENEER SUPPORT BY WOOD CONSTRUCTION

CHAPTER 8
ROOF-CEILING CONSTRUCTION

SECTION R801
GENERAL

R801.1 Application. The provisions of this chapter shall control the design and construction of the roof-ceiling system for all buildings.

R801.2 Requirements. Roof and ceiling construction shall be capable of accommodating all loads imposed according to Section R301 and of transmitting the resulting loads to the supporting structural elements.

R801.3 Roof drainage. In areas where expansive or collapsible soils are known to exist, all dwellings shall have a controlled method of water disposal from roofs that will collect and discharge all roof drainage to the ground surface at least 5 feet (1524 mm) from foundation walls or to an approved drainage system.

SECTION R802
WOOD ROOF FRAMING

R802.1 Identification and grade. Load-bearing dimension lumber for rafters, trusses and ceiling joists shall be identified by a grade mark of a lumber grading or inspection agency that has been approved by an accreditation body that complies with DOC PS 20. In lieu of a grade mark, a certificate of inspection issued by a lumber grading or inspection agency meeting the requirements of this section shall be accepted.

R802.1.1 Blocking. Blocking shall be a minimum of utility grade lumber.

R802.1.2 End-jointed lumber. Approved end-jointed lumber identified by a grade mark conforming to Section R802.1 may be used interchangeably with solid-sawn members of the same species and grade.

R802.1.3 Fire-retardant-treated lumber. The allowable unit stresses for fire-retardant-treated lumber, including fastener values, shall be developed from an approved method of investigation that considers the effects of anticipated temperature and humidity to which the fire-retardant lumber will be subjected, the type of treatment and redrying process.

R802.1.3.1 Labeling and grade marking. Fire-retardant-treated lumber and wood structural panels shall bear the identification mark of an approved agency. Such identification marks shall indicate conformance with appropriate standards. Additionally, fire-retardant-treated lumber and wood structural panels shall be identified in accordance with Section R802.1 and Section R803.2.1 respectively.

R802.1.3.2 Moisture content. Fire-retardant-treated wood shall be dried in accordance with AWPA C20 to a moisture content of 19 percent or less for lumber and 15 percent or less in accordance with AWPA C27 for wood structural panels before use.

R802.2 Design and construction. Roof-ceilings shall be designed and constructed in accordance with the provisions of this chapter and Figures R606.10(1), R606.10(2) and R606.10(3) or in accordance with AFPA/NDS. Components of roof-ceilings shall be fastened in accordance with Table R602.3(1).

R802.3 Framing details. Rafters shall be framed to ridge board or to each other with a gusset plate as a tie. Ridge board shall be at least 1-inch (25.4 mm) nominal thickness and not less in depth than the cut end of the rafter. At all valleys and hips there shall be a valley or hip rafter not less than 2-inch (51 mm) nominal thickness and not less in depth than the cut end of the rafter. Hip and valley rafters shall be supported at the ridge by a brace to a bearing partition or be designed to carry and distribute the specific load at that point. Where the roof pitch is less than three units vertical in 12 units horizontal (25-percent slope), structural members that support rafters and ceiling joists, such as ridge beams, hips and valleys, shall be designed as beams.

R802.3.1 Ceiling joist and rafter connections. Ceiling joists and rafters shall be nailed to each other in accordance with Tables R602.3(1) and R802.5.1(9), and the assembly shall be nailed to the top wall plate in accordance with Table R602.3(1). Ceiling joists shall be continuous or securely joined where they meet over interior partitions and nailed to adjacent rafters to provide a continuous tie across the building when such joists are parallel to the rafters.

Where ceiling joists are not parallel to rafters, subflooring or metal straps attached to the ends of the rafters shall be installed in a manner to provide a continuous tie across the building, or rafters shall be tied to 1-inch by 4-inch (25.4 mm by 102 mm) (nominal) minimum-size crossties. The connections shall be in accordance with Table R602.3(1) or connections of equivalent capacities shall be provided. Where ceiling joists or rafter ties are not provided at the top plate, the ridge formed by these rafters shall also be supported by a girder designed in accordance with accepted engineering practice.

Rafter ties shall be spaced not more than 4 feet (1219 mm) on center.

R802.3.2 Ceiling joists lapped. Ends of ceiling joists shall be lapped a minimum of 3 inches (76 mm) or butted over bearing partitions or beams and toenailed to the bearing member. When ceiling joists are used to provide resistance to rafter thrust, lapped joists shall be nailed together in accordance with Table R602.3(1) and butted joists shall be tied together in a manner to resist such thrust.

R802.4 Allowable ceiling joist spans. Spans for ceiling joists shall be in accordance with Tables R802.4(1) and R802.4(2). For other grades and species and for other loading conditions, refer to the AF&PA Span Tables for Joists and Rafters.

TABLE R802.4(1)
CEILING JOIST SPANS FOR COMMON LUMBER SPECIES
(Uninhabitable attics without storage, live load = 20 psf, L/Δ = 240)

CEILING JOIST SPACING (inches)	SPECIES AND GRADE		DEAD LOAD = 5 psf			
			2x4	2x6	2x8	2x10
			Maximum ceiling joist spans			
			(feet - inches)	(feet - inches)	(feet - inches)	(feet - inches)
12	Douglas fir-larch	SS	13-2	20-8	Note a	Note a
	Douglas fir-larch	#1	12-8	19-11	Note a	Note a
	Douglas fir-larch	#2	12-5	19-6	25-8	Note a
	Douglas fir-larch	#3	10-10	15-10	20-1	24-6
	Hem-fir	SS	12-5	19-6	25-8	Note a
	Hem-fir	#1	12-2	19-1	25-2	Note a
	Hem-fir	#2	11-7	18-2	24-0	Note a
	Hem-fir	#3	10-10	15-10	20-1	24-6
	Southern pine	SS	12-11	20-3	Note a	Note a
	Southern pine	#1	12-8	19-11	Note a	Note a
	Southern pine	#2	12-5	19-6	25-8	Note a
	Southern pine	#3	11-6	17-0	21-8	25-7
	Spruce-pine-fir	SS	12-2	19-1	25-2	Note a
	Spruce-pine-fir	#1	11-10	18-8	24-7	Note a
	Spruce-pine-fir	#2	11-10	18-8	24-7	Note a
	Spruce-pine-fir	#3	10-10	15-10	20-1	24-6
16	Douglas fir-larch	SS	11-11	18-9	24-8	Note a
	Douglas fir-larch	#1	11-6	18-1	23-10	Note a
	Douglas fir-larch	#2	11-3	17-8	23-0	Note a
	Douglas fir-larch	#3	9-5	13-9	17-5	21-3
	Hem-fir	SS	11-3	17-8	23-4	Note a
	Hem-fir	#1	11-0	17-4	22-10	Note a
	Hem-fir	#2	10-6	16-6	21-9	Note a
	Hem-fir	#3	9-5	13-9	17-5	21-3
	Southern pine	SS	11-9	18-5	24-3	Note a
	Southern pine	#1	11-6	18-1	23-1	Note a
	Southern pine	#2	11-3	17-8	23-4	Note a
	Southern pine	#3	10-0	14-9	18-9	22- 2
	Spruce-pine-fir	SS	11-0	17-4	22-10	Note a
	Spruce-pine-fir	#1	10-9	16-11	22- 4	Note a
	Spruce-pine-fir	#2	10-9	16-11	22- 4	Note a
	Spruce-pine-fir	#3	9-5	13-9	17-5	21-3
19.2	Douglas fir-larch	SS	11-3	17-8	23-3	Note a
	Douglas fir-larch	#1	10-10	17-0	22-5	Note a
	Douglas fir-larch	#2	10-7	16-7	21-0	25-8
	Douglas fir-larch	#3	8-7	12-6	15-10	19-5
	Hem-fir	SS	10-7	16-8	21-11	Note a
	Hem-fir	#1	10-4	16-4	21-6	Note a
	Hem-fir	#2	9-11	15-7	20-6	25-3
	Hem-fir	#3	8-7	12-6	15-10	19-5
	Southern -pine	SS	11-0	17-4	22-10	Note a
	Southern pine	#1	10-10	17-0	22-5	Note a
	Southern pine	#2	10-7	16-8	21-11	Note a
	Southern pine	#3	9-1	13-6	17-2	20-3
	Spruce-pine-fir	SS	10-4	16-4	21-6	Note a
	Spruce-pine-fir	#1	10-2	15-11	21-0	25-8
	Spruce-pine-fir	#2	10-2	15-11	21-0	25-8
	Spruce-pine-fir	#3	8-7	12-6	15-10	19-5

(continued)

TABLE R802.4(1)—continued
CEILING JOIST SPANS FOR COMMON LUMBER SPECIES
(Uninhabitable attics without storage, live load = 20 psf, L/Δ = 240)

CEILING JOIST SPACING (inches)	SPECIES AND GRADE		DEAD LOAD = 5 psf			
			2x4	2x6	2x8	2x10
			Maximum ceiling joist spans			
			(feet - inches)	(feet - inches)	(feet - inches)	(feet - inches)
24	Douglas fir-larch	SS	10-5	16-4	21-7	Note a
	Douglas fir-larch	#1	10-0	15-9	20-1	24-6
	Douglas fir-larch	#2	9-10	14-10	18-9	22-11
	Douglas fir-larch	#3	7-8	11-2	14-2	17-4
	Hem-fir	SS	9-10	15-6	20-5	Note a
	Hem-fir	#1	9-8	15-2	19-7	23-11
	Hem-fir	#2	9-2	14-5	18-6	22-7
	Hem-fir	#3	7-8	11-2	14-2	17-4
	Southern pine	SS	10-3	16-1	21-2	Note a
	Southern pine	#1	10-0	15-9	20-10	Note a
	Southern pine	#2	9-10	15-6	20-1	23-11
	Southern pine	#3	8-2	12-0	15-4	18-1
	Spruce-pine-fir	SS	9-8	15-2	19-11	25-5
	Spruce-pine-fir	#1	9-5	14-9	18-9	22-11
	Spruce-pine-fir	#2	9-5	14-9	18-9	22-11
	Spruce-pine-fir	#3	7-8	11-2	14-2	17-4

Check sources for availability of lumber in lengths greater than 20 feet.

For SI: 1 inch = 25.4 mm, 1 foot = 304.8 mm, 1 pound per square foot = 0.0479 kN/m^2.

a. Span exceeds 26 feet in length

TABLE R802.4(2)
CEILING JOIST SPANS FOR COMMON LUMBER SPECIES
(Uninhabitable attics with limited storage, live load = 20 psf, L/Δ = 240)

CEILING JOIST SPACING (inches)	SPECIES AND GRADE		DEAD LOAD = 10 psf			
			2x4	2x6	2x8	2x10
			Maximum ceiling joist spans			
			(feet - inches)	(feet - inches)	(feet - inches)	(feet - inches)
12	Douglas fir-larch	SS	10-5	16-4	21-7	Note a
	Douglas fir-larch	#1	10-0	15-9	20-1	24-6
	Douglas fir-larch	#2	9-10	14-10	18-9	22-11
	Douglas fir-larch	#3	7-8	11-2	14-2	17-4
	Hem-fir	SS	9-10	15-6	20-5	Note a
	Hem-fir	#1	9-8	15-2	19-7	23-11
	Hem-fir	#2	9-2	14-5	18-6	22-7
	Hem-fir	#3	7-8	11-2	14-2	17-4
	Southern pine	SS	10-3	16-1	21-2	Note a
	Southern pine	#1	10-0	15-9	20-10	Note a
	Southern pine	#2	9-10	15-6	20-1	23-11
	Southern pine	#3	8-2	12-0	15-4	18-1
	Spruce-pine-fir	SS	9-8	15-2	19-11	25-5
	Spruce-pine-fir	#1	9-5	14-9	18-9	22-11
	Spruce-pine-fir	#2	9-5	14-9	18-9	22-11
	Spruce-pine-fir	#3	7-8	11-2	14-2	17-4
16	Douglas fir-larch	SS	9-6	14-11	19-7	25-0
	Douglas fir-larch	#1	9-1	13-9	17-5	21-3
	Douglas fir-larch	#2	8-9	12-10	16-3	19-10
	Douglas fir-larch	#3	6-8	9-8	12-4	15-0
	Hem-fir	SS	8-11	14-1	18-6	23-8
	Hem-fir	#1	8-9	13-5	16-10	20-8
	Hem-fir	#2	8-4	12-8	16-0	19-7
	Hem-fir	#3	6-8	9-8	12-4	15-0
	Southern pine	SS	9-4	14-7	19-3	24-7
	Southern pine	#1	9-1	14-4	18-11	23-1
	Southern pine	#2	8-11	13-6	17-5	20-9
	Southern pine	#3	7-1	10-5	13-3	15-8
	Spruce-pine-fir	SS	8-9	13-9	18-1	23-1
	Spruce-pine-fir	#1	8-7	12-10	16-3	19-10
	Spruce-pine-fir	#2	8-7	12-10	16-3	19-10
	Spruce-pine-fir	#3	6-8	9-8	12-4	15-0
19.2	Douglas fir-larch	SS	8-11	14-0	18-5	23-4
	Douglas fir-larch	#1	8-7	12-6	15-10	19-5
	Douglas fir-larch	#2	8-0	11-9	14-10	18-2
	Douglas fir-larch	#3	6-1	8-10	11-3	13-8
	Hem-fir	SS	8-5	13-3	17-5	22-3
	Hem-fir	#1	8-3	12-3	15-6	18-11
	Hem-fir	#2	7-10	11-7	14-8	17-10
	Hem-fir	#3	6-1	8-10	11-3	13-8
	Southern pine	SS	8-9	13-9	18-1	23-1
	Southern pine	#1	8-7	13-6	17-9	21-1
	Southern pine	#2	8-5	12-3	15-10	18-11
	Southern pine	#3	6-5	9-6	12-1	14-4
	Spruce-pine-fir	SS	8-3	12-11	17-1	21-8
	Spruce-pine-fir	#1	8-0	11-9	14-10	18-2
	Spruce-pine-fir	#2	8-0	11-9	14-10	18-2
	Spruce-pine-fir	#3	6-1	8-10	11-3	13-8

(continued)

TABLE R802.4(2)—continued
CEILING JOIST SPANS FOR COMMON LUMBER SPECIES
(Uninhabitable attics with limited storage, live load = 20 psf, L/Δ = 240)

CEILING JOIST SPACING (inches)	SPECIES AND GRADE		DEAD LOAD = 10 psf			
			2x4	2x6	2x8	2x10
			Maximum Ceiling Joist Spans			
			(feet - inches)	(feet - inches)	(feet - inches)	(feet - inches)
24	Douglas fir-larch	SS	8-3	13-0	17-1	20-11
	Douglas fir-larch	#1	7-8	11-2	14-2	17-4
	Douglas fir-larch	#2	7-2	10-6	13-3	16-3
	Douglas fir-larch	#3	5-5	7-11	10-0	12-3
	Hem-fir	SS	7-10	12-3	16-2	20-6
	Hem-fir	#1	7-6	10-11	13-10	16-11
	Hem-fir	#2	7-1	10-4	13-1	16-0
	Hem-fir	#3	5-5	7-11	10-0	12-3
	Southern pine	SS	8-1	12-9	16-10	21-6
	Southern pine	#1	8-0	12-6	15-10	18-10
	Southern pine	#2	7-8	11-0	14-2	16-11
	Southern pine	#3	5-9	8-6	10-10	12-10
	Spruce-pine-fir	SS	7-8	12-0	15-10	19-5
	Spruce-pine-fir	#1	7-2	10-6	13-3	16-3
	Spruce-pine-fir	#2	7-2	10-6	13-3	16-3
	Spruce-pine-fir	#3	5-5	7-11	10-0	12-3

For SI: 1 inch = 25.4 mm, 1 foot = 304.8 mm, 1 pound per square foot = 0.0479 kN/m^2.

a. Check sources for availability of lumber in lengths greater than 20 feet.

R802.5 Allowable rafter spans. Spans for rafters shall be in accordance with Tables R802.5.1(1) through R802.5.1(8). For other grades and species and for other loading conditions, refer to the AF&PA Span Tables for Joists and Rafters. The span of each rafter shall be measured along the horizontal projection of the rafter.

R802.5.1 Purlins. Purlins are permitted to be installed to reduce the span of rafters as shown in Figure R802.5.1. Purlins shall be sized no less than the required size of the rafters that they support. Purlins shall be continuous and shall be supported by 2-inch by 4-inch (51 mm by 102 mm) braces installed to bearing walls at a slope not less than 45 degrees from the horizontal. The braces shall be spaced not more than 4 feet (1219 mm) on center and the unbraced length of braces shall not exceed 8 feet (2438 mm).

R802.6 Bearing. The ends of each rafter or ceiling joist shall have not less than 1$^1/_2$ inches (38 mm) of bearing on wood or metal and not less than 3 inches (76 mm) on masonry or concrete.

R802.6.1 Finished ceiling material. If the finished ceiling material is installed on the ceiling prior to the attachment of the ceiling to the walls, such as in construction at a factory, a compression strip of the same thickness as the finish ceiling material shall be installed directly above the top plate of bearing walls if the compressive strength of the finish ceiling material is less than the loads it will be required to withstand. The compression strip shall cover the entire length of such top plate and shall be at least one-half the width of the top plate. It shall be of material capable of transmitting the loads transferred through it.

R802.7 Cutting and notching. Structural roof members shall not be cut, bored or notched in excess of the limitations specified in this section.

R802.7.1 Sawn lumber. Notches in solid lumber joists, rafters and beams shall not exceed one-sixth of the depth of the member, shall not be longer than one-third of the depth of the member and shall not be located in the middle one-third of the span. Notches at the ends of the member shall not exceed one-fourth the depth of the member. The tension side of members 4 inches (102 mm) or greater in nominal thickness shall not be notched except at the ends of the members. The diameter of the holes bored or cut into members shall not exceed one-third the depth of the member. Holes shall not be closer than 2 inches (51 mm) to the top or bottom of the member, or to any other hole located in the member. Where the member is also notched, the hole shall not be closer than 2 inches (51 mm) to the notch.

Exception: Notches on cantilevered portions of rafters are permitted provided the dimension of the remaining portion of the rafter is not less than 4-inch nominal (102 mm) and the length of the cantilever does not exceed 24 inches (610 mm).

R802.7.2 Engineered wood products. Cuts, notches and holes bored in laminated veneer lumber, glue-laminated members or I-joists are not permitted unless the effect of such penetrations are specifically considered in the design of the member.

TABLE R802.5.1(1)
RAFTER SPANS FOR COMMON LUMBER SPECIES
(Roof live load=20 psf, ceiling not attached to rafters, L/Δ=180)

RAFTER SPACING (inches)	SPECIES AND GRADE		DEAD LOAD = 10 psf					DEAD LOAD = 20 psf				
			2x4	2x6	2x8	2x10	2x12	2x4	2x6	2x8	2x10	2x12
			Maximum rafter spans[a]									
			(feet - inches)	(feet - inches)	(feet - inches)	(feet - inches)	(feet - inches)	(feet - inches)	(feet - inches)	(feet - inches)	(feet - inches)	(feet - inches)
12	Douglas fir-larch	SS	11-6	18-0	23-9	Note b	Note b	11-6	18-0	23-5	Note b	Note b
	Douglas fir-larch	#1	11-1	17-4	22-5	Note b	Note b	10-6	15-4	19-5	23-9	Note b
	Douglas fir-larch	#2	10-10	16-7	21-0	25-8	Note b	9-10	14-4	18-2	22-3	25-9
	Douglas fir-larch	#3	8-7	12-6	15-10	19-5	22-6	7-5	10-10	13-9	16-9	19-6
	Hem-fir	SS	10-10	17-0	22-5	Note b	Note b	10-10	17-0	22-5	Note b	Note b
	Hem-fir	#1	10-7	16-8	21-10	Note b	Note b	10-3	14-11	18-11	23-2	Note b
	Hem-fir	#2	10-1	15-11	20-8	25-3	Note b	9-8	14-2	17-11	21-11	25-5
	Hem-fir	#3	8-7	12-6	15-10	19-5	22-6	7-5	10-10	13-9	16-9	19-6
	Southern pine	SS	11-3	17-8	23-4	Note b	Note b	11-3	17-8	23-4	Note b	Note b
	Southern pine	#1	11-1	17-4	22-11	Note b	Note b	11-1	17-3	21-9	25-10	Note b
	Southern pine	#2	10-10	17-0	22-5	Note b	Note b	10-6	15-1	19-5	23-2	Note b
	Southern pine	#3	9-1	13-6	17-2	20-3	24-1	7-11	11-8	14-10	17-6	20-11
	Spruce-pine-fir	SS	10-7	16-8	21-11	Note b	Note b	10-7	16-8	21-9	Note b	Note b
	Spruce-pine-fir	#1	10-4	16-3	21-0	25-8	Note b	9-10	14-4	18-2	22-3	25-9
	Spruce-pine-fir	#2	10-4	16-3	21-0	25-8	Note b	9-10	14-4	18-2	22-3	25-9
	Spruce-pine-fir	#3	8-7	12-6	15-10	19-5	22-6	7-5	10-10	13-9	16-9	19-6
16	Douglas fir-larch	SS	10-5	16-4	21-7	Note b	Note b	10-5	16-0	20-3	24-9	Note b
	Douglas fir-larch	#1	10-0	15-4	19-5	23-9	Note b	9-1	13-3	16-10	20-7	23-10
	Douglas fir-larch	#2	9-10	14-4	18-2	22-3	25-9	8-6	12-5	15-9	19-3	22-4
	Douglas fir-larch	#3	7-5	10-10	13-9	16-9	19-6	6-5	9-5	11-11	14-6	16-10
	Hem-fir	SS	9-10	15-6	20-5	Note b	Note b	9-10	15-6	19-11	24-4	Note b
	Hem-fir	#1	9-8	14-11	18-11	23-2	Note b	8-10	12-11	16-5	20-0	23-3
	Hem-fir	#2	9-2	14-2	17-11	21-11	25-5	8-5	12-3	15-6	18-11	22-0
	Hem-fir	#3	7-5	10-10	13-9	16-9	19-6	6-5	9-5	11-11	14-6	16-10
	Southern pine	SS	10-3	16-1	21-2	Note b	Note b	10-3	16-1	21-2	Note b	Note b
	Southern pine	#1	10-0	15-9	20-10	25-10	Note b	10-0	15-0	18-10	22-4	Note b
	Southern pine	#2	9-10	15-1	19-5	23-2	Note b	9-1	13-0	16-10	20-1	23-7
	Southern pine	#3	7-11	11-8	14-10	17-6	20-11	6-10	10-1	12-10	15-2	18-1
	Spruce-pine-fir	SS	9-8	15-2	19-11	25-5	Note b	9-8	14-10	18-10	23-0	Note b
	Spruce-pine-fir	#1	9-5	14-4	18-2	22-3	25-9	8-6	12-5	15-9	19-3	22-4
	Spruce-pine-fir	#2	9-5	14-4	18-2	22-3	25-9	8-6	12-5	15-9	19-3	22-4
	Spruce-pine-fir	#3	7-5	10-10	13-9	16-9	19-6	6-5	9-5	11-11	14-6	16-10
19.2	Douglas fir-larch	SS	9-10	15-5	20-4	25-11	Note b	9-10	14-7	18-6	22-7	Note b
	Douglas fir-larch	#1	9-5	14-0	17-9	21-8	25-2	8-4	12-2	15-4	18-9	21-9
	Douglas fir-larch	#2	8-11	13-1	16-7	20-3	23-6	7-9	11-4	14-4	17-7	20-4
	Douglas fir-larch	#3	6-9	9-11	12-7	15-4	17-9	5-10	8-7	10-10	13-3	15-5
	Hem-fir	SS	9-3	14-7	19-2	24-6	Note b	9-3	14-4	18-2	22-3	25-9
	Hem-fir	#1	9-1	13-8	17-4	21-1	24-6	8-1	11-10	15-0	18-4	21-3
	Hem-fir	#2	8-8	12-11	16-4	20-0	23-2	7-8	11-2	14-2	17-4	20-1
	Hem-fir	#3	6-9	9-11	12-7	15-4	17-9	5-10	8-7	10-10	13-3	15-5
	Southern pine	SS	9-8	15-2	19-11	25-5	Note b	9-8	15-2	19-11	25-5	Note b
	Southern pine	#1	9-5	14-10	19-7	23-7	Note b	9-3	13-8	17-2	20-5	24-4
	Southern pine	#2	9-3	13-9	17-9	21-2	24-10	8-4	11-11	15-4	18-4	21-6
	Southern pine	#3	7-3	10-8	13-7	16-0	19-1	6-3	9-3	11-9	13-10	16-6
	Spruce-pine-fir	SS	9-1	14-3	18-9	23-11	Note b	9-1	13-7	17-2	21-0	24-4
	Spruce-pine-fir	#1	8-10	13-1	16-7	20-3	23-6	7-9	11-4	14-4	17-7	20-4
	Spruce-pine-fir	#2	8-10	13-1	16-7	20-3	23-6	7-9	11-4	14-4	17-7	20-4
	Spruce-pine-fir	#3	6-9	9-11	12-7	15-4	17-9	5-10	8-7	10-10	13-3	15-5

(continued)

TABLE R802.5.1(1)—continued
RAFTER SPANS FOR COMMON LUMBER SPECIES
(Roof live load=20 psf, ceiling not attached to rafters, L/Δ=180)

RAFTER SPACING (inches)	SPECIES AND GRADE		DEAD LOAD = 10 psf					DEAD LOAD = 20 psf				
			2x4	2x6	2x8	2x10	2x12	2x4	2x6	2x8	2x10	2x12
			Maximum rafter spans[a]									
			(feet - inches)	(feet - inches)	(feet - inches)	(feet - inches)	(feet - inches)	(feet - inches)	(feet - inches)	(feet - inches)	(feet - inches)	(feet - inches)
24	Douglas fir-larch	SS	9-1	14-4	18-10	23-4	23-4	8-11	13-1	16-7	20-3	23-5
	Douglas fir-larch	#1	8-7	12-6	15-10	19-5	19-5	7-5	10-10	13-9	16-9	19-6
	Douglas fir-larch	#2	8-0	11-9	14-10	18-2	18-2	6-11	10-2	12-10	15-8	18-3
	Douglas fir-larch	#3	6-1	8-10	11-3	13-8	13-8	5-3	7-8	9-9	11-10	13-9
	Hem-fir	SS	8-7	13-6	17-10	22-9	22-9	8-7	12-10	16-3	19-10	23-0
	Hem-fir	#1	8-4	12-3	15-6	18-11	18-11	7-3	10-7	13-5	16-4	19-0
	Hem-fir	#2	7-11	11-7	14-8	17-10	17-10	6-10	10-0	12-8	15-6	17-11
	Hem-fir	#3	6-1	8-10	11-3	13-8	13-8	5-3	7-8	9-9	11-10	13-9
	Southern pine	SS	8-11	14-1	18-6	23-8	23-8	8-11	14-1	18-6	22-11	Note b
	Southern pine	#1	8-9	13-9	17-9	21-1	21-1	8-3	12-3	15-4	18-3	21-9
	Southern pine	#2	8-7	12-3	15-10	18-11	18-11	7-5	10-8	13-9	16-5	19-3
	Southern pine	#3	6-5	9-6	12-1	14-4	14-4	5-7	8-3	10-6	12-5	14-9
	Spruce-pine-fir	SS	8-5	13-3	17-5	21-8	21-8	8-4	12-2	15-4	18-9	21-9
	Spruce-pine-fir	#1	8-0	11-9	14-10	18-2	18-2	6-11	10-2	12-10	15-8	18-3
	Spruce-pine-fir	#2	8-0	11-9	14-10	18-2	18-2	6-11	10-2	12-10	15-8	18-3
	Spruce-pine-fir	#3	6-1	8-10	11-3	13-8	13-8	5-3	7-8	9-9	11-10	13-9

Check sources for availability of lumber in lengths greater than 20 feet.

For SI: 1 inch = 25.4 mm, 1 foot = 304.8 mm, 1 pound per square foot = 0.0479 kN/m^2.

a. The tabulated rafter spans assume that ceiling joists are located at the bottom of the attic space or that some other method of resisting the outward push of the rafters on the bearing walls, such as rafter ties, is provided at that location. When ceiling joists or rafter ties are located higher in the attic space, the rafter spans shall be multiplied by the factors given below:

H_C/H_R	Rafter Span Adjustment Factor
2/3 or greater	0.50
1/2	0.58
1/3	0.67
1/4	0.76
1/5	0.83
1/6	0.90
1/7.5 and less	1.00

where: H_C = Height of ceiling joists or rafter ties measured vertically above the top of the rafter support walls.

H_R = Height of roof ridge measured vertically above the top of the rafter support walls.

b. Span exceeds 26 feet in length.

TABLE R802.5.1(2)
RAFTER SPANS FOR COMMON LUMBER SPECIES
(Roof live load=20 psf, ceiling attached to rafters, L/Δ=240)

RAFTER SPACING (inches)	SPECIES AND GRADE		DEAD LOAD = 10 psf					DEAD LOAD = 20 psf				
			2x4	2x6	2x8	2x10	2x12	2x4	2x6	2x8	2x10	2x12
			Maximum rafter spans[a]									
			(feet - inches)	(feet - inches)	(feet - inches)	(feet - inches)	(feet - inches)	(feet - inches)	(feet - inches)	(feet - inches)	(feet - inches)	(feet - inches)
12	Douglas fir-larch	SS	10-5	16-4	21-7	Note b	Note b	10-5	16-4	21-7	Note b	Note b
	Douglas fir-larch	#1	10-0	15-9	20-10	Note b	Note b	10-0	15-4	19-5	23-9	Note b
	Douglas fir-larch	#2	9-10	15-6	20-5	25-8	Note b	9-10	14-4	18-2	22-3	25-9
	Douglas fir-larch	#3	8-7	12-6	15-10	19-5	22-6	7-5	10-10	13-9	16-9	19-6
	Hem-fir	SS	9-10	15-6	20-5	Note b	Note b	9-10	15-6	20-5	Note b	Note b
	Hem-fir	#1	9-8	15-2	19-11	25-5	Note b	9-8	14-11	18-11	23-2	Note b
	Hem-fir	#2	9-2	14-5	19-0	24-3	Note b	9-2	14-2	17-11	21-11	25-5
	Hem-fir	#3	8-7	12-6	15-10	19-5	22-6	7-5	10-10	13-9	16-9	19-6
	Southern pine	SS	10-3	16-1	21-2	Note b	Note b	10-3	16-1	21-2	Note b	Note b
	Southern pine	#1	10-0	15-9	20-10	Note b	Note b	10-0	15-9	20-10	25-10	Note b
	Southern pine	#2	9-10	15-6	20-5	Note b	Note b	9-10	15-1	19-5	23-2	Note b
	Southern pine	#3	9-1	13-6	17-2	20-3	24-1	7-11	11-8	14-10	17-6	20-11
	Spruce-pine-fir	SS	9-8	15-2	19-11	25-5	Note b	9-8	15-2	19-11	25-5	Note b
	Spruce-pine-fir	#1	9-5	14-9	19-6	24-10	Note b	9-5	14-4	18-2	22-3	25-9
	Spruce-pine-fir	#2	9-5	14-9	19-6	24-10	Note b	9-5	14-4	18-2	22-3	25-9
	Spruce-pine-fir	#3	8-7	12-6	15-10	19-5	22-6	7-5	10-10	13-9	16-9	19-6
16	Douglas fir-larch	SS	9-6	14-11	19-7	25-0	Note b	9-6	14-11	19-7	24-9	Note b
	Douglas fir-larch	#1	9-1	14-4	18-11	23-9	Note b	9-1	13-3	16-10	20-7	23-10
	Douglas fir-larch	#2	8-11	14-1	18-2	22-3	25-9	8-6	12-5	15-9	19-3	22-4
	Douglas fir-larch	#3	7-5	10-10	13-9	16-9	19-6	6-5	9-5	11-11	14-6	16-10
	Hem-fir	SS	8-11	14-1	18-6	23-8	Note b	8-11	14-1	18-6	23-8	Note b
	Hem-fir	#1	8-9	13-9	18-1	23-1	Note b	8-9	12-11	16-5	20-0	23-3
	Hem-fir	#2	8-4	13-1	17-3	21-11	25-5	8-4	12-3	15-6	18-11	22-0
	Hem-fir	#3	7-5	10-10	13-9	16-9	19-6	6-5	9-5	11-11	14-6	16-10
	Southern pine	SS	9-4	14-7	19-3	24-7	Note b	9-4	14-7	19-3	24-7	Note b
	Southern pine	#1	9-1	14-4	18-11	24-1	Note b	9-1	14-4	18-10	22-4	Note b
	Southern pine	#2	8-11	14-1	18-6	23-2	Note b	8-11	13-0	16-10	20-1	23-7
	Southern pine	#3	7-11	11-8	14-10	17-6	20-11	6-10	10-1	12-10	15-2	18-1
	Spruce-pine-fir	SS	8-9	13-9	18-1	23-1	Note b	8-9	13-9	18-1	23-0	Note b
	Spruce-pine-fir	#1	8-7	13-5	17-9	22-3	25-9	8-6	12-5	15-9	19-3	22-4
	Spruce-pine-fir	#2	8-7	13-5	17-9	22-3	25-9	8-6	12-5	15-9	19-3	22-4
	Spruce-pine-fir	#3	7-5	10-10	13-9	16-9	19-6	6-5	9-5	11-11	14-6	16-10
19.2	Douglas fir-larch	SS	8-11	14-0	18-5	23-7	Note b	8-11	14-0	18-5	22-7	Note b
	Douglas fir-larch	#1	8-7	13-6	17-9	21-8	25-2	8-4	12-2	15-4	18-9	21-9
	Douglas fir-larch	#2	8-5	13-1	16-7	20-3	23-6	7-9	11-4	14-4	17-7	20-4
	Douglas fir-larch	#3	6-9	9-11	12-7	15-4	17-9	5-10	8-7	10-10	13-3	15-5
	Hem-fir	SS	8-5	13-3	17-5	22-3	Note b	8-5	13-3	17-5	22-3	25-9
	Hem-fir	#1	8-3	12-11	17-1	21-1	24-6	8-1	11-10	15-0	18-4	21-3
	Hem-fir	#2	7-10	12-4	16-3	20-0	23-2	7-8	11-2	14-2	17-4	20-1
	Hem-fir	#3	6-9	9-11	12-7	15-4	17-9	5-10	8-7	10-10	13-3	15-5
	Southern pine	SS	8-9	13-9	18-1	23-1	Note b	8-9	13-9	18-1	23-1	Note b
	Southern pine	#1	8-7	13-6	17-9	22-8	Note b	8-7	13-6	17-2	20-5	24-4
	Southern pine	#2	8-5	13-3	17-5	21-2	24-10	8-4	11-11	15-4	18-4	21-6
	Southern pine	#3	7-3	10-8	13-7	16-0	19-1	6-3	9-3	11-9	13-10	16-6
	Spruce-pine-fir	SS	8-3	12-11	17-1	21-9	Note b	8-3	12-11	17-1	21-0	24-4
	Spruce-pine-fir	#1	8-1	12-8	16-7	20-3	23-6	7-9	11-4	14-4	17-7	20-4
	Spruce-pine-fir	#2	8-1	12-8	16-7	20-3	23-6	7-9	11-4	14-4	17-7	20-4
	Spruce-pine-fir	#3	6-9	9-11	12-7	15-4	17-9	5-10	8-7	10-10	13-3	15-5

(continued)

TABLE R802.5.1(2)—continued
RAFTER SPANS FOR COMMON LUMBER SPECIES
(Roof live load=20 psf, ceiling attached to rafters, L/Δ=240)

RAFTER SPACING (inches)	SPECIES AND GRADE		DEAD LOAD = 10 psf					DEAD LOAD = 20 psf				
			2x4	2x6	2x8	2x10	2x12	2x4	2x6	2x8	2x10	2x12
			Maximum rafter spans[a]									
			(feet - inches)	(feet - inches)	(feet - inches)	(feet - inches)	(feet - inches)	(feet - inches)	(feet - inches)	(feet - inches)	(feet - inches)	(feet - inches)
24	Douglas fir-larch	SS	8-3	13-0	17-2	21-10	Note b	8-3	13-0	16-7	20-3	23-5
	Douglas fir-larch	#1	8-0	12-6	15-10	19-5	22-6	7-5	10-10	13-9	16-9	19-6
	Douglas fir-larch	#2	7-10	11-9	14-10	18-2	21-0	6-11	10-2	12-10	15-8	18-3
	Douglas fir-larch	#3	6-1	8-10	11-3	13-8	15-11	5-3	7-8	9-9	11-10	13-9
	Hem-fir	SS	7-10	12-3	16-2	20-8	25-1	7-10	12-3	16-2	19-10	23-0
	Hem-fir	#1	7-8	12-0	15-6	18-11	21-11	7-3	10-7	13-5	16-4	19-0
	Hem-fir	#2	7-3	11-5	14-8	17-10	20-9	6-10	10-0	12-8	15-6	17-11
	Hem-fir	#3	6-1	8-10	11-3	13-8	15-11	5-3	7-8	9-9	11-10	13-9
	Southern pine	SS	8-1	12-9	16-10	21-6	Note b	8-1	12-9	16-10	21-6	Note b
	Southern pine	#1	8-0	12-6	16-6	21-1	25-2	8-0	12-3	15-4	18-3	21-9
	Southern pine	#2	7-10	12-3	15-10	18-11	22-2	7-5	10-8	13-9	16-5	19-3
	Southern pine	#3	6-5	9-6	12-1	14-4	17-1	5-7	8-3	10-6	12-5	14-9
	Spruce-pine-fir	SS	7-8	12-0	15-10	20-2	24-7	7-8	12-0	15-4	18-9	21-9
	Spruce-pine-fir	#1	7-6	11-9	14-10	18-2	21-0	6-11	10-2	12-10	15-8	18-3
	Spruce-pine-fir	#2	7-6	11-9	14-10	18-2	21-0	6-11	10-2	12-10	15-8	18-3
	Spruce-pine-fir	#3	6-1	8-10	11-3	13-8	15-11	5-3	7-8	9-9	11-10	13-9

Check sources for availability of lumber in lengths greater than 20 feet.

For SI: 1 inch = 25.4 mm, 1 foot = 304.8 mm, 1 pound per square foot = 0.0479 kN/m^2.

a. The tabulated rafter spans assume that ceiling joists are located at the bottom of the attic space or that some other method of resisting the outward push of the rafters on the bearing walls, such as rafter ties, is provided at that location. When ceiling joists or rafter ties are located higher in the attic space, the rafter spans shall be multiplied by the factors given below:

H_C/H_R	Rafter Span Adjustment Factor
2/3 or greater	0.50
1/2	0.58
1/3	0.67
1/4	0.76
1/5	0.83
1/6	0.90
1/7.5 and less	1.00

where: H_C = Height of ceiling joists or rafter ties measured vertically above the top of the rafter support walls.

H_R = Height of roof ridge measured vertically above the top of the rafter support walls.

b. Span exceeds 26 feet in length.

TABLE R802.5.1(3)
RAFTER SPANS FOR COMMON LUMBER SPECIES
(Ground snow load=30 psf, ceiling not attached to rafters, L/Δ=180)

RAFTER SPACING (inches)	SPECIES AND GRADE		DEAD LOAD = 10 psf					DEAD LOAD = 20 psf				
			2x4	2x6	2x8	2x10	2x12	2x4	2x6	2x8	2x10	2x12
			Maximum rafter spans[a]									
			(feet - inches)	(feet - inches)	(feet - inches)	(feet - inches)	(feet - inches)	(feet - inches)	(feet - inches)	(feet - inches)	(feet - inches)	(feet - inches)
12	Douglas fir-larch	SS	10-0	15-9	20-9	Note b	Note b	10-0	15-9	20-1	24-6	Note b
	Douglas fir-larch	#1	9-8	14-9	18-8	22-9	Note b	9-0	13-2	16-8	20-4	23-7
	Douglas fir-larch	#2	9-5	13-9	17-5	21-4	24-8	8-5	12-4	15-7	19-1	22-1
	Douglas fir-larch	#3	7-1	10-5	13-2	16-1	18-8	6-4	9-4	11-9	14-5	16-8
	Hem-fir	SS	9-6	14-10	19-7	25-0	Note b	9-6	14-10	19-7	24-1	Note b
	Hem-fir	#1	9-3	14-4	18-2	22-2	25-9	8-9	12-10	16-3	19-10	23-0
	Hem-fir	#2	8-10	13-7	17-2	21-0	24-4	8-4	12-2	15-4	18-9	21-9
	Hem-fir	#3	7-1	10-5	13-2	16-1	18-8	6-4	9-4	11-9	14-5	16-8
	Southern pine	SS	9-10	15-6	20-5	Note b	Note b	9-10	15-6	20-5	Note b	Note b
	Southern pine	#1	9-8	15-2	20-0	24-9	Note b	9-8	14-10	18-8	22-2	Note b
	Southern pine	#2	9-6	14-5	18-8	22-3	Note b	9-0	12-11	16-8	19-11	23-4
	Southern pine	#3	7-7	11-2	14-3	16-10	20-0	6-9	10-0	12-9	15-1	17-11
	Spruce-pine-fir	SS	9-3	14-7	19-2	24-6	Note b	9-3	14-7	18-8	22-9	Note b
	Spruce-pine-fir	#1	9-1	13-9	17-5	21-4	24-8	8-5	12-4	15-7	19-1	22-1
	Spruce-pine-fir	#2	9-1	13-9	17-5	21-4	24-8	8-5	12-4	15-7	19-1	22-1
	Spruce-pine-fir	#3	7-1	10-5	13-2	16-1	18-8	6-4	9-4	11-9	14-5	16-8
16	Douglas fir-larch	SS	9-1	14-4	18-10	23-9	Note b	9-1	13-9	17-5	21-3	24-8
	Douglas fir-larch	#1	8-9	12-9	16-2	19-9	22-10	7-10	11-5	14-5	17-8	20-5
	Douglas fir-larch	#2	8-2	11-11	15-1	18-5	21-5	7-3	10-8	13-6	16-6	19-2
	Douglas fir-larch	#3	6-2	9-0	11-5	13-11	16-2	5-6	8-1	10-3	12-6	14-6
	Hem-fir	SS	8-7	13-6	17-10	22-9	Note b	8-7	13-6	17-1	20-10	24-2
	Hem-fir	#1	8-5	12-5	15-9	19-3	22-3	7-7	11-1	14-1	17-2	19-11
	Hem-fir	#2	8-0	11-9	14-11	18-2	21-1	7-2	10-6	13-4	16-3	18-10
	Hem-fir	#3	6-2	9-0	11-5	13-11	16-2	5-6	8-1	10-3	12-6	14-6
	Southern pine	SS	8-11	14-1	18-6	23-8	Note b	8-11	14-1	18-6	23-8	Note b
	Southern pine	#1	8-9	13-9	18-1	21-5	25-7	8-8	12-10	16-2	19-2	22-10
	Southern pine	#2	8-7	12-6	16-2	19-3	22-7	7-10	11-2	14-5	17-3	20-2
	Southern pine	#3	6-7	9-8	12-4	14-7	17-4	5-10	8-8	11-0	13-0	15-6
	Spruce-pine-fir	SS	8-5	13-3	17-5	22-1	25-7	8-5	12-9	16-2	19-9	22-10
	Spruce-pine-fir	#1	8-2	11-11	15-1	18-5	21-5	7-3	10-8	13-6	16-6	19-2
	Spruce-pine-fir	#2	8-2	11-11	15-1	18-5	21-5	7-3	10-8	13-6	16-6	19-2
	Spruce-pine-fir	#3	6-2	9-0	11-5	13-11	16-2	5-6	8-1	10-3	12-6	14-6
19.2	Douglas fir-larch	SS	8-7	13-6	17-9	21-8	25-2	8-7	12-6	15-10	19-5	22-6
	Douglas fir-larch	#1	7-11	11-8	14-9	18-0	20-11	7-1	10-5	13-2	16-1	18-8
	Douglas fir-larch	#2	7-5	10-11	13-9	16-10	19-6	6-8	9-9	12-4	15-1	17-6
	Douglas fir-larch	#3	5-7	8-3	10-5	12-9	14-9	5-0	7-4	9-4	11-5	13-2
	Hem-fir	SS	8-1	12-9	16-9	21-4	24-8	8-1	12-4	15-7	19-1	22-1
	Hem-fir	#1	7-9	11-4	14-4	17-7	20-4	6-11	10-2	12-10	15-8	18-2
	Hem-fir	#2	7-4	10-9	13-7	16-7	19-3	6-7	9-7	12-2	14-10	17-3
	Hem-fir	#3	5-7	8-3	10-5	12-9	14-9	5-0	7-4	9-4	11-5	13-2
	Southern pine	SS	8-5	13-3	17-5	22-3	Note b	8-5	13-3	17-5	22-0	25-9
	Southern pine	#1	8-3	13-0	16-6	19-7	23-4	7-11	11-9	14-9	17-6	20-11
	Southern pine	#2	7-11	11-5	14-9	17-7	20-7	7-1	10-2	13-2	15-9	18-5
	Southern pine	#3	6-0	8-10	11-3	13-4	15-10	5-4	7-11	10-1	11-11	14-2
	Spruce-pine-fir	SS	7-11	12-5	16-5	20-2	23-4	7-11	11-8	14-9	18-0	20-11
	Spruce-pine-fir	#1	7-5	10-11	13-9	16-10	19-6	6-8	9-9	12-4	15-1	17-6
	Spruce-pine-fir	#2	7-5	10-11	13-9	16-10	19-6	6-8	9-9	12-4	15-1	17-6
	Spruce-pine-fir	#3	5-7	8-3	10-5	12-9	14-9	5-0	7-4	9-4	11-5	13-2

(continued)

TABLE R802.5.1(3)—continued
RAFTER SPANS FOR COMMON LUMBER SPECIES
(Ground snow load=30 psf, ceiling not attached to rafters, L/Δ=180)

RAFTER SPACING (inches)	SPECIES AND GRADE	DEAD LOAD = 10 psf					DEAD LOAD = 20 psf				
		2x4	2x6	2x8	2x10	2x12	2x4	2x6	2x8	2x10	2x12
		Maximum rafter spans[a]									
		(feet - inches)	(feet - inches)	(feet - inches)	(feet - inches)	(feet - inches)	(feet - inches)	(feet - inches)	(feet - inches)	(feet - inches)	(feet - inches)
24	Douglas fir-larch SS	7-11	12-6	15-10	19-5	22-6	7-8	11-3	14-2	17-4	20-1
	Douglas fir-larch #1	7-1	10-5	13-2	16-1	18-8	6-4	9-4	11-9	14-5	16-8
	Douglas fir-larch #2	6-8	9-9	12-4	15-1	17-6	5-11	8-8	11-0	13-6	15-7
	Douglas fir-larch #3	5-0	7-4	9-4	11-5	13-2	4-6	6-7	8-4	10-2	11-10
	Hem-fir SS	7-6	11-10	15-7	19-1	22-1	7-6	11-0	13-11	17-0	19-9
	Hem-fir #1	6-11	10-2	12-10	15-8	18-2	6-2	9-1	11-6	14-0	16-3
	Hem-fir #2	6-7	9-7	12-2	14-10	17-3	5-10	8-7	10-10	13-3	15-5
	Hem-fir #3	5-0	7-4	9-4	11-5	13-2	4-6	6-7	8-4	10-2	11-10
	Southern pine SS	7-10	12-3	16-2	20-8	25-1	7-10	12-3	16-2	19-8	23-0
	Southern pine #1	7-8	11-9	14-9	17-6	20-11	7-1	10-6	13-2	15-8	18-8
	Southern pine #2	7-1	10-2	13-2	15-9	18-5	6-4	9-2	11-9	14-1	16-6
	Southern pine #3	5-4	7-11	10-1	11-11	14-2	4-9	7-1	9-0	10-8	12-8
	Spruce-pine-fir SS	7-4	11-7	14-9	18-0	20-11	7-1	10-5	13-2	16-1	18-8
	Spruce-pine-fir #1	6-8	9-9	12-4	15-1	17-6	5-11	8-8	11-0	13-6	15-7
	Spruce-pine-fir #2	6-8	9-9	12-4	15-1	17-6	5-11	8-8	11-0	13-6	15-7
	Spruce-pine-fir #3	5-0	7-4	9-4	11-5	13-2	4-6	6-7	8-4	10-2	11-10

Check sources for availability of lumber in lengths greater than 20 feet.

For SI: 1 inch = 25.4 mm, 1 foot = 304.8 mm, 1 pound per square foot = 0.0479 kN/m^2.

a. The tabulated rafter spans assume that ceiling joists are located at the bottom of the attic space or that some other method of resisting the outward push of the rafters on the bearing walls, such as rafter ties, is provided at that location. When ceiling joists or rafter ties are located higher in the attic space, the rafter spans shall be multiplied by the factors given below:

H_C/H_R	Rafter Span Adjustment Factor
2/3 or greater	0.50
1/2	0.58
1/3	0.67
1/4	0.76
1/5	0.83
1/6	0.90
1/7.5 and less	1.00

where: H_C = Height of ceiling joists or rafter ties measured vertically above the top of the rafter support walls.

H_R = Height of roof ridge measured vertically above the top of the rafter support walls.

b. Span exceeds 26 feet in length.

TABLE R802.5.1(4)
RAFTER SPANS FOR COMMON LUMBER SPECIES
(Ground snow load=50 psf, ceiling not attached to rafters, L/Δ=180)

RAFTER SPACING (inches)	SPECIES AND GRADE	DEAD LOAD = 10 psf					DEAD LOAD = 20 psf				
		2x4	2x6	2x8	2x10	2x12	2x4	2x6	2x8	2x10	2x12
		Maximum rafter spans[a]									
		(feet - inches)	(feet - inches)	(feet - inches)	(feet - inches)	(feet - inches)	(feet - inches)	(feet - inches)	(feet - inches)	(feet - inches)	(feet - inches)
12	Douglas fir-larch SS	8- 5	13- 3	17- 6	22- 4	26- 0	8- 5	13- 3	17- 0	20- 9	24- 0
	Douglas fir-larch #1	8- 2	12- 0	15- 3	18- 7	21- 7	7- 7	11- 2	14- 1	17- 3	20- 0
	Douglas fir-larch #2	7- 8	11- 3	14- 3	17- 5	20- 2	7- 1	10- 5	13- 2	16- 1	18- 8
	Douglas fir-larch #3	5-10	8- 6	10- 9	13- 2	15- 3	5- 5	7-10	10- 0	12- 2	14- 1
	Hem-fir SS	8- 0	12- 6	16- 6	21- 1	25- 6	8- 0	12- 6	16- 6	20- 4	23- 7
	Hem-fir #1	7-10	11- 9	14-10	18- 1	21- 0	7- 5	10-10	13- 9	16- 9	19- 5
	Hem-fir #2	7- 5	11- 1	14- 0	17- 2	19-11	7- 0	10- 3	13- 0	15-10	18- 5
	Hem-fir #3	5-10	8- 6	10- 9	13- 2	15- 3	5- 5	7-10	10- 0	12- 2	14- 1
	Southern pine SS	8- 4	13- 0	17- 2	21-11	Note b	8- 4	13- 0	17- 2	21-11	Note b
	Southern pine #1	8- 2	12-10	16-10	20- 3	24- 1	8- 2	12- 6	15- 9	18- 9	22- 4
	Southern pine #2	8- 0	11- 9	15- 3	18- 2	21- 3	7- 7	10-11	14- 1	16-10	19- 9
	Southern pine #3	6- 2	9- 2	11- 8	13- 9	16- 4	5- 9	8- 5	10- 9	12- 9	15- 2
	Spruce-pine-fir SS	7-10	12- 3	16- 2	20- 8	24- 1	7-10	12- 3	15- 9	19- 3	22- 4
	Spruce-pine-fir #1	7- 8	11- 3	14- 3	17- 5	20- 2	7- 1	10- 5	13- 2	16- 1	18- 8
	Spruce-pine-fir #2	7- 8	11- 3	14- 3	17- 5	20- 2	7- 1	10- 5	13- 2	16- 1	18- 8
	Spruce-pine-fir #3	5-10	8- 6	10- 9	13- 2	15- 3	5- 5	7-10	10- 0	12- 2	14- 1
16	Douglas fir-larch SS	7- 8	12- 1	15-10	19- 5	22- 6	7- 8	11- 7	14- 8	17-11	20-10
	Douglas fir-larch #1	7- 1	10- 5	13- 2	16- 1	18- 8	6- 7	9- 8	12- 2	14-11	17- 3
	Douglas fir-larch #2	6- 8	9- 9	12- 4	15- 1	17- 6	6- 2	9- 0	11- 5	13-11	16- 2
	Douglas fir-larch #3	5- 0	7- 4	9- 4	11- 5	13- 2	4- 8	6-10	8- 8	10- 6	12- 3
	Hem-fir SS	7- 3	11- 5	15- 0	19- 1	22- 1	7- 3	11- 5	14- 5	17- 8	20- 5
	Hem-fir #1	6-11	10- 2	12-10	15- 8	18- 2	6- 5	9- 5	11-11	14- 6	16-10
	Hem-fir #2	6- 7	9- 7	12- 2	14-10	17- 3	6- 1	8-11	11- 3	13- 9	15-11
	Hem-fir #3	5- 0	7- 4	9- 4	11- 5	13- 2	4- 8	6-10	8- 8	10- 6	12- 3
	Southern pine SS	7- 6	11-10	15- 7	19-11	24- 3	7- 6	11-10	15- 7	19-11	23-10
	Southern pine #1	7- 5	11- 7	14- 9	17- 6	20-11	7- 4	10-10	13- 8	16- 2	19- 4
	Southern pine #2	7- 1	10- 2	13- 2	15- 9	18- 5	6- 7	9- 5	12- 2	14- 7	17- 1
	Southern pine #3	5- 4	7-11	10- 1	11-11	14- 2	4-11	7- 4	9- 4	11- 0	13- 1
	Spruce-pine-fir SS	7- 1	11- 2	14- 8	18- 0	20-11	7- 1	10- 9	13- 8	15-11	19- 4
	Spruce-pine-fir #1	6- 8	9- 9	12- 4	15- 1	17- 6	6- 2	9- 0	11- 5	13-11	16- 2
	Spruce-pine-fir #2	6- 8	9- 9	12- 4	15- 1	17- 6	6- 2	9- 0	11- 5	13-11	16- 2
	Spruce-pine-fir #3	5- 0	7- 4	9- 4	11- 5	13- 2	4- 8	6-10	8- 8	10- 6	12- 3
19.2	Douglas fir-larch SS	7- 3	11- 4	14- 6	17- 8	20- 6	7- 3	10- 7	13- 5	16- 5	19- 0
	Douglas fir-larch #1	6- 6	9- 6	12- 0	14- 8	17- 1	6- 0	8-10	11- 2	13- 7	15- 9
	Douglas fir-larch #2	6- 1	8-11	11- 3	13- 9	15-11	5- 7	8- 3	10- 5	12- 9	14- 9
	Douglas fir-larch #3	4- 7	6- 9	8- 6	10- 5	12- 1	4- 3	6- 3	7-11	9- 7	11- 2
	Hem-fir SS	6-10	10- 9	14- 2	17- 5	20- 2	6-10	10- 5	13- 2	16- 1	18- 8
	Hem-fir #1	6- 4	9- 3	11- 9	14- 4	16- 7	5-10	8- 7	10-10	13- 3	15- 5
	Hem-fir #2	6- 0	8- 9	11- 1	13- 7	15- 9	5- 7	8- 1	10- 3	12- 7	14- 7
	Hem-fir #3	4- 7	6- 9	8- 6	10- 5	12- 1	4- 3	6- 3	7-11	9- 7	11- 2
	Southern pine SS	7- 1	11- 2	14- 8	18- 9	22-10	7- 1	11- 2	14- 8	18- 7	21- 9
	Southern pine #1	7- 0	10- 8	13- 5	16- 0	19- 1	6- 8	9-11	12- 5	14-10	17- 8
	Southern pine #2	6- 6	9- 4	12- 0	14- 4	16-10	6- 0	8- 8	11- 2	13- 4	15- 7
	Southern pine #3	4-11	7- 3	9- 2	10-10	12-11	4- 6	6- 8	8- 6	10- 1	12- 0
	Spruce-pine-fir SS	6- 8	10- 6	13- 5	16- 5	19- 1	6- 8	9-10	12- 5	15- 3	17- 8
	Spruce-pine-fir #1	6- 1	8-11	11- 3	13- 9	15-11	5- 7	8- 3	10- 5	12- 9	14- 9
	Spruce-pine-fir #2	6-1	8-11	11-3	13-9	15-11	5-7	8-3	10-5	12-9	14-9
	Spruce-pine-fir #3	4- 7	6- 9	8- 6	10- 5	12- 1	4- 3	6- 3	7-11	9- 7	11- 2

(continued)

TABLE R802.5.1(4)—continued
RAFTER SPANS FOR COMMON LUMBER SPECIES
(Ground snow load=50 psf, ceiling not attached to rafters, L/Δ=180)

RAFTER SPACING (inches)	SPECIES AND GRADE		DEAD LOAD = 10 psf					DEAD LOAD = 20 psf				
			2x4	2x6	2x8	2x10	2x12	2x4	2x6	2x8	2x10	2x12
			Maximum rafter spans[a]									
			(feet - inches)	(feet - inches)	(feet - inches)	(feet - inches)	(feet - inches)	(feet - inches)	(feet - inches)	(feet - inches)	(feet - inches)	(feet - inches)
24	Douglas fir-larch	SS	6- 8	10- 3	13- 0	15-10	18- 4	6- 6	9- 6	12- 0	14- 8	17- 0
	Douglas fir-larch	#1	5-10	8- 6	10- 9	13- 2	15- 3	5- 5	7-10	10- 0	12- 2	14- 1
	Douglas fir-larch	#2	5- 5	7-11	10- 1	12- 4	14- 3	5- 0	7- 4	9- 4	11- 5	13- 2
	Douglas fir-larch	#3	4- 1	6- 0	7- 7	9- 4	10- 9	3-10	5- 7	7- 1	8- 7	10- 0
	Hem-fir	SS	6- 4	9-11	12- 9	15- 7	18- 0	6- 4	9- 4	11- 9	14- 5	16- 8
	Hem-fir	#1	5- 8	8- 3	10- 6	12-10	14-10	5- 3	7- 8	9- 9	11-10	13- 9
	Hem-fir	#2	5- 4	7-10	9-11	12- 1	14- 1	4-11	7- 3	9- 2	11- 3	13- 0
	Hem-fir	#3	4- 1	6- 0	7- 7	9- 4	10- 9	3-10	5- 7	7- 1	8- 7	10- 0
	Southern pine	SS	6- 7	10- 4	13- 8	17- 5	21- 0	6- 7	10- 4	13- 8	16- 7	19- 5
	Southern pine	#1	6- 5	9- 7	12- 0	14- 4	17- 1	6- 0	8-10	11- 2	13- 3	15- 9
	Southern pine	#2	5-10	8- 4	10- 9	12-10	15- 1	5- 5	7- 9	10- 0	11-11	13-11
	Southern pine	#3	4- 4	6- 5	8- 3	9- 9	11- 7	4- 1	6- 0	7- 7	9- 0	10- 8
	Spruce-pine-fir	SS	6- 2	9- 6	12- 0	14- 8	17- 1	6- 0	8-10	11- 2	13- 7	15- 9
	Spruce-pine-fir	#1	5- 5	7-11	10- 1	12- 4	14- 3	5- 0	7- 4	9- 4	11- 5	13- 2
	Spruce-pine-fir	#2	5- 5	7-11	10- 1	12- 4	14- 3	5- 0	7- 4	9- 4	11- 5	13- 2
	Spruce-pine-fir	#3	4- 1	6- 0	7- 7	9- 4	10- 9	3-10	5- 7	7- 1	8- 7	10- 0

Check sources for availability of lumber in lengths greater than 20 feet.

For SI: 1 inch = 25.4 mm, 1 foot = 304.8 mm, 1 pound per square foot = 0.0479 kN/m^2.

a. The tabulated rafter spans assume that ceiling joists are located at the bottom of the attic space or that some other method of resisting the outward push of the rafters on the bearing walls, such as rafter ties, is provided at that location. When ceiling joists or rafter ties are located higher in the attic space, the rafter spans shall be multiplied by the factors given below:

H_C/H_R	Rafter Span Adjustment Factor
2/3 or greater	0.50
1/2	0.58
1/3	0.67
1/4	0.76
1/5	0.83
1/6	0.90
1/7.5 and less	1.00

where: H_C = Height of ceiling joists or rafter ties measured vertically above the top of the rafter support walls.

H_R = Height of roof ridge measured vertically above the top of the rafter support walls.

b. Span exceeds 26 feet in length.

TABLE R802.5.1(5)
RAFTER SPANS FOR COMMON LUMBER SPECIES
(Ground snow load=30 psf, ceiling attached to rafters, L/Δ=240)

RAFTER SPACING (inches)	SPECIES AND GRADE	DEAD LOAD = 10 psf					DEAD LOAD = 20 psf				
		2x4	2x6	2x8	2x10	2x12	2x4	2x6	2x8	2x10	2x12
		Maximum rafter spans[a]									
		(feet - inches)	(feet - inches)	(feet - inches)	(feet - inches)	(feet - inches)	(feet - inches)	(feet - inches)	(feet - inches)	(feet - inches)	(feet - inches)
12	Douglas fir-larch SS	9- 1	14- 4	18-10	24- 1	Note b	9- 1	14- 4	18-10	24- 1	Note b
	Douglas fir-larch #1	8- 9	13- 9	18- 2	22- 9	Note b	8- 9	13- 2	16- 8	20- 4	23- 7
	Douglas fir-larch #2	8- 7	13- 6	17- 5	21- 4	24- 8	8- 5	12- 4	15- 7	19- 1	22- 1
	Douglas fir-larch #3	7- 1	10- 5	13- 2	16- 1	18- 8	6- 4	9- 4	11- 9	14- 5	16- 8
	Hem-fir SS	8- 7	13- 6	17-10	22- 9	Note b	8- 7	13- 6	17-10	22- 9	Note b
	Hem-fir #1	8- 5	13- 3	17- 5	22- 2	25- 9	8- 5	12-10	16- 3	19-10	23- 0
	Hem-fir #2	8- 0	12- 7	16- 7	21- 0	24- 4	8- 0	12- 2	15- 4	18- 9	21- 9
	Hem-fir #3	7- 1	10- 5	13- 2	16- 1	18- 8	6- 4	9- 4	11- 9	14- 5	16- 8
	Southern pine SS	8-11	14- 1	18- 6	23- 8	Note b	8-11	14- 1	18- 6	23- 8	Note b
	Southern pine #1	8- 9	13- 9	18- 2	23- 2	Note b	8- 9	13- 9	18- 2	22- 2	Note b
	Southern pine #2	8- 7	13- 6	17-10	22- 3	Note b	8- 7	12-11	16- 8	19-11	23- 4
	Southern pine #3	7- 7	11- 2	14- 3	16-10	20- 0	6- 9	10- 0	12- 9	15- 1	17-11
	Spruce-pine-fir SS	8- 5	13- 3	17- 5	22- 3	Note b	8- 5	13- 3	17- 5	22- 3	Note b
	Spruce-pine-fir #1	8- 3	12-11	17- 0	21- 4	24- 8	8- 3	12- 4	15- 7	19- 1	22- 1
	Spruce-pine-fir #2	8- 3	12-11	17- 0	21- 4	24- 8	8- 3	12- 4	15- 7	19- 1	22- 1
	Spruce-pine-fir #3	7- 1	10- 5	13- 2	16- 1	18- 8	6- 4	9- 4	11- 9	14- 5	16- 8
16	Douglas fir-larch SS	8- 3	13- 0	17- 2	21-10	Note b	8- 3	13- 0	17- 2	21- 3	24- 8
	Douglas fir-larch #1	8- 0	12- 6	16- 2	19- 9	22-10	7-10	11- 5	14- 5	17- 8	20- 5
	Douglas fir-larch #2	7-10	11-11	15- 1	18- 5	21- 5	7- 3	10- 8	13- 6	16- 6	19- 2
	Douglas fir-larch #3	6- 2	9- 0	11- 5	13-11	16- 2	5- 6	8- 1	10- 3	12- 6	14- 6
	Hem-fir SS	7-10	12- 3	16- 2	20- 8	25- 1	7-10	12- 3	16- 2	20- 8	24- 2
	Hem-fir #1	7- 8	12- 0	15- 9	19- 3	22- 3	7- 7	11- 1	14- 1	17- 2	19-11
	Hem-fir #2	7- 3	11- 5	14-11	18- 2	21- 1	7- 2	10- 6	13- 4	16- 3	18-10
	Hem-fir #3	6- 2	9- 0	11- 5	13-11	16- 2	5- 6	8- 1	10- 3	12- 6	14- 6
	Southern pine SS	8- 1	12- 9	16-10	21- 6	Note b	8- 1	12- 9	16-10	21- 6	Note b
	Southern pine #1	8- 0	12- 6	16- 6	21- 1	25- 7	8- 0	12- 6	16- 2	19- 2	22-10
	Southern pine #2	7-10	12- 3	16- 2	19- 3	22- 7	7-10	11- 2	14- 5	17- 3	20- 2
	Southern pine #3	6- 7	9- 8	12- 4	14- 7	17- 4	5-10	8- 8	11- 0	13- 0	15- 6
	Spruce-pine-fir SS	7- 8	12- 0	15-10	20- 2	24- 7	7- 8	12- 0	15-10	19- 9	22-10
	Spruce-pine-fir #1	7- 6	11- 9	15- 1	18- 5	21- 5	7- 3	10- 8	13- 6	16- 6	19- 2
	Spruce-pine-fir #2	7- 6	11- 9	15- 1	18- 5	21- 5	7- 3	10- 8	13- 6	16- 6	19- 2
	Spruce-pine-fir #3	6- 2	9- 0	11- 5	13-11	16- 2	5- 6	8- 1	10- 3	12- 6	14- 6
19.2	Douglas fir-larch SS	7- 9	12- 3	16- 1	20- 7	25- 0	7- 9	12- 3	15-10	19- 5	22- 6
	Douglas fir-larch #1	7- 6	11- 8	14- 9	18- 0	20-11	7- 1	10- 5	13- 2	16- 1	18- 8
	Douglas fir-larch #2	7- 4	10-11	13- 9	16-10	19- 6	6- 8	9- 9	12- 4	15- 1	17- 6
	Douglas fir-larch #3	5- 7	8- 3	10- 5	12- 9	14- 9	5- 0	7- 4	9- 4	11- 5	13- 2
	Hem-fir SS	7- 4	11- 7	15- 3	19- 5	23- 7	7- 4	11- 7	15- 3	19- 1	22- 1
	Hem-fir #1	7- 2	11- 4	14- 4	17- 7	20- 4	6-11	10- 2	12-10	15- 8	18- 2
	Hem-fir #2	6-10	10- 9	13- 7	16- 7	19- 3	6- 7	9- 7	12- 2	14-10	17- 3
	Hem-fir #3	5- 7	8- 3	10- 5	12- 9	14- 9	5- 0	7- 4	9- 4	11- 5	13- 2
	Southern pine SS	7- 8	12- 0	15-10	20- 2	24- 7	7- 8	12- 0	15-10	20- 2	24- 7
	Southern pine #1	7- 6	11- 9	15- 6	19- 7	23- 4	7- 6	11- 9	14- 9	17- 6	20-11
	Southern pine #2	7- 4	11- 5	14- 9	17- 7	20- 7	7- 1	10- 2	13- 2	15- 9	18- 5
	Southern pine #3	6- 0	8-10	11- 3	13- 4	15-10	5- 4	7-11	10- 1	11-11	14- 2
	Spruce-pine-fir SS	7- 2	11- 4	14-11	19- 0	23- 1	7- 2	11- 4	14- 9	18- 0	20-11
	Spruce-pine-fir #1	7- 0	10-11	13- 9	16-10	19- 6	6- 8	9- 9	12- 4	15- 1	17- 6
	Spruce-pine-fir #2	7- 0	10-11	13- 9	16-10	19- 6	6- 8	9- 9	12- 4	15- 1	17- 6
	Spruce-pine-fir #3	5- 7	8- 3	10- 5	12- 9	14- 9	5- 0	7- 4	9- 4	11- 5	13- 2

(continued)

TABLE R802.5.1(5)—continued
RAFTER SPANS FOR COMMON LUMBER SPECIES
(Ground snow load=30 psf, ceiling attached to rafters, L/Δ=240)

RAFTER SPACING (inches)	SPECIES AND GRADE		DEAD LOAD = 10 psf					DEAD LOAD = 20 psf				
			2x4	2x6	2x8	2x10	2x12	2x4	2x6	2x8	2x10	2x12
			Maximum rafter spans[a]									
			(feet - inches)	(feet - inches)	(feet - inches)	(feet - inches)	(feet - inches)	(feet - inches)	(feet - inches)	(feet - inches)	(feet - inches)	(feet - inches)
24	Douglas fir-larch	SS	7- 3	11- 4	15- 0	19- 1	22- 6	7- 3	11- 3	14- 2	17- 4	20- 1
	Douglas fir-larch	#1	7- 0	10- 5	13- 2	16- 1	18- 8	6- 4	9- 4	11- 9	14- 5	16- 8
	Douglas fir-larch	#2	6- 8	9- 9	12- 4	15- 1	17- 6	5-11	8- 8	11- 0	13- 6	15- 7
	Douglas fir-larch	#3	5- 0	7- 4	9- 4	11- 5	13- 2	4- 6	6- 7	8- 4	10- 2	11-10
	Hem-fir	SS	6-10	10- 9	14- 2	18- 0	21-11	6-10	10- 9	13-11	17- 0	19- 9
	Hem-fir	#1	6- 8	10- 2	12-10	15- 8	18- 2	6- 2	9- 1	11- 6	14- 0	16- 3
	Hem-fir	#2	6- 4	9- 7	12- 2	14-10	17- 3	5-10	8- 7	10-10	13- 3	15- 5
	Hem-fir	#3	5- 0	7- 4	9- 4	11- 5	13- 2	4- 6	6- 7	8- 4	10- 2	11-10
	Southern pine	SS	7- 1	11- 2	14- 8	18- 9	22-10	7- 1	11- 2	14- 8	18- 9	22-10
	Southern pine	#1	7- 0	10-11	14- 5	17- 6	20-11	7- 0	10- 6	13- 2	15- 8	18- 8
	Southern pine	#2	6-10	10- 2	13- 2	15- 9	18- 5	6- 4	9- 2	11- 9	14- 1	16- 6
	Southern pine	#3	5- 4	7-11	10- 1	11-11	14- 2	4- 9	7- 1	9- 0	10- 8	12- 8
	Spruce-pine-fir	SS	6- 8	10- 6	13-10	17- 8	20-11	6- 8	10- 5	13- 2	16- 1	18- 8
	Spruce-pine-fir	#1	6- 6	9- 9	12- 4	15- 1	17- 6	5-11	8- 8	11- 0	13- 6	15- 7
	Spruce-pine-fir	#2	6- 6	9- 9	12- 4	15- 1	17- 6	5-11	8- 8	11- 0	13- 6	15- 7
	Spruce-pine-fir	#3	5- 0	7- 4	9- 4	11- 5	13- 2	4- 6	6- 7	8- 4	10- 2	11-10

Check sources for availability of lumber in lengths greater than 20 feet.

For SI: 1 inch = 25.4 mm, 1 foot = 304.8 mm, 1 pound per square foot = 0.0479 kN/m^2.

a. The tabulated rafter spans assume that ceiling joists are located at the bottom of the attic space or that some other method of resisting the outward push of the rafters on the bearing walls, such as rafter ties, is provided at that location. When ceiling joists or rafter ties are located higher in the attic space, the rafter spans shall be multiplied by the factors given below:

H_C/H_R	Rafter Span Adjustment Factor
2/3 or greater	0.50
1/2	0.58
1/3	0.67
1/4	0.76
1/5	0.83
1/6	0.90
1/7.5 and less	1.00

where: H_C = Height of ceiling joists or rafter ties measured vertically above the top of the rafter support walls.

H_R = Height of roof ridge measured vertically above the top of the rafter support walls.

b. Span exceeds 26 feet in length.

TABLE R802.5.1(6)
RAFTER SPANS FOR COMMON LUMBER SPECIES
(Ground snow load=50 psf, ceiling attached to rafters, L/Δ=240)

RAFTER SPACING (inches)	SPECIES AND GRADE		DEAD LOAD = 10 psf					DEAD LOAD = 20 psf				
			2x4	2x6	2x8	2x10	2x12	2x4	2x6	2x8	2x10	2x12
			Maximum rafter spans[a]									
			(feet - inches)	(feet - inches)	(feet - inches)	(feet - inches)	(feet - inches)	(feet - inches)	(feet - inches)	(feet - inches)	(feet - inches)	(feet - inches)
12	Douglas fir-larch	SS	7-8	12-1	15-11	20-3	24-8	7-8	12-1	15-11	20-3	24-0
	Douglas fir-larch	#1	7-5	11-7	15-3	18-7	21-7	7-5	11-2	14-1	17-3	20-0
	Douglas fir-larch	#2	7-3	11-3	14-3	17-5	20-2	7-1	10-5	13-2	16-1	18-8
	Douglas fir-larch	#3	5-10	8-6	10-9	13-2	15-3	5-5	7-10	10-0	12-2	14-1
	Hem-fir	SS	7-3	11-5	15-0	19-2	23-4	7-3	11-5	15-0	19-2	23-4
	Hem-fir	#1	7-1	11-2	14-8	18-1	21-0	7-1	10-10	13-9	16-9	19-5
	Hem-fir	#2	6-9	10-8	14-0	17-2	19-11	6-9	10-3	13-0	15-10	18-5
	Hem-fir	#3	5-10	8-6	10-9	13-2	15-3	5-5	7-10	10-0	12-2	14-1
	Southern pine	SS	7-6	11-10	15-7	19-11	24-3	7-6	11-10	15-7	19-11	24-3
	Southern pine	#1	7-5	11-7	15-4	19-7	23-9	7-5	11-7	15-4	18-9	22-4
	Southern pine	#2	7-3	11-5	15-0	18-2	21-3	7-3	10-11	14-1	16-10	19-9
	Southern pine	#3	6-2	9-2	11-8	13-9	16-4	5-9	8-5	10-9	12-9	15-2
	Spruce-pine-fir	SS	7-1	11-2	14-8	18-9	22-10	7-1	11-2	14-8	18-9	22-4
	Spruce-pine-fir	#1	6-11	10-11	14-3	17-5	20-2	6-11	10-5	13-2	16-1	18-8
	Spruce-pine-fir	#2	6-11	10-11	14-3	17-5	20-2	6-11	10-5	13-2	16-1	18-8
	Spruce-pine-fir	#3	5-10	8-6	10-9	13-2	15-3	5-5	7-10	10-0	12-2	14-1
16	Douglas fir-larch	SS	7-0	11-0	14-5	18-5	22-5	7-0	11-0	14-5	17-11	20-10
	Douglas fir-larch	#1	6-9	10-5	13-2	16-1	18-8	6-7	9-8	12-2	14-11	17-3
	Douglas fir-larch	#2	6-7	9-9	12-4	15-1	17-6	6-2	9-0	11-5	13-11	16-2
	Douglas fir-larch	#3	5-0	7-4	9-4	11-5	13-2	4-8	6-10	8-8	10-6	12-3
	Hem-fir	SS	6-7	10-4	13-8	17-5	21-2	6-7	10-4	13-8	17-5	20-5
	Hem-fir	#1	6-5	10-2	12-10	15-8	18-2	6-5	9-5	11-11	14-6	16-10
	Hem-fir	#2	6-2	9-7	12-2	14-10	17-3	6-1	8-11	11-3	13-9	15-11
	Hem-fir	#3	5-0	7-4	9-4	11-5	13-2	4-8	6-10	8-8	10-6	12-3
	Southern pine	SS	6-10	10-9	14-2	18-1	22-0	6-10	10-9	14-2	18-1	22-0
	Southern pine	#1	6-9	10-7	13-11	17-6	20-11	6-9	10-7	13-8	16-2	19-4
	Southern pine	#2	6-7	10-2	13-2	15-9	18-5	6-7	9-5	12-2	14-7	17-1
	Southern pine	#3	5-4	7-11	10-1	11-11	14-2	4-11	7-4	9-4	11-0	13-1
	Spruce-pine-fir	SS	6-5	10-2	13-4	17-0	20-9	6-5	10-2	13-4	16-8	19-4
	Spruce-pine-fir	#1	6-4	9-9	12-4	15-1	17-6	6-2	9-0	11-5	13-11	16-2
	Spruce-pine-fir	#2	6-4	9-9	12-4	15-1	17-6	6-2	9-0	11-5	13-11	16-2
	Spruce-pine-fir	#3	5-0	7-4	9-4	11-5	13-2	4-8	6-10	8-8	10-6	12-3
19.2	Douglas fir-larch	SS	6-7	10-4	13-7	17-4	20-6	6-7	10-4	13-5	16-5	19-0
	Douglas fir-larch	#1	6-4	9-6	12-0	14-8	17-1	6-0	8-10	11-2	13-7	15-9
	Douglas fir-larch	#2	6-1	8-11	11-3	13-9	15-11	5-7	8-3	10-5	12-9	14-9
	Douglas fir-larch	#3	4-7	6-9	8-6	10-5	12-1	4-3	6-3	7-11	9-7	11-2
	Hem-fir	SS	6-2	9-9	12-10	16-5	19-11	6-2	9-9	12-10	16-1	18-8
	Hem-fir	#1	6-1	9-3	11-9	14-4	16-7	5-10	8-7	10-10	13-3	15-5
	Hem-fir	#2	5-9	8-9	11-1	13-7	15-9	5-7	8-1	10-3	12-7	14-7
	Hem-fir	#3	4-7	6-9	8-6	10-5	12-1	4-3	6-3	7-11	9-7	11-2
	Southern pine	SS	6-5	10-2	13-4	17-0	20-9	6-5	10-2	13-4	17-0	20-9
	Southern pine	#1	6-4	9-11	13-1	16-0	19-1	6-4	9-11	12-5	14-10	17-8
	Southern pine	#2	6-2	9-4	12-0	14-4	16-10	6-0	8-8	11-2	13-4	15-7
	Southern pine	#3	4-11	7-3	9-2	10-10	12-11	4-6	6-8	8-6	10-1	12-0
	Spruce-pine-fir	SS	6-1	9-6	12-7	16-0	19-1	6-1	9-6	12-5	15-3	17-8
	Spruce-pine-fir	#1	5-11	8-11	11-3	13-9	15-11	5-7	8-3	10-5	12-9	14-9
	Spruce-pine-fir	#2	5-11	8-11	11-3	13-9	15-11	5-7	8-3	10-5	12-9	14-9
	Spruce-pine-fir	#3	4-7	6-9	8-6	10-5	12-1	4-3	6-3	7-11	9-7	11-2

(continued)

TABLE R802.5.1(6)—continued
RAFTER SPANS FOR COMMON LUMBER SPECIES
(Ground snow load=50 psf, ceiling attached to rafters, L/Δ=240)

RAFTER SPACING (inches)	SPECIE AND GRADE		DEAD LOAD = 10 psf					DEAD LOAD = 20 psf				
			2x4	2x6	2x8	2x10	2x12	2x4	2x6	2x8	2x10	2x12
			Maximum rafter spans[a]									
			(feet - inches)	(feet - inches)	(feet - inches)	(feet - inches)	(feet - inches)	(feet - inches)	(feet - inches)	(feet - inches)	(feet - inches)	(feet - inches)
24	Douglas fir-larch	SS	6- 1	9- 7	12- 7	15-10	18- 4	6- 1	9- 6	12- 0	14- 8	17- 0
	Douglas fir-larch	#1	5-10	8- 6	10- 9	13- 2	15- 3	5- 5	7-10	10- 0	12- 2	14- 1
	Douglas fir-larch	#2	5- 5	7-11	10- 1	12- 4	14- 3	5- 0	7- 4	9- 4	11- 5	13- 2
	Douglas fir-larch	#3	4- 1	6- 0	7- 7	9- 4	10- 9	3-10	5- 7	7- 1	8- 7	10- 0
	Hem-fir	SS	5- 9	9- 1	11-11	15- 2	18- 0	5- 9	9- 1	11- 9	14- 5	15-11
	Hem-fir	#1	5- 8	8- 3	10- 6	12-10	14-10	5- 3	7- 8	9- 9	11-10	13- 9
	Hem-fir	#2	5- 4	7-10	9-11	12- 1	14- 1	4-11	7- 3	9- 2	11- 3	13- 0
	Hem-fir	#3	4- 1	6- 0	7- 7	9- 4	10- 9	3-10	5- 7	7- 1	8- 7	10- 0
	Southern pine	SS	6- 0	9- 5	12- 5	15-10	19- 3	6- 0	9- 5	12- 5	15-10	19- 3
	Southern pine	#1	5-10	9- 3	12- 0	14- 4	17- 1	5-10	8-10	11- 2	13- 3	15- 9
	Southern pine	#2	5- 9	8- 4	10- 9	12-10	15- 1	5- 5	7- 9	10- 0	11-11	13-11
	Southern pine	#3	4- 4	6- 5	8- 3	9- 9	11- 7	4- 1	6- 0	7- 7	9- 0	10- 8
	Spruce-pine-fir	SS	5- 8	8-10	11- 8	14- 8	17- 1	5- 8	8-10	11- 2	13- 7	15- 9
	Spruce-pine-fir	#1	5- 5	7-11	10- 1	12- 4	14- 3	5- 0	7- 4	9- 4	11- 5	13- 2
	Spruce-pine-fir	#2	5- 5	7-11	10- 1	12- 4	14- 3	5- 0	7- 4	9- 4	11- 5	13- 2
	Spruce-pine-fir	#3	4- 1	6- 0	7- 7	9- 4	10- 9	3-10	5- 7	7- 1	8- 7	10- 0

Check sources for availability of lumber in lengths greater than 20 feet.

For SI: 1 inch = 25.4 mm, 1 foot = 304.8 mm, 1 pound per square foot = 0.0479 kN/m^2.

a. The tabulated rafter spans assume that ceiling joists are located at the bottom of the attic space or that some other method of resisting the outward push of the rafters on the bearing walls, such as rafter ties, is provided at that location. When ceiling joists or rafter ties are located higher in the attic space, the rafter spans shall be multiplied by the factors given below:

H_C/H_R	Rafter Span Adjustment Factor
2/3 or greater	0.50
1/2	0.58
1/3	0.67
1/4	0.76
1/5	0.83
1/6	0.90
1/7.5 and less	1.00

where: H_C = Height of ceiling joists or rafter ties measured vertically above the top of the rafter support walls.

H_R = Height of roof ridge measured vertically above the top of the rafter support walls.

b. Span exceeds 26 feet in length.

TABLE R802.5.1(7)
RAFTER SPANS FOR 70 PSF GROUND SNOW LOAD
(Ceiling not attached to rafters, L/Δ=180)

RAFTER SPACING (inches)	SPECIES AND GRADE	DEAD LOAD = 10 psf					DEAD LOAD = 20 psf				
		2x4	2x6	2x8	2x10	2x12	2x4	2x6	2x8	2x10	2x12
		Maximum Rafter Spans[a]									
		(feet - inches)	(feet - inches)	(feet - inches)	(feet - inches)	(feet - inches)	(feet - inches)	(feet - inches)	(feet - inches)	(feet - inches)	(feet - inches)
12	Douglas fir-larch SS	7- 7	11-10	15- 8	19- 5	22- 6	7- 7	11-10	15- 0	18- 3	21- 2
	Douglas fir-larch #1	7- 1	10- 5	13- 2	16- 1	18- 8	6- 8	9-10	12- 5	15- 2	17- 7
	Douglas fir-larch #2	6- 8	9- 9	12- 4	15- 1	17- 6	6- 3	9- 2	11- 8	14- 2	16- 6
	Douglas fir-larch #3	5- 0	7- 4	9- 4	11- 5	13- 2	4- 9	6-11	8- 9	10- 9	12- 5
	Hem-fir SS	7- 2	11- 3	14- 9	18-10	22- 1	7- 2	11- 3	14- 8	18- 0	20-10
	Hem-fir #1	6-11	10- 2	12-10	15- 8	18- 2	6- 6	9- 7	12- 1	14-10	17- 2
	Hem-fir #2	6- 7	9- 7	12- 2	14-10	17- 3	6- 2	9- 1	11- 5	14- 0	16- 3
	Hem-fir #3	5- 0	7- 4	9- 4	11- 5	13- 2	4- 9	6-11	8- 9	10- 9	12- 5
	Southern pine SS	7- 5	11- 8	15- 4	19- 7	23-10	7- 5	11- 8	15- 4	19- 7	23-10
	Southern pine #1	7- 3	11- 5	14- 9	17- 6	20-11	7- 3	11- 1	13-11	16- 6	19- 8
	Southern pine #2	7- 1	10- 2	13- 2	15- 9	18- 5	6- 8	9- 7	12- 5	14-10	17- 5
	Southern pine #3	5- 4	7-11	10- 1	11-11	14- 2	5- 1	7- 5	9- 6	11- 3	13- 4
	Spruce-pine-fir SS	7- 0	11- 0	14- 6	18- 0	20-11	7- 0	11- 0	13-11	17- 0	19- 8
	Spruce-pine-fir #1	6- 8	9- 9	12- 4	15- 1	17- 6	6- 3	9- 2	11- 8	14- 2	16- 6
	Spruce-pine-fir #2	6- 8	9- 9	12- 4	15- 1	17- 6	6- 3	9- 2	11- 8	14- 2	16- 6
	Spruce-pine-fir #3	5- 0	7- 4	9- 4	11- 5	13- 2	4- 9	6-11	8- 9	10- 9	12- 5
16	Douglas fir-larch SS	6-10	10- 9	13- 9	16-10	19- 6	6-10	10- 3	13- 0	15-10	18- 4
	Douglas fir-larch #1	6- 2	9- 0	11- 5	13-11	16- 2	5-10	8- 6	10- 9	13- 2	15- 3
	Douglas fir-larch #2	5- 9	8- 5	10- 8	13- 1	15- 2	5- 5	7-11	10- 1	12- 4	14- 3
	Douglas fir-larch #3	4- 4	6- 4	8- 1	9-10	11- 5	4- 1	6- 0	7- 7	9- 4	10- 9
	Hem-fir SS	6- 6	10- 2	13- 5	16- 6	19- 2	6- 6	10- 1	12- 9	15- 7	18- 0
	Hem-fir #1	6- 0	8- 9	11- 2	13- 7	15- 9	5- 8	8- 3	10- 6	12-10	14-10
	Hem-fir #2	5- 8	8- 4	10- 6	12-10	14-11	5- 4	7-10	9-11	12- 1	14- 1
	Hem-fir #3	4- 4	6- 4	8- 1	9-10	11- 5	4- 1	6- 0	7- 7	9- 4	10- 9
	Southern pine SS	6- 9	10- 7	14- 0	17-10	21- 8	6- 9	10- 7	14- 0	17-10	21- 0
	Southern pine #1	6- 7	10- 2	12- 9	15- 2	18- 1	6- 5	9- 7	12- 0	14- 4	17- 1
	Southern pine #2	6- 2	8-10	11- 5	13- 7	16- 0	5-10	8- 4	10- 9	12-10	15- 1
	Southern pine #3	4- 8	6-10	8- 9	10- 4	12- 3	4- 4	6- 5	8- 3	9- 9	11- 7
	Spruce-pine-fir SS	6- 4	10- 0	12- 9	15- 7	18- 1	6- 4	9- 6	12- 0	14- 8	17- 1
	Spruce-pine-fir #1	5- 9	8- 5	10- 8	13- 1	15- 2	5- 5	7-11	10- 1	12- 4	14- 3
	Spruce-pine-fir #2	5- 9	8- 5	10- 8	13- 1	15- 2	5- 5	7-11	10- 1	12- 4	14- 3
	Spruce-pine-fir #3	4- 4	6- 4	8- 1	9-10	11- 5	4- 1	6- 0	7- 7	9- 4	10- 9
19.2	Douglas fir-larch SS	6- 5	9-11	12- 7	15- 4	17- 9	6- 5	9- 4	11-10	14- 5	16- 9
	Douglas fir-larch #1	5- 7	8- 3	10- 5	12- 9	14- 9	5- 4	7- 9	9-10	12- 0	13-11
	Douglas fir-larch #2	5- 3	7- 8	9- 9	11-11	13-10	5- 0	7- 3	9- 2	11- 3	13- 0
	Douglas fir-larch #3	4- 0	5-10	7- 4	9- 0	10- 5	3- 9	5- 6	6-11	8- 6	9-10
	Hem-fir SS	6- 1	9- 7	12- 4	15- 1	17- 4	6- 1	9- 2	11- 8	14- 2	15-5
	Hem-fir #1	5- 6	8- 0	10- 2	12- 5	14- 5	5- 2	7- 7	9- 7	11- 8	13- 7
	Hem-fir #2	5- 2	7- 7	9- 7	11- 9	13- 7	4-11	7- 2	9- 1	11- 1	12-10
	Hem-fir #3	4- 0	5-10	7- 4	9- 0	10- 5	3- 9	5- 6	6-11	8- 6	9-10
	Southern pine SS	6- 4	10- 0	13- 2	16- 9	20- 4	6- 4	10- 0	13- 2	16- 5	19- 2
	Southern pine #1	6- 3	9- 3	11- 8	13-10	16- 6	5-11	8- 9	11- 0	13- 1	15- 7
	Southern pine #2	5- 7	8- 1	10- 5	12- 5	14- 7	5- 4	7- 7	9-10	11- 9	13- 9
	Southern pine #3	4- 3	6- 3	8- 0	9- 5	11- 2	4- 0	5-11	7- 6	8-10	10- 7
	Spruce-pine-fir SS	6- 0	9- 2	11- 8	14- 3	16- 6	5-11	8- 8	11- 0	13- 5	15- 7
	Spruce-pine-fir #1	5- 3	7- 8	9- 9	11-11	13-10	5- 0	7- 3	9- 2	11- 3	13- 0
	Spruce-pine-fir #2	5- 3	7- 8	9- 9	11-11	13-10	5- 0	7- 3	9- 2	11- 3	13- 0
	Spruce-pine-fir #3	4- 0	5-10	7- 4	9- 0	10- 5	3- 9	5- 6	6-11	8- 6	9-10

(continued)

TABLE R802.5.1(7)—continued
RAFTER SPANS FOR 70 PSF GROUND SNOW LOAD
(Ceiling not attached to rafters, L/Δ=180)

RAFTER SPACING (inches)	SPECIES AND GRADE		DEAD LOAD = 10 psf					DEAD LOAD = 20 psf				
			2x4	2x6	2x8	2x10	2x12	2x4	2x6	2x8	2x10	2x12
			Maximum rafter spans[a]									
			(feet - inches)	(feet - inches)	(feet - inches)	(feet - inches)	(feet - inches)	(feet - inches)	(feet - inches)	(feet - inches)	(feet - inches)	(feet - inches)
24	Douglas fir-larch	SS	6- 0	8-10	11- 3	13- 9	15-11	5- 9	8- 4	10- 7	12-11	15- 0
	Douglas fir-larch	#1	5- 0	7- 4	9- 4	11- 5	13- 2	4- 9	6-11	8- 9	10- 9	12- 5
	Douglas fir-larch	#2	4- 8	6-11	8- 9	10- 8	12- 4	4- 5	6- 6	8- 3	10- 0	11- 8
	Douglas fir-larch	#3	3- 7	5- 2	6- 7	8- 1	9- 4	3- 4	4-11	6- 3	7- 7	8-10
	Hem-fir	SS	5- 8	8- 8	11- 0	13- 6	13-11	5- 7	8- 3	10- 5	12- 4	12-4
	Hem-fir	#1	4-11	7- 2	9- 1	11- 1	12-10	4- 7	6- 9	8- 7	10- 6	12- 2
	Hem-fir	#2	4- 8	6- 9	8- 7	10- 6	12- 2	4- 4	6- 5	8- 1	9-11	11- 6
	Hem-fir	#3	3- 7	5- 2	6- 7	8- 1	9- 4	3- 4	4-11	6- 3	7- 7	8-10
	Southern pine	SS	5-11	9- 3	12- 2	15- 7	18- 2	5-11	9- 3	12- 2	14- 8	17- 2
	Southern pine	#1	5- 7	8- 3	10- 5	12- 5	14- 9	5- 3	7-10	9-10	11- 8	13-11
	Southern pine	#2	5- 0	7- 3	9- 4	11- 1	13- 0	4- 9	6-10	8- 9	10- 6	12- 4
	Southern pine	#3	3- 9	5- 7	7- 1	8- 5	10- 0	3- 7	5- 3	6- 9	7-11	9- 5
	Spruce-pine-fir	SS	5- 6	8- 3	10- 5	12- 9	14- 9	5- 4	7- 9	9-10	12- 0	12-11
	Spruce-pine-fir	#1	4- 8	6-11	8- 9	10- 8	12- 4	4- 5	6- 6	8- 3	10- 0	11- 8
	Spruce-pine-fir	#2	4- 8	6-11	8- 9	10- 8	12- 4	4- 5	6- 6	8- 3	10- 0	11- 8
	Spruce-pine-fir	#3	3- 7	5- 2	6- 7	8- 1	9- 4	3- 4	4-11	6- 3	7- 7	8-10

Check sources for availability of lumber in lengths greater than 20 feet.

For SI: 1 inch = 25.4 mm, 1 foot = 304.8 mm, 1 pound per square foot = 0.0479 kN/m^2.

a. The tabulated rafter spans assume that ceiling joists are located at the bottom of the attic space or that some other method of resisting the outward push of the rafters on the bearing walls, such as rafter ties, is provided at that location. When ceiling joists or rafter ties are located higher in the attic space, the rafter spans shall be multiplied by the factors given below:

H_C/H_R	Rafter Span Adjustment Factor
2/3 or greater	0.50
1/2	0.58
1/3	0.67
1/4	0.76
1/5	0.83
1/6	0.90
1/7.5 and less	1.00

where: H_C = Height of ceiling joists or rafter ties measured vertically above the top of the rafter support walls.

H_R = Height of roof ridge measured vertically above the top of the rafter support walls.

TABLE R802.5.1(8)
RAFTER SPANS FOR 70 PSF GROUND SNOW LOAD
(Ceiling attached to rafters, L/Δ=240)

RAFTER SPACING (inches)	SPECIES AND GRADE		DEAD LOAD = 10 psf					DEAD LOAD = 20 psf				
			2x4	2x6	2x8	2x10	2x12	2x4	2x6	2x8	2x10	2x12
			Maximum rafter spans[a]									
			(feet - inches)	(feet - inches)	(feet - inches)	(feet - inches)	(feet - inches)	(feet - inches)	(feet - inches)	(feet - inches)	(feet - inches)	(feet - inches)
12	Douglas fir-larch	SS	6-10	10- 9	14- 3	18- 2	22- 1	6-10	10- 9	14- 3	18- 2	21- 2
	Douglas fir-larch	#1	6- 7	10- 5	13- 2	16- 1	18- 8	6- 7	9-10	12- 5	15- 2	17- 7
	Douglas fir-larch	#2	6- 6	9- 9	12- 4	15- 1	17- 6	6- 3	9- 2	11- 8	14- 2	16- 6
	Douglas fir-larch	#3	5- 0	7- 4	9- 4	11- 5	13- 2	4- 9	6-11	8- 9	10- 9	12- 5
	Hem-fir	SS	6- 6	10- 2	13- 5	17- 2	20-10	6- 6	10- 2	13- 5	17- 2	20-10
	Hem-fir	#1	6- 4	10- 0	12-10	15- 8	18- 2	6- 4	9- 7	12- 1	14-10	17- 2
	Hem-fir	#2	6- 1	9- 6	12- 2	14-10	17- 3	6- 1	9- 1	11- 5	14- 0	16- 3
	Hem-fir	#3	5- 0	7- 4	9- 4	11- 5	13- 2	4- 9	6-11	8- 9	10- 9	12- 5
	Southern pine	SS	6- 9	10- 7	14- 0	17-10	21- 8	6- 9	10- 7	14- 0	17-10	21- 8
	Southern pine	#1	6- 7	10- 5	13- 8	17- 6	20-11	6- 7	10- 5	13- 8	16- 6	19- 8
	Southern pine	#2	6- 6	10- 2	13- 2	15- 9	18- 5	6- 6	9- 7	12- 5	14-10	17- 5
	Southern pine	#3	5- 4	7-11	10- 1	11-11	14- 2	5- 1	7- 5	9- 6	11- 3	13- 4
	Spruce-pine-fir	SS	6- 4	10- 0	13- 2	16- 9	20- 5	6- 4	10- 0	13- 2	16- 9	19- 8
	Spruce-pine-fir	#1	6- 2	9- 9	12- 4	15- 1	17- 6	6- 2	9- 2	11- 8	14- 2	16- 6
	Spruce-pine-fir	#2	6- 2	9- 9	12- 4	15- 1	17- 6	6- 2	9- 2	11- 8	14- 2	16- 6
	Spruce-pine-fir	#3	5- 0	7- 4	9- 4	11- 5	13- 2	4- 9	6-11	8- 9	10- 9	12- 5
16	Douglas fir-larch	SS	6- 3	9-10	12-11	16- 6	19- 6	6- 3	9-10	12-11	15-10	18- 4
	Douglas fir-larch	#1	6- 0	9- 0	11- 5	13-11	16- 2	5-10	8- 6	10- 9	13- 2	15- 3
	Douglas fir-larch	#2	5- 9	8- 5	10- 8	13- 1	15- 2	5- 5	7-11	10- 1	12- 4	14- 3
	Douglas fir-larch	#3	4- 4	6- 4	8- 1	9-10	11- 5	4- 1	6- 0	7- 7	9- 4	10- 9
	Hem-fir	SS	5-11	9- 3	12- 2	15- 7	18-11	5-11	9- 3	12- 2	15- 7	18- 0
	Hem-fir	#1	5- 9	8- 9	11- 2	13- 7	15- 9	5- 8	8- 3	10- 6	12-10	14-10
	Hem-fir	#2	5- 6	8- 4	10- 6	12-10	14-11	5- 4	7-10	9-11	12- 1	14- 1
	Hem-fir	#3	4- 4	6- 4	8- 1	9-10	11- 5	4- 1	6- 0	7- 7	9- 4	10- 9
	Southern pine	SS	6- 1	9- 7	12- 8	16- 2	19- 8	6- 1	9- 7	12- 8	16- 2	19- 8
	Southern pine	#1	6- 0	9- 5	12- 5	15- 2	18- 1	6- 0	9- 5	12- 0	14- 4	17- 1
	Southern pine	#2	5-11	8-10	11- 5	13- 7	16- 0	5-10	8- 4	10- 9	12-10	15- 1
	Southern pine	#3	4- 8	6-10	8- 9	10- 4	12- 3	4- 4	6- 5	8- 3	9- 9	11- 7
	Spruce-pine-fir	SS	5- 9	9- 1	11-11	15- 3	18- 1	5- 9	9- 1	11-11	14- 8	17- 1
	Spruce-pine-fir	#1	5- 8	8- 5	10- 8	13- 1	15- 2	5- 5	7-11	10- 1	12- 4	14- 3
	Spruce-pine-fir	#2	5- 8	8- 5	10- 8	13- 1	15- 2	5- 5	7-11	10- 1	12- 4	14- 3
	Spruce-pine-fir	#3	4- 4	6- 4	8- 1	9-10	11- 5	4- 1	6- 0	7- 7	9- 4	10- 9
19.2	Douglas fir-larch	SS	5-10	9- 3	12- 2	15- 4	17- 9	5-10	9- 3	11-10	14- 5	16- 9
	Douglas fir-larch	#1	5- 7	8- 3	10- 5	12- 9	14- 9	5- 4	7- 9	9-10	12- 0	13-11
	Douglas fir-larch	#2	5- 3	7- 8	9- 9	11-11	13-10	5- 0	7- 3	9- 2	11- 3	13- 0
	Douglas fir-larch	#3	4- 0	5-10	7- 4	9- 0	10- 5	3- 9	5- 6	6-11	8- 6	9-10
	Hem-fir	SS	5- 6	8- 8	11- 6	14- 8	17- 4	5- 6	8- 8	11- 6	14- 2	15-5
	Hem-fir	#1	5- 5	8- 0	10- 2	12- 5	14- 5	5- 2	7- 7	9- 7	11- 8	13- 7
	Hem-fir	#2	5- 2	7- 7	9- 7	11- 9	13- 7	4-11	7- 2	9- 1	11- 1	12-10
	Hem-fir	#3	4- 0	5-10	7- 4	9- 0	10- 5	3- 9	5- 6	6-11	8- 6	9-10
	Southern pine	SS	5- 9	9- 1	11-11	15- 3	18- 6	5- 9	9- 1	11-11	15- 3	18- 6
	Southern pine	#1	5- 8	8-11	11- 8	13-10	16- 6	5- 8	8- 9	11- 0	13- 1	15- 7
	Southern pine	#2	5- 6	8- 1	10- 5	12- 5	14- 7	5- 4	7- 7	9-10	11- 9	13- 9
	Southern pine	#3	4- 3	6- 3	8- 0	9- 5	11- 2	4- 0	5-11	7- 6	8-10	10- 7
	Spruce-pine-fir	SS	5- 5	8- 6	11- 3	14- 3	16- 6	5- 5	8- 6	11- 0	13- 5	15- 7
	Spruce-pine-fir	#1	5- 3	7- 8	9- 9	11-11	13-10	5- 0	7- 3	9- 2	11- 3	13- 0
	Spruce-pine-fir	#2	5- 3	7- 8	9- 9	11-11	13-10	5- 0	7- 3	9- 2	11- 3	13- 0
	Spruce-pine-fir	#3	4- 0	5-10	7- 4	9- 0	10- 5	3- 9	5- 6	6-11	8- 6	9-10

(continued)

TABLE R802.5.1(8)—continued
RAFTER SPANS FOR 70 PSF GROUND SNOW LOAD[a]
(Ceiling attached to rafters, L/Δ=240)

RAFTER SPACING (inches)	SPECIES AND GRADE		DEAD LOAD = 10 psf					DEAD LOAD = 20 psf				
			2x4	2x6	2x8	2x10	2x12	2x4	2x6	2x8	2x10	2x12
			Maximum rafter spans[a]									
			(feet - inches)	(feet - inches)	(feet - inches)	(feet - inches)	(feet - inches)	(feet - inches)	(feet - inches)	(feet - inches)	(feet - inches)	(feet - inches)
24	Douglas fir-larch	SS	5- 5	8- 7	11- 3	13- 9	15-11	5- 5	8- 4	10- 7	12-11	15- 0
	Douglas fir-larch	#1	5- 0	7- 4	9- 4	11- 5	13- 2	4- 9	6-11	8- 9	10- 9	12- 5
	Douglas fir-larch	#2	4- 8	6-11	8- 9	10- 8	12- 4	4- 5	6- 6	8- 3	10- 0	11- 8
	Douglas fir-larch	#3	3- 7	5- 2	6- 7	8- 1	9- 4	3- 4	4-11	6- 3	7- 7	8-10
	Hem-fir	SS	5- 2	8- 1	10- 8	13- 6	13-11	5- 2	8- 1	10- 5	12- 4	12-4
	Hem-fir	#1	4-11	7- 2	9- 1	11- 1	12-10	4- 7	6- 9	8- 7	10- 6	12- 2
	Hem-fir	#2	4- 8	6- 9	8- 7	10- 6	12- 2	4- 4	6- 5	8- 1	9-11	11- 6
	Hem-fir	#3	3- 7	5- 2	6- 7	8- 1	9- 4	3- 4	4-11	6- 3	7- 7	8-10
	Southern pine	SS	5- 4	8- 5	11- 1	14- 2	17- 2	5- 4	8- 5	11- 1	14- 2	17- 2
	Southern pine	#1	5- 3	8- 3	10- 5	12- 5	14- 9	5- 3	7-10	9-10	11- 8	13-11
	Southern pine	#2	5- 0	7- 3	9- 4	11- 1	13- 0	4- 9	6-10	8- 9	10- 6	12- 4
	Southern pine	#3	3- 9	5- 7	7- 1	8- 5	10- 0	3- 7	5- 3	6- 9	7-11	9- 5
	Spruce-pine-fir	SS	5- 0	7-11	10- 5	12- 9	14- 9	5- 0	7- 9	9-10	12- 0	12-11
	Spruce-pine-fir	#1	4- 8	6-11	8- 9	10- 8	12- 4	4- 5	6- 6	8- 3	10- 0	11- 8
	Spruce-pine-fir	#2	4- 8	6-11	8- 9	10- 8	12- 4	4- 5	6- 6	8- 3	10- 0	11- 8
	Spruce-pine-fir	#3	3- 7	5- 2	6- 7	8- 1	9- 4	3- 4	4-11	6- 3	7- 7	8-10

Check sources for availability of lumber in lengths greater than 20 feet.

For SI: 1 inch = 25.4 mm, 1 foot = 304.8 mm, 1 pound per square foot = 0.0479 kN/m^2.

a. The tabulated rafter spans assume that ceiling joists are located at the bottom of the attic space or that some other method of resisting the outward push of the rafters on the bearing walls, such as rafter ties, is provided at that location. When ceiling joists or rafter ties are located higher in the attic space, the rafter spans shall be multiplied by the factors given below:

H_C/H_R	Rafter Span Adjustment Factor
2/3 or greater	0.50
1/2	0.58
1/3	0.67
1/4	0.76
1/5	0.83
1/6	0.90
1/7.5 and less	1.00

where: H_C = Height of ceiling joists or rafter ties measured vertically above the top of the rafter support walls.

H_R = Height of roof ridge measured vertically above the top of the rafter support walls.

TABLE R802.5.1(9)
RAFTER/CEILING JOIST HEEL JOINT CONNECTIONS[a,b,c,d,e,f]

RAFTER SLOPE	RAFTER SPACING (inches)	GROUND SNOW LOAD (psf)											
		30 psf				50 psf				70 psf			
		Roof span (feet)											
		12	20	28	36	12	20	28	36	12	20	28	36
		Required number of 16d common nails[a,b] per heel joint splices[c,d,e,f]											
3:12	12	4	6	8	11	5	8	12	15	6	11	15	20
	16	5	8	11	14	6	11	15	20	8	14	20	26
	24	7	11	16	21	9	16	23	30	12	21	30	39
4:12	12	3	5	6	8	4	6	9	11	5	8	12	15
	16	4	6	8	11	5	8	12	15	6	11	15	20
	24	5	9	12	16	7	12	17	22	9	16	23	29
5:12	12	3	4	5	7	3	5	7	9	4	7	9	12
	16	3	5	7	9	4	7	9	12	5	9	12	16
	24	4	7	10	13	6	10	14	18	7	13	18	23
7:12	12	3	3	4	5	3	4	5	7	3	5	7	9
	16	3	4	5	6	3	5	7	9	4	6	9	11
	24	3	5	7	9	4	7	10	13	5	9	13	17
9:12	12	3	3	3	4	3	3	4	5	3	4	5	7
	16	3	3	4	5	3	4	5	7	3	5	7	9
	24	3	4	6	7	3	6	8	10	4	7	10	13
12:12	12	3	3	3	3	3	3	3	4	3	3	4	5
	16	3	3	3	4	3	3	4	5	3	4	5	7
	24	3	3	4	6	3	4	6	8	3	6	8	10

For SI: 1 inch = 25.4 mm, 1 foot = 304.8 mm, 1 pound per square foot = 0.0479 kN/m^2.

a. 40d box nails shall be permitted to be substituted for 16d common nails.

b. Nailing requirements shall be permitted to be reduced 25 percent if nails are clinched.

c. Heel joint connections are not required when the ridge is supported by a load-bearing wall, header or ridge beam.

d. When intermediate support of the rafter is provided by vertical struts or purlins to a loadbearing wall, the tabulated heel joint connection requirements shall be permitted to be reduced proportionally to the reduction in span.

e. Equivalent nailing patterns are required for ceiling joist to ceiling joist lap splices.

f. When rafter ties are substituted for ceiling joists, the heel joint connection requirement shall be taken as the tabulated heel joint connection requirement for two-thirds of the actual rafter-slope.

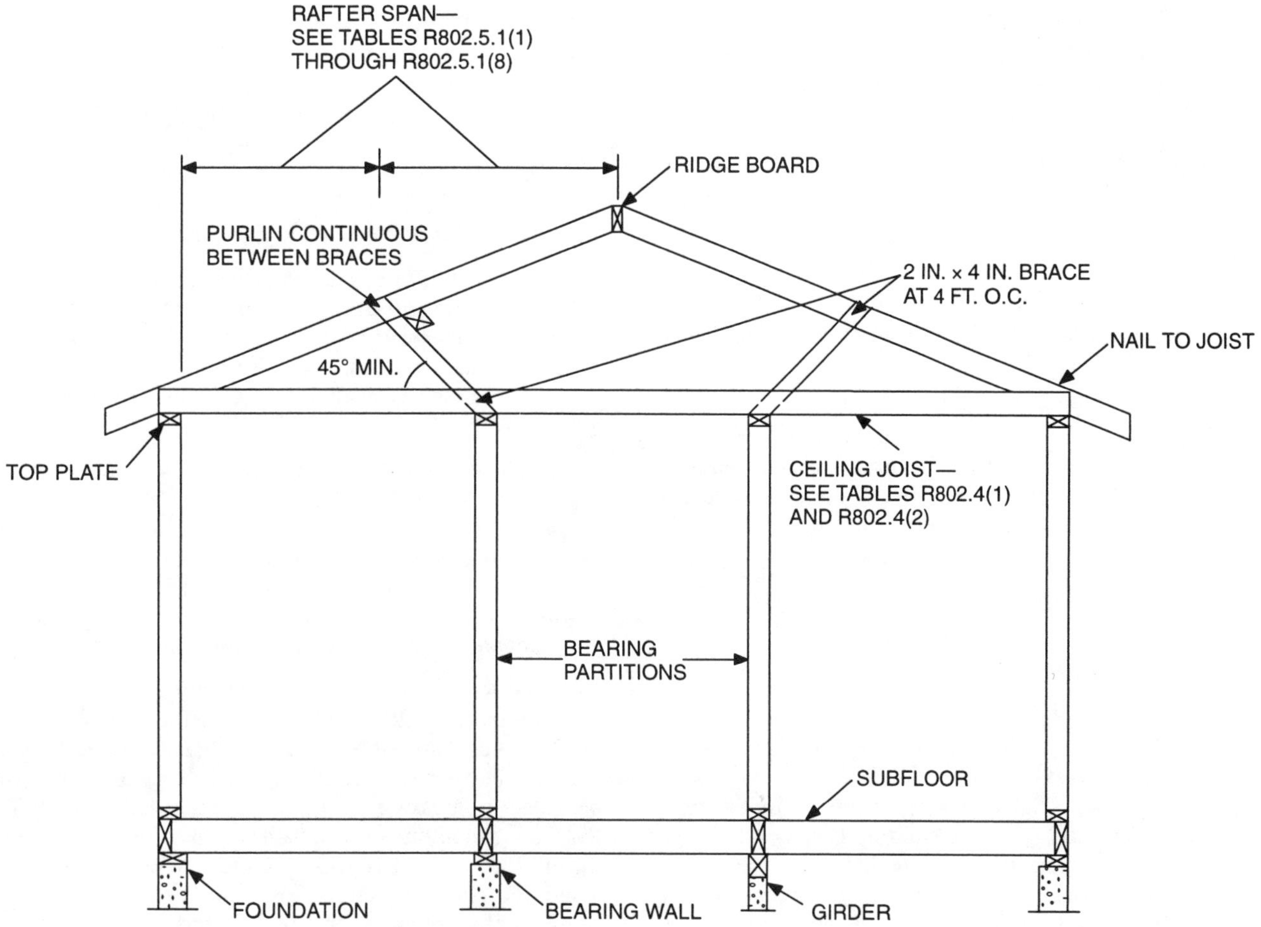

For SI: 1 inch = 25.4 mm, 1 foot = 304.8 mm, 1 degree = 0.018 rad.

Note: Where ceiling joints run perpendicular to the rafters, rafter ties shall be nailed to the rafter near the plate line and spaced not more than 4 feet on center.

FIGURE R802.5.1
BRACED RAFTER CONSTRUCTION

R802.8 Lateral support. Rafters and ceiling joists having a depth-to-thickness ratio exceeding 5 to 1 based on nominal dimensions shall be provided with lateral support at points of bearing to prevent rotation.

R802.8.1 Bridging. Rafters and ceiling joists having a depth-to-thickness ratio exceeding 6 to 1 based on nominal dimensions shall be supported laterally by solid blocking, diagonal bridging (wood or metal) or a continuous 1-inch by 3-inch (25.4 mm by 76 mm) wood strip nailed across the rafters or ceiling joists at intervals not exceeding 8 feet (2438 mm).

R802.9 Framing of openings. Openings in roof and ceiling framing shall be framed with header and trimmer joists. When the header joist span does not exceed 4 feet (1219 mm), the header joist may be a single member the same size as the ceiling joist or rafter. Single trimmer joists may be used to carry a single header joist that is located within 3 feet (914 mm) of the trimmer joist bearing. When the header joist span exceeds 4 feet (1219 mm), the trimmer joists and the header joist shall be doubled and of sufficient cross section to support the ceiling joists or rafter framing into the header. Approved hangers shall be used for the header joist to trimmer joist connections when the header joist span exceeds 6 feet (1829 mm). Tail joists over 12 feet (3658 mm) long shall be supported at the header by framing anchors or on ledger strips not less than 2 inches by 2 inches (51 mm by 51 mm).

R802.10 Wood trusses.

R802.10.1 Truss design drawings. Truss design drawings, prepared in conformance with Section R802.10.1, shall be provided to the building official and approved prior to installation. Truss design drawings shall include, at a minimum, the information specified below. Truss design drawing shall be provided with the shipment of trusses delivered to the jobsite.

1. Slope or depth, span and spacing.
2. Location of all joints.

3. Required bearing widths.
4. Design loads as applicable.
   4.1. Top chord live load (including snow loads).
   4.2. Top chord dead load.
   4.3. Bottom chord live load.
   4.4. Bottom chord dead load.
   4.5. Concentrated loads and their points of application.
   4.6. Controlling wind and earthquake loads.
5. Adjustments to lumber and joint connector design values for conditions of use.
6. Each reaction force and direction.
7. Joint connector type and description (e.g., size, thickness or gauge) and the dimensioned location of each joint connector except where symmetrically located relative to the joint interface.
8. Lumber size, species and grade for each member.
9. Connection requirements for:
   9.1. Truss to truss girder.
   9.2. Truss ply to ply.
   9.3. Field splices.
10. Calculated deflection ratio and/or maximum description for live and total load.
11. Maximum axial compression forces in the truss members to enable the building designer to design the size, connections and anchorage of the permanent continuous lateral bracing. Forces shall be shown on the truss design drawing or on supplemental documents.
12. Required permanent truss member bracing location.

R802.10.2 Design. Wood trusses shall be designed in accordance with accepted engineering practice. The design and manufacture of metal plate connected wood trusses shall comply with ANSI/TPI 1. The truss design drawings shall be prepared by a registered professional where required by the statutes of the jurisdiction in which the project is to be constructed in accordance with Section R106.1.

R802.10.3 Bracing. Trusses shall be braced to prevent rotation and provide lateral stability in accordance with the requirements specified in the construction documents for the building and on the individual truss design drawings. In the absence of specific bracing requirements, trusses shall be braced in accordance with TPI/HIB.

R802.10.4 Alterations to trusses. Truss members shall not be cut, notched, drilled, spliced or otherwise altered in any way without the approval of a registered design professional. Alterations resulting in the addition of load (e.g., HVAC equipment, water heater) that exceeds the design load for the truss shall not be permitted without verification that the truss is capable of supporting such additional loading.

R802.11 Roof tie-down. Roof assemblies subject to wind uplift pressures of 20 pounds per square foot (0.958 kN/m^2) or greater, as established in Table R301.2(2), adjusted for height and exposure per Table R301.2(3), shall have rafter or truss ties provided at bearing locations in accordance with Table R802.11. A continuous load path shall be provided to transmit the uplift forces from the rafter or truss ties to the foundation. Wind uplift pressure on roof assemblies shall be determined using an effective wind area of 100 square feet (9.3 m^2) and Zone 1 in Table R301.2(2).

TABLE R802.11
REQUIRED STRENGTH OF TRUSS OR RAFTER TIE-DOWN CONNECTIONS TO RESIST WIND UPLIFT FORCES[a,b,c,d]
(Pounds per tie-down connection)

DESIGN WIND LOAD (psf)[d]	TOTAL ROOF WIDTH INCLUDING OVERHANG (feet)				
	24	28	32	36	40
20	192	224	256	288	320
30	432	504	576	648	720
40	672	784	895	1,008	1,120
50	912	1,064	1,216	1,368	1,520
60	1,152	1,344	1,536	1,728	1,920

For SI: 1 inch = 25.4 mm, 1 foot = 304.8 mm, 1 pound per square foot = 0.0479 kN/m^2, 1 pound = 0.454 kg.

a. Wind uplift forces are based on 24-inch spacing of roof trusses or rafters. For spacing other than 24 inches, forces shall be adjusted accordingly.

b. Interpolation is permitted for intermediate values of wind uplift pressures and roof widths.

c. The rated capacity of approved tie-down devices is permitted to include up to a 60-percent increase for wind effects where allowed by material specifications.

d. Figure R301.2(4) and Table R301.2(2) shall be used in determining the design wind load.

SECTION R803
ROOF SHEATHING

R803.1 Lumber sheathing. Allowable spans for lumber used as roof sheathing shall conform to Table R803.1. Spaced lumber sheathing for wood shingle and shake roofing shall conform to the requirements of Sections R905.7 and R905.8. Spaced lumber sheathing is not allowed in Seismic Design Category D_2.

TABLE R803.1
MINIMUM THICKNESS OF LUMBER ROOF SHEATHING

RAFTER OR BEAM SPACING (inches)	MINIMUM NET THICKNESS (inches)
24	$^5/_8$
48[a]	$1^1/_2$ T & G
60[b]	
72[c]	

For SI: 1 inch = 25.4 mm.
a. Minimum 270 F_b, 340,000 E.
b. Minimum 420 F_b, 660,000 E.
c. Minimum 600 F_b, 1,150,000 E.

R803.2 Wood structural panel sheathing.

R803.2.1 Identification and grade. Wood structural panels shall conform to DOC PS 1, DOC PS 2 or, when manufactured in Canada, CSA 0437, and shall be identified by grade mark or certificate of inspection issued by an approved agency. Wood structural panels shall comply with the grades specified in Table R503.2.1.1(1).

R803.2.1.1 Exposure durability. All wood structural panels, when designed to be permanently exposed in outdoor applications, shall be of an exterior exposure durability. Wood structural panel roof sheathing exposed to the underside may be of interior type bonded with exterior glue, identified as Exposure 1.

R803.2.1.2 Fire-retardant-treated plywood. The allowable unit stresses for fire-retardant-treated plywood, including fastener values, shall be developed from an approved method of investigation that considers the effects of anticipated temperature and humidity to which the fire-retardant-treated plywood will be subjected, the type of treatment and redrying process. The fire-retardant-treated plywood shall be graded by an approved agency.

R803.2.2 Allowable spans. The maximum allowable spans for wood structural panel roof sheathing shall not exceed the values set forth in Table R503.2.1.1(1).

R803.2.3 Installation. Wood structural panel used as roof sheathing shall be installed with joints staggered or nonstaggered in accordance with Table R602.3(1), or APA E30 for wood roof framing or with Table R804.3 for steel roof framing.

SECTION R804
STEEL ROOF FRAMING

R804.1 General. Elements shall be straight and free of any defects that would significantly affect their structural performance. Cold-formed steel roof framing members shall comply with the requirements of this section.

R804.1.1 Applicability limits. The provisions of this section shall control the construction of steel roof framing for buildings not greater than 60 feet (18 288 mm) in length perpendicular to the joist, rafter or truss span, not greater than 36 feet (10 973 mm) in width parallel to the joist span or truss, not greater than two stories in height with each story not greater than 10 feet (3048 mm) high, and roof slopes not smaller than 3:12 (25-percent slope) or greater than 12:12 (100-percent slope). Steel roof framing constructed in accordance with the provisions of this section shall be limited to sites subjected to a maximum design wind speed of 130 miles per hour (209 km/h) Exposure A, B or C and a maximum ground snow load of 70 psf (3.35 kN/m^2).

R804.1.2 In-line framing. Steel roof framing constructed in accordance with Section R804 shall be located directly in-line with load-bearing studs below with a maximum tolerance of $^3/_4$ inch (19.1 mm) between the centerline of the stud and roof joist/rafter.

R804.2 Structural framing. Load-bearing steel roof framing members shall comply with Figure R804.2(1) and the dimensional and minimum thickness requirements specified in Tables R804.2(1) and R804.2(2). Tracks shall comply with Figure R804.2(2) and shall have a minimum flange width of $1^1/_4$ inches (32 mm). The maximum inside bend radius for load-bearing members shall be the greater of $^3/_{32}$ inch (2.4 mm) or twice the uncoated steel thickness. Holes in roof framing members shall not exceed 1.5 inches (38 mm) in width or 4 inches (102 mm) in length as shown in Figure R804.2(3). Holes shall be permitted only along the centerline of the web of the framing member. Holes shall not be less than 24 inches (610 mm) center-to-center and shall not be located less than 10 inches (254 mm) from the edge of the hole to the edge of the bearing surface or support unless patched in accordance with Section R804.3.6.

R804.2.1 Material. Load-bearing steel framing members shall be cold-formed to shape from structural quality sheet steel complying with the requirements of one of the following:

1. ASTM A 653; Grades 33, 37, 40 and 50 (Classes 1 and 3).
2. ASTM A 792; Grades 33, 37, 40 and 50A.
3. ASTM A 875; Grades 33, 37, 40 and 50 (Classes 1 and 3).
4. Steels that comply with ASTM A 653, except for tensile and elongation, shall be permitted provided the ratio of tensile strength to yield point is at least 1.08 and the total elongation is at least 10 percent for a 2-inch (51 mm) gage length or 7 percent for an 8-inch (203 mm) gage length.

R804.2.2 Identification. Load-bearing steel framing members shall have a legible label, stencil, stamp or embossment with the following information as a minimum:

1. Manufacturer's identification.
2. Minimum uncoated steel thickness in inches (mm).
3. Minimum coating designation.
4. Minimum yield strength, in kips per square inch (ksi).

TABLE R804.2(1)
LOAD-BEARING COLD-FORMED STEEL MEMBER SIZES

NOMINAL MEMBER SIZE MEMBER DESIGNATION[a]	WEB DEPTH (inches)	MINIMUM FLANGE WIDTH (inches)	MAXIMUM FLANGE WIDTH (inches)	MINIMUM LIP SIZE (inches)
350S162-t	3.5	1.625	2	0.5
550S162-t	5.5	1.625	2	0.5
800S162-t	8	1.625	2	0.5
1000S162-t	10	1.625	2	0.5
1200S162-t	12	1.625	2	0.5

For SI: 1 inch = 25.4 mm.

a. The member designation is defined by the first number representing the member depth in $^1/_{100}$ inch, the letter "s" representing a stud or joist member, the second number representing the flange width in $^1/_{100}$ inch, and the letter "t" shall be a number representing the minimum base metal thickness in mils [see Table R804.2(2)].

TABLE R804.2(2)
MINIMUM THICKNESS OF COLD-FORMED STEEL ROOF FRAMING MEMBERS

DESIGNATION (mils)	MINIMUM UNCOATED THICKNESS (inches)	REFERENCED GAGE NUMBER
33	0.033	20
43	0.043	18
54	0.054	16
68	0.068	14

For SI: 1 inch = 25.4 mm, 1 mil = 0.0254 mm.

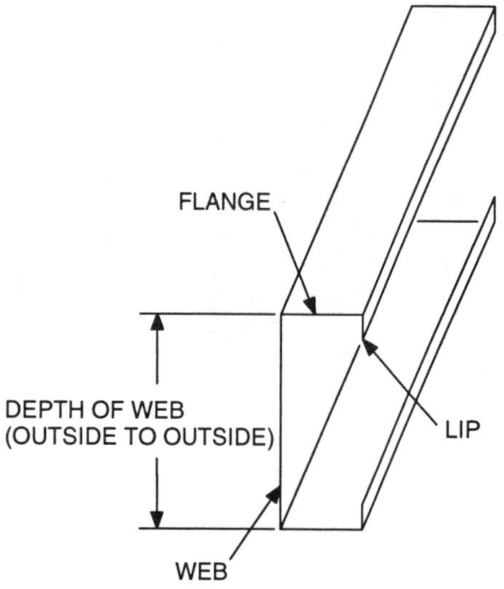

FIGURE R804.2(1)
C-SECTION

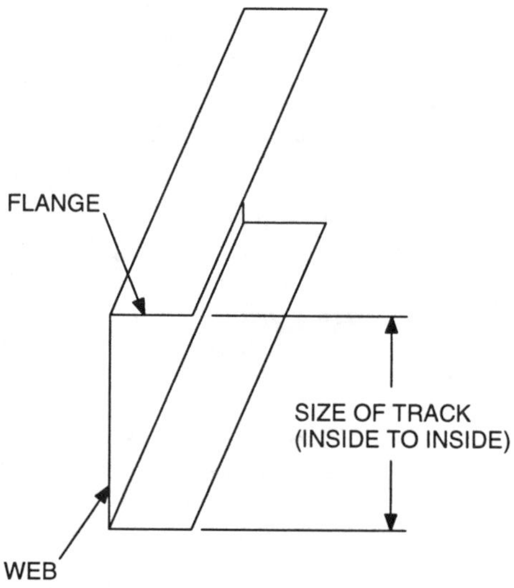

FIGURE R804.2(2)
TRACK SECTION

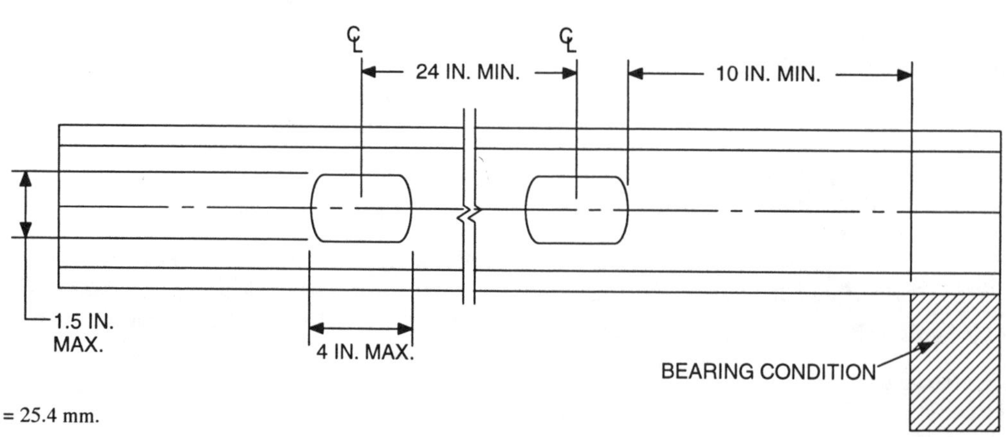

For SI: 1 inch = 25.4 mm.

FIGURE R804.2(3)
WEB HOLES

R804.2.3 Corrosion protection. Load-bearing steel framing shall have a metallic coating complying with one of the following:

1. A minimum of G 60 in accordance with ASTM A 653.
2. A minimum of AZ 50 in accordance with ASTM A 792.
3. A minimum of GF 60 in accordance with ASTM A 875.

R804.2.4 Fastening requirements. Screws for steel-to-steel connections shall be installed with a minimum edge distance and center-to-center spacing of $^1/_2$ inch (12.7 mm), shall be self-drilling tapping, and shall conform to SAE J78. Structural sheathing shall be attached to roof rafters with minimum No. 8 self-drilling tapping screws that conform to SAE J78. Screws for attaching structural sheathing to steel roof framing shall have a minimum head diameter of 0.292 inch (7.4 mm) with countersunk heads and shall be installed with a minimum edge distance of $^3/_8$ inch (9.5 mm). Gypsum board ceilings shall be attached to steel joists with minimum No. 6 screws conforming to ASTM C 954 and shall be installed in accordance with Section R805. For all connections, screws shall extend through the steel a minimum of three exposed threads. All self-drilling tapping screws conforming to SAE J78 shall have a minimum Type II coating in accordance with ASTM B 633.

Where No. 8 screws are specified in a steel-to-steel connection, the required number of screws in the connection is permitted to be reduced in accordance with the reduction factors in Table R804.2.4 when larger screws are used or when one of the sheets of steel being connected is thicker that 33 mils (0.83 mm). When applying the reduction factor, the resulting number of screws shall be rounded up.

TABLE R804.2.4
SCREW SUBSTITUTION FACTOR

SCREW SIZE	THINNEST CONNECTED STEEL SHEET (mils)	
	33	43
#8	1.0	0.67
#10	0.93	0.62
#12	0.86	0.56

For SI: 1 mil = 0.0254 mm.

R804.3 Roof construction. Steel roof systems constructed in accordance with the provisions of this section shall consist of both ceiling joists and rafters in accordance with Figure R804.3 and fastened in accordance with Table R804.3.

R804.3.1 Allowable ceiling joist spans. The clear span of cold-formed steel ceiling joists shall not exceed the limits set forth in Table R804.3.1(1) or R804.3.1(2). Ceiling joists shall have a minimum bearing length of 1.5 inches (38 mm) and shall be connected to rafters (heel joint) in accordance with Figure R804.3.1(1) and Table R804.3.1(3). When continuous joists are framed across interior bearing supports, the interior bearing supports shall be located within 24 inches (610 mm) of midspan of the ceiling joist, and the individual spans shall not exceed the applicable spans in Table R804.3.1(1) or R804.3.1(2). Where required in Table R804.3.1(1) or R804.3.1(2), bearing stiffeners shall be installed at each bearing location in accordance with Section R804.3.8 and Figure R804.3.8. When the attic is to be used as an occupied space, the ceiling joists shall be designed in accordance with Section R505.

TABLE R804.3
ROOF FRAMING FASTENING SCHEDULE[a,b]

DESCRIPTION OF BUILDING ELEMENTS	NUMBER AND SIZE OF FASTENERS	SPACING OF FASTENERS
Ceiling joist to top track of load-bearing wall	2 No. 10 screws	Each joist
Roof sheathing (oriented strand board or plywood) to rafters	No. 8 screws	6″ o.c. on edges and 12″ o.c. at interior supports. 6″ o.c. at gable end truss
Truss to bearing wall[a]	2 No. 10 screws	Each truss
Gable end truss to endwall top track	No. 10 screws	12″ o.c.
Rafter to ceiling joist	Minimum No. 10 screws, per Table R804.3.1(3)	Evenly spaced, less than ½″ from all edges.

For SI: 1 inch = 25.4 mm, 1 foot = 304.8 mm, 1 pound per square foot = 0.0479 kN/m^2, 1 mil = 0.0254 mm.

a. Screws shall be applied through the flanges of the truss or ceiling joist or a 54 mil clip angle shall be used with two No. 10 screws in each leg. See Section R804.4 for additional requirements to resist uplift forces.

b. Spacing of fasteners on roof sheathing panel edges applies to panel edges supported by framing members and at all roof plane perimeters. Blocking of roof sheathing panel edges perpendicular to the framing members shall not be required except at the intersection of adjacent roof planes. Roof perimeter shall be supported by framing members or cold-formed blocking of the same depth and gauge as the floor members.

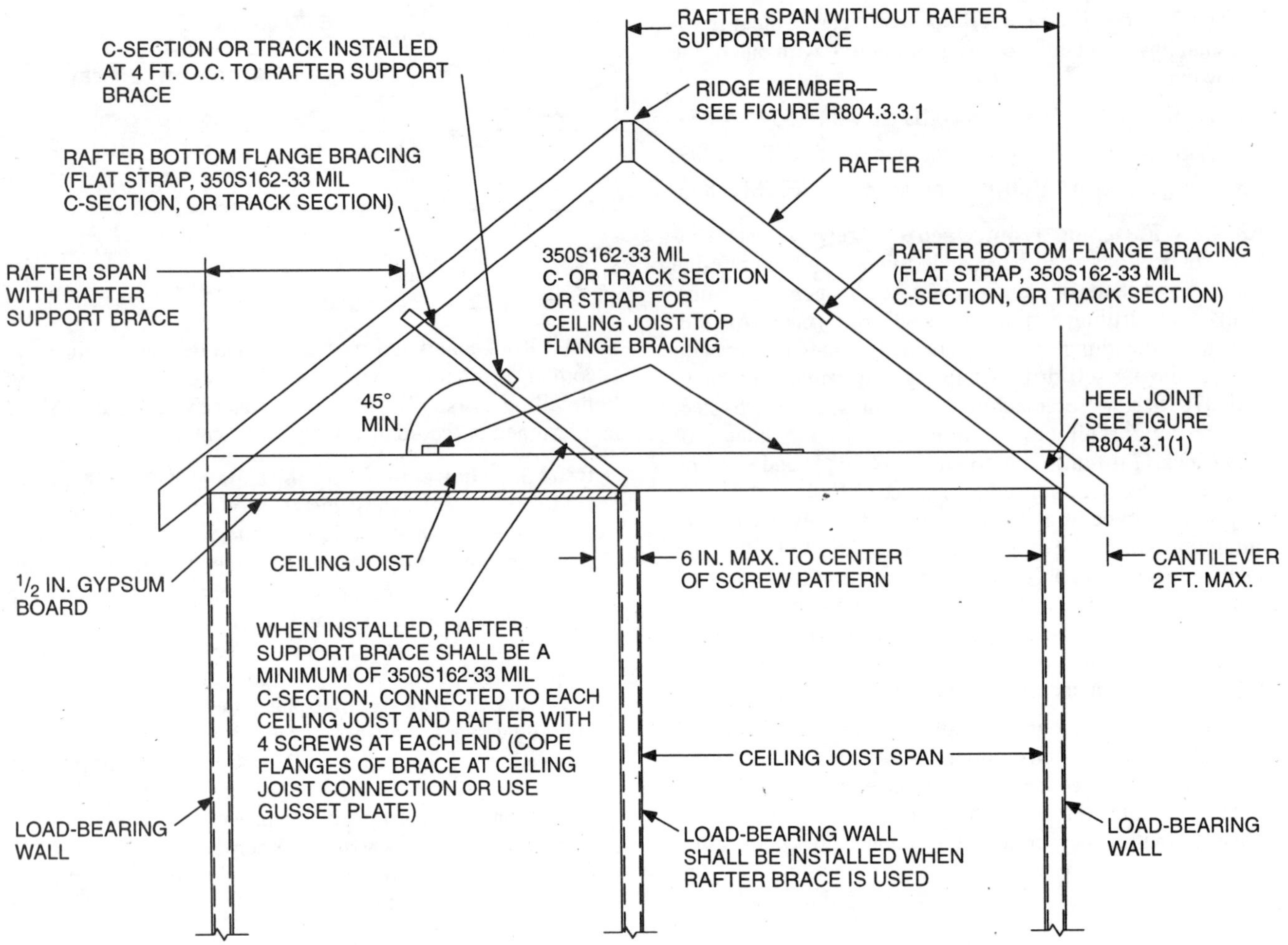

For SI: 1 inch = 25.4 mm, 1 foot = 304.8 mm, 1 mil = 0.0254 mm.

FIGURE R804.3
STEEL ROOF CONSTRUCTION

TABLE R804.3.1(1)
ALLOWABLE SPANS FOR COLD-FORMED STEEL CEILING JOISTS[a,b,c] 10 psf live load (no attic storage)

NOMINAL JOIST SIZE	LATERAL SUPPORT OF TOP (COMPRESSION) FLANGE					
	Unbraced		Mid-span bracing		Third-point bracing	
	Spacing (inches)		Spacing (inches)		Spacing (inches)	
	16	24	16	24	16	24
350S162-33	9′-2″	8′-3″	11′-9″	10′-1″	11′-9″	10′-4″
350S162-43	9′-11″	8′-10″	12′-10″	11′-2″	12′-10″	11′-2″
350S162-54	10′-8″	9′-6″	13′-9″	12′-0″	13′-9″	12′-0″
350S162-68	11′-7″	10′-4″	14′-8″	12′-10″	14′-8″	12′-10″
550S162-33	10′-5″	9′-5″	14′-5″	12′-8″[c]	16′-4″	13′-10″[c]
550S162-43	11′-2″	10′-1″	15′-7″	13′-10″	18′-0″	15′-5″
550S162-54	12′-0″	10′-9″	16′-7″	14′-9″	19′-5″	16′-8″
550S162-68	12′-11″	11′-7″	17′-8″	15′-10″	20′-11″	18′-1″
800S162-33	11′-8″[c]	10′-6″[c]	16′-5″[c]	14′-9″[c]	19′-5″[c]	16′-7″[c]
800S162-43	12′-6″	11′-3″	17′-6″	15′-10″	21′-2″	18′-7″
800S162-54	13′-4″	11′-11″	18′-7″	16′-9″	22′-7″	20′-0″
800S162-68	14′-3″	12′-9″	19′-8″	17′-8″	23′-11″	21′-4″
1000S162-43	13′-4″[c]	12′-1″[c]	18′-9″[c]	16′-11″[c]	22′-11″[c]	20′-6″[c]
1000S162-54	14′-2″	12′-9″	19′-10″	17′-10″	24′-2″	21′-9″
1000S162-68	15′-2″	13′-7″	21′-0″	18′-11″	25′-6″	23′-0″
1200S162-43	14′-1″[c]	12′-8″[c]	19′-10″[c]	17′-11″[c]	24′-3″[c]	21′-6″[c]
1200S162-54	15′-0″[c]	13′-5″[c]	20′-11″[c]	18′-11″[c]	25′-7″[c]	23′-1″[c]
1200S162-68	15′-11″	14′-4″	22′-2″	19′-11″	27′-0″	24′-4″

For SI: 1 inch = 25.4 mm, 1 foot = 304.8 mm, 1 pound per square foot = 0.0479 kN/m^2.

a. Deflection criteria: *L*/240 for total loads.

b. Ceiling dead load = 5 psf.

c. Bearing stiffeners are required at all bearing points and concentrated load locations.

TABLE R804.3.1(2)
ALLOWABLE SPANS FOR COLD-FORMED STEEL CEILING JOISTS[a,b,c]
20 psf live load (Limited attic storage where development of future rooms is not possible)

NOMINAL JOIST SIZE	LATERAL SUPPORT OF TOP FLANGE					
	Unbraced		Mid-span bracing		Third-point bracing	
	Spacing (inches)		Spacing (inches)		Spacing (inches)	
	16	24	16	24	16	24
350S162-33	8′-0″	6′-0″	9′-8″	6′-0″	9′-0″	6′-0″
350S162-43	8′-8″	7′-8″	10′-9″	9′-1″	10′-10″	9′-5″
350S162-54	9′-3″	8′-3″	11′-7″	9′-11″	11′-7″	10′-1″
350S162-68	10′-0″	8′-11″	12′-5″	10′-10″	12′-5″	10′-10″
550S162-33	9′-2″	6′-0″	12′-2″[c]	10′-5″[c]	13′-3″	11′-0″[c]
550S162-43	9′-10″	8′-10″	13′-4″	11′-6″	14′-9″	12′-5″
550S162-54	10′-5″	9′-5″	14′-4″	12′-6″	16′-1″	13′-7″
550S162-68	11′-3″	10′-0″	15′-4″	13′-5″	17′-5″	14′-10″
800S162-33	10′-3″[c]	9′-3″[c]	14′-4″[c]	12′-5″[c]	15′-11″[c]	13′-4″[c]
800S162-43	10′-11″	9′-10″	15′-5″	13′-8″[c]	17′-11″[c]	15′-5″[c]
800S162-54	11′-8″	10′-6″	16′-3″	14′-7″	19′-3″	16′-8″
800S162-68	12′-5″	11′-2″	17′-3″	15′-6″	20′-7″	18′-0″
1000S162-43	11′-9″	10′-7″[c]	16′-6″[c]	14′-10″[c]	19′-10″[c]	17′-1″[c]
1000S162-54	12′-5″	11′-2″	17′-5″	15′-8″	21′-1″	18′-7″
1000S162-68	13′-3″	11′-10″	18′-5″	16′-7″	22′-4″	19′-11″
1200S162-43	12′-5″[c]	11′-2″[c]	17′-5″[c]	15′-8″[c]	20′-9″[c]	18′-0″[c]
1200S162-54	13′-1″[c]	11′-9″[c]	18′-5″[c]	16′-7″[c]	22′-5″[c]	20′-1″[c]
1200S162-68	13′-11″	12′-6″	19′-5″	17′-6″	23′-8″	21′-3″

For SI: 1 inch = 25.4 mm, 1 foot = 304.8 mm, 1 pound per square foot = 0.0479 kN/m^2.

a. Deflection criteria: *L*/240 for total loads.

b. Ceiling dead load = 5 psf.

c. Bearing stiffeners are required at all bearing points and concentrated load locations.

TABLE R804.3.1(3)
NUMBER OF SCREWS REQUIRED FOR CEILING JOIST TO RAFTER CONNECTION[a]

ROOF SLOPE	BUILDING WIDTH (feet)															
	24				28				32				36			
	Ground snow load (psf)				Ground snow load (psf)				Ground snow load (psf)				Ground snow load (psf)			
	20	30	50	70	20	30	50	70	20	30	50	70	20	30	50	70
3/12	5	6	9	12	6	7	10	13	7	8	12	15	8	9	13	17
4/12	4	5	7	9	5	6	8	10	6	6	9	12	6	7	10	13
5/12	4	4	6	7	4	5	7	9	5	5	8	10	5	6	9	11
6/12	3	4	5	7	4	4	6	8	4	5	7	9	4	5	7	10
7/12	3	3	5	6	3	4	5	7	4	4	6	8	4	5	7	9
8/12	3	3	4	5	3	3	5	6	3	4	5	7	4	4	6	8
9/12	2	3	4	5	3	3	4	6	3	4	5	6	3	4	6	7
10/12	2	3	4	5	3	3	4	5	3	3	5	6	3	4	5	7
11/12	2	3	4	4	3	3	4	5	3	3	5	6	3	4	5	6
12/12	2	3	3	4	2	3	4	5	3	3	4	6	3	4	5	6

For SI: 1 foot = 304.8 mm, 1 pound per square fot = 0.0479 kN/m^2.

a. Screws shall be No. 10 minimum.

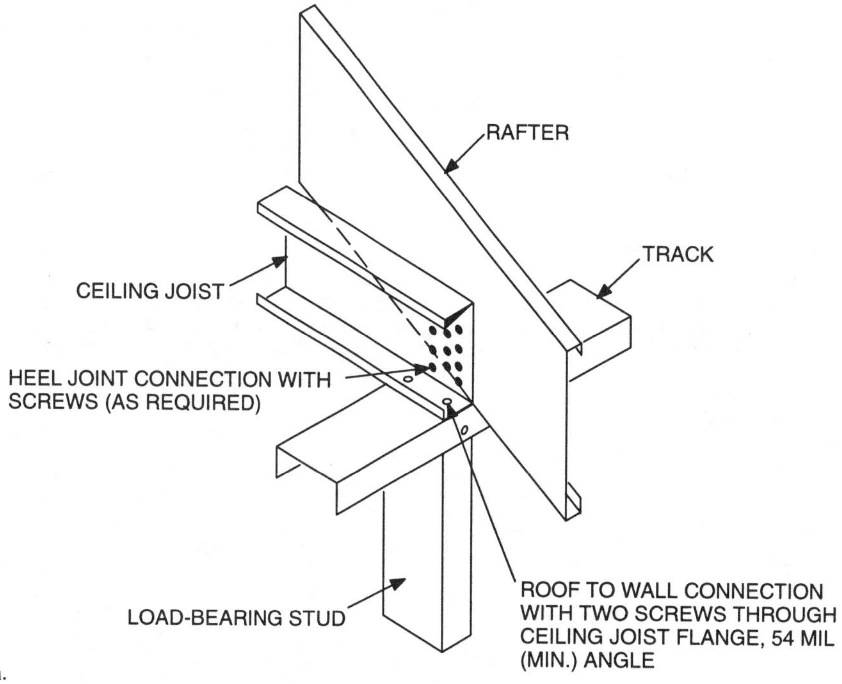

For SI: 1 mil = 0.0254 mm.

FIGURE R804.3.1(1)
JOIST TO RAFTER CONNECTION

R804.3.2 Ceiling joist bracing. The bottom flanges of steel ceiling joists shall be laterally braced in accordance with Section R702. The top flanges of steel ceiling joists shall be laterally braced with a minimum of 33 mil (0.84 mm) C-section, 33 mil (0.84 mm) track section, or $1^1/_2$ inch by 33 mil (38 mm by 0.84 mm) continuous steel strapping as required in Table R804.3.1(1) or R804.3.1(2). Lateral bracing shall be installed in accordance with Figure R804.3. C-section, tracks or straps shall be fastened to the top flange at each joist with at least one No. 8 screw and shall be fastened to blocking with at least two No. 8 screws. Blocking or bridging (X-bracing) shall be installed between joists in-line with strap bracing at a maximum spacing of 12 feet (3658 mm) measured perpendicular to the joists, and at the termination of all straps. The third point bracing span values from Table R804.3.1(1) or R804.3.1(2) shall be used for straps installed at closer spacings than third point bracing, or when sheathing is applied to the top of the ceiling joists.

R804.3.3 Allowable rafter spans. The horizontal projection of the rafter span, as shown in Figure R804.3, shall not exceed the limits set forth in Table R804.3.3(1). Wind speeds shall be converted to equivalent ground snow loads in accordance with Table R804.3.3(2). Rafter spans shall be selected based on the higher of the ground snow load or the equivalent snow load converted from the wind speed. When required, a rafter support brace shall be a minimum of 350S162-33 C-section with maximum length of 8 feet (2438 mm) and shall be connected to a ceiling joist and rafter with four No. 10 screws at each end.

R804.3.3.1 Rafter framing. Rafters shall be connected to a parallel ceiling joist to form a continuous tie between exterior walls in accordance with Figures R804.3 and R804.3.1(1) and Table R804.3.1(3). Rafters shall be connected to a ridge member with a minimum 2-inch by 2-inch (51 mm by 51 mm) clip angle fastened with minimum No. 10 screws to the ridge member in accordance with Figure R804.3.3.1 and Table R804.3.3.1. The clip angle shall have a minimum steel thickness as the rafter member and shall extend the full depth of the rafter member. The ridge member shall be fabricated from a C-section and a track section, which shall be of a minimum size and steel thickness as the adjacent rafters and shall be installed in accordance with Figure R804.3.3.1.

TABLE R804.3.3(1)
ALLOWABLE HORIZONTAL RAFTER SPANS[a,b]

NOMINAL RAFTER SIZE	GROUND SNOW LOAD							
	20 psf		30 psf		50 psf		70 psf	
	Spacing (inches-feet)		Spacing (inches-feet)		Spacing (inches-feet)		Spacing (inches-feet)	
	16	24	16	24	16	24	16	24
550S162-33	12′-8″	10′-4″	11′-9″	9′-7″	9′-11″	8′-1″	8′-10″	7′-2″
550S162-43	15′-5″	12′-7″	14′-3″	11′-8″	12′-1″	9′-10″	10′-8″	8′-9″
550S162-54	13′-0″	14′-2″	16′-1″	13′-1″	13′-8″	11′-2″	12′-1″	9′-10″
550S162-68	18′-1″	15′-10″	17′-3″	14′-9″	15′-4″	12′-6″	13′-6″	11′-1″
800S162-33	15′-5″	11′-5″	14′-4″	9′-10″	10′-7″	7′-1″	8′-3″	5′-6″
800S162-43	19′-1″	15′-7″	17′-9″	14′-6″	15′-1″	12′-3″	13′-3″	10′-9″
800S162-54	22′-7″	18′-5″	21′-0″	17′-1″	17′-9″	14′-6″	15′-9″	12′-10″
800S162-68	24′-7″	20′-9″	23′-4″	19′-3″	20′-0″	16′-4″	17′-8″	14′-5″
1000S162-43	21′-2″	17′-3″	19′-8″	16′-0″	16′-8″	13′-1″	14′-9″	10′-3″
1000S162-54	25′-1″	20′-6″	23′-3″	19′-0″	19′-9″	16′-1″	17′-5″	14′-3″
1000S162-68	29′-6″	24′-6″	27′-9″	22′-9″	23′-8″	19′-3″	21′-0″	17′-1″
1200S162-43	23′-0″	18′-2″	21′-4″	15′-7″	16′-9″	11′-3″	13′-2″	8′-9″
1200S162-54	27′-3″	22′-3″	25′-3″	20′-7″	21′-5″	17′-6″	18′-11″	15′-5″
1200S162-68	32′-1″	26′-2″	29′-9″	24′-3″	25′-3″	20′-7″	22′-4″	18′-2″

For SI: 1 inch = 25.4 mm, 1 foot = 304.8 mm, 1 pound per square foot = 0.0479 kN/m^2.

a. Deflection criteria: *L*/240 for live loads and *L*/180 for total loads.

b. Roof dead load = 12 pounds per square foot.

TABLE R804.3.3(2)
BASIC WIND SPEED TO EQUIVALENT SNOW LOAD CONVERSION[a]

BASIC WIND SPEED AND EXPOSURE		EQUIVALENT GROUND SNOW LOAD (psf)									
		Roof slope									
Exp. A/B	Exp. C	3:12	4:12	5:12	6:12	7:12	8:12	9:12	10:12	11:12	12:12
85 mph	—	20	20	20	20	20	20	30	30	30	30
100 mph	85 mph	20	20	20	20	30	30	30	30	50	50
110 mph	100 mph	20	20	20	20	30	50	50	50	50	50
120 mph	110 mph	30	30	30	50	50	50	70	70	70	—
130 mph	120 mph	30	50	50	50	70	70	70	—	—	—
—	130 mph	50	50	50	70	70	—	—	—	—	—

For SI: 1 mile per hour = 0.447 m/s, 1 pound per square foot = 0.0479 kN/m^2.

a. In areas where the basic wind speed equals or exceeds 110 miles per hour the equivalent snow load shall be used only to determine the size of members. Connections of rafters to the ridge and the roof members to walls shall comply with Sections R804.3.3.1.1 and R804.4.1.

R804.3.3.1.1 High wind ridge tension connections. In areas where the basic wind speed is 110 miles per hour (177 km/h) or greater, roof rafters shall be provided with a connection at the ridge line to transfer tension loads. The ridge connection shall be capable of resisting the unit loads listed in Table R804.3.3.1.1(1) multiplied by the appropriate spacing multiplier. Alternatively, a steel ridge strap shall be provided with minimum No. 8 screws on each end of the strap as required in Table R804.3.3.1.1(1). The number of screws shall be increased to account for the spacing multipliers shown in the table. The width and thickness of the steel ridge strap shall be as shown in Table R804.3.3.1.1(2), based upon the required number of screws on one side of the strap.

TABLE R804.3.3.1
NUMBER OF SCREWS REQUIRED AT EACH LEG OF CLIP ANGLE FOR RAFTER TO RIDGE MEMBER CONNECTION[a]

BUILDING WIDTH (feet)	GROUND SNOW LOAD (psf)			
	0 to 20	21 to 30	31 to 50	51 to 70
24	2	3	4	4
28	2	3	4	5
32	3	3	4	5
36	3	4	5	6

For SI: 1 foot = 304.8 mm, 1 pound per square foot = 0.0479 kN/m^2.

a. Screws shall be No. 10 minimum.

TABLE R804.3.3.1.1(1)
RIDGE TENSION STRAP CONNECTION REQUIREMENTS PER FOOT OF RIDGE SPAN[c]

Roof pitch	Roof span (feet)	BASIC WIND SPEED (mph)			BASIC WIND SPEED (mph)		
		110	120	130	110	120	130
		Number of #8 screws in each end of a steel ridge strap [Table R804.3.3.1.1(2)]			Required ridge construction capacity[a,b] (plf)		
3:12	24	2	2	3	475	624	788
	28	2	3	3	554	727	919
	32	2	3	3	633	831	1051
	36	3	3	4	712	935	1182
4:12	24	2	2	3	378	495	765
	28	2	2	3	441	577	727
	32	2	2	3	504	659	831
	36	2	4	3	567	742	935
5:12	24	1	2	2	289	380	480
	28	1	2	2	338	443	560
	32	2	2	2	386	507	640
	36	2	2	3	434	570	720
6:12	24	1	1	2	262	342	431
	28	1	2	2	306	399	503
	32	2	2	2	350	457	575
	36	2	2	2	393	514	647
7:12-12:12	24	1	2	2	246	320	401
	28	1	2	2	287	373	468
	32	1	2	2	328	426	535
	36	2	3	2	369	480	602
Framing spacing		12 in.	16 in.	24 in.	19.2 in.	—	—
Multiplier		1.0	1.33	2.0	1.6	—	—

For SI: 1 inch = 25.4 mm, 1 foot = 304.8 mm, 1 pound force = 4.45 N, 1 mile per hour = 0.447 m/s, 1 pound per square foot = 0.0479 kN/m^2, 1 pound per linear foot = 14.60 N/m.

a. Connection requirements are based on a foot assembly dead load of 10 pounds per square foot.

b. Connection capacities shown in the table are based on a 12-inch ridge strap spacing. For spacing greater than 12 inches, capacity values shall be increased using the multipliers in the table above.

c. The required number of screws shown in the table are based on a 12-inch strap spacing. For spacing other than 12 inches, the appropriate connection capacity in the table shall be increased using the multipliers above and dividing by the screw shear value of 263 pounds = 350 pounds/screw.

TABLE R804.3.3.1.1(2)
MINIMUM SIZE OF RIDGE STRAP

STRAP WIDTH (inches)	MINIMUM THICKNESS OF RIDGE STRAP (inches) REQUIRED NUMBER OF RIDGE STRAP SCREWS[a]					
	4 or less	5	6	7	8	9
1.25	0.043	0.054	0.054	—	—	—
1.5	0.043	0.043	0.054	0.054	—	—
1.75	0.043	0.043	0.043	0.054	0.054	—
2	0.043	0.043	0.043	0.043	0.054	0.054

For SI: 1 inch = 25.4 mm.

a. Required number of screws per Table R804.3.3.1.1(1) on each end of the ridge strap.

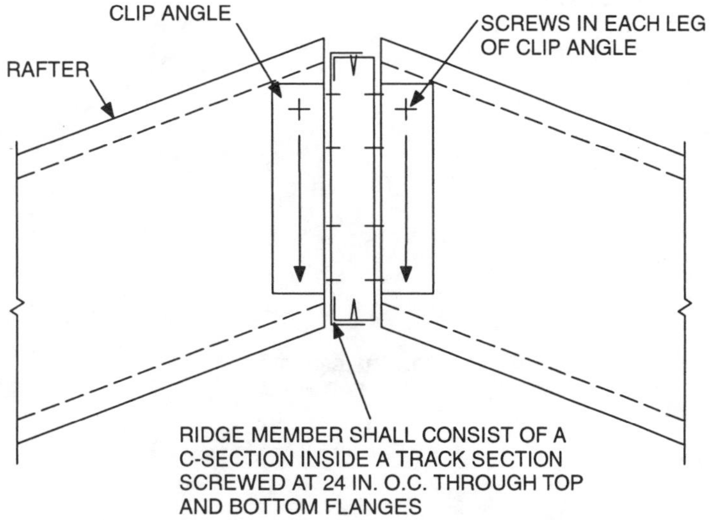

For SI: 1 inch = 25.4 mm.

FIGURE R804.3.3.1
RIDGE BOARD CONNECTION

R804.3.3.2 Roof cantilevers. Roof cantilevers shall not exceed 24 inches (610 mm) in accordance with Figure R804.3. Roof cantilevers shall be supported by a header in accordance with Section R603.6 or shall be supported by the floor framing in accordance with Section R505.3.7.

R804.3.4 Rafter bottom flange bracing. The bottom flanges of steel rafters shall be continuously braced with a minimum 33-mil (0.84 mm) C-section, 33-mil (0.84 mm) track section, or a $1^1/_2$-inch by 33-mil (38 mm by 0.84 mm) steel strapping at a maximum spacing of 8 feet (2438 mm) as measured parallel to the rafters. Bracing shall be installed in accordance with Figure R804.3. The C-section, track section, or straps shall be fastened to blocking with at least two No. 8 screws. Blocking or bridging (X-bracing) shall be installed between rafters in-line with the continuous bracing at a maximum spacing of 12 feet (3658 mm) measured perpendicular to the rafters and at the termination of all straps. The ends of continuous bracing shall be fastened to blocking with at least two No. 8 screws.

R804.3.5 Cutting and notching. Flanges and lips of load-bearing steel roof framing members shall not be cut or notched. Holes in webs shall be in accordance with Section R804.2.

R804.3.6 Hole patching. Holes in ceiling joist and rafters with dimensions conforming to Section R804.2 that are closer than 10 inches (254 mm) from the edge of the hole to the edge of bearing surface shall be patched with a solid steel plate, C-section or track section in accordance with Figure R804.3.6. The steel patch shall be of a minimum thickness as the receiving member and shall extend at least 1 inch (25.4 mm) beyond all edges of the hole. The steel patch shall be fastened to the web with No. 8 screws (minimum) spaced no greater than 1 inch (25.4 mm) center-to-center along the edges of the patch, with a minimum edge distance of $^1/_2$ inch (12.7 mm).

R804.3.7 Splicing. Rafters and other structural members, except ceiling joists, shall not be spliced. Splices in ceiling joists shall only be permitted at interior bearing points and shall be constructed in accordance with Figure R804.3.7(1). Spliced ceiling joists shall be connected with the same number and size of screws on connection. Splicing of tracks shall conform with Figure R804.3.7(2).

R804.3.8 Bearing stiffener. A bearing stiffener shall be fabricated from a minimum 33-mil (0.84 mm) C-section or track section. Each stiffener shall be fastened to the web of the ceiling joist with a minimum of four No. 8 screws equally spaced as shown in Figure R804.3.8. Stiffeners shall extend across the full depth of the web and shall be installed on either side of the web.

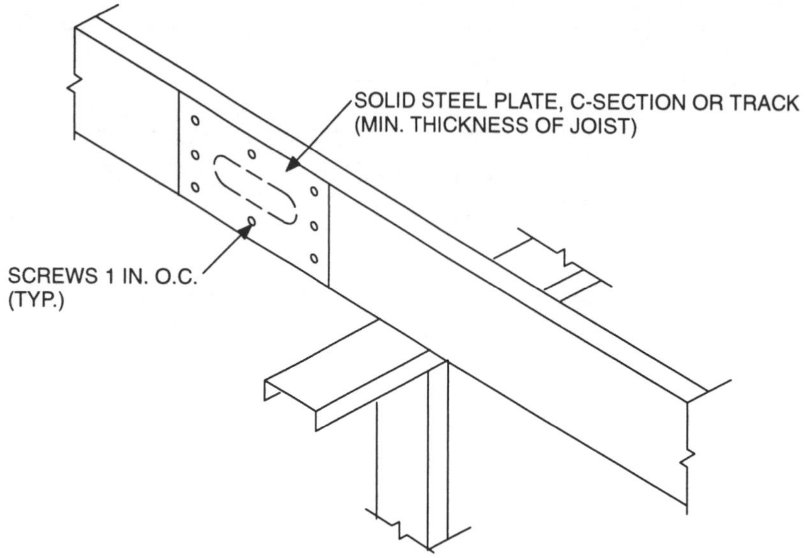

For SI: 1 inch = 25.4 mm.

FIGURE R804.3.6
HOLE PATCHING

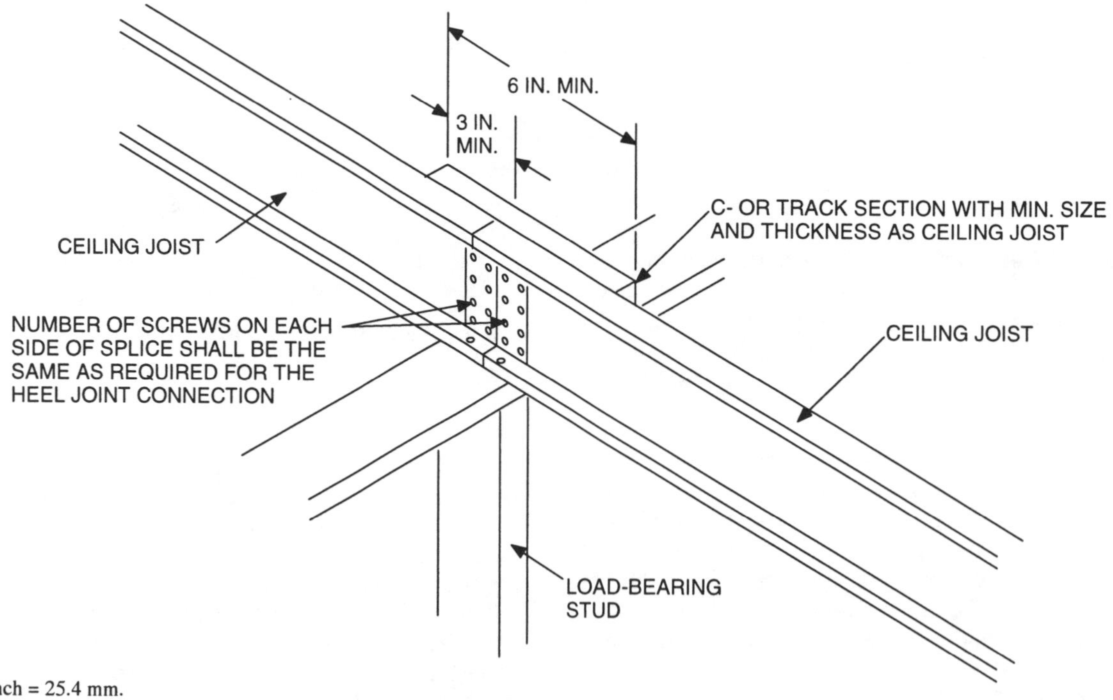

For SI: 1 inch = 25.4 mm.

FIGURE R804.3.7(1)
SPLICED CEILING JOISTS

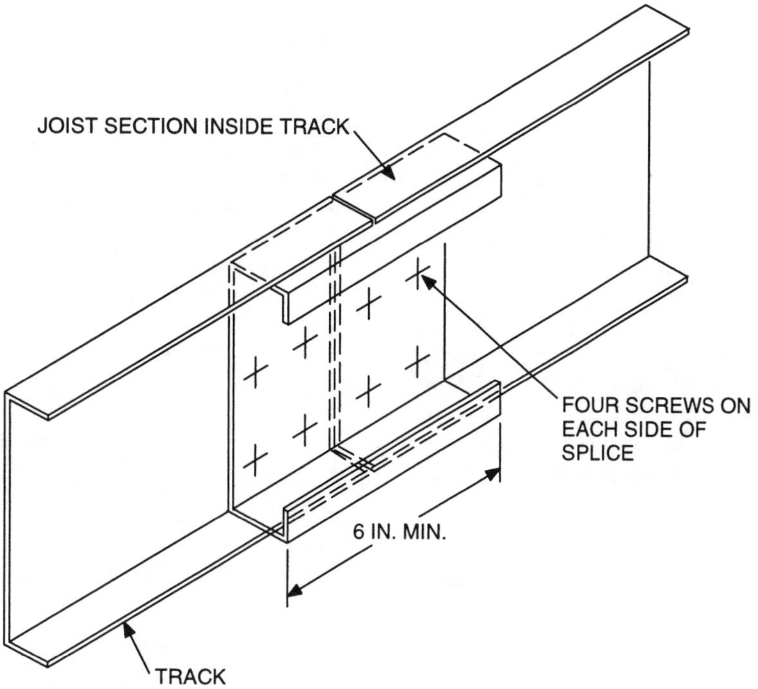

For SI: 1 inch = 25.4 mm.

FIGURE R804.3.7(2)
TRACK SPLICE

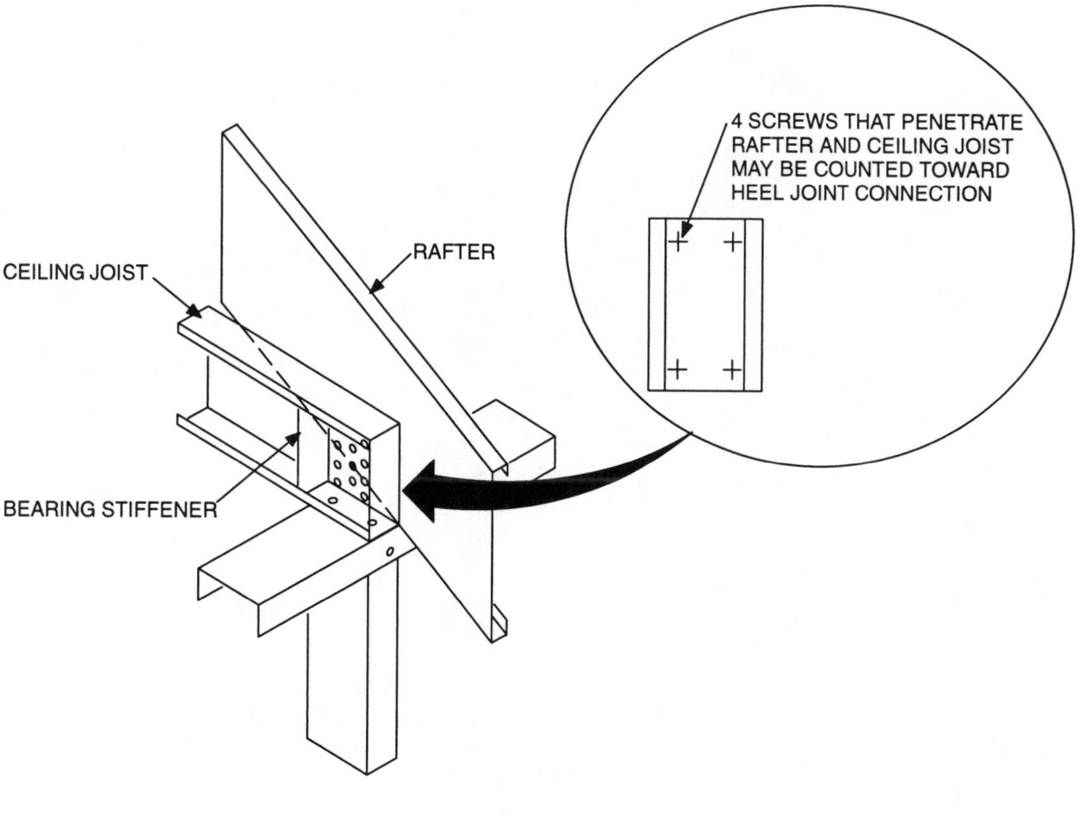

FIGURE R804.3.8
BEARING STIFFENER

R804.3.9 Headers. Roof-ceiling framing above wall openings shall be supported on headers. The allowable spans for headers in bearing walls shall not exceed the values set forth in Table R603.6(1).

R804.3.10 Framing of opening. Openings in roof and ceiling framing shall be framed with headers and trimmers between ceiling joists or rafters. Header joist spans shall not exceed 4 feet (1219 mm). Header and trimmer joists shall be fabricated from joist and track sections, which shall be of a minimum size and thickness in accordance with Figures R804.3.10(1) and R804.3.10(2). Each header joist shall be connected to trimmer joist with a minimum of four 2-inch by 2-inch (51 by 51 mm) clip angles. Each clip angle shall be fastened to both the header and trimmer joists with four No. 8 screws, evenly spaced, through each leg of the clip angle. The clip angles shall have a steel thickness not less than that of the floor joist.

R804.4 Roof tie-down. Roof assemblies subject to wind uplift pressures of 20 pounds per square foot (0.96 kN/m^2) or greater, as established in Table R301.2(2), shall have rafter-to-bearing wall ties provided in accordance with Table R802.11.

R804.4.1 High wind roof tie-down. In areas where the basic wind speed is 110 miles per hour (177 km/h) or greater, roof rafters or trusses shall be attached to their supporting wall assemblies by connections capable of resisting the uplift loads listed in Table R603.8.3.2.1(1). Alternatively, a 1$^1/_4$-inch (32 mm) by minimum 33-mil (0.84 mm) steel uplift strap connecting the rafter or truss to the in-line framing stud below shall be permitted. Each end of the uplift strap shall be fastened with minimum No. 8 screws as required by Table R603.8.3.2.1(2).

SECTION R805
CEILING FINISHES

R805.1 Ceiling installation. Ceilings shall be installed in accordance with the requirements for interior wall finishes as provided in Section R702.

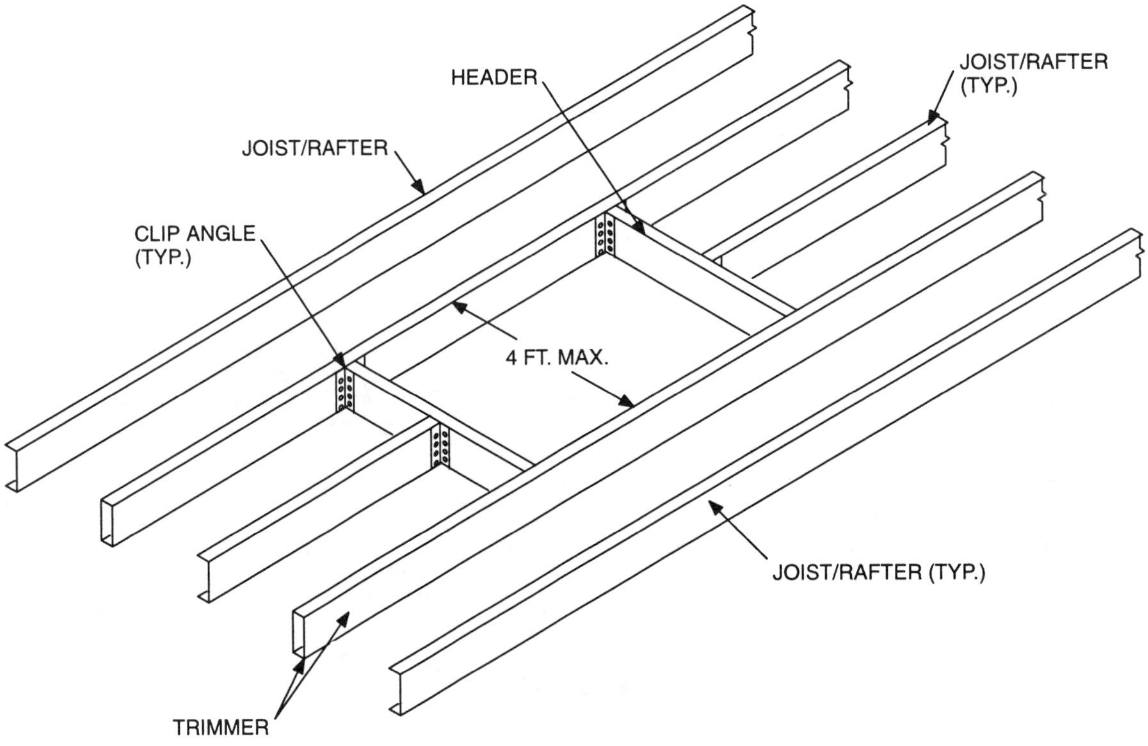

For SI: 1 foot = 304.8 mm.

FIGURE R804.3.10(1)
ROOF OPENING

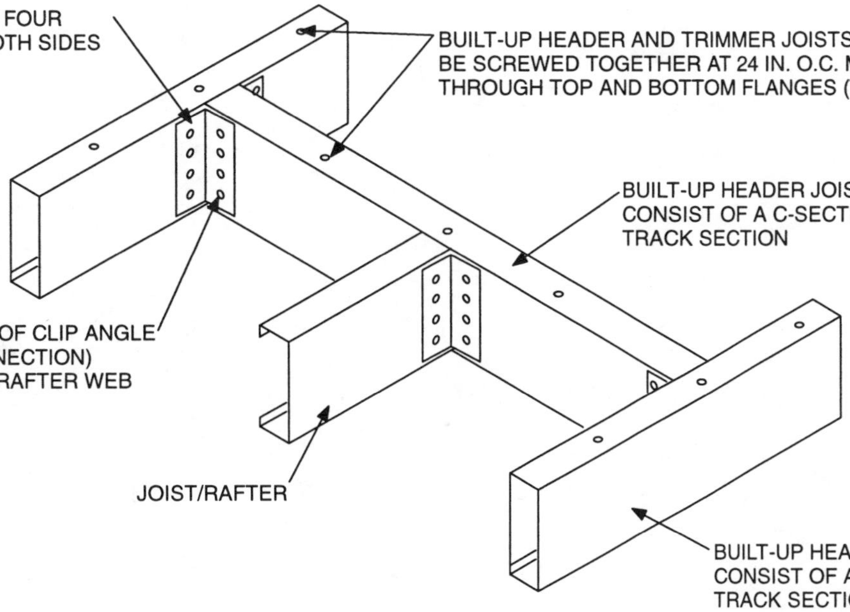

For SI: 1 inch = 25.4 mm.

FIGURE R804.3.10(2)
HEADER TO TRIMMER CONNECTION

SECTION R806 ROOF VENTILATION

R806.1 Ventilation required. Enclosed attics and enclosed rafter spaces formed where ceilings are applied directly to the underside of roof rafters shall have cross ventilation for each separate space by ventilating openings protected against the entrance of rain or snow. Ventilating openings shall be provided with corrosion-resistant wire mesh, with $^1/_8$ inch (3.2 mm) minimum to $^1/_4$ inch (6.35 mm) maximum openings.

R806.2 Minimum area. The total net free ventilating area shall not be less than 1 to 150 of the area of the space ventilated except that the total area is permitted to be reduced to 1 to 300, provided at least 50 percent and not more than 80 percent of the required ventilating area is provided by ventilators located in the upper portion of the space to be ventilated at least 3 feet (914 mm) above eave or cornice vents with the balance of the required ventilation provided by eave or cornice vents. As an alternative, the net free cross-ventilation area may be reduced to 1 to 300 when a vapor barrier having a transmission rate not exceeding 1 perm (57.4 mg/s · m^2 · Pa) is installed on the warm side of the ceiling.

R806.3 Vent clearance. Where eave or cornice vents are installed, insulation shall not block the free flow of air. A minimum of a 1-inch (25.4 mm) space shall be provided between the insulation and the roof sheathing at the location of the vent.

SECTION R807 ATTIC ACCESS

R807.1 Attic access. In buildings with combustible ceiling or roof construction, an attic access opening shall be provided to attic areas that exceed 30 square feet (2.8 m^2) and have a vertical height of 30 inches (762 mm) or greater.

The rough-framed opening shall not be less than 22 inches by 30 inches (559 mm by 762 mm) and shall be located in a hallway or other readily accessible location. A 30-inch (762 mm) minimum unobstructed headroom in the attic space shall be provided at some point above the access opening. See Section M1305.1.3 for access requirements where mechanical equipment is located in attics.

SECTION R808 INSULATION CLEARANCE

R808.1 Combustible insulation. Combustible insulation shall be separated a minimum of 3 inches (76 mm) from recessed lighting fixtures, fan motors and other heat-producing devices.

Exception: When heat-producing devices are listed for lesser clearances, combustible insulation complying with the listing requirements shall be separated in accordance with the conditions stipulated in the listing.

Recessed lighting fixtures installed in the building thermal envelope shall meet the requirements of Section N1101.3.

CHAPTER 9
ROOF ASSEMBLIES

SECTION R901
GENERAL

R901.1 Scope. The provisions of this chapter shall govern the design, materials, construction and quality of roof assemblies.

SECTION R902
ROOF CLASSIFICATION

R902.1 Roofing covering materials. Roofs shall be covered with materials as set forth in Sections R904 and R905. Class A, B or C roofing shall be installed in areas designated by law as requiring their use or when the edge of the roof is less than 3 feet (914 mm) from a property line. Classes A, B and C roofing required to be listed by this section shall be tested in accordance with UL 790 or ASTM E 108. Roof assemblies with coverings of brick, masonry, slate, clay or concrete roof tile, exposed concrete roof deck, ferrous or copper shingles or sheets, and metal sheets and shingles, shall be considered Class A roof coverings.

R902.2 Wood shingles and shakes. When testing wood shingles and shakes in accordance with ASTM E 108 (including the rain test) and ASTM D 2898, the fire tests shall include the intermittent flame test, spread of flame test, burning brand test and flying brand test; additionally, at the conclusion of the rain test, test panels shall be subjected to the intermittent flame test, burning brand test and flying brand test.

R902.2.1 Fire-retardant-treated shingles and shakes. Fire-retardant-treated wood shakes and shingles shall be treated by impregnation with chemicals by the full-cell vacuum-pressure process, in accordance with AWPA C1. Each bundle shall be marked to identify the manufactured unit and the manufacturer, and shall also be labeled to identify the classification of the material in accordance with the testing required in Section R902.2, the treating company and the quality control agency.

SECTION R903
WEATHER PROTECTION

R903.1 General. Roof decks shall be covered with approved roof coverings secured to the building or structure in accordance with the provisions of this chapter. Roof assemblies shall be designed and installed in accordance with this code and the approved manufacturer's installation instructions such that the roof assembly shall serve to protect the building or structure.

R903.2 Flashing. Flashings shall be installed in such a manner as to prevent moisture entering the wall through the joints in the coping, through moisture permeable material, at intersections with the roof plane or at parapet wall penetrations.

R903.2.1 Locations. Flashings shall be installed at wall and roof intersections; wherever there is a change in roof slope or direction; and around roof openings. Where flashing is of metal, the metal shall be corrosion-resistant with a thickness of not less than 0.019 inch (No. 26 galvanized sheet).

R903.3 Coping. Parapet walls shall be properly coped with noncombustible, weatherproof materials of a width no less than the thickness of the parapet wall.

R903.4 Roof drainage. Unless roofs are sloped to drain over roof edges, roof drains shall be installed at each low point of the roof. Where required for roof drainage, scuppers shall be placed level with the roof surface in a wall or parapet. The scupper shall be located as determined by the roof slope and contributing roof area.

R903.4.1 Overflow drains and scuppers. Where roof drains are required, overflow drains having the same size as the roof drains shall be installed with the inlet flow line located 2 inches (51 mm) above the low point of the roof, or overflow scuppers having three times the size of the roof drains and having a minimum opening height of 4 inches (102 mm) may be installed in the adjacent parapet walls with the inlet flow located 2 inches (51 mm) above the low point of the adjacent roof. The installation and sizing of overflow drains, leaders and conductors shall comply with the *International Plumbing Code*.

Overflow drains shall discharge to an approved location and shall not be connected to roof drain lines.

SECTION R904
MATERIALS

R904.1 Scope. The requirements set forth in this section shall apply to the application of roof covering materials specified herein. Roof assemblies shall be applied in accordance with this chapter and the manufacturer's installation instructions. Installation of roof assmblies shall comply with the applicable provisions of Section R905.

R904.2 Compatibility of materials. Roof assemblies shall be of materials that are compatible with each other and with the building or structure to which the materials are applied.

R904.3 Material specifications and physical characteristics. Roof covering materials shall conform to the applicable standards listed in this chapter. In the absence of applicable standards or where materials are of questionable suitability, testing by an approved testing agency shall be required by the building offical to determine the character, quality and limitations of application of the materials.

R904.4 Product identification. Roof covering materials shall be delivered in packages bearing the manufacturer's identifying marks and approved testing agency labels when required. Bulk shipments of materials shall be accompanied with the same information issued in the form of a certificate or on a bill of lading by the manufacturer.

SECTION R905
REQUIREMENTS FOR ROOF COVERINGS

R905.1 Roof covering application. Roof coverings shall be applied in accordance with the applicable provisions of this section and the manufacturer's installation instructions.

R905.2 Asphalt shingles. The installation of asphalt shingles shall comply with the provisions of this section.

R905.2.1 Sheathing requirements. Asphalt shingles shall be fastened to solidly sheathed decks.

R905.2.2 Slope. Asphalt shingles shall only be used on roof slopes of two units vertical in 12 units horizontal (2:12) or greater. For roof slopes from two units vertical in 12 units horizontal (2:12) up to four units vertical in 12 units horizontal (4:12), double underlayment application is required in accordance with Section R905.2.7.

R905.2.3 Underlayment. Unless otherwise noted, required underlayment shall conform with ASTM D 226, Type I, or ASTM D 4869, Type I.

Self-adhering polymer modified bitumen sheet shall comply with ASTM D 1970.

R905.2.4 Asphalt shingles. Asphalt shingles shall have self-seal strips or be interlocking, and comply with ASTM D 225 or D 3462.

R905.2.5 Fasteners. Fasteners for asphalt shingles shall be galvanized steel, stainless steel, aluminum or copper roofing nails, minimum 12 gage [0.105 inch (2.67 mm)] shank with a minimum $^{3}/_{8}$-inch (9.5 mm) diameter head, ASTM F 1667, of a length to penetrate through the roofing materials and a minimum of $^{3}/_{4}$ inch (19.1 mm) into the roof sheathing. Where the roof sheathing is less than $^{3}/_{4}$ inch (19.1 mm) thick, the fasteners shall penetrate through the sheathing. Fasteners shall comply with ASTM F 1667.

R905.2.6 Attachment. Asphalt shingles shall have the minimum number of fasteners required by the manufacturer. For normal application, asphalt shingles shall be secured to the roof with not less than four fasteners per strip shingle or two fasteners per individual shingle. Where the roof slope exceeds 20 units vertical in 12 units horizontal (20:12), special methods of fastening are required.

Exception: Asphalt strip shingles shall have a minimum of six fasteners per shingle where the roof is in one of the following categories:

1. The basic wind speed per Figure R301.2(4) is 110 miles per hour (177 km/h) or greater and the eave is 20 feet (6096 mm) or higher above grade.
2. The basic wind speed per Figure R301.2(4) is 120 miles per hour (193 km/h) or greater.
3. Special wind zones per Figure R301.2(4).

R905.2.7 Underlayment application. For roof slopes from two units vertical in 12 units horizontal (17-percent slope), up to four units vertical in 12 units horizontal (33-percent slope), underlayment shall be two layers applied in the following manner. Apply a 19-inch (483 mm) strip of underlayment felt parallel with and starting at the eaves, fastened sufficiently to hold in place. Starting at the eave, apply 36-inch-wide (914 mm) sheets of underlayment, overlapping successive sheets 19 inches (483 mm), and fastened sufficiently to hold in place. For roof slopes of four units vertical in 12 units horizontal (33-percent slope) or greater, underlayment shall be one layer applied in the following manner. Underlayment shall be applied shingle fashion, parallel to and starting from the eave and lapped 2 inches (51 mm), fastened sufficiently to hold in place. End laps shall be offset by 6 feet (1829 mm).

R905.2.7.1 Ice protection. In areas where the average daily temperature in January is 25°F (-4°C) or less, an ice barrier that consists of a least two layers of underlayment cemented together or of a self-adhering polymer modified bitumen sheet, shall be used in lieu of normal underlayment and extend from the eave's edge to a point at least 24 inches (610 mm) inside the exterior wall line of the building.

R905.2.7.2 Underlayment and high wind. Underlayment applied in areas subject to high winds [greater than 110 mph (177km/h) per Figure R301.2(4)] shall be applied with corrosion-resistant fasteners in accordance with manufacturer's installation instructions. Fasteners are to be applied along the overlap not farther apart than 36 inches (914 mm) on center.

R905.2.8 Flashing. Flashing for asphalt shingles shall comply with this section.

R905.2.8.1 Base and cap flashing. Base and cap flashing shall be installed in accordance with manufacturer's installation instructions. Base flashing shall be of either corrosion-resistant metal of minimum nominal 0.019-inch (0.483 mm) thickness or mineral surface roll roofing weighing a minimum of 77 pounds per 100 square feet (3.76 kg/m^{2}). Cap flashing shall be corrosion-resistant metal of minimum nominal 0.019-inch (0.483 mm) thickness.

R905.2.8.2 Valleys. Valley linings shall be installed in accordance with manufacturer's installation instructions before applying shingles. Valley linings of the following types shall be permitted:

1. For open valley (valley lining exposed) lined with metal, the valley lining shall be at least 24 inches (610 mm) wide and of any of the corrosion-resistant metals in Table R905.2.8.2.
2. For open valleys, valley lining of two plies of mineral surface roll roofing, complying with ASTM D 249, shall be permitted. The bottom layer shall be 18 inches (457 mm) and the top layer a minimum of 36 inches (914 mm) wide.
3. For closed valleys (valley covered with shingles), valley lining of one ply of smooth roll roofing complying with ASTM D 224 Type II or Type III and at least 36 inches (914 mm) wide or valley lining as described in Items 1 and 2 above shall be permitted. Specialty underlayment complying with ASTM D 1970 may be used in lieu of the lining material.

TABLE R905.2.8.2
VALLEY LINING MATERIAL

MATERIAL	MINIMUM THICKNESS (inches)	GAGE	WEIGHT (pounds)
Copper	—	—	1
Aluminum	0.024	—	—
Stainless steel	—	28	—
Galvanized steel	0.0179	26 (zinc coated G90)	—
Zinc alloy	0.027	—	
Lead	—	—	$2^1/_2$
Painted terne	—	—	20

For SI: 1 inch = 25.4 mm, 1 pound = 0.454 kg.

R905.2.8.3 Crickets and saddles. A cricket or saddle shall be installed on the ridge side of any chimney greater than 30 inches (762 mm) wide. Cricket or saddle coverings shall be sheet metal or of the same material as the roof covering.

R905.2.8.4 Sidewall flashing. Flashing against a vertical sidewall shall be by the step-flashing method.

R905.2.8.5 Other flashing. Flashing against a vertical front wall, as well as soil stack, vent pipe and chimney flashing, shall be applied according to asphalt shingle manufacturer's printed instructions.

R905.3 Clay and concrete tile. The installation of clay and concrete shall comply with the provisions of this section. Clay roof tile shall comply with ASTM C1167.

R905.3.1 Deck requirements. Concrete and clay tile shall be installed only over solid sheathing or spaced structural sheathing boards.

R905.3.2 Deck slope. Clay and concrete roof tile shall be installed on roof slopes of two and one-half units vertical in 12 units horizontal ($2^1/_2$:12) or greater. For roof slopes from two and one-half units vertical in 12 units horizontal ($2^1/_2$:12) to four units vertical in 12 units horizontal (4:12), double underlayment application is required in accordance with Section R905.3.3.

R905.3.3 Underlayment. Unless otherwise noted, required underlayment shall conform with ASTM D 226, Type II; ASTM D 2626, Type I; or ASTM D 249 mineral surfaced roll roofing.

R905.3.3.1 Low slope roofs. For roof slopes from two and one half units vertical in 12 units horizontal ($2^1/_2$:12), up to four units vertical in 12 units horizontal (4:12), underlayment shall be a minimum of two layers underlayment applies as follows:

1. Starting at the eave, a 19-inch (483 mm) strip of underlayment shall be applied parallel with the eave and fastened sufficiently in place.
2. Starting at the eave, 36-inch-wide (914 mm) strips of underlayment felt shall be applied, overlapping successive sheets 19 inches (483 mm), and fastened sufficiently in place.

R905.3.3.2 High slope roofs. For roof slopes of four units vertical in 12 units horizontal (4:12) or greater, underlayment shall be a minimum of one layer of underlayment felt applied shingle fashion, parallel to and starting from the eaves and lapped 2 inches (51 mm), fastened sufficiently in place.

R905.3.3.3 Underlayment and high wind. Underlayment applied in areas subject to high wind [greater than 110 miles per hour (177 km/h) per Figure R301.2(4)] shall be applied with corrosion-resistant fasteners in accordance with manufacturer's installation instructions. Fasteners are to be applied along the overlap not farther apart than 36 inches (914 mm) on center.

R905.3.4 Tile. Clay roof tile shall comply with ASTM C 1167.

R905.3.5 Concrete tile. Concrete roof tiles shall be in accordance with the physical test requirements as follows:

1. The transverse strength of tiles shall be determined according to Section 5.3 of ASTM C 1167 and in accordance with Table R905.3.5
2. The absorption of concrete roof tiles shall be according to Section 8 of ASTM C 140. Roof tiles shall absorb not more than 15 percent of the dry weight of the tile during a 24-hour immersion test.
3. Roof tiles shall be tested for freeze/thaw resistance according to Section 8 of ASTM C 67. Roof tiles shall show no breakage and not have more than 1-percent loss in dry weight of any individual concrete roof tile.

TABLE R905.3.5
TRANSVERSE BREAKING STRENGTH OF CONCRETE ROOF TILE

TILE PROFILE	DRY (pounds)	
	AVERAGE OF 5 TILES	INDIVIDUAL TILE
High profile	400	350
Medium profile	300	250
Flat profile	300	250

For SI: 1 pound = 4.45 N.

R905.3.6 Fasteners. Nails shall be corrosion-resistant and not less than 11 gage, $^5/_{16}$-inch (10.6 mm) head, and of sufficient length to penetrate the deck a minimum of $^3/_4$ inch (19.1 mm) or through the thickness of the deck, whichever is less. Attaching wire for clay or concrete tile shall not be smaller than 0.083 inch (2.1 mm). Perimeter fastening areas include three tile courses but not less than 36 inches (914 mm) from either side of hips or ridges and edges of eaves and gable rakes.

R905.3.7 Application. Tile shall be applied in accordance with this chapter and the manufacturer's installation instructions, based on the following:

1. Climatic conditions.
2. Roof slope.
3. Underlayment system.
4. Type of tile being installed.

Clay and concrete roof tiles shall be fastened in accordance with this section and the manufacturer's installation instructions. Perimeter tiles shall be fastened with a minimum of one fastener per tile. Tiles with installed weight less than 9 pounds per square foot (0.43 kN/m^2) require a minimum of one fastener per tile regardless of roof slope. Clay and concrete roof tile attachment shall be in accordance with the manufacturer's installation instructions where applied in areas where the wind speed exceeds 100 miles per hour (161 km/h) and on buildings where the roof is located more than 40 feet (12 192 mm) above grade. In areas subject to snow, a minimum of two fasteners per tile is required. In all other areas, clay and concrete roof tiles shall be attached in accordance with Table R905.3.7.

TABLE R905.3.7
CLAY AND CONCRETE TILE ATTACHMENT

SHEATHING	ROOF SLOPE	NUMBER OF FASTENERS
Solid without battens	All	One per tile
Spaced or solid with battens and slope < 5:12	Fasteners not required	—
Spaced sheathing without battens	5:12 ≤ slope < 12:12	One per tile/every other row
	12:12 ≤ slope < 24:12	One per tile

R905.3.8 Flashing. At the juncture of roof vertical surfaces, flashing and counter flashing shall be provided in accordance with this chapter and the manufacturer's installation instructions and, where of metal, shall not be less than 0.019 inch (0.48 mm) (No. 26 galvanized sheet gage) corrosion-resistant metal. The valley flashing shall extend at least 11 inches (279 mm) from the center line each way and have a splash diverter rib not less than 1 inch (25.4 mm) high at the flow line formed as part of the flashing. Sections of flashing shall have an end lap of not less than 4 inches (102 mm). For roof slopes of 3:12 and over, valley flashing shall have a 36-inch-wide (914 mm) underlayment of one layer of Type I underlayment running the full length of the valley, in addition to other required underlayment. In areas where the average daily temperature in January is 25°F (-4°C) or less, metal valley flashing underlayment shall be solid cemented to the roofing underlayment for slopes under 7:12.

R905.4 Metal roof shingles. The installation of metal roof shingles shall comply with the provisions of this section.

R905.4.1 Deck requirements. Metal roof shingles shall be applied to a solid or closely fitted deck, except where the roof covering is specifically designed to be applied to spaced sheathing.

R905.4.2 Deck slope. Metal roof shingles shall not be installed on roof slopes below three units vertical in 12 units horizontal (25-percent slope).

R905.4.3 Underlayment. In areas where the average daily temperature in January is 25°F (-4°C) or less, an ice barrier that consists of at least two layers of underlayment cemented together or of a self-adhering polymer modified bitumen sheet shall be used in lieu of normal underlayment and extend from the eave's edge to a point at least 24 inches (610 mm) inside the exterior wall line of the building. Underlayment shall comply with ASTM D 226, Type I.

R905.4.4 Material standards. Metal roof shingle roof coverings of galvanized steel shall be 0.013 inch (0.378 mm) minimum thickness. Metal roof shingle roof coverings of aluminum shall be of 0.024 inch (0.610 mm) minimum thickness.

R905.4.5 Application. Metal roof shingles shall be secured to the roof in accordance with this chapter and the approved manufacturer's installation instructions.

R905.4.6 Flashing. Roof valley flashing shall be provided of corrosion-resistant metal of the same material as the roof covering or shall comply with the standards in Table R905.10.3. The valley flashing shall extend at least 8 inches (203 mm) from the center line each way and shall have a splash diverter rib not less than $^3/_4$ inch (19.1 mm) high at the flow line formed as part of the flashing. Sections of flashing shall have an end lap of not less than 4 inches (102 mm). The metal valley flashing shall have a 36-inch-wide (914 mm) underlayment directly under it consisting of one layer of underlayment running the full length of the valley, in addition to underlayment required for metal roof shingles. In areas where the average daily temperature in January is 25°F (-4°C) or less , the metal valley flashing underlayment shall be solid cemented to the roofing underlayment for roof slopes under seven units vertical in 12 units horizontal (58.3-percent slope) or self-adhering polymer modified bitumen sheet.

R905.5 Mineral-surfaced roll roofing. The installation of mineral-surfaced rolling roofing shall comply with this section.

R905.5.1 Deck requirements. Mineral-surfaced roll roofing shall be fastened to solidly sheathed roofs.

R905.5.2 Deck slope. Mineral-surfaced roll roofing shall not be applied on roof slopes below one unit vertical in 12 units horizontal (8-percent slope).

R905.5.3 Underlayment. In areas where the average daily temperature in January is 25°F (-4°C) or less, an ice barrier that consists of at least two layers of underlayment cemented together or of a self-adhering polymer modified bitumen sheet shall extend from the eave's edge to a point at least 24 inches (610 mm) inside the exterior wall line of the building. Underlayment shall conform with ASTM D 226, Type I.

R905.5.4 Material standards. Mineral-surfaced roll roofing shall conform to ASTM D 224, D 249, D 371 or D 3909.

R905.5.5 Application. Mineral-surfaced roll roofing shall be installed in accordance with this chapter and the manufacturer's installation instructions.

R905.6 Slate and slate-type shingles. The installation of slate and slate-type shingles shall comply with the provisions of this section.

R905.6.1 Deck requirements. Slate shingles shall be fastened to solidly sheathed roofs.

R905.6.2 Deck slope. Slate shingles shall only be used on slopes of four units vertical in 12 units horizontal (33-percent slope) or greater.

R905.6.3 Underlayment. In areas where the average daily temperature in January is 25°F (-4°C) or less, an ice barrier that consists of at least two layers of underlayment cemented together or of a self-adhering polymer modified bitumen sheet shall extend from the eave's edge to a point at least 24 inches (610 mm) inside the exterior wall line of the building. Underlayment shall comply with ASTM D 226, Type II.

R905.6.4 Material standards. Slate shingles shall comply with ASTM C 406.

R905.6.5 Application. Minimum headlap for slate shingles shall be in accordance with Table R905.6.5. Slate shingles shall be secured to the roof with two fasteners per slate. Slate shingles shall be installed in accordance with this chapter and the manufacturer's installation instructions.

TABLE R905.6.5
SLATE SHINGLE HEADLAP

SLOPE	HEADLAP (inches)
4:12 ≤ slope < 8:12	4
8:12 ≤ slope < 20:12	3
slope ≥ 20:12	2

For SI: 1 inch = 25.4 mm.

R905.6.6 Flashing. Flashing and counterflashing shall be made with sheet metal. Valley flashing shall be a minimum of 15 inches (381 mm) wide. Valley and flashing metal shall be a minimum uncoated thickness of 0.0179-inch (0.455 mm) zinc coated G90. Chimneys, stucco or brick walls shall have a minimum of two plies of felt for a cap flashing consisting of a 4-inch-wide (102 mm) strip of felt set in plastic cement and extending 1 inch (25.4 mm) above the first felt and a top coating of plastic cement. The felt shall extend over the base flashing 2 inches (51 mm).

R905.7 Wood shingles. The installation of wood shingles shall comply with the provisions of this section.

R905.7.1 Deck requirements. Wood shingles shall be installed on solid or spaced sheathing. Where spaced sheathing is used, sheathing boards shall not be less than 1-inch by 4-inch (25.4 mm by 102 mm) nominal dimensions and shall be spaced on centers equal to the weather exposure to coincide with the placement of fasteners.

R905.7.1.1 Solid sheathing required. In areas where the average daily temperature in January is 25°F (-4°C) or less, solid sheathing is required on that portion of the roof requiring the application of an ice shield.

R905.7.2 Deck slope. Wood shingles shall be installed on slopes of three units vertical in 12 units horizontal (25-percent slope) or greater.

R905.7.3 Underlayment. In areas where the average daily temperature in January is 25°F (-4°C) or less, an ice barrier that consists of at least two layers of underlayment cemented together or of a self-adhering polymer modified bitumen sheet shall extend from the eave's edge to a point at least 24 inches (610 mm) inside the exterior wall line of the building. Underlayment shall comply with ASTM D 226, Type I.

R905.7.4 Material standards. Wood shingles shall be of naturally durable wood and comply with the requirements of Table R905.7.4.

TABLE R905.7.4
WOOD SHINGLE MATERIAL REQUIREMENTS

MATERIAL	MINIMUM GRADES	APPLICABLE GRADING RULES
Wood shingles of naturally durable wood	1, 2 or 3	Cedar Shake and Shingle Bureau

R905.7.5 Application. Wood shingles shall be installed according to this chapter and the manufacturer's installation instructions. Wood shingles shall be laid with a side lap not less than $1^1/_2$ inches (38 mm) between joints in courses, and no two joints in any three adjacent courses shall be in direct alignment. Spacing between shingles shall not be less than $^1/_4$ inch to $^3/_8$ inch (6.4 mm to 9.5 mm). Weather exposure for wood shingles shall not exceed those set in Table R905.7.5. Fasteners for wood shingles shall be corrosion-resistant with a minimum penetration of $^1/_2$ inch (12.7 mm) into the sheathing. For sheathing less than $^1/_2$ inch (12.7 mm) in thickness, the fasteners shall extend through the sheathing. Wood shingles shall be attached to the roof with two fasteners per shingle, positioned no more than $^3/_4$ inch (19.1 mm) from each edge and no more than 1 inch (25.4 mm) above the exposure line.

TABLE R905.7.5
WOOD SHINGLE WEATHER EXPOSURE AND ROOF SLOPE

ROOFING MATERIAL	LENGTH (inches)	GRADE	EXPOSURE (inches)	
			3:12 pitch to < 4:12	4:12 pitch or steeper
Shingles of naturally durable wood	16	No. 1	$3^3/_4$	5
		No. 2	$3^1/_2$	4
		No. 3	3	$3^1/_2$
	18	No. 1	$4^1/_4$	$5^1/_2$
		No. 2	4	$4^1/_2$
		No. 3	$3^1/_2$	4
	24	No. 1	$5^3/_4$	$7^1/_2$
		No. 2	$5^1/_2$	$6^1/_2$
		No. 3	5	$5^1/_2$

For SI: 1 inch = 25.4 mm.

R905.7.6 Valley flashing. Roof flashing shall be not less than No. 26 gage [0.019 inches (0.48 mm)] corrosion-resistant sheet metal and shall extend 10 inches (254 mm) from the centerline each way for roofs having slopes less than 12 units vertical in 12 units horizontal (100-percent slope), and 7 inches (178 mm) from the centerline each way for slopes of 12 units vertical in 12 units horizontal and greater. Sections of flashing shall have an end lap of not less than 4 inches (102 mm).

R905.7.7 Label required. Each bundle of shingles shall be identified by a label of an approved grading or inspection bureau or agency.

R905.8 Wood shakes. The installation of wood shakes shall comply with the provisions of this section.

R905.8.1 Deck requirements. Wood shakes shall only be used on solid or spaced sheathing. Where spaced sheathing is used, sheathing boards shall not be less than 1-inch by 4-inch (25 mm by 102 mm) nominal dimensions and shall be spaced on centers equal to the weather exposure to coincide with the placement of fasteners. Where 1-inch by 4-inch (25 mm by 102 mm) spaced sheathing is installed at 10 inches (254 mm) on center, additional 1-inch by 4-inch (25 mm by 102 mm) boards shall be installed between the sheathing boards.

R905.8.1.1 Solid sheathing required. In areas where the average daily temperature in January is 25°F (-4°C) or less, solid sheathing is required on that portion of the roof requiring an ice barrier.

R905.8.2 Deck slope. Wood shakes shall only be used on slopes of three units vertical in 12 units horizontal (25-percent slope) or greater.

R905.8.3 Underlayment. In areas where the average daily temperature in January is 25°F (-4°C) or less, an ice barrier that consists of at least two layers of underlayment cemented together or a self-adhering polymer modified bitumen sheet shall extend from the edge of the eave to a point at least 24 inches (610 mm) inside the exterior wall line of the building. Underlayment shall comply with ASTM D 226, Type I.

R905.8.4 Interlayment. Interlayment shall comply with ASTM D 226, Type I.

R905.8.5 Material standards. Wood shakes shall comply with the requirements of Table R905.8.5.

TABLE R905.8.5
WOOD SHAKE MATERIAL REQUIREMENTS

MATERIAL	MINIMUM GRADES	APPLICABLE GRADING RULES
Wood shakes of naturally durable wood	1	Cedar Shake and Shingle Bureau
Taper sawn shakes of naturally durable wood	1 or 2	Cedar Shake and Shingle Bureau
Preservative-treated shakes and shingles of naturally durable wood	1	Cedar Shake and Shingle Bureau
Fire-retardant-treated shakes and shingles of naturally durable wood	1	Cedar Shake and Shingle Bureau
Preservative-treated tapersawn shakes of southern yellow pine treated in accordance with AWPA Standard C2	1 or 2	Forest Products Laboratory of the Texas Forest Services

R905.8.6 Application. Wood shakes shall be installed according to this chapter and the manufacturer's installation instructions. Wood shakes shall be laid with a side lap not less than $1^1/_2$ inches (38 mm) between joints in adjacent courses. Spacing between shakes in the same course shall be $^1/_8$ inch to $^5/_8$ inch (3.2 mm to 15.9 mm) for shakes and tapersawn shakes of naturally durable wood and shall be $^1/_4$ inch to $^3/_8$ inch (6.4 mm to 9.5 mm) for preservative tapersawn shakes. Weather exposure for wood shakes shall not exceed those set forth in Table R905.8.6. Fasteners for wood shakes shall be corrosion-resistant, with a minimum penetration of $^1/_2$ inch (12.7 mm) into the sheathing. For sheathing less than $^1/_2$ inch (12.7 mm) in thickness, the fasteners shall extend through the sheathing. Wood shakes shall be attached to the roof with two fasteners per shake, positioned no more than 1 inch (25.4 mm) from each edge and no more than 2 inches (51 mm) above the exposure line.

TABLE R905.8.6
WOOD SHAKE WEATHER EXPOSURE AND ROOF SLOPE

ROOFING MATERIAL	LENGTH (inches)	GRADE	EXPOSURE (inches)
			4:12 pitch or steeper
Shakes of naturally durable wood	18	No. 1	$7^1/_2$
	24	No. 1	10[a]
Preservative-treated taper sawn shakes of southern yellow pine	18	No. 1	$7^1/_2$
	24	No. 1	10
	18	No. 2	$5^1/_2$
	24	No. 2	$7^1/_2$
Taper-sawn shakes of naturally durable wood	18	No. 1	$7^1/_2$
	24	No. 1	10
	18	No. 2	$5^1/_2$
	24	No. 2	$7^1/_2$

For SI: 1 inch = 25.4 mm.

a. For 24-inch by $^3/_8$-inch handsplit shakes, the maximum exposure is $7^1/_2$ inches.

R905.8.7 Shake placement. The starter course at the eaves shall be doubled and the bottom layer shall be either 15-inch (381 mm), 18-inch (457 mm) or 24-inch (610 mm) wood shakes or wood shingles. Fifteen-inch (381 mm) or 18-inch (457 mm) wood shakes may be used for the final course at the ridge. Shakes shall be interlaid with 18-inch-wide (457 mm) strips of not less than No. 30 felt shingled between each course in such a manner that no felt is exposed to the weather by positioning the lower edge of each felt strip above the butt end of the shake it covers a distance equal to twice the weather exposure.

R905.8.8 Valley flashing. Roof valley flashing shall not be less than No. 26 gage [0.019 inches (0.48 mm)] corrosion-resistant sheet metal and shall extend at least 11 inches (279 mm) from the centerline each way. Sections of flashing shall have an end lap of not less than 4 inches (102 mm).

R905.8.9 Label required. Each bundle of shakes shall be identified by a label of an approved grading or inspection bureau or agency.

R905.9 Built-up roofs. The installation of built-up roofs shall comply with the provisions of this section.

R905.9.1 Slope. Built-up roofs shall have a design slope of a minimum of one-fourth unit vertical in 12 units horizontal (2-percent slope) for drainage, except for coal-tar built-up roofs, which shall have a design slope of a minimum one-eighth unit vertical in 12 units horizontal (1-percent slope).

R905.9.2 Material standards. Built-up roof covering materials shall comply with the standards in Table R905.9.2.

R905.9.3 Application. Built-up roofs shall be installed according to this chapter and the manufacturer's installation instructions.

R905.10 Metal roof panels. The installation of metal roof panels shall comply with the provisions of this section.

R905.10.1 Deck requirements. Metal roof panel roof coverings shall be applied to a solid or spaced sheathing, except where the roof covering is specifically designed to be applied to spaced supports.

R905.10.2 Slope. The minimum slope for lapped, nonsoldered seam metal roofs without applied lap sealant shall be three units vertical in 12 units horizontal (25-percent slope). The minimum slope for lapped, nonsoldered seam metal roofs with applied lap sealant shall be one-half vertical unit in 12 units horizontal (4-percent slope). The minimum slope for standing seam roof systems shall be one-fourth unit vertical in 12 units horizontal (2-percent slope).

R905.10.3 Material standards. Metal-sheet roof covering systems that incorporate supporting structural members shall be designed in accordance with the *International Building Code*. Metal-sheet roof coverings installed over structural decking shall comply with Table R905.10.3.

TABLE R905.9.2
BUILT-UP ROOFING MATERIAL STANDARDS

MATERIAL STANDARD	STANDARD
Acrylic coatings used in roofing	ASTM D 6083
Aggregate surfacing	ASTM D 1863
Asphalt adhesive used in roofing	ASTM D 3747
Asphalt cements used in roofing	ASTM D 3019; D 2822; D 4586
Asphalt-coated glass fiber base sheet	ASTM D 4601
Asphalt coatings used in roofing	ASTM D 1227; D 2823; D 2824; D 4479
Asphalt glass felt	ASTM D 2178
Asphalt primer used in roofing	ASTM D 41
Asphalt-saturated and asphalt-coated organic felt base sheet	ASTM D 2626
Asphalt-saturated organic felt (perforated)	ASTM D 226
Asphalt used in roofing	ASTM D 312
Coal tar cements used in roofing	ASTM D 4022; D 5643
Coal-tar saturated organic felt	ASTM D 227
Coal-tar used in roofing	ASTM D 450, Types I or II
Glass mat, coal tar	ASTM D 4990
Glass mat, venting type	ASTM D 4897
Mineral-surfaced inorganic cap sheet	ASTM D 3909
Thermoplastic fabrics used in roofing	ASTM D 5665; D 5726

TABLE R905.10.3
METAL ROOF COVERINGS STANDARDS AND INSTALLATION

ROOF COVERING TYPE	STANDARD APPLICATION RATE/THICKNESS
Galvanized steel	ASTM A 653 G-90 zinc coated, 0.013 inch thick minimum
Prepainted steel	ASTM A 755
Aluminum zinc alloy coated steel	ASTM A 792 AZ 50
Lead-coated copper	ASTM B 101
Copper	ASTM B 370, 16 oz. per sq. ft. for metal sheet roof covering systems; 12 oz. per sq. ft. for preformed metal shingle systems.CDA 4115
Hard lead	2 lbs. per sq. ft.
Soft lead	3 lbs. per sq. ft.
Aluminum	0.024 inch minimum thickness
Terne (tin) and terne coated stainless	Terne coating of 40 lb. per double base box, field painted where applicable in accordance with manufacturer's installation instructions.

For SI: 1 inch = 25.4 mm, 1 square foot = 0.0929 m^2, 1 pound = 0.454 kg, 1 ounce per square foot = 305 g/m^2, 1 pound per square foot = 0.0479 kN/m^2.

R905.10.4 Attachment. Metal roofing shall be installed in accordance with this chapter and the manufacturer's installation instructions. Metal roofing fastened directly to steel framing shall be attached by approved fasteners. The following fasteners shall be used:

1. Galvanized fasteners shall be used for galvanized roofs.
2. Hard copper or copper alloy shall be used for copper roofs.
3. Stainless steel fasteners are acceptable for metal roofs.

R905.11 Modified bitumen roofing. The installation of modified bitumen roofing shall comply with the provisions of this section.

R905.11.1 Slope. Modified bitumen membrane roofs shall have a design slope of a minimum of one-fourth unit vertical in 12 units horizontal (2-percent slope) for drainage.

R905.11.2 Material standards. Modified bitumen roof coverings shall comply with the standards in Table R905.11.2.

TABLE R905.11.2
MODIFIED BITUMEN ROOFING MATERIAL STANDARDS

MATERIAL	STANDARD
Modified bitumen roof membrane	ASTM D 6162; D 6163; D 6164; D 6222; D 6223; CGSB 37-56M
Asphalt primer	ASTM D 41
Asphalt cement	ASTM D 3019
Asphalt adhesive	ASTM D 3747
Asphalt coating	ASTM D 1227; D 2824
Acrylic coating	ASTM D 6083

R905.11.3 Application. Modified bitumen roofs shall be installed according to this chapter and the manufacturer's installation instructions.

R905.12 Thermoset single-ply roofing. The installation of thermoset single-ply roofing shall comply with the provisions of this section.

R905.12.1 Slope. Thermoset single-ply membrane roofs shall have a design slope of a minimum of one-fourth unit vertical in 12 units horizontal (2-percent slope) for drainage.

R905.12.2 Materials standards. Thermoset single-ply roof coverings shall comply with RMA RP-1, RP-2 or RP-3, or ASTM D 4637 or CGSB 37-52M.

R905.12.3 Application. Thermoset single-ply roofs shall be installed according to this chapter and the manufacturer's installation instructions.

R905.13 Thermoplastic single-ply roofing. The installation of thermoplastic single-ply roofing shall comply with the provisions of this section.

R905.13.1 Slope. Thermoplastic single-ply membrane roofs shall have a design slope of a minimum of one-fourth unit vertical in 12 units horizontal (2-percent slope).

R905.13.2 Material standards. Thermoplastic single-ply roof coverings shall comply with ASTM D 4434 or CGSB 37-54M.

R905.13.3 Application. Thermoplastic single-ply roofs shall be installed according to this chapter and the manufacturer's installation instructions.

R905.14 Sprayed polyurethane foam roofing. The installation of sprayed polyurethane foam roofing shall comply with the provisions of this section.

R905.14.1 Slope. Sprayed polyurethane foam roofs shall have a design slope of a minimum of one-fourth unit vertical in 12 units horizontal (2-percent slope) for drainage.

R905.14.2 Material standards. Spray-applied polyurethane-foam insulation shall comply with ASTM C 1029.

R905.14.3 Application. Foamed in place roof insulation shall be installed in accordance with this chapter and the manufacturer's installation instructions. A liquid-applied protective coating that complies with Section R905.15 shall be applied no less than 2 hours nor more than 72 hours following the application of the foam.

R905.14.4 Foam plastics. Foam plastic materials and installation shall comply with Section R318.

R905.15 Liquid-applied coatings. The installation of liquid-applied coatings shall comply with the provisions of this section.

R905.15.1 Slope. Liquid-applied roofs shall have a design slope of a minimum of one-fourth unit vertical in 12 units horizontal (2-percent slope).

R905.15.2 Material standards. Liquid-applied roof coatings shall comply with ASTM C 836, C 957, D 1227, D 3468 or D 6083.

R905.15.3 Application. Liquid-applied roof coatings shall be installed according to this chapter and the manufacturer's installation instructions.

SECTION R906
ROOF INSULATION

R906.1 General. The use of above deck thermal insulation shall be permitted provided such insulation is covered with an approved roof covering and passes FM 4450 or UL 1256.

SECTION R907
REROOFING

R907.1 General. Materials and methods of application used for recovering or replacing an existing roof covering shall comply with the requirements of this chapter. Roof repairs to existing roofs and roof coverings shall comply with the provisions of Chapter 34 of the *International Building Code*, but more than 25 percent of the roof covering of any building shall not be removed and replaced within a 12-month period unless the entire roof covering is made to conform to the requirements for new roofing.

Exception: Reroofing shall not be required to meet the minimum design slope requirement of one-fourth vertical in 12 units horizontal (2-percent slope) in Section R905 for roofs that provide positive roof drainage.

R907.2 Structural and construction loads. The structural roof components shall be capable of supporting the roof covering system and the material and equipment loads that will be encountered during installation of the roof covering system.

R907.3 Recovering versus replacement. New roof coverings shall not be installed without first removing existing roof coverings where any of the following conditions occur:

1. Where the existing roof or roof covering is water-soaked or has deteriorated to the point that the existing roof or roof covering is not adequate as a base for additional roofing.
2. Where the existing roof covering is wood shake, slate, clay, cement or asbestos-cement tile.
3. Where the existing roof has two or more applications of any type of roof covering.
4. For asphalt shingles, when the building is located in an area subject to severe hail damage according to Figure R907.3.

Exceptions:

1. Complete and separate roofing systems, such as standing-seam metal roof systems, that are designed to transmit the roof loads directly to the building's structural system and that do not rely on existing roofs and roof coverings for support shall not require the removal of existing roof coverings.
2. Metal panel, metal shingle, and concrete and clay tile roof coverings shall be permitted to be installed over existing wood shake roofs when applied in accordance with Section R907.4.

R907.4 Roof recovering. Where the application of a new roof covering over wood shingle or shake roofs creates a combustible concealed space, the entire existing surface shall be covered with gypsum board, mineral fiber, glass fiber or other approved materials securely fastened in place.

R907.5 Reinstallation of materials. Existing slate, clay or cement tile shall be permitted for reinstallation, except that damaged, cracked or broken slate or tile shall not be reinstalled. Existing vent flashing, metal edgings, drain outlets, collars and metal counterflashings shall not be reinstalled where rusted, damaged or deteriorated. Aggregate surfacing materials shall not be reinstalled.

R907.6 Flashings. Flashings shall be reconstructed in accordance with approved manufacturer's installation instructions. Metal flashing to which bituminous materials are to be adhered shall be primed prior to installation.

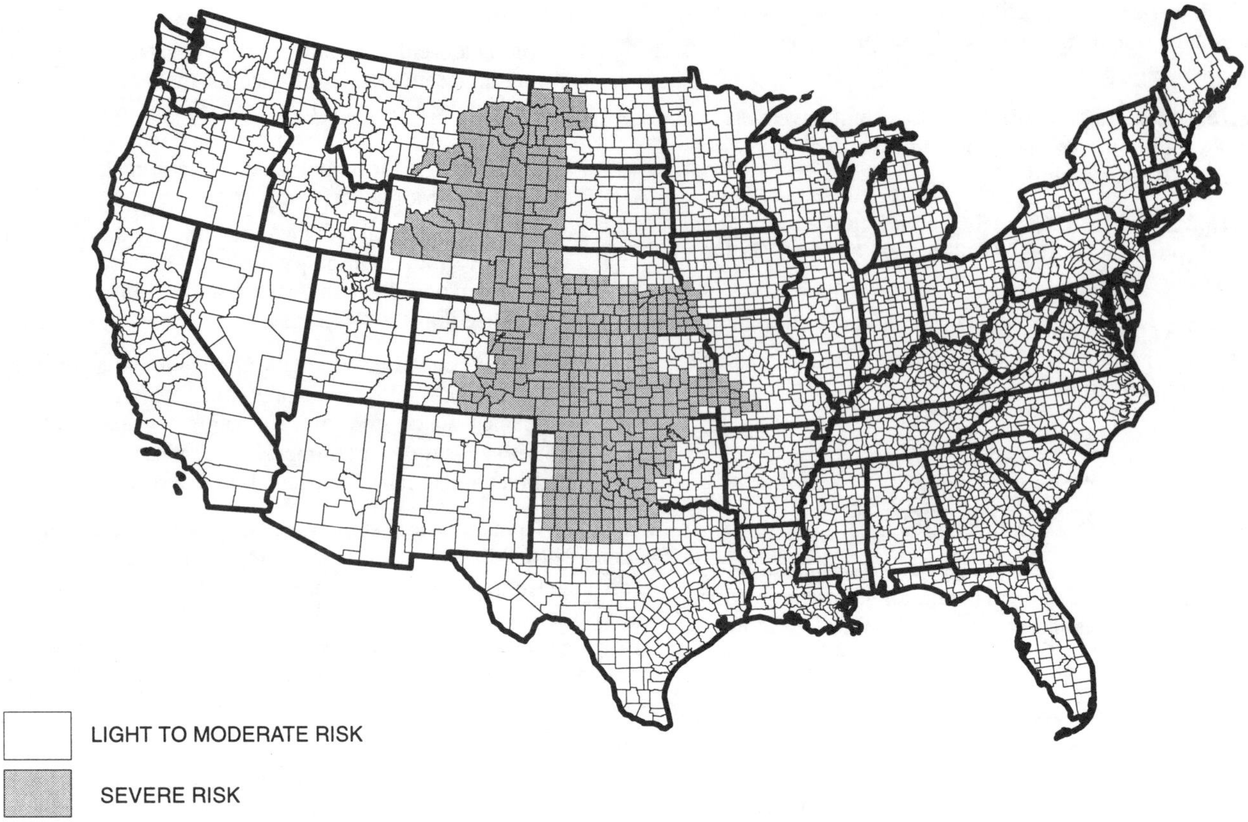

For SI: 1 inch = 25.4 mm.

a. Shaded portion of map denotes locations with a high probability of severe hail. A severe classification is where weather conditions frequently produce damaging hailstorms.

b. Areas with greater than a 50/50 chance of a 1-inch or larger diameter hailstone occurring at a point at least once in a 5-year period.

FIGURE R907.3
HAIL RISK MAP[a,b]

CHAPTER 10
CHIMNEYS AND FIREPLACES

SECTION R1001
MASONRY CHIMNEYS

R1001.1 General. Masonry chimneys shall be constructed, anchored, supported and reinforced as required in this chapter and the applicable provisions of Chapters 3, 4 and 6. In Seismic Design Categories D_1 and D_2, masonry and concrete chimneys shall be reinforced and anchored as detailed in Section R1003 for chimneys serving fireplaces. In Seismic Design Category A, B or C, reinforcement and seismic anchorage is not required. Chimneys shall be structurally sound, durable, smoke-tight and capable of conveying flue gases to the exterior safely.

R1001.1.1 Support. Masonry chimneys shall be supported on foundations of solid masonry or concrete at least 12 inches (305 mm) thick and at least 6 inches (152 mm) beyond each side of the exterior dimensions of the chimney. Footings shall be founded on natural, undisturbed earth below the frostline. In areas not subject to freezing, footings shall be located a minimum of 12 inches (305 mm) below finished grade.

R1001.2 Corbeling. Masonry chimneys shall not be corbeled more than one-half of the chimney's wall thickness from a wall or foundation, nor shall a chimney be corbeled from a wall or foundation that is less than 12 inches (305 mm) in thickness unless it projects equally on each side of the wall, except that on the second story of a two-story dwelling, corbeling of chimneys on the exterior of the enclosing walls may equal the wall thickness. The projection of a single course shall not exceed one-half the unit height or one-third of the unit bed depth, whichever is less.

R1001.3 Changes in dimension. The chimney wall or chimney flue lining shall not change in size or shape within 6 inches (152 mm) above or below where the chimney passes through floor components, ceiling components or roof components.

R1001.4 Offsets. Where a masonry chimney is constructed with a fireclay flue liner surrounded by one wythe of masonry, the maximum offset shall be such that the centerline of the flue above the offset does not extend beyond the center of the chimney wall below the offset. Where the chimney offset is supported by masonry below the offset in an approved manner, the maximum offset limitations shall not apply. Each individual corbeled masonry course of the offset shall not exceed the projection limitations specified in Section R1001.2.

R1001.5 Additional load. Chimneys shall not support loads other than their own weight unless they are designed and constructed to support the additional load. Masonry chimneys shall be permitted to be constructed as part of the masonry walls or reinforced concrete walls of the building.

R1001.6 Termination. Chimneys shall extend at least 2 feet (610 mm) higher than any portion of a building within 10 feet (3048 mm), but shall not be less than 3 feet (914 mm) above the point where the chimney passes through the roof.

R1001.7 Wall thickness. Masonry chimney walls shall be constructed of solid masonry units or hollow masonry units grouted solid with not less than a 4-inch (102 mm) nominal thickness.

R1001.8 Flue lining (material). Masonry chimneys shall be lined. The lining material shall be appropriate for the type of appliance connected, according to the terms of the appliance listing and manufacturer's instructions.

R1001.8.1 Residential-type appliances (general). Flue lining systems shall comply with one of the following:

1. Clay flue lining complying with the requirements of ASTM C 315 or equivalent.
2. Listed chimney lining systems complying with UL 1777.
3. Factory-built chimneys or chimney units listed for installation within masonry chimneys.
4. Other approved materials that will resist, without cracking, softening or corrosion, and flue gases and condensate at temperatures up to 1,800°F (982°C).

R1001.8.2 Flue linings for specific appliances. Flue linings other than these covered in Section R1001.8.1, intended for use with specific types of appliances, shall comply with Sections R1001.8.3 through R1001.8.6.

R1001.8.3 Gas appliances. Flue lining systems for gas appliances shall be in accordance with Chapter 24.

R1001.8.4 Pellet fuel-burning appliances. Flue lining and vent systems for use in masonry chimneys with pellet fuel burning appliances shall be limited to the following:

1. Flue lining systems complying with Section R1001.8.1.
2. Pellet vents listed for installation within masonry chimneys. (See Section R1001.8.6 for marking.)

R1001.8.5 Oil-fired appliances approved for use with Type L vent. Flue lining and vent systems for use in masonry chimneys with oil-fired appliances approved for use with Type L vent shall be limited to the following:

1. Flue lining systems complying with Section R1001.8.1.
2. Listed chimney liners complying with UL 641. (See Section R1001.8.6 for marking.)

R1001.8.6 Notice of usage. When a flue is relined with a material not complying with Section R1001.8.1, the chimney shall be plainly and permanently identified by a label attached to a wall, ceiling or other conspicuous location adjacent to where the connector enters the chimney. The label shall include the following message or equivalent language:

THIS CHIMNEY FLUE IS FOR USE ONLY WITH [TYPE OR CATEGORY OF APPLIANCE] APPLIANCES THAT BURN [TYPE OF FUEL]. DO NOT CONNECT OTHER TYPES OF APPLIANCES.

R1001.9 Flue lining (installation). Flue liners shall be installed in accordance with ASTM C 1283 and extend from a point not less than 8 inches (203 mm) below the lowest inlet or, in the case of fireplaces, from the top of the smoke chamber to a point above the enclosing walls. The lining shall be carried up vertically, with a maximum slope no greater than 30 degrees from the vertical.

Fireclay flue liners shall be laid in medium-duty refractory mortar conforming to ASTM C 199 with tight mortar joints left smooth on the inside and installed to maintain an air space or insulation not to exceed the thickness of the flue liner separating the flue liners from the interior face of the chimney masonry walls. Flue lining shall be supported on all sides. Only enough mortar shall be placed to make the joint and hold the liners in position.

R1001.9.1 Listed materials. Listed materials used as flue linings shall be installed in accordance with the terms of their listings and manufacturer's instructions.

R1001.9.2 Space around lining. The space surrounding a chimney lining system or vent installed within a masonry chimney shall not be used to vent any other appliance.

Exception: This shall not prevent the installation of a separate flue lining in accordance with the manufacturer's installation instructions.

R1001.10 Multiple flues. When two or more flues are located in the same chimney, masonry wythes shall be built between adjacent flue linings. The masonry wythes shall be at least 4 inches (102 mm) thick and bonded into the walls of the chimney.

Exception: When venting only one appliance, two flues may adjoin each other in the same chimney with only the flue lining separation between them. The joints of the adjacent flue linings shall be staggered at least 4 inches (102 mm).

R1001.11 Flue area (appliance). Chimney flues shall not be smaller in area than that of the area of the connector from the appliance [see Tables R1001.11(1) and R1001.11(2)]. The sizing of a chimney flue to which multiple-appliance venting systems are connected shall be in accordance with Section M1805.3.

TABLE R1001.11(1)
NET CROSS-SECTIONAL AREA OF ROUND FLUE SIZES[a]

FLUE SIZE, INSIDE DIAMETER (inches)	CROSS-SECTIONAL AREA (square inches)
6	28
7	38
8	50
10	78
$10^3/_4$	90
12	113
15	176
18	254

For SI: 1 inch = 25.4 mm, 1 square inch = 645.16 mm^2.

a. Flue sizes are based on ASTM C 315.

TABLE R1001.11(2)
NET CROSS-SECTIONAL AREA OF SQUARE AND RECTANGULAR FLUE SIZES[a]

FLUE SIZE, OUTSIDE DIMENSIONS (inches)	CROSS-SECTIONAL AREA (square inches)
$4^1/_2 \times 13$	34
$7^1/_2 \times 7^1/_2$	37
$8^1/_2 \times 8^1/_2$	47
$7^1/_2 \times 11^1/_2$	58
$8^1/_2 \times 13$	74
$7^1/_2 \times 15^1/_2$	82
$11^1/_2 \times 11^1/_2$	91
$8^1/_2 \times 17^1/_2$	101
13×13	122
$11^1/_2 \times 15^1/_2$	124
$13 \times 17^1/_2$	165
$15^1/_2 \times 15^1/_2$	168
$15^1/_2 \times 19^1/_2$	214
$17^1/_2 \times 17^1/_2$	226
$19^1/_2 \times 19^1/_2$	269
20×20	286

For SI: 1 inch = 25.4 mm, 1 square inch = 645.16 mm^2.

a. Flue sizes are based on ASTM C 315.

R1001.12 Flue area (masonry fireplace). Flue sizing for chimneys serving fireplaces shall be in accordance with Section R1001.12.1 or Section R1001.12.2.

R1001.12.1 Option 1. Round chimney flues shall have a minimum net cross-sectional area of at $^1/_{12}$ of the fireplace opening. Square chimney flues shall have a minimum net cross-sectional area of $^1/_{10}$ of the fireplace opening. Rectangular chimney flues with an aspect ratio less than 2 to 1 shall have a minimum net cross-sectional area of $^1/_{10}$ of the fireplace opening. Rectangular chimney flues with an aspect ratio of 2 to 1 or more shall have a minimum net cross-sectional area of $^1/_8$ of the fireplace opening.

R1001.12.2 Option 2. The minimum net cross-sectional area of the chimney flue shall be determined in accordance with Figure R1001.12.2. A flue size providing at least the equivalent net cross-sectional area shall be used. Cross-sectional areas of clay flue linings are provided in Tables R1001.11(1) and R1001.11(2) or as provided by the manufacturer or as measured in the field. The height of the chimney shall be measured from the firebox floor to the top of the chimney flue.

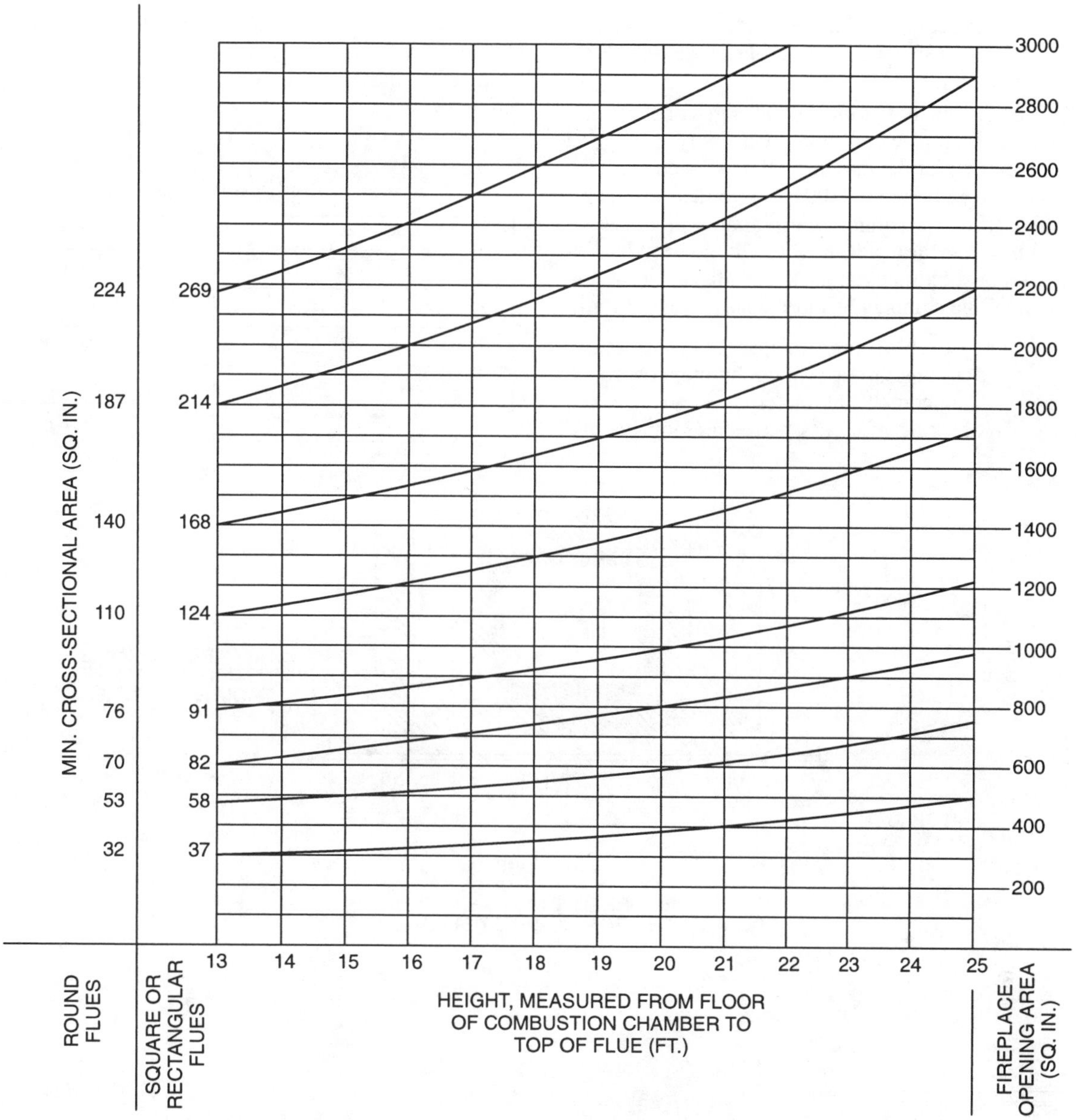

For SI: 1 foot = 304.8 mm, 1 square inch = 645.16 mm^2.

FIGURE R1001.12.2
FLUE SIZES FOR MASONRY CHIMNEYS

R1001.13 Inlet. Inlets to masonry chimneys shall enter from the side. Inlets shall have a thimble of fireclay, rigid refractory material or metal that will prevent the connector from pulling out of the inlet or from extending beyond the wall of the liner.

R1001.14 Masonry chimney cleanout openings. Cleanout openings shall be provided within 6 inches (152 mm) of the base of each flue within every masonry chimney. The upper edge of the cleanout shall be located at least 6 inches (152 mm) below the lowest chimney inlet opening. The height of the opening shall be at least 6 inches (152 mm). The cleanout shall be provided with a noncombustible cover.

Exception: Chimney flues serving masonry fireplaces where cleaning is possible through the fireplace opening.

R1001.15 Chimney clearances. Any portion of a masonry chimney located in the interior of the building or within the exterior wall of the building shall have a minimum air space clearance to combustibles of 2 inches (51 mm). Chimneys located entirely outside the exterior walls of the building, including chimneys that pass through the soffit or cornice, shall have a minimum air space clearance of 1 inch (25.4 mm). The air space shall not be filled, except to provide fire blocking in accordance with Section R1001.16.

Exceptions:

1. Masonry chimneys equipped with a chimney lining system listed and labeled for use in chimneys in contact with combustibles in accordance with UL 1777 and installed in accordance with the manufacturer's installation instructions are permitted to have combustible material in contact with their exterior surfaces.
2. When masonry chimneys are constructed as part of masonry or concrete walls, combustible materials shall

not be in contact with the masonry or concrete wall less than 12 inches (306 mm) from the inside surface of the nearest flue lining.

3. Exposed combustible trim and the edges of sheathing materials, such as wood siding and flooring, shall be permitted to abut the masonry chimney side walls, in accordance with Figure R1001.15 provided such combustible trim or sheathing is a minimum of 12 inches (306 mm) from the inside surface of the nearest flue lining. Combustible material and trim shall not overlap the corners of the chimney by more than 1 inch (25 mm).

R1001.16 Chimney fireblocking. All spaces between chimneys and floors and ceilings through which chimneys pass shall be fireblocked with noncombustible material securely fastened in place. The fireblocking of spaces between chimneys and wood joists, beams or headers shall be to a depth of 1 inch (25 mm) and shall only be placed on strips of metal or metal lath laid across the spaces between combustible material and the chimney.

TABLE R1001.17
CRICKET DIMENSIONS

ROOF SLOPE	H
12 - 12	$^1/_2$ of W
8 - 12	$^1/_3$ of W
6 - 12	$^1/_4$ of W
4 - 12	$^1/_6$ of W
3 - 12	$^1/_8$ of W

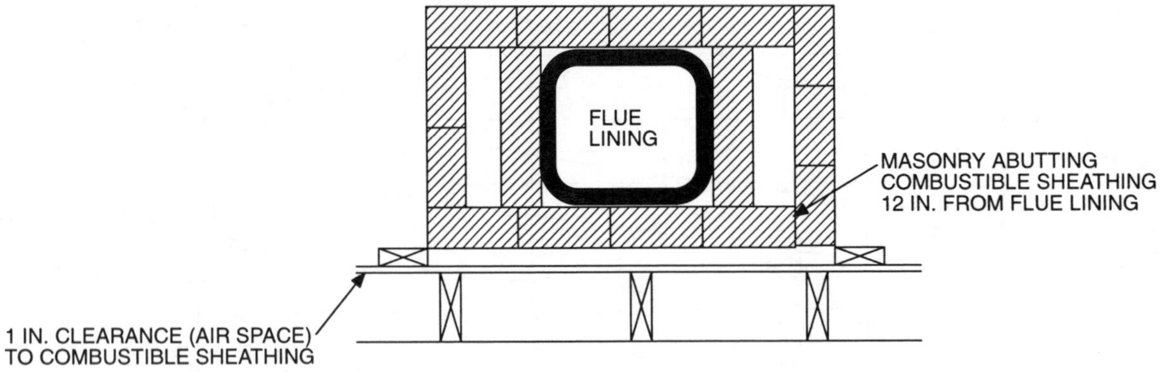

For SI: 1 inch = 25.4 mm.

FIGURE R1001.15
CLEARANCE FROM COMBUSTIBLES

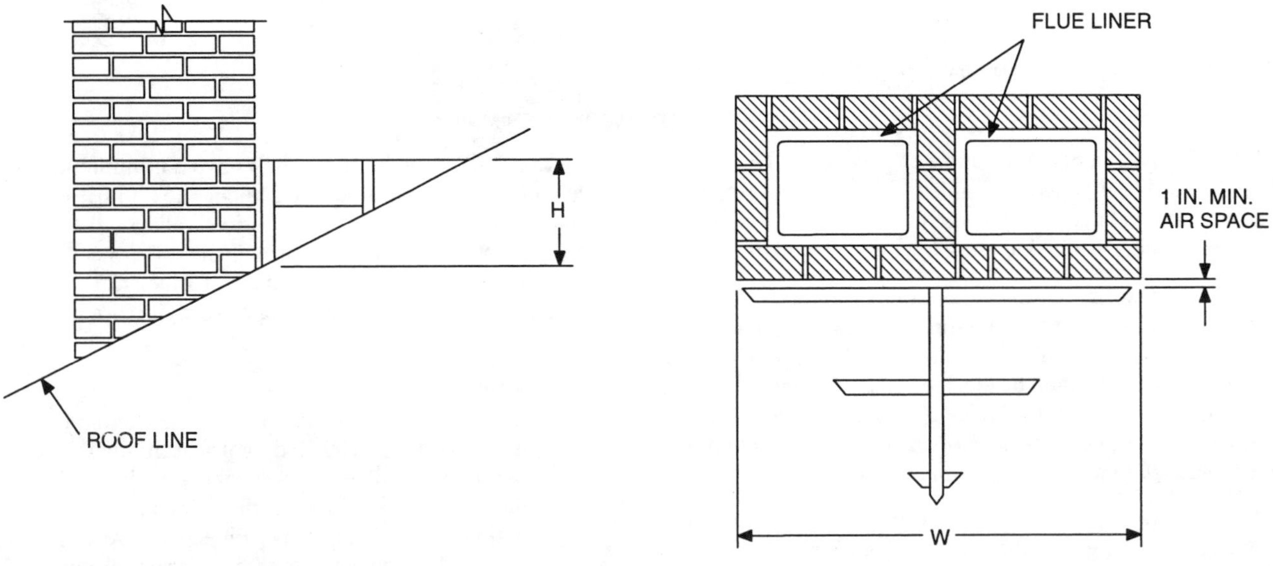

For SI: 1 inch = 25.4 mm.

FIGURE R1001.17
CHIMNEY CRICKET

R1001.17 Chimney crickets. Chimneys shall be provided with crickets when the dimension parallel to the ridgeline is greater than 30 inches (762 mm) and does not intersect the ridgeline. The intersection of the cricket and the chimney shall be flashed and counterflashed in the same manner as normal roof-chimney intersections. Crickets shall be constructed in compliance with Figure R1001.17 and Table R1001.17.

SECTION R1002 FACTORY-BUILT CHIMNEYS

R1002.1 Listing. Factory-built chimneys shall be listed and labeled and shall be installed and terminated in accordance with the manufacturer's installation instructions.

R1002.2 Decorative shrouds. Decorative shrouds shall not be installed at the termination of factory-built chimneys except where such shrouds are listed and labeled for use with the specific factory-built chimney system and installed in accordance with the manufacturer's installation instructions.

R1002.3 Solid fuel appliances. Factory-built chimneys for use with solid fuel-burning appliances shall comply with the Type HT requirements of UL 103.

Exception: Chimneys for use with fireplace stoves listed only to UL 737 shall comply with the requirements of UL 103.

R1002.4 Factory-built fireplaces. Chimneys for use with factory-built fireplaces shall comply with the requirements of UL 127.

R1002.5 Support. Where factory-built chimneys are supported by structural members, such as joists and rafters, such members shall be designed to support the additional load.

R1002.6 Medium-heat appliances. Factory-built chimneys for medium-heat appliances producing flue gases having a temperature above 1,000°F (538°C), measured at the entrance to the chimney shall comply with UL 959.

SECTION R1003 MASONRY FIREPLACES

R1003.1 General. Masonry fireplaces shall be constructed in accordance with this section and the applicable provisions of Chapters 3 and 4.

R1003.2 Footings and foundations. Foundations for masonry fireplaces and their chimneys shall be constructed of concrete or solid masonry at least 12 inches (305 mm) thick and shall extend at least 6 inches (152 mm) beyond the face of the fireplace or support wall on all sides. Footings shall be founded on natural, undisturbed earth or engineered fill below frost depth. In areas not subjected to freezing, footings shall be at least 12 inches (305 mm) below finished grade.

R1003.3 Seismic reinforcing. Masonry or concrete chimneys in Seismic Design Categories D_1 and D_2 shall be reinforced. Reinforcing shall conform to the requirements set forth in Table R1003.1 and Section R609, Grouted Masonry.

R1003.3.1 Vertical reinforcing. For chimneys up to 40 inches (1016 mm) wide, four No. 4 continuous vertical bars shall be placed between wythes of solid masonry or within the cells of hollow unit masonry and grouted in accordance with Section R609. Grout shall be prevented from bonding with the flue liner so that the flue liner is free to move with thermal expansion. For chimneys greater than 40 inches (1016 mm) wide, two additional No. 4 vertical bars shall be provided for each additional flue incorporated in the chimney or for each additional 40 inches (1016 mm) in width or fraction thereof.

R1003.3.2 Horizontal reinforcing. Vertical reinforcement shall be placed within $^1/_4$-inch (6.4 mm) ties, or other reinforcing of equivalent net cross-sectional area, placed in the bed joints according to Section R607 at a minimum of every 18 inches (457 mm) of vertical height. Two such ties shall be provided at each bend in the vertical bars.

R1003.4 Seismic anchorage. Masonry and concrete chimneys in Seismic Design Categories D_1 and D_2 shall be anchored at each floor, ceiling or roof line more than 6 feet (1829 mm) above grade, except where constructed completely within the exterior walls. Anchorage shall conform to the requirements of Section R1003.4.1.

R1003.4.1 Anchorage. Two $^3/_{16}$-inch by 1-inch (4.8 mm by 25.4 mm) straps shall be embedded a minimum of 12 inches (305 mm) into the chimney. Straps shall be hooked around the outer bars and extend 6 inches (152 mm) beyond the bend. Each strap shall be fastened to a minimum of four floor ceiling or floor joists or rafters with two $^1/_2$-inch (12.7 mm) bolts.

R1003.5 Fireplace walls. Masonry fireplaces shall be constructed of solid masonry units, hollow masonry units grouted solid, stone or reinforced concrete. When a lining of firebrick at least 2 inches (51 mm) in thickness or other approved lining is provided, the total minimum thickness of back and side walls shall be 8 inches (203 mm) of solid masonry, including the lining. The width of joints between firebricks shall not be greater than $^1/_4$ inch (6.4 mm). When no lining is provided, the total minimum thickness of back and side walls shall be 10 inches (254 mm) of solid masonry. Firebrick shall conform to ASTM C 27 or C 1261 and shall be laid with medium-duty refractory mortar conforming to ASTM C 199.

R1003.6 Steel fireplace units. Steel fireplace units incorporating a firebox liner of not less than $^1/_4$ inch (6.4 mm) in thickness and an air chamber may be installed with masonry to provide a total thickness at the back and sides of not less than 8 inches (203 mm), of which not less than 4 inches (102 mm) shall be of solid masonry. Warm-air ducts employed with steel fireplace units of the circulating air type shall be constructed of metal or masonry.

R1003.7 Lintel and throat. Masonry over a fireplace opening shall be supported by a lintel of noncombustible material. The minimum required bearing length on each end of the fireplace opening shall be 4 inches (102 mm). The fireplace throat or damper shall be located a minimum of 8 inches (203 mm) above the lintel.

TABLE R1003.1
SUMMARY OF REQUIREMENTS FOR MASONRY FIREPLACES AND CHIMNEYS

ITEM	LETTER[a]	REQUIREMENTS
Hearth slab thickness	A	4″
Hearth extension (each side of opening)	B	8″ fireplace opening < 6 sq. ft. 12″ fireplace opening ≥ 6 sq. ft.
Hearth extension (front of opening)	C	16″ fireplace opening < 6 sq. ft. 20″ fireplace opening ≥ 6 sq. ft.
Hearth slab reinforcing	D	Reinforced to carry its own weight and all imposed loads.
Thickness of wall of firebox	E	10″ solid brick or 8″ where a firebrick lining is used. Joints in firebrick $^1/_4$″ max.
Distance from top of opening to throat	F	8″
Smoke chamber wall thickness unlined walls	G	6″ 8″
Chimney Vertical reinforcing[b]	H	Four No. 4 full-length bars for chimney up to 40″ wide. Add two No. 4 bars for each additional 40″ or fraction of width or each additional flue.
Horizontal reinforcing	J	$^1/_4$″ ties at 18″ and two ties at each bend in vertical steel.
Bond beams	K	No specified requirement
Fireplace lintel	L	Noncombustible material.
Chimney walls with flue lining	M	Solid masonry units or hollow masonry units grouted solid with at least 4 inch nominal thickness.
Distances between adjacent flues	N	See Section R1001.10.
Effective flue area (based on area of fireplace opening)	P	See Section R1001.12.
Clearances: Combustible material Mantel and trim Above roof	R	 See Sections R1001.15 and R1003.12. See Section R1001.13. 3′ at roofline and 2′ at 10′.
Anchorage[b] Strap Number Embedment into chimney Fasten to Bolts	S	 $^3/_{16}$″ × 1″ Two 12″ hooked around outer bar with 6″ extension 4 joists Two $^1/_2$″ diameter.
Footing Thickness Width	T	 12″ min. 6″ each side of fireplace wall.

For SI: 1 inch = 25.4 mm, 1 foot = 304.8 mm, 1 square foot = 0.0929 m^2.

NOTE: This table provides a summary of major requirements for the construction of masonry chimneys and fireplaces. Letter references are to Figure R1003.1, which shows examples of typical construction. This table does not cover all requirements, nor does it cover all aspects of the indicated requirements. For the actual mandatory requirements of the code, see the indicated section of text.

a. The letters refer to Figure R1003.1.

b. Not required in Seismic Design Category A, B or C.

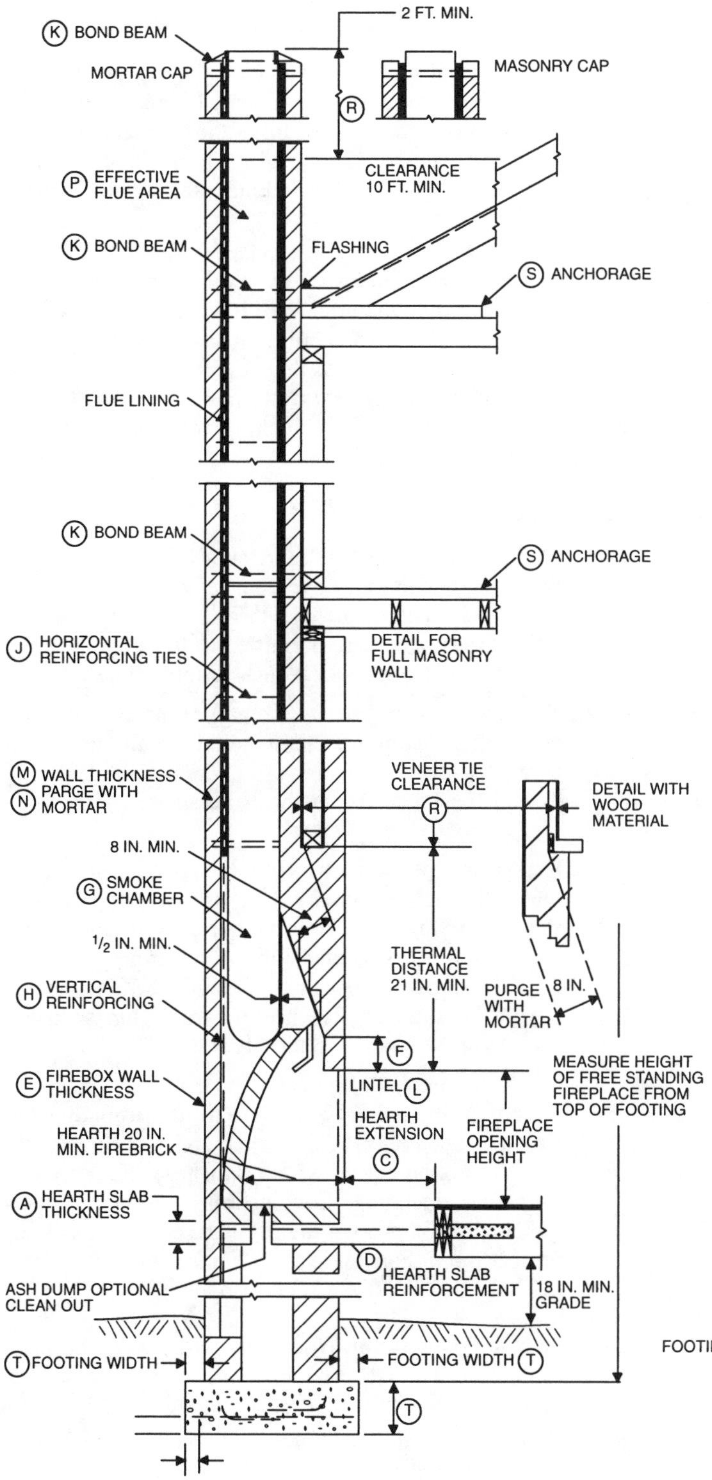

BRICK FIREBOX AND CHIMNEY—
SECTIONAL SIDE VIEW ON WOOD FLOOR

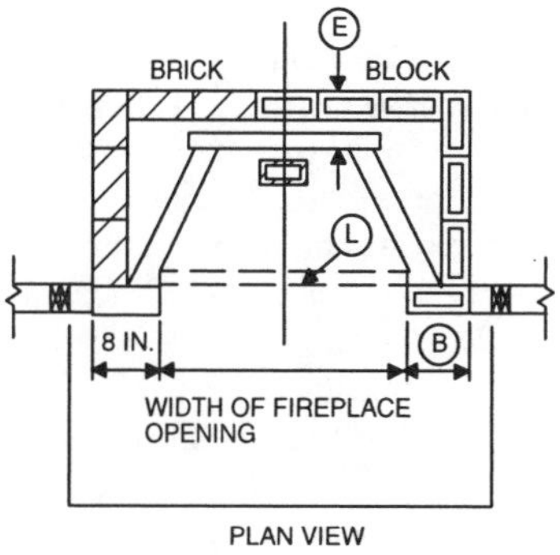

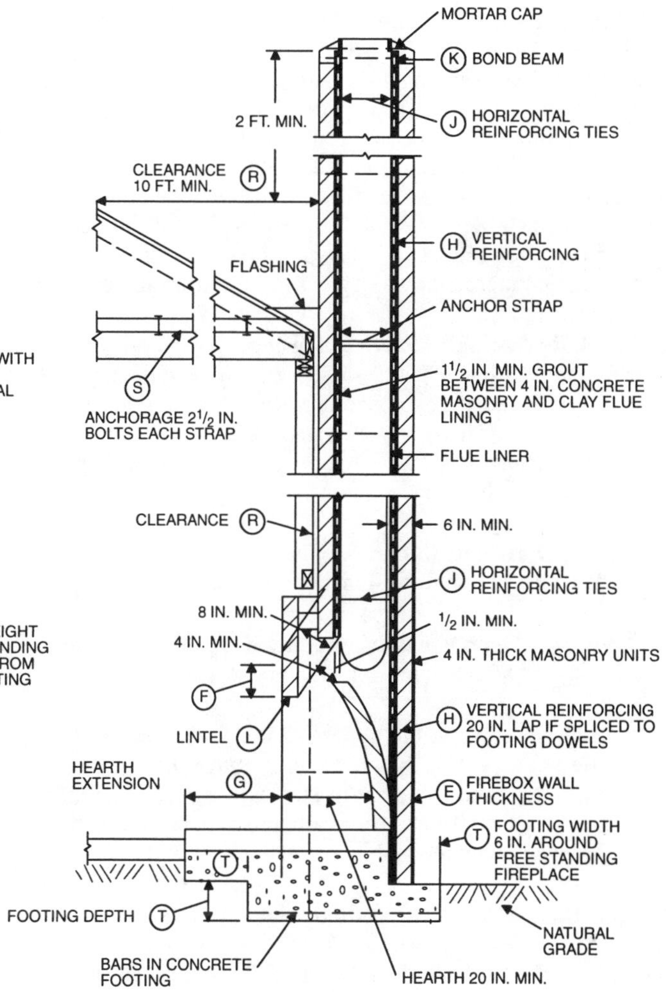

BRICK FIREBOX AND BLOCK CHIMNEY—
SECTIONAL SIDE VIEW ON CONCRETE SLAB

For SI: 1 inch = 25.4 mm, 1 foot = 304.8 mm.

FIGURE R1003.1
FIREPLACE AND CHIMNEY DETAILS

R1003.8 Smoke chamber. Smoke chamber walls shall be constructed of solid masonry units, hollow masonry units grouted solid, stone or reinforced concrete. Corbelling of masonry units shall not leave unit cores exposed to the inside of the smoke chamber. When a lining of firebrick at least 2 inches (51 mm) thick, or a lining of vitrified clay at least $^5/_8$ inch (15.9 mm) thick, is provided, the total minimum thickness of front, back and side walls shall be 6 inches (152 mm) of solid masonry, including the lining. Firebrick shall conform to ASTM C 27 or C 1261 and shall be laid with medium-duty refractory mortar conforming to ASTM C 199. Where no lining is provided, the total minimum thickness of front, back and side walls shall be 8 inches (203 mm) of solid masonry. When the inside surface of the smoke chamber is formed by corbeled masonry, the inside surface shall be parged smooth.

R1003.8.1 Smoke chamber dimensions. The inside height of the smoke chamber from the fireplace throat to the beginning of the flue shall not be greater than the inside width of the fireplace opening. The inside surface of the smoke chamber shall not be inclined more than 45 degrees (0.39 rad) from vertical when prefabricated smoke chamber linings are used or when the smoke chamber walls are rolled or sloped rather than corbeled. When the inside surface of the smoke chamber is formed by corbeled masonry, the walls shall not be corbeled more than 30 degrees (0.26 rad) from vertical.

R1003.9 Hearth and hearth extension. Masonry fireplace hearths and hearth extensions shall be constructed of concrete or masonry, supported by noncombustible materials, and reinforced to carry their own weight and all imposed loads. No combustible material shall remain against the underside of hearths and hearth extensions after construction.

R1003.9.1 Hearth thickness. The minimum thickness of fireplace hearths shall be 4 inches (102 mm).

R1003.9.2 Hearth extension thickness. The minimum thickness of hearth extensions shall be 2 inches (51 mm).

Exception: When the bottom of the firebox opening is raised at least 8 inches (203 mm) above the top of the hearth extension, a hearth extension of not less than $^3/_8$-inch-thick (9.5 mm) brick, concrete, stone, tile or other approved noncombustible material is permitted.

R1003.10 Hearth extension dimensions. Hearth extensions shall extend at least 16 inches (406 mm) in front of and at least 8 inches (203 mm) beyond each side of the fireplace opening. Where the fireplace opening is 6 square feet (0.557 m^2) or larger, the hearth extension shall extend at least 20 inches (508 mm) in front of and at least 12 inches (305 mm) beyond each side of the fireplace opening.

R1003.11 Firebox dimensions. The firebox of a concrete or masonry fireplace shall have a minimum depth of 20 inches (508 mm). The throat shall not be less than 8 inches (203 mm) above the fireplace opening. The throat opening shall not be less than 4 inches (102 mm) in depth. The cross-sectional area of the passageway above the firebox, including the throat, damper and smoke chamber, shall not be less than the cross-sectional area of the flue.

Exception: Rumford fireplaces shall be permitted provided that the depth of the fireplace is at least 12 inches (305 mm) and at least one-third of the width of the fireplace opening, that the throat is at least 12 inches (305 mm) above the lintel and is at least $^1/_{20}$ the cross-sectional area of the fireplace opening.

R1003.12 Fireplace clearance. All wood beams, joists, studs and other combustible material shall have a clearance of not less than 2 inches (51 mm) from the front faces and sides of masonry fireplaces and not less than 4 inches (102 mm) from the back faces of masonry fireplaces. The air space shall not be filled, except to provide fire blocking in accordance with Section R1003.14.

Exceptions:

1. Masonry fireplaces listed and labeled for use in contact with combustibles in accordance with UL 127 and installed in accordance with the manufacturer's installation instructions are permitted to have combustible material in contact with their exterior surfaces.
2. When masonry fireplaces are part of masonry or concrete walls, combustible materials shall not be in contact with the masonry or concrete walls less than 12 inches (305 mm) from the inside surface of the nearest firebox lining.
3. Exposed combustible trim and the edges of sheathing materials such as wood siding, flooring and drywall shall be permitted to abut the masonry fireplace side walls and hearth extension in accordance Figure R1003.12, provided such combustible trim or sheathing is a minimum of 12 inches (305 mm) from the inside surface of the nearest firebox lining.
4. Exposed combustible mantels or trim may be placed directly on the masonry fireplace front surrounding the fireplace opening providing such combustible materials shall not be placed within 6 inches (152 mm) of a fireplace opening. Combustible material within 12 inches (306 mm) of the fireplace opening shall not project more than $^1/_8$ inch (3.2 mm) for each 1-inch (25 mm) distance from such an opening.

R1003.13 Mantel and trim. Woodwork or other combustible materials shall not be placed within 6 inches (152 mm) of a fireplace opening. Combustible material within 12 inches (305 mm) of the fireplace opening shall not project more than $^1/_8$ inch (3.2 mm) for each 1-inch (25.4 mm) distance from such opening.

R1003.14 Fireplace fireblocking. See Section R602.8.

R1003.15 Ash dump cleanout. Cleanout openings, when provided, shall be equipped with ferrous metal doors and frames constructed to remain tightly closed except when in use. Cleanouts shall be accessible and located so that ash removal will not create a hazard to combustible materials.

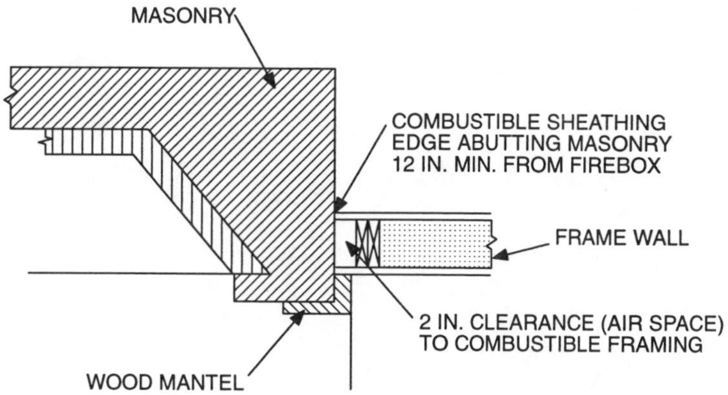

For SI: 1 inch = 25.4 mm.

FIGURE R1003.12
CLEARANCE FROM COMBUSTIBLES

SECTION R1004
FACTORY-BUILT FIREPLACES

R1004.1 General. Factory-built fireplaces shall be listed and labeled and shall be installed in accordance with the conditions of the listing. Factory-built fireplaces shall be tested in accordance with UL 127.

R1004.2 Hearth extensions. Hearth extensions of approved factory-built fireplaces shall be installed in accordance with the listing of the fireplace. The hearth extension shall be readily distinguishable from the surrounding floor area.

R1004.3 Decorative shrouds. Decorative shrouds shall not be installed at the termination of chimneys for factory-built fireplaces except where such shrouds are listed and labeled for use with the specific factory-built fireplace system and installed in accordance with the manufacturer's installation instructions.

R1004.4 Unvented gas log heaters. An unvented gas log heater shall not be installed in a factory-built fireplace unless the fireplace system has been specifically tested, listed and labeled for such use in accordance with UL 127.

SECTION R1005
EXTERIOR AIR SUPPLY

R1005.1 Exterior air. Factory-built or masonry fireplaces covered in this chapter shall be equipped with an exterior air supply to assure proper fuel combustion unless the room is mechanically ventilated and controlled so that the indoor pressure is neutral or positive.

R1005.1.1 Factory-built fireplaces. Exterior combustion air ducts for factory-built fireplaces shall be a listed component of the fireplace and shall be installed according to the fireplace manufacturer's instructions.

R1005.1.2 Masonry fireplaces. Listed combustion air ducts for masonry fireplaces shall be installed according to the terms of their listing and manufacturer's instructions.

R1005.2 Exterior air intake. The exterior air intake shall be capable of providing all combustion air from the exterior of the dwelling or from spaces within the dwelling ventilated with outside air such as crawl or attic spaces. The exterior air intake shall not be located within the garage or basement of the dwelling nor shall the air intake be located at an elevation higher than the firebox. The exterior air intake shall be covered with a corrosion-resistant screen of $^1/_4$-inch (6.4 mm) mesh.

R1005.3 Clearance. Unlisted combustion air ducts shall be installed with a minimum 1-inch (25.4 mm) clearance to combustibles for all parts of the duct within 5 feet (1524 mm) of the duct outlet.

R1005.4 Passageway. The combustion air passageway shall be a minimum of 6 square inches (3870 mm^2) and not more than 55 square inches (0.035 m^2), except that combustion air systems for listed fireplaces shall be constructed according to the fireplace manufacturer's instructions.

R1005.5 Outlet. The exterior air outlet is permitted to be located in the back or sides of the firebox chamber or within 24 inches (610 mm) of the firebox opening on or near the floor. The outlet shall be closable and designed to prevent burning material from dropping into concealed combustible spaces.

Part IV — Energy Conservation

CHAPTER 11

ENERGY EFFICIENCY

SECTION N1101
GENERAL

N1101.1 Scope. This chapter sets forth energy-efficiency-related requirements for the design and construction of buildings regulated by this code.

Exception: Portions of the building thermal envelope that do not enclose conditioned space.

N1101.2 Compliance. Compliance with this chapter shall be demonstrated by meeting the requirements of the applicable sections and tables of this chapter. Where applicable, provisions are based on the climate zone where the building is located. The climate zone where the building is located shall be based on zone assignments in Table N1101.2 for the county and state in which the building is being constructed. Alternatively, the climate zone shall be permitted to be determined by the Heating Degree Days assigned by the building official.

N1101.2.1 Residential buildings, Type A-1. Compliance shall be demonstrated by either:

1. Meeting the requirements of this chapter for buildings with a glazing area that does not exceed 15 percent of the gross area of exterior walls; or
2. Meeting the requirements of the *International Energy Conservation Code* for residential buildings, Type A-1.

N1101.2.2 Residential buildings, Type A-2. Compliance shall be demonstrated by either:

1. Meeting the requirements of this chapter for buildings with a glazing area that does not exceed 25 percent of the gross area of exterior walls; or
2. Meeting the requirements of the *International Energy Conservation Code* for residential buildings, Type A-2.

N1101.3 Materials and equipment. Materials and equipment shall be identified in a manner that will allow a determination of their compliance with the provisions of this chapter. Materials and equipment used to conform to the applicable provisions of this chapter shall be installed in accordance with the manufacturer's installation instructions.

N1101.3.1 Insulation. The thermal resistance (*R*-value) shall be indicated on all insulation and the insulation installed such that the *R*-value can be verified during inspection, or a certification of the installed *R*-value shall be provided at the job site by the insulation installer. Where blown-in or sprayed insulation is applied in walls, the installer shall provide a certification of the installed density and *R*-value. Where blown-in or sprayed insulation is applied in the roof-ceiling assembly, the installer shall provide a certification of the initial installed thickness, settled thickness, coverage area, and number of bags of insulating material installed. Markers shall be provided for every 300 square feet (28 m^2) of attic area, attached to the trusses, rafters, or joists, and indicate in 1-inch-high (25.4 mm) numbers the installed thickness of the insulation.

N1101.3.2 Fenestration. The *U*-factor of fenestration shall be determined in accordance with NFRC 100 by an accredited, independent laboratory, and labeled and certified by the manufacturer. The solar heat gain coefficient (SHGC) of fenestration shall be determined in accordance with NFRC 200 by an accredited, independent laboratory, and labeled and certified by the manufacturer.

N1101.3.2.1 Default fenestration performance. When a manufacturer has not determined a fenestration product's *U*-factor in accordance with NFRC 100, compliance shall be determined by assigning such products a default *U*-factor from Tables 102.5.2(1) and 102.5.2(2) in the *International Energy Conservation Code.* When a manufacturer has not determined a fenestration product's SHGC in accordance with NFRC 200, compliance shall be determined by assigning such products a default SHGC from Table 102.5.2(3) of the *International Energy Conservation Code.*

N1101.3.2.2 Air leakage. The air leakage of prefabricated fenestration shall be determined in accordance with AAMA/WDMA 101/I.S.2 by an accredited, independent laboratory, and labeled and certified by the manufacturer. Alternatively, the manufacturer shall certify that the fenestration is installed in accordance with Section 502.1.4 of the *International Energy Conservation Code.*

N1101.3.3 Maintenance. Where mechanical or plumbing system components require preventive maintenance for efficient operation, regular maintenance requirements shall be clearly stated and affixed to the component, or the source for such information shall be shown on a label attached to the component.

TABLE N1101.2
CLIMATE ZONES BY STATES AND COUNTIES

County	Zone
ALABAMA	
Autauga[a]	6
Baldwin[a]	4
Barbour[a]	5
Bibb[a]	6
Blount	7
Bullock[a]	5
Butler[a]	5
Calhoun[a]	6
Chambers[a]	6
Cherokee	7
Chilton[a]	6
Choctaw[a]	5
Clarke[a]	5
Clay	7
Cleburne	7
Coffee[a]	4
Colbert	8
Conecuh[a]	5
Coosa[a]	6
Covington[a]	4
Crenshaw[a]	5
Cullman	7
Dale[a]	4
Dallas[a]	5
De Kalb	8
Elmore[a]	6
Escambia[a]	4
Etowah	7
Fayette	7
Franklin	8
Geneva[a]	4
Greene[a]	5
Hale[a]	5
Henry[a]	4
Houston[a]	4
Jackson	8
Jefferson[a]	6
Lamar	7
Lauderdale	8
Lawrence	8
Lee[a]	6
Limestone	8
Lowndes[a]	5
Macon[a]	6
Madison	8
Marengo[a]	5
Marion	7
Marshall	8
Mobile[a]	4
Monroe[a]	5
Montgomery[a]	6
Morgan	8
Perry[a]	5
Pickens[a]	6
Pike[a]	5
Randolph	7
Russell[a]	5
Shelby[a]	6
St Clair[a]	6
Sumter[a]	5
Talladega[a]	6
Tallapoosa[a]	6
Tuscaloosa[a]	6
Walker[a]	6
Washington[a]	5
Wilcox[a]	5
Winston	7
ALASKA BOROUGHS AND REAA's	
Adak Region	16
Alaska Gateway	19
Aleutian Region	17
Aleutians East	17
Anchorage	17
Annette Island	15
Bering Straits	19
Bristol Bay	17
Chatham	16
Chugach	17
Copper River	18
Delta/Greely	18
Denali	18
Fairbanks N. Star	18
Haines	16
Iditarod Area	19
Juneau	16
Kashunamiut	18
Kenai Peninsula	17
Ketchikan Gateway	15
Kodiak Island	16
Kuspuk	18
Lake & Peninsula	17
Lower Kuskokwim	18
Lower Yukon	18
Matanuska-Susitna	17
North Slope	19
Northwest Arctic	19
Pribilof Islands	17
Sitka	15
Southeast Island	15
Southwest Region	17
Yakutat	17
Yukon Flats	19
Yukon-Koyukuk	19
Yupiit	18
ARIZONA	
Apache	13
Cochise	6
Coconino	14
Gila	8
Graham	6
Greenlee	6
La Paz	3
Maricopa	3
Mohave	7
Navajo	10
Pima	4
Pinal	4
Santa Cruz	6
Yavapai	10
Yuma	3
ARKANSAS	
Arkansas[a]	6
Ashley[a]	6
Baxter	9
Benton	9
Boone	9
Bradley[a]	6
Calhoun[a]	6
Carroll	9
Chicot[a]	6
Clark[a]	6
Clay	8
Cleburne	8
Cleveland[a]	6
Columbia[a]	6
Conway[a]	7
Craighead	8
Crawford	8
Crittenden[a]	7
Cross[a]	7
Dallas[a]	6
Desha[a]	6
Drew[a]	6
Faulkner[a]	7
Franklin	8
Fulton	8
Garland[a]	7
Grant[a]	6
Greene	8
Hempstead[a]	7
Hot Spring[a]	7
Howard[a]	7
Independence	8
Izard	8
Jackson	8
Jefferson[a]	6
Johnson	8
Lafayette[a]	6
Lawrence	8
Lee[a]	7
Lincoln[a]	6
Little River[a]	6
Logan[a]	7
Lonoke[a]	7
Madison	9
Marion	9
Miller[a]	6
Mississippi	8
Monroe[a]	7
Montgomery	8
Nevada[a]	6
Newton	9
Ouachita[a]	6
Perry[a]	7
Phillips[a]	7
Pike[a]	7
Poinsett	8
Polk	8
Pope	8
Prairie[a]	7
Pulaski[a]	7
Randolph	8
Saline[a]	7
Scott[a]	7
Searcy	9
Sebastian	8
Sevier[a]	7
Sharp	8
St Francis[a]	7
Stone	9
Union[a]	6
Van Buren	8
Washington	9
White[a]	7
Woodruff[a]	7
Yell[a]	7
CALIFORNIA	
Alameda	6
Alpine	15
Amador	8
Butte	6
Calaveras	8
Colusa	6
Contra Costa	6
Del Norte	9
El Dorado	8
Fresno	6
Glenn	6
Humboldt	9
Imperial	3
Inyo	9
Kern	5
Kings	6
Lake	8
Lassen	13
Los Angeles	4
Madera	6
Marin	6
Mariposa	8
Mendocino	8
Merced	6
Modoc	15
Mono	15
Monterey	6
Napa	6
Nevada	11
Orange	4
Placer	8
Plumas	13
Riverside	4
Sacramento	6
San Benito	6
San Bernardino	4
San Diego	3
San Francisco	6
San Joaquin	6
San Luis Obispo	6
San Mateo	6
Santa Barbara	5
Santa Clara	6
Santa Cruz	6
Shasta	6
Sierra	11
Siskiyou	11
Solano	6
Sonoma	6
Stanislaus	6
Sutter	6
Tehama	6
Trinity	9
Tulare	6
Tuolume	8
Ventura	4
Yolo	6
Yuba	6
COLORADO	
Adams	13
Alamosa	16
Arapahoe	13
Archuleta	16
Baca	11
Bent	11
Boulder	13
Chaffee	16
Cheyenne	13
Clear Creek	17
Conejos	16
Costilla	16
Crowley	11
Custer	16
Delta	13
Denver	13
Dolores	15
Douglas	13
Eagle	15
El Paso	13
Elbert	13
Fremont	11
Garfield	15
Gilpin	13
Grand	17
Gunnison	17
Hinsdale	17
Huerfano	11
Jackson	17
Jefferson	13
Kiowa	13
Kit Carson	13
La Plata	15
Lake	17
Larimer	13
Las Animas	11
Lincoln	13
Logan	13
Mesa	13
Mineral	17
Moffat	15
Montezuma	15
Montrose	13
Morgan	13
Otero	11
Ouray	15
Park	17
Phillips	13
Pitkin	17
Prowers	11
Pueblo	11
Rio Blanco	15
Rio Grande	17
Routt	17
Saguache	16
San Juan	17
San Miguel	15
Sedgwick	13
Summit	17
Teller	13
Washington	13
Weld	13
Yuma	13
CONNECTICUT	
Fairfield	12
Hartford	13
Litchfield	14
Middlesex	12
New Haven	12
New London	12
Tolland	14
Windham	14
DELAWARE	
Kent	9
New Castle	10
Sussex	9
DC	
Washington	10
FLORIDA	
Alachua[a]	3
Baker[a]	3
Bay[a]	4
Bradford[a]	3
Brevard[a]	2
Broward[a]	1
Calhoun[a]	4
Charlotte[a]	2
Citrus[a]	2
Clay[a]	3
Collier[a]	1
Columbia[a]	3
Dade[a]	1
De Soto[a]	2
Dixie[a]	3
Duval[a]	3
Escambia[a]	4
Flagler[a]	3
Franklin[a]	4
Gadsden[a]	4

(continued)

TABLE N1101.2—continued
CLIMATE ZONES BY STATES AND COUNTIES

County	Zone
Gilchrist[a]	3
Glades[a]	1
Gulf[a]	4
Hamilton[a]	3
Hardee[a]	2
Hendry[a]	1
Hernando[a]	2
Highlands[a]	2
Hillsborough[a]	2
Holmes[a]	4
Indian River[a]	2
Jackson[a]	4
Jefferson[a]	4
Lafayette[a]	3
Lake[a]	2
Lee[a]	1
Leon[a]	4
Levy[a]	2
Liberty[a]	4
Madison[a]	3
Manatee[a]	2
Marion[a]	2
Martin[a]	1
Monroe[a]	1
Nassau[a]	3
Okaloosa[a]	4
Okeechobee[a]	2
Orange[a]	2
Osceola[a]	2
Palm Beach[a]	1
Pasco[a]	2
Pinellas[a]	2
Polk[a]	2
Putnam[a]	3
Santa Rosa[a]	4
Sarasota[a]	2
Seminole[a]	2
St Johns[a]	3
St Lucie[a]	2
Sumter[a]	2
Suwannee[a]	3
Taylor[a]	3
Union[a]	3
Volusia[a]	2
Wakulla[a]	4
Walton[a]	4
Washington[a]	4
GEORGIA	
Appling[a]	4
Atkinson[a]	4
Bacon[a]	4
Baker[a]	4
Baldwin[a]	6
Banks	7
Barrow	7
Bartow	7
Ben Hill[a]	5
Berrien[a]	4
Bibb[a]	5
Bleckley[a]	5
Brantley[a]	4
Brooks[a]	4
Bryan[a]	4
Bulloch[a]	5
Burke[a]	6
Butts	7
Calhoun[a]	5
Camden[a]	4
Candler[a]	5
Carroll	7
Catoosa	8
Charlton[a]	4
Chatham[a]	4
Chattahoochee[a]	5
Chattooga	8
Cherokee	8
Clarke	7
Clay[a]	5
Clayton	7
Clinch[a]	4
Cobb	7
Coffee[a]	5
Colquitt[a]	4
Columbia[a]	6
Cook[a]	4
Coweta	7
Crawford[a]	5
Crisp[a]	5
Dade	8
Dawson	8
De Kalb	7
Decatur[a]	4
Dodge[a]	5
Dooly[a]	5
Dougherty[a]	5
Douglas	7
Early[a]	5
Echols[a]	4
Effingham[a]	4
Elbert	7
Emanuel[a]	5
Evans[a]	4
Fannin	8
Fayette	7
Floyd	7
Forsyth	8
Franklin	7
Fulton	7
Gilmer	8
Glascock[a]	6
Glynn[a]	4
Gordon	8
Grady[a]	4
Greene[a]	6
Gwinnett	7
Habersham	8
Hall	7
Hancock[a]	6
Haralson	7
Harris[a]	6
Hart	7
Heard[a]	6
Henry	7
Houston[a]	5
Irwin[a]	5
Jackson	7
Jasper[a]	6
Jeff Davis[a]	4
Jefferson[a]	6
Jenkins[a]	5
Johnson[a]	5
Jones[a]	6
Lamar[a]	6
Lanier[a]	4
Laurens[a]	5
Lee[a]	5
Liberty[a]	4
Lincoln[a]	6
Long[a]	4
Lowndes[a]	4
Lumpkin	8
Macon[a]	5
Madison	7
Marion[a]	5
Mcduffie[a]	6
Mcintosh[a]	4
Meriwether[a]	6
Miller[a]	4
Mitchell[a]	4
Monroe[a]	6
Montgomery[a]	5
Morgan[a]	6
Murray	8
Muscogee[a]	5
Newton	7
Oconee	7
Oglethorpe	7
Paulding	7
Peach[a]	5
Pickens	8
Pierce[a]	4
Pike[a]	6
Polk	7
Pulaski[a]	5
Putnam[a]	6
Quitman[a]	5
Rabun	8
Randolph[a]	5
Richmond[a]	6
Rockdale	7
Schley[a]	5
Screven[a]	5
Seminole[a]	4
Spalding	7
Stephens	7
Stewart[a]	5
Sumter[a]	5
Talbot[a]	5
Taliaferro[a]	6
Tattnall[a]	4
Taylor[a]	5
Telfair[a]	5
Terrell[a]	5
Thomas[a]	4
Tift[a]	5
Toombs[a]	4
Towns	8
Treutlen[a]	5
Troup[a]	6
Turner[a]	5
Twiggs[a]	5
Union	8
Upson[a]	5
Walker	8
Walton	7
Ware[a]	4
Warren[a]	6
Washington[a]	6
Wayne[a]	4
Webster[a]	5
Wheeler[a]	5
White	8
Whitfield	8
Wilcox[a]	5
Wilkes	7
Wilkinson[a]	5
Worth[a]	5
HAWAII	
Hawaii	1
Honolulu	1
Kalawao	1
Kauai	1
Maui	1
IDAHO	
Ada	12
Adams	15
Bannock	15
Bear Lake	15
Benewah	14
Bingham	15
Blaine	15
Boise	15
Bonner	15
Bonneville	15
Boundary	15
Butte	16
Camas	15
Canyon	12
Caribou	15
Cassia	14
Clark	15
Clearwater	12
Custer	16
Elmore	13
Franklin	15
Fremont	16
Gem	13
Gooding	13
Idaho	15
Jefferson	16
Jerome	14
Kootenai	14
Latah	14
Lemhi	15
Lewis	15
Lincoln	15
Madison	16
Minidoka	15
Nez Perce	12
Oneida	15
Owyhee	12
Payette	12
Power	15
Shoshone	14
Teton	16
Twin Falls	14
Valley	16
Washington	13
ILLINOIS	
Adams	12
Alexander	10
Bond	11
Boone	14
Brown	12
Bureau	13
Calhoun	11
Carroll	14
Cass	12
Champaign	12
Christian	11
Clark	12
Clay	11
Clinton	10
Coles	12
Cook	14
Crawford	11
Cumberland	12
De Kalb	14
De Witt	12
Douglas	12
Du Page	14
Edgar	12
Edwards	11
Effingham	11
Fayette	11
Ford	13
Franklin	10
Fulton	13
Gallatin	10
Greene	11
Grundy	13
Hamilton	10
Hancock	13
Hardin	10
Henderson	13
Henry	13
Iroquois	13
Jackson	10
Jasper	11
Jefferson	11
Jersey	10
Jo Daviess	14
Johnson	10
Kane	14
Kankakee	13
Kendall	13
Knox	13
La Salle	13
Lake	14
Lawrence	11
Lee	14
Livingston	13
Logan	12
Macon	12
Macoupin	11
Madison	10
Marion	11
Marshall	13
Mason	12
Massac	10
Mcdonough	13
Mchenry	14
Mclean	12
Menard	12
Mercer	13
Monroe	10
Montgomery	11
Morgan	12
Moultrie	12
Ogle	14
Peoria	13
Perry	10
Piatt	12
Pike	12
Pope	10
Pulaski	10
Putnam	13
Randolph	10
Richland	11
Rock Island	13
Saline	10
Sangamon	12
Schuyler	12
Scott	12
Shelby	11
St Clair	10
Stark	13
Stephenson	14
Tazewell	12
Union	10
Vermilion	12
Wabash	11
Warren	13
Washington	10
Wayne	11
White	10
Whiteside	14
Will	13
Williamson	10
Winnebago	14
Woodford	13

(continued)

TABLE N1101.2—continued
CLIMATE ZONES BY STATES AND COUNTIES

County	Zone
INDIANA	
Adams	13
Allen	13
Bartholomew	11
Benton	13
Blackford	13
Boone	12
Brown	11
Carroll	13
Cass	13
Clark	10
Clay	12
Clinton	13
Crawford	11
Daviess	11
De Kalb	13
Dearborn	11
Decatur	12
Delaware	13
Dubois	11
Elkhart	13
Fayette	12
Floyd	10
Fountain	12
Franklin	12
Fulton	14
Gibson	10
Grant	13
Greene	11
Hamilton	12
Hancock	12
Harrison	10
Hendricks	12
Henry	12
Howard	13
Huntington	14
Jackson	11
Jasper	13
Jay	13
Jefferson	10
Jennings	11
Johnson	12
Knox	11
Kosciusko	14
La Porte	13
Lagrange	14
Lake	13
Lawrence	11
Madison	13
Marion	12
Marshall	13
Martin	11
Miami	14
Monroe	11
Montgomery	12
Morgan	12
Newton	13
Noble	14
Ohio	11
Orange	11
Owen	12
Parke	12
Perry	10
Pike	11
Porter	13
Posey	10
Pulaski	13
Putnam	12
Randolph	13
Ripley	11
Rush	12
Scott	11
Shelby	12
Spencer	10
St Joseph	13
Starke	13
Steuben	14
Sullivan	11
Switzerland	10
Tippecanoe	13
Tipton	13
Union	12
Vanderburgh	10
Vermillion	12
Vigo	12
Wabash	14
Warren	12
Warrick	10
Washington	11
Wayne	12
Wells	13
White	13
Whitley	14
IOWA	
Adair	14
Adams	13
Allamakee	15
Appanoose	13
Audubon	14
Benton	14
Black Hawk	15
Boone	14
Bremer	15
Buchanan	15
Buena Vista	15
Butler	15
Calhoun	15
Carroll	14
Cass	14
Cedar	14
Cerro Gordo	15
Cherokee	15
Chickasaw	15
Clarke	13
Clay	15
Clayton	15
Clinton	13
Crawford	14
Dallas	14
Davis	13
Decatur	13
Delaware	15
Des Moines	13
Dickinson	15
Dubuque	14
Emmet	15
Fayette	15
Floyd	15
Franklin	15
Fremont	13
Greene	14
Grundy	15
Guthrie	14
Hamilton	15
Hancock	15
Hardin	15
Harrison	14
Henry	13
Howard	15
Humboldt	15
Ida	15
Iowa	14
Jackson	14
Jasper	14
Jefferson	13
Johnson	13
Jones	14
Keokuk	13
Kossuth	15
Lee	13
Linn	14
Louisa	13
Lucas	13
Lyon	15
Madison	14
Mahaska	13
Marion	13
Marshall	14
Mills	13
Mitchell	15
Monona	14
Monroe	13
Montgomery	13
Muscatine	13
Obrien	15
Osceola	15
Page	13
Palo Alto	15
Plymouth	15
Pocahontas	15
Polk	14
Pottawattamie	14
Poweshiek	14
Ringgold	13
Sac	15
Scott	13
Shelby	14
Sioux	15
Story	14
Tama	14
Taylor	13
Union	13
Van Buren	13
Wapello	13
Warren	14
Washington	13
Wayne	13
Webster	15
Winnebago	15
Winneshiek	15
Woodbury	15
Worth	15
Wright	15
KANSAS	
Allen	10
Anderson	10
Atchison	11
Barber	9
Barton	11
Bourbon	10
Brown	11
Butler	10
Chase	10
Chautauqua	9
Cherokee	9
Cheyenne	13
Clark	10
Clay	11
Cloud	12
Coffey	10
Comanche	9
Cowley	9
Crawford	9
Decatur	13
Dickinson	11
Doniphan	11
Douglas	10
Edwards	11
Elk	9
Ellis	12
Ellsworth	11
Finney	11
Ford	11
Franklin	10
Geary	11
Gove	12
Graham	12
Grant	11
Gray	11
Greeley	12
Greenwood	10
Hamilton	11
Harper	9
Harvey	11
Haskell	11
Hodgeman	11
Jackson	11
Jefferson	11
Jewell	12
Johnson	11
Kearny	11
Kingman	10
Kiowa	10
Labette	9
Lane	12
Leavenworth	11
Lincoln	11
Linn	10
Logan	12
Lyon	11
Marion	11
Marshall	12
Mcpherson	11
Meade	10
Miami	10
Mitchell	12
Montgomery	9
Morris	11
Morton	10
Nemaha	11
Neosho	9
Ness	12
Norton	13
Osage	10
Osborne	12
Ottawa	11
Pawnee	11
Phillips	12
Pottawatomie	11
Pratt	10
Rawlins	13
Reno	11
Republic	12
Rice	11
Riley	11
Rooks	12
Rush	11
Russell	11
Saline	11
Scott	12
Sedgwick	10
Seward	10
Shawnee	11
Sheridan	12
Sherman	13
Smith	12
Stafford	11
Stanton	11
Stevens	10
Sumner	9
Thomas	13
Trego	12
Wabaunsee	11
Wallace	12
Washington	12
Wichita	12
Wilson	9
Woodson	10
Wyandotte	11
KENTUCKY	
Adair	9
Allen	9
Anderson	10
Ballard	9
Barren	9
Bath	11
Bell	10
Boone	11
Bourbon	10
Boyd	11
Boyle	10
Bracken	11
Breathitt	10
Breckenridge	9
Bullitt	10
Butler	9
Caldwell	9
Calloway	9
Campbell	11
Carlisle	9
Carroll	10
Carter	11
Casey	10
Christian	9
Clark	10
Clay	10
Clinton	10
Crittenden	9
Cumberland	9
Daviess	9
Edmonson	9
Elliot	11
Estill	10
Fayette	10
Fleming	11
Floyd	10
Franklin	10
Fulton	9
Gallatin	11
Garrard	10
Grant	11
Graves	9
Grayson	9
Green	9
Greenup	11
Hancock	9
Hardin	9
Harlan	10
Harrison	11
Hart	9
Henderson	9
Henry	10
Hickman	9
Hopkins	9
Jackson	10
Jefferson	10
Jessamine	10
Johnson	11
Kenton	11
Knott	10
Knox	10
Larue	9

(continued)

TABLE N1101.2—continued
CLIMATE ZONES BY STATES AND COUNTIES

County	Zone
Laurel	10
Lawrence	11
Lee	10
Leslie	10
Letcher	10
Lewis	11
Lincoln	10
Livingston	9
Logan	9
Lyon	9
Madison	10
Magoffin	10
Marion	10
Marshall	9
Martin	11
Mason	11
Mccracken	9
Mccreary	10
Mclean	9
Meade	9
Menifee	10
Mercer	10
Metcalfe	9
Monroe	9
Montgomery	10
Morgan	10
Muhlenberg	9
Nelson	10
Nicholas	11
Ohio	9
Oldham	10
Owen	10
Owsley	10
Pendleton	11
Perry	10
Pike	10
Powell	10
Pulaski	10
Robertson	11
Rockcastle	10
Rowan	11
Russell	10
Scott	11
Shelby	10
Simpson	9
Spencer	10
Taylor	9
Todd	9
Trigg	9
Trimble	10
Union	9
Warren	9
Washington	10
Wayne	10
Webster	9
Whitley	10
Wolfe	10
Woodford	10
LOUISIANA	
Acadia[a]	4
Allen[a]	4
Ascension[a]	4
Assumption[a]	3
Avoyelles[a]	5
Beauregard[a]	4
Bienville[a]	6
Bossier[a]	6
Caddo[a]	6
Calcasieu[a]	4
Caldwell[a]	6
Cameron[a]	4
Catahoula[a]	5
Claiborne[a]	6
Concordia[a]	5
De Soto[a]	5
East Baton Rouge[a]	4
East Carroll[a]	6
East Feliciana[a]	4
Evangeline[a]	4
Franklin[a]	6
Grant[a]	5
Iberia[a]	4
Iberville[a]	4
Jackson[a]	6
Jefferson[a]	3
Jefferson Davis[a]	4
La Salle[a]	5
Lafayette[a]	4
Lafourche[a]	3
Lincoln[a]	6
Livingston[a]	4
Madison[a]	6
Morehouse[a]	6
Natchitoches[a]	5
Orleans[a]	3
Ouachita[a]	6
Plaquemines[a]	3
Pointe Coupee[a]	4
Rapides[a]	5
Red River[a]	5
Richland[a]	6
Sabine[a]	5
St Bernard[a]	3
St Charles[a]	3
St Helena[a]	4
St James[a]	3
St John The Baptist[a]	3
St Landry[a]	4
St Martin[a]	4
St Mary[a]	3
St Tammany[a]	4
Tangipahoa[a]	4
Tensas[a]	5
Terrebonne[a]	3
Union[a]	6
Vermilion[a]	4
Vernon[a]	5
Washington[a]	4
Webster[a]	6
West Baton Rouge[a]	4
West Carroll[a]	6
West Feliciana[a]	4
Winn[a]	5
MAINE	
Androscoggin	15
Aroostook	17
Cumberland	15
Franklin	16
Hancock	15
Kennebec	15
Knox	15
Lincoln	15
Oxford	16
Penobscot	15
Piscataquis	17
Sagadahoc	15
Somerset	17
Waldo	15
Washington	15
York	15
MARYLAND	
Allegany	12
Anne Arundel	9
Baltimore	10
Baltimore City	9
Calvert	9
Caroline	9
Carroll	11
Cecil	10
Charles	9
Dorchester	9
Frederick	11
Garrett	13
Harford	10
Howard	10
Kent	10
Montgomery	10
Prince Georges	10
Queen Annes	9
Somerset	9
St Marys	9
Talbot	9
Washington	11
Wicomico	9
Worcester	9
MASSACHUSETTS	
Barnstable	12
Berkshire	14
Bristol	12
Dukes	12
Essex	13
Franklin	14
Hampden	14
Hampshire	14
Middlesex	13
Nantucket	12
Norfolk	13
Plymouth	12
Suffolk	13
Worcester	14
MICHIGAN	
Alcona	15
Alger	16
Allegan	14
Alpena	15
Antrim	15
Arenac	15
Baraga	17
Barry	14
Bay	15
Benzie	15
Berrien	14
Branch	14
Calhoun	14
Cass	14
Charlevoix	15
Cheboygan	15
Chippewa	16
Clare	15
Clinton	14
Crawford	15
Delta	16
Dickinson	16
Eaton	14
Emmet	15
Genesee	14
Gladwin	15
Gogebic	17
Grand Traverse	15
Gratiot	14
Hillsdale	14
Houghton	17
Huron	14
Ingham	14
Ionia	14
Iosco	15
Iron	17
Isabella	15
Jackson	14
Kalamazoo	14
Kalkaska	15
Kent	14
Keweenaw	17
Lake	15
Lapeer	14
Leelanau	15
Lenawee	14
Livingston	14
Luce	16
Mackinac	16
Macomb	14
Manistee	15
Marquette	16
Mason	15
Mecosta	15
Menominee	16
Midland	15
Missaukee	15
Monroe	13
Montcalm	14
Montmorency	15
Muskegon	14
Newaygo	15
Oakland	14
Oceana	15
Ogemaw	15
Ontonagon	17
Osceola	15
Oscoda	15
Otsego	15
Ottawa	14
Presque Isle	15
Roscommon	15
Saginaw	14
Sanilac	14
Schoolcraft	16
Shiawassee	14
St Clair	14
St Joseph	14
Tuscola	14
Van Buren	14
Washtenaw	13
Wayne	13
Wexford	15
MINNESOTA	
Aitkin	17
Anoka	16
Becker	17
Beltrami	17
Benton	16
Big Stone	16
Blue Earth	15
Brown	15
Carlton	17
Carver	15
Cass	17
Chippewa	16
Chisago	16
Clay	17
Clearwater	17
Cook	17
Cottonwood	15
Crow Wing	17
Dakota	15
Dodge	15
Douglas	16
Faribault	15
Fillmore	15
Freeborn	15
Goodhue	15
Grant	16
Hennepin	15
Houston	15
Hubbard	17
Isanti	16
Itasca	17
Jackson	15
Kanabec	16
Kandiyohi	16
Kittson	17
Koochiching	17
Lac Qui Parle	15
Lake	17
Lake Of The Woods	17
Le Sueur	15
Lincoln	15
Lyon	15
Mahnomen	17
Marshall	17
Martin	15
Mcleod	15
Meeker	16
Mille Lacs	16
Morrison	16
Mower	15
Murray	15
Nicollet	15
Nobles	15
Norman	17
Olmsted	15
Otter Tail	17
Pennington	17
Pine	16
Pipestone	15
Polk	17
Pope	16
Ramsey	15
Red Lake	17
Redwood	15
Renville	15
Rice	15
Rock	15
Roseau	17
Scott	15
Sherburne	16
Sibley	15
St Louis	17
Stearns	16
Steele	15
Stevens	16
Swift	16
Todd	16
Traverse	16
Wabasha	15
Wadena	17
Waseca	15
Washington	15
Watonwan	15
Wilkin	17
Winona	15
Wright	16
Yellow Medicine	15
MISSISSIPPI	
Adams[a]	5
Alcorn	7
Amite[a]	4
Attala[a]	6
Benton	7

(continued)

TABLE N1101.2—continued
CLIMATE ZONES BY STATES AND COUNTIES

County	Zone
Bolivar[a]	6
Calhoun[a]	6
Carroll[a]	6
Chickasaw[a]	6
Choctaw[a]	6
Claiborne[a]	5
Clarke[a]	5
Clay[a]	6
Coahoma	7
Copiah[a]	5
Covington[a]	5
De Soto	7
Forrest[a]	5
Franklin[a]	5
George[a]	4
Greene[a]	5
Grenada[a]	6
Hancock[a]	4
Harrison[a]	4
Hinds[a]	6
Holmes[a]	6
Humphreys[a]	6
Issaquena[a]	6
Itawamba	7
Jackson[a]	4
Jasper[a]	5
Jefferson[a]	5
Jefferson Davis[a]	5
Jones[a]	5
Kemper[a]	6
Lafayette	7
Lamar[a]	4
Lauderdale[a]	6
Lawrence[a]	5
Leake[a]	6
Lee	7
Leflore[a]	6
Lincoln[a]	5
Lowndes[a]	6
Madison[a]	6
Marion[a]	4
Marshall	7
Monroe[a]	6
Montgomery[a]	6
Neshoba[a]	6
Newton[a]	6
Noxubee[a]	6
Oktibbeha[a]	6
Panola	7
Pearl River[a]	4
Perry[a]	5
Pike[a]	4
Pontotoc	7
Prentiss	7
Quitman	7
Rankin[a]	6
Scott[a]	6
Sharkey[a]	6
Simpson[a]	5
Smith[a]	5
Stone[a]	4
Sunflower[a]	6
Tallahatchie	7
Tate	7
Tippah	7
Tishomingo	7
Tunica	7
Union	7
Walthall[a]	4
Warren[a]	6
Washington[a]	6
Wayne[a]	5
Webster[a]	6
Wilkinson[a]	4
Winston[a]	6
Yalobusha	7
Yazoo[a]	6
MISSOURI	
Adair	12
Andrew	12
Atchison	13
Audrain	12
Barry	9
Barton	10
Bates	11
Benton	11
Bollinger	10
Boone	11
Buchanan	12
Butler	9
Caldwell	12
Callaway	11
Camden	11
Cape Girardeau	9
Carroll	12
Carter	10
Cass	11
Cedar	11
Chariton	12
Christian	10
Clark	13
Clay	11
Clinton	12
Cole	11
Cooper	11
Crawford	10
Dade	10
Dallas	10
Daviess	12
De Kalb	12
Dent	10
Douglas	10
Dunklin	9
Franklin	10
Gasconade	11
Gentry	13
Greene	10
Grundy	12
Harrison	13
Henry	11
Hickory	11
Holt	12
Howard	11
Howell	9
Iron	10
Jackson	11
Jasper	9
Jefferson	10
Johnson	11
Knox	12
Laclede	10
Lafayette	11
Lawrence	10
Lewis	12
Lincoln	11
Linn	12
Livingston	12
Macon	12
Madison	10
Maries	11
Marion	12
Mcdonald	9
Mercer	13
Miller	11
Mississippi	9
Moniteau	11
Monroe	12
Montgomery	11
Morgan	11
New Madrid	9
Newton	9
Nodaway	13
Oregon	9
Osage	11
Ozark	9
Pemiscot	9
Perry	10
Pettis	11
Phelps	10
Pike	12
Platte	11
Polk	10
Pulaski	10
Putnam	13
Ralls	12
Randolph	12
Ray	11
Reynolds	10
Ripley	9
Saline	11
Schuyler	13
Scotland	13
Scott	9
Shannon	10
Shelby	12
St Charles	10
St Clair	11
St Francois	10
St Louis	10
St Louis City	10
Ste Genevieve	10
Stoddard	9
Stone	9
Sullivan	12
Taney	9
Texas	10
Vernon	11
Warren	11
Washington	10
Wayne	10
Webster	10
Worth	13
Wright	10
MONTANA	
Beaverhead	15
Big Horn	15
Blaine	16
Broadwater	15
Carbon	15
Carter	15
Cascade	15
Chouteau	15
Custer	15
Daniels	16
Dawson	15
Deer Lodge	16
Fallon	15
Fergus	15
Flathead	16
Gallatin	15
Garfield	15
Glacier	16
Golden Valley	15
Granite	16
Hill	16
Jefferson	15
Judith Basin	15
Lake	15
Lewis And Clark	15
Liberty	16
Lincoln	15
Madison	15
McCone	15
Meagher	15
Mineral	15
Missoula	15
Musselshell	15
Park	15
Petroleum	15
Phillips	16
Pondera	16
Powder River	15
Powell	16
Prairie	15
Ravalli	15
Richland	15
Roosevelt	16
Rosebud	15
Sanders	15
Sheridan	16
Silver Bow	16
Stillwater	15
Sweet Grass	15
Teton	15
Toole	16
Treasure	15
Valley	16
Wheatland	15
Wibaux	15
Yellowstone	15
Yellowstone National Park	15
NEBRASKA	
Adams	13
Antelope	14
Arthur	14
Banner	14
Blaine	14
Boone	14
Box Butte	15
Boyd	14
Brown	14
Buffalo	13
Burt	14
Butler	13
Cass	13
Cedar	14
Chase	13
Cherry	14
Cheyenne	14
Clay	13
Colfax	13
Cuming	14
Custer	14
Dakota	14
Dawes	15
Dawson	13
Deuel	14
Dixon	14
Dodge	13
Douglas	13
Dundy	13
Fillmore	13
Franklin	13
Frontier	13
Furnas	13
Gage	13
Garden	14
Garfield	14
Gosper	13
Grant	14
Greeley	14
Hall	13
Hamilton	13
Harlan	13
Hayes	13
Hitchcock	13
Holt	14
Hooker	14
Howard	14
Jefferson	13
Johnson	13
Kearney	13
Keith	14
Keya Paha	14
Kimball	14
Knox	14
Lancaster	13
Lincoln	14
Logan	14
Loup	14
Madison	14
Mcpherson	14
Merrick	13
Morrill	14
Nance	13
Nemaha	13
Nuckolls	13
Otoe	13
Pawnee	13
Perkins	13
Phelps	13
Pierce	14
Platte	13
Polk	13
Red Willow	13
Richardson	13
Rock	14
Saline	13
Sarpy	13
Saunders	13
Scotts Bluff	14
Seward	13
Sheridan	15
Sherman	14
Sioux	15
Stanton	14
Thayer	13
Thomas	14
Thurston	14
Valley	14
Washington	13
Wayne	14
Webster	13
Wheeler	14
York	13
NEVADA	
Carson City	12
Churchill	12
Clark	5
Douglas	13
Elko	15
Esmeralda	12
Eureka	15
Humboldt	13
Lander	13
Lincoln	12
Lyon	13
Mineral	12
Nye	12
Pershing	12

(continued)

TABLE N1101.2—continued
CLIMATE ZONES BY STATES AND COUNTIES

County	Zone
Storey	12
Washoe	12
White Pine	15
NEW HAMPSHIRE	
Belknap	15
Carroll	15
Cheshire	15
Coos	16
Grafton	15
Hillsborough	15
Merrimack	15
Rockingham	15
Strafford	15
Sullivan	15
NEW JERSEY	
Atlantic	10
Bergen	12
Burlington	11
Camden	10
Cape May	10
Cumberland	10
Essex	11
Gloucester	10
Hudson	11
Hunterdon	12
Mercer	11
Middlesex	11
Monmouth	11
Morris	12
Ocean	11
Passaic	12
Salem	10
Somerset	12
Sussex	13
Union	11
Warren	12
NEW MEXICO	
Bernalillo	9
Catron	11
Chaves	7
Cibola	12
Colfax	13
Curry	9
De Baca	9
Dona Ana	7
Eddy	7
Grant	9
Guadalupe	9
Harding	11
Hidalgo	7
Lea	7
Lincoln	9
Los Alamos	13
Luna	7
Mckinley	13
Mora	15
Otero	7
Quay	8
Rio Arriba	12
Roosevelt	8
San Juan	12
San Miguel	12
Sandoval	13
Santa Fe	13
Sierra	8
Socorro	9
Taos	15
Torrance	11
Union	11
Valencia	10
NEW YORK	
Albany	14
Allegheny	15
Bronx	11
Broome	15
Cattaraugus	15
Cayuga	14
Chautauqua	13
Chemung	15
Chenango	15
Clinton	15
Columbia	13
Cortland	15
Delaware	15
Dutchess	13
Erie	14
Essex	16
Franklin	16
Fulton	15
Genesee	14
Greene	14
Hamilton	16
Herkimer	15
Jefferson	15
Kings	10
Lewis	15
Livingston	14
Madison	14
Monroe	14
Montgomery	14
Nassau	11
New York	10
Niagara	14
Oneida	15
Onondaga	14
Ontario	14
Orange	12
Orleans	14
Oswego	14
Otsego	15
Putnam	12
Queens	10
Rensselaer	14
Richmond	11
Rockland	12
Saratoga	14
Schenectady	14
Schoharie	15
Schuyler	15
Seneca	14
St Lawrence	15
Steuben	15
Suffolk	11
Sullivan	15
Tioga	15
Tompkins	15
Ulster	15
Warren	15
Washington	15
Wayne	14
Westchester	12
Wyoming	14
Yates	14
NORTH CAROLINA	
Alamance	8
Alexander	8
Allegheny	11
Anson	7
Ashe	11
Avery	11
Beaufort[a]	6
Bertie	7
Bladen[a]	6
Brunswick[a]	6
Buncombe	9
Burke	8
Cabarrus	7
Caldwell	8
Camden	7
Carteret[a]	6
Caswell	8
Catawba	8
Chatham	8
Cherokee	9
Chowan	7
Clay	9
Cleveland	7
Columbus[a]	6
Craven[a]	6
Cumberland	7
Currituck	7
Dare[a]	6
Davidson	8
Davie	8
Duplin[a]	6
Durham	8
Edgecombe	7
Forsyth	8
Franklin	8
Gaston	7
Gates	7
Graham	9
Granville	8
Greene	7
Guilford	8
Halifax	7
Harnett	7
Haywood	9
Henderson	9
Hertford	7
Hoke	7
Hyde[a]	6
Iredell	8
Jackson	9
Johnston	7
Jones[a]	6
Lee	7
Lenoir	7
Lincoln	7
Macon	9
Madison	9
Martin	7
Mcdowell	8
Mecklenburg	7
Mitchell	11
Montgomery	7
Moore	7
Nash	7
New Hanover[a]	6
Northampton	7
Onslow[a]	6
Orange	8
Pamlico[a]	6
Pasquotank	7
Pender[a]	6
Perquimans	7
Person	8
Pitt	7
Polk	7
Randolph	8
Richmond	7
Robeson	7
Rockingham	8
Rowan	7
Rutherford	7
Sampson[a]	6
Scotland	7
Stanly	7
Stokes	9
Surry	9
Swain	9
Transylvania	9
Tyrrell[a]	6
Union	7
Vance	8
Wake	7
Warren	8
Washington	7
Watauga	11
Wayne	7
Wilkes	9
Wilson	7
Yadkin	8
Yancey	11
NORTH DAKOTA	
Adams	16
Barnes	17
Benson	17
Billings	16
Bottineau	17
Bowman	16
Burke	17
Burleigh	16
Cass	17
Cavalier	17
Dickey	16
Divide	17
Dunn	16
Eddy	17
Emmons	16
Foster	17
Golden Valley	16
Grand Forks	17
Grant	16
Griggs	17
Hettinger	16
Kidder	17
La Moure	16
Logan	16
Mchenry	17
Mcintosh	16
Mckenzie	16
Mclean	17
Mercer	16
Morton	16
Mountrail	17
Nelson	17
Oliver	16
Pembina	17
Pierce	17
Ramsey	17
Ransom	16
Renville	17
Richland	16
Rolette	17
Sargent	16
Sheridan	17
Sioux	16
Slope	16
Stark	16
Steele	17
Stutsman	17
Towner	17
Traill	17
Walsh	17
Ward	17
Wells	17
Williams	17
OHIO	
Adams	11
Allen	13
Ashland	13
Ashtabula	13
Athens	11
Auglaize	13
Belmont	12
Brown	11
Butler	12
Carroll	13
Champaign	13
Clark	13
Clermont	11
Clinton	12
Columbiana	13
Coshocton	12
Crawford	13
Cuyahoga	13
Darke	13
Defiance	14
Delaware	13
Erie	13
Fairfield	12
Fayette	12
Franklin	12
Fulton	14
Gallia	11
Geauga	13
Greene	12
Guernsey	12
Hamilton	11
Hancock	13
Hardin	13
Harrison	13
Henry	14
Highland	11
Hocking	12
Holmes	13
Huron	13
Jackson	11
Jefferson	13
Knox	13
Lake	13
Lawrence	11
Licking	12
Logan	13
Lorain	13
Lucas	14
Madison	12
Mahoning	13
Marion	13
Medina	13
Meigs	11
Mercer	13
Miami	13
Monroe	12
Montgomery	12
Morgan	12
Morrow	13
Muskingum	12
Noble	12
Ottawa	13
Paulding	14
Perry	12
Pickaway	12
Pike	11
Portage	13
Preble	12
Putnam	13
Richland	13

(continued)

TABLE N1101.2—continued
CLIMATE ZONES BY STATES AND COUNTIES

County	Zone
Ross	12
Sandusky	13
Scioto	11
Seneca	13
Shelby	13
Stark	13
Summit	13
Trumbull	13
Tuscarawas	13
Union	13
Van Wert	13
Vinton	11
Warren	12
Washington	11
Wayne	13
Williams	14
Wood	14
Wyandot	13
OKLAHOMA	
Adair	8
Alfalfa	9
Atoka	7
Beaver	10
Beckham	8
Blaine	8
Bryan	7
Caddo	8
Canadian	8
Carter[a]	6
Cherokee	8
Choctaw[a]	6
Cimarron	10
Cleveland	7
Coal	7
Comanche	7
Cotton	7
Craig	9
Creek	8
Custer	8
Delaware	8
Dewey	9
Ellis	9
Garfield	8
Garvin	7
Grady	7
Grant	9
Greer	7
Harmon	7
Harper	9
Haskell	7
Hughes	7
Jackson	7
Jefferson[a]	6
Johnston[a]	6
Kay	9
Kingfisher	8
Kiowa	7
Latimer	7
Le Flore	7
Lincoln	7
Logan	8
Love[a]	6
Major	9
Marshall[a]	6
Mayes	8
Mcclain	7
Mccurtain	7
Mcintosh	7
Murray	7
Muskogee	7
Noble	8
Nowata	9
Okfuskee	7
Oklahoma	8
Okmulgee	8
Osage	8
Ottawa	9
Pawnee	8
Payne	8
Pittsburg	7
Pontotoc	7
Pottawatomie	7
Pushmataha[a]	6
Roger Mills	9
Rogers	9
Seminole	7
Sequoyah	7
Stephens	7
Texas	10
Tillman	7
Tulsa	8
Wagoner	8
Washington	9
Washita	8
Woods	9
Woodward	9
OREGON	
Baker	15
Benton	10
Clackamas	10
Clatsop	11
Columbia	11
Coos	9
Crook	14
Curry	9
Deschutes	14
Douglas	9
Gilliam	12
Grant	15
Harney	15
Hood River	12
Jackson	11
Jefferson	13
Josephine	9
Klamath	14
Lake	15
Lane	10
Lincoln	11
Linn	10
Malheur	12
Marion	10
Morrow	12
Multnomah	10
Polk	10
Sherman	13
Tillamook	11
Umatilla	12
Union	13
Wallowa	15
Wasco	13
Washington	10
Wheeler	13
Yamhill	10
PENNSYLVANIA	
Adams	11
Allegheny	12
Armstrong	13
Beaver	12
Bedford	13
Berks	12
Blair	13
Bradford	15
Bucks	11
Butler	14
Cambria	13
Cameron	15
Carbon	13
Centre	13
Chester	11
Clarion	14
Clearfield	15
Clinton	13
Columbia	13
Crawford	14
Cumberland	12
Dauphin	12
Delaware	10
Elk	15
Erie	14
Fayette	12
Forest	15
Franklin	11
Fulton	12
Greene	12
Huntingdon	12
Indiana	13
Jefferson	15
Juniata	12
Lackawanna	14
Lancaster	11
Lawrence	14
Lebanon	12
Lehigh	12
Luzerne	13
Lycoming	13
Mckean	15
Mercer	14
Mifflin	12
Monroe	13
Montgomery	11
Montour	13
Northampton	12
Northumberland	13
Perry	12
Philadelphia	10
Pike	13
Potter	15
Schuylkill	13
Snyder	13
Somerset	13
Sullivan	14
Susquehanna	15
Tioga	15
Union	13
Venango	14
Warren	14
Washington	12
Wayne	15
Westmoreland	13
Wyoming	14
York	11
RHODE ISLAND	
Bristol	12
Kent	12
Newport	12
Providence	14
Washington	12
SOUTH CAROLINA	
Abbeville	7
Aiken[a]	6
Allendale[a]	5
Anderson	7
Bamberg[a]	5
Barnwell[a]	5
Beaufort[a]	5
Berkeley[a]	5
Calhoun[a]	6
Charleston[a]	5
Cherokee	7
Chester	7
Chesterfield	7
Clarendon[a]	6
Colleton[a]	5
Darlington[a]	6
Dillon[a]	6
Dorchester[a]	5
Edgefield[a]	6
Fairfield	7
Florence[a]	6
Georgetown[a]	5
Greenville	7
Greenwood	7
Hampton[a]	5
Horry[a]	5
Jasper[a]	5
Kershaw	7
Lancaster	7
Laurens	7
Lee[a]	6
Lexington[a]	6
Marion[a]	6
Marlboro[a]	6
Mccormick[a]	6
Newberry[a]	6
Oconee	7
Orangeburg[a]	6
Pickens	7
Richland[a]	6
Saluda[a]	6
Spartanburg	7
Sumter[a]	6
Union	7
Williamsburg[a]	6
York	7
SOUTH DAKOTA	
Aurora	15
Beadle	15
Bennett	14
Bon Homme	14
Brookings	16
Brown	16
Brule	15
Buffalo	15
Butte	15
Campbell	15
Charles Mix	14
Clark	16
Clay	14
Codington	16
Corson	15
Custer	15
Davison	15
Day	16
Deuel	16
Dewey	15
Douglas	14
Edmunds	15
Fall River	15
Faulk	15
Grant	16
Gregory	14
Haakon	15
Hamlin	16
Hand	15
Hanson	15
Harding	15
Hughes	15
Hutchinson	14
Hyde	15
Jackson	14
Jerauld	15
Jones	15
Kingsbury	15
Lake	15
Lawrence	15
Lincoln	15
Lyman	15
Marshall	16
Mccook	15
Mcpherson	16
Meade	15
Mellette	14
Miner	15
Minnehaha	15
Moody	15
Pennington	15
Perkins	15
Potter	15
Roberts	16
Sanborn	15
Shannon	15
Spink	15
Stanley	15
Sully	15
Todd	14
Tripp	14
Turner	15
Union	14
Walworth	15
Yankton	14
Ziebach	15
TENNESSEE	
Anderson	9
Bedford	8
Benton	9
Bledsoe	8
Blount	8
Bradley	8
Campbell	10
Cannon	9
Carroll	9
Carter	10
Cheatham	9
Chester	8
Claiborne	10
Clay	9
Cocke	9
Coffee	8
Crockett	8
Cumberland	9
Davidson	8
De Kalb	9
Decatur	8
Dickson	9
Dyer	8
Fayette[a]	7
Fentress	10
Franklin	8
Gibson	9
Giles	8
Grainger	9
Greene	9
Grundy	9
Hamblen	9
Hamilton	8
Hancock	10
Hardeman	8
Hardin	8
Hawkins	9

(continued)

TABLE N1101.2—continued
CLIMATE ZONES BY STATES AND COUNTIES

County	Zone
Haywood	8
Henderson	8
Henry	9
Hickman	9
Houston	9
Humphreys	9
Jackson	9
Jefferson	9
Johnson	10
Knox	8
Lake	9
Lauderdale	8
Lawrence	8
Lewis	8
Lincoln	8
Loudon	8
Macon	9
Madison	8
Marion	8
Marshall	8
Maury	9
Mcminn	8
Mcnairy	8
Meigs	8
Monroe	8
Montgomery	9
Moore	8
Morgan	10
Obion	9
Overton	9
Perry	8
Pickett	10
Polk	8
Putnam	9
Rhea	8
Roane	9
Robertson	9
Rutherford	8
Scott	10
Sequatchie	8
Sevier	9
Shelby[a]	7
Smith	9
Stewart	9
Sullivan	9
Sumner	9
Tipton	8
Trousdale	9
Unicoi	10
Union	9
Van Buren	9
Warren	9
Washington	9
Wayne	8
Weakley	9
White	9
Williamson	8
Wilson	9
TEXAS	
Anderson[a]	5
Andrews	6
Angelina[a]	4
Aransas[a]	3
Archer	6
Armstrong	9
Atascosa[a]	3
Austin[a]	4
Bailey	9
Bandera[a]	4
Bastrop[a]	4
Baylor	7
Bee[a]	3
Bell[a]	5
Bexar[a]	4
Blanco[a]	5
Borden	7
Bosque[a]	5
Bowie	6
Brazoria[a]	3
Brazos[a]	4
Brewster[a]	5
Briscoe	8
Brooks[a]	3
Brown[a]	5
Burleson[a]	4
Burnet[a]	5
Caldwell[a]	4
Calhoun[a]	3
Callahan	6
Cameron[a]	2
Camp	6
Carson	9
Cass	6
Castro	9
Chambers[a]	4
Cherokee[a]	5
Childress	7
Clay	6
Cochran	8
Coke	6
Coleman[a]	5
Collin	5
Collingsworth	7
Colorado[a]	4
Comal[a]	4
Comanche[a]	5
Concho[a]	5
Cooke	6
Coryell[a]	5
Cottle	7
Crane[a]	5
Crockett[a]	5
Crosby	7
Culberson	6
Dallam	9
Dallas[a]	5
Dawson	7
De Witt[a]	3
Deaf Smith	9
Delta	6
Denton	5
Dickens	7
Dimmit[a]	3
Donley	8
Duval[a]	3
Eastland	6
Ector	6
Edwards[a]	4
El Paso	6
Ellis[a]	5
Erath	5
Falls[a]	5
Fannin	6
Fayette[a]	4
Fisher	6
Floyd	8
Foard	7
Fort Bend[a]	4
Franklin	6
Freestone[a]	5
Frio[a]	3
Gaines	7
Galveston[a]	3
Garza	7
Gillespie[a]	5
Glasscock	6
Goliad[a]	3
Gonzales[a]	4
Gray	9
Grayson	6
Gregg	6
Grimes[a]	4
Guadalupe[a]	4
Hale	8
Hall	8
Hamilton[a]	5
Hansford	9
Hardeman	7
Hardin[a]	4
Harris[a]	4
Harrison	6
Hartley	9
Haskell	6
Hays[a]	4
Hemphill	8
Henderson[a]	5
Hidalgo[a]	2
Hill[a]	5
Hockley	8
Hood[a]	5
Hopkins	6
Houston[a]	5
Howard	6
Hudspeth	6
Hunt	6
Hutchinson	9
Irion[a]	5
Jack	6
Jackson[a]	3
Jasper[a]	4
Jeff Davis	6
Jefferson[a]	4
Jim Hogg[a]	3
Jim Wells[a]	3
Johnson[a]	5
Jones	6
Karnes[a]	3
Kaufman	5
Kendall[a]	5
Kenedy[a]	2
Kent	7
Kerr[a]	5
Kimble[a]	5
King	7
Kinney[a]	4
Kleberg[a]	2
Knox	7
La Salle[a]	3
Lamar	6
Lamb	8
Lampasas[a]	5
Lavaca[a]	4
Lee[a]	4
Leon[a]	5
Liberty[a]	4
Limestone[a]	5
Lipscomb	9
Live Oak[a]	3
Llano[a]	5
Loving	6
Lubbock	7
Lynn	7
Madison[a]	4
Marion	6
Martin	6
Mason[a]	5
Matagorda[a]	3
Maverick[a]	3
Mcculloch[a]	5
Mclennan[a]	5
Mcmullen[a]	3
Medina[a]	4
Menard[a]	5
Midland	6
Milam[a]	4
Mills[a]	5
Mitchell	6
Montague	6
Montgomery[a]	4
Moore	9
Morris	6
Motley	7
Nacogdoches[a]	5
Navarro[a]	5
Newton[a]	4
Nolan	6
Nueces[a]	3
Ochiltree	9
Oldham	9
Orange[a]	4
Palo Pinto	6
Panola[a]	5
Parker	5
Parmer	9
Pecos[a]	5
Polk[a]	4
Potter	9
Presidio[a]	5
Rains	6
Randall	9
Reagan[a]	5
Real[a]	4
Red River	6
Reeves	6
Refugio[a]	3
Roberts	9
Robertson[a]	4
Rockwall	5
Runnels[a]	5
Rusk[a]	5
Sabine[a]	5
San Augustine[a]	5
San Jacinto[a]	4
San Patricio[a]	3
San Saba[a]	5
Schleicher[a]	5
Scurry	7
Shackelford	6
Shelby[a]	5
Sherman	9
Smith[a]	5
Somervell[a]	5
Starr[a]	2
Stephens	6
Sterling	6
Stonewall	7
Sutton[a]	5
Swisher	9
Tarrant[a]	5
Taylor	6
Terrell[a]	5
Terry	7
Throckmorton	6
Titus	6
Tom Green[a]	5
Travis[a]	4
Trinity[a]	4
Tyler[a]	4
Upshur	6
Upton[a]	5
Uvalde[a]	4
Val Verde[a]	4
Van Zandt	5
Victoria[a]	3
Walker[a]	4
Waller[a]	4
Ward	6
Washington[a]	4
Webb[a]	3
Wharton[a]	3
Wheeler	8
Wichita	7
Wilbarger	7
Willacy[a]	2
Williamson[a]	4
Wilson[a]	4
Winkler	6
Wise	5
Wood	6
Yoakum	8
Young	6
Zapata[a]	2
Zavala[a]	3
UTAH	
Beaver	14
Box Elder	12
Cache	15
Carbon	14
Daggett	15
Davis	12
Duchesne	15
Emery	14
Garfield	14
Grand	10
Iron	12
Juab	12
Kane	10
Millard	13
Morgan	15
Piute	13
Rich	15
Salt Lake	12
San Juan	13
Sanpete	14
Sevier	13
Summit	15
Tooele	12
Uintah	15
Utah	12
Wasatch	15
Washington	10
Wayne	14
Weber	12
VERMONT	
Addison	15
Bennington	15
Caledonia	16
Chittenden	15
Essex	16
Franklin	15
Grand Isle	15
Lamoille	16
Orange	16
Orleans	16
Rutland	15
Washington	16
Windham	15
Windsor	15
VIRGINIA	
Accomack	8
Albemarle	9
Alleghany	10

(continued)

TABLE N1101.2—continued
CLIMATE ZONES BY STATES AND COUNTIES

County	Zone
Amelia	9
Amherst	9
Appomattox	9
Arlington	10
Augusta	11
Bath	11
Bedford	9
Bland	11
Botetourt	9
Brunswick	8
Buchanan	10
Buckingham	9
Campbell	9
Caroline	9
Carroll	11
Charles City	8
Charlotte	9
Chesterfield	9
Clarke	11
Craig	10
Culpeper	10
Cumberland	9
Dickenson	10
Dinwiddie	8
Essex	9
Fairfax	10
Fauquier	10
Floyd	11
Fluvanna	9
Franklin	10
Frederick	11
Fredericksburg Giles	10
Gloucester	8
Goochland	9
Grayson	11
Greene	10
Greensville	8
Halifax	9
Hampton	8
Hanover	9
Henrico	8
Henry	10
Highland	11
Isle Of Wight	8
James City	8
King And Queen	9
King George	9
King William	9
Lancaster	8
Lee	10
Loudoun	10
Louisa	9
Lunenburg	9
Madison	11
Mathews	8
Mecklenburg	9
Middlesex	8
Montgomery	11
Nansemond	8
Nelson	9
New Kent	8
Newport News	8
Norfolk	8
Northampton	8
Northumberland	8
Nottoway	9
Orange	10
Page	11
Patrick	10
Pittsylvania	9
Powhatan	9
Prince Edward	9
Prince George	8
Prince William	10
Pulaski	11
Rappahannock	11
Richmond	8
Roanoke	9
Rockbridge	9
Rockingham	11
Russell	10
Scott	10
Shenandoah	11
Smyth	11
Southampton	8
Spotsylvania	10
Stafford	10
Surry	8
Sussex	8
Tazewell	11
Virginia Beach	8
Warren	11
Washington	11
Westmoreland	8
Wise	10
Wythe	11
York	8
VIRGINIA INDEPENDENT CITIES	
Alexandria	10
Bedford	9
Bristol	11
Buena Vista	9
Charlottesville	9
Chesapeake	8
Clifton Forge	10
Colonial Hts	9
Covington	10
Danville	9
Emporia	8
Fairfax	10
Falls Church	10
Franklin	8
Fredericksburg	10
Galax	11
Hampton	8
Harrisonburg	11
Hopewell	8
Lexington	9
Lynchburg	9
Manassas	10
Manassas Park	10
Martinsville	10
Newport News	8
Norfolk	8
Norton	10
Petersburg	8
Poquoson	8
Portsmouth	8
Radford	11
Richmond	8
Roanoke	9
Salem	9
South Boston	9
Staunton	11
Suffolk	8
Virginia Beach	8
Waynesboro	11
Williamsburg	8
Winchester	11
WASHINGTON	
Adams	12
Asotin	12
Benton	11
Chelan	12
Clallam	12
Clark	11
Columbia	12
Cowlitz	11
Douglas	14
Ferry	15
Franklin	11
Garfield	12
Grant	12
Grays Harbor	11
Island	12
Jefferson	11
King	10
Kitsap	11
Kittitas	14
Klickitat	12
Lewis	11
Lincoln	15
Mason	11
Okanogan	15
Pacific	11
Pend Oreille	15
Pierce	11
San Juan	12
Skagit	11
Skamania	11
Snohomish	11
Spokane	14
Stevens	15
Thurston	11
Wahkiakum	11
Walla Walla	11
Whatcom	12
Whitman	14
Yakima	12
WEST VIRGINIA	
Barbour	13
Berkeley	11
Boone	10
Braxton	11
Brooke	12
Cabell	10
Calhoun	11
Clay	11
Doddridge	12
Fayette	12
Gilmer	11
Grant	13
Greenbrier	12
Hampshire	11
Hancock	12
Hardy	12
Harrison	12
Jackson	11
Jefferson	11
Kanawha	10
Lewis	12
Lincoln	10
Logan	10
Marion	12
Marshall	12
Mason	11
Mcdowell	11
Mercer	11
Mineral	12
Mingo	10
Monongalia	12
Monroe	11
Morgan	11
Nicholas	12
Ohio	12
Pendleton	13
Pleasants	11
Pocahontas	13
Preston	13
Putnam	10
Raleigh	12
Randolph	13
Ritchie	11
Roane	11
Summers	12
Taylor	12
Tucker	13
Tyler	11
Upshur	12
Wayne	10
Webster	12
Wetzel	12
Wirt	11
Wood	11
Wyoming	11
WISCONSIN	
Adams	15
Ashland	17
Barron	16
Bayfield	17
Brown	15
Buffalo	15
Burnett	17
Calumet	15
Chippewa	15
Clark	15
Columbia	15
Crawford	15
Dane	15
Dodge	15
Door	15
Douglas	17
Dunn	15
Eau Claire	15
Florence	17
Fond Du Lac	15
Forest	17
Grant	15
Green	15
Green Lake	15
Iowa	15
Iron	17
Jackson	15
Jefferson	15
Juneau	15
Kenosha	15
Kewaunee	15
La Crosse	15
Lafayette	15
Langlade	17
Lincoln	17
Manitowoc	15
Marathon	15
Marinette	15
Marquette	15
Menominee	15
Milwaukee	15
Monroe	15
Oconto	15
Oneida	17
Outagamie	15
Ozaukee	15
Pepin	15
Pierce	15
Polk	16
Portage	15
Price	17
Racine	15
Richland	15
Rock	15
Rusk	16
Sauk	15
Sawyer	17
Shawano	15
Sheboygan	15
St Croix	15
Taylor	17
Trempealeau	15
Vernon	15
Vilas	17
Walworth	15
Washburn	17
Washington	15
Waukesha	15
Waupaca	15
Waushara	15
Winnebago	15
Wood	15
WYOMING	
Albany	16
Big Horn	15
Campbell	15
Carbon	16
Converse	15
Crook	15
Fremont	15
Goshen	14
Hot Springs	15
Johnson	15
Laramie	15
Lincoln	17
Natrona	15
Niobrara	15
Park	15
Platte	14
Sheridan	15
Sublette	17
Sweetwater	16
Teton	17
Uinta	16
Washakie	15
Weston	15

a. A moisture vapor retarder is not required in these counties. See Section R322.1.

SECTION N1102
BUILDING ENVELOPE

N1102.1 Thermal performance criteria. The minimum required insulation *R*-value or maximum required *U*-factor for each element in the building thermal envelope (fenestration, roof/ceiling, opaque wall, floor, slab edge, crawl space wall and basement wall) shall be in accordance with the criteria in Table N1102.1.

Residential buildings, Type A-1, with greater than 15-percent glazing area; residential buildings, Type A-2, with greater than 25-percent glazing area; and any residential building in climates with HDD equal to or greater than 13,000; shall determine compliance using the building envelope requirements of the *International Energy Conservation Code.*

N1102.1.1 Exterior walls. The minimum required *R*-value in Table N1102.1 shall be met by the sum of the *R*-values of the insulation materials installed in framing cavities and/or insulating sheathing applied, and not by framing, drywall, structural sheathing, or exterior siding materials. Insulation separated from the conditioned space by a vented space shall not be counted towards the required *R*-value.

N1102.1.1.1 Mass walls. Mass walls shall be permitted to meet the Mass Wall criteria in Table N1102.1.1.1(1), based on the insulation position and the climate zone where the building is located. Other mass walls shall meet the frame wall criteria for the building type and the climate zone where the building is located, based on the sum of interior and exterior insulation. Walls with exterior insulation position have the entire effective mass layer interior to an insulation layer. Walls with integral insulation position have either insulation and mass materials well mixed as in wood (logs); or substantially equal amounts of mass material on the interior and exterior of insulation as in concrete masonry units with insulated cores or masonry cavity walls. Walls with interior insulation position have the mass material located exterior to the insulating material. Walls not meeting either of the above descriptions for exterior or integral positions shall meet the requirements for "Other mass walls" in Table N1102.1.1.1(1). The *R*-value of the mass assembly for typical masonry constructions shall be taken from Table N1102.1.1.1(2). The *R*-value for a solid concrete wall with a thickness of 4 inches (102 mm) or greater is R-1.1. *R*-values for other assemblies are permitted to be based on hot box tests referenced in ASTM C 236 or ASTM C 976, two dimensional calculations or isothermal planes calculations.

N1102.1.1.2 Steel-frame walls. The minimum required *R*-values for steel-frame walls shall be in accordance with Table N1102.1.1.2.

TABLE N1102.1
SIMPLIFIED PRESCRIPTIVE BUILDING ENVELOPE THERMAL COMPONENT CRITERIA
MINIMUM REQUIRED THERMAL PERFORMANCE (*U*-FACTOR AND *R*-VALUE)

HDD	MAXIMUM GLAZING U-FACTOR [Btu / (hr · ft^2 · °F)]	MINIMUM INSULATION *R*-VALUE [(hr·ft^2·°F) / Btu]					
		Ceilings	Walls	Floors	Basement walls	Slab perimeter *R*-value and depth	Crawl space walls
0-499	Any	R-13	R-11	R-11	R-0	R-0	R-0
500-999	0.90	R-19	R-11	R-11	R-0	R-0	R-4
1,000-1,499	0.75	R-19	R-11	R-11	R-0	R-0	R-5
1,500-1,999	0.75	R-26	R-13	R-11	R-5	R-0	R-5
2,000-2,499	0.65	R-30	R-13	R-11	R-5	R-0	R-6
2,500-2,999	0.60	R-30	R-13	R-19	R-6	R-4, 2 ft.	R-7
3,000-3,499	0.55	R-30	R-13	R-19	R-7	R-4, 2 ft.	R-8
3,500-3,999	0.50	R-30	R-13	R-19	R-8	R-5, 2 ft.	R-10
4,000-4,499	0.45	R-38	R-13	R-19	R-8	R-5, 2 ft.	R-11
4,500-4,999	0.45	R-38	R-16	R-19	R-9	R-6, 2 ft.	R-17
5,000-5,499	0.45	R-38	R-18	R-19	R-9	R-6, 2 ft.	R-17
5,500-5,999	0.40	R-38	R-18	R-21	R-10	R-9, 4 ft.	R-19
6,000-6,499	0.35	R-38	R-18	R-21	R-10	R-9, 4 ft.	R-20
6,500-6,999	0.35	R-49	R-21	R-21	R-11	R-11, 4 ft.	R-20
7,000-8,499	0.35	R-49	R-21	R-21	R-11	R-13, 4 ft.	R-20
8,500-8,999	0.35	R-49	R-21	R-21	R-18	R-14, 4 ft.	R-20
9,000-12,999	0.35	R-49	R-21	R-21	R-19	R-18, 4 ft.	R-20

For SI: 1 Btu/(hr · ft^2 · °F) = 5.68W/m^2 · K, 1 (hr · ft^2 · °F)/Btu = 0.176m^2 · K/W.

TABLE N1102.1.1.1(1)
MASS WALL PRESCRIPTIVE BUILDING ENVELOPE REQUIREMENTS

BUILDING LOCATION		MASS WALL ASSEMBLY *R*-VALUE[a] (hr · ft^2 · °F) / Btu	
Zone	HDD	Exterior or integral insulation	Other mass walls
1	0-499	R-3.8	R-9.7
2	500-999	R-4.8	R-9.7
3	1,000-1,499	R-4.8	R-9.7
4	1,500-1,999	R-8.1	R-10.8
5	2,000-2,499	R-8.9	R-10.8
6	2,500-2,999	R-8.9	R-10.8
7	3,000-3,499	R-8.9	R-10.8
8	3,500-3,999	R-8.9	R-10.8
9	4,000-4,499	R-8.9	R-10.9
10	4,500-4,999	R-10.4	R-12.3
11	5,000-5,499	R-11.9	R-15.2
12	5,500-5,999	R-11.9	R-15.2
13	6,000-6,499	R-11.9	R-15.2
14	6,500-6,999	R-15.5	R-18.4
15	7,000-8,499	R-15.5	R-18.4
16	8,500-8,999	R-18.4	R-18.4
17	9,000-12,999	R-18.4	R-18.4

For SI: 1 (hr · ft^2 · °F)/Btu = 0.176 m^2 · K/W.

a. The sum of the value in Table N1102.1.1.1(2) and additional insulation layers.

TABLE N1102.1.1.1(2)
MASS ASSEMBLY *R*-VALUES

ASSEMBLY TYPE	UNGROUTED CELLS NOT INSULATED (hr · ft^2 · °F) / Btu	UNGROUTED CELLS INSULATED (hr · ft^2 · °F) / Btu		
		No grout	Vertical cells grouted at 10 ft. o.c. or greater	Vertical cells grouted at less than 10 ft. o.c.
6″ Light weight concrete block	2.3	5.0	4.5	3.8
6″ Medium weight concrete block	2.1	4.2	3.8	3.2
6″ Normal weight concrete block	1.9	3.3	3.1	2.7
8″ Light weight concrete block	2.6	6.7	5.9	4.8
8″ Medium weight concrete block	2.3	5.3	4.8	4.0
8″ Normal weight concrete block	2.1	4.2	3.8	3.3
12″ Light weight concrete block	2.9	9.1	7.9	6.3
12″ Medium weight concrete block	2.6	7.1	6.4	5.2
12″ Normal weight concrete block	2.3	5.6	5.1	4.3
Brick cavity wall	3.7	6.7	6.2	5.4
Hollow clay brick	2.0	2.7	2.6	2.4

For SI: 1 inch = 25.4 mm, 1 foot = 304.8 mm, 1 (hr · ft^2 · °F)/Btu = 0.176 m^2 · K/W.

TABLE N1102.1.1.2
STEEL-FRAME WALL MINIMUM PERFORMANCE REQUIREMENTS (*R*-VALUE)

CLIMATE ZONES	HDD	EQUIVALENT STEEL-FRAME WALL CAVITY AND SHEATHING *R*-VALUE[a] (hr·ft^2·°F) / Btu
1–4	0-1999	R-11+R-5, R-15+R-4, R-21+R-3
5–8	2,000-3,999	R-11+R-5, R-15+R-4, R-21+R-3
9–12	4,000-5,999	R-11+R-9, R-15+R-8, R-21+R-7
13–15	6,000-8,499	R-13+R-10, R-19+R-9, R-25+R-8
16 and 17	8,500-12,999	R-13+R-10, R-19+R-9, R-25+R-8

For SI: 1 (hr · ft^2 · °F)/Btu = 0.176 m^2 · K/W.

a. The cavity insulation *R*-value requirement is listed first, followed by the sheathing *R*-value requirement.

N1102.1.2 Ceilings. The required "Ceiling *R*-value" in Table N1102.1 assumes standard truss or rafter construction and shall apply to all roof/ceiling portions of the building thermal envelope, including cathedral ceilings. Where the construction technique allows the required *R*-value of ceiling insulation to be obtained over the wall top plate, R-30 shall be permitted to be used where R-38 is required and R-38 shall be permitted to be used where R-49 is required.

N1102.1.3 Opaque doors. Opaque doors separating conditioned and unconditioned space shall have a maximum *U*-factor of 0.35. One opaque door shall be permitted to be exempt from this *U*-factor requirement.

N1102.1.4 Floors. The required *R*-value in Table N1102.1 shall apply to all floors, except any individual floor assembly with over 25 percent of its conditioned floor area exposed directly to outside air shall meet the *R*-value requirement in Table N1102.1 for "Ceilings."

N1102.1.5 Basement walls. When the basement is a conditioned space, the basement walls shall be insulated in accordance with Table N1102.1. When the basement is not a conditioned space, either the basement walls or the ceilings separating the basement from conditioned space shall be insulated in accordance with Table N1102.1. When insulating basement walls, the required *R*-value shall be applied from the top of the basement wall to a depth of 10 feet (3048 mm) below grade or to the top of the basement floor, whichever is less.

N1102.1.6 Slab-on-grade floors. For slabs with a top edge above finished grade of 12 inches (305 mm) or less below finished grade, the required *R*-value in Table N1102.1 shall be applied to the outside of the foundation or the inside of the foundation wall. The insulation shall extend downward from the top of the slab, or downward to the bottom of the slab and then horizontally in either direction, until the distance listed in Table N1102.1 is reached.

When installed between the exterior wall and the edge of the interior slab, the top edge of the insulation shall be permitted to be cut at a 45-degree (0.79 rad) angle away from the exterior wall. Insulation extending horizontally away from the building shall be protected by pavement or by a minimum of 10 inches (254 mm) of soil.

In locations of 500 HDD or greater, R-2 shall be added to the values in Table N1102.1 where uninsulated hot water pipes, air distribution ducts or electric heating cables are installed within or under the slab.

Exception: Slab perimeter insulation is not required for unheated slabs in areas of very heavy termite infestation probability as shown in Figure R301.2(6). Where this exception is used, building envelope compliance shall be demonstrated by (a) using *International Energy Conservation Code* Section 502.2.2 or *International Energy Conservation Code* Chapter 4 with the actual slab insulation *R*-value in Table N1102.1, or (b) using *International Energy Conservation Code* Section 502.2.4.

N1102.1.7 Crawl space walls. Where the floor above the crawl space is uninsulated, insulation shall be installed on crawl space walls when the crawl space is not vented to outside air. The required *R*-value in Table N1102.1 shall be applied inside of the crawl space wall, downward from the sill plate to the exterior finished grade level and then vertically and/or horizontally for at least an additional 24 inches (610 mm). The exposed earth in all crawl space foundations shall be covered with a continuous vapor retarder having a maximum permeance rating of 1.0 perm [(57 mg)/(s · m^2 · Pa)], when tested in accordance with ASTM E 96.

N1102.1.8 Masonry veneer. For exterior foundation insulation, that horizontal portion of the foundation that supports a masonry veneer shall not be required to be insulated.

N1102.1.9 Protection. Exposed insulating materials applied to the exterior of foundation walls shall be protected. The protection shall extend at least 6 inches (152 mm) below finished grade level.

N1102.1.10 Air leakage. All joints, seams, penetrations; site-built windows, doors, and skylights; openings between window and door assemblies and their respective jambs and framing; and other sources of air leakage (infiltration and exfiltration) through the building thermal envelope shall be caulked, gasketed, weatherstripped, wrapped, or otherwise sealed to limit uncontrolled air movement.

N1102.2 Maximum solar heat gain coefficient for fenestration products. The area-weighted-average solar heat gain coefficient (SHGC) for glazed fenestration installed in locations with 3,500 or fewer heating degree days shall not exceed 0.40.

N1102.3 Fenestration exemption. Up to 1 percent of the total glazing area shall be exempt from *U*-factor requirements.

N1102.4 Replacement fenestration. Where an entire fenestration product, including frame, sash and glazed portion, is being replaced in an existing building, the replacement fenestration product shall have a *U*-factor that does not exceed the "Maximum fenestration *U*-factor" in Table 502.2.5 of the 2000 *International Energy Conservation Code* applicable to the climate zone (HDD) where the building is located. Replacement skylights and roof windows shall be permitted to have a maximum *U*-factor of 0.50 when installed in any location above 1,999 HDD. The replacement fenestration products must also satisfy the SHGC and air leakage requirements of Sections N1102.2 and N1101.3.2.2, respectively.

SECTION N1103 MECHANICAL SYSTEMS

N1103.1 Heating and air conditioning appliance and equipment performance. Performance of equipment listed in Table N1103.1 is covered by preemptive Federal law. Appliances and equipment not listed in Table N1103.1 shall meet the minimum efficiency requirements of Section 503.2 of the *International Energy Conservation Code.*

N1103.2 Controls. At least one thermostat shall be provided for each separate heating, cooling, or combination heating and cooling system. Heat pumps shall have controls that prevent supplementary electric resistance heater operation when the heating load can be met by the heat pump alone. Supplementary heater operation shall be permitted during outdoor coil defrost cycles not exceeding 15 minutes.

N1103.3 Duct insulation. All portions of the air distribution system shall be installed in accordance with Section M1601 and be insulated to an installed R-5 when system components are located within the building but outside of conditioned space, and R-8 when located outside of the building. When located within a building envelope assembly, at least R-8 shall be applied between the duct and that portion of the assembly furthest from conditioned space.

Exceptions: Exhaust air ducts and portions of the air distribution system within appliances or equipment.

N1103.4 Duct sealing. All ducts shall be sealed in accordance with Section M1601.3.1.

N1103.5 Piping insulation. All mechanical system piping shall be insulated in accordance with Table N1103.5.

Exceptions: Piping installed within appliances and equipment or piping serving fluids between 55°F (13°C) and 120°F (49°C).

SECTION N1104 SERVICE WATER HEATING

N1104.1 Water heating appliance and equipment performance. Performance of equipment listed in Table N1104.1 is covered by preemptive Federal law. Appliances and equipment not listed in Table N1104.1 shall meet the minimum efficiency requirements of Section 504.2 of the *International Energy Conservation Code.*

TABLE N1103.1 MINIMUM EQUIPMENT PERFORMANCE

EQUIPMENT CATEGORY	SUB-CATEGORY[e]	REFERENCED STANDARD	MINIMUM PERFORMANCE
Air-cooled heat pumps heating mode < 65,000 Btu/h cooling capacity	Split systems	ARI 210/240	6.8 HSPF[a,b]
	Single package		6.6 HSPF[a,b]
Gas-fired or oil-fired furnace < 225,000 Btu/h	—	DOE 10 CFR Part 430, Subpart B, Appendix N	AFUE 78%[b] *Et* 80%[c]
Gas-fired or oil-fired steam and hot-water boilers < 300,000 Btu/h	—	DOE 10 CFR Part 430, Subpart B, Appendix N	AFUE 78%[b,d]
Air-cooled air conditioners and heat pumps cooling mode < 65,000 Btu/h cooling capacity	Split systems	ARI 210/240	10.0 SEER[b]
	Single package		9.7 SEER[b]

For SI: 1 Btu/h = 0.2931 W.

a. For multicapacity equipment, the minimum performance shall apply to each capacity step provided. Multicapacity refers to manufacturer-published ratings for more than one capacity mode allowed by the product's controls.

b. This is used to be consistent with the National Appliance Energy Conservation Act (NAECA) of 1987 (Public Law 100-12).

c. These requirements apply to combination units not covered by NAECA (three-phase power or cooling capacity 65,000 Btu/h).

d. Except for gas-fired steam boilers, for which the minimum AFUE shall be 75 percent.

e. Seasonal rating.

TABLE N1103.5
MINIMUM HVAC PIPING INSULATION THICKNESSES[a]

PIPING SYSTEM TYPES	FLUID TEMP RANGE (°F)	INSULATION THICKNESS inches[b]
Heating systems		
Low pressure/temperature	201-250	1.5
Low temperature	120-200	1.0
Steam condensate (for feed water)	Any	1.5
Cooling systems		
Chilled water, refrigerant or brine	40-55	0.75
	Below 40	1.25

For SI: 1 inch = 25.4 mm, °C = [(°F)-32]/1.8.

a. The pipe insulation thicknesses specified in this table are based on insulation *R*-values ranging from R-4 to R-4.6 per inch of thickness. For materials with an *R*-value greater than R-4.6, the insulation thickness specified in this table may be reduced as follows:

$$\text{New Minimum Thickness} = \frac{4.6 \times \text{Table Thickness}}{\text{Actual } R\text{-Value}}$$

For materials with an *R*-value less than R-4, the minimum insulation thickness shall be increased as follows:

$$\text{New Minimum Thickness} = \frac{4.0 \times \text{Table Thickness}}{\text{Actual } R\text{-Value}}$$

b. For piping exposed to outdoor air, increase thickness by 0.5 inch.

TABLE N1104.1
REQUIRED PERFORMANCE OF DOMESTIC HOT WATER HEATING EQUIPMENT SUBJECT TO MINIMUM FEDERAL STANDARDS

CATEGORY	MAXIMUM INPUT RATING	MINIMUM EFFICIENCY
Electric; storage or instantaneous	12 kW	0.93 - 0.00132 × *V*[a]
Gas; storage	75,000 Btu/h	0.62 - 0.0019 × *V*[a]
Gas; instantaneous	200,000 Btu/h	0.62 - 0.0019 × *V*[a]
Oil; storage	105,000 Btu/h	0.59 - 0.0019 × *V*[a]
Oil; instantaneous	210,000 Btu/h	0.59 - 0.0019 × *V*[a]

For SI: 1 Btu/h = 0.2931 W, 1 gallon = 3.785 L.

a. *V* is the rated storage volume in gallons as specified by the manufacturer.

Part V — Mechanical

CHAPTER 12
MECHANICAL ADMINISTRATION

SECTION M1201
GENERAL

M1201.1 Scope. The provisions of Chapters 12 through 24 shall regulate the design, installation, maintenance, alteration and inspection of mechanical systems that are permanently installed and utilized to provide control of environmental conditions within buildings. These chapters shall also regulate those mechanical systems, system components, equipment and appliances specifically addressed in this code.

M1201.2 Application. In addition to the general administration requirements of Chapter 1, the administrative provisions of this chapter shall also apply to the mechanical requirements of Chapters 13 through 24.

M1201.3 Conformity. Conformity to the standards specified in Chapter 43 shall indicate that the equipment is safe for installation under this code.

SECTION M1202
EXISTING MECHANICAL SYSTEMS

M1202.1 Additions, alterations or repairs. Additions, alterations, renovations or repairs to a mechanical system shall conform to that required for a new mechanical system without requiring the existing mechanical system to comply with all of the requirements of this code. Additions, alterations or repairs shall not cause an existing mechanical system to become unsafe, hazardous or overloaded. Minor additions, alterations or repairs to existing mechanical systems shall meet the provisions for new construction, unless such work is done in the same manner and arrangement as was in the existing system, is not hazardous, and is approved.

M1202.2 Existing installations. Except as otherwise provided for in this code, a provision in this code shall not require the removal, alteration or abandonment of, nor prevent the continued utilization and maintenance of, an existing mechanical system lawfully in existence at the time of the adoption of this code.

M1202.3 Maintenance. Mechanical systems, both existing and new, and parts thereof shall be maintained in proper operating condition in accordance with the original design and in a safe and sanitary condition. Devices or safeguards that are required by this code shall be maintained in compliance with the code edition under which installed. The owner or the owner's designated agent shall be responsible for maintenance of the mechanical systems. To determine compliance with this provision, the building official shall have the authority to require a mechanical system to be reinspected.

CHAPTER 13
GENERAL MECHANICAL SYSTEM REQUIREMENTS

SECTION M1301
GENERAL

M1301.1 Scope. The provisions of this chapter shall govern the installation of mechanical systems not specifically covered in other chapters applicable to mechanical systems. Installations of mechanical appliances, equipment and systems not addressed by this code shall comply with the applicable provisions of the *International Mechanical Code* and the *International Fuel Gas Code.*

SECTION M1302
APPROVAL

M1302.1 General. All appliances shall be listed and bear the label of an approved agency unless otherwise approved in accordance with Section R104.11.

SECTION M1303
LABELING OF APPLIANCES

M1303.1 Label information. A permanent factory-applied nameplate(s) shall be affixed to appliances on which shall appear, in legible lettering, the manufacturer's name or trademark, the model number, serial number, and the seal or mark of the testing agency. A label shall also include the following:

1. Electrical appliances. Electrical rating in volts, amperes and motor phase; identification of individual electrical components in volts, amperes or watts, and motor phase; and in Btu/h (W) output and required clearances.
2. Absorption units. Hourly rating in Btu/h (W), minimum hourly rating for units having step or automatic modulating controls, type of fuel, type of refrigerant, cooling capacity in Btu/h (W), and required clearances.
3. Fuel-burning units. Hourly rating in Btu/h (W), type of fuel approved for use with the appliance and required clearances.
4. Electric comfort heating appliances. Name and trademark of the manufacturer; the model number or equivalent; the electric rating in volts, ampacity and phase; Btu/h (W) output rating; individual marking for each electrical component in amperes or watts, volts and phase; required clearances from combustibles; and a seal indicating approval of the appliance by an approved agency.

SECTION M1304
TYPE OF FUEL

M1304.1 Fuel types. Fuel-fired appliances shall be designed for use with the type of fuel to which they will be connected and the altitude at which they are installed. Appliances that comprise parts of the building mechanical system shall not be converted for the usage of a different fuel, except where approved and converted in accordance with the manufacturer's instructions. The fuel input rate shall not be increased or decreased beyond the limit rating for the altitude at which the appliance is installed.

SECTION M1305
APPLIANCE ACCESS

M1305.1 Appliance access for inspection service, repair and replacement. Appliances shall be accessible for inspection, service, repair and replacement without removing permanent construction. Thirty inches (762 mm) of working space shall be provided in front of the control side to service an appliance. Room heaters shall be permitted to be installed with at least an 18-inch (457 mm) working space. A platform shall not be required for room heaters.

M1305.1.1 Central furnaces. Central furnaces within compartments or alcoves shall have a minimum working space clearance of 3 inches (76 mm) along the sides, back and top with a total width of the enclosing space being at least 12 inches (305 mm) wider than the furnace. Furnaces having a firebox open to the atmosphere shall have at least a 6-inch (152 mm) working space along the front combustion chamber side. Combustion air openings at the rear or side of the compartment shall comply with the requirements of Chapter 17.

Exception: This section shall not apply to replacement appliances installed in existing compartments and alcoves where the working space clearances are in accordance with the equipment or appliance manufacturer's installation instructions.

M1305.1.2 Appliances in rooms. Appliances installed in a compartment, alcove, basement or similar space shall be accessed by an opening or door and an unobstructed passageway measuring not less than 24 inches (610 mm) wide and large enough to allow removal of the largest appliance in the space, provided that a level service space of not less than 30 inches (762 mm) deep and the height of the appliance, but not less than 30 inches (762 mm), is present at the front or service side of the appliance with the door open.

M1305.1.3 Appliances in attics. Attics containing appliances requiring access shall be provided with an opening and a clear and unobstructed passageway large enough to allow removal of the largest appliance, but not less than 30 inches (762 mm) high and 22 inches (559 mm) wide and not more than 20 feet (6096 mm) in length when measured along the centerline of the passageway from the opening to the appliance. The passageway shall have continuous solid flooring in accordance with Chapter 5 not less than 24 inches (610 mm) wide. A level service space at least 30 inches (762 mm) deep and 30 inches (762 mm) wide shall be present along all sides of the appliance where access is required. The clear access opening dimensions shall be a minimum of 20 inches by 30 inches (508 mm by 762 mm), where such dimensions are large enough to allow removal of the largest appliance.

Exception: The passageway and level service space are not required where the appliance is capable of being serviced and removed through the required opening.

M1305.1.3.1 Electrical requirements. A lighting fixture controlled by a switch located at the required passageway opening and a receptacle outlet shall be provided at or near the appliance location in accordance with Chapter 38.

M1305.1.4 Appliances under floors. Underfloor spaces containing appliances requiring access shall be provided with an unobstructed passageway large enough to remove the largest appliance, but not less than 30 inches (762 mm) high and 22 inches (559 mm) wide, nor more than 20 feet (6096 mm) in length when measured along the centerline of the passageway from the opening to the appliance. A level service space at least 30 inches (762 mm) deep and 30 inches (762 mm) wide shall be present at the front or service side of the appliance. If the depth of the passageway or the service space exceeds 12 inches (305 mm) below the adjoining grade, the walls of the passageway shall be lined with concrete or masonry extending 4 inches (102 mm) above the adjoining grade in accordance with Chapter 4. The rough-framed access opening dimensions shall be a minimum of 22 inches by 30 inches (559 mm by 762 mm), where the dimensions are large enough to remove the largest appliance.

Exception: The passageway is not required where the level service space is present when the access is open, and the appliance is capable of being serviced and removed through the required opening.

M1305.1.4.1 Ground clearance. Appliances supported from the ground shall be level and firmly supported on a concrete slab or other approved material extending above the adjoining ground. Appliances suspended from the floor shall have a clearance of not less than 6 inches (152 mm) from the ground.

M1305.1.4.2 Excavations. Excavations for appliance installations shall extend to a depth of 6 inches (152 mm) below the appliance and 12 inches (305 mm) on all sides, except that the control side shall have a clearance of 30 inches (762 mm).

M1305.1.4.3 Electrical requirements. A lighting fixture controlled by a switch located at the required passageway opening and a receptacle outlet shall be provided at or near the appliance location in accordance with Chapter 38.

SECTION M1306
CLEARANCES FROM COMBUSTIBLE CONSTRUCTION

M1306.1 Appliance clearance. Appliances shall be installed with the clearances from unprotected combustible materials as indicated on the appliance label and in the manufacturer's installation instructions.

M1306.2 Clearance reduction. Reduction of clearances shall be in accordance with the appliance manufacturer's instructions and Table M1306.2. Forms of protection with ventilated air space shall conform to the following requirements:

1. Not less than 1-inch (25.4 mm) air space shall be provided between the protection and combustible wall surface.
2. Air circulation shall be provided by having edges of the wall protection open at least 1 inch (25.4 mm).
3. If the wall protection is mounted on a single flat wall away from corners, air circulation shall be provided by having the bottom and top edges, or the side and top edges open at least 1 inch (25.4 mm).
4. Wall protection covering two walls in a corner shall be open at the bottom and top edges at least 1 inch (25.4 mm).

M1306.2.1 Solid fuel appliances. Table M1306.2 shall not be used to reduce the clearance required for appliances listed for installation with minimum clearances of 12 inches (305 mm) or less. For appliances listed for installation with minimum clearances greater than 12 inches (305 mm), Table M1306.2 shall not be used to reduce the clearance to less than 12 inches (305 mm).

SECTION M1307
APPLIANCE INSTALLATION

M1307.1 General. Installation of appliances shall conform to the conditions of their listing and label and the manufacturer's installation instructions. The manufacturer's operating and installation instructions shall remain attached to the appliance.

M1307.2 Anchorage of appliances. Appliances designed to be fixed in position shall be fastened or anchored in an approved manner. In Seismic Design Categories D_1 and D_2, water heaters shall be anchored or strapped to resist horizontal displacement due to earthquake motion. Strapping shall be at points within the upper one-third and lower one-third of the appliance's vertical dimensions. At the lower point, the strapping shall maintain a minimum distance of 4 inches (102 mm) above the controls.

M1307.3 Elevation of ignition source. Appliances having an ignition source shall be elevated such that the source of ignition is not less than 18 inches (457 mm) above the floor in garages. For the purpose of this section, rooms or spaces that are not part of the living space of a dwelling unit and that communicate with a private garage through openings shall be considered to be part of the garage.

M1307.3.1 Protection from impact. Appliances located in a garage or carport shall be protected from impact by automobiles.

M1307.4 Electrical appliances. Electrical appliances shall be installed in accordance with Chapters 14, 21 and 33 through 42 of this code.

SECTION M1308
MECHANICAL SYSTEMS INSTALLATION

M1308.1 Drilling and notching. Wood-framed structural members shall be drilled, notched or altered in accordance with the provisions of Sections R502.8, R602.6, R602.6.1 and R802.7. Holes in cold-formed, steel-framed, load-bearing members shall only be permitted in accordance with Sections R505.2, R603.2 and R804.2. Cutting and notching of flanges and lips of cold-formed, steel-framed, load-bearing members shall not be permitted.

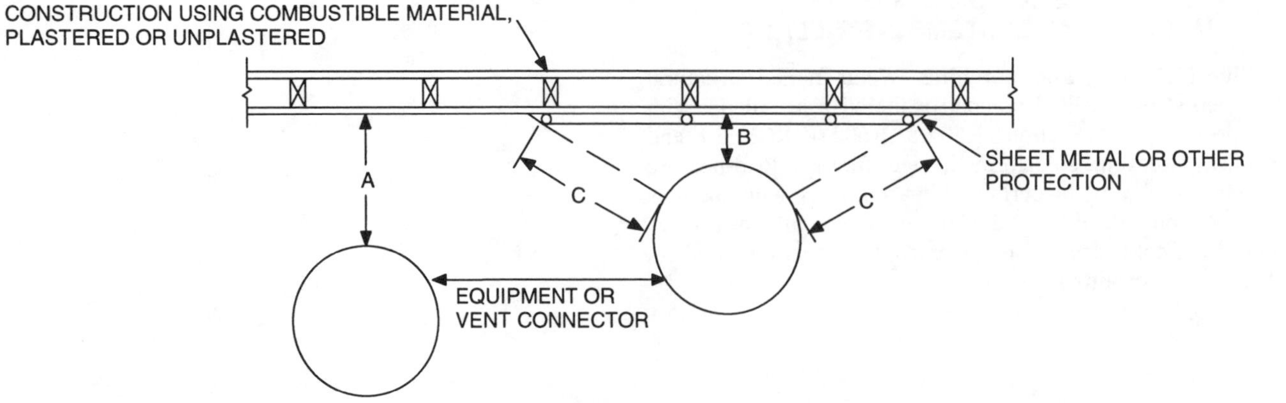

Note: "A" equals the required clearance with no protection. "B" equals the reduced clearance permitted in accordance with Table M1306.2. The protection applied to the construction using combustible material shall extend far enough in each direction to make "C" equal to "A."

FIGURE M1306.1
REDUCED CLEARANCE DIAGRAM

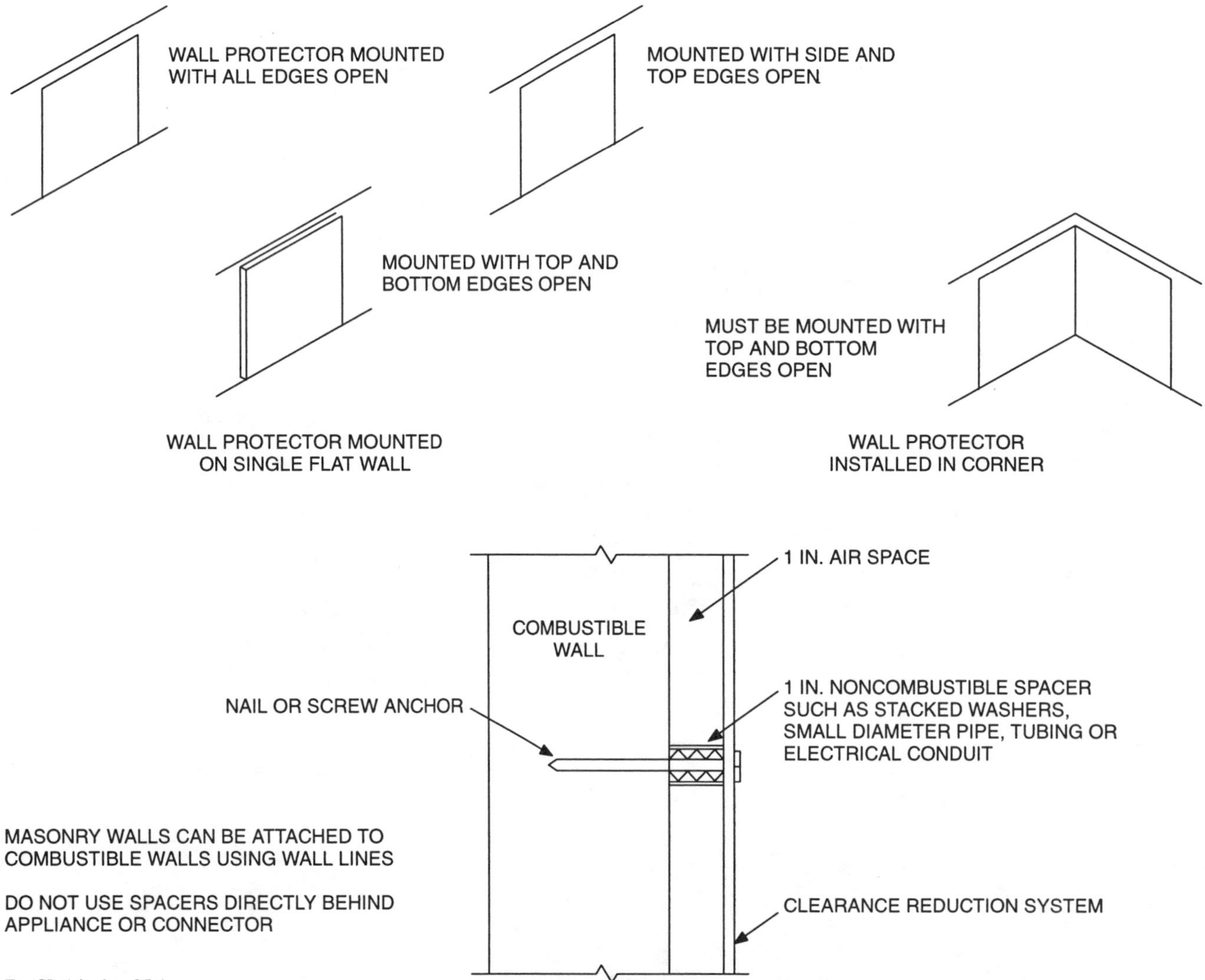

For SI: 1 inch = 25.4 mm.

FIGURE M1306.2
WALL PROTECTOR CLEARANCE REDUCTION SYSTEM

TABLE M1306.2
REDUCTION OF CLEARANCES WITH SPECIFIED FORMS OF PROTECTION[a,b,c,d,e,f,g,h,i,j,k]

TYPE OF PROTECTION APPLIED TO AND COVERING ALL SURFACES OF COMBUSTIBLE MATERIAL WITHIN THE DISTANCE SPECIFIED AS THE REQUIRED CLEARANCE WITH NO PROTECTION [SEE FIGURES M1306.1 AND M1306.2]	WHERE THE REQUIRED CLEARANCE WITH NO PROTECTION FROM APPLIANCE, VENT CONNECTOR, OR SINGLE WALL METAL PIPE IS:									
	36 inches		18 inches		12 inches		9 inches		6 inches	
	Allowable clearances with specified protection (Inches)[b]									
	Use column 1 for clearances above an appliance or horizontal connector. Use column 2 for clearances from an appliance, vertical connector and single-wall metal pipe.									
	Above column 1	Sides and rear column 2	Above column 1	Sides and rear column 2	Above column 1	Sides and rear column 2	Above column 1	Sides and rear column 2	Above column 1	Sides and rear column 2
$3\frac{1}{2}$-in. thick masonry wall without ventilated air space	—	24	—	12	—	9	—	6	—	5
$\frac{1}{2}$-in. insulation board over 1-in. glass fiber or mineral wool batts	24	18	12	9	9	6	6	5	4	3
0.024 sheet metal over 1-in. glass fiber or mineral wool batts reinforced with wire on rear face with ventilated air space	18	12	9	6	6	4	5	3	3	3
$3\frac{1}{2}$-in. thick masonry wall with ventilated air space	—	12	—	6	—	6	—	6	—	6
0.024 sheet metal with ventilated air space	18	12	9	6	6	4	5	3	3	2
$\frac{1}{2}$-in. thick insulation board with ventilated air space	18	12	9	6	6	4	5	3	3	3
0.024 sheet metal with ventilated air space over 0.024 sheet metal with ventilated air space	18	12	9	6	6	4	5	3	3	3
1-in. glass fiber or mineral wool batts sandwiched between two sheets 0.024 sheet metal with ventilated air space.	18	12	9	6	6	4	5	3	3	3

For SI: 1 inch = 25.4 mm, 1 pound per cubic foot = 16.019 kg/m^3, °C = [(°F)-32/1.8], 1 Btu/(n · ft^2 · °F/in) = 0.001442299 (w/cm^2 · °C/cm).

a. Reduction of clearances from combustible materials shall not interfere with combustion air, draft hood clearance and relief, and accessibility of servicing.

b. Clearances shall be measured from the surface of the heat producing appliance or equipment to the outer surface of the combustible material or combustible assembly.

c. Spacers and ties shall be of noncombustible material. No spacer or tie shall be used directly opposite appliance or connector.

d. Where all clearance reduction systems use a ventilated air space, adequate provision for air circulation shall be provided as described. (See Figures M1306.1 and M1306.2.)

e. There shall be at least 1 inch between clearance reduction systems and combustible walls and ceilings for reduction systems using ventilated air space.

f. If a wall protector is mounted on a single flat wall away from corners, adequate air circulation shall be permitted to be provided by leaving only the bottom and top edges or only the side and top edges open with at least a 1-inch air gap.

g. Mineral wool and glass fiber batts (blanket or board) shall have a minimum density of 8 pounds per cubic foot and a minimum melting point of 1,500°F.

h. Insulation material used as part of a clearance reduction system shall have a thermal conductivity of 1.0 Btu inch per square foot per hour °F or less. Insulation board shall be formed of noncombustible material.

i. There shall be at least 1 inch between the appliance and the protector. In no case shall the clearance between the appliance and the combustible surface be reduced below that allowed in Table M1306.2.

j. All clearances and thicknesses are minimum; larger clearances and thicknesses are acceptable.

k. Listed single-wall connectors shall be permitted to be installed in accordance with the terms of their listing and the manufacturer's instructions.

CHAPTER 14
HEATING AND COOLING EQUIPMENT

SECTION M1401
GENERAL

M1401.1 Installation. Heating and cooling equipment shall be installed in accordance with the manufacturer's installation instructions and the requirements of this code.

M1401.2 Access. Heating and cooling equipment shall be located with respect to building construction and other equipment to permit maintenance, servicing and replacement. Clearances shall be maintained to permit cleaning of heating and cooling surfaces; replacement of filters, blowers, motors, controls and vent connections; lubrication of moving parts; and adjustments.

M1401.3 Sizing. Heating and cooling equipment shall be sized based on building loads calculated in accordance with ACCA Manual J or other approved heating and cooling calculation methodologies.

M1401.4 Exterior installations. Equipment installed outdoors shall be listed and labeled for outdoor installation. Supports and foundations shall prevent excessive vibration, settlement or movement of the equipment. Supports and foundations shall be level and conform to the manufacturer's installation instructions.

SECTION M1402
CENTRAL FURNACES

M1402.1 General. Central furnaces shall conform to ANSI/UL 727.

M1402.2 Clearances. Clearances shall be provided in accordance with the listing and the manufacturer's installation instructions.

M1402.3 Combustion air. Combustion air shall be supplied in accordance with Chapter 17. Combustion air openings shall be unobstructed for a distance of not less than 6 inches (152 mm) in front of the openings.

M1402.4 Electric furnaces. Electric furnaces shall conform to UL 1995 and shall be installed in compliance with their listing and the manufacturer's installation instructions.

SECTION M1403
HEAT PUMP EQUIPMENT

M1403.1 Heat pumps. The minimum unobstructed total area of the outside and return air ducts or openings to a heat pump shall be not less than 6 square inches per 1,000 Btu/h (13 208 mm^2/kW) output rating or as indicated by the conditions of the listing of the heat pump. Electric heat pumps shall conform to UL 559 or UL 1995.

M1403.2 Foundations and supports. Supports and foundations for the outdoor unit of a heat pump shall be raised at least 3 inches (76 mm) above the ground to permit free drainage of defrost water, and shall conform to the manufacturer's installation instructions.

SECTION M1404
REFRIGERATION COOLING EQUIPMENT

M1404.1 Compliance. Refrigeration cooling equipment shall comply with Section M1411.

SECTION M1405
BASEBOARD CONVECTORS

M1405.1 General. Electric baseboard convectors shall be installed in accordance with the manufacturer's installation instructions and Chapters 33 through 42 of this code.

SECTION M1406
RADIANT HEATING SYSTEMS

M1406.1 General. Electric radiant heating systems shall be installed in accordance with the manufacturer's installation instructions and Chapters 33 through 42 of this code.

M1406.2 Clearances. Clearances for radiant heating panels or elements to any wiring, outlet boxes and junction boxes used for installing electrical devices or mounting lighting fixtures shall comply with Chapters 33 through 42 of this code.

M1406.3 Installation of radiant panels. Radiant panels installed on wood framing shall conform to the following requirements:

1. Heating panels shall be installed parallel to framing members and secured to the surface of framing members or mounted between framing members.
2. Panels shall be nailed or stapled only through the unheated portions provided for this purpose and shall not be fastened at any point closer than $^1/_4$ inch (6.4 mm) from an element.
3. Unless listed and labeled for field cutting, heating panels shall be installed as complete units.

M1406.4 Installation in concrete or masonry. Radiant heating systems installed in concrete or masonry shall conform to the following requirements:

1. Radiant heating systems shall be identified as being suitable for the installation, and shall be secured in place, as specified in the manufacturer's installation instructions.
2. Radiant heating panels or radiant heating panel sets shall not be installed where they bridge expansion joints unless protected from expansion and contraction.

M1406.5 Gypsum panels. Where radiant heating systems are used on gypsum assemblies, operating temperatures shall not exceed 125°F (52°C).

M1406.6 Finish surfaces. Finish materials installed over radiant heating panels or systems shall be installed in accordance with the manufacturer's installation instructions. Surfaces shall be secured so that nails or other fastenings do not pierce the radiant heating elements.

SECTION M1407 DUCT HEATERS

M1407.1 General. Electric duct heaters shall be installed in accordance with the manufacturer's installation instructions and Chapters 33 through 42 of this code. Electric furnaces shall be tested in accordance with UL 1995.

M1407.2 Installation. Electric duct heaters shall be installed so that they will not create a fire hazard. Class 1 ducts, duct coverings and linings shall be interrupted at each heater to provide the clearances specified in the manufacturer's installation instructions. Such interruptions are not required for duct heaters listed and labeled for zero clearance to combustible materials. Insulation installed in the immediate area of each heater shall be classified for the maximum temperature produced on the duct surface.

M1407.3 Installation with heat pumps and air conditioners. Duct heaters located within 4 feet (1219 mm) of a heat pump or air conditioner shall be listed and labeled for such installations. The heat pump or air conditioner shall additionally be listed and labeled for such duct heater installations.

M1407.4 Access. Duct heaters shall be accessible for servicing, and clearance shall be maintained to permit adjustment, servicing and replacement of controls and heating elements.

M1407.5 Fan interlock. The fan circuit shall be provided with an interlock to prevent heater operation when the fan is not operating.

SECTION M1408 VENTED FLOOR FURNACES

M1408.1 General. Vented floor furnaces shall conform to ANSI/UL 729 and be installed in accordance with their listing, the manufacturer's installation instructions and the requirements of this code.

M1408.2 Clearances. Vented floor furnaces shall be installed in accordance with their listing and the manufacturer's installation instructions.

M1408.3 Location. Location of floor furnaces shall conform to the following requirements:

1. Floor registers of floor furnaces shall be installed not less than 6 inches (152 mm) from a wall.
2. Wall registers of floor furnaces shall be installed not less than 6 inches (152 mm) from the adjoining wall at inside corners.
3. The furnace register shall be located not less than 12 inches (305 mm) from doors in any position, draperies or similar combustible objects.
4. The furnace register shall be located at least 5 feet (1524 mm) below any projecting combustible materials.
5. The floor furnace burner assembly shall not project into an occupied under-floor area.
6. The floor furnace shall not be installed in concrete floor construction built on grade.
7. The floor furnace shall not be installed where a door can swing within 12 inches (305 mm) of the grill opening.

M1408.4 Access. An opening in the foundation not less than 18 inches by 24 inches (457 mm by 610 mm), or a trap door not less than 22 inches by 30 inches (559 mm by 762 mm) shall be provided for access to a floor furnace. The opening and passageway shall be large enough to allow replacement of any part of the equipment.

M1408.5 Installation. Floor furnace installations shall conform to the following requirements:

1. Thermostats controlling floor furnaces shall be located in the room in which the register of the floor furnace is located.
2. Floor furnaces shall be supported independently of the furnace floor register.
3. Floor furnaces shall be installed not closer than 6 inches (152 mm) to the ground. Clearance may be reduced to 2 inches (51 mm), provided that the lower 6 inches (152 mm) of the furnace is sealed to prevent water entry.
4. Where excavation is required for a floor furnace installation, the excavation shall extend 30 inches (762 mm) beyond the control side of the floor furnace and 12 inches (305 mm) beyond the remaining sides. Excavations shall slope outward from the perimeter of the base of the excavation to the surrounding grade at an angle not exceeding 45 degrees (0.39 rad) from horizontal.
5. Floor furnaces shall not be supported from the ground.

SECTION M1409 VENTED WALL FURNACES

M1409.1 General. Vented wall furnaces shall conform to ANSI/UL 730 and be installed in accordance with their listing, the manufacturer's installation instructions and the requirements of this code.

M1409.2 Location. The location of vented wall furnaces shall conform to the following requirements:

1. Vented wall furnaces shall be located so as not to cause a fire hazard to walls, floors, combustible furnishings or doors. Vented wall furnaces installed between bathrooms and adjoining rooms shall not circulate air from bathrooms to other parts of the building.
2. Vented wall furnaces shall not be located where a door can swing within 12 inches (305 mm) of the furnace air inlet or outlet measured at right angles to the opening. Doorstops or door closers shall not be installed to obtain this clearance.

M1409.3 Installation. Vented wall furnace installations shall conform to the following requirements:

1. Required wall thicknesses shall be in accordance with the manufacturer's installation instructions.
2. Ducts shall not be attached to a wall furnace. Casing extensions or boots shall only be installed when listed as part of a listed and labeled appliance.
3. A manual shut off valve shall be installed ahead of all controls.

M1409.4 Access. Vented wall furnaces shall be provided with access for cleaning of heating surfaces; removal of burners; replacement of sections, motors, controls, filters and other working parts; and for adjustments and lubrication of parts requiring such attention. Panels, grilles and access doors that must be removed for normal servicing operations shall not be attached to the building construction.

SECTION M1410 VENTED ROOM HEATERS

M1410.1 General. Vented room heaters shall be tested in accordance with UL 1482 or UL 896 and installed in accordance with their listing, the manufacturer's installation instructions and the requirements of this code.

M1410.2 Floor mounting. Room heaters shall be installed on noncombustible floors or approved assemblies constructed of noncombustible materials that extend at least 18 inches (457 mm) beyond the appliance on all sides.

Exceptions:

1. Listed room heaters shall be installed on noncombustible floors, assemblies constructed of noncombustible materials or listed floor protectors with materials and dimensions in accordance with the appliance manufacturer's instructions.
2. Room heaters listed for installation on combustible floors without floor protection shall be installed in accordance with the appliance manufacturer's instructions.

SECTION M1411 REFRIGERATION COOLING EQUIPMENT

M1411.1 Approved refrigerants. Refrigerants used in direct refrigerating systems shall conform to the applicable provisions of ANSI/ASHRAE 34.

M1411.2 Refrigeration coils in warm-air furnaces. Where a cooling coil is located in the supply plenum of a warm-air furnace, the furnace blower shall be rated at not less than 0.5-inch water column (124 Pa) static pressure unless the furnace is listed and labeled for use with a cooling coil. Cooling coils shall not be located upstream from heat exchangers unless listed and labeled for such use. Conversion of existing furnaces for use with cooling coils shall be permitted provided the furnace will operate within the temperature rise specified for the furnace.

M1411.3 Condensate disposal. Condensate from all cooling coils or evaporators shall be conveyed from the drain pan outlet to an approved place of disposal. Condensate shall not discharge into a street, alley or other areas so as to cause a nuisance.

M1411.3.1 Auxiliary and secondary drain systems. In addition to the requirements of Section M1411.3, a secondary drain or auxiliary drain pan shall be required for each cooling or evaporator coil where damage to any building components will occur as a result of overflow from the equipment drain pan or stoppage in the condensate drain piping. Drain piping shall be a minimum of $^3/_4$-inch (19.1 mm) nominal pipe size. One of the following methods shall be used:

1. An auxiliary drain pan with a separate drain shall be provided under the coils on which condensation will occur. The auxiliary pan drain shall discharge to a conspicuous point of disposal to alert occupants in the event of a stoppage of the primary drain. The pan shall have a minimum depth of 1.5 inches (38 mm), shall not be less than 3 inches (76 mm) larger than the unit or the coil dimensions in width and length and shall be constructed of corrosion resistant material. Metallic pans shall have a minimum thickness of not less than 0.0276-inch (0.7 mm) galvanized sheet metal. Nonmetallic pans shall have a minimum thickness of not less than 0.0625 inch (1.6 mm).
2. A separate overflow drain line shall be connected to the drain pan provided with the equipment. Such overflow drain shall discharge to a conspicuous point of disposal to alert occupants in the event of a stoppage of the primary drain. The overflow drain line shall connect to the drain pan at a higher level than the primary drain connection.
3. An auxiliary drain pan without a separate drain line shall be provided under the coils on which condensate will occur. Such pan shall be equipped with a water level detection device that will shut off the equipment served prior to overflow of the pan. The auxiliary drain pan shall be constructed in accordance with Item 1 of this section.

M1411.4 Insulation of refrigerant piping. Piping and fittings for refrigerant vapor (suction) lines shall be insulated with insulation having a thermal resistivity of at least $R = 4.0\ hr \cdot ft^2 \cdot {}^\circ F/Btu$ and having external surface permeance not exceeding 0.05 perms [2.87 ng/(s·m^2·Pa)] when tested in accordance with ASTM E 96.

SECTION M1412 ABSORPTION COOLING EQUIPMENT

M1412.1 Approval of equipment. Absorption systems shall be installed in accordance with the manufacturer's installation instructions.

M1412.2 Condensate disposal. Condensate from the cooling coil shall be disposed of as provided in Section M1411.3.

M1412.3 Insulation of piping. Refrigerant piping, brine piping and fittings within a building shall be insulated to prevent condensation from forming on piping.

M1412.4 Pressure-relief protection. Absorption systems shall be protected by a pressure-relief device. Discharge from the pressure-relief device shall be located so as not to create a hazard to persons or property.

SECTION M1413
EVAPORATIVE COOLING EQUIPMENT

M1413.1 General. Cooling equipment that utilizes evaporation of water for cooling shall be installed in accordance with the manufacturer's installation instructions. Evaporative coolers shall be installed on a level platform or base not less than 3 inches (76 mm) above the adjoining ground and secured to prevent displacement. Openings in exterior walls shall be flashed in accordance with Section R703.8.

M1413.2 Protection of potable water. The potable water system shall be protected from backflow in accordance with the provisions in Section P2902.

SECTION 1414
FIREPLACE STOVES

M1414.1 General. Fireplace stoves shall be listed, labeled and installed in accordance with the terms of the listing. Fireplace stoves shall be tested in accordance with UL 737.

M1414.2 Hearth extensions. Hearth extensions for fireplace stoves shall be installed in accordance with the listing of the fireplace stove. The supporting structure for a hearth extension for a fireplace stove shall be at the same level as the supporting structure for the fireplace unit. The hearth extension shall be readily distinguishable from the surrounding floor area.

CHAPTER 15
EXHAUST SYSTEMS

SECTION M1501
CLOTHES DRYERS EXHAUST

M1501.1 General. Dryer exhaust systems shall be independent of all other systems, shall convey the moisture to the outdoors and shall terminate on the outside of the building. Exhaust duct terminations shall be in accordance with the dryer manufacturer's installation instructions. Screens shall not be installed at the duct termination. Exhaust ducts shall not be connected with sheet-metal screws or fastening means which extend into the duct. Exhaust ducts shall be equipped with a backdraft damper. Exhaust ducts shall be constructed of minimum 0.016-inch-thick (0.406 mm) rigid metal ducts, having smooth interior surfaces with joints running in the direction of air flow. Flexible transition ducts used to connect the dryer to the exhaust duct system shall be limited to single lengths, not to exceed 8 feet (2438 mm) in length and shall be listed and labeled in accordance with UL 2158A. Transition ducts shall not be concealed within construction.

M1501.2 Exhaust duct size. The minimum diameter of the exhaust duct shall be as recommended by the manufacturer and shall be at least the diameter of the appliance outlet.

M1501.3 Length limitation. The maximum length of a clothes dryer exhaust duct shall not exceed 25 feet (7620 mm) from the dryer location to the wall or roof termination. The maximum length of the duct shall be reduced 2.5 feet (762 mm) for each 45-degree (0.79 rad) bend and 5 feet (1524 mm) for each 90-degree (1.6 rad) bend. The maximum length of the exhaust duct does not include the transition duct.

> **Exception:** Where the make and model of the clothes dryer to be installed is known and the manufacturer's installation instructions for such dryer are provided to the building official, the maximum length of the exhaust duct, including any transition duct, shall be permitted to be in accordance with the dryer manufacturer's installation instructions.

SECTION M1502
RANGE HOODS

M1502.1 General. Range hoods shall discharge to the outdoors through a single-wall duct. The duct serving the hood shall have a smooth interior surface, shall be air tight and shall be equipped with a backdraft damper. Ducts serving range hoods shall not terminate in an attic or crawl space or areas inside the building.

> **Exception:** Where installed in accordance with the manufacturer's installation instructions, and where mechanical or natural ventilation is otherwise provided, listed and labeled ductless range hoods shall not be required to discharge to the outdoors.

M1502.2 Duct material. Single-wall ducts serving range hoods shall be constructed of galvanized steel, stainless steel or copper.

> **Exception:** Ducts for domestic kitchen cooking appliances equipped with down draft exhaust systems shall be permitted to be constructed of schedule 40 PVC pipe provided that the installation complies with all of the following:
>
> 1. The duct shall be installed under a concrete slab poured on grade,
> 2. The underfloor trench in which the duct is installed shall be completely backfilled with sand or gravel,
> 3. The PVC duct shall extend not greater than 1 inch (25 mm) above the indoor concrete floor surface,
> 4. The PVC duct shall extend not greater than 1 inch (25 mm) above grade outside of the building, and
> 5. The PVC ducts shall be solvent cemented.

SECTION M1503
INSTALLATION OF MICROWAVE OVENS

M1503.1 Installation of microwave oven over a cooking appliance. The installation of a listed and labeled cooking appliance or microwave oven over a listed and labeled cooking appliance shall conform to the terms of the upper appliance's listing and label and the manufacturer's installation instructions.

SECTION M1504
OVERHEAD EXHAUST HOODS

M1504.1 General. Domestic open-top broiler units shall be provided with a metal exhaust hood, not less than 28 gage, with a clearance of not less than 0.25 inch (6.4 mm) between the hood and the underside of combustible material or cabinets. A clearance of at least 24 inches (610 mm) shall be maintained between the cooking surface and the combustible material or cabinet. The hood shall be at least as wide as the broiler unit and shall extend over the entire unit. Such exhaust hood shall discharge to the outdoors and shall be equipped with a backdraft damper or other means to control infiltration/exfiltration when not in operation. Broiler units incorporating an integral exhaust system, and listed and labeled for use without an exhaust hood, need not be provided with an exhaust hood.

SECTION M1505
EXHAUST DUCTS

M1505.1 Ducts. Where exhaust duct construction is not specified in this chapter, such construction shall comply with Chapter 16.

CHAPTER 16
DUCT SYSTEMS

SECTION M1601
DUCT CONSTRUCTION

M1601.1 Materials. Duct systems serving heating, cooling and ventilation equipment shall be fabricated in accordance with the provisions of this section.

M1601.1.1 Above-ground duct systems. Above-ground duct systems shall conform to the following:

1. Equipment connected to duct systems shall be designed to limit discharge air temperature to a maximum of 250°F (121°C).
2. Factory-made air ducts shall be constructed of Class 0 or Class 1 materials as designated in Table M1601.1.1(1).
3. Fibrous duct construction shall conform to the SMACNA "Fibrous Glass Duct Construction Standards" or NAIMA "Fibrous Glass Duct Construction Standards."
4. Minimum thicknesses of metal duct material shall be as listed in Table M1601.1.1(2). Galvanized steel shall conform to ASTM A 525.
5. Gypsum products are permitted to be used to construct return air ducts or plenums, provided that the air temperature does not exceed 125°F (52°C) and exposed surfaces are not subject to condensation.
6. Duct systems shall be constructed of materials having a flame-spread index not greater than 200.
7. Cavities between studs or partitions to be used as return ducts shall be isolated from unused spaces with tight-fitting stops of sheet metal, or with wood not less than a 2-inch (51 mm) nominal thickness. Such cavities or spaces shall not be utilized as a plenum for supply air. Such cavities or spaces shall not be part of a required fire-resistance-rated assembly. Stud wall cavities shall not convey air from more than one floor level.

TABLE M1601.1.1(1)
CLASSIFICATION OF FACTORY-MADE AIR DUCTS

DUCT CLASS	MAXIMUM FLAME-SPREAD RATING
0	0
1	25

M1601.1.2 Underground duct systems. Underground duct systems shall be constructed of approved concrete, clay, metal or plastic. The maximum duct temperature for plastic ducts shall not be greater than 150°F (66°C). Metal ducts shall be protected from corrosion in an approved manner or shall be completely encased in concrete not less than 2 inches (51 mm) in thickness. Nonmetallic ducts shall be installed in accordance with the manufacturer's installation instructions. Plastic pipe and fitting materials shall conform to cell classification 12454-B of ASTM D 1248 or ASTM D 1784 and external loading properties of ASTM D 2412. All ducts shall slope to an accessible point for drainage. Where encased in concrete, ducts shall be sealed and secured prior to pouring any concrete. Metallic ducts having an approved protective coating and nonmetallic ducts shall be installed in accordance with the manufacturer's installation instructions.

M1601.2 Factory-made ducts. Factory-made air ducts or duct material shall be approved for the use intended, and shall be installed in accordance with the manufacturer's installation instructions. Each portion of a factory-made air duct system shall bear a listing and label indicating compliance with UL 181 and UL 181A or UL 181B.

TABLE M1601.1.1(2)
GAGES OF METAL DUCTS AND PLENUMS USED FOR HEATING OR COOLING

TYPE OF DUCT	SIZE (inches)	MINIMUM THICKNESS (inch)	EQUIVALENT GALVANIZED SHEET GAGE	APPROXIMATE ALUMINUM B & S GAGE
Round ducts and enclosed rectangular ducts	14 or less over 14	0.013 0.016	30 28	26 24
Exposed rectangular ducts	14 or less over 14	0.016 0.019	28 26	24 22

For SI: 1 inch = 25.4 mm.

M1601.2.1 Duct insulation materials. Duct insulation materials shall conform to the following requirements:

1. Duct coverings and linings shall have a flame-spread index not greater than 25, and a smoke-developed index not greater than 50 when tested in accordance with ASTM E 84.
2. Duct coverings and linings shall not flame, glow, smolder or smoke when tested in accordance with ASTM C 411 at the temperature to which they are exposed in service. The test temperature shall not fall below 250°F (121°C).
3. External duct insulation and factory-insulated flexible ducts shall be legibly printed or identified at intervals not greater than 36 inches (914 mm) with the name of the manufacturer; the thermal resistance *R*-value at the specified installed thickness; and the flame spread and smoke-developed indexes of the composite materials. All duct insulation product *R*-values shall be based on insulation only, excluding air films, vapor retarders or other duct components, and shall be based on tested *C*-values at 75°F (24°C) mean temperature at the installed thickness, in accordance with recognized industry procedures. The installed thickness of duct insulation used to determine its *R*-values shall be determined as follows:
   - 3.1. For duct board, duct liner and factory-made rigid ducts not normally subjected to compression, the nominal insulation thickness shall be used.
   - 3.2. For duct wrap, the installed thickness shall be assumed to be 75 percent (25-percent compression) of nominal thickness.
   - 3.3. For factory-made flexible air ducts, the installed thickness shall be determined by dividing the difference between the actual outside diameter and nominal inside diameter by two.

M1601.2.2 Vibration isolators. Vibration isolators installed between mechanical equipment and metal ducts shall be fabricated from approved materials and shall not exceed 10 inches (254 mm) in length.

M1601.3 Installation. Duct installation shall comply with Sections M1601.3.1 through M1601.3.6.

M1601.3.1 Joints and seams. Joints of duct systems shall be made substantially airtight by means of tapes, mastics, gasketing or other approved closure systems. Closure systems used with rigid fibrous glass ducts shall comply with UL 181A and shall be marked "181A-P" for pressure-sensitive tape, "181 A-M" for mastic or "181 A-H" for heat-sensitive tape. Closure systems used with flexible air ducts and flexible air connectors shall comply with UL 181B and shall be marked "181B-FX" for pressure-sensitive tape or "181B-M" for mastic. Duct connections to flanges of air distribution system equipment or sheet metal fittings shall be mechanically fastened. Crimp joints for round ducts shall have a contact lap of at least 1.5 inches (38 mm) and shall be mechanically fastened by means of at least three sheet metal screws or rivets equally spaced around the joint.

M1601.3.2 Support. Metal ducts shall be supported by 0.5-inch (12.7 mm) wide 18-gage metal straps or 12-gage galvanized wire at intervals not exceeding 10 feet (3048 mm) or other approved means. Nonmetallic ducts shall be supported in accordance with the manufacturer's installation instructions.

M1601.3.3 Fireblocking. Duct installations shall be fireblocked in accordance with Section R602.8.

M1601.3.4 Duct insulation. Duct insulation shall be installed in accordance with the following requirements:

1. A vapor retarder having a maximum permeance of 0.05 perm [(2.87 ng/(s·m^2·Pa)] in accordance with ASTM E 96, or aluminum foil with a minimum thickness of 2 mils (0.051 mm), shall be installed on the exterior of insulation on cooling supply ducts that pass through nonconditioned spaces conducive to condensation.
2. Exterior duct systems shall be protected against the elements.
3. Duct coverings shall not penetrate a fireblocked wall or floor.

M1601.3.5 Factory-made air ducts. Factory-made air ducts shall not be installed in or on the ground, in tile or metal pipe, or within masonry or concrete.

M 1601.3.6 Duct separation. Ducts shall be installed with at least 4 inches (102 mm) separation from earth except where they meet the requirements of Section M1601.1.2.

M1601.4 Under-floor plenums. An under-floor space used as a supply plenum shall conform to the requirements of this section. Fuel gas lines and plumbing waste cleanouts shall not be located within the space.

M1601.4.1 General. The space shall be cleaned of loose combustible materials and scrap, and shall be tightly enclosed. The ground surface of the space shall be covered with a moisture barrier having a minimum thickness of 4 mils (0.102 mm).

M1601.4.2 Materials. The under-floor space, including the sidewall insulation, shall be formed by materials having flame-spread ratings not greater than 200 when tested in accordance with ASTM E 84.

M1601.4.3 Furnace connections. A duct shall extend from the furnace supply outlet to not less than 6 inches (152 mm) below the combustible framing. This duct shall comply with the provisions of Section M1601.1. A noncombustible receptacle shall be installed below any floor opening into the plenum in accordance with the following requirements:

1. The receptacle shall be securely suspended from the floor members and shall not be more than 18 inches (457 mm) below the floor opening.
2. The area of the receptacle shall extend 3 inches (76 mm) beyond the opening on all sides.
3. The perimeter of the receptacle shall have a vertical lip at least 1 inch (25.4 mm) high at the open sides.

M1601.4.4 Access. Access to an under-floor plenum shall be provided through an opening in the floor with minimum dimensions of 18 inches by 24 inches (457 mm by 610 mm).

M1601.4.5 Furnace controls. The furnace shall be equipped with an automatic control that will start the air-circulating fan when the air in the furnace bonnet reaches a temperature not greater than 150°F (66°C). The furnace shall additionally be equipped with an approved automatic control that limits the outlet air temperature to 200°F (93°C).

SECTION M1602
RETURN AIR

M1602.1 Return air. Return air shall be taken from inside the dwelling. Dilution of return air with outdoor air shall not be prohibited.

M1602.2 Required area. The total unobstructed area of return ducts or openings to a warm-air furnace shall be in accordance with the manufacturer's installation instructions, but not less than 2 square inches (1290 mm^2) for each 1,000 Btu/h (293 W) output rating of the furnace. The minimum unobstructed total area of the return air ducts or openings to a central air-conditioning unit and/or heat pump shall be in accordance with the manufacturer's installation instructions, but shall not be less than 6 square inches (3870 mm^2) for each 1,000 Btu/h (293 W) nominal cooling output rating.

M1602.3 Prohibited sources. Outside or return air for a forced-air heating or cooling system shall not be taken from the following locations:

1. Closer than 10 feet (3048 mm) from an appliance vent outlet, a vent opening from a plumbing drainage system or the discharge outlet of an exhaust fan, unless the outlet is 3 feet (914 mm) above the outside air inlet.
2. Where there is the presence of flammable vapors; or where located less than 10 feet (3048 mm) above the surface of any abutting public way or driveway; or where located at grade level by a sidewalk, street, alley or driveway.
3. A room or space, the volume of which is less than 25 percent of the entire volume served by such system. Where connected by a permanent opening having an area sized in accordance with Sections M1602.2 and M1603.1, adjoining rooms or spaces shall be considered as a single room or space for the purpose of determining the volume of such rooms or spaces.

   Exception: The minimum volume requirement shall not apply where the amount of return air taken from a room or space is less than or equal to the amount of supply air delivered to such room or space.
4. A closet, bathroom, toilet room, kitchen, garage, mechanical room, furnace room or other dwelling unit.
5. A room or space containing a fuel-burning appliance where such room or space serves as the sole source of return air.

   Exceptions:

   1. The fuel-burning appliance is a direct-vent appliance or an appliance not requiring a vent in accordance with Section M1801.1 or Chapter 24.
   2. The room or space complies with the following requirements:
      2.1. The return air shall be taken from a room or space having a volume exceeding 1 cubic foot for each 10 Btu/h (9.6 L/W) of combined input rating of all fuel-burning appliances therein.
      2.2. The volume of supply air discharged back into the same space shall be approximately equal to the volume of return air taken from the space.
      2.3. Return-air inlets shall not be located within 10 feet (3048 mm) of any appliance firebox or draft hood in the same room or space.
   3. Rooms or spaces containing solid-fuel burning appliances, provided that return-air inlets are located not less than 10 feet (3048 mm) from the firebox of such appliances.

M1602.4 Inlet opening protection. Outdoor air inlets shall be covered with screens having not less than $^1/_4$-inch (6.4 mm) openings and not greater than $^1/_2$-inch (12.7 mm) openings.

SECTION M1603
SUPPLY AIR

M1603.1 General. The minimum unobstructed total area of supply ducts from a warm-air furnace shall be in accordance with the manufacturer's installation instructions, but shall not be less than 2 square inches (1290 mm^2) for each 1,000 Btu/h (293 W) output rating of the furnace. The minimum unobstructed total area of the supply air ducts from a central air-conditioning unit and/or heat pump shall be in accordance with the manufacturer's installation instructions, but shall not be less than 6 square inches (3870 mm^2) for each 1,000 Btu/h (293 W) nominal cooling output rating. Dampers, grilles or registers installed for the purpose of controlling the supply airflow shall not be considered as obstructions.

CHAPTER 17
COMBUSTION AIR

SECTION M1701
GENERAL

M1701.1 Air supply. Liquid and solid fuel-burning appliances shall be provided with a supply of air for fuel combustion, draft hood dilution and ventilation of the space in which the appliance is installed, in accordance with Section M1702 or Section M1703. The methods of providing combustion air in this chapter do not apply to fireplaces, fireplace stoves and direct-vent appliances.

M1701.1.1 Buildings of unusually tight construction. In buildings of unusually tight construction, combustion air shall be obtained from outside the sealed thermal envelope. In buildings of ordinary tightness, insofar as infiltration is concerned, all or a portion of the combustion air for fuel-burning appliances may be obtained from infiltration when the room or space has a volume of 50 cubic feet per 1,000 Btu/h (4.83 L/W) input.

M1701.2 Exhaust and ventilation system. Air requirements for the operation of exhaust fans, kitchen ventilation systems, clothes dryers and fireplaces shall be considered in determining the adequacy of a space to provide combustion air.

M1701.3 Volume dampers prohibited. Volume dampers shall not be installed in combustion air openings.

M1701.4 Prohibited sources. Combustion air ducts and openings shall not connect appliance enclosures with space in which the operation of a fan may adversely affect the flow of combustion air. Combustion air shall not be obtained from an area in which flammable vapors present a hazard. Fuel-fired appliances shall not obtain combustion air from any of the following rooms or spaces:

1. Sleeping rooms.
2. Bathrooms.
3. Toilet rooms.

Exception: The following appliances shall be permitted to obtain combustion air from sleeping rooms, bathrooms and toilet rooms:

1. Solid fuel-fired appliances provided that the room is not a confined space and the building is not of unusually tight construction.
2. Appliances installed in an enclosure in which all combustion air is taken from the outdoors and the enclosure is equipped with a solid weatherstripped door and self-closing device.

M1701.5 Opening area. The free area of each opening shall be used for determining combustion air. Unless otherwise specified by the manufacturer or determined by actual measurement, the free area shall be considered 75 percent of the gross area for metal louvers and 25 percent of the gross area for wood louvers.

SECTION M1702
ALL AIR FROM INSIDE THE BUILDING

M1702.1 Required volume. Where the volume of the space in which fuel-burning appliances are installed is greater than 50 cubic feet per 1,000 Btu/h (4.83 L/W) of aggregate input rating in buildings of ordinary tightness, insofar as infiltration is concerned, normal infiltration shall be regarded as adequate to provide combustion air. Rooms communicating directly with the space in which the appliances are installed through openings not furnished with doors shall be considered part of the required volume.

M1702.2 Confined space. Where the space in which the appliance is located does not meet the criterion specified in Section M1702.1, two permanent openings to adjacent spaces shall be provided so that the combined volume of all spaces meets the criterion. One opening shall be within 12 inches (305 mm) of the top and one within 12 inches (305 mm) of the bottom of the space, as illustrated in Figure M1702.2. Each opening shall have a free area equal to a minimum of 1 square inch per 1,000 Btu/h (2.20 mm^2/W) input rating of all appliances installed within the space, but not less than 100 square inches (0.064 m^2).

M1702.3 Unusually tight construction. Where the space is of adequate volume in accordance with Section M1702.1 or Section M1702.2, but is within a building sealed so tightly that infiltration air is not adequate for combustion, combustion air shall be obtained from outdoors or from spaces freely communicating with the outdoors in accordance with Section M1703.

SECTION M1703
ALL AIR FROM OUTDOORS

M1703.1 Outdoor air. Where the space in which fuel-burning appliances are located does not meet the criterion for indoor air specified in Section M1702, outside combustion air shall be supplied as specified in Section M1703.2 or Section M1703.3.

M1703.2 Two openings or ducts. Outside combustion air shall be supplied through openings or ducts, as illustrated in Figures M1703.2(1), M1703.2(2), M1703.2(3) and M1703.2(4). One opening shall be within 12 inches (305 mm) of the top of the enclosure, and one within 12 inches (305 mm) of the bottom of the enclosure. Openings are permitted to connect to spaces directly communicating with the outdoors, such as ventilated crawl spaces or ventilated attic spaces. The same duct or opening shall not serve both combustion air openings. The duct serving the upper opening shall be level or extend upward from the appliance space.

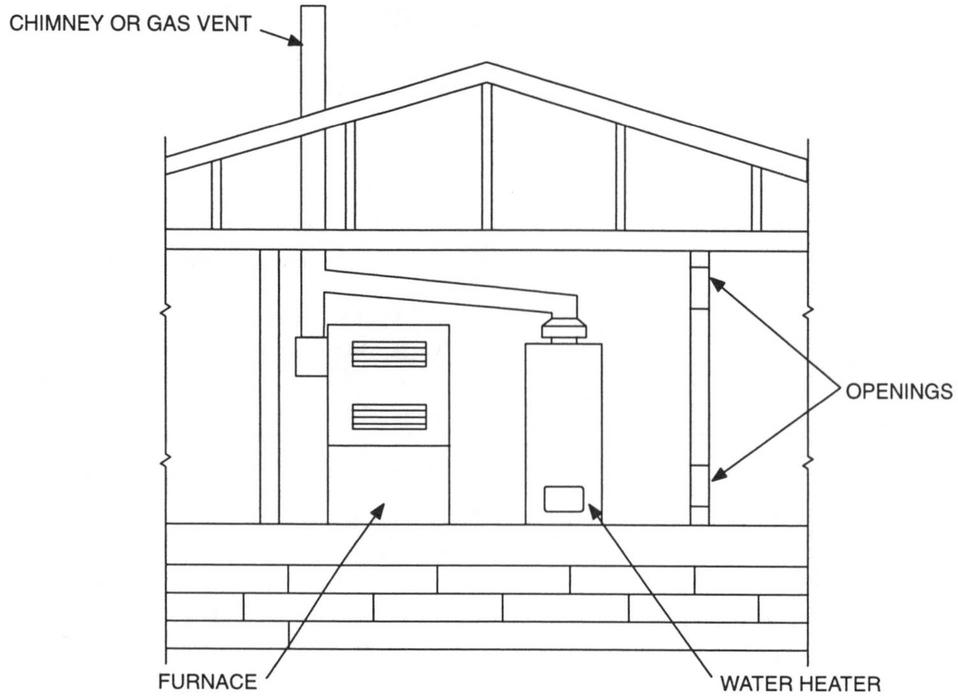

For SI: 1 square inch = 645.16 mm^2, 1 British thermal unit per hour = 0.2931 W.

Note: Each opening shall have a free area of not less than 1 square inch per 1,000 Btu/h of the total input rating of all appliances in the enclosure, but not less than 100 square inches.

FIGURE M1702.2
APPLIANCES LOCATED IN CONFINED SPACES—ALL AIR TAKEN FROM ADJACENT SPACES WITHIN THE BUILDING

M1703.2.1 Size of openings. Where communicating with the outdoors by means of vertical ducts, each opening shall have a free area of at least 1 square inch per 4,000 Btu/h (0.550 mm^2/W) of total input rating of all appliances in the space. Where horizontal ducts are used, each opening shall have a free area of at least 1 square inch per 2,000 Btu/h (0.275 mm^2/W) of total input of all appliances in the space. Ducts shall be of the same minimum cross-sectional area as the required free area of the openings to which they connect. The minimum cross-sectional dimension of rectangular air ducts shall be 3 inches (76 mm).

M1703.3 Single opening or duct. For an appliance with a minimum clearance of 1 inch (25.4 mm) on the sides and back and 6 inches (152 mm) on the front, outside combustion air shall be supplied through one opening or duct, as illustrated in Figures M1703.3(1), M1703.3(2) and M1703.3(3). The opening or duct shall be within 12 inches (305 mm) of the top of the enclosure. The opening is permitted to connect to spaces directly communicating with the outdoors, such as ventilated attic spaces. A duct shall be level or extend upward from the appliance space.

M1703.3.1 Size of opening. Where communicating with the outdoors by means of a single opening or duct, the opening or duct shall have a free area of at least 1 square inch per 3,000 Btu/h (0.413 mm^2/W) of total input rating of all appliances in the space, but not smaller than the vent flow area. A duct shall be of the same minimum cross-sectional area as the required free area of the opening to which it connects. The minimum cross-sectional dimension of a rectangular air duct shall be 3 inches (76 mm).

M1703.4 Attic combustion air. Combustion air obtained from an attic area, as illustrated in Figure M1703.2(3), shall be in accordance with the following:

1. The attic ventilation shall be sufficient to provide the required volume of combustion air.
2. The combustion air opening shall be provided with a metal sleeve extending from the appliance enclosure to at least 6 inches (152 mm) above the top of the ceiling joists and ceiling insulation.
3. An inlet air duct within an outlet air duct shall be an acceptable means of supplying attic combustion air to an appliance room provided that the inlet duct extends at least 12 inches (305 mm) above the top of the outlet duct in the attic space, as illustrated in Figure M1703.3(4).
4. The end of ducts that terminate in an attic shall not be screened.

M1703.5 Under-floor combustion air. Combustion air obtained from under-floor areas, as illustrated in Figure M1703.2(4), shall have free opening areas to the outside equivalent to not less than twice the required combustion air opening.

M1703.6 Opening requirements. Outside combustion air openings shall be covered with corrosion-resistant screen or equivalent protection having not less than 0.25-inch (6.4 mm) openings, and not greater than 0.5-inch (12.7 mm) openings.

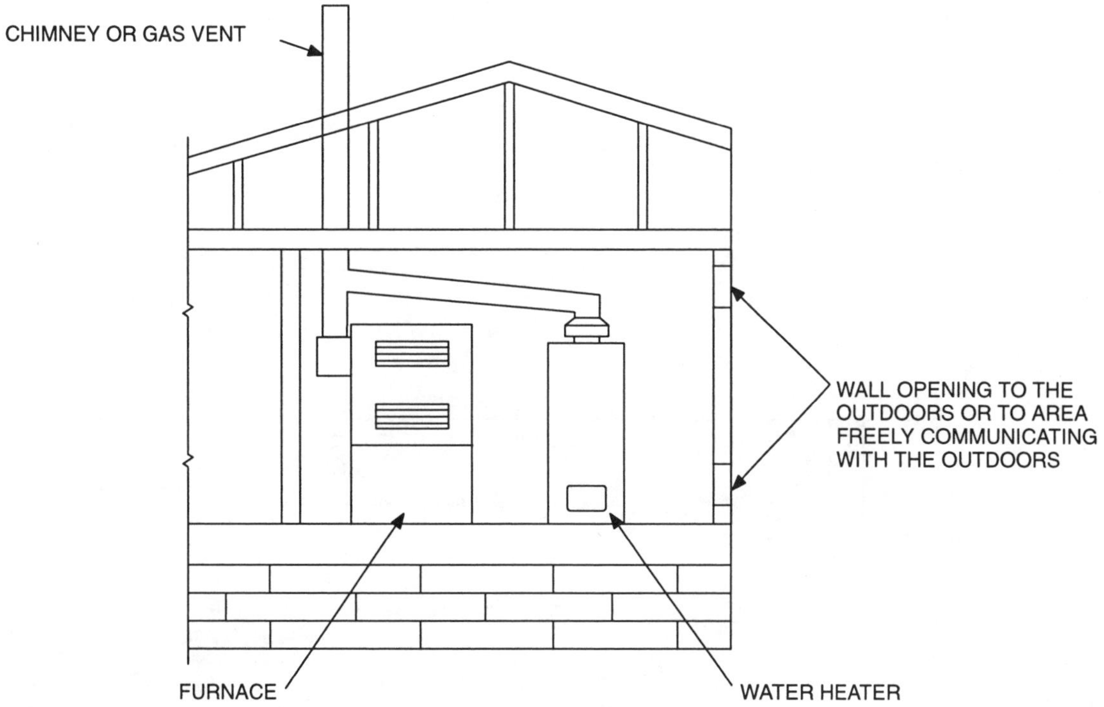

For SI: 1 square inch = 645.16 mm^2, 1 British thermal unit per hour = 0.2931 W.

Note: Each opening shall have a free area of not less than 1 square inch per 4,000 Btu/h of the total input rating of all appliances in the enclosure.

FIGURE M1703.2(1)
APPLIANCES LOCATED IN CONFINED SPACES—ALL AIR TAKEN FROM OUTDOORS THROUGH TWO OPENINGS

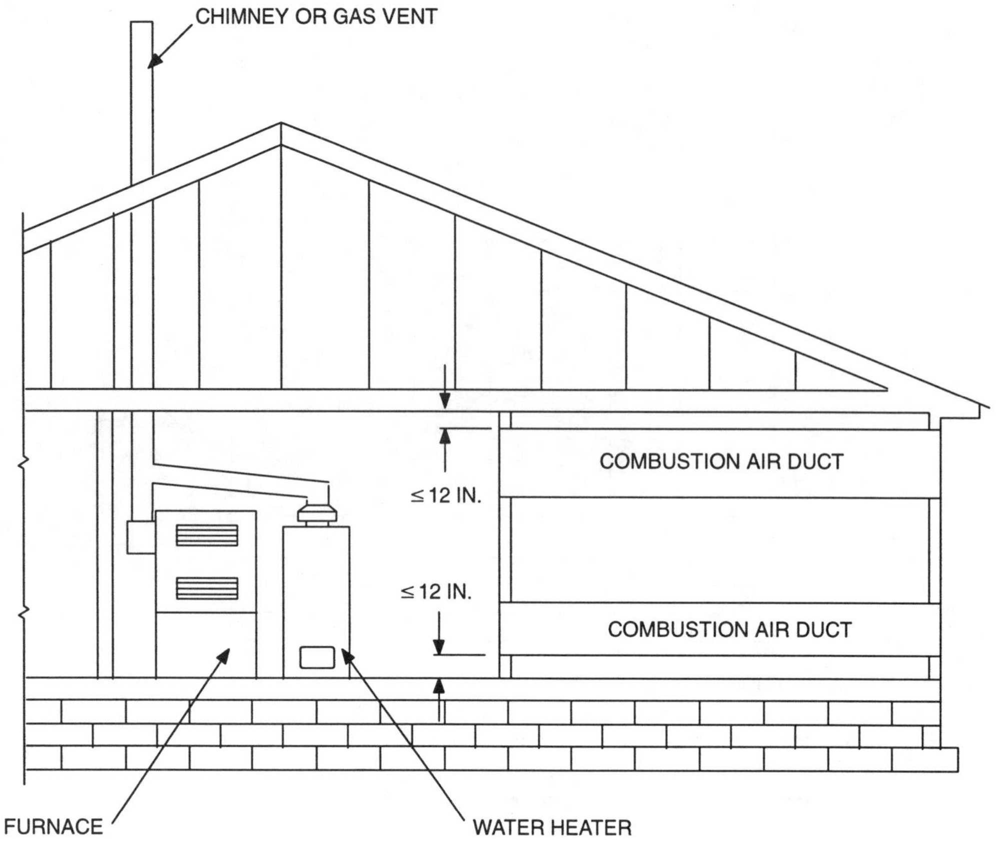

For SI: 1 inch = 25.4 mm, 1 square inch = 645.16 mm^2, 1 British thermal unit per hour = 0.2931 W.

Note: Each opening shall have a free area of at least 1 square inch per 2,000 Btu/h of the total input of all appliances in the space.

FIGURE M1703.2(2)
APPLIANCES LOCATED IN CONFINED SPACES—ALL AIR TAKEN FROM OUTDOORS THROUGH HORIZONTAL DUCTS

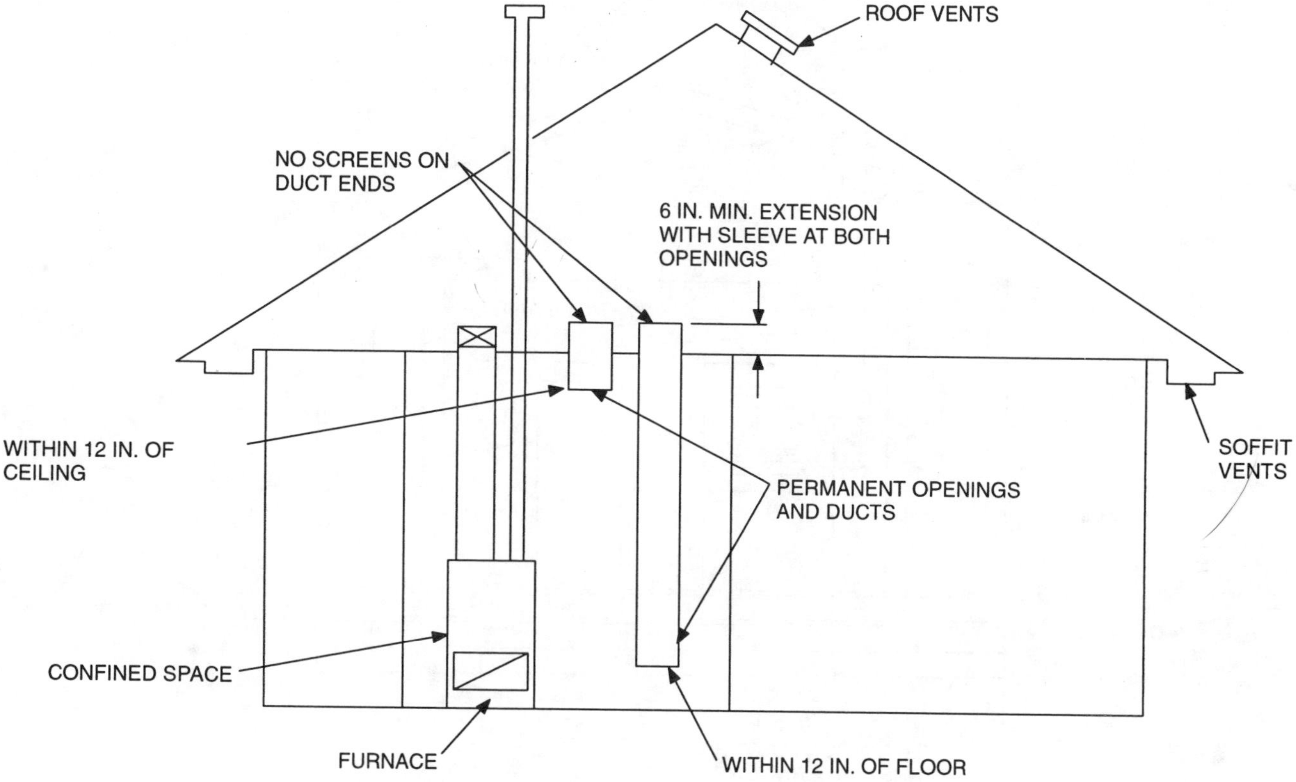

For SI: 1 inch = 25.4 mm, 1 British thermal unit per hour = 0.2931 W, 1 square inch = 645.16 mm^2.

Note: Each opening shall have a free area of at least 1 square inch per 4,000 Btu/h of the total input of all appliances in the space. The attic must be sufficiently vented for combustion air to be taken from the attic.

FIGURE M1703.2(3)
APPLIANCES LOCATED IN CONFINED SPACES—ALL AIR TAKEN FROM OUTDOORS THROUGH VENTILATED ATTIC

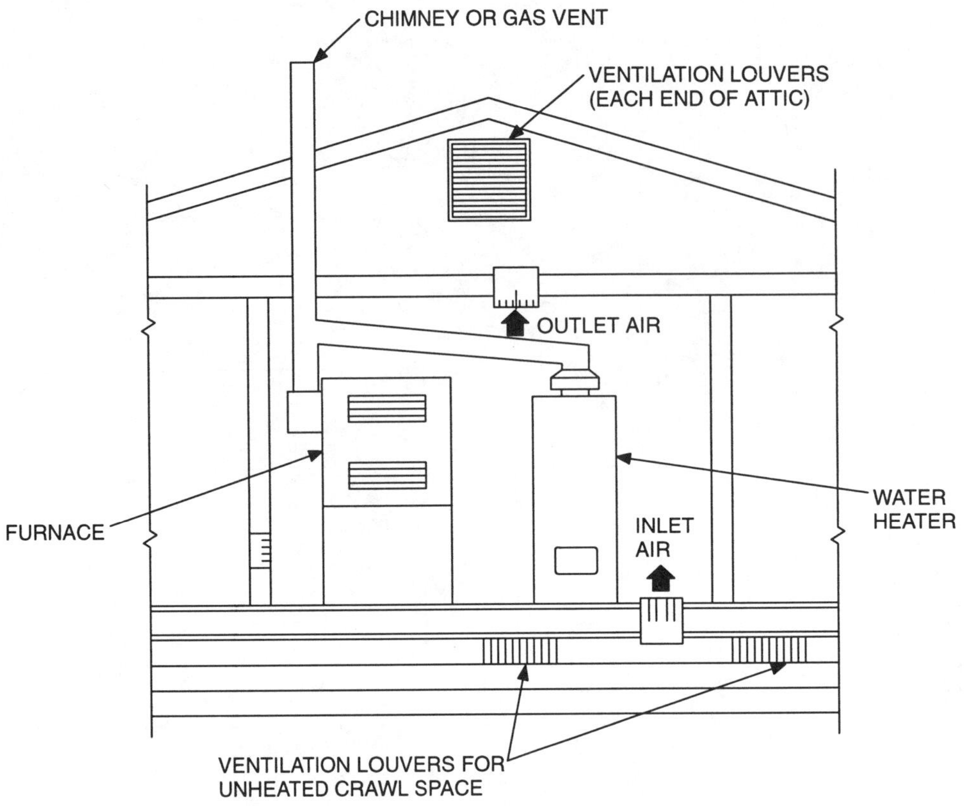

For SI: 1 square inch = 645.16 mm^2, 1 British thermal unit per hour = 0.2931 W.

Note: The inlet and outlet air openings shall have a free area of not less than 1 square inch per 4,000 Btu/h of the total input rating of all appliances in the enclosure.

FIGURE M1703.2(4)
APPLIANCES LOCATED IN CONFINED SPACES—INLET AIR TAKEN FROM VENTILATED CRAWL SPACE AND OUTLET AIR TO VENTILATED ATTIC

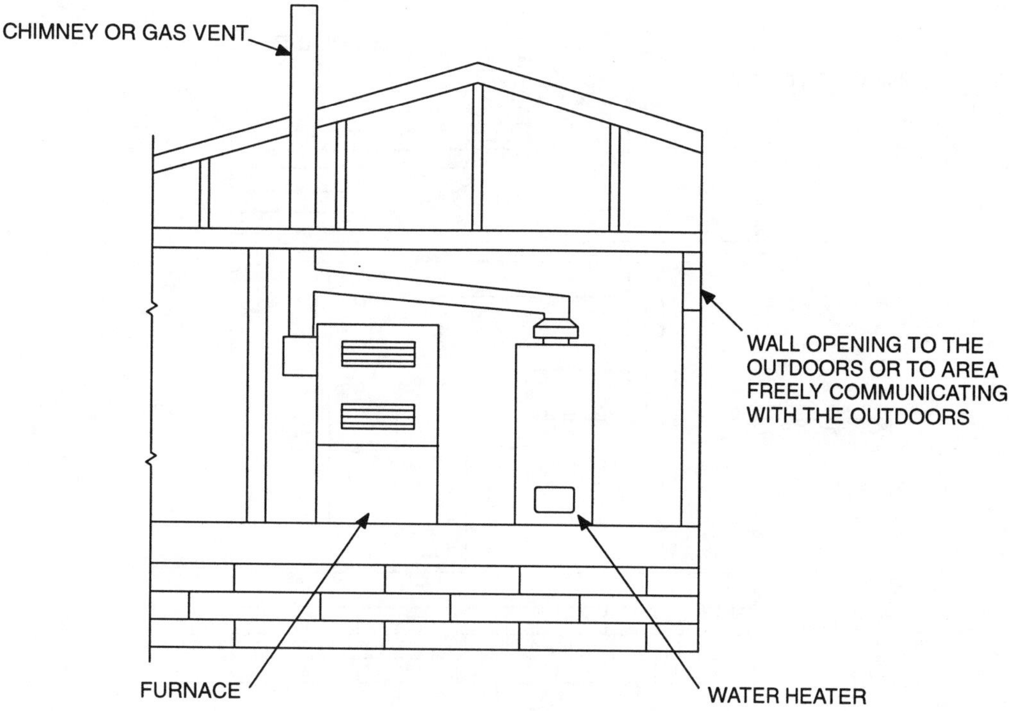

For SI: 1 square inch = 645.16 mm^2, 1 British thermal unit per hour = 0.2931 W.

Note: The air opening shall have a free area of not less than 1 square inch per 3,000 Btu/h of the total input rating of all appliances in the enclosure but not smaller than the vent flow area.

FIGURE M1703.3(1)
APPLIANCES LOCATED IN CONFINED SPACES—ALL AIR TAKEN FROM OUTDOORS THROUGH SINGLE OPENING

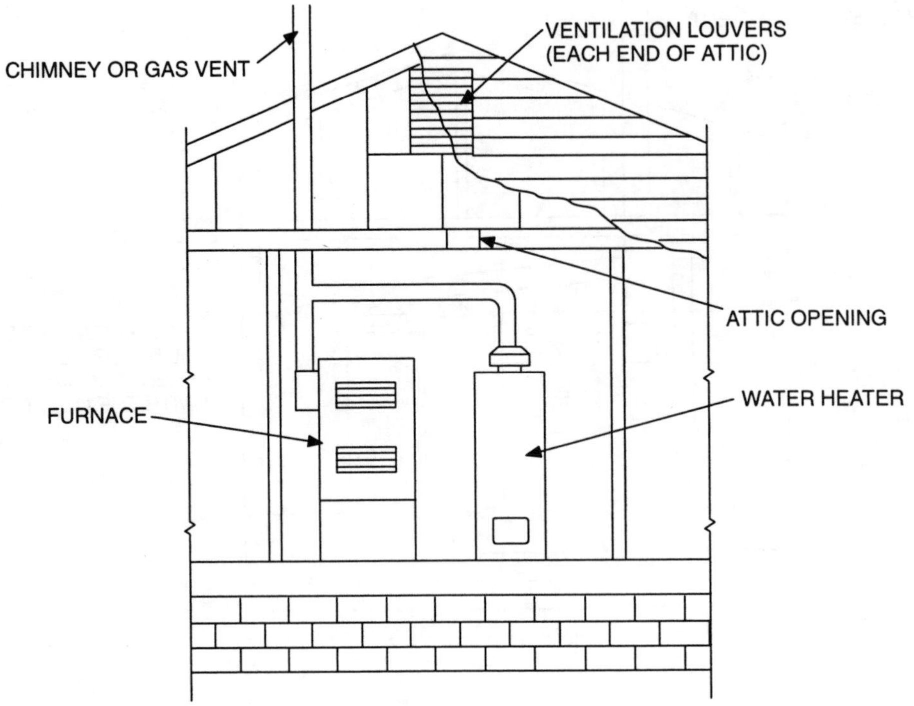

For SI: 1 square inch = 645.16 mm^2, 1 British thermal unit per hour = 0.2931 W.

Note: The air opening and duct shall have a free area of not less than 1 square inch per 3,000 Btu/h of the total input rating of all appliances in the enclosure but not smaller than the vent flow area.

FIGURE M1703.3(2)
APPLIANCES LOCATED IN CONFINED SPACES—ALL AIR TAKEN FROM OUTDOORS THROUGH SINGLE OPENING OR DUCT

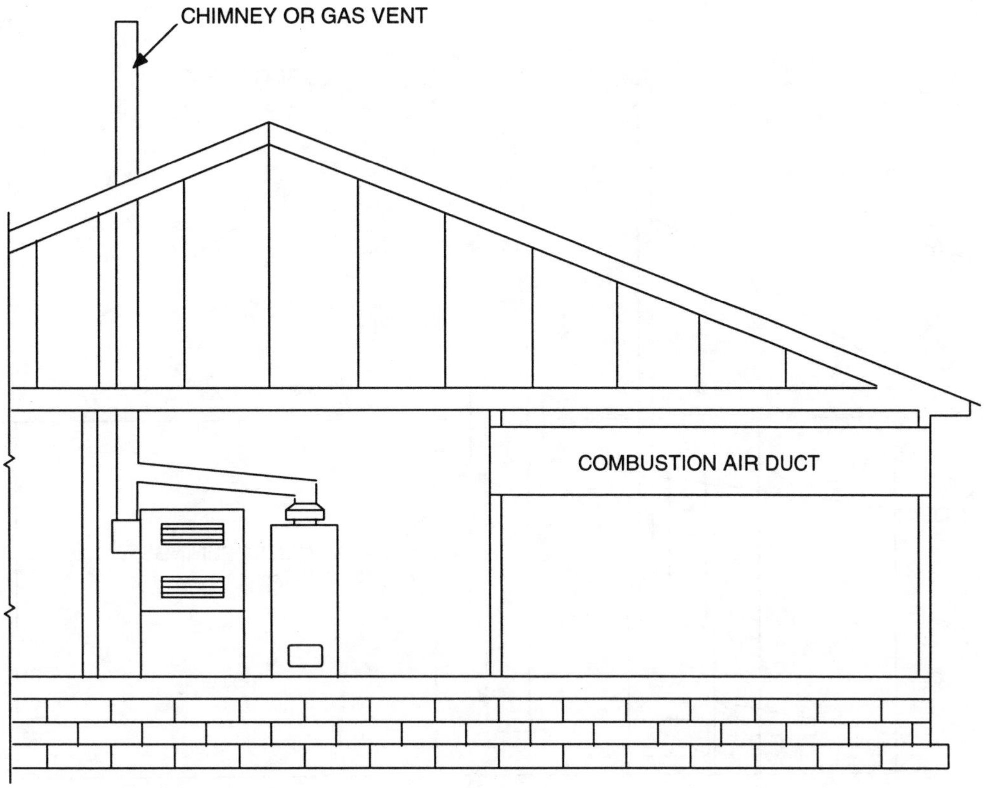

For SI: 1 square inch = 645.16 mm^2, 1 British thermal unit per hour = 0.2931 W.

Note: The air opening and duct shall have a free area of not less than 1 square inch per 3,000 Btu/h of the total input rating of all appliances in the enclosure but not smaller than the vent flow area.

FIGURE M1703.3(3)
APPLIANCES LOCATED IN CONFINED SPACES—ALL AIR TAKEN FROM OUTDOORS THROUGH SINGLE OPENING OR DUCT

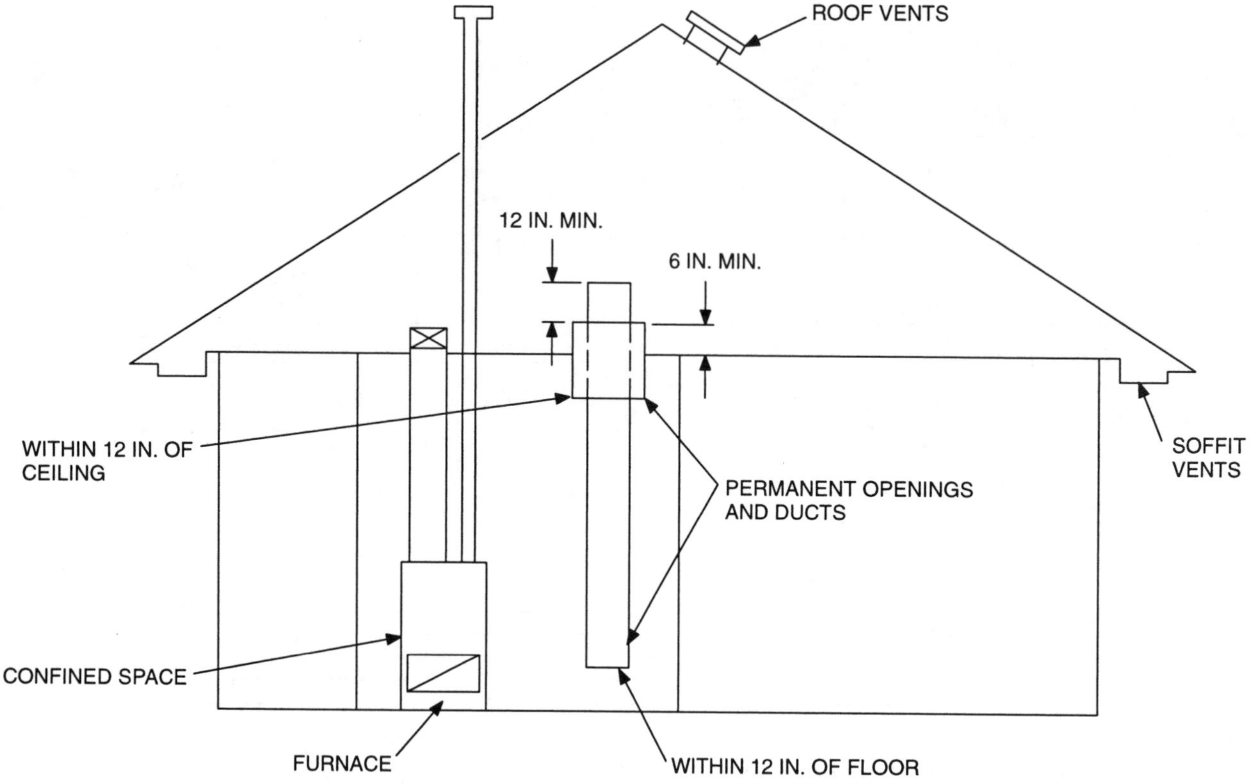

For SI: 1 inch = 25.4 mm, 1 British thermal unit per hour = 0.2931 W, 1 square inch = 645.16 mm^2.

Note: Each opening shall have a free area of at least 1 square inch per 4,000 Btu/h of the total input of all appliances in the space. The attic must be sufficiently vented for combustion air to be taken from the attic.

FIGURE M1703.3(4)
APPLIANCES LOCATED IN CONFINED SPACES—ALL AIR TAKEN FROM OUTDOORS THROUGH VENTILATED ATTIC, INLET DUCT WITHIN OUTLET DUCT

CHAPTER 18
CHIMNEYS AND VENTS

SECTION M1801
GENERAL

M1801.1 Venting required. Fuel-burning appliances shall be vented to the outside in accordance with their listing and label and manufacturer's installation instructions except appliances listed and labeled for unvented use. Venting systems shall consist of approved chimneys or vents, or venting assemblies that are integral parts of labeled appliances. Gas-fired appliances shall be vented in accordance with Chapter 24.

M1801.2 Draft requirements. A venting system shall satisfy the draft requirements of the appliance in accordance with the manufacturer's installation instructions, and shall be constructed and installed to develop a positive flow to convey combustion products to the outside atmosphere.

M1801.3 Existing chimneys and vents. Where an appliance is permanently disconnected from an existing chimney or vent, or where an appliance is connected to an existing chimney or vent during the process of a new installation, the chimney or vent shall comply with Sections M1801.3.1 through M1801.3.4.

M1801.3.1 Size. The chimney or vent shall be resized as necessary to control flue gas condensation in the interior of the chimney or vent and to provide the appliance, or appliances served, with the required draft. For the venting of oil-fired appliances to masonry chimneys, the resizing shall be done in accordance with NFPA 31.

M1801.3.2 Flue passageways. The flue gas passageway shall be free of obstructions and combustible deposits and shall be cleaned if previously used for venting a solid or liquid fuel-burning appliance or fireplace. The flue liner, chimney inner wall or vent inner wall shall be continuous and free of cracks, gaps, perforations, or other damage or deterioration that would allow the escape of combustion products, including gases, moisture and creosote.

M1801.3.3 Cleanout. Masonry chimneys shall be provided with a cleanout opening complying with Section R1001.12.

M1801.3.4 Clearances. Chimneys and vents shall have air-space clearance to combustibles in accordance with this code and the chimney or vent manufacturer's installation instructions.

Exception: Masonry chimneys equipped with a chimney lining system tested and listed for installation in chimneys in contact with combustibles in accordance with UL 1777, and installed in accordance with the manufacturer's instruction, shall not be required to have a clearance between combustible materials and exterior surfaces of the masonry chimney. Noncombustible firestopping shall be provided in accordance with this code.

M1801.4 Space around lining. The space surrounding a flue lining system or other vent installed within a masonry chimney shall not be used to vent any other appliance. This shall not prevent the installation of a separate flue lining in accordance with the manufacturer's installation instructions and this code.

M1801.5 Mechanical draft systems. A mechanical draft system shall be used only with appliances listed and labeled for such use. Provisions shall be made to prevent the flow of fuel to the equipment when the draft system is not operating. Forced draft systems and all portions of induced draft systems under positive pressure during operation shall be designed and installed so as to prevent leakage of flue gases into a building.

M1801.6 Direct-vent appliances. Direct-vent appliances shall be installed in accordance with the manufacturer's installation instructions.

M1801.7 Support. Venting systems shall be adequately supported for the weight of the material used.

M1801.8 Duct penetrations. Chimneys, vents and vent connectors shall not extend into or through supply and return air ducts or plenums.

M1801.9 Fireblocking. Vent and chimney installations shall be fireblocked in accordance with Section R602.8.

M1801.10 Unused openings. Unused openings in any venting system shall be closed or capped.

M1801.11 Multiple-appliance venting systems. Two or more listed and labeled appliances connected to a common natural draft venting system shall comply with the following requirements:

1. Appliances that are connected to common venting systems shall be located on the same floor of the dwelling.

   Exception: Engineered systems as provided for in Section G2426.
2. Inlets to common venting systems shall be offset such that no portion of an inlet is opposite another inlet.
3. Connectors serving appliances operating under a natural draft shall not be connected to any portion of a mechanical draft system operating under positive pressure.
4. Solid-fuel-burning appliances shall not be connected to a vent or chimney flue serving other appliances.

SECTION M1802
VENT COMPONENTS

M1802.1 Draft hoods. Draft hoods shall be located in the same room or space as the combustion air openings for the appliances.

M1802.2 Vent dampers. Vent dampers shall comply with Sections M1802.2.1 and M1802.2.2.

M1802.2.1 Manually operated. Manually operated dampers shall not be installed except in connectors or chimneys serving solid-fuel-burning appliances.

M1802.2.2 Automatically operated. Automatically operated dampers shall conform to UL 17 and be installed in accordance with the terms of their listing and label. The installation shall prevent firing of the burner when the damper is not opened to a safe position.

M1802.3 Draft regulators. Draft regulators shall be provided for oil-fired appliances required to be connected to a chimney. Draft regulators provided for solid-fuel-burning appliances to reduce draft intensity shall be installed and set in accordance with the manufacturer's installation instructions.

M1802.3.1 Location. Where required, draft regulators shall be installed in the same room or enclosure as the appliance so that no difference in pressure will exist between the air at the regulator and the combustion air supply.

SECTION M1803 CHIMNEY AND VENT CONNECTORS

M1803.1 General. Connectors shall be used to connect fuel-burning appliances to a vertical chimney or vent except where the chimney or vent is attached directly to the appliance.

M1803.2 Connectors for oil and solid fuel appliances. Connectors for oil and solid-fuel-burning appliances shall be constructed of factory-built chimney material, Type L vent material or single-wall metal pipe having resistance to corrosion and heat and thickness not less than that of galvanized steel as specified in Table M1803.2.

TABLE M1803.2 THICKNESS FOR SINGLE-WALL METAL PIPE CONNECTORS

DIAMETER OF CONNECTOR (inches)	GALVANIZED SHEET METAL GAGE NUMBER	MINIMUM THICKNESS (inch)
Less than 6	26	0.019
6 to 10	24	0.024
Over 10 through 16	22	0.029

For SI: 1 inch = 25.4 mm.

M1803.3 Installation. Vent and chimney connectors shall be installed in accordance with the manufacturer's installation instructions and within the space that the appliance is located. Appliances shall be located as close as practical to the vent or chimney. Connectors shall be as short and straight as possible and installed with a slope of not less than $^1/_4$ inch (6.4 mm) rise per foot of run. Connectors shall be securely supported and joints shall be fastened with sheet metal screws or rivets. Devices that obstruct the flow of flue gases shall not be installed in a connector unless listed and labeled or approved for such installations.

M1803.3.1 Floor, ceiling and wall penetrations. A chimney connector or vent connector shall not pass through any floor or ceiling. A chimney connector or vent connector shall not pass through a wall or partition unless the connector is listed and labeled for wall pass-through, or is routed through a device listed and labeled for wall pass-through and is installed in accordance with the conditions of its listing and label. Connectors for oil-fired appliances listed and labeled for Type L vents, passing through walls or partitions shall be in accordance with the following:

1. Type L vent material for oil appliances shall be installed with not less than listed and labeled clearances to combustible material.
2. Single-wall metal pipe shall be guarded by a ventilated metal thimble not less than 4 inches (102 mm) larger in diameter than the vent connector. A minimum 6 inches (152 mm) of clearance shall be maintained between the thimble and combustibles.

M1803.3.2 Length. The horizontal run of an uninsulated connector to a natural draft chimney shall not exceed 75 percent of the height of the vertical portion of the chimney above the connector. The horizontal run of a listed connector to a natural draft chimney shall not exceed 100 percent of the height of the vertical portion of the chimney above the connector.

M1803.3.3 Size. A connector shall not be smaller than the flue collar of the appliance.

Exception: Where installed in accordance with the appliance manufacturer's installation instructions

M1803.3.4 Clearance. Connectors shall be installed with clearance to combustibles as set forth in Table M1803.3.4. Reduced clearances to combustible materials shall be in accordance with Table M1306.2 and Figure M1306.1.

TABLE M1803.3.4 CHIMNEY AND VENT CONNECTOR CLEARANCES TO COMBUSTIBLE MATERIALS[a]

TYPE OF CONNECTOR	MINIMUM CLEARANCE (inches)
Single-wall metal pipe connectors:	
Oil and solid-fuel appliances	18
Oil appliances listed for use with type L vents	9
Type L vent piping connectors:	
Oil and solid-fuel appliances	9
Oil appliances listed for use with type L vents	3[b]

For SI: 1 inch = 25.4 mm.

a. These minimum clearances apply to unlisted single-wall chimney and vent connectors. Reduction of required clearances are permitted as in Table M1306.2.

b. When listed Type L vent piping is used, the clearance shall be in accordance with the vent listing.

M1803.3.5 Access. The entire length of a connector shall be accessible for inspection, cleaning and replacement.

M1803.4 Connection to fireplace flue. Connection of appliances to chimney flues serving fireplaces shall comply with Sections M1803.4.1 through M1803.4.4.

M1803.4.1 Closure and accessibility. A noncombustible seal shall be provided below the point of connection to prevent entry of room air into the flue. Means shall be provided for access to the flue for inspection and cleaning.

M1803.4.2 Connection to factory-built fireplace flue. A different appliance shall not be connected to a flue serving a factory-built fireplace unless the appliance is specifically listed for such an installation. The connection shall be made in conformance with the appliance manufacturer's instructions.

M1803.4.3 Connection to masonry fireplace flue. A connector shall extend from the appliance to the flue serving a masonry fireplace such that the flue gases are conveyed directly into the flue. The connector shall be accessible or removable for inspection and cleaning of both the connector and the flue. Listed direct connection devices shall be installed in accordance with their listing.

M1803.4.4 Size of flue. The size of the fireplace flue shall be in accordance with Section M1805.3.1.

SECTION M1804
VENTS

M1804.1 Type of vent required. Appliances shall be provided with a listed and labeled venting system as set forth in Table M1804.1.

TABLE M1804.1
VENT SELECTION CHART

TYPE L OIL	PELLET VENTS
Oil-burning appliances listed and labeled for use with Type L vents.	Pellet fuel burning
	Pellet fuel-burning appliances listed and labeled for use with pellet vents

M1804.2 Termination. Vent termination shall comply with Sections M1804.2.1 through M1804.2.6.

M1804.2.1 Through the roof. Vents passing through a roof shall extend through flashing and terminate in accordance with the manufacturer's installation requirements.

M1804.2.2 Decorative shrouds. Decorative shrouds shall not be installed at the termination of vents except where such shrouds are listed and labeled for use with the specific venting system and are installed in accordance with the manufacturer's installation instructions.

M1804.2.3 Natural draft appliances. Vents for natural draft appliances shall terminate at least 5 feet (1524 mm) above the highest connected appliance outlet, and natural draft gas vents serving wall furnaces shall terminate at an elevation at least 12 feet (3658 mm) above the bottom of the furnace.

M1804.2.4 Type L vent. Type L venting systems shall conform to UL 641. Such vents shall terminate with a listed and labeled cap in accordance with the vent manufacturer's installation instructions not less than 2 feet (610 mm) above the roof and not less than 2 feet (610 mm) above any portion of the building within 10 feet (3048 mm).

M1804.2.5 Direct vent appliances. Vent terminals for direct vent appliances shall be located in accordance with Sections M1804.2.5.1 through M1804.2.5.3. The bottom of the vent terminal and air intake shall be located not less than 12 inches (305 mm) above grade.

M1804.2.5.1 Input 10,000 Btu/h or less. The vent terminal of a direct vent appliance with an input of 10,000 Btu/h (2931 W) or less shall be located not less than 6 inches (153 mm) from any opening through which flue gases could enter a building.

M1804.2.5.2 Input greater than 10,000 Btu/h and not greater than 50,000 Btu/h. The vent terminal of a direct vent appliance with an input of greater than 10,000 Btu/h (2931 W) and not greater than 50,000 Btu/h (14 655 W) shall be located not less than 9 inches (229 mm) from any opening through which flue gases could enter a building.

M1804.2.5.3 Input greater than 50,000 Btu/h. The vent terminal of a direct vent appliance with an input over 50,000 Btu/h (14 655 W) shall be located not less than 12 inches (305 mm) from any opening through which flue gases could enter a building.

1804.2.6 Mechanical draft systems. Mechanical draft systems other than direct vent systems shall be installed in accordance with their listing, the manufacturer's installation instructions and the following requirements:

1. The vent terminal shall be located not less than 3 feet (914 mm) above a forced air inlet located within 10 feet (3048 mm).
2. The vent terminal shall be located not less than 4 feet (1219 mm) below, 4 feet (1219 mm) horizontally from, or 1 foot (305 mm) above a door, window or gravity air inlet into a dwelling.
3. The vent terminal shall be located not less than 2 feet (610 mm) from an adjacent building.
4. The bottom of the vent terminal shall be located at least 12 inches (305 mm) above finished ground level.

M1804.3 Installation. Type L and pellet vents shall be installed in accordance with the terms of their listing and label and the manufacturer's installation instructions.

M1804.3.1 Size of single appliance venting systems. An individual vent for a single appliance shall have a cross-sectional area equal to or greater than the area of the connector to the appliance, but not less than 7 square inches (4515 mm^2) except where the vent is an integral part of a listed and labeled appliance.

SECTION M1805
MASONRY AND FACTORY-BUILT CHIMNEYS

M1805.1 General. Masonry and factory-built chimneys shall be built and installed in accordance with Sections R1001 and R1002, respectively. Flue lining for masonry chimneys shall comply with Section R1001.7.

M1805.2 Masonry chimney connection. A chimney connector shall enter a masonry chimney not less than 6 inches (152 mm) above the bottom of the chimney. Where it is not-possible to locate the connector entry at least 6 inches (152 mm) above the bottom of the chimney flue, a cleanout shall be provided by installing a capped tee in the connector next to the chimney. A connector entering a masonry chimney shall extend through, but not beyond the wall and shall be flush with the inner face of the liner. Connectors, or thimbles where used, shall be firmly cemented into the masonry.

M1805.3 Size of chimney flues. The effective area of a natural draft chimney flue for one appliance shall be not less than the area of the connector to the appliance. Chimney flues connected to more than one appliance shall be not less than the area of the largest connector plus 50 percent of the areas of additional chimney connectors.

Exception: Chimney flues serving oil-fired appliances sized in accordance with NFPA 31.

M1805.3.1 Size of chimney flue for solid fuel appliance. Except where otherwise specified in the manufacturer's installation instructions, the cross-sectional area of a flue connected to a solid-fuel-burning appliance shall be not less than the area of the flue collar or connector, and not larger than three times the area of the flue collar.

CHAPTER 19
SPECIAL FUEL-BURNING EQUIPMENT

SECTION M1901
RANGES AND OVENS

M1901.1 Clearances. Freestanding or built-in ranges shall have a vertical clearance above the cooking top of not less than 30 inches (762 mm) to unprotected combustible material. Reduced clearances are permitted in accordance with the listing and labeling of the range hoods or appliances.

M1901.2 Cooking appliances. Household cooking appliances shall be listed and labeled and shall be installed in accordance with the manufacturer's installation instructions. The installation shall not interfere with combustion air or accessibility for operation and servicing.

SECTION M1902
SAUNA HEATERS

M1902.1 Locations and protection. Sauna heaters shall be protected from accidental contact by persons with a guard of material having a low thermal conductivity, such as wood. The guard shall have no substantial effect on the transfer of heat from the heater to the room.

M1902.2 Installation. Sauna heaters shall be installed in accordance with the manufacturer's installation instructions.

M1902.3 Combustion air. Combustion air and venting for a nondirect vent-type heater shall be provided in accordance with Chapters 17 and 18, respectively.

M1902.4 Controls. Sauna heaters shall be equipped with a thermostat that will limit room temperature to not greater than 194°F (90°C). Where the thermostat is not an integral part of the heater, the heat-sensing element shall be located within 6 inches (152 mm) of the ceiling.

CHAPTER 20
BOILERS/WATER HEATERS

SECTION M2001
BOILERS

M2001.1 Installation. In addition to the requirements of this code, the installation of boilers shall conform to the manufacturer's instructions. The manufacturer's rating data, the nameplate and operating instructions of a permanent type shall be attached to the boiler. Boilers shall have all controls set, adjusted and tested by the installer. A complete control diagram together with complete boiler operating instructions shall be furnished by the installer. Solid and liquid-fuel-burning boilers shall be provided with combustion air as required by Chapter 17.

M2001.1.1 Standards. Oil-fired boilers and their control systems shall be listed and labeled in accordance with UL 726. Electric boilers and their control systems shall be listed in accordance with UL 834. Boilers shall be designed and constructed in accordance with the requirements of ASME CSD-1 and as applicable, the ASME *Boiler and Pressure Vessel Code*, Sections I and IV. Gas-fired boilers shall conform to the requirements listed in Chapter 24.

M2001.2 Clearance. Boilers shall be installed in accordance with their listing and label.

SECTION M2002
OPERATING AND SAFETY CONTROLS

M2002.1 Safety controls. Electrical and mechanical operating and safety controls for boilers shall be listed and labeled.

M2002.2 Boiler gauges. Boilers shall be equipped with pressure and temperature gauges that indicate pressure and temperature within the normal range of operation.

M2002.3 Pressure-relief valve. Boilers shall be equipped with pressure-relief valves with minimum rated capacities for the equipment served. Pressure-relief valves shall be set at the maximum rating of the boiler. Discharge shall be piped to drains by gravity to within 18 inches (457 mm) of the floor or to an open receptor.

SECTION M2003
EXPANSION TANKS

M2003.1 General. Boilers shall be provided with expansion tanks. Nonpressurized expansion tanks shall be securely fastened to the structure or boiler and supported to carry twice the weight of the tank filled with water. Provisions shall be made for draining nonpressurized tanks without emptying the system.

M2003.1.1 Pressurized expansion tanks. Pressurized expansion tanks shall be consistent with the volume and capacity of the system. Tanks shall be capable of withstanding a hydrostatic test pressure of two and one-half times the allowable working pressure of the system.

M2003.2 Minimum capacity. The minimum capacity of expansion tanks shall be determined from Table M2003.2.

SECTION M2004
WATER HEATERS USED FOR SPACE HEATING

M2004.1 General. Water heaters utilized both to supply potable hot water and to provide hot water for space heating shall be installed in accordance with this chapter, Chapter 24 and the manufacturer's installation instructions.

TABLE M2003.2
EXPANSION TANK MINIMUM CAPACITY[a] FOR FORCED HOT-WATER SYSTEMS

SYSTEM VOLUME[b] (gallons)	PRESSURIZED DIAPHRAGM TYPE	NONPRESSURIZED TYPE
10	1.0	1.5
20	1.5	3.0
30	2.5	4.5
40	3.0	6.0
50	4.0	7.5
60	5.0	9.0
70	6.0	10.5
80	6.5	12.0
90	7.5	13.5
100	8.0	15.0

For SI: 1 gallon = 3.785 L, 1 pound per square inch gauge = 6.895 kPa, °C = [(°F)-32]/1.8.

a. Based on average water temperature of 195°F, fill pressure of 12 psig and a maximum operating pressure of 30 psig.

b. System volume includes volume of water in boiler, convectors and piping, not including the expansion tank.

SECTION M2005
WATER HEATERS

M2005.1 General. Water heaters shall be installed in accordance with the manufacturer's installation instructions and the requirements of this code. Water heaters installed in an attic shall conform to the requirements of Section M1305.1.3. Gas-fired water heaters shall conform to the requirements in Chapter 24. Domestic electric water heaters shall conform to UL 174 or UL 1453. Commercial electric water heaters shall conform to UL 1453. Oiled-fired water heaters shall conform to UL 732.

M2005.2 Prohibited locations. Fuel-fired water heaters shall not be installed in a room used as a storage closet. Water heaters located in a bedroom or bathroom shall be installed in a sealed enclosure so that combustion air will not be taken from the living space. Direct-vent water heaters are not required to be installed within an enclosure.

M2005.2.1 Water heater access. Access to water heaters that are located in an attic or underfloor crawl space is permitted to be through a closet located in a sleeping room or bathroom where ventilation of those spaces is in accordance with this code.

M2005.3 Electric water heaters. Electric water heaters shall also be installed in accordance with the applicable provisions of Chapters 33 through 42.

M2005.4 Supplemental water-heating devices. Potable water heating devices that utilize refrigerant-to-water heat exchangers shall be approved and installed in accordance with the manufacturer's installation instructions.

SECTION M2006
POOL HEATERS

M2006.1 General. Pool and spa heaters shall be installed in accordance with the manufacturer's installation instructions. Oil-fired pool heaters shall be tested in accordance with UL 726.

M2006.2 Location. Pool heaters shall be located or protected to guard against accidental contact of hot surfaces by persons.

M2006.3 Clearances. In no case shall the clearances be such as to interfere with combustion air, draft hood or flue terminal relief, or accessibility for servicing.

M2006.4 Temperature and pressure-limiting devices. Pool heaters shall have temperature and pressure-relief valves.

M2006.5 Bypass valves. Where an integral bypass system is not provided as a part of the pool heater, a bypass line and valve shall be installed between the inlet and outlet piping for use in adjusting the flow of water through the heater.

CHAPTER 21
HYDRONIC PIPING

SECTION M2101
HYDRONIC PIPING SYSTEMS INSTALLATION

M2101.1 General. Hydronic piping shall conform to Table M2101.1. Approved piping, valves, fittings and connections shall be installed in accordance with the manufacturer's installation instructions. Pipe and fittings shall be rated for use at the operating temperature and pressure of the hydronic system. Used pipe, fittings, valves or other materials shall be free of foreign materials.

M2101.2 System drain down. Hydronic piping systems shall be installed to permit the system to be drained. When the system drains to the plumbing drainage system, the installation shall conform to the requirements of Chapters 25 through 32 of this code.

M2101.3 Protection of potable water. The potable water system shall be protected from backflow in accordance with the provisions listed in Section P2902.

M2101.4 Pipe penetrations. Openings through concrete or masonry building elements shall be sleeved.

M2101.5 Contact with building material. A hydronic piping system shall not be in direct contact with any building material that causes the piping material to degrade or corrode.

M2101.6 Drilling and notching. Wood-framed structural members shall be drilled, notched or altered in accordance with the provisions of Sections R502.6, R602.6, R602.6.1 and R802.6. Holes in cold-formed, steel-framed, load-bearing members shall only be permitted in accordance with Sections R506.2, R603.2 and R804.2. In accordance with the provisions of Sections R505.3.5, R603.3.4 and R804.3.5, cutting and notching of flanges and lips of cold-formed, steel-framed, load-bearing members shall not be permitted.

M2101.7 Prohibited tee applications. Fluid in the supply side of a hydronic system shall not enter a tee fitting through the branch opening.

M2101.8 Expansion, contraction and settlement. Piping shall be installed so that piping, connections and equipment shall not be subjected to excessive strains or stresses. Provisions shall be made to compensate for expansion, contraction, shrinkage and structural settlement.

M2101.9 Piping support. Hangers and supports shall be of material of sufficient strength to support the piping, and shall be fabricated from materials compatible with the piping material. Piping shall be supported at intervals not exceeding the spacing specified in Table M2101.9.

SECTION M2102
BASEBOARD CONVECTORS

M2102.1 General. Baseboard convectors shall be installed in accordance with the manufacturer's installation instructions. Convectors shall be supported independently of the hydronic piping.

SECTION M2103
FLOOR HEATING SYSTEMS

M2103.1 Piping materials. Piping for embedment in concrete or gypsum materials shall be standard-weight steel pipe, copper tubing, chlorinated polyvinyl chloride (CPVC) or polybutylene with a minimum rating of 100 psi at 180°F (68.9 kPa at 146°C).

M2103.2 Piping joints. Piping joints that are embedded shall be installed in accordance with the following requirements:

1. Steel pipe joints shall be welded.
2. Copper tubing shall be joined with brazing material having a melting point exceeding 1,000°F (538°C).
3. Polybutylene pipe and tubing joints shall be installed with socket-type heat-fused polybutylene fittings.
4. CPVC tubing shall be joined using solvent cement joints.

M2103.3 Testing. Piping or tubing to be embedded shall be tested by applying a hydrostatic pressure of not less than 100 psi (68.9 kPa). The pressure shall be maintained for 30 minutes, during which all joints shall be visually inspected for leaks.

SECTION M2104
LOW TEMPERATURE PIPING

M2104.1 Piping materials. Low temperature piping for embedment in concrete or gypsum materials shall be as indicated in Table M2101.1.

M2104.2 Piping joints. Piping joints (other than those in Section M2103.2) that are embedded shall be installed in accordance with the following requirements:

1. Crosslinked Polyethylene (PEX) tubing shall follow manufacturer's instructions.
2. Polyethylene tubing shall be installed with heat fusion joints.

M2104.2.1 Polyethylene plastic pipe and tubing for ground source heat pump loop systems. Joints between polyethylene plastic pipe and tubing or fittings for ground source heat pump loop systems shall be heat fusion joints conforming to Section M2104.2.1.1, electrofusion joints conforming to Section M2104.2.1.2 or stab type insertion joints conforming to Section M2104.2.1.3.

M2104.2.1.1 Heat-fusion joints. Joints shall be of the socket-fusion, saddle-fusion or butt-fusion type, fabricated in accordance with the piping manufacturer's instructions. Joint surfaces shall be clean and free of moisture. Joint surfaces shall be heated to melt temperatures and joined. The joint shall be undisturbed until cool. Fittings shall be manufactured in accordance with ASTM D 2683.

M2104.2.1.2 Electrofusion joints. Joints shall be of the electrofusion type. Joint surfaces shall be clean and free of moisture, and scoured to expose virgin resin. Joint surfaces shall be heated to melt temperatures for the period of time specified by the manufacturer. The joint shall be undisturbed until cool. Fittings shall be manufactured in accordance with ASTM F 1055.

M2104.2.1.3 Stab type insert fittings. Joint surfaces shall be clean and free of moisture. Pipe ends shall be chamfered and inserted into the fitting to full depth. Fittings shall be manufactured in accordance with ASTM D 2513.

SECTION M2105 GROUND SOURCE HEAT PUMP SYSTEM LOOP PIPING

M2105.1 Testing. The assembled loop system shall be pressure tested with water at 100 psi (690 kPa) for 30 minutes with no observed leaks before connection (header) trenches are backfilled. Flow rates and pressure drops shall be compared tocalculated values. If actual flow rate or pressure drop figures differ from calculated values by more than 10 percent, the problem shall be identified and corrected.

**TABLE M2101.1
HYDRONIC PIPING MATERIALS**

MATERIAL	USE CODE[a]	STANDARD[b]	JOINTS	NOTES
Brass pipe	1	ASTM B 43	Brazed, welded, threaded, mechanical and flanged fittings	
Brass tubing	1	ASTM B 135	Brazed, soldered and mechanical fittings	
Chlorinated poly (vinyl chloride) (CPVC) pipe and tubing	1, 2, 3	ASTM D 2846	Solvent cement joints, compression joints and threaded adapters	
Copper pipe	1	ASTM B 42, B 302	Brazed, soldered and mechanical fittings threaded, welded and flanged	
Copper tubing (type K, L or M)	1, 2	ASTM B 75, B 88, B 251, B 306	Brazed, soldered and flared mechanical fittings	Joints embedded in concrete
Crosslinked polyethylene (PEX) tubing	2, 3	ASTM F 876, F 877	Mechanical compression	Install in accordance with manufacturer's instructions.
Crosslinked polyethylene/ aluminum/crosslinked polyethylene-(PEX-AL-PEX) pressure pipe	1, 2	ASTM F 1281 or CAN/ CSA B137.10	Mechanical, crimp/insert	Install in accordance with manufacturer's instructions.
Plastic fittings PEX		ASTM F 1807		
Polybutylene (PB) pipe and tubing	1, 2, 3	ASTM D 3309	Heat-fusion, crimp/insert and compression	Joints in concrete shall be heat-fused.
Polyethylene (PE) pipe, tubing and fittings (for ground source heat pump loop systems)	1, 2, 4	ASTM D 2513; ASTM D 3350; ASTM D 2513; ASTM D 3035; ASTM D 2447; ASTM D 2683; ASTM F 1055; ASTM D 2837; ASTM D 3350; ASTM D 1693	Heat-fusion	
Soldering fluxes	1	ASTM B 813	Copper tube joints	
Steel pipe	1, 2	ASTM A 53; A 106	Brazed, welded, threaded, flanged and mechanical fittings	Joints in concrete shall be welded. Galvanized pipe shall not be welded or brazed.
Steel tubing	1	ASTM A 254	Mechanical fittings, welded	

For SI: °C = [(°F)-32]/1.8.

a. Use code:

1. Above ground.
2. Embedded in radiant system.
3. Temperatures below 180°F only.
4. Low temperature (below 130°F) applications only.

b. Standards as listed in Chapter 43.

TABLE M2101.9
HANGER SPACING INTERVALS

PIPING MATERIAL	MAXIMUM HORIZONTAL SPACING (feet)	MAXIMUM VERTICAL SPACING (feet)
ABS	4	10
CPVC ≤ 1 inch pipe or tubing	3	5
CPVC ≥ $1^1/_4$ inch	4	10
Copper or copper alloy pipe	12	10
Copper or copper alloy tubing	6	10
PB pipe or tubing	2.67	4
PE pipe or tubing	2.67	4
PEX tubing	2.67	4
PVC	4	10
Steel pipe	12	15
Steel tubing	8	10

For SI: 1 inch = 25.4 mm, 1 foot = 304.8 mm.

CHAPTER 22
SPECIAL PIPING AND STORAGE SYSTEMS

SECTION M2201
OIL TANKS

M2201.1 Materials. Supply tanks shall be listed and labeled and shall conform to UL 58 for underground tanks and UL 80 for inside tanks.

M2201.2 Above-ground tanks. The maximum amount of fuel oil stored above ground or inside of a building shall be 660 gallons (2498 L). The supply tank shall be supported on rigid noncombustible supports to prevent settling or shifting.

M2201.2.1 Tanks within buildings. Supply tanks for use inside of buildings shall be of such size and shape to permit installation and removal from dwellings as whole units. Supply tanks larger than 10 gallons (38 L) shall be placed not less than 5 feet (1524 mm) from any fire or flame either within or external to any fuel-burning appliance.

M2201.2.2 Outside above-ground tanks. Tanks installed outside above ground shall be a minimum of 5 feet (1524 mm) from an adjoining property line. Such tanks shall be suitably protected from the weather and from physical damage.

M2201.3 Underground tanks. Excavations for underground tanks shall not undermine the foundations of existing structures. The clearance from the tank to the nearest wall of a basement, pit or property line shall not be less than 1 foot (305 mm). Tanks shall be set on and surrounded with noncorrosive inert materials such as clean earth, sand or gravel well tamped in place. Tanks shall be covered with not less than 1 foot (305 mm) of earth. Corrosion protection shall be provided in accordance with Section M2203.7.

M2201.4 Multiple tanks. Cross connection of two supply tanks shall be permitted in accordance with Section M2203.6.

M2201.5 Oil gauges. Inside tanks shall be provided with a device to indicate when the oil in the tank has reached a predetermined safe level. Glass gauges or a gauge subject to breakage that could result in the escape of oil from the tank shall not be used.

SECTION M2202
OIL PIPING, FITTING AND CONNECTIONS

M2202.1 Materials. Piping shall consist of steel pipe, copper tubing or steel tubing conforming to ASTM A 539. Aluminum tubing shall not be used between the fuel-oil tank and the burner units.

M2202.2 Joints and fittings. Piping shall be connected with standard fittings compatible with the piping material. Cast iron fittings shall not be used for oil piping. Unions requiring gaskets or packings, right or left couplings, and sweat fittings employing solder having a melting point less than 1,000°F (538°C) shall not be used for oil piping. Threaded joints and connections shall be made tight with a lubricant or pipe thread compound.

M2202.3 Flexible connectors. Flexible metal hose used where rigid connections are impractical or to reduce the effect of jarring and vibration shall be listed and labeled and shall be installed in compliance with its label and the manufacturer's installation instructions. Connectors made from combustible materials shall not be used inside of buildings or above ground outside of buildings.

SECTION M2203
INSTALLATION

M2203.1 General. Piping shall be installed in a manner to avoid placing stresses on the piping, and to accommodate expansion and contraction of the piping system.

M2203.2 Supply piping. Supply piping used in the installation of oil burners and appliances shall be not smaller than $^3/_8$-inch (9.5 mm) pipe or $^3/_8$-inch (9.5 mm) outside diameter tubing. Copper tubing and fittings shall be a minimum Type L.

M2203.3 Fill piping. Fill piping shall terminate outside of buildings at a point at least 2 feet (610 mm) from any building opening at the same or lower level. Fill openings shall be equipped with a tight metal cover.

M2203.4 Vent piping. Vent piping shall be not smaller than $1^1/_4$-inch (32 mm) pipe. Vent piping shall be laid to drain toward the tank without sags or traps in which the liquid can collect. Vent pipes shall not be cross connected with fill pipes, lines from burners or overflow lines from auxiliary tanks. The lower end of a vent pipe shall enter the tank through the top and shall extend into the tank not more than 1 inch (25.4 mm).

M2203.5 Vent termination. Vent piping shall terminate outside of buildings at a point not less than 2 feet (610 mm), measured vertically or horizontally, from any building opening. Outer ends of vent piping shall terminate in a weather-proof cap or fitting having an unobstructed area at least equal to the cross-sectional area of the vent pipe, and shall be located sufficiently above the ground to avoid being obstructed with snow and ice.

M2203.6 Cross connection of tanks. Cross connection of two supply tanks, not exceeding 660 gallons (2498 L) aggregate capacity, with gravity flow from one tank to another, shall be acceptable providing that the two tanks are on the same horizontal plane.

M2203.7 Corrosion protection. Underground tanks and buried piping shall be protected by corrosion-resistant coatings or special alloys or fiberglass-reinforced plastic.

SECTION M2204
OIL PUMPS AND VALVES

M2204.1 Pumps. Oil pumps shall be positive displacement types that automatically shut off the oil supply when stopped. Automatic pumps shall be listed and labeled and shall be installed in compliance with their listing and label.

M2204.2 Shutoff valves. A readily accessible manual shutoff valve shall be installed between the oil supply tank and the burner. Where the shutoff valve is installed in the discharge line of an oil pump, a pressure-relief valve shall be incorporated to bypass or return surplus oil.

M2204.3 Maximum pressure. Pressure at the oil supply inlet to an appliance shall be not greater than 3 psi (20.7 kPa).

M2204.4 Relief valves. Fuel-oil lines incorporating heaters shall be provided with relief valves that will discharge to a return line when excess pressure exists.

CHAPTER 23
SOLAR SYSTEMS

SECTION M2301
SOLAR ENERGY SYSTEMS

M2301.1 General. This section provides for construction, installation, alteration, and repair of equipment and systems utilizing solar energy to provide space heating or cooling, hot water heating, and swimming pool heating.

M2301.2 Installation. Installation of solar energy systems shall comply with Sections M2301.2.1 through M2301.2.9.

M2301.2.1 Access. Solar energy collectors, controls, dampers, fans, blowers and pumps shall be accessible for inspection, maintenance, repair and replacement.

M2301.2.2 Roof-mounted collectors. The roof shall be constructed to support the loads imposed by roof-mounted solar collectors. Roof-mounted solar collectors that serve as a roof covering shall conform to the requirements for roof coverings in Chapter 9 of this code. Where mounted on or above the roof coverings, the collectors and supporting structure shall be constructed of noncombustible materials or fire-retardant-treated wood equivalent to that required for the roof construction.

M2301.2.3 Pressure and temperature relief. System components containing fluids shall be protected with pressure- and temperature-relief valves. Relief devices shall be installed in sections of the system so that a section cannot be valved off or isolated from a relief device.

M2301.2.4 Vacuum relief. System components that might be subjected to pressure drops below atmospheric pressure during operation or shutdown shall be protected by a vacuum-relief valve.

M2301.2.5 Protection from freezing. System components shall be protected from damage resulting from freezing of heat-transfer liquids at the winter design temperature provided in Table R301.2(1). Freeze protection shall be provided by heating, insulation, thermal mass and heat transfer fluids with freeze points lower than the winter design temperature, heat tape or other approved methods, or combinations thereof.

Exception: Where the winter design temperature is greater than 32°F (0°C).

M2301.2.6 Expansion tanks. Expansion tanks in solar energy systems shall be installed in accordance with Section M2003 in closed fluid loops that contain heat transfer fluid.

M2301.2.7 Roof penetrations. Roof penetrations shall be flashed and waterproofed in accordance with Chapter 9 of this code.

M2301.2.8 Solar loop isolation. Valves shall be installed to allow the solar collectors to be isolated from the remainder of the system. Each isolation valve shall be labeled with the open and closed position.

M2301.2.9 Maximum temperature limitation. Systems shall be equipped with means to limit the maximum water temperature of the system fluid entering or exchanging heat with any pressurized vessel inside the dwelling to 180°F (82°C). This protection is in addition to the required temperature and pressure relief valves required by Section M2301.2.3.

M2301.3 Labeling. Labeling shall comply with Sections M2301.3.1 and M2301.3.2.

M2301.3.1 Collectors. Collectors shall be listed and labeled to show the manufacturer's name, model number, serial number, collector weight, collector maximum allowable temperatures and pressures, and the type of heat transfer fluids that are compatible with the collector. The label shall clarify that these specifications apply only to the collector.

M2301.3.2 Thermal storage units. Pressurized thermal storage units shall be listed and labeled to show the manufacturer's name, model number, serial number, storage unit maximum and minimum allowable operating temperatures and pressures, and the type of heat transfer fluids that are compatible with the storage unit. The label shall clarify that these specifications apply only to the thermal storage unit.

M2301.4 Prohibited heat transfer fluids. Flammable gases and liquids shall not be used as heat transfer fluids.

M2301.5 Backflow protection. All connections from the potable water supply to solar systems shall comply with Section P2902.4.5.

Part VI — Fuel Gas

CHAPTER 24
FUEL GAS

The text of this chapter is excerpted from the 2000 edition of the *International Fuel Gas Code* and has been modified where necessary to make such text conform to the scope of application of the *International Residential Code for One- and Two-Family Dwellings*. The section numbers appearing in parentheses after each section number represent the location of the corresponding text in the *International Fuel Gas Code.*

SECTION G2401 (101)
GENERAL

G2401.1 (101.2) Application. This chapter covers those fuel gas piping systems, fuel gas utilization equipment and related accessories, venting systems and combustion air configurations most commonly encountered in the construction of one- and two-family dwellings and structures regulated by this code.

Coverage of piping systems shall extend from the point of delivery to the connections with gas utilization equipment. (See "point of delivery.") Piping systems requirements shall include design, materials, components, fabrication, assembly, installation, testing, inspection, operation, and maintenance. Requirements for gas utilization equipment and related accessories shall include installation, combustion, and ventilation air and venting.

The omission from this chapter of any material or method of installation provided for in the *International Fuel Gas Code* shall not be construed as prohibiting the use of such material or method of installation. Fuel gas piping systems, fuel gas utilization equipment and related accessories, venting systems and combustion air configurations not specifically covered in these chapters shall comply with the applicable provisions of the *International Fuel Gas Code.*

This chapter shall not apply to the following:

1. Liquified natural gas (LNG) installations.
2. Temporary LP-Gas piping for buildings under construction or renovation that is not to become part of the permanent piping system.
3. Except as provided in Section 401.1.1, gas piping, meters, gas presure regulators, and other appurtenances used by the serving gas supplier in the distribution of gas, other than undiluted LP-Gas.

SECTION G2402 (201)
GENERAL

G2402.1 (201.1) Scope. Unless otherwise expressly stated, the following words and terms shall, for the purposes of this chapter, have the meanings indicated in this chapter.

G2402.2 (201.2) Interchangeability. Words used in the present tense include the future; words in the masculine gender include the feminine and neuter; the singular number includes the plural and the plural, the singular.

G2402.3 (201.3) Terms defined in other codes. Where terms are not defined in this code and are defined in the ICC *Electrical Code, International Building Code, International Fire Code, International Mechanical Code* or *International Plumbing Code*, such terms shall have meanings ascribed to them as in those codes.

SECTION G2403 (202)
GENERAL DEFINITIONS

AIR CONDITIONING, GAS FIRED. A gas-burning, automatically operated appliance for supplying cooled and/or dehumidified air or chilled liquid.

AIR, EXHAUST. Air being removed from any space or piece of equipment and conveyed directly to the atmosphere by means of openings or ducts.

AIR-HANDLING UNIT. A blower or fan used for the purpose of distributing supply air to a room, space or area.

AIR, MAKEUP. Air that is provided to replace air being exhausted.

ALTERATION. A change in a system that involves an extension, addition or change to the arrangement, type or purpose of the original installation.

ANODELESS RISER. A transition assembly in which plastic piping is installed and terminated aboveground outside of a building.

APPLIANCE (EQUIPMENT). Any apparatus or equipment that utilizes gas as a fuel or raw material to produce light, heat, power, refrigeration or air conditioning.

APPLIANCE, FAN-ASSISTED COMBUSTION. An appliance equipped with an integral mechanical means to either draw or force products of combustion through the combustion chamber or heat exchanger.

APPLIANCE, AUTOMATICALLY CONTROLLED. Appliances equipped with an automatic burner ignition and safety shut-off device and other automatic devices, which accomplish complete turn-on and shut-off of the gas to the main burner or burners, and graduate the gas supply to the burner or burners, but do not affect complete shut-off of the gas.

APPLIANCE, UNVENTED. An appliance designed or installed in such a manner that the products of combustion are not conveyed by a vent or chimney directly to the outside atmosphere.

APPLIANCE, VENTED. An appliance designed and installed in such a manner that all of the products of combustion are conveyed directly from the appliance to the outside atmosphere through an approved chimney or vent system.

ATMOSPHERIC PRESSURE. The pressure of the weight of air and water vapor on the surface of the earth, approximately 14.7 pounds per square inch (psia) (101 kPa absolute) at sea level.

AUTOMATIC IGNITION. Ignition of gas at the burner(s) when the gas controlling device is turned on, including reignition if the flames on the burner(s) have been extinguished by means other than by the closing of the gas controlling device.

BAROMETRIC DRAFT REGULATOR. A balanced damper device attached to a chimney, vent connector, breeching or flue gas manifold to protect combustion equipment by controlling chimney draft. A double-acting barometric draft regulator is one whose balancing damper is free to move in either direction to protect combustion equipment from both excessive draft and backdraft.

BOILER, LOW-PRESSURE. A self-contained gas-fired appliance for supplying steam or hot water.

Hot water heating boiler. A boiler in which no steam is generated, from which hot water is circulated for heating purposes and then returned to the boiler, and that operates at water pressures not exceeding 160 psig (1100 kPa gauge) and at water temperatures not exceeding 250°F (121°C) at or near the boiler outlet.

Hot water supply boiler. A boiler, completely filled with water, which furnishes hot water to be used externally to itself, and that operates at water pressures not exceeding 160 psig (1100 kPa gauge) and at water temperatures not exceeding 250°F (121°C) at or near the boiler outlet.

Steam heating boiler. A boiler in which steam is generated and that operates at a steam pressure not exceeding 15 psig (100 kPa gauge).

BRAZING. A metal joining process wherein coalescence is produced by the use of a nonferrous filler metal having a melting point above 1,000°F (538°C), but lower than that of the base metal being joined. The filler material is distributed between the closely fitted surfaces of the joint by capillary action.

BTU. Abbreviation for British thermal unit, which is the quantity of heat required to raise the temperature of 1 pound (454 g) of water 1°F (0.56°C) (1 Btu = 1055 J).

BURNER. A device for the final conveyance of the gas, or a mixture of gas and air, to the combustion zone.

Induced-draft. A burner that depends on draft induced by a fan that is an integral part of the appliance and is located downstream from the burner.

Power. A burner in which gas, air or both are supplied at pressures exceeding, for gas, the line pressure, and for air, atmospheric pressure, with this added pressure being applied at the burner.

CHIMNEY. A primarily vertical structure containing one or more flues, for the purpose of carrying gaseous products of combustion and air from an appliance to the outside atmosphere.

Factory-built chimney. A listed and labeled chimney composed of factory-made components, assembled in the field in accordance with manufacturer's instructions and the conditions of the listing.

Masonry chimney. A field-constructed chimney composed of solid masonry units, bricks, stones or concrete.

CLEARANCE. The minimum distance through air measured between the heat-producing surface of the mechanicalappliance, device or equipment and the surface of the combustible material or assembly.

CLOTHES DRYER. An appliance used to dry wet laundry by means of heated air.

Type 1. Factory-built package, multiple production. Primarily used in the family living environment. Usually the smallest unit physically and in function output.

CODE. These regulations, subsequent amendments thereto, or any emergency rule or regulation that the administrative authority having jurisdiction has lawfully adopted.

CODE OFFICIAL. The officer or other designated authority charged with the administration and enforcement of this code, or a duly authorized representative.

COMBUSTION. In the context of this code, refers to the rapid oxidation of fuel accompanied by the production of heat or heat and light.

COMBUSTION AIR. Air necessary for complete combustion of a fuel, including theoretical air and excess air.

COMBUSTION CHAMBER. The portion of an appliance within which combustion occurs.

COMBUSTION PRODUCTS. Constituents resulting from the combustion of a fuel with the oxygen of the air, including the inert gases, but excluding excess air.

CONCEALED PIPING. Piping that is located in a concealed location. (See "concealed location.")

CONCEALED LOCATION. A location that cannot be accessed without damaging permanent parts of the building structure or finish surface. Spaces above, below or behind readily removable panels or doors shall not be considered as concealed.

CONDENSATE. The liquid that condenses from a gas (including flue gas) caused by a reduction in temperature or increase in pressure.

CONFINED SPACES. A space having a volume less than 50 cubic feet per 1,000 British thermal units per hour (Btu/h) (4.8 m^3/kW) of the aggregate input rating of all appliances installed in that space.

CONNECTOR. The pipe that connects an approved appliance to a chimney, flue or vent.

CONTROL. A manual or automatic device designed to regulate the gas, air, water or electrical supply to, or operation of, a mechanical system.

CONVERSION BURNER. A unit consisting of a burner and its controls for installation in an appliance originally utilizing another fuel.

CUBIC FOOT. The amount of gas that occupies 1 cubic foot (0.02832 m^3) when at a temperature of 60°F (16°C), saturated with water vapor and under a pressure equivalent to that of 30 inches of mercury (101 kPa).

DAMPER. A manually or automatically controlled device to regulate draft or the rate of flow of air or combustion gases.

DECORATIVE GAS APPLIANCE, VENTED. A vented appliance wherein the primary function lies in the aesthetic effect of the flames.

DECORATIVE GAS APPLIANCES FOR INSTALLATION IN VENTED FIREPLACES. A vented appliance designed for installation within the fire chamber of a vented fireplace, wherein the primary function lies in the aesthetic effect of the flames.

DEMAND. The maximum amount of gas input required per unit of time, usually expressed in cubic feet per hour, or Btu/h (1 Btu/h = 0.2931 W).

DILUTION AIR. Air that is introduced into a draft hood and is mixed with the flue gases.

DIRECT-VENT APPLIANCES. Appliances that are constructed and installed so that all air for combustion is derived directly from the outside atmosphere and all flue gases are discharged directly to the outside atmosphere.

DRAFT. The pressure difference existing between the equipment or any component part and the atmosphere, that causes a continuous flow of air and products of combustion through the gas passages of the appliance to the atmosphere.

Mechanical or induced draft. The pressure difference created by the action of a fan, blower or ejector that is located between the appliance and the chimney or vent termination.

Natural draft. The pressure difference created by a vent or chimney because of its height, and the temperature difference between the flue gases and the atmosphere.

DRAFT HOOD. A nonadjustable device built into an appliance, or made as part of the vent connector from an appliance, that is designed to (1) provide for ready escape of the flue gases from the appliance in the event of no draft, backdraft, or stoppage beyond the draft hood, (2) prevent a backdraft from entering the appliance, and (3) neutralize the effect of stack action of the chimney or gas vent upon operation of the appliance.

DRAFT REGULATOR. A device that functions to maintain a desired draft in the appliance by automatically reducing the draft to the desired value.

DRIP. The container placed at a low point in a system of piping to collect condensate and from which the condensate is removable.

DUCT FURNACE. A warm-air furnace normally installed in an air-distribution duct to supply warm air for heating. This definition shall apply only to a warm-air heating appliance that depends for air circulation on a blower not furnished as part of the furnace.

EQUIPMENT. See "Appliance."

FIREPLACE. A fire chamber and hearth constructed of noncombustible material for use with solid fuels and provided with a chimney.

Masonry fireplace. A hearth and fire chamber of solid masonry units such as bricks, stones, listed masonry units or reinforced concrete, provided with a suitable chimney.

Factory-built fireplace. A fireplace composed of listed factory-built components assembled in accordance with the terms of listing to form the completed fireplace.

FLAME SAFEGUARD. A device that will automatically shut off the fuel supply to a main burner or group of burners when the means of ignition of such burners become inoperative, and when flame failure occurs on the burner or group of burners.

FLOOR FURNACE. A completely self-contained furnace suspended from the floor of the space being heated, taking air for combustion from outside such space and with means for observing flames and lighting the appliance from such space.

FLUE, APPLIANCE. The passage(s) within an appliance through which combustion products pass from the combustion chamber of the appliance to the draft hood inlet opening on an appliance equipped with a draft hood or to the outlet of the appliance on an appliance not equipped with a draft hood.

FLUE COLLAR. That portion of an appliance designed for the attachment of a draft hood, vent connector or venting system.

FLUE GASES. Products of combustion plus excess air in appliance flues or heat exchangers.

FLUE LINER (LINING). A system or material used to form the inside surface of a flue in a chimney or vent, for the purpose of protecting the surrounding structure from the effects of combustion products and for conveying combustion products without leakage to the atmosphere.

FUEL GAS. A natural, manufactured, liquefied petroleum or a mixture of these.

FUEL GAS UTILIZATION EQUIPMENT. See "Appliance."

FURNACE. A completely self-contained heating unit that is designed to supply heated air to spaces remote from or adjacent to the appliance location.

FURNACE, CENTRAL FURNACE. A self-contained appliance for heating air by transfer of heat of combustion through metal to the air, and designed to supply heated air through ducts to spaces remote from or adjacent to the appliance location.

GAS CONVENIENCE OUTLET. A permanently mounted, manually operated device that provides the means for connecting an appliance to, and disconnecting an appliance from, the gas supply piping. The device includes an integral, manually operated valve with a nondisplaceable valve member and is designed so that disconnection of an appliance only occurs when the manually operated valve is in the closed position.

GAS PIPING. An installation of pipe, valves or fittings installed on a premises or in a building and utilized to convey fuel gas.

GAS UTILIZATION EQUIPMENT. An appliance that utilizes gas as a fuel or raw material or both.

HAZARDOUS LOCATION. Any location considered to be a fire hazard for flammable vapors, dust, combustible fibers or other highly combustible substances. The location is not necessarily categorized in the building code as a high-hazard use group classification.

HOUSE PIPING. See "Piping system."

IGNITION PILOT. A pilot that operates during the lighting cycle and discontinues during main burner operation.

IGNITION SOURCE. A flame spark or hot surface capable of igniting flammable vapors or fumes. Such sources include appliance burners, burner ignitors and electrical switching devices.

INFRARED RADIANT HEATER. A heater which directs a substantial amount of its energy output in the form of infrared radiant energy into the area to be heated. Such heaters are of either the vented or unvented type.

JOINT, FLARED. A metal-to-metal compression joint in which a conical spread is made on the end of a tube that is compressed by a flare nut against a mating flare.

JOINT, MECHANICAL. A general form of gas-tight joints obtained by the joining of metal parts through a positive-holding mechanical construction, such as flanged joint, threaded joint, flared joint or compression joint.

JOINT, PLASTIC ADHESIVE. A joint made in thermoset plastic piping by the use of an adhesive substance which forms a continuous bond between the mating surfaces without dissolving either one of them.

LIQUEFIED PETROLEUM GAS or LPG (LP-GAS). Liquefied petroleum gas composed predominately of propane, propylene, butanes or butylenes, or mixtures thereof that is gaseous under normal atmospheric conditions, but is capable of being liquefied under moderate pressure at normal temperatures.

LIVING SPACE. Space within a dwelling unit utilized for living, sleeping, eating, cooking, bathing, washing and sanitation purposes.

LOG LIGHTER, GAS-FIRED. A manually operated solid-fuel ignition appliance for installation in a vented solid-fuel-burning fireplace.

MAIN BURNER. A device or group of devices essentially forming an integral unit for the final conveyance of gas or a mixture of gas and air to the combustion zone, and on which combustion takes place to accomplish the function for which the appliance is designed.

MECHANICAL EXHAUST SYSTEM. Equipment installed in and made a part of the vent, which will provide a positive induced draft.

METER. The instrument installed to measure the volume of gas delivered through it.

MODULATING. Modulating or throttling is the action of a control from its maximum to minimum position in either predetermined steps or increments of movement as caused by its actuating medium.

OFFSET (VENT). A combination of approved bends that make two changes in direction bringing one section of the vent out of line, but into a line parallel with the other section.

OUTLET. A threaded connection or bolted flange in a pipe system to which a gas-burning appliance is attached.

OXYGEN DEPLETION SAFETY SHUTOFF SYSTEM (ODS). A system designed to act to shut off the gas supply to the main and pilot burners if the oxygen in the surrounding atmosphere is reduced below a predetermined level.

PILOT. A small flame that is utilized to ignite the gas at the main burner or burners.

PIPING. Where used in this code, "piping" refers to either pipe or tubing, or both.

Pipe. A rigid conduit of iron, steel, copper, brass or plastic.

Tubing. Semirigid conduit of copper, aluminum, plastic or steel.

PIPING SYSTEM. All fuel piping, valves, and fittings from the outlet of the point of delivery to the connections with the gas utilization equipment.

PLASTIC, THERMOPLASTIC. A plastic that is capable of being repeatedly softened by increase of temperature and hardened by decrease of temperature.

POINT OF DELIVERY. The point of delivery is the outlet of the service meter assembly, or the outlet of the service regulator or service shutoff valve where a meter is not provided. For undiluted liquefied petroleum gas systems, the point of delivery shall be considered the outlet of the first stage pressure regulator that provides utilization pressure, exclusive of line gas regulators, in the system.

PRESSURE DROP. The loss in pressure due to friction or obstruction in pipes, valves, fittings, regulators and burners.

PRESSURE TEST. An operation performed to verify the gas-tight integrity of gas piping following its installation or modification.

READY ACCESS (TO). That which enables a device, appliance or equipment to be directly reached, without requiring the removal or movement of any panel, door or similar obstruction. (See "Access.")

REGULATOR. A device for controlling and maintaining a uniform gas supply pressure, either pounds-to-inches water column (MP regulator) or inches-to-inches water column (appliance regulator).

REGULATOR, GAS APPLIANCE. A pressure regulator for controlling pressure to the manifold of gas equipment.

REGULATOR, LINE GAS PRESSURE. A device placed in a gas line between the service pressure regulator and the equipment for controlling, maintaining or reducing the pressure in that portion of the piping system downstream of the device.

REGULATOR, PRESSURE. A device placed in a gas line for reducing, controlling and maintaining the pressure in that portion of the piping system downstream of the device.

REGULATOR, SERVICE PRESSURE. A device installed by the serving gas supplier to reduce and limit the service line gas pressure to delivery pressure.

RELIEF OPENING. The opening provided in a draft hood to permit the ready escape to the atmosphere of the flue products from the draft hood in the event of no draft, back draft or stoppage beyond the draft hood, and to permit air into the draft hood in the event of a strong chimney updraft.

RELIEF VALVE (DEVICE). A safety valve designed to forestall the development of a dangerous condition by relieving either pressure, temperature or vacuum in the hot water supply system.

RELIEF VALVE, PRESSURE. An automatic valve which opens and closes a relief vent, depending on whether the pressure is above or below a predetermined value.

RELIEF VALVE, TEMPERATURE

Reseating or self-closing type. An automatic valve which opens and closes a relief vent, depending on whether the temperature is above or below a predetermined value.

Manual reset type. A valve which automatically opens a relief vent at a predetermined temperature and which must be manually returned to the closed position.

RELIEF VALVE, VACUUM. A valve that automatically opens and closes a vent for relieving a vacuum within the hot water supply system, depending on whether the vacuum is above or below a predetermined value.

RISER, GAS. A vertical pipe supplying fuel gas.

ROOM HEATER, UNVENTED. See "Unvented room heater."

ROOM HEATER, VENTED. A free-standing gas-fired heating unit used for direct heating of the space in and adjacent to that in which the unit is located. [see also "Vented room heater."]

SAFETY SHUTOFF DEVICE. See "Flame safeguard."

SHAFT. An enclosed space extending through one or more stories of a building, connecting vertical openings in successive floors, or floors and the roof.

SPECIFIC GRAVITY. As applied to gas, specific gravity is the ratio of the weight of a given volume to that of the same volume of air, both measured under the same condition.

THERMOSTAT

Electric switch type. A device that senses changes in temperature and controls electrically, by means of separate components, the flow of gas to the burner(s) to maintain selected temperatures.

Integral gas valve type. An automatic device, actuated by temperature changes, designed to control the gas supply to the burner(s) in order to maintain temperatures between predetermined limits, and in which the thermal actuating element is an integral part of the device.

1. Graduating thermostat. A thermostat in which the motion of the valve is approximately in direct proportion to the effective motion of the thermal element induced by temperature change.
2. Snap-acting thermostat. A thermostat in which the thermostatic valve travels instantly from the closed to the open position, and vice versa.

TRANSITION FITTINGS, PLASTIC TO STEEL. An adapter for joining plastic pipe to steel pipe. The purpose of this fitting is to provide a permanent, pressure-tight connection between two materials that cannot be joined directly one to another.

UNIT HEATER

High-static pressure type. A self-contained, automatically controlled, vented appliance having integral means for circulation of air against 0.2 inch (15 mm H_2O) or greater static pressure. Such appliance is equipped with provisions for attaching an outlet air duct and, where the appliance is for indoor installation remote from the space to be heated, is also equipped with provisions for attaching an inlet air duct.

Low-static pressure type. A self-contained, automatically controlled, vented appliance, intended for installation in the space to be heated without the use of ducts, having integral means for circulation of air. Such units are allowed to be equipped with louvers or face extensions made in accordance with the manufacturer's specifications.

UNVENTED ROOM HEATER. An unvented heating appliance designed for stationary installation and utilized to provide comfort heating. Such appliances provide radiant heat or convection heat by gravity or fan circulation directly from the heater and do not utilize ducts.

VALVE. A device used in piping to control the gas supply to any section of a system of piping or to an appliance.

Automatic. An automatic or semiautomatic device consisting essentially of a valve and operator that control the gas supply to the burner(s) during operation of an appliance. The operator shall be actuated by application of gas pressure on a flexible diaphragm, by electrical means, by mechanical means or by other approved means.

Automatic gas shutoff. A valve used in conjunction with an automatic gas shutoff device to shut off the gas supply to a water heating system. It shall be constructed integrally with the gas shutoff device or shall be a separate assembly.

Equipment shutoff. A valve located in the piping system, used to isolate individual equipment for purposes such as service or replacement.

Individual main burner. A valve that controls the gas supply to an individual main burner.

Main burner control. A valve that controls the gas supply to the main burner manifold.

Manual main gas-control. A manually operated valve in the gas line for the purpose of completely turning on or shutting off the gas supply to the appliance, except to pilot or pilots that are provided with independent shutoff.

Manual reset. An automatic shutoff valve installed in the gas supply piping and set to shut off when unsafe conditions occur. The device remains closed until manually reopened.

Service shutoff. A valve, installed by the serving gas supplier between the service meter or source of supply and the customer piping system, to shut off the entire piping system.

VENT. A pipe or other conduit composed of factory-made components, containing a passageway for conveying combustion products and air to the atmosphere, listed and labeled for use with a specific type or class of appliance.

Special gas vent. A vent listed and labeled for use with listed Category II, III and IV gas appliances.

Type B vent. A vent listed and labeled for use with appliances with draft hoods and other Category I appliances that are listed for use with Type B vents.

Type BW vent. A vent listed and labeled for use with wall furnaces.

Type L vent. A vent listed and labeled for use with appliances that are listed for use with Type L or Type B vents.

VENT CONNECTOR. (See "Connector.")

VENTED GAS APPLIANCE CATEGORIES. Appliances that are categorized for the purpose of vent selection are classified into the following four categories:

Category I. An appliance that operates with a nonpositive vent static pressure and with a vent gas temperature that avoids excessive condensate production in the vent.

Category II. An appliance that operates with a nonpositive vent static pressure and with a vent gas temperature that is capable of causing excessive condensate production in the vent.

Category III. An appliance that operates with a positive vent static pressure and with a vent gas temperature that avoids excessive condensate production in the vent.

Category IV. An appliance that operates with a positive vent static pressure and with a vent gas temperature that is capable of causing excessive condensate production in the vent.

VENTED ROOM HEATER. A vented self-contained, free-standing, nonrecessed appliance for furnishing warm air to the space in which it is installed, directly from the heater without duct connections.

VENTED WALL FURNACE. A self-contained vented appliance complete with grilles or equivalent, designed for incorporation in or permanent attachment to the structure of a building, mobile home or travel trailer, and furnishing heated air circulated by gravity or by a fan directly into the space to be heated through openings in the casing. This definition shall exclude floor furnaces, unit heaters and central furnaces as herein defined.

VENTING SYSTEM. A continuous open passageway from the flue collar or draft hood of an appliance to the outside atmosphere for the purpose of removing flue or vent gases. A venting system is usually composed of a vent or a chimney and vent connector, if used, assembled to form the open passageway.

WALL HEATER, UNVENTED TYPE. A room heater of the type designed for insertion in or attachment to a wall or partition. Such heater does not incorporate concealed venting arrangements in its construction and discharges all products of combustion through the front into the room being heated.

WATER HEATER. Any heating appliance or equipment that heats potable water and supplies such water to the potable hot water distribution system.

SECTION G2404 (301)
GENERAL

G2404.1 (301.1) Scope. This chapter shall govern the approval and installation of all equipment and appliances that comprise parts of the installations regulated by this code in accordance with Section G2401.

G2404.2 (301.1.1) Other fuels. The requirements for combustion and dilution air for gas-fired appliances shall be governed by Section G2407. The requirements for combustion and dilution air for appliances operating with fuels other than fuel gas shall be regulated by Chapter 17.

G2404.3 (301.3) Listed and labeled. Appliances regulated by this code shall be listed and labeled unless otherwise approved in accordance with Section R104.11. The approval of unlisted appliances in accordance with Section R104.11 shall be based upon approved engineering evaluation.

G2404.4 (301.8) Vibration isolation. Where means for isolation of vibration of an appliance is installed, an approved means for support and restraint of that appliance shall be provided.

G2404.5 (301.9) Repair. Defective material or parts shall be replaced or repaired in such a manner so as to preserve the original approval or listing.

G2404.6 (301.10) Wind resistance. Appliances and supports that are exposed to wind shall be designed and installed to resist the wind pressures determined in accordance with this code.

G2404.7 (301.11) Flood hazard. For structures located in special flood-hazard areas, the appliance, equipment and system installations regulated by this code shall comply with the flood-resistant construction requirements of this code.

G2404.8 (301.12) Seismic resistance. When earthquake loads are applicable in accordance with this code, the supports shall be designed and installed for the seismic forces in accordance with this code

G2404.9 (301.14) Rodent proofing. Buildings or structures and the walls enclosing habitable or occupiable rooms and spaces in which persons live, sleep or work, or in which feed, food or foodstuffs are stored, prepared, processed, served or sold, shall be constructed to protect against the entry of rodents.

SECTION G2405 (302) STRUCTURAL SAFETY

G2405.1 (302.1) Structural safety. The building shall not be weakened by the installation of any gas piping. In the process of installing or repairing any gas piping, the finished floors, walls, ceilings, tile work or any other part of the building or premises, which are required to be changed or replaced shall be left in a safe structural condition in accordance with the requirements of this code.

SECTION G2406 (303) APPLIANCE LOCATION

G2406.1 (303.1) General. Appliances shall be located as required by this section, specific requirements elsewhere in this code and the conditions of the equipment and appliance listing.

G2406.2 (303.3) Prohibited locations. Fuel-fired appliances shall not be located in, or obtain combustion air from, any of the following rooms or spaces:

1. Sleeping rooms.
2. Bathrooms.
3. Toilet rooms.
4. Storage closets.

Exceptions:

1. Direct-vent appliances that obtain all combustion air directly from the outdoors.
2. Vented room heaters, wall furnaces, vented decorative appliances and decorative appliances for installation in vented solid fuel-burning fireplaces, provided that the room is not a confined space and the building is not of unusually tight construction.
3. A single listed wall-mounted unvented room heater equipped with an oxygen depletion safety shutoff system and installed in a bathroom provided that the input rating does not exceed 6,000 Btu per hour (1760 W/hr) and the bathroom is not a confined space.
4. A single listed wall-mounted unvented room heater equipped with an oxygen depletion safety shutoff system and installed in a bedroom provided that the input rating does not exceed 10,000 Btu per hour (2930 W/hr) and the bedroom is not a confined space.
5. Appliances installed in a dedicated enclosure in which all combustion air is taken directly from the outdoors, in accordance with Section G2407.11. Access to such an enclosure shall be through a weather-stripped solid door equipped with an approved self-closing device.

G2406.3 (303.6) Outdoor locations. Equipment installed in outdoor locations shall be either listed for outdoor installation or provided with protection from outdoor environmental factors that influence the operability, durability and safety of the equipment.

SECTION G2407 (304) COMBUSTION, VENTILATION AND DILUTION AIR

G2407.1 (304.1) General. The provisions of Section G2407 shall apply to gas utilization equipment installed in buildings and that requires air for combustion, ventilation, and dilution of flue gases.

Exceptions:

1. Direct-vent equipment that is constructed and installed so that all air for combustion is obtained directly from the outdoors and all flue gases are discharged to the outdoors.
2. Enclosed furnaces that incorporate an integral total enclosure and use only outdoor air for combustion and dilution of flue gases.

G2407.2 (304.2) Appliance/equipment location. Equipment shall be located so as not to interfere with proper circulation of combustion, ventilation and dilution air.

G2407.3 (304.3) Outdoor air required. Where normal infiltration does not provide the necessary air, outdoor air shall be introduced in accordance with Section G2407.11 or G2407.13

G2407.4 (304.4) Process air. In addition to air needed for combustion, process air shall be provided as required for cooling of equipment or material, controlling dew point, heating, drying, oxidation, dilution, safety exhaust, odor control, and air for compressors.

G2407.5 (304.5) Ventilation air. In addition to air needed for combustion, air shall be supplied for ventilation, including all air required for comfort and proper working conditions for personnel.

G2407.6 (304.6) Draft head/regulator location. A draft hood or a barometric draft regulator shall be installed in the same room or enclosure as the equipment served so as to prevent any difference in pressure between the hood or regulator and the combustion air supply.

G2407.7 (304.7) Makeup air provisions. Air requirements for the operation of exhaust fans, kitchen ventilation systems, kitchen ventilation systems, clothes dryers and fireplaces shall be considered in determining the adequacy of a space to provide combustion air requirements.

G2407.8 (304.8) Combustion air methods. Air for combustion, ventilation and dilution of flue gases for gas utilization equipment vented by natural draft shall be obtained by application of one of the methods covered in Sections G2407.10 through G2407.13.

G2407.9 (304.9) Unusually tight construction. Equipment located in buildings of unusually tight construction shall be provided with air for combustion, ventilation and dilution of flue gases using one of the methods described in Section G2407.11 or Section G2407.13.

G2407.10 (304.10) All air from inside the building. A confined space shall be provided with two permanent openings communicating directly with other spaces of sufficient volume so that the combined volume of all spaces meets the criteria for an unconfined space. The total input of all equipment installed in the combined spaces shall be used to determine the required minimum volume. Each opening shall have a minimum free area of not less than 1 square inch per 1,000 Btu per hour (22 cm^2 per kW) of the total input rating of all gas utilization equipment in the confined space, but not less than 100 square inches (64 415 mm^2). One opening shall commence within 12 inches (305 mm) of the top, and one opening shall commence within 12 inches (305 mm) of the bottom of the enclosure (see Figure G2407.10). The minimum dimension of air openings shall be not less than 3 inches (76 mm).

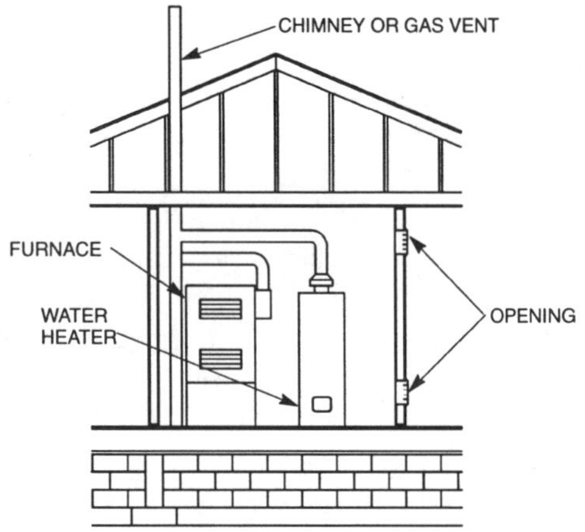

FIGURE G2407.10 (304.10)
APPLIANCES LOCATED IN CONFINED SPACES; ALL AIR FROM INSIDE THE BUILDING. SEE SECTION G2407.10.

G2407.11 (304.11) All air from outdoors. The confined space shall communicate with the outdoors in accordance with Section G2407.11.1 or Section G2407.11.2. The minimum dimension of air openings shall not be less than 3 inches (76 mm). Where ducts are used, they shall be of the same cross-sectional area as the free area of the openings to which they connect.

G2407.11.1 (304.11.1) Two opening method. Two permanent openings, one commencing within 12 inches (305 mm) of the top, and one commencing within 12 inches (305 mm) of the bottom, of the enclosure shall be provided. The openings shall communicate directly, or by ducts, with the outdoors or spaces (crawl or attic) that freely communicate with the outdoors.

Where directly communicating with the outdoors, or where communicating to the outdoors through vertical ducts, each opening shall have a minimum free area of 1 square inch per 4,000 Btu per hour (5.5 cm^2 per kW) of total input rating of all equipment in the enclosure. [See Figures G2407.11(1) and G2407.11(2).]

Where communicating with the outdoors through horizontal ducts, each opening shall have a minimum free area of 1 square inch per 2,000 Btu per hour (11 cm^2 per kW) of total input rating of all equipment in the enclosure. [See Figure G2407.11(3).]

G2407.11.2 (304.11.2) One opening method. One permanent opening, commencing within 12 inches (305 mm) of the top of the enclosure, shall be provided. The equipment shall have clearances of at least 1 inch (25.4 mm) from the sides and back and 6 inches (152 mm) from the front of the appliance. The opening shall directly communicate with the outdoors or shall communicate through a vertical or horizontal duct to the outdoors or spaces (crawl or attic) that freely communicate with the outdoors [see Figure G2407.11(4)], and shall have a minimum free area of 1 square inch per 3,000 Btu per hr (7 cm^2 per kW) of the total input rating of all equipment located in the enclosure, and not less than the sum of the areas of all vent connectors in the confined space.

G2407.12 (304.12) Combination of air from inside and from outdoors. Where the building in which the fuel-burning appliances are located is not unusually tight construction and the communicating interior spaces containing the fuel-burning appliances comply with all of the requirements of Section G2407.10, except the volumetric requirement of Section G2407.10, required combustion and dilution air shall be obtained by opening the room to the outdoors utilizing a combination of inside and outdoor air prorated in accordance with Section G2407.12.6. Openings connecting the interior spaces shall comply with Section G2407.10. The ratio of interior spaces shall comply with Section G2407.12.5. The number, location and ratios of openings connecting the space with the outdoor air shall comply with Sections G2407.12.1 through G2407.12.4.

G2407.12.1 (304.12.1) Number and location of openings. At least two openings shall be provided, one within 1 foot (305 mm) of the ceiling of the room and one within 1 foot (305 mm) of the floor.

G2407.12.2 (304.12.2) Ratio of direct openings. Where direct openings to the outdoors are provided in accordance with Section G2407.11.1, the ratio of direct openings shall be the sum of the net free areas of both direct openings to the outdoors, divided by the sum of the required areas for both such openings as determined in accordance with Section G2407.11.1.

G2407.12.3 (304.12.3) Ratio of horizontal openings. Where openings connected to the outdoors through horizontal ducts are provided in accordance with Section G2407.11.1, the ratio of horizontal openings shall be the sum of the net free areas of both such openings, divided by the sum of the required areas for both such openings as determined in accordance with Section G2407.11.1.

G2407.12.4 (304.12.4) Ratio of vertical openings. Where openings connected to the outdoors through vertical ducts are provided in accordance with Section G2407.11.1, the ratio of vertical openings shall be the sum of the net free areas of both such openings, divided by the sum of the required areas for both such openings as determined in accordance with Section G2407.11.1.

G2407.12.5 (304.12.5) Ratio of interior spaces. The ratio of interior spaces shall be the available volume of all communicating spaces, divided by the required volume as determined in accordance with Section G2407.10.

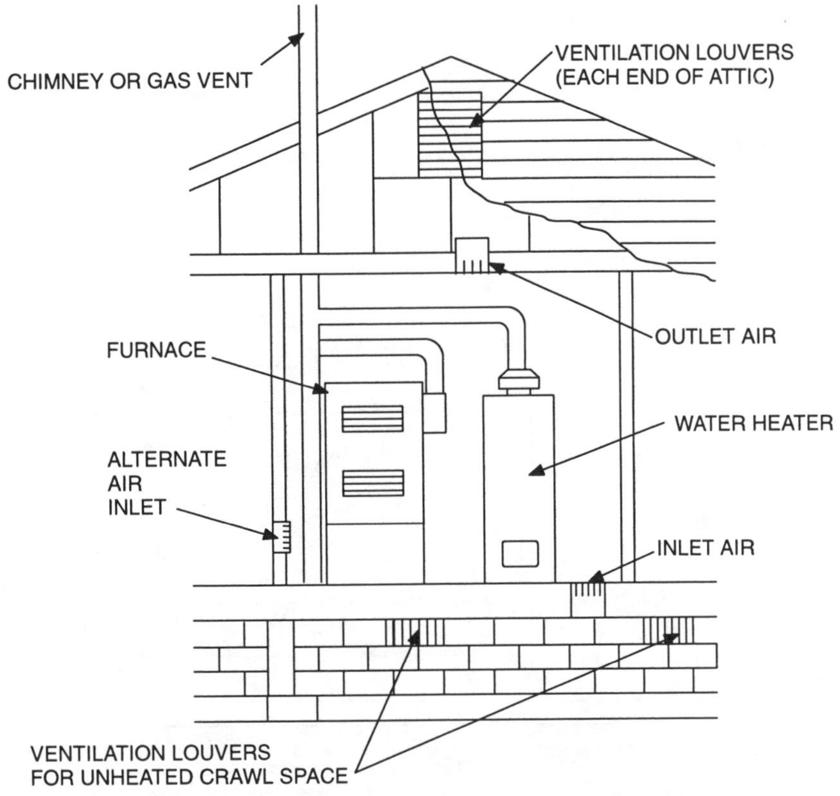

FIGURE G2407.11(1) [304.11(1)]
APPLIANCES LOCATED IN CONFINED SPACES; ALL AIR FROM OUTDOOR—INLET AIR FROM VENTILATED CRAWL SPACE AND OUTLET AIR TO VENTILATED ATTIC. SEE SECTION G2407.11.1.

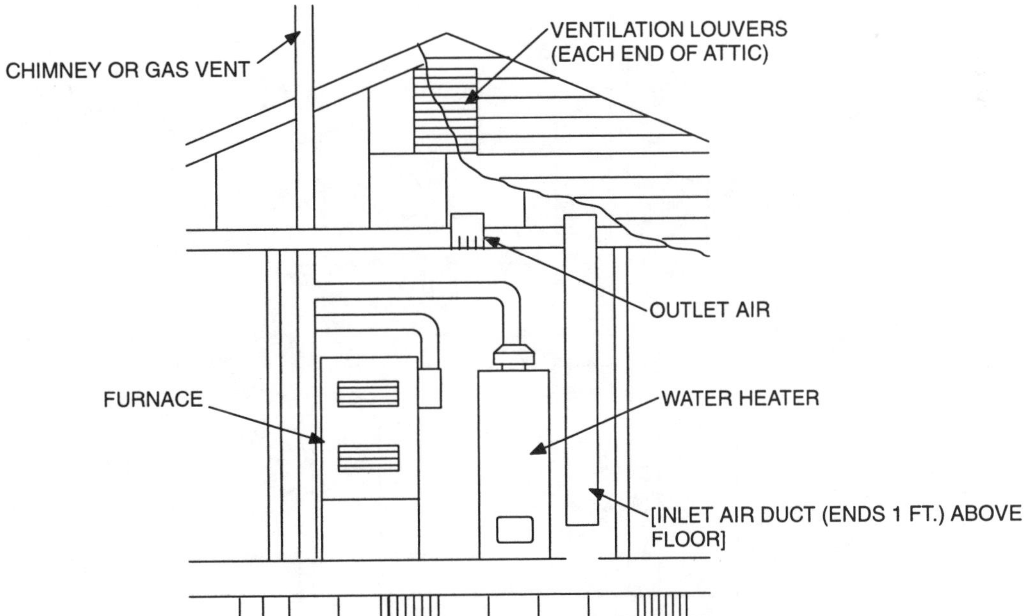

For SI: 1 foot = 304.8 mm.

FIGURE G2407.11(2) [304.11(2)]
APPLIANCES LOCATED IN CONFINED SPACES; ALL AIR FROM OUTDOORS THROUGH VENTILATED ATTIC. SEE SECTION G2407.11.

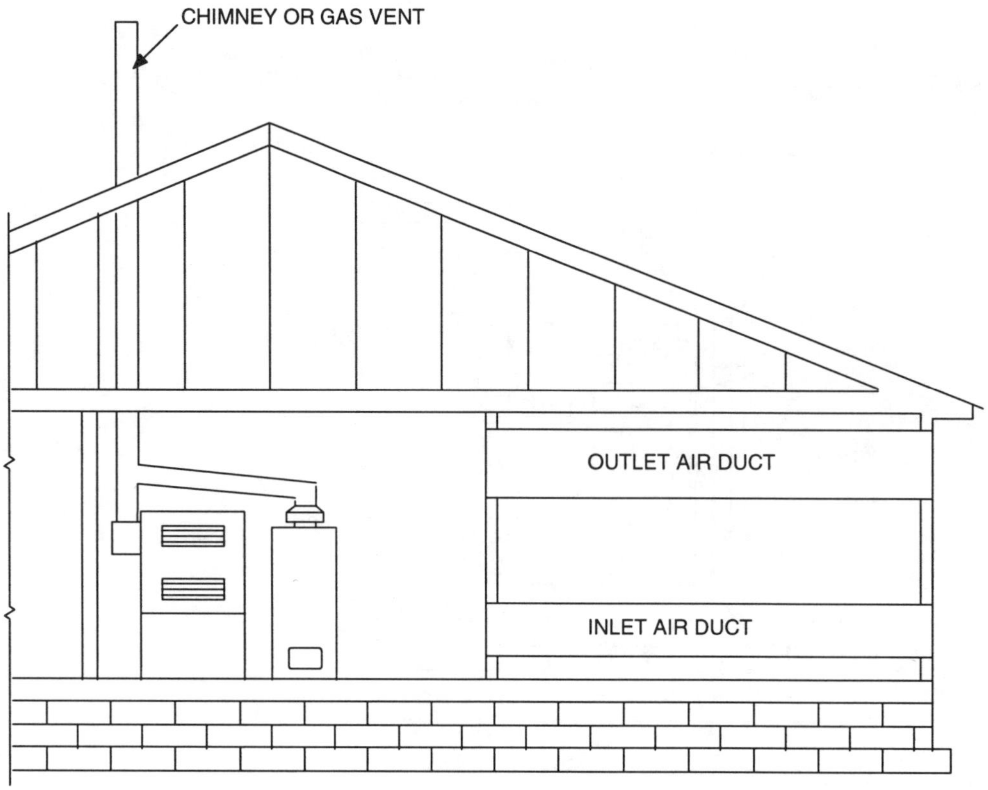

FIGURE G2407.11(3) [304.11(3)]
APPLIANCES LOCATED IN CONFINED SPACES; ALL AIR FROM OUTDOORS.
SEE SECTION G2407.11.1.

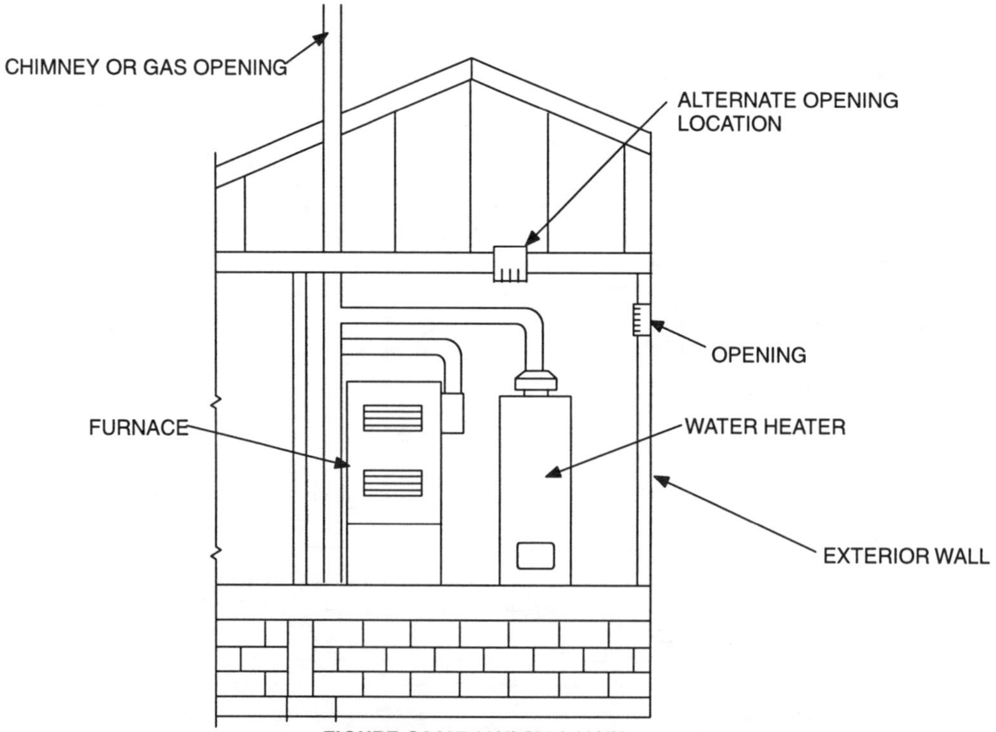

FIGURE G2407.11(4) [304.11(4)]
APPLIANCES LOCATED IN CONFINED SPACES; SINGLE COMBUSTION AIR OPENING.
SEE SECTION G2407.11.2.

G2407.12.6 (304.12.6) Prorating of inside and outdoor air. In spaces that utilize a combination of inside and outdoor air, the sum of the ratios of all direct openings, horizontal openings, vertical openings and interior spaces shall equal or exceed 1.

G2407.13 (304.13) Specially engineered installations. As an alternative to the provisions of Sections G2407.10, G2407.11 and G2407.12, the necessary supply of air for combustion, ventilation and dilution of flue gases shall be provided by an approved engineered system.

G2407.14 (304.14) Louvers and grilles. In calculating free area in Sections G2407.10, G2407.11 and G2407.12, the required size of openings for combustion, ventilation and dilution air shall be based on the net free area of each opening. If the free area through a design of louver or grille is known, it shall be used in calculating the size opening required to provide the free area specified. If the design and free area are not known, it shall be assumed that wood louvers will have 20- to 25-percent free area and metal louvers and grilles will have 60- to 75-percent free area. Louvers and grilles shall be fixed in the open position.

Exception: Louvers interlocked with the equipment so that they are proven in the full open position prior to main burner ignition and during main burner operation. Means shall be provided to prevent the main burner from igniting if the louvers fail to open during burner startup and to shut down the main burner if the louvers close during operation.

G2407.15 (304.15) Combustion air ducts. Combustion air ducts shall comply with all of the following:

1. Ducts shall be of galvanized steel complying with Chapter 16 of this code or of equivalent corrosion-resistant material approved for this application.

   Exception: Unobstructed stud and joist spaces shall not be prohibited from conveying combustion air, provided that not more than one required fireblock is removed.

2. Ducts shall terminate in an unobstructed space allowing free movement of combustion air to the appliances.
3. Ducts shall serve a single enclosure.
4. Ducts shall not serve both upper and lower combustion air openings where both such openings are used. The separation between ducts serving upper and lower combustion air openings shall be maintained to the source of combustion air.
5. Ducts shall not be screened where terminating in an attic space.
6. Horizontal upper combustion air ducts shall not slope downward toward the source of combustion air.

SECTION G2408 (305) INSTALLATION

G2408.1 (305.1) General. Equipment and appliances shall be installed as required by the terms of their approval. Equipment and appliances shall be installed in accordance with the conditions of listing, the manufacturer's installation instructions and this code. Manufacturer's installation instructions shall be available on the job site at the time of inspection.

Approved unlisted appliances shall be limited to uses recommended by the manufacturer and shall be installed in accordance with the manufacturer's installation instructions, the provisions of this code and the requirements determined by the code official.

G2408.2 (305.2) Elevation of ignition source. Equipment and appliances having an ignition source shall be elevated such that the source of ignition is not less than 18 inches (457 mm) above the floor in hazardous locations and private garages. For the purpose of this section, rooms or spaces that are not part of the living space of a dwelling unit and that communicate directly with a private garage through openings shall be considered to be part of the private garage.

G2408.3 (305.4) Private garages. Appliances located in private garages shall be installed with a minimum clearance of 6 feet (1829 mm) above the floor.

Exception: The requirements of this section shall not apply where the appliances are protected from motor vehicle impact and installed in accordance with Section 305.2.

SECTION G2409 (308) CLEARANCE REDUCTION

G2409.1 (308.1) Scope. This section shall govern the reduction in required clearances to combustible materials and combustible assemblies for chimneys, vents, kitchen exhaust equipment, fuel gas appliances, and fuel gas devices and equipment.

G2409.2 (308.2) Reduction table. The allowable clearance reduction shall be based on one of the methods specified in Table G2409.2 or shall utilize an assembly listed for such application. Where required clearances are not listed in Table 308.2, the reduced clearances shall be determined by linear interpolation between the distances listed in the table. Reduced clearances shall not be derived by extrapolation below the range of the table. The reduction of the required clearances to combustibles for listed and labeled appliances and equipment shall be in accordance with the requirements of this section except that such clearances shall not be reduced where reduction is specifically prohibited by the terms of the appliance or equipment listing. See Figures G2409.2(1) and G2409.2(2).

G2409.3 (308.3) Clearances for Indoor Air-Conditioning Equipment. Clearance requirements for indoor air-conditioning equipment shall comply with Sections G2409.3.1 through G2409.3.5.

G2409.3.1 (308.3.1) Equipment installed in rooms that are large in comparison with the size of the equipment. Air-conditioning equipment installed in rooms that are large in comparison with the size of the equipment shall be installed with clearances per the terms of their listing and the manufacturer's instructions.

TABLE G2409.2 (308.2) REDUCTION OF CLEARANCES WITH SPECIFIED FORMS OF PROTECTION[a, b, c, d, e, f, g, h, i, j, k]

Type of protection applied to and covering all surfaces of combustible material within the distance specified as the required clearance with no protection [*see* Figures G2409.2(1), G2409(2) and G2409(3)]	WHERE THE REQUIRED CLEARANCE WITH NO PROTECTION FROM APPLIANCE, VENT CONNECTOR OR SINGLE WALL METAL PIPE IS: Allowable clearances with specified protection (inches). Use column 1 for clearances above appliance or horizontal connector. Use column 2 for clearances from appliance, vertical connector and single-wall metal pipe.									
	36 inches		18 inches		12 inches		9 inches		6 inches	
	Above column 1	Sides and rear column 2	Above column 1	Sides and rear column 2	Above column 1	Sides and rear column 2	Above column 1	Sides and rear column 2	Above column 1	Sides and rear column 2
3.5-in. thick masonry wall without ventilated air space	—	24	—	12	—	9	—	6	—	5
0.5-in. insulation board over 1-in. glass fiber or mineral wool batts	24	18	12	9	9	6	6	5	4	3
0.024 sheet metal over 1-in. glass fiber or mineral wool batts reinforced with wire on rear face with ventilated air space	18	12	9	6	6	4	5	3	3	3
3.5-in. thick masonry wall with ventilated air space	—	12	—	6	—	6	—	6	—	6
0.024 sheet metal with ventilated air space	18	12	9	6	6	4	5	3	3	2
0.5-in. thick insulation board with ventilated air space	18	12	9	6	6	4	5	3	3	3
0.024 sheet metal with ventilated air space over 0.024 sheet metal with ventilated air space	18	12	9	6	6	4	5	3	3	3
1-in. glass fiber or mineral wool batts sandwiched between two sheets 0.024 sheet metal with ventilated air space	18	12	9	6	6	4	5	3	3	3

For SI: 1 inch = 25.4 mm.

NOTES:

a. Reduction of clearances from combustible materials shall not interfere with combustion air, draft hood clearance and relief, and accessibility of servicing.

b. All clearances shall be measured from the outer surface of the combustible material to the nearest point on the surface of the appliance, disregarding any intervening protection applied to the combustible material.

c. Spacers and ties shall be of noncombustible material. No spacer or tie shall be used directly opposite appliance or connector.

d. Where clearance reduction systems use ventilated air space, adequate provision for air circulation shall be provided as described. [See Figure G2409.2(2).]

e. There shall be at least 1 inch between clearance reduction systems and combustible walls and ceilings for reduction systems using ventilated air space.

f. If a wall protector is mounted on a single flat wall away from corners, adequate air circulation shall be permitted to be provided by leaving only the bottom and top edges or only the side and top edges open with at least 1-inch air gap.

g. Mineral wool batts (blanket or board) shall have a minimum density of 8 pounds per cubic foot (128 kg per m) and a minimum melting point of 1,500°F (816°C).

h. Insulation material used as part of clearance reduction system shall have a thermal conductivity of 1.0 Btu inch per square foot per hour-°F (0.144 Watts per meter-Kelvin) or less.

i. There shall be at least 1 inch between the appliance and the protector. In no case shall the clearance between the appliance and the combustible surface be reduced below that allowed in Table G2409.2.

j. All clearances and thicknesses are minimum; larger clearances and thicknesses are acceptable.

k. Listed single-wall connectors shall be permitted to be installed in accordance with the terms of their listing and the manufacturer's instructions.

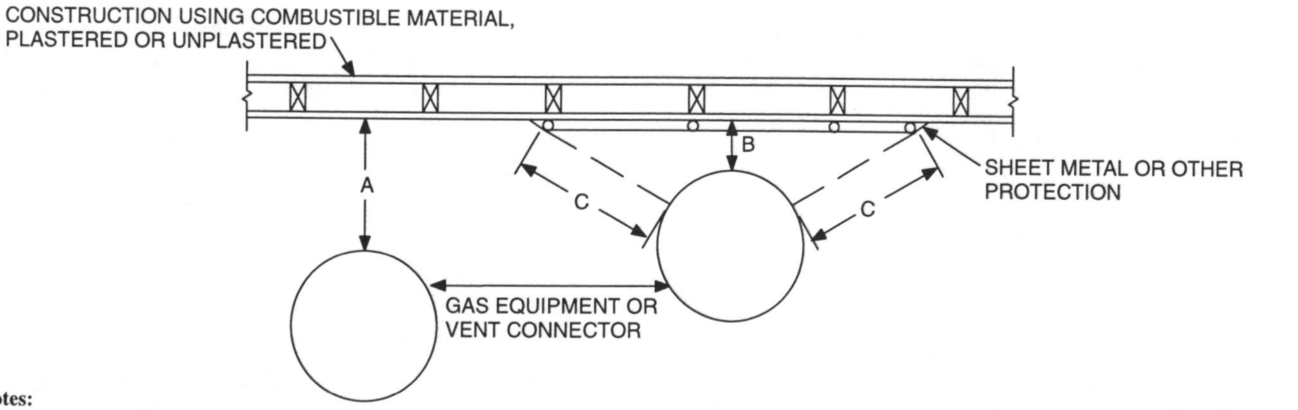

Notes:

"A" equals the clearance with no protection.

"B" equals the reduced clearance permitted in accordance with Table G2409.2. The protection applied to the construction using combustible material shall extend far enough in each direction to make "C" equal to "A."

FIGURE G2409.2(1) [308.2(1)]
EXTENT OF PROTECTION NECESSARY TO REDUCE CLEARANCES FROM GAS EQUIPMENT OR VENT CONNECTORS

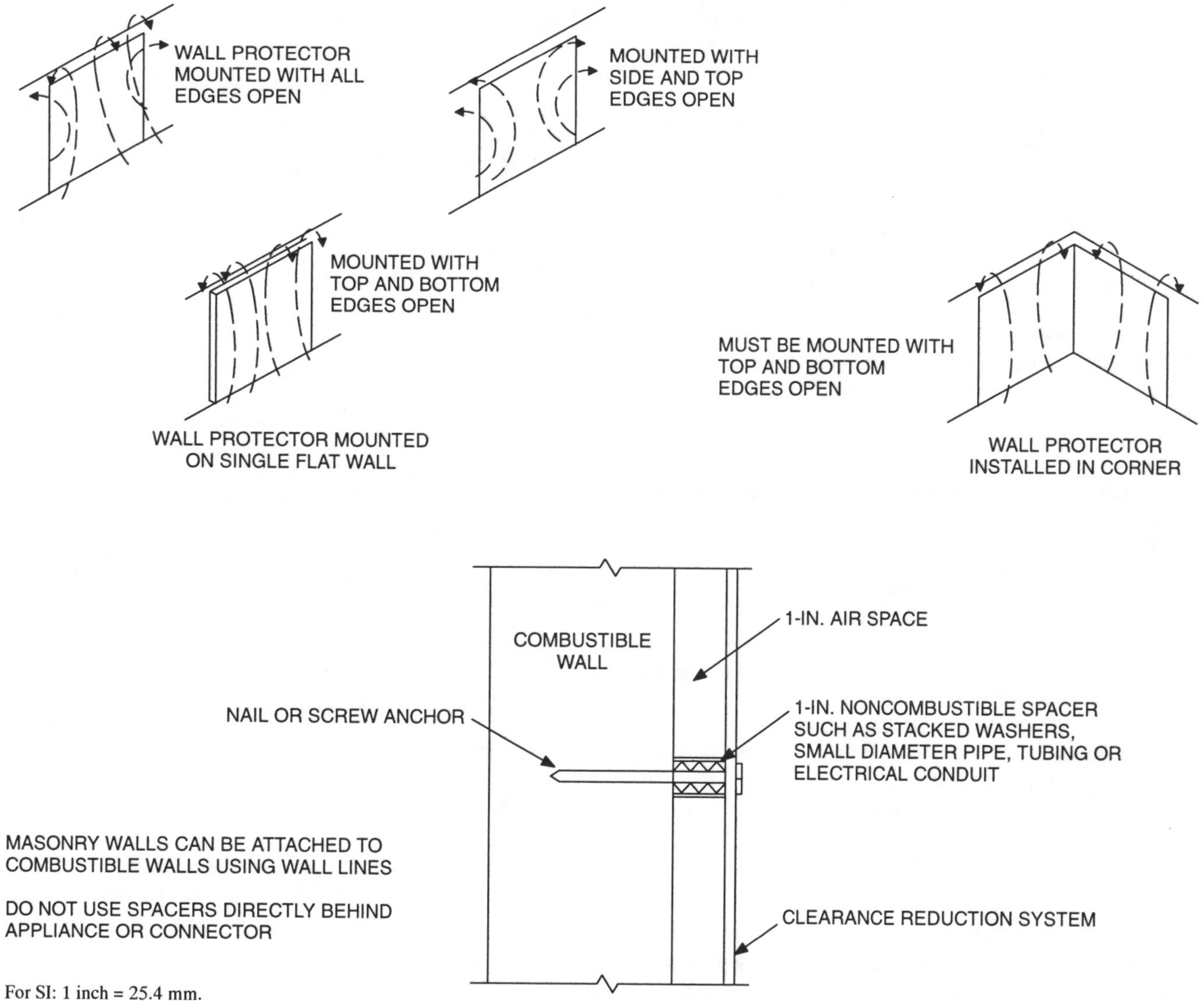

For SI: 1 inch = 25.4 mm.

FIGURE G2409.2(2) [308.2(2)]
WALL PROTECTOR CLEARANCE REDUCTION SYSTEM

G2409.3.2 (308.3.2) Equipment installed in rooms that are not large in comparison with the size of the equipment. Air-conditioning equipment installed in rooms that are not large (such as alcoves and closets) in comparison with the size of the equipment shall be listed for such installations and installed in accordance with the manufacturer's instructions. Listed clearances shall not be reduced by the protection methods described in Table G2409.2, regardless of whether the enclosure is of combustible or noncombustible material.

G2409.3.3 (308.3.4) Clearance reduction. Air-conditioning equipment installed in rooms that are large in comparison with the size of the equipment shall be permitted to be installed with reduced clearances to combustible material provided the combustible material or equipment is protected as described in Table G2409.2

G2409.3.4 (308.3.5) Plenum clearances. Where the plenum is adjacent to plaster on metal lath or noncombustible material attached to combustible material, the clearance shall be measured to the surface of the plaster or other noncombustible finish where the clearance specified is 2 inches (51 mm) or less.

G2409.3.5 (308.3.6) Clearance from supply ducts. Air-conditioning equipment shall have the clearance from supply ducts within 3 feet (914 mm) of the plenum be not less than that specified from the plenum. No clearance is necessary beyond this distance.

G2409.4 (308.4) Central heating boilers and furnaces. Clearance requirements for central heating boilers and furnaces shall comply with Sections G2409.4.1 through G2409.4.7. The clearance to this equipment shall not interfere with combustion air, draft hood clearance and relief, and accessibility for servicing.

G2409.4.1 (308.4.1) Equipment installed in rooms that are large in comparison with the size of the equipment. Central heating furnaces and low-pressure boilers installed in rooms large in comparison with the size of the equipment shall be installed with clearances per terms of their listing and the manufacturer's instructions.

G2409.4.2 (308.4.2) Equipment installed in rooms that are not large in comparison with the size of the equipment. Central heating furnaces and low-pressure boilers installed in rooms that are NOT large (such as alcoves and closets) in comparison with the size of the equipment shall be listed for such installations. Listed clearances shall not be reduced by the protection methods described in Table G2409.2 and illustrated in Figures G2409.2(1) and G2409.2(2), regardless of whether the enclosure is of combustible or noncombustible material.

G2409.4.3 (308.4.3) Clearance reduction. Central heating furnaces and low-pressure boilers installed in rooms that are large in comparison with the size of the equipment shall be permitted to be installed with reduced clearances to combustible material provided the combustible material or equipment is protected as described in Table G2409.2.

G2409.4.4 (308.4.5) Plenum clearances. Where the plenum is adjacent to plaster on metal lath or noncombustible material attached to combustible material, the clearance shall be measured to the surface of the plaster or other noncombustible finish where the clearance specified is 2 inches (51 mm) or less.

G2409.4.5 (308.4.6) Clearance from supply ducts. Central heating furnaces shall have the clearance from supply ducts within 3 feet (914 mm) of the plenum be not less than that specified from the plenum. No clearance is necessary beyond this distance.

G2409.4.6 (308.4.7) Unlisted central heating furnaces. Unlisted central heating furnaces with temperature limit controls that cannot be set higher than 250°F (121°C) shall have the clearance from supply duct with 6 feet (1829 mm) of the plenum be not less than 6 inches (152 mm). No clearance is necessary beyond this distance.

G2409.4.7 (308.4.4) Clearance for servicing equipment. Front clearance shall be sufficient for servicing the burner and the furnace or boiler.

SECTION G2410 (309) ELECTRICAL

G2410.1 (309.1) Grounding. Gas piping shall not be used as a grounding electrode.

G2410.2 (309.2) Connections. Electrical connections between gas utilization equipment and the building wiring, including the grounding of the equipment, shall conform to Chapters 33 through 42.

SECTION G2411 (401) GENERAL

G2411.1 (401.1) Scope. This chapter shall govern the design, installation, modification and maintenance of piping systems. The applicability of this code to piping systems extends from the point of delivery to the connections with the equipment and includes the design, materials, components, fabrication, assembly, installation, testing, inspection, operation and maintenance of such piping systems.

G2411.1.1 (401.1.1) Utility piping systems located within buildings. Utility service piping located within buildings shall be installed in accordance with the structural safety and fire protection provisions of this code.

G2411.2 (401.2) Liquefied petroleum gas storage. The storage system for liquefied petroleum gas shall be designed and installed in accordance with the *International Fire Code* and NFPA 58.

G2411.3 (401.3) Modifications to existing systems. In modifying or adding to existing piping systems, sizes shall be maintained in accordance with this chapter.

G2411.4 (401.4) Additional appliances. Where an additional appliance is to be served, the existing piping shall be checked to determine if it has adequate capacity for all appliances served. If inadequate, the existing system shall be enlarged as required or separate piping of adequate capacity shall be provided.

G2411.5 (401.5) Identification. For other than black steel pipe, exposed piping shall be identified by a yellow label marked "Gas" in black letters. The marking shall be spaced at intervals not exceeding 5 feet (1524 mm). The marking shall not be required on pipe located in the same room as the equipment served.

G2411.6 (401.6) Interconnections. Where two or more meters are installed on the same premises, but supply separate consumers, the piping systems shall not be interconnected on the outlet side of the meters.

G2411.7 (401.7) Piping meter identification. Piping from multiple meter installations shall be marked with an approved permanent identification by the installer so that the piping system supplied by each meter is readily identifiable.

G2411.8 (401.8) Minimum sizes. All pipe utilized for the installation, extension and alteration of any piping system shall be sized to supply the full number of outlets for the intended purpose and shall be sized in accordance with Section G2412.

SECTION G2412 (402)
PIPE SIZING

G2412.1 (402.1) General considerations. Piping systems shall be of such size and so installed as to provide a supply of gas sufficient to meet the maximum demand without undue loss of pressure between the point of delivery and the gas utilization equipment.

G2412.2 (402.2) Maximum gas demand. The volume of gas to be provided in cubic feet per hour shall be determined directly from the manufacturer's input ratings of the gas utilization equipment served. Where input rating is not indicated, the gas supplier, equipment manufacturer or a qualified agency shall be contacted for estimating the volume of gas to be supplied.

The total connected hourly load shall be used as the basis for piping sizing assuming that all equipment could be operating at full capacity simultaneously. Where a diversity of load can be established, pipe sizing shall be permitted to be based on such loads.

G2412.3 (402.3) Sizing. Gas piping shall be sized in accordance with Tables G2412.3(1) through G2412.3(6), or other approved engineering methods (see Appendix A).

G2412.4 (402.4) Allowable pressure drop. The design pressure loss in any piping system under maximum probable flow conditions, from the point of delivery to the inlet connection of the equipment, shall be such that the supply pressure at the equipment is greater than the minimum pressure required for proper equipment operation.

G2412.5 (402.5) Maximum design operating pressure. The maximum design operating pressure for piping systems located inside buildings shall not exceed 5 psig (34 kPa gauge) except where one or more of the following conditions are met:

1. The piping system is welded.
2. The piping is located in a ventilated chase or otherwise enclosed for protection against accidental gas accumulation.
3. The piping is a temporary installation for buildings under construction.

G2412.5.1 (402.5.1) Liquefied petroleum gas systems. The operating pressure for undiluted LP- Gas systems shall not exceed 20 psig (140 kPa gauge). Buildings having systems designed to operate below -5°F (-21°C) or with butane or a propane-butane mix shall be designed to either accommodate liquid LP-Gas or prevent LP-Gas vapor from condensing back into a liquid.

TABLE G2412.3(1) [402.3(1)]
MAXIMUM CAPACITY OF PIPE IN CUBIC FEET OF GAS PER HOUR FOR GAS PRESSURES OF 0.5 PSI OR LESS AND A PRESSURE DROP OF 0.5 INCH WATER COLUMN
(Based on a 0.60 Specific Gravity Gas)

NOMINAL IRON PIPE SIZE (inches)	INTERNAL DIAMETER (inches)	LENGTH OF PIPE (feet) 10	20	30	40	50	60	70	80	90	100	125	150
1/4	.364	43	29	24	20	18	16	15	14	13	12	11	10
3/8	.493	95	65	52	45	40	36	33	31	29	27	24	22
1/2	.622	175	120	97	82	73	66	61	57	53	50	44	40
3/4	.824	360	250	200	170	151	138	125	118	110	103	93	84
1	1.049	680	465	375	320	285	260	240	220	205	195	175	160
1 1/4	1.380	1,400	950	770	660	580	530	490	460	430	400	360	325

For SI: 1 inch = 24.4 mm, 1 cubic foot = 0.028 m^3, 1 foot = 304.8 mm, 1-inch water column = 0.249 kPa, 1 psig = 6.895 kPa, 1,000 Btu per hour = 0.293 kw, 90 degrees = 1.6 rad.

TABLE G2412.3(2) [402.3(2)]
PIPE SIZING TABLE FOR 2 PSI PRESSURE CAPACITY OF PIPES OF DIFFERENT DIAMETERS AND LENGTHS IN CUBIC FEET PER HOUR FOR AN INITIAL PRESSURE OF 2.0 PSI WITH A 1.0 PSI PRESSURE DROP AND A GAS OF 0.6 SPECIFIC GRAVITY

PIPE SIZE OF SCHEDULE 40 STANDARD PIPE (inches)	INTERNAL DIAMETER (inches)	TOTAL EQUIVALENT LENGTH OF PIPE (feet)											
		10	20	30	40	50	60	70	80	90	100	125	150
1/2	.622	1,506	1,065	869	753	673	615	569	532	502	462	414	372
3/4	.824	3,041	2,150	1,756	1,521	1,360	1,241	1,150	1,075	1,014	934	836	751
1	1.049	5,561	3,932	3,211	2,781	2,487	2,270	2,102	1,966	1,854	1,708	1,528	1,373
1 1/4	1.380	11,415	8,072	6,591	5,708	5,105	4,660	4,315	4,036	3,805	3,508	3,138	2,817

For SI: 1 inch = 25.4 mm, 1 cubic foot = 0.028 m^3, 1 foot = 304.8 mm, 1-inch water column = 0.249 kPa, 1 psig = 6.895 kPa, 1,000 Btu per hour = 0.293 kw, 90 degrees = 1.6 rad.

TABLE G2412.3(3) [402.3(14)]
MAXIMUM CAPACITY OF SEMI-RIGID TUBING IN CUBIC FEET OF GAS PER HOUR FOR GAS PRESSURES OF 0.5 PSI OR LESS AND A PRESSURE DROP OF 0.5 INCH WATER COLUMN (Based on a 0.60 specific gravity gas)

OUTSIDE DIAMETER (inch)	LENGTH OF TUBING (feet)											
	10	20	30	40	50	60	70	80	90	100	125	150
3/8	27	18	15	13	11	10	9	9	8	8	7	6
1/2	56	38	31	26	23	21	19	18	17	16	14	13
5/8	113	78	62	53	47	43	39	37	34	33	29	26
3/4	197	136	109	93	83	75	69	64	60	57	50	46
7/8	280	193	155	132	117	106	98	91	85	81	71	65

For SI: 1 inch = 25.4 mm, 1 cubic foot = 0.028 m^3, 1 foot = 304.8 mm, 1-inch water column = 0.249 kPa, 1 psig = 6.895 kPa, 1,000 Btu per hour = 0.293 kw, 90 degrees = 1.6 rad.

TABLE G2412.3(4) [402.3(17)]
MAXIMUM CAPACITY OF SEMI-RIGID TUBING IN CUBIC FEET OF GAS PER HOUR FOR A GAS PRESSURE OF 2.0 PSI OR LESS AND A PRESSURE DROP OF 1.0 PSI (Based on a 0.6 specific gravity gas)

NOMINAL TUBING DIAMETER INSIDE (inches)	INTERNAL DIAMETER (inches)	LENGTH OF TUBING (feet)													
		5	10	15	20	30	40	50	60	70	80	90	100	125	150
1/4	0.315	459	306	242	204	163	139	122	110	100	93	87	82	72	65
3/8	0.430	1,071	722	569	484	382	323	285	255	234	217	204	191	168	151
1/2	0.545	2,040	1,385	1,088	918	731	620	548	493	450	416	391	365	323	289
5/8	0.666	3,527	2,363	1,827	1,581	1,258	1,062	935	850	773	722	671	629	552	497
3/4	0.785	5,524	3,697	2,932	2,507	1,955	1,700	1,487	1,326	1,215	1,130	1,045	986	871	782
1	1.025	8,923	6,459	5,269	4,589	3,739	3,229	2,847	2,592	2,380	2,252	2,125	1,997	1,785	1,615

For SI: 1 inch = 25.4 mm, 1 cubic foot = 0.028 m^3, 1 foot = 304.8 mm, 1-inch water column = 0.249 kPa, 1 psig = 6.895 kPa, 1,000 Btu per hour = 0.293 kw, 90 degrees = 1.6 rad.

TABLE G2412.3(5) [402.3(19)]
MAXIMUM CAPACITY OF CSST IN CUBIC FEET PER HOUR FOR GAS PRESSURE OF 0.5 PSI OR LESS AND PRESSURE DROP OF 0.5 INCH WATER COLUMN[a] (Based on a 0.60 specific gravity gas)

EHD[b] FLOW DESIGNATION	TUBING LENGTH (feet)															
	5	10	15	20	25	30	40	50	60	70	80	90	100	150	200	250	300
13	46	32	25	22	19	18	15	13	12	11	10	10	9	7	6	5	5
14	63	44	35	31	27	25	21	19	17	16	15	14	13	10	9	8	7
18	115	82	66	58	52	47	41	37	34	31	29	28	26	20	18	16	15
19	134	95	77	67	60	55	47	42	38	36	33	32	30	23	21	19	17
23	225	161	132	116	104	96	83	75	68	63	60	57	54	42	38	34	32
25	270	192	157	137	122	112	97	87	80	74	69	65	62	48	44	39	36
30	471	330	267	231	206	188	162	144	131	121	113	107	101	78	71	63	57
31	546	383	310	269	240	218	188	168	153	141	132	125	118	91	82	74	67

For SI: 1 inch = 25.4 mm, 1 cubic foot = 0.028 m^3, 1 foot = 304.8 m m, 1-inch water column = 0.249 kPa, 1 psig = 6.895 kPa, 1,000 Btu per hour = 0.293 kw, 90 degrees = 1.6 rad.

a. Table includes losses for four 90-degree bends and two end fittings. Tubing runs with larger numbers of bends and/or fittings shall be increased by an equivalent length of tubing to the following equation: $L = 1.3n$ where L is additional length (ft) of tubing and n is the number of additional fittings and/or bends.

b. EHD—Equivalent Hydraulic Diameter—A measure of the relative hydraulic efficiency between different tubing sizes. The greater the value of EHD, the greater the gas capacity of the tubing.

TABLE G2412.3(6) [402.3(22)]
MAXIMUM CAPACITY OF CSST IN CUBIC FEET PER HOUR FOR GAS PRESSURE OF 2 PSI AND A PRESSURE DROP OF 1 PSI
(Based on 0.60 specific gravity gas)[a]

EHD[b] FLOW DESIGNATION	TUBING LENGTH (feet)													
	10	25	30	40	50	75	80	100	150	200	250	300	400	500
13	270	166	151	129	115	93	89	79	64	55	49	44	38	34
14	353	220	200	172	154	124	120	107	87	75	67	61	52	46
18	587	374	342	297	266	218	211	189	155	135	121	110	96	86
19	700	444	405	351	314	257	249	222	182	157	141	129	111	100
23	1,098	709	650	567	510	420	407	366	302	263	236	217	189	170
25	1,372	876	801	696	624	512	496	445	364	317	284	260	225	202
30	2,592	1,620	1,475	1,273	1,135	922	892	795	646	557	497	453	390	348
31	2,986	1,869	1,703	1,470	1,311	1,066	1,031	920	748	645	576	525	453	404

For SI: 1 inch = 25.4 mm, 1 cubic foot = 0.028 m^3, 1 foot = 304.8 mm, 1-inch water column = 0.249 kPa, 1 psig = 6.895 kPa, 1,000 Btu per hour = 0.293 kw, 90 degrees = 1.6 rad.

NOTE: Table does not include effect of pressure drop across the line regulator. If regulator loss exceeds 0.75 psi (based on 7 in. water column outlet pressure), DO NOT USE THIS TABLE. Consult with regulator manufacturer for pressure drops and capacity factors. Pressure drops across a regulator may vary with flow rate.

CAUTION: Capacities shown in table may exceed maximum capacity for a selected regulator. Consult with regulator or tubing manufacturer for guidance.

a. Table includes losses for four 90-degree bends and two end fittings. Tubing runs with larger number of bends and/or fittings shall be increased by an equivalent length of tubing according to the following equation: $L = 1.3n$ where L is additional length (ft) of tubing and n is the number of additional fittings and/or bends.

b. EHD—Equivalent Hydraulic Diameter—A measure of the relative hydraulic efficiency between different tubing sizes. The greater the value of EHD, the greater the gas capacity of the tubing.

SECTION G2413 (403)
PIPING MATERIALS

G2413.1 (403.1) Material application. Materials and components conforming to standards or specifications listed herein and those approached by the code official shall be permitted to be used for the appropriate applications, as prescribed and imited by this code.

G2413.2 (403.2) Used materials. Pipe, fittings, valves or other materials shall not be used again unless they are free of foreign materials and have been ascertained to be adequate for the service intended.

G2413.3 (403.3) Other materials. Material not covered by the standards specifications listed herein shall be investigated and tested to determine that it is safe and suitable for the proposed service, and, in addition, shall be recommended for that service by the manufacturer and shall be approved by the code official.

G2413.4 (403.4) Metallic pipe. Metallic pipe shall comply with Sections G2413.4.1 and G2413.4.2.

G2413.4.1 (403.4.1) Cast iron. Cast-iron pipe shall not be used.

G2413.4.2 (403.4.2) Steel. Steel and wrought-iron pipe shall be at least of standard weight (Schedule 40) and shall comply with one of the following:

1. ASME B36.10, 10M; or
2. ASTM A 53; or
3. ASTM A 106.

G2413.5 (403.5) Metallic tubing. Seamless copper, aluminum alloy or steel tubing shall be permitted to be used with gases not corrosive to such material.

G2413.5.1 (403.5.1) Steel tubing shall comply with ASTM A 539 or ASTM A 254.

G2413.5.2 (403.5.2) Copper tubing. Copper tubing shall comply with standard Type K or L of ASTM B 88 or ASTM B 280.

Copper and brass tubing shall not be used if the gas contains more than an average of 0.3 grains of hydrogen sulfide per 100 standard cubic feet of gas (0.7 milligrams per 100 liters).

G2413.5.3 (403.5.4) Corrugated stainless steel tubing. Corrugated stainless steel tubing shall be tested and listed in compliance with the construction, installation and performance requirements of ANSI/AGA LC 1.

G2413.6 (403.6) Plastic pipe, tubing and fittings. Plastic pipe, tubing and fittings shall be used outside underground only and shall conform with ASTM D 2513. Pipe to be used shall be marked "gas" and "ASTM D 2513."

G2413.6.1 (403.6.1) Anodeless risers. Anodeless risers shall comply with the following:

1. Factory-assembled anodeless risers shall be recommended by the manufacturer for the gas used and shall be leak-tested by the manufacturer in accordance with written procedures.
2. Service head adapters and field-assembled anodeless risers incorporating service head adapters shall be recommended by the manufacturer for the gas used by the manufacturer and shall be designed certified to meet the requirements of Category I of ASTM D 2513, and U.S. Department of Transportation, Code of Federal Regulations, Title 49, Part 192.281(e). The manufacturer shall provide the user qualified installation instructions as prescribed by the U.S. Department of Transportation, Code of Federal Regulations, Title 49, Part 192.283(b).

G2413.6.2 (403.6.2) LP-Gas systems. The use of plastic pipe, tubing and fittings in undiluted liquefied petroleum gas piping systems shall be in accordance with NFPA 58.

G2413.7 (403.7) Workmanship and defects. Pipe or tubing and fittings shall be clear and free from cutting burrs and defects in structure or threading, and shall be thoroughly brushed, and chip and scale blown.

Defects in pipe or tubing or fittings shall not be repaired. Defective pipe, tubing or fittings shall be replaced. (See Section G2416.1.2.)

G2413.8 (403.8) Protective coating. Where in contact with material or atmosphere exerting a corrosive action, metallic piping and fittings coated with a corrosion-resistant material shall be used. External or internal coatings or linings used on piping or components shall not be considered as adding strength.

G2413.9 (403.9) Metallic pipe threads. Metallic pipe and fitting threads shall be taper pipe threads and shall comply with ASME B1.20.1.

G2413.9.1 (403.9.1) Damaged threads. Pipe with threads that are stripped, chipped, corroded or otherwise damaged shall not be used. If a weld opens during the operation of cutting or threading, that portion of the pipe shall not be used.

G2413.9.2 (403.9.2) Number of threads. Field threading of metallic pipe shall be in accordance with Table G2413.9.2.

TABLE G2413.9.2 (403.9.2)
SPECIFICATIONS FOR THREADING METALLIC PIPE

IRON PIPE SIZE (inches)	APPROXIMATE LENGTH OF THREADED PORTION (inch)	APPROXIMATE NO. OF THREADS TO BE CUT
$^1/_2$	$^3/_4$	10
$^3/_4$	$^3/_4$	10
1	$^7/_8$	10
$1^1/_4$	1	11
$1^1/_2$	1	11

For SI: 1 inch = 25.4 mm.

G2413.9.3 (403.9.3) Thread compounds. Thread (joint) compounds (pipe dope) shall be resistant to the action of liquefied petroleum gas or to any other chemical constituents of the gases to be conducted through the piping.

G2413.10 (403.10) Metallic piping joints and fittings. The type of piping joint used shall be suitable for the pressure-temperature conditions and shall be selected giving consideration to joint tightness and mechanical strength under the service conditions. The joint shall be able to sustain the maximum end force due to the internal pressure and any additional forces due to temperature expansion or contraction, vibration, fatigue, or to the weight of the pipe and its contents.

G2413.10.1 (403.10.1) Pipe joints. Pipe joints shall be threaded, flanged or welded, and nonferrous pipe shall be permitted to also be brazed with materials having a melting point in excess of 1,000°F (538°C). Brazing alloys shall not contain more than 0.05-percent phosphorus.

G2413.10.2 (403.10.2) Tubing joints. Tubing joints shall either be made with approved gas tubing fittings or be brazed with a material having a melting point in excess of 1,000°F (538°C). Brazing alloys shall not contain more than 0.05-percent phosphorus.

G2413.10.3 (403.10.3) Flared joints. Flared joints shall be used only in systems constructed from nonferrous pipe and tubing where experience or tests have demonstrated that the joint is suitable for the conditions and where provisions are made in the design to prevent separation of the joints.

G2413.10.4 (403.10.4) Metallic fittings. Metallic fittings, including valves, strainers and filters shall comply with the following:

1. Fittings used with steel or wrought-iron pipe shall be steel, brass, bronze, malleable iron, ductile iron or cast iron.
2. Fittings used with copper or brass pipe shall be copper, brass or bronze.
3. Brass, bronze or copper fittings. Fittings, if exposed to soil, shall have a minimum 80-percent copper content.
4. Special fittings. Fittings such as couplings, proprietary-type joints, saddle tees, gland-type compression fittings, flared, flareless or compression-type tubing fittings shall be permitted to be used provided they are used within the fitting manufacturer's pressure-temperature recommendations; used within the service conditions anticipated with respect to vibration, fatigue, thermal expansion or contraction; installed or braced to prevent separation of the joint by gas pressure or external physical damage; and approved.

G2413.11 (403.11) Plastic piping, joints and fittings. Plastic pipe, tubing and fittings shall be joined in accordance with the manufacturers' instructions. Such joints shall comply with the following:

1. The joints shall be designed and installed so that the longitudinal pull-out resistance of the joints will be at least equal to the tensile strength of the plastic piping material.
2. Heat-fusion joints shall be made in accordance with qualified procedures that have been established and proven by test to produce gas-tight joints at least as strong as the pipe or tubing being joined. Joints shall be made with the joining method recommended by the pipe manufacturer. Heat fusion fittings shall be marked "ASTM D2513."
3. Where compression type mechanical joints are used, the gasket material in the fitting shall be compatible with the plastic piping and with the gas distributed by the system. An internal tubular rigid stiffener shall be used in conjunction with the fitting. The stiffener shall be flush with the end of the pipe or tubing and shall extend at least to the outside end of the pipe or tubing and shall extend at least to the outside end of the compression fitting when installed. The stiffener shall be free of rough or sharp edges and shall not be a force fit in the plastic. Split tubular stiffeners shall not be used.

4. Plastic piping joints and fittings for use in liquefied petroleum gas piping systems shall be in accordance with NFPA 58.

SECTION G2414 (404) PIPING SYSTEM INSTALLATION

G2414.1 (404.1) Prohibited locations. Piping shall not be installed in or through a circulating air duct, clothes chute, chimney or gas vent, ventilating duct, dumbwaiter or elevator shaft.

G2414.2 (404.2) Piping in solid partitions and walls. Concealed piping shall not be located in solid partitions and solid walls, unless installed in a chase or casing.

G2414.3 (404.3) Piping in concealed locations. Portions of a piping system installed in concealed locations shall not have unions, tubing fittings, right and left couplings, bushings, compression couplings and swing joints made by combinations of fittings.

Exceptions:

1. Tubing joined by brazing.
2. Fittings listed for use in concealed locations.

G2414.4 (404.4) Piping through foundation wall. Underground piping, where installed below grade through the outer foundation or basement wall of a building, shall be encased in a protective pipe sleeve. The annular space between the gas piping and the sleeve shall be sealed.

G2414.5 (404.5) Protection against physical damage. In concealed locations, where piping other than black or galvanized steel is installed through holes or notches in wood studs, joists, rafters or similar members less than 1 inch (25.4 mm) from the nearest edge of the member, the pipe shall be protected by shield plates. Shield plates shall be a minimum of 0.0625-inch-thick (1.6 mm) steel, shall cover the area of the pipe where the member is notched or bored, and shall extend a minimum of 4 inches (102 mm) above sole plates, below top plates and to each side of a stud, joist or rafter.

G2414.6 (404.6) Piping in solid floors. Piping in solid floors shall be laid in channels in the floor and covered in a manner that will allow access to the piping with a minimum amount of damage to the building. Where such piping is subject to exposure to excessive moisture or corrosive substances, the piping shall be protected in an approved manner. As an alternative to installation in channels, the piping shall be installed in a casing of schedule 40 steel, wrought iron, PVC or ABS pipe with tightly sealed ends and joints. Both ends of such casing shall extend not less than 2 inches (51 mm) beyond the point where the pipe emerges from the floor.

G2414.7 (404.7) Above-ground piping outdoors. Piping installed above ground outdoors shall be securely supported and located where it will be protected from physical damage. Where passing through an outside wall, the piping shall also be protected against corrosion by coating or wrapping with an inert material. Where piping is encased in a protective pipe sleeve, the annular space between the gas piping and the sleeve shall be sealed.

G2414.8 (404.8) Protection against corrosion. Metallic pipe or tubing exposed to corrosive action, such as soil condition or moisture, shall be protected in an approved manner. Zinc coatings (galvanizing) shall not be deemed adequate protection for gas piping underground. Ferrous metal exposed in exterior locations shall be protected from corrosion in a manner satisfactory to the code official. Where dissimilar metals are joined underground, an insulating coupling or fitting shall be used. Piping shall not be laid in contact with cinders.

G2414.8.1 (404.8.1) Prohibited use. Uncoated threaded or socket welded joints shall not be used in piping in contact with soil or where internal or external crevice corrosion is known to occur.

G2414.8.2 (404.8.2) Protective coatings and wrapping. Pipe protective coatings and wrappings shall be approved for the application and shall be factory applied.

Exception: Where installed in accordance with the manufacturer's installation instructions, field application of coatings and wrappings shall be permitted for pipe nipples, fittings and locations where the factory coating or wrapping has been damaged or necessarily removed at joints.

G2414.9 (404.9) Minimum burial depth. Underground piping systems shall be installed a minimum depth of 12 inches (305 mm) below grade, except as provided for in Section G2414.9.1.

G2414.9.1 (404.9.1) Individual outside appliances. Individual lines to outside lights, grills or other appliances shall be installed a minimum of 8 inches (203 mm) below finished grade, provided that such installation is approved and is installed in locations not susceptible to physical damage.

G2414.10 (404.10) Trenches. The trench shall be graded so that the pipe has a firm, substantially continuous bearing on the bottom of the trench.

G2414.11 (404.11) Piping underground beneath buildings. Piping installed underground beneath buildings is prohibited except where the piping is encased in a conduit of wrought iron, plastic pipe, or steel pipe designed to withstand the superimposed loads. Such conduit shall extend into an occupiable portion of the building and, at the point where the conduit terminates in the building, the space between the conduit and the gas piping shall be sealed to prevent the possible entrance of any gas leakage. If the end sealing is capable of withstanding the full pressure of the gas pipe, the conduit shall be designed for the same pressure as the pipe. Such conduit shall extend not less than 4 inches (102 mm) outside the building, shall be vented above grade to the outdoors, and shall be installed so as prevent the entrance of water and insects. The conduit shall be protected from corrosion in accordance with Section G2414.8.

G2414.12 (404.12) Outlet closures. Gas outlets that do not connect to appliances shall be capped gas tight.

Exception: Listed and labeled flush-mounted-type quick-disconnect devices and listed and labeled gas convenience outlets shall be installed in accordance with the manufacturer's installation instructions.

G2414.13 (404.13) Location of outlets. The unthreaded portion of piping outlets shall extend not less than 1 inch (25.4 mm) through finished ceilings and walls and where extending through floors, outdoor patios and slabs, shall not be less than 2 inches (51 mm) above them. The outlet fitting or piping shall be securely supported. Outlets shall not be placed behind doors. Outlets shall be located in the room or space where the appliance is installed.

Exception: Listed and labeled flush-mounted-type quick-disconnect devices and listed and labeled gas convenience outlets shall be installed in accordance with the manufacturer's installation instructions.

G2414.14 (404.14) Plastic pipe. The installation of plastic pipe shall comply with Sections G2414.14.1 through G2414.14.3.

G2414.14.1 (404.14.1) Limitations. Plastic pipe shall be installed outside underground only. Plastic pipe shall not be used within or under any building or slab or be operated at pressures greater than 100 psig (689 kPa) for natural gas or 30 psig (207 kPa) for LP gas.

Exceptions:

1. Plastic pipe shall be permitted to terminate aboveground outside of buildings where installed in premanufactured anodeless risers or service head adapter risers that are installed in accordance with that manufacturer's installation instructions.
2. Plastic pipe shall be permitted to terminate with a wall head adapter within buildings where the plastic pipe is inserted in a piping material for fuel gas use in buildings.

G2414.14.2 (404.14.2) Connections. Connections made outside and underground between metallic and plastic piping shall be made only with transition fittings categorized as Category I in accordance with ASTM D 2513.

G2414.14.3 (404.14.3) Tracer. A yellow insulated copper tracer wire or other approved conductor shall be installed adjacent to underground nonmetallic piping. Access shall be provided to the tracer wire or the tracer wire shall terminate above ground at each end of the nonmetallic piping. The tracer wire size shall not be less than 18 AWG and the insulation type shall be suitable for direct burial.

G2414.15 (404.15) Prohibited devices. A device shall not be placed inside the piping or fittings that will reduce the cross-sectional area or otherwise obstruct the free flow of gas.

Exception: Approved gas filters.

G2414.16 (404.16) Testing of piping. Before any system of piping is put in service or concealed, it shall be tested to ensure that it is gas tight. Testing, inspection and purging of piping systems shall comply with Section G2416.

SECTION G2415 (405)
PIPING BENDS AND CHANGES IN DIRECTION

G2415.1 (405.1) General. Changes in direction of pipe shall be permitted to be made by the use of fittings, factory bends or field bends.

G2415.2 (405.2) Metallic pipe. Metallic pipe bends shall comply with the following:

1. Bends shall be made only with bending equipment and procedures especially intended for that purpose.
2. All bends shall be smooth and free from buckling, cracks or other evidence of mechanical damage.
3. The longitudinal weld of the pipe shall be near the neutral axis of the bend.
4. Pipe shall not be bent through an arc of more than 90 degrees (1.6 rad).
5. The inside radius of a bend shall be not less than six times the outside diameter of the pipe.

G2415.3 (405.3) Plastic pipe. Plastic pipe bends shall comply with the following:

1. The pipe shall not be damaged and the internal diameter of the pipe shall not be effectively reduced.
2. Joints shall not be located in pipe bends.
3. The radius of the inner curve of such bends shall not be less than 25 times the inside diameter of the pipe.
4. Where the piping manufacturer specifies the use of special bending equipment or procedures, such equipment or procedures shall be used.

SECTION G2416 (406)
INSPECTION, TESTING AND PURGING

G2416.1 (406.1) General. Prior to acceptance and initial operation, all piping installations shall be inspected and pressure tested to determine that the materials, design, fabrication, and installation practices comply with the requirements of this code.

G2416.1.1 (406.1.1) Inspections. Inspection shall consist of visual examination, during or after manufacture, fabrication, assembly or pressure tests as appropriate.

G2416.1.2 (406.1.2) Repairs and additions. In the event repairs or additions are made following the pressure test, the affected piping shall be tested.

Exception: Minor repairs or additions, provided the work is inspected and connections are tested with a noncorrosive leak-detecting fluid or other leak-detecting methods approved by the code official.

G2416.1.3 (406.1.3) Section testing. A piping system shall be permitted to be tested as a complete unit or in sections. Under no circumstances shall a valve in a line be used as a bulkhead between gas in one section of the piping system and test medium in an adjacent section, unless two valves are installed in series with a valved "tell-tale" located between these valves. A valve shall not be subjected to the test pressure unless it can be determined that the valve, including the valve closing mechanism, is designed to safely withstand the test pressure.

G2416.1.4 (406.1.4) Regulators and valve assemblies. Regulator and valve assemblies fabricated independently of the piping system in which they are to be installed shall be permitted to be tested with inert gas or air at the time of fabrication.

G2416.2 (406.2) Test medium. The test medium shall be air or an inert gas. Oxygen shall not be used.

G2416.3 (406.3) Test preparation. Pipe joints, including welds, shall be left exposed for examination during the test. If the pipe end joints have been previously tested in accordance with this code, they shall be permitted to be covered or concealed.

G2416.3.1 (406.3.1) Expansion joints. Expansion joints shall be provided with temporary restraints, if required, for the additional thrust load under test.

G2416.3.2 (406.3.2) Equipment isolation. Equipment that is not to be included in the test shall be either disconnected from the piping or isolated by blanks, blind flanges or caps.

G2416.3.3 (406.3.3) Equipment disconnection. Where the piping system is connected to equipment or components designed for operating pressures of less than the test pressure, such equipment or equipment components shall be isolated from the piping system by disconnecting them and capping the outlet(s).

G2416.3.4 (406.3.4) Valve isolation. Where the piping system is connected to equipment or components designed for operating pressures equal to or greater than the test pressure, such equipment shall be isolated from the piping system by closing the individual equipment shutoff valve(s).

G2416.3.5 (406.3.5) Testing precautions. All testing of piping systems shall be done with due regard for the safety of employees and the public during the test. Prior to testing, the interior of the pipe shall be cleared of all foreign material.

G2416.4 (406.4) Test pressure measurement. Test pressure shall be measured with a manometer or with a pressure measuring device designed and calibrated to read, record or indicate a pressure loss due to leakage during the pressure test period. The source of pressure shall be isolated before the pressure tests are made.

G2416.4.1 (406.4.1) Test pressure. The test pressure to be used shall be not less than one and one-half times the proposed maximum working pressure, but not less than 3 psig (20 kPa gauge), irrespective of design pressure. Where the test pressure exceeds 125 psig (862 kPa gauge), the test pressure shall not exceed a value that produces a hoop stress in the piping greater than 50 percent of the specified minimum yield strength of the pipe.

G2416.4.2 (406.4.2) Test duration. The test duration shall be not less than 10 minutes.

G2416.5 (406.5) Detection of leaks and defects. The piping system shall withstand the test pressure specified without showing any evidence of leakage or other defects. Any reduction of test pressures as indicated by pressure gauges shall be deemed to indicate the presence of a leak unless such reduction can be readily attributed to some other cause.

G2416.5.1 (406.5.1) Detection methods. The leakage shall be located by means of an approved combustible gas detector, a noncorrosive leak detection fluid or an equivalent nonflammable solution. Matches, candles, open flames or other methods that could provide a source of ignition shall not be used.

G2416.5.2 (406.5.2) Corrections. Where leakage or other defects are located, the affected portion of the piping system shall be repaired or replaced and retested.

G2416.6 (406.6) System and equipment leakage test. Leakage testing of systems and equipment shall be in accordance with Sections G2416.6.1 through G2416.6.4.

G2416.6.1 (406.6.1) Test gases. Fuel gas shall be permitted to be used for leak checks in piping systems that have been tested in accordance with Section G2416.

G2416.6.2 (406.6.2) Before turning gas on. Before gas is introduced into a system of new gas piping, the entire system shall be inspected to determine that there are no open fittings or ends and that all manual valves at outlets on equipment are closed and all unused valves at outlets are closed and plugged or capped.

G2416.6.3 (406.6.3) Test for leakage. Immediately after the gas is turned on into a new system or into a system that has been initially restored after an interruption of service, the piping system shall be tested for leakage. If leakage is indicated, the gas supply shall be shut off until the necessary repairs have been made.

G2416.6.4 (406.6.4) Placing equipment in operation. Gas utilization equipment shall be permitted to be placed in operation after the piping system has been tested and determined to be free of leakage and purged in accordance with Section G2416.7.2.

G2416.7 (406.7) Purging. Purging of piping shall comply with Sections G2416.7.1 through G2416.7.4.

G2416.7.1 (406.7.1) Removal from service. When gas piping is to be opened for servicing, addition or modification, the section to be worked on shall be turned off from the gas supply at the nearest convenient point, and the line pressure vented to the outdoors, or to ventilated areas of sufficient size to prevent accumulation of flammable mixtures.

G2416.7.2 (406.7.2) Placing in operation. When piping full of air is placed in operation, the air in the piping shall be displaced with fuel gas. The air can be safely displaced with fuel gas provided that a moderately rapid and continuous flow of fuel gas is introduced at one end of the line and air is vented out at the other end. The fuel gas flow should be continued without interruption until the vented gas is free of air. The point of discharge shall not be left unattended during purging. After purging, the vent shall then be closed.

G2416.7.3 (406.7.3) Discharge of purged gases. The open end of piping systems being purged shall not discharge into confined spaces or areas where there are sources of ignition unless precautions are taken to perform this operation in a safe manner by ventilation of the space, control or purging rate, and elimination of all hazardous conditions.

G2416.7.4 (406.7.4) Placing equipment in operation. After the piping has been placed in operation, all equipment shall be purged and then placed in operation, as necessary.

SECTION G2417 (407)
PIPING SUPPORT

G2417.1 (407.1) General. Piping shall be provided with support in accordance with Section G2417.2.

G2417.2 (407.2) Design and installation. Piping shall be supported with pipe hooks, metal pipe straps, bands, brackets or hangers suitable for the size of piping, of adequate strength and quality, and located at intervals so as to prevent or damp out excessive vibration. Piping shall be anchored to prevent undue strains on connected equipment and shall not be supported by other piping. Pipe hangers and supports shall conform to the requirements of MSS SP-58 and shall be spaced in accordance with Section G2423. Supports, hangers, and anchors shall be installed so as not to interfere with the free expansion and contraction of the piping between anchors. All parts of the supporting equipment shall be designed and installed so they will not be disengaged by movement of the supported piping.

SECTION G2418 (408)
DRIPS AND SLOPED PIPING

G2418.1 (408.1) Slopes. Piping for other than dry gas conditions shall be sloped not less than 0.25 inch in 15 feet (6.4 mm in 4572 mm) to prevent traps.

G2418.2 (408.2) Drips. Where wet gas exists, a drip shall be provided at any point in the line of pipe where condensate could collect. A drip shall also be provided at the outlet of the meter and shall be installed so as to constitute a trap wherein an accumulation of condensate will shut off the flow of gas before the condensate will run back into the meter.

G2418.3 (408.3) Location of drips. Drips shall be provided with ready access to permit cleaning or emptying. A drip shall not be located where the condensate is subject to freezing.

G2418.4 (408.4) Sediment trap. Where a sediment trap is not incorporated as a part of the gas utilization equipment, a sediment trap shall be installed as close to the inlet of the equipment as practical. The sediment trap shall be either a tee fitting with a capped nipple in the bottom opening of the run of the tee or other device approved as an effective sediment trap. Illuminating appliances, ranges, clothes dryers, and outdoor grills need not be so equipped.

SECTION G2419 (409)
GAS SHUTOFF VALVES

G2419.1 (409.1) General. Piping systems shall be provided with shutoff valves in accordance with this section.

G2419.1.1 (409.1.1) Valve approval. Shutoff valves shall be of an approved type. Shutoff valves shall be constructed of materials compatible with the gas piping. Shutoff valves installed in a portion of a piping system operating above 0.5 psig shall comply with ANSI/ASME B16.33. Shutoff valves installed in a portion of a piping system operating at 0.5 psig or less shall comply with ANSI Z21.15 or ASME B16.33.

G2419.1.2 (409.1.2) Prohibited locations. Shutoff valves shall be prohibited in concealed locations and spaces used as plenums.

G2419.1.3 (409.1.3) Access to shutoff valves. Shutoff valves shall be located in places so as to provide access for operation and shall be installed so as to be protected from damage.

G2419.2 (409.2) Meter valve. Every meter shall be equipped with a shutoff valve located on the supply side of the meter.

G2419.3 (409.3.2) Individual buildings. In a common system serving more than one building, shutoff valves shall be installed outdoors at each building.

G2419.4 (409.4) MP regulator valves. A listed shutoff valve shall be installed immediately ahead of each MP regulator.

G2419.5 (409.5) Equipment shutoff valve. Each appliance shall be provided with a shutoff valve separate from the appliance. The shutoff valve shall be located in the same room as the appliance, not further than 6 feet (1829 mm) from the appliance, and shall be installed upstream from the union, connector or quick disconnect device it serves. Such shutoff valves shall be provided with ready access.

Exception: Shutoff valves for vented decorative appliances and decorative appliances for installation in vented fireplaces shall not be prohibited from being installed in an area remote from the appliance where such valves are provided with ready access. Such valves shall be permanently identified and shall serve no other equipment.

G2419.5.1 (409.5.1) Shutoff valve in fireplace. Equipment shutoff valves located in the firebox of a fireplace shall be installed in accordance with the appliance manufacturer's instructions.

SECTION G2420 (410)
FLOW CONTROLS

G2420.1 (410.1) Pressure regulators. A line pressure regulator shall be installed where the appliance is designed to operate at a lower pressure than the supply system. Access shall be provided to pressure regulators. Pressure regulators shall be protected from physical damage. Regulators installed on the exterior of the building shall be approved for outdoor installation.

G2420.2 (410.2) MP regulators. MP pressure regulators shall comply with the following:

1. The MP regulator shall be approved and shall be suitable for the inlet and outlet gas pressures for the application.
2. The MP regulator shall maintain a reduced outlet pressure under lockup (no-flow) conditions.

3. The capacity of the MP regulator, determined by published ratings of its manufacturer, shall be adequate to supply the appliances served.
4. The MP pressure regulator shall be provided with access. Where located indoors, the regulator shall be vented to the outdoors or shall be equipped with a leak-limiting device, in either case complying with Section G2420.3.
5. A tee fitting with one opening capped or plugged shall be installed between the MP regulator and its upstream shutoff valve. Such tee fitting shall be positioned to allow connection of a pressure measuring instrument and to serve as a sediment trap.
6. A tee fitting with one opening capped or plugged shall be installed not less than 10 pipe diameters downstream of the MP regulator outlet. Such tee fitting shall be positioned to allow connection of a pressure measuring instrument.

G2420.3 (410.3) Venting of regulators. Pressure regulators that require a vent shall have an independent vent to the outside of the building. The vent shall be designed to prevent the entry of water or foreign objects.

Exception: A vent to the outside of the building is not required for regulators equipped with and labeled for utilization with approved vent-limiting devices installed in accordance with the manufacturer's instructions.

SECTION G2421 (411) APPLIANCE CONNECTIONS

G2421.1 (411.1) Connecting appliances. Appliances shall be connected to the piping system by one of the following:

1. Rigid metallic pipe and fittings.
2. Listed and labeled appliance connectors installed in accordance with the manufacturer's installation instructions and located entirely in the same room as the appliance.
3. Listed and labeled quick-disconnect devices used in conjunction with listed and labeled appliance connectors.
4. Listed and labeled gas convenience outlets used in conjunction with listed and labeled appliance connectors.

G2421.1.1 (411.1.1) Protection from damage. Connectors and tubing shall be installed so as to be protected against physical damage.

G2421.1.2 (411.1.2) Appliance fuel connectors. Connectors shall have an overall length not to exceed 3 feet (914 mm), except for range and domestic clothes dryer connectors, which shall not exceed 6 feet (1829 mm) in length. Connectors shall not be concealed within, or extended through, walls, floors, partitions, ceilings or appliance housings. A shutoff valve not less than the nominal size of the connector shall be installed ahead of the connector in accordance with Section G2419.5. Connectors shall be sized to provide the total demand of the connected appliance.

Exception: Fireplace inserts factory equipped with grommets, sleeves or other means of protection in accordance with the listing of the appliance.

G2421.1.3 (411.1.3) Movable appliances. Where appliances are equipped with casters or are otherwise subject to periodic movement or relocation for purposes such as routine cleaning and maintenance, such appliances shall be connected to the supply system piping by means of an approved flexible connector designed and labeled for the application. Such flexible connectors shall be installed and protected against physical damage in accordance with the manufacturer's installation instructions.

SECTION G2422 (413) CNG GAS-DISPENSING SYSTEMS

G2422.1 (413.1) General. Service stations for CNG fuel shall be in accordance with this section and the *International Fire Code.* The operation of CNG service stations shall be regulated by the *International Fire Code.*

SECTION G2423 (415) PIPING SUPPORT INTERVALS

G2423.1 (415.1) Interval of support. Piping shall be supported at intervals not exceeding the spacing specified in Table G2423.1.

TABLE G2423.1 (415.1) SUPPORT OF PIPING

STEEL PIPE, NOMINAL SIZE OF PIPE (inches)	SPACING OF SUPPORTS (feet)	NOMINAL SIZE OF TUBING (inch O.D.)	SPACING OF SUPPORTS (feet)
1/2	6	1/2	4
3/4 or 1	8	5/8 or 3/4	6
1 1/4 or larger (horizontal)	10	7/8 or 1	8
1 1/4 or larger (vertical)	every floor level		

For SI: 1 inch = 25.4 mm, 1 foot = 304.8 mm.

SECTION G2424 (501) GENERAL

G2424.1 (501.1) Scope. This chapter shall govern the installation, maintenance, repair and approval of factory-built and masonry chimneys, chimney liners, vents and connectors serving gas-fired appliances.

G2424.2 (501.2) General. Every appliance shall discharge the products of combustion to the outdoors, except for appliances exempted by Section G2424.8.

G2424.3 (501.3) Masonry chimneys. Masonry chimneys shall be constructed in accordance with Section G2426.5.3 and Chapter 10.

G2424.4 (501.4) Minimum size of chimney or vent. Chimneys and vents shall be sized in accordance with Section G2427.

G2424.5 (501.5) Abandoned inlet openings. Abandoned inlet openings in chimneys and vents shall be closed by an approved method.

G2424.6 (501.6) Positive pressure. Where an appliance equipped with a mechanical forced draft system creates a positive pressure in the venting system, the venting system shall be designed for positive pressure applications.

G2424.7 (501.7) Connection to fireplace. Connection of appliances to chimney flues serving fireplaces shall be in accordance with Sections G2424.7.1 through G2424.7.3.

G2424.7.1 (501.7.1) Closure and access. A noncombustible seal shall be provided below the point of connection to prevent entry of room air into the flue. Means shall be provided for access to the flue for inspection and cleaning.

G2424.7.2 (501.7.2) Connection to factory-built fireplace flue. An appliance shall not be connected to a flue serving a factory-built fireplace unless the appliance is specifically listed for such installation. The connection shall be made in accordance with the appliance manufacturer's installation instructions.

G2424.7.3 (501.7.3) Connection to masonry fireplace flue. A connector shall extend from the appliance to the flue serving a masonry fireplace such that the flue gases are exhausted directly into the flue. The connector shall be accessible or removable for inspection and cleaning of both the connector and the flue. Listed direct connection devices shall be installed in accordance with their listing.

G2424.8 (501.8) Equipment not required to be vented. The following appliances shall not be required to be vented:

1. Ranges.
2. Built-in domestic cooking units listed and marked for optional venting.
3. Hot plates and laundry stoves.
4. Type 1 Clothes dryers (Type 1 clothes dryers shall be exhausted in accordance with the requirements of Section G2437).
5. Refrigerators.
6. Counter appliances.
7. Room heaters listed for unvented use.

Where the appliances and equipment listed in Items 1 through 7 above are installed so that the aggregate input rating exceeds 20 Btu per hour per cubic foot (207 watts per m^3) of volume of the room or space in which such appliances and equipment are installed, one or more shall be provided with venting systems or other approved means for conveying the vent gases to the outdoor atmosphere so that the aggregate input rating of the remaining unvented appliances and equipment does not exceed the 20 Btu per hour per cubic foot (207 watts per m^3) figure. Where the room or space in which the equipment is installed is directly connected to another room or space by a doorway, archway or other opening of comparable size that cannot be closed, the volume of such adjacent room or space shall be permitted to be included in the calculations.

G2424.9 (501.9) Chimney entrance. Connectors shall connect to a masonry chimney flue at a point not less than 12 inches (305 mm) above the lowest portion of the interior of the chimney flue.

G2424.10 (501.10) Connections to exhauster. Appliance connections to a chimney or vent equipped with a power exhauster shall be made on the inlet side of the exhauster. Joints on the positive pressure side of the exhauster shall be sealed to prevent flue-gas leakage as specified by the manufacturer's installation instructions for the exhauster.

G2424.11 (501.11) Masonry chimneys. Masonry chimneys utilized to vent appliances shall be located, constructed and sized as specified in the manufacturer's installation instructions for the appliances being vented and Section G2426.

G2424.12 (501.12) Residential and low-heat appliances flue lining systems. Flue lining systems for use with residential-type and low-heat appliances shall be limited to the following:

1. Clay flue lining complying with the requirements of ASTM C 315 or equivalent. Clay flue lining shall be installed in accordance with Chapter 10.
2. Listed chimney lining systems complying with UL 1777.
3. Other approved materials that will resist, without cracking, softening or corrosion, flue gases and condensate at temperatures up to 1,800°F (982°C).

G2424.13 (501.13) Category I appliance flue lining systems. Flue lining systems for use with Category I appliances shall be limited to the following:

1. Flue lining systems complying with Section G2424.12.
2. Chimney lining systems listed and labeled for use with appliances with draft hoods and other Category I gas appliances listed and labeled for use with Type B vents.

G2424.14 (501.14) Category II, III and IV appliance venting systems. The design, sizing and installation of vents for Category II, III and IV appliances shall be in accordance with the appliance manufacturer's installation instructions.

G2424.15 (501.15) Existing chimneys and vents. Where an appliance is permanently disconnected from an existing chimney or vent, or where an appliance is connected to an existing chimney or vent during the process of a new installation, the chimney or vent shall comply with Sections G2424.15.1 through G2424.15.4.

G2424.15.1 (501.15.1) Size. The chimney or vent shall be resized as necessary to control flue gas condensation in the interior of the chimney or vent and to provide the appliance or appliances served with the required draft. For Category I appliances, the resizing shall be in accordance with Section G2425.

G2424.15.2 (501.15.2) Flue passageways. The flue gas passageway shall be free of obstructions and combustible deposits and shall be cleaned if previously used for venting a solid or liquid fuel-burning appliance or fireplace. The flue liner, chimney inner wall or vent inner wall shall be continuous and shall be free of cracks, gaps, perforations, or other damage or deterioration that would allow the escape of combustion products, including gases, moisture and creosote.

G2424.15.3 (501.15.3) Cleanout. Masonry chimney flues shall be provided with a cleanout opening having a minimum height of 6 inches (152 mm). The upper edge of the opening shall be located not less than 6 inches (152 mm) below the lowest chimney inlet opening. The cleanout shall be provided with a tight-fitting, noncombustible cover.

G2424.15.4 (501.15.4) Clearances. Chimneys and vents shall have airspace clearance to combustibles in accordance with Chapter 10 and the chimney or vent manufacturer's installation instructions. Noncombustible firestopping or fireblocking shall be provided in accordance with Chapter 10.

> **Exception:** Masonry chimneys equipped with a chimney lining system tested and listed for installation in chimneys in contact with combustibles in accordance with UL 1777, and installed in accordance with the manufacturer's instructions, shall not be required to have clearance between combustible materials and exterior surfaces of the masonry chimney.

SECTION G2425 (502) VENTS

G2425.1 (502.1) General. All vents, except as provided in Section G2426.7, shall be listed and labeled. Type B and BW vents shall be tested in accordance with UL 441. Type L vents shall be tested in accordance with UL 641. Vents for Category II and III appliances shall be tested in accordance with UL 1738. Plastic vents for Category IV appliances shall not be required to be listed and labeled where such vents are as specified by the appliance manufacturer and are installed in accordance with the appliance manufacturer's installation instructions.

G2425.2 (502.2) Connectors required. Connectors shall be used to connect appliances to the vertical chimney or vent, except where the chimney or vent is attached directly to the appliance. Vent connector size, material, construction and installation shall be in accordance with Section G2426.

G2425.3 (502.3) Vent application. The application of vents shall be in accordance with Table G2426.4.

G2425.4 (502.4) Insulation shield. Where vents pass through insulated assemblies, an insulation shield constructed of not less than 26 gage sheet (0.016 inch) (0.4 mm) metal shall be installed to provide clearance between the vent and the insulation material. The clearance shall not be less than the clearance to combustibles specified by the vent manufacturer's installation instructions. Where vents pass through attic space, the shield shall terminate not less than 2 inches (51 mm) above the insulation materials and shall be secured in place to prevent displacement. Insulation shields provided as part of a listed vent system shall be installed in accordance with the manufacturer's installation instructions.

G2425.5 (502.5) Installation. Vent systems shall be sized, installed and terminated in accordance with the vent and appliance manufacturer's installation instructions and Section G2426.

G2425.6 (502.6) Support of vents. All portions of vents shall be adequately supported for the design and weight of the materials employed.

SECTION G2426 (503) VENTING OF EQUIPMENT

G2426.1 (503.1) General. This section recognizes that the choice of venting materials and the methods of installation of venting systems are dependent on the operating characteristics of the equipment being vented. The operating characteristics of vented equipment can be categorized with respect to (1) positive or negative pressure within the venting system; and (2) whether or not the equipment generates flue or vent gases that may condense in the venting system. See Section G2403 for the definition of these vented appliance categories.

G2426.2 (503.2) Venting systems required. Except as permitted in Section G2426.2.1, Section G2426.2.2 and Section G2424.8, all equipment shall be connected to venting systems.

G2426.2.1 (503.2.3) Direct-vent equipment. Listed direct-vent equipment shall be considered properly vented where installed in accordance with the terms of its listing, the manufacturer's instructions and Section G2426.8 Item 3.

G2426.2.2 (503.2.4) Equipment with integral vents. Equipment incorporating integral venting means shall be considered properly vented when installed in accordance with its listing, the manufacturer's instructions and Sections G2426.8(1) and G2426.8(2).

G2426.3 (503.3) Design and construction. A venting system shall be designed and constructed so as to develop a positive flow adequate to remove flue or vent gases to the outdoor atmosphere.

G2426.3.1 (503.3.1) Equipment draft requirements. A venting system shall satisfy the draft requirements of the equipment in accordance with the manufacturer's instructions.

G2426.3.2 (503.3.2) Design and construction. Gas utilization equipment required to be vented shall be connected to a venting system designed and constructed in accordance with the provisions of Sections G2426.4 through G2426.15.

G2426.3.3 (503.3.3) Mechanical draft systems. Mechanical draft systems shall comply with the following:

1. Equipment, except incinerators, requiring venting shall be permitted to be vented by means of mechanical draft systems of either forced or induced draft design.
2. Forced draft systems and all portions of induced draft systems under positive pressure during operation shall be designed and installed so as to prevent leakage of flue or vent gases into a building.
3. Vent connectors serving equipment vented by natural draft shall not be connected into any portion of mechanical draft systems operating under positive pressure.
4. When a mechanical draft system is employed, provision shall be made to prevent the flow of gas to the main burners when the draft system is not performing so as to

satisfy the operating requirements of the equipment for safe performance.

5. The exit terminals of mechanical draft systems shall be not less than 7 feet (2134 mm) above grade where located adjacent to public walkways and shall be located as specified in Section G2426.8, Items 1 and 2.
6. Mechanical draft systems shall be installed in accordance with the terms of their listing and the manufacturers' instructions.

G2426.3.4 (503.3.5) Circulating air ducts and plenums. No portion of a venting system shall extend into or pass through any circulating air duct or plenum.

G2426.4 (503.4) Type of venting system to be used. The type of venting system to be used shall be in accordance with Table G2426.4.

G2426.4.1 (503.4.1) Plastic piping. Approved plastic piping shall be permitted to be used for venting equipment listed for use with such venting materials.

G2426.4.2 (503.4.2) Special gas vent. Special gas vents shall be listed and installed in accordance with the terms of the special gas vent listing and the manufacturers' instructions.

G2426.5 (503.5) Masonry, metal, and factory-built chimneys. Masonry, metal and factory-built chimneys shall comply with Sections G2426.5.1 through G2426.5.10.

G2426.5.1 (503.5.1) Factory-built chimneys. Factory-built chimneys shall be installed in accordance with their listing and the manufacturers' instructions. Factory-built chimneys used to vent appliances that operate at positive vent pressure shall be listed for such application.

G2426.5.2 (503.5.3) Masonry chimneys. Masonry chimneys shall be built and installed in accordance with this code and shall be lined with approved clay flue lining, a listed chimney lining system, or other approved material that will resist corrosion, erosion, softening or cracking from vent gases at temperatures up to 1,800°F (982°C).

Exception: Masonry chimney flues serving listed gas appliances with draft hoods, Category I appliances, and other gas appliances listed for use with Type B vent shall be permitted to be lined with a chimney lining system specifically listed for use only with such appliances. The liner shall be installed in accordance with the liner manufacturers' instructions and the terms of the listing. A permanent identifying label shall be attached at the point where the connection is to be made to the liner. The label shall read: "This chimney liner is for appliances that burn gas only. Do not connect to solid or liquid fuel-burning appliances or incinerators."

G2426.5.3 (503.5.4) Chimney termination. Chimneys shall extend at least 3 feet (914 mm) above the highest point where it passes through a roof of a building and at least 2 feet (610 mm) higher than any portion of a building within a horizontal distance of 10 feet (3048 mm). (See Figure G2426.5.) Chimneys shall extend at least 5 feet (1524 mm) above the highest connected equipment draft hood outlet or flue collar. Decorative shrouds shall not be installed at the termination of factory-built chimneys except where such shrouds are listed and labeled for use with the specific factory-built chimney system and are installed in accordance with the manufacturers' installation instructions.

TABLE G2426.4 (503.4)
TYPE OF VENTING SYSTEM TO BE USED

GAS UTILIZATION EQUIPMENT	TYPE OF VENTING SYSTEM
Listed Category I equipment Listed equipment equipped with draft hood Equipment listed for use with Type B gas vent	Type B gas vent (G2426.6) Chimney (G2426.5) Single-wall metal pipe (G2426.7) Listed chimney lining system for gas venting (G2426.5.3). Special gas vent listed for this equipment (G2426.4.2)
Listed vented wall furnaces	Type B-W gas vent (G2426.6, G2434)
Category II equipment	As specified or furnished by manufacturers of listed equipment (G2426.4.1, G2426.4.2)
Category III equipment	As specified or furnished by manufacturers of listed equipment (G2426.4.1, G2426.4.2)
Category IV equipment	As specified or furnished by manufacturers of listed equipment (G2426.4.1, G2426.4.2)
Unlisted equipment	Chimney (G2426.5)
Decorative appliance in vented fireplace	Chimney
Direct-vent equipment	See G2426.2.1
Equipment with integral vent	See G2426.2.2

G2426.5.4 (503.5.5) Size of chimneys. The effective area of a chimney venting system serving listed appliances with draft hoods, Category I appliances, and other appliances listed for use with Type B vents shall be in accordance with Section G2427 or other approved engineering methods.

Exceptions:

1. As an alternate method of sizing an individual chimney venting system for a single appliance with a draft hood, the effective areas of the vent connector and chimney flue shall be not less than the area of the appliance flue collar or draft hood outlet, nor greater than seven times the draft hood outlet area.
2. As an alternate method for sizing a chimney venting system connected to two appliances with draft hoods, the effective area of the chimney flue shall be not less than the area of the larger draft hood outlet plus 50 percent of the area of the smaller draft hood outlet, nor greater than seven times the smallest draft hood outlet area.

G2426.5.5 (503.5.6) Inspection of chimneys. Before replacing an existing appliance or connecting a vent connector to a chimney, the chimney passageway shall be examined to ascertain that it is clear and free of obstructions and it shall be cleaned if previously used for venting solid or liquid fuel-burning appliances or fireplaces.

Exception: Existing chimneys shall be permitted to have their use continued when an appliance is replaced by an appliance of similar type, input rating and efficiency.

G2426.5.5.1 (503.5.6.1) Chimney lining. Chimneys shall be lined in accordance with Chapter 10.

G2426.5.5.2 (503.5.6.2) Cleanouts. Cleanouts shall be examined to determine they will remain tightly closed when not in use.

G2426.5.5.3 (503.5.6.3) Unsafe chimneys. Where inspection reveals that an existing chimney is not safe for the intended application, it shall be repaired, rebuilt, lined, relined, or replaced with a vent or chimney to conform to M1805 and Chapter 10, and it shall be suitable for the equipment to be vented.

G2426.5.6 (503.5.7) Chimney serving equipment burning other fuels. Chimneys serving equipment burning other fuels shall comply with Sections G2426.5.7.1 through G2426.5.7.4.

G2426.5.6.1 (503.5.7.1) Solid fuel-burning appliances. Gas utilization equipment shall not be connected to a chimney flue serving a separate appliance designed to burn solid fuel.

G2426.5.6.2 (503.5.7.2) Liquid fuel-burning appliances. Gas utilization equipment and equipment burning liquid fuel shall be permitted to be connected to one chimney flue through separate openings or shall be permitted to be connected through a single opening if joined by a suitable fitting located as close as practical to the chimney. If two or more openings are provided into one chimney flue, they shall be at different levels. If the gas utilization equipment is automatically controlled, it shall be equipped with a safety shutoff device.

G2426.5.6.3 (503.5.7.3) Combination gas- and solid fuel-burning appliances. A combination gas- and solid fuel-burning appliance equipped with a manual reset device to shut off gas to the main burner in the event of sustained backdraft or flue gas spillage shall be permitted to be connected to a single chimney flue. The chimney flue shall be sized to properly vent the appliance.

G2426.5.6.4 (503.5.7.4) Combination gas- and oil fuel-burning appliances. A listed combination gas- and oil-burning appliance shall be permitted to be connected to a single chimney flue. The chimney flue shall be sized to properly vent the appliance.

G2426.5.7 (503.5.8) Support of chimneys. All portions of chimneys shall be supported for the design and weight of the materials employed. Factory-built chimneys shall be supported and spaced in accordance with their listings and the manufacturers' instructions.

G2426.5.8 (503.5.9) Cleanouts. Where a chimney that formerly carried flue products from liquid or solid fuel-burning appliances is used with an appliance using fuel gas, an accessible cleanout shall be provided. The cleanout shall have a tight-fitting cover and be installed so its upper edge is at least 6 inches (152 mm) below the lower edge of the lowest chimney inlet opening.

G2426.5.9 (503.5.10) Space surrounding lining or vent. The remaining space surrounding a chimney liner, gas vent, special gas vent or plastic piping installed within a masonry chimney flue shall not be used to vent another appliance.

Exception: The insertion of another liner or vent within the chimney as provided in this code and the liner or vent manufacturer's instructions.

G2426.6 (503.6) Gas vents. Gas vents shall comply with Sections G2426.6.1 through G2426.6.12. (See Section G2403, Definitions.)

G2426.6.1 (503.6.1) Installation, general. Gas vents shall be installed in accordance with the terms of their listings and the manufacturer's instructions.

G2426.6.2 (503.6.2) Type B-W vent capacity. A Type B-W gas vent shall have a listed capacity not less than that of the listed vented wall furnace to which it is connected.

G2426.6.3 (503.6.3) Roof penetration. A gas vent passing through a roof shall extend through the roof flashing, roof jack or roof thimble and shall be terminated by a listed termination cap.

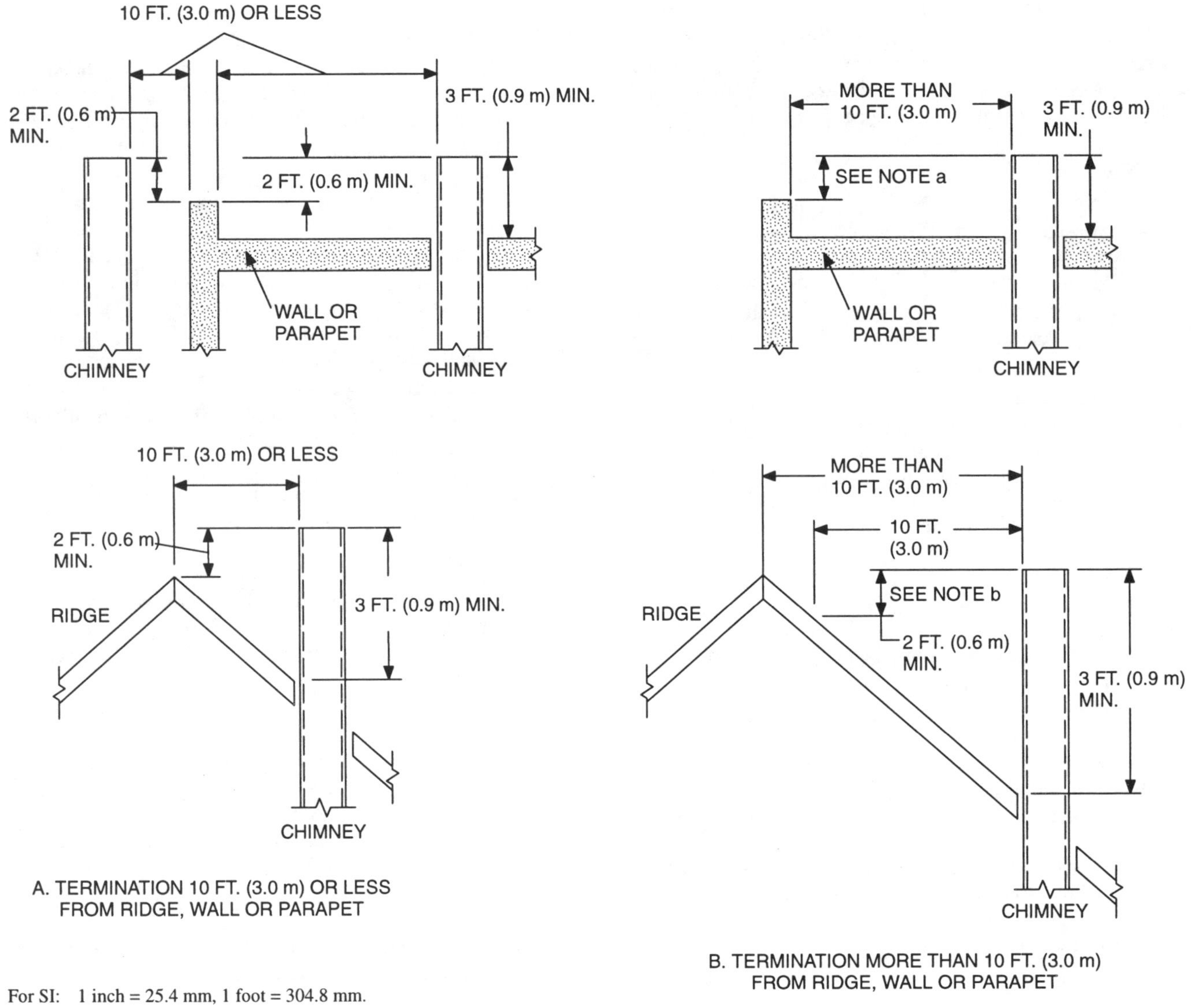

For SI: 1 inch = 25.4 mm, 1 foot = 304.8 mm.

Notes:

a. No height above parapet required when distance from walls or parapet is more than 10 feet.

b. Height above any roof surface within 10 feet horizontally.

FIGURE G2426.5 (503.5)
TYPICAL TERMINATION LOCATIONS FOR CHIMNEYS AND SINGLE-WALL METAL PIPES

G2426.6.4 (503.6.4) Offsets. Type B or Type L vents shall extend in a generally vertical direction with offsets not exceeding 45 degrees, except that a vent system having not more than one 60-degree offset shall be permitted. Any angle greater than 45 degrees from the vertical is considered horizontal. The total horizontal distance of a vent plus the horizontal vent connector serving drafthood-equipped appliances shall not be greater than 75 percent of the vertical height of the vent.

Exception: Systems designed and sized as provided in Section G2427 or in accordance with other approved engineering methods.

Vents serving Category I fan-assisted appliances shall be installed in accordance with the appliance manufacturer's instructions and Section G2427 or other approved engineering methods.

G2426.6.5 (503.6.6) Gas vent termination. A gas vent shall terminate above the roof surface with a listed cap or listed roof assembly. Gas vents 12 inches (305 mm) in size or smaller with listed caps shall be permitted to be terminated in accordance with Figure G2426.6.5, provided that such vents are at least 8 feet (2438 mm) from a vertical wall or similar obstruction. All other gas vents shall terminate not less than 2 feet (610 mm) above the highest point where they pass through the roof and at least 2 feet (610 mm) higher than any portion of a building within 10 feet (3048 mm).

Exceptions:

1. Direct vent systems as provided in Section G2426.2.3.
2. Equipment with integral vents as provided in Section G2426.2.4.
3. Mechanical draft systems as provided in Section G2426.3.3.

G2426.6.6 (503.6.7) Minimum height. A Type B or a Type L gas vent shall terminate at least 5 feet (1524 mm) in vertical height above the highest connected equipment draft hood or flue collar. A Type B-W gas vent shall terminate at least 12 feet (3658 mm) in vertical height above the bottom of the wall furnace.

G2426.6.7 (503.6.8) Exterior wall penetrations. A gas vent extending through an exterior wall shall not terminate adjacent to the wall or below eaves or parapets, except as provided in Sections G2426.2.4 and G2426.3.4.

G2426.6.8 (503.6.9) Size of gas vents. Venting systems shall be sized and constructed in accordance with Section G2427 or other approved engineering methods and the gas vent and gas equipment manufacturers' instructions.

G2426.6.8.1 (503.6.9.1) Category I appliances. The sizing of natural draft venting systems serving one or more listed appliances equipped with a draft hood or appliances listed for use with Type B gas vent, installed in a single story of a building, shall be in accordance with Section G2427 or in accordance with sound engineering practice. Category I appliances are either draft hood-equipped or fan-assisted combustion system in design. Different vent design methods are required for draft hood-equipped and fan-assisted combustion system appliances.

Exceptions:

1. As an alternate method for sizing an individual gas vent for a single, draft hood-equipped appliance, the effective area of the vent connector and the gas vent shall be not less than the area of the appliance draft hood outlet, nor greater than seven times the draft hood outlet area. Vents serving fan-assisted combustion system appliances shall be sized in accordance with Section G2427 or other approved engineering methods.
2. As an alternate method for sizing a gas vent connected to two appliances with draft hoods, the effective area of the vent shall be not less than the area of the larger draft hood plus 50 percent of the smaller draft hood outlets area, nor greater than seven times the smallest draft hood outlet area. Vents serving fan-assisted combustion system appliances, or combinations of fan-assisted combustion system and draft hood-equipped appliances, shall be sized in accordance with Section G2427 or other approved engineering methods.

G2426.6.8.2 (503.6.9.2) Category II, Category III, and Category IV appliances. The sizing of gas vents for Category II, Category III and Category IV equipment shall be in accordance with the equipment manufacturer's instructions.

G2426.6.9 (503.6.11) Support of gas vents. Gas vents shall be supported and spaced in accordance with their listings and the manufacturers' instructions.

G2426.6.10 (503.6.12) Marking. In those localities where solid and liquid fuels are used extensively, gas vents shall be permanently identified by a label attached to the wall or ceiling at a point where the vent connector enters the gas vent. The determination of where such localities exist shall be made by the code official. The label shall read:

"This gas vent is for appliances that burn gas. Do not connect to solid or liquid fuel-burning appliances or incinerators."

G2426.7 (503.7) Single-wall metal pipe. Single-wall metal pipe vents shall comply with Sections G2426.7.1 through G2426.7.12.

G2426.7.1 (503.7.1) Construction. Single-wall metal pipe shall be constructed of galvanized sheet steel not less than 0.0304 inch (0.7 mm) thick, or other approved, noncombustible, corrosion-resistant material.

G2426.7.2 (503.7.2) Cold climate. Uninsulated single-wall metal pipe shall not be used outdoors in cold climates for venting gas utilization equipment.

G2426.7.3 (503.7.3) Termination. Single-wall metal pipe shall terminate at least 5 feet (1524 mm) in vertical height above the highest connected equipment draft hood outlet or flue collar. Single-wall metal pipe shall extend at least 2 feet (610 mm) above the highest point where it passes through a roof of a building and at least 2 feet (610 mm) higher than any portion of a building within a horizontal distance of 10 feet (3048 mm). (See Figure G2426.5.) An approved cap or roof assembly shall be attached to the terminus of a single-wall metal pipe. [Also see Section 2426.7.8, Item 3.]

G2426.7.4 (503.7.4) Limitations of use. Single-wall metal pipe shall be used only for runs directly from the space in which the equipment is located through the roof or exterior wall to the outdoor atmosphere.

G2426.7.5 (503.7.5) Roof penetrations. A pipe passing through a roof shall extend without interruption through the roof flashing, roof jacket or roof thimble. Where a single-wall metal pipe passes through a roof constructed of combustible material, a noncombustible, nonventilating thimble shall be used at the point of passage. The thimble shall extend at least 18 inches (457 mm) above and 6 inches (152 mm) below the roof with the annular space open at the bottom and closed only at the top. The thimble shall be sized in accordance with Section G2426.10.16.

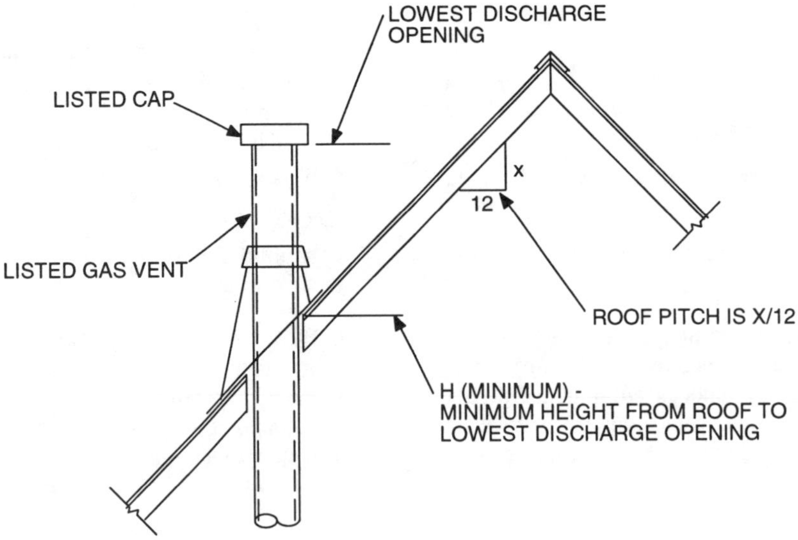

ROOF PITCH	H (minimum) ft	m
Flat to 6/12	1.0	0.30
6/12 to 7/12	1.25	0.38
Over 7/12 to 8/12	1.5	0.46
Over 8/12 to 9/12	2.0	0.61
Over 9/12 to 10/12	2.5	0.76
Over 10/12 to 11/12	3.25	0.99
Over 11/12 to 12/12	4.0	1.22
Over 12/12 to 14/12	5.0	1.52
Over 14/12 to 16/12	6.0	1.83
Over 16/12 to 18/12	7.0	2.13
Over 18/12 to 20/12	7.5	2.27
Over 20/12 to 21/12	8.0	2.44

For SI: 1 foot = 304.8 mm.

FIGURE G2426.6.5 (503.6.6)
GAS VENT TERMINATION LOCATIONS FOR LISTED CAPS 12 INCHES OR LESS IN SIZE AT LEAST 8 FEET FROM A VERTICAL WALL

G2426.7.6 (503.7.6) Installation. Single-wall metal pipe shall not originate in any unoccupied attic or concealed space and shall not pass through any attic, inside wall, concealed space or floor. The installation of a single-wall metal pipe through an exterior combustible wall shall comply with Section G2426.10.16.

G2426.7.7 (503.7.7) Clearances. Minimum clearances from single-wall metal pipe to combustible material shall be in accordance with Table G2426.7.7. The clearance from single-wall metal pipe to combustible material shall be permitted to be reduced where the combustible material is protected as specified for vent connectors in Table G2409.2.

G2426.7.8 (503.7.8) Size of single-wall metal pipe. A venting system constructed of single-wall metal pipe shall be sized in accordance with one of the following methods and the equipment manufacturer's instructions:

1. For a draft hood-equipped appliance, in accordance with Section G2427.
2. For a venting system for a single appliance with a draft hood, the areas of the connector and the pipe each shall not be less than the area of the appliance flue collar or draft hood outlet, whichever is smaller. The vent area shall not be greater than seven times the draft hood outlet area.
3. Other approved engineering methods.

G2426.7.9 (503.7.9) Pipe geometry. Any shaped single-wall metal pipe shall be permitted to be used, provided its equivalent effective area is equal to the effective area of the round pipe for which it is substituted and provided the minimum internal dimension of the pipe is not less than 2 inches (51 mm).

G2426.7.10 (503.7.10) Termination capacity. The vent cap or a roof assembly shall have a venting capacity not less than that of the pipe to which it is attached.

TABLE G2426.7.7 (503.7.7)
CLEARANCES FOR CONNECTORS

	MINIMUM DISTANCE FROM COMBUSTIBLE MATERIAL			
Equipment	**Listed Type B gas vent material**	**Listed Type L vent material**	**Single-wall metal pipe**	**Factory-built chimney sections**
Listed equipment with draft hoods and equipment listed for use with Type B Gas Vents	as listed	as listed	6 inches	as listed
Residential boilers and furnaces with listed gas conversion burner and with draft hood	6 inches	6 inches	9 inches	as listed
Residential appliances listed for use with Type L vents	not permitted	as listed	9 inches	as listed
Unlisted residential appliances with draft hood	not permitted	6 inches	9 inches	as listed
Residential and low-heat equipment other than those above	not permitted	9 inches	18 inches	as listed

For SI: 1 inch = 25.4 mm.
NOTE: These clearances shall apply unless the listing of an appliance or connector specifies different clearances, in which case the listed clearances shall apply.

G2426.7.11 (503.7.11) Support of single-wall metal pipe. All portions of single-wall metal pipe shall be supported for the design and weight of the material employed.

G2426.7.12 (503.7.12) Marking. Single-wall metal pipe shall comply with the marking provisions of Section G2426.6.12.

G2426.8 (503.8) Venting system location. The location of venting system terminations shall comply with following (see Appendix C):

1. A mechanical draft venting system shall terminate at least 3 feet (914 mm) above any forced air inlet located within 10 feet (3048 mm).

   Exception: This provision shall not apply to the combustion air intake of a direct vent appliance.

2. The mechanical draft venting system, excluding direct-vent appliances, shall terminate at least 4 feet (1219 mm) below, 4 feet (1219 mm) horizontally from, or 1 foot (305 mm) above any door, window, or gravity air inlet into any building. The bottom of the vent terminal shall be located at least 12 inches (305 mm) above grade.
3. The vent terminal of a direct-vent appliance with an input of 10,000 Btu per hour (3 kW) or less shall be located at least 6 inches (152 mm) from any air opening into a building, and such an appliance with an input over 10,000 Btu per hour (3 kW) but not over 50,000 Btu per hour (14.7 kW) shall be installed with a 9 inch (230 mm) vent termination clearance, and an appliance with an input over 50,000 btu/hr shall have at least a 12-inch (305 mm) vent termination clearance. The bottom of the vent terminal and the air intake shall be located at least 12 inches (305 mm) above grade.
4. Through-the-wall vents for Category II and Category IV appliances and noncategorized condensing appliances shall not terminate over public walkways or over an area where condensate or vapor could create a nuisance or hazard or could be detrimental to the operation of regulators, relief valves, or other equipment. Where local experience indicates that condensate is a problem with Category I and Category III appliances, this provision shall also apply.

G2426.9 (503.9) Condensation drain. Provision shall be made to collect and dispose of condensate from venting systems serving Category II and Category IV equipment and noncategorized condensing appliances in accordance with Item G2426.8(4). Where local experience indicates that condensation is a problem, provision shall be made to drain off and dispose of condensate from venting systems serving Category I and Category III equipment in accordance with Item G2426(4).

G2426.10 (503.10) Vent connectors for Category I gas utilization equipment. Vent connectors for Category I equipment shall comply with Sections G2426.10.1 through G2426.10.16.

G2426.10.1 (503.10.1) Where required. A vent connector shall be used to connect equipment to a gas vent, chimney or single-wall metal pipe, except where the gas vent, chimney or single-wall metal pipe is directly connected to the equipment.

G2426.10.2 (503.10.2) Materials. Vent connectors shall be constructed in accordance with Sections G2426.10.2.1 through G2426.10.2.5.

G2426.10.2.1 (503.10.2.1) General. A vent connector shall be made of noncombustible corrosion-resistant material capable of withstanding the vent gas temperature produced by the equipment and of sufficient thickness to withstand physical damage.

G2426.10.2.2 (503.10.2.2) Vent connectors located in unconditioned areas. Where the vent connector used for equipment having a draft hood or a Category I appliance is located in or passes through an attic space or other unconditioned area, that portion of the vent connector shall be listed Type B or Type L or listed vent material or listed material having equivalent insulation qualities.

G2426.10.2.3 (503.10.2.3) Residential-type appliance connectors. Where vent connectors for residential-type appliances are not installed in attics or other unconditioned spaces, connectors for listed appliances having draft hoods and for appliances having draft hoods and equipped with listed conversion burners shall be one of the following:

1. Type B or Type L vent material; or
2. Galvanized sheet steel not less than 0.018 inch (0.46 mm) thick; or
3. Aluminum (1100 or 3003 alloy or equivalent) sheet not less than 0.027 inch (0.69 mm) thick; or
4. Stainless steel sheet not less than 0.012 inch (0.31 mm) thick; or
5. Smooth interior wall metal pipe having resistance to heat and corrosion equal to or greater than that of Items 2, 3 or 4 above; or
6. A listed vent connector.

G2426.10.2.4 (503.10.2.4) Low-heat equipment. A vent connector for low-heat equipment shall be a factory-built chimney section or steel pipe having resistance to heat and corrosion equivalent to that for the appropriate galvanized pipe as specified in Table G2426.10.2.4. Factory-built chimney sections shall be joined together in accordance with the chimney manufacturers' instructions.

TABLE G2426.10.2.4 (503.10.2.4)
MINIMUM THICKNESS FOR GALVANIZED STEEL VENT CONNECTORS FOR LOW-HEAT APPLIANCES

DIAMETER OF CONNECTOR (inches)	MINIMUM THICKNESS (inch)
less than 6	0.019
6 to less than 10	0.023
10 to 12 inclusive	0.029
14 to 16 inclusive	0.034
over 16	0.056

For SI: 1 inch = 25.4 mm, 1 square inch = 645.16 mm^2.

G2426.10.3 (503.10.3) Size of vent connector. Vent connectors shall be sized in accordance with Sections G2426.10.3.1 through G2426.10.3.6.

G2426.10.3.1 (503.10.3.1) Single draft hood and fan-assisted. A vent connector for equipment with a single draft hood or for a Category I fan-assisted combustion system appliance shall be sized and constructed in accordance with Section G2427 and other approved engineering methods.

G2426.10.3.2 (503.10.3.2) Multiple draft hood. For a single appliance having more than one draft hood outlet or flue collar, the manifold shall be constructed according to the instructions of the appliance manufacturer. If there are no instructions, the manifold shall be designed and constructed in accordance with approved engineering practices. As an alternate method, the effective area of the manifold shall equal the combined area of the flue collars or draft hood outlets and the vent connectors shall have a minimum 1 foot (305 mm) rise.

G2426.10.3.3 (503.10.3.3) Multiple appliances. Where two or more appliances are connected to a common vent or chimney, each vent connector shall be sized in accordance with Section G2427 or other approved engineering methods.

As an alternative method applicable only where all of the appliances are draft hood-equipped, each vent connector shall have an effective area not less than the area of the draft hood outlet of the appliance to which it is connected.

G2426.10.3.4 (503.10.3.4) Common connector/manifold. Where two or more gas appliances are vented through a common vent connector or vent manifold, the common vent connector or vent manifold shall be located at the highest level consistent with available headroom and the required clearance to combustible materials and shall be sized in accordance with Section G2427 or other approved engineering methods.

As an alternate method applicable only where there are two draft-hood equipped appliances, the effective area of the common vent connector or vent manifold and all junction fittings shall be not less than the area of the larger vent connector plus 50 percent of the area of the smaller flue collar outlet.

G2426.10.3.5 (503.10.3.5) Size increase. Where the size of a vent connector is increased to overcome installation limitations and obtain connector capacity equal to the equipment input, the size increase shall be made at the equipment draft hood outlet.

G2426.10.3.6 (503.10.3.6) Approved engineering practices. The effective area of the vent connector, where connected to one or more appliances requiring draft for operation, shall be obtained by the application of approved engineering practices to perform as specified in Sections G2426.3 and G2426.3.1.

G2426.10.4 (503.10.4) Two or more appliances connected to a single vent. Where two or more vent connectors enter a common gas vent, chimney flue or single-wall metal pipe, the smaller connector shall enter at the highest level consistent with the available headroom or clearance to combustible material. Vent connectors serving Category I appliances shall not be connected to any portion of a mechanical draft system operating under positive static pressure, such as those serving Category III or Category IV appliances.

G2426.10.5 (503.10.5) Clearance. Minimum clearances from vent connectors to combustible material shall be in accordance with Table G2426.7.7.

Exception: The clearance between a vent connector and combustible material shall be permitted to be reduced where the combustible material is protected as specified for vent connectors in Table G2409.2.

G2426.10.6 (503.10.6) Flow resistance. A vent connector shall be installed so as to avoid turns or other construction features that create excessive resistance to flow of vent gases.

G2426.10.7 (503.10.7) Joints. Joints between sections of connector piping and connections to flue collars or hood outlets shall be fastened by sheet-metal screws or other approved means.

Exception: Vent connectors of listed vent material, assembled and connected to flue collars and draft hood outlets in accordance with the manufacturers' instructions.

G2426.10.8 (503.10.8) Slope. A vent connector shall be installed without any dips or sags and shall slope upward toward the vent or chimney at least 0.25 inch per foot (6.35 mm/305 mm).

G2426.10.9 (503.10.9) Length of vent connector. A vent connector shall be as short as practical and the equipment located as close as practical to the chimney or vent. Except as provided for in Section G2426.10.3, the maximum horizontal length of a single-wall connector shall be 75 percent of the height of the chimney or vent. Except as provided for in Section G2426.10.3, the maximum horizontal length of a Type B double-wall connector shall be 100 percent of the height of the chimney or vent. For a chimney or vent system serving multiple appliances, the maximum length of an individual connector, from the appliance outlet to the junction with the common vent or another connector, shall be 100 percent of the height of the chimney or vent.

G2426.10.10 (503.10.10) Support. A vent connector shall be supported for the design and weight of the material employed to maintain clearances and prevent physical damage and separation of joints.

G2426.10.11 (503.10.11) Location. Where the vent connector used for equipment having a draft hood or for Category I appliances is located in or passes through an attic, crawl space, or other unconditioned area subject to low ambient temperatures, that portion of the vent connector shall be of listed double-wall Type B or Type L vent material or listed material having equivalent insulation qualities.

G2426.10.12 (503.10.12) Chimney connection. Where entering a flue in a masonry or metal chimney, the vent connector shall be installed above the extreme bottom to avoid stoppage. A thimble or slip joint shall be permitted to be used to facilitate removal of the connector. The connector shall be firmly attached to or inserted into the thimble or slip joint to prevent the connector from falling out. Means shall be employed to prevent the connector from entering so far as to restrict the space between its end and the opposite wall of the chimney flue. (See Section G2402.9.)

G2426.10.13 (503.10.13) Inspection. The entire length of a vent connector shall be provided with ready access for inspection, cleaning and replacement.

G2426.10.14 (503.10.14) Fireplaces. A vent connector shall not be connected to a chimney flue serving a fireplace unless the fireplace flue opening is permanently sealed.

G2426.10.15 (503.10.15) Passage through ceilings, floors or walls. A vent connector shall not pass through any ceiling, floor or fire-resistance-rated wall. A single-wall metal pipe connector shall not pass through any interior wall.

Exception: Vent connectors made of listed Type B or Type L vent material and serving listed equipment with draft hoods and other equipment listed for use with Type B gas vents shall be permitted to pass through walls or partitions constructed of combustible material if the connectors are installed with not less than the listed clearance to combustible material.

G2426.10.16 (503.10.16) Single-wall connector penetrations of combustible walls. A vent connector made of a single-wall metal pipe shall not pass through a combustible exterior wall unless guarded at the point of passage by a ventilated metal thimble not smaller than the following:

1. For listed equipment equipped with draft hoods and equipment listed for use with Type B gas vents, the thimble shall be not less than 4 inches (102 mm) larger in diameter than the vent connector. Where there is a run of not less than 6 feet (1829 mm) of vent connector in the open between the draft hood outlet and the thimble, the thimble shall be permitted to be 2 inches (51 mm) larger in diameter than the vent connector.
2. For unlisted equipment having draft hoods, the thimble shall be not less than 6 inches (152 mm) larger in diameter than the vent connector.
3. For residential incinerators and all other residential and low-heat equipment, the thimble shall be not less than 12 inches (305 mm) larger in diameter than the vent connector.

Exception: In lieu of thimble protection, all combustible material in the wall shall be removed from the vent connector a sufficient distance to provide the specified clearance from such vent connector to combustible material. Any material used to close up such opening shall be noncombustible.

G2426.11 (503.11) Vent connectors for Category II, Category III and Category IV gas utilization equipment. Vent connectors for category II, III and IV appliances shall be as specified for the venting system in accordance with Section G2426.4.

G2426.12 (503.12) Draft hoods and draft controls. The installation of draft hoods and draft controls shall comply with Sections G2426.12.1 through G2426.12.8.

G2426.12.1 (503.12.1) Equipment requiring draft hoods. Vented equipment shall be installed with draft hoods.

Exception: Dual oven type combination ranges; incinerators; direct-vent equipment; fan-assisted combustion system appliances; equipment requiring chimney draft for operation; single firebox boilers equipped with conversion burners with inputs greater than 400,000 Btu per hour (117 kw); equipment equipped with blast, power or pressure burners that are not listed for use with draft hoods; and equipment designed for forced venting.

G2426.12.2 (503.12.2) Installation. A draft hood supplied with or forming a part of listed vented equipment shall be installed without alteration, exactly as furnished and specified by the equipment manufacturer. If a draft hood is not supplied by the equipment manufacturer where one is required, a draft hood shall be installed, shall be of a listed or approved type and, in the absence of other instructions, shall be of the same size as the equipment flue collar. Where a draft hood is required with a conversion burner, it shall be of a listed or approved type.

Exception: Where it is determined that a draft hood of special design is needed or preferable for a particular installation, the installation shall be in accordance with the recommendations of the equipment manufacturer and shall be approved.

G2426.12.3 (503.12.3) Draft control devices. Where a draft control device is part of the equipment or is supplied by the equipment manufacturer, it shall be installed in accordance with the manufacturer's instructions. In the absence of manufacturers' instructions, the device shall be attached to the fluecollar of the equipment or as near to the equipment as practical.

G2426.12.4 (503.12.4) Additional devices. Equipment requiring controlled chimney draft shall be permitted to be equipped with a listed double acting barometric draft regulator installed and adjusted in accordance with the manufacturer's instructions.

G2426.12.5 (503.12.6) Location. Draft hoods and draft regulators shall be installed in the same room or enclosure as the equipment in such a manner as to prevent any difference in pressure between the hood or regulator and the combustion air supply.

G2426.12.6 (503.12.7) Positioning. Draft hoods and draft regulators shall be installed in the position for which they were designed with reference to the horizontal and vertical planes and shall be located so that the relief opening is not obstructed by any part of the equipment or adjacent construction. The equipment and its draft hood shall be located so that the relief opening is accessible for checking vent operation.

G2426.12.7 (503.12.8) Clearance. A draft hood shall be located so its relief opening is not less than 6 inches (152 mm) from any surface except that of the equipment it serves and the venting system to which the draft hood is connected. Where a greater or lesser clearance is indicated on the equipment label, the clearance shall be not less than that specified on the label. Such clearances shall not be reduced.

G2426.13 (503.13) Manually operated dampers. A manually operated damper shall not be placed in the vent connector for any equipment. Fixed baffles shall not be classified as manually operated dampers.

G2426.14 (503.14) Automatically operated vent dampers. An automatically operated vent damper shall be of a listed type.

G2426.15 (503.15) Obstructions. A device that retards the flow of vent gases shall not be installed in a vent connector, chimney or vent. The tables in Section G2427 shall not apply where the devices covered in this section are installed in the vent. Other approved engineering methods shall be used to size such vents.

Exceptions:

1. Draft regulators and safety controls specifically listed for installation in venting systems and installed in accordance with the terms of their listing.
2. Draft regulators and safety controls that are designed and installed in accordance with approved engineering methods and that are approved.
3. Listed heat reclaimers and automatically operated vent dampers installed in accordance with the terms of their listing.
4. Approved economizers, heat reclaimers, and recuperators installed in venting systems of equipment not required to be equipped with draft hoods, provided the gas utilization equipment manufacturer's instructions cover the installation of such a device in the venting system and performance in accordance with Section G2426.3 and Section G2426.3.1 is obtained.

SECTION G2427 (504) SIZING OF CATEGORY I APPLIANCE VENTING SYSTEMS

G2427.1 (504.1) Definitions. The following definitions apply to tables in this section.

APPLIANCE CATEGORIZED VENT DIAMETER/AREA. The minimum vent area/diameter permissible for Category I appliances to maintain a nonpositive vent static pressure when tested in accordance with nationally recognized standards.

FAN-ASSISTED COMBUSTION SYSTEM. An appliance equipped with an integral mechanical means to either draw or force products of combustion through the combustion chamber or heat exchanger.

FAN MIN. The minimum input rating of a Category I fan-assisted appliance attached to a vent or connector.

FAN MAX. The maximum input rating of a Category I fan-assisted appliance attached to a vent or connector.

NAT MAX. The maximum input rating of a Category I draft-hood equipped appliance attached to a vent or connector.

FAN + FAN. The maximum combined appliance input rating of two or more Category I fan-assisted appliances attached to the common vent.

FAN + NAT. The maximum combined appliance input rating of one or more Category I fan-assisted appliances and one or more Category I draft hood-equipped appliances attached to the common vent.

NA. Vent configuration is not permitted due to potential for condensate formation or pressurization of the venting system, or not applicable due to physical or geometric restraints.

NAT + NAT. The maximum combined appliance input rating of two or more Category I draft hood- equipped appliances attached to the common vent.

G2427.2 (504.2) Application of single appliance vent Tables G2427.2(1) through G2427.2(5). The application of

Tables G2427.2(1) through G2427.2(5) shall be subject to the requirements of Sections G2427.2.1 through G2427.2.13.

G2427.2.1 (504.2.1) Vent obstructions. These venting tables shall not be used where obstructions, as described in the exceptions to Section G2427.15, are installed in the venting system. The installation of vents serving listed appliances with vent dampers shall be in accordance with the appliance manufacturer's instructions or in accordance with the following:

1. The maximum capacity of the vent system shall be determined using the "NAT Max" column.
2. The minimum capacity shall be determined as if the appliance were a fan-assisted appliance, using the "FAN Min" column to determine the minimum capacity of the vent system. Where the corresponding "FAN Min" is "NA," the vent configuration shall not be permitted and an alternative venting configuration shall be utilized.

G2427.2.2 (504.2.2) Minimum size. Where the vent size determined from the tables is smaller than the appliance draft hood outlet or flue collar, the smaller size shall be permitted to be used provided all of the following are met:

1. The total vent height (H) is at least 10 feet (3048 mm).
2. Vents for appliance draft hood outlets or flue collars 12 inches (305 mm) in diameter or smaller are not reduced more than one table size.
3. Vents for appliance draft hood outlets or flue collars larger than 12 inches (305 mm) in diameter are not reduced more than two table sizes.
4. The maximum capacity listed in the tables for a fan-assisted appliance is reduced by 10 percent (0.90 by maximum table capacity).
5. The draft hood outlet is greater than 4 inches (102 mm) in diameter. Do not connect a 3-inch-diameter (76 mm) vent to a 4-inch-diameter (102 mm) draft hood outlet. This provision shall not apply to fan-assisted appliances.

G2427.2.3 (504.2.3) Vent offsets. Single-appliance venting configurations with zero lateral lengths in Tables G2427.2(1) and G2427.2(2) shall have no elbows in the venting system. For vent configurations with lateral lengths, the venting tables include allowance for two 90-degree turns. For each additional 90-degree (1.6 rad) turn, or equivalent, the maximum capacity listed in the venting tables shall be reduced by 10 percent (0.90 by maximum table capacity). Two or more turns, the combined angles of which equal 90 degrees, shall be considered equivalent to one 90-degree (1.6 rad) turn.

G2427.2.4 (504.2.4) Zero lateral. Zero (0) lateral (L) shall apply only to a straight vertical vent attached to a top outlet draft hood or flue collar.

G2427.2.5 (504.2.5) High altitude installations. Sea level input ratings shall be used when determining maximum capacity for high altitude installation. Actual input, derated for altitude, shall be used for determining minimum capacity for high altitude installation.

G2427.2.6 (504.2.6) Multiple input rate appliances. For appliances with more than one input rate, the minimum vent capacity (FAN Min) determined from the tables shall be less than the lowest appliance input rating, and the maximum vent capacity (FAN Max/NAT Max) determined from the tables shall be greater than the highest appliance rating input.

G2427.2.7 (504.2.7) Liner system sizing. Listed corrugated metallic chimney liner systems in masonry chimneys shall be sized by using Table G2427.2(1) or G2427.2(2) for Type B vents with the maximum capacity reduced by 20 percent (0.80 × maximum capacity) and the minimum capacity as shown in Table G2427.2(1) or G2427.2(2). Corrugated metallic liner systems installed with bends or offsets shall have their maximum capacity further reduced in accordance with Section G2427.2.3.

G2427.2.8 (504.2.8) Vent area and diameter Where the vertical vent has a larger diameter than the vent connector, the vertical vent diameter shall be used to determine the minimum vent capacity, and the connector diameter shall be used to determine the maximum vent capacity. The flow area of the vertical vent shall not exceed seven times the flow area of the listed appliance categorized vent area, flue collar area, or draft hood outlet area unless designated in accordance with approved engineering methods.

G2427.2.9 (504.2.9) Chimney and vent locations. Tables G2427.2(1) and G2427.2(2) shall be used for chimneys and vents not exposed to the outdoors below the roof line. A Type B vent or listed chimney lining system passing through an unused masonry chimney flue shall not be considered to be exposed to the outdoors.

Exception: The installation of vents serving listed appliances shall be permitted to be in accordance with the appliance manufacturer's instructions and the terms of the listing.

G2427.2.10 (504.2.10) Vent connector size limitation. Vent connectors shall not be increased in size more than two sizes greater than the listed appliance categorized vent diameter, flue collar diameter or draft hood outlet diameter.

G2427.2.11 (502.2.11) Component commingling. In a single run of vent or vent connector, different diameters and types of vent and connector components shall be permitted to be used, provided that all such sizes and types are permitted by the tables.

TABLE G2427.2(1) [504.2(1)]
CAPACITY OF TYPE B DOUBLE-WALL GAS VENTS WHEN CONNECTED DIRECTLY TO A SINGLE CATEGORY I APPLIANCE

HEIGHT H (feet)	LATERAL L (feet)	VENT DIAMETER — D (inches)																				
		3			4			5			6			7			8			9		
		Appliance input rating in thousands of Btu/h																				
		FAN		NAT	FAN		NAT	FAN		NAT	FAN		NAT	FAN		NAT	FAN		NAT	FAN		NAT
		Min	Max	Max	Min	Max	Max	Min	Max	Max	Min	Max	Max	Min	Max	Max	Min	Max	Max	Min	Max	Max
6	0	0	78	46	0	152	86	0	251	141	0	375	205	0	524	285	0	698	370	0	897	470
	2	13	51	36	18	97	67	27	157	105	32	232	157	44	321	217	53	425	285	63	543	370
	4	21	49	34	30	94	64	39	153	103	50	227	153	66	316	211	79	419	279	93	536	362
	6	25	46	32	36	91	61	47	149	100	59	223	149	78	310	205	93	413	273	110	530	354
8	0	0	84	50	0	165	94	0	276	155	0	415	235	0	583	320	0	780	415	0	1,006	537
	2	12	57	40	16	109	75	25	178	120	28	263	180	42	365	247	50	483	322	60	619	418
	5	23	53	38	32	103	71	42	171	115	53	255	173	70	356	237	83	473	313	99	607	407
	8	28	49	35	39	98	66	51	164	109	64	247	165	84	347	227	99	463	303	117	596	396
10	0	0	88	53	0	175	100	0	295	166	0	447	255	0	631	345	0	847	450	0	1,096	585
	2	12	61	42	17	118	81	23	194	129	26	289	195	40	402	273	48	533	355	57	684	457
	5	23	57	40	32	113	77	41	187	124	52	280	188	68	392	263	81	522	346	95	671	446
	10	30	51	36	41	104	70	54	176	115	67	267	175	88	376	245	104	504	330	122	651	427
15	0	0	94	58	0	191	112	0	327	187	0	502	285	0	716	390	0	970	525	0	1,263	682
	2	11	69	48	15	136	93	20	226	150	22	339	225	38	475	316	45	633	414	53	815	544
	5	22	65	45	30	130	87	39	219	142	49	330	217	64	463	300	76	620	403	90	800	529
	10	29	59	41	40	121	82	51	206	135	64	315	208	84	445	288	99	600	386	116	777	507
	15	35	53	37	48	112	76	61	195	128	76	301	198	98	429	275	115	580	373	134	755	491
20	0	0	97	61	0	202	119	0	349	202	0	540	307	0	776	430	0	1,057	575	0	1,384	752
	2	10	75	51	14	149	100	18	250	166	20	377	249	33	531	346	41	711	470	50	917	612
	5	21	71	48	29	143	96	38	242	160	47	367	241	62	519	337	73	697	460	86	902	599
	10	28	64	44	38	133	89	50	229	150	62	351	228	81	499	321	95	675	443	112	877	576
	15	34	58	40	46	124	84	59	217	142	73	337	217	94	481	308	111	654	427	129	853	557
	20	48	52	35	55	116	78	69	206	134	84	322	206	107	464	295	125	634	410	145	830	537
30	0	0	100	64	0	213	128	0	374	220	0	587	336	0	853	475	0	1,173	650	0	1,548	855
	2	9	81	56	13	166	112	14	283	185	18	432	280	27	613	394	33	826	535	42	1,072	700
	5	21	77	54	28	160	108	36	275	176	45	421	273	58	600	385	69	811	524	82	1,055	688
	10	27	70	50	37	150	102	48	262	171	59	405	261	77	580	371	91	788	507	107	1,028	668
	15	33	64	NA	44	141	96	57	249	163	70	389	249	90	560	357	105	765	490	124	1,002	648
	20	56	58	NA	53	132	90	66	237	154	80	374	237	102	542	343	119	743	473	139	977	628
	30	NA	NA	NA	73	113	NA	88	214	NA	104	346	219	131	507	321	149	702	444	171	929	594
50	0	0	101	67	0	216	134	0	397	232	0	633	363	0	932	518	0	1,297	708	0	1,730	952
	2	8	86	61	11	183	122	14	320	206	15	497	314	22	715	445	26	975	615	33	1,276	813
	5	20	82	NA	27	177	119	35	312	200	43	487	308	55	702	438	65	960	605	77	1,259	798
	10	26	76	NA	35	168	114	45	299	190	56	471	298	73	681	426	86	935	589	101	1,230	773
	15	59	70	NA	42	158	NA	54	287	180	66	455	288	85	662	413	100	911	572	117	1,203	747
	20	NA	NA	NA	50	149	NA	63	275	169	76	440	278	97	642	401	113	888	556	131	1,176	722
	30	NA	NA	NA	69	131	NA	84	250	NA	99	410	259	123	605	376	141	844	522	161	1,125	670

For SI: 1 inch = 25.4 mm, 1 foot = 304.8 mm, 1,000 Btu = 0.293 kW, 1 square inch = 645.16 mm^2.

TABLE G2427.2(2) [504.2(2)] CAPACITY OF TYPE B DOUBLE-WALL VENTS WITH SINGLE-WALL METAL CONNECTORS SERVING A SINGLE CATEGORY I APPLIANCE

HEIGHT H (feet)	LATERAL L (feet)	VENT DIAMETER — D (inches)																										
		3			4			5			6			7			8			9			10			12		
		Appliance input rating in thousands of Btu/h																										
		FAN		NAT	FAN		NAT	FAN		NAT	FAN		NAT	FAN		NAT	FAN		NAT	FAN		NAT	FAN		NAT	FAN		NAT
		Min	Max	Max	Min	Max	Max	Min	Max	Max	Min	Max	Max	Min	Max	Max	Min	Max	Max	Min	Max	Max	Min	Max	Max	Min	Max	Max
6	0	38	77	45	59	151	85	85	249	140	126	373	204	165	522	284	211	695	369	267	894	469	371	1,118	569	537	1,639	849
	2	39	51	36	60	96	66	85	156	104	123	231	156	159	320	213	201	423	284	251	541	368	347	673	453	498	979	648
	4	NA	NA	33	74	92	63	102	152	102	146	225	152	187	313	208	237	416	277	295	533	360	409	664	443	584	971	638
	6	NA	NA	31	83	89	60	114	147	99	163	220	148	207	307	203	263	409	271	327	526	352	449	656	433	638	962	627
8	0	37	83	50	58	164	93	83	273	154	123	412	234	161	580	319	206	777	414	258	1,002	536	360	1,257	658	521	1,852	967
	2	39	56	39	59	108	75	83	176	119	121	261	179	155	363	246	197	482	321	246	617	417	339	768	513	486	1,120	743
	5	NA	NA	37	77	102	69	107	168	114	151	252	171	193	352	235	245	470	311	305	604	404	418	754	500	598	1,104	730
	8	NA	NA	33	90	95	64	122	161	107	175	243	163	223	342	225	280	458	300	344	591	392	470	740	486	665	1,089	715
10	0	37	87	53	57	174	99	82	293	165	120	444	254	158	628	344	202	844	449	253	1,093	584	351	1,373	718	507	2,031	1,057
	2	39	61	41	59	117	80	82	193	128	119	287	194	153	400	272	193	531	354	242	681	456	332	849	559	475	1,242	848
	5	52	56	39	76	111	76	105	185	122	148	277	186	190	388	261	241	518	344	299	667	443	409	834	544	584	1,224	825
	10	NA	NA	34	97	100	68	132	171	112	188	261	171	237	369	241	296	497	325	363	643	423	492	808	520	688	1,194	788
15	0	36	93	57	56	190	111	80	325	186	116	499	283	153	713	388	195	966	523	244	1,259	681	336	1,591	838	488	2,374	1,237
	2	38	69	47	57	136	93	80	225	149	115	337	224	148	473	314	187	631	413	232	812	543	319	1,015	673	457	1,491	983
	5	51	63	44	75	128	86	102	216	140	144	326	217	182	459	298	231	616	400	287	795	526	392	997	657	562	1,469	963
	10	NA	NA	39	95	116	79	128	201	131	182	308	203	228	438	284	284	592	381	349	768	501	470	966	628	664	1,433	928
	15	NA	NA	NA	NA	NA	72	158	186	124	220	290	192	272	418	269	334	568	367	404	742	484	540	937	601	750	1,399	894
20	0	35	96	60	54	200	118	78	346	201	114	537	306	149	772	428	190	1,053	573	238	1,379	750	326	1,751	927	473	2,631	1,346
	2	37	74	50	56	148	99	78	248	165	113	375	248	144	528	344	182	708	468	227	914	611	309	1,146	754	443	1,689	1,098
	5	50	68	47	73	140	94	100	239	158	141	363	239	178	514	334	224	692	457	279	896	596	381	1,126	734	547	1,665	1,074
	10	NA	NA	41	93	129	86	125	223	146	177	344	224	222	491	316	277	666	437	339	866	570	457	1,092	702	646	1,626	1,037
	15	NA	NA	NA	NA	NA	80	155	208	136	216	325	210	264	469	301	325	640	419	393	838	549	526	1,060	677	730	1,587	1,005
	20	NA	NA	NA	NA	NA	NA	186	192	126	254	306	196	309	448	285	374	616	400	448	810	526	592	1,028	651	808	1,550	973
30	0	34	99	63	53	211	127	76	372	219	110	584	334	144	849	472	184	1,168	647	229	1,542	852	312	1,971	1056	454	2,996	1,545
	2	37	80	56	55	164	111	76	281	183	109	429	279	139	610	392	175	823	533	219	1,069	698	296	1,346	863	424	1,999	1,308
	5	49	74	52	72	157	106	98	271	173	136	417	271	171	595	382	215	806	521	269	1,049	684	366	1,324	846	524	1,971	1,283
	10	NA	NA	NA	91	144	98	122	255	168	171	397	257	213	570	367	265	777	501	327	1,017	662	440	1,287	821	620	1,927	1,243
	15	NA	NA	NA	115	131	NA	151	239	157	208	377	242	255	547	349	312	750	481	379	985	638	507	1,251	794	702	1,884	1,205
	20	NA	NA	NA	NA	NA	NA	181	223	NA	246	357	228	298	524	333	360	723	461	433	955	615	570	1,216	768	780	1,841	1,166
	30	NA	NA	NA	NA	NA	NA	NA	NA	NA	NA	NA	NA	389	477	305	461	670	426	541	895	574	704	1,147	720	937	1,759	1,101
50	0	33	99	66	51	213	133	73	394	230	105	629	361	138	928	515	176	1,292	704	220	1,724	948	295	2,223	1,189	428	3,432	1,818
	2	36	84	61	53	181	121	73	318	205	104	495	312	133	712	443	168	971	613	209	1,273	811	280	1,615	1,007	401	2,426	1,509
	5	48	80	NA	70	174	117	94	308	198	131	482	305	164	696	435	204	953	602	257	1,252	795	347	1,591	991	496	2,396	1,490
	10	NA	NA	NA	89	160	NA	118	292	186	162	461	292	203	671	420	253	923	583	313	1,217	765	418	1,551	963	589	2,347	1,455
	15	NA	NA	NA	112	148	NA	145	275	174	199	441	280	244	646	405	299	894	562	363	1,183	736	481	1,512	934	668	2,299	1,421
	20	NA	NA	NA	NA	NA	NA	176	257	NA	236	420	267	285	622	389	345	866	543	415	1,150	708	544	1,473	906	741	2,251	1,387
	30	NA	NA	NA	NA	NA	NA	NA	NA	NA	315	376	NA	373	573	NA	442	809	502	521	1,086	649	674	1,399	848	892	2,159	1,318

For SI: 1 inch = 25.4 mm, 1 foot = 304.8 mm, 1,000 Btu = 0.293 kW.

G2427.2.12 (504.2.12) Table interpolation. Interpolation shall be permitted in calculating capacities for vent dimensions that fall between the table entries. (See Example 3, Appendix B.)

G2427.2.13 (504.2.13) Extrapolation prohibited. Extrapolation beyond the table entries shall not be permitted.

G2427.3 (504.3) Application of multiple appliance vent Tables 504.3(3) through 504.3(6). The application of Tables G2427.3(1) through G2427.3(4) shall be subject to the requirements of Sections G2427.3.1 through G2427.3.21.

G2427.3.1 (504.3.1) Vent obstructions. These venting tables shall not be used where obstructions, as described in the exceptions to Section 503.15, are installed in the venting system. The installation of vents serving listed appliances with vent dampers shall be in accordance with the appliance manufacturer's instructions or in accordance with the following:

1. The maximum capacity of the vent connector shall be determined using the NAT Max column.
2. The maximum capacity of the vertical vent or chimney shall be determined using the FAN+NAT column when the second appliance is a fan-assisted appliance, or the NAT+NAT column when the second appliance is equipped with a draft hood.

3. The minimum capacity shall be determined as if the appliance were a fan-assisted appliance.
   3.1. The minimum capacity of the vent connector shall be determined using the FAN Min column.
   3.2. The FAN+FAN column shall be used when the second appliance is a fan-assisted appliance, and the FAN+NAT column shall be used when the second appliance is equipped with a draft hood, to determine whether the vertical vent or chimney configuration is not permitted (NA). Where the vent configuration is NA, the vent configuration shall not be permitted and an alternative venting configuration shall be utilized.

G2427.3.2 (504.3.2) Connector length limit. The vent connector shall be routed to the vent utilizing the shortest possible route. Except as provided in Section G2427.3.3, the maximum vent connector horizontal length shall be 1.5 feet (457 mm) for each inch (18 mm per mm) of connector diameter as shown in Table G2427.3.2.

TABLE G2427.3.2 (504.3.2)
MAXIMUM VENT CONNECTOR LENGTH

CONNECTOR DIAMETER	CONNECTOR HORIZONTAL
Maximum (inches)	Length (feet)
3	4.5
4	6
5	7.5
6	9
7	10.5
8	12
9	13.5

For SI: 1 inch = 25.4 mm, 1 foot = 304.8 mm.

G2427.3.3 (504.3.3) Connectors with longer lengths. Connectors with longer horizontal lengths than those listed in Section G2427.3.2 are permitted under the following conditions:

1. The maximum capacity (FAN Max or NAT Max) of the vent connector shall be reduced 10 percent for each additional multiple of the length listed above. For example, the maximum length listed above for a 4-inch (102 mm) connector is 6 feet (1829 mm). With a connector length greater than 6 feet (1829 mm), but not exceeding 12 feet (3658 mm), the maximum capacity must be reduced by 10 percent (0.90 maximum vent connector capacity). With a connector length greater than 12 feet (3658 mm), but not exceeding 18 feet (5486 mm), the maximum capacity must be reduced by 20 percent (0.80 maximum vent capacity).
2. For a connector serving a fan-assisted appliance, the minimum capacity (FAN Min) of the connector shall be determined by referring to the corresponding single appliance table. For Type B double-wall connectors, Table G2427.2(1) shall be used. For single-wall connectors, Table G2427.2(2) shall be used. The height (H) and lateral (L) shall be measured according to the procedures for a single appliance vent, as if the other appliances were not present.

G2427.3.4 (504.3.4) Combined connectors. Where the vent connectors are combined prior to entering the common vent, the maximum common vent capacity listed in the common venting tables shall be reduced by 10 percent (0.90 maximum common vent capacity). The length of the common vent connector manifold (L_M) shall not exceed 1.5 feet for each inch (18 mm per mm) of common vent connector manifold diameter (D). (See Figure B-11).

G2427.3.5 (504.3.5) Common vertical vent offset. Where the common vertical vent is offset as shown in Figure B-12, the maximum common vent capacity listed in the common venting tables shall be reduced by 20 percent (0.80 maximum common vent capacity), the equivalent of two 90-degree (1.6 rad) turns. The horizontal length of the common vent offset (L_M) shall not exceed 1.5 feet for each inch (18 cm per cm) of common vent diameter (D).

G2427.3.6 (504.3.6) Additional capacity reduction. Excluding elbows counted in Section G2427.3.5, for each additional 90-degree turn in excess of two, the maximum capacity of that portion of the venting system shall be reduced by 10 percent (0.90 maximum common vent capacity). Two or more turns, the combined angles of which equal 90 degrees, shall be considered equivalent to one 90-degree turn.

G2427.3.7 (504.3.7) Common vent minimum size. The cross-sectional area of the common vent shall be equal to or greater than the cross-sectional area of the largest connector.

G2427.3.8 (504.3.8) Common vent fittings. Interconnection fittings shall be the same size as the common vent.

G2427.3.9 (504.3.9) High altitude installations. Sea level input ratings shall be used when determining maximum capacity for high altitude installation. Actual input, derated for altitude, shall be used for determining minimum capacity for high altitude installation.

G2427.3.10 (504.3.10) Connector rise measurement. Connector rise (R) for each appliance connector shall be measured from the draft hood outlet or flue collar to the centerline where the vent gas streams come together.

G2427.3.11 (504.3.11) Vent height measurement. For multiple units of equipment all located on one floor, available total height (H) shall be measured from the highest draft hood outlet or flue collar up to the level of the outlet of the common vent.

TABLE G2427.3(1) [504.3(1)] CAPACITY OF TYPE B DOUBLE-WALL VENTS WITH TYPE B DOUBLE-WALL CONNECTORS SERVING TWO OR MORE CATEGORY I APPLIANCES

Vent Connector Capacity

VENT HEIGHT H (feet)	CONNECTOR RISE R (feet)	TYPE B DOUBLE-WALL VENT AND CONNECTOR DIAMETER — D (inches)																							
		3			4			5			6			7			8			9			10		
		Appliance input rating limits in thousands of Btu/h																							
		FAN		NAT	FAN		NAT	FAN		NAT	FAN		NAT	FAN		NAT	FAN		NAT	FAN		NAT	FAN		NAT
		Min	Max	Max	Min	Max	Max	Min	Max	Max	Min	Max	Max	Min	Max	Max	Min	Max	Max	Min	Max	Max	Min	Max	Max
6	1	22	37	26	35	66	46	46	106	72	58	164	104	77	225	142	92	296	185	109	376	237	128	466	289
	2	23	41	31	37	75	55	48	121	86	60	183	124	79	253	168	95	333	220	112	424	282	131	526	345
	3	24	44	35	38	81	62	49	132	96	62	199	139	82	275	189	97	363	248	114	463	317	134	575	386
8	1	22	40	27	35	72	48	49	114	76	64	176	109	84	243	148	100	320	194	118	408	248	138	507	303
	2	23	44	32	36	80	57	51	128	90	66	195	129	86	269	175	103	356	230	121	454	294	141	564	358
	3	24	47	36	37	87	64	53	139	101	67	210	145	88	290	198	105	384	258	123	492	330	143	612	402
10	1	22	43	28	34	78	50	49	123	78	65	189	113	89	257	154	106	341	200	125	436	257	146	542	314
	2	23	47	33	36	86	59	51	136	93	67	206	134	91	282	182	109	374	238	128	479	305	149	596	372
	3	24	50	37	37	92	67	52	146	104	69	220	150	94	303	205	111	402	268	131	515	342	152	642	417
15	1	21	50	30	33	89	53	47	142	83	64	220	120	88	298	163	110	389	214	134	493	273	162	609	333
	2	22	53	35	35	96	63	49	153	99	66	235	142	91	320	193	112	419	253	137	532	323	165	658	394
	3	24	55	40	36	102	71	51	163	111	68	248	160	93	339	218	115	445	286	140	565	365	167	700	444
20	1	21	54	31	33	99	56	46	157	87	62	246	125	86	334	171	107	436	224	131	552	285	158	681	347
	2	22	57	37	34	105	66	48	167	104	64	259	149	89	354	202	110	463	265	134	587	339	161	725	414
	3	23	60	42	35	110	74	50	176	116	66	271	168	91	371	228	113	486	300	137	618	383	164	764	466
30	1	20	62	33	31	113	59	45	181	93	60	288	134	83	391	182	103	512	238	125	649	305	151	802	372
	2	21	64	39	33	118	70	47	190	110	62	299	158	85	408	215	105	535	282	129	679	360	155	840	439
	3	22	66	44	34	123	79	48	198	124	64	309	178	88	423	242	108	555	317	132	706	405	158	874	494
50	1	19	71	36	30	133	64	43	216	101	57	349	145	78	477	197	97	627	257	120	797	330	144	984	403
	2	21	73	43	32	137	76	45	223	119	59	358	172	81	490	234	100	645	306	123	820	392	148	1,014	478
	3	22	75	48	33	141	86	46	229	134	61	366	194	83	502	263	103	661	343	126	842	441	151	1,043	538

Common Vent Capacity

VENT HEIGHT H (feet)	TYPE B DOUBLE-WALL COMMON VENT DIAMETER — D (inches)																				
	4			5			6			7			8			9			10		
	Combined appliance input rating in thousands of Btu/h																				
	FAN +FAN	FAN +NAT	NAT +NAT	FAN +FAN	FAN +NAT	NAT +NAT	FAN +FAN	FAN +NAT	NAT +NAT	FAN +FAN	FAN +NAT	NAT +NAT	FAN +FAN	FAN +NAT	NAT +NAT	FAN +FAN	FAN +NAT	NAT +NAT	FAN +FAN	FAN +NAT	NAT +NAT
6	92	81	65	140	116	103	204	161	147	309	248	200	404	314	260	547	434	335	672	520	410
8	101	90	73	155	129	114	224	178	163	339	275	223	444	348	290	602	480	378	740	577	465
10	110	97	79	169	141	124	243	194	178	367	299	242	477	377	315	649	522	405	800	627	495
15	125	112	91	195	164	144	283	228	206	427	352	280	556	444	365	753	612	465	924	733	565
20	136	123	102	215	183	160	314	255	229	475	394	310	621	499	405	842	688	523	1,035	826	640
30	152	138	118	244	210	185	361	297	266	547	459	360	720	585	470	979	808	605	1,209	975	740
50	167	153	134	279	244	214	421	353	310	641	547	423	854	706	550	1,164	977	705	1,451	1,188	860

For SI: 1 inch = 25.4 mm, 1 foot = 304.8 mm, 1,000 Btu = 0.293 kW.

TABLE G2427.3(2) [504.3(2)]
CAPACITY OF TYPE B DOUBLE-WALL VENT WITH SINGLE-WALL CONNECTORS SERVING TWO OR MORE CATEGORY I APPLIANCES

Vent Connector Capacity

VENT HEIGHT H (feet)	CONNECTOR RISE R (feet)	SINGLE-WALL METAL VENT CONNECTOR DIAMETER — D (inches)																							
		3			4			5			6			7			8			9			10		
		Appliance input rating in thousands of Btu/h																							
		FAN		NAT	FAN		NAT	FAN		NAT	FAN		NAT	FAN		NAT	FAN		NAT	FAN		NAT	FAN		NAT
		Min	Max	Max	Min	Max	Max	Min	Max	Max	Min	Max	Max	Min	Max	Max	Min	Max	Max	Min	Max	Max	Min	Max	Max
6	1	NA	NA	26	NA	NA	46	NA	NA	71	NA	NA	102	207	223	140	262	293	183	325	373	234	447	463	286
	2	NA	NA	31	NA	NA	55	NA	NA	85	168	182	123	215	251	167	271	331	219	334	422	281	458	524	344
	3	NA	NA	34	NA	NA	62	121	131	95	175	198	138	222	273	188	279	361	247	344	462	316	468	574	385
8	1	NA	NA	27	NA	NA	48	NA	NA	75	NA	NA	106	226	240	145	285	316	191	352	403	244	481	502	299
	2	NA	NA	32	NA	NA	57	125	126	89	184	193	127	234	266	173	293	353	228	360	450	292	492	560	355
	3	NA	NA	35	NA	NA	64	130	138	100	191	208	144	241	287	197	302	381	256	370	489	328	501	609	400
10	1	NA	NA	28	NA	NA	50	119	121	77	182	186	110	240	253	150	302	335	196	372	429	252	506	534	308
	2	NA	NA	33	84	85	59	124	134	91	189	203	132	248	278	183	311	369	235	381	473	302	517	589	368
	3	NA	NA	36	89	91	67	129	144	102	197	217	148	257	299	203	320	398	265	391	511	339	528	637	413
15	1	NA	NA	29	79	87	52	116	138	81	177	214	116	238	291	158	312	380	208	397	482	266	556	596	324
	2	NA	NA	34	83	94	62	121	150	97	185	230	138	246	314	189	321	411	248	407	522	317	568	646	387
	3	NA	NA	39	87	100	70	127	160	109	193	243	157	255	333	215	331	438	281	418	557	360	579	690	437
20	1	49	56	30	78	97	54	115	152	84	175	238	120	233	325	165	306	425	217	390	538	276	546	664	336
	2	52	59	36	82	103	64	120	163	101	182	252	144	243	346	197	317	453	259	400	574	331	558	709	403
	3	55	62	40	87	107	72	125	172	113	190	264	164	252	363	223	326	476	294	412	607	375	570	750	457
30	1	47	60	31	77	110	57	112	175	89	169	278	129	226	380	175	296	497	230	378	630	294	528	779	358
	2	51	62	37	81	115	67	117	185	106	177	290	152	236	397	208	307	521	274	389	662	349	541	819	425
	3	54	64	42	85	119	76	122	193	120	185	300	172	244	412	235	316	542	309	400	690	394	555	855	482
50	1	46	69	34	75	128	60	109	207	96	162	336	137	217	460	188	284	604	245	364	768	314	507	951	384
	2	49	71	40	79	132	72	114	215	113	170	345	164	226	473	223	294	623	293	376	793	375	520	983	458
	3	52	72	45	83	136	82	119	221	123	178	353	186	235	486	252	304	640	331	387	816	423	535	1,013	518

Common Vent Capacity

VENT HEIGHT H (feet)	TYPE B DOUBLE-WALL VENT AND DIAMETER — D (inches)																				
	4			5			6			7			8			9			10		
	Combined appliance input rating in thousands of Btu/h																				
	FAN +FAN	FAN +NAT	NAT +NAT	FAN +FAN	FAN +NAT	NAT +NAT	FAN +FAN	FAN +NAT	NAT +NAT	FAN +FAN	FAN +NAT	NAT +NAT	FAN +FAN	FAN +NAT	NAT +NAT	FAN +FAN	FAN +NAT	NAT +NAT	FAN +FAN	FAN +NAT	NAT +NAT
6	NA	78	64	NA	113	99	200	158	144	304	244	196	398	310	257	541	429	332	665	515	407
8	NA	87	71	NA	126	111	218	173	159	331	269	218	436	342	285	592	473	373	730	569	460
10	NA	94	76	163	137	120	237	189	174	357	292	236	467	369	309	638	512	398	787	617	487
15	121	108	88	189	159	140	275	221	200	416	343	274	544	434	357	738	599	456	905	718	553
20	131	118	98	208	177	156	305	247	223	463	383	302	606	487	395	824	673	512	1,013	808	626
30	145	132	113	236	202	180	350	286	257	533	446	349	703	570	459	958	790	593	1,183	952	723
50	159	145	128	268	233	208	406	337	296	622	529	410	833	686	535	1,139	954	689	1,418	1,157	838

For SI: 1 inch = 25.4 mm, 1 foot = 304.8 mm, 1,000 Btu = 0.293 kW.

TABLE G2427.3(3) [504.3(3)] CAPACITY OF MASONRY CHIMNEY WITH TYPE B DOUBLE-WALL CONNECTORS SERVING TWO OR MORE CATEGORY I APPLIANCES

Vent Connector Capacity

VENT HEIGHT H (feet)	CONNECTOR RISE R (feet)	TYPE B DOUBLE-WALL VENT CONNECTOR DIAMETER — D (inches)																							
		3			4			5			6			7			8			9			10		
		Appliance input rating limits in thousands of Btu/h																							
		FAN		NAT	FAN		NAT	FAN		NAT	FAN		NAT	FAN		NAT	FAN		NAT	FAN		NAT	FAN		NAT
		Min	Max	Max	Min	Max	Max	Min	Max	Max	Min	Max	Max	Min	Max	Max	Min	Max	Max	Min	Max	Max	Min	Max	Max
6	1	24	33	21	39	62	40	52	106	67	65	194	101	87	274	141	104	370	201	124	479	253	145	599	319
	2	26	43	28	41	79	52	53	133	85	67	230	124	89	324	173	107	436	232	127	562	300	148	694	378
	3	27	49	34	42	92	61	55	155	97	69	262	143	91	369	203	109	491	270	129	633	349	151	795	439
8	1	24	39	22	39	72	41	55	117	69	71	213	105	94	304	148	113	414	210	134	539	267	156	682	335
	2	26	47	29	40	87	53	57	140	86	73	246	127	97	350	179	116	473	240	137	615	311	160	776	394
	3	27	52	34	42	97	62	59	159	98	75	269	145	99	383	206	119	517	276	139	672	358	163	848	452
10	1	24	42	22	38	80	42	55	130	71	74	232	108	101	324	153	120	444	216	142	582	277	165	739	348
	2	26	50	29	40	93	54	57	153	87	76	261	129	103	366	184	123	498	247	145	652	321	168	825	407
	3	27	55	35	41	105	63	58	170	100	78	284	148	106	397	209	126	540	281	147	705	366	171	893	463
15	1	24	48	23	38	93	44	54	154	74	72	277	114	100	384	164	125	511	229	153	658	297	184	824	375
	2	25	55	31	39	105	55	56	174	89	74	299	134	103	419	192	128	558	260	156	718	339	187	900	432
	3	26	59	35	41	115	64	57	189	102	76	319	153	105	448	215	131	597	292	159	760	382	190	960	486
20	1	24	52	24	37	102	46	53	172	77	71	313	119	98	437	173	123	584	239	150	752	312	180	943	397
	2	25	58	31	39	114	56	55	190	91	73	335	138	101	467	199	126	625	270	153	805	354	184	1011	452
	3	26	63	35	40	123	65	57	204	104	75	353	157	104	493	222	129	661	301	156	851	396	187	1067	505
30	1	24	54	25	37	111	48	52	192	82	69	357	127	96	504	187	119	680	255	145	883	337	175	1,115	432
	2	25	60	32	38	122	58	54	208	95	72	376	145	99	531	209	122	715	287	149	928	378	179	1,171	484
	3	26	64	36	40	131	66	56	221	107	74	392	163	101	554	233	125	746	317	152	968	418	182	1,220	535
50	1	23	51	25	36	116	51	51	209	89	67	405	143	92	582	213	115	798	294	140	1,049	392	168	1,334	506
	2	24	59	32	37	127	61	53	225	102	70	421	161	95	604	235	118	827	326	143	1,085	433	172	1,379	558
	3	26	64	36	39	135	69	55	237	115	72	435	180	98	624	260	121	854	357	147	1,118	474	176	1,421	611

Common Vent Capacity

VENT HEIGHT H (feet)	MINIMUM INTERNAL AREA OF MASONRY CHIMNEY FLUE (square inches)																							
	12			19			28			38			50			63			78			113		
	Combined appliance input rating in thousands of Btu/h																							
	FAN +FAN	FAN +FAN	NAT +NAT	FAN +FAN	FAN +FAN	NAT +NAT	FAN +FAN	FAN +FAN	NAT +NAT	FAN +FAN	FAN +FAN	NAT +NAT	FAN +FAN	FAN +FAN	NAT +NAT	FAN +FAN	FAN +FAN	NAT +NAT	FAN +FAN	FAN +FAN	NAT +NAT	FAN +FAN	FAN +FAN	NAT +NAT
6	NA	74	25	NA	119	46	NA	178	71	NA	257	103	NA	351	143	NA	458	188	NA	582	246	1,041	853	NA
8	NA	80	28	NA	130	53	NA	193	82	NA	279	119	NA	384	163	NA	501	218	724	636	278	1,144	937	408
10	NA	84	31	NA	138	56	NA	207	90	NA	299	131	NA	409	177	606	538	236	776	686	302	1,226	1,010	454
15	NA	NA	36	NA	152	67	NA	233	106	NA	334	152	523	467	212	682	611	283	874	781	365	1,374	1,156	546
20	NA	NA	41	NA	NA	75	NA	250	122	NA	368	172	565	508	243	742	668	325	955	858	419	1,513	1,286	648
30	NA	NA	NA	NA	NA	NA	NA	270	137	NA	404	198	615	564	278	816	747	381	1,062	969	496	1,702	1,473	749
50	NA	NA	NA	NA	NA	NA	NA	NA	NA	NA	NA	NA	NA	620	328	879	831	461	1,165	1,089	606	1,905	1,692	922

For SI: 1 inch = 25.4 mm, 1 foot = 304.8 mm, 1,000 Btu = 0.293 kW, 1 square inch = 645.16 mm^2.

TABLE G2427.3(4) [504.3(4)] CAPACITY OF MASONRY CHIMNEY WITH SINGLE-WALL CONNECTORS SERVING TWO OR MORE CATEGORY I APPLIANCES

Vent Connector Capacity

VENT HEIGHT H (feet)	CONNECTOR RISE R (feet)	SINGLE-WALL METAL VENT CONNECTOR DIAMETER — D (inches)																							
		3			4			5			6			7			8			9			10		
		Appliance input rating limits in thousands of Btu/h																							
		FAN		NAT	FAN		NAT	FAN		NAT	FAN		NAT	FAN		NAT	FAN		NAT	FAN		NAT	FAN		NAT
		Min	Max	Max	Min	Max	Max	Min	Max	Max	Min	Max	Max	Min	Max	Max	Min	Max	Max	Min	Max	Max	Min	Max	Max
6	1	NA	NA	21	NA	NA	39	NA	NA	66	179	191	100	231	271	140	292	366	200	362	474	252	499	594	316
	2	NA	NA	28	NA	NA	52	NA	NA	84	186	227	123	239	321	172	301	432	231	373	557	299	509	696	376
	3	NA	NA	34	NA	NA	61	134	153	97	193	258	142	247	365	202	309	491	269	381	634	348	519	793	437
8	1	NA	NA	21	NA	NA	40	NA	NA	68	195	208	103	250	298	146	313	407	207	387	530	263	529	672	331
	2	NA	NA	28	NA	NA	52	137	139	85	202	240	125	258	343	177	323	465	238	397	607	309	540	766	391
	3	NA	NA	34	NA	NA	62	143	156	98	210	264	145	266	376	205	332	509	274	407	663	356	551	838	450
10	1	NA	NA	22	NA	NA	41	130	151	70	202	225	106	267	316	151	333	434	213	410	571	273	558	727	343
	2	NA	NA	29	NA	NA	53	136	150	86	210	255	128	276	358	181	343	489	244	420	640	317	569	813	403
	3	NA	NA	34	97	102	62	143	166	99	217	277	147	284	389	207	352	530	279	430	694	363	580	880	459
15	1	NA	NA	23	NA	NA	43	129	151	73	199	271	112	268	376	161	349	502	225	445	646	291	623	808	366
	2	NA	NA	30	92	103	54	135	170	88	207	295	132	277	411	189	359	548	256	456	706	334	634	884	424
	3	NA	NA	34	96	112	63	141	185	101	215	315	151	286	439	213	368	586	289	466	755	378	646	945	479
20	1	NA	NA	23	87	99	45	128	167	76	197	303	117	265	425	169	345	569	235	439	734	306	614	921	387
	2	NA	NA	30	91	111	55	134	185	90	205	325	136	274	455	195	355	610	266	450	787	348	627	986	443
	3	NA	NA	35	96	119	64	140	199	103	213	343	154	282	481	219	365	644	298	461	831	391	639	1,042	496
30	1	NA	NA	24	86	108	47	126	187	80	193	347	124	259	492	183	338	665	250	430	864	330	600	1,089	421
	2	NA	NA	31	91	119	57	132	203	93	201	366	142	269	518	205	348	699	282	442	908	372	613	1,145	473
	3	NA	NA	35	95	127	65	138	216	105	209	381	160	277	540	229	358	729	312	452	946	412	626	1,193	524
50	1	NA	NA	24	85	113	50	124	204	87	188	392	139	252	567	208	328	778	287	417	1,022	383	582	1,302	492
	2	NA	NA	31	89	123	60	130	218	100	196	408	158	262	588	230	339	806	320	429	1,058	425	596	1,346	545
	3	NA	NA	35	94	131	68	136	231	112	205	422	176	271	607	255	349	831	351	440	1,090	466	610	1,386	597

Common Vent Capacity

VENT HEIGHT H (feet)	MINIMUM INTERNAL AREA OF MASONRY CHIMNEY FLUE (square inches)																							
	12			19			28			38			50			63			78			113		
	Combined appliance input rating in thousands of Btu/h																							
	FAN +FAN	FAN +NAT	NAT +NAT	FAN +FAN	FAN +NAT	NAT +NAT	FAN +FAN	FAN +NAT	NAT +NAT	FAN +FAN	FAN +NAT	NAT +NAT	FAN +FAN	FAN +NAT	NAT +NAT	FAN +FAN	FAN +NAT	NAT +NAT	FAN +FAN	FAN +NAT	NAT +NAT	FAN +FAN	FAN +NAT	NAT +NAT
6	NA	NA	25	NA	118	45	NA	176	71	NA	255	102	NA	348	142	NA	455	187	NA	579	245	NA	846	NA
8	NA	NA	28	NA	128	52	NA	190	81	NA	276	118	NA	380	162	NA	497	217	NA	633	277	1,136	928	405
10	NA	NA	31	NA	136	56	NA	205	89	NA	295	129	NA	405	175	NA	532	234	771	680	300	1,216	1,000	450
15	NA	NA	36	NA	NA	66	NA	230	105	NA	335	150	NA	400	210	677	602	280	866	772	360	1,359	1,139	540
20	NA	NA	NA	NA	NA	74	NA	247	120	NA	362	170	NA	503	240	765	661	321	947	849	415	1,495	1,264	640
30	NA	NA	NA	NA	NA	NA	NA	NA	135	NA	398	195	NA	558	275	808	739	377	1,052	957	490	1,682	1,447	740
50	NA	NA	NA	NA	NA	NA	NA	NA	NA	NA	NA	NA	NA	612	325	NA	821	456	1,152	1,076	600	1,879	1,672	910

For SI: 1 inch = 25.4 mm, 1 foot = 304.8 mm, 1,000 Btu = 0.293 kW, 1 square inch = 645.16 mm^2.

G2427.3.12 (504.3.15) Vertical vent maximum size. Where two or more appliances are connected to a vertical vent or chimney, the flow area of the largest section of vertical vent or chimney shall not exceed seven times the smallest listed appliance categorized vent areas, flue collar area, or draft hood outlet area unless designed in accordance with approved engineering methods.

G2427.3.13 (504.3.16) Multiple input rate appliances. For appliances with more than one input rate, the minimum vent connector capacity (FAN Min) determined from the tables shall be less than the lowest appliance input rating, and the maximum vent connector capacity (FAN Max or NAT Max) determined from the tables shall be greater than the highest appliance input rating.

G2427.3.14 (504.3.17) Liner system sizing. Listed, corrugated metallic chimney liner systems in masonry chimneys shall be sized by using Table G2427.3(1) or G2427.3(2) for Type B vents, with the maximum capacity reduced by 20 percent (0.80 maximum capacity) and the minimum capacity as shown in Table G2427.3(1) or G2427.3(2). Corrugated metallic liner systems installed with bends or offsets shall have their maximum capacity further reduced in accordance with Sections G2427.3.5 and G2427.3.6.

G2427.3.15 (504.3.18) Chimney and vent location. Tables G2427.3(1), G2427.3(2), G2427.3(3) and G2427.3(4) shall be used for chimneys and vents not exposed to the outdoors below the roof line. A Type B vent or listed chimney lining system passing through an unused masonry chimney flue shall not be considered to be exposed to the outdoors.

Exception: The installation of vents serving listed appliances shall be permitted to be in accordance with the appliance manufacturer's instructions and the terms of the listing.

G2427.3.16 (504.3.19) Connector maximum size. Vent connectors shall not be increased in size more than two sizes greater than the listed appliance categorized vent diameter, flue collar diameter, or draft hood outlet diameter. Vent connectors for draft hood-equipped appliances shall not be smaller than the draft hood outlet diameter. Where a vent connector size(s) determined from the tables for a fan-assisted appliance(s) is smaller than the flue collar diameter, the smaller size(s) shall be permitted to be used provided all of the following conditions are met:

1. Vent connectors for fan-assisted appliance flue collars 12 inches (305 mm) in diameter or smaller are not reduced by more than one table size [e.g., 12 inches to 10 inches (305 mm to 254 mm) is a one-size reduction].
2. The fan-assisted appliance(s) is common vented with a draft hood-equipped appliance(s).

G2427.3.17 (504.3.20) Component commingling. All combination of pipe sizes, single-wall, and double-wall metal pipe shall be allowed within any connector run(s) or within the common vent, provided all of the appropriate tables permit all of the desired sizes and types of pipe, as if they were used for the entire length of the subject connector or vent. Where single-wall and Type B double-wall metal pipes are used for vent connectors, the common vent must be sized using Table G2427.3(2) or G2427.3(4) as appropriate.

G2427.3.18 (504.3.21) Multiple sizes permitted. Where a table permits more than one diameter of pipe to be used for a connector or vent, all the permitted sizes shall be permitted to be used.

G2427.3.19 (504.3.22) Table interpolation. Interpolation shall be permitted in calculating capacities for vent dimensions that fall between table entries. (See Example 3, Appendix B.)

G2427.3.20 (504.3.23) Extrapolation prohibited. Extrapolation beyond the table entries shall not be permitted.

G2427.3.21 (504.3.24) Engineering calculations. For vent heights lower than 6 feet (1829 mm) and greater than shown in the tables, engineering methods shall be used to calculate vent capacities.

SECTION G2428 (505) DIRECT-VENT, INTEGRAL VENT, MECHANICAL VENT AND VENTILATION/EXHAUST HOOD VENTING

G2428.1 (505.1) General. The installation of direct-vent and integral vent appliances shall be in accordance with Section G2426. Mechanical venting systems shall be designed and installed in accordance with Section G2426.

SECTION G2429 (506) FACTORY-BUILT CHIMNEYS

G2429.1 (506.1) Listing. Factory-built chimneys for building heating appliances producing flue gases having a temperature not greater than 1,000° F (538° C), measured at the entrance to the chimney, shall be listed and labeled in accordance with UL 103 and shall be installed and terminated in accordance with the manufacturer's installation instructions.

G2429.2 (506.2) Support. Where factory-built chimneys are supported by structural members, such as joists and rafters, such members shall be designed to support the additional load.

SECTION G2430 (601) GENERAL

G2430.1 (601.1) Scope. This chapter shall govern the approval, design, installation, construction, maintenance, alteration and repair of the appliances and equipment specifically identified herein.

SECTION G2431 (602) DECORATIVE APPLIANCES FOR INSTALLATION IN FIREPLACES

G2431.1 (602.1) General. Decorative appliances for installation in approved solid fuel burning fireplaces shall be tested in accordance with ANSI Z21.60 and shall be installed in accordance with the manufacturer's installation instructions.

Manually lighted natural gas decorative appliances shall be tested in accordance with an approved method.

G2431.2 (602.2) Flame safeguard device. Decorative appliances for installation in approved solid fuel-burning fireplaces, with the exception of those tested in accordance with an approved method, shall utilize a direct ignition device, an ignitor or a pilot flame to ignite the fuel at the main burner, and shall be equipped with a flame safeguard device. The flame safeguard device shall automatically shut off the fuel supply to a main burner or group of burners when the means of ignition of such burners becomes inoperative.

G2431.3 (602.3) Prohibited installations. Decorative appliances for installation in fireplaces shall not be installed where prohibited by Section G2406.2.

SECTION G2432 (603) LOG LIGHTERS

G2432.1 (603.1) General. Log lighters shall be tested in accordance with IAS 8 and shall be installed in accordance with the manufacturer's installation instructions.

SECTION G2433 (604) VENTED DECORATIVE APPLIANCES

G2433.1 (604.1) General. Vented gas-fired decorative appliances shall be tested in accordance with ANSI Z21.50, shall be installed in accordance with the manufacturer's installation instructions and shall be designed and equipped as specified in Section G2431.2.

G2433.2 (604.2) Access. Panels, grilles, and access doors that are required to be removed for normal servicing operations shall not be attached to the building.

SECTION G2434 (607) VENTED WALL FURNACES

G2434.1 (607.1) General. Vented wall furnaces shall be tested in accordance with ANSI Z21.49 and shall be installed in accordance with the manufacturer's installation instructions.

G2434.2 (607.2) Venting. Vented wall furnaces shall be vented in accordance with Section G2426.

G2434.3 (607.3) Location. Vented wall furnaces shall be located so as not to cause a fire hazard to walls, floors, combustible furnishings or doors. Vented wall furnaces installed between bathrooms and adjoining rooms shall not circulate air from bathrooms to other parts of the building.

G2434.4 (607.4) Door swing. Vented wall furnaces shall be located so that a door cannot swing within 12 inches (305 mm) of an air inlet or air outlet of such furnace measured at right angles to the opening. Doorstops or door closers shall not be installed to obtain this clearance.

G2434.5 (607.5) Ducts prohibited. Ducts shall not be attached to wall furnaces. Casing extension boots shall not be installed unless listed as part of the appliance.

G2434.6 (607.6) Access. Vented wall furnaces shall be provided with access for cleaning of heating surfaces, removal of burners, replacement of sections, motors, controls, filters and other working parts, and for adjustments and lubrication of parts requiring such attention. Panels, grilles and access doors that are required to be removed for normal servicing operations shall not be attached to the building construction.

SECTION G2435 (608) FLOOR FURNACES

G2435.1 (608.1) General. Floor furnaces shall be tested in accordance with ANSI Z21.48 and shall be installed in accordance with the manufacturer's installation instructions.

G2435.2 (608.2) Placement. The following provisions apply to floor furnaces:

1. Floors. Floor furnaces shall not be installed in the floor of any doorway, stairway landing, aisle or passageway of any enclosure, public or private or in an exitway from any such room or space.
2. Walls and corners. The register of a floor furnace with a horizontal warm air outlet shall not be placed closer than 6 inches (152 mm) to the nearest wall. A distance of at least 18 inches (457 mm) from two adjoining sides of the floor furnace register to walls shall be provided to eliminate the necessity of occupants walking over the warm air discharge. The remaining sides shall be permitted to be placed not closer than 6 inches (152 mm) to a wall. Wall-register models shall not be placed closer than 6 inches (152 mm) to a corner.
3. Draperies. The furnace shall be placed so that a door, drapery, or similar object cannot be nearer than 12 inches (305 mm) to any portion of the register of the furnace.
4. Floor construction. Floor furnaces shall not be installed in concrete floor construction built on grade.
5. Thermostat. The controlling thermostat for a floor furnace shall be located within the same room or space as the floor furnace or shall be located in an adjacent room or space that is permanently open to the room or space containing the floor furnace.

G2435.3 (608.3) Bracing. The floor around the furnace shall be braced and headed with a support framework designed in accordance with Chapter 5.

G2435.4 (608.4) Clearance. The lowest portion of the floor furnace shall have not less than a 6-inch (152 mm) clearance from the grade level; except where the lower 6-inch (152 mm) portion of the floor furnace is sealed by the manufacturer to prevent entrance of water, the minimum clearance shall be reduced to not less than 2 inches (51 mm). Where these clearances cannot be provided, the ground below and to the sides shall be excavated to form a pit under the furnace so that the required clearance is provided beneath the lowest portion of the furnace. A 12-inch (305 mm) minimum clearance shall be provided on all sides except the control side, which shall have an 18-inch (457 mm) minimum clearance.

G2435.5 (608.5) First floor installation. Where the basement story level below the floor in which a floor furnace is installed is utilized as habitable space, such floor furnaces shall be enclosed as specified in Section G2435.6 and shall project into a nonhabitable space.

G2435.6 (608.6) Upper floor installations. Floor furnaces installed in upper stories of buildings shall project below into nonhabitable space and shall be separated from the nonhabitable space by an enclosure constructed of noncombustible materials. The floor furnace shall be provided with access, clearance to all sides and bottom of not less than 6 inches (152 mm) and combustion air in accordance with Section G2407.

SECTION G2436 (612) CLOTHES DRYERS

G2436.1 (612.1) General. Clothes dryers shall be tested in accordance with ANSI Z21.5.1 and shall be installed in accordance with the manufacturer's installation instructions.

SECTION G2437 (613) CLOTHES DRYER EXHAUST

G2437.1 (613.1) Installation. Clothes dryers shall be exhausted in accordance with the manufacturer's instructions. Dryer exhaust systems shall be independent of all other systems and shall convey the moisture and any products of combustion to the outside of the building.

G2437.2 (613.2) Duct penetrations. Ducts that exhaust clothes dryers shall not penetrate or be located within any fireblocking, draftstopping or any wall, floor/ceiling or other assembly required by this code to be fire-resistance-rated, unless such duct is constructed of galvanized steel or aluminum of the thickness specified in the mechanical provisions of this code and the fire-resistance rating is maintained in accordance with this code.

G2437.3 (613.4) Exhaust material. Dryer exhaust ducts for clothes dryers shall terminate on the outside of the building and shall be equipped with a backdraft damper. Screens shall not be installed at the duct termination. Ducts shall not be connected or installed with sheet metal screws or other fasteners that will obstruct the flow. Clothes dryer exhaust ducts shall not be connected to a vent connector, vent or chimney. Clothes dryer exhaust ducts shall not extend into or through ducts or plenums.

G2437.4 (613.5) Makeup air. Installations exhausting more than 200 cfm (0.09 m^3/s) shall be provided with makeup air. Where a closet is designed for the installation of a clothes dryer, an opening having an area of not less than 100 square inches (0.0645 m^2) for makeup air shall be provided in the closet enclosure, or makeup air shall be provided by other approved means.

G2437.5 (613.6) Clothes dryer ducts. Exhaust ducts for domestic clothes dryers shall have a smooth interior finish. The exhaust duct shall be a minimum nominal size of 4 inches (102 mm) in diameter. The entire exhaust system shall be supported and secured in place. The male end of the duct at overlapped duct joints shall extend in the direction of airflow. Clothes dryer transition ducts used to connect the appliance to the exhaust duct system shall be metal and limited to a single length not to exceed 8 feet (2438 mm) in length and shall be listed and labeled for the application. Transition ducts shall not be concealed within construction.

G2437.5.1 (613.6.1) Maximum length. The maximum length of a clothes dryer exhaust duct shall not exceed 25 feet (7620 mm) from the dryer location to the outlet terminal. The maximum length of the duct shall be reduced $2^1/_2$ feet (762 mm) for each 45 degree (0.79 rad) bend and 5 feet (1524 mm) for each 90 degree (1.6 rad) bend.

Exception: Where the make and model of the clothes dryer to be installed is known and the manufacturer's installation instructions for such dryer are provided to the code official, the maximum length of the exhaust duct, including any transition duct, shall be permitted to be in accordance with the dryer manufacturer's installation instructions.

G2437.5.2 (613.6.2) Rough-in-required. Where a compartment or space for a clothes dryer is provided, an exhaust duct system shall be installed.

SECTION G2438 (614) SAUNA HEATERS

G2438.1 (614.1) General. Sauna heaters shall be installed in accordance with the manufacturer's installation instructions.

G2438.2 (614.2) Location and protection. Sauna heaters shall be located so as to minimize the possibility of accidental contact by a person in the room.

G2438.2.1 (614.2.1) Guards. Sauna heaters shall be protected from accidental contact by an approved guard or barrier of material having a low coefficient of thermal conductivity. The guard shall not substantially affect the transfer of heat from the heater to the room.

G2438.3 (614.3) Access. Panels, grilles and access doors that are required to be removed for normal servicing operations, shall not be attached to the building.

G2438.4 (614.4) Combustion and dilution air intakes. Sauna heaters of other than the direct-vent type shall be installed with the draft hood and combustion air intake located outside the sauna room. Where the combustion air inlet and the draft hood are in a dressing room adjacent to the sauna room, there shall be provisions to prevent physically blocking the combustion air inlet and the draft hood inlet, and to prevent physical contact with the draft hood and vent assembly, or warning notices shall be posted to avoid such contact. Any warning notice shall be easily readable, shall contrast with its background, and the wording shall be in letters not less than 0.25 inch (6.4 mm) high.

G2438.5 (614.5) Combustion and ventilation air. Combustion air shall not be taken from inside the sauna room. Combustion and ventilation air for a sauna heater not of the direct-vent type shall be provided to the area in which the combustion air inlet and draft hood are located in accordance with Section G2407.

G2438.6 (614.6) Heat and time controls. Sauna heaters shall be equipped with a thermostat which will limit room temperature to 194°F (90°C). If the thermostat is not an integral part of the sauna heater, the heat-sensing element shall be located within 6 inches (152 mm) of the ceiling. If the heat-sensing element is a capillary tube and bulb, the assembly shall be attached to the wall or other support, and shall be protected against physical damage.

G2438.6.1 (614.6.1) Timers. A timer, if provided to control main burner operation, shall have a maximum operating time of 1 hour. The control for the timer shall be located outside the sauna room.

G2438.7 (614.7) Sauna room. A ventilation opening into the sauna room shall be provided. The opening shall be not less than 4 inches by 8 inches (102 mm by 203 mm) located near the top of the door into the sauna room.

SECTION G2439 (616) POOL AND SPA HEATERS

G2439.1 (616.1) General. Pool and spa heaters shall be tested in accordance with ANSI Z21.56 and shall be installed in accordance with the manufacturer's installation instructions.

SECTION G2440 (617) FORCED-AIR WARM-AIR FURNACES

G2440.1 (617.1) General. Forced-air warm-air furnaces shall be tested in accordance with ANSI Z21.47 or UL 795 and shall be installed in accordance with the manufacturer's installation instructions.

G2440.2 (617.2) Forced-air furnaces. The minimum unobstructed total area of the outside and return air ducts or openings to a forced air warm-air furnace shall be not less than 2 square inches for each 1,000 Btu/h (4402 mm^2/W) output rating capacity of the furnace and not less than that specified in the furnace manufacturer's installation instructions. The minimum unobstructed total area of supply ducts from a forced-air warm-air furnace shall be not less than 2 square inches for each 1,000 Btu/h (4402 mm^2/W) output rating capacity of the furnace and not less than that specified in the furnace manufacturer's installation instructions.

Exception: The total area of the supply air ducts and outside and return air ducts shall not be required to be larger than the minimum size required by the furnace manufacturer's installation instructions.

G2440.3 (617.3) Dampers. Volume dampers shall not be placed in the air inlet to a furnace in a manner that will reduce the required air to the furnace.

G2440.4 (617.4) Circulating air ducts for forced-air warm-air furnaces. Circulating air for forced-air-type, warm-air furnaces shall be conducted into the blower housing from outside the furnace enclosure by continuous air-tight ducts.

G2440.5 (617.5) Prohibited sources. Outside or return air for a forced-air heating system shall not be taken from the following locations:

1. Closer than 10 feet (3048 mm) from an appliance vent outlet, a vent opening from a plumbing drainage system or the discharge outlet of an exhaust fan, unless the outlet is 3 feet (914 mm) above the outside air inlet.
2. Where there is the presence of objectionable odors, fumes or flammable vapors; or where located less than 10 feet (3048 mm) above the surface of any abutting public way or driveway; or where located at grade level by a sidewalk, street, alley or driveway.
3. A room or space, the volume of which is less than 25 percent of the entire volume served by such system. Where connected by a permanent opening having an area sized in accordance with Section G2440.2, adjoining rooms or spaces shall be considered as a single room or space for the purpose of determining the volume of such rooms or spaces.

   Exception: The minimum volume requirement shall not apply where the amount of return air taken from a room or space is less than or equal to the amount of supply air delivered to such room or space.
4. A room or space containing an appliance where such a room or space serves as the sole source of return air.

   Exception: This shall not apply where:

   1. The appliance is a direct-vent appliance or an appliance not requiring a vent in accordance with Section G2424.8.
   2. The room or space complies with the following requirements:
      - 2.1. The return air shall be taken from a room or space having a volume exceeding 1 cubic foot (28 316.85 mm^3) for each 10 Btu/h (9.6 L/W) of combined input rating of all fuel-burning appliances therein.
      - 2.2. The volume of supply air discharged back into the same space shall be approximately equal to the volume of return air taken from the space.
      - 2.3. Return-air inlets shall not be located within 10 feet (3048 mm) of any appliance firebox or draft hood in the same room or space.
   3. Rooms or spaces containing solid-fuel burning appliances, provided that return-air inlets are located not less than 10 feet (3048 mm) from the firebox of such appliances.
5. A closet, bathroom, toilet room, kitchen, garage, mechanical room, boiler room or furnace room.

G2440.6 (617.6) Screen. Required outdoor air inlets shall be covered with a screen having $^1/_4$ inch (6.4 mm) openings. Required outdoor air inlets serving a nonresidential portion of a building shall be covered with screen having openings larger than $^1/_4$ inch (6.4 mm) and not larger than 1 inch (25.4 mm).

G2440.7 (617.7) Return-air limitation. Return air from one dwelling unit shall not be discharged into another dwelling unit.

SECTION G2441 (618) CONVERSION BURNERS

G2441.1 (618.1) Conversion burners. The installation of conversion burners shall conform to ANSI Z21.8.

SECTION G2442 (619) UNIT HEATERS

G2442.1 (619.1) General. Unit heaters shall be tested in accordance with ANSI Z83.8 and shall be installed in accordance with the manufacturer's installation instructions.

G2442.2 (619.2) Support. Suspended-type unit heaters shall be supported by elements that are designed and constructed to accommodate the weight and dynamic loads. Hangers and brackets shall be of noncombustible material.

G2442.3 (619.3) Ductwork. Ducts shall not be connected to a unit heater unless the heater is listed for such installation.

G2442.4 (619.4) Clearance. Suspended-type unit heaters shall be installed with clearances to combustible materials of not less than 18 inches (457 mm) at the sides, 12 inches (305 mm) at the bottom and 6 inches (152 mm) above the top where the unit heater has an internal draft hood or 1 inch (25.4 mm) above the top of the sloping side of the vertical draft hood.

Floor-mounted type unit heaters shall be installed with clearances to combustible materials at the back and one side only of not less than 6 inches (152 mm). Where the flue gases are vented horizontally, the 6-inch (152 mm) clearance shall be measured from the draft hood of vent instead of the rear wall of the unit heater. Floor-mounted-type unit heaters shall not be installed on combustible floors unless listed for such installation.

Clearance for servicing all unit heaters shall be in accordance with the manufacturer's installation instructions.

Exception: Unit heaters listed for reduced clearance shall be permitted to be installed with such clearances in accordance with their listing and the manufacturer's instructions.

SECTION G2443 (620) UNVENTED ROOM HEATERS

G2443.1 (620.1) General. Unvented room heaters shall be tested in accordance with ANSI Z21.11.2 and shall be installed in accordance with the conditions of the listing and the manufacturer's installation instructions.

G2443.2 (620.2) Prohibited use. One or more unvented room heaters shall not be used as the sole source of comfort heating in a dwelling unit.

G2443.3 (620.3) Input rating. Unvented room heaters shall not have an input rating in excess of 40,000 Btu/h (11.7 kW).

G2443.4 (620.4) Prohibited locations. The location of unvented room heaters shall comply with Section G2406.2.

G2443.5 (620.5) Room or space volume. The aggregate input rating of all unvented appliances installed in a room or space shall not exceed 20 Btu/h per cubic foot (0.21 kW/m^3) of volume of such room or space. Where the room or space in which the equipment is installed is directly connected to another room or space by a doorway, archway or other opening of comparable size that cannot be closed, the volume of such adjacent room or space shall be permitted to be included in the calculations.

G2443.6 (620.6) Oxygen-depletion safety system. Unvented room heaters shall be equipped with an oxygen-depletion-sensitive safety shutoff system. The system shall shut off the gas supply to the main and pilot burners when the oxygen in the surrounding atmosphere is depleted to the percent concentration specified by the manufacturer, but not lower than 18 percent. The system shall not incorporate field adjustment means capable of changing the set point at which the system acts to shut off the gas supply to the room heater.

G2443.7 (620.7) Unvented log heaters. An unvented log heater shall not be installed in a factory-built fireplace unless the fireplace system has been specifically tested, listed and labeled for such use in accordance with UL 127.

SECTION G2444 (621) VENTED ROOM HEATERS

G2444.1 (621.1) General. Vented room heaters shall be tested in accordance with ANSI Z21.11.1, shall comply with Section G2431.2 and shall be installed in accordance with the manufacturer's installation instructions.

SECTION G2445 (622) COOKING APPLIANCES

G2445.1 (622.1) Cooking appliances. Cooking appliances that are designed for permanent installation, including ranges, ovens, stoves, broilers, grills, fryers, griddles, hot plates and barbecues, shall be tested in accordance with ANSI Z21.1 or ANSI Z21.58 and shall be installed in accordance with the manufacturer's installation instructions.

G2445.2 (622.2) Prohibited location. Cooking appliances designed, tested, listed and labeled for use in commercial occupancies shall not be installed within dwelling units or within any area where domestic cooking operations occur.

G2445.3 (622.3) Domestic appliances. Cooking appliances installed within dwelling units and within areas where domestic cooking operations occur shall be listed and labeled as household-type appliances for domestic use.

G2445.4 (622.4) Range installation. Ranges installed on combustible floors shall be set on their own bases or legs and shall be installed with clearances of not less than that shown on the label.

SECTION G2446 (623) WATER HEATERS

G2446.1 (623.1) General. Water heaters shall be tested in accordance with ANSI Z21.10.1 and ANSI Z21.10.3 and shall be installed in accordance with the manufacturer's installation instructions.

G2446.1.1 (623.1.1) Installation requirements. The requirements for water heaters relative to sizing, relief valves, drain pans and scald protection shall be in accordance with this code.

G2446.2 (623.2) Water heaters utilized for space heating. Water heaters utilized both to supply potable hot water and provide hot water for space-heating applications shall be listed and labeled for such applications by the manufacturer and shall be installed in accordance with the manufacturer's installation instructions and this code.

SECTION G2447 (626) AIR CONDITIONING EQUIPMENT

G2447.1 (626.1) General. Air conditioning equipment shall be tested in accordance with ANSI Z21.40.1 or ANSI Z21.40.2 and shall be installed in accordance with the manufacturer's installation instructions.

G2447.2 (626.2) Independent piping. Gas piping serving heating equipment shall be permitted to also serve cooling equipment where such heating and cooling equipment cannot be operated simultaneously. (See Section G2412.)

G2447.3 (626.3) Connection of gas engine-powered air conditioners. To protect against the effects of normal vibration in service, gas engines shall not be rigidly connected to the gas supply piping.

G2447.4 (626.6) Installation. Air conditioning equipment shall be installed in accordance with the manufacturer's instructions. Unless the equipment is listed for installation on a combustible surface such as a floor or roof, or unless the surface is protected in an approved manner, equipment shall be installed on a surface of noncombustible construction with noncombustible material and surface finish and with no combustible material against the underside thereof.

SECTION G2448 (627) ILLUMINATING APPLIANCES

G2448.1 (627.1) General. Illuminating appliances shall be tested in accordance with ANSI Z21.42 and shall be installed in accordance with the manufacturer's installation instructions.

G2448.2 (627.2) Mounting on buildings. Illuminating appliances designed for wall or ceiling mounting shall be securely attached to substantial structures in such a manner that they are not dependent on the gas piping for support.

G2448.3 (627.3) Mounting on posts. Illuminating appliances designed for post mounting shall be securely and rigidly attached to a post. Posts shall be rigidly mounted. The strength and rigidity of posts greater than 3 feet (914 mm) in height shall be at least equivalent to that of a 2.5-inch (64 mm) diameter post constructed of 0.064-inch (1.6 mm) thick steel or a 1-inch (25.4 mm) Schedule 40 steel pipe. Posts 3 feet (914 mm) or less in height shall not be smaller than a $^3/_4$-inch (19 mm) Schedule 40 steel pipe. Drain openings shall be provided near the base of posts where there is a possibility of water collecting inside them.

G2448.4 (627.4) Appliance pressure regulators. Where an appliance pressure regulator is not supplied with an illuminating appliance and the service line is not equipped with a service pressure regulator, an appliance pressure regulator shall be installed in the line to the illuminating appliance. For multiple installations, one regulator of adequate capacity shall be permitted to serve more than one illuminating appliance.

SECTION G2449 (629) INFRARED RADIANT HEATERS

G2449.1 (629.1) General. Infrared radiant heaters shall be tested in accordance with ANSI Z 83.6 and shall be installed in accordance with the manufacturer's installation instructions.

G2449.2 (629.2) Support. Infrared radiant heaters shall be safely and adequately fixed in an approved position independent of gas and electric supply lines. Hanger and brackets shall be of noncombustible material.

SECTION G2450 (630) BOILERS

G2450.1 (630.1) General. Boilers shall be listed in accordance with the requirements of ANSI Z21.13 or UL 795. When required by the code official, the boiler shall be designed and constructed in accordance with the requirements of ASME CSD-1 and as applicable, the ASME Boiler and Pressure Vessel Code Sections I, II, IV, V and IX; NFPA 8501; NFPA 8502; and NFPA 8504.

G2450.2 (630.2) Installation. In addition to the requirements of this code, the installation of boilers shall be in accordance with the manufacturer's instructions and this code. Operating instructions of a permanent type shall be attached to the boiler. Boilers shall have all controls set, adjusted and tested by the installer. A complete control diagram together with complete boiler operating instructions shall be furnished by the installer. The manufacturer's rating data and the nameplate shall be attached to the boiler.

G2450.3 (630.3) Clearance to combustible material. Clearances to combustible materials shall be in accordance with Section G2409.4.

SECTION G2451 (632) CHIMNEY DAMPER OPENING AREA

G2451.1 (632.1) Free opening area of chimeny dampers. Where an unlisted decorative appliance for installation in a vented fireplace is installed, the fireplace damper shall have a permanent free opening equal to or greater than specified in Table G2451.1.

**TABLE G2451.1 (632.1)
FREE OPENING AREA OF CHIMNEY DAMPER FOR VENTING FLUE GASES
FROM UNLISTED DECORATIVE APPLIANCES FOR INSTALLATION IN VENTED FIREPLACES**

CHIMNEY HEIGHT (feet)	MINIMUM PERMANENT FREE OPENING (square inches)[a]						
	8	13	20	29	39	51	64
	Appliance input rating (Btu per hour)						
6	7,800	14,000	23,200	34,000	46,400	62,400	80,000
8	8,400	15,200	25,200	37,000	50,400	68,000	86,000
10	9,000	16,800	27,600	40,400	55,800	74,400	96,400
15	9,800	18,200	30,200	44,600	62,400	84,000	108,800
20	10,600	20,200	32,600	50,400	68,400	94,000	122,200
30	11,200	21,600	36,600	55,200	76,800	105,800	138,600

For SI: 1 foot = 304.8 mm, 1 square inch = 645.16 mm^2, 1,000 Btu per hour = 0.293 kW.

a. The first six minimum permanent free openings (8 square inches to 51 square inches) correspond approximately to the cross-sectional areas of chimneys having diameters of 3 inches through 8 inches, respectively. The 64-square inch opening corresponds to the cross-sectional area of standard 8-inch by 8-inch chimney tile.

Part VII — Plumbing

CHAPTER 25
PLUMBING ADMINISTRATION

SECTION P2501 GENERAL

P2501.1 Scope. The provisions of this chapter shall establish the general administrative requirements applicable to plumbing systems and inspection requirements of this code.

P2501.2 Application. In addition to the general administration requirements of Chapter 1, the administrative provisions of this chapter shall also apply to the plumbing requirements of Chapters 25 through 32.

SECTION P2502 EXISTING PLUMBING SYSTEMS

P2502.1 Existing building sewers and drains. Existing building sewers and drains shall be used in connection with new systems when found by examination and/or test to conform to the requirements prescribed by this document.

P2502.2 Additions, alterations or repairs. Additions, alterations, renovations or repairs to any plumbing system shall conform to that required for a new plumbing system without requiring the existing plumbing system to comply with all the requirements of this code. Additions, alterations or repairs shall not cause an existing system to become unsafe, insanitary or overloaded.

Minor additions, alterations, renovations and repairs to existing plumbing systems shall be permitted in the same manner and arrangement as in the existing system, provided that such repairs or replacement are not hazardous and are approved.

SECTION P2503 INSPECTION AND TESTS

P2503.1 Inspection required. New plumbing work and parts of existing systems affected by new work or alterations shall be inspected by the building official to ensure compliance with the requirements of this code.

P2503.2 Concealment. A plumbing or drainage system, or part thereof, shall not be covered, concealed or put into use until it has been tested, inspected and approved by the building official.

P2503.3 Responsibility of permittee. Test equipment, materials and labor shall be furnished by the permittee.

P2503.4 Building sewer testing. The building sewer shall be tested by insertion of a test plug at the point of connection with the public sewer and filling the building sewer with water, testing with not less than a 10-foot (3048 mm) head of water and be able to maintain such pressure for 15 minutes.

P2503.5 DWV systems testing. Rough and finished plumbing installations shall be tested in accordance with Sections P2503.5.1 and P2503.5.2.

P2503.5.1 Rough plumbing. DWV systems shall be tested on completion of the rough piping installation by water or air with no evidence of leakage. Either test shall be applied to the drainage system in its entirety or in sections after rough piping has been installed, as follows:

1. Water test. Each section shall be filled with water to a point not less than 10 feet (3048 mm) above the highest fitting connection in that section, or to the highest point in the completed system. Water shall be held in the section under test for 15 minutes before inspection. The system shall prove leak free by visual inspection.
2. Air test. The portion under test shall be maintained at a gauge pressure of 5 psi (34 kPa) or 10 inches of mercury column (2488 Pa). This pressure shall be held without introduction of additional air for a period of 15 minutes.

P2503.5.2 Finished plumbing. After the plumbing fixtures have been set and their traps filled with water, their connections shall be tested and proved gas tight and/or water tight as follows:

1. Water tightness. Each fixture shall be filled and then drained. Traps and fixture connections shall be proven water tight by visual inspection.
2. Gas tightness. When required by the local administrative authority, a final test for gas tightness of the DWV system shall be made by the smoke or peppermint test as follows:
   - 2.1. Smoke test. Introduce a pungent, thick smoke into the system. When the smoke appears at vent terminals, such terminals shall be sealed and a pressure equivalent to a 1-inch water column (249 Pa) shall be applied and maintained for the period of inspection.
   - 2.2. Peppermint test. Introduce 2 ounces (59 mL) of oil of peppermint into the system. Add 10 quarts (9464 mL) of hot water and seal all vent terminals. The odor of peppermint shall not be detected at any trap or other point in the system.

P2503.6 Water-supply system testing. Upon completion of the water-supply system or a section thereof, the system, or portion completed, shall be tested and proved tight under a water pressure not less than the working pressure of the system; or, for piping systems other than plastic, by an air test of not less than 50 psi (344 kPa). The water used for tests shall be obtained from a potable water source.

P2503.7 Inspection and testing of backflow prevention devices. Inspection and testing of backflow prevention devices shall comply with Section P2503.7.1 and Section P2503.7.2.

P2503.7.1 Inspections. Inspections shall be made of all backflow prevention assemblies to determine whether they are operable.

P2503.7.2 Testing. Reduced pressure principle backflow preventers, double check valve assemblies, double-detector check valve assemblies and pressure vacuum breaker assemblies shall be tested at the time of installation, immediately after repairs or relocation and at least annually.

CHAPTER 26
GENERAL PLUMBING REQUIREMENTS

SECTION P2601
GENERAL

P2601.1 Scope. The provisions of this chapter shall govern the installation of plumbing not specifically covered in other chapters applicable to plumbing systems.

P2601.2 Connection. Plumbing fixtures, drains and appliances used to receive or discharge liquid wastes or sewage shall be connected to the drainage system of the building or premises in accordance with the requirements of this code. This section shall not be construed to prevent indirect waste systems.

SECTION P2602
INDIVIDUAL WATER SUPPLY AND SEWAGE DISPOSAL

P2602.1 General. The water-distribution and drainage system of any building or premises where plumbing fixtures are installed shall be connected to a public water-supply or sewer system, respectively, if available. When either a public water-supply or sewer system, or both, are not available, or connection thereto is not feasible, an individual water-supply or individual (private) sewage-disposal system, or both, shall be provided.

SECTION P2603
STRUCTURAL AND PIPING PROTECTION

P2603.1 General. In the process of installing or repairing any part of a plumbing and drainage installation, the finished floors, walls, ceilings, tile work or any other part of the building or premises that must be changed or replaced shall be left in a safe structural condition in accordance with the requirements of the building portion of this code.

P2603.2 Drilling and notching. Wood-framed structural members shall not be drilled, notched or altered in any manner except as provided in Sections R502.6, R602.5, R602.6, R802.7 and R802.7.1. Holes in cold-formed steel-framed load-bearing members shall only be permitted in accordance with Sections R506.2, R603.2 and R804.2. In accordance with the provisions of Sections R603.3.4 and R804.3.5 cutting and notching of flanges and lips of cold-formed steel-framed load-bearing members shall not be permitted.

P2603.2.1 Protection against physical damage. In concealed locations, where piping, other than cast-iron or galvanized steel, is installed through holes or notches in studs, joists, rafters or similar members less than 1.5 inches (38 mm) from the nearest edge of the member, the pipe shall be protected by shield plates. Protective shield plates shall be a minimum of 0.062-inch-thick (1.6 mm) steel, shall cover the area of the pipe where the member is notched or bored and shall extend a minimum of 2 inches (51 mm) above sole plates and below top plates.

P2603.3 Breakage and corrosion. Pipes passing through or under walls shall be protected from breakage. Pipes passing through concrete or cinder walls and floors, cold-formed steel framing or other corrosive material shall be protected against external corrosion by a protective sheathing or wrapping or other means that will withstand any reaction from lime and acid of concrete, cinder or other corrosive material. Sheathing or wrapping shall allow for expansion and contraction of piping to prevent any rubbing action. Minimum wall thickness of material shall be 0.025 inch (0.64 mm).

P2603.4 Sleeves. Annular spaces between sleeves and pipes shall be filled or tightly caulked as approved by the building official. Annular spaces between sleeves and pipes in fire-rated assemblies shall be filled or tightly caulked in accordance with the building portion of this code.

P2603.5 Pipes through footings or foundation walls. A soil or waste pipe, or building drain passing under a footing or through a foundation wall shall be provided with a relieving arch; or there shall be built into the masonry wall a pipe sleeve two pipe sizes greater than the pipe passing through.

P2603.6 Freezing. In localities having a winter design temperature of 32°F (0°C) or lower as shown in Table R301.2(1) of this code, a water, soil or waste pipe shall not be installed outside of a building, in exterior walls, in attics or crawl spaces, or in any other place subjected to freezing temperature unless adequate provision is made to protect it from freezing by insulation or heat or both. Water service pipe shall be installed not less than 12 inches (305 mm) deep or less than 6 inches (152 mm) below the frost line.

P2603.7 Depth. Piping installed deeper than and parallel to footings or bearing walls shall be located above the foot bearing plane. The footing bearing plane shall extend downward from the bottom of the footing at an angle of 45 degrees (0.79 rad) from the horizontal (see Figure P2603.7).

SECTION P2604
TRENCHING AND BACKFILLING

P2604.1 Trenching and bedding. Piping shall be installed in trenches so that the piping rests on solid and continuous bearing. When overexcavated, the trench shall be backfilled to the proper grade with compacted earth, sand, fine gravel or similar granular material. Piping may not be supported on rocks or blocks at any point. Rocky or unstable soil shall be overexcavated by two or more pipe diameters and brought to the proper grade with suitable compacted granular material.

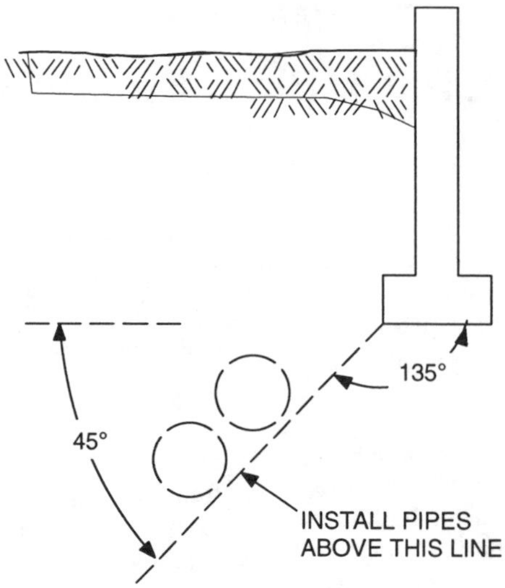

For SI: 1 degree = 0.018 rad.

FIGURE P2603.7
PIPE LOCATION WITH RESPECT TO FOOTINGS

P2604.2 Common trench. See Section P2904.4.1.

P2604.3 Backfilling. Care shall be exercised in backfilling trenches to avoid rocks, broken concrete, frozen chunks and other rubble until the pipe is covered by at least 12 inches (305 mm) of tamped earth. Backfill shall be placed evenly on both sides of the pipe and tamped to retain proper alignment. Loose earth shall be carefully placed in the trench in 6-inch (152 mm) layers and tamped in place.

P2604.4 Protection of footings. Trenching installed parallel to footings shall not extend below the 45-degree (0.79 rad) bearing plane of the bottom edge of a wall or footing.

P2604.5 Inspection. Excavations required for the installation of a building drainage system shall be open trench work and shall be kept open until the piping has been inspected, tested and approved.

SECTION P2605
SUPPORT

P2605.1 General. Support for piping shall be provided in accordance with the following:

1. Piping shall be supported so as to ensure alignment and prevent sagging, and allow movement associated with the expansion and contraction of the piping system.
2. Piping in the ground shall be laid on a firm bed for its entire length, except where support is otherwise provided.
3. Hangers and anchors shall be of sufficient strength to maintain their proportional share of the weight of pipe and contents and of sufficient width to prevent distortion to the pipe. Hangers and strapping shall be of approved material that will not promote galvanic action. Rigid support sway bracing shall be provided at changes in direction greater than 45 degrees (0.39 rad) for pipe sizes 4 inches (102 mm) and larger.
4. Piping shall be supported at distances not to exceed those indicated in Table P2605.1.

SECTION P2606
WATERPROOFING OF OPENINGS

P2606.1 General. Roof and exterior wall penetrations shall be made water tight. Joints at the roof, around vent pipes, shall be water tight by the use of lead, copper or galvanized iron flashings or an approved elastomeric material. Counterflashing shall not restrict the required internal cross-sectional area of any vent.

SECTION P2607
WORKMANSHIP

P2607.1 General. Valves, pipes and fittings shall be installed in correct relationship to the direction of the flow. Burred ends shall be reamed to the full bore of the pipe.

SECTION P2608
MATERIALS EVALUATION AND LISTING

P2608.1 General. Plastic pipes, plastic plumbing pipe components and related plastic materials shall be evaluated and listed as conforming to ANSI/NSF 14 by an approved agency. The approved agency shall maintain periodic evaluations of production of listed products or materials, and their listing shall state that the products or materials meet ANSI/NSF 14. All plastic pipe, fittings and components shall be identified with the mark of an approved agency as conforming to ANSI/NSF 14.

P2608.2 Water-supply systems. All water-service pipes, water-distribution pipes and the necessary connecting pipes, fittings, control valves, faucets and all appurtenances used to dispense water intended for human ingestion shall be evaluated and listed as conforming to the requirements of ANSI/NSF 61.

TABLE P2605.1
PIPING SUPPORT

PIPE MATERIAL	MAXIMUM HORIZONTAL SPACING	MAXIMUM VERTICAL SPACING
Cast-iron soil pipe	5′ except may be 10′ where 10′ lengths of pipe are installed.	Base and each story height but not to exceed 15′ 0″
Threaded-steel pipe	3/4″ diameter and under—10′ 0” 1″ diameter and over—12′ 0”	15′ 0″
Copper tube and copper pipe	1 1/4″ diameter and under—6′ 0” 1 1/2″ diameter and over—10′ 0”	Each story height but not to exceed 10′ 0″
Lead pipe	Continuous support	4′ 0″
Plastic parallel water distribution manifold bundles	4′0″	Each story height and piping shall have a midstory guide.
Plastic pipe (DWV)	4′ 0″	Each story height and piping shall have a midstory guide.
Plastic pipe and tube, hot- and cold-water, rigid	3′ 0″	Each story height and piping shall have a midstory guide.
Plastic pipe and tube, hot- and cold-water, flexible	32″	Each story height and piping shall have a midstory guide.

For SI: 1 inch = 25.4 mm, 1 foot = 304.8 mm.

CHAPTER 27
PLUMBING FIXTURES

SECTION P2701
FIXTURES, FAUCETS AND FIXTURE FITTINGS

P2701.1 General. Plumbing fixtures, faucets and fixture fittings shall be constructed of approved materials, shall have smooth impervious surfaces, shall be free from defects and concealed fouling surfaces, and except as permitted elsewhere in this code, shall conform to the standards specified in Table P2701.1. Plumbing fixtures shall be provided with an adequate supply of potable water to flush and keep the fixtures in a clean and sanitary condition without danger of backflow or cross connection.

SECTION P2702
FIXTURE ACCESSORIES

P2702.1 Plumbing fixtures. Plumbing fixtures, other than water closets, shall be provided with approved strainers.

P2702.2 Material for tail pieces and traps. Continuous wastes, waste and overflow fittings, traps and tail pieces constructed of seamless drawn brass shall be of No. 20 gage (0.8 mm) minimum thickness.

P2702.3 Plastic tubular fittings. Plastic tubular fittings shall conform to ASTM F 409 listed in Table P2701.1.

SECTION P2703
TAIL PIECES

P2703.1 Minimum size. Fixture tail pieces shall be not less than 1.5 inches (38 mm) in diameter for sinks, dishwashers, laundry tubs, bathtubs and similar fixtures, and not less than 1.25 inches (32 mm) in diameter for bidets lavatories and similar fixtures.

SECTION P2704
ACCESS TO CONNECTIONS

P2704.1 General. Slip jonts shall be made with an approved elastomeric gasket and shall only be installed in the trap inlet and within trap seal. Fixtures with concealed slip-joint connections shall be provided with an access panel or utility space at least 12 inches (305 mm) in its smallest dimension or other approved arrangement so as to provide access to the slip connections for inspection and repair.

SECTION P2705
INSTALLATION

P2705.1 General. The installation of fixtures shall conform to the following:

1. Floor-outlet or floor-mounted fixtures shall be secured to the drainage connection and to the floor, when so designed, by screws, bolts, washers, nuts and similar fasteners of copper, brass or other corrosion-resistant material.
2. Wall-hung fixtures shall be rigidly supported so that strain is not transmitted to the plumbing system.
3. Where fixtures come in contact with walls, the contact area shall be water tight.
4. Plumbing fixtures shall be functionally accessible.
5. The center line of water closets or bidets shall not be less than 15 inches (381 mm) from adjacent walls or partitions or not less than 30 inches (762 mm) center to center from an adjacent water closet or bidet. There shall be at least 21 inches (533 mm) clearance in front of the water closet, bidet or lavatory to any wall, fixture or door.
6. The location of piping, fixtures or equipment shall not interfere with the operation of windows or doors.

SECTION P2706
WASTE RECEPTORS

P2706.1 General. Every waste receptor shall be of an approved type. Plumbing fixtures or other receptors receiving the discharge of indirect waste pipes shall be shaped and have a capacity to prevent splashing or flooding and shall be readily accessible for inspection and cleaning. Waste receptors and standpipes shall be trapped and vented and shall connect to the building drainage system. A removable strainer or basket shall cover the waste outlet of waste receptors. Waste receptors shall be installed in ventilated spaces. Waste receptors shall not be installed in bathrooms or in any inaccessible or unventilated space such as a closet. Ready access shall be provided to waste receptors.

Exception: Open hub waste receptors shall be permitted in the form of a hub or pipe extending not less than 2 inches (51 mm) above a water-impervious floor, and are not required to have a strainer.

P2706.2 Standpipes. Standpipes shall extend a minimum of 18 inches (457 mm) and a maximum of 42 inches (1067 mm) above the trap. Access shall be provided to all standpipe traps and drains for rodding.

P2706.2.1 Laundry tray connection. A laundry tray waste line is permitted to connect into a standpipe for the automatic clothes washer drain. The standpipes shall not be less than 30 inches (762 mm) as measured from the crown weir. The outlet of the laundry tray shall be a maximum horizontal distance of 30 inches (762 mm) from the standpipe trap.

TABLE P2701.1
PLUMBING FIXTURES, FAUCETS AND FIXTURE FITTINGS

MATERIAL	STANDARD
Diverters for faucets with hose spray anti-syphon type, residential application	ASSE/ANSI 1025
Enameled cast-iron plumbing fixtures	ASME/ANSI A112.19.1
Floor drains	ASME/ANSI A112.21.1
Handheld showers	ASSE/ANSI 1014
Home laundry equipment	ASSE/ANSI 1007
Hose connection vacuum breaker	ASSE 1052
Hot water dispensers, household storage type, electrical	ASSE/ANSI 1023
Household dishwashing machines	ASSE/ANSI 1006
Household disposers	ASSE/ANSI 1008
Hydraulic performance for water closets and urinals	ASME/ANSI A112.19.6
Individual pressure balancing valves for individual fixture fittings	ASSE 1066
Individual shower control valves anti-scald	ASSE/ANSI 1016
Nonvitreous ceramic plumbing fixtures	ANSI A112.19.9
Plastic bathtub units	ANSI Z124.1
Plastic lavatories	ANSI Z124.3
Plastic shower receptors and shower stall	ANSI Z124.2
Plastic sinks	ANSI Z124.6
Plastic water closet bowls and tanks	ANSI Z124.4
Plumbing fixture fittings	ANSI/ASME A112.18.1
Porcelain enameled formed steel plumbing fixtures	ASME/ANSI A112.19.4M
Pressurized flushing devices for plumbing fixtures	ASSE/ANSI 1037
Specification for copper sheet and strip for building construction	ASTM B 370
Suction fittings for use in swimming pools, wading pools, spas, hot tubs and whirlpool bathtub appliances	ASME/ANSI A112.19.8M
Stainless steel plumbing fixtures (residential)	ASME/ANSI A112.19.3M
Thermoplastic accessible and replaceable plastic tube and tubular fittings	ASTM F 409
Trim for water closet bowls, tanks and urinals	ASME/ANSI A112.19.5
Vacuum breaker wall hydrant-frost resistant automatic draining type	ASSE 1019
Vitreous china plumbing fixtures	ASME/ANSI A112.19.2M
Water closet flush tank ball cocks	ASSE/ANSI 1002
Whirlpool bathtub appliances	ASME/ANSI A112.19.7M

P2706.3 Prohibited waste receptors. Plumbing fixtures that are used for domestic or culinary purposes shall not be used to receive the discharge of an indirect waste.

Exceptions:

1. A kitchen sink trap is acceptable for use as a receptor for a dishwasher.
2. A laundry tray is acceptable for use as a receptor for a clothes washing machine.

SECTION P2707 DIRECTIONAL FITTINGS

P2707.1 Directional fitting required. Approved directional-type branch fittings shall be installed in fixture tail pieces receiving the discharge from food waste disposal units or dishwashers.

SECTION P2708 SHOWERS

P2708.1 General. Shower compartments shall have at least 900 square inches (0.581 m^2) of floor area and be of sufficient size to inscribe a circle with a diameter not less than 30 inches (762 mm). Hinged shower doors shall open outward. The wall area above built-in tubs having installed shower heads and in-shower compartments shall be constructed as per Section R307.2. Such walls shall form a water-tight joint with each other and with either the tub, receptor or shower floor.

Exception: Fold-down seats shall be permitted in the shower, provided the required 900-square-inch (0.581 m^2) dimension is maintained when the seat is in the folded-up position.

P2708.2 Water-supply riser. The water supply riser from the shower valve to the shower head outlet shall be secured to the permanent structure.

P2708.3 Shower control valves. Showers and tub/shower combinations shall be equipped with control valves of the pressure balance, the thermostatic mixing or the combination pressure balance/thermostatic mixing valve types with high limit stops in accordance with ASSE/ANSI 1016. The high limit stops shall be set to limit water temperature to a maximum 120°F (49°C).

SECTION P2709
SHOWER RECEPTORS

P2709.1 Construction. Shower receptors shall have a finished curb threshold not less than 1 inch (25.4 mm) below the sides and back of the receptor. The curb shall be not less than 2 inches (51 mm) and not more than 9 inches (229 mm) in depth when measured from the top of the curb to the top of the drain. The finished floor shall slope uniformly toward the drain not less than one-fourth unit vertical in 12 units horizontal (2-percent slope) nor more than 0.5 inch (12.7 mm), and floor drains shall be flanged to provide a water-tight joint in the floor.

P2709.2 Lining required. The adjoining walls and floor framing, enclosing on-site built-up shower receptors shall be lined with sheet lead, copper or a plastic liner material that complies with ASTM D 4068. The lining material shall extend not less than 3 inches (76 mm) beyond or around the rough jambs and not less than 3 inches (76 mm) above finished thresholds. Hot-mopping shall be permitted in accordance with Section P2709.2.1.

P2709.2.1 Hot-mopping. Shower receptors lined by hot-mopping shall be built-up with not less than three layers of standard grade Type 15 asphalt-impregnated roofing felt. The bottom layer shall be fitted to the formed subbase and each succeeding layer thoroughly hot-mopped to that below. All corners shall be carefully fitted and shall be made strong and water tight by folding or lapping, and each corner shall be reinforced with suitable webbing hot-mopped in place. All folds, laps and reinforcing webbing shall extend at least 4 inches (102 mm) in all directions from the corner and all webbing shall be of approved type and mesh, producing a tensile strength of not less than 50 pounds per inch (893 kg/m) in either direction.

P2709.3 Installation. Lining materials shall be pitched one-fourth unit vertical in 12 units horizontal (2-percent slope) to weep holes in the subdrain by means of a smooth, solidly formed subbase, shall be properly recessed and fastened to approved backing so as not to occupy the space required for the wall covering, and shall not be nailed or perforated at any point less than 1 inch (25.4 mm) above the finished threshold.

P2709.3.1 Materials. Lead and copper linings shall be insulated from conducting substances other than the connecting drain by 15-pound (6.80 kg) asphalt felt or its equivalent. Sheet lead liners shall weigh not less than 4 pounds per square foot (19.5 kg/m^2). Sheet copper liners shall weigh not less than 12 ounces per square foot (3.82 kg/m^2). Joints in lead and copper pans or liners shall be burned or silver brazed, respectively. Joints in plastic liner materials shall be jointed per the manufacturer's recommendations.

2709.4 Receptor drains. An approved flanged drain shall be installed with shower subpans or linings. The flange shall be placed flush with the subbase and be equipped with a clamping ring or other device to make a water-tight connection between the lining and the drain. The flange shall have weep holes into the drain.

SECTION P2710
SHOWER WALLS

P2710.1 Finished. Shower walls shall be finished in accordance with Section R702.4.

SECTION P2711
LAVATORIES

P2711.1 Lavatory waste outlets. Lavatories shall have waste outlets not less than 1.25 inch (32 mm) in diameter. A strainer, pop-up stopper, crossbar or other device shall be provided to restrict the clear opening of the waste outlet.

SECTION P2712
WATER CLOSETS

P2712.1 Prohibited water closets. Water closets that have an invisible seal and unventilated space or walls that are not thoroughly washed at each discharge shall be prohibited. Water closets that permit backflow of the contents of the bowl into the flush tank shall be prohibited.

P2712.2 Flushing devices required. Water closets shall be provided with a flush tank, flushometer tank or flushometer valve designed and installed to supply water in sufficient quantity and flow to flush the contents of the fixture, to cleanse the fixture and refill the fixture trap in accordance with ASME/ANSI A112.19.2 and ASME/ANSI A112.19.6.

P2712.3 Water supply for flushing devices. An adequate quantity of water shall be provided to flush and clean the fixture served. The water supply to flushing devices equipped for manual flushing shall be controlled by a float valve or other automatic device designed to refill the tank after each discharge and to completely shut off the water flow to the tank when the tank is filled to operational capacity. Provision shall be made to automatically supply water to the fixture so as to refill the trap after each flushing.

P2712.4 Flush valves in flush tanks. Flush valve seats in tanks for flushing water closets shall be at least 1 inch (25.4 mm) above the flood-level rim of the bowl connected thereto, except an approved water closet and flush tank combination designed so that when the tank is flushed and the fixture is clogged or partially clogged, the flush valve will close tightly so that water will not spill continuously over the rim of the bowl or backflow from the bowl to the tank.

P2712.5 Overflows in flush tanks. Flush tanks shall be provided with overflows discharging to the water closet connected thereto and such overflow shall be of sufficient size to prevent flooding the tank at the maximum rate at which the tanks are supplied with water.

P2712.6 Access. All parts in a flush tank shall be accessible for repair and replacement.

P2712.7 Water closet seats. Water closets shall be equipped with seats of smooth, nonabsorbent material and shall be properly sized for the water closet bowl type.

P2712.8 Flush tank lining. Sheet copper used for flush tank linings shall have a minimum weight of 10 ounces per square foot (3.18 kg/m^2).

SECTION P2713 BATHTUBS

P2713.1 Bathtub waste outlets and overflows. Bathtubs shall have outlets and overflows at least 1.5 inches (38 mm) in diameter, and the waste outlet shall be equipped with an approved stopper.

SECTION P2714 SINKS

P2714.1 Sink waste outlets. Sinks shall be provided with waste outlets not less than 1.5 inches (38 mm) in diameter. A strainer, crossbar or other device shall be provided to restrict the clear opening of the waste outlet. Sinks on which a waste grinder is installed shall have a waste opening not less than 3.5 inches (89 mm) in diameter.

SECTION P2715 LAUNDRY TUBS

P2715.1 Laundry tub waste outlet. Each compartment of a laundry tub shall be provided with a waste outlet not less than 1.5 inches (38 mm) in diameter and a strainer or crossbar to restrict the clear opening of the waste outlet.

SECTION P2716 FOOD-WASTE GRINDER

P2716.1 Food-waste grinder waste outlets. Food waste grinders shall be connected to a drain of not less than 1.5 inches (38 mm) in diameter.

P2716.2 Water supply required. Food waste grinders shall be provided with an adequate supply of water at a sufficient flow rate to ensure proper functioning of the unit.

SECTION P2717 DISHWASHING MACHINES

P2717.1 Protection of water supply. The water supply for dishwashers shall be protected by an air gap or integral backflow preventor.

P2717.2 Sink and dishwasher. A sink and dishwasher are permitted to discharge through a single 1.5-inch (38 mm) trap. The discharge pipe from the dishwasher shall be increased to a minimum of 0.75 inch (19.1 mm) in diameter and shall be connected with a wye fitting to the sink tail piece. The dishwasher waste line shall rise and be securely fastened to the underside of the counter before connecting to the sink tail piece.

P2717.3 Sink, dishwasher and food grinder. The combined discharge from a sink, dishwasher, and waste grinder is permitted to discharge through a single 1.5 inch (38 mm) trap. The discharge pipe from the dishwasher shall be increased to a minimum of 0.75 inch (19.1 mm) in diameter and shall connect with a wye fitting between the discharge of the food-waste grinder and the trap inlet or to the head of the food grinder. The dishwasher waste line shall rise and be securely fastened to the underside of the counter before connecting to sink tail piece or food grinder.

SECTION P2718 CLOTHES WASHING MACHINE

P2718.1 Backflow protection. The discharge from a clothes washing machine shall be through an air break.

SECTION P2719 FLOOR DRAINS

P2719.1 Minimum size. Floor drains shall have waste outlets not less than 2 inches (51 mm) in diameter and shall be provided with a removable strainer with an open area of at least two-thirds of the cross-sectional area of the drain line to which it connects.

SECTION P2720 WHIRLPOOL BATHTUBS

P2720.1 Access panel. A door or panel of sufficient size shall be installed to provide access to the pump for repair and/or replacement.

P2720.2 Piping drainage. The circulation pump shall be accessibly located above the crown weir of the trap. The pump drain line shall be properly graded to ensure minimum water retention in the volute after fixture use. The circulation piping shall be installed to be self-draining.

P2720.3 Leak testing. Leak testing and pump operation shall be performed in accordance with the manufacturer's installation instructions.

P2720.4 Manufacturer's instructions. The product shall be installed in accordance with the manufacturer's installation instructions.

SECTION P2721
BIDET INSTALLATIONS

P2721.1 Water supply. The bidet shall be equipped with either an air-gap-type or vacuum-breaker-type fixture supply fitting.

SECTION P2722
FIXTURE FITTING

P2722.1 General. Fixture supply valves and faucets shall comply with ASME/ANSI A112.18.1 as listed in Table P2701.1. Faucets and fixture fittings that supply drinking water for human ingestion shall conform to the requirements of ANSI/NSF 61, Section 9.

P2722.2 Hot water. Fixture fittings, faucets and diverters shall be installed and adjusted so that the flow of hot water from the fittings corresponds to the left-hand side of the fitting.

Exception: Shower or tub/shower mixing valves conforming to ASSE/ANSI 1016, where the flow of hot water corresponds to the markings on the device.

SECTION P2723
MACERATING TOILET SYSTEMS

P2723.1 General. Macerating toilet systems shall be installed in accordance with manufacturer's installation instructions.

P2723.2 Drain. The minimum size of the drain from the macerating toilet system shall be 0.75 inch (19.1 mm) in diameter.

CHAPTER 28
WATER HEATERS

SECTION P2801
GENERAL

P2801.1 Required. Each dwelling shall have an approved automatic water heater or other type of domestic water-heating system sufficient to supply hot water to plumbing fixtures and appliances intended for bathing, washing or culinary purposes. Storage tanks shall be constructed of noncorrosive metal or shall be lined with noncorrosive material.

P2801.2 Installation. Water heaters shall be installed in accordance with this chapter and Chapters 20 and 24.

P2801.3 Location. Water heaters and storage tanks shall be located and connected so as to provide access for observation, maintenance, servicing and replacement.

P2801.4 Prohibited locations. Water heaters shall be located in accordance with Chapter 20.

Exception: Direct-vent water heaters.

P2801.5 Required pan. Where water heaters or hot water storage tanks are installed in locations where leakage of the tanks or connections will cause damage, the tank or water heater shall be installed in a galvanized steel pan having a minimum thickness of 24 gage (0.016 inch) (0.4 mm) or other pans listed for such use.

P2801.5.1 Pan size and drain. The pan shall be not less than 1.5 inches (38 mm) deep and shall be of sufficient size and shape to receive all dripping and condensate from the tank or water heater. The pan shall be drained by an indirect waste pipe having a minimum diameter of 1 inch (25.4 mm) or the outlet diameter of the relief valve, whichever is larger.

P2801.5.2 Pan drain termination. The pan drain shall extend full-size and terminate over a suitably located indirect waste receptor or shall extend to the exterior of the building and terminate not less than 6 inches (152 mm) and not more than 24 inches (610 mm) above the adjacent ground surface.

P2801.6 Water heaters installed in garages. Water heaters having an ignition source shall be elevated such that the source of ignition is not less than 18 inches (457 mm) above the garage floor.

SECTION P2802
WATER HEATERS USED FOR SPACE HEATING

P2802.1 Protection of potable water. Piping and components connected to a water heater for space heating applications shall be suitable for use with potable water in accordance with Chapter 29. Water heaters that will be used to supply potable water shall not be connected to a heating system or components previously used with nonpotable water heating appliances. Chemicals for boiler treatment shall not be introduced into the water heater.

P2802.2 Scald protection. Where a combination water heater-space heating system requires water for space heating at temperatures exceeding 140°F (60°C), a means such as a mixing valve shall be installed to temper the water for domestic uses.

SECTION P2803
RELIEF VALVES

P2803.1 Relief valves required. Appliances and equipment used for heating water or storing hot water shall be protected by:

1. A separate pressure-relief valve and a separate temperature-relief valve; or
2. A combination pressure- and temperature-relief valve.

P2803.2 Rating. Relief valves shall have a minimum rated capacity for the equipment served and shall conform to ANSI Z 21.22.

P2803.3 Pressure relief valves. Pressure-relief valves shall have a relief rating adequate to meet the pressure conditions for the appliances or equipment protected. In tanks, they shall be installed directly into a tank tapping or in a water line close to the tank. They shall be set to open at least 25 psi (172 kPa) above the system pressure but not over 150 psi (1034 kPa). The relief-valve setting shall not exceed the tank's rated working pressure.

P2803.4 Temperature relief valves. Temperature-relief valves shall have a relief rating compatible with the temperature conditions of the appliances or equipment protected. The valves shall be installed such that the temperature-sensing element monitors the water within the top 6 inches (152 mm) of the tank. The valve shall be set to open at a maximum temperature of 210°F (99°C).

P2803.5 Combination pressure/temperature relief valves. Combination pressure/temperature-relief valves shall comply with all the requirements of separate pressure- and temperature-relief valves.

P2803.6 Installation of relief valves. A check or shutoff valve shall not be installed in the following locations:

1. Between a relief valve and the termination point of the relief valve discharge pipe;
2. Between a relief valve and a tank; or
3. Between a relief valve and heating appliances or equipment.

P2803.6.1 Requirements of discharge pipe. The outlet of a pressure relief valve, temperature relief valve or combination thereof, shall not be directly connected to the drainage system. The discharge from the relief valve shall be piped full size separately to the outside of the building or to an indirect waste receptor located inside the building. In areas subject to freezing, the relief valve shall discharge through an air gap into an indirect waste receptor located within a heated space, or by other approved means. The discharge shall be installed in a manner that does not cause personal injury or property damage and that is readily observable by the building occupants. The discharge from a relief valve shall not be trapped. The diameter of the discharge piping shall not be less than the diameter of the relief valve outlet. The discharge pipe shall be installed so as to drain by gravity flow and shall terminate atmospherically not more than 6 inches (152 mm) above the floor. The end of the discharge pipe shall not be threaded.

P2803.6.2 Relief valve drains. Relief valve drains shall comply with Section P2904.5 or ASME/ANSI A112.4.1.

CHAPTER 29

WATER SUPPLY AND DISTRIBUTION

SECTION P2901 GENERAL

P2901.1 Potable water required. Dwelling units shall be provided with a supply of potable water in the amounts and pressures specified in this chapter. In a building where both a potable and nonpotable water-distribution system are installed, each system shall be identified by color marking, metal tag or other appropriate method. Any nonpotable outlet that could inadvertently be used for drinking or domestic purposes shall be posted.

SECTION P2902 PROTECTION OF POTABLE WATER SUPPLY

P2902.1 General. A potable water supply system shall be designed and installed in such a manner as to prevent contamination from nonpotable liquids, solids or gases being introduced into the potable water supply. Connections shall not be made to a potable water supply in a manner that could contaminate the water supply or provide a cross-connection between the supply and source of contamination unless an approved backflow-prevention device is provided. Cross-connections between an individual water supply and a potable public water supply shall be prohibited.

P2902.2 Backflow protection. A means of protection against backflow shall be provided in accordance with Sections P2902.2.1 through P2902.2.6.

P2902.2.1 Air gaps. Air gaps shall comply with ASME/ANSI A112.1.2 or the requirements of this section. The minimum air gap shall be measured vertically from the lowest end of a water supply outlet to the flood level rim of the fixture or receptor into which such potable water outlets discharge. The minimum required air gap shall be twice the diameter of the effective opening of the outlet, but in no case less than the values specified in Table P2902.2.1. An air gap is required at the discharge point of a relief valve or piping. Air gap devices shall be incorporated in dishwashing and clothes washing appliances.

P2902.2.2 Atmospheric vacuum breakers. Pipe applied atmospheric vacuum breakers shall conform to ASSE/ANSI 1001. Hose connection vacuum breakers shall conform to ASSE/ANSI 1011, or ASSE/ANSI 1019. These devices shall operate under normal atmospheric pressure. Except deck-mounted and integral vacuum breakers, the critical level of an atmospheric-type vacuum breaker shall be set a minimum of 6 inches (152 mm) above the flood level rim of the fixture, device or highest point of usage. The vacuum breaker shall be installed in accordance with the manufacturer's installation instructions.

P2902.2.3 Backflow preventer with intermediate atmospheric vent. Backflow preventers with intermediate atmospheric vents shall conform to ASSE/ANSI 1012. These devices shall be permitted to be installed where subject to continuous pressure conditions. The relief opening shall discharge by air gap and shall be prevented from being submerged.

P2902.2.4 Hose connection backflow preventer. Hose connection backflow preventers shall conform to ASSE/ANSI 1052. The backflow preventer shall be installed on the discharge side of a hose threaded outlet.

P2902.2.5 Pressure-type vacuum breakers. Pressure-type vacuum breakers shall conform to ASSE/ANSI 1020 for outdoor use and ASSE/ANSI 1056 for spill-proof indoor use. These devices shall be designed for installation under continuous pressure conditions when the critical level is installed at the required height.

TABLE P2902.2.1 MINIMUM AIR GAPS

FIXTURE	MINIMUM AIR GAP	
	Away from a wall[a] (inches)	Close to a wall (inches)
Lavatories and other fixtures with effective opening not greater than $^1/_2$ inch in diameter	1	1.5
Sink, laundry trays, gooseneck back faucets and other fixtures with effective openings not greater than $^3/_4$ inch in diameter	1.5	2.5
Over-rim bath fillers and other fixtures with effective openings not greater than 1 inch in diameter	2	3
Effective openings greater than 1 inch	Two times the diameter of the effective opening	Three times the diameter of the effective opening

For SI: 1 inch = 25.4 mm.

a. Applicable where walls or obstructions are spaced from the nearest inside edge of the spout opening a distance greater than three times the diameter of the effective opening for a single wall, or a distance greater than four times the diameter of the effective opening for two intersecting walls.

P2902.2.6 Reduced pressure principle backflow preventer. Reduced pressure principle backflow preventers shall conform to ASSE/ANSI 1013. These devices are permitted to be installed where subject to continuous pressure conditions. The relief opening shall discharge by air gap and shall be prevented from being submerged.

P2902.3 Protection of potable water outlets. All potable water openings and outlets shall be protected by an air gap, atmospheric-type vacuum breaker, pressure-type vacuum breaker or hose connection backflow preventer.

P2902.3.1 Flush tanks. Flush tanks shall be equipped with an antisiphon ball cock conforming to ASSE/ANSI 1002. The ball cock shall be installed a minimum of 1 inch (25.4 mm) above the full opening of the overflow pipe.

P2902.3.2 Deck-mounted and integral vacuum breakers. Approved deck-mounted vacuum breakers and faucets with integral atmospheric or spill-proof vacuum breakers shall be installed in accordance with the manufacturer's installation instructions with the critical level not less than 1 inch (25.4 mm) above the flood level rim.

P2902.3.3 Hose connection. Sillcocks, hose bibbs, wall hydrants and other openings with a hose connection shall be protected by an atmospheric-type or pressure-type vacuum breaker or a permanently attached hose connection vacuum breaker complying with ASSE/ANSI 1001, ASSE/ANSI 1011, ASSE/ANSI 1012, ASSE/ANSI 1019, ASSE/ANSI 1020, ASSE/ANSI 1052 or ASSE/ANSI 1056.

Exceptions:

1. This section shall not apply to water heater and boiler drain valves that are provided with hose connection threads and that are intended only for tank or vessel draining.
2. This section shall not apply to water supply valves intended for connection of clothes washing machines where backflow prevention is otherwise provided or is integral with the machine.

P2902.4 Protection of potable water connections. All connections to the potable water shall conform to Sections P2902.4.1 through P2902.4.5.

P2902.4.1 Connections to boilers. The potable supply to the boiler shall be equipped with a backflow preventer with an intermediate atmospheric vent complying with ASSE/ANSI 1012. When boilers have conditioning chemicals introduced into the system, the potable water connection shall be protected by a reduced pressure principle backflow preventer complying with ASSE/ANSI 1013.

P2902.4.2 Heat exchangers. Heat exchangers utilizing an essentially toxic transfer fluid shall be separated from the potable water by double-wall construction. An air gap open to the atmosphere shall be provided between the two walls. Heat exchangers utilizing an essentially nontoxic transfer fluid are permitted to be of single-wall construction.

Exception: Heat exchangers utilizing an essentially toxic transfer fluid with a pressure on the transfer fluid side a minimum of 10 psi (69 kPa) lower than the pressure on the potable water side protected with a pressure gradient monitor are permitted to be of single-wall construction.

P2902.4.3 Lawn irrigation systems. The potable water supply to lawn irrigation systems shall be protected against backflow by an atmospheric-type vacuum breaker, a pressure-type vacuum breaker or a reduced pressure principle backflow preventer. A valve shall not be installed downstream from an atmospheric vacuum breaker. All devices shall be installed in accordance with manufacturer's installation instructions.

P2902.4.4 Connections to automatic fire sprinkler systems. The potable water supply to automatic fire sprinkler systems shall be protected against backflow by a double check-valve assembly or a reduced pressure principle backflow preventer.

Exception: Where systems are installed as a portion of the water distribution system in accordance with the requirements of this code and are not provided with a fire department connection, isolation of the water supply system shall not be required.

P2902.4.4.1 Additives or nonpotable source. Where systems contain chemical additives or antifreeze, or where systems are connected to a nonpotable secondary water supply, the potable water supply shall be protected against backflow by a reduced pressure principle backflow preventer. Where chemical additives or antifreeze are added to only a portion of an automatic fire sprinkler or standpipe system, the reduced pressure principle backflow preventer shall be permitted to be located so as to isolate that portion of the system.

P2902.4.5 Solar systems. The potable water supply to a solar system shall be equipped with a backflow preventer with intermediate atmospheric vent complying with ASSE/ANSI 1012 or a reduced pressure principle backflow preventer complying with ASSE/ANSI 1013. Where chemicals are utilized, the potable water supply shall be protected by a reduced pressure principle backflow preventer.

P2902.5 Access. All backflow prevention devices shall be accessible for inspection and servicing.

SECTION P2903 WATER-SUPPLY SYSTEM

P2903.1 Water supply system design criteria. The water service and water distribution systems shall be designed and pipe sizes shall be selected such that under conditions of peak demand, the capacities at the point of outlet discharge shall not be less than shown in Table P2903.1.

P2903.2 Maximum flow and water consumption. The maximum water consumption flow rates and quantities for all plumbing fixtures and fixture fittings shall be in accordance with Table P2903.2.

TABLE P2903.1
REQUIRED CAPACITIES AT POINT OF OUTLET DISCHARGE

FIXTURE AT POINT OF OUTLET	FLOW RATE (gpm)	FLOW PRESSURE (psi)
Bathtub	4	8
Bidet	2	4
Dishwasher	2.75	8
Laundry tub	4	8
Lavatory	2	8
Shower	3	8
Shower, temperature controlled	3	20
Sillcock, hose bibb	5	8
Sink	2.5	8
Water closet, flushometer tank	1.6	15
Water closet, tank, close coupled	3	8
Water closet, tank, one-piece	6	20

For SI: 1 gallon per minute = 3.785 L/m, 1 pound per square inch = 6.895 kPa.

TABLE P2903.2
MAXIMUM FLOW RATES AND CONSUMPTION FOR PLUMBING FIXTURES AND FIXTURE FITTINGS[b]

PLUMBING FIXTURE OR FIXTURE FITTING	PLUMBING FIXTURE OR FIXTURE FITTING
Lavatory faucet	2.2 gpm at 60 psi
Shower head[a]	2.5 gpm at 80 psi
Sink faucet	2.2 gpm at 60 psi
Water closet	1.6 gallons per flushing cycle

For SI: 1 gallon per minute = 3.785 L/m, 1 pound per square inch = 6.895 kPa.

a. A handheld shower spray is also a shower head.

b. Consumption tolerances shall be determined from referenced standards.

P2903.3 Minimum pressure. Minimum static pressure (as determined by the local water authority) at the building entrance for either public or private water service shall be 40 psi (276 kPa).

P2903.3.1 Maximum pressure. Maximum static pressure shall be 80 psi (551 kPa). When main pressure exceeds 80 psi (551 kPa), an approved pressure-reducing valve conforming to ASSE/ANSI 1003 shall be installed on the domestic water branch main or riser at the connection to the water-service pipe.

P2903.4 Thermal expansion. In addition to the required pressure relief valve, an approved device for thermal expansion control shall be installed on any water supply system utilizing storage water heating equipment whenever the building supply pressure is greater than the required relief valve pressure setting or when any device, such as a pressure reducing valve, backflow preventer or check valve, is installed that prevents pressure relief through the building supply. The thermal expansion control device shall be sized in accordance with the manufacturer's installation instructions.

P2903.5 Water hammer. The flow velocity of the water distribution system shall be controlled to reduce the possibility of water hammer. A water-hammer arrestor shall be installed where quick-closing valves are utilized, unless otherwise approved. Water hammer arrestors shall be installed in accordance with manufacturer's installation instructions. Water-hammer arrestors shall conform to ASSE/ANSI 1010.

P2903.6 Determining water-supply fixture units. Supply load in the building water-distribution system shall be determined by total load on the pipe being sized, in terms of water-supply fixture units (w.s.f.u.), as shown in Table P2903.6. For fixtures not listed, choose a w.s.f.u. value of a fixture with similar flow characteristics.

P2903.7 Size of water-service mains, branch mains and risers. The minimum size water service pipe shall be $^3/_4$ inch (19.1 mm). The size of water service mains, branch mains and risers shall be determined according to water supply demand [gpm (L/m)], available water pressure [psi (kPa)] and friction loss due to the water meter and developed length of pipe [feet (m)], including equivalent length of fittings. The size of each water distribution system shall be determined according to the procedure outlined in this section or by other design methods conforming to acceptable engineering practice and approved by the administrative authority:

1. Obtain the minimum daily static service pressure [psi (kPa)] available (as determined by the local water authority) at the water meter or other source of supply at the installation location. Adjust this minimum daily static pressure [psi (kPa)] for the following conditions:
   1.1. Determine the difference in elevation between the source of supply and the highest water supply outlet. Where the highest water supply outlet is located above the source of supply, deduct 0.5 psi (3.4 kPa) for each foot (0.3 m) of difference in elevation. Where the highest water supply outlet is located below the source of supply, add 0.5 psi (3.4 kPa) for each foot (0.3 m) of difference in elevation.
   1.2. Where a water pressure reducing valve is installed in the water distribution system, the minimum daily static water pressure available is 80 percent of the minimum daily static water pressure at the source of supply or the set pressure downstream of the pressure reducing valve, whichever is smaller.
   1.3. Deduct all pressure losses due to special equipment such as a backflow preventer, water filter and water softener. Pressure loss data for each piece of equipment shall be obtained through the manufacturer of such devices.
   1.4. Deduct the pressure in excess of 8 psi (55 kPa) due to installation of the special plumbing fixture, such as temperature controlled shower and flushometer tank water closet.

   Using the resulting minimum available pressure, find the corresponding pressure range in Table P2903.7.
2. The maximum developed length for water piping is the actual length of pipe between the source of supply and the most remote fixture, including either hot (through the water heater) or cold water branches multiplied by a factor of 1.2 to compensate for pressure loss through fittings.

Select the appropriate column in Table P2903.7 equal to or greater than the calculated maximum developed length.

3. To determine the size of water service pipe, meter and main distribution pipe to the building using the appropriate table, follow down the selected "maximum developed length" column to a fixture unit equal to, or greater than the total installation demand calculated by using the "combined" water supply fixture unit column of Table P2903.6. Read the water service pipe and meter sizes in the first left-hand column and the main distribution pipe to the building in the second left-hand column on the same row.
4. To determine the size of each water distribution pipe, start at the most remote outlet on each branch (either hot or cold branch) and, working back toward the main distribution pipe to the building, add up the water supply fixture unit demand passing through each segment of the distribution system using the related hot or cold column of Table P2903.6. Knowing demand, the size of each segment shall be read from the second left-hand column of the same table and maximum developed length column selected in Steps 1 and 2, under the same or next smaller size meter row. In no case does the size of any branch or main need to be larger that the size of the main distribution pipe to the building established in Step 3.

P2903.8 Parallel water-distribution system manifolds. Hot and cold parallel water-distribution system manifolds with individual distribution lines to each fixture or fixture fitting shall be sized and installed in accordance with Sections P2903.8.1 through P2903.8.7

P2903.8.1 Sizing of manifolds. Manifold shall be sized in accordance with Table P2903.8.1. Total gallons per minute is the demand for all outlets.

P2903.8.2 Minimum size. The minimum size of individual distribution lines shall be $^3/_8$ inch (9.5 mm). Certain fixtures such as one-piece water closets and whirlpool bathtubs shall require a larger size where specified by the manufacturer. If a water heater is fed from the end of a cold water manifold, the manifold shall be one size larger than the water heater feed.

P2903.8.3 Maximum length. The maximum length of individual distribution lines shall be 60 feet (18 288 mm) nominal.

P2903.8.4 Orientation. Manifolds shall be permitted to be installed in a horizontal or vertical position.

P2903.8.5 Support and protection. Plastic piping bundles shall be secured in accordance with the manufacturer's installation instructions and supported in accordance with Section P2605. Bundles that have a change in direction equal to or greater than 45 degrees (0.79 rad) shall be protected from chaffing at the point of contact with framing members by sleeving or wrapping.

P2903.8.6 Valving. Fixture valves, when installed, shall be located either at the fixture or at the manifold. If valves are installed at the manifold, they shall be labeled indicating the fixture served.

P2903.8.7 Hose bibb bleed. A readily accessible air bleed shall be installed in hose bibb supplies at the manifold or at the hose bibb exit point.

TABLE P2903.6
WATER-SUPPLY FIXTURE-UNIT VALUES FOR VARIOUS PLUMBING FIXTURES AND FIXTURE GROUPS

TYPE OF FIXTURES OR GROUP OF FIXTURES	WATER-SUPPLY FIXTURE-UNIT VALUE (w.s.f.u.)		
	Hot	Cold	Combined
Bathtub (with/without overhead shower head)	1.0	1.0	1.4
Clothes washer	1.0	1.0	1.4
Dishwasher	1.4	—	1.4
Hose bibb (sillcock)[a]	—	2.5	2.5
Kitchen sink	1.0	1.0	1.4
Lavatory	0.5	0.5	0.7
Laundry tub	1.0	1.0	1.4
Shower stall	1.0	1.0	1.4
Water closet (tank type)	—	2.2	2.2
Full-bath group with bathtub (with/without shower head) or shower stall	1.5	2.7	3.6
Half-bath group (water closet and lavatory)	0.5	2.5	2.6
Kitchen group (dishwasher and sink with/without garbage grinder)	1.9	1.0	2.5
Laundry group (clothes washer standpipe and laundry tub)	1.8	1.8	2.5

For SI: 1 gallon per minute = 3.785 L/m.

a. The fixture unit value 2.5 assumes a flow demand of 2.5 gpm, such as for an individual lawn sprinkler device. If a hose bibb/sill cock will be required to furnish a greater flow rate, the equivalent fixture-unit value may be obtained from Table P2903.6 or Table P2903.7.

TABLE P2903.7
MINIMUM SIZE OF WATER METERS, MAINS AND DISTRIBUTION PIPING BASED ON WATER SUPPLY FIXTURE UNIT VALVES (w.s.f.u.)

Pressure Range—30 to 39 psi

METER AND SERVICE PIPE (inches)	DISTRIBUTION PIPE (inches)	MAXIMUM DEVELOPMENT LENGTH (feet)									
		40	60	80	100	150	200	250	300	400	500
3/4	1/2[a]	2.5	2	1.5	1.5	1	1	.5	.5	0	0
3/4	3/4	9.5	7.5	6	5.5	4	3.5	3	2.5	2	1.5
3/4	1	32	25	20	16.5	11	9	7.5	6.5	5.5	4.5
1	1	32	32	27	21	13.5	10	8	7	5.5	5
3/4	1 1/4	32	32	32	32	30	24	20	17	13	10.5
1	1 1/4	80	80	70	61	45	34	27	22	16	12
1 1/2	1 1/4	80	80	80	75	54	40	31	25	17.5	13
1	1 1/2	87	87	87	87	84	73	74	56	45	36
1 1/2	1 1/2	151	151	151	151	117	92	79	69	54	43

a. Minimum size for building supply is 3/4-inch pipe.

Pressure Range—40 to 49 psi

METER AND SERVICE PIPE (inches)	DISTRIBUTION PIPE (inches)	MAXIMUM DEVELOPMENT LENGTH (feet)									
		40	60	80	100	150	200	250	300	400	500
3/4	1/2[a]	3	2.5	2	1.5	1.5	1	1	.5	.5	.5
3/4	3/4	9.5	9.5	8.5	7	5.5	4.5	3.5	3	2.5	2
3/4	1	32	32	32	26	18	13.5	10.5	9	7.5	6
1	1	32	32	32	32	21	15	11.5	9.5	7.5	6.5
3/4	1 1/4	32	32	32	32	32	32	32	27	21	16.5
1	1 1/4	80	80	80	80	65	52	42	35	26	20
1 1/2	1 1/4	80	80	80	80	75	59	48	39	28	21
1	1 1/2	87	87	87	87	87	87	87	78	65	55
1 1/2	1 1/2	151	151	151	151	151	130	109	93	75	63

a. Minimum size for building supply is 3/4-inch pipe.

Pressure Range—50 to 60 psi

METER AND SERVICE PIPE (inches)	DISTRIBUTION PIPE (inches)	MAXIMUM DEVELOPMENT LENGTH (feet)									
		40	60	80	100	150	200	250	300	400	500
3/4	1/2[a]	3	3	2.5	2	1.5	1	1	1	.5	.5
3/4	3/4	9.5	9.5	9.5	8.5	6.5	5	4.5	4	3	2.5
3/4	1	32	32	32	32	25	18.5	14.5	12	9.5	8
1	1	32	32	32	32	30	22	16.5	13	10	8
3/4	1 1/4	32	32	32	32	32	32	32	32	29	24
1	1 1/4	80	80	80	80	80	68	57	48	35	28
1 1/2	1 1/4	80	80	80	80	80	75	63	53	39	29
1	1 1/2	87	87	87	87	87	87	87	87	82	70
1 1/2	1 1/2	151	151	151	151	151	151	139	120	94	79

a. Minimum size for building supply is 3/4-inch pipe.

(continued)

TABLE P2903.7—continued
MINIMUM SIZE OF WATER METERS, MAINS AND DISTRIBUTION PIPING BASED ON WATER SUPPLY FIXTURE UNIT VALVES (w.s.f.u.)

Pressure Range—over 60 psi

METER AND SERVICE PIPE (inches)	DISTRIBUTION PIPE (inches)	MAXIMUM DEVELOPMENT LENGTH (feet)									
		40	60	80	100	150	200	250	300	400	500
3/4	1/2[a]	3	3	3	2.5	2	1.5	1.5	1	1	.5
3/4	3/4	9.5	9.5	9.5	9.5	7.5	6	5	4.5	3.5	3
3/4	1	32	32	32	32	32	24	19.5	15.5	11.5	9.5
1	1	32	32	32	32	32	28	22	17	12	9.5
3/4	1 1/4	32	32	32	32	32	32	32	32	32	30
1	1 1/4	80	80	80	80	80	80	69	60	46	36
1 1/2	1 1/4	80	80	80	80	80	80	76	65	50	38
1	1 1/2	87	87	87	87	87	87	87	87	87	84
1 1/2	1 1/2	151	151	151	151	151	151	151	144	114	94

For SI: 1 inch = 25.4 mm, 1 foot = 304.8 mm.

a. Minimum size for building supply is 3/4-inch pipe.

TABLE P2903.8.1
MANIFOLD SIZING

PLASTIC		METALLIC	
Nominal Size ID (inches)	Maximum[a] gpm	Nominal Size ID (inches)	Maximum[a] gpm
3/4	17	3/4	11
1	29	1	20
1 1/4	46	1 1/4	31
1 1/2	66	1 1/2	44

For SI: 1 inch = 25.4 mm, 1 gallon per minute = 3.785 L/m, 1 foot per second = 0.3048 m/s.

NOTE: See Table P2903.7 for w.s.f.u.

a. Based on velocity limitation: plastic—12 fps; metal—8 fps.

P2903.9 Valves. Valves shall be installed in accordance with Sections P2903.9.1 through P2903.9.3.

P2903.9.1 Service valve. Each dwelling unit shall be provided with an accessible main shutoff valve near the entrance of the water service. The valve shall be of a full-open type having nominal restriction to flow, with provision for drainage such as a bleed orifice or installation of a separate drain valve. Additionally, the water service shall be valved at the curb or property line in accordance with local requirements.

P2903.9.2 Water heater valve. A readily accessible full-open valve shall be installed in the cold-water supply pipe to each water heater at or near the water heater.

P2903.9.3 Valve requirements. Valves serving individual fixtures, appliances, risers and branches shall be provided with access. An individual shutoff valve shall be required on the water supply pipe to each water closet.

P2903.10 Hose bibb. Hose bibbs subject to freezing, including the "frost-proof" type, shall be equipped with an accessible stop-and-waste-type valve inside the building so that they may be controlled and/or drained during cold periods.

Exception: Frost-proof hose bibbs installed such that the stem extends through the building insulation into an open heated or semi-conditioned space need not be separately valved. (See Figure P2903.10.)

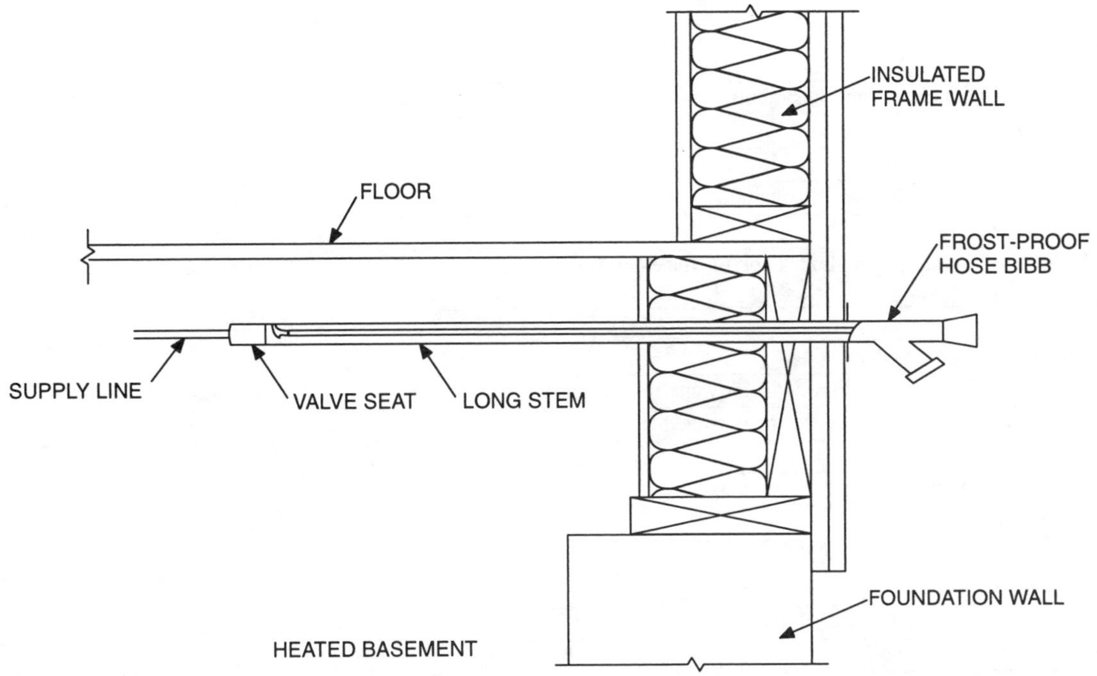

FIGURE P2903.10
TYPICAL FROST-PROOF HOSE BIBB INSTALLATION NOT REQUIRING SEPARATE VALUE

SECTION P2904
MATERIALS, JOINTS AND CONNECTIONS

P2904.1 Soil and groundwater. The installation of water service piping, fittings, valves, appurtenances and gaskets shall be prohibited in soil and groundwater that is contaminated with solvents, fuels, organic compounds or other detrimental materials that cause permeation, corrosion, degradation or structural failure of the water service material.

P2904.1.1 Investigation required. Where detrimental conditions are suspected by or brought to the attention of the building official, a chemical analysis of the soil and groundwater conditions shall be required to ascertain the acceptability of the water service material for the specific installation.

P2904.1.2 Detrimental condition. When a detrimental condition exists, approved alternate materials or alternate routing shall be required.

P2904.2 Lead content. Pipe and fittings utilized in the water-supply system shall have a maximum of 8 percent lead.

P2904.3 Polyethylene plastic piping installation. Polyethylene pipe shall be cut square, using a cutter designed for plastic pipe. Except when joined by heat fusion, pipe ends shall be chamfered to remove sharp edges. Pipe that has been kinked shall not be installed. For bends, the installed radius of pipe curvature shall be greater than 30 pipe diameters or the coil radius when bending with the coil. Coiled pipe shall not be bent beyond straight. Bends shall not be permitted within ten pipe diameters of any fitting or valve. Stiffener inserts used with compression-type fittings shall not extend beyond the clamp or nut of the fitting. Flared joints shall be permitted where recommended by the manufacturer and made by the use of a tool designed for that operation.

P2904.4 Water service pipe. Water service pipe shall conform to one of the standards listed in Table P2904.4.1. Water service pipe or tubing, installed underground and outside of the structure, shall have a minimum working pressure rating of 160 psi at 73°F (1100 kPa at 23°C).

P2904.4.1 Water service installation. Trenching, pipe installation and backfilling shall be in accordance with Section P2604. Water-service pipe is permitted to be located in the same trench with a building sewer provided such sewer is constructed of materials listed for underground use within a building in Section P3002.1. If the building sewer is not constructed of materials listed in Section P3002.1, the water-service pipe shall be separated from the building sewer by a minimum of 5 feet (1524 mm), measured horizontally, of undisturbed or compacted earth or placed on a solid ledge at least 12 inches (305 mm) above and to one side of the highest point in the sewer line.

TABLE P2904.4.1[a]
WATER SERVICE, SUPPLY AND DISTRIBUTION PIPING

MATERIAL	STANDARD
ABS plastic pipe (SDR-PR)	ASTM D 2282
ABS plastic pipe schedule 40 and 80	ASTM D 1527
CPVC plastic hot and cold water distribution systems	ASTM D 2846
CPVC plastic pipe schedule 40 and 80	ASTM F 441
CPVC plastic pipe schedule (SDR-PR)	ASTM F 442
Crosslinked P.E. plastic hot and cold water distribution	ASTM F 877
Crosslinked polyethylene (PEX) tubing	ASTM F 876
Crosslinked polyethylene/aluminum/crosslinked polybutylene (PEX-AL-PEX) for water service and water distribution piping	ASTM F 1281
Ductile iron pressure pipe	ASTM A 377
P.B. plastic hot water distribution systems	ASTM D 3309
Polybutylene/aluminum/polyethylene (PE-AL-PE) for water service piping	ASTM F 1282
Polybutylene (PB) plastic pipe (SDR-PR) based on outside diameter	ASTM D 3000
Polybutylene (PB) plastic tubing	ASTM D 2666
Polybutylene (PE) plastic tubing	ASTM D 2737
Polyethylene (PE) plastic pipe (SDR-PR)	ASTM D 3000
Polyethylene (PE) plastic pipe controlled OD	ASTM D 2447
Polyethylene plastic pipe, schedule 40	ASTM D 2104
Polyethylene plastic pipe (SDR-PR) controlled ID	ASTM D 2239
PVC plastic pipe schedule 40, 80 and 120	ASTM D 1785
PVC pressure rated pipe (SDR Series)	ASTM D 2241
Seamless brass type	ASTM B 135
Seamless copper tube	ASTM B 75
Seamless copper watertube Type K, L and M	ASTM B 88
Seamless red brass pipe, standard sizes	ASTM B 43
Specification for polybutylene (PB) plastic pipe (SIDR-PR) based on controlled inside diameter	ASTM D 2662
Welded copper water tube (WK, WL, WM)	ASTM B 447

a. Please note that Table P2904.4.1 includes standards for materials for use with cold water only.

P2904.5 Water-distribution pipe. Water-distribution piping within dwelling units shall conform to one of the standards in Table P2904.4.1. ABS, PVC, PE-AL-PE or PE plastic pipe shall not be used for water distribution piping. CPVC, Crosslinked PEX, PEX-AL-PEX or PB plastic pipe is permitted to be used for both hot and cold water. All hot-water-distribution pipe and tubing shall have a minimum pressure rating of 100 psi at 180°F (689 kPa at 82°C). Water distribution pipe shall conform to NSF 61.

P2904.5.1 Under concrete slabs. Inaccessible water distribution piping under slabs shall be copper water tube minimum Type M, brass, ductile iron pressure pipe, galvanized steel pipe, chlorinated polyvinyl chloride (CPVC) or polybutylene (PB) or crosslinked polyethylene (PEX) plastic pipe or tubing—all to be installed with approved fittings or bends. The minimum pressure rating for plastic pipe or tubing installed under slabs shall be 100 psi at 180°F (689 kPa at 82°C).

P2904.6 Fittings. Pipe fittings shall be approved for installation with the piping material installed, and shall conform to the respective pipe standards listed in Table P2904.4.1 or one of the standards listed in Table P2904.6.

TABLE P2904.6
PIPE FITTINGS

MATERIAL	STANDARD
ABS plastic fittings, schedule 40	ASTM D 2468
Butt heat fusion P.E. fittings for plastic pipe and fittings	ASTM D 3261
Cast copper alloy solder-joint pressure fittings	ASME B 16.18
Metal insert fittings for P.B. tubing	ASTM F 1380
Metal insert fittings utilizing a copper crimp ring SDR9 (PEX) tubing	ASTM F 1807
Plastic insert fittings for P.B. tubing	ASTM F 845
Plastic insert fittings for P.E. plastic pipe	ASTM D 2609
PVC plastic pipe fittings, schedule 40	ASTM D 2466
Socket bell for PVC plastic pipe	ASTM D 2672
Socket Type CPVC plastic pipe fittings, schedule 40	ASTM F 438
Socket Type CPVC plastic pipe fittings, schedule 80	ASTM F 439
Socket Type PVC plastic pipe fittings sch. 80	ASTM D 2467

P2904.7 Joint and connection tightness. Joints and connections in the plumbing system shall be gas tight and water tight for the intended use or required test pressure.

P2904.8 Plastic pipe joints. Joints in plastic piping shall be made with approved fittings by solvent cementing, heat fusion, corrosion-resistant metal clamps with insert fittings or compression connections. Flared joints for polyethylene pipe are permitted in accordance with Section P2904.3.

P2904.8.1 Solvent cementing. Solvent cemented joints shall comply with Sections P2904.8.1.1 through P2904.8.1.3.

P2904.8.1.1 ABS plastic pipe. Solvent cement for ABS plastic pipe conforming to ASTM D 2235 shall be applied to all joint surfaces.

P2904.8.1.2 CPVC plastic pipe. Solvent cement for CPVC plastic pipe conforming to ASTM F 493 shall be applied to all joint surfaces. The joint shall be made while the cement is wet, and in accordance with ASTM D 2846 or ASTM F 493. Solvent-cement joints shall be permitted above or below ground.

P2904.8.1.3 PVC plastic pipe. A primer complying with ASTM F 656 shall be applied to all PVC solvent cemented joints. Solvent cement for PVC plastic pipe conforming to ASTM D 2564 shall be applied to all joint surfaces.

P2904.8.1.4 Cross-linked polyethylene plastic (PEX). Joints between cross-linked polyethylene plastic tubing or fittings shall comply with Section P2904.8.1.4.1 or Section P2904.8.1.4.2.

P2904.8.1.4.1 Flared joints. Flared pipe ends shall be made by a tool designed for that operation.

P2904.8.1.4.2 Mechanical joints. Mechanical joints shall be installed in accordance with the manufacturer's instructions. Metallic lock rings and insert fittings complying with ASTM F 1807 shall be installed in accordance with the manufacturer's instructions.

P2904.9 Threaded pipe joints. Threaded joints shall conform to American National Taper Pipe Thread specifications. Pipe ends shall be deburred and chips removed. Pipe joint compound shall be used only on male threads.

P2904.10 Soldered joints. Soldered joints in tubing shall be made with fittings approved for water piping and shall conform to ASTM B 828. Surfaces to be soldered shall be cleaned bright. The joints shall be properly fluxed and made with approved solder. Solders and fluxes used in potable water-supply systems shall have a maximum of 0.2 percent lead. Fluxes shall conform to ASTM B 813.

P2904.11 Flared joints. Flared joints in water tubing shall be made with approved fittings. The tubing shall be reamed and then expanded with a flaring tool.

P2904.12 Underground joints. Joints in polybutylene (PB) plastic pipe or tubing underground or under a concrete floor slab shall be installed using heat fusion, in accordance with the manufacturer's installation instructions. Joints in copper pipe or tube installed in a concrete floor slab or under a concrete floor slab on grade shall be installed using wrought-copper fittings and brazed joints.

P2904.13 Above-ground joints. Joints within the building between copper pipe, polybutylene tubing or CPVC tubing, in any combination with compatible outside diameters, are permitted to be made with the use of approved push-in mechanical fittings of a pressure-lock design.

P2904.14 Joints between different materials. Joints between different piping materials shall be made in accordance with Sections P2904.14.1 and P2904.14.2 or with a mechanical joint of the compression or mechanical sealing type having an elastomeric seal conforming to ASTM D 1869 or ASTM F 477. Joints shall be installed in accordance with the manufacturer's instructions.

P2904.14.1 Copper or copper-alloy tubing to galvanized steel pipe. Joints between copper or copper-alloy tubing and galvanized steel pipe shall be made with a brass converter fitting or dielectric fitting. The copper tubing shall be joined to the fitting in an approved manner, and the fitting shall be screwed to the threaded pipe.

P2904.14.2 Plastic pipe or tubing to other piping material. Joints between different grades of plastic pipe or between plastic pipe and other piping material shall be made with an approved adapter fitting. Joints between plastic pipe and cast-iron hub pipe shall be made by a caulked joint or a mechanical compression joint.

SECTION P2905
CHANGES IN DIRECTION

P2905.1 Bends. Changes in direction in copper tube are permitted to be made with bends having a radius of not less than four diameters of the tube, providing such bends are made by use of forming equipment that does not deform or create loss in cross-sectional area of the tube.

SECTION P2906
SUPPORT

P2906.1 General. Pipe and tubing support shall conform to Section P2605.

SECTION P2907
DRINKING WATER TREATMENT UNITS

P2907.1 Design. Drinking water treatment units shall meet the requirements of either NSF 42, NSF 44 or NSF 53.

P2907.2 Reverse osmosis drinking water treatment units. Point-of-use reverse osmosis drinking water treatment units, designed for residential use, shall meet requirements of NSF 58. Waste or discharge from reverse osmosis drinking water treatment units shall enter the drainage system through an air gap or an air gap device that meets the requirements of NSF 58.

P2907.3 Connection tubing. The tubing to and from drinking water treatment units shall be of a size and material as recommended by the manufacturer. The tubing shall comply with NSF 14, NSF 42, NSF 44, NSF 53, NSF 58 or NSF 61.

CHAPTER 30
SANITARY DRAINAGE

SECTION P3001 GENERAL

P3001.1 Scope. The provisions of this chapter shall govern the materials, design, construction and installation of sanitary drainage systems. Plumbing materials shall conform to the requirements of this chapter. The drainage waste and vent (DWV) system shall consist of all piping for conveying wastes from plumbing fixtures, appliances and appurtenances, including fixture traps; above-grade drainage piping; below-grade drains within the building (building drain); below- and above-grade venting systems; and piping to the public sewer or private septic system.

P3001.2 Protection from freezing. No portion of the above-grade DWV system other than vent terminals shall be located outside of a building, in attics or crawl spaces, concealed in outside walls, or in any other place subjected to freezing temperatures unless adequate provision is made to protect them from freezing by insulation or heat or both, except in localities having a winter design temperature above 32°F (0°C) (ASHRAE 97.5 percent column, winter, see Chapter 3).

SECTION P3002 MATERIALS

P3002.1 Piping within buildings. Drain, waste and vent (DWV) piping in buildings shall be as shown in Table P3002.1 except that galvanized wrought-iron or galvanized steel pipe shall not be used underground and shall be maintained not less than 6 inches (152 mm) above ground. Allowance shall be made for the thermal expansion and contraction on plastic piping.

P3002.2 Building sewer. Building sewer piping shall be as shown in Table P3002.2.

P3002.3 Fittings. Fittings shall be approved and compatible with the type of piping being used and shall be of a sanitary or DWV design for drainage and venting. Water-pipe fittings shall be permitted for use in engineer designed systems where the design indicates compliance with Section P3101.2.1.

P3002.3.1 Drainage. Drainage fittings shall have a smooth interior waterway of the same diameter as the piping served. All fittings shall conform to the type of pipe used. Drainage fittings shall have no ledges, shoulders or reductions which can retard or obstruct drainage flow in the piping. Threaded drainage pipe fittings shall be of the recessed drainage type, black or galvanized. Drainage fittings shall be designed to maintain one-fourth unit vertical in 12 units horizontal (2-percent slope) grade.

P3002.4 Other materials. Sheet lead, lead bends, lead traps and sheet copper shall comply with Sections P3002.4.1 through P3002.4.3.

P3002.4.1 Sheet lead. Sheet lead for the following uses shall weigh not less than indicated below:

1. Flashing of vent terminals, 3 psf (0.144 kN/m^2).
2. Prefabricated flashing for vent pipes, $2^1/_2$ psf (0.120 kN/m^2).

P3002.4.2 Lead bends and traps. Lead bends and lead traps shall not be less than 0.125-inch (3.2 mm) wall thickness.

P3002.4.3 Sheet copper. Sheet copper for the following uses shall weigh not less than indicated below:

1. General use, 12 ounces per square feet (3.82 L/m^2).
2. Flashing for vent pipes, 8 ounces per square feet (2.55 L/m^2).

SECTION P3003 JOINTS AND CONNECTIONS

P3003.1 Tightness. Joints and connections in the DWV system shall be gas tight and water tight for the intended use or pressure required by test.

P3003.2 Prohibited joints. Running threads and bands shall not be used in the drainage system. Drainage and vent piping shall not be drilled, tapped, burned or welded.

The following types of joints and connections shall be prohibited:

1. Cement or concrete.
2. Mastic or hot-pour bituminous joints.
3. Joints made with fittings not approved for the specific installation.
4. Joints between different diameter pipes made with elastomeric rolling O-rings.
5. Solvent-cement joints between different types of plastic pipe.
6. Saddle-type fittings.

P3003.3 Joint requirements, similar piping materials. Joints between similar piping materials shall be made in accordance with Sections P3003.3.1 through P3003.3.7.

P3003.3.1 Cast-iron pipe, caulked joints. Lead-caulked joints for cast-iron hub and spigot soil pipe shall be firmly packed with oakum and filled with molten lead not less than 1 inch (25.4 mm) deep and shall not extend more than 0.125 inch (3.2 mm) below the rim of the hub. Lead shall be poured in one operation and shall be caulked tight.

P3003.3.2 Cast-iron pipe, mechanical joints. Mechanical joints used with cast-iron pipe shall comply with Section P3003.3.2.1 or P3003.3.2.2.

P3003.3.2.1 Hubless pipe. Joints for hubless cast-iron soil pipe shall be made with an approved elastomeric sealing sleeve and stainless steel-retaining sleeve.

TABLE P3002.1
DRAIN, WASTE AND VENT PIPING AND FITTING MATERIALS

MATERIAL	STANDARD
Welded and seamless steel pipe (black or galvanized)	ASTM A 53
Cast-iron soil pipe and fittings (hub and spigot)	ASTM A 74
Cast-iron fittings (threaded)	ASTM A 126
Malleable iron fittings (threaded)	ASTM A 197
Seamless copper pipe, standard sizes	ASTM B 42
Seamless red brass pipe, standard sizes	ASTM B 43
Seamless copper tube	ASTM B 75
Seamless copper water tube Type K, L and M	ASTM B 88
ABS-DWV pipe and fittings	ASTM D 2661
PVC-DWV pipe and fittings	ASTM D 2665
3.25-inch O.D. PVC-DWV pipe and fittings	ASTM D 2949
ABS-DWV sch.40 pipe with cellular core	ASTM F 628
Co-extruded PVC plastic pipe with cellular core	ASTM F 891
Co-extended composite ABS-DWV pipe and fittings	ASTM F 1488
Cast-iron soil pipe and fittings (hubless)	CISPI 301
Copper drainage tube (DWV)	ASTM B 306
Mechanical couplings for drain, waste and vent pipe and sewer pipe	CSA B602M
Solvent cement for ABS-DWV pipe and fittings	ASTM D 2235
Solvent cement for PVC-DWV pipe and fittings	ASTM D 2564
Socket bell for PVC-DWV pipe and fittings	ASTM D 2672
Primers for solvent cemented PVC-DWV pipe and fittings	ASTM F 656
Hubless cast-iron soil pipe and fittings	ASTM A 888
Couplings for use in connection with hubless cast-iron soil pipe and fittings	CISPI 310-95
Shielded couplings joining hubless cast-iron soil pipe and fittings	ASTM C 1277
Co-extruded composite ABS DWV schedule 40 IPS pipe (solid or cellular core) solvent cement fittings	ASTM F 1488; ASTM D 2235; ASTM D 2661; ASTM F 628
Co-extruded composite PVC DWV schedule 40 IPS pipe (solid or cellular core) solvent cement fittings	ASTM F 1488; ASTM D 2564; ASTM D 2665; ASTM F 891
Co-extruded composite PVC DWV IPS pipe-DR-PS140 PS200 solvent cement fittings	ASTM F 1488; ASTM D 2564; ASTM D 2665; ASTM F 891

For SI: 1 inch = 25.4 mm.

TABLE P3002.2
BUILDING SEWER PIPING

MATERIAL	STANDARD
Cast-iron pipe and fittings	ASTM A 74
Cast-iron soil pipe and fittings for hubless sanitary system	CISPI 301
Seamless copper tube	ASTM B 75
Copper water tube	ASTM B 88
Concrete sewer, storm drain and culvert pipe	ASTM C 14
Compression joints for vitrified clay pipe and fittings	ASTM C 425
Vitrified clay pipe and fittings	ASTM C 700
Bitumenized fiber drain and sewer pipe	ASTM D 1861
ABS-DWV pipe and fittings	ASTM D 2661
PVC/DWV pipe and fittings	ASTM D 2665
ABS Sewer pipe and fittings	ASTM D 2751
3.25-inch O.D. PVC/DWV pipe and fittings	ASTM D 2949
Type PSM/PVC sewer pipe and fittings	ASTM D 3034
Joints for drain and sewer plastic pipe using flexible elastomeric seals	ASTM D 3212
ABS Schedule 40 DWV pipe with cellular core	ASTM F 628
Co-extruded PVC schedule 40, PS 50 or PS 100 plastic pipe with cellular core	ASTM F 891
Copper drainage tube (DWV)	ASTM B 306
Mechanical couplings for drain waste and vent pipe and sewer pipe	CSA B602M
Solvent cement for ABS-DWV pipe and fittings	ASTM D 2235
Solvent cement for PVC-DWV pipe and fittings	ASTM D 2564
Socket bell for PVC-DWV pipe and fittings	ASTM D 2672
Primers for solvent cemented PVC-DWV pipe and fittings	ASTM F 656
Couplings for hubless cast-iron soil pipe and fittings	CISPI 310
Shielded couplings joining cast-iron soil pipe and fittings	ASTM C 1277
Cast-iron soil pipe and fittings for hubless sanitary system	ASTM A 888
Co-extruded composite ABS DWV schedule 40 IPS pipe (solid or cellular core) solvent cement fittings	ASTM F 1488; ASTM D 2235; ASTM D 2661; ASTM F 628
Co-extruded composite PVC DWV schedule 40 IPS pipe (solid or cellular core) solvent cement fittings	ASTM F 1488; ASTM D 2564; ASTM D 2665; ASTM F 891
Co-extruded composite PVC DWV IPS-DR-PS in PS35, PS50, PS100, PS140, PS 200 solvent cement fittings	ASTM F 1488; ASTM D 2564; ASTM D 2665; ASTM F 891
Co-extruded composite ABS sewer and drain pipe DR-PS in PS35, PS50, PS100, PS140, PS200 solvent cement fittings	ASTM F 1488; ASTM D 2235; ASTM D 2751
Co-extruded composite PVC sewer and drain pipe DR-PS in PS35, PS50, PS100, PS140, PS200 solvent cement fittings	ASTM F 1488; ASTM D 2564; ASTM D 3034; ASTM F 789

For SI: 1 inch = 25.4 mm.

P3003.3.2.2 Hub and spigot joints. An approved positive-seal one-piece elastomeric compression-type gasket that is placed in the hub before the spigot is inserted shall be permitted to be used for joining hub and spigot cast-iron soil piping and fittings as an alternate for lead and oakum joints.

P3003.3.3 Threaded pipe joints. Threaded joints shall conform to American National Taper Pipe Thread. Pipe ends shall be reamed or filed to size and all chips removed. Pipe-joint compound or tape shall be applied on the male threads only.

P3003.3.4 Soldered joints. Soldered joints in tubing shall be made with fittings approved for DWV piping. Surfaces to be soldered shall be cleaned bright. The joints shall be properly fluxed and made with approved solder. Fluxes shall conform to ASTM B 813. The joint shall be soldered with a solder conforming to ASTM B 32.

P3003.3.5 Clay or cement soil pipe joints. Joints between concrete pipe or fittings shall be made with an elastomeric seal conforming to ASTM C 443, ASTM C 1173, CSA-A257.3 or CSA CAN/CSA-B602. Joints between vitrified clay pipe or fittings shall be made with an elastomeric seal conforming to ASTM C 445, ASTM C 1173 or CSA CAN/CSA-B602.

P3003.3.6 Plastic pipe joints. Joints in plastic piping shall be made with approved fittings by solvent cementing, elastomeric gaskets or other approved manufactured system.

P3003.3.7 Slip joints. Slip joints shall be made using approved gaskets or compression washers. Ground joint connections that allow adjustment of tubing but provide a rigid joint when made up shall not be considered slip joints. When a ground joint connection is used and the assembled joint does not permit free movement, the joint need not be accessible.

P3003.4 Joints between different piping materials. Joints between different piping material shall comply with Sections P3003.4.1 through P3003.4.5.

P3003.4.1 Hub-type cast-iron or vitrified clay to other piping materials. Joints between hub-type drainage piping and other materials, including steel, plastic and copper may be made with a lead and oakum joint and an approved caulking ferrule of the other material or by a lead and oakum joint directly with the other material omitting the adapter ferrule. Where the outside diameter of the other material matches that of the hub pipe, an elastomeric gasket designed for use with the hub pipe may be used.

P3003.4.2 Hubless pipe to other piping materials. Joints between hubless drainage piping and any other materials shall be joined with an approved elastomeric sleeve and stainless steel clamp.

P3003.4.3 Threaded pipe to cast-iron soil pipe. Joints between threaded pipe and cast-iron soil pipe shall be made with approved adapter fittings.

P3003.4.4 Threaded pipe to copper or plastic pipe. Joints from threaded pipe to copper or plastic piping shall utilize approved adapter fittings.

P3003.4.5 Joints between drainage piping and water closets. Joints between drainage piping and water closets or similar fixtures shall be made by means of a closet flange compatible with the drainage system material, securely fastened to a structurally firm base. The inside diameter of the drainage pipe shall not be used as a socket fitting for a four by three closet flange. The joint shall be bolted, with an approved gasket or setting compound between the fixture and the closet flange.

SECTION P3004
DETERMINING DRAINAGE FIXTURE UNITS

P3004.1 DWV system load. The load on DWV-system piping shall be computed in terms of drainage fixture unit (d.f.u.) values in accordance with Table P3004.1.

SECTION P3005
DRAINAGE SYSTEM

P3005.1 Drainage fittings and connections. Changes in direction in drainage piping shall be made by the appropriate use of sanitary tees, wyes, sweeps, bends or by a combination of these drainage fittings in accordance with Table P3005.1. Change in direction by combination fittings, heel or side inlets or increasers shall be installed in accordance with Table P3005.1 and Sections P3005.1.1 through P3005.1.4. based on the pattern of flow created by the fitting.

P3005.1.1 Horizontal to vertical (multiple connection fittings). Double fittings such as double sanitary tees and tee-wyes or approved multiple connection fittings and back-to-back fixture arrangements that connect two or more branches at the same level shall be permitted as long as directly opposing connections are the same size and the discharge into directly opposing connections is from similar fixture types or fixture groups.

Exception: Double sanitary tee patterns shall not receive the discharge of water closets located within 30 inches (762 mm) in developed length of the fitting or any fixture with pumping action discharge.

P3005.1.2 Heel- or side-inlet quarter bends, drainage. Heel-inlet quarter bends shall be an acceptable means of connection, except where the quarter bends serves a water closet. A low-heel inlet shall not be used as a wet-vented connection. Side-inlet quarter bends shall be an acceptable means of connection for both drainage, wet venting and stack venting arrangements.

P3005.1.3 Heel- or side-inlet quarter bends, venting. Heel-inlet or side-inlet quarter bends, or any arrangement of pipe and fittings producing a similar effect, shall be acceptable as a dry vent where the inlet is placed in a vertical position. The inlet is permitted to be placed in a horizontal position only where the entire fitting is part of a dry vent arrangement.

TABLE P3004.1
DRAINAGE FIXTURE UNIT (d.f.u.) VALUES FOR VARIOUS PLUMBING FIXTURES

TYPE OF FIXTURE OR GROUP OF FIXTURES	DRAINAGE FIXTURE UNIT VALUE (d.f.u.)[a]
Bar sink	1
Bathtub (with or without shower head and/or whirlpool attachments)	2
Bidet	1
Clothes washer standpipe	2
Dishwasher	2
Floor drain[b]	0
Kitchen sink	2
Lavatory	1
Laundry tub	2
Shower stall	2
Water closet (1.6 gallons per flush)	3
Water closet (greater than 1.6 gallons per flush)	4
Full-bath group with bathtub (with 1.6 gallon per flush water closet, and with or without shower head and/or whirlpool attachment on the bathtub or shower stall)	5
Full-bath group with bathtub (water closet greater than 1.6 gallon per flush, and with or without shower head and/or whirlpool attachment on the bathtub or shower stall)	6
Half-bath group (1.6 gallon per flush water closet plus lavatory)	4
Half-bath group (water closet greater than 1.6 gallon per flush plus lavatory)	5
Kitchen group (dishwasher and sink with or without garbage grinder)	3
Laundry group (clothes washer standpipe and laundry tub)	3
Multiple-bath groups[c]: 1.5 baths 2 baths 2.5 baths 3 baths 3.5 baths	 7 8 9 10 11

For SI: 1 gallon = 3.785 L.

a. For a continuous or semicontinuous flow into a drainage system, such as from a pump or similar device, 1.5 fixture units shall be allowed per gpm of flow. For a fixture not listed, use the highest d.f.u. value for a similar listed fixture.

b. A floor drain itself adds no hydraulic load. However, where used as a receptor, the fixture unit value of the fixture discharging into the receptor shall be applicable.

c. Add 2 d.f.u. for each additional full bath.

P3005.1.4 Water closet connection between flange and pipe. One-quarter bends 3 inches (76 mm) in diameter shall be acceptable for water closet or similar connections, provided a 4-inch-by-3-inch (102 mm by 76 mm) flange is installed to receive the closet fixture horn. Alternately, a 4-inch-by-3-inch (102 mm by 76 mm) elbow shall be acceptable with a 4-inch (102 mm) flange.

P3005.1.5 Dead ends. Dead ends shall be prohibited except where necessary to extend a cleanout or as an approved part of a rough-in more than 2 feet (610 mm) in length.

P3005.1.6 Provisions for future fixtures. Where drainage has been roughed-in for future fixtures, the drainage unit values of the future fixtures shall be considered in determining the required drain sizes. Such future installations shall be terminated with an accessible permanent plug or cap fitting.

P3005.2 Drainage pipe cleanouts. Drainage pipe cleanouts shall comply with Sections P3005.2.1 through P3005.2.11.

P3005.2.1 Materials. Cleanouts shall be liquid and gas tight. Cleanout plugs shall be brass or plastic.

TABLE P3005.1
FITTINGS FOR CHANGE IN DIRECTION

TYPE OF FITTING PATTERN	CHANGE IN DIRECTION		
	Horizontal to vertical[c]	Vertical to horizontal	Horizontal to horizontal
Sixteenth bend	X	X	X
Eighth bend	X	X	X
Sixth bend	X	X	X
Quarter bend	X	X[a]	X[a]
Short sweep	X	X[a,b]	X[a]
Long sweep	X	X	X
Sanitary tee	X[c]		
Wye	X	X	X
Combination wye and eighth bend	X	X	X

For SI: 1 inch = 25.4 mm.

a The fittings shall only be permitted for a 2-inch or smaller fixture drain.

b Three inches and larger.

c For a limitation on multiple connection fittings, see Section P3005.1.1.

P3005.2.2 Spacing. Cleanouts shall be installed not more than 100 feet (30 480 mm) apart in horizontal drainage lines.

P3005.2.3 Underground drainage cleanouts. When installed in underground drains, cleanouts shall be extended vertically to or above finished grade either inside or outside the building.

P3005.2.4 Change of direction. Cleanouts shall be installed at each change of direction of the drainage system greater than 45 degrees, except not more than one cleanout shall be required in each 40 feet (12 192 mm) of run regardless of change in direction.

P3005.2.5 Accessibility. Cleanouts shall be accessible. Minimum clearance in front of cleanouts shall be 18 inches on 3 inches (457 mm on 76 mm) and larger pipes, and 12 inches (305 mm) on smaller pipes. Concealed cleanouts shall be provided with access of sufficient size to permit removal of the cleanout plug and rodding of the system. Cleanout plugs shall not be concealed by permanent finishing material.

P3005.2.6 Base of stacks. Accessible cleanouts shall be provided near the base of each vertical waste or soil stack. Alternatively, such cleanouts may be installed outside the building within 3 feet (914 mm) of the building wall.

P3005.2.7 Building drain and building sewer junction. There shall be a cleanout near the junction of the building drain and building sewer. This cleanout may be either inside or outside the building wall, provided it is brought up to finish grade or to the lowest floor level. An accessible interior building drain cleanout or test tee within close proximity to the building drain exit point shall fulfill this requirement.

P3005.2.8 Direction of flow. Cleanouts shall be installed so that the cleanout opens to allow cleaning in the direction of the flow of the drainage line.

P3005.2.9 Cleanout size. Cleanouts shall be sized in accordance with Table P3005.2.9.

TABLE P3005.2.9
CLEANOUTS

PIPE SIZE (inches)	CLEANOUT SIZE (inches)
$1^1/_2$	$1^1/_2$
2	$1^1/_2$
3	$2^1/_2$
4 and larger	$3^1/_2$

For SI: 1 inch = 25.4 mm.

P3005.2.10 Cleanout equivalent. A fixture trap or a fixture with integral trap, readily removable without disturbing concealed piping shall be acceptable as a cleanout equivalent.

P3005.2.11 Connections to cleanouts prohibited. Cleanout openings shall not be used for the installation of new fixtures or floor drains except where approved and an acceptable alternate cleanout is provided.

P3005.3 Horizontal drainage piping slope. Horizontal drainage piping shall be installed in uniform alignment at uniform slopes not less than one-fourth unit vertical in 12 units horizontal (2-percent slope) for $2^1/_2$-inch (64 mm) diameter and less, and not less than one-eighth unit vertical in 12 units horizontal (1-percent slope) for diameters of 3 inches (76 mm) or more.

P3005.4 Drain pipe sizing. Drain pipes shall be sized according to drainage fixture unit (d.f.u.) loads. The following general procedure is permitted to be used:

1. Draw an isometric layout or riser diagram denoting fixtures on the layout.
2. Assign d.f.u. values to each fixture group plus individual fixtures using Table P3004.1.
3. Starting with the top floor or most remote fixtures, work downstream toward the building drain accumulating d.f.u. values for fixture groups plus individual fixtures for each branch. Where multiple bath groups are being added, use the reduced d.f.u. values in Table P3004.1, which take into account probability factors of simultaneous use.
4. Size branches and stacks by equating the assigned d.f.u. values to pipe sizes shown in Table P3005.4.1.
5. Determine the pipe diameter and slope of the building drain and building sewer based on the accumulated d.f.u. values, using Table P3005.4.2.

P3005.4.1 Fixture branch and stack sizing.

1. Branches and stacks shall be sized according to Table P3005.4.1. Below grade drain pipes shall not be less than $1^1/_2$ inches (38 mm) in diameter.
2. Minimum stack size. Drain stacks shall not be smaller than the largest horizontal branch connected, with the following exceptions:
   2.1. A 4-inch-by-3-inch (102 mm by 76 mm) closet bend or flange or a 4-inch (102 mm) closet bend into a 3-inch (76 mm) stack tee shall be acceptable (see Section P3005.1.4).

TABLE P3005.4.1
MAXIMUM FIXTURE UNITS ALLOWED TO BE CONNECTED TO BRANCHES AND STACKS

NOMINAL PIPE SIZE (inches)	ANY HORIZONTAL FIXTURE BRANCH	ANY ONE VERTICAL STACK OR DRAIN
$1^1/_4$[a]	—	—
$1^1/_2$[b]	3	4
2[b]	6	10
$2^1/_2$[b]	12	20
3	20	48
4	160	240

For SI: 1 inch = 25.4 mm.

a. $1^1/_4$-inch pipe size limited to a single-fixture drain or trap arm. See Table P3201.7.

b. No water closets.

P3005.4.2 Building drain and sewer size and slope. Pipe sizes and slope shall be determined from Table P3005.4.2 on the basis of drainage load in fixture units (d.f.u.) computed from Table P3004.1.

TABLE P3005.4.2
MAXIMUM NUMBER OF FIXTURE UNITS ALLOWED TO BE CONNECTED TO THE BUILDING DRAIN, BUILDING DRAIN BRANCHES OR THE BUILDING SEWER

DIAMETER OF PIPE (inches)	SLOPE PER FOOT		
	1/8 inch	1/4 inch	1/2 inch
1 1/2 [a,b]	—	—[a]	—[a]
2[b]	—	21	27
2 1/2[b]	—	24	31
3	36	42	50
4	180	216	250

For SI: 1 inch = 25.4 mm, 1 foot = 304.8 mm.

a. 1 1/2-inch pipe size limited to a building drain branch serving not more than two waste fixtures, or not more than one waste fixture if serving a pumped discharge fixture or garbage grinder discharge.

b. No water closets.

SECTION P3006
SIZING OF DRAIN PIPE OFFSETS

P3006.1 Vertical offsets. An offset in a vertical drain, with a change of direction of 45 degrees (0.79 rad) or less from the vertical, shall be sized as a straight vertical drain.

P3006.2 Horizontal offsets above the lowest branch. A stack with an offset of more than 45 degrees (0.79 rad) from the vertical shall be sized as follows:

1. The portion of the stack above the offset shall be sized as for a regular stack based on the total number of fixture units above the offset.
2. The offset shall be sized as for a building drain in accordance with Table P3005.4.2.
3. The portion of the stack below the offset shall be sized as for the offset or based on the total number of fixture units on the entire stack, whichever is larger.
4. Drainage branch connections shall not be made within five pipe diameters above or below the offset unless a relief vent is installed at the offset.

P3006.3 Horizontal offsets below the lowest branch. In soil or waste stacks below the lowest horizontal branch, there shall be no change in diameter required if the offset is made at an angle not greater than 45 degrees (0.79 rad) from the vertical. If an offset greater than 45 degrees (0.79 rad) from the vertical is made, the offset and stack below it shall be sized as a building drain (see Table P3005.4.2).

SECTION P3007
SUMPS AND EJECTORS

P3007.1 Sewage ejectors or sewage pumps. A sewage ejector, sewage pump, or grinder pump receiving discharge from a water closet shall have minimum discharge velocity of 1.9 feet per second (0.579 m/s) throughout the discharge piping to the point of connection with a gravity building drain, gravity sewer or pressure sewer system. A nongrinding pump or ejector shall be capable of passing a 1.5-inch-diameter (38 mm) solid ball, and the discharge piping shall be not less than 2 inches (51 mm) in diameter. The discharge piping of grinding pumps shall be not less than 1.25 inches (32 mm) in diameter. A check valve and a gate valve located on the discharge side of the check valve shall be installed in the pump or ejector discharge piping between the pump or ejector and the drainage system. Access shall be provided to such valves. Such valves shall be located above the sump cover or, when the discharge pipe from the ejector is below grade, the valves shall be accessibly located outside the sump below grade in an access pit with a removable access cover.

Exception: Macerating toilet systems shall be permitted to have the discharge pipe sized in accordance with manufacturer's instruction, but not less than 0.75 inch (19.1 mm) in diameter.

P3007.2 Building drains below sewer (building subdrains). Building drains which cannot be discharged to the sewer by gravity flow shall be discharged into a tightly covered and vented sump from which the contents shall be lifted and discharged into the building gravity drainage system by automatic pumping equipment.

P3007.2.1 Drainage piping. The system of drainage piping below the sewer level shall be installed and vented in a manner similar to that of the gravity system. Only such drains that must be lifted for discharge shall be discharged into sumps. All other drains shall be discharged by gravity.

Exception: Macerating toilet systems shall be permitted as an alternate to the sewage pump or ejector system. The macerating toilet system shall be installed in accordance with manufacturer's instructions.

SECTION P3008
BACKWATER VALVES

P3008.1 General. Fixtures that have flood level rims located below the elevation of the next upstream manhole cover of the public sewer serving such fixtures shall be protected from backflow of sewage by installing an approved backwater valve. Fixtures having flood level rims above the elevation of the next upstream manhole shall not discharge through the backwater valve. Backwater valves shall be provided with access.

P3008.2 Construction. Backwater valves shall have noncorrosive bearings, seats and self-aligning discs, and shall be constructed to ensure a positive mechanical seal. Valve access covers shall be water tight.

CHAPTER 31
VENTS

SECTION P3101
VENT SYSTEMS

P3101.1 General. This chapter shall govern the selection and installation of piping, tubing and fittings for vent systems. This chapter shall control the minimum diameter of vent pipes, circuit vents, branch vents and individual vents, and the size and length of vents and various aspects of vent stacks and stack vents. Additionally, this chapter regulates vent grades and connections, height above fixtures and relief vents for stacks and fixture traps, and the venting of sumps and sewers.

P3101.2 Trap seal protection. The plumbing system shall be provided with a system of vent piping that will permit the admission or emission of air so that the seal of any fixture trap shall not be subjected to a pneumatic pressure differential of more than 1 inch of water (249 Pa).

P3101.2.1 Venting required. Every trap and trapped fixture shall be vented in accordance with one of the venting methods specified in this chapter.

P3101.3 Use limitations. The plumbing vent system shall not be utilized for purposes other than the venting of the plumbing system.

P3101.4 Extension outside a structure. In climates where the 97.5-percent value for outside design temperature is less than 0°F (-18°C) (ASHRAE 97.5-percent column, winter, see Chapter 3), vent pipes installed on the exterior of the structure shall be protected against freezing by insulation, heat or both. Vent terminals shall be protected from frost closure in accordance with Section P3103.2.

SECTION P3102
VENT STACKS AND STACK VENTS

P3102.1 Main vent required. Every building shall have a main vent that is either a vent stack or a stack vent. Such vent shall run undiminished in size and as directly as possible from the building drain through to the open air above the roof.

P3102.2 Vent connection to drainage system. Every vent stack shall connect to the base of the drainage stack. The vent stack shall connect at or below the lowest horizontal branch. Where the vent stack connects to the building drain, the connection shall be located within 10 pipe diameters downstream of the drainage stack. A stack vent shall be a vertical extension of the drainage stack.

P3102.3 Vent termination. Every vent stack or stack vent shall extend outdoors and terminate to the open air or terminate to an air admittance valve.

SECTION P3103
VENT TERMINALS

P3103.1 Roof extension. All open vent pipes which extend through a roof shall be terminated at least [number] inches above the roof or [number] inches above the anticipated snow accumulation, except that where a roof is to be used for any purpose other than weather protection, the vent extensions shall be run at least 7 feet (2134 mm) above the roof.

P3103.2 Frost closure. Where the 97.5-percent value for outside design temperature is 0°F (-18°C) or less, (ASHRAE 97.5-percent column, winter, see Chapter 3), every vent extension through a roof or wall shall be a minimum of 3 inches (76 mm) in diameter. Any increase in the size of the vent shall be made inside the structure a minimum of 1 foot (305 mm) below the roof or inside the wall.

P3103.3 Flashings and sealing. The juncture of each vent pipe with the roof line shall be made water tight by an approved flashing. Vent extensions in walls and soffits shall be made weather tight by caulking.

P3103.4 Prohibited use. Vent terminals shall not be used as a flag pole or to support flag poles, TV aerials, or similar items, except when the piping has been anchored in an approved manner.

P3103.5 Location of vent terminal. An open vent terminal from a drainage system shall not be located less than 4 feet (1219 mm) directly beneath any door, openable window, or other air intake opening of the building or of an adjacent building, nor shall any such vent terminal be within 10 feet (3048 mm) horizontally of such an opening unless it is at least 2 feet (610 mm) above the top of such opening.

P3103.6 Extension through the wall. Vent terminals extending through the wall shall terminate a minimum of 10 feet (3048 mm) from the lot line and 10 feet (3048 mm) above the highest adjacent grade within 10 feet (3048 mm) horizontally of the vent terminal. Vent terminals shall not terminate under the overhang of a structure with soffit vents. Side wall vent terminals shall be protected to prevent birds or rodents from entering or blocking the vent opening.

SECTION P3104
VENT CONNECTIONS AND GRADES

3104.1 Connection. All individual branch and circuit vents shall connect to a vent stack, stack vent or extend to the open air.

Exception: Individual, branch and circuit vents shall be permitted to terminate at an air admittance valve in accordance with Section P3114.

P3104.2 Grade. All vent and branch vent pipes shall be so graded, connected and supported as to allow moisture and condensate to drain back to the soil or waste pipe by gravity.

P3104.3 Vent connection to drainage system. Every dry vent connecting to a horizontal drain shall connect above the centerline of the horizontal drain pipe.

P3104.4 Vertical rise of vent. Every dry vent shall rise vertically to a minimum of 6 inches (152 mm) above the flood level rim of the highest trap or trapped fixture being vented.

P3104.5 Height above fixtures. A connection between a vent pipe and a vent stack or stack vent shall be made at least 6 inches (152 mm) above the flood level rim of the highest fixture served by the vent. Horizontal vent pipes forming branch vents shall be at least 6 inches (152 mm) above the flood level rim of the highest fixture served.

P3104.6 Vent for future fixtures. Where the drainage piping has been roughed-in for future fixtures, a rough-in connection for a vent shall be installed a minimum of one-half the diameter of the drain. The vent rough-in shall connect to the vent system or shall be vented by other means as provided in this chapter. The connection shall be identified to indicate that the connection is a vent.

SECTION P3105
FIXTURE VENTS

P3105.1 Distance of trap from vent. Each fixture trap shall have a protecting vent located so that the slope and the developed length in the fixture drain from the trap weir to the vent fitting are within the requirements set forth in Table P3105.1.

TABLE P3105.1
MAXIMUM DISTANCE OF FIXTURE TRAP FROM VENT

SIZE OF TRAP (inches)	SLOPE (inch per foot)	DISTANCE FROM TRAP (feet)
$1^1/_4$	$^1/_4$	5
$1^1/_2$	$^1/_4$	6
2	$^1/_4$	8
3	$^1/_4$	12
4	$^1/_8$	16

For SI: 1 inch = 25.4 mm, 1 foot = 304.8 mm, 1 inch per foot = 0.0833 mm/m.

P3105.2 Fixture drains. The total fall in a fixture drain due to pipe slope shall not exceed one pipe diameter, nor shall the vent pipe connection to a fixture drain, except for water closets, be below the weir of the trap, except as provided in Section P3105.3.

P3105.3 Vertical leg for waste fixture drains. A vertical leg (see Figure P3105.3) is permitted within a fixture drain of a waste fixture in accordance with the following criteria:

1. Minimum trap diameter shall be in accordance with Table P3201.7.
2. The diameter of Section A shall be equal to the diameter of the trap.
3. The length of Section A shall not be less than 8 inches (203 mm) and in accordance with Table P3105.1.
4. The diameter of Section B shall be one pipe size larger than the diameter of Section A.
5. The length of Section B shall not be more than 36 inches (914 mm).
6. The diameter of Section C shall be one pipe size larger than the diameter of Section B.
7. The total length of Section A and Section C shall not exceed the distance allowed in Table P3105.1.
8. Bends shall be the diameter of the largest connected section.

P3105.4 Crown vent. A vent shall not be installed within two pipe diameters of the trap weir.

SECTION P3106
INDIVIDUAL VENT

P3106.1 Individual vent permitted. Each trap and trapped fixture is permitted to be provided with an individual vent. The individual vent shall connect to the fixture drain of the trap or trapped fixture being vented.

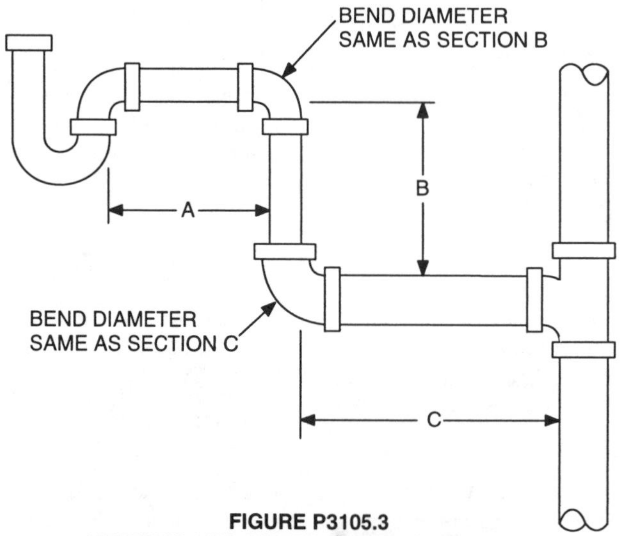

FIGURE P3105.3
VERTICAL LEG FIXTURE DRAIN SCHEMATIC

SECTION P3107
COMMON VENT

P3107.1 Individual vent as common vent. An individual vent is permitted to vent two traps or trapped fixtures as a common vent. The traps or trapped fixtures being common vented shall be located on the same floor level.

P3107.2 Connection at the same level. Where the fixture drains being common vented connect at the same level, the vent connection shall be at the interconnection of the fixture drains or downstream of the interconnection.

P3107.3 Connection at different levels. Where the fixture drains connect at different levels, the vent shall connect as a vertical extension of the vertical drain. The vertical drain pipe connecting the two fixture drains shall be considered the vent for the lower fixture drain, and shall be sized in accordance with Table P3107.3. The upper fixture shall not be a water closet.

TABLE P3107.3
COMMON VENT SIZES

PIPE SIZE (inches)	MAXIMUM DISCHARGE FROM UPPER FIXTURE DRAIN (d.f.u.)
$1^1/_2$	1
2	4
$2^1/_2$ to 3	6

For SI: 1 inch = 25.4 mm.

SECTION P3108
WET VENTING

P3108.1 Wet vent permitted. Any combination of fixtures within two bathroom groups located on the same floor level are permitted to be vented by a wet vent. The wet vent shall be considered the vent for the fixtures and shall extend from the connection of the dry vent along the direction of the flow in the drain pipe to the most downstream fixture drain connection to the horizontal branch drain. Only the fixtures within the bathroom groups shall connect to the wet vented horizontal branch drain. Any additional fixtures shall discharge downstream of the wet vent. [See Figures P3108.1(1), P3108.1(2) and P3108.1(3) for typical wet vent configurations.]

P3108.2 Vent connections. The dry vent connection to the wet vent shall be an individual vent or common vent to the lavatory, bidet, shower or bathtub.

P3108.3 Size. Horizontal and vertical wet vents shall be of a minimum size as specified in Table P3108.3, based on the fixture unit discharge to the wet vent.

P3108.4 Vertical wet vent. A combination of fixtures located on the same floor level are permitted to be vented by a vertical wet vent. The vertical wet vent shall extend from the connection to the dry vent down to the lowest fixture drain connection. Each fixture shall connect independently to the vertical wet vent. All water closet drains shall connect at the same elevation. Other fixture drains shall connect above or at the same elevation as the water closet fixture drains. The dry vent connection to the vertical wet vent shall be an individual or common vent serving one or two fixtures. See Figures P3108.2(1) and P3108.2(2) for typical vertical wet vent configurations.

TABLE P3108.3
WET VENT SIZE

WET VENT PIPE SIZE (inches)	FIXTURE UNIT LOAD (d.f.u.)
$1^1/_2$	1
2	4
$2^1/_2$	6
3	12
4	32

For SI: 1 inch = 25.4 mm.

SECTION P3109
WASTE STACK VENT

P3109.1 Waste stack vent permitted. A waste stack shall be considered a vent for all of the fixtures discharging to the stack where installed in accordance with the requirements of this section.

P3109.2 Stack installation. The waste stack shall be vertical, and both horizontal and vertical offsets shall be prohibited. Every fixture drain shall connect separately to the waste stack. The stack shall not receive the discharge of water closets or urinals. See Figure P3109.2 for a typical waste stack serving as a wet vent configuration.

P3109.3 Stack vent. A stack vent shall be provided for the waste stack. The size of the stack vent shall be equal to the size of the waste stack.

P3109.4 Waste stack size. The waste stack shall be sized based on the total discharge to the stack and the discharge within a branch interval in accordance with Table P3109.4. The waste stack shall be the same size throughout the length of the waste stack.

TABLE P3109.4
WASTE STACK VENT SIZE

STACK SIZE (inches)	MAXIMUM NUMBER OF FIXTURE UNITS (d.f.u.)	
	Total discharge into one branch interval	Total discharge for stack
$1^1/_2$	1	2
2	2	4
$2^1/_2$	No limit	8
3	No limit	24
4	No limit	50

For SI: 1 inch = 25.4 mm.

SECTION P3110
CIRCUIT VENTING

P3110.1 Circuit vent permitted. A maximum of eight fixtures connected to a horizontal branch drain shall be permitted to be circuit vented. Each fixture drain shall connect horizontally to the horizontal branch being circuit vented. The horizontal branch drain shall be classified as a vent from the most downstream fixture drain connection to the most upstream fixture drain connection to the horizontal branch.

P3110.2 Vent connection. The circuit vent connection shall be located between the two most upstream fixture drains. The vent shall connect to the horizontal branch and shall be installed in accordance with Section P3104. The circuit vent pipe shall not receive the discharge of any soil or waste.

P3110.3 Slope and size of horizontal branch. The maximum slope of the vent section of the horizontal branch drain shall be one unit vertical in 12 units horizontal (8-percent slope). The entire length of the vent section of the horizontal branch drain shall be sized for the total drainage discharge to the branch in accordance with Table P3005.4.1.

P3110.4 Additional fixtures. Fixtures, other than the circuit-vented fixtures, are permitted to discharge to the horizontal branch drain. Such fixtures shall be located on the same floor as the circuit-vented fixtures and shall be either individually or common vented. See Figure P3110.4 for a typical circuit vent with additional noncircuit vented fixtures

SECTION P3111 COMBINATION WASTE AND VENT SYSTEM

P3111.1 Type of fixtures. A combination waste and vent system shall not serve fixtures other than floor drains, standpipes, sinks and lavatories.

P3111.2 Installation. The only vertical pipe of a combination drain and vent system shall be the connection between the fixture drain of a sink, lavatory or standpipe, and the horizontal combination waste and vent pipe. The maximum vertical distance shall be 36 inches (914 mm).

P3111.2.1 Slope. The horizontal combination waste and vent pipe shall have a maximum slope of one-half unit vertical in 12 units horizontal (4-percent slope). The minimum slope shall be in accordance with Section P3005.3.

P3111.2.2 Connection. The combination waste and vent pipe shall connect to a horizontal drain that is vented or a vent shall connect to the combination waste and vent. The vent connecting to the combination waste and vent pipe shall extend vertically a minimum of 6 inches (152 mm) above the flood level rim of the highest fixture being vented before offsetting horizontally.

P3111.2.3 Vent size. The vent shall be sized for the total fixture unit load in accordance with Section P3113.1.

P3111.2.4 Fixture branch or drain. The fixture branch or fixture drain shall connect to the combination waste and vent within a distance specified in Table P3105.1. The combination waste and vent pipe shall be considered the vent for the fixture.

P3111.3 Size. The minimum size of a combination waste and vent pipe shall be in accordance with Table P3111.3.

TABLE P3111.3
SIZE OF COMBINATION WASTE AND VENT PIPE

DIAMETER PIPE (inches)	MAXIMUM NUMBER OF FIXTURE UNITS (d.f.u.)	
	Connecting to a horizontal branch or stack	Connecting to a building drain or building subdrain
2	3	4
$2^1/_2$	6	26
3	12	31
4	20	50

For SI: 1 inch = 25.4 mm.

SECTION P3112 ISLAND FIXTURE VENTING

P3112.1 Limitation. Island fixture venting shall not be permitted for fixtures other than sinks and lavatories. Kitchen sinks with a dishwasher waste connection, a food waste grinder, or both, in combination with the kitchen sink waste, shall be permitted to be vented in accordance with this section.

P3112.2 Vent connection. The island fixture vent shall connect to the fixture drain as required for an individual or common vent. The vent shall rise vertically to above the drainage outlet of the fixture being vented before offsetting horizontally or vertically downward. The vent or branch vent for multiple island fixture vents shall extend to a minimum of 6 inches (152 mm) above the highest island fixture being vented before connecting to the outside vent terminal.

P3112.3 Vent installation below the fixture flood level rim. The vent located below the flood level rim of the fixture being vented shall be installed as required for drainage piping in accordance with Chapter 30, except for sizing. The vent shall be sized in accordance with Section P3113.1. The lowest point of the island fixture vent shall connect full size to the drainage system. The connection shall be to a vertical drain pipe or to the top half of a horizontal drain pipe. Cleanouts shall be provided in the island fixture vent to permit rodding of all vent piping located below the flood level rim of the fixtures. Rodding in both directions shall be permitted through a cleanout.

SECTION P3113 VENT PIPE SIZING

P3113.1 Size of vents. The minimum required diameter of individual vents, branch vents, circuit vents, vent stacks and stack vents shall be at least one-half the required diameter of the drain served. The required size of the drain shall be determined in accordance with Chapter 30. Vent pipes shall be not less than $1^1/_4$ inch (32 mm) in diameter. Vents exceeding 40 feet (12 192 mm) in developed length shall be increased by one nominal pipe size for the entire developed length of the vent pipe.

P3113.2 Developed length. The developed length of individual, branch, and circuit vents shall be measured from the farthest point of vent connection to the drainage system, to the point of connection to the vent stack, stack vent or termination outside of the building.

P3113.3 Multiple branch vents. Where multiple branch vents are connected to a common branch vent, the common branch vent shall be sized in accordance with this section, based on the size of the common horizontal drainage branch that is or would be required to serve the total drainage fixture unit (dfu) load being vented.

P3113.4 Sump vents. Sump vent sizes shall be determined in accordance with Sections P3113.4.1 and P3113.4.2.

P3113.4.1 Sewage pumps and sewage ejectors other than pneumatic. Drainage piping below sewer level shall be vented in a similar manner to that of a gravity system. Building sump vent sizes for sumps with sewage pumps or sewage ejectors, other than pneumatic, shall be determined in accordance with Table P3113.4.1.

P3113.4.2 Pneumatic sewage ejectors. The air pressure relief pipe from a pneumatic sewage ejector shall be connected to an independent vent stack terminating as required for vent extensions through the roof. The relief pipe shall be sized to relieve air pressure inside the ejector to atmospheric pressure, but shall not be less than $1^1/_4$ inches (32 mm) in size.

SECTION P3114 AIR ADMITTANCE VALVES

P3114.1 General. Vent systems utilizing air admittance valves shall comply with this section. Individual- and branch-type air admittance valves shall conform to ASSE 1051. Stack type air admittance valves shall conform to ASSE 1050.

P3114.2 Installation. The valves shall be installed in accordance with the requirements of this section and the manufacturer's installation instructions. Air admittance valves shall be installed after the DWV testing required by Section P2503.5.1 or P2503.5.2 has been performed.

P3114.3 Where permitted. Individual vents, branch vents, circuit vents and stack vents shall be permitted to terminate with a connection to an air admittance valve.

P3114.4 Location. The air admittance valve shall be located a minimum of 4 inches (102 mm) above the horizontal branch drain or fixture drain being vented. The air admittance valve shall be located within the maximum developed length permitted for the vent. The air admittance valve shall be installed a minimum of 6 inches (152 mm) above insulation materials where installed in attics.

P3114.5 Access and ventilation. Access shall be provided to all air admittance valves. The valve shall be located within a ventilated space that allows air to enter the valve.

P3114.6 Size. The air admittance valve shall be rated for the size of the vent to which the valve is connected.

P3114.7 Vent required. Within each plumbing system, a minimum of one stack vent or vent stack shall extend outdoors to the open air.

TABLE P3113.4.1 SIZE AND LENGTH OF SUMP VENTS

DISCHARGE CAPACITY OF PUMP (gpm)	MAXIMUM DEVELOPED LENGTH OF VENT (feet)[a]				
	Diameter of vent (inches)				
	$1^1/_4$	$1^1/_2$	2	$2^1/_2$	3
10	No limit[b]	No limit	No limit	No limit	No limit
20	270	No limit	No limit	No limit	No limit
40	72	160	No limit	No limit	No limit
60	31	75	270	No limit	No limit

For SI: 1 inch = 25.4 mm, 1 foot = 304.8 mm, 1 gpm = 3.785 L/m.

a. Developed length plus an appropriate allowance for entrance losses and friction due to fittings, changes in direction and diameter. Suggested allowances shall be obtained from NBS Monograph 31 or other approved sources. An allowance of 50 percent of the developed length shall be assumed if a more precise value is not available.

b. Actual values greater than 500 feet.

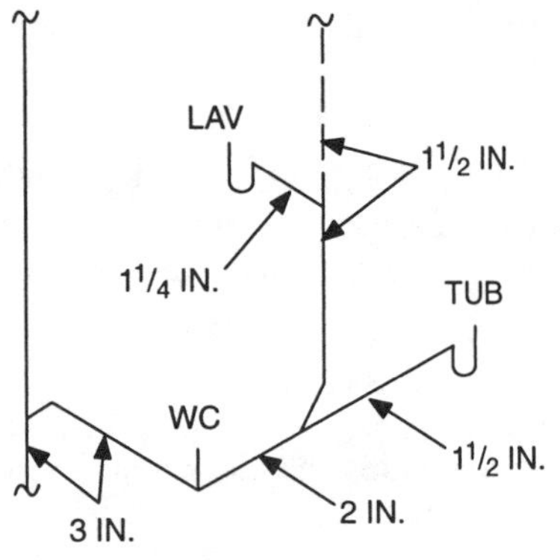

A. TYPICAL SINGLE-BATH ARRANGEMENT

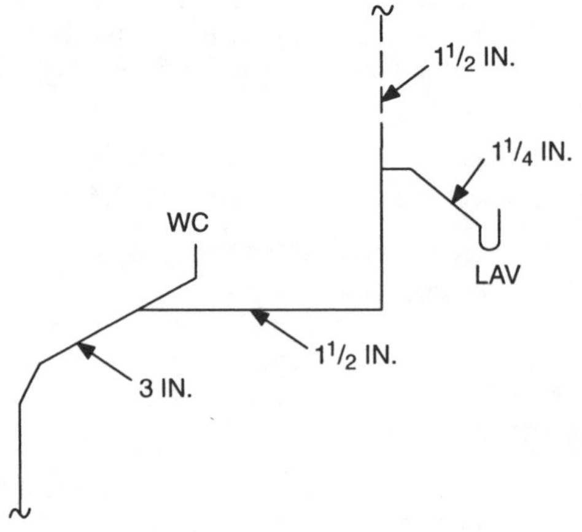

B. TYPICAL POWDER ROOM

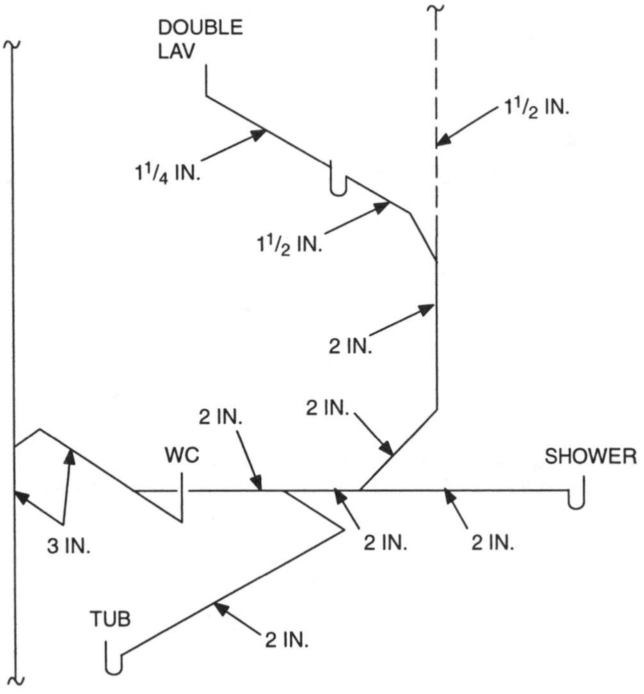

C. MORE ELABORATE SINGLE-BATH ARRANGEMENT

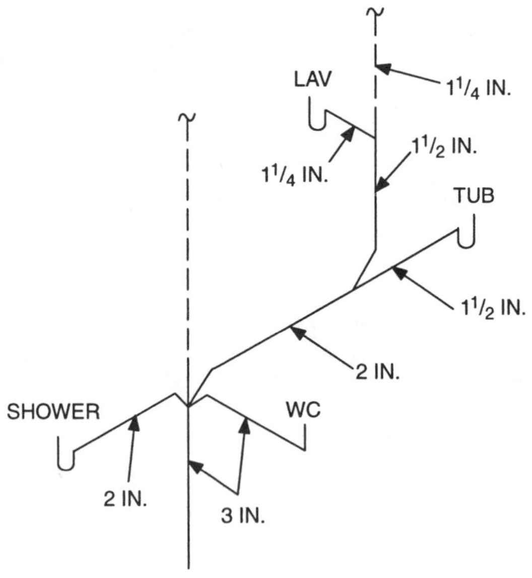

D. COMBINATION WET- AND STACK-VENTING WITH STACK FITTING

For SI: 1 inch = 25.4 mm.

Note: This figure is for illustration purposes only and the written text shall apply.

FIGURE P3108.1(1)
TYPICAL SINGLE-BATH WET-VENT ARRANGEMENTS

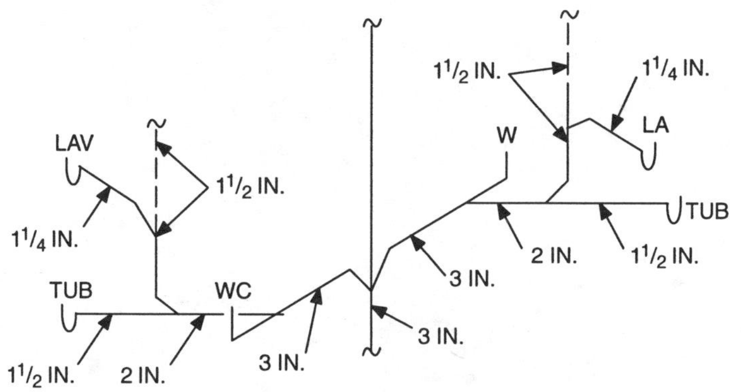

A. TYPICAL BACK-TO-BACK BATHS

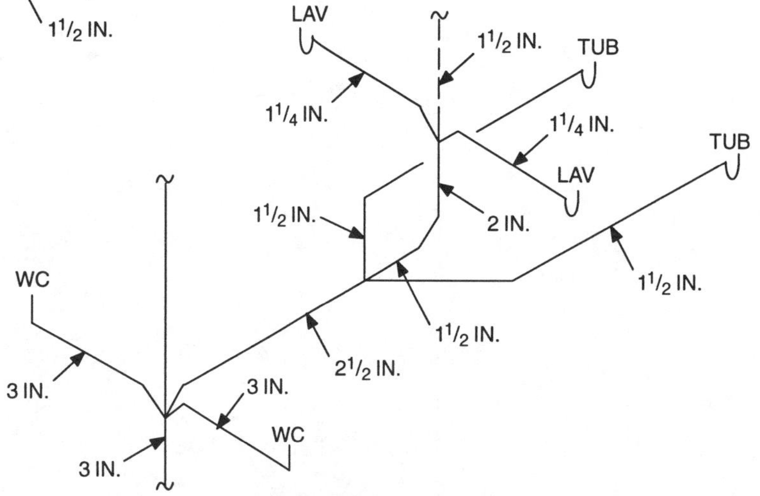

B. DOUBLE BATHS WITH WASTE FIXTURES ON COMMON HORIZONTAL BRANCH, COMMON WET VENT

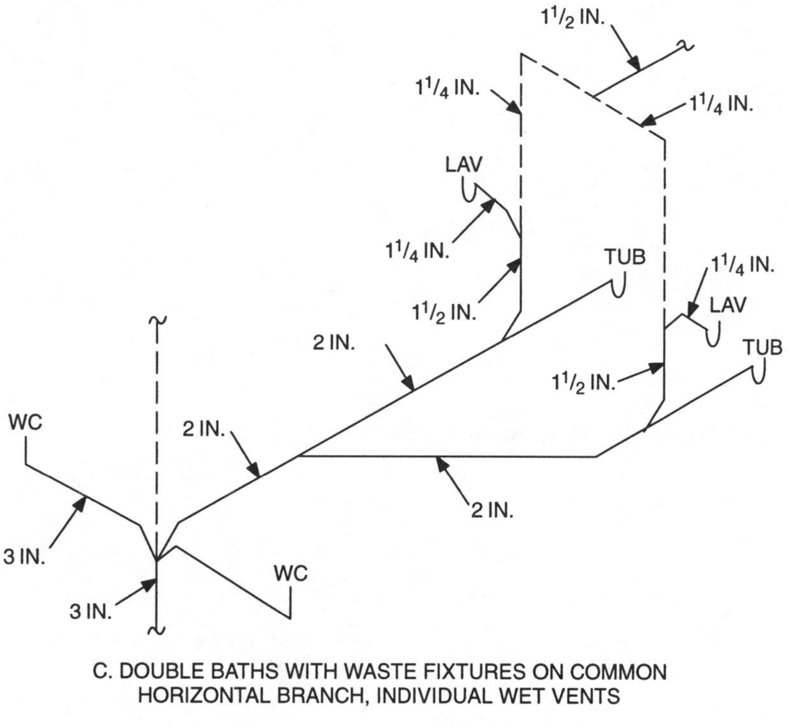

C. DOUBLE BATHS WITH WASTE FIXTURES ON COMMON HORIZONTAL BRANCH, INDIVIDUAL WET VENTS

For SI: 1 inch = 25.4 mm.

Note: This figure is for illustration purposes only and the written text shall apply.

FIGURE P3108.1(2)
TYPICAL DOUBLE-BATH WET-VENT ARRANGEMENTS

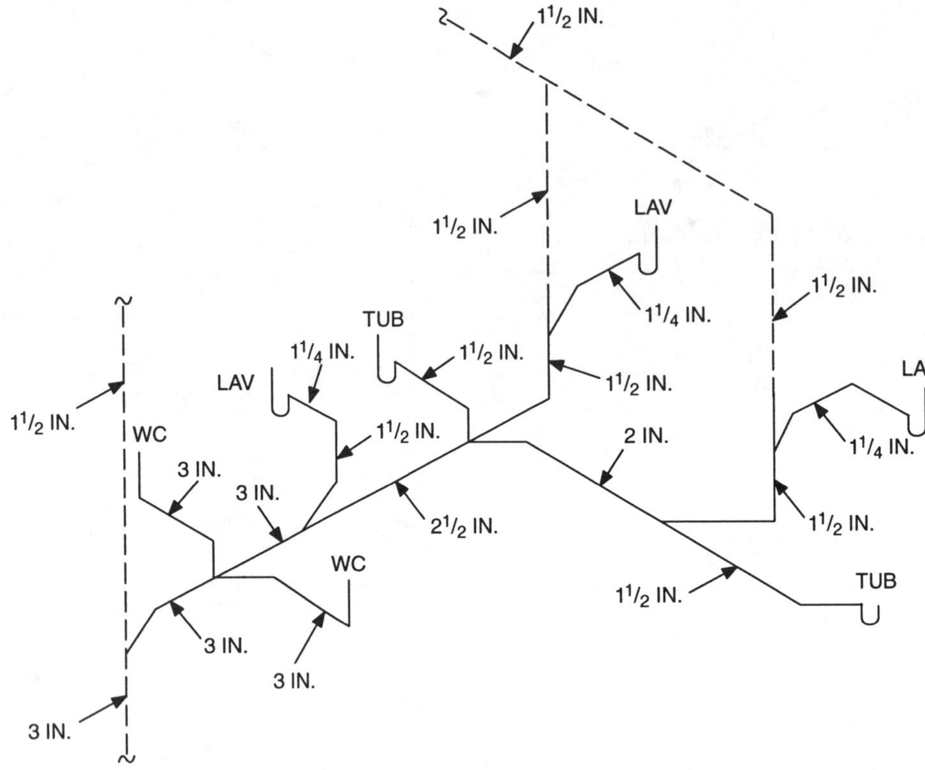

For SI: 1 inch = 25.4 mm.

Note: This figure is for illustration purposes only and the written text shall apply.

FIGURE P3108.1(3)
TYPICAL HORIZONTAL WET VENTING

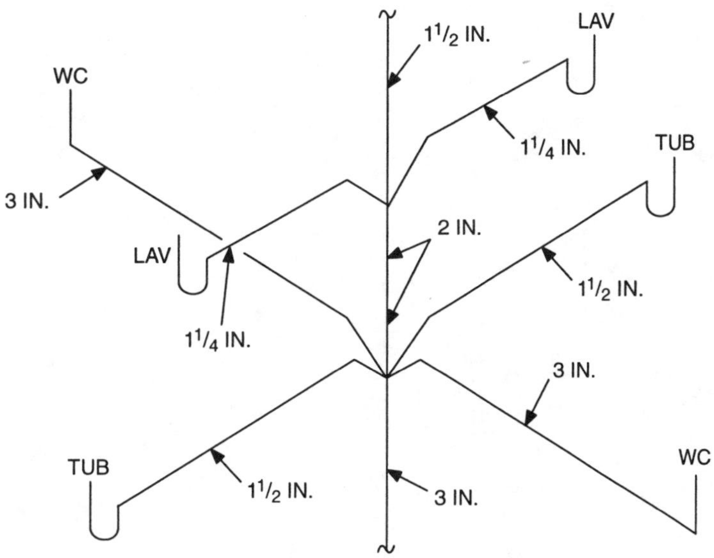

A. VERTICAL WET VENTING

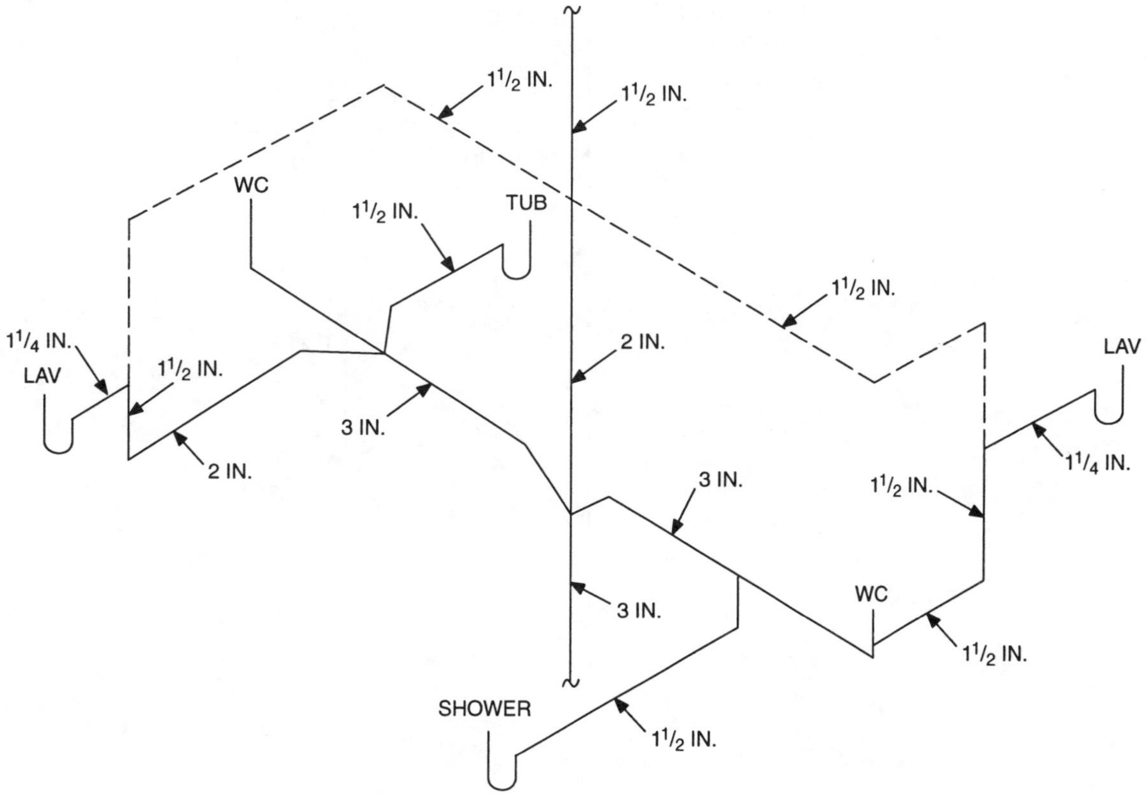

B. COMBINATION VERTICAL AND HORIZONTAL WET VENTING

For SI: 1 inch = 25.4 mm.

Note: This figure is for illustration purposes only and the written text shall apply.

FIGURE P3108.2(1)
TYPICAL METHODS OF WET VENTING

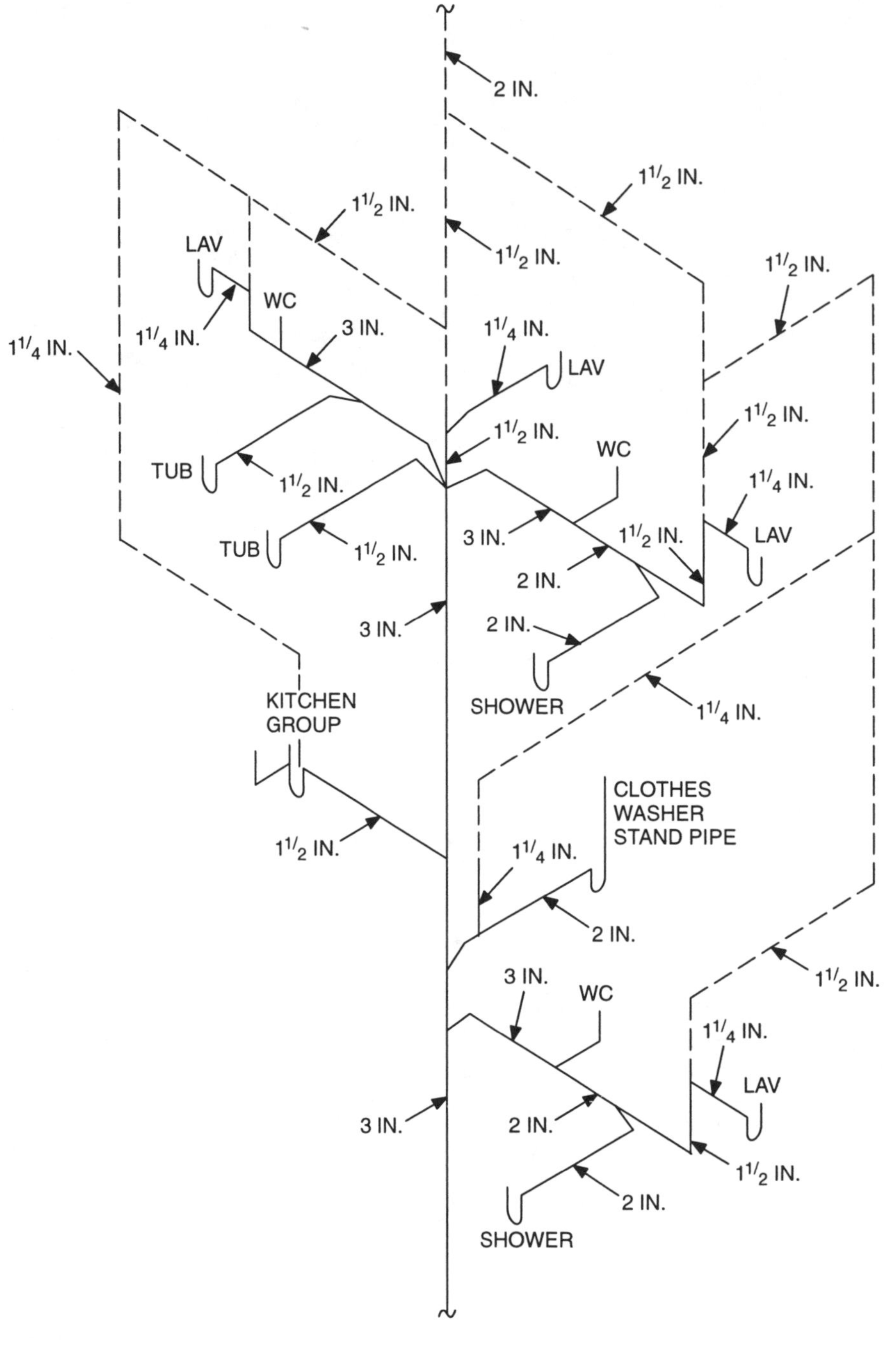

For SI: 1 inch = 25.4 mm.

Note: This figure is for illustration purposes only and the written text shall apply.

FIGURE P3108.2(2)
TYPICAL SINGLE-STACK SYSTEM FOR A TWO-STORY DWELLING

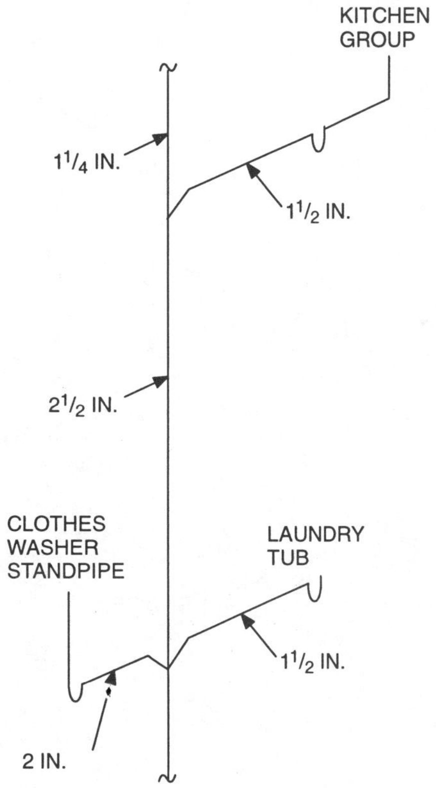

For SI: 1 inch = 25.4 mm.

Note: This figure is for illustration purposes only and the written text shall apply.

FIGURE P3109.2
WASTE STACK SERVING AS WET VENT FOR LAUNDRY GROUP

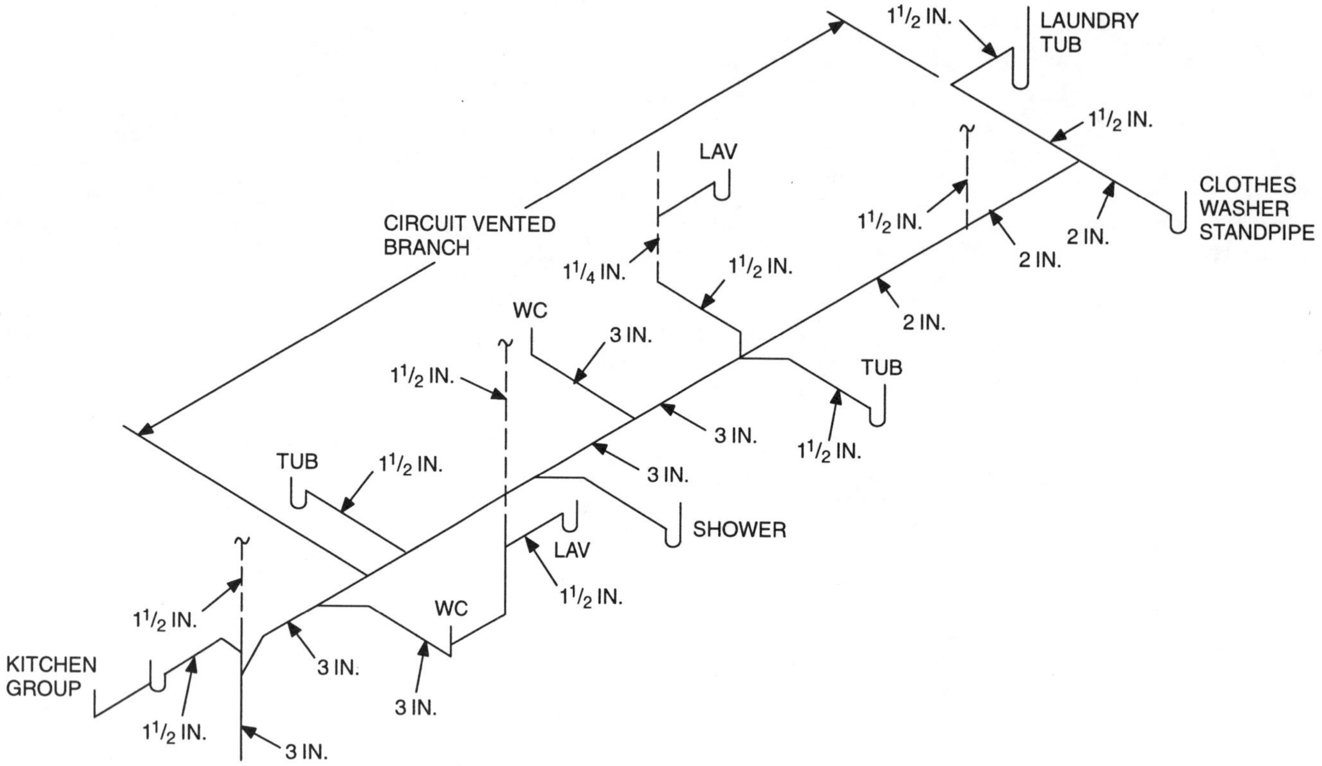

For SI: 1 inch = 25.4 mm.

Note: This figure is for illustration purposes only and the written text shall apply.

FIGURE P3110.4
CIRCUIT VENT WITH ADDITIONAL NONCIRCUIT VENTED FIXTURE—CIRCUIT VENTED BRANCH

CHAPTER 32
TRAPS

SECTION P3201
FIXTURE TRAPS

P3201.1 Design of traps. Traps shall be of standard design, shall have smooth uniform internal waterways, shall be self-cleaning and shall not have interior partitions except where integral with the fixture. Traps shall be constructed of lead, cast iron, cast or drawn brass or approved plastic. Tubular brass traps shall be not less than No. 20 gage (0.8 mm) thickness. Solid connections, slip joints and couplings are permitted to be used on the trap inlet, trap outlet, or within the trap seal. Slip joints shall be accessible.

P3201.2 Trap seals and trap seal protection. Traps shall have a liquid seal not less than 2 inches (51 mm) and not more than 4 inches (102 mm).

Exception: Traps for floor drains shall be fitted with a trap primer or shall be of the deep seal design.

P3201.3 Trap setting and protection. Traps shall be set level with respect to their water seals and shall be protected from freezing. Trap seals shall be protected from siphonage, aspiration or back pressure by an approved system of venting (see Section P3101).

P3201.4 Building traps. Building traps shall not be installed, except in special cases where sewer gases are extremely corrosive or noxious, as directed by the building official.

P3201.5 Prohibited trap designs. The following types of traps are prohibited:

1. Bell traps.
2. Separate fixture traps with interior partitions, except those lavatory traps made of plastic, stainless steel or other corrosion-resistant material.
3. "S" traps except as permitted under Section P3102.3.
4. Drum traps.
5. Trap designs with moving parts.

P3201.6 Number of fixtures per trap. Each plumbing fixture shall be separately trapped by a water seal trap placed as close as possible to the fixture outlet. The vertical distance from the fixture outlet to the trap weir shall not exceed 24 inches (610 mm). Fixtures shall not be double trapped. Exceptions to the separate trapping requirements are as follows:

1. Fixtures that have integral traps.
2. A single trap shall be permitted to serve two or three like fixtures limited to kitchen sinks, laundry tubs and lavatories. Such fixtures shall be adjacent to each other and located in the same room with a continuous waste arrangement. The trap shall be installed at the center fixture where three such fixtures are installed. Common trapped fixture outlets shall not be more than 30 inches (762 mm) apart.
3. The connection of a laundry tray waste line is permitted to be made into a standpipe for the automatic clothes-washer drain in accordance with Section P2706.2.1.

P3201.7 Size of fixture traps. Fixture trap size (nominal diameter) shall be sufficient to drain the fixture rapidly and in no case less than given in Table P3201.7. Traps, including integral traps, shall not be larger than the drainage pipe into which it discharges.

TABLE P3201.7
SIZE OF TRAPS AND TRAP ARMS FOR PLUMBING FIXTURES

PLUMBING FIXTURE	TRAP SIZE MINIMUM (inches)
Bathtub (with or without shower head and/or whirlpool attachments)	$1^1/_2$
Bidet	$1^1/_4$
Clothes washer standpipe	2
Dishwasher (on separate trap)	$1^1/_2$
Floor drain	2
Kitchen sink (one or two traps, with or without dishwasher and garbage grinder)	$1^1/_2$
Laundry tub (one or more compartments)	$1^1/_2$
Lavatory	$1^1/_4$
Shower	2
Water closet	a

For SI: 1 inch = 25.4 mm.

a. Consult fixture standards for trap dimensions of specific bowls.

Part VIII — Electrical

CHAPTER 33
GENERAL REQUIREMENTS

 See Appendix L, International Residential Code Electrical Provisions/National Electrical Code Cross Reference.

SECTION E3301
GENERAL

E3301.1 Applicability. The provisions of Chapters 33 through 42 shall establish the general scope of the electrical system and equipment requirements of this code. Chapters 33 through 42 cover those wiring methods and materials most commonly encountered in the construction of one- and two-family dwellings and structures regulated by this code. Other wiring methods, materials and subject matter covered in the NFPA 70 are also allowed by this code.

E3301.2 Scope. Chapters 33 through 42 shall cover the installation of electrical systems, equipment and components indoors and outdoors that are within the scope of this code, including services, power distribution systems, fixtures, appliances, devices and appurtenances. Services within the scope of this code shall be limited to 120/240-volt, 0- to 400-ampere, single-phase systems. These chapters specifically cover the equipment, fixtures, appliances, wiring methods and materials that are most commonly used in the construction or alteration of one- and two-family dwellings and accessory structures regulated by this code. The omission from these chapters of any material or method of construction provided for in the referenced standard NFPA 70 shall not be construed as prohibiting the use of such material or method of construction. Electrical systems, equipment or components not specifically covered in these chapters shall comply with the applicable provisions of the NFPA 70.

E3301.3 Not covered. Chapters 33 through 42 do not cover the following:

1. Installations, including associated lighting, under the exclusive control of communications utilities and electric utilities.
2. Services over 400 amperes.

E3301.4 Additions and alterations. Any addition or alteration to an existing electrical system shall be made in conformity with the provisions of Chapters 33 through 42. Where additions subject portions of existing systems to loads exceeding those permitted herein, such portions shall be made to comply with Chapters 33 through 42.

SECTION E3302
BUILDING STRUCTURE PROTECTION

E3302.1 Drilling and notching. Wood-framed structural members shall not be drilled, notched or altered in any manner except as provided for in this code.

E3302.2 Penetrations of fire-resistance-rated assemblies. Electrical installations in hollow spaces, vertical shafts, and ventilation or air-handling ducts shall be made so that the possible spread of fire or products of combustion will not be substantially increased. Electrical penetrations through fire-resistance-rated walls, partitions, floors or ceilings shall be protected by approved methods to maintain the fire-resistance rating of the element penetrated. Penetrations of fire-resistance-rated walls shall be limited as specified in Section R320.2.

E3302.3 Penetrations of firestops and draftstops. Penetrations through fire blocking and draftstopping shall be protected in an approved manner to maintain the integrity of the element penetrated.

SECTION E3303
INSPECTION AND APPROVAL

E3303.1 Approval. Electrical materials, components and equipment shall be approved.

E3303.2 Inspection required. New electrical work and parts of existing systems affected by new work or alterations shall be inspected by the building official to ensure compliance with the requirements of Chapters 33 through 42.

E3303.3 Listing and labeling. Electrical materials, components, devices, fixtures and equipment shall be listed for the application, shall bear the label of an approved agency and shall be installed, and used, or both, in accordance with the manufacturer's installation instructions.

SECTION E3304
GENERAL EQUIPMENT REQUIREMENTS

E3304.1 Voltages. Throughout Chapters 33 through 42, the voltage considered shall be that at which the circuit operates.

E3304.2 Interrupting rating. Equipment intended to interrupt current at fault levels shall have a minimum interrupting rating of 10,000 amperes. Equipment intended to break interrupt current at levels other than fault levels shall have an interrupting rating at nominal circuit voltage sufficient for the current that must be interrupted.

E3304.3 Circuit characteristics. The overcurrent protective devices, total impedance, component short-circuit current ratings and other characteristics of the circuit to be protected shall be so selected and coordinated as to permit the circuit protective devices that are used to clear a fault to do so without extensive damage to the electrical components of the circuit. This fault shall be assumed to be either between two or more of the circuit conductors or between any circuit conductor and the grounding conductor or enclosing metal raceway. Listed products applied in accordance with their listing shall be considered to meet the requirements of this section.

E3304.4 Protection of equipment. Equipment identified only as "dry locations," "Type 1," or "indoor use only" shall be protected against permanent damage from the weather during building construction.

E3304.5 Unused openings. Unused openings in boxes, cabinets, equipment cases or housings shall be effectively closed to afford protection equivalent to the wall of the equipment.

E3304.6 Integrity of electrical equipment. Internal parts of electrical equipment, including busbars, wiring terminals, insulators and other surfaces, shall not be damaged or contaminated by foreign materials such as paint, plaster, cleaners or abrasives, and corrosive residues. There shall not be any damaged parts that might adversely affect safe operation or mechanical strength of the equipment such as parts that are broken; bent; cut; deteriorated by corrosion, chemical action, or overheating. Foreign debris shall be removed from equipment.

E3304.7 Mounting. Electric equipment shall be firmly secured to the surface on which it is mounted.

E3304.8 Energized parts guarded against accidental contact. Approved enclosures shall guard energized parts that are operating at 50 volts or more against accidental contact.

E3304.9 Prevent physical damage. In locations where electrical equipment is likely to be exposed to physical damage, enclosures or guards shall be so arranged and of such strength as to prevent such damage.

E3304.10 Equipment identification. The manufacturer's name, trademark or other descriptive marking by which the organization responsible for the product can be identified shall be placed on all electric equipment. Other markings shall be provided that indicate voltage, current, wattage or other ratings as specified elsewhere in Chapters 33 through 42. The marking shall have the durability to withstand the environment involved.

E3304.11 Identification of disconnecting means. Each required disconnecting means for motors and appliances, and each service, feeder and branch circuit at the point where it originates, shall be legibly marked to indicate its purpose, except where located and arranged so that the purpose is evident. The marking shall have the durability to withstand the environment involved.

SECTION E3305
EQUIPMENT LOCATION AND CLEARANCES

E3305.1 Working space and clearances. Access and working space shall be provided and maintained around all electrical equipment to permit ready and safe operation and maintenance of such equipment in accordance with this section and Figure E3305.1.

E3305.2 Working clearances for energized equipment and panelboards. Except as otherwise specified in Chapters 33 through 42, the dimension of the working space in the direction of access to panelboards and live parts likely to require examination, adjustment, servicing or maintenance while energized shall be not less than 36 inches (914 mm) in depth. Distances shall be measured from the energized parts where such parts are exposed or from the enclosure front or opening where such parts are enclosed. In addition to the 36-inch dimension (914 mm), the work space shall not be less than 30 inches (762 mm) wide in front of the electrical equipment and not less than the width of such equipment. The work space shall be clear and shall extend from the floor or platform to a height of 6.5 feet (1981 mm). In all cases, the work space shall allow at least a 90-degree opening of equipment doors or hinged panels. Equipment associated with the electrical installation located above or below the electrical equipment shall be permitted to extend not more than 6 inches (152 mm) beyond the front of the electrical equipment.

E3305.3 Clearances over panelboards. A dedicated space directly over a panelboard that extends from the panelboard to the structural ceiling or to a height of 6 feet (1829 mm) above the panelboard, whichever is lower, and has a width and depth equal to the equipment shall be dedicated and kept clear of equipment unrelated to the electrical equipment. Piping, ducts or equipment unrelated to the electrical equipment shall not be installed in such dedicated space.

E3305.4 Location of clear spaces. Required working space shall not be designated for storage. Panelboards and overcurrent protection devices shall not be located in clothes closets or bathrooms.

E3305.5 Access and entrance to working space. Access shall be provided to the required working space.

E3305.6 Illumination. Artificial illumination shall be provided for all working spaces for service equipment and panelboards installed indoors.

E3305.7 Headroom. The minimum headroom for working spaces for service equipment and panelboards shall be 6.5 feet (1981mm).

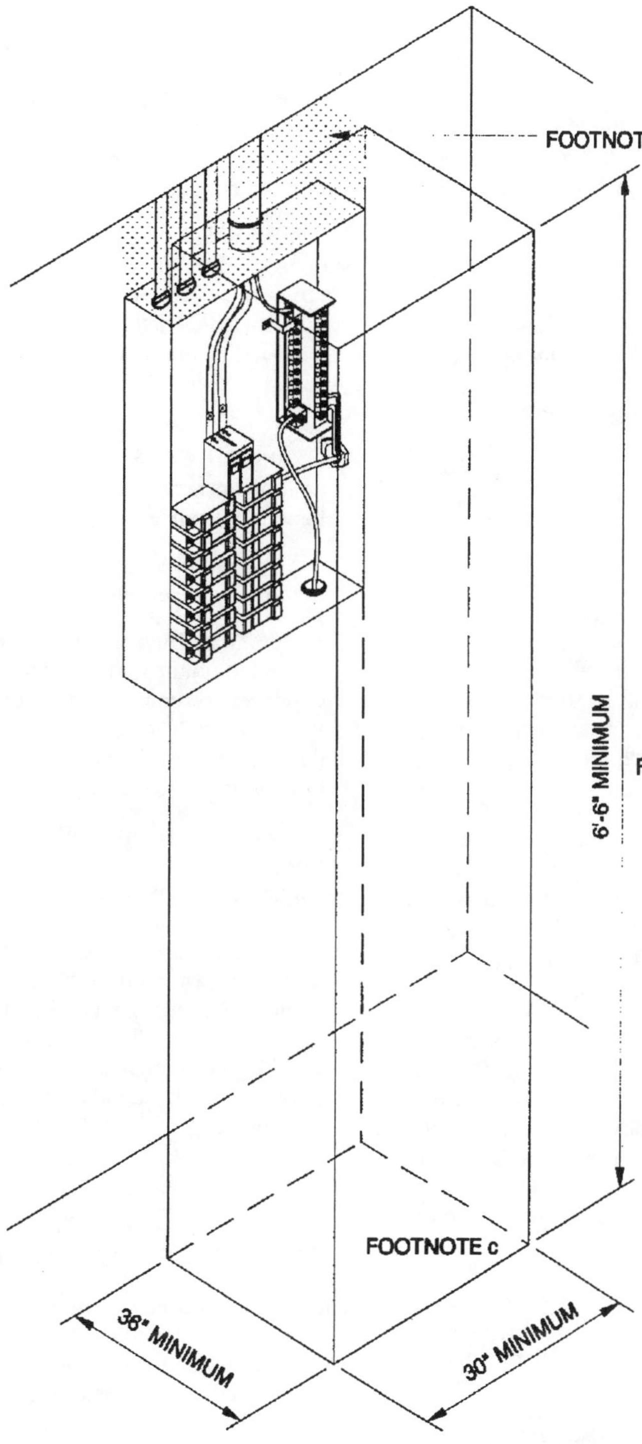

FIGURE E3305.1[a,b,c,d,e]
WORKING SPACE AND CLEARANCES

For SI: 1 inch = 25.4 mm, 1 foot = 304.8 mm.

a. Equipment, piping and ducts shall not be placed in the shaded area extending directly above the top of the panelboard from the panelboard to the ceiling above.

b. The working space shall be clear and unobstructed from the floor to a height of 6.5 feet.

c. The working space shall not be designated for storage.

d. Panelboards, service equipment and similar enclosures shall not be located in bathrooms, toilet rooms and clothes closets.

e. Such work spaces shall be provided with artificial lighting where located indoors.

SECTION E3306
ELECTRICAL CONDUCTORS AND CONNECTIONS

E3306.1 General. This section provides general requirements for conductors, connections and splices. These requirements do not apply to conductors that form an integral part of equipment, such as motors, appliances and similar equipment, or to conductors specifically provided for elsewhere in Chapters 33 through 42.

E3306.2 Conductor material. Conductors used to conduct current shall be of copper except as otherwise provided in Chapters 33 through 42. Where the conductor material is not specified, the material and the sizes given in these chapters shall apply to copper conductors. Where other materials are used, the conductor sizes shall be changed accordingly.

E3306.3 Minimum size of conductors. The minimum size of conductors for feeders and branch circuits shall be No. 14 copper and No. 12 aluminum. The minimum size of service conductors shall be as specified in Chapter 35. The minimum size of class 2 remote control, signaling and power-limited circuits conductors shall be as specified in Chapter 42.

E3306.4 Stranded conductors. Where installed in raceways, conductors of size No. 8 and larger shall be stranded. A solid No. 8 conductor shall be permitted to be installed in a raceway only to meet the requirements of Section E4104.

E3306.5 Individual conductor insulation. Except where otherwise permitted in Sections E3505.1 and E3808.9, and E4203, current-carrying conductors shall be insulated. Insulated conductors shall have insulation types identified as RH, RHW, RHW-2, THHN, THHW,THW, THW-2, THWN, THWN-2, TW, UF, USE, USE-2, XHHW or XHHW-2. Insulation types shall be approved for the application.

E3306.6 Conductors in parallel. Circuit conductors that are electrically joined at each end to form a single conductor shall be limited to sizes No. 1/0 and larger. Conductors in parallel shall be of the same length, same conductor material, same circular mil area and same insulation type. Conductors in parallel shall be terminated in the same manner. Where run in separate raceways or cables, the raceway or cables shall have the same physical characteristics.

E3306.7 Conductors of the same circuit. All conductors of the same circuit and, where used, the grounded conductor and all equipment grounding conductors shall be contained within the same raceway, cable or cord.

E3306.8 Aluminum and copper connections. Terminals and splicing connectors shall be approved for the material of the conductors joined. Conductors of dissimilar metals shall not be joined in a terminal or splicing connector where physical contact occurs between dissimilar conductors such as copper and aluminum, copper and copper-clad aluminum, or aluminum and copper-clad aluminum, except where the device is listed for the purpose and conditions of application. Materials such as inhibitors and compounds shall be suitable for the application and shall be of a type that will not adversely affect the conductors, installation or equipment.

E3306.9 Terminals. Connection of conductors to terminal parts shall be made without damaging the conductors and shall be made by means of pressure connectors, including set-screw type, by means of splices to flexible leads, or for conductor sizes of No. 10 and smaller, by means of wire binding screws or studs and nuts having upturned lugs or the equivalent. Terminals for more than one conductor and terminals for connecting aluminum conductors shall be identified for the application.

E3306.10 Splices. Conductors shall be spliced or joined with splicing devices listed for the purpose. Splices and joints and the free ends of conductors shall be covered with an insulation equivalent to that of the conductors or with an insulating device listed for the purpose. Wire connectors or splicing means installed on conductors for direct burial shall be listed for such use.

E3306.10.1 Continuity. Conductors in raceways shall be continuous between outlets, devices and junctions and shall be without splices or taps in the raceway.

Exception: Splices shall be permitted within surface-mounted raceways that have a removable cover.

E3306.10.2 Device connections. The continuity of a grounded conductor in multiwire branch circuits shall not be dependent on connection to devices such as receptacles and lampholders. The continuity of equipment grounding conductors shall not be dependent on such connections in any type of branch circuit.

E3306.10.3 Length of conductor for splice or termination. Where conductors are to be spliced, terminated or connected to fixtures or devices, a minimum length of 6 inches (152 mm) of free conductor, shall be provided at each outlet, junction or switch point. The required length shall be measured from the point in the box where the conductor emerges from its raceway or cable sheath. Where the opening to an outlet, junction, or switch point is less than 8 inches (203 mm) in any dimension, each conductor shall be long enough to extend at least 3 inches (76 mm) outside of such opening.

SECTION E3307
CONDUCTOR AND TERMINAL IDENTIFICATION

E3307.1 Grounded conductors. Insulated grounded conductors of sizes No. 6 and smaller shall be identified by a continuous white or natural gray outer finish or by three continuous white stripes on other than green insulation along the entire length of the conductors. Conductors of sizes larger than No. 6 shall be identified by a continuous white or natural gray outer finish or by three continuous white stripes on other than green insulation along the entire length of the conductor, or shall be identified by a distinctive white marking that is applied at the time of installation and that encircles the conductor at all terminations.

E3307.2 Equipment grounding conductors. Equipment grounding conductors shall be identified by a continuous green color or a continuous green color with one or more yellow stripes on the insulation or covering, except where bare.

E3307.3 Ungrounded conductors. Insulation on the ungrounded conductors shall be a continuous color other than white, gray or green.

Exception: An insulated conductor that is part of a cable or flexible cord assembly and that has a white or gray finish or a finish with three continuous white stripes shall be permitted to be used as an ungrounded conductor where it is permanently re-identified to indicate its use as an ungrounded conductor at all terminations and at each location where the conductor is visible and accessible.

E3307.4 Identification of terminals. Terminals for attachment to conductors shall be identified in accordance with Sections E3307.4.1 and E3307.4.2.

E3307.4.1 Device terminals. All devices, excluding lighting and appliance branch-circuit panelboards, provided with terminals for the attachment of conductors and intended for connection to more than one side of the circuit shall have terminals properly marked for identification, except where the terminal intended to be connected to the grounded conductor is clearly evident.

Exception: For other than polarized attachment plugs and polarized receptacles for attachment plugs, terminal identification shall not be required for devices that have a normal current rating of over 30 amperes.

E3307.4.2 Receptacles, plugs, and connectors. Receptacles, polarized attachment plugs and cord connectors for plugs and polarized plugs shall have the terminal intended for connection to the grounded (white) conductor identified.

Identification shall be by a metal or metal coating substantially white in color or by the word "white" or the letter "W" located adjacent to the identified terminal.

Where the terminal is not visible, the conductor entrance hole for the connection shall be colored white or marked with the word "white or the letter "W".

E3307.5 Tag marking. AC cable shall be marked by means of a printed tag attached to the coil, reel or carton.

Chapter 34
ELECTRICAL DEFINITIONS

SECTION E3401
GENERAL

E3401.1 Scope. This chapter contains definitions that shall apply only to the electrical requirements of Chapters 33 through 42. Unless otherwise expressly stated, the following terms shall, for the purpose of this code, have the meanings indicated in this chapter. Words used in the present tense include the future; the singular number includes the plural and the plural the singular. Where terms are not defined in this section and are defined in Sections R202, M1202 and P3002 of this code, such terms shall have the meanings ascribed to them in those sections. Where terms are not defined in these sections, they shall have their ordinarily accepted meanings or such as the context implies.

ACCESSIBLE. (As applied to wiring methods.) Capable of being removed or exposed without damaging the building structure or finish, or not permanently closed in by the structure or finish of the building.

ACCESSIBLE. (As applied to equipment.) Admitting close approach; not guarded by locked doors, elevation or other effective means.

ACCESSIBLE, READILY. Capable of being reached quickly for operation, renewal or inspections, without requiring those to whom ready access is requisite to climb over or remove obstacles or to resort to portable ladders, chairs, etc.

AMPACITY. The current in amperes that a conductor can carry continuously under the conditions of use without exceeding its temperature rating.

APPLIANCE. Utilization equipment, normally built in standardized sizes or types, that is installed or connected as a unit to perform one or more functions such as clothes washing, air conditioning, food mixing, deep frying, etc.

APPROVED. Acceptable to the authority having jurisdiction.

ARC-FAULT CIRCUIT INTERRUPTER. A device intended to provide protection from the effects of arc-faults by recognizing characteristics unique to arcing and by functioning to de-energize the circuit when an arc-fault is detected.

ATTACHMENT PLUG (PLUG CAP)(PLUG). A device that, by insertion into a receptacle, establishes connection between the conductors of the attached flexible cord and the conductors connected permanently to the receptacle.

AUTOMATIC. Self-acting, operating by its own mechanism when actuated by some impersonal influence as, for example, a change in current strength, pressure, temperature or mechanical configuration.

BATHROOM. An area, including a basin, with one or more of the following: a toilet, a tub or a shower.

BONDING. The permanent joining of metallic parts to form an electrically conductive path that will ensure electrical continuity and the capacity to conduct safely any current likely to be imposed.

BONDING JUMPER. A reliable conductor to ensure the required electrical conductivity between metal parts required to be electrically connected.

BONDING JUMPER, CIRCUIT. The connection between portions of a conductor in a circuit to maintain the required ampacity of the circuit.

BONDING JUMPER, MAIN. The connection between the grounded circuit conductor and the equipment grounding conductor at the service.

BRANCH CIRCUIT. The circuit conductors between the final overcurrent device protecting the circuit and the outlet(s).

BRANCH CIRCUIT, APPLIANCE. A branch circuit supplying energy to one or more outlets to which appliances are to be connected, and that has no permanently connected lighting fixtures that are not a part of an appliance.

BRANCH CIRCUIT, GENERAL PURPOSE. A branch circuit that supplies a number of outlets for lighting and appliances.

BRANCH CIRCUIT, INDIVIDUAL. A branch circuit that supplies only one utilization equipment.

BRANCH CIRCUIT, MULTIWIRE. A branch circuit consisting of two or more ungrounded conductors having a potential difference between them, and a grounded conductor having equal potential difference between it and each ungrounded conductor of the circuit, and that is connected to the neutral or grounded conductor of the system.

CABINET. An enclosure designed either for surface or flush mounting and provided with a frame, mat or trim in which a swinging door or doors are or may be hung.

CIRCUIT BREAKER. A device designed to open and close a circuit by nonautomatic means and to open the circuit automatically on a predetermined overcurrent without damage to itself when properly applied within its rating.

CONCEALED. Rendered inaccessible by the structure or finish of the building. Wires in concealed raceways are considered concealed, even though they may become accessible by withdrawing them. [See "Accessible (As applied to wiring methods)."]

CONDUCTOR

Bare. A conductor having no covering or electrical insulation whatsoever.

Covered. A conductor encased within material of composition or thickness that is not recognized by this code as electrical insulation.

Insulated. A conductor encased within material of composition and thickness that is recognized by this code as electrical insulation.

CONDUIT BODY. A separate portion of a conduit or tubing system that provides access through a removable cover(s) to the interior of the system at a junction of two or more sections of the system or at a terminal point of the system. Boxes such as FS and FD or larger cast or sheet metal boxes are not classified as conduit bodies.

CONNECTOR, PRESSURE (SOLDERLESS). A device that establishes a connection between two or more conductors or between one or more conductors and a terminal by means of mechanical pressure and without the use of solder.

CONTINUOUS LOAD. A load where the maximum current is expected to continue for 3 hours or more.

COOKING UNIT, COUNTER-MOUNTED. A cooking appliance designed for mounting in or on a counter and consisting of one or more heating elements, internal wiring, and built-in or separately mountable controls.

COPPER-CLAD ALUMINUM CONDUCTORS. Conductors drawn from a copper-clad aluminum rod with the copper metallurgically bonded to an aluminum core. The copper forms a minimum of 10 percent of the cross-sectional area of a solid conductor or each strand of a stranded conductor.

CUTOUT BOX. An enclosure designed for surface mounting and having swinging doors or covers secured directly to and telescoping with the walls of the box proper. (See "Cabinet.")

DEAD FRONT. Without live parts exposed to a person on the operating side of the equipment.

DEMAND FACTOR. The ratio of the maximum demand of a system, or part of a system, to the total connected load of a system or the part of the system under consideration.

DEVICE. A unit of an electrical system that is intended to carry but not utilize electric energy.

DISCONNECTING MEANS. A device, or group of devices, or other means by which the conductors of a circuit can be disconnected from their source of supply.

DWELLING

Dwelling unit. One or more rooms for the use of one or more persons as a housekeeping unit with space for eating, living and sleeping, and permanent provisions for cooking and sanitation.

One-family dwelling. A building consisting solely of one dwelling unit.

Two-family dwelling. A building consisting solely of two dwelling units.

ENCLOSED. Surrounded by a case, housing, fence or walls that will prevent persons from accidentally contacting energized parts.

ENCLOSURE. The case or housing of apparatus, or the fence or walls surrounding an installation, to prevent personnel from accidentally contacting energized parts or to protect the equipment from physical damage.

ENERGIZED. Electrically connected to a source of potential difference.

EQUIPMENT. A general term including material, fittings, devices, appliances, fixtures, apparatus and the like used as a part of, or in connection with, an electrical installation.

EXPOSED. (As applied to live parts.) Capable of being inadvertently touched or approached nearer than a safe distance by a person. It is applied to parts not suitably guarded, isolated or insulated.

EXPOSED. (As applied to wiring methods.) On or attached to the surface or behind panels designed to allow access.

EXTERNALLY OPERABLE. Capable of being operated without exposing the operator to contact with live parts.

FEEDER. All circuit conductors between the service equipment, or the source of a separately derived system, or other power supply source and the final branch-circuit overcurrent device.

FITTING. An accessory such as a locknut, bushing or other part of a wiring system that is intended primarily to perform a mechanical rather than an electrical function.

GROUND. A conducting connection, whether intentional or accidental, between an electrical circuit or equipment and the earth, or to some conducting body that serves in place of the earth.

GROUNDED. Connected to earth or to some conducting body that serves in place of the earth.

GROUNDED, EFFECTIVELY. Intentionally connected to earth through a ground connection or connections of sufficiently low impedance and having sufficient current-carrying capacity to prevent the buildup of voltages that may result in undue hazards to connected equipment or to persons.

GROUNDED CONDUCTOR. A system or circuit conductor that is intentionally grounded.

GROUNDING CONDUCTOR. A conductor used to connect equipment or the grounded circuit of a wiring system to a grounding electrode or electrodes.

GROUNDING CONDUCTOR, EQUIPMENT. The conductor used to connect the noncurrent-carrying metal parts of equipment, raceways and other enclosures to the system grounded conductor, the grounding electrode conductor or both, at the service equipment or at the source of a separately derived system.

GROUNDING ELECTRODE CONDUCTOR. The conductor used to connect the grounding electrode to the equipment grounding conductor, the grounded conductor or both, of the circuit at the service equipment or at the source of a separately derived system.

GROUND-FAULT CIRCUIT-INTERRUPTER. A device intended for the protection of personnel that functions to de-energize a circuit or portion thereof within an established period of time when a current to ground exceeds some predetermined value that is less than that required to operate the overcurrent protective device of the supply circuit.

GUARDED. Covered, shielded, fenced, enclosed or otherwise protected by means of suitable covers, casings, barriers, rails, screens, mats or platforms to remove the likelihood of approach or contact by persons or objects to a point of danger.

IDENTIFIED. (As applied to Equipment.) Recognizable as suitable for the specific purpose, function, use, environment, application, etc., where described in a particular code requirement.

INTERRUPTING RATING. The highest current at rated voltage that a device is intended to interrupt under standard test conditions.

ISOLATED. Not readily accessible to persons unless special means for access are used.

LABELED. Equipment or materials to which has been attached a label, symbol or other identifying mark of an organization acceptable to the authority having jurisdiction and concerned with product evaluation that maintains periodic inspection of production of labeled equipment or materials and by whose labeling the manufacturer indicates compliance with appropriate standards or performance in a specified manner.

LIGHTING OUTLET. An outlet intended for the direct connection of a lampholder, a lighting fixture or a pendant cord terminating in a lampholder.

LISTED. Equipment, materials, or services included in a list published by an organization that is acceptable to the authority having jurisdiction and concerned with product evaluation of products or services, that maintains periodic inspection of production of listed equipment or materials or periodic evaluation of services, and whose listing states either that the equipment, material, or services meets identified standards or has been tested and found suitable for a specified purpose.

LIVE PARTS. Electric conductors, buses, terminals or components that are uninsulated or exposed and a shock hazard exists.

LOCATION

Damp location. Partially protected locations under canopies, marquees, roofed open porches and like locations, and interior locations subject to moderate degrees of moisture, such as some basements, some barns and some cold-storage warehouses.

Dry location. A location not normally subject to dampness or wetness. A location classified as dry may be temporarily subject to dampness or wetness, as in the case of a building under construction.

Wet location. Installations underground or in concrete slabs or masonry in direct contact with the earth and locations subject to saturation with water or other liquids, such as vehicle-washing areas, and locations exposed to weather and unprotected.

MULTIOUTLET ASSEMBLY. A type of surface, or flush, or freestanding raceway; designed to hold conductors and receptacles, assembled in the field or at the factory.

OUTLET. A point on the wiring system at which current is taken to supply utilization equipment.

OVEN, WALL-MOUNTED. An oven for cooking purposes and consisting of one or more heating elements, internal wiring, and built-in or separately mountable controls.

OVERCURRENT. Any current in excess of the rated current of equipment or the ampacity of a conductor. Such current results from overload (see definition), short circuit or ground fault.

OVERLOAD. Operation of equipment in excess of normal, full-load rating, or of a conductor in excess of rated ampacity that, when it persists for a sufficient length of time, would cause damage or dangerous overheating. A fault, such as a short circuit or ground fault, is not an overload.

PANELBOARD. A single panel or group of panel units designed for assembly in the form of a single panel, including buses and automatic overcurrent devices, and equipped with or without switches for the control of light, heat or power circuits, designed to be placed in a cabinet or cutout box placed in or against a wall or partition and accessible only from the front.

PLENUM. A compartment or chamber to which one or more air ducts are connected and that forms part of the air distribution system.

POWER OUTLET. An enclosed assembly that may include receptacles, circuit breakers, fuseholders, fused switches, buses and watt-hour meter mounting means, intended to supply and control power to mobile homes, recreational vehicles or boats, or to serve as a means for distributing power required to operate mobile or temporarily installed equipment.

PREMISES WIRING (SYSTEM). That interior and exterior wiring, including power, lighting, control and signal circuit wiring together with all of their associated hardware, fittings and wiring devices, both permanently and temporarily installed, that extends from the service point of utility conductors or source of power such as a battery, a solarphotovoltaic system, or a generator, transformer, or converter winding, to the outlet(s). Such wiring does not include wiring internal to appliances, fixtures, motors, controllers, motor control centers and similar equipment.

RACEWAY. An enclosed channel of metal or nonmetallic materials designed expressly for holding wires, cables, or busbars, with additional functions as permitted in this code. Raceways include, but are not limited to, rigid metal conduit, rigid nonmetallic conduit, intermediate metal conduit, liquid-tight flexible conduit, flexible metallic tubing, flexible metal conduit, electrical nonmetallic tubing, electrical metallic tubing, underfloor raceways, cellular concrete floor raceways, cellular metal floor raceways, surface raceways, wireways and busways.

RAINPROOF. Constructed, protected or treated so as to prevent rain from interfering with the successful operation of the apparatus under specified test conditions.

RAIN TIGHT. Constructed or protected so that exposure to a beating rain will not result in the entrance of water under specified test conditions.

RECEPTACLE. A receptacle is a contact device installed at the outlet for the connection of an attachment plug. A single receptacle is a single contact device with no other contact device on the same yoke. A multiple receptacle is two or more contact devices on the same yoke.

RECEPTACLE OUTLET. An outlet where one or more receptacles are installed.

SERVICE. The conductors and equipment for delivering energy from the serving utility to the wiring system of the premises served.

SERVICE CABLE. Service conductors made up in the form of a cable.

SERVICE CONDUCTORS. The conductors from the service point or other source of power to the service disconnecting means.

SERVICE DROP. The overhead service conductors from the last pole or other aerial support to and including the splices, if any, connecting to the service-entrance conductors at the building or other structure.

SERVICE-ENTRANCE CONDUCTORS, OVERHEAD SYSTEM. The service conductors between the terminals of the service equipment and a point usually outside the building, clear of building walls, where joined by tap or splice to the service drop.

SERVICE-ENTRANCE CONDUCTORS, UNDERGROUND SYSTEM. The service conductors between the terminals of the service equipment and the point of connection to the service lateral.

SERVICE EQUIPMENT. The necessary equipment, usually consisting of a circuit breaker(s) or switch(es) and fuse(s), and their accessories, connected to the load end of the service conductors to a building or other structure, or an otherwise designated area, and intended to constitute the main control and cutoff of the supply.

SERVICE LATERAL. The underground service conductors between the street main, including any risers at a pole or other structure or from transformers, and the first point of connection to the service-entrance conductors in a terminal box or meter or other enclosure, inside or outside the building wall. Where there is no terminal box, meter or other enclosure with adequate space, the point of connection shall be considered to be the point of entrance of the service conductors into the building.

SERVICE POINT. Service point is the point of connection between the facilities of the serving utility and the premises wiring.

SWITCHES

General-use snap switch. A form of general-use switch constructed so that it can be installed in device boxes or on box covers or otherwise used in conjunction with wiring systems recognized by this code.

General-use switch. A switch intended for use in general distribution and branch circuits. It is rated in amperes and is capable of interrupting its rated current at its rated voltage.

Isolating switch. A switch intended for isolating an electric circuit from the source of power. It has no interrupting rating and is intended to be operated only after the circuit has been opened by some other means.

Motor-circuit switch. A switch, rated in horsepower, capable of interrupting the maximum operating overload current of a motor of the same horsepower rating as the switch at the rated voltage.

UTILIZATION EQUIPMENT. Equipment that utilizes electric energy for electronic, electromechanical, chemical, heating, lighting or similar purposes.

VENTILATED. Provided with a means to permit circulation of air sufficient to remove an excess of heat, fumes or vapors.

VOLTAGE (OF A CIRCUIT). The greatest root-mean-square (rms) (effective) difference of potential between any two conductors of the circuit concerned.

VOLTAGE, NOMINAL. A nominal value assigned to a circuit or system for the purpose of conveniently designating its voltage class (e.g., 120/240). The actual voltage at which a circuit operates can vary from the nominal within a range that permits satisfactory operation of equipment.

VOLTAGE TO GROUND. For grounded circuits, the voltage between the given conductor and that point or conductor of the circuit that is grounded. For ungrounded circuits, the greatest voltage between the given conductor and any other conductor of the circuit.

WATER TIGHT. So constructed that moisture will not enter the enclosure under specified test conditions.

WEATHERPROOF. So constructed or protected that exposure to the weather will not interfere with successful operation.

Chapter 35
SERVICES

SECTION E3501
GENERAL SERVICES

E3501.1 Scope. This chapter covers service conductors and equipment for the control and protection of services and their installation requirements.

E3501.2 Number of services. A dwelling unit shall be supplied by only one service.

E3501.3 One building or other structure not to be supplied through another. Service conductors supplying a building or other structure shall not pass through the interior of another building or other structure.

E3501.4 Other conductors in raceway or cable. Conductors other than service conductors shall not be installed in the same service raceway or service cable.

Exceptions:

1. Grounding conductors and bonding jumpers.
2. Load management control conductors having overcurrent protection.

E3501.5 Raceway seal. Where a service raceway enters from an underground distribution system, it shall be sealed in accordance with Section E3703.6.

E3501.6 Service disconnect required. A means shall be provided to disconnect all conductors from the service conductors.

E3501.6.1 Marking of service equipment and disconnects. Service disconnects shall be permanently marked as a service disconnect. Service equipment shall be listed for the purpose. Individual meter socket enclosures shall not be considered service equipment.

E3501.6.2 Service disconnect location. The service disconnecting means shall be installed at a readily accessible location either outside of a building or inside nearest the point of entrance of the service conductors. Service disconnecting means shall not be installed in bathrooms. Each occupant shall have access to the disconnect serving the dwelling unit in which they reside.

E3501.7 Maximum number of disconnects. The service disconnecting means shall consist of not more than six switches or six circuit breakers mounted in a single enclosure or in a group of separate enclosures.

SECTION E3502
SERVICE SIZE AND RATING

E3502.1 Rating of ungrounded conductors. Ungrounded service conductors shall have an ampacity of not less than the load served. For one-family dwellings, the rating of the ungrounded conductors shall be not less than 100 amperes, 3 wire. For all other installations, the rating of the ungrounded conductors shall be not less than 60 amperes:

E3502.2 Service load. The minimum load for ungrounded service conductors and service devices that serve 100 percent of the dwelling unit load shall be computed in accordance with Table E3502.2. Ungrounded service conductors and service devices that serve less than 100 percent of the dwelling unit load shall be computed as required for feeders in accordance with Chapter 36.

TABLE E3502.2
MINIMUM SERVICE LOAD CALCULATION

LOADS AND PROCEDURE
3 volt-amperes per square foot of floor area for general lighting and convenience receptacle outlets.
Plus
1,500 volt-amperes × total number of 20-ampere-rated small appliance and laundry circuits.
Plus
The nameplate volt-ampere rating of all fastened-in-place, permanently connected or dedicated circuit-supplied appliances such as ranges, ovens, cooking units, clothes dryers and water heaters.
Apply the following demand factors to the above subtotal:
The minimum subtotal for the loads above shall be 100 percent of the first 10,000 volt-amperes of the sum of the above loads plus 40 percent of any portion of the sum that is in excess of 10,000 volt-amperes.
Plus the largest of the following:
Nameplate rating(s) of air-conditioning and cooling equipment including heat pump compressors.
Nameplate rating of the electric thermal storage and other heating systems where the usual load is expected to be continuous at the full nameplate value. Systems qualifying under this selection shall not be figured under any other category in this table.
Sixty-five percent of nameplate rating of central electric space-heating equipment, including integral supplemental heating for heat pump systems.
Sixty-five percent of nameplate rating(s) of electric space-heating units if less than four separately controlled units.
Forty percent of nameplate rating(s) of electric space-heating units of four or more separately controlled units.
The minimum total load in amperes shall be the volt-ampere sum calculated above divided by 240 volts.

E3502.2.1 Services under 100 amperes. Services that are not required to be 100 amperes shall be sized in accordance with Chapter 36.

E3502.3 Rating of service disconnect. The combined rating of all individual service disconnects serving a single dwelling unit shall not be less than the load determined from Table 3502.2 and shall not be less than as specified in Section 3502.1.

E3502.4 Voltage rating. Systems shall be three-wire, 120/240-volt, single-phase with a grounded neutral.

SECTION E3503
SERVICE, FEEDER AND GROUNDING ELECTRODE CONDUCTOR SIZING

E3503.1 Grounded and ungrounded service conductor size. Conductors used as ungrounded service entrance conductors, service lateral conductors, and feeder conductors that serve as the main power feeder to a dwelling unit shall be those listed in Table E3503.1. Ungrounded service conductors shall have a minimum size in accordance with Table E3503.1. The grounded conductor size shall not be less than the maximum unbalance of the load and shall not be smaller than the required minimum grounding electrode conductor size specified in Table E3503.1.

E3503.2 Ungrounded service conductors for accessory buildings and structures. Ungrounded conductors for other than dwelling units shall have a minimum rating of 60 amperes and shall be sized as required for feeders in Chapter 36.

Exceptions:

1. For loads consisting of not more than two two-wire branch circuits, the minimum size shall be No. 8 copper or No. 6 aluminum or copper-clad aluminum.
2. For loads limited by demand or by the source of supply, the minimum size shall be No. 8 copper or No. 6 aluminum or copper-clad aluminum.
3. For limited loads of a single branch circuit, the minimum size shall be No. 12 copper or No. 10 aluminum or copper-clad aluminum, but in no case smaller than the branch-circuit conductors.

E3503.3 Overload protection. Each ungrounded service conductor shall have overload protection.

E3503.3.1 Ungrounded conductor. Overload protection shall be provided by an overcurrent device in series with each ungrounded service conductor having a rating or setting not higher than the allowable ampacity specified in Table E3503.1.

Exception: Not more than six circuit breakers or six sets of fuses shall serve as the overcurrent device to provide the overload protection. Where the sum of the ratings of the circuit breakers or fuses exceeds the ampacity of the service conductors, the load calculated in accordance with Section E3502 shall not exceed the allowable ampacity of the service conductors.

TABLE E3503.1
SERVICE CONDUCTOR AND GROUNDING ELECTRODE CONDUCTOR SIZING

CONDUCTOR TYPES AND SIZES—THHW, THW, THWN, USE, XHHW (Parallel sets of 1/0 and larger conductors are permitted in either a single raceway or in separate raceways)		ALLOWABLE AMPACITY	MINIMUM GROUNDING ELECTRODE CONDUCTOR SIZE[a]	
Copper (AWG)	Aluminum and copper-clad aluminum (AWG)	Maximum load (amps)	Copper (AWG)	Aluminum (AWG)
4	2	100	8[b]	6[c]
3	1	110	8[b]	6[c]
2	1/0	125	8[b]	6[c]
1	2/0	150	6[c]	4
1/0	3/0	175	6[c]	4
2/0	4/0 or two sets of 1/0	200	4[d]	2[d]
3/0	250 kcmil or two sets of 2/0	225	4[d]	2[d]
4/0 or two sets of 1/0	300 kcmil or two sets of 3/0	250	2[d]	1/0[d]
250 kcmil or two sets of 2/0	350 kcmil or two sets of 4/0	300	2[d]	1/0[d]
350 kcmil or two sets of 3/0	500 kcmil or two sets of 250 kcmil	350	2[d]	1/0[d]
400 kcmil or two sets of 4/0	600 kcmil or two sets of 300 kcmil	400	1/0[d]	3/0[d]

For SI: 1 inch = 25.4 mm.

a. Where protected by a metal raceway, grounding electrode conductors shall be electrically bonded to the metal raceway at both ends.

b. No. 8 grounding electrode conductors shall be protected with metal conduit or nonmetallic conduit.

c. Where not protected, No. 6 grounding electrode conductors shall closely follow a structural surface for physical protection. The supports shall be spaced not more than 24 inches on center and shall be within 12 inches of any enclosure or termination.

d. Where the sole grounding electrode system is a ground rod or pipe as covered in Section E3508.2, the grounding electrode conductor shall not be required to be larger than No. 6 copper or No. 4 aluminum. Where the sole grounding electrode system is the footing steel as covered in Section E3508.1.2, the grounding electrode conductor shall not be required to be larger than No. 4 copper conductor.

A set of fuses shall be considered all the fuses required to protect all the ungrounded conductors of a circuit. Single-pole circuit breakers, grouped with "handle ties" or a "master handle" to disconnect all conductors of the service, with not more than six operations of the hand, shall be considered as the protective device.

E3503.3.2 Not in grounded conductor. Overcurrent devices shall not be connected in series with a grounded service conductor except where a circuit breaker is used that simultaneously opens all conductors of the circuit.

E3503.4 Grounding electrode conductor size. The grounding electrode conductors shall be sized based on the size of the service entrance conductors as required in Table E3503.1.

SECTION E3504 OVERHEAD SERVICE-DROP AND SERVICE CONDUCTOR INSTALLATION

E3504.1 Clearance from building openings. Open conductors and multiconductor cables without an overall outer jacket shall have a clearance of not less than 3 feet (914 mm) from the sides of doors, porches, decks, stairs, ladders, fire escapes, and balconies, and from the sides and bottom of windows that open. See Figure E3504.1.

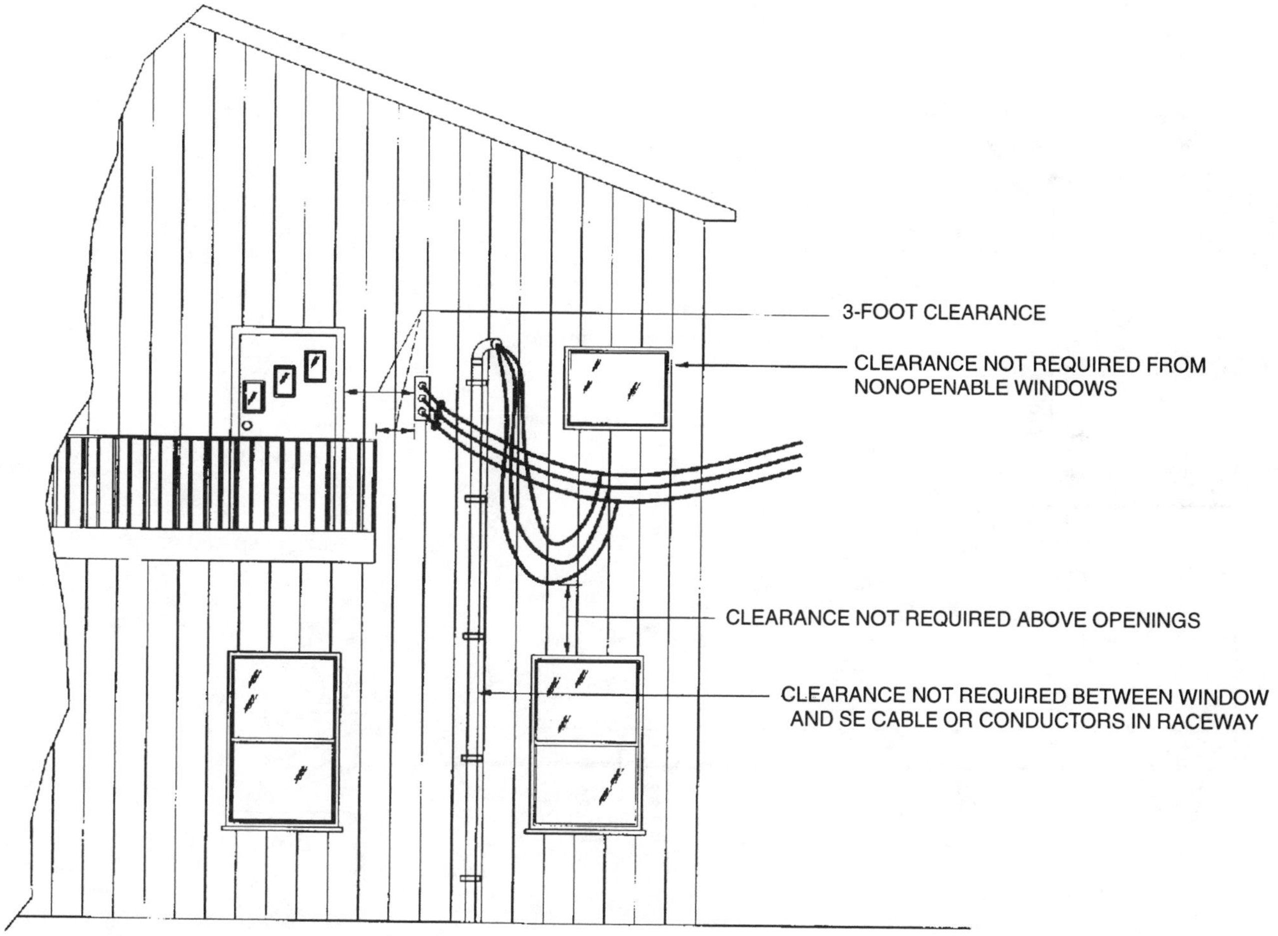

For SI: 1 foot = 304.8 mm.

FIGURE E3504.1
CLEARANCES FROM BUILDING OPENINGS

E3504.2 Vertical clearances. The vertical clearances of all service-drop conductors shall be based on a conductor temperature of 60°F (15°C), with no wind and with final unloaded sag in the wire, conductor or cable. Service-drop conductors shall not have ready access and shall comply with Sections E3504.2.1 and E3504.2.2.

E3504.2.1 Above roofs. Conductors shall have a vertical clearance of not less than 8 feet (2438 mm) above the roof surface. The vertical clearance above the roof level shall be maintained for a distance of not less than 3 feet (914 mm) in all directions from the edge of the roof. See Figure E3504.2.1.

Exceptions:

1. Conductors above a roof surface subject to pedestrian traffic shall have a vertical clearance from the roof surface in accordance with Section E3504.2.2.
2. Where the roof has a slope of 4 inches (102 mm) in 12 inches (305 mm), or greater, the minimum clearance shall be 3 feet (914 mm).
3. The minimum clearance above only the overhanging portion of the roof shall not be less than 18 inches (457 mm) where not more than 6 feet (1829 mm) of conductor length passes over 4 feet (1219 mm) or less of roof surface measured horizontally and such conductors are terminated at a through-the-roof raceway or approved support.
4. The requirement for maintaining the vertical clearance for a distance of 3 feet (914 mm) from the edge of the roof shall not apply to the final conductor span where the service drop is attached to the side of a building.

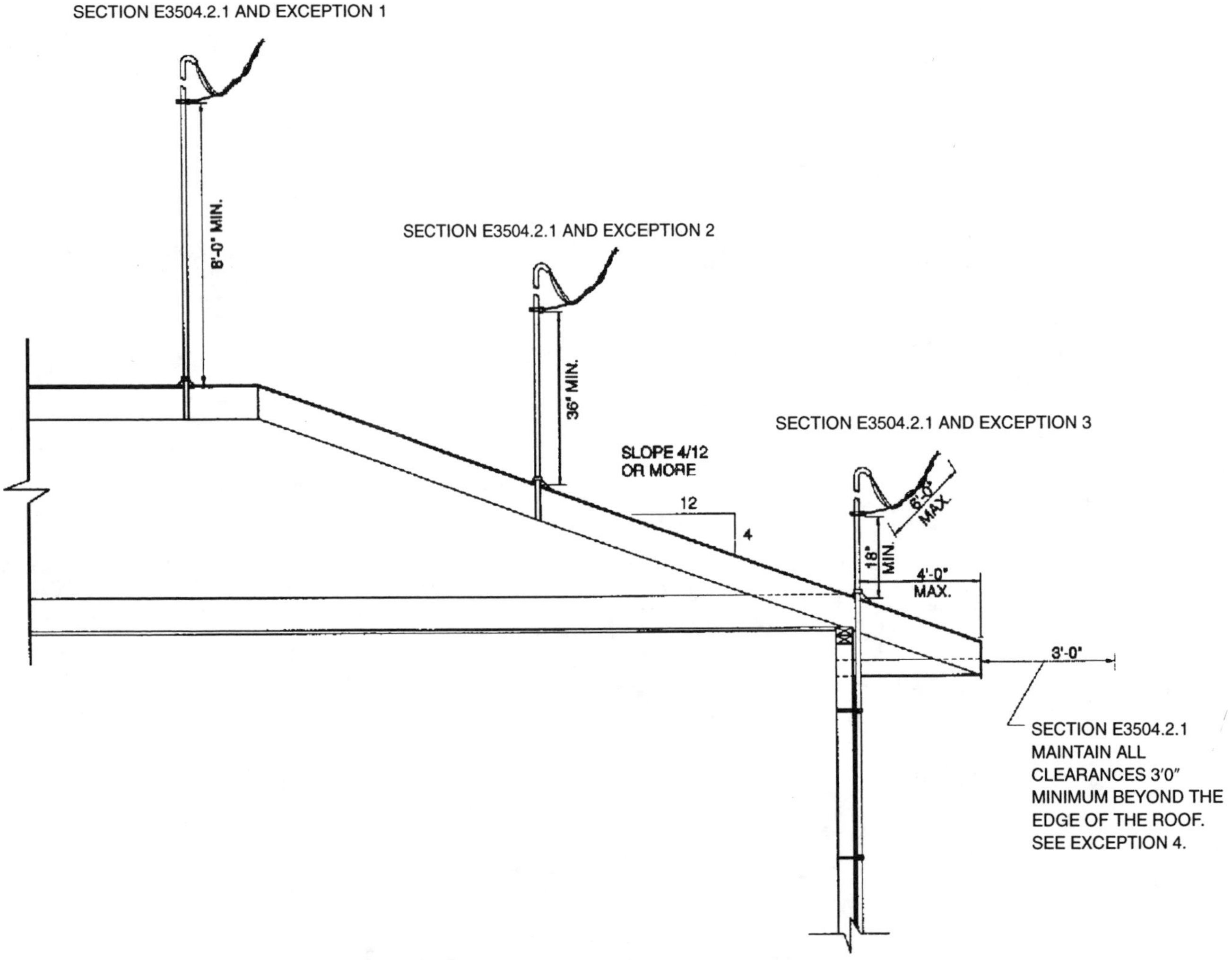

For SI: 1 inch = 25.4 mm, 1 foot = 304.8 mm.

FIGURE E3504.2.1
CLEARANCES FROM ROOFS

E3504.2.2 Vertical clearance from grade. Service-drop conductors shall have the following minimum clearances from final grade:

1. For service-drop cables supported on and cabled together with a grounded bare messenger wire, the minimum vertical clearance shall be 10 feet (3048 mm) at the electric service entrance to buildings, at the lowest point of the drip loop of the building electric entrance, and above areas or sidewalks accessed by pedestrians only. Such clearance shall be measured from final grade or other accessible surfaces.
2. Twelve feet (3658 mm)—over residential property and driveways.
3. Eighteen feet (5486 mm)—over public streets, alleys, roads or parking areas subject to truck traffic.

E3504.3 Point of attachment. The point of attachment of the service-drop conductors to a building or other structure shall provide the minimum clearances as specified in Sections E3504.1 through E3504.2.2. In no case shall the point of attachment be less than 10 feet (3048 mm) above finished grade.

E3504.4 Means of attachment. Multiconductor cables used for service drops shall be attached to buildings or other structures by fittings approved for the purpose.

E3504.5 Service masts as supports. Where a service mast is used for the support of service-drop conductors, it shall be of adequate strength or be supported by braces or guys to withstand the strain imposed by the service drop. Where raceway-type service masts are used, all equipment shall be approved. Only power service drop conductors shall be permitted to be attached to a service mast.

E3504.6 Supports over buildings. Service-drop conductors passing over a roof shall be securely supported. Where practicable, such supports shall be independent of the building.

SECTION E3505
SERVICE-ENTRANCE CONDUCTORS

E3505.1 Insulation of service-entrance conductors. Service-entrance conductors entering or on the exterior of buildings or other structures shall be insulated in accordance with Section E3306.5.

Exceptions:

1. A copper grounded conductor shall not be required to be insulated where it is:
   1.1. In a raceway or part of a service cable assembly,
   1.2. Directly buried in soil of suitable condition or
   1.3. Part of a cable assembly listed for direct burial without regard to soil conditions.
2. An aluminum or copper-clad aluminum grounded conductor shall not be required to be insulated where part of a cable or where identified for direct burial or utilization in underground raceways.

E3505.2 Wiring methods for services. Service-entrance wiring methods shall be installed in accordance with the applicable requirements in Chapter 37.

E3505.3 Spliced conductors. Service-entrance conductors shall permitted to be spliced or tapped by clamped or bolted connections. Splices shall be made in enclosures or, if directly buried, with listed underground splice kits. Conductor splices shall be made in accordance with Chapters 33, 36, 37 and 38.

E3505.4 Protection against physical damage. Underground service-entrance conductors shall be protected against physical damage in accordance with Chapter 37.

E3505.5 Protection of service cables against damage. Above-ground service-entrance cables, where subject to physical damage, such as where installed in exposed areas near sidewalks, walkways and driveways or where subject to contact with awnings, shutters or similar objects, shall be protected by one or more of the following: rigid metal conduit, intermediate metal conduit, rigid nonmetallic conduit suitable for the location, electrical metallic tubing or other approved means.

E3505.6 Mounting supports. Service cables shall be supported by straps or other approved means within 12 inches (305 mm) of every service head, gooseneck or connection to a raceway or enclosure and at intervals not exceeding 30 inches (762 mm).

E3505.7 Raceways to drain. Where exposed to the weather, raceways enclosing service-entrance conductors shall be raintight and arranged to drain. Where embedded in masonry, raceways shall be arranged to drain.

E3505.8 Overhead service locations. Connections at service heads shall be in accordance with Sections E3505.8.1 through E3505.8.7.

E3505.8.1 Rain-tight service head. Service raceways shall be equipped with a rain-tight service head at the point of connection to service-drop conductors.

E3505.8.2 Service cable, service head or gooseneck. Service cable shall be equipped with a rain-tight service head or shall be formed into a gooseneck in an approved manner.

E3505.8.3 Service head location. Service heads, and goosenecks in service-entrance cables, shall be located above the point of attachment of the service-drop conductors to the building or other structure.

Exception: Where it is impracticable to locate the service head above the point of attachment, the service head location shall not be more than 24 inches (610 mm) from the point of attachment.

E3505.8.4 Separately bushed openings. Service heads shall have conductors of different potential brought out through separately bushed openings.

E3505.8.5 Drip loops. Drip loops shall be formed on individual conductors. To prevent the entrance of moisture, service-entrance conductors shall be connected to the service-drop conductors either below the level of the service head or below the level of the termination of the service-entrance cable sheath.

E3505.8.6 Conductor arrangement. Service-drop conductors and service-entrance conductors shall be arranged so that water will not enter service raceways or equipment.

E3505.8.7 Secured. Service cables shall be held securely in place.

SECTION E3506 SERVICE EQUIPMENT—GENERAL

E3506.1 Service equipment enclosures. Energized parts of service equipment shall be enclosed.

E3506.2 Working space. In no case shall the working space in the vicinity of service equipment be less than that specified in Chapter 33.

E3506.3 Available short-circuit current. Service equipment shall be suitable for the maximum fault current available at its supply terminals, but not less than 10,000 amperes.

E3506.4 Marking. Service equipment shall be marked to identify it as being suitable for use as service equipment. Individual meter socket enclosures shall not be considered service equipment.

SECTION E3507 SYSTEM GROUNDING

E3507.1 System service ground. The premises wiring system shall be grounded at the service with a grounding electrode conductor connected to a grounding electrode system as required by this code. Grounding electrode conductors shall be sized in accordance with Table E3503.1.

E3507.2 Location of grounding electrode conductor connection. The grounding electrode conductor shall be connected to the grounded service conductor at any accessible point from the load end of the service drop or service lateral to and including the terminal or bus to which the grounded service conductor is connected at the service disconnecting means. A grounding connection shall not be made to any grounded circuit conductor on the load side of the service disconnecting means, except as provided in Section E3507.3.

E3507.3 Two or more buildings or structures supplied from a common service. Where two or more buildings or structures are supplied from a common service, the grounding electrode(s) required in Section E3508 at each building or structure shall be connected in a manner specified in Section E3507.3.1 or E3507.3.2. Where there is no existing grounding electrode, the grounding electrode(s) required in Section E3508 shall be installed.

Exception: A grounding electrode at separate buildings or structures shall not be required where only one branch circuit supplies the building or structure and the branch circuit includes an equipment grounding conductor for grounding the noncurrent-carrying parts of all equipment.

E3507.3.1 Equipment grounding conductor. An equipment grounding conductor as described in Section 3808 shall be run with the supply conductors and connected to the building or structure disconnecting means and to the grounding electrode(s). The equipment grounding conductor shall be used for grounding or bonding of equipment, structures, or frames required to be grounded or bonded. Any installed grounded conductor shall not be connected to the equipment grounding conductor or to the grounding electrode(s).

E3507.3.2 Grounded conductor. Where an equipment grounding conductor is not run with the supply conductors to the building or structure, and there are no continuous metallic paths bonded to the grounding system in both buildings or structures involved, and ground-fault protection of equipment has not been installed on the common ac service, the grounded circuit conductor run with the supply conductors to the building or structure shall be connected to the building or structure disconnecting means and to the grounding electrode(s) and shall be used for grounding or bonding of equipment, structures, or frames required to be grounded or bonded.

E3507.4 Grounding electrode conductor. A grounding electrode conductor shall be used to connect the equipment grounding conductors, the service equipment enclosures, and the grounded service conductor to the grounding electrode(s).

E3507.5 Main bonding jumper. An unspliced main bonding jumper shall be used to connect the equipment grounding conductor(s) and the service-disconnect enclosure to the grounded conductor of the system within the enclosure for each service disconnect.

E3507.6 Common grounding electrode. The service grounding electrode in or at a building shall be used to ground conductor enclosures and equipment in or on that building. Two or more grounding electrodes that are effectively bonded together shall be considered as a single grounding electrode system.

SECTION E3508 GROUNDING ELECTRODE SYSTEM

E3508.1 Grounding electrode system. Where available on the premises at each building or structure served, electrodes specified in Sections E3508.1.1, E3508.1.2 and E3508.1.3, and any made electrodes specified in Section E3508.2, shall be bonded together to form the grounding electrode system. Interior metal water piping located more than 5 feet (1524 mm) from the point of entrance to the building shall not be used as part of the grounding electrode system or as a conductor to interconnect the electrodes that are part of the grounding electrode system. The bonding jumper shall be sized and installed in accordance with Section E3509. The unspliced grounding electrode conductor shall run to any convenient grounding electrode available in the grounding electrode system. Such conductor shall be sized based on the largest grounding electrode conductor required for any of the available electrodes.

E3508.1.1 Metal underground water pipe. A metal underground water pipe in direct contact with the earth for

10 feet (3048 mm) or more and electrically continuous to the points of connection of the grounding electrode conductor and the bonding conductors shall be considered as a grounding electrode. Continuity of the grounding path or the bonding connection to interior piping shall not rely on water meters or filtering devices and similar equipment. A metal underground water pipe shall be supplemented by an additional electrode of a type specified in this section or in Section E3508.2. The supplemental electrode shall be bonded to the grounding electrode conductor, the grounded service-entrance conductor, the grounded service raceway or any grounded service enclosure.

Where the supplemental electrode is a made electrode in accordance with Section E3508.2, that portion of the bonding jumper that is the sole connection to the supplemental grounding electrode shall not be required to be larger than No. 6 copper wire or No. 4 aluminum wire.

E3508.1.2 Concrete-encased electrode. An electrode encased by at least 2 inches (51 mm) of concrete, located within and near the bottom of a concrete foundation or footing that is in direct contact with the earth, consisting of at least 20 feet (6096 mm) of one or more bare or zinc-galvanized or other electrically conductive coated steel reinforcing bars or rods of not less than $^{1}/_{2}$ inch (12.7 mm) diameter, or consisting of at least 20 feet (6096 mm) of bare copper conductor not smaller than No. 4 shall be considered as a grounding electrode. Reinforcing bars shall be permitted to be bonded together by the usual tie wires or other effective means.

E3508.1.3 Ground rings. A ground ring encircling the building or structure, in direct contact with the earth at a depth below the earth's surface of not less than 2.5 feet (762 mm), consisting of at least 20 feet (6096 mm) of bare copper conductor not smaller than No. 2 shall be considered as a grounding electrode.

E3508.2 Made and other electrodes. Where none of the electrodes specified in Section E3508.1 is available, one or more of the electrodes specified in Section E3508.2.1 shall be used. Where practicable, made electrodes shall be embedded below permanent moisture level. Made electrodes shall be free from nonconductive coatings such as paint or enamel. Where more than one electrode is used, each electrode of one grounding system shall be not less than 6 feet (1829 mm) from any other electrode of another grounding system. Two or more grounding electrodes that are effectively bonded together shall be considered as a single grounding electrode system.

E3508.2.1 Rod and pipe electrodes. Rod and pipe electrodes shall be not less than 8 feet (2438 mm) in length and shall consist of the following materials:

1. Electrodes of pipe or conduit shall not be smaller than $^{3}/_{4}$-inch (19.1 mm) trade size and, where of iron or steel, shall have the outer surface galvanized or otherwise metal-coated for corrosion protection.
2. Electrodes of rods of iron or steel shall be at least $^{5}/_{8}$ inch (15.9 mm) in diameter. Stainless steel rods less than $^{5}/_{8}$ inch (15.9 mm) in diameter, nonferrous rods or their equivalent shall be listed and shall be not less than $^{1}/_{2}$ inch (12.7 mm) in diameter.

E3508.2.2 Installation. The electrode shall be installed such that at least 8 feet (2438 mm) of length is in contact with the soil. It shall be driven to a depth of not less than 8 feet (2438 mm) except that, where rock bottom is encountered, the electrode shall be driven at an oblique angle not to exceed 45 degrees from the vertical or shall be buried in a trench that is at least 2.5 feet (762 mm) deep. The upper end of the electrode shall be flush with or below ground level unless the aboveground end and the grounding electrode conductor attachment are protected against physical damage.

E3508.2.3 Aluminum electrodes. Aluminum electrodes shall not be permitted.

E3508.3 Resistance of made electrodes. A single electrode consisting of a rod or pipe that does not have a resistance to ground of 25 ohms or less shall be augmented by one additional electrode of any of the types specified in Sections E3508.1 or E3508.2. Where multiple rod or pipe electrodes are installed to meet the requirements of this section, they shall be not less than 6 feet (1829 mm) apart.

E3508.4 Metal underground gas piping system. A metal underground gas piping system shall not be used as a grounding electrode.

SECTION E3509 BONDING

E3509.1 General. Bonding shall be provided where necessary to ensure electrical continuity and the capacity to conduct safely any fault current likely to be imposed.

E3509.2 Bonding of services. The noncurrent-carrying metal parts of the following equipment shall be effectively bonded together:

1. The service raceways or service cable armor.
2. All service enclosures containing service conductors, including meter fittings, and boxes, interposed in the service raceway or armor.
3. Any metallic raceways or armor enclosing a grounding electrode conductor. Bonding shall apply at each end and to all intervening raceways, boxes and enclosures between the service equipment and the grounding electrode.

E3509.3 Bonding to other systems. An accessible means external to enclosures for connecting intersystem bonding and grounding conductors shall be provided at the service by at least one of the following means:

1. Exposed nonflexible metallic service raceways.
2. Exposed grounding electrode conductor.
3. Approved means for the external connection of a copper or other corrosion-resistant bonding or grounding conductor to the service raceway or equipment.

For the purposes of providing an accessible means for intersystem bonding, the disconnecting means at a separate building or structure as permitted in Section E3507.3 shall be considered the service equipment.

E3509.4 Method of bonding at the service. Electrical continuity at service equipment, service raceways, and service conductor enclosures shall be ensured by one or more of the methods specified in Sections E3509.4.1 through E3509.4.4.

Bonding jumpers meeting the other requirements of this code shall be used around concentric or eccentric knockouts that are punched or otherwise formed so as to impair the electrical connection to ground. Standard locknuts or bushings shall not be the sole means for the bonding required by this section.

E3509.4.1 Grounded service conductor. Equipment shall be bonded to the grounded service conductor in a manner provided in this code.

E3509.4.2 Threaded connections. Equipment shall be bonded by connections using threaded couplings or threaded bosses on enclosures. Such connections shall be made wrench tight.

E3509.4.3 Threadless couplings and connectors. Equipment shall be bonded by threadless couplings and connectors for metal raceways and metal-clad cables. Such couplings and connectors shall be made wrench tight. Standard locknuts or bushings shall not be used for the bonding required by this section.

E3509.4.4 Other devices. Equipment shall be bonded by approved devices, such as bonding-type locknuts and bushings.

E3509.5 Sizing bonding jumper on supply side of service and main bonding jumper. The bonding jumper shall not be smaller than the sizes shown in Table E3503.1 for grounding electrode conductors. Where the service-entrance conductors are paralleled in two or more raceways or cables, the equipment bonding jumper, where routed with the raceways or cables, shall be run in parallel. The size of the bonding jumper for each raceway or cable shall be based on the size of the service-entrance conductors in each raceway or cable. The bonding jumper for a grounding electrode conductor raceway shall be the same size or larger than the required enclosed grounding electrode conductor.

E3509.6 Metal water piping bonding. The interior metal water piping system shall be bonded to the service equipment enclosure, the grounded conductor at the service, the grounding electrode conductor where of sufficient size, or to the one or more grounding electrodes used. The bonding jumper shall be sized in accordance with Table E3503.1. The points of attachment of the bonding jumper(s) shall be accessible.

E3509.7 Metal gas piping bonding. Each aboveground portion of a gas piping system upstream from the equipment shutoff valve shall be electrically continuous and bonded to the grounding electrode system.

E3509.8 Bonding other metal piping. Interior metal piping capable of becoming energized shall be bonded to the service equipment enclosure, the grounded conductor at the service, the grounding electrode conductor where of sufficient size, or to the one or more grounding electrodes used. The bonding jumper shall be sized in accordance with Table E3808.12 using the rating of the circuit capable of energizing the piping. A piping system shall be considered as bonded where connected to the equipment grounding conductor for the circuit capable of energizing such piping system.

SECTION E3510 GROUNDING ELECTRODE CONDUCTORS

E3510.1 Installation. A grounding electrode conductor or its enclosure shall be securely fastened to the surface on which it is carried. A No. 4 or larger conductor shall be protected where exposed to severe physical damage. A No. 6 grounding conductor that is free from exposure to physical damage shall be permitted to be run along the surface of the building construction without metal covering or protection where it is and securely fastened to the construction; otherwise, it shall be in rigid metal conduit, intermediate metal conduit, rigid nonmetallic conduit, electrical metallic tubing or cable armor. Grounding conductors smaller than No. 6 shall be in rigid metal conduit, intermediate metal conduit, rigid nonmetallic conduit, electrical metallic tubing or cable armor.

Insulated or bare aluminum or copper-clad aluminum grounding conductors shall not be used where in direct contact with masonry or the earth or where subject to corrosive conditions. Where used outside, aluminum or copper-clad aluminum grounding conductors shall not be installed within 18 inches (457 mm) of the earth.

E3510.2 Enclosures for grounding electrode conductors. Metal enclosures for grounding electrode conductors shall be electrically continuous from the point of attachment to cabinets or equipment to the grounding electrode, and shall be securely fastened to the ground clamp or fitting. Metal enclosures that are not physically continuous from cabinet or equipment to the grounding electrode shall be made electrically continuous by bonding each end to the grounding conductor. Where a raceway is used as protection for a grounding conductor, the installation shall comply with the requirements of Chapter 37.

SECTION E3511 GROUNDING ELECTRODE CONDUCTOR CONNECTION TO THE GROUNDING ELECTRODES

E3511.1 Methods of grounding conductor connection to electrodes. The grounding conductor shall be connected to the grounding electrode by listed lugs, listed pressure connectors, listed clamps or other listed means. Ground clamps shall be listed for the materials of the grounding electrode and the grounding electrode conductor and, where used on pipe, rod or other buried electrodes, shall also be listed for direct soil burial. Not more than one conductor shall be connected to the grounding electrode by a single clamp or fitting unless the clamp or fitting is listed for multiple conductors. One of the methods indicated in the following items shall be used:

1. Bolted clamp. A listed bolted clamp of cast bronze or brass, or plain or malleable iron.
2. Pipe fitting, pipe plug, etc. A pipe fitting, pipe plug or other approved device screwed into a pipe or pipe fitting.
3. Sheet-metal-strap-type ground clamp. For indoor telecommunications purposes only. A listed sheet-metal-

strap-type ground clamp having a rigid metal base that seats on the electrode and having a strap of such material and dimensions that is not likely to stretch during or after installation.

4. Other means. Equally substantial approved means.

E3511.2 Accessibility. The connection of the grounding electrode conductor to the grounding electrodes that are not buried or concrete encased shall be accessible.

E3511.3 Effective grounding path. The connection of the grounding electrode conductor shall be made in a manner that will ensure a permanent and effective grounding path. Where necessary to ensure effective grounding for a metal piping system used as a grounding electrode, effective bonding shall be provided around insulated joints and sections and around any equipment that is likely to be disconnected for repairs or replacement. Bonding conductors shall be of sufficient length to permit removal of such equipment while retaining the integrity of the bond.

E3511.4 Protection of ground clamps and fittings. Ground clamps or other fittings shall be approved for applications without protection or shall be protected from physical damage with approved enclosures.

E3511.5 Clean surfaces. Nonconductive coatings on equipment to be grounded shall be removed from threads and other contact surfaces to ensure good electrical continuity or shall be connected by fittings that make such removal unnecessary.

Chapter 36
BRANCH CIRCUIT AND FEEDER REQUIREMENTS

SECTION E3601 GENERAL

E3601.1 Scope. This chapter covers branch circuits and feeders and specifies the minimum required branch circuits, the allowable loads and the required overcurrent protection for branch circuits and feeders that serve less than 100 percent of the total dwelling unit load. Feeder circuits that serve 100 percent of the dwelling unit load shall be sized in accordance with the procedures in Chapter 35.

E3601.2 Branch-circuit and feeder ampacity. Branch-circuit and feeder conductors shall have ampacities not less than the maximum load to be served. Where a branch circuit or a feeder supplies continuous loads or any combination of continuous and noncontinuous loads, the minimum branch-circuit or feeder conductor size, before the application of any adjustment or correction factors, shall have an allowable ampacity equal to or greater than the noncontinuous load plus 125 percent of the continuous load.

E3601.3 Multi-outlet branch circuits. Conductors of multi-outlet branch circuits supplying more than one receptacle for cord- and plug-connected portable loads shall have ampacities of not less than the rating of the branch circuit.

E3601.4 Multiwire branch circuits. All conductors for multiwire branch circuits shall originate from the same panelboard. Where two or more devices on the same yoke or strap are supplied by a multiwire branch circuit, a means shall be provided at the panelboard where the circuit originates to simultaneously disconnect all ungrounded conductors of the multiwire circuit. Except where all ungrounded conductors are opened simultaneously by the branch-circuit overcurrent device, multiwire branch circuits shall supply only line-to-neutral loads or only one appliance.

SECTION E3602 BRANCH CIRCUIT RATINGS

E3602.1 Branch-circuit voltage limitations. The voltage rating of branch circuits that supply lighting fixtures, receptacles for cord- and plug-connected loads of up to 1,440 volt-amperes, and loads of less than 0.25 horsepower shall be limited to a maximum rating of 120 volts, nominal, between conductors.

Branch circuits that supply cord- and plug-connected or permanently connected utilization equipment and appliances rated at 1,440 volt-amperes and higher or more than $^1/_4$ horsepower shall be rated at 120 volts or 240 volts, nominal.

E3602.2 Branch-circuit ampere rating. Branch circuits shall be rated in accordance with the maximum allowable ampere rating or setting of the overcurrent protection device. The rating for other than individual branch circuits shall be 15, 20, 30, 40 and 50 amperes. Where conductors of higher ampacity are used, the ampere rating or setting of the specified overcurrent device shall determine the circuit rating.

E3602.3 Fifteen- and 20-ampere branch circuits. A 15- or 20-ampere branch circuit shall be permitted to supply lighting units, or other utilization equipment, or a combination of both. The rating of any one cord- and plug-connected utilization equipment shall not exceed 80 percent of the branch-circuit ampere rating. The total rating of utilization equipment fastened in place, other than lighting fixtures, shall not exceed 50 percent of the branch-circuit ampere rating where lighting units, cord- and plug-connected utilization equipment not fastened in place, or both, are also supplied.

E3602.4 Thirty-ampere branch circuits. A 30-ampere branch circuit shall be permitted to supply fixed utilization equipment. A rating of any one cord- and plug-connected utilization equipment shall not exceed 80 percent of the branch-circuit ampere rating.

E3602.5 Branch circuits serving multiple loads or outlets. General-purpose branch circuits shall supply lighting outlets, appliances, equipment or receptacle outlets, and combinations of such. The rating of a fastened-in-place appliance or equipment, where used in combination on the same branch circuit with light fixtures, receptacles, and/or other appliances or equipment not fastened in place, shall not exceed 50 percent of the branch-circuit rating. Multioutlet branch circuits serving lighting or receptacles shall be limited to a maximum branch-circuit rating of 20 amperes.

E3602.6 Branch circuits serving a single motor. Branch-circuit conductors supplying a single motor shall have an ampacity not less than 125 percent of the motor full-load current rating.

E3602.7 Branch circuits serving motor-operated and combination loads. For circuits supplying loads consisting of motor-operated utilization equipment that is fastened in place and that has a motor larger than 0.125 horsepower in combination with other loads, the total computed load shall be based on 125 percent of the largest motor load plus the sum of the other loads.

E3602.8 Branch-circuit inductive lighting loads. For circuits supplying lighting units having ballasts, the computed load shall be based on the total ampere ratings of such units and not on the total watts of the lamps.

E3602.9 Branch-circuit load for ranges and cooking appliances. It shall be permissible to compute the branch-circuit load for one range in accordance with Table E3604.3(2). The branch-circuit load for one wall-mounted oven or one counter-mounted cooking unit shall be the nameplate rating of the appliance. The branch-circuit load for a counter-mounted cooking unit and not more than two wall-mounted ovens all supplied from a single branch circuit and located in the same room shall be computed by adding the nameplate ratings of the individual appliances and treating the total as equivalent to one range.

E3602.9.1 Minimum branch circuit for ranges. Ranges with a rating of 8.75 kVA or more shall be supplied by a branch circuit having a minimum rating of 40 amperes.

E3602.10 Branch circuits serving heating loads. Branch circuit ratings for space-heating and water-heating appliances shall be 125 percent of the nameplate rating of the appliance. Branch circuits supplying two or more outlets for fixed electric space-heating equipment shall be rated 15, 20 or 30 amperes.

E3602.11 Branch circuits for air-conditioning and heat pump equipment. The ampacity of the conductors supplying multimotor and combination load equipment shall not be less than the minimum circuit ampacity marked on the equipment. The branch-circuit overcurrent device rating shall be the size and type marked on the appliance and shall be listed for the specific purpose.

E3602.12 Branch circuits serving room air conditioners. A room air conditioner shall be considered as a single motor unit in determining its branch-circuit requirements where all the following conditions are met:

1. It is cord- and attachment plug-connected.
2. The rating is not more than 40 amperes and 250 volts; single phase.
3. Total rated-load current is shown on the room air-conditioner nameplate rather than individual motor currents.
4. The rating of the branch-circuit short-circuit and ground-fault protective device does not exceed the ampacity of the branch-circuit conductors, or the rating of the branch-circuit conductors, or the rating of the receptacle, whichever is less.

E3602.12.1 Where no other loads are supplied. The total marked rating of a cord- and attachment plug-connected room air conditioner shall not exceed 80 percent of the rating of a branch circuit where no other appliances are also supplied.

E3602.12.2 Where lighting units or other appliances are also supplied. The total marked rating of a cord- and attachment plug-connected room air conditioner shall not exceed 50 percent of the rating of a branch circuit where lighting or other appliances are also supplied.

E3602.13 Branch-circuit requirement—summary. The requirements for circuits having two or more outlets, or receptacles, other than the receptacle circuits of Section E3603.2, are summarized in Table E3602.13. Branch circuits in dwelling units shall supply only loads within that dwelling unit or loads associated only with that dwelling unit. Branch circuits required for the purpose of lighting, central alarm, signal, communications, or other needs for public or common areas of a two-family dwelling shall not be supplied from equipment that supplies an individual dwelling unit.

TABLE E3602.13
BRANCH-CIRCUIT REQUIREMENTS—SUMMARY[a,b]

	CIRCUIT RATING		
	15 amp	**20 amp**	**30 amp**
Conductors: Minimum size (AWG) circuit conductors	14	12	10
Maximum overcurrent-protection device rating Ampere rating	15	20	30
Outlet devices: Lampholders permitted Receptacle rating (amperes)	 Any type 15 maximum	 Any type 15 or 20	 N/A 30
Maximum load (amperes)	15	20	30

a. These gages are for copper conductors.
b. N/A means not allowed.

SECTION E3603 REQUIRED BRANCH CIRCUITS

E3603.1 Branch circuits for heating. Central heating equipment other than fixed electric space heating shall be supplied by an individual branch circuit. Auxiliary equipment, such as pumps, motorized valves, humidifiers and air cleaners directly associated with the central heating equipment, shall not be prohibited from connecting to the same branch circuit as the central heating equipment.

E3603.2 Kitchen and dining area receptacles. A minimum of two 20-ampere-rated branch circuits shall be provided to serve receptacles located in the kitchen, pantry, breakfast area and dining area. The kitchen countertop receptacles shall be served by a minimum of two 20-ampere-rated branch circuits, either or both of which shall also be permitted to supply other receptacle outlets in the kitchen, pantry, breakfast area and dining area.

E3603.3 Laundry circuit. A minimum of one 20-ampere-rated branch circuit shall be provided for receptacles located in the laundry area and shall serve only receptacle outlets located in the laundry area.

E3603.4 Bathroom branch circuits. A minimum of one 20-ampere branch circuit shall be provided to supply the bathroom receptacle outlet(s). Such circuits shall have no other outlets.

Exception: Where the 20-ampere circuit supplies a single bathroom, outlets for other equipment within the same bathroom shall be permitted to be supplied in accordance with Section E3602.

E3603.5 Number of branch circuits. The minimum number of branch circuits shall be determined from the total computed load and the size or rating of the circuits used. The number of circuits shall be sufficient to supply the load served. In no case shall the load on any circuit exceed the maximum specified by Section E3602.

E3603.6 Branch-circuit load proportioning. Where the branch-circuit load is computed on a volt-amperes-per-square-foot (m^2) basis, the wiring system shall have the capacity to serve not less than the calculated load. This load shall be evenly proportioned among multioutlet branch circuits.

SECTION E3604
FEEDER REQUIREMENTS

E3604.1 Conductor size. Feeder conductors that do not serve 100 percent of the dwelling unit load and branch-circuit conductors shall be of a size sufficient to carry the load as determined by this chapter. Feeder conductors shall not be required to be larger than the service-entrance conductors that supply the dwelling unit. The load for feeder conductors that serve as the main power feeder to a dwelling unit shall be determined as specified in Chapter 35 for services.

E3604.2 Minimum feeder conductor size. The size of feeder conductors shall not be less than No. 10 copper or No. 8 aluminum where the load supplied consists of any of the following number and types of circuits: (1) two or more two-wire branch circuits supplied by a two-wire feeder; (2) three or more two-wire branch circuits supplied by a three-wire feeder; or (3) two or more three-wire branch circuits supplied by a three-wire feeder.

E3604.3 Feeder loads. The minimum load in volt-amperes shall be computed in accordance with the load calculation procedure prescribed in Table E3604.3(1). The associated table demand factors shall be applied to the actual load to determine the minimum load for feeders.

E3604.4 Feeder neutral load. The feeder neutral load shall be the maximum unbalance of the load determined in accordance with this chapter. The maximum unbalanced load shall be the maximum net computed load between the neutral and any one ungrounded conductor. For feeders supplying electric ranges, wall-mounted ovens, counter-mounted cooking units and electric dryers, the maximum unbalanced load shall be 70 percent of the load on the ungrounded conductors.

TABLE E3604.3(1)
FEEDER LOAD CALCULATION

LOAD CALCULATION PROCEDURE	APPLIED DEMAND FACTOR
Lighting and receptacles: A unit load of not less than 3 VA per square foot of total floor area shall constitute the lighting and 120-volt, 15- and 20-ampere convenience receptacle load. 1,500 VA shall be added for each 20-ampere branch circuit serving receptacles in the kitchen, dining room, pantry, breakfast area and laundry area.	100 percent of first 3,000 VA or less and 35 percent of that in excess of 3,000 VA.
Plus	
Appliances and motors: The nameplate rating load of all fastened-in-place appliances other than dryers, ranges, air-conditioning and space-heating equipment.	100 percent of load for three or less appliances. 75 percent of load for four or more appliances.
Plus	
Fixed motors: Full-load current of motors plus 25 percent of the full load current of the largest motor.	
Plus	
Electric clothes dryer: The dryer load shall be 5,000 VA for each dryer circuit or the nameplate rating load of each dryer, whichever is greater.	
Plus	
Cooking appliances: The nameplate rating of ranges, wall-mounted ovens, counter-mounted cooking units, and other cooking appliances rated in excess of 1.75 kVA shall be summed.	Demand factors shall be as allowed by Table E3604.3(2).
Plus the largest of either the heating or cooling load	
Largest of the following two selections: 1. 100 percent of the nameplate rating(s) of the air conditioning and cooling, including heat pump compressors. 2. 100 percent of the fixed electric space heating.	

For SI: 1 square foot = 0.0929 m^2.

TABLE E3604.3(2)
DEMAND LOADS FOR ELECTRIC RANGES, WALL-MOUNTED OVENS, COUNTER-MOUNTED COOKING UNITS AND OTHER COOKING APPLIANCES OVER $1^3/_4$ kVA RATING[a,b]

NUMBER OF APPLIANCES	MAXIMUM DEMAND[b,c] Column A maximum 12 kVA rating	DEMAND FACTORS (percent)[d] Column B less than $3^1/_2$ kVA rating	Column C $3^1/_2$ to $8^3/_4$ kVA rating
1	8 kVA	80	80
2	11 kVA	75	65

a. Column A shall be used in all cases except as provided for in Footnote d.

b. For ranges all having the same rating and individually rated more than 12 kVA but not more than 27 kVA, the maximum demand in Column A shall be increased 5 percent for each additional kVA of rating or major fraction thereof by which the rating of individual ranges exceeds 12 kVA.

c. For ranges of unequal ratings and individually rated more than 8.75 kVA, but none exceeding 27 kVA, an average value of rating shall be computed by adding together the ratings of all ranges to obtain the total connected load (using 12 kVA for any ranges rated less than 12 kVA) and dividing by the total number of ranges; and then the maximum demand in Column A shall be increased 5 percent for each kVA or major fraction thereof by which this average value exceeds 12 kVA.

d. Over 1.75 kVA through 8.75 kVA. As an alternative to the method provided in Column A, the nameplate ratings of all ranges rated more than 1.75 kVA but not more than 8.75 kVA shall be added and the sum shall be multiplied by the demand factor specified in Column B or C for the given number of appliances.

E3604.5 Lighting and convenience receptacle load. A unit load of not less than three volt-amperes shall constitute the minimum lighting and convenience receptacle load for each square foot (m^2) of floor area. The floor area for each floor shall be computed from the outside dimensions of the building. The computed floor area shall not include open porches, garages, or unused or unfinished spaces not adaptable for future use.

E3604.6 Ampacity and computed loads. The computed load of a feeder shall be not less than the sum of the loads on the branch circuits supplied, as determined by Section E3604, after any applicable demand factors permitted by Section E3604 have been applied.

Feeder conductors shall have sufficient ampacity to supply the load served. In no case shall the computed load of a feeder be less than the sum of the loads on the branch circuits supplied as determined by this chapter after any permitted demand factors have been applied.

E3604.7 Feeder and branch-circuit conductors. Feeder and branch-circuit conductors shall be protected by overcurrent-protective devices connected at the point the conductors receive their supply.

SECTION E3605 CONDUCTOR SIZING AND OVERCURRENT PROTECTION

E3605.1 General. Ampacities for conductors shall be determined based in accordance with Table E3605.1 and Sections E3605.2 and E3605.3.

E3605.2 Correction factor for ambient temperatures. For ambient temperatures other than 30°C (86°F), multiply the allowable ampacities specified in Table E3605.1 by the appropriate correction factor shown in Table E3605.2.

TABLE E3605.1 ALLOWABLE AMPACITIES

CONDUCTOR SIZE	CONDUCTOR TEMPERATURE RATING						CONDUCTOR SIZE
	60°C	75°C	90°C	60°C	75°C	90°C	
	Types TW, UF	Types RHW, THHW, THW, THWN, USE, XHHW	Types RHW-2, THHN, THHW, THW-2, THWN-2, XHHW, XHHW-2	Types TW, UF	Types RHW, THHW, THW, THWN, USE, XHHW	Types RHW-2, THHN, THHW, THW-2, THWN-2, XHHW, XHHW-2	
AWG kcmil	Copper			Aluminum or copper-clad aluminum			AWG kcmil
18	—	—	14	—	—	—	—
16	—	—	18	—	—	—	—
14	20	20	25	—	—	—	—
12	25	25	30	20	20	25	12
10	30	35	40	25	30	35	10
8	40	50	55	30	40	45	8
6	55	65	75	40	50	60	6
4	70	85	95	55	65	75	4
3	85	100	110	65	75	85	3
2	95	115	130	75	90	100	2
1	110	130	150	85	100	115	1
1/0	125	150	170	100	120	135	1/0
2/0	145	175	195	115	135	150	2/0
3/0	165	200	225	130	155	175	3/0
4/0	195	230	260	150	180	205	4/0

For SI: °C = [(°F)-32]/1.8.

TABLE E3605.2 AMBIENT TEMPERATURE CORRECTION FACTORS

AMBIENT TEMP. °C	FOR AMBIENT TEMPERATURES OTHER THAN 30°C (86°F), MULTIPLY THE ALLOWABLE AMPACITIES SPECIFIED IN TABLE E3605.1 BY THE APPROPRIATE FACTOR SHOWN BELOW						AMBIENT TEMP. °F
21-25	1.08	1.05	1.04	1.08	1.05	1.04	70-77
26-30	1.00	1.00	1.00	1.00	1.00	1.00	78-86
31-35	0.91	0.94	0.96	0.91	0.94	0.96	87-95
36-40	0.82	0.88	0.91	0.82	0.88	0.91	96-104
41-45	0.71	0.82	0.87	0.71	0.82	0.87	105-113
46-50	0.58	0.75	0.82	0.58	0.75	0.82	114-122
51-55	0.41	0.67	0.76	0.41	0.67	0.76	123-131
56-60	—	0.58	0.71	—	0.58	0.71	132-140
61-70	—	0.33	0.58	—	0.33	0.58	141-158
71-80	—	—	0.41	—	—	0.41	159-176

For SI: °C = [(°F)-32]/1.8.

E3605.3 Adjustment factor for conductor proximity. Where the number of current-carrying conductors in a raceway or cable exceeds three, or where single conductors or multiconductor cables are stacked or bundled for distances greater than 24 inches (610 mm) without maintaining spacing and are not installed in raceways, the allowable ampacity of each conductor shall be reduced as shown in Table E3605.3.

Exceptions:

1. Adjustment factors shall not apply to conductors in nipples having a length not exceeding 24 inches (610 mm).
2. Adjustment factors shall not apply to underground conductors entering or leaving an outdoor trench if those conductors have physical protection in the form of rigid metal conduit, intermediate metal conduit, or rigid nonmetallic conduit having a length not exceeding 10 feet (3048 mm) and the number of conductors does not exceed four.

TABLE E3605.3
CONDUCTOR PROXIMITY ADJUSTMENT FACTORS

NUMBER OF CURRENT-CARRYING CONDUCTORS IN CABLE OR RACEWAY	PERCENT OF VALUES IN TABLE E3605.1
4-6	80
7-9	70
10-20	50
21-30	45
31-40	40
41 and above	35

E3605.4 Temperature limitations. The temperature rating associated with the ampacity of a conductor shall be so selected and coordinated to not exceed the lowest temperature rating of any connected termination, conductor or device. Conductors with temperature ratings higher than specified for terminations shall be permitted to be used for ampacity adjustment, correction, or both.

E3605.4.1 Conductors rated 60°C. Termination provisions of equipment for circuits rated 100 amperes or less, or marked for No. 14 through No. 1 conductors, shall be used only for one of the following:

1. Conductors rated 60°C (140°F), or
2. Conductors with higher temperature ratings, provided that the ampacity of such conductors is determined based on the 60°C (140°F) ampacity of the conductor size used, or
3. Conductors with higher temperature ratings where the equipment is listed and identified for use with such conductors, or
4. For motors marked with design letters B, C, D, or E, conductors having an insulation rating of 75°C (167°F) or higher shall be permitted to be used provided that the ampacity of such conductors does not exceed the 75°C (167°F) ampacity.

E3605.4.2 Conductors rated 75°C. Termination provisions of equipment for circuits rated over 100 amperes, or marked for conductors larger than No. 1, shall be used only for:

1. Conductors rated 75°C (167°F).
2. Conductors with higher temperature ratings provided that the ampacity of such conductors does not exceed the 75°C (167°F) ampacity of the conductor size used, or provided that the equipment is listed and identified for use with such conductors.

E3605.4.3 Separately installed pressure connectors. Separately installed pressure connectors shall be used with conductors at the ampacities not exceeding the ampacity at the listed and identified temperature rating of the connector.

E3605.4.4 Conductors of Type NM cable. Conductors in NM cable assemblies shall be rated at 90°C (194°F). Types NM, NMC, and NMS cable identified by the markings NM-B, NMC-B, and NMS-B meet this requirement. The ampacity of Types NM, NMC, and NMS cable shall be at 60°C (140°F) conductors and shall comply with Tables E3605.1 and E3605.7. The 90°C (194°F) rating shall be permitted to be used for ampacity correction and adjustment purposes provided that the final corrected or adjusted ampacity does not exceed that for a 60°C (140°F) rated conductor.

E3605.5 Overcurrent protection required. All ungrounded branch-circuit and feeder conductors shall be protected against overcurrent by an overcurrent device installed at the point where the conductors receive their supply. Overcurrent devices shall not be connected in series with a grounded conductor. Overcurrent protection and allowable loads for branch circuits and feeders that do not serve as the main power feeder to the dwelling unit load shall be in accordance with this chapter.

Branch-circuit conductors and equipment shall be protected by overcurrent protective devices having a rating or setting not exceeding the allowable ampacity specified in Table E3605.1 and Sections E3605.2, E3605.3 and E3605.4 except where otherwise permitted or required in Sections E3605.5.1 through E3605.5.3.

E3605.5.1 Cords. Cords shall be protected in accordance with Section E3809.2.

E3605.5.2 Overcurrent devices of the next higher rating. The next higher standard overcurrent device rating, above the ampacity of the conductors being protected, shall be permitted to be used, provided that all of the following conditions are met:

1. The conductors being protected are not part of a multi-outlet branch circuit supplying receptacles for cord- and plug-connected portable loads.
2. The ampacity of conductors does not correspond with the standard ampere rating of a fuse or a circuit breaker without overload trip adjustments above its rating (but that shall be permitted to have other trip or rating adjustments).
3. The next higher standard device rating does not exceed 400 amperes.

E3605.5.3 Small conductors. Except as specifically permitted by Section E3605.5.4, the rating of overcurrent protection devices shall not exceed the ratings shown in Table E3605.5.3 for the conductors specified therein.

TABLE E3605.5.3
OVERCURRENT-PROTECTION RATING

COPPER		ALUMINUM OR COPPER-CLAD ALUMINUM	
Size (AWG)	Maximum overcurrent-protection-device rating[a] (amps)	Size (AWG)	Maximum overcurrent-protection-device rating[a] (amps)
14	15	12	15
12	20	10	25
10	30	8	30

a. The maximum overcurrent-protection-device rating shall not exceed the conductor allowable ampacity determined by the application of the correction and adjustment factors in accordance with Sections E3605.2 and E3605.3.

E3605.5.4 Air conditioning and heat pump equipment. Air-conditioning and heat pump equipment circuit conductors shall be permitted to be protected against overcurrent in accordance with Section E3602.11.

E3605.6 Fuses and fixed trip circuit breakers. The standard ampere ratings for fuses and inverse time circuit breakers shall be considered 15, 20, 25, 30, 35, 40, 45, 50, 60, 70, 80, 90, 100, 110, 125, 150, 175, 200, 225, 250, 300, 350 and 400 amperes.

E3605.7 Location of overcurrent devices in or on premises. Overcurrent devices shall:

1. Be readily accessible.
2. Not be located where they will be exposed to physical damage.
3. Not be located where they will be in the vicinity of easily ignitible material such as in clothes closets.
4. Not be located in bathrooms.

Exceptions:

1. This section shall not apply to supplementary overcurrent protection that is integral to utilization equipment.
2. Overcurrent devices installed adjacent to the utilization equipment that they supply shall be permitted to be accessible by portable means.

E3605.8 Ready access for occupants. Each occupant shall have ready access to all overcurrent devices protecting the conductors supplying that occupancy.

E3605.9 Enclosures for overcurrent devices. Overcurrent devices shall be enclosed in cabinets or cutout boxes except where an overcurrent device is part of an assembly that provides equivalent protection. The operating handle of a circuit breaker shall be permitted to be accessible without opening a door or cover.

SECTION E3606
PANELBOARDS

E3606.1 Panelboard rating. All panelboards shall have a rating not less than that of the minimum service entrance or feeder capacity required for the computed load.

E3606.2 Panelboard circuit identification. The use or purpose of all panelboard circuits shall be legibly marked and identified on a circuit directory located on the face or inside the door of the enclosure.

E3606.3 Panelboard overcurrent protection. Panelboards shall be protected on the supply side by not more than two main circuit breakers or two sets of fuses having a combined rating not greater than that of the panelboard.

Exception: Individual protection for a panelboard shall not be required if the panelboard feeder has overcurrent protection not greater than the rating of the panelboard.

Chapter 37
WIRING METHODS

SECTION E3701
GENERAL REQUIREMENTS

E3701.1 Scope. This chapter covers the wiring methods for services, feeders and branch circuits for electrical power and distribution.

E3701.2 Allowable wiring methods. The allowable wiring methods for electrical installations shall be those listed in Table E3701.2. Single conductors shall be used only where part of one of the recognized wiring methods listed in Table E3701.2. As used in this code, abbreviations of the wiring-method types shall be as indicated in Table E3701.2.

TABLE E3701.2
ALLOWABLE WIRING METHODS

ALLOWABLE WIRING METHOD	DESIGNATED ABBREVIATION
Armored cable	AC
Electrical metallic tubing	EMT
Electrical nonmetallic tubing	ENT
Flexible metal conduit	FMC
Intermediate metal conduit	IMC
Liquidtight flexible conduit	LFC
Metal-clad cable	MC
Nonmetallic sheathed cable	NM
Rigid nonmetallic conduit	RNC
Rigid metallic conduit	RMC
Service entrance cable	SE
Surface raceways	SR
Underground feeder cable	UF
Underground service cable	USE

E3701.3 Circuit conductors. All conductors of a circuit, including equipment grounding conductors, shall be contained in the same raceway, trench, cable or cord.

E3701.4 Wiring method applications. Wiring methods shall be applied in accordance with Table E3701.4.

SECTION E3702
ABOVE-GROUND INSTALLATION REQUIREMENTS

E3702.1 Installation and support requirements. Wiring methods shall be installed and supported in accordance with Table E3702.1.

E3702.2 Cables in accessible attics. Cables in attics or roof spaces provided with access shall be installed as specified in Sections E3702.2.1 and E3702.2.2.

E3702.2.1 Across structural members. Where run across the top of floor joists, or run within 7 feet (2134 mm) of floor or floor joists across the face of rafters or studding, in attics and roof spaces that are provided with access, the cable shall be protected by substantial guard strips that are at least as high as the cable. Where such spaces are not provided with access by permanent stairs or ladders, protection shall only be required within 6 feet (1829 mm) of the nearest edge of the attic entrance.

E3702.2.2 Cable installed parallel to framing members. Where cables are installed parallel to the sides of rafters, studs or floor joists, guard strips and running boards shall not be required, and the installation shall comply with Table E3702.1.

E3702.3 Exposed cable. In exposed work, except as provided for in Sections E3702.2 and E3702.4, cable assemblies shall be installed as specified in Sections E3702.3.1 and E3702.3.2.

E3702.3.1 Surface installation. Cables shall closely follow the surface of the building finish or running boards.

E3702.3.2 Protection from physical damage. Where subject to physical damage, cables shall be protected by conduit, electrical metallic tubing, Schedule 80 PVC rigid nonmetallic conduit, pipe, guard strips or other approved means. Where passing through a floor, the cable shall be enclosed in rigid metal conduit, intermediate metal conduit, electrical metallic tubing, Schedule 80 PVC rigid nonmetallic conduit or other metal pipe extending not less than 6 inches (152 mm) above the floor.

E3702.4 In unfinished basements. Where the cable is run at angles with joists in unfinished basements, cable assemblies containing two or more conductors of sizes No. 6 and larger and assemblies containing three or more conductors of sizes No. 8 and larger shall not require additional protection where attached directly to the bottom of the joists.

E3702.5 Bends. Bends shall be made so as not to damage the wiring method or reduce the internal diameter of raceways.

For types NM and SE cable, bends shall be so made, and other handling shall be such that the cable will not be damaged and the radius of the curve of the inner edge of any bend shall be not less than five times the diameter of the cable.

SECTION E3703
UNDERGROUND INSTALLATION REQUIREMENTS

E3703.1 Minimum cover requirements. Direct buried cable or raceways shall be installed in accordance with the minimum cover requirements of Table E3703.1.

E3703.2 Grounding. All underground installations shall be grounded and bonded in accordance with this code.

TABLE E3701.4
ALLOWABLE APPLICATIONS FOR WIRING METHODS[a,b,c,d,e,f,g,h,i]

ALLOWABLE APPLICATIONS (application allowed where marked with an "A")	AC	EMT	ENT	FMC	IMC RMC RNC	LFC[a]	MC	NM	SR	SE	UF	USE
Services	—	A	A[h]	A[i]	A	A[i]	A	—	—	A	—	A
Feeders	A	A	A	A	A	A	A	A	—	A[b]	A	A[b]
Branch circuits	A	A	A	A	A	A	A	A	A	A[c]	A	—
Inside a building	A	A	A	A	A	A	A	A	A	A	A	—
Wet locations exposed to sunlight	—	A	A[h]	A[d]	A	A	A	—	—	A	A[e]	A[e]
Damp locations	—	A	A	A[d]	A	A	A	—	—	A	A	A
Embedded in noncinder concrete in dry location	—	A	A	—	A	—	—	—	—	—	—	—
In noncinder concrete in contact with grade	—	A[f]	A	—	A[f]	—	—	—	—	—	—	—
Embedded in plaster not exposed to dampness	A	A	A	A	A	A	A	—	—	A	A	—
Embedded in masonry	—	A	A	—	A[f]	A	A	—	—	—	—	—
In masonry voids and cells exposed to dampness or below grade line	—	A[f]	A	A[d]	A[f]	A	A	—	—	A	A	—
Fished in masonry voids	A	—	—	A	—	A	A	A	—	A	A	—
In masonry voids and cells not exposed to dampness	A	A	A	A	A	A	A	A	—	A	A	—
Run exposed	A	A	A	A	A	A	A	A	A	A	A	A
Run exposed and subject to physical damage	—	—	—	—	A[g]	—	—	—	—	—	—	—
For direct burial	—	A[f]	—	—	A[f]	A	A[f]	—	—	—	A	A

For SI: 1 foot = 304.8 mm.

a. Liquid-tight flexible nonmetallic conduit without integral reinforcement within the conduit wall shall not exceed 6 feet in length.

b. The grounded conductor shall be insulated except where used to supply other buildings on the same premises. Type USE cable shall not be used inside buildings.

c. The grounded conductor shall be insulated.

d. Conductors shall be a type approved for wet locations and the installation shall prevent water from entering other raceways.

e. Shall be listed as sunlight resistant.

f. Metal raceways shall be protected from corrosion and approved for the application.

g. RNC shall be Schedule 80.

h. Shall be listed as "Sunlight Resistant" where exposed to the direct rays of the sun.

i. Conduit shall not exceed 6 feet in length.

TABLE E3702.1
GENERAL INSTALLATION AND SUPPORT REQUIREMENTS FOR WIRING METHODS[a,b,c,d,e,f,g,h,i]

INSTALLATION REQUIREMENTS (requirement applicable only to wiring methods marked "A")	AC MC	EMT IMC RMC	ENT	FMC LFC	NM UF	RNC	SE	SR[a]	USE
Where run parallel with the framing member, the wiring shall be 1.25 inches from the edge of a framing member such as a joist, rafter or stud or shall be physically protected.	A	—	A	A	A	—	A	—	—
Bored holes in studs and vertical framing members for wiring shall be located 1.25 inches from the edge or shall be protected with a minimum 0.0625-inch steel plate or sleeve or other physical protection.	A	—	A	A	A	—	A	—	—
Where installed in grooves, to be covered by wallboard, siding, paneling, carpeting, or similar finish, wiring methods shall be protected by 0.0625-inch-thick steel plate, sleeve, or equivalent or by not less than 1.25-inch free space for the full length of the groove in which the cable or raceway is installed.	A	—	A	A	A	—	A	A	A
Bored holes in joists, rafters, beams and other horizontal framing members shall be 2 inches from the edge of the structural framing member.	A	A	A	A	A	A	A	—	—
Securely fastened bushings or grommets shall be provided to protect wiring run through openings in metal framing members.	—	—	A	—	A	—	A	—	—
The maximum number of 90-degree bends shall not exceed four between junction boxes.	—	A	A	A	—	A	—	—	—
Bushings shall be provided where entering a box, fitting or enclosure unless the box or fitting is designed to afford equivalent protection.	A	A	A	A	—	A	—	A	—
Ends of raceways shall be reamed to remove rough edges.	—	A	A	A	—	A	—	A	—
Maximum allowable on center support spacing for the wiring method in feet.	4.5[b, c]	10	3[b]	4.5[b]	4.5[i]	3[d]	2.5[e]	—	2.5[e]
Maximum support distance in inches from box or other terminations.	12[b, f]	36	36	12[b, g]	12[h,i]	36	12	—	12

For SI: 1 inch = 25.4 mm, 1 foot = 304.8 mm, 1 degree = 0.009 rad.

a. Installed in accordance with listing requirements.

b. Supports not required in accessible ceiling spaces between light fixtures where lengths do not exceed 6 feet.

c. Six feet for MC cable.

d. Five feet for trade sizes greater than 1 inch.

e. Two and one-half feet where used for service or outdoor feeder and 4.5 feet where used for branch circuit or indoor feeder.

f. Twenty-four inches where flexibility is necessary.

g. Thirty-six inches where flexibility is necessary.

h. Within 8 inches of boxes without cable clamps.

i. Flat cables shall not be stapled on edge.

TABLE E3703.1
MINIMUM COVER REQUIREMENTS, BURIAL IN INCHES[a,b,c,d]

LOCATION OF WIRING METHOD OR CIRCUIT	TYPE OF WIRING METHOD OR CIRCUIT				
	1 Direct burial cables or conductors	2 Rigid metal conduit or intermediate metal conduit	3 Nonmetallic raceways listed for direct burial without concrete encasement or other approved raceways	4 Residential branch circuits rated 120 volts or less with GFCI protection and maximum overcurrent protection of 20 amperes	5 Circuits for control of irrigation and landscape lighting limited to not more than 30 volts and installed with type UF or in other identified cable or raceway
All locations not specified below	24	6	18	12	6
In trench below 2-inch-thick concrete or equivalent	18	6	12	6	6
Under a building	0 (In raceway only)	0	0	0 (In raceway only)	0 (In raceway only)
Under minimum of 4-inch-thick concrete exterior slab with no vehicular traffic and the slab extending not less than 6 inches beyond the underground installation	18	4	4	6 (Direct burial) 4 (In raceway)	6 (Direct burial) 4 (In raceway)
Under streets, highways, roads, alleys, driveways and parking lots	24	24	24	24	24
One- and two-family dwelling driveways and outdoor parking areas, and used only for dwelling-related purposes	18	18	18	12	18
In solid rock where covered by minimum of 2 inches concrete extending down to rock	2 (In raceway only)	2	2	2 (In raceway only)	2 (In raceway only)

For SI: 1 inch = 25.4 mm.

a. Raceways approved for burial only where encased concrete shall require concrete envelope not less than 2 inches thick.

b. Lesser depths shall be permitted where cables and conductors rise for terminations or splices or where access is otherwise required.

c. Where one of the wiring method types listed in columns 1 to 3 is combined with one of the circuit types in columns 4 and 5, the shallower depth of burial shall be permitted.

d. Where solid rock prevents compliance with the cover depths specified in this table, the wiring shall be installed in metal or nonmetallic raceway permitted for direct burial. The raceways shall be covered by a minimum of 2 inches of concrete extending down to the rock.

E3703.3 Protection from damage. Direct buried conductors and cables emerging from the ground shall be protected by enclosures or raceways extending from the minimum cover distance below grade required by Section 3703.1 to a point at least 8 feet (2438 mm) above finished grade. In no case shall the protection be required to exceed 18 inches (457 mm) below finished grade. Service laterals that are not encased in concrete and that are buried 18 inches (457 mm) or more below grade shall have their location identified by a warning ribbon that is placed in the trench at least 12 inches (305 mm) above the underground installation. Conductors entering a building shall be protected to the point of entrance. Where the enclosure or raceway is subject to physical damage, the conductors shall be installed in rigid metal conduit, intermediate metal conduit, Schedule 80 rigid nonmetallic conduit or the equivalent.

E3703.4 Splices and taps. Direct buried conductors or cables shall be permitted to be spliced or tapped without the use of splice boxes. The splices or taps shall be made by approved methods with materials listed for the application.

E3703.5 Backfill. Backfill containing large rock, paving materials, cinders, large or sharply angular substances, or corrosive material shall not be placed in an excavation where such materials cause damage to raceways, cables or other substructures or prevent adequate compaction of fill or contribute to corrosion of raceways, cables or other substructures. Where necessary to prevent physical damage to the raceway or cable, protection shall be provided in the form of granular or selected material, suitable boards, suitable sleeves or other approved means.

E3703.6 Raceway seals. Conduits or raceways shall be sealed or plugged at either or both ends where moisture will enter and contact energized parts.

E3703.7 Bushing. A bushing, or terminal fitting, with an integral bushed opening shall be installed on the end of a conduit or other raceway that terminates underground where the conductors or cables emerge as a direct burial wiring method.

A seal incorporating the physical protection characteristics of a bushing shall be considered equivalent to a bushing.

E3703.8 Single conductors. All conductors of the same circuit and, where present, the grounded conductor and all equipment grounding conductors shall be installed in the same raceway or shall be installed in close proximity in the same trench.

Exception: Where conductors are installed in parallel in raceways, each raceway shall contain all conductors of the same circuit including grounding conductors.

E3703.9 Ground movement. Where direct buried conductors, raceways, or cables are subject to movement by settlement or frost, direct buried conductors, raceways, or cables shall be arranged to prevent damage to the enclosed conductors or to equipment connected to the raceways.

Chapter 38
POWER AND LIGHTING DISTRIBUTION

SECTION E3801
RECEPTACLE OUTLETS

E3801.1 General. Receptacle outlets shall be provided in accordance with Sections E3801.2 through E3801.11. Receptacle outlets required by this section shall be in addition to any receptacle that is part of a lighting fixture or appliance, that is located within cabinets or cupboards, or that is located over 5.5 feet (1676 mm) above the floor.

Permanently installed electric baseboard heaters equipped with factory-installed receptacle outlets, or outlets provided as a separate assembly by the baseboard manufacturer shall be permitted as the required outlet or outlets for the wall space utilized by such permanently installed heaters. Such receptacle outlets shall not be connected to the heater circuits.

E3801.2 Convenience receptacle distribution. In every kitchen, family room, dining room, living room, parlor, library, den, sun room, bedroom, recreation room, or similar room or area of dwelling units, receptacle outlets shall be installed in accordance with the general provisions specified in Sections E3801.2.1 through E3801.2.3. See Figure E3801.2.

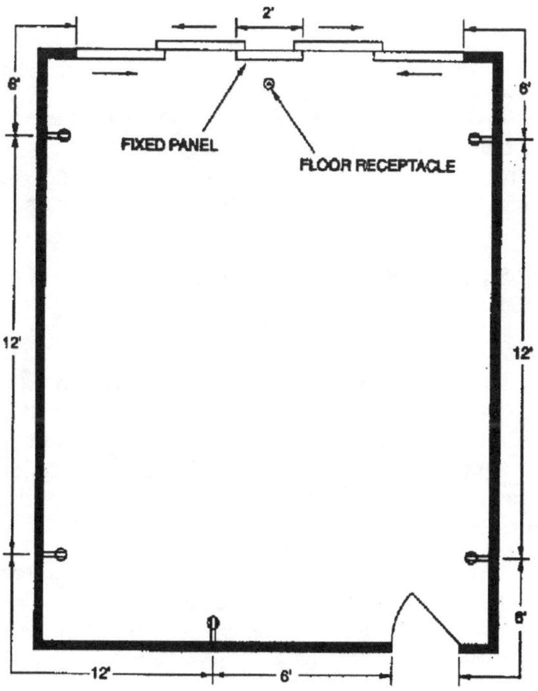

For SI: 1 foot = 304.8 mm.

FIGURE E3801.2
GENERAL USE RECEPTACLE DISTRIBUTION

E3801.2.1 Spacing. Receptacles shall be installed so that no point along the floor line in any wall space is more than 6 feet (1829 mm), measured horizontally, from an outlet in that space. Receptacles shall, insofar as practicable, be spaced equal distances apart.

E3801.2.2 Wall space. As used in this section, a wall space shall include the following:

1. Any space that is 2 feet (610 mm) or more in width, (including space measured around corners), and that is unbroken along the floor line by doorways, fireplaces, and similar openings.
2. The space occupied by fixed panels in exterior walls, excluding sliding panels.
3. The space created by fixed room dividers such as railings and freestanding bar-type counters.

E3801.2.3 Floor receptacles. Receptacle outlets in floors shall not be counted as part of the required number of receptacle outlets except where located within 18 inches (457 mm) of the wall.

E3801.3 Small appliance receptacles. In the kitchen, pantry, breakfast room, dining room, or similar area of a dwelling unit, the two or more 20-ampere small-appliance branch circuits required by Section E3603.2, shall serve all receptacle outlets covered by Sections E3801.2 and E3801.4 and those receptacle outlets provided for refrigeration appliances.

Exceptions:

1. In addition to the required receptacles specified by Sections E3801.1 and E3801.2, switched receptacles supplied from a general-purpose branch circuit as defined in Section E3803.2, Exception 1, shall be permitted.
2. The receptacle outlet for refrigeration appliances shall be permitted to be supplied from an individual branch circuit rated at 15 amperes or greater.

E3801.3.1 Other outlets prohibited. The two or more small-appliance branch circuits specified in Section 3801.3 shall serve no other outlets.

Exceptions:

1. A receptacle installed solely for the electrical supply to and support of an electric clock in any of the rooms specified in Section E3801.3.
2. Receptacles installed to provide power for supplemental equipment and lighting on gas-fired ranges, ovens, and counter-mounted cooking units.

E3801.3.2 Limitations. Receptacles installed in a kitchen to serve countertop surfaces shall be supplied by not less than two small-appliance branch circuits, either or both of which shall also be permitted to supply receptacle outlets in the same kitchen and in other rooms specified in Section E3801.3. Additional small-appliance branch circuits shall be permitted to supply receptacle outlets in the kitchen and other rooms specified in Section E3801.3. A small-appliance branch circuit shall not serve more than one kitchen.

E3801.4 Countertop receptacles. In kitchens and dining rooms of dwelling units, receptacle outlets for counter spaces shall be installed in accordance with Sections E3801.4.1 through E3801.4.5. (See Figure E3801.4.)

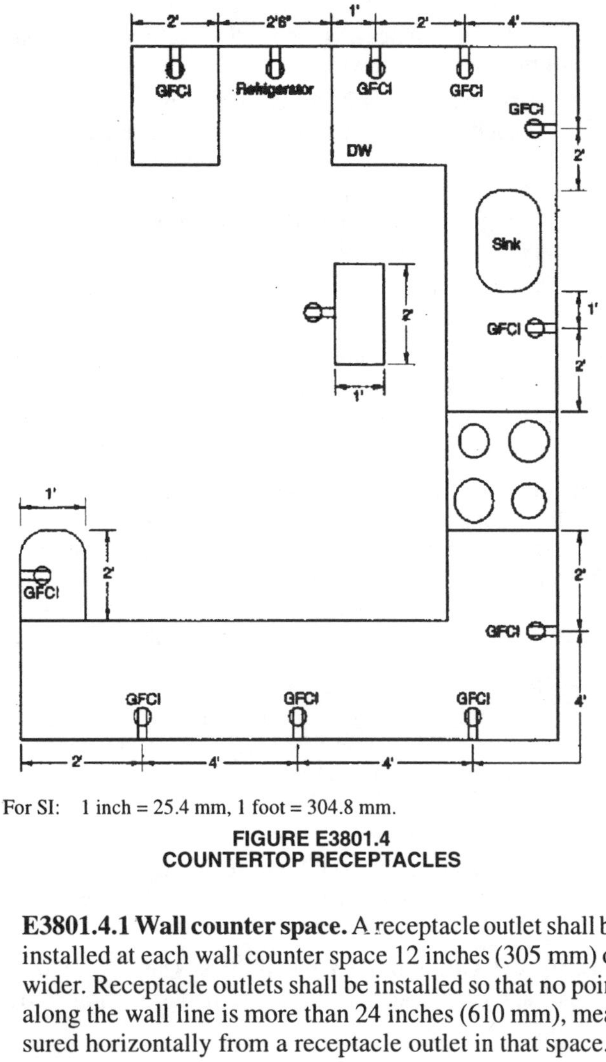

For SI: 1 inch = 25.4 mm, 1 foot = 304.8 mm.

FIGURE E3801.4
COUNTERTOP RECEPTACLES

E3801.4.1 Wall counter space. A receptacle outlet shall be installed at each wall counter space 12 inches (305 mm) or wider. Receptacle outlets shall be installed so that no point along the wall line is more than 24 inches (610 mm), measured horizontally from a receptacle outlet in that space.

E3801.4.2 Island counter spaces. At least one receptacle outlet shall be installed at each island counter space with a long dimension of 24 inches (610 mm) or greater and a short dimension of 12 inches (305 mm) or greater.

E3801.4.3 Peninsular counter space. At least one receptacle outlet shall be installed at each peninsular counter space with a long dimension of 24 inches (610 mm) or greater and a short dimension of 12 inches (305 mm) or greater. A peninsular countertop is measured from the connecting edge.

E3801.4.4 Separate spaces. Countertop spaces separated by range tops, refrigerators, or sinks shall be considered as separate countertop spaces in applying the requirements of Sections E3801.4.1, E3801.4.2 and E3801.4.3.

E3801.4.5 Receptacle outlet location. Receptacle outlets shall be located not more than 18 inches (458 mm) above the countertop. Receptacle outlets shall not be installed in a face-up position in the work surfaces or countertops. Receptacle outlets rendered not readily accessible by appliances fastened in place or appliances occupying dedicated space shall not be considered as these required outlets.

Exception: Receptacle outlets shall be permitted to be mounted not more than 12 inches (305 mm) below the countertop in construction designed for the physically impaired and for island and peninsular countertops where the countertop is flat across its entire surface and there are no means to mount a receptacle within 18 inches (458 mm) above the countertop, such as in an overhead cabinet. Receptacles mounted below the countertop in accordance with this exception shall not be located where the countertop extends more than 6 inches (152 mm) beyond its support base.

E3801.5 Appliance outlets. Appliance receptacle outlets installed for specific appliances, such as laundry equipment, shall be installed within 6 feet (1829 mm) of the intended location of the appliance.

E3801.6 Bathroom. At least one wall receptacle outlet shall be installed in bathrooms and such outlet shall be located within 36 inches (914 mm) of the outside edge of each lavatory basin. The receptacle outlet shall be located on a wall that is adjacent to the lavatory basin location.

Receptacle outlets shall not be installed in a face-up position in the work surfaces or countertops in a bathroom basin location.

E3801.7 Outdoor outlets. At least one receptacle outlet accessible at grade level and not more than 6 feet, 6 inches (1981 mm) above grade, shall be installed outdoors at the front and back of each dwelling unit having direct access to grade.

E3801.8 Laundry areas. At least one receptacle outlet shall be installed to serve laundry appliances.

E3801.9 Basements and garages. At least one receptacle outlet, in addition to any provided for laundry equipment, shall be installed in each basement and in each attached garage, and in each detached garage that is provided with electrical power. Where a portion of the basement is finished into a habitable room(s), the receptacle outlet required by this section shall be installed in the unfinished portion.

E3801.10 Hallways. Hallways of 10 feet (3048 mm) or more in length shall have at least one receptacle outlet. The hall length shall be considered the length measured along the centerline of the hall without passing through a doorway.

E3801.11 HVAC outlet. A convenience receptacle outlet shall be installed for the servicing of heating, air-conditioning and refrigeration equipment located in attics and crawl spaces. The receptacle shall be accessible and shall be located on the same level and within 25 feet (7620 mm) of the heating, air-conditioning and refrigeration equipment. The receptacle outlet shall not be connected to the load side of the HVAC equipment disconnecting means.

SECTION E3802 GROUND-FAULT AND ARC-FAULT CIRCUIT-INTERRUPTER PROTECTION

E3802.1 Bathroom receptacles. All 125-volt, single-phase, 15- and 20-ampere receptacles installed in bathrooms shall have ground-fault circuit-interrupter protection for personnel.

E3802.2 Garage receptacles. All 125-volt, single-phase, 15- or 20-ampere receptacles installed in garages and grade-level portions of unfinished accessory buildings used for storage or work areas shall have ground-fault circuit-interrupter protection for personnel.

Exceptions:

1. Receptacles that are not readily accessible.
2. A single receptacle or a duplex receptacle for two appliances located within dedicated space for each appliance that in normal use is not easily moved from one place to another, and that is cord- and plug-connected.

E3802.3 Outdoor receptacles. All 125-volt, single-phase, 15- and 20-ampere receptacles installed outdoors shall have ground-fault circuit-interrupter protection for personnel.

Exception: Receptacles as covered in Section E4001.7.

E3802.4 Crawl space receptacles. Where a crawl space is at or below grade level, all 125-volt, single-phase, 15- and 20-ampere receptacles installed in such spaces shall have ground-fault circuit-interrupter protection for personnel.

E3802.5 Unfinished basement receptacles. All 125-volt, single-phase, 15- and 20-ampere receptacles installed in unfinished basements shall have ground-fault circuit-interrupter protection for personnel. For purposes of this section, unfinished basements are defined as portions or areas of the basement not intended as habitable rooms and limited to storage areas, work areas, and the like.

Exceptions:

1. Receptacles that are not readily accessible.
2. A single receptacle or duplex receptacle for two appliances located within dedicated space for each appliance that in normal use is not easily moved from one place to another, and that is cord- and plug-connected.

E3802.6 Kitchen receptacles. All 125-volt, single-phase, 15- and 20-ampere receptacles that serve countertop surfaces shall have ground-fault circuit-interrupter protection for personnel.

E3802.7 Bar sink receptacles. All 125-volt, single-phase, 15- and 20-ampere receptacles that serve a countertop surface, and are located within 6 feet (1829 mm) of the outside edge of a wet bar sink shall have ground-fault circuit-interrupter protection for personnel. Receptacle outlets shall not be installed in a face-up position in the work surfaces or countertops.

E3802.8 Exempt receptacles. Receptacles installed under exceptions to Sections E3802.2 and E3802.5 shall not be considered as meeting the requirements of Section E3801.10.

SECTION E3803 LIGHTING OUTLETS

E3803.1 General. Lighting outlets shall be provided in accordance with Sections E3803.2 through E3803.3.

E3803.2 Habitable rooms. At least one wall switch-controlled lighting outlet shall be installed in every habitable room and bathroom.

Exceptions:

1. In other than kitchens and bathrooms, one or more receptacles controlled by a wall switch shall be considered equivalent to the required lighting outlet.
2. Lighting outlets shall be permitted to be controlled by occupancy sensors that are in addition to wall switches, or that are located at a customary wall switch location and equipped with a manual override that will allow the sensor to function as a wall switch.

E3803.3 Additional locations. At least one wall switch-controlled lighting outlet shall be installed in hallways, stairways, attached garages, and detached garages with electric power. At least one wall-switch-controlled lighting outlet shall be installed to provide illumination on the exterior side of each outdoor egress door having grade level access. A vehicle door in a garage shall not be considered as an outdoor egress door. Where lighting outlets are installed in interior stairways, there shall be a wall switch at each floor level to control the lighting outlet where the stairway has 6 or more risers.

Exception: In hallways, stairways, and at outdoor egress doors, remote, central, or automatic control of lighting shall be permitted.

E3803.4 Storage or equipment spaces. In attics, under-floor spaces, utility rooms and basements, at least one lighting outlet shall be installed where these spaces are used for storage or contain equipment requiring servicing. Such lighting outlet shall be controlled by a wall switch or shall have an integral switch. At least one point of control shall be at the usual point of entry to these spaces. The lighting outlet shall be provided at or near the equipment requiring servicing.

SECTION E3804 GENERAL INSTALLATION REQUIREMENTS

E3804.1 Electrical continuity of metal raceways and enclosures. Metal raceways, cable armor and other metal enclosures for conductors shall be mechanically joined together into a continuous electric conductor and shall be connected to all boxes, fittings and cabinets so as to provide effective electrical continuity. Raceways and cable assemblies shall be mechanically secured to boxes, fittings cabinets and other enclosures.

Exceptions:

1. As provided in Section E3805.3.2 for nonmetallic boxes used with Type NM cable.
2. Short sections of raceway used to provide cable assemblies with support or protection against physical damage.

E3804.2 Mechanical continuity—raceways and cables. Metal or nonmetallic raceways, cable armors and cable sheaths shall be continuous between cabinets, boxes, fittings or other enclosures or outlets.

Exception: Short sections of raceway used to provide cable assemblies with support or protection against physical damage.

E3804.3 Securing and supporting. Raceways, cable assemblies, boxes, cabinets and fittings shall be securely fastened in place.

E3804.4 Raceways as means of support. Raceways shall be used as a means of support for other raceways, cables or nonelectric equipment only under the following conditions:

1. Where the raceway or means of support is identified for the purpose; or
2. Where the raceway contains power supply conductors for electrically controlled equipment and is used to support Class 2 circuit conductors or cables that are solely for the purpose of connection to the control circuits of the equipment served by such raceway; or
3. Where the raceway is used to support boxes or conduit bodies in accordance with Sections E3806.8.4 and E3806.8.5.

E3804.5 Raceway installations. Raceways shall be installed complete between outlet, junction or splicing points prior to the installation of conductors.

E3804.6 Conduit and tubing fill. The maximum number of conductors installed in conduit or tubing shall be in accordance with Tables E3804.6(1) through E3804.6(10).

E3804.7 Air handling—stud cavity and joist spaces. Where wiring methods having a nonmetallic covering pass through stud cavities and joist spaces used for air handling, such wiring shall pass through such spaces perpendicular to the long dimension of the spaces.

SECTION E3805
BOXES, CONDUIT BODIES AND FITTINGS

E3805.1 Box, conduit body or fitting—where required. A box or conduit body shall be installed at each conductor splice point, outlet, switch point, junction point and pull point except as otherwise permitted in Sections E3805.1.1 through E3805.1.7.

Fittings and connectors shall be used only with the specific wiring methods for which they are are designed and listed.

E3805.1.1 Equipment. An integral junction box or wiring compartment that is part of listed equipment shall be permitted to serve as a box or conduit body.

E3805.1.2 Protection. A box or conduit body shall not be required where cables enter or exit from conduit or tubing that is used to provide cable support or protection against physical damage. A fitting shall be provided on the end(s) of the conduit or tubing to protect the cable from abrasion.

E3805.1.3 Integral enclosure. A wiring device with integral enclosure identified for the use, having brackets that securely fasten the device to walls or ceilings of conventional on-site frame construction, for use with nonmetallic-sheathed cable, shall be permitted in lieu of a box or conduit body.

E3805.1.4 Fitting. A fitting identified for the use shall be permitted in lieu of a box or conduit body where such fitting is accessible after installation and does not contain spliced or terminated conductors.

E3805.1.5 Buried conductors. Splices and taps in buried conductors and cables shall not be required to be enclosed in a box or conduit body where installed in accordance with Section 3703.4.

E3805.1.6 Fixtures. Where a fixture is listed to be used as a raceway, a box or conduit body shall not be required for wiring installed therein.

E3805.1.7 Closed loop. Where a device identified and listed as suitable for installation without a box is used with a closed-loop power-distribution system, a box or conduit body shall not be required.

E3805.2 Metal boxes. All metal boxes shall be grounded.

E3805.3 Nonmetallic boxes. Nonmetallic boxes shall be used only with nonmetallic-sheathed cable and nonmetallic raceways.

Exceptions:

1. Where internal bonding means are provided between all entries, nonmetallic boxes shall be permitted to be used with metal raceways and metal-armored cables.
2. Where integral bonding means with a provision for attaching an equipment grounding jumper inside the box are provided between all threaded entries in nonmetallic boxes listed for the purpose, nonmetallic boxes shall be permitted to be used with metal raceways and metal-armored cables.

TABLE E3804.6(1)
MAXIMUM NUMBER OF CONDUCTORS IN ELECTRICAL METALLIC TUBING (EMT)[a]

TYPE LETTERS	CONDUCTOR SIZE AWG/kcmil	TRADE SIZES (inches)					
		1/2	3/4	1	1 1/4	1 1/2	2
RHW, RHW-2	14	4	7	11	20	27	46
	12	3	6	9	17	23	38
	10	2	5	8	13	18	30
	8	1	2	4	7	9	16
	6	1	1	3	5	8	13
	4	1	1	2	4	6	10
	3	1	1	1	4	5	9
	2	1	1	1	3	4	7
	1	0	1	1	1	3	5
	1/0	0	1	1	1	2	4
	2/0	0	1	1	1	2	4
	3/0	0	0	1	1	1	3
	4/0	0	0	1	1	1	3
TW	14	8	15	25	43	58	96
	12	6	11	19	33	45	74
	10	5	8	14	24	33	55
	8	2	5	8	13	18	30
RHW[a], RHW-2[a], THHW, THW, THW-2	14	6	10	16	28	39	64
	12	4	8	13	23	31	51
	10	3	6	10	18	24	40
	8	1	4	6	10	14	24
RHW[a], RHW-2[a], TW, THW, THHW, THW-2	6	1	3	4	8	11	18
	4	1	1	3	6	8	13
	3	1	1	3	5	7	12
	2	1	1	2	4	6	10
	1	1	1	1	3	4	7
	1/0	0	1	1	2	3	6
	2/0	0	1	1	1	3	5
	3/0	0	1	1	1	2	4
	4/0	0	0	1	1	1	3
THHN, THWN, THWN-2	14	12	22	35	61	84	138
	12	9	16	26	45	61	101
	10	5	10	16	28	38	63
	8	3	6	9	16	22	36
	6	2	4	7	12	16	26
	4	1	2	4	7	10	16
	3	1	1	3	6	8	13
	2	1	1	3	5	7	11
	1	1	1	1	4	5	8
	1/0	1	1	1	3	4	7
	2/0	0	1	1	2	3	6
	3/0	0	1	1	1	3	5
	4/0	0	1	1	1	2	4
XHHW, XHHW-2	14	8	15	25	43	58	96
	12	6	11	19	33	45	74
	10	5	8	14	24	33	55
	8	2	5	8	13	18	30
	6	1	3	6	10	14	22
	4	1	2	4	7	10	16
	3	1	1	3	6	8	14
	2	1	1	3	5	7	11
	1	1	1	1	4	5	8
	1/0	1	1	1	3	4	7
	2/0	0	1	1	2	3	6
	3/0	0	1	1	1	3	5
	4/0	0	1	1	1	2	4

For SI: 1 inch = 25.4 mm.

a. Types RHW, and RHW-2 without outer covering.

TABLE E3804.6(2)
MAXIMUM NUMBER OF CONDUCTORS IN ELECTRICAL NONMETALLIC TUBING (ENT)[a]

TYPE LETTERS	CONDUCTOR SIZE AWG/kcmil	TRADE SIZES (inches)					
		1/2	3/4	1	1 1/4	1 1/2	2
RHW, RHW-2	14	3	6	10	19	26	43
	12	2	5	9	16	22	36
	10	1	4	7	13	17	29
	8	1	1	3	6	9	15
	6	1	1	3	5	7	12
	4	1	1	2	4	6	9
	3	1	1	1	3	5	8
	2	0	1	1	3	4	7
	1	0	1	1	1	3	5
	1/0	0	0	1	1	2	4
	2/0	0	0	1	1	1	3
	3/0	0	0	1	1	1	3
	4/0	0	0	1	1	1	2
TW	14	7	13	22	40	55	92
	12	5	10	17	31	42	71
	10	4	7	13	23	32	52
	8	1	4	7	13	17	29
RHW[a], RHW-2[a], THHW, THW, THW-2	14	4	8	15	27	37	61
	12	3	7	12	21	29	49
	10	3	5	9	17	23	38
	8	1	3	5	10	14	23
RHW[a], RHW-2[a], TW, THW, THHW, THW-2	6	1	2	4	7	10	17
	4	1	1	3	5	8	13
	3	1	1	2	5	7	11
	2	1	1	2	4	6	9
	1	0	1	1	3	4	6
	1/0	0	1	1	2	3	5
	2/0	0	1	1	1	3	5
	3/0	0	0	1	1	2	4
	4/0	0	0	1	1	1	3
THHN, THWN, THWN-2	14	10	18	32	58	80	132
	12	7	13	23	42	58	96
	10	4	8	15	26	36	60
	8	2	5	8	15	21	35
	6	1	3	6	11	15	25
	4	1	1	4	7	9	15
	3	1	1	3	5	8	13
	2	1	1	2	5	6	11
	1	1	1	1	3	5	8
	1/0	0	1	1	3	4	7
	2/0	0	1	1	2	3	5
	3/0	0	1	1	1	3	4
	4/0	0	0	1	1	2	4
XHHW, XHHW-2	14	7	13	22	40	55	92
	12	5	10	17	31	42	71
	10	4	7	13	23	32	52
	8	1	4	7	13	17	29
	6	1	3	5	9	13	21
	4	1	1	4	7	9	15
	3	1	1	3	6	8	13
	2	1	1	2	5	6	11
	1	1	1	1	3	5	8
	1/0	0	1	1	3	4	7
	2/0	0	1	1	2	3	6
	3/0	0	1	1	1	3	5
	4/0	0	0	1	1	2	4

For SI: 1 inch = 25.4 mm.

a. Types RHW, and RHW-2 without outer covering.

TABLE E3804.6(3)
MAXIMUM NUMBER OF CONDUCTORS IN FLEXIBLE METALLIC CONDUIT (FMC)[a]

TYPE LETTERS	CONDUCTOR SIZE AWG/kcmil	TRADE SIZES (inches)					
		1/2	3/4	1	1 1/4	1 1/2	2
RHW, RHW-2	14	4	7	11	17	25	44
	12	3	6	9	14	21	37
	10	3	5	7	11	17	30
	8	1	2	4	6	9	15
	6	1	1	3	5	7	12
	4	1	1	2	4	5	10
	3	1	1	1	3	5	7
	2	1	1	1	3	4	7
	1	0	1	1	1	2	5
	1/0	0	1	1	1	2	4
	2/0	0	1	1	1	1	3
	3/0	0	0	1	1	1	3
TW	14	9	15	23	36	53	94
	12	7	11	18	28	41	72
	10	5	8	13	21	30	54
	8	3	5	7	11	17	30
RHW[a], RHW-2[a], THHW, THW, THW-2	14	6	10	15	24	35	62
	12	5	8	12	19	28	50
	10	4	6	10	15	22	39
	8	1	4	6	9	13	23
RHW[a], RHW-2[a], TW, THW, THHW, THW-2	6	1	3	4	7	10	18
	4	1	1	3	5	7	13
	3	1	1	3	4	6	11
	2	1	1	2	4	5	10
	1	1	1	1	2	4	7
	1/0	0	1	1	1	3	6
	2/0	0	1	1	1	3	5
	3/0	0	1	1	1	2	4
	4/0	0	0	1	1	1	3
	4/0	0	0	1	1	1	2
THHN, THWN, THWN-2	14	13	22	33	52	76	134
	12	9	16	24	38	56	98
	10	6	10	15	24	35	62
	8	3	6	9	14	20	35
	6	2	4	6	10	14	25
	4	1	2	4	6	9	16
	3	1	1	3	5	7	13
	2	1	1	3	4	6	11
	1	1	1	1	3	4	8
	1/0	1	1	1	2	4	7
	2/0	0	1	1	1	3	6
	3/0	0	1	1	1	2	5
	4/0	0	1	1	1	1	4
XHHW, XHHW-2	14	9	15	23	36	53	94
	12	7	11	18	28	41	72
	10	5	8	13	21	30	54
	8	3	5	7	11	17	30
	6	1	3	5	8	12	22
	4	1	2	4	6	9	16
	3	1	1	3	5	7	13
	2	1	1	3	4	6	11
	1	1	1	1	3	5	8
	1/0	1	1	1	2	4	7
	2/0	0	1	1	2	3	6
	3/0	0	1	1	1	3	5
	4/0	0	1	1	1	2	4

For SI: 1 inch = 25.4 mm.

a. Types RHW, and RHW-2 without outer covering.

TABLE E3804.6(4)
MAXIMUM NUMBER OF CONDUCTORS IN INTERMEDIATE METALLIC CONDUIT (IMC)[a]

TYPE LETTERS	CONDUCTOR SIZE AWG/kcmil	TRADE SIZES (inches)					
		$^1/_2$	$^3/_4$	1	$1^1/_4$	$1^1/_2$	2
RHW, RHW-2	14	4	8	13	22	30	49
	12	4	6	11	18	25	41
	10	3	5	8	15	20	33
	8	1	3	4	8	10	17
	6	1	1	3	6	8	14
	4	1	1	3	5	6	11
	3	1	1	2	4	6	9
	2	1	1	1	3	5	8
	1	0	1	1	2	3	5
	1/0	0	1	1	1	3	4
	2/0	0	1	1	1	2	4
	3/0	0	0	1	1	1	3
	4/0	0	0	1	1	1	3
TW	14	10	17	27	47	64	104
	12	7	13	21	36	49	80
	10	5	9	15	27	36	59
	8	3	5	8	15	20	33
RHW[a], RHW-2[a], THHW, THW, THW-2	14	6	11	18	31	42	69
	12	5	9	14	25	34	56
	10	4	7	11	19	26	43
	8	2	4	7	12	16	26
RHW[a], RHW-2[a], TW, THW, THHW, THW-2	6	1	3	5	9	12	20
	4	1	2	4	6	9	15
	3	1	1	3	6	8	13
	2	1	1	3	5	6	11
	1	1	1	1	3	4	7
	1/0	1	1	1	3	4	6
	2/0	0	1	1	2	3	5
	3/0	0	1	1	1	3	4
	4/0	0	1	1	1	2	4
THHN, THWN, THWN-2	14	14	24	39	68	91	149
	12	10	17	29	49	67	109
	10	6	11	18	31	42	68
	8	3	6	10	18	24	39
	6	2	4	7	13	17	28
	4	1	3	4	8	10	17
	3	1	2	4	6	9	15
	2	1	1	3	5	7	12
	1	1	1	2	4	5	9
	1/0	1	1	1	3	4	8
	2/0	1	1	1	3	4	6
	3/0	0	1	1	2	3	5
	4/0	0	1	1	1	2	4
XHHW, XHHW-2	14	10	17	27	47	64	104
	12	7	13	21	36	49	80
	10	5	9	15	27	36	59
	8	3	5	8	15	20	33
	6	1	4	6	11	15	24
	4	1	3	4	8	11	18
	3	1	2	4	7	9	15
	2	1	1	3	5	7	12
	1	1	1	2	4	5	9
	1/0	1	1	1	3	5	8
	2/0	1	1	1	3	4	6
	3/0	0	1	1	2	3	5
	4/0	0	1	1	1	2	4

For SI: 1 inch = 25.4 mm.

a. Types RHW, and RHW-2 without outer covering.

TABLE E3804.6(5)
MAXIMUM NUMBER OF CONDUCTORS IN LIQUID-TIGHT FLEXIBLE NONMETALLIC CONDUIT (FNMC-B)[a]

TYPE LETTERS	CONDUCTOR SIZE AWG/kcmil	TRADE SIZES (inches)						
		3/8	1/2	3/4	1	1 1/4	1 1/2	2
RHW, RHW-2	14	2	4	7	12	21	27	44
	12	1	3	6	10	17	22	36
	10	1	3	5	8	14	18	29
	8	1	1	2	4	7	9	15
	6	1	1	1	3	6	7	12
	4	0	1	1	2	4	6	9
	3	0	1	1	1	4	5	8
	2	0	1	1	1	3	4	7
	1	0	0	1	1	1	3	5
	1/0	0	0	1	1	1	2	4
	2/0	0	0	1	1	1	1	3
	3/0	0	0	0	1	1	1	3
	4/0	0	0	0	1	1	1	2
TW	14	5	9	15	25	44	57	93
	12	4	7	12	19	33	43	71
	10	3	5	9	14	25	32	53
	8	1	3	5	8	14	18	29
RHW[a], RHW-2[a], THHW, THW, THW-2	14	3	6	10	16	29	38	62
	12	3	5	8	13	23	30	50
	10	1	3	6	10	18	23	39
	8	1	1	4	6	11	14	23
RHW[a], RHW-2[a], TW, THW, THHW, THW-2	6	1	1	3	5	8	11	18
	4	1	1	1	3	6	8	13
	3	1	1	1	3	5	7	11
	2	0	1	1	2	4	6	9
	1	0	1	1	1	3	4	7
	1/0	0	0	1	1	2	3	6
	2/0	0	0	1	1	2	3	5
	3/0	0	0	1	1	1	2	4
	4/0	0	0	0	1	1	1	3
THHN, THWN, THWN-2	14	8	13	22	36	63	81	133
	12	5	9	16	26	46	59	97
	10	3	6	10	16	29	37	61
	8	1	3	6	9	16	21	35
	6	1	2	4	7	12	15	25
	4	1	1	2	4	7	9	15
	3	1	1	1	3	6	8	13
	2	1	1	1	3	5	7	11
	1	0	1	1	1	4	5	8
	1/0	0	1	1	1	3	4	7
	2/0	0	0	1	1	2	3	6
	3/0	0	0	1	1	1	3	5
	4/0	0	0	1	1	1	2	4
XHHW, XHHW-2	14	5	9	15	25	44	57	93
	12	4	7	12	19	33	43	71
	10	3	5	9	14	25	32	53
	8	1	3	5	8	14	18	29
	6	1	1	3	6	10	13	22
	4	1	1	2	4	7	9	16
	3	1	1	1	3	6	8	13
	2	1	1	1	3	5	7	11
	1	0	1	1	1	4	5	8
	1/0	0	1	1	1	3	4	7
	2/0	0	0	1	1	2	3	6
	3/0	0	0	1	1	1	3	5
	4/0	0	0	1	1	1	2	4

For SI: 1 inch = 25.4 mm.

a. Types RHW, and RHW-2 without outer covering.

TABLE E3804.6(6)
MAXIMUM NUMBER OF CONDUCTORS IN LIQUID-TIGHT FLEXIBLE NONMETALLIC CONDUIT (FNMC-A)[a]

TYPE LETTERS	CONDUCTOR SIZE AWG/kcmil	TRADE SIZES (inches)						
		3/8	1/2	3/4	1	1 1/4	1 1/2	2
RHW, RHW-2	14	2	4	7	11	20	27	45
	12	1	3	6	9	17	23	38
	10	1	3	5	8	13	18	30
	8	1	1	2	4	7	9	16
	6	1	1	1	3	5	7	13
	4	0	1	1	2	4	6	10
	3	0	1	1	1	4	5	8
	2	0	1	1	1	3	4	7
	1	0	0	1	1	1	3	5
	1/0	0	0	1	1	1	2	4
	2/0	0	0	1	1	1	1	4
	3/0	0	0	0	1	1	1	3
	4/0	0	0	0	1	1	1	3
TW	14	5	9	15	24	43	58	96
	12	4	7	12	19	33	44	74
	10	3	5	9	14	24	33	55
	8	1	3	5	8	13	18	30
RHW[a], RHW-2[a], THHW, THW, THW-2	14	3	6	10	16	28	38	64
	12	3	4	8	13	23	31	51
	10	1	3	6	10	18	24	40
	8	1	1	4	6	10	14	24
RHW[a], RHW-2[a], TW, THW, THHW, THW-2	6	1	1	3	4	8	11	18
	4	1	1	1	3	6	8	13
	3	1	1	1	3	5	7	11
	2	0	1	1	2	4	6	10
	1	0	1	1	1	3	4	7
	1/0	0	0	1	1	2	3	6
	2/0	0	0	1	1	1	3	5
	3/0	0	0	1	1	1	2	4
	4/0	0	0	0	1	1	1	3
THHN, THWN, THWN-2	14	8	13	22	35	62	83	137
	12	5	9	16	25	45	60	100
	10	3	6	10	16	28	38	63
	8	1	3	6	9	16	22	36
	6	1	2	4	6	12	16	26
	4	1	1	2	4	7	9	16
	3	1	1	1	3	6	8	13
	2	1	1	1	3	5	7	11
	1	0	1	1	1	4	5	8
	1/0	0	1	1	1	3	4	7
	2/0	0	0	1	1	2	3	6
	3/0	0	0	1	1	1	3	5
	4/0	0	0	1	1	1	2	4
XHHW, XHHW-2	14	5	9	15	24	43	58	96
	12	4	7	12	19	33	44	74
	10	3	5	9	14	24	33	55
	8	1	3	5	8	13	18	30
	6	1	1	3	5	10	13	22
	4	1	1	2	4	7	10	16
	3	1	1	1	3	6	8	14
	2	1	1	1	3	5	7	11
	1	0	1	1	1	4	5	8
	1/0	0	1	1	1	3	4	7
	2/0	0	0	1	1	2	3	6
	3/0	0	0	1	1	1	3	5
	4/0	0	0	1	1	1	2	4

For SI: 1 inch = 25.4 mm.

a. Types RHW, and RHW-2 without outer covering.

TABLE E3804.6(7)
MAXIMUM NUMBER OF CONDUCTORS IN LIQUID-TIGHT FLEXIBLE METAL CONDUIT (LFMC)[a]

TYPE LETTERS	CONDUCTOR SIZE AWG/kcmil	TRADE SIZES (inches)					
		1/2	3/4	1	1 1/4	1 1/2	2
RHW, RHW-2	14	4	7	12	21	27	44
	12	3	6	10	17	22	36
	10	3	5	8	14	18	29
	8	1	2	4	7	9	15
	6	1	1	3	6	7	12
	4	1	1	2	4	6	9
	3	1	1	1	4	5	8
	2	1	1	1	3	4	7
	1	0	1	1	1	3	5
	1/0	0	1	1	1	2	4
	2/0	0	1	1	1	1	3
	3/0	0	0	1	1	1	3
	4/0	0	0	1	1	1	2
TW	14	9	15	25	44	57	93
	12	7	12	19	33	43	71
	10	5	9	14	25	32	53
	8	3	5	8	14	18	29
RHW[a], RHW-2[a], THHW, THW, THW-2	14	6	10	16	29	38	62
	12	5	8	13	23	30	50
	10	3	6	10	18	23	39
	8	1	4	6	11	14	23
RHW[a], RHW-2[a], TW, THW, THHW, THW-2	6	1	3	5	8	11	18
	4	1	1	3	6	8	13
	3	1	1	3	5	7	11
	2	1	1	2	4	6	9
	1	1	1	1	3	4	7
	1/0	0	1	1	2	3	6
	2/0	0	1	1	2	3	5
	3/0	0	1	1	1	2	4
	4/0	0	0	1	1	1	3
THHN, THWN, THWN-2	14	13	22	36	63	81	133
	12	9	16	26	46	59	97
	10	6	10	16	29	37	61
	8	3	6	9	16	21	35
	6	2	4	7	12	15	25
	4	1	2	4	7	9	15
	3	1	1	3	6	8	13
	2	1	1	3	5	7	11
	1	1	1	1	4	5	8
	1/0	1	1	1	3	4	7
	2/0	0	1	1	2	3	6
	3/0	0	1	1	1	3	5
	4/0	0	1	1	1	2	4
XHHW, XHHW-2	14	9	15	25	44	57	93
	12	7	12	19	33	43	71
	10	5	9	14	25	32	53
	8	3	5	8	14	18	29
	6	1	3	6	10	13	22
	4	1	2	4	7	9	16
	3	1	1	3	6	8	13
	2	1	1	3	5	7	11
	1	1	1	1	4	5	8
	1/0	1	1	1	3	4	7
	2/0	0	1	1	2	3	6
	3/0	0	1	1	1	3	5
	4/0	0	1	1	1	2	4

For SI: 1 inch = 25.4 mm.

a. Types RHW, and RHW-2 without outer covering.

TABLE E3804.6(8)
MAXIMUM NUMBER OF CONDUCTORS IN RIGID METAL CONDUIT (RMC)[a]

TYPE LETTERS	CONDUCTOR SIZE AWG/kcmil	TRADE SIZES (inches)					
		1/2	3/4	1	1 1/4	1 1/2	2
RHW, RHW-2	14	4	7	12	21	28	46
	12	3	6	10	17	23	38
	10	3	5	8	14	19	31
	8	1	2	4	7	10	16
	6	1	1	3	6	8	13
	4	1	1	2	4	6	10
	3	1	1	2	4	5	9
	2	1	1	1	3	4	7
	1	0	1	1	1	3	5
	1/0	0	1	1	1	2	4
	2/0	0	1	1	1	2	4
	3/0	0	0	1	1	1	3
	4/0	0	0	1	1	1	3
TW	14	9	15	25	44	59	98
	12	7	12	19	33	45	75
	10	5	9	14	25	34	56
	8	3	5	8	14	19	31
RHW[a], RHW-2[a], THHW, THW, THW-2	14	6	10	17	29	39	65
	12	5	8	13	23	32	52
	10	3	6	10	18	25	41
	8	1	4	6	11	15	24
RHW[a], RHW-2[a], TW, THW, THHW, THW-2	6	1	3	5	8	11	18
	4	1	1	3	6	8	14
	3	1	1	3	5	7	12
	2	1	1	2	4	6	10
	1	1	1	1	3	4	7
	1/0	0	1	1	2	3	6
	2/0	0	1	1	2	3	5
	3/0	0	1	1	1	2	4
	4/0	0	0	1	1	1	3
THHN, THWN, THWN-2	14	13	22	36	63	85	140
	12	9	16	26	46	62	102
	10	6	10	17	29	39	64
	8	3	6	9	16	22	37
	6	2	4	7	12	16	27
	4	1	2	4	7	10	16
	3	1	1	3	6	8	14
	2	1	1	3	5	7	11
	1	1	1	1	4	5	8
	1/0	1	1	1	3	4	7
	2/0	0	1	1	2	3	6
	3/0	0	1	1	1	3	5
	4/0	0	1	1	1	2	4
XHHW, XHHW-2	14	9	15	25	44	59	98
	12	7	12	19	33	45	75
	10	5	9	14	25	34	56
	8	3	5	8	14	19	31
	6	1	3	6	10	14	23
	4	1	2	4	7	10	16
	3	1	1	3	6	8	14
	2	1	1	3	5	7	12
	1	1	1	1	4	5	9
	1/0	1	1	1	3	4	7
	2/0	0	1	1	2	3	6
	3/0	0	1	1	1	3	5
	4/0	0	1	1	1	2	4

For SI: 1 inch = 25.4 mm.

a. Types RHW, and RHW-2 without outer covering.

TABLE E3804.6(9)
MAXIMUM NUMBER OF CONDUCTORS IN RIGID PVC CONDUIT, SCHEDULE 80 (PVC-80)[a]

TYPE LETTERS	CONDUCTOR SIZE AWG/kcmil	TRADE SIZES (inches)					
		1/2	3/4	1	1 1/4	1 1/2	2
RHW, RHW-2	14	3	5	9	17	23	39
	12	2	4	7	14	19	32
	10	1	3	6	11	15	26
	8	1	1	3	6	8	13
	6	1	1	2	4	6	11
	4	1	1	1	3	5	8
	3	0	1	1	3	4	7
	2	0	1	1	3	4	6
	1	0	1	1	1	2	4
	1/0	0	0	1	1	1	3
	2/0	0	0	1	1	1	3
	3/0	0	0	1	1	1	3
	4/0	0	0	0	1	1	2
TW	14	6	11	20	35	49	82
	12	5	9	15	27	38	63
	10	3	6	11	20	28	47
	8	1	3	6	11	15	26
RHW[a], RHW-2[a], THHW, THW, THW-2	14	4	8	13	23	32	55
	12	3	6	10	19	26	44
	10	2	5	8	15	20	34
	8	1	3	5	9	12	20
RHW[a], RHW-2[a], TW, THW, THHW, THW-2	6	1	1	3	7	9	16
	4	1	1	3	5	7	12
	3	1	1	2	4	6	10
	2	1	1	1	3	5	8
	1	0	1	1	2	3	6
	1/0	0	1	1	1	3	5
	2/0	0	1	1	1	2	4
	3/0	0	0	1	1	1	3
	4/0	0	0	1	1	1	3
THHN, THWN, THWN-2	14	9	17	28	51	70	118
	12	6	12	20	37	51	86
	10	4	7	13	23	32	54
	8	2	4	7	13	18	31
	6	1	3	5	9	13	22
	4	1	1	3	6	8	14
	3	1	1	3	5	7	12
	2	1	1	2	4	6	10
	1	0	1	1	3	4	7
	1/0	0	1	1	2	3	6
	2/0	0	1	1	1	3	5
	3/0	0	1	1	1	2	4
	4/0	0	0	1	1	1	3
XHHW, XHHW-2	14	6	11	20	35	49	82
	12	5	9	15	27	38	63
	10	3	6	11	20	28	47
	8	1	3	6	11	15	26
	6	1	2	4	8	11	19
	4	1	1	3	6	8	14
	3	1	1	3	5	7	12
	2	1	1	2	4	6	10
	1	0	1	1	3	4	7
	1/0	0	1	1	2	3	6
	2/0	0	1	1	1	3	5
	3/0	0	1	1	1	2	4
	4/0	0	0	1	1	1	3

For SI: 1 inch = 25.4 mm.

a. Types RHW, and RHW-2 without outer covering.

**TABLE E3804.6(10)
MAXIMUM NUMBER OF CONDUCTORS IN RIGID PVC CONDUIT SCHEDULE 40 (PVC-40)[a]**

TYPE LETTERS	CONDUCTOR SIZE AWG/kcmil	TRADE SIZES (inches)					
		1/2	3/4	1	1 1/4	1 1/2	2
RHW, RHW-2	14	4	7	11	20	27	45
	12	3	5	9	16	22	37
	10	2	4	7	13	18	30
	8	1	2	4	7	9	15
	6	1	1	3	5	7	12
	4	1	1	2	4	6	10
	3	1	1	1	4	5	8
	2	1	1	1	3	4	7
	1	0	1	1	1	3	5
	1/0	0	1	1	1	2	4
	2/0	0	0	1	1	1	3
	3/0	0	0	1	1	1	3
	4/0	0	0	1	1	1	2
TW	14	8	14	24	42	57	94
	12	6	11	18	32	44	72
	10	4	8	13	24	32	54
	8	2	4	7	13	18	30
RHW[a], RHW-2[a], THHW, THW, THW-2	14	5	9	16	28	38	63
	12	4	8	12	22	30	50
	10	3	6	10	17	24	39
	8	1	3	6	10	14	23
RHW[a], RHW-2[a], TW, THW, THHW, THW-2	6	1	2	4	8	11	18
	4	1	1	3	6	8	13
	3	1	1	3	5	7	11
	2	1	1	2	4	6	10
	1	0	1	1	3	4	7
	1/0	0	1	1	2	3	6
	2/0	0	1	1	1	3	5
	3/0	0	1	1	1	2	4
	4/0	0	0	1	1	1	3
THHN, THWN, THWN-2	14	11	21	34	60	82	135
	12	8	15	25	43	59	99
	10	5	9	15	27	37	62
	8	3	5	9	16	21	36
	6	1	4	6	11	15	26
	4	1	2	4	7	9	16
	3	1	1	3	6	8	13
	2	1	1	3	5	7	11
	1	1	1	1	3	5	8
	1/0	1	1	1	3	4	7
	2/0	0	1	1	2	3	6
	3/0	0	1	1	1	3	5
	4/0	0	1	1	1	2	4
XHHW, XHHW-2	14	8	14	24	42	57	94
	12	6	11	18	32	44	72
	10	4	8	13	24	32	54
	8	2	4	7	13	18	30
	6	1	3	5	10	13	22
	4	1	2	4	7	9	16
	3	1	1	3	6	8	13
	2	1	1	3	5	7	11
	1	1	1	1	3	5	8
	1/0	1	1	1	3	4	7
	2/0	0	1	1	2	3	6
	3/0	0	1	1	1	3	5
	4/0	0	1	1	1	2	4

For SI: 1 inch = 25.4 mm.

a. Types RHW, and RHW-2 without outer covering.

E3805.3.1 Nonmetallic-sheathed cable and nonmetallic boxes. Where nonmetallic-sheathed cable is used, the cable assembly, including the sheath, shall extend into the box not less than $^1/_4$ inch (6.4 mm) through a nonmetallic-sheathed cable knockout opening.

E3805.3.2 Securing to box. All permitted wiring methods shall be secured to the boxes.

Exception: Where nonmetallic-sheathed cable is used with boxes not larger than a nominal size of $2^1/_4$ inches by 4 inches (57 mm by 102 mm) mounted in walls or ceilings, and where the cable is fastened within 8 inches (203 mm) of the box measured along the sheath, and where the sheath extends through a cable knockout not less than $^1/_4$ inch (6.4 mm), securing the cable to the box shall not be required.

E3805.3.3 Conductor rating. Nonmetallic boxes shall be suitable for the lowest temperature-rated conductor entering the box.

E3805.4 Minimum depth of outlet boxes. Boxes shall have an internal depth of not less than 0.5 inch (12.7 mm). Boxes designed to enclose flush devices shall have an internal depth of not less than 0.938 inch (2.4 mm).

E3805.5 Boxes enclosing flush-mounted devices. Boxes enclosing flush-mounted devices shall be of such design that the devices are completely enclosed at the back and all sides and shall provide support for the devices. Screws for supporting the box shall not be used for attachment of the device contained therein.

E3805.6 Boxes at lighting fixture outlets. Boxes for lighting fixture outlets shall be approved for the purpose and shall be installed to allow a lighting fixture to be attached.

E3805.7 Floor boxes. Where outlet boxes for receptacles are installed in the floor, such boxes shall be listed specifically for that application.

E3805.8 Boxes at fan outlets. Outlet boxes shall not be depended on as the sole support for ceiling (paddle) fans except where such boxes are listed for that application.

E3805.9 Conduit bodies and junction, pull and outlet boxes to be accessible. Conduit bodies and junction, pull and outlet boxes shall be installed so that the wiring therein can be accessed without removing any part of the building or, in underground circuits, without excavating sidewalks, paving, earth or other substance used to establish the finished grade.

Exception: Boxes covered by gravel, light aggregate or noncohesive granulated soil shall be listed for the application, and the box locations shall be effectively identified and access shall be provided for excavation.

E3805.10 Damp or wet locations. In damp or wet locations, boxes, conduit bodies and fittings shall be placed or equipped so as to prevent moisture from entering or accumulating within the box, conduit body or fitting. Boxes, conduit bodies and fittings installed in wet locations shall be listed for use in wet locations.

E3805.11 Number of conductors in outlet, device, and junction boxes, and conduit bodies. Boxes and conduit bodies shall be of sufficient size to provide free space for all enclosed conductors. In no case shall the volume of the box, as calculated in Section E3805.11.1, be less than the box fill calculation as calculated in Section E3805.11.2. The minimum volume for conduit bodies shall be as calculated in Section E3805.11.3. The provisions of this section shall not apply to terminal housings supplied with motors.

E3805.11.1 Box volume calculations. The volume of a wiring enclosure (box) shall be the total volume of the assembled sections, and, where used, the space provided by plaster rings, domed covers, extension rings, etc., that are marked with their volume in cubic inches or are made from boxes the dimensions of which are listed in Table E3805.11.1.

E3805.11.1.1 Standard boxes. The volumes of standard boxes that are not marked with a cubic-inch capacity shall be as given in Table E3805.11.1.

E3805.11.1.2 Other boxes. Boxes 100 cubic inches (1640 cm^3) or less, other than those described in Table E3805.11.1, and nonmetallic boxes shall be durably and legibly marked by the manufacturer with their cubic-inch capacity. Boxes described in Table E3805.11.1 that have a larger cubic inch capacity than is designated in the table shall be permitted to have their cubic-inch capacity marked as required by this section.

E3805.11.2 Box fill calculations. The volumes in Section E3805.11.2.1 through Section E3805.11.2.5, as applicable, shall be added together. No allowance shall be required for small fittings such as locknuts and bushings.

E3805.11.2.1 Conductor fill. Each conductor that originates outside the box and terminates or is spliced within the box shall be counted once, and each conductor that passes through the box without splice or termination shall be counted once. The conductor fill, in cubic inches, shall be computed using Table E3805.11.2.1. A conductor, no part of which leaves the box, shall not be counted.

Exception: An equipment grounding conductor or not more than four fixture wires smaller than No. 14, or both, shall be permitted to be omitted from the calculations where such conductors enter a box from a domed fixture or similar canopy and terminate within that box.

E3805.11.2.2 Clamp fill. Where one or more internal cable clamps, whether factory or field supplied, are present in the box, a single volume allowance in accordance with Table E3805.11.2.1 shall be made based on the largest conductor present in the box. No allowance shall be required for a cable connector with its clamping mechanism outside the box.

E3805.11.2.3 Support fittings fill. Where one or more fixture studs or hickeys are present in the box, a single volume allowance in accordance with Table E3805.11.2.1 shall be made for each type of fitting based on the largest conductor present in the box.

TABLE E3805.11.1
MAXIMUM NUMBER OF CONDUCTORS IN METAL BOXES[a]

BOX DIMENSIONS (inches trade size and type)	MAXIMUM CAPACITY (cubic inches)	MAXIMUM NUMBER OF CONDUCTORS[a]						
		No. 18	No. 16	No. 14	No. 12	No. 10	No. 8	No. 6
4 × 1 1/4 round or octagonal	12.5	8	7	6	5	5	4	2
4 × 1 1/2 round or octagonal	15.5	10	8	7	6	6	5	3
4 × 2 1/8 round or octagonal	21.5	14	12	10	9	8	7	4
4 × 1 1/4 square	18.0	12	10	9	8	7	6	3
4 × 1 1/2 square	21.0	14	12	10	9	8	7	4
4 × 2 1/8 square	30.3	20	17	15	13	12	10	6
4 11/16 × 1 1/4 square	25.5	17	14	12	11	10	8	5
4 11/16 × 1 1/2 square	29.5	19	16	14	13	11	9	5
4 11/16 × 2 1/8 square	42.0	28	24	21	18	16	14	8
3 × 2 × 1 1/2 device	7.5	5	4	3	3	3	2	1
3 × 2 × 2 device	10.0	6	5	5	4	4	3	2
3 × 2 × 2 1/4 device	10.5	7	6	5	4	4	3	2
3 × 2 × 2 1/2 device	12.5	8	7	6	5	5	4	2
3 × 2 × 2 3/4 device	14.0	9	8	7	6	5	4	2
3 × 2 × 3 1/2 device	18.0	12	10	9	8	7	6	3
4 × 2 1/8 × 1 1/2 device	10.3	6	5	5	4	4	3	2
4 × 2 1/8 × 1 7/8 device	13.0	8	7	6	5	5	4	2
4 × 2 1/8 × 2 1/8 device	14.5	9	8	7	6	5	4	2
3 3/4 × 2 × 2 1/2 masonry box/gang	14.0	9	8	7	6	5	4	2
3 3/4 × 2 × 3 1/2 masonry box/gang	21.0	14	12	10	9	8	7	4

For SI: 1 inch = 25.4 mm, 1 cubic inch = 16.4 cm^3.

a. Where no volume allowances are required by Sections E3805.11.2.2 through E3805.11.2.5.

TABLE E3805.11.2.1
VOLUME ALLOWANCE REQUIRED PER CONDUCTOR

SIZE OF CONDUCTOR (AWG)	FREE SPACE WITHIN BOX FOR EACH CONDUCTOR (cubic inches)
No. 18	1.50
No. 16	1.75
No. 14	2.00
No. 12	2.25
No. 10	2.50
No. 8	3.00
No. 6	5.00

For SI: 1 cubic inch = 16.4 cm^3.

E3805.11.2.4 Device or equipment fill. For each yoke or strap containing one or more devices or equipment, a double volume allowance in accordance with Table E3805.11.2.1 shall be made for each yoke or strap based on the largest conductor connected to a device(s) or equipment supported by that yoke or strap.

E3805.11.2.5 Equipment grounding conductor fill. Where one or more equipment grounding conductors or equipment bonding jumpers enters a box, a single volume allowance in accordance with Table E3805.11.2.1 shall be made based on the largest equipment grounding conductor or equipment bonding jumper present in the box.

E3805.11.3 Conduit bodies. Conduit bodies enclosing No. 6 conductors or smaller, other than short radius conduit bodies, shall have a cross-sectional area not less than twice the cross-sectional area of the largest conduit or tubing to which it is attached. The maximum number of conductors permitted shall be the maximum number permitted by Table E3804.6 for the conduit to which it is attached.

E3805.11.3.1 Splices, taps or devices. Only those conduit bodies that are durably and legibly marked by the manufacturer with their cubic inch capacity shall be permitted to contain splices, taps or devices. The maximum number of conductors shall be computed using the same procedure for similar conductors in other than standard boxes.

SECTION E3806
INSTALLATION OF BOXES, CONDUIT BODIES AND FITTINGS

E3806.1 Conductors entering boxes, conduit bodies or fittings. Conductors entering boxes, conduit bodies or fittings shall be protected from abrasion.

E3806.1.1 Insulated fittings. Where raceways containing ungrounded conductors No. 4 or larger enter a cabinet, box enclosure, or raceway, the conductors shall be protected by a substantial fitting providing a smoothly rounded insulating surface, unless the conductors are separated from the fitting or raceway by substantial insulating material securely fastened in place.

Exception: Where threaded hubs or bosses that are an integral part of a cabinet, box enclosure, or raceway provide a smoothly rounded or flared entry for conductors.

Conduit bushings constructed wholly of insulating material shall not be used to secure a fitting or raceway. The insulating fitting or insulating material shall have a temperature rating not less than the insulation temperature rating of the installed conductors.

E3806.2 Openings. Openings through which conductors enter shall be adequately closed.

E3806.3 Metal boxes, conduit bodies and fittings. Where raceway or cable is installed with metal boxes, or conduit bodies, the raceway or cable shall be secured to such boxes and conduit bodies.

E3806.4 Unused openings. Unused cable or raceway openings in boxes and conduit bodies shall be effectively closed to afford protection substantially equivalent to that of the wall of the box or conduit body. Metal plugs or plates used with nonmetallic boxes or conduit bodies shall be recessed at least 0.25 inch (6.4 mm) from the outer surface of the box or conduit body.

E3806.5 In wall or ceiling. In walls or ceilings of concrete, tile or other noncombustible material, boxes shall be installed so that the front edge of the box will not be set back from the finished surface more than 0.25 inch (6.4 mm). In walls and ceilings constructed of wood or other combustible material, outlet boxes shall be flush with the finished surface or project therefrom.

E3806.6 Plaster, gypsum board and plasterboard. Openings in plaster, gypsum board or plasterboard surfaces that accommodate boxes shall be made so that there are no gaps or open spaces greater than 0.125 inch (3.2 mm) around the edge of the box or fitting.

E3806.7 Exposed surface extensions. Surface extensions from an outlet box of a concealed wiring system shall be made by mounting and mechanically securing a box or extension ring over the concealed box.

Exception: A surface extension shall be permitted to be made from the cover of a concealed box where the cover is designed so it is unlikely to fall off, or be removed if its securing means becomes loose. The wiring method shall be flexible and arranged so that any required grounding continuity is independent of the connection between the box and cover.

E3806.8 Supports. Boxes and enclosures shall be supported in accordance with one or more of the provisions in Sections E3806.8.1 through E3806.8.6.

E3806.8.1 Surface mounting. An enclosure mounted on a building or other surface shall be rigidly and securely fastened in place. If the surface does not provide rigid and secure support, additional support in accordance with other provisions of Section E3806.8 shall be provided.

E3806.8.2 Structural mounting. An enclosure supported from a structural member of a building or from grade shall be rigidly supported either directly, or by using a metal, polymeric or wood brace.

E3806.8.2.1 Nails. Nails, where used as a fastening means, shall be attached by using brackets on the outside of the enclosure, or they shall pass through the interior within 0.25 inch (6.4 mm) of the back or ends of the enclosure.

E3806.8.2.2 Braces. Metal braces shall be protected against corrosion and formed from metal that is not less than 0.020 inch (.508 mm) thick uncoated. Wood braces shall have a cross section not less than nominal 1 inch by 2 inches (25 mm by 51 mm). Wood braces in wet locations shall be treated for the conditions. Polymeric braces shall be identified as being suitable for the use.

E3806.8.3 Mounting in finished surfaces. An enclosure mounted in a finished surface shall be rigidly secured thereto by clamps, anchors, or fittings identified for the application.

E3806.8.4 Raceway supported enclosures without devices or fixtures. An enclosure that does not contain a device(s) or support a fixture(s) or other equipment, and that is supported by entering raceways shall not exceed 100 cubic inches (1640 cm^3) in size. The enclosure shall have threaded entries or have hubs identified for the purpose. The enclosure shall be supported by two or more conduits threaded wrench tight into the enclosure or hubs. Each conduit shall be secured within 3 feet (914 mm) of the enclosure, or within 18 inches (457 mm) of the enclosure if all entries are on the same side of the enclosure.

Exception: Rigid metal, intermediate metal, or rigid nonmetallic conduit or electrical metallic tubing shall be permitted to support a conduit body of any size, provided that the conduit body is not larger in trade size than the largest trade size of the supporting conduit or electrical metallic tubing.

E3806.8.5 Raceway supported enclosures, with devices or fixtures. An enclosure that contains a device(s) or supports a fixture(s) or other equipment and is supported by entering raceways shall not exceed 100 cubic inches (1640 cm^3) in size. The enclosure shall have threaded entries or have hubs identified for the purpose. The enclosure shall be supported by two or more conduits threaded wrench tight into the enclosure or hubs. Each conduit shall be secured within 18 inches (457 mm) of the enclosure.

Exception: Rigid metal or intermediate metal conduit shall be permitted to support a conduit body of any size, provided that the conduit bodies are not larger in trade size than the largest trade size of the supporting conduit.

E3806.8.6 Enclosures in concrete or masonry. An enclosure supported by embedment shall be identified as being suitably protected from corrosion and shall be securely embedded in concrete or masonry.

E3806.9 Covers and canopies. Outlet boxes shall be effectively closed with a cover, faceplate or fixture canopy.

E3806.10 Metal covers and plates. Metal covers and plates shall be grounded.

E3806.11 Exposed combustible finish. Combustible wall or ceiling finish exposed between the edge of a fixture canopy or pan and the outlet box shall be covered with noncombustible material.

SECTION E3807 CABINETS AND PANELBOARDS

E3807.1 Damp or wet locations. In damp or wet locations, cabinets and panelboards of the surface type shall be placed or equipped so as to prevent moisture or water from entering and accumulating within the cabinet, and shall be mounted to provide an airspace not less than 0.25 inch (6.4 mm) between the enclosure and the wall or other supporting surface. Cabinets installed in wet locations shall be weatherproof.

E3807.2 Position in wall. In walls of concrete, tile or other noncombustible material, cabinets and panelboards shall be installed so that the front edge of the cabinet will not set back of the finished surface more than 0.25 inch (6.4 mm). In walls constructed of wood or other combustible material, cabinets shall be flush with the finished surface or shall project therefrom.

E3807.3 Unused openings. Unused openings in cabinets and panelboards shall be effectively closed to afford protection equivalent to that of the wall of the cabinet. Metal plugs or plates used with nonmetallic cabinets shall be recessed at least 0.25 inch (6.4 mm) from the outer surface.

E3807.4 Conductors entering cabinets. Conductors entering cabinets and panelboards shall be protected from abrasion and shall comply with Section E3806.1.1.

E3807.5 Openings to be closed. Openings through which conductors enter cabinets, panelboards and meter sockets shall be adequately closed.

E3807.6 Cables. Where cables are used, each cable shall be secured to the cabinet or panelboard.

SECTION E3808 GROUNDING

E3808.1 Metal enclosures. Metal enclosures of conductors, devices and equipment shall be grounded.

Exceptions:

1. Short sections of metal enclosures or raceways used to provide cable assemblies with support or protection against physical damage.
2. A metal elbow that is installed in an underground installation of rigid nonmetallic conduit and is isolated from possible contact by a minimum cover of 18 inches (457 mm) to any part of the elbow.

E3808.2 Equipment fastened in place or connected by permanent wiring methods (fixed). Exposed noncurrent-carrying metal parts of fixed equipment likely to become energized shall be grounded where any of the following conditions apply:

1. Where within 8 feet (2438 mm) vertically or 5 feet (1524 mm) horizontally of earth or grounded metal objects and subject to contact by persons,
2. Where located in a wet or damp location and not isolated, or
3. Where in electrical contact with metal.

E3808.3 Specific equipment fastened in place or connected by permanent wiring methods. Exposed noncurrent-carrying metal parts of the following equipment and enclosures shall be grounded:

1. Lighting fixtures as provided in Chapter 39.
2. Motor-operated water pumps, including submersible types. Where a submersible pump is used in a metal well casing, the well casing shall be bonded to the pump circuit equipment grounding conductor.

E3808.4 Performance of fault current path. The fault current path shall be permanent and electrically continuous, shall be capable of safely conducting the maximum fault current likely to be imposed on it, and shall have sufficiently low impedance to facilitate the operation of overcurrent devices under fault conditions.

E3808.5 Earth as a grounding conductor. The earth shall not be used as the sole equipment grounding conductor or fault current path.

E3808.6 Load-side neutral. A grounding connection shall not be made to any grounded circuit conductor on the load side of the service disconnecting means.

Exception: A grounding conductor connection shall be made at each separate building where required by Section E3507.3.

E3808.7 Load-side equipment. A grounded circuit conductor shall not be used for grounding noncurrent-carrying metal parts of equipment on the load side of the service disconnecting means.

Exception: For separate buildings, in accordance with Section E3507.3.

E3808.8 Types of equipment grounding conductors. The equipment grounding conductor run with or enclosing the circuit conductors shall be one or more or a combination of the following:

1. A copper or other corrosion-resistant conductor. This conductor shall be solid or stranded; insulated, covered or bare; in the form of a wire or a busbar of any shape.
2. Rigid metal conduit.
3. Intermediate metal conduit.
4. Electrical metallic tubing.
5. Flexible metal conduit, where both the conduit and fittings are listed for grounding.
6. Armor of Type AC cable.
7. Surface metal raceway.
8. Metal-clad cable, where both the cable and fittings are listed for grounding.

E3808.8.1 Flexible metal conduit. Flexible metal conduit that is not listed for grounding shall not be used as a grounding conductor except where all of the following conditions are met:

1. The conduit is terminated in fittings listed for grounding.
2. The circuit conductors contained in the conduit are protected by overcurrent devices rated at 20 amperes or less.
3. The combined length of flexible metal conduit and flexible metallic tubing and liquid-tight flexible metal conduit in the same ground return path does not exceed 6 feet (1829 mm).
4. The conduit is not installed to provide flexibility.

E3808.8.2 Liquid-tight flexible metal conduit. Liquid-tight flexible metal conduit shall not be used as a grounding conductor except where all of the following conditions are met:

1. The conduit is terminated in fittings listed for grounding.
2. For trade sizes $^3/_8$ inch through $^1/_2$ inch (9.5 mm through 12.7 mm), the circuit conductors contained in the conduit are protected by overcurrent devices rated at 20 amperes or less.
3. For trade sizes $^3/_4$ inch through $1^1/_4$ inches (19.1 mm through 32 mm), the circuit conductors contained in the conduit are protected by overcurrent devices rated at not more than 60 amperes and there is no flexible metal conduit, flexible metallic tubing, or liquid-tight flexible metal conduit in trade sizes $^3/_8$ inch or $^1/_2$ inch (9.5 mm through 12.7 mm) in the grounding path.
4. The combined length of flexible metal conduit and flexible metallic tubing and liquidtight flexible metal conduit in the same ground return path does not exceed 6 feet (1829 mm).
5. The conduit is not installed for flexibility.

E3808.8.3 Nonmetallic sheathed cable (Type NM). In addition to the insulated conductors, the cable shall be permitted to have an insulated or bare conductor for equipment grounding purposes only. Equipment grounding conductors shall be sized in accordance with Table E3808.12.

E3808.9 Equipment fastened in place or connected by permanent wiring methods. Noncurrent-carrying metal parts of equipment, raceways and other enclosures, where required to be grounded, shall be grounded by one of the following methods:

1. By any of the equipment grounding conductors permitted by Sections E3808.8 through E3808.8.3.
2. By an equipment grounding conductor contained within the same raceway, cable or cord, or otherwise run with the circuit conductors. Individually, covered or insulated equipment grounding conductors shall have a continuous outer finish that is either green or green with one or more yellow stripes.

E3808.10 Methods of equipment grounding. Fixtures and equipment shall be considered grounded where mechanically connected to an equipment grounding conductor as specified in Sections E3808.8 through E3808.8.3 and sized in accordance with Section E3808.12.

E3808.11 Equipment grounding conductor installation. Where an equipment grounding conductor consists of a raceway, cable armor or cable sheath or where such conductor is a wire within a raceway or cable, it shall be installed in accordance with the provisions of this chapter and Chapters 33 and 37 using fittings for joints and terminations approved for installation with the type of raceway or cable used. All connections, joints and fittings shall be made tight using suitable tools.

E3808.12 Equipment grounding conductor size. Copper, aluminum and copper-clad aluminum equipment grounding conductors shall be not smaller than shown in Table E3808.12, but shall not be required to be larger than the circuit conductors supplying the equipment. Where a raceway or a cable armor or sheath is used as the equipment grounding conductor, as provided in Section E3808.8, it shall comply with Section E3808.4.

TABLE E3808.12
EQUIPMENT GROUNDING CONDUCTOR SIZING

RATING OR SETTING OF AUTOMATIC OVERCURRENT DEVICE IN CIRCUIT AHEAD OF EQUIPMENT, CONDUIT, ETC., NOT EXCEEDING THE FOLLOWING RATINGS (amperes)	MINIMUM SIZE	
	Copper wire No. (AWG)	Aluminum or copper-clad aluminum wire No. (AWG)
15	14	12
20	12	10
30	10	8
40	10	8
60	10	8
100	8	6
200	6	4
300	4	2
400	3	1

E3808.12.1 Multiple circuits. Where a single equipment grounding conductor is run with multiple circuits in the same raceway or cable, it shall be sized for the largest overcurrent device protecting conductors in the raceway or cable.

E3808.13 Continuity and attachment of equipment grounding conductors to boxes. Where more than one equipment grounding conductor enters a box, all such conductors shall be spliced or joined within the box or to the box with devices suitable for the use. Splices shall be made in accordance with Section E3306.10 except that insulation shall not be required. The arrangement of grounding connections shall be such that the disconnection or removal of a receptacle, fixture or other device fed from the box will not interfere with or interrupt the grounding continuity.

E3808.14 Connecting receptacle grounding terminal to box. An equipment bonding jumper shall be used to connect the grounding terminal of a grounding-type receptacle to a grounded box except where grounded in accordance with one of the following:

1. Surface mounted box. Where the box is mounted on or at the surface, direct metal-to-metal contact between the device yoke and the box shall be permitted to ground the receptacle to the box. This provision shall not apply to cover-mounted receptacles unless the box and cover combination are listed as providing satisfactory ground continuity between the box and the receptacle.
2. Contact devices or yokes. Contact devices or yokes designed and listed for the purpose shall be permitted in conjunction with the supporting screws to establish the grounding circuit between the device yoke and flush-type boxes.
3. Floor boxes. The receptacle is installed in a floor box designed for and listed as providing satisfactory ground continuity between the box and the device.

E3808.15 Metal boxes. A connection shall be made between the one or more equipment grounding conductors and a metal box by means of a grounding screw that shall be used for no other purpose, or by means of a listed grounding device. Sheet-metal screws shall not be used to connect grounding conductors to boxes.

E3808.16 Nonmetallic boxes. One or more equipment grounding conductors brought into a nonmetallic outlet box shall be arranged to allow connection to fittings or devices installed in that box.

E3808.17 Clean surfaces. Nonconductive coatings such as paint, lacquer and enamel on equipment to be grounded shall be removed from threads and other contact surfaces to ensure electrical continuity or the equipment shall be connected by means of fittings designed so as to make such removal unnecessary.

E3808.18 Bonding other enclosures. Metal raceways, cable armor, cable sheath, enclosures, frames, fittings and other metal noncurrent-carrying parts that serve as grounding conductors, with or without the use of supplementary equipment grounding conductors, shall be effectively bonded where necessary to ensure electrical continuity and the capacity to conduct safely any fault current likely to be imposed on them. Any nonconductive paint, enamel and similar coating shall be removed at threads, contact points and contact surfaces, or connections shall be made by means of fittings designed so as to make such removal unnecessary.

E3808.19 Size of equipment bonding jumper on load side of service. The equipment bonding jumper on the load side of the service overcurrent devices shall be sized, as a minimum, in accordance with Table E3808.12, but shall not be required to be larger than the circuit conductors supplying the equipment. An equipment grounding conductor shall be not smaller than No. 14.

A single common continuous equipment bonding jumper shall be permitted to bond two or more raceways or cables where the bonding jumper is sized in accordance with Table 3808.12 for the largest overcurrent device supplying circuits therein.

E3808.20 Installation—equipment bonding jumper. The equipment bonding jumper shall be permitted to be installed inside or outside of a raceway or enclosure. Where installed on the outside, the length of the equipment bonding jumper shall not exceed 6 feet (1829 mm) and shall be routed with the raceway or enclosure. Where installed inside of a raceway, the equipment bonding jumper shall comply with the requirements of Sections E3808.9, Item 2; E3808.13; E3808.15; and E3808.16.

SECTION E3809 FLEXIBLE CORDS

E3809.1 Where permitted. Flexible cords shall be used only for the connection of appliances where the fastening means and mechanical connections of such appliances are designed to permit ready removal for maintenance and repair, and the appliance is listed for flexible cord connection. Flexible cords shall not be installed as a substitute for the fixed wiring of a structure; shall not be run through holes in walls, ceilings or floors; shall not be concealed behind building walls, ceilings or floors; and shall not be installed in raceways.

E3809.2 Loading and protection. The ampere load and overcurrent protection of flexible cords serving fixed appliances shall be in accordance with Table E3809.2. This table shall be used in conjunction with applicable end use product standards to ensure selection of the proper size and type.

TABLE E3809.2
MAXIMUM AMPERE LOAD AND OVERCURRENT-PROTECTION-DEVICE RATING FOR FLEXIBLE CORDS

CORD SIZE (AWG)	CORD TYPES S, SE, SEO, SJ, SJE, SJEO, SJO, SJOO, SJT, SJTO, SJTOO, SO, SOO, SRD, SRDE, SRDT, ST, STD, SV, SVO, SVOO, SVTO, SVTOO — Maximum ampere load: Three current-carrying conductors	Maximum ampere load: Two current-carrying conductors	MAXIMUM AMPERE RATING OF BRANCH-CIRCUIT OVERCURRENT-PROTECTIVE DEVICE[a]
18	7	10	20
16	10	13	30
14	15	18	30
12	20	25	40

a. This column applies only where the cord is used in accordance with the appliance listing.

E3809.3 Splices. Flexible cord shall be used only in continuous lengths without splices or taps.

E3809.4 Attachment plugs. Where used in accordance with Section E3809.1, each flexible cord shall be equipped with an attachment plug and shall be energized from a receptacle outlet.

Chapter 39

DEVICES AND LIGHTING FIXTURES

SECTION E3901 SWITCHES

E3901.1 Rating and application of snap switches. General-use snap switches shall be used within their ratings and shall control only the following loads:

1. Resistive and inductive loads, including electric-discharge lamps, not exceeding the ampere rating of the switch at the voltage involved.
2. Tungsten-filament lamp loads not exceeding the ampere rating of the switch at 120 volts.
3. Motor loads not exceeding 80 percent of the ampere rating of the switch at its rated voltage.

E3901.2 CO/ALR snap switches. Snap switches rated 20 amperes or less directly connected to aluminum conductors shall be marked CO/ALR.

E3901.3 Indicating. General-use and motor-circuit switches and circuit breakers shall clearly indicate whether they are in the open OFF or closed ON position. Where single-throw switches or circuit breaker handles are operated vertically rather than rotationally or horizontally, the up position of the handle shall be the ON position.

E3901.4 Time switches and similar devices. Time switches and similar devices shall be of the enclosed type or shall be mounted in cabinets or boxes or equipment enclosures. A barrier shall be used around energized parts to prevent operator exposure when making manual adjustments or switching.

E3901.5 Grounding of enclosures. Metal enclosures for switches or circuit breakers shall be grounded. Where nonmetallic enclosures are used with metal raceways or metal-armored cables, provisions shall be made for maintaining grounding continuity.

Metal boxes for switches shall be effectively grounded. Nonmetallic boxes for switches shall be installed with a wiring method that provides or includes an equipment grounding conductor.

E3901.6 Access. All switches and circuit breakers used as switches shall be located to allow operation from a readily accessible location. Such devices shall be installed so that the center of the grip of the operating handle of the switch or circuit breaker, when in its highest position, will not be more than 6 feet 7 inches (2007 mm) above the floor or working platform.

E3901.7 Wet locations. A switch or circuit breaker located in a wet location or outside of a building shall be enclosed in a weatherproof enclosure or cabinet. Switches shall not be installed within wet locations in tub or shower spaces unless installed as part of a listed tub or shower assembly.

E3901.8 Grounded conductors. Switches or circuit breakers shall not disconnect the grounded conductor of a circuit except where the switch or circuit breaker simultaneously disconnects all conductors of the circuit.

E3901.9 Switch connections. Three- and four-way switches shall be wired so that all switching occurs only in the ungrounded circuit conductor. Color coding of switch connection conductors shall comply with Section E3307.4. Where in metal raceways or metal-jacketed cables, wiring between switches and outlets shall be in accordance with Section 3306.7.

Exception: Switch loops do not require a grounded conductor.

E3901.10 Box mounted. Flush-type snap switches mounted in boxes that are recessed from the wall surfaces as covered in Section 3806.5 shall be installed so that the extension plaster ears are seated against the surface of the wall. Flush-type snap switches mounted in boxes that are flush with the wall surface or project therefrom shall be installed so that the mounting yoke or strap of the switch is seated against the box.

E3901.11 Snap switch faceplates. Snap switches mounted in boxes shall have faceplates installed so as to completely cover the opening and seat against the finished surface.

E3901.11.1 Faceplate grounding. Snap switches, including dimmer switches, shall be effectively grounded and shall provide a means to ground metal faceplates, whether or not a metal faceplate is installed. Snap switches shall be considered effectively grounded where either of the following conditions are met:

1. The switch is mounted with metal screws to a metal box or to a nonmetallic box with integral means for grounding devices.
2. An equipment grounding conductor or equipment bonding jumper is connected to an equipment grounding termination of the snap switch.

SECTION E3902 RECEPTACLES

E3902.1 Rating and type. Receptacles installed for the attachment of portable cords shall be rated at not less than 15 amperes, 125 volts, or 15 amperes, 250 volts, and shall not be a lampholder type. Receptacles shall be rated in accordance with this section.

E3902.1.1 Single receptacle. A single receptacle installed on an individual branch circuit shall have an ampere rating not less than that of the branch circuit.

E3902.1.2 Two or more receptacles. Where connected to a branch circuit supplying two or more receptacles or outlets, receptacles shall conform to the values listed in Table E3902.1.2.

TABLE E3902.1.2
RECEPTACLE RATINGS FOR VARIOUS SIZE MULTIOUTLET CIRCUITS

CIRCUIT RATING (amperes)	RECEPTACLE RATING (amperes)
15	15
20	15 or 20
30	30
40	40 or 50
50	50

E3902.2 Grounding type. Receptacles installed on 15- and 20-ampere-rated branch circuits shall be of the grounding type.

E3902.3 CO/ALR receptacles. Receptacles rated at 20 amperes or less and directly connected to aluminum conductors shall be marked CO/ALR.

E3902.4 Faceplates. Metal faceplates shall be grounded.

E3902.5 Position of receptacle faces. After installation, receptacle faces shall be flush with or project from faceplates of insulating material and shall project a minimum of 0.015 inch (0.381 mm) from metal faceplates. Faceplates shall be installed so as to completely cover the opening and seat against the mounting surface.

E3902.6 Receptacle mounted in boxes. Receptacles mounted in boxes that are set back of the wall surface shall be installed so that the mounting yoke or strap of the receptacle is held rigidly at the surface of the wall. Receptacles mounted in boxes that are flush with the wall surface or project therefrom shall be so installed that the mounting yoke or strap is seated against the box or raised cover.

E3902.7 Receptacles mounted on covers. Receptacles mounted to and supported by a cover shall be secured by more than one screw or shall be a device assembly or box cover listed and identified for securing by a single screw.

E3902.8 Damp locations. A receptacle installed outdoors in a location protected from the weather or in other damp locations shall have an enclosure for the receptacle that is weatherproof when the receptacle cover(s) is closed and an attachment plug cap is not inserted. An installation suitable for wet locations shall also be considered suitable for damp locations. A receptacle shall be considered to be in a location protected from the weather where located under roofed open porches, canopies and similar structures and not subject to rain or water runoff.

E3902.9 Wet locations. Where a receptacle is installed in a wet location and where the product intended to be plugged into it is not attended while in use, the receptacle shall have an enclosure that is weatherproof both when the attachment plug cap is inserted and when it is removed. Where a receptacle is installed in a wet location and where the product intended to be plugged into it will be attended while in use, the receptacle shall have an enclosure that is weatherproof when the attachment plug cap is removed.

E3902.10 Bathtub and shower space. A receptacle shall not be installed within a bathtub or shower space.

E3902.11 Flush mounting with faceplate. In damp or wet locations, the enclosure for a receptacle installed in an outlet box flush-mounted on a wall surface shall be made weatherproof by means of a weatherproof faceplate assembly that provides a water-tight connection between the plate and the wall surface.

E3902.12 Outdoor installation. A receptacle outlet installed outdoors shall be located so that water accumulation is not likely to contact the outlet cover or plate.

SECTION E3903 FIXTURES

E3903.1 Energized parts. Fixtures, lampholders, lamps and receptacles shall not have energized parts normally exposed to contact.

E3903.2 Fixtures near combustible material. Fixtures shall be installed so that combustible material will not be subjected to temperatures in excess of 90°C (194°F).

E3903.3 Exposed conductive parts. The exposed conductive parts of lighting fixtures and equipment directly wired or attached to outlets shall be grounded. Lamp tie wires, mounting screws, clips and decorative bands on glass spaced at least 1.5 inches (38 mm) from lamp terminals shall not be required to be grounded.

E3903.4 Screw-shell type. Lampholders of the screw-shell type shall be installed for use as lampholders only.

E3903.5 Recessed incandescent fixtures. Recessed incandescent fixtures shall have thermal protection and shall be listed as thermally protected.

Exceptions:

1. Thermal protection shall not be required in recessed fixtures listed for the purpose and installed in poured concrete.
2. Thermal protection shall not be required in recessed fixtures having design, construction, and thermal performance characteristics equivalent to that of thermally protected fixtures, and such fixtures are identified as inherently protected.

E3903.6 Thermal protection. The ballast of a fluorescent fixtures installed indoors shall have integral thermal protection. Replacement ballasts shall also have thermal protection integral with the ballast. A simple reactance ballast in a fluorescent fixture with straight tubular lamps shall not be required to be thermally protected.

E3903.7 High-intensity discharge fixtures. Recessed high-intensity fixtures designed to be installed in wall or ceiling cavities shall have thermal protection and be identified as thermally protected. Thermal protection shall not be required in recessed high-intensity fixtures having design, construction and thermal performance characteristics equivalent to that of thermally protected fixtures, and such fixtures are identified as inherently protected. Thermal protection shall not be required in recessed high-intensity discharge fixtures installed in and identified for use in poured concrete.

A recessed remote ballast for a high-intensity discharge fixture shall have thermal protection that is integral with the ballast and shall be identified as thermally protected.

E3903.8 Wet or damp locations. Fixtures installed in wet or damp locations shall be installed so that water cannot enter or accumulate in wiring compartments, lampholders or other electrical parts. All fixtures installed in wet locations shall be marked SUITABLE FOR WET LOCATIONS. All fixtures installed in damp locations shall be marked SUITABLE FOR WET LOCATIONS or SUITABLE FOR DAMP LOCATIONS.

E3903.9 Lampholders in wet or damp locations. Lampholders installed in wet or damp locations shall be of the weatherproof type.

E3903.10 Bathtub and shower areas. Cord-connected fixtures, hanging fixtures, lighting track, pendants, and ceiling-suspended paddle fans shall not have any parts located within a zone measured 3 feet (914 mm) horizontally and 8 feet (2438 mm) vertically from the top of a bathtub rim or shower stall threshold. This zone is all encompassing and includes the zone directly over the tub or shower.

E3903.11 Fixtures in clothes closets. For the purposes of this section, storage space shall be defined as a volume bounded by the sides and back closet walls and planes extending from the closet floor vertically to a height of 6 feet (1829 mm) or the highest clothes-hanging rod and parallel to the walls at a horizontal distance of 24 inches (610 mm) from the sides and back of the closet walls respectively, and continuing vertically to the closet ceiling parallel to the walls at a horizontal distance of 12 inches (305 mm) or the width of the shelf, whichever is greater. For a closet that permits access to both sides of a hanging rod, the storage space shall include the volume below the highest rod extending 12 inches (305 mm) on either side of the rod on a plane horizontal to the floor extending the entire length of the rod (see Figure E3903.11).

The types of fixtures installed in clothes closets shall be limited to surface-mounted or recessed incandescent fixtures with completely enclosed lamps, and surface-mounted or recessed fluorescent fixtures. Incandescent fixtures with open or partially enclosed lamps and pendant fixtures or lampholders shall be prohibited. Fixture installations shall be in accordance with one or more of the following:

1. Surface-mounted incandescent fixtures shall be installed on the wall above the door or on the ceiling, provided there is a minimum clearance of 12 inches (305 mm) between the fixture and the nearest point of a storage space.
2. Surface-mounted fluorescent fixtures shall be installed on the wall above the door or on the ceiling, provided there is a minimum clearance of 6 inches (152 mm) between the fixture and the nearest point of a storage space.
3. Recessed incandescent fixtures with a completely enclosed lamp shall be installed in the wall or the ceiling provided there is a minimum clearance of 6 inches (152 mm) between the fixture and the nearest point of a storage space.
4. Recessed fluorescent fixtures shall be installed in the wall or on the ceiling provided there is a minimum clearance of 6 inches (152 mm) between the fixture and the nearest point of a storage space.

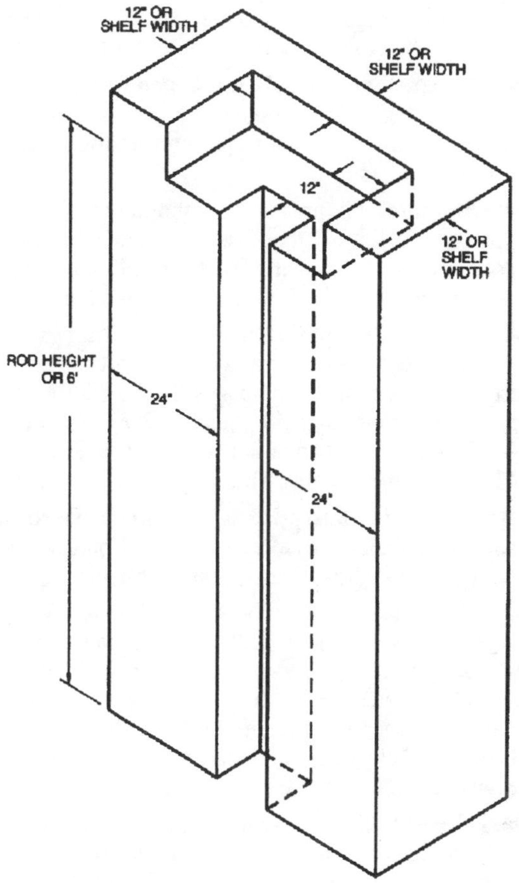

For SI: 1 inch = 25.4 mm, 1 foot = 304.8 mm.

FIGURE E3903.11
CLOSET STORAGE SPACE

E3903.12 Fixture wiring—general. Wiring within fixtures or fixture canopies shall not be exposed to physical damage. Conductors shall be arranged so that they are not subjected to temperatures above those for which the conductors are rated.

E3903.12.1 Polarization of fixtures. Fixtures shall be wired so that the screw shells of lampholders will be connected to the same fixture or circuit conductor or terminal. The grounded conductor shall be connected to the screw shell.

E3903.12.2 Fixtures as raceways. Fixtures shall not be used as raceways for circuit conductors except where such fixtures are listed for the purpose.

SECTION E3904
FIXTURE INSTALLATION

E3904.1 Outlet box covers. In a completed installation, each outlet box shall be provided with a cover except where covered by means of a fixture canopy, lampholder or device with a faceplate.

E3904.2 Combustible material at outlet boxes. Combustible wall or ceiling finish exposed between the inside edge of a fixture canopy or pan and the outlet box to which the fixture connects shall be covered with a noncombustible material.

E3904.3 Access. Fixtures shall be installed so that the connections between the fixture conductors and the circuit conductors can be accessed without requiring the disconnection of any part of the wiring.

E3904.4 Supports. Fixtures and lampholders shall be securely supported. A fixture that weighs more than 6 pounds (2.72 kg) or exceeds 16 inches (406 mm) in any dimension shall not be supported by the screw shell of a lampholder.

E3904.5 Means of support. Outlet boxes or fittings installed as required by Sections E3805 and E3806 shall be permitted to support fixtures weighing 50 pounds (22.7 kg) or less. A fixture that weighs more than 50 pounds (22.7 kg) shall be supported independently of an outlet box except where the outlet box is listed for the weight to be supported.

E3904.6 Exposed ballasts. Fixtures having exposed ballasts or transformers shall be installed so that such ballasts or transformers are not in contact with combustible material.

E3904.7 Combustible low-density cellulose fiberboard. Where a surface-mounted fixture containing a ballast is installed on combustible low-density cellulose fiberboard, the fixture shall be listed for the purpose or it shall be spaced not less than 1.5 inches (38 mm) from the surface of the fiberboard. Where such fixtures are partially or wholly recessed, the provisions of Sections E3904.8 and E3904.9 shall apply.

E3904.8 Recessed fixture clearance. A recessed fixture that is not identified for contact with insulation shall have all recessed parts spaced at least 0.5 inch (12.7 mm) from combustible materials. The points of support and the finish trim parts at the opening in the ceiling or wall surface shall be permitted to be in contact with combustible materials. A recessed fixture that is identified for contact with insulation, Type IC, shall be permitted to be in contact with combustible materials at recessed parts, points of support, and portions passing through the building structure and at finish trim parts at the opening in the ceiling or wall.

E3904.9 Recessed fixture installation. Thermal insulation shall not be installed above a recessed fixture or within 3 inches (76 mm) of the recessed fixture's enclosure, wiring compartment or ballast except where such fixture is identified for contact with insulation, Type IC.

SECTION E3905
TRACK LIGHTING

E3905.1 Fastening. Lighting track shall be securely mounted so that each fastening will be suitable for supporting the maximum weight of fixtures that can be installed. Except where identified for supports at greater intervals, a single section 4 feet (1219 mm) or shorter in length shall have two supports and, where installed in a continuous row, each individual section of not more than 4 feet (1219 mm) in length shall have one additional support.

E3905.2 Grounding. Lighting track shall be grounded in accordance with Chapter 38, and the track sections shall be securely coupled to maintain continuity of the circuitry, polarization and grounding throughout.

Chapter 40
APPLIANCE INSTALLATION

SECTION E4001
GENERAL

E4001.1 Scope. This section covers installation requirements for appliances and fixed heating equipment.

E4001.2 Installation. Appliances and equipment shall be installed in accordance with the manufacturer's installation instructions. Electrically heated appliances and equipment shall be installed with the required clearances to combustible materials.

E4001.3 Flexible cords. Cord- and plug-connected appliances shall use cords suitable for the environment and physical conditions likely to be encountered. Flexible cords shall only be used where the appliance is listed to be connected with a flexible cord. Specific appliances have additional requirements as specified in Table E4001.3. (See Section E3809.)

TABLE E4001.3
FLEXIBLE CORD LENGTH

APPLIANCE	MINIMUM CORD LENGTH (inches)	MAXIMUM CORD LENGTH (inches)
Kitchen waste disposal	18	36
Built-in dishwasher	36	48
Trash compactor	36	48

For SI: 1 inch = 25.4 mm.

E4001.4 Overcurrent protection. Each appliance shall be protected against overcurrent in accordance with the rating of the appliance and its listing.

E4001.4.1 Single nonmotor-operated appliance. The overcurrent protection for a branch circuit that supplies a single nonmotor-operated appliance shall not exceed that marked on the appliance. Where the overcurrent protection rating is not marked and the appliance is rated at over 13.3 amperes, the overcurrent protection shall not exceed 150 percent of the appliance rated current. Where 150 percent of the appliance rating does not correspond to a standard overcurrent device ampere rating, the next higher standard rating shall be permitted. Where the overcurrent protection rating is not marked and the appliance is rated at 13.3 amperes or less, the overcurrent protection shall not exceed 20 amperes.

E4001.5 Disconnecting means. Each appliance shall be provided with a means to disconnect all ungrounded supply conductors. For fixed electric space-heating equipment, means shall be provided to disconnect the heater and any motor controller(s) and supplemental overcurrent-protective devices. Switches and circuit breakers used as a disconnecting means shall be of the indicating type. Disconnecting means shall be as set forth in Table E4001.5.

E4001.6 Support of ceiling-suspended paddle fans. Ceiling-suspended paddle fans that do not exceed 35 pounds (15.88 kg) in weight, with or without accessories, shall be permitted to be supported by outlet boxes identified for such use and supported in accordance with Sections E3805 and E3806. Ceiling-suspended paddle fans exceeding 35 pounds (15.88 kg) in weight, with or without accessories, shall be supported independently of the outlet box (see Section 3806).

Exception: Listed outlet boxes or outlet box systems that are identified for the purpose shall be permitted to support ceiling-suspended paddle fans, with or without accessories, that weigh 70 pounds (31.76 kg) or less.

E4001.7 Snow-melting and deicing equipment protection. Outdoor receptacles that are not readily accessible and are supplied from a dedicated branch circuit for electric snow-melting or deicing equipment shall be permitted to be installed without ground-fault circuit-interrupter protection for personnel. However, ground-fault protection of equipment shall be provided for fixed outdoor electric deicing and snow-melting equipment.

TABLE E4001.5
DISCONNECTING MEANS

DESCRIPTION	ALLOWED DISCONNECTING MEANS
Permanently connected appliance rated at not over 300 volt-amperes or 0.125 horsepower.	Branch-circuit overcurrent device.
Permanently connected appliances rated in excess of 300 volt-amperes or 0.125 horsepower.	Branch circuit breaker or switch located within sight of appliance or such devices in any location that are capable of being locked in the open position.
Appliances listed for cord-and-plug connection.	A separable connector or attachment plug and receptacle provided with access.
Permanently installed heating equipment with motors rated at not over 0.125 horsepower with supplementary overcurrent protection.	Disconnect, on the supply side of fuses, in sight from the supplementary overcurrent device, and in sight of the heating equipment or, in any location, if capable of being locked in the open position.
Heating equipment containing motors rated over 0.125 horsepower with supplementary overcurrent protection.	Disconnect permitted to serve as required disconnect for both the heating equipment and the controller where, on the supply side of fuses, and in sight from the supplementary overcurrent devices, if the disconnecting means is also in sight from the controller, or is capable of being locked off.
Heating equipment containing no motor rated over 0.125 horsepower without supplementary overcurrent protection.	Branch-circuit switch or circuit breaker where within sight from the heating equipment or capable of being locked off.
Heating equipment containing motors rated over 0.125 horsepower without supplementary overcurrent protection.	Disconnecting means in sight from motor controller or as provided for heating equipment with motor rated over 0.125 horsepower with supplementary overcurrent protection.
Air-conditioning condensing units and heat pump units.	Disconnect within sight from unit.
Appliances and fixed heating equipment with unit switches having a marked OFF position.	Unit switch where an additional individual switch or circuit breaker serves as a redundant disconnecting means.
Thermostatically controlled fixed heating equipment.	Thermostats with a marked OFF position that directly open all ungrounded conductors, which when manually placed in the OFF position are designed so that the circuit cannot be energized automatically and that are located within sight of the equipment controlled.

Chapter 41
SWIMMING POOLS

SECTION E4101
GENERAL

E4101.1 Scope. The provisions of this chapter shall apply to the construction and installation of electric wiring and equipment associated with all swimming pools, wading pools, hot tubs and spas, and hydromassage bathtubs, whether permanently installed or storable, and shall apply to metallic auxiliary equipment, such as pumps, filters and similar equipment. Sections E4102 through E4106 provide general rules for permanent pools, spas and hot tubs. Section E4107 provides specific rules for storable pools. Section E4108 provides specific rules for spas and hot tubs. Section E4109 provides specific rules for hydromassage bathtubs.

E4101.2 Definitions.

CORD- AND PLUG-CONNECTED LIGHTING ASSEMBLY. A lighting assembly consisting of a cord and plug-connected transformer and a lighting fixture intended for installation in the wall of a spa, hot tub, or storable pool.

DRY-NICHE LIGHTING FIXTURE. A lighting fixture intended for installation in the wall of a pool or fountain in a niche that is sealed against the entry of pool water.

FORMING SHELL. A structure designed to support a wet-niche lighting fixture assembly and intended for mounting in a pool or fountain structure.

HYDROMASSAGE BATHTUB. A permanently installed bathtub equipped with a recirculating piping system, pump, and associated equipment. It is designed so it can accept, circulate and discharge water upon each use.

NO-NICHE LIGHTING FIXTURE. A lighting fixture intended for installation above or below the water without a niche.

PACKAGED SPA OR HOT TUB EQUIPMENT ASSEMBLY. A factory-fabricated unit consisting of water-circulating, heating and control equipment mounted on a common base, intended to operate a spa or hot tub. Equipment may include pumps, air blowers, heaters, lights, controls and sanitizer generators.

PERMANENTLY INSTALLED SWIMMING AND WADING POOLS. Those that are constructed in the ground or partially in the ground, and all others capable of holding water with a depth greater than 42 inches (1067 mm), and all pools installed inside of a building, regardless of water depth, whether or not served by electrical circuits of any nature.

POOLCOVER, ELECTRICALLY OPERATED. Motor-driven equipment designed to cover and uncover the water surface of a pool by means of a flexible sheet or rigid frame.

SELF-CONTAINED SPA OR HOT TUB. A factory-fabricated unit consisting of a spa or hot tub vessel with all water-circulating, heating and control equipment integral to the unit. Equipment may include pumps, air blowers, heaters, lights, controls and sanitizer generators.

SPA OR HOT TUB. A hydromassage pool, or tub for recreational or therapeutic use, not located in health care facilities, designed for immersion of users, and usually having a filter, heater, and motor-driven blower. They are installed indoors or outdoors, on the ground or supporting structure, or in the ground or supporting structure. Generally, a spa or hot tub is not designed or intended to have its contents drained or discharged after each use.

STORABLE SWIMMING OR WADING POOL. Those that are constructed on or above the ground and are capable of holding water with a maximum depth of 42 inches (1067 mm), or a pool with nonmetallic, molded polymeric walls or inflatable fabric walls regardless of dimension.

WET-NICHE LIGHTING FIXTURE. A lighting fixture intended for installation in a forming shell mounted in a pool or fountain structure where the fixture will be completely surrounded by water.

SECTION E4102
WIRING METHODS FOR POOLS, SPAS, HOT TUBS AND HYDROMASSAGE BATHTUBS

E4102.1 General. Wiring methods used in conjunction with permanently installed swimming pools, spas, hot tubs or hydromassage bathtubs shall be installed in accordance with Table E4102.1 and Chapter 37 except as otherwise stated in this section. Storable swimming pools shall comply with Section E4107.

E4102.2 Flexible cords. Flexible cords used in conjunction with a pool, spa, hot tub or hydromassage bathtub shall be installed in accordance with the following:

1. For other than underwater lighting fixtures, fixed or stationary equipment, rated at 20 amperes or less shall be permitted to be connected with a flexible cord to facilitate the removal or disconnection for maintenance or repair. For other than storable pools, the flexible cord shall not exceed 3 feet (914 mm) in length. Cords that supply swimming pool equipment, shall have a copper equipment grounding conductor not smaller than No. 12 and shall be provided with a grounding-type attachment plug.
2. Flexible cord that is supplied as part of a listed underwater swimming pool lighting fixture shall be permitted to be installed in any of the permitted wiring methods from the fixture to a deck box or other enclosure. Splices shall not be made within a raceway. The equipment grounding conductor shall be an insulated copper conductor that is not smaller than the supply conductors and not smaller than No.16 AWG.

3. A listed packaged spa or hot tub installed outdoors that is GFCI protected, shall be permitted to be cord and plug connected provided that such cord does not exceed 15 feet (4572 mm) in length.
4. A listed packaged spa or hot tub rated at 20 amperes or less and installed indoors shall be permitted to be cord and plug connected to facilitate maintenance and repair.
5. For other than underwater and storable pool lighting fixtures, the requirements of Item 1 shall apply to any cord equipped lighting fixture that is located within 16 feet (4877 mm) radially from any point on the water surface.

SECTION E4103
EQUIPMENT LOCATION AND CLEARANCES

E4103.1 Receptacle outlets. Receptacles outlets shall be installed and located in accordance with Sections E4103.1.1 through E4103.1.5. Distances shall be measured as the shortest path that an appliance supply cord connected to the receptacle would follow without penetrating a floor, wall, ceiling, doorway with hinged or sliding door, window opening, or other effective permanent barrier.

E4103.1.1 Location. Receptacles that provide power for water-pump motors or other loads directly related to the circulation and sanitation system shall be permitted to be located between 5 feet and 10 feet (1524 mm and 3048 mm) from the inside walls of pools and outdoor spas and hot tubs, and, where so located, shall be single and of the locking and grounding type and shall be protected by ground-fault circuit interrupters.

Other receptacles on the property shall be located not less than 10 feet (3048 m) from the inside walls of pools and outdoor spas and hot tubs.

E4103.1.2 Where required. At least one 125-volt 15- or 20-ampere receptacle supplied by a general-purpose branch circuit shall be located a minimum of 10 feet (3048 mm) from and not more than 20 feet (6096 mm) from the inside wall of pools and outdoor spas and hot tubs. This receptacle shall be located not more than 6 feet, 6 inches (1981 mm) above the floor, platform or grade level serving the pool, spa or hot tub.

E4103.1.3 GFCI protection. All 125-volt receptacles located within 20 feet (6096 mm) of the inside walls of pools and outdoor spas and hot tubs shall be protected by a ground-fault circuit-interrupter.

E4103.1.4 Indoor locations. Receptacles shall be located not less than 5 feet (1524 mm) from the inside walls of indoor spas and hot tubs. A minimum of one 125-volt receptacle shall be located between 5 feet (1524 mm) and 10 feet (3048 mm) from the inside walls of indoor spas or hot tubs.

E4103.1.5 Indoor GFCI protection. One hundred twenty-five-volt receptacles located within 10 feet (3048 mm) of the inside walls of spas and hot tubs installed indoors shall be protected by ground-fault circuit-interrupters. One hundred twenty-five-volt receptacles located within 5 feet (1524 mm) of the inside walls of hydromassage bathtubs shall be protected by a ground-fault circuit-interrupter.

E4103.2 Switching devices. Switching devices shall be located not less than 5 feet (1524 mm) horizontally from the inside walls of pools, spas and hot tubs except where separated from the pool, spa or hot tub by a solid fence, wall, or other permanent barrier. Switching devices located in a room or area containing a hydromassage bathtub shall be located in accordance with the general requirements of this code.

E4103.3 Disconnecting means. An accessible disconnecting means shall be provided and located within sight from all pools, spas, and hot tub equipment, and shall be located not less than 5 feet (1524 mm) from the inside walls of the pool, spa or hot tub.

E4103.4 Lighting fixtures and ceiling fans. Lighting outlets, fixtures, and ceiling-suspended paddle fans shall be installed and located in accordance with Sections E4103.4.1 through E4103.4.5.

E4103.4.1 Outdoor location. In outdoor pool, outdoor spas and outdoor hot tubs areas, lighting fixtures, lighting outlets, and ceiling-suspended paddle fans shall not be installed over the pool or over the area extending 5 feet (1524 mm) horizontally from the inside walls of a pool except where no part of the lighting fixture or ceiling-suspended paddle fan is less than 12 feet (3658 mm) above the maximum water level.

E4103.4.2 Indoor locations. In indoor pool areas, the limitations of Section E4103.4.1 shall apply except where the fixtures, lighting outlets and ceiling-suspended paddle fans comply with all of the following conditions:

1. The fixtures are of a totally enclosed type, and
2. A ground-fault circuit interrupter is installed in the branch circuit supplying the fixture(s) or ceiling-suspended (paddle) fans, and
3. The distance from the bottom of the fixture or ceiling-suspended (paddle) fan to the maximum water level is not less than 7 feet, 6 inches (2286 mm).

E4103.4.3 Existing lighting outlets and fixtures. Existing lighting outlets and fixtures that are located within 5 feet (1524 mm) horizontally from the inside walls of pools and outdoor spas and hot tubs shall be permitted to be located not less than 5 feet (1524 mm) vertically above the maximum water level, provided that such fixtures and outlets are rigidly attached to the existing structure and ground-fault circuit-interrupter protection is provided for the branch circuit that supplies such lighting fixtures and outlets.

TABLE E4102.1
ALLOWABLE APPLICATIONS FOR WIRING METHODS[a,b,c,d,e,f,g]

WIRING LOCATION OR PURPOSE	AC, FMC, NM, SR, SE[b]	EMT[c]	ENT[b]	IMC, RMC, RNC	LFMC	LFNMC	UF	MC	Flex Cord[g]
Panelboard(s) that supply pool equipment: from service equipment to panelboard	A[e]	A	—	A	—	A	A[e]	A[e]	—
Wet niche and no niche fixtures: from branch circuit OCPD to deck or junction box	—	A	A	A	—	A	—	—	—
Wet niche and no niche fixtures: from deck or junction box to forming shell	—	—	—	A[d]	—	A	—	—	A
Dry niche: from branch circuit OCPD to fixture	—	A	A	A	—	A	—	—	—
Pool-associated motors: from branch circuit OCPD to motor	A	A	A	A	A[f]	A[f]	A[b]	A	A
Packaged or self-contained outdoor spas and hot tubs with underwater lighting fixture: from branch circuit OCPD to spa or hot tub	—	A	A	A	—	A[f]	—	—	A
Packaged or self-contained outdoor spas and hot tubs without underwater lighting fixture: from branch circuit OCPD to spa or hot tub	A	A	A	A	A[f]	A[f]	A[b]	A	A
Indoor spa and hot tubs, hydromassage bathtubs, and other pool, spa or hot tub associated equipment: from branch circuit OCPD to equipment	A	A	A	A	A	A	A	A	A

For SI: 1 foot = 304.8 mm.

a. For all wiring methods, see Section E4105.2 for equipment grounding conductor requirements.

b. Limited to use within buildings.

c. Limited to use on or within buildings.

d. Metal conduit shall be constructed of brass or other approved corrosion resistant metal

e. Permitted only for existing feeder panelboards where the equipment grounding conductor is insulated or covered.

f. Limited to use in lengths not to exceed 3 feet at pool, spa or hot tub equipment where flexibility is necessary.

g. Flexible cord shall be installed in accordance with Section E4102.2.

E4103.4.4 Indoor spas and hot tubs.

1. Lighting fixtures, lighting outlets, and ceiling-suspended paddle fans located over the spa or hot tub or within 5 feet (1524 mm) from the inside walls of the spa or hot tub shall be a minimum of 7 feet, 6 inches (2286 mm) above the maximum water level and shall be protected by a ground-fault circuit interrupter.

   Lighting fixtures, lighting outlets, and ceiling-suspended paddle fans that are located 12 feet (3658 mm) or more above the maximum water level shall not require ground-fault circuit interrupter protection.

2. Lighting fixtures protected by a ground-fault circuit interrupter and complying with Item 2.1. or 2.2. shall be permitted to be installed less than 7 feet, 6 inches (2286 mm) over a spa or hot tub.

   2.1. Recessed fixtures shall have a glass or plastic lens and nonmetallic or electrically isolated metal trim, and shall be suitable for use in damp locations.

   2.2. Surface-mounted fixtures shall have a glass or plastic globe and a nonmetallic body or a metallic body isolated from contact. Such fixtures shall be suitable for use in damp locations.

E4103.4.5 GFCI protection. Lighting fixtures and outlets that are installed in the area extending between 5 feet (1524 mm) and 10 feet (3048 mm) from the inside walls of pools and outdoor spas and hot tubs shall be protected by ground-fault circuit-interrupters except where such fixtures and outlets are installed not less than 5 feet (1524 mm) above the maximum water level and are rigidly attached to the structure.

E4103.5 Overhead conductor clearances. Except where installed with the clearances specified in Table E4103.5, the following parts of pools and outdoor spas and hot tubs shall not be placed under existing service-drop conductors or any other open overhead wiring; nor shall such wiring be installed above the following:

1. Pools and the areas extending 10 feet (3048 mm) horizontally from the inside of the walls of the pool;
2. Diving structures; or
3. Observation stands, towers, and platforms.

Utility-owned, -operated and -maintained communications conductors, community antenna system coaxial cables and the supporting messengers shall be permitted at a height of not less than 10 feet (3048 mm) above swimming and wading pools, diving structures, and observation stands, towers, and platforms.

E4103.6 Underground wiring. Underground wiring shall not be installed under or within the area extending 5 feet (1524 mm) horizontally from the inside walls of pools and outdoor hot tubs and spas except where the wiring is installed to supply pool, spa or hot tub equipment or where space limitations prevent wiring from being routed 5 feet (1524 mm) or more horizontally from the inside walls. Where installed within 5 feet (1524 mm) of the inside walls, the wiring method shall be rigid metal conduit, intermediate metal conduit or a nonmetallic raceway system. Metal conduit shall be corrosion resistant and suitable for the location. The minimum raceway burial depth shall be in accordance with Table E4103.6.

SECTION E4104 BONDING

E4104.1 Bonded parts. The following parts shall be bonded together:

1. All metallic parts of pool, spa and hot tub structures, including the reinforcing metal of pool, spa and hot tub shells, coping stones, and decks. The usual steel tie wires shall be considered suitable for bonding the reinforcing steel together, and welding or special clamping shall not be required. Such tie wires shall be made tight. Where reinforcing steel is effectively insulated by a listed encapsulating nonconductive compound, at the time of manufacture and installation, it shall not be required to be bonded.
2. All forming shells and mounting brackets of no-niche fixtures except where a listed low-voltage lighting system is used that does not require bonding.
3. All metal fittings within or attached to pool, spa and hot tub structures. Isolated parts that are not over 4 inches (102 mm) in any dimension and do not penetrate into the pool structure more than 1 inch (25.4 mm) shall not require bonding. The metal bands or hoops used to secure wooden staves for a hot tub or spa shall not be required to be bonded.
4. Metal parts of electrical equipment associated with pool, spa and hot tub water circulating systems, including pump motors and metal parts of equipment associated with pool covers, including electric motors. Metal parts of listed equipment incorporating an approved system of double insulation and providing a means for grounding internal nonaccessible, noncurrent-carrying metal parts shall not be bonded.
5. Metal-sheathed cables and raceways, metal piping and all fixed metal parts that are within 5 feet (1524 mm) horizontally of the inside walls of the pool, spa or hot tub and that are within 12 feet (3658 mm) above the maximum water level of the pool or any observation stands, towers or platforms, or from any diving structures, and that are not separated from the pool by a permanent barrier.

For pool water heaters rated at more than 50 amperes and having specific instructions regarding bonding and grounding, only those parts designated to bonded shall be bonded and only those parts designated to be grounded shall be grounded.

E4104.2 Parts not required to be bonded. Small conductive surfaces not likely to become energized, such as towel bars, mirror frames, and air and water jets and drain fittings that are not connected to metallic piping, and similar equipment installed on or within indoor spas and hot tubs shall not be required to be bonded.

E4104.3 Methods of bonding. It shall not be the intent to require that the No. 8 or larger solid copper bonding conductor be extended or attached to any remote panelboard, service equipment, or any electrode, but only that it shall be employed to eliminate voltage gradients in the pool area as prescribed. Bonding shall be accomplished by one or more of the following methods:

1. Common Bonding Grid. The parts specified in Section E4104.1 above shall be connected to a common bonding grid with a solid copper conductor, insulated, covered, or bare, not smaller than No. 8. Connection shall be made by exothermic welding or by pressure connectors or clamps that are labeled as being suitable for the purpose and that are made of stainless steel, brass, copper or copper alloy.

   The common bonding grid shall be permitted to be any of the following:

   1.1. The structural reinforcing steel of a concrete pool where the reinforcing rods are bonded together by the usual steel tie wires made up tight or the equivalent; or

   1.2. The wall of a bolted or welded metal pool; or

   1.3. A solid copper conductor, insulated, covered, or bare, not smaller than No. 8.
2. For hot tubs and spas, metal to metal mounting on a common frame or base
3. The interconnection of threaded metal piping and fittings.

TABLE E4103.5
OVERHEAD CONDUCTOR CLEARANCES

	INSULATED SUPPLY OR SERVICE DROP CABLES, 0-750 VOLTS TO GROUND, SUPPORTED ON AND CABLED TOGETHER WITH AN EFFECTIVELY GROUNDED BARE MESSENGER OR EFFECTIVELY GROUNDED NEUTRAL CONDUCTOR (feet)	ALL OTHER SUPPLY OR SERVICE DROP CONDUCTORS (feet)	
		Voltage to ground	
		0-15 kV	Greater than 15 to 50 kV
A. Clearance in any direction to the water level, edge of water surface, base of diving platform, or permanently-anchored raft	22	25	27
B. Clearance in any direction to the diving platform	14	17	18
C. Horizontal limit of clearance measured from inside wall of the pool	This limit shall extend to the outer edge of the structures listed in Rows (A) and (B) above but not less than 10 feet.		

For SI: 1 foot = 304.8 mm.

TABLE E4103.6
MINIMUM BURIAL DEPTHS

WIRING METHOD	MINIMUM BURIAL DEPTH (inches)
Rigid metal conduit	6
Intermediate metal conduit	6
Nonmetallic raceways listed for direct burial without concrete encasement	18
Other approved raceways[a]	18

For SI: 1 inch = 25.4 mm.
a. Raceways approved for burial only where concrete-encased shall require a concrete envelope not less than 2 inches in thickness.

SECTION E4105
GROUNDING

E4105.1 Equipment to be grounded. The following equipment shall be grounded:

1. Wet-niche, dry-niche and no-niche underwater lighting fixtures other than those low-voltage systems listed for the application without a grounding conductor.
2. All electrical equipment located within 5 feet (1524 mm) of the inside wall of the pool, spa or hut tub.
3. All electrical equipment associated with the recirculating system of the pool, spa or hot tub.
4. Junction boxes.
5. Transformer enclosures.
6. Ground-fault circuit-interrupters.
7. Panelboards that are not part of the service equipment and that supply any electrical equipment associated with the pool, spa or hot tub.

E4105.2 Light fixtures and related equipment. Wet-niche, dry-niche, or no-niche lighting fixtures shall be connected to an insulated copper equipment grounding conductor sized in accordance with Table E3808.12 but not smaller than No. 12. The equipment grounding conductor between the wiring chamber of the secondary winding of a transformer and a junction box shall be sized in accordance with the overcurrent device in such circuit. The junction box, transformer enclosure, or other enclosure in the supply circuit to a wet-niche or no-niche lighting fixture and the field-wiring chamber of a dry-niche lighting fixture shall be grounded to the equipment grounding terminal of the panelboard. The equipment grounding terminal shall be directly connected to the panelboard enclosure. The equipment grounding conductor shall be installed without joint or splice.

Exceptions:

1. Where more than one underwater lighting fixture is supplied by the same branch circuit, the equipment grounding conductor, installed between the junction boxes, transformer enclosures, or other enclosures in the supply circuit to wet-niche fixtures, or between the field-wiring compartments of dry-niche fixtures, shall be permitted to be terminated on grounding terminals.
2. Where an underwater lighting fixture is supplied from a transformer, ground-fault circuit-interrupter, clock-operated switch, or a manual snap switch that is located between the panelboard and a junction box connected to the conduit that extends directly to the underwater lighting fixture, the equipment grounding conductor shall be permitted to terminate on grounding terminals on the transformer, ground-fault circuit-interrupter, clock-operated switch enclosure, or an outlet box used to enclose a snap switch.

E4105.3 Nonmetallic conduit. Where a nonmetallic conduit is installed between a wet-niche fixture and a junction box, transformer enclosure, or other enclosure, a No. 8 insulated copper conductor shall be installed in this conduit with provisions for terminating in the forming shell, junction box or transformer enclosure, or ground-fault circuit-interrupter enclosure. The termination of the No. 8 conductor in the forming shell shall be covered with, or encapsulated in, a listed potting compound to protect such connection from the possible deteriorating effect of pool water.

E4105.4 Flexible cords. Wet-niche or no-niche lighting fixtures that are supplied by a flexible cord or cable shall have all exposed noncurrent-carrying metal parts grounded by an insulated copper equipment grounding conductor that is an integral part of the cord or cable. This grounding conductor shall be connected to a grounding terminal in the supply junction box, transformer enclosure, or other enclosure. The grounding conductor shall not be smaller than the supply conductors and not smaller than No. 16.

E4105.5 Motors. Pool-associated motors shall be connected to a copper equipment grounding conductor sized in accordance with Table E3808.12, but not smaller than No. 12.

E4105.6 Panelboards. A panelboard that is not part of the service equipment, or source of a separately derived system shall have an equipment grounding conductor installed between its grounding terminal and the grounding terminal of the applicable service equipment or source of a separately derived system. This conductor shall be sized in accordance with Table E3808.12, but not smaller than No. 12.

E4105.7 Cord-connected equipment. Where fixed or stationary equipment is connected with a flexible cord to facilitate removal or disconnection for maintenance, repair, or storage, as provided in Section E4102.2, the equipment grounding conductors shall be connected to a fixed metal part of the assembly. The removable part shall be mounted on or bonded to the fixed metal part.

E4105.8 Other equipment. Other electrical equipment shall be grounded in accordance with Section E3808.

SECTION E4106
EQUIPMENT INSTALLATION

E4106.1 Transformers. Transformers used for the supply of underwater fixtures, together with the transformer enclosure, shall be identified for the purpose. Such transformers shall be of an isolated winding type having a grounded metal barrier between the primary and secondary windings.

E4106.2 Ground-fault circuit-interrupters. Ground-fault circuit-interrupters shall be self-contained units, circuit-breaker types, receptacle types or other approved types.

E4106.3 Wiring on load side of ground-fault circuit-interrupters and transformers. For other than grounding conductors, conductors installed on the load side of a ground-fault circuit-interrupter or transformer used to comply with the provisions of Section E4106.4, shall not occupy raceways, boxes, or enclosures containing other conductors except where the other conductors are protected by ground-fault circuit interrupters or are grounding conductors. Supply conductors to a feed-through type ground-fault circuit interrupter shall be permitted in the same enclosure. Ground-fault circuit interrupters shall be permitted in a panelboard that contains circuits protected by other than ground-fault circuit interrupters.

E4106.4 Underwater lighting fixtures. The design of an underwater lighting fixture supplied from a branch circuit either directly or by way of a transformer meeting the requirements of Section E4106.1, shall be such that, where the fixture is properly installed without a ground-fault circuit-interrupter, there is no shock hazard with any likely combination of fault conditions during normal use (not relamping). In addition, a ground-fault circuit-interrupter shall be installed in the branch circuit supplying fixtures operating at more than 15 volts, so that there is no shock hazard during relamping. The installation of the ground-fault circuit-interrupter shall be such that there is no shock hazard with any likely fault-condition combination that involves a person in a conductive path from any ungrounded part of the branch circuit or the fixture to ground. Compliance with this requirement shall be obtained by the use of a listed underwater lighting fixture and by installation of a listed ground-fault circuit-interrupter in the branch circuit. Fixtures that depend on submersion for safe operation shall be inherently protected against the hazards of overheating when not submerged.

E4106.4.1 Maximum voltage. Lighting fixtures shall not be installed for operation on supply circuits over 150 volts between conductors.

E4106.4.2 Fixture location. Lighting fixtures mounted in walls shall be installed with the top of the fixture lens not less than 18 inches (457 mm) below the normal water level of the pool, except where the lighting fixture is listed and identified for use at a depth of not less than 4 inches (102 mm) below the normal water level of the pool. A lighting fixture facing upward shall have the lens adequately guarded to prevent contact by any person.

E4106.5 Wet-niche fixtures. Forming shells shall be installed for the mounting of all wet-niche underwater fixtures and shall be equipped with provisions for conduit entries. Conduit shall extend from the forming shell to a suitable junction box or other enclosure located as provided in Section E4106.8. Metal parts of the fixture and forming shell in contact with the pool water shall be of brass or other approved corrosion-resistant metal.

The end of flexible-cord jackets and flexible-cord conductor terminations within a fixture shall be covered with, or encapsulated in, a suitable potting compound to prevent the entry of water into the fixture through the cord or its conductors. In addition, the grounding connection within a fixture shall be similarly treated to protect such connection from the deteriorating effect of pool water in the event of water entry into the fixture.

Fixtures shall be bonded to and secured to the forming shell by a positive locking device that ensures a low-resistance contact and requires a tool to remove the fixture from the forming shell.

E4106.6 Dry-niche fixtures. Dry-niche lighting fixtures shall be provided with provisions for drainage of water and means for accommodating one equipment grounding conductor for each conduit entry. Junction boxes shall not be required but, if used, shall not be required to be elevated or located as specified in Section E4106.8 if the fixture is specifically identified for the purpose.

E4106.7 No-niche fixtures. No-niche fixtures shall be listed for the purpose and shall be installed in accordance with the requirements of Section E4106.5. Where connection to a forming shell is specified, the connection shall be to the mounting bracket.

E4106.8 Junction boxes and enclosures for transformers or ground-fault circuit interrupters. Junction boxes for underwater lighting fixtures and enclosures for transformers and ground-fault circuit-interrupters that supply underwater lighting fixtures shall comply with the following:

E4106.8.1 Junction boxes. A junction box connected to a conduit that extends directly to a forming shell or mounting bracket of a no-niche fixture shall be:

1. Listed and labled for the purpose; and
2. Equipped with threaded entries or hubs or a nonmetallic hub listed for the purpose; and
3. Constructed of copper, brass, suitable plastic, or other approved corrosion-resistant material; and
4. Provided with electrical continuity between every connected metal conduit and the grounding terminals by means of copper, brass, or other approved corrosion-resistant metal that is integral with the box; and
5. Located not less than 4 inches (102 mm), measured from the inside of the bottom of the box, above the ground level, or pool deck, or not less than 8 inches (203 mm) above the maximum pool water level, whichever provides the greatest elevation, and shall be located not less than 4 feet (1219 mm) from the inside wall of the pool, unless separated from the pool by a solid fence, wall or other permanent barrier. Where used on a lighting system operating at 15 volts or less, a flush deck box shall be permitted provided that an approved potting compound is used to fill the box to prevent the entrance of moisture; and the flush deck box is located not less than 4 feet (1219 mm) from the inside wall of the pool.

E4106.8.2 Other enclosures. An enclosure for a transformer, ground-fault circuit-interrupter or a similar device connected to a conduit that extends directly to a forming shell or mounting bracket of a no-niche fixture shall be:

1. Listed and labeled for the purpose; and
2. Equipped with threaded entries or hubs or a nonmetallic hub listed for the purpose; and
3. Provided with an approved seal, such as duct seal at the conduit connection, that prevents circulation of air between the conduit and the enclosures; and
4. Provided with electrical continuity between every connected metal conduit and the grounding terminals by means of copper, brass or other approved corrosion-resistant metal that is integral with the enclosures; and
5. Located not less than 4 inches (102 mm), measured from the inside bottom of the enclosure, above the ground level or pool deck, or not less than 8 inches (203 mm) above the maximum pool water level, whichever provides the greater elevation, and shall be located not less than 4 feet (1219 mm) from the inside wall of the pool, except where separated from the pool by a solid fence, wall or other permanent barrier.

E4106.8.3 Protection of junction boxes and enclosures. Junction boxes and enclosures mounted above the grade of the finished walkway around the pool shall not be located in the walkway unless afforded additional protection, such as by location under diving boards or adjacent to fixed structures.

E4106.8.4 Grounding terminals. Junction boxes, transformer enclosures, and ground-fault circuit-interrupter enclosures connected to a conduit that extends directly to a forming shell or mounting bracket of a no-niche fixture shall be provided with grounding terminals in a quantity not less than the number of conduit entries plus one.

E4106.8.5 Strain relief. The termination of a flexible cord of an underwater lighting fixture within a junction box, transformer enclosure, ground-fault circuit-interrupter, or other enclosure shall be provided with a strain relief.

E4106.9 Underwater audio equipment. Underwater audio equipment shall be identified for the purpose.

E4106.9.1 Speakers. Each speaker shall be mounted in an approved metal forming shell, the front of which is enclosed by a captive metal screen, or equivalent, that is bonded to and secured to the forming shell by a positive locking device that ensures a low-resistance contact and requires a tool to open for installation or servicing of the speaker. The forming shell shall be installed in a recess in the wall or floor of the pool.

E4106.9.2 Wiring methods. Rigid metal conduit or intermediate metal conduit of brass or other identified corrosion-resistant metal or rigid nonmetallic conduit shall extend from the forming shell to a suitable junction box or other enclosure as provided in Section E4106.8. Where rigid nonmetallic conduit is used, a No. 8 insulated copper conductor shall be installed in this conduit with provisions for terminating in the forming shell and the junction box. The termination of the No. 8 conductor in the forming shell shall be covered with, or encapsulated in, a suitable potting compound to protect such connection from the possible deteriorating effect of pool water.

E4106.9.3 Forming shell and metal screen. The forming shell and metal screen shall be of brass or other approved corrosion-resistant metal.

E4106.10 Electrically operated pool covers. The electric motors, controllers, and wiring for pool covers shall be located not less than 5 feet (1524 mm) from the inside wall of the pool except where separated from the pool by a wall, cover, or other permanent barrier. Electric motors installed below grade level shall be of the totally enclosed type. The electric motor and controller shall be connected to a circuit protected by a ground-fault circuit-interrupter.

E4106.11 Electric pool water heaters. All electric pool water heaters shall have the heating elements subdivided into loads not exceeding 48 amperes and protected at not more than 60 amperes. The ampacity of the branch-circuit conductors and the rating or setting of overcurrent protective devices shall be not less than 125 percent of the total nameplate load rating.

E4106.12 Pool area heating. The provisions of Sections E4106.12.1 through E4106.12.3 shall apply to all pool deck areas, including a covered pool, where electrically operated comfort heating units are installed within 20 feet (6096 mm) of the inside wall of the pool.

E4106.12.1 Unit heaters. Unit heaters shall be rigidly mounted to the structure and shall be of the totally enclosed or guarded types. Unit heaters shall not be mounted over the pool or within the area extending 5 feet (1524 mm) horizontally from the inside walls of a pool.

E4106.12.2 Permanently wired radiant heaters. Electric radiant heaters shall be suitably guarded and securely fastened to their mounting devices. Heaters shall not be installed over a pool or within the area extending 5 feet (1524 mm) horizontally from the inside walls of the pool and shall be mounted not less than 12 feet (3658 mm) vertically above the pool deck.

E4106.12.3 Radiant heating cables prohibited. Radiant heating cables embedded in or below the deck shall be prohibited.

E4106.13 Double insulated pool pumps. A permanently installed pool shall be permitted to be provided with listed cord- and plug-connected pool pumps incorporating an approved system of double insulation that provides a means for grounding only the internal and nonaccessible, noncurrent-carrying metal parts of the pump.

SECTION E4107
STORABLE SWIMMING POOLS

E4107.1 Pumps. A cord-connected pool filter pump for use with storable pools shall incorporate an approved system of double insulation or its equivalent and shall be provided with means for grounding only the internal and nonaccessible non-current-carrying metal parts of the appliance.

The means for grounding shall be an equipment grounding conductor run with the power-supply conductors in a flexible cord that is properly terminated in a grounding-type attachment plug having a fixed grounding contact.

E4107.2 Ground-fault circuit-interrupters required. Electrical equipment, including power-supply cords, used with storable pools shall be protected by ground-fault circuit-interrupters.

E4107.3 Lighting fixtures. Lighting fixtures for storable pools shall not have exposed metal parts and shall be listed for the purpose as an assembly. In addition, lighting fixtures for storable pools shall comply with the requirements of Section E4107.3.1 or E4107.3.2.

E4107.3.1 Fifteen (15) volts or less. A lighting fixture installed in or on the wall of a storable pool shall be part of a cord- and plug-connected lighting assembly. The assembly shall:

1. Have a fixture lamp that operates at 15 volts or less; and
2. Have an impact-resistant polymeric lens, fixture body, and transformer enclosure; and
3. Have a transformer meeting the requirements of Section E4106.1 with a primary rating not over 150 volts.

E4107.3.2 Not over 150 volts. A lighting assembly without a transformer, and with the fixture lamp(s) operating at not over 150 volts, shall be permitted to be cord- and plug-connected where the assembly complies with all of the following:

1. It has an impact-resistant polymeric lens and fixture body.
2. A ground-fault circuit interrupter with open neutral protection is provided as an integral part of the assembly.
3. The fixture lamp is permanently connected to the ground-fault circuit interrupter with open-neutral protection.
4. It complies with the requirements of Section E4106.4.

SECTION E4108
SPAS AND HOT TUBS

E4108.1 Ground-fault circuit-interrupters. The outlet(s) that supplies a self-contained spa or hot tub, or a packaged spa or hot tub equipment assembly, or a field-assembled spa or hot tub with a heater load of 50 amperes or less, shall be protected by a ground-fault circuit-interrupter.

A listed self-contained unit or listed packaged equipment assembly marked to indicate that integral ground-fault circuit-interrupter protection is provided for all electrical parts within the unit or assembly, including pumps, air blowers, heaters, lights, controls, sanitizer generators and wiring, shall not require that the outlet supply be protected by a ground-fault circuit interrupter.

A combination pool/hot tub or spa assembly commonly bonded need not be protected by a ground-fault circuit interrupter.

E4108.2 Electric water heaters. Electric spa and hot tub water heaters shall be listed and shall have the heating elements subdivided into loads not exceeding 48 amperes and protected at not more than 60 amperes. The ampacity of the branch-circuit conductors, and the rating or setting of overcurrent protective devices, shall be not less than 125 percent of the total nameplate load rating.

E4108.3 Underwater audio equipment. Underwater audio equipment used with spas and hot tubs shall comply with the provisions of Section E4106.9.

SECTION E4109
HYDROMASSAGE BATHTUBS

E4109.1 Ground-fault circuit-interrupters. Hydromassage bathtubs and their associated electrical components shall be protected in accordance with Section E4108.

E4109.2 Other electric equipment. Lighting fixtures, switches, receptacles, and other electrical equipment located in the same room, and not directly associated with a hydromassage bathtub, shall be installed in accordance with the requirements of this code relative to the installation of electrical equipment in bathrooms.

E4109.3 Accessibility. Hydromassage bathtub electrical equipment shall be accessible without damaging the building structure or building finish.

E4109.4 Bonding. All metal piping systems, metal parts of electrical equipment, and pump motors associated with the hydromassage tub shall be bonded together using a copper bonding jumper, insulated, covered, or bare, not smaller than No. 8 solid.

Metal parts of listed equipment incorporating an approved system of double insulation and providing a means for grounding internal nonaccessible, noncurrent-carrying metal parts shall not be bonded.

Chapter 42

CLASS 2 REMOTE-CONTROL, SIGNALING AND POWER-LIMITED CIRCUITS

SECTION E4201 GENERAL

E4201.1 Scope. This chapter contains requirements for power supplies and wiring methods associated with Class 2 remote-control, signaling, and power-limited circuits that are not an integral part of a device or appliance. Other classes of remote-control, signaling and power-limited conductors shall comply with Article 725 of the *National Electrical Code®*.

E4201.2 Definitions.

CLASS 2 CIRCUIT. That portion of the wiring system between the load side of a Class 2 power source and the connected equipment. Due to its power limitations, a Class 2 circuit considers safety from a fire initiation standpoint and provides acceptable protection from electric shock.

REMOTE-CONTROL CIRCUIT. Any electrical circuit that controls any other circuit through a relay or an equivalent device.

SIGNALING CIRCUIT. Any electrical circuit that energizes signaling equipment.

SECTION E4202 POWER SOURCES

E4202.1 Power sources for Class 2 circuits. The power source for a Class 2 circuit shall be one of the following:

1. A listed Class 2 transformer.
2. A listed Class 2 power supply.
3. Other listed equipment marked to identify the Class 2 power source.
4. Listed computer equipment limited power circuits.
5. A dry cell battery provided that the voltage is 30 volts or less and the capacity is equal to or less than that available from series connected No. 6 carbon zinc cells.

E4202.2 Interconnection of power sources. A Class 2 power source shall not have its output connections paralleled or otherwise interconnected with another Class 2 power source except where listed for such interconnection.

SECTION E4203 WIRING METHODS

E4203.1 Wiring methods on supply side of Class 2 power source. Conductors and equipment on the supply side of the power source shall be installed in accordance with the appropriate requirements of Chapters 33 through 40. Transformers or other devices supplied from electric light or power circuits shall be protected by an overcurrent device rated at not over 20 amperes. The input leads of a transformer or other power source supplying Class 2 circuits shall be permitted to be smaller than No. 14, if not over 12 inches (305 mm) long and if the conductor insulation is rated at not less than 600 volts. In no case shall such leads be smaller than No. 18.

E4203.2 Wiring methods and materials on load side of the Class 2 power source. Class 2 cables installed as wiring within buildings shall be listed as being resistant to the spread of fire and listed as meeting the criteria specified in Sections E4203.2.1 through E4203.2.3. Cables shall be marked in accordance with Section E4203.2.4. Cable substitutions as described in Table E4203.2 and wiring methods covered in Chapter 37 shall also be permitted.

TABLE E4203.2
CABLE USES AND PERMITTED SUBSTITUTIONS

CABLE TYPE	USE	PERMITTED SUBSTITUTIONS[a]
CL2P	Class 2 Plenum Cable	MPP, CMP, FPLP, CL3P
CL2	Class 2 Cable	MPP, CMP, FPLP, CL3P, CL2P, MPR, CMR, FPLR, CL3R, CL2R MPG, MP, CMG, CM, FPL, CL3
CL2X	Class 2 Cable, Limited Use	MPP, CMP, FPLP, CL3P CL2P, MPR, CMR, FPLR, CL3R, CL2R, MPG, MP, CMG, CM, FPL, CL3, CL2, CMX, CL3X

a. For identification of cables other than Class 2 cables, see the *National Electrical Code®*.

E4203.2.1 Type CL2P cables. Cables installed in ducts, plenums and other spaces used to convey environmental air shall be Type CL2P cables listed as being suitable for the use and listed as having adequate fire-resistant and low smoke-producing characteristics.

E4203.2.2 Type CL2 cables. Cables for general-purpose use, shall be listed as being resistant to the spread of fire and listed for the use.

E4203.2.3 Type CL2X cables. Type CL2X limited-use cable shall be listed as being suitable for use in dwellings and for the use and in raceways and shall also be listed as being flame retardant. Cables with a diameter of 0.25 inches (6.4 mm) or less shall be permitted to be installed as open cable without a raceway.

E4203.2.4 Marking. Cables shall be marked in accordance with Table E4203.2. Voltage ratings shall not be marked on cables.

SECTION E4204
INSTALLATION REQUIREMENTS

E4204.1 Separation from other conductors. In cables, compartments, enclosures, outlet boxes, device boxes, and raceways, conductors of Class 2 circuits shall not be placed in any cable, compartment, enclosure, outlet box, device box, raceway, or similar fitting with conductors of electric light, power, Class 1 and nonpower-limited fire alarm circuits.

Exceptions:

1. Where the conductors of the electric light, power, Class 1 and nonpower-limited fire alarm circuits are separated by a barrier from the Class 2 circuits. In enclosures, Class 2 circuits shall be permitted to be installed in a raceway within the enclosure to separate them from Class 1, electric light, power and nonpower-limited fire alarm circuits
2. Class 2 conductors in compartments, enclosures, device boxes, outlet boxes and similar fittings where electric light, power, Class 1 or nonpower-limited fire alarm circuit conductors are introduced solely to connect to the equipment connected to the Class 2 circuits. The electric light, power, Class 1 and nonpower-limited fire alarm circuit conductors shall be routed to maintain a minimum of 0.25 inch (6.4 mm) separation from the conductors and cables of the Class 2 circuits; or the electric light power, Class 1 and nonpower-limited fire alarm circuit conductors operate at 150 volts or less to ground and the Class 2 circuits are installed using Types CL3, CL3R, or CL3P or permitted substitute cables, and provided that these Class 3 cable conductors extending beyond their jacket are separated by a minimum of 0.25 inch (6.4 mm) or by a nonconductive sleeve or nonconductive barrier from all other conductors.

E4204.2 Other applications. Conductors of Class 2 circuits shall be separated by not less than 2 inches (51 mm) from conductors of any electric light, power, Class 1 or nonpower-limited fire alarm circuits except where one of the following conditions is met:

1. All of the electric light, power, Class 1 and nonpower-limited fire alarm circuit conductors are in raceways or in metal-sheathed, metal-clad, nonmetallic-sheathed or Type UF cables.
2. All of the Class 2 circuit conductors are in raceways or in metal-sheathed, metal-clad, nonmetallic-sheathed or Type UF cables.

E4204.3 Class 2 circuits with communications circuits. Where Class 2 circuit conductors are in the same cable as communications circuits, the Class 2 circuits shall be classified as communications circuits and shall meet the requirements of Article 800 of the *National Electrical Code®*. The cables shall be listed as communications cables or multipurpose cables.

Exception: Cables constructed of individually listed Class 2 and communications cables under a common jacket shall not be required to be classified as communications cables. The fire-resistance rating of the composite cable shall be determined by the performance of the composite cable.

E4204.4 Class 2 cables with other circuit cables. Jacketed cables of Class 2 circuits shall be permitted in the same enclosure or raceway with jacketed cables of any of the following:

1. Power-limited fire alarm systems in compliance with Article 760 of the *National Electrical Code®*.
2. Nonconductive and conductive optical fiber cables in compliance with Article 770 of the *National Electrical Code®*.
3. Communications circuits in compliance with Article 800 of the *National Electrical Code®*.
4. Community antenna television and radio distribution systems in compliance with Article 820 of the *National Electrical Code®*.

E4204.5 Installation of conductors and cables. Cables shall be supported by the building structure in such a manner that the cable will not be damaged by normal building use. The installation of wires and cables shall not prevent access to equipment nor prevent removal of panels, including suspended ceiling panels. Raceways shall not be used as a means of support for Class 2 circuit conductors.

Exception: Raceways containing power supply conductors for electrically controlled equipment shall be permitted to support Class 2 circuit conductors or cables that are used only to control such equipment.

Part IX — Referenced Standards

CHAPTER 43
REFERENCED STANDARDS

This chapter lists the standards that are referenced in various sections of this document. The standards are listed herein by the promulgating agency of the standard, the standard identification, the effective date and title, and the section or sections of this document that reference the standard. The application of the referenced standard shall be as specified in Section 102.4.

AAMA

American Architectural Manufacturers Association
1827 Walden Office Square, Suite 104
Schaumburg, IL 60173

Standard reference number	Title	Referenced in code section number
101—97	Voluntary Specifications for Aluminum Vinyl PVC and Wood Windows and Glass Doors	R613.3

ACI

American Concrete Institute
38800 Country Club Drive
Farmington Hills, MI 48331

Standard reference number	Title	Referenced in code section number
318-99	Building Code Requirements for Structural Concrete	R402.2, R404.1, R404.4.6.1, Table R404.4(1), Table R404.4(2), Table R404.4(3), Table R404.4(4), R611.1, R611.5, Table R611.3, Table R611.4(1), Table R611.5, Table R611.7(1), Table R611.7(2), Table R611.7(3), Table R611.7(4), Table R611.7(5), Table R611.7(6), Table R611.7(7), Table R611.7(8), R611.7.1.1, R612.1
ACI 530-99/ASCE 5-96/ TMS 0402-95	Building Code Requirements for Masonry Structures	R404.1, R606.1, R606.1.1, R606.11.1, R606.11.2.2.1, R606.11.2.2.2, R606.11.3.1

ACCA

Air Conditioning Contractors of America
1228 17th Street, NW
Washington DC 20046

Standard reference number	Title	Referenced in code section number
Manual J—86	Load Calculations for Residential Winter and Summer Air Conditioning	M1401.3

AFPA

American Forest and Paper Association
111 19th Street, NW, #800
Washington, DC 20036

Standard reference number	Title	Referenced in code section number
AFPA NDS-97	Wood Construction-Design Values for Wood Construction	R404.2.2, R502.2, Table R503.1, R602.3, R802.2
WFCM—96	Wood Frame Construction Manual for One- and Two-family Dwellings	R301.2.1.1
AFPA—93	Span Tables for Joists and Rafters	R502.3, R802.4, R802.5
TR7—87	Technical Report No. 7-The Permanent Wood Foundation System, Basic Requirements	R401.1

AGA

American Gas Association
1515 Wilson Boulevard
Arlington, VA 22209

Standard reference number	Title	Referenced in code section number
7—90	Requirements for Gas Convenience Outlets	Chapter 24

AHA

American Hardboard Association
1210 West Northwest Highway
Palatine, IL 60067

Standard reference number	Title	Referenced in code section number
A135.4—95	Basic Hardware	Table R602.3(1)
A135.5—95	Prefinished Hardboard Paneling	R702.5
A135.6—98	Hardboard Siding	Table R703.4
A194.1—85	Cellulosic Fiber Board	Table R602.3(1)

ANSI

American National Standards Institute
25 West 43rd Street
New York, NY 10036

Standard reference number	Title	Referenced in code section number
A108.1A&B—92	Glazed Wall Tile, Ceramic Mosaic Tile, Quarry Tile and Paver Tile Installed with Portland Cement Mortar	R702.4.1, M2103.9
A108.4—92	Ceramic Tile Installed with Organic Adhesives or Water-cleanable Tile Setting Epoxy Adhesive	R702.4.1, M2103.9.7
A108.5—92	Ceramic Tile Installed with Dry-Set Portland Cement Mortar or Latex Portland Cement Mortar	R702.4.1, M2103.9.1, M2103.9.2, M2103.9.3, M2103.9.5
A108.6—92	Ceramic Tile Installed with Chemical-Resistant, Water-cleanable Tile-Setting-and-Grout Epoxy	R702.4.1, M2103.9.4
A108.11—92	Standard Specification for Installation of Ceramic Tile	R702.4.1
A118.1—92	Standard Specification for Dry-Set Portland Cement Mortar	R702.4.1
A118.3—92	Standard Specification for Chemical Resistant, Water-cleanable Tile-Setting and Grouting Epoxy and Water-cleanable Tile Setting Epoxy Adhesive	R702.4.1
A136.1—92	Organic Adhesives for Installation of Ceramic Tile	R702.4.1
A137.1—88	Ceramic Tile	R702.4.1
A208.1—93	Particleboard	R503.3.1, R605.1
ANSI/AGA LC-1—91	Interior Fuel Gas Piping Systems Using Corrugated Stainless Steel Tubing	G2413.5.3, Chapter 24
Z21.1—96	Household Cooking Gas Appliances	G2436.1
Z21.5.1—95	Gas Clothes Dryers—Volume I—Type I Clothes Dryers	Chapter 24
Z21.8—94	Installation of Domestic Gas Conversion Burners	G2441.1
Z21.10.1—98	Gas Water Heaters—Volume I—Storage, Water Heaters with Input Ratings of 75,000 Btu per Hour or Less	G2446.1
Z21.10.3—98	Gas Water Heaters—Volume III—Storage, Water Heaters	G2446.1
Z21.11.1—91	Gas-Fired Room Heaters—Volume I—Vented Room Heaters	G2444.1
Z21.11.2—96	Gas-Fired Room Heaters—Volume II—Unvented Room Heaters	G2445.1
Z21.13—91	Gas-Fired low Pressure Steam and Hot Water Boilers	G2450.1
Z21.15—92	Manually Operated Gas Valves for Appliances, Appliance Connector Valves and Hose End Valves	G2419.1.1
Z21.22—86	Relief Valves and Automatic Gas Shutoff Devices for Hot Water Supply Systems	P2803.2
Z21.40.1—96	Gas-Fired Absorption Summer Air Conditioning Appliances	G2447.1
Z21.40.2—96	Gas-Fired Work Activated Air Conditioning and Heat Pump Appliances (Internal Combustion)	G2447.1
Z21.42—93	Gas-Fired Illuminating Appliances	G2448.1
Z21.47—93	Gas-Fired Central Furnaces (except Direct-Vent Central Furnaces)	G2440.1
Z21.48—92	Gas-Fired Gravity and Fan-Type Floor Furnace	G2435.1
Z21.49—92	Gas-Fired Gravity and Fan-Type Vented Wall Furnaces	G2434.1
Z21.50—98	Vented Gas Fireplaces	G2433.1
Z21.56—94	Gas-Fired Pool Heaters	G2439.1
Z21.58—91	Outdoor Cooking Gas Appliances	G2445.1
Z21.60—96	Decorative Gas Appliances for Installation in Vented Fireplaces	G2431.1
Z21.64—90	Direct Vent Central Furnaces	M1402.1
Z83.6—87	Gas-Fired Infrared Heaters—with 1989 Addenda	G2442.1, G2449.1
Z83.11—96	Gas Food Service Equipment (Ranges and Unit Broilers), Baking and Roasting Ovens, Fat Fryers, Counter Appliances and Kettles, Steam Cookers, and Steam Generators	G2445.1
Z97.1—94	Safety Performance Specifications and Methods of Test for Safety Glazing Materials Used in Buildings	R308.3
Z124.1—95	Plastic Bathtub Units	Table P2701.1
Z124.2—95	Plastic Shower Receptors and Shower Stalls	Table P2701.1
Z124.3—95	Plastic Lavatories	Table P2701.1
Z124.4—96	Plastic Water Closet Bowls and Tanks	Table P2701.1
Z124.6—97	Plastic Sinks	Table P2701.1

ASCE

American Society of Civil Engineers
1801 Alexander Bell Drive
Reston, VA 20191

Standard reference number	Title	Referenced in code section number
ASCE 7—98	Minimum Design Loads for Buildings and Other Structures	R301.2.1.1

ASHRAE

American Society of Heating, Refrigerating
and Air-Conditioning Engineers, Inc.
1791 Tullie Circle, NE
Atlanta, GA 30329

Standard reference number	Title	Referenced in code section number
ASNI/ASHRAE/34—92	Number Designation and Safety Classification of Refrigerants—with Addenda through 1997	M1411.1
ASHRAE—97	Handbook of Fundamentals	P3001.2, P3002.3, P3101.4, P3103.2, Appendix B

ASME

American Society of Mechanical Engineers
Three Park Avenue
New York, NY 10016-5990

Standard reference number	Title	Referenced in code section number
A112.1.2—91	Air Gaps in Plumbing Systems	P2902.2.1
A112.4.1—93	Water Heater Relief Valve Drain Tubes	P2803.6.2
A112.18.1M—96	Brass Plumbing Fixture Units	Table P2701.1, P2722.1
A112.19.1M—94	Enameled Cast Iron Plumbing Fixtures	Table P2701.1
A112.19.2M—98	Vitreous China Plumbing Fixtures	Table P2701.1, Table P2712.2
A112.19.3M—87	Stainless Steel Plumbing Fixtures (Designed for Residential Use)	Table P2701.1
A112.19.4M—94	Porcelain Enameled Formed Steel Plumbing Fixtures	Table P2701.1
A112.19.5—79	Trim for Water-Closet Bowls, Tanks, and Urinals	Table P2701.1
A112.19.6—95	Hydraulic Performance Requirements for Water Closets and Urinals	Table P2701.1, P2712.2
A112.19.7M—95	Whirlpool Bathtub Appliances	Table P2701.1
A112.19.8M—87	Suction Fittings for Use in Swimming Pools, Wading Pools, Spas, Hot Tubs, and Whirlpool Bathtub Appliances	Table P2701.1
A112.19.9—91	Nonvitreous Ceramic Plumbing Fixtures	Table P2701.1
A112.21.1—91	Floor Drains	Table P2701.1
B1.20.1—83	Pipe Threads, General Purpose (Inch)	G2413.9
B16.18-R94	Cast Copper Alloy Solder Joint Pressure Fittings	Table P2904.6
B16.33—90	Manually Operated Metallic Gas Valves for Use in Gas Piping Systems up to 125 psig (Sizes $^1/_2$ through 2)	G2419.1
B36.10—85	Welded and Seamless Wrought-Steel Pipe	G2413.4.2
CSD-1—98	Controls and Safety Devices for Automatically Fired Boilers	G2450.1, M2001.1.1
ASME—98	Boiler and Pressure Vessel Code, Section VIII, Divisions 1, 2 and 3	M2001.1.1

ASSE

American Society of Sanitary Engineering
28901 Clemens Road, #100
Westlake, OH 44145

Standard reference number	Title	Referenced in code section number
1001—90	Performance Requirements for Pipe Applied Atmospheric Type Vacuum Breakers	P2902.2.2, P2902.3.3
1002-79—99	Performance Requirements for Water Closet Flush Tank Ball Cocks	Table P2701.1, P2702.3.1
1003—95	Performance Requirements for Water Pressure	P2903.3.1
1006—89	Performance Requirements for Residential Use (Household) Dishwashers	Table P2701.1
1007—92	Performance Requirements for Home Laundry Equipment	Table P2701.1
1008—89	Performance Requirements for Household Food Waste Disposer Units	Table P2701.1

ASSE—continued

1010—98	Performance Requirements for Water Hammer Arresters—with 1982 Revision	P2903.5
1011—95	Performance Requirements for Hose Connection Vacuum Breakers	P2902.2.2, P2902.3.3
1012—95	Performance Requirements for Backflow Preventers with Intermediate Atmospheric Vent	P2902.2.3, P2902.3.3, P2902.4.1, P2902.4.5
1013—93	Performance Requirements for Reduced Pressure Principle Backflow Preventers	P2902.2.6, P2902.4.1, P2902.4.5
1014—90	Performance Requirements for Handheld Showers Reducing Valves for Domestic Water Supply Systems	Table P2701.1
1016—96	Performance Requirements for Individual Thermostatic, Pressure Balancing and Combination Control Valves for Bathing Facilities	Table P2701.1, P2708.3, P2709.4, P2722.2, P2723.2
1019—97	Performance Requirements for Wall Hydrants, Freezeless, Automatic Draining, Anti-Backflow Types	Table P2701.1, P2902.2.2, P2902.3.3
1020—98	Performance Requirements for Pressure Vacuum Breaker Assembly	P2902.2.5, P2902.3.3
1023—79	Hot Water Dispensers Household Storage Type Electrical	Table P2701.1
1025—78	Performance Requirements for Diverters for Plumbing Faucets with Hose Spray, Anti-Siphon Type, Residential Applications	Table P2701.1
1037—90	Performance Requirements for Pressurized Flushing Devices for Plumbing Fixtures	Table P2701.1
1050—91	Performance Requirements for Air Admittance Valves for Plumbing Drainage Systems	P3114.1
1051—98	Performance Requirements for Air Admittance Valves for Plumbing Drainage Systems, Fixture and Branch Devices	P3114.1
1052—94	Performance Requirements for Hose Connection Backflow Preventers	Table P2701.1, P2902.2.4, P2902.3.3
1056—95	Back Siphonage Vacuum Breaker	P2902.2.5, P2902.3.3

ASTM

American Society for Testing and Materials
100 Barr Harbor Drive
West Conshohocken, PA 19428

Standard reference number	Title	Referenced in code section number
A 36—97a	Standard Specification for Carbon Structural Steel	R606.14
A 53—98	Standard Specification for Pipe, Steel, Black and Hot-Dipped, Zinc-Coated Welded and Seamless	Table M2101.1, Table P2903.8(1); Table P3002.1, G2413.4.2
A 74—98	Standard Specification for Cast Iron Soil Pipe and Fittings	Table P3002.1,Table P3002.2
A 106—97a	Standard Specification for Seamless Carbon Steel Pipe for High-Temperature Service	M2101.1, G2413.4.2
A 126—95e01	Standard Specification for Gray Iron Castings for Valves, Flanges and Pipe Fittings	Table P3002.1
A 153M—95	Standard Specification for Zinc Coating (Hot Dip) on Iron and Steel Hardware	Table R606.14.1
A 167—96	Standard Specification for Stainless and Heat-Resisting Chromium-Nickel Steel Plate, Sheet, and Strip	R606.14, Table R606.14.1
A 197—92	Standard Specification for Cupola Malleable Iron	Table P3002.1
A 254—97	Standard Specification for Copper Brazed Steel Tubing	Table M2101.1, G2413.5.1
A 377—95e1	Standard Index of Standard Specification for Ductile Iron Pressure Pipe	Table P2904.4.1
A 510—96	Standard Specification for General Requirements for Wire Rods and Coarse Round Wire, Carbon Steel	R606.14
A 525	**(Discontinued 1994, replaced by A 653)**	Table R606.14.1, M1601.1.1
A 539—96	Standard Specification for Electronic-Resistance-Welded Coiled Steel Tubing for Gas and Fuel Oil Lines	M2202.1, G2413.5.1
A 615/A 615M—96a	Standard Specification for Deformed and Plain Billet Steel Bars for Concrete Reinforcement	R404.4.6.1, R611.6.2
A 616M—96a	Standard Specification for Rail Steel Deformed and Plain Bars for Concrete Reinforcement **(Discontinued 1999, replaced by A 996)**	R404.4.6.1, R611.6.2
A 617/A 617M—96	Standard Specification for Axle Steel Deformed and Plain Bars for Concrete Reinforcement **(Discontinued 1999, replaced by A 996)**	R404.4.6.1, R611.6.2
A 641M—98	Standard Specification for Zinc-Coated (Galvanized) Carbon Steel Wire	Table R606.14.1
A 653M—97	Standard Specification for Steel Sheet, Zinc-Coated Galvanized) or Zinc-Iron Alloy-Coated (Galvannealed) by the Hot-Dip Process	R505.2.1, R505.2.3, R603.2.1, R603.2.3, R804.2.1, R804.2.3, Table R905.10.3
A 706M—98	Standard Specification for Low-Alloy Steel Deformed Plain Bars for Concrete Reinforcement	R404.4.6.1, R611.6.2
A 755M—96	Standard Specification for Steel Sheet, Metallic Coated by the Hot Dip Process and Prepainted by the Coil Coating Process for Exterior Exposed Building Products	Table R905.10.3
A 792M—97a	Standard Specification for Steel Sheet, 55% Aluminum-Zinc Alloy-Coated by the Hot-Dip Process	R505.2.1, R603.2.1, R603.2.3, R804.2.1, R804.2.3
A 875M—97a	Standard Specification for Steel Sheet, Zinc-5%, Aluminum Alloy-Coated by the Hot-Dip Process	R505.2.1, R505.2.3, R603.2.1, R603.2.3, R804.2.1, R804.2.3
A 888—98e1	Standard Specification for Hubless Cast Iron Soil Pipe and Fittings for Sanitary and Storm Drain, Waste, and Vent Piping Application	Table P3002.1, Table P3002.2
A 951—98	Standard Specification for Masonry Joint Reinforcement	R606.14
B 32—96	Standard Specification for Solder Metal	P3003.3.4
B 42—98	Standard Specification for Seamless Copper Pipe, Standard Sizes	Table M2101.1, Table P3002.1
B 43—98	Standard Specification for Seamless Red Brass Pipe, Standard Sizes	Table M2101.1, G2413.5.2, Table P3002.1, Table P2904.4.4.1

ASTM—continued

B 75—97	Standard Specification for Seamless Copper Tube [Metric]	Table M2101.1, Table P2904.4.1, Table P3002.1, Table P3002.2
B 88—96	Standard Specification for Seamless Copper Water Tube	Table M2101.1, G2413.5.2, Table P2904.4.1, Table P3002.1, Table P3002.2, Table P2903.8(1)
B 101—96	Standard Specification for Lead-Coated Copper Sheets and Strip for Building Construction	Table R905.10.3
B 135—96	Standard Specification for Seamless Brass Tube [Metric]	Table P2904.4.1, Table M2101.1
B 227—88(1993)e[1]	Standard Specification for Hard-Drawn Copper-Clad Steel Wire	R606.14
B 251—97	Standard Specification for General Requirements for Wrought Seamless Copper and Copper-Alloy Tube	Table M2101.1
B 280—98	Standard Specification for Seamless Copper Tube for Air Conditioning and Refrigeration Field Service	G2413.5.2
B 302—98	Standard Specification for Threadless Copper Pipe	Table M2101.1
B 306—96	Standard Specification for Copper Drainage Tube (DWV)	Table M2101.1, Table P3002.1, Table P3002.2
B 370—98	Standard Specification for Copper Sheet and Strip for Building Construction	Table R905.10.3, Table P2701.1
B 447—97	Standard Specification for Welded Copper Tube	Table P2904.4.1
B 633—98 (Re-approved 1994)e[1]	Standard Specification for Electrodeposited Coatings of Zinc on Iron and Steel	R505.2.4, R603.2.4, R804.2.4
B 813—93	Standard Standard Specification for Liquid and Paste Fluxes for Soldering Applications of Copper and Copper Alloy Tube	Table M2101.1, P3004.10, P3003.3.4
B 828—92e[1]	Standard Practice for Making Capillary Joints by Soldering of Copper and Copper Alloy Tube and Fittings	P3004.10
C 5—98	Standard Specification for Quicklime for Structural Purposes	R702.2
C 14—95	Standard Specification for Concrete Sewer, Storm Drain, and Culvert Pipe	Table P3002.2
C 27—98	Standard Classification of Fireclay and High-Alumina Refractory Brick	R1003.5, R1003.8
C 28—96	Standard Specification for Gypsum Plasters	R702.2
C 34—96	Standard Specification for Structural Clay Load-Bearing Wall Tile	Table R301.2(1)
C 35—95	Standard Specification for Inorganic Aggregates for Use in Gypsum Plaster	R702.2
C 36—97	Standard Specification for Gypsum Wallboard	R702.3.1
C 37—95	Standard Specification for Gypsum Lath	R702.2
C 55—97	Standard Specification for Concrete Brick	R202, Table R301.2(1)
C 59—95	Standard Specification for Gypsum Casting Plaster and Gypsum Molding Plaster	R702.2
C 61—95	Standard Specification for Gypsum Keene's Cement	R702.2
C 62—97a	Standard Specification for Building Brick (Solid Masonry Units Made from Clay or Shale)	R202, Table R301.2(1)
C 67—98	Standard Test Methods of Sampling and Testing Brick and Structural Clay Tile	R905.3.5
C 73—97a	Standard Specification for Calcium Silicate Face Brick (Sand Lime Brick)	R202, Table R301.2(1)
C 79M—97	Standard Specification for Treated Core and Nontreated Core Gypsum Sheathing Board	Table R602.3(1), R702.3.1
C 90—99(98)	Standard Specification for Load-Bearing Concrete Masonry Units	Table R301.2(1)
C 129—97	Standard Specification for Nonload-bearing Concrete Masonry Units	Table R301.2(1)
C 140—98	Standard Test Methods of Sampling and Testing Concrete Masonry Units	R905.3.5
C 143/C 143M—97	Standard Test Method for Slump or Hydraulic Cement Concrete	R404.4.5, R611.6.1
C 145—85	Standard Specification for Solid Load-bearing Concrete Masonry Units **(Discontinued, replaced by C 90)**	R202, Table R301.2(1)
C 199—84 (Reapproved 199A)e[1]	Standard Test Method for Pier Test for Refractory Mortar	R1001.9, R1003.5, R1003.8
C 208—95	Standard Specification for Cellulosic Fiber Insulating Board	Table R602.3(1)
C 216—98	Standard Specification for Facing Brick (Solid Masonry Units Made from Clay or Shale)	R202, Table R301.2(1)
C 236—89 (1993)e1	Standard Test Method for Steady-State Thermal Performance of Building Assemblies by Means of a Guarded Hot Box	N1102.1.1.1
C 270—99	Standard Specification for Mortar for Unit Masonry	AE602, R607.1
C 315—98b	Standard Specification for Clay Flue Linings	Table R1001.11(1), Table R1001.11(2), R1001.8.1
C 411—97	Standard Test Method for Hot-Surface Performance of High-Temperature Thermal Insulation	M1601.2.1
C 425—98	Standard Specification for Compression Joints for Vitrified Clay Pipe and Fittings	Table P3002.2
C 443—98	Standard Specification for Joints for Circular Concrete Sewer and Culvert Pipe, Using Rubber Gaskets	P3003.3.5
C 445	Discontinued 1980, Replaced by no Replacement	P3003.3.5
C 475—94	Standard Specification for Joint Compound and Joint Tape for Finishing Gypsum Board	R702.3.1
C 476—99	Standard Specification for Grout for Masonry	R609.1.1
C 514—96	Standard Specification for Nails for the Application of Gypsum Wallboard	R702.3.1
C 557—93a	Standard Specification for Adhesives for Fastening Gypsum Wallboard to Wood Framing	R702.3.1
C 578—95	Standard Specification for Rigid, Cellular Polystyrene Thermal Insulation	R403.3
C 587—97	Standard Specification for Gypsum Veneer Plaster	R702.2
C 588—95a	Standard Specification for Gypsum Base for Veneer Plasters	R702.2
C 630/630M—96a	Standard Specification for Water-Resistant Gypsum Backing Board	R702.3.1, R702.4.2
C 631—95a	Standard Specification for Bonding Compounds for Interior Gypsum Plastering	R702.2
C 645—99	Standard Specification for Nonstructural Steel Framing Members	R702.3.3
C 652—97	Standard Specification for Hollow Brick (Hollow Masonry Units Made from Clay or Shale)	R202, Table R301.2(1)

ASTM—continued

C 700—97	Standard Specification for Vitrified Clay Pipe, Extra Strength, Standard Strength, and Perforated	Table P3002.2
C 836—95	Standard Specification for High Solids Content, Cold Liquid-Applied Elastomeric Waterproofing Membrane for Use with Separate Wearing Course	R905.15.2
C 843—98	Standard Specification for Application of Gypsum Veneer Plaster	R702.2
C 844—98a	Standard Specification for Application of Gypsum Base to Receive Gypsum Veneer Plaster	R702.2
C 847—95	Standard Specification for Metal Lath	R702.2
C 887—79a(89)ae[01]	Standard Specification for Packaged, Dry, Combined Materials for Surface Bonding Mortar	R406.1
C 897—96	Standard Specification for Aggregate for Job-Mixed Portland Cement-Based Plasters	R702.2
C 933—96a	Standard Specification for Welded Wire Lath	R702.2
C 954—98	Standard Specification for Steel Drill Screws for the Application of Gypsum Panel Products or Metal Plaster Bases to Steel Studs From 0.033 in. (0.84 mm) to 0.112 in. (2.84 mm) in Thickness	R505.2.4, R603.2.4, R702.3.6, R804.2.4
C 955—98	Standard Specification for Load-Bearing (Transverse and Axial) Steel Studs, Runners (Tracks), and Bracing or Bridging for Screw Application of Gypsum Panel Products and Metal Plaster Bases	R702.3.3
C 957—93(1998)	Standard Specification for High-Solids Content, Cold-Liquid Applied Elastomeric Waterproofing Membrane for Use with Integral Wearing Surface	R905.15.2
C 960/C 960M—97	Standard Specification for Predecorated Gypsum Board	R702.3.1
C 976—90(1996)e	Standard Test Method for Thermal Performance of Building Assemblies by Means of a Guarded Hot Box	N1102.1.1.1
C 1002—98	Standard Specification for Steel Drill Screws for the Application of Gypsum Panel Products or Metal Plaster Bases	R702.3.1, R702.3.6, Table R702.3.4
C 1029—96	Standard Specification for Spray-Applied Rigid Cellular Polyurethane Thermal Insulation	R905.14.2
C 1032—96	Standard Specification for Woven Wire Plaster Base	R702.2
C 1047—98	Standard Specification for Accessories for Gypsum Wallboard and Gypsum Veneer Base	R702.2, R702.3.1
C 1063—98	Standard Specification for Installation of Lathing and Furring to Receive Interior and Exterior Portland Cement-Based Plaster	R702.2
C 1157—98a	Standard Performance Specification for Hydraulic Cements	R402.2
C 1167—96	Standard Specification for Clay Roof Tiles	R905.3, R905.3.4, R905.3.5
C 1173—97	Standard Specification for Flexible Transition Couplings for Underground Piping Systems	P3003.3.5
C 1177/C 1177M—96	Standard Specification for Glass Mat Gypsum Substrate for Use as Sheathing	R702.3.1
C 1178/C 1178M—96	Standard Specification for Glass Mat Water-Resistant Gypsum Backing Panel	R702.3.1, R702.4.2
C 1261—98	Standard Specification for Firebox Brick for Residential Fireplaces	R1003.5, R1003.8
C 1277—97	Standard Specification for Shielded Couplings Joining Hubless Cast Iron Soil Pipe and Fittings	Table P3002.1, Table P3002.2
C 1278/C 1278M—96	Standard Specification for Fiber-Reinforced Gypsum Panels	R702.3.1
C 1283—97	Standard Specification for Practice for Installing Clay Flue Lining	R1001.9
C 1395/C 1395M—98	Standard Specification for Gypsum Ceiling Board	R702.3.1
D 41—94	Standard Specification for Asphalt Primer Used in Roofing, Dampproofing, and Waterproofing	Table R905.9.2, Table R905.11.2
D 224—89(1996)	Standard Specification for Smooth-Surfaced Asphalt Roll Roofing(Organic Felt)	R905.2.8.2
D 225—95	Standard Specification for Asphalt Shingles (Organic Felt) Surfaced with Mineral Granules	R905.2.4
D 226—97a	Standard Specification for Asphalt-Saturated (Organic Felt) Used in Roofing and Waterproofing	R703.2, R703.9.1, Table R905.2.3, R905.4.3, R905.5.3, R905.5.4, R905.6.3, R905.7.3, R905.8.3, R905.8.4, Table 905.9.2
D 227—97a	Standard Specification for Coal Tar Saturated (Organic Felt) Used in Roofing and Waterproofing	Table R905.9.2
D 249—89(96)	Standard Specification for Asphalt Roll Roofing (Organic Felt) Surfaced with Mineral Granules	Table R905.9.2, R905.2.8.2, R905.3.3, R905.5.4
D 312—95a	Standard Specification for Asphalt Used in Roofing	R905.9.2
D 371—89(1996)	Standard Specification for Asphalt Roll Roofing (Organic Felt) Surfaced with Mineral Granules: Wide-Selvage	R905.5.4
D 422—63	Standard Test Method for Particle-Size Analysis of Soils	R403.1.7.5.1
D 406	Discontinued 1961, Replaced by No Replacement	R905.6.4
D 449—89	Standard Specification for Asphalt Used in Dampproofing and Waterproofing	R406.2
D 450—96	Standard Specification for Coal-Tar Pitch Used in Roofing , Dampproofing and Waterproofing	Table R905.9.2
D 1227—95	Standard Specification for Emulsified Asphalt Used as a Protective Coating for Roofing	Table R905.9.2, Table R905.11.2, R905.15.2
D 1248—84(89)e[01]	Standard Specification for Polyethylene Plastics Molding and Extrusion Materials	M1601.1.2
D 1527—96a	Standard Specification for Acrylonitrile-Butadiene-Styrene (ABS) Plastic Pipe, Schedules 40 and 80	Table P2904.4.1
D 1693—97a	Standard Test Method for Environmental Stress-Cracking of Ethylene Plastics	Table M2101.1
D 1784—97	Standard Specification for Rigid Poly (Vinyl Chloride) (PVC) Compounds and Chlorinated Poly (Vinyl Chloride) (CPVC) Compounds	M1601.1.2
D 1785—96b	Standard Specification for Poly (Vinyl Chloride) (PVC) Plastic Pipe, Schedules 40, 80 and 120	Table P2904.4.1
D 1861	**(Discontinued, 1992 Specification for Homogeneous Bituminized Fiber Drain and Sewer Pipe, Replaced by no replacement)**	Table P3002.2
D 1863—93(96)	Standard Specification for Mineral Aggregate Used in Built-Up Roofs	Table R905.9.2, Table R906.3.2
D 1869—95	Standard Specification for Rubber Rings for Asbestos-Cement Pipe	P2904.14
D 1970—97	Standard Specification for Self-Adhering Polymer Modified Bitumen Sheet Materials Used as Steep Roofing Underlayment for Ice Dam Protection	R905.2.3, R905.2.8.3

ASTM—continued

D 2104—96	Standard Specification for Polyethylene (PE) Plastic Pipe, Schedule 40	Table P2904.4.1
D 2178—97a	Standard Specification for Asphalt Glass Felt Used in Roofing and Waterproofing	Table R905.9.2
D 2235—96a	Standard Specification for Solvent Cement for Acrylonitrile-Butadiene-Styrene (ABS) Plastic Pipe and Fittings	P2904.8.1.1, Table P3002.1, Table P3002.2
D 2239—96a	Standard Specification for Polyethylene (PE) Plastic Pipe (SIDR-PR) Based on Controlled Inside Diameter	Table P2904.4.1
D 2241—96b	Standard Specification for Poly (Vinyl Chloride) (PVC) Pressure-Rated Pipe (SDR-Series)	Table P2904.4.1
D 2282—96a	Standard Specification for Acrylonitrile-Butadiene-Styrene (ABS) Plastic Pipe (SDR-PR)	Table P2904.4.1
D 2412—96a	Standard Test Method for Determination of External Loading Characteristics of Plastic Pipe by Parallel-Plate Loading	M1601.1.2
D 2447—99	Standard Specification for Polyethylene (PE) Plastic Pipe Schedules 40 and 80, Based on Outside Diameter	Table P2903.8(1), Table 2904.4.1, Table M2101.1
D 2466—97	Standard Specification for Poly(Vinyl Chloride) (PVC) Plastic Pipe Fittings, Schedule 40	P2904.6
D 2467—96a	Standard Specification for Poly (Vinyl Chloride) (PVC) Plastic Pipe Fittings, Schedule 80	Table P2904.6
D 2468—96a	Standard Specification for Acrylonitrile-Butadiene-Styrene (ABS) Plastic Pipe Fittings, Schedule 40	Table P2904.6
D 2513—98b	Standard Specification for Thermoplastic Gas Pressure Pipe, Tubing, and Fittings	G2413.6, G2413.6.1, G2413.11 G2414.14.2, Table M2101.1, M2104.2.1.3
D 2564—96a	Standard Specification for Solvent Cements for Poly (Vinyl Chloride) (PVC) Plastic Piping Systems	P2904.8.1.3, Table P3002.1, Table P3002.2
D 2609—97	Standard Specification for Plastic Insert Fittings for Polyethylene (PE) Plastic Pipe	Table P2904.6
D 2626—97b	Standard Specification for Asphalt-Saturated and Coated Organic Felt Base Sheet Used in Roofing	R905.3.3, Table R905.9.2
D 2661—97a	Standard Specification for Acrylonitrile-Butadiene-Styrene (ABS) Schedule 40 Plastic Drain, Waste, and Vent Pipe and Fittings	Table P3002.1, Table P3002.2
D 2662—96a	Standard Specification for Polybutylene (PB) Plastic Pipe (SIDR-PR) Based on Controlled Inside Diameter	Table P2904.4.1
D 2665—98	Standard Specification for Poly (Vinyl Chloride) (PVC) Plastic Drain, Waste, and Vent Pipe and Fittings	Table P3002.1, Table P3002.2
D 2666—96a	Standard Specification for Polybutylene (PB) Plastic Tubing	Table P2904.4.1
D 2672—96a	Standard Specification for Joints for IPS PVC Pipe Using Solvent Cement	Table P3002.1, Table, P3002.2, Table P2904.6
D 2683—98	Standard Specification for Socket-Type Polyethylene Fittings for Outside Diameter-Controlled Polyethylene Pipe and Tubing	Table M2101.1, M2104.2.1.1
D 2737—96a	Standard Specification for Polyethylene (PE) Plastic Tubing	Table P2904.4.1
D 2751—96a	Standard Specification for Acrylonitrile-Butadiene-Styrene (ABS) Sewer Pipe and Fittings	Table P3002.2
D 2822—91(1997)e[1]	Standard Standard Specification for Asphalt Roof Cement	Table R905.9.2, Table R905.11.2
D 2823—90(1997)e[1]	Standard Specification for Asphalt Roof Coatings	Table R905.9.2, Table R905.11.2
D 2824—94	Standard Specification for Aluminum-Pigmented Asphalt Roof Coatings, Non-Fibered, Asbestos Fibered, and Fibered without Asbestos	Table R905.9.2, Table R905.11.2
D 2837—98	Standard Test Method for Obtaining Hydrostatic Design Basis for Thermoplastic Pipe Materials	Table M2101.1
D 2846/D 2846M—97	Standard Specification for Chlorinated Poly (Vinyl Chloride) (CPVC) Plastic Hot- and Cold-Water Distribution Systems	Table P2903.8(1), Table P2904.4.1, P2904.8.1.2, M2101.1
D 2898—94(1999)	Standard Test Methods for Accelerated Weathering of Fire-Retardant-Treated Wood for Fire Testing	R902.2
D 2949—98	Standard Specification for 3.25-in. Outside Diameter Poly (Vinyl Chloride) (PVC) Plastic Drain, Waste, and Vent Pipe and Fittings	Table P3002.1, Table P3002.2
D 3000—95ae[1]	Standard Specification for Polybutylene (PB) Plastic Pipe (SDR-PR) Based on Outside Diameter	Table P2904.4.1
D 3019—94	Standard Specification for Lap Cement Used with Asphalt Roll Roofing, Nonfibered, Asbestos Fibered, and Nonasbestos Fibered	Table R905.9.2, Table R905.11.2
D 3034—98	Standard Specification for Type PSM Poly (Vinyl Chloride) (PVC) Sewer Pipe and Fittings	Table P3002.2
D 3035—95	Standard Specification fro Polyethylene (PE) Plastic Pipe (DR-PR) Based On Controlled Outside Diameter	Table M2101.1
D 3212—96a	Standard Specification for Joints for Drain and Sewer Plastic Pipes Using Flexible Elastomeric Seals	Table P3002.2
D 3261—97	Standard Specification for Butt Heat Fusion Polyethylene (PE) Plastic Fittings for Polyethylene (PE) Plastic Pipe and Tubing	Table P2904.6
D 3309—97a	Standard Specification for Polybutylene (PB) Plastic Hot- and Cold-Water Distribution Systems	Table P2903.8(1), Table P2904.4.1, Table M2101.1
D 3350—93	Standard Specification for Polyethylene Plastic Pipe and Fitting Materials	Table M2101.1
D 3462—96	Standard Specification for Asphalt Shingles Made From Glass Felt and Surfaced with Mineral Granules	R905.2.4
D 3468—97	Standard Specification for Liquid-Applied Neoprene and Chlorosulfanated Polyethylene Used in Roofing and Waterproofings	Table R905.15.2
D 3679—96a	Standard Specification for Rigid Poly (Vinyl Chloride) (PVC) Siding	Table R703.4
D 3747—79(1995)	Standard Specification for Emulsified Asphalt Adhesive for Adhering Roof Insulation	Table R905.9.2, Table R905.11.2
D 3909—97b	Standard Specification for Asphalt Roll Roofing (Glass Felt) Surfaced with Mineral Granules	R905.5.4, Table R905.9.2, Table R906.3.2
D 4022—94	Standard Specification for Coal Tar Roof Cement, Asbestos Containing	Table R905.9.2
D 4068—96	Standard Specification for Chlorinated Polyethylene (CPE) Sheeting for Concealed Water-Containment Membrane	P2709.1, P2709.2
D 4318—98	Standard Test Method for Liquid Limit, Plastic Limit and Plasticity Index of Soils	R403.1.7.5.1
D 4434—96	Standard Specification for Poly (Vinyl Chloride) Sheet Roofing	R905.13.2

ASTM—continued

D 4479—93	Standard Specification for Asphalt Roof Coatings-Asbestos-Free	Table R905.9.2
D 4586—93	Standard Specification for Asphalt Roof Cement, Asbestos-Free	Table R905.9.2
D 4601—98	Standard Specification for Asphalt-Coated Glass Fiber Base Sheet Used in Roofing	Table R905.9.2
D 4637—96	Standard Specification for EPDM Sheet Used in Single-Ply Roof Membrane	R905.12.2
D 4869—88(1993)e1	Standard Specification for Asphalt-Saturated (Organic Felt) Shingle Underlayment Used in Roofing	R905.2.3
D 4897—98	Standard Specification for Asphalt Coated Glass-Fiber Venting Base Sheet Used in Roofing	Table R905.9.2
D 4990—97a	Specication for Coal Tar Glass Felt Used in Roofing and Waterproofing	Table R905.9.2
D 5055—97e01	Standard Specification for Establishing and Monitoring Structural Capacities of Prefabricated Wood I-Joists	R502.1.4
D 5643—94	Standard Specification for Coal Tar Cement Used in Roofing	Table R905.9.2
D 5665—99a	Standard Specification for Thermoplastic Fabrics Used in Cold-Applied Roofing and Waterproofing	Table R905.9.2
D 5726—98	Standard Specification for Thermoplastic Fabrics Used in Hot-Applied Roofing and Waterproofing	Table R905.9.2
D 6083—97a	Standard Specification for Liquid Applied Acyrlic Coating Used in Roofing	Table R905.9.2, Table R905.11.2, Table R905.15.2
D 6162—98	Standard Specification for Styrene Butadiene Styrene (SBS) Modified Bituminous Sheet Materials Using a Combination of Polyester and Glass Fiber Reinforcements	Table R905.11.2
D 6163—98	Standard Specification for Styrene Butadiene Styrene (SBS) Modified Bituminous Sheet Materials Using Glass Fiber Reinforcements	Table R905.11.2
D 6164—98	Standard Specification for Styrene Butadiene Styrene (SBS) Modified Bituminous Sheet Materials Using Polyester Reinforcements	Table R905.11.2
D 6221—98	Standard Specification for Reinforced Bituminous Flashing Sheets for Roofing and Waterproofing	Table R905.11.2
D 6222—98	Standard Specification for Atactic Polypropelene (APP) Modified Bituminous Sheet Materials Using Polyester Reinforcement	Table R905.11.2
D 6223—98	Standard Specification for Atactic Polypropelene (APP) Modified Bituminous Sheet Materials Using a Combination of Polyester and Glass Fiber Reinforcement	Table R905.11.2
E 84—98e01	Standard Test Method for Surface Burning Characteristics of Building Materials	R202, R318.1.1, R318.2.6, R318.3, R319.3, R320.1, R320.2, M1601.2.1, M1601.4.2
E 90—97	Standard Test Method for Laboratory Measurement of Airborne Sound Transmission Loss of Building Partitions and Elements	AK102, R320.3.1
E 96—95	Standard Test Method for Water Vapor Transmission of Materials	R202, R321.1, M1411.4, M1601.3.4, N1102.1.7
E 108—96	Standard Test Method for Fire Tests of Roof Coverings	R902.1, R902.2
E 119—98	Standard Test Method for Fire Tests of Building Construction and Materials	R318.1.2, R321.1, R321.3.1
E 136—98e01	Standard Test Method for Behavior of Materials in a Vertical Tube Furnace at 750° C	R202
E 152—95	Standard Specification for Method of Fire Tests of Door Assemblies **(Replaced by no replacement)**	R318.3
E 330—97	Standard Test Method for Structural Performance of Exterior Windows, Curtain Walls, and Doors by Uniform Static Air Pressure Difference	R613.3
E 492—90(1996)e	Standard Specification for Laboratory Measurement of Impact Sound Transmission through Floor-Ceiling Assemblies using the Tapping Machine	AK103
E 814—97	Standard Test Method for Fire Tests of Through-Penetration Fire Stops	R321.3.1.2
E 970—98	Standard Test Method for Critical Radiant Flux of Exposed Attic Floor Insulation Using a Radiant Heat Energy Source	R320.5
E 1886—97	Standard Test Method for Performance of Exterior Windows, Curtain Walls, Doors and Storm Shutters Impacted by Missle(s) and Exposed to Cyclic Pressure Differentials	R301.2.1.2
E 1996—99	Specification for Performance of Exterior Windows, Glazed Curtain Walls, Doors and Storm Shutters Impacted by Windborne Debris in Hurricanes	R301.2.1.2
F 409—98	Standard Specification for Thermoplastic Accessible and Replaceable Plastic Tube and Tubular Fittings	Table P2701.1, P2702.3, Table P3002.1
F 438—97	Standard Specification for Socket-Type Chlorinated Poly (Vinyl Chloride) (CPVC) Plastic Pipe Fittings, Schedule 40	Table P2904.6
F 439—97	Standard Specification for Socket-Type Chlorinated Poly (Vinyl Chloride) (CPVC) Plastic Pipe Fittings, Schedule 80	Table P2904.6
F 441—97	Standard Specification for Chlorinated Poly (Vinyl Chloride) (CPVC) Plastic Pipe, Schedules 40 and 80	Table P2903.8(1), Table P2904.4.1
F 442—97	Standard Specification for Chlorinated Poly (Vinyl Chloride) (CPVC) Plastic Pipe (SDR-PR)	Table P2904.4.1
F 477—96a	Standard Specification for Elastomeric Seals (Gaskets) for joining Plastic Pipe	P2904.14
F 493—97	Standard Specification for Solvent Cements for Chlorinated Poly (Vinyl Chloride) (CPVC) Plastic Pipe and Fittings	P2904.8.1.2
F 628—97	Standard Specification for Acrylonitrile-Butadiene-Styrene (ABS) Schedule 40 Plastic Drain, Waste, and Vent Pipe with a Cellular Core	Table P3002.1, Table P3002.2
F 656—96a	Standard Specification for Primers for Use in Solvent Cement Joints of Poly (Vinyl Chloride) (PVC) Plastic Pipe and Fittings	P2904.8.1.3, Table P3002.1, Table P3002.2
F 789—95a	Standard Specification for Type PS-46 and Type PS-115 Poly (Vinyl Chloride) (PVC) Plastic Gravity Flow Sewer Pipe and Fittings	Table P3002.2
F 845—96	Standard Specification for Plastic Insert Fittings for Polybutylene (PB) Tubing	Table P2904.6
F 876—97	Standard Specification for Crosslinked Polyethylene (PEX) Tubing	Table M2101.1, Table P2904.4.1
F 877—99	Standard Specification for Crosslinked Polyethylene (PEX) Plastic Hot- and Cold-Water Distribution Systems	Table M2101.1, Table P2904.4.1
F 891—98	Standard Specification for Coextruded Poly (Vinyl Chloride) (PVC) Plastic Pipe with a Cellular Core	Table P3002.1, Table P3002.2

ASTM—continued

F 1055—95a	Standard Specification for Electrofusion Type Polyethylene Fittings for Outside Diameter Controlled Polyethylene Pipe and Fittings	Table M2101.1, M2104.2.1.2
F 1281—98	Crosslinked Polyethylene/Aluminum/Crosslinked Polyethylene (PEX-AL-PEX) Pipe	Table P2904.4.1
F 1282—98	Polyethylene/Aluminum/Polyethylene (PE-AL-PE) Pipe	Table P2904.4.1
F 1346—91	Standard Performance Standard Specification for Safety Covers and Labeling Requirements for All Covers for Swimming Pools, Spas and Hot Tubs	AG105.2, AG105.5, AG107
F 1380—95a	Standard Specification for Metal Insert Fittings for Polybutylene (PB) Tubing	Table P2904.6
F 1488—98	Standard Specification for Coextruded Composite Pipe	Table P3002.1, P3002.2
F 1667—97	Standard Specification for Driven Fasteners, Nails, Spikes, and Staples	R905.2.5
F 1807—98	Standard Specification for Metal Insert Fittings Utilizing a Copper Crimp Ring for SDR9 Cross-linked Polyethylene (PEX) Tubing	Table P2904.6, P2904.8.1.4.2, Table M2101.1

AWPA

American Wood-Preservers' Association
P.O. Box 5690
Granbury, Texas 76049

Standard reference number	Title	Referenced in code section number
C1—98	All Timber Products — Preservative Treatment by Pressure Processes	R323.1, R902.2.1
C2—98	Lumber, Timbers, Bridge Ties and Mine Ties — Preservative Treatment by Pressure Processes	R323.1, R327.1.7, Table R905.8.5
C3—97	Piles — Preservative Treatment by Pressure Processes	R323.1, R327.1.7
C4—90	Poles — Preservative Treatment by Pressure Processes	R323.1, R327.1.7
C9—97	Plywood — Preservative Treatment by Pressure Processes	R323.1, R327.1.7
C15—98	Wood for Commercial-Residential Construction Preservative Treatment by Pressure Process	R323.1, R327.1.7
C18—90	Standard for Pressure Treated Material in Marine Construction	R323.1, R327.1.7
C20—93	Structural Lumber-Fire-Retardant Treatment by Pressure Processes	R802.1.3.2
C22—96	Lumber and Plywood for Permanent Wood Foundations — Preservative Treatment by Pressure Processes	R323.1, R402.1.2, R504.3
C23—84	Round Poles and Posts Used in Building Construction-Preservative Treatment by Pressure Processes	R323.1, R327.1.7
C24—96	Sawn Timber Piles Used for Residential and Commercial Building	R323.1, R327.1.7
C27—88	Plywood-Fire-Retardant Treatment by Pressure Process	R802.1.3.2
C28—90	Standard for Preservative Treatment of Structural Glued Laminated Members and Laminations Before Gluing of Southern Pine, Pacific Coast Douglas Fir, Hem Fir and Western Hemlock by Pressure Process	R323.1, R327.1.7
P1—98	Standard for Coal Tar Creosote for Land and Fresh Water and Marine (Coastal Water) Use	R323.1, R327.1.7
P2—98	Standard for Creosote Solutions	R323.1, R327.1.7
P3—89	Standard for Creosote- Petroleum Oil Solution	R323.1, R327.1.7
M4—88	Standard for the Care of Preservative -Treated Wood Products	R324.3.1

CDA

The Copper Development Association, Inc.
260 Madison Avenue
New York, NY 10016

Standard reference number	Title	Referenced in code section number
4115—1929	Copper in Architecture—Design Handbook	Table R905.10.3

CGSB

Canadian General Standards Board
222 Queens Street
14th Floor, Suite 1402
Ottawa, Ontario, Canada KIA 1G6

Standard reference number	Title	Referenced in code section number
37-52M—84	Roofing and Waterproofing Membrane, Sheet Applied, Elastomeric	R905.12.2
37-54M—79	Roofing and Waterproofing Membrane, Sheet Applied, Flexible, Polyvinyl Chloride	R905.13.2
37-56M—80	Membrane, Modified Bituminous, Prefabricated and Reinforced for Roofing—with December 1985 Amendment	Table R905.11.2

CISPI

Cast Iron Soil Pipe Institute
Suite 419
5959 Shallowford Road
Chattanooga, TN 37421

Standard reference number	Title	Referenced in code section number
301—97	Standard Standard Specification for Hubless Cast Iron Soil Pipe and Fittings for Sanitary and Storm Drain, Waste and Vent Piping Applications	Table P3002.1, Table P3002.2
310—97	Standard Specification for Coupling for Use in Connection with Hubless Cast Iron Soil Pipe and Fittings for Sanitary and Storm Drain, Waste, and Vent Piping Applications	Table P3002.1, Table P3002.2

CPSC

Consumer Product Safety Commission
4330 East West Highway
Bethesda, MD 20814-4408

Standard reference number	Title	Referenced in code section number
16 CFR 1201—77	Safety Standard for Architectural Glazing	R308.1.1, R308.3
16 CFR 1209—79	Interim Safety Standard for Cellulose Insulation	R320.3
16 CFR 1404	Cellulose Insulation	R320.3

CSA

Canadian Standards Association
178 Rexdale Blvd.
Rexdale (Toronto), Ontario, Canada M9W IR3

Standard reference number	Title	Referenced in code section number
0325.0—92	Construction Sheathing	R503.2.1
0437-Series—93	Standards on OSB and Waferboard	R503.2.1, R803.2.1
A257.3—M92	Joints for Circular Concrete Sewer and Culvert Pipe, Manhole Sections, and Fittings Using Rubber Gaskets	P3003.3.5
B137.10M—98	Crosslinked Polyethylene/Aluminum/Polyethylene Composite Pressure Pipe Systems	Table M2101.1
B602M—90	Mechanical Couplings for Drain, Waste, and Vent Pipe and Sewer Pipe	Table P3002.1, Table P3002.2

CSSB

Cedar Shake & Shingle Bureau
515 116th Avenue, NE, #275
Bellevue, WA 98004-5294

Standard reference number	Title	Referenced in code section number
CSSB—97	Grading Rules for Wood Shakes and Shingles	Table R905.7.4, R702.6, R703.5
CSSB—97	Grading and Packing Rules for Certigrade Red Cedar Shingles	Table R905.7.4, R703.5

DOC

United States Department of Commerce
National Institute of Standards and Technology
Gaithersburg, MD 20899

Standard reference number	Title	Referenced in code section number
PS 1—95	Construction and Industrial Plywood	R404.2.1, Table R404.2.3, R503.2.1, R604.1, R803.2.1
PS 2—92	Performance Standard for Wood-Based Structural-Use Panels	R404.2.1, Table R404.2.3, R503.2.1, R604.1, R803.2.1
PS 20—94	American Softwood Lumber Standard	R404.2.1, R502.1, R602.1, R802.1

EWA (formerly APA)

APA-The Engineered Wood Association
PO Box 11700
Tacoma, WA 98411-0700

Standard reference number	Title	Referenced in code section number
EWA E30—98	APA Design and Construction Guide: Residential and Commercial	R803.2.3

FEMA

Federal Emergency Management Agency
500 C Street, SW
Washington, DC 20472

Standard reference number	Title	Referenced in code section number
FEMA/FIA-TB-2—93	Flood-Resistant Materials Requirements For Buildings Located in Special Flood Hazard Areas in Accordance with the National Flood Insurance Progam	R327.1.7

FM

Factory Mutual
Standards Laboratories Department
1151 Boston Providence Turnpike
Norwood, MA 02062

Standard reference number	Title	Referenced in code section number
4450—90	Approval Standard for Class I Insulated Steel Deck Roofs—Supplements thru 7/92	R906.1
4880—94	Approval Standard for Class 1: A) Insulated Wall or Wall & Roof/Ceiling Panels, B) Plastic Interior Finish Materials, C) Plastic Exterior Building Panels, D) Wall/Ceiling Coating Systems and E) Interior or Exterior Finish Systems	R318.3

GA

Gypsum Association
810 First Street, Northeast, Suite 510
Washington, DC 20002-4268

Standard reference number	Title	Referenced in code section number
253—99	Recommended Standard Specification for the Application of Gypsum Sheathing	Table R602.3(1)

HPVA

Hardwood Plywood & Veneer Association
1825 Michael Faraday Drive
Reston, Virginia 20190-5350

Standard reference number	Title	Referenced in code section number
ANSI/HPVA HP-1—1994	The American National Standard for Hardwood and Decorative Plywood	R702.5

ICC

International Code Council
5203 Leesburg Pike, Suite 708
Falls Church, VA 22041

Standard reference number	Title	Referenced in code section number
IBC—2000	International Building Code®	R110.2, R1001.8.2
ICCEC—2000	ICC Electrical Code—Administrative Provisions	R107.3
IECC—2000	International Energy Conservation Code™	R104.11
IFC—2000	International Fire Code	R102.7
IFGC—2000	International Fuel Gas Code®	R104.11
[Note: See Chapter 7 of the International Fuel Gas Code® for all referenced standards listed in Chapter 24.]		
IMC—2000	International Mechanical Code®	R104.11, M2106.1
IPC—2000	International Plumbing Code®	R104.11
IPMC—2000	International Property Maintenance Code	R102.7
ICC Standard 18-1	Standard for Expansive Soil Tests	R403.1.7.5.1
SBCCI SSTD 10-99	Standard for Hurricane Resistant Construction	R301.2.1.1

NCMA

National Concrete Masonry Association
2302 Horse Pen Road
Herndon, VA 20171-3499

Standard reference number	Title	Referenced in code section number
TR 68-A	Discontinued by NCMA	R404.1

NFPA

National Fire Protection Association
Batterymarch Park
Quincy, MA 02269

Standard reference number	Title	Referenced in code section number
13—96	Installation of Sprinkler Systems	R321.1
31—97	Installation of Oil-Burning Equipment	M1801.3.1, M1805.3
58—98	Liquefied Petroleum Gas Code	G2411.2, G2413.6.2
70—99	National Electrical Code®	E3301.1, E3301.2
72—96	National Fire Alarm Code	R317.1
259—97	Standard Test Method for Potential Heat of Building Materials	R318.2.5
501A—92	Standard on Manufactured Housing	AE201, R202
8501—97	Single Burner Boiler Operation	G2450.1
8502—99	Prevention of Furmace Explosions/Implosions in Multiple Burner Boiler-Furnaces	G2450.1
8504—96	Atmospheric Fluidized-Bed Boiler Operation	G2450.1

NFRC

National Fenestration Rating Council, Inc.
Suite 120
1300 Spring Park
Silver Spring, MD 20910

Standard reference number	Title	Referenced in code section number
100—97	Prodedure for Determining Fenestration Product U-Factors	N1101.3.2, N1101.3.2.1
200—97	Prodedure for Determining Fenestration product Solar Heat Gain Coefficients at Normal Incidence	N1101.3.2, N1101.3.2.1

NSF

National Sanitation Foundation
3475 Plymouth Road
Ann Arbor, MI 48105

Standard reference number	Title	Referenced in code section number
ANSI/NSF 14—96	Plastic Piping System Components and Related Materials	P2608.1, P2907.3
ANSI/NSF 42—98	Drinking Water Treatment Units—Anesthetic Effects	P2907.1, P2907.3
ANSI/NSF 44—98	Cajon Exchange Water Softners	P2907.1, P2907.3
ANSI/NSF 53—98	Drinking Water Treatment Units—Health Effects	P2907.1, P2907.3
ANSI/NSF 58—97	Reverse Osmosis Drinking Water Treatment Systems	P2907.2, P2907.3
ANSI/NSF 61—99	Drinking Water System Components—Health Effects	P2608.2, P2722.1, P2904.5, P2907.3

NSPI

National Spa & Pool Institute
2111 Eisenhower Avenue
Alexandria, VA 22314

Standard reference number	Title	Referenced in code section number
ANSI/NSPI-3—1992	Standard for Permanently Installed Residential Spas	AG104.1

NSPI—continued

ANSI/NSPI-4—1992	Standard for Aboveground/Onground Residential Swimming Pools	AG103.2
ANSI/NSPI-5—1995	Standard for Residential Inground Swimming Pools	AG103.1
ANSI/NSPI-6—1992	Standard for Residential Portable Spas	AG104.2

RMA

Rubber Manufacturer's Association
1400 K Street N.W.
Washington, DC 20005

Standard reference number	Title	Referenced in code section number
RP1—90	Minimum Requirements for Non-Reinforced Black EPDM Rubber Sheets	R905.12.2
RP2—90	Minimum Requirements for Fabric-Reinforced Black EPDM Rubber Sheets	R905.12.2
RP3—85	Minimum Requirements for Fabric-Reinforced Black Polychloroprene Rubber Sheets	R905.12.2

SAE

Society of Automotive Engineers
4700 West Lake Avenue
Glenview, IL 60025

Standard reference number	Title	Referenced in code section number
J 78—79	Steel Self Drilling Tapping Screws	R505.2.4, R603.2.4, R804.2.4

SMACNA

Sheet Metal & Air Conditioning Contractors National Assoc., Inc.
4021 Lafayette Center Road
Chantilly, VA 22021

Standard reference number	Title	Referenced in code section number
SMACNA—92	Fibrous Glass Duct Construction Standards	M1601.1.1

TPI

Truss Plate Institute
583 D'Onofrio Drive, #200
Madison, WI 53719

Standard reference number	Title	Referenced in code section number
1—95	National Design Standard for Metal-Plate-Connected Wood Truss Construction	R502.11.1, R502.11.2 R802.10.2, R802.10.3, R802.11.1

UL

Underwriters Laboratories, Inc.
333 Pfingsten Road
Northbrook, IL 60062

Standard reference number	Title	Referenced in code section number
17—94	Vent or Chimney Connector Dampers for Oil-fired Appliances	M1802.2.2
58—96	Steel Underground Tanks for Flammable and Combustible Liquids	M2201.1
80—80	Steel Inside Tanks for Oil-Burner Fuel	M2201.1
103—98	Chimneys, Factory-Built, Residential Type and Building Heating Appliance—with Revision thru March 1999	G2429.1, R202, R1002.3
127—99	Factory-Built Fireplaces	G2443.7, R1002.4, R1003.12, R1004.1, R1004.4
174—98	Household Electric Storage Tank Water Heaters	M2005.1
181—98	Factory-made Air Ducts and Air Connectors—with Revisions thru December 1998	M1601.2, M1601.3.1
181A—98	Closure Systems for Use with Rigid Air Ducts and Air Connectors	M1601.2, M1601.3.1
181B—98	Closure Systems for Use with Flexible Air Ducts and Air Connectors—with Revisions thru December 1998	M1601.2, M1601.3.1

UL—continued

441—98	Gas Vents	G2425.1
559—85	Heat Pumps	M1403.1
641—98	Low-Temperature Venting Systems, Type L	G2425.1, M1804.2.4, R202, R1001.8.5
726—98	Oil-Fired Boiler Furnaces Assemblies—with Revisions thru January 1999	G2425.1, M2001.1.1, M2006.1
727—98	Oil-Fired Central Furnaces—with Revisions thru January 1999	M1402.1
729—98	Oil-Fired Floor Furnaces—with Revisions thru January 1999	M1408.1
730—98	Oil-Fired Wall Furnaces—with Revisions thru January 1999	M1409.1
732—95	Oil-Fired Storage Tank Water Heaters	M2005.1
737—96	Fireplaces Stoves—with Revisions thru May 1998	R1002.3, M1414.1
790—98	Tests for Fire Resistance of Roof Covering Materials	R902.1
795—98	Commercial-Industrial Gas Heating Equipment	G2440.1, G2450.1
834—98	Heating, Water Supply, and Power-Electric—with Revisions thru November 1995	M2001.1.1
896—98	Oil-Burning Stoves	M1410.1
959—95	Medium Heat Appliance Factory Built Chimneys—with Revisions thru April 1998	R1002.6
1040—98	Fire Test of Insulated Wall Construction	R318.3
1256—98	Fire Test of Roof Deck Construction	R906.1
1453—95	Electronic Booster and Commercial Storage Tank Water Heaters	M2005.1
1482—98	Room Heaters, Solid-Fuel Type—with Revisions thru September 1998	M1410.1
1715—97	Fire Test of Interior Finish Material	R318.3
1738—93	Venting Systems for Gas-Burning Appliances, Categories II, III and IV	G2425.1
1777—98	Chimney Liners—with Revisions thru July 1998	R1001.8.1, R1001.15, M1801.3.4, G2424.15.4
1995—98	Heating and Cooling Equipment—with Revisions thru July 1998	M1402.4, M1403.1, M1407.1
2158A—96	Clothes Dryer Transition Duct	M1501.1

ULC

Underwriters' Laboratories of Canada
7 Crouse Road
Scarborough, Ontario, Canada MIR 3A9

Standard reference number	Title	Referenced in code section number
CAN/ULC-S 102—88	Surface Burning Characteristics of Building Materials and Assemblies	R320.2

WDMA

Window & Door Manufacturers Association
1400 East Touhy Avenue, #470
Des Plaines, IL 60018

Standard reference number	Title	Referenced in code section number
ANSI/AAMA/NWWDA 101/I.S. 2—97	Units Voluntary Specifications for Aluminum Vinyl (PVC) and Wood Windows and Glass Doors	R613.3

This appendix is a direct excerpt from the 2000 *Interantional Fuel Gas Code* and no attempt has been made to coordinate with Chapter 24 of the *International Residential Code.*

APPENDIX A (IFGS)
SIZING AND CAPACITIES OF GAS PIPING

(This Appendix is informative and is not part of the Code.)

In order to determine the size of piping to be used in designing a gas piping system, the following factors must be considered:

(a) Allowable loss in pressure from point of delivery to equipment

(b) Maximum gas demand

(c) Length of piping and number of fittings

(d) Specific gravity of the gas

(e) Diversity factor.

For any gas piping system, for special gas utilization equipment, or for conditions other than those covered by Tables 402.3(1) through 402.3(4), or Tables 402.3(15), 402.3(16), or 402.3(17) such as longer runs, greater gas demands, or greater pressure drops, the size of each gas piping system should be determined by standard engineering practices acceptable to the authority having jurisdiction.

Description of Tables

(a) The quantity of gas to be provided at each outlet should be determined, whenever possible, directly from the manufacturer's Btu input rating of the equipment which will be installed. In case the ratings of the equipment to be installed are not known, Table A-1 shows the approximate consumption of average appliances of certain types in Btu per hour.

To obtain the cubic feet per hour of gas required, divide the total Btu input of all equipment by the average Btu heating value per cubic foot of the gas. The average Btu per cubic foot of the gas in the area of the installation may be obtained from the serving gas supplier.

(b) Capacities for gas at low pressure [0.5 psig (35 kPa gauge) or less] in cubic feet per hour of 0.60 specific gravity gas for different sizes and lengths are shown in Tables 402.3(1) and 402.3(2) for iron pipe or equivalent rigid pipe, in Tables 402.3(3) and 402.3(4) for smooth wall semirigid tubing, and Tables 402.3(18), 402.3(19), and 402.3(20) for corrugated stainless steel tubing. Tables 402.3(1) and 402.3(3) are based upon a pressure drop of 0.3 inch (75 pa) water column, whereas Tables 402.3(2), 402.3(4), and 402.3(18) are based upon a pressure drop of 0.5 inch (125 pa) water column. Tables 402.3(19) and 402.3(20) are special low-pressure applications based upon pressure drops greater than 0.5 inch water column (125 pa). In using these tables, no additional allowance is necessary for an ordinary number of fittings.

(c) Capacities in thousands of Btu per hour of undiluted liquefied petroleum gases based on a pressure drop of 0.5 inch (125 pa) water column for different sizes and lengths are shown in Table 402.3(14) for iron pipe or equivalent rigid pipe and in Table 402.3(15) for smooth wall semirigid tubing, and in Table 402.3(23) for corrugated stainless steel tubing. Tables 402.3(24) and 402.3(25) for corrugated stainless steel tubing are based on pressure drops greater than 0.5 inches water column (125 pa). In using these tables, no additional allowance is necessary for an ordinary number of fittings.

(d) Gas piping systems that are to be supplied with gas of a specific gravity of 0.70 or less can be sized directly from Tables 402.3(1) through 402.3(4), unless the authority having jurisdiction specifies that a gravity factor be applied. Where the specific gravity of the gas is greater than 0.70, the gravity factor should be applied.

Application of the gravity factor converts the figures given in Tables 402.3(1) through 402.3(4) to capacities with another gas of different specific gravity. Such application is accomplished by multiplying the capacities given in Tables 402.3(1) through 402.3(4) by the multipliers shown in Table 402.3(13). In case the exact specific gravity does not appear in the table, choose the next higher value specific gravity shown.

(e) Capacities for gas at pressures greater than 0.5 psig (3.5 kPa gauge) in cubic feet per hour of 0.60 specific gravity gas for different sizes and lengths are shown in Tables 402.3(5) to 402.3(12) for iron pipe or equivalent rigid pipe and Tables 402.3(23) and 402.3(24) for corrugated stainless steel tubing.

Use of Capacity Tables

To determine the size of each section of gas piping in a system within the range of the capacity tables, proceed as follows: (Also see sample calculation at end of Appendix A.)

1. Determine the gas demand of each appliance to be attached to the piping system. Where Tables 402.3(1) through 402.3(23) are to be used to select the piping size, calculate the gas demand in terms of

cubic feet per hour for each piping system outlet. Where Tables 402.3(25) through 402.3(34) are to be used to select the piping size, calculate the gas demand in terms of thousands of Btu per hour for each piping system outlet.

2. Where the piping system is for use with other than undiluted liquefied petroleum gases, determine the design system pressure, the allowable loss in pressure (pressure drop), and the specific gravity of the gas to be used in the piping system.
3. Measure the length of piping from the point of delivery to the most remote outlet in the building. Where a multipressure gas piping system is used, gas piping shall be sized for the maximum length of pipe measured from the gas pressure regulator to the most remote outlet of each similarly pressured section.
4. In the appropriate capacity table, select the column showing the measured length, or the next longer length if the table does not give the exact length. This is the only length used in determining the size of any section of gas piping. If the gravity factor is to be applied, the values in the selected column of the table are multiplied by the appropriate multiplier from Table 402.3(24).

Capacities of smooth wall pipe or tubing may also be determined by using the following formulae*:

High Pressure [1.5 psig (10.3 kPa) and above]:

$$Q = 181.6 \sqrt{\frac{D^5 \cdot (P_1^2 - P_2^2) \cdot Y}{Cr \cdot fba \cdot L}}$$

$$= 2237\, D^{2.623} \left[\frac{(P_1^2 - P_2^2) \cdot Y}{Cr.L}\right]^{0.541}$$

Low Pressure [Less than 1.5 psig (10.3 kPa)]:

$$Q = 187.3 \sqrt{\frac{D^5 \cdot \Delta H}{Cr \cdot fba \cdot L}}$$

$$= 2313 D^{2.623} \left(\frac{\Delta H}{Cr.L}\right)^{0.541}$$

where

Q = Rate, cu ft per hr at 60°F (15.6°C) and 30 in. mercury column

D = Inside diameter of pipe, in.

P_1 = Upstream pressure, psia

P_2 = Downstream pressure, psia

Y = Superexpansibility factor = 1/supercompressibility factor

Cr = Factor for viscosity, density, and temperature

$$= 0.00354\, ST\left(\frac{Z}{S}\right)^{0.152}$$

S = Specific gravity of gas at 60°F and 30 in. mercury column

T = Absolute temperature, °F or = $t + 460$

t = Temperature, °F

Z = Viscosity of gas, centipoise (0.012 for natural gas, 0.008 for propane), or = 1488μ

m = Viscosity, pounds per second ft

fba = Base friction factor for air at 60°F ($CF = 1$)

L = Length of pipe, ft

ΔH = Pressure drop, in. water column (27.7 in. H_2O = 1 psi)

$$CF = \text{Factor } CF = \left(\frac{fb}{fba}\right)$$

fb = Base friction factor for any fluid at a given temperature, °F

*For further details on the formulae, refer to "Polyflo Flow Computer," available from Polyflo Company, 3412 High Bluff, Dallas, Texas 75234.

†For values for natural gas, refer to Manual for Determination of Supercompressibility Factors for Natural Gas, available from American Gas Association, 1515 Wilson Boulevard, Arlington, Virginia 22209. For values for liquefied petroleum gases, refer to *Engineering Data Book*, available from Gas Processors Association, 1812 First Place, Tulsa, Oklahoma 74102.

5. Use this vertical column to locate ALL gas demand figures for this particular system of piping.
6. Starting at the most remote outlet, find in the vertical column just selected the gas demand for that outlet. If the exact figure of demand is not shown, choose the next larger figure below in the column.
7. Opposite this demand figure, in the first column at the left, will be found the correct size of gas piping.
8. Proceed in a similar manner for each outlet and each section of gas piping. For each section of piping, determine the total gas demand supplied by that section.

**TABLE A-1
APPROXIMATE GAS INPUT
FOR TYPICAL APPLIANCES**

Appliance	Input Btu per hr. (Approx.)
Range, Free Standing, Domestic	65,000
Built-in Oven or Broiler Unit, Domestic	25,000
Built-in Top Unit, Domestic	40,000
Water Heater, Automatic Storage 30 to 40 Gal. Tank	45,000
Water Heater, Automatic Storage 50 Gal. Tank	55,000
Water Heater, Automatic Instantaneous	
Capacity 2 gal. per minute	142,800
Capacity 4 gal. per minute	285,000
Capacity 6 gal. per minute	428,400
Water Heater, Domestic, Circulating or Side-Arm	35,000
Refrigerator	3,000
Clothes Dryer, Type 1 (Domestic)	35,000
Gas Light	2,500
Incinerator, Domestic	35,000

For SI: 1 Btu per hour = 0.293 W

For specific appliances or appliances not shown above, the input should be determined from the manufacturer's rating.

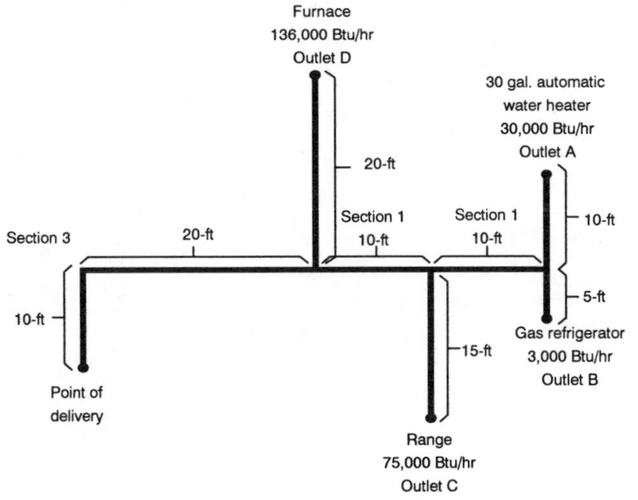

For SI units: 1 foot = 0.305 m; 1 gallon = 3.785 L;.1000 Btu per hour = 0.293 kw

EXHIBIT 1

Example of Piping System Design:

Determine the required pipe size of each section and out-let of the piping system shown in Exhibit 1, with a designated pressure drop of 0.50 inch water column (125 pa). The gas to be used has 0.65 specific gravity and a heating value of 1,000 Btu per cubic foot (37.5 Mj/m^3).

Exhibit 1

Solution:

(1)Maximum gas demand for outlet A:

$$\frac{\text{Consumption (rating plate input,or Table A-1 if necessary)} =}{\text{Btu of gas}}$$

$$\frac{\text{30,000 Btu per hour rating}}{\text{1,000 Btu per cubic foot}} = \frac{\text{30 cubic feet per hour}}{\text{(or 30 cfh)}}$$

Maximum gas demand for outlet B:

$$\frac{\text{Consumption}}{\text{Btu of gas}} = \frac{3{,}000}{1{,}000} = 3 \text{ cfh}$$

Maximum gas demand for outlet C:

$$\frac{\text{Consumption}}{\text{Btu of gas}} = \frac{75{,}000}{1{,}000} = 75 \text{ cfh}$$

Maximum gas demand for outlet D:

$$\frac{\text{Consumption}}{\text{Btu of gas}} = \frac{136{,}000}{1{,}000} = 136 \text{ cfh}$$

(2) The length of pipe from the point of delivery to the most remote Outlet (A) is 60 feet (18.3 m). This is the only distance used.

(3) Using the column marked 60 feet (18.3 m) in Table 402.3(2):

Outlet A supplying 30 cfh (0.8 m^3/h), requires 3/8 inch pipe.

Outlet B, supplying 3 cfh (0.08 m^3/h), requires 1/4 inch pipe.

Section 1, supplying Outlets A and B, or 33 cfh (0.9 m^3/hr), requires 3/8 inch pipe.

Outlet C, supplying 75 cfh (2.1 m^3/hr), requires 3/4 inch pipe.

Section 2, supplying Outlets A, B, and C, or 108 cfh (3.0 m^3/hr), requires $^3/_4$ inch pipe.

Outlet D, supplying 136 cfh (3.8 m^3/hr), requires 3/4 inch pipe.

Section 3, supplying Outlets A, B, and C, or 244 cfh (6.8 m^3/hr), requires 1 inch pipe.

(4) If the gravity factor [see (d) under "Description of Tables"] is applied to this example, the values in the column marked 60 feet (18.3 m) of Table 402.3(2) would be multiplied by the multiplier (0.962) from Table 402.3(13) and the resulting cubic feet per hour values would be used to size the piping.

APPENDIX B (IFGS)

SIZING OF VENTING SYSTEMS SERVING APPLIANCES EQUIPPED WITH DRAFT HOODS, CATEGORY I APPLIANCES, AND APPLIANCES LISTED FOR USE AND TYPE B VENTS

(This appendix is informative and is not part of the code.)

EXAMPLES USING SINGLE APPLIANCE VENTING TABLES

Example 1: Single Draft Hood-Equipped Appliance

An installer has a 120,000 Btu per hour input appliance with a 5-inch (127 mm) diameter draft hood outlet that needs to be vented into a 10-foot high Type B vent system. What size vent should be used assuming (a) a 5-foot (152 cm) lateral single-wall metal vent connector is used with two 90-degree elbows, or (b) a 5-foot (152 cm) lateral single-wall metal vent connector is used with three 90-degree elbows in the vent system?

Solution:

Table 502(2) should be used to solve this problem, because single-wall metal vent connectors are being used with a Type B vent.

(a) Read down the first column in Table 502(2) until the row associated with a 10-foot height and 5-foot (152 cm) lateral is found. Read across this row until a vent capacity greater than 120,000 Btu per hour is located in the shaded columns labeled "NAT Max" for draft hood-equipped appliances. In this case, a 5-inch (127 mm) diameter vent has a capacity of 122,000 Btu per hour and may be used for this application.

(b) If three 90-degree elbows are used in the vent system, then the maximum vent capacity listed in the tables must be reduced by 10 percent (*see Note 2 for Single Appliance Vents*). This implies that the 5-inch diameter vent has an adjusted capacity of only 110,000 Btu per hour. In this case, the vent system must be increased to 6 inches in diameter. See calculations below.

122,000 (.90) = 110,000 for 5-inch Vent

From Table 502(2), Select 6-inch Vent

186,000 (.90) = 167,000; This is greater than the required 120,000. Therefore, use a 6-inch vent and connector where three elbows are used.

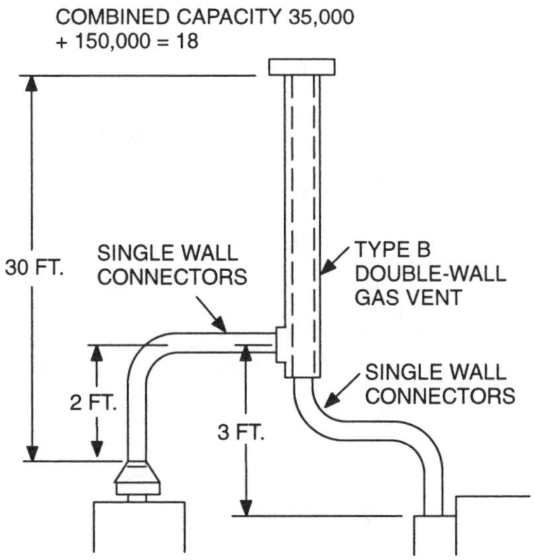

For SI: 1 foot = 304.8 mm.

Table 502(1) is used when sizing a Type B double-wall gas vent connected directly to the appliance.

Note: The appliance may either be Category I draft hood-equipped or fan-assisted types.

FIGURE B-1
TYPE B DOUBLE-WALL VENT SYSTEM SERVING A SINGLE APPLIANCE WITH A TYPE B DOUBLE-WALL VENT

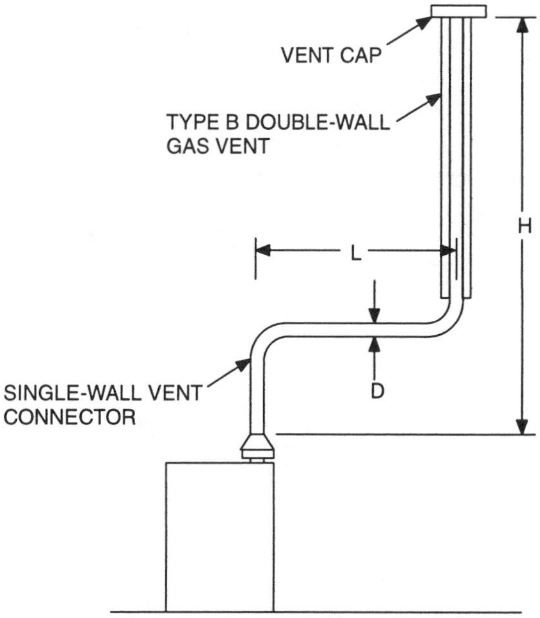

Table 502(2) is used when sizing a single-wall metal vent connector attached to a Type B double-wall gas vent.

Note: The appliance may either be Category I draft hood-equipped or fan-assisted types.

FIGURE B-2
TYPE B DOUBLE-WALL VENT SYSTEM SERVING A SINGLE APPLIANCE WITH A SINGLE-WALL METAL VENT CONNECTOR

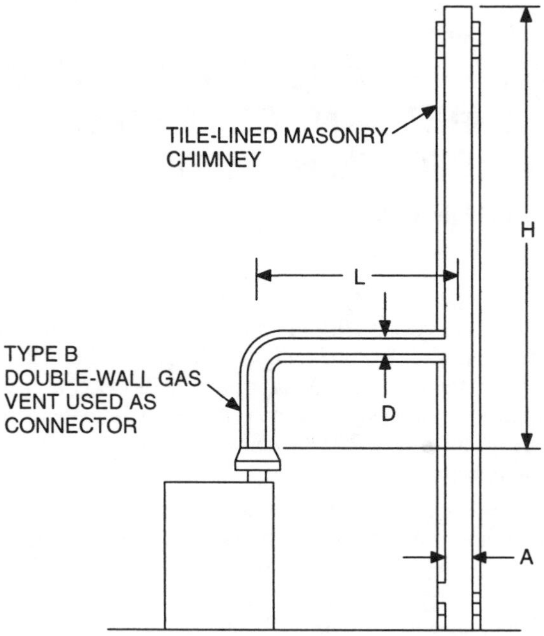

Table 502(3) is used when sizing Type B double-wall gas vent connector attached to a tile-lined masonry chimney.

Notes: "A" is the equivalent cross-sectional area of the tile liner. The appliance may either be Category I draft hood-equipped or fan-assisted type.

FIGURE B-3
VENT SYSTEM SERVING A SINGLE APPLIANCE WITH A MASONRY CHIMNEY OF TYPE B DOUBLE-WALL VENT CONNECTOR

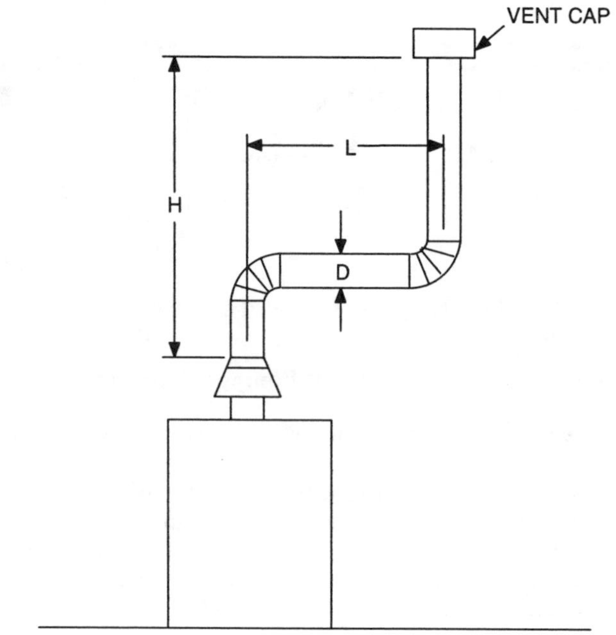

Asbestos cement Type B or single-wall metal vent serving a single draft hood-equipped appliance. [See Table 502.(5).]

FIGURE B-5
ASBESTOS CEMENT TYPE B OR SINGLE-WALL METAL VENT SYSTEM SERVING A SINGLE DRAFT HOOD-EQUIPPED APPLIANCE

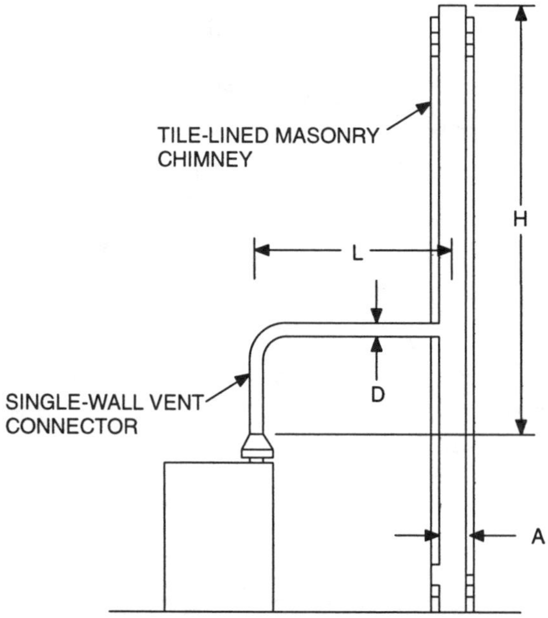

Table 502(4) is used when sizing a single-wall vent connector attached to a tile-lined masonry chimney.

Notes: "A" is equivalent cross-sectional area of the tile liner. The appliance may be either Category I draft hood-equipped or fan-assisted types.

FIGURE B-4
VENT SYSTEM SERVING A SINGLE APPLIANCE USING A MASONRY CHIMNEY AND A SINGLE-WALL METAL VENT CONNECTOR

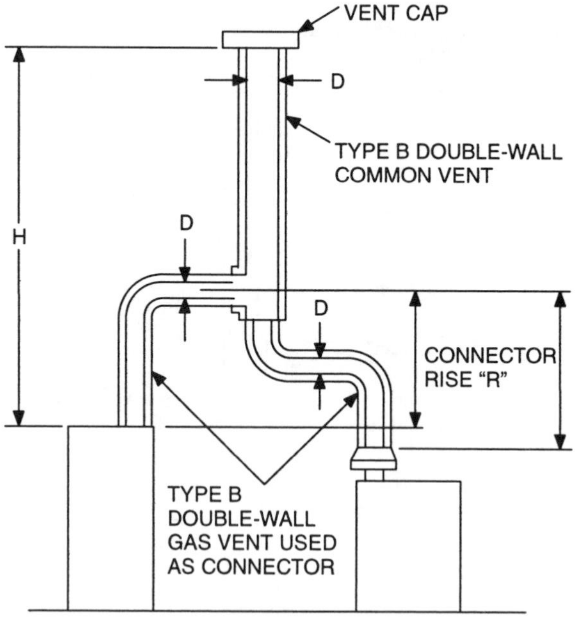

Table 502(6) is used when sizing Type B double-wall vent connectors attached to a Type B double-wall common vent.

Note: Each appliance may be either Category I draft hood-equipped or fan-assisted type.

FIGURE B-6
VENT SYSTEM SERVING TWO OR MORE APPLIANCES WITH TYPE B DOUBLE-WALL VENT AND TYPE B DOUBLE-WALL VENT CONNECTOR

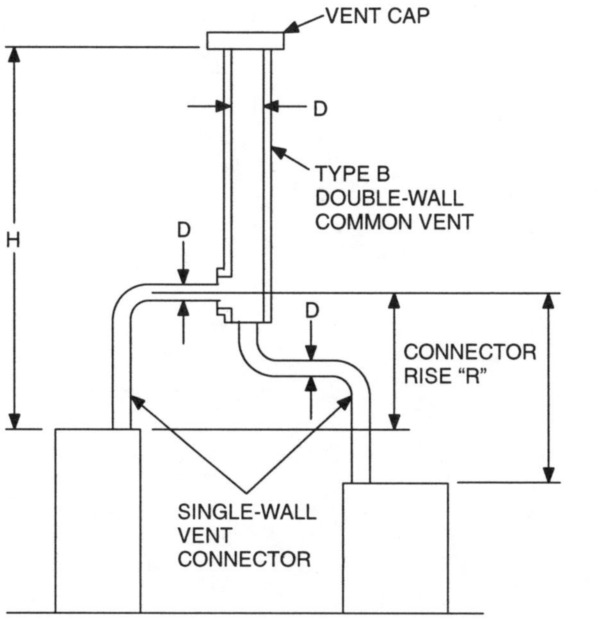

Table 502(7) is used when sizing single-wall vent connectors attached to a Type B double-wall common vent.

Note: Each appliance may either be Category I draft hood-equipped or fan-assisted type.

FIGURE B-7
VENT SYSTEM SERVING TWO OR MORE APPLIANCES WITH TYPE B DOUBLE-WALL VENT AND SINGLE-WALL METAL VENT CONNECTORS

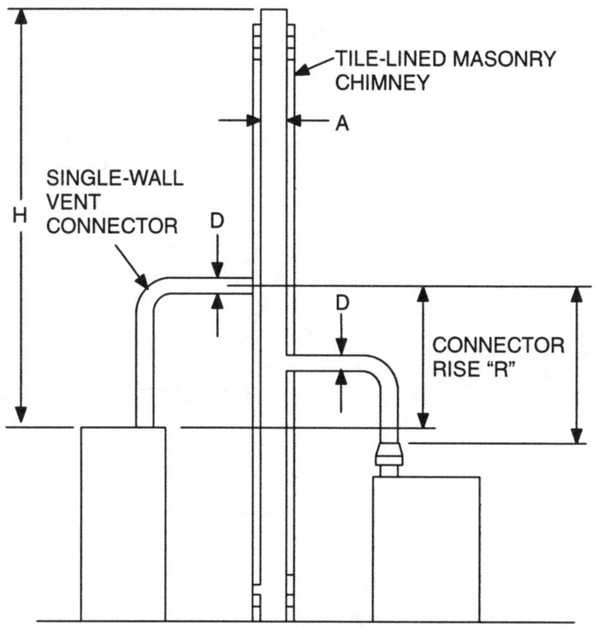

Table 502(9) is used when sizing single-wall vent connectors attached to a tile-lined masonry chimney.

Notes: "A" is the equivalent cross-sectional area of the tile liner. Each appliance may either be Category I draft hood-equipped or fan-assisted type.

FIGURE B-9
MASONRY CHIMNEY SERVING TWO OR MORE APPLIANCES WITH SINGLE-WALL METAL VENT CONNECTORS

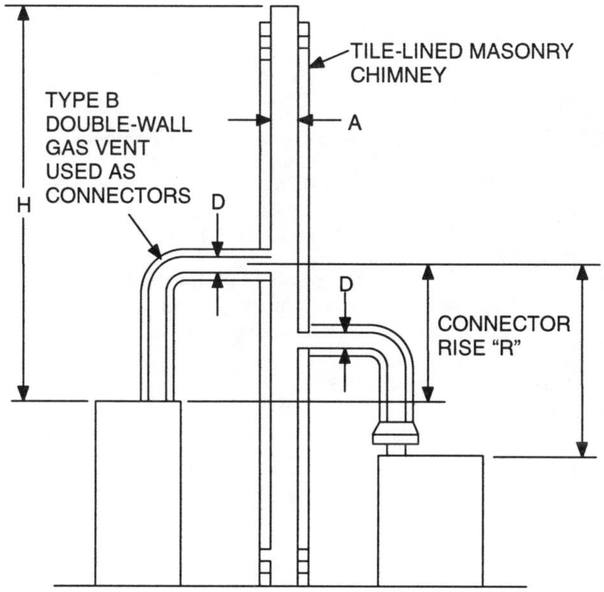

Table 502(8) is used when sizing Type B double-wall vent connectors attached to a tile-lined masonry chimney.

Notes: "A" is equivalent cross-sectional area of the tile liner. Each appliance may either be Category I draft hood-equipped or fan-assisted types.

FIGURE B-8
MASONRY CHIMNEY SERVING TWO OR MORE APPLIANCES WITH TYPE B DOUBLE-WALL VENT CONNECTOR

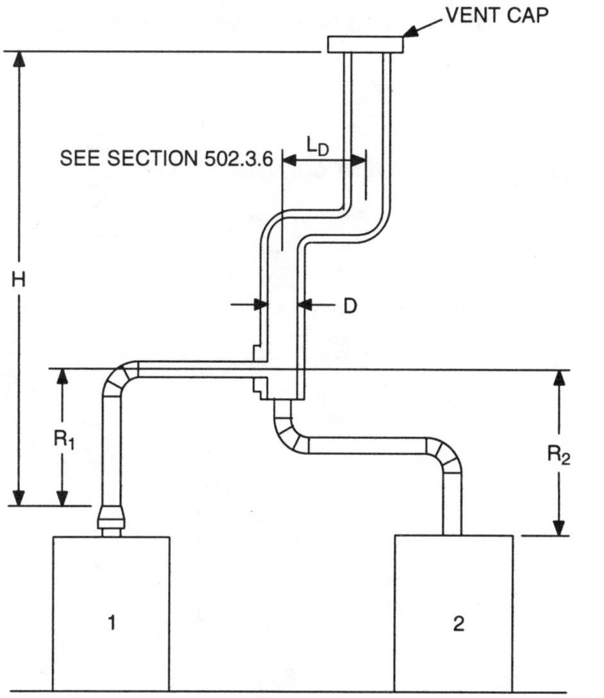

Asbestos cement Type B or single-wall metal pipe vent serving two or more draft hood-equipped appliances [see Table 502(10)].

FIGURE B-10
ASBESTOS CEMENT TYPE B OR SINGLE-WALL METAL VENT SYSTEM SERVING TWO OR MORE DRAFT HOOD-EQUIPPED APPLIANCES

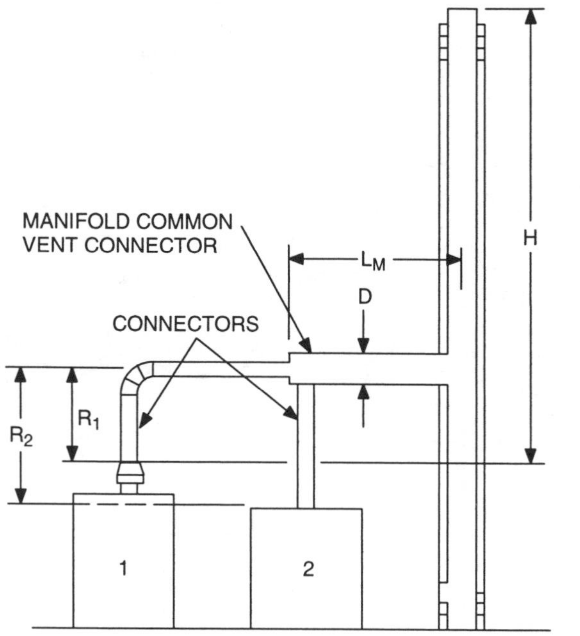

Example: Manifold Common Vent Connector L_M shall be no greater than 18 times the common vent connector manifold inside diameter: i.e., a 4 in. inside diameter common vent connector manifold shall not exceed 72 in. in length. (See Note 504.3.4).

Note: This is an illustration of a typical manifolded vent connector. Different appliance, vent connector or common vent types are possible. Consult Section 504.3.

For SI: 1 inch = 25.4 mm.

FIGURE B-11
USE OF MANIFOLD COMMON VENT CONNECTOR

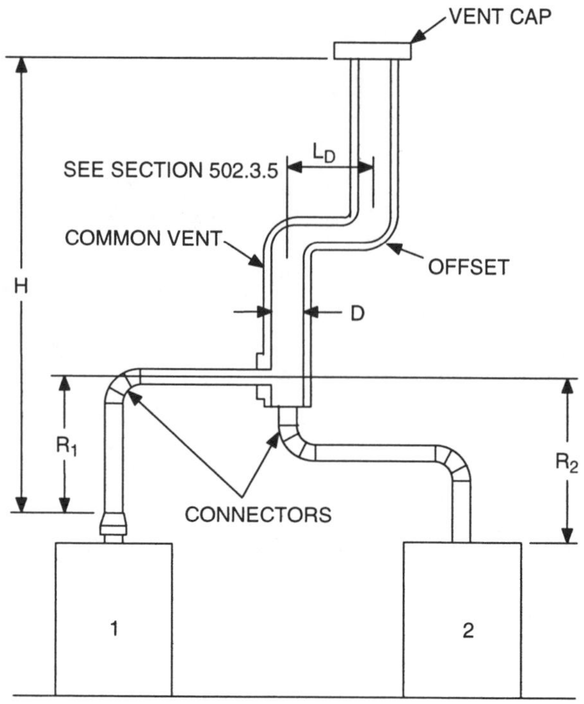

Example: Offset Common Vent

Note: This is an illustration of a typical offset vent. Different appliance, vent connector or vent types are possible. Consult Sections 502.2 and 502.3

FIGURE B-12
USE OF OFFSET COMMON VENT

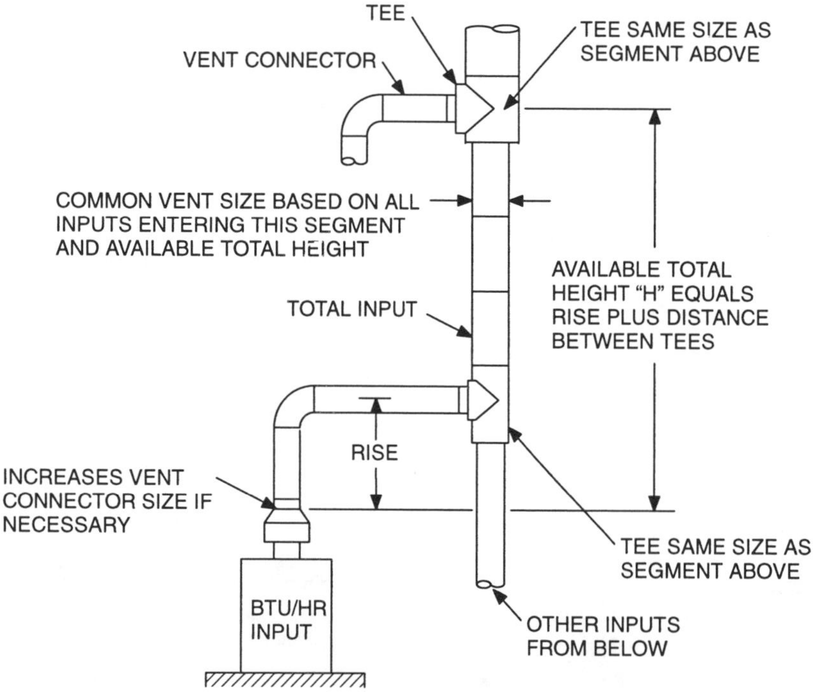

Vent connector size depends on:
- Input
- Rise
- Available total height "H"
- Table 502(6) connectors

Common vent size depends on:
- Combined inputs
- Available total height "H"
- Table 502(6), Common Vent

FIGURE B-13
MULTISTORY GAS VENT DESIGN PROCEDURE FOR EACH SEGMENT OF SYSTEM

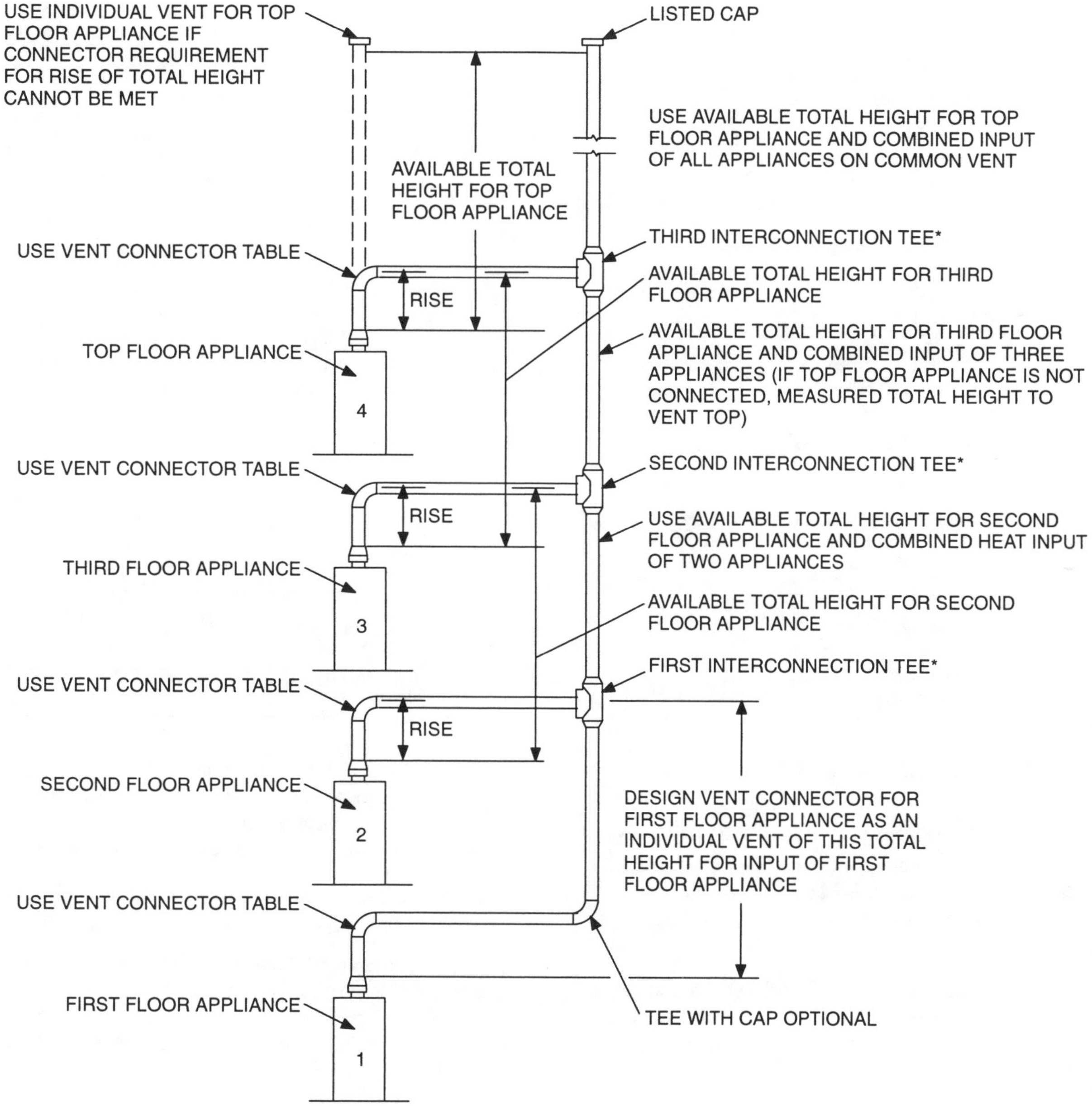

*EACH INTERCONNECTION TEE IS SAME SIZE AS THE SEGMENT OF COMMON VENT DIRECTLY ABOVE

Note: Principles of design of multistory vents using vent connector and common vent design tables. (See 504.3.10 through 504.3.15.)

FIGURE B-14
MULTISTORY VENT SYSTEMS

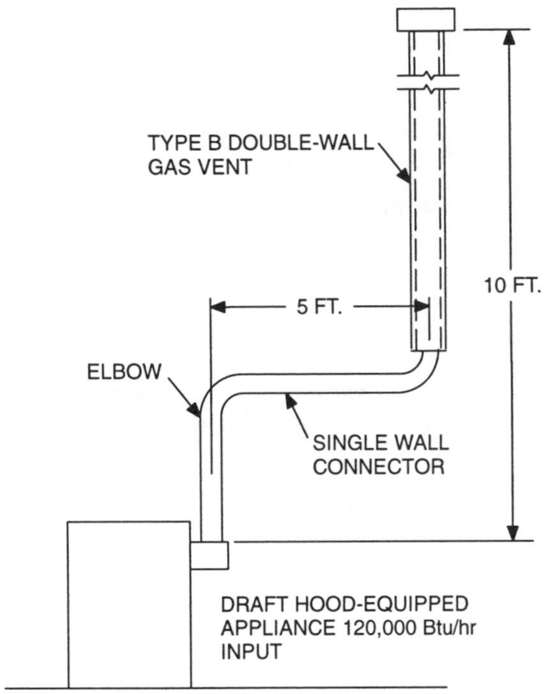

For SI: 1 foot = 304.8 mm, 1 Btu/hr = 0.2931 W.

FIGURE B-15 (EXAMPLE 1)
SINGLE DRAFT HOOD-EQUIPPED APPLIANCE

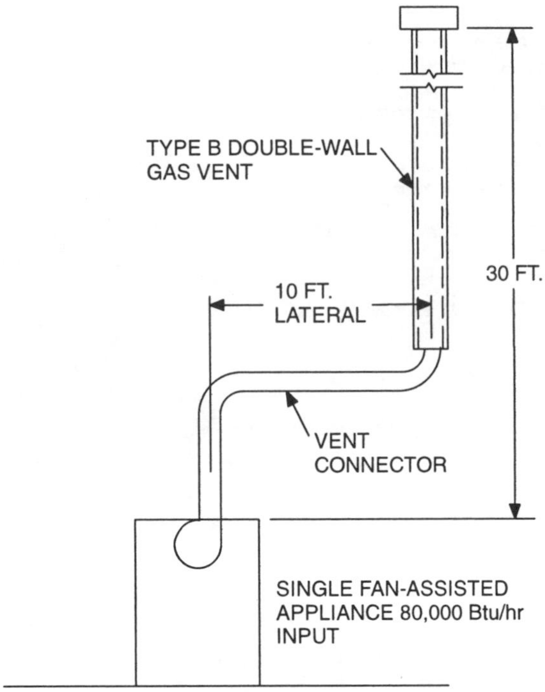

For SI: 1 foot = 304.8 mm, 1 Btu/hr = 0.2931 W.

FIGURE B-15 (EXAMPLE 2)
SINGLE FAN-ASSISTED APPLIANCE

Example 2: Single Fan-Assisted Appliance

An installer has an 80,000 Btu per hour input fan-assisted appliance that must be installed using 10 feet of lateral connector attached to a 30-foot high Type B vent. Two 90-degree elbows are needed for the installation. Can a single-wall metal vent connector be used for this application?

Solution:

Table 502(2) refers to the use of single-wall metal vent connectors with Type B vent. In the first column find the row associated with a 30-foot height and a 10-foot lateral. Read across this row, looking at the FAN Min and FAN Max columns, to find that a 3-inch diameter single-wall metal vent connector is not recommended. Moving to the next larger size single wall connector (4 inches) we find that a 4-inch diameter single-wall metal connector has a recommended minimum vent capacity of 91,000 Btu per hour and a recommended maximum vent capacity of 144,000 Btu per hour. The 80,000 Btu per hour fan-assisted appliance is outside this range, so the conclusion is that a single-wall metal vent connector cannot be used to vent this appliance using 10 feet of lateral for the connector.

However, if the 80,000 Btu per hour input appliance could be moved to within 5 feet of the vertical vent, then a 4-inch single-wall metal connector could be used to vent the appliance. Table 502(2) shows the acceptable range of vent capacities for a 4-inch vent with 5 feet of lateral to be between 72,000 Btu per hour and 157,000 Btu per hour.

If the appliance cannot be moved closer to the vertical vent, then Type B vent could be used as the connector material. In this case, Table 502(1) shows that for a 30-foot high vent with 10 feet of lateral, the acceptable range of vent capacities for a 4-inch diameter vent attached to a fan-assisted appliance is between 37,000 Btu per hour and 150,000 Btu per hour.

Example 3: Interpolating Between Table Values

An installer has an 80,000 Btu per hour input appliance with a 4-inch diameter draft hood outlet that needs to be vented into a 12-foot high Type B vent. The vent connector has a 5-foot lateral length and is also Type B. Can this appliance be vented using a 4-inch diameter vent?

Solution:

Table 502(1) is used in the case of an all Type B vent system. However, since there is no entry in Table 502(1) for a height of 12 feet, interpolation must be used. Read down the 4-inch diameter NAT Max column to the row associated with 10-foot height and 5-foot lateral to find the capacity value of 77,000 Btu per hour. Read further down to the 15-foot height, 5-foot lateral row to find the capacity value of 87,000 Btu per hour. The difference between the 15-foot height capacity value and the 10-foot height capacity value is 10,000 Btu per hour. The capacity for a vent system with a 12-foot height is equal to the capacity for a 10-foot height plus $^2/_5$ of the difference between the 10-foot and 15-foot height values, or 77,000 + $^2/_5$ (10,000 = 81,000 Btu per hour.) Therefore, a 4-inch diameter vent may be used in the installation.

EXAMPLES USING COMMON VENTING TABLES

Example 4: Common Venting Two Draft Hood-Equipped Appliances

A 35,000 Btu per hour water heater is to be common vented with a 150,000 Btu per hour furnace using a common vent with a total height of 30 feet. The connector rise is 2 feet for the water heater with a horizontal length of 4 feet. The connector rise for the furnace is 3 feet with a horizontal length of 8 feet. Assume single-wall metal connectors will be used with Type B vent. What size connectors and combined vent should be used in this installation?

Solution:

Table 502(7) should be used to size single-wall metal vent connectors attached to Type B vertical vents. In the vent connector capacity portion of Table 502(7), find the row associated with a 30-foot vent height. For a 2-foot rise on the vent connector for the water heater, read the shaded columns for draft hood-equipped appliances to find that a 3-inch diameter vent connector has a capacity of 37,000 Btu per hour. Therefore, a 3-inch single-wall metal vent connector may be used with the water heater. For a draft hood-equipped furnace with a 3-foot rise, read across the appropriate row to find that a 5-inch diameter vent connector has a maximum capacity of 120,000 Btu per hour (which is too small for the furnace) and a 6-inch diameter vent connector has a maximum vent capacity of 172,000 Btu per hour. Therefore, a 6-inch diameter vent connector should be used with the 150,000 Btu per hour furnace. Since both vent connector horizontal lengths are less than the maximum lengths listed in Notes for Multiple Appliance Vents, Tables 502(6) through 502(10), Note 2, the table values may be used without adjustments.

In the common vent capacity portion of Table 502(7), find the row associated with a 30-foot vent height and read over to the NAT + NAT portion of the 6-inch diameter column to find a maximum combined capacity of 257,000 Btu per hour. Since the two appliances total only 185,000 Btu per hour, a 6-inch common vent may be used.

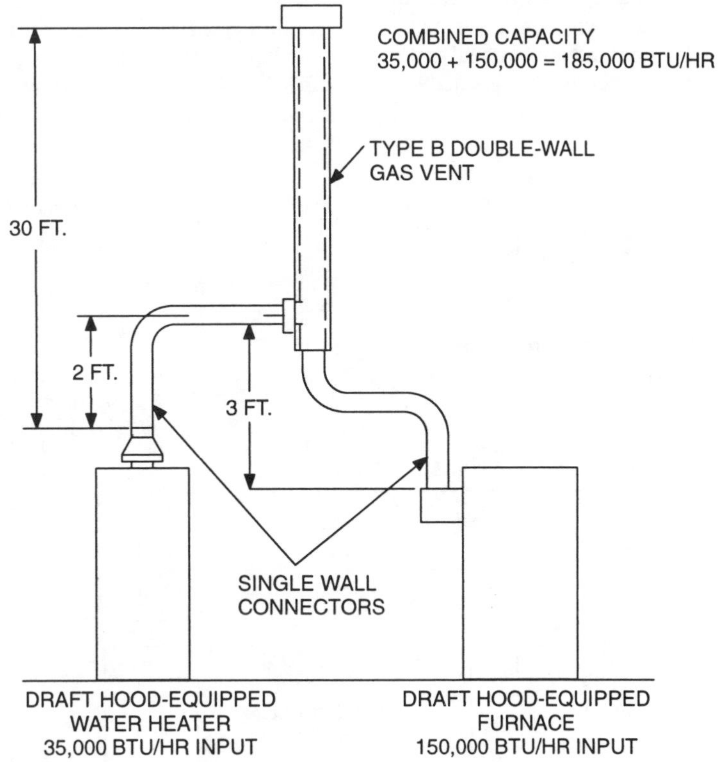

For SI: 1 foot = 304.8 mm, 1 Btu/hr = 0.2931 W.

FIGURE B-17 (EXAMPLE 4)
COMMON VENTING TWO DRAFT HOOD-EQUIPPED APPLIANCES

Example 5a: Common Venting a Draft Hood-Equipped Water Heater with a Fan-Assisted Furnace into a Type B Vent

In this case, a 35,000 Btu per hour input draft hood-equipped water heater with a 4-inch diameter draft hood outlet, 2 feet of connector rise, and 4 feet of horizontal length is to be common vented with a 100,000 Btu per hour fan-assisted furnace with a 4-inch diameter flue collar, 3 feet of connector rise, and 6 feet of horizontal length. The common vent consists of a 30-foot height of Type B vent. What are the recommended vent diameters for each connector and the common vent? The installer would like to use a single-wall metal vent connector.

Solution: [Table 502(7)]

Water Heater Vent Connector Diameter. Since the water heater vent connector horizontal length of 4 feet is less than the maximum value listed in Notes for Multiple Appliance Vents, Tables 502(6) through 502(10), Note 2, the venting table values may be used without adjustments. Using the Vent Connector Capacity portion of Table 502(7), read down the Total Vent Height (H) column to 30 feet and read across the 2-foot Connector Rise (R) row to the first Btu per hour rating in the NAT Max column that is equal to or greater than the water heater input rating. The table shows that a 3-inch vent connector has a maximum input rating of 37,000 Btu per hour. Although this is greater than the water heater input rating, a 3-inch vent connector is prohibited by Note 17. A 4-inch vent connector has a maximum input rating of 67,000 Btu per hour and is equal to the draft hood outlet diameter. A 4-inch vent connector is selected. Since the water heater is equipped with a draft hood, there are no minimum input rating restrictions.

Furnace Vent Connector Diameter. Using the Vent Connector Capacity portion of Table, read down the Total Vent Height (H) column to 30 feet and across the 3-foot Connector Rise (R) row. Since the furnace has a fan-assisted combustion system, find the first FAN Max column with a Btu per hour rating greater than the furnace input rating. The 4-inch vent connector has a maximum input rating of 119,000 Btu per hour and a minimum input rating of 85,000 Btu per hour. The 100,000 Btu per hour furnace in this example falls within this range, so a 4-inch connector is adequate. Since the furnace vent connector horizontal length of 6 feet is less than the maximum value listed in Notes for Multiple Appliance Vents, Tables 504(6) through 504(10), Note 2, the venting table values may be used without adjustment. If the furnace had an input rating of 80,000 Btu per hour, then a Type B vent connector [*see Table 504(6)*] would be needed in order to meet the minimum capacity limit.

Common Vent Diameter. The total input to the common vent is 135,000 Btu per hour. Using the Common Vent Capacity portion of Table 502(7), read down the Total Vent Height (H) column to 30 feet and across this row to find the smallest vent diameter in the FAN + NAT column that has a Btu per hour rating equal to or greater than 135,000 Btu per hour. The 4-inch common vent has a capacity of 132,000 Btu per hour and the 5-inch common vent has a capacity of 202,000 Btu per hour. Therefore, the 5-inch common vent should be used in this example.

Summary. In this example, the installer may use a 4-inch diameter, single-wall metal vent connector for the water heater and a 4-inch diameter, single-wall metal vent connector for the furnace. The common vent should be a 5-inch diameter Type B vent.

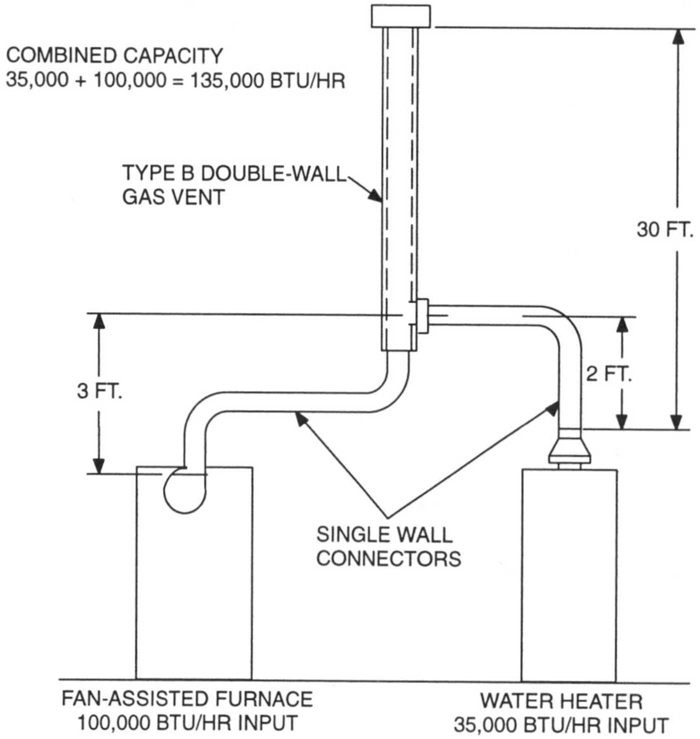

For SI: 1 foot = 304.8 mm, 1 Btu/hr = 0.2931 W.

FIGURE B-18 (EXAMPLE 4)
COMMON VENTING A DRAFT HOOD WITH A FAN-ASSISTED FURNACE INTO A TYPE B DOUBLE-WALL COMMON VENT

Example 5b: Common Venting Into a Masonry Chimney

In this case, the water heater and fan-assisted furnace of Example 5a are to be common vented into a clay tile-lined masonry chimney with a 30-foot height. The chimney is not exposed to the outdoors below the roof line. The internal dimensions of the clay tile liner are nominally 8 inches by 12 inches Assuming the same vent connector heights, laterals, and materials found in Example 5a, what are the recommended vent connector diameters, and is this an acceptable installation?

Solution:

Table 502(9) is used to size common venting installations involving single-wall connectors into masonry chimneys.

Water Heater Vent Connector Diameter. Using Table 502(9), Vent Connector Capacity, read down the Total Vent Height (H) column to 30 feet, and read across the 2-foot Connector Rise (R) row to the first Btu per hour rating in the NAT Max column that is equal to or greater than the water heater input rating. The table shows that a 3-inch vent connector has a maximum input of only 31,000 Btu per hour while a 4-inch vent connector has a maximum input of 57,000 Btu per hour. A 4-inch vent connector must therefore be used.

Furnace Vent Connector Diameter. Using the Vent Connector Capacity portion of Table 502(9), read down the Total Vent Height (H) column to 30 feet and across the 3-foot Connector Rise (R) row. Since the furnace has a fan-assisted combustion system, find the first FAN Max column with a Btu per hour rating greater than the furnace input rating. The 4-inch vent connector has a maximum input rating of 127,000 Btu per hour and a minimum input rating of 95,000 Btu per hour. The 100,000 Btu per hour furnace in this example falls within this range, so a 4-inch connector is adequate.

Masonry Chimney. From Table B-1, the Equivalent Area for a Nominal Liner size of 8 inches (12 inches is 63.6 square inches). Using Table 501.13.2(9), Common Vent Capacity, read down the FAN + NAT column under the Minimum Internal Area of Chimney value of 63 to the row for 30-foot height to find a capacity value of 739,000 Btu per hour. The combined input rating of the furnace and water heater, 135,000 Btu per hour, is less than the table value, so this is an acceptable installation.

Section 502.3.15 requires the common vent area to be no greater than seven times the smallest listed appliance categorized vent area, flue collar area, or draft hood outlet area. Both appliances in this installation have 4-inch diameter outlets. From Table B-1, the Equivalent Area for an Inside Diameter of 4 inches is 12.2 square inches. Seven times 12.2 equals 85.4, which is greater than 63.6, so this configuration is acceptable.

Example 5c: Common Venting into an Exterior Masonry Chimney

In this case, the water heater and fan-assisted furnace of Examples 5a and 5b are to be common vented into an exterior masonry chimney. The chimney height, clay tile liner dimensions, and vent connector heights and laterals are the same as in Example 5b. This system is being installed in Charlotte, North Carolina. Does this exterior masonry chimney need to be relined? If so, what corrugated metallic liner size is recommended? What vent connector diameters are recommended?

Solution:

According to 502.3.18, Type B vent connectors are required to be used with exterior masonry chimneys. Table 502(13) to size FAN+NAT common venting installations involving Type-B double wall connectors into exterior masonry chimneys.

The local 99-percent winter design temperature needed to use Table 502(13) can be found in ASHRAE Handbook—Fundamentals. For Charlotte, North Carolina, this design temperature is 19°F (-7.2°C).

Chimney Liner Requirement. As in Example 5b, use the 63 square inches Internal Area columns for this size clay tile liner. Read down the 63 square inches column of Table 50.2(13a) to the 30-feet height row to find that the combined Appliance Maximum Input is 747,000 Btu per hour. The combined input rating of the appliances in this installation, 135,000 Btu per hour, is less than the maximum value, so this criterion is satisfied. Table 502(13b), at a 19°F (-7.2°C) Design Temperature, and at the same Vent Height and Internal Area used above, shows that the minimum allowable input rating of a space-heating appliance is 470,000 Btu per hour. The furnace input rating of 100,000 Btu per hour is less than this minimum value. So this criterion is not satisfied, and an alternative venting design needs to be used, such as a Type B vent shown in Example 5a or a listed chimney liner system shown in the remainder of the example.

According to 502.3.17, Table 502(6) or Table 501.13.2(7) are used for sizing corrugated metallic liners in masonry chimneys, with the maximum common vent capacities reduced by 20 percent. This example will be continued assuming Type B vent connectors.

Water Heater Vent Connector Diameter. Using Table 502(6), Vent Connector Capacity, read down the Total Vent Height (H) column to 30 feet, and read across the 2-feet Connector Rise (R) row to the first Btu per hour rating in the NAT Max column that is equal to or greater than the water heater input rating. The table shows that a 3-inch vent connector has a maximum capacity of 39,000 Btu per hour. So the 35,000 Btu per hour water heater in this example can use a 3-inch connector.

Furnace Vent Connector Diameter. Using Table 502(6), Vent Connector Capacity, read down the Total Vent height (H) column to 30 feet, and read across the 3-feet Connector Rise (R) row to the first Btu per hour rating in the FAN Max column that is equal to or greater than the furnace input rating. The 100,000 Btu per hour furnace in this example falls within this range, so a 4-inch connector is adequate.

Chimney Liner Diameter. The total input to the common vent is 135,000 Btu per hour. Using the Common Vent Capacity Portion of Table 502(6), read down the Total Vent Height (H) column to 30 feet and across this row to find the smallest vent diameter in the FAN+NAT column that has a Btu per hour rating greater than 135,000 Btu per hour. The 4-inch common vent has a capacity of 138,000 Btu per hour. Reducing the maximum capacity by 20 percent (502.3.17) results in a maximum capacity for a 4-inch corrugated liner of 110,000 Btu per hour, less than the total input of 135,000 Btu per hour. So a larger liner is needed. The 5-inch common vent capacity listed in Table 502(6) is 210,000 Btu per hour, and after reducing by 20 percent is 168,000 Btu per hour. Therefore, a 5-inch corrugated metal liner should be used in this example.

Single Wall Connectors. Once it has been established that relining the chimney is necessary, Type B double wall vent connectors are not specifically required. This example could be redone using Table 502(7) for single wall vent connectors. For this case, the vent connector and liner diameters would be the same as found above with Type B double wall connectors.

TABLE B-1
MASONRY CHIMNEY LINER DIMENSIONS WITH CIRCULAR EQUIVALENTS

NOMINAL LINER SIZE (inches)	INSIDE DIMENSIONS OF LINER (inches)	INSIDE DIAMETER OR EQUIVALENT DIAMETER (inches)	EQUIVALENT AREA (square inches)
4 × 8	2 1/2 × 6 1/2	4	12.2
		5	19.6
		6	28.3
		7	38.3
8 × 8	6 3/4 × 6 3/4	7.4	42.7
		8	50.3
8 × 12	6 1/2 × 10 1/2	9	63.6
		10	78.5
12 × 12	9 3/4 × 9 3/4	10.4	83.3
		11	95
12 × 16	9 1/2 × 13 1/2	11.8	107.5
		12	113.0
		14	153.9
16 × 16	13 1/4 × 13 1/4	14.5	162.9
		15	176.7
16 × 20	13 × 17	16.2	206.1
		18	254.4
20 × 20	16 3/4 × 16 3/4	18.2	260.2
		20	314.1
20 × 24	16 1/2 × 20 1/2	20.1	314.2
		22	380.1
24 × 24	20 1/4 × 20 1/4	22.1	380.1
		24	452.3
24 × 28	20 1/4 × 24 1/4	24.1	456.2
28 × 28	24 1/4 × 24 1/4	26.4	543.3
		27	572.5
30 × 30	25 1/2 × 25 1/2	27.9	607
		30	706.8
30 × 36	25 1/2 × 31 1/2	30.9	749.9
		33	855.3
36 × 36	31 1/2 × 31 1/2	34.4	929.4
		36	1017.9

For SI: 1 inch = 25.4 mm, 1 square inch = 0.0929 m^2.

NOTE: When liner sizes differ dimensionally from those shown in Table B-7, equivalent diameters may be determined from published tables for square and rectangular ducts of equivalent carrying capacity or by other engineering methods.

APPENDIX C (IFGS)

EXIT TERMINALS OF MECHANICAL DRAFT AND DIRECT-VENT VENTING SYSTEMS

(This appendix is informative and is not part of the code.)

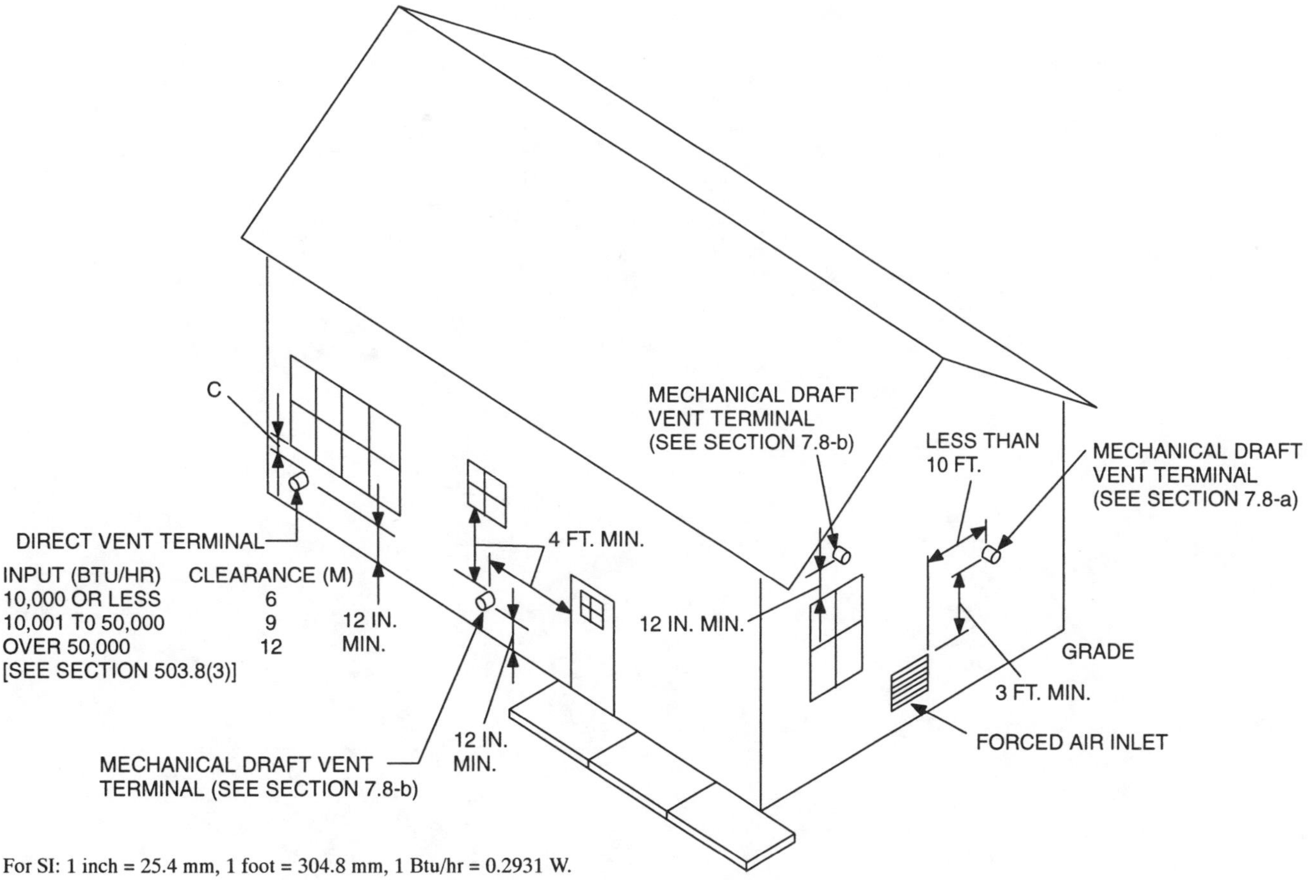

For SI: 1 inch = 25.4 mm, 1 foot = 304.8 mm, 1 Btu/hr = 0.2931 W.

APPENDIX C
EXIT TERMINALS OF MECHANICAL DRAFT AND DIRECT-VENT SYSTEMS

APPENDIX D (IFGS)

RECOMMENDED PROCEDURE FOR SAFETY INSPECTION OF AN EXISTING APPLIANCE INSTALLATION

(This appendix is informative and is not part of the code.)

The following procedure is intended as a guide to aid in determining that an appliance is properly installed and is in a safe condition for continuing use.

This procedure is predicated on central furnace and boiler installations, and it should be recognized that generalized procedures cannot anticipate all situations. Accordingly, in some cases, deviation from this procedure is necessary to determine safe operation of the equipment.

(a) This procedure should be performed prior to any attempt at modification of the appliance or of the installation.

(b) If it is determined there is a condition that could result in unsafe operation, the appliance should be shut off and the owner advised of the unsafe condition.

The following steps should be followed in making the safety inspection:

1. Conduct a test for gas leakage.
2. Visually inspect the venting system for proper size and horizontal pitch and determine there is no blockage or restriction, leakage, corrosion and other deficiencies that could cause an unsafe condition.
3. Shut off all gas to the appliance and shut off any other fuel-gas-burning appliance within the same room. **Use the shut-off valve in the supply line to each appliance.**
4. Inspect burners and crossovers for blockage and corrosion.
5. **Applicable only to furnaces.** Inspect the heat exchanger for cracks, openings or excessive corrosion.
6. **Applicable only to boilers.** Inspect for evidence of water or combustion product leaks.
7. Insofar as is practical, close all building doors and windows and all doors between the space in which the appliance is located and other spaces of the building. Turn on clothes dryers. Turn on any exhaust fans, such as range hoods and bathroom exhausts, so they will operate at maximum speed. Do not operate a summer exhaust fan. Close fireplace dampers. If, after completing Steps 8 through 13, it is believed sufficient combustion air is not available, refer to Section 304 of this code for guidance.
8. Place the appliance being inspected in operation. **Follow the lighting instructions.** Adjust the thermostat so appliance will operate continuously.
9. Determine that the pilot(s), where provided, is burning properly and that the main burner ignition is satisfactory by interrupting and reestablishing the electrical supply to the appliance in any convenient manner. If the appliance is equipped with a continuous pilot(s), test the pilot safety device(s) to determine if it is operating properly by extinguishing the pilot(s) when the main burner(s) is off and determining, after 3 minutes, that the main burner gas does not flow upon a call for heat. If the appliance is not provided with a pilot(s), test for proper operation of the ignition system in accordance with the appliance manufacturer's lighting and operating instructions.
10. Visually determine that the main burner gas is burning properly (i.e., no floating, lifting, or flashback). Adjust the primary air shutter(s) as required.

    If the appliance is equipped with high and low flame controlling or flame modulation, check for proper main burner operation at low flame.
11. Test for spillage at the draft hood relief opening after 5 minutes of main burner operation. Use a flame of a match or candle or smoke from a cigarette, cigar, or pipe.
12. Turn on all other fuel-gas-burning appliances within the same room so they will operate at their full inputs. **Follow lighting instructions for each appliance.**
13. Repeat Steps 10 and 11 on the appliance being inspected.
14. Return doors, windows, exhaust fans, fireplace dampers and any other fuel-gas-burning appliance to their previous conditions of use.
15. **Applicable only to furnaces.** Check both the limit control and the fan control for proper operation. Limit control operation can be checked by blocking the circulating air inlet or temporarily disconnecting the electrical supply to the blower motor and determining that the limit control acts to shut off the main burner gas.
16. **Applicable only to boilers.** Determine that the water pumps are in operating condition. Test low water cutoffs, automatic feed controls, pressure and temperature limit controls, and relief valves in accordance with the manufacturer's recommendations to determine that they are in operating condition.

APPENDIX E
MANUFACTURED HOUSING USED AS DWELLINGS

SECTION AE101
SCOPE

AE101.1 General. These provisions shall be applicable only to a manufactured home used as a single dwelling unit installed on privately owned (nonrental) lots and shall apply to the following:

1. Construction, alteration and repair of any foundation system which is necessary to provide for the installation of a manufactured home unit.
2. Construction, installation, addition, alteration, repair or maintenance of the building service equipment which is necessary for connecting manufactured homes to water, fuel, or power supplies and sewage systems.
3. Alterations, additions or repairs to existing manufactured homes. The construction, alteration, moving, demolition, repair and use of accessory buildings and structures and their building service equipment shall comply with the requirements of the codes adopted by this jurisdiction.

These provisions shall not be applicable to the design and construction of manufactured homes and shall not be deemed to authorize either modifications or additions to manufactured homes where otherwise prohibited.

Exception: In addition to these provisions, new and replacement manufactured homes to be located in flood hazard areas as established in Table R301.2(1) of the *International Residential Code* shall meet the applicable requirements of Sections R327 of the *International Residential Code*.

SECTION AE102
APPLICATION TO EXISTING MANUFACTURED HOMES AND BUILDING SERVICE EQUIPMENT

AE102.1 General. Manufactured homes and their building service equipment to which additions, alterations or repairs are made shall comply with all the requirements of these provisions for new facilities, except as specifically provided in this section.

AE102.2 Additions, alterations or repairs. Additions made to a manufactured home shall conform to one of the following:

1. Be certified under the National Manufactured Housing Construction and Safety Standards Act of 1974 (42 U.S.C. Section 5401, et seq.).
2. Be designed and constructed to conform with the applicable provisions of the National Manufactured Housing Construction and Safety Standards Act of 1974 (42 U.S.C. Section 5401, et seq.).
3. Be designed and constructed in conformance with the code adopted by this jurisdiction.

Additions shall be structurally separated from the manufactured home.

Exception: A structural separation need not be provided when structural calculations are provided to justify the omission of such separation.

Alterations or repairs may be made to any manufactured home or to its building service equipment without requiring the existing manufactured home or its building service equipment to comply with all the requirements of these provisions, provided the alteration or repair conforms to that required for new construction, and provided further that no hazard to life, health or safety will be created by such additions, alterations or repairs.

Alterations or repairs to an existing manufactured home which are nonstructural and do not adversely affect any structural member or any part of the building or structure having required fire protection may be made with materials equivalent to those of which the manufactured home structure is constructed, subject to approval by the building official.

Exception: The installation or replacement of glass shall be required for new installations.

Minor additions, alterations and repairs to existing building service equipment installations may be made in accordance with the codes in effect at the time the original installation was made subject to approval of the building official, and provided such additions, alterations and repairs will not cause the existing building service equipment to become unsafe, insanitary or overloaded.

AE102.3 Existing installations. Building service equipment lawfully in existence at the time of the adoption of the applicable codes may have their use, maintenance or repair continued if the use, maintenance or repair is in accordance with the original design and no hazard to life, health or property has been created by such building service equipment.

AE102.4 Existing occupancy. Manufactured homes which are in existence at the time of the adoption of these provisions may have their existing use or occupancy continued if such use or occupancy was legal at the time of the adoption of these provisions, provided such continued use is not dangerous to life, health and safety.

The use or occupancy of any existing manufactured home shall not be changed unless evidence satisfactory to the building official is provided to show compliance with all applicable provisions of the codes adopted by this jurisdiction. Upon any change in use or occupancy, the manufactured home shall cease to be classified as such within the intent of these provisions.

AE102.5 Maintenance. All manufactured homes and their building service equipment, existing and new, and all parts thereof shall be maintained in a safe and sanitary condition. All device or safeguards which are required by applicable codes or by the Manufactured Home Standards shall be maintained in conformance with the code or standard under which it was installed. The owner or the owner's designated agent shall be responsible for the maintenance of manufactured homes, accessory buildings, structures and their building service equipment. To determine compliance with this subsection, the building official may cause any manufactured home, accessory building or structure to be reinspected.

AE102.6 Relocation. Manufactured homes which are to be relocated within this jurisdiction shall comply with these provisions.

SECTION AE201 DEFINITIONS

AE201.1 General. For the purpose of these provisions, certain abbreviations, terms, phrases, words and their derivatives shall be construed as defined or specified herein.

ACCESSORY BUILDING. Any building or structure, or portion thereto, located on the same property as a manufactured home which does not qualify as a manufactured home as defined herein.

BUILDING SERVICE EQUIPMENT. Refers to the plumbing, mechanical and electrical equipment including piping, wiring, fixtures and other accessories which provide sanitation, lighting, heating ventilation, cooling, fire protection and facilities essential for the habitable occupancy of a manufactured home or accessory building or structure for its designated use and occupancy.

MANUFACTURED HOME. A structure transportable in one or more sections which, in the traveling mode, is 8 body feet (2438 body mm) or more in width or 40 body feet (12 192 body mm) or more in length or, when erected on site, is 320 or more square feet (30 m^2), and which is built on a permanent chassis and designed to be used as a dwelling with or without a permanent foundation when connected to the required utilities, and includes the plumbing, heating, air-conditioning and electrical systems contained therein; except that such term shall include any structure which meets all the requirements of this paragraph except the size requirements and with respect to which the manufacturer voluntarily files a certification required by the secretary (HUD) and complies with the standards established under this title.

For mobile homes built prior to June 15, 1976, a label certifying compliance to the Standard for Mobile Homes, NFPA 501, ANSI 119.1, in effect at the time of manufacture is required. For the purpose of these provisions, a mobile home shall be considered a manufactured home.

MANUFACTURED HOME INSTALLATION. Construction which is required for the installation of a manufactured home, including the construction of the foundation system, required structural connections thereto and the installation of on-site water, gas, electrical and sewer systems and connections thereto which are necessary for the normal operation of the manufactured home.

MANUFACTURED HOME STANDARDS. The Manufactured Home Construction and Safety Standards as promulgated by the United States Department of Housing and Urban Development.

PRIVATELY OWNED (NONRENTAL) LOT. A parcel of real estate outside of a manufactured home rental community (park) where the land and the manufactured home to be installed thereon are held in common ownership.

SECTION AE301 PERMITS

AE301.1 Initial installation. A manufactured home shall not be installed on a foundation system reinstalled or altered without first obtaining a permit from the building official. A separate permit shall be required for each manufactured home installation. When approved by the building official, such permit may include accessory buildings and structures and their building service equipment when the accessory buildings or structures will be constructed in conjunction with the manufactured home installation.

AE301.2 Additions, alterations and repairs to a manufactured home. A permit shall be obtained to alter, remodel, repair or add accessory buildings or structures to a manufactured home subsequent to its initial installation. Permit issuance and fees therefor shall be in conformance with the codes applicable to the type of work involved.

An addition made to a manufactured home as defined in these provisions shall comply with these provisions.

AE301.3 Accessory buildings. Except as provided in Section AE301.1, permits shall be required for all accessory buildings and structures and their building service equipment. Permit issuance and fees therefor shall be in conformance with the codes applicable to the types of work involved.

AE301.4 Exempted work. A permit shall not be required for the types of work specifically exempted by the applicable codes. Exemption from the permit requirements of any of said codes shall not be deemed to grant authorization for any work to be done in violation of the provisions of said codes or any other laws or ordinances of this jurisdiction.

SECTION AE302 APPLICATION FOR PERMIT

AE302.1 Application. To obtain a manufactured home installation permit, the applicant shall first file an application inwriting on a form furnished by the building official for that purpose. At the option of the building official, every such application shall:

1. Identify and describe the work to be covered by the permit for which application is made.
2. Describe the land on which the proposed work is to be done by legal description, street address or similar description that will readily identify and definitely locate the proposed building or work.
3. Indicate the use or occupancy for which the proposed work is intended.
4. Be accompanied by plans, diagrams, computations and specifications and other data as required in Section AE302.2.

5. Be accompanied by a soil investigation when required by Section AE502.2.
6. State the valuation of any new building or structure or any addition, remodeling or alteration to an existing building.
7. Be signed by permittee, or permittee's authorized agent, who may be required to submit evidence to indicate such authority.
8. Give such other data and information as may be required by the building official.

AE302.2 Plans and specifications. Plans, engineering calculations, diagrams and other data as required by the building official shall be submitted in not less than two sets with each application for a permit. The building official may require plans, computations and specifications to be prepared and designed by an engineer or architect licensed by the state to practice as such.

Where no unusual site conditions exist, the building official may accept approved standard foundation plans and details in conjunction with the manufacturer's approved installation instructions without requiring the submittal of engineering calculations.

AE302.3 Information on plans and specifications. Plans and specifications shall be drawn to scale on substantial paper or cloth and shall be of sufficient clarity to indicate the location, nature and extent of the work proposed and shown in detail that it will conform to the provisions of these provisions and all relevant laws, ordinances, rules and regulations. The building official shall determine what information is required on plans and specifications to ensure compliance.

SECTION AE303 PERMITS ISSUANCE

AE303.1 Issuance. The application, plans and specifications and other data filed by an applicant for permit shall be reviewed by the building official. Such plans may be reviewed by other departments of this jurisdiction to verify compliance with any applicable laws under their jurisdiction. If the building official finds that the work described in an application for a permit and the plans, specifications and other data filed therewith conform to the requirements of these provisions and other data filed therewith conform to the requirements of these provisions and other pertinent codes, laws and ordinances, and that the fees specified in Section AE304 have been paid, the building official shall issue a permit therefor to the applicant.

When the building official issues the permit where plans are required, the building official shall endorse in writing or stamp the plans and specifications APPROVED. Such approved plans and specifications shall not be changed, modified or altered without authorization from the building official, and all work shall be done in accordance with the approved plans.

AE303.2 Retention of plans. One set of approved plans and specifications shall be returned to the applicant and shall be kept on the site of the building or work at all times during which the work authorized thereby is in progress. One set of approved plans, specification and computations shall be retained by the building official until final approval of the work.

AE303.3 Validity of permit. The issuance of a permit or approval of plans and specifications shall shall not be construed to be a permit for, or an approval of, any violation of any of the provisions of these provisions or other pertinent codes of any other ordinance of the jurisdiction. No permit presuming to give authority to violate or cancel these provisions shall be valid.

The issuance of a permit based on plans, specifications and other data shall not prevent the building official from thereafter requiring the correction of errors in said plans, specifications and other data, or from preventing building operations being carried on thereunder when in violation of these provisions or of any other ordinances of this jurisdiction.

AE303.4 Expiration. Every permit issued by the building official under these provisions shall expire by limitation and become null and void if the work authorized by such permit is not commenced within 180 days from the date of such permit, or if the work authorized by such permit is suspended or abandoned at any time after the work is commenced for a period of 180 days. Before such work can be recommenced, a new permit shall be first obtained, and the fee therefor shall be one-half the amount required for a new permit for such work, provided no changes have been made or will be made in the original plans and specifications for such work, and provided further that such suspension or abandonment has not exceeded one year. In order to renew action on a permit after expiration, the permittee shall pay a new full permit fee.

Any permittee holding an unexpired permit may apply for an extension of the time within which work may commence under that permit when the permittee is unable to commence work within the time required by this section for good and satisfactory reasons. The building official may extend the time for action by the permittee for a period not exceeding 180 days upon written request by the permittee showing that circumstances beyond the control of the permittee have prevented action from being taken. No permit shall be extended more than once.

AE303.5 Suspension or revocation. The building official may, in writing, suspend or revoke a permit issued under these provisions whenever the permit is issued in error or on the basis of incorrect information supplied, or in violation of any ordinance or regulation or any of these provisions.

SECTION AE304 FEES

AE304.1 Permit fees. The fee for each manufactured home installation permit shall be established by the building official.

When permit fees are to be based on the value or valuation of the work to be performed, the determination of value or valuation under these provisions shall be made by the building official. The value to be used shall be the total value of all work required for the manufactured home installation plus the total value of all work required for the construction of accessory buildings and structures for which the permit is issued as well as all finish work, painting, roofing, electrical, plumbing, heating, air conditioning, elevators, fire-extinguishing systems and any other permanent equipment which is a part of the accessory building or structure. The value of the manufactured home itself shall not be included.

AE304.2 Plan review fees. When a plan or other data are required to be submitted by Section AE302.2, a plan review fee shall be paid at the time of submitting plans and specifications for review. Said plan review fee shall be as established by the building official. Where plans are incomplete or changed so as to require additional plan review, an additional plan review fee shall be charged at a rate as established by the building official.

AE304.3 Other provisions.

AE304.3.1 Expiration of plan review. Applications for which no permit is issued within 180 days following the date of application shall expire by limitation, and plans and other data submitted for review may thereafter be returned to the applicant or destroyed by the building official. The building official may extend the time for action by the applicant for a period not exceeding 180 days upon request by the applicant showing that circumstances beyond the control of the applicant have prevented action from being taken. No application shall be extended more than once. In order to renew action on an application after expiration, the applicant shall resubmit plans and pay a new plan review fee.

AE304.3.2 Investigation fees: work without a permit.

AE304.3.2.1 Investigation. Whenever any work for which a permit is required by these provisions has been commenced without first obtaining said permit, a special investigation shall be made before a permit may be issued for such work.

AE304.3.2.2 Fee. An investigation fee, in addition to the permit fee, shall be collected whether or not a permit is then or subsequently issued. The investigation fee shall be equal to the amount of the permit fee required. The minimum investigation fee shall be the same as the minimum fee established by the building official. The payment of such investigation fee shall not exempt any person from compliance with all other provisions of either these provisions or other pertinent codes or from any penalty prescribed by law.

E304.3.3 Fee refunds.

AE304.3.3.1 Permit fee erroneously paid or collected. The building official may authorize the refunding of any fee paid hereunder which was erroneously paid or collected.

AE304.3.3.2 Permit fee paid when no work done. The building official may authorize the refunding of not more than 80 percent of the permit fee paid when no work has been done under a permit issued in accordance with these provisions.

AE304.3.3.3 Plan review fee. The building official may authorize the refunding of not more than 80 percent of the plan review fee paid when an application for a permit for which a plan review fee has been paid is withdrawn or canceled before any plan reviewing is done.

The building official shall not authorize the refunding of any fee paid except upon written application by the original permittee not later than 180 days after the date of the fee payment.

SECTION AE305 INSPECTIONS

AE305.1 General. All construction or work for which a manufactured home installation permit is required shall be subject to inspection by the building official, and certain types of construction shall have continuous inspection by special inspectors as specified in Section AE306. A survey of the lot may be required by the building official to verify that the structure is located in accordance with the approved plans.

It shall be the duty of the permit applicant to cause the work to be accessible and exposed for inspection purposes. Neither the building official nor this jurisdiction shall be liable for expense entailed in the removal or replacement of any material required to allow inspection.

AE305.2 Inspection requests. It shall be the duty of the person doing the work authorized by a manufactured home installation permit to notify the building official that such work is ready for inspection. The building official may require that every request for inspection be filed at least one working day before such inspection is desired. Such request may be in writing or by telephone at the option of the building official.

It shall be the duty of the person requesting any inspections required either by these provisions or other applicable codes to provide access to and means for proper inspection of such work.

AE305.3 Inspection record card. Work requiring a manufactured home installation permit shall not be commenced until the permit holder or the permit holder's agent shall have posted an inspection record card in a conspicuous place on the premises and in such position as to allow the building official conveniently to make the required entries thereon regarding inspection of the work. This card shall be maintained in such position by the permit holder until final approval has been issued by the building official.

AE305.4 Approval required. Work shall not be done on any part of the manufactured home installation beyond the point indicated in each successive inspection without first obtaining the approval of the building official. Such approval shall be given only after an inspection has been made of each successive step in the construction as indicated by each of the inspections required in Section AE305.5. There shall be a final inspection and approval of the manufactured home installation, including connections to its building service equipment, when completed and ready for occupancy or use.

AE305.5 Required inspections.

AE305.5.1 Structural inspections for the manufactured home installation. Reinforcing steel or structural framework of any part of any manufactured home foundation system shall not be covered or concealed without first obtaining the approval of the building official. The building official, upon notification from the permit holder or the permit holder's agent, shall make the following inspections and shall either approve that portion of the construction as completed or shall notify the permit holder or the permit holder's agent wherein the same fails to comply with these provisions or other applicable codes:

1. Foundation inspection: To be made after excavations for footings are completed and any required reinforcing steel is in place. For concrete foundations, any required forms shall be in place prior to inspection. All materials for the foundation shall be on the job, except where concrete from a central mixing plant (commonly termed "transit mixed") is to be used, the concrete materials need not be on the job. Where the foundation is to be constructed of approved treated wood, additional framing inspections as required by the building official may be required.
2. Concrete slab or under-floor inspection: To be made after all in-slab or underfloor building service equipment, conduit, piping accessories and other ancillary equipment items are in place but before any concrete is poured or the manufactured home is installed.
3. Anchorage inspection: To be made after the manufactured home has been installed and permanently anchored.

AE305.5.2 Structural inspections for accessory building and structures. Inspections for accessory buildings and structures shall be made as set forth in this code.

AE305.5.3 Building service equipment inspections. All building service equipment which is required as a part of a manufactured home installation, including accessory buildings and structures authorized by the same permit, shall be inspected by the building official. Building service equipment shall be inspected and tested as required by the applicable codes. Such inspections and testing shall be limited to site construction and shall not include building service equipment which is a part of the manufactured home itself. No portion of any building service equipment intended to be concealed by any permanent portion of the construction shall be concealed until inspected and approved. Building service equipment shall not be connected to the water, fuel or power supply or sewer system until authorized by the building official.

AE305.5.4 Final inspection. When finish grading and the manufactured home installation, including the installation of all required building service equipment, is completed and the manufactured home is ready for occupancy, a final inspection shall be made.

AE305.6 Other inspections. In addition to the called inspections specified above, the building official may make or require other inspections of any construction work to ascertain compliance with these provisions or other codes and laws which are enforced by the code enforcement agency.

SECTION AE306
SPECIAL INSPECTIONS

AE306.1 General. In addition to the inspections required by Section AE305, the building official may require the owner to employ a special inspector during construction of specific types of work as described in this code.

SECTION AE307
UTILITY SERVICE

AE307.1 General. Utility service shall not be provided to any building service equipment which is regulated by these provisions or other applicable codes and for which a manufactured home installation permit is required by these provisions until approved by the building official.

SECTION AE401
OCCUPANCY CLASSIFICATION

AE401.1 Manufactured homes. A manufactured home shall be limited in use to use as a single dwelling unit.

AE401.2 Accessory buildings. Accessory buildings shall be classified as to occupancy by the building official as set forth in this code.

SECTION AE402
LOCATION ON PROPERTY

AE402.1 General. Manufactured homes and accessory buildings shall be located on the property in accordance with applicable codes and ordinances of this jurisdiction.

SECTION AE501
DESIGN

AE501.1 General. A manufactured home shall be installed on a foundation system which is designed and constructed to sustain within the stress limitations specified in this code and all loads specified in this code.

> **Exception:** When specifically authorized by the building official, foundation and anchorage systems which are constructed in accordance with the methods specified in Section AE600 of these provisions, or in the United States Department of Housing and Urban Development Handbook, *Permanent Foundations for Manufactured Housing,* 1984 Edition, Draft, shall be deemed to meet the requirements of this Appendix E.

AE501.2 Manufacturer's installation instructions. The installation instructions as provided by the manufacturer of the manufactured home shall be used to determine permissible points of support for vertical loads and points of attachment for anchorage systems used to resist horizontal and uplift forces.

AE501.3 Rationality. Any system or method of construction to be used shall admit to a rational analysis in accordance with well-established principles of mechanics.

SECTION AE502
FOUNDATION SYSTEMS

AE502.1 General. Foundation systems designed and constructed in accordance with this section may be considered as a permanent installation.

AE502.2 Soil classification. The classification of the soil at each manufactured home site shall be determined when required by the building official. The building official may require that the determination be made by an engineer or architect licensed by the state to conduct soil investigations.

The classification shall be based on observation and any necessary tests of the materials disclosed by borings or excavations made in appropriate locations. Additional studies may be necessary to evaluate soil strength, the effect of moisture variation on soil-bearing capacity, compressibility and expansiveness.

When required by the building official, the soil classification design bearing capacity and lateral pressure shall be shown on the plans.

AE502.3 Footings and foundations. Footings and foundations, unless otherwise specifically provided, shall be constructed of materials specified by this code for the intended use and in all cases shall extend below the frost line. Footings of concrete and masonry shall be of solid material. Foundations supporting untreated wood shall extend at least 8 inches (203 mm) above the adjacent finish grade. Footings shall have a minimum depth below finished grade of 12 inches (305 mm) unless a greater depth is recommended by a foundation investigation.

Piers and bearing walls shall be supported on masonry or concrete foundations or piles, or other approved foundation systems which shall be of sufficient capacity to support all loads.

AE502.4 Foundation design. When a design is provided, the foundation system shall be designed in accordance with the applicable structural provisions of this code and shall be designed to minimize differential settlement. Where a design is not provided, the minimum foundation requirements shall be as set forth in this code.

AE502.5 Drainage. Provisions shall be made for the control and drainage of surface water away from the manufactured home.

AE502.6 Under-floor clearances—ventilation and access. A minimum clearance of 12 inches (305 mm) shall be maintained beneath the lowest member of the floor support framing system. Clearances from the bottom of wood floor joists or perimeter joists shall be as specified in this code.

Under-floor spaces shall be ventilated with openings as specified in this code. If combustion air for one or more heat-producing appliances is taken from within the under-floor spaces, ventilation shall be adequate for proper appliance operation.

Under-floor access openings shall be provided. Such openings shall be not less than 18 inches (457 mm) in any dimension and not less than 3 square feet (0.279 m^2) in area and shall be located so that any water supply and sewer drain connections located under the manufactured home are accessible.

SECTION AE503
SKIRTING AND PERIMETER ENCLOSURES

AE503.1 Skirting and permanent perimeter enclosures. Skirting and permanent perimeter enclosures shall be installed only where specifically required by other laws or ordinances. Skirting, when installed, shall be of material suitable for exterior exposure and contact with the ground. Permanent perimeter enclosures shall be constructed of materials as required by this code for regular foundation construction.

Skirting shall be installed in accordance with the skirting manufacturer's installation instructions. Skirting shall be adequately secured to assure stability, to minimize vibration and susceptibility to wind damage, and to compensate for possible frost heave.

AE503.2 Retaining walls. Where retaining walls are used as a permanent perimeter enclosure, they shall resist the lateral displacements of soil or other materials and shall conform to this code as specified for foundation walls. Retaining walls and foundation walls shall be constructed of approved treated wood, concrete, masonry or other approved materials or combination of materials as for foundations as specified in this code. Siding materials shall extend below the top of the exterior of the retaining or foundation wall or the joint between siding and enclosure wall shall be flashed in accordance with this code.

SECTION AE504
STRUCTURAL ADDITIONS

AE504.1 General. Accessory buildings shall not be structurally supported by or attached to a manufactured home unless engineering calculations are submitted to substantiate any proposed structural connection.

> **Exception:** The building official may waive the submission of engineering calculations if it is found that the nature of the work applied for is such that engineering calculations are not necessary to show conformance to these provisions.

SECTION AE505
BUILDING SERVICE EQUIPMENT

AE505.1 General. The installation, alteration, repair, replacement, addition to or maintenance of the building service equipment within the manufactured home shall conform to regulations set forth in the Manufactured Home Standards. Such work which is located outside the manufactured home shall comply with the applicable codes adopted by this jurisdiction.

SECTION AE506
EXITS

AE506.1 Site development. Exterior stairways and ramps which provide egress to the public way shall comply with applicable provisions of this code.

AE506.2 Accessory buildings. Every accessory building or portion thereof shall be provided with exits as required by this code.

SECTION AE507
OCCUPANCY, FIRE SAFETY AND ENERGY CONSERVATION STANDARDS

AE507.1 General. Alterations made to a manufactured home subsequent to its initial installation shall conform to the occupancy, fire-safety and energy conservation requirements set forth in the Manufactured Home Standards.

SECTION AE600
SPECIAL REQUIREMENTS FOR FOUNDATION SYSTEMS

AE600.1 General. Section AE600 is applicable only when specifically authorized by the building official.

SECTION AE601
FOOTINGS AND FOUNDATIONS

AE601.1 General. The capacity of individual load-bearing piers and their footings shall be sufficient to sustain all loads specified in this code within the stress limitations specified in this code. Footings, unless otherwise approved by the building official, shall be placed level on firm, undisturbed soil or an engineered fill which is free of organic material, such as weeds and grasses. Where used, an engineered fill shall provide a minimum load-bearing capacity of not less than 1,000 psf (48 kN/m^2). Continuous footings shall conform to the requirements of this code. Section AE502 of these provisions shall apply to footings and foundations constructed under the provisions of this section.

SECTION AE602
PIER CONSTRUCTION

AE602.1 General. Piers shall be designed and constructed to distribute loads evenly. Multiple section homes may have concentrated roof loads which will require special consideration. Load-bearing piers may be constructed utilizing one of the methods listed below. Such piers shall be considered to resist only vertical forces acting in a downward direction. They shall not be considered as providing any resistance to horizontal loads induced by wind or earthquake forces.

1. A prefabricated load-bearing device that is listed and labeled for the intended use.
2. Mortar shall comply with ASTM C 270 Type M, S or N; this may consist of one part portland cement, one-half part hydrated lime and four parts sand by volume. Lime shall not be used with plastic or waterproof cement.
3. A cast-in-place concrete pier with concrete having specified compressive strength at 28 days of 2,500 psi (17 225 kPa).

Alternate materials and methods of construction may be used for piers which have been designed by an engineer or architect licensed by the state to practice as such.

Caps and leveling spacers may be used for leveling of the manufactured home. Spacing of piers shall be as specified in the manufacturer's installation instructions, if available, or by an approved designer.

SECTION AE603
HEIGHT OF PIERS

AE603.1 General. Piers constructed as indicated in Section AE602 may have heights as follows:

1. Except for corner piers, piers 36 inches (914 mm) or less in height may be constructed of masonry units, placed with cores or cells vertically. Piers shall be installed with their long dimension at right angles to the main frame member they support and shall have a minimum cross-sectional area of 128 square inches (82 560 mm^2). Piers shall be capped with minimum 4-inch (102 mm) solid masonry units or equivalent.
2. Piers between 36 and 80 inches (914 mm and 2032 mm) in height and all corner piers over 24 inches (610 mm) in height shall be at least 16 inches by 16 inches (406 mm by 406 mm) consisting of interlocking masonry units and shall be fully capped with minimum 4-inch (102 mm) solid masonry units or equivalent.
3. Piers over 80 inches (2032 mm) in height may be constructed in accordance with the provisions of Item 2 above, provided the piers shall be filled solid with grout and reinforced with four continuous No. 5 bars. One bar shall be placed in each corner cell of hollow masonry unit piers or in each corner of the grouted space of piers constructed of solid masonry units.
4. Cast-in-place concrete piers meeting the same size and height limitations of Items 1, 2 and 3 above may be substituted for piers constructed of masonry units.

SECTION AE604
ANCHORAGE INSTALLATIONS

AE604.1 Ground anchors. Ground anchors shall be designed and installed to transfer the anchoring loads to the ground. The load-carrying portion of the ground anchors shall be installed to the full depth called for by the manufacturer's installation directions and shall extend below the established frost line into undisturbed soil.

Manufactured ground anchors shall be listed and installed in accordance with the terms of their listing and the anchor manufacturer's instructions and shall include means of attachment of ties meeting the requirements of Section AE605. Ground anchor manufacturer's installation instructions shall include the amount of preload required and load capacity in various types of soil. These instructions shall include tensioning adjustments which may be needed to prevent damage to the manufactured home, particularly damage that can be caused by frost heave. Each ground anchor shall be marked with the manufacturer's identification and listed model identification number which shall be visible after installation. Instructions shall accompany each listed ground anchor specifying the types of soil for which the anchor is suitable under the requirements of this section.

Each approved ground anchor, when installed, shall be capable of resisting an allowable working load at least equal to

3,150 pounds (14 kN) in the direction of the tie plus a 50 percent overload [4,725 pounds (21 kN) total] without failure. Failure shall be considered to have occurred when the anchor moves more than 2 inches (51 mm) at a load of 4,725 pounds (21 kN) in the direction of the tie installation. Those ground anchors which are designed to be installed so that loads on the anchor are other than direct withdrawal shall be designed and installed to resist an applied design load of 3,150 pounds (14 kN) at 40 to 50 degrees from vertical or within the angle limitations specified by the home manufacturer without displacing the tie end of the anchor more than 4 inches (102 mm) horizontally. Anchors designed for connection of multiple ties shall be capable of resisting the combined working load and overload consistent with the intent expressed herein.

When it is proposed to use ground anchors and the building official has reason to believe that the soil characteristics at a given site are such as to render the use of ground anchors i advisable, or when there is doubt regarding the ability of the ground anchors to obtain their listed capacity, the building official may require that a representative field installation be made at the site in question and tested to demonstrate ground anchor capacity. The building official shall approve the test procedures.

AE604.2 Anchoring equipment. Anchoring equipment, when installed as a permanent installation, shall be capable of resisting all loads as specified within these provisions. When the stabilizing system is designed by an engineer or architect licensed by the state to practice as such, alternative designs may be used, providing the anchoring equipment to be used is capable of withstanding a load equal to 1.5 times the calculated load. All anchoring equipment shall be listed and labeled as being capable of meeting the requirements of these provisions. Anchors as specified in this code may be attached to the main frame of the manufactured home by an approved $^3/_{16}$-inch-thick (4.76 mm) slotted steel plate anchoring device. Other anchoring devices or methods meeting the requirements of these provisions may be permitted when approved by the building official.

Anchoring systems shall be so installed as to be permanent. Anchoring equipment shall be so designed to prevent self-disconnection with no hook ends used.

AE604.3 Resistance to weather deterioration. All anchoring equipment, tension devices and ties shall have a resistance to deterioration as required by this code.

AE604.4 Tensioning devices. Tensioning devices, such as turnbuckles or yoke-type fasteners, shall be ended with clevis or welded eyes.

SECTION AE605
TIES, MATERIALS AND INSTALLATION

AE605.1 General. Steel strapping, cable, chain or other approved materials shall be used for ties. All ties shall be fastened to ground anchors and drawn tight with turnbuckles or other adjustable tensioning devices or devices supplied with the ground anchor. Tie materials shall be capable of resisting an allowable working load of 3,150 pounds (14 kN) with no more than 2 percent elongation and shall withstand a 50 percent overload [4,750 pounds (21 kN)]. Ties shall comply with the weathering requirements of Section AE604.3. Ties shall connect the ground anchor and the main structural frame. Ties shall not connect to steel outrigger beams which fasten to and intersect the main structural frame unless specifically stated in the manufacturer's installation instructions. Connection of cable ties to main frame members shall be $^5/_8$-inch (15.9 mm) closed-eye bolts affixed to the frame member in an approved manner. Cable ends shall be secured with at least two U-bolt cable clamps with the "U" portion of the clamp installed on the short (dead) end of the cable to assure strength equal to that required by this section.

Wood floor support systems shall be fixed to perimeter foundation walls in accordance with provisions of this code. The minimum number of ties required per side shall be sufficient to resist the wind load stated in this code. Ties shall be evenly spaced as practicable along the length of the manufactured home with the distance from each end of the home and the tie nearest that end not exceeding 8 feet (2438 mm). When continuous straps are provided as vertical ties, such ties shall be positioned at rafters and studs. Where a vertical tie and diagonal tie are located at the same place, both ties may be connected to a single anchor, provided the anchor used is capable of carrying both loadings. Multisection manufactured homes require diagonal ties only. Diagonal ties shall be installed on the exterior main frame and slope to the exterior at an angle of 40 to 50 degrees from the vertical or within the angle limitations specified by the home manufacturer. Vertical ties which are not continuous over the top of the manufactured home shall be attached to the main frame.

APPENDIX F
RADON CONTROL METHODS

SECTION AF101
SCOPE

AF101.1 General. This appendix contains requirements for new construction in jurisdictions where radon-resistant construction is required.

Inclusion of this appendix by jurisdictions shall be determined through the use of locally available data or determination of Zone 1 designation in Figure AF101.

SECTION AF102
DEFINITIONS

AF102.1 General. For the purpose of these requirements, the terms used shall be defined as follows:

SUB-SLAB DEPRESSURIZATION SYSTEM (Passive). A system designed to achieve lower sub-slab air pressure relative to indoor air pressure by use of a vent pipe routed through the conditioned space of a building and connecting the sub-slab area with outdoor air, thereby relying on the convective flow of air upward in the vent to draw air from beneath the slab.

SUB-SLAB DEPRESSURIZATION SYSTEM (Active). A system designed to achieve lower sub-slab air pressure relative to indoor air pressure by use of a fan-powered vent drawing air from beneath the slab.

DRAIN TILE LOOP. A continuous length of drain tile o perforated pipe extending around all or part of the internal or external perimeter of a basement or crawl space footing.

RADON GAS. A naturally-occurring, chemically inert, radioactive gas that is not detectable by human senses. As a gas, it can move readily through particles of soil and rock and can accumulate under the slabs and foundations of homes where it can easily enter into the living space through construction cracks and openings.

SOIL-GAS-RETARDER. A continuous membrane of 6-mil (0.15 mm) polyethylene or other equivalent material used to retard the flow of soil gases into a building.

SUB-MEMBRANE DEPRESSURIZATION SYSTEM. A system designed to achieve lower-sub-membrane air pressure relative to crawl space air pressure by use of a vent drawing air from beneath the soil-gas-retarder membrane.

SECTION AF103
REQUIREMENTS

AF103.1 General. The following construction techniques are intended to resist radon entry and prepare the building for post-construction radon mitigation, if necessary (see Figure AF102). These techniques are required in areas where designated by the jurisdiction.

AF103.2 Subfloor preparation. A layer of gas-permeable material shall be placed under all concrete slabs and other floor systems that directly contact the ground and are within the walls of the living spaces of the building, to facilitate future installation of a sub-slab depressurization system, if needed. The gas-permeable layer shall consist of one of the following:

1. A uniform layer of clean aggregate, a minimum of 4 inches (102 mm) thick. The aggregate shall consist of material that will pass through a 2-inch (51 mm) sieve and be retained by a $^1/_4$-inch (6.4 mm) sieve.
2. A uniform layer of sand (native or fill), a minimum of 4 inches (102 mm) thick, overlain by a layer or strips of geotextile drainage matting designed to allow the lateral flow of soil gases.
3. Other materials, systems or floor designs with demonstrated capability to permit depressurization across the entire sub-floor area.

AF103.3 Soil-gas-retarder. A minimum 6-mil (0.15 mm) [or 3-mil (0.075 mm) cross-laminated] polyethylene or equivalent flexible sheeting material shall be placed on top of the gas-permeable layer prior to casting the slab or placing the floor assembly to serve as a soil-gas-retarder by bridging any cracks that develop in the slab or floor assembly and to prevent concrete from entering the void spaces in the aggregate base material. The sheeting shall cover the entire floor area with separate sections of sheeting lapped at least 12 inches (305 mm). The sheeting shall fit closely around any pipe, wire or other penetrations of the material. All punctures or tears in the material shall be sealed or covered with additional sheeting.

AF103.4 Entry routes. Potential radon entry routes shall be closed in accordance with Sections AF103.4.1 through AF103.4.10.

AF103.4.1 Floor openings. Openings around bathtubs, showers, water closets, pipes, wires or other objects that penetrate concrete slabs or other floor assemblies shall be filled with a polyurethane caulk or equivalent sealant applied in accordance with the manufacturer's recommendations.

AF103.4.2 Concrete joints. All control joints, isolation joints, construction joints and any other joints in concrete slabs or between slabs and foundation walls shall be sealed with a caulk or sealant. Gaps and joints shall be cleared of loose material and filled with polyurethane caulk or other elastomeric sealant applied in accordance with the manufacturer's recommendations.

AF103.4.3 Condensate drains. Condensate drains shall be trapped or routed through nonperforated pipe to daylight.

AF103.4.4 Sumps. Sump pits open to soil or serving as the termination point for sub-slab or exterior drain tile loops shall be covered with a gasketed or otherwise sealed lid. Sumps used as the suction point in a sub-slab depressurization system shall have a lid designed to accommodate the vent pipe. Sumps used as a floor drain shall have a lid equipped with a trapped inlet.

AF103.4.5 Foundation walls. Hollow block masonry foundation walls shall be constructed with either a continuous course of solid masonry, one course of masonry grouted solid, or a solid concrete beam at or above finished ground surface to prevent passage of air from the interior of the wall into the living space. Where a brick veneer or other masonry ledge is installed, the course immediately below that ledge shall be sealed. Joints, cracks or other openings around all penetrations of both exterior and interior surfaces of masonry block or wood foundation walls below the ground surface shall be filled with polyurethane caulk or equivalent sealant. Penetrations of concrete walls shall be filled.

AF103.4.6 Dampproofing. The exterior surfaces of portions of concrete and masonry block walls below the ground surface shall be dampproofed in accordance with Section R406 of this code.

AF103.4.7 Air-handling units. Air-handling units in crawl spaces shall be sealed to prevent air from being drawn into the unit.

Exception: Units with gasketed seams or units that are otherwise sealed by the manufacturer to prevent leakage.

AF103.4.8 Ducts. Ductwork passing through a crawl space or beneath a slab shall be of seamless material unless the air-handling system is designed to maintain continuous positive pressure within such ducting. Joints in such ductwork shall be sealed to prevent air leakage.

AF103.4.9 Crawl space floors. Openings around all penetrations through floors above crawl spaces shall be caulked or otherwise filled to prevent air leakage.

AF103.4.10 Crawl space access. Access doors and other openings or penetrations between basements and adjoining crawl spaces shall be closed, gasketed or otherwise filled to prevent air leakage.

AF103.5 Passive sub-membrane depressurization system. In buildings with crawl space foundations, the following components of a passive sub-membrane depressurization system shall be installed during construction.

Exception: Buildings in which an approved mechanical crawl space ventilation system or other equivalent system is installed.

AF103.5.1 Ventilation. Crawl spaces shall be provided with vents to the exterior of the building. The minimum net area of ventilation openings shall comply with Section R408.1 of this code.

AF103.5.2 Soil-gas-retarder. The soil in crawl spaces shall be covered with a continuous layer of minimum 6-mil (0.15 mm) polyethylene soil-gas-retarder. The ground cover shall be lapped a minimum of 12 inches (305 mm) at joints and shall extend to all foundation walls enclosing the crawl space area.

AF103.5.3 Vent pipe. A plumbing tee or other approved connection shall be inserted horizontally beneath the sheeting and connected to a 3- or 4-inch-diameter (76 mm or 102 mm) fitting with a vertical vent pipe installed through the sheeting. The vent pipe shall be extended up through the building floors, terminate at least 12 inches (305 mm) above the roof in a location at least10 feet (3048 mm) away from any window or other opening into the conditioned spaces of the building that is less than 2 feet (610 mm) below the exhaust point, and 10 feet (3048 mm) from any window or other opening in adjoining or adjacent buildings.

AF103.6 Passive sub-slab depressurization system. In basement or slab-on-grade buildings, the following components of a passive sub-slab depressurization system shall be installed during construction.

AF103.6.1 Vent pipe. A minimum 3-inch-diameter (76 mm) ABS, PVC or equivalent gas-tight pipe shall be embedded vertically into the sub-slab aggregate or other permeable material before the slab is cast. A "T" fitting or equivalent method shall be used to ensure that the pipe opening remains within the sub-slab permeable material. Alternatively, the 3-inch (76 mm) pipe shall be inserted directly into an interior perimeter drain tile loop or through a sealed sump cover where the sump is exposed to the sub-slab aggregate or connected to it through a drainage system.

The pipe shall be extended up through the building floors, terminate at least 12 inches (305 mm) above the surface of the roof in a location at least 10 feet (3048 mm) away from any window or other opening into the conditioned spaces of the building that is less than 2 feet (610 mm) below the exhaust point, and 10 feet (3048 mm) from any window or other opening in adjoining or adjacent buildings.

AF103.6.2 Multiple vent pipes. In buildings where interior footings or other barriers separate the sub-slab aggregate or other gas-permeable material, each area shall be fitted with an individual vent pipe. Vent pipes shall connect to a single vent that terminates above the roof or each individual vent pipe shall terminate separately above the roof.

AF103.7 Vent pipe drainage. All components of the radon vent pipe system shall be installed to provide positive drainage to the ground beneath the slab or soil-gas-retarder.

AF103.8 Vent pipe accessibility. Radon vent pipes shall be accessible for future fan installation through an attic or other area outside the habitable space.

Exception: The radon vent pipe need not be accessible in an attic space where an approved roof-top electrical supply is provided for future use.

AF103.9 Vent pipe identification. All exposed and visible interior radon vent pipes shall be identified with at least one label on each floor and in accessible attics. The label shall read: "Radon Reduction System."

AF103.10 Combination foundations. Combination basement/crawl space or slab-on-grade/crawl space foundations shall have separate radon vent pipes installed in each type of foundation area. Each radon vent pipe shall terminate above the roof or shall be connected to a single vent that terminates above the roof.

AF103.11 Building depressurization. Joints in air ducts and plenums in unconditioned spaces shall meet the requirements of Section M1601. Thermal envelope air infiltration requirements shall comply with the energy conservation provisions in Chapter 11. Firestopping shall meet the requirements contained in Section R602.8.

AF103.12 Power source. To provide for future installation of an active sub-membrane or sub-slab depressurization system, an electrical circuit terminated in an approved box shall be installed during construction in the attic or other anticipated location of vent pipe fans. An electrical supply shall also be accessible in anticipated locations of system failure alarms.

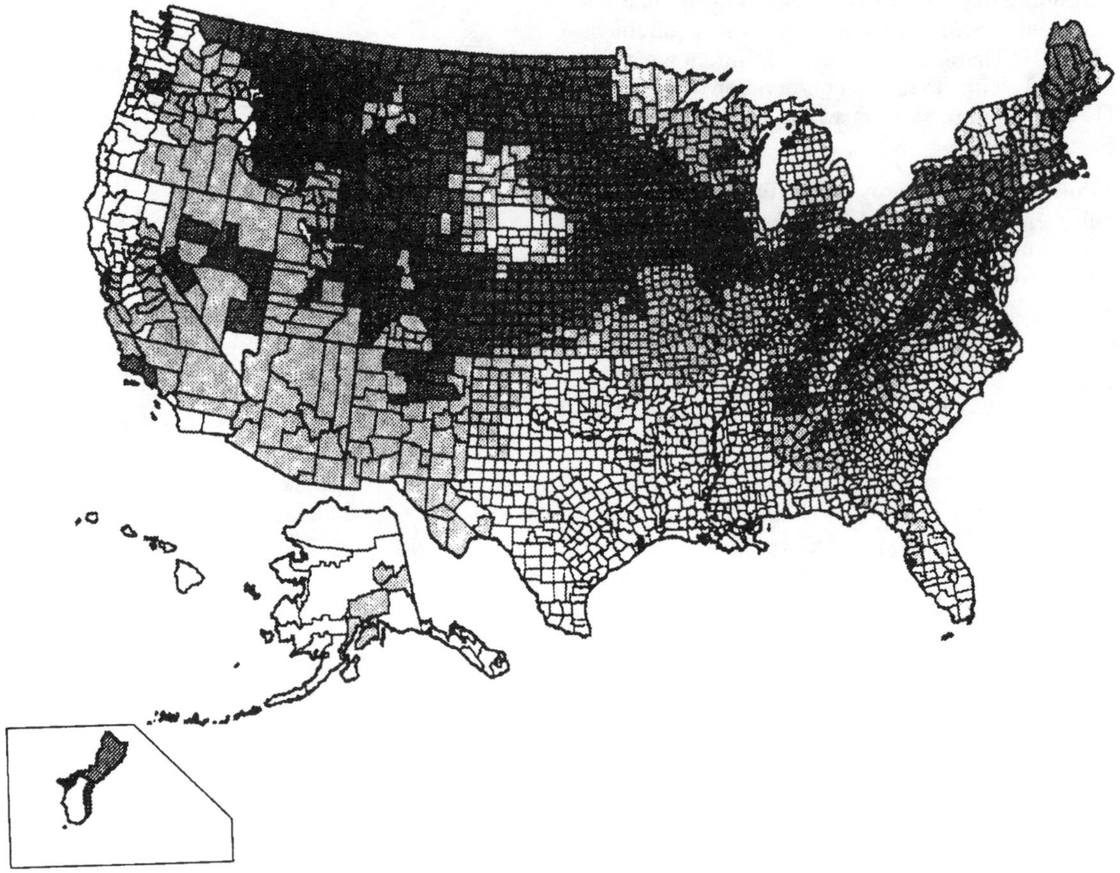

LEGEND

ZONE 1 HIGH POTENTIAL (GREATER THAN 4 pCi/L[a])

ZONE 2 MODERATE POTENTIAL (FROM 2 TO 4 pCi/L)

ZONE 3 LOW POTENTIAL (LESS THAN 2 pCi/L)

a. pCi/L standard for picocuries per liter of radon gas. EPA recommends that all homes that measure 4 pCi/L and greater be mitigated.

The United States Environmental Protection Agency and the United States Geological Survey have evaluated the radon potential in the United States and have developed a map of radon zones designed to assist building building officials in deciding whether radon-resistant features are applicable in new construction.

The map assigns each of the 3,141 counties in the United States to one of three zones based on radon potential. Each zone designation reflects the average short-term radon measurement that can be expected to be measured in a building without the implementation of radon control methods. The radon zone designation of highest priority is Zone 1. Table 1 of this appendix lists the Zone 1 counties illustrated on the map. More detailed information can be obtained from state-specific booklets (EPA-402-R-93-021 through 070) available through State Radon Offices or from U.S. EPA Regional Offices.

FIGURE AF101
EPA MAP OF RADON ZONES

TABLE AF101(1) HIGH RADON POTENTIAL (ZONE 1) COUNTIES[a]

ALABAMA
- Calhoun
- Clay
- Cleburne
- Colbert
- Coosa
- Franklin
- Jackson
- Lauderdale
- Lawrence
- Limestone
- Madison
- Morgan
- Talladega

CALIFORNIA
- Santa Barbara
- Ventura

COLORADO
- Adams
- Arapahoe
- Baca
- Bent
- Boulder
- Chaffee
- Cheyenne
- Clear Creek
- Crowley
- Custer
- Delta
- Denver
- Dolores
- Douglas
- El Paso
- Elbert
- Fremont
- Garfield
- Gilpin
- Grand
- Gunnison
- Huerfano
- Jackson
- Jefferson
- Kiowa
- Kit Carson
- Lake
- Larimer
- Las Animas
- Lincoln
- Logan
- Mesa
- Moffat
- Montezuma
- Montrose
- Morgan
- Otero
- Ouray
- Park
- Phillips
- Pitkin
- Prowers
- Pueblo
- Rio Blanco
- San Miguel
- Summit
- Teller
- Washington
- Weld
- Yuma

CONNECTICUT
- Fairfield
- Middlesex
- New Haven
- New London

GEORGIA
- Cobb
- De Kalb
- Fulton
- Gwinnett

IDAHO
- Benewah
- Blaine
- Boise
- Bonner
- Boundary
- Butte
- Camas
- Clark
- Clearwater
- Custer
- Elmore
- Fremont
- Gooding
- Idaho
- Kootenai
- Latah
- Lemhi
- Shoshone
- Valley

ILLINOIS
- Adams
- Boone
- Brown
- Bureau
- Calhoun
- Carroll
- Cass
- Champaign
- Coles
- De Kalb
- De Witt
- Douglas
- Edgar
- Ford
- Fulton
- Greene
- Grundy
- Hancock
- Henderson
- Henry
- Iroquois
- Jersey
- Jo Daviess
- Kane
- Kendall
- Knox
- La Salle
- Lee
- Livingston
- Logan
- Macon
- Marshall
- Mason
- McDonough
- McLean
- Menard
- Mercer
- Morgan
- Moultrie
- Ogle
- Peoria
- Piatt
- Pike
- Putnam
- Rock Island
- Sangamon
- Schuyler
- Scott
- Stark
- Stephenson
- Tazewell
- Vermilion
- Warren
- Whiteside
- Winnebago
- Woodford

INDIANA
- Adams
- Allen
- Bartholomew
- Benton
- Blackford
- Boone
- Carroll
- Cass
- Clark
- Clinton
- De Kalb
- Decatur
- Delaware
- Elkhart
- Fayette
- Fountain
- Fulton
- Grant
- Hamilton
- Hancock
- Harrison
- Hendricks
- Henry
- Howard
- Huntington
- Jay
- Jennings
- Johnson
- Kosciusko
- Lagrange
- Lawrence
- Madison
- Marion
- Marshall
- Miami
- Monroe
- Montgomery
- Noble
- Orange
- Putnam
- Randolph
- Rush
- Scott
- Shelby
- Steuben
- St. Joseph
- Tippecanoe
- Tipton
- Union
- Vermillion
- Wabash
- Warren
- Washington
- Wayne
- Wells
- White
- Whitley

IOWA
- All Counties

KANSAS
- Atchison
- Barton
- Brown
- Cheyenne
- Clay
- Cloud
- Decatur
- Dickinson
- Douglas
- Ellis
- Ellsworth
- Finney
- Ford
- Geary
- Gove
- Graham
- Grant
- Gray
- Greeley
- Hamilton
- Haskell
- Hodgeman
- Jackson
- Jewell
- Johnson
- Kearny
- Kingman
- Kiowa
- Lane
- Leavenworth
- Lincoln
- Logan
- Marion
- Marshall
- McPherson
- Meade
- Mitchell
- Nemaha
- Ness
- Norton
- Osborne
- Ottawa
- Pawnee
- Phillips
- Pottawatomie
- Pratt
- Rawlins
- Republic
- Rice
- Riley
- Rooks
- Rush
- Russell
- Saline
- Scott
- Sheridan
- Sherman
- Smith
- Stanton
- Thomas
- Trego
- Wallace
- Washington
- Wichita
- Wyandotte

KENTUCKY
- Adair
- Allen
- Barren
- Bourbon
- Boyle
- Bullitt
- Casey
- Clark
- Cumberland
- Fayette
- Franklin
- Green
- Harrison
- Hart
- Jefferson
- Jessamine
- Lincoln
- Marion
- Mercer
- Metcalfe
- Monroe
- Nelson
- Pendleton
- Pulaski
- Robertson
- Russell
- Scott
- Taylor
- Warren
- Woodford

MAINE
- Androscoggin
- Aroostook
- Cumberland
- Franklin
- Hancock
- Kennebec
- Lincoln
- Oxford
- Penobscot
- Piscataquis
- Somerset
- York

MARYLAND
- Baltimore
- Calvert
- Carroll
- Frederick
- Harford
- Howard
- Montgomery
- Washington

MASS.
- Essex
- Middlesex
- Worcester

MICHIGAN
- Branch
- Calhoun
- Cass
- Hillsdale
- Jackson
- Kalamazoo
- Lenawee
- St. Joseph
- Washtenaw

MINNESOTA
- Becker
- Big Stone
- Blue Earth
- Brown
- Carver
- Chippewa
- Clay
- Cottonwood
- Dakota
- Dodge
- Douglas
- Faribault
- Fillmore
- Freeborn
- Goodhue
- Grant
- Hennepin
- Houston
- Hubbard
- Jackson
- Kanabec
- Kandiyohi
- Kittson
- Lac Qui Parle
- Le Sueur
- Lincoln
- Lyon
- Mahnomen
- Marshall
- Martin
- McLeod
- Meeker
- Mower
- Murray
- Nicollet
- Nobles
- Norman
- Olmsted
- Otter Tail
- Pennington
- Pipestone
- Polk
- Pope
- Ramsey
- Red Lake
- Redwood
- Renville
- Rice
- Rock
- Roseau
- Scott
- Sherburne
- Sibley
- Stearns
- Steele
- Stevens
- Swift
- Todd
- Traverse
- Wabasha
- Wadena
- Waseca
- Washington
- Watonwan
- Wilkin
- Winona
- Wright
- Yellow Medicine

MISSOURI
- Andrew
- Atchison
- Buchanan
- Cass
- Clay
- Clinton
- Holt
- Iron
- Jackson
- Nodaway
- Platte

MONTANA
- Beaverhead
- Big Horn
- Blaine
- Broadwater
- Carbon
- Carter
- Cascade
- Chouteau
- Custer
- Daniels
- Dawson
- Deer Lodge
- Fallon
- Fergus
- Flathead
- Gallatin
- Garfield
- Glacier
- Granite
- Hill
- Jefferson
- Judith Basin
- Lake
- Lewis and Clark
- Liberty
- Lincoln
- Madison
- McCone
- Meagher
- Mineral
- Missoula
- Park
- Phillips
- Pondera
- Powder River
- Powell
- Prairie
- Ravalli
- Richland
- Roosevelt
- Rosebud
- Sanders
- Sheridan
- Silver Bow
- Stillwater
- Teton
- Toole
- Valley
- Wibaux
- Yellowstone National Park

NEBRASKA
- Adams
- Boone
- Boyd
- Burt
- Butler
- Cass
- Cedar
- Clay
- Colfax
- Cuming
- Dakota
- Dixon
- Dodge
- Douglas
- Fillmore
- Franklin
- Frontier
- Furnas
- Gage
- Gosper
- Greeley
- Hamilton
- Harlan
- Hayes
- Hitchcock
- Hurston
- Jefferson
- Johnson
- Kearney
- Knox
- Lancaster
- Madison
- Nance

a. EPA recommends that this county listing be supplemented with other available State and local data to further understand the radon potential of Zone 1 areas.

(continued)

TABLE AF101(1)—continued
HIGH RADON POTENTIAL (ZONE 1) COUNTIES[a]

Nemaha
Nuckolls
Otoe
Pawnee
Phelps
Pierce
Platte
Polk
Red Willow
Richardson
Saline
Sarpy
Saunders
Seward
Stanton
Thayer
Washington
Wayne
Webster
York

NEW HAMPSHIRE
Carroll

NEW JERSEY
Hunterdon
Mercer
Monmouth
Morris
Somerset
Sussex
Warren

NEW MEXICO
Bernalillo
Colfax
Mora
Rio Arriba
San Miguel
Santa Fe
Taos

NEVADA
Carson City
Douglas
Eureka
Lander
Lincoln
Lyon
Mineral
Pershing
White Pine

NEW YORK
Albany
Allegheny
Broome
Cattaraugus
Cayuga
Chautauqua
Chemung
Chenango
Columbia
Cortland
Delaware
Dutchess
Erie
Genesee
Greene
Livingston
Madison
Onondaga
Ontario
Orange
Otsego
Putnam
Rensselaer
Schoharie
Schuyler
Seneca
Steuben
Sullivan
Tioga
Tompkins
Ulster
Washington
Wyoming
Yates

N. CAROLINA
Allegheny
Buncombe
Cherokee
Henderson
Mitchell
Rockingham
Transylvania
Watauga

N. DAKOTA
All Counties

OHIO
Adams
Allen
Ashland
Auglaize
Belmont
Butler
Carroll
Champaign
Clark
Clinton
Columbiana
Coshocton
Crawford
Darke
Delaware
Fairfield
Fayette
Franklin
Greene
Guernsey
Hamilton
Hancock
Hardin
Harrison
Holmes
Huron
Jefferson
Knox
Licking
Logan
Madison
Marion
Mercer
Miami
Montgomery
Morrow
Muskingum
Perry
Pickaway
Pike
Preble
Richland
Ross
Seneca
Shelby
Stark
Summit
Tuscarawas
Union
Van Wert
Warren
Wayne
Wyandot

PENNSYLVANIA
Adams
Allegheny
Armstrong
Beaver
Bedford
Berks
Blair
Bradford
Bucks
Butler
Cameron
Carbon
Centre
Chester
Clarion
Clearfield
Clinton
Columbia
Cumberland
Dauphin
Delaware
Franklin
Fulton
Huntingdon
Indiana
Juniata
Lackawanna
Lancaster
Lebanon
Lehigh
Luzerne
Lycoming
Mifflin
Monroe
Montgomery
Montour
Northampton
Northumberland
Perry
Schuylkill
Snyder
Sullivan
Susquehanna
Tioga
Union
Venango
Westmoreland
Wyoming
York

RHODE ISLAND
Kent
Washington

S. CAROLINA
Greenville

S. DAKOTA
Aurora
Beadle
Bon Homme
Brookings
Brown
Brule
Buffalo
Campbell
Charles Mix
Clark
Clay
Codington
Corson
Davison
Day
Deuel
Douglas
Edmunds
Faulk
Grant
Hamlin
Hand
Hanson
Hughes
Hutchinson
Hyde
Jerauld
Kingsbury
Lake
Lincoln
Lyman
Marshall
McCook
McPherson
Miner
Minnehaha
Moody
Perkins
Potter
Roberts
Sanborn
Spink
Stanley
Sully
Turner
Union
Walworth
Yankton

TENNESSEE
Anderson
Bedford
Blount
Bradley
Claiborne
Davidson
Giles
Grainger
Greene
Hamblen
Hancock
Hawkins
Hickman
Humphreys
Jackson
Jefferson
Knox
Lawrence
Lewis
Lincoln
Loudon
Marshall
Maury
McMinn
Meigs
Monroe
Moore
Perry
Roane
Rutherford
Smith
Sullivan
Trousdale
Union
Washington
Wayne
Williamson
Wilson

UTAH
Carbon
Duchesne
Grand
Piute
Sanpete
Sevier
Uintah

VIRGINIA
Allegheny
Amelia
Appomattox
Augusta
Bath
Bland
Botetourt
Bristol
Brunswick
Buckingham
Buena Vista
Campbell
Chesterfield
Clarke
Clifton Forge
Covington
Craig
Cumberland
Danville
Dinwiddie
Fairfax
Falls Church
Fluvanna
Frederick
Fredericksburg
Giles
Goochland
Harrisonburg
Henry
Highland
Lee
Lexington
Louisa
Martinsville
Montgomery
Nottoway
Orange
Page
Patrick
Pittsylvania
Powhatan
Pulaski
Radford
Roanoke
Rockbridge
Rockingham
Russell
Salem
Scott
Shenandoah
Smyth
Spotsylvania
Stafford
Staunton
Tazewell
Warren
Washington
Waynesboro
Winchester
Wythe

WASHINGTON
Berkeley
Brooke
Clark
Ferry
Grant
Greenbrier
Hampshire
Hancock
Hardy
Jefferson
Marshall
Mercer
Mineral
Monongalia
Monroe
Morgan
Ohio
Okanogan
Pend Oreille
Pendleton
Pocahontas
Preston
Skamania
Spokane
Stevens
Summers
Wetzel

WISCONSIN
Buffalo
Crawford
Dane
Dodge
Door
Fond du Lac
Grant
Green
Green Lake
Iowa
Jefferson
Lafayette
Langlade
Marathon
Menominee
Pepin
Pierce
Portage
Richland
Rock
Shawano
St. Croix
Vernon
Walworth
Washington
Waukesha
Waupaca
Wood

WYOMING
Albany
Big Horn
Campbell
Carbon
Converse
Crook
Fremont
Goshen
Hot Springs
Johnson
Laramie
Lincoln
Natrona
Niobrara
Park
Sheridan
Sublette
Sweetwater
Teton
Uinta
Washakie

a. EPA recommends that this county listing be supplemented with other available State and local data to further understand the radon potential of Zone 1 areas.

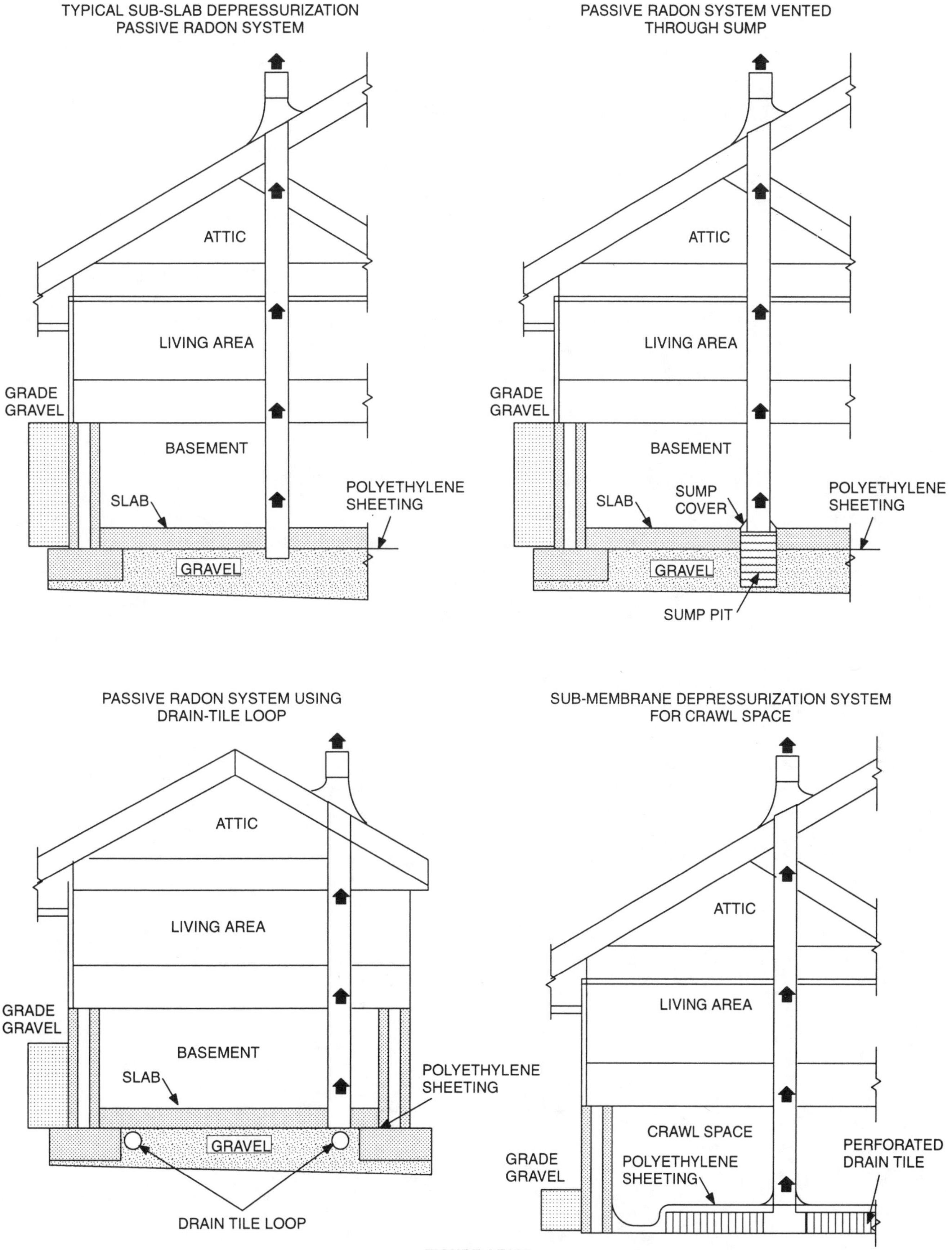

FIGURE AF102
RADON-RESISTANT CONSTRUCTION DETAILS FOR FOUR FOUNDATION TYPES

APPENDIX G
SWIMMING POOLS, SPAS AND HOT TUBS

SECTION AG101 GENERAL

AG101.1 General. The provisions of this appendix shall control the design and construction of swimming pools, spas and hot tubs installed in or on the lot of a one- and two-family dwelling.

SECTION AG102 DEFINITIONS

AG102.1 General. For the purposes of these requirements, the terms used shall be defined as follows and as set forth in Chapter 2.

ABOVE-GROUND/ON-GROUND POOL. See "Swimming pool."

BARRIER. A fence, wall, building wall or combination thereof which completely surrounds the swimming pool and obstructs access to the swimming pool.

HOT TUB. See "Swimming pool."

IN-GROUND POOL. See "Swimming pool."

RESIDENTIAL. That which is situated on the premises of a detached one- or two-family dwelling or a one-family townhouse not more than three stories in height.

SPA, NONPORTABLE. See "Swimming pool."

SPA, PORTABLE. A nonpermanent structure intended for recreational bathing, in which all controls, water-heating and water-circulating equipment are an integral part of the product.

SWIMMING POOL. Any structure intended for swimming or recreational bathing that contains water over 24 inches (610 mm) deep. This includes in-ground, aboveground and on-ground swimming pools, hot tubs and spas.

SWIMMING POOL, INDOOR. A swimming pool which is totally contained within a structure and surrounded on all four sides by walls of said structure.

SWIMMING POOL, OUTDOOR. Any swimming pool which is not an indoor pool.

SECTION AG103 SWIMMING POOLS

AG103.1 In-ground pools. In-ground pools shall be designed and constructed in conformance with ANSI/NSPI-5 as listed in Section AG107.

AG103.2 Above-ground and on-ground pools. Above-ground and on-ground pools shall be designed and constructed in conformance with ANSI/NSPI-4 as listed in Section AG107.

SECTION AG104 SPAS AND HOT TUBS

AG104.1 Permanently installed spas and hot tubs. Permanently installed spas and hot tubs shall be designed and constructed in conformance with ANSI/NSPI-3 as listed in Section AG107.

AG104.2 Portable spas and hot tubs. Portable spas and hot tubs shall be designed and constructed in conformance with ANSI/NSPI-6 as listed in Section AG107.

SECTION AG105 BARRIER REQUIREMENTS

AG105.1 Application. The provisions of this chapter shall control the design of barriers for residential swimming pools, spas and hot tubs. These design controls are intended to provide protection against potential drownings and near-drownings by restricting access to swimming pools, spas and hot tubs.

AG105.2 Outdoor swimming pool. An outdoor swimming pool, including an in-ground, aboveground or on-ground pool, hot tub or spa shall be provided with a barrier which shall comply with the following:

1. The top of the barrier shall be at least 48 inches (1219 mm) above grade measured on the side of the barrier which faces away from the swimming pool. The maximum vertical clearance between grade and the bottom of the barrier shall be 2 inches (51 mm) measured on the side of the barrier which faces away from the swimming pool. Where the top of the pool structure is above grade, such as an aboveground pool, the barrier may be at ground level, such as the pool structure, or mounted on top of the pool structure. Where the barrier is mounted on top of the pool structure, the maximum vertical clearance between the top of the pool structure and the bottom of the barrier shall be 4 inches (102 mm).
2. Openings in the barrier shall not allow passage of a 4-inch-diameter (102 mm) sphere.
3. Solid barriers which do not have openings, such as a masonry or stone wall, shall not contain indentations or protrusions except for normal construction tolerances and tooled masonry joints.
4. Where the barrier is composed of horizontal and vertical members and the distance between the tops of the horizontal members is less than 45 inches (1143 mm), the horizontal members shall be located on the swimming pool side of the fence. Spacing between vertical members shall not exceed 1.75 inches (44 mm) in width. Where there are decorative cutouts within vertical members, spacing within the cutouts shall not exceed 1.75 inches (44 mm) in width.
5. Where the barrier is composed of horizontal and vertical members and the distance between the tops of the horizontal members is 45 inches (1143 mm) or more, spacing

between vertical members shall not exceed 4 inches (102 mm). Where there are decorative cutouts within vertical members, spacing within the cutouts shall not exceed 1.75 inches (44 mm) in width.

6. Maximum mesh size for chain link fences shall be a 1.25-inch (32 mm) square unless the fence is provided with slats fastened at the top or the bottom which reduce the openings to not more than 1.75 inches (44 mm).
7. Where the barrier is composed of diagonal members,such as a lattice fence, the maximum opening formed by the diagonal members shall not be more than 1.75 inches (44 mm).
8. Access gates shall comply with the requirements of Section AG105.2, Items 1 through 7, and shall be equipped to accommodate a locking device. Pedestrian access gates shall open outward away from the pool and shall be self-closing and have a self-latching device. Gates other than pedestrian access gates shall have a self-latching device. Where the release mechanism of the self-latching device is located less than 54 inches (1372 mm) from the bottom of the gate, the release mechanism and openings shall comply with the following:
   - 8.1. The release mechanism shall be located on the pool side of the gate at least 3 inches (76 mm) below the top of the gate, and
   - 8.2. The gate and barrier shall have no opening greater than 0.5 inch (12.7 mm) within 18 inches (457 mm) of the release mechanism.
9. Where a wall of a dwelling serves as part of the barrier one of the following conditions shall be met:
   - 9.1. The pool shall be equipped with a powered safety cover in compliance with ASTM F1346; or
   - 9.2. All doors with direct access to the pool through that wall shall be equipped with an alarm which produces an audible warning when the door and its screen, if present, are opened. The alarm shall sound continuously for a minimum of 30 seconds immediately after the door is opened and be capable of being heard throughout the house during normal household activities. The alarm shall automatically reset under all conditions. The alarm system shall be equipped with a manual means, such as touchpad or switch, to temporarily deactivate the alarm for a single opening. Such deactivation shall last for not more than 15 seconds. The deactivation switch(es) shall be located at least 54 inches (1372 mm) above the threshold of the door; or
   - 9.3. Other means of protection, such as self-closing doors with self-latching devices, which are approved by the governing body, shall be acceptable so long as the degree of protection afforded is not less than the protection afforded by Item 9.1 or 9.2 described above.
10. Where an aboveground pool structure is used as a barrier or where the barrier is mounted on top of the pool structure, and the means of access is a ladder or steps, then:
   - 10.1. The ladder or steps shall be capable of being secured, locked or removed to prevent access, or
   - 10.2. The ladder or steps shall be surrounded by a barrier which meets the requirements of Section AG105.2, Items 1 through 9. When the ladder or steps are secured, locked or removed, any opening created shall not allow the passage of a 4-inch-diameter (102 mm) sphere.

AG105.3 Indoor swimming pool. All walls surrounding an indoor swimming pool shall comply with Section AG105.2, Item 9.

AG105.4 Prohibited locations. Barriers shall be located so as to prohibit permanent structures, equipment or similar objects from being used to climb the barriers.

AG105.5 Barrier exceptions. Spas or hot tubs with a safety cover which complies with ASTM F 1346, as listed in Section AG107, shall be exempt from the provisions of this appendix.

SECTION AG106 ABBREVIATIONS

AG106.1 General.

ANSI—American National Standards Institute
11 West 42nd Street, New York, NY 10036

ASTM—American Society for Testing and Materials
1916 Race Street, Philadelphia, PA 19103

NSPI—National Spa and Pool Institute
2111 Eisenhower Avenue, Alexandria, VA 22314

SECTION AG107 STANDARDS

AG107.1 General.

ANSI/NSPI

ANSI/NSPI-3 Standard for Permanently Installed Residential Spas AG104.1

ANSI/NSPI-4 Standard for Above-ground/On-ground Residential Swimming Pools AG103.2

ANSI/NSPI-5 Standard for Residential In-ground Swimming Pools AG103.1

ANSI/NSPI-6 Standard for Residential Portable Spas AG104.2

ASTM

ASTM F 1346-91 Standard Performance Specifications for Safety Covers and Labeling Requirements for All Covers for Swimming Pools, Spas and Hot Tubs AG105.2, AG105.5

APPENDIX H
PATIO COVERS

SECTION AH101
GENERAL

AH101.1 Scope. Patio covers shall conform to the requirements of this appendix chapter.

SECTION AH102
DEFINITION

Patio covers. One-story structures not exceeding 12 feet (3657 mm) in height. Enclosure walls shall be permitted to be of any configuration, provided the open or glazed area of the longer wall and one additional wall is equal to at least 65 percent of the area below a minimum of 6 feet 8 inches (2032 mm) of each wall, measured from the floor. Openings shall be permitted to be enclosed with (1) insect screening, (2) approved translucent or transparent plastic not more than 0.125 inch (3.2 mm) in thickness, (3) glass conforming to the provisions of Section R308, or (4) any combination of the foregoing.

SECTION AH103
PERMITTED USES

AH103.1 General. Patio covers shall be permitted to be detached from or attached to dwelling units. Patio covers shall be used only for recreational, outdoor living purposes and not as carports, garages, storage rooms or habitable rooms.

SECTION AH104
DESIGN LOADS

AH104.1 General. Patio covers shall be designed and constructed to sustain, within the stress limits of this code, all dead loads plus a minimum vertical live load of 10 pounds per square foot (0.48 kN/m^2) except that snow loads shall be used where such snow loads exceed this minimum. Such covers shall be designed to resist the minimum wind loads set forth in Table R301.2(1).

SECTION AH105
LIGHT AND VENTILATION/EMERGENCY EGRESS

AH105.1 General. Exterior openings required for light and ventilation shall be permitted to open into a patio structure conforming to Section AH101, provided that the patio structure shall be unenclosed if such openings are serving as emergency egress or rescue openings from sleeping rooms. Where such exterior openings serve as an exit from the dwelling unit, the patio structure, unless unenclosed, shall be provided with exits conforming to the provisions of Section R310 of this code.

SECTION AH106
FOOTINGS

AH106.1 General. In areas with a frostline depth of zero as specified in Table R301.2(1), a patio cover shall be permitted to be supported on a slab on grade without footings, provided the slab conforms to the provisions of Section R505 of this code, is not less than 3.5 inches (89 mm) thick and the columns do not support live and dead loads in excess of 750 pounds (3.34 kN) per column.

APPENDIX I
PRIVATE SEWAGE DISPOSAL

SECTION AI101
GENERAL

AI101.1 Scope. Private sewage disposal systems shall conform to the *International Private Sewage Disposal Code.*

APPENDIX J
EXISTING BUILDINGS AND STRUCTURES

SECTION AJ101
PURPOSE AND INTENT

AJ101.1 General. The purpose of these provisions is to encourage the continued use or reuse of legally existing buildings and structures. These provisions are intended to permit work in existing buildings that is consistent with the purpose of the *International Residential Code*. Compliance with these provisions shall be deemed to meet the requirements of the *International Residential Code*.

AJ101.2 Classification of work. For purposes of this appendix, all work in existing buildings shall be classified into the categories of repair, renovation, alteration and reconstruction. Specific requirements are established for each category of work in these provisions.

AJ101.3 Multiple categories of work. Work of more than one category may be part of a single work project. All related work permitted within a 12-month period shall be considered a single work project. Where a project includes one category of work in one building area and another category of work in a separate and unrelated area of the building, each project area shall comply with the requirements of the respective category of work. Where a project with more than one category of work is performed in the same area or in related areas of the building, the project shall comply with the requirements of the more stringent category of work.

SECTION AJ102
COMPLIANCE

AJ102.1 General. Regardless of the category of work being performed, the work shall not cause the structure to become unsafe or adversely affect the performance of the building; shall not cause an existing mechanical or plumbing system to become unsafe, hazardous, insanitary or overloaded; and unless expressly permitted by these provisions, shall not make the building any less conforming to this code or to any previously approved alternative arrangements than it was before the work was undertaken.

AJ102.2 Requirements by category of work. Repairs shall conform to the requirements of Section AJ301. Renovations shall conform to the requirements of Section AJ401. Alterations shall conform to the requirements of Section AJ501 and the requirements for renovations. Reconstructions shall conform to the requirements of Section AJ601 and the requirements for alterations and renovations.

AJ102.3 Smoke detectors. Regardless of the category of work, smoke detectors shall be provided where required by Section R317.1.1.

AJ102.4 Replacement windows. Regardless of the category of work, when an entire existing window, including frame, sash and glazed portion is replaced, the replacement window shall comply with the requirements of Section N1102.4.

AJ102.5 Flood Hazard areas. Work performed in existing buildings located in a flood hazard area as established by Table R301.2(1) shall be subject to the provisions of R105.3.1.1.

AJ102.6 Equivalent alternatives. These provisions are not intended to prevent the use of any alternate material, alternate design or alternate method of construction not specifically prescribed herein, provided any alternate has been deemed to be equivalent and its use authorized by the building official.

AJ102.7 Other alternatives. Where compliance with these provisions or with this code as required by these provisions is technically infeasible or would impose disproportionate costs because of structural, construction or dimensional difficulties, other alternatives may be accepted by the building official. These alternatives may include materials, design features and/or operational features.

AJ102.8 More restrictive requirements. Buildings or systems in compliance with the requirements of this code for new construction shall not be required to comply with any more restrictive requirement of these provisions.

AJ102.9 Features exceeding International Residential Code requirements. Elements, components and systems of existing buildings with features that exceed the requirements of this code for new construction, and are not otherwise required as part of approved alternative arrangements or deemed by the building official to be required to balance other building elements not complying with this code for new construction, shall not be prevented by these provisions from being modified as long as they remain in compliance with the applicable requirements for new construction.

SECTION AJ103
PRELIMINARY MEETING

AJ103.1 General. If a building permit is required at the request of the prospective permit applicant, the building official or his designee shall meet with the prospective applicant to discuss plans for any proposed work under these provisions prior to the application for the permit. The purpose of this preliminary meeting is for the building official to gain an understanding of the prospective applicant's intentions for the proposed work, and to determine, together with the prospective applicant, the specific applicability of these provisions.

SECTION AJ104
EVALUATION OF AN EXISTING BUILDING

AJ104.1 General. The building official may require an existing building to be investigated and evaluated by a registered design professional in the case of proposed reconstruction of any portion of a building. The evaluation shall determine the existence of any potential nonconformities with these provisions, and shall provide a basis for determining the impact of the proposed changes on the performance of the building. The evaluation shall utilize the following sources of information, as applicable:

1. Available documentation of the existing building.
   1.1. Field surveys.
   1.2. Tests (nondestructive and destructive).
   1.3. Laboratory analysis.

Exception: Detached one- or two-family dwellings that are not irregular buildings under Section R301.2.2.6 and are not undergoing an extensive reconstruction shall not be required to be evaluated.

SECTION AJ105 PERMIT

AJ105.1 Identification of work area. The work area shall be clearly identified on all permits issued under these provisions.

SECTION AJ201 DEFINITIONS

AJ201.1 General. For purposes of this appendix, the terms used shall be defined as follows.

ALTERATION. The reconfiguration of any space, the addition or elimination of any door or window, the reconfiguration or extension of any system, or the installation of any additional equipment.

CATEGORIES OF WORK. The nature and extent of construction work undertaken in an existing building. The categories of work covered in this Appendix, listed in increasing order of stringency of requirements, are repair, renovation, alteration and reconstruction.

DANGEROUS. Where the stresses in any member; the condition of the building, or any of its components or elements or attachments; or other condition that results in an overload exceeding 150 percent of the stress allowed for the member or material in this code.

EQUIPMENT OR FIXTURE. Any plumbing, heating, electrical, ventilating, air conditioning, refrigerating and fire protection equipment, and elevators, dumb waiters, boilers, pressure vessels, and other mechanical facilities or installations that are related to building services.

LOAD-BEARING ELEMENT. Any column, girder, beam, joist, truss, rafter, wall, floor or roof sheathing that supports any vertical load in addition to its own weight, and/or any lateral load.

MATERIALS AND METHODS REQUIREMENTS. Those requirements in this code that specify material standards, details of installation and connection, joints, penetrations, and continuity of any element, component or system in the building. The required quantity, fire resistance, flame spread, acoustic or thermal performance, or other performance attribute is specifically excluded from materials and methods requirements.

RECONSTRUCTION. The reconfiguration of a space that affects an exit, a renovation and/or alteration when the work area is not permitted to be occupied because existing means of egress and fire protection systems, or their equivalent, are not in place or continuously maintained; and/or there are extensive alterations as defined in Section AJ501.3.

REHABILITATION. Any repair, renovation, alteration or reconstruction work undertaken in an existing building.

RENOVATION. The change, strengthening or addition of load-bearing elements; and/or the refinishing, replacement, bracing, strengthening, upgrading or extensive repair of existing materials, elements, components, equipment and/or fixtures. Renovation involves no reconfiguration of spaces. Interior and exterior painting are not considered refinishing for purposes of this definition, and are not renovation.

REPAIR. The patching, restoration and/or minor replacement of materials, elements, components, equipment and/or fixtures for the purposes of maintaining such materials, elements, components, equipment and/or fixtures in good or sound condition.

WORK AREA. That portion of a building affected by any renovation, alteration or reconstruction work as initially intended by the owner and indicated as such in the permit. Work area excludes other portions of the building where incidental work entailed by the intended work must be performed, and portions of the building where work not initially intended by the owner is specifically required by these provisions for a renovation, alteration or reconstruction.

SECTION AJ301 REPAIRS

AJ301.1 Materials. Except as otherwise required herein, work shall be done using like materials or materials permitted by this code for new construction.

AJ301.1.1 Hazardous materials. Hazardous materials no longer permitted, such as asbestos and lead-based paint, shall not be used.

AJ301.1.2 Plumbing materials and supplies. The following plumbing materials and supplies shall not be used:

1. All-purpose solvent cement, unless listed for the specific application;
2. Flexible traps and tailpieces, unless listed for the specific application; and
3. Solder having more than 0.2 percent lead in the repair of potable water systems.

AJ301.2 Water closets. When any water closet is replaced with a newly manufactured water closet, the replacement water closet shall comply with the requirements of Section P2903.2.

AJ301.3 Safety glazing. Replacement glazing in hazardous locations shall comply with the safety glazing requirements of Section R308.1.

AJ301.4 Electrical. Existing electrical wiring and equipment undergoing repair shall be permitted to be repaired or replaced with like material.

Exceptions:

1. Replacement of electrical receptacles shall comply with the requirements of Chapters 33 through 42.
2. Plug fuses of the Edison-base type shall be used for replacements only where there is no evidence of overfusing or tampering per the applicable requirements of Chapters 33 through 42.
3. For replacement of nongrounding-type receptacles with grounding-type receptacles and for branch circuits that do not have an equipment grounding conductor in the branch circuitry, the grounding conductor of a grounding type receptacle outlet shall be permitted to be grounded to any accessible point on the grounding electrode system, or to any accessible point on the grounding electrode conductor, as allowed and described in Chapters 33 through 42.

SECTION AJ401 RENOVATIONS

AJ401.1 Materials and methods. The work shall comply with the materials and methods requirements of this code.

AJ401.2 Door and window dimensions. Minor reductions in the clear opening dimensions of replacement doors and windows that result from the use of different materials shall be allowed, whether or not they are permitted by this code.

AJ401.3 Interior finish. Wood paneling and textile wall coverings used as an interior finish shall comply with the flame spread requirements of Section R319.

AJ401.4 Structural. Unreinforced masonry buildings located in Seismic Design Categories D_2 or E shall have parapet bracing and wall anchors installed at the roofline whenever a reroofing permit is issued. Such parapet bracing and wall anchors shall be of an approved design.

SECTION AJ501 ALTERATIONS

AJ501.1 Newly constructed elements. All newly constructed elements, components and systems shall comply with the requirements of this code.

Exceptions:

1. Openable windows may be added without requiring compliance with the light and ventilation requirements of Section R303.
2. Newly installed electrical equipment shall comply with the requirements of Section AJ501.5.

AJ501.2 Nonconformities. The work shall not increase the extent of noncompliance with the requirements of Section AJ601, or create nonconformity with those requirements which did not previously exist.

AJ501.3 Extensive alterations. When the total area of all the work areas included in an alteration exceeds 50 percent of the area of the dwelling unit, the work shall be considered as a reconstruction and shall comply with the requirements of these provisions for reconstruction work.

Exception: Work areas in which the alteration work is exclusively plumbing, mechanical or electrical shall not be included in the computation of total area of all work areas.

AJ501.4 Structural. The minimum design loads for the structure shall be the loads applicable at the time the building was constructed, provided that no dangerous condition is created. Structural elements that are uncovered during the course of the alteration and that are found to be unsound or dangerous shall be made to comply with the applicable requirements of this code.

AJ501.5 Electrical equipment and wiring.

AJ501.5.1 Materials and methods. All newly installed electrical equipment and wiring relating to work done in any work area shall comply with the materials and methods requirements of Chapters 33 through 42.

Exception: Electrical equipment and wiring in newly installed partitions and ceilings shall comply with all applicable requirements of Chapters 33 through 42.

AJ501.5.2 Electrical service. Service to the dwelling unit shall be a minimum of 100 ampere, three-wire capacity and service equipment shall be dead front having no live parts exposed whereby accidental contact could be made. Type "S" fuses shall be installed when fused equipment is used.

Exception: Existing service of 60 ampere, three-wire capacity, and feeders of 30 ampere or larger two- or three-wire capacity shall be accepted if adequate for the electrical load being served.

AJ501.5.3 Additional electrical requirements. When the work area includes any of the following areas within a dwelling unit, the requirements of Sections AJ501.5.3.1 through AJ501.5.3.5 shall apply.

AJ501.5.3.1 Enclosed areas. All enclosed areas other than closets, kitchens, basements, garages, hallways, laundry areas and bathrooms shall have a minimum of two duplex receptacle outlets, or one duplex receptacle outlet and one ceiling or wall type lighting outlet.

AJ501.5.3.2 Kitchen and laundry areas. Kitchen areas shall have a minimum of two duplex receptacle outlets. Laundry areas shall have a minimum of one duplex receptacle outlet located near the laundry equipment and installed on an independent circuit.

AJ501.5.3.3 Ground-fault circuit-interruption. Ground fault circuit interruption shall be provided on newly installed receptacle outlets if required by Chapters 33 through 42.

AJ501.5.3.4 Lighting outlets. At least one lighting outlet shall be provided in every bathroom, hallway, stairway, attached garage and detached garage with electric power to illuminate outdoor entrances and exits, and in utility rooms and basements where these spaces are used for storage or contain equipment requiring service.

AJ501.5.3.5 Clearance. Clearance for electrical service equipment shall be provided in accordance with Chapters 33 through 42.

AJ501.6 Ventilation. All reconfigured spaces intended for occupancy and all spaces converted to habitable or occupiable space in any work area shall be provided with ventilation in accordance with Section R303.

SECTION AJ601
RECONSTRUCTION

AJ601.1 Stairways, handrails and guards.

AJ601.1.1 Stairways. Stairways within the work area shall be provided with illumination in accordance with Section R303.4.

AJ601.1.2 Handrails. Every required exit stairway that has three or more risers; is part of the means of egress for any work area; and is not provided with at least one handrail, or in which the existing handrails are judged to be in danger of collapsing, shall be provided with handrails designed and installed in accordance with Section R315 for the full length of the run of steps on at least one side.

AJ601.1.3 Guards. Every open portion of a stair, landing or balcony that is more than 30 inches (762 mm) above the floor or grade below; is part of the egress path for any work area; and is not provided with guards or in which the existing guards are judged to be in danger of collapsing, shall be provided with guards designed and installed in accordance with Section R316.

AJ601.2 Wall and ceiling finish. The interior finish of walls and ceilings in any work area shall comply with the requirements of Section R319. All existing interior finish materials that do not comply with those requirements shall be removed or shall be treated with an approved fire retardant coating in accordance with the manufacturer's instructions to secure compliance with the requirements of this section.

AJ601.3 Separation walls. Where the work area is in an attached dwelling unit, all walls separating dwelling units that are not continuous from the foundation to the underside of the roof sheathing shall be constructed to provide a continuous fire separation using construction materials consistent with the existing wall or complying with the requirements for new structures. Work shall only be required to be performed on the side of the wall of the dwelling unit that is part of the work area.

APPENDIX K
SOUND TRANSMISSION

SECTION AK101
GENERAL

AK 101.1 General. Wall and floor-ceiling assemblies separating dwelling units shall provide airborne sound insulation for walls, and both airborne and impact sound insulation for floor-ceiling assemblies.

SECTION AK102
AIRBORNE SOUND

AK102.1 General. Airborne sound insulation for wall and floor-ceiling assemblies shall meet a Sound Transmission Class (STC) rating of 45 when tested in accordance with ASTM E 90.

SECTION AK103
STRUCTURAL-BORNE SOUND

AK103.1 General. Floor/ceiling assemblies between dwelling units or between a dwelling unit and a public or service area within a structure shall have an Impact Insulation Class (IIC) rating of not less than 45 when tested in accordance with ASTM E 492.

APPENDIX L

INTERNATIONAL RESIDENTIAL CODE ELECTRICAL PROVISIONS/ NATIONAL ELECTRICAL CODE CROSS REFERENCE

This table is a cross-reference of the *International Residential Code,* Chapters 33 through 42, and the 1999 *National Electrical Code* (NFPA 70-1999).

International Residential Code		National Electrical Code
CHAPTER 33	**GENERAL REQUIREMENTS**	
SECTION E3301	**GENERAL**	
E3301.1	Applicability	None
E3301.2	Scope	Sections 90-1 and 90-2
E3301.3	Not covered	Section 90-2
E3301.4	Additions and alterations	None
SECTION E3302	**BUILDING STRUCTURE PROTECTION**	
E3302.1	Drilling and notching	None
E3302.2	Penetrations of fire-resistance-rated assemblies	Section 300-21
E3302.3	Penetrations of firestops and draftstops	Section 300-21
SECTION E3303	**INSPECTION AND APPROVAL**	
E3303.1	Approval	Section 110-2
E3303.2	Inspection required	None
E3303.3	Listing and labeling	Section 110-3
SECTION E3304	**GENERAL EQUIPMENT REQUIREMENTS**	
E3304.1	Voltages	Section 110-4
E3304.2	Interrupting rating	Section 110-9
E3304.3	Circuit characteristics	Section 110-10
E3304.4	Protection of equipment	Section 110-11
E3304.5	Unused openings	Section 110-12(a)
E3304.6	Integrity of electrical equipment	Section 110-12(c)
E3304.7	Mounting	Section 110-13(a)
E3304.8	Energized parts guarded against accidental contact	Section 110-27(a)
E3304.9	Prevent physical damage	Section 110-27(b)
E3304.10	Equipment identification	Section 110-21
E3304.11	Identification of disconnecting means	Section 110-22
SECTION E3305	**EQUIPMENT LOCATION AND CLEARANCES**	
E3305.1	Working space and clearances	Section 110-26
Figure E3305.1	Working space and clearances	Sections 110-26
	Footnote a.	Section 110-26(f)
	Footnote b.	Section 110-26(e)
	Footnote c.	Section 110-26(b)
	Footnote d.	Sections 230-70(a), 240-24(d) and 240-24(e)
	Footnote e.	Section 110-26(d)
E3305.2	Working clearances for energized equipment and panelboards	Sections 110-26(a), 110-26(d), and 110-26(f)
E3305.3	Clearances over panelboards	Section 110-26(f)(1)(a)
E3305.4	Location of clear spaces	Sections 110-26(b), 230-70(a) and 240-24
E3305.5	Access and entrance to working space	Section 110-26(c)
E3305.6	Illumination	Section 110-26(d)
E3305.7	Headroom	Section 110-26(e)
SECTION E3306	**ELECTRICAL CONDUCTORS AND CONNECTIONS**	
E3306.1	General	Article 110 and Section 310-1
E3306.2	Conductor material	Section 110-5

International Residential Code		National Electrical Code
E3306.3	Minimum size of conductors	Section 310-5
E3306.4	Stranded conductors	Section 310-3
E3306.5	Individual conductor insulation	Section 310-2(a)
E3306.6	Conductors in parallel	Section 310-4
E3306.7	Conductors of the same circuit	Section 300-3(b)
E3306.8	Aluminum and copper connections	Section 110-14
E3306.9	Terminals	Section 110-14(a)
E3306.10	Splices	Section 110-14(b)
E3306.10.1	Continuity	Section 300-13(a)
	Exception	Section 300-13(a)
E3306.10.2	Device connections	Sections 250-148 and 300-13(b)
E3306.10.3	Length of conductor for splice or termination	Section 300-14
SECTION E3307	**CONDUCTOR AND TERMINAL IDENTIFICATION**	
E3307.1	Grounded conductors	Sections 200-6(a), 200-6(b) and 310-12(a)
E3307.2	Equipment grounding conductors	Sections 250-119 and 310-12(b)
E3307.3	Ungrounded conductors	Section 310-12(c)
	Exception	Section 200-7(c)(2)
E3307.4	Identification of terminals	Section 200-10
E3307.4.1	Device terminals	Section 200-10(a)
E3307.4.2	Receptacles, plugs, and connectors	Section 200-10(b)
E3307.5	Tag marking	Section 310-11(b)(3)
CHAPTER 34	**ELECTRICAL DEFINITIONS**	
SECTION E3401	**GENERAL**	**Article 100, Definitions**
	Arc-fault circuit-interrupter	210-12(a)
CHAPTER 35	**SERVICES**	
SECTION 3501	**GENERAL SERVICES**	
E3501.1	Scope	Section 230-1
E3501.2	Number of services	Section 230-2
E3501.3	One building or other structure not to be supplied through another	Section 230-3
E3501.4	Other conductors in raceway or cable	Section 230-7
E3501.5	Raceway seal	Section 230-8
E3501.6	Service disconnect required	Section 230-70
E3501.6.1	Marking of service equipment and disconnects	Sections 230-66 and 230-70(b)
E3501.6.2	Service disconnect location	Sections 230-70(a) and 230-72(c)
E3501.7	Maximum number of disconnects	Section 230-71(a)
SECTION E3502	**SERVICE SIZE AND RATING**	
E3502.1	Rating of ungrounded conductors	Section 230-79(c) and (d)
E3502.2	Service load	Section 220-30
Table E3502.2	Minimum service load calculation	Table 220-30
E3502.2.1	Services under 100 amperes	None
E3502.3	Rating of service disconnect	Section 230-79
E3502.4	Voltage rating	Section 220-30(a)
SECTION E3503	**SERVICE, FEEDER AND GROUNDING ELECTRODE CONDUCTOR SIZING**	
E3503.1	Grounded and ungrounded service conductor size	310-15(b)(6)
Table E3503.1	Service conductor and grounding electrode conductor sizing	Table 310-16 and Table 250-66
	Footnote a.	Section 250-64(e)
	Footnotes b. and c.	Section 250-64(b)
	Footnote d.	Section 250-66(a) and (b)
E3503.2	Ungrounded service conductors for accessory buildings and structures	Section 230-42(b)

International Residential Code		National Electrical Code
	Exceptions 1,2, and 3	Section 230-42(b) and 230-79
E3503.3	Overload protection	Section 230-90
E3503.3.1	Ungrounded conductor	Section 230-90(a)
	Exception	Section 230-90(a), Exception No. 3
E3503.3.2	Not in grounded conductor	Section 230-90(b)
E3503.4	Grounding electrode conductor size	Section 250-66
SECTION E3504	**OVERHEAD SERVICE-DROP AND SERVICE CONDUCTOR INSTALLATION**	
E3504.1	Clearance from building openings	Section 230-9
Figure E3504.1	Clearances from building openings	Section 230-9
E3504.2	Vertical clearances	Section 230-24
E3504.2.1	Above roofs	Section 230-24(a)
	Exception 1	Section 230-24(a), Exception No. 1
	Exception 2	Section 230-24(a), Exception No. 2
	Exception 3	Section 230-24(a), Exception No. 3
	Exception 4	Section 230-24(a), Exception No. 4
Figure E3504.2.1	Clearances from roofs	Section 230-24
E3504.2.2	Vertical clearance from grade	Section 230-24(b)
	Items 1, 2, and 3	Section 230-24(b)
E3504.3	Point of attachment	Section 230-26
E3504.4	Means of attachment	Section 230-27
E3504.5	Service masts as supports	Section 230-28
E3504.6	Supports over buildings	Section 230-29
SECTION E3505	**SERVICE-ENTRANCE CONDUCTORS**	
E3505.1	Insulation of service-entrance conductors	Section 230-41
	Exceptions 1 and 2	Section 230-41, Exception
E3505.2	Wiring methods for services	Section 230-43
E3505.3	Spliced conductors	Section 230-46
E3505.4	Protection against physical damage	Section 230-49
E3505.5	Protection of service cables against damage	Section 230-50(a)
E3505.6	Mounting supports	Section 230-51
E3505.7	Raceways to drain	Section 230-53
E3505.8	Overhead service locations	Section 230-54
E3505.8.1	Raintight service head	Section 230-54(a)
E3505.8.2	Service cable, service head or gooseneck	Section 230-54(b)
E3505.8.3	Service head location	Section 230-54(c)
	Exception	Section 230-54(c), Exception
E3505.8.4	Separately bushed openings	Section 230-54(e)
E3505.8.5	Drip loops	Section 230-54(f)
E3505.8.6	Conductor arrangement	Section 230-54(g)
E3505.8.7	Secured	Section 230-54(d)
E3505.9	Termination at service equipment	Section 230-55
SECTION E3506	**SERVICE EQUIPMENT—GENERAL**	
E3506.1	Service equipment enclosures	Section 230-62
E3506.2	Working space	Section 110-26
E3506.3	Available short-circuit current	none
E3506.4	Marking	Section 230-66
SECTION E3507	**SYSTEM GROUNDING**	
E3507.1	System service ground	Sections 250-20(b)(1) and 250-24(a)
E3507.2	Location of grounding electrode conductor connection	Section 250-24(a)(1) and (a)(5)
E3507.3	Two or more buildings or structures supplied from a common service	250-32(a)
	Exception	Section 250-32(a), Exception
E3507.3.1	Equipment grounding conductor	Section 250-32(b)(1) and Table 250-122

International Residential Code		National Electrical Code
E3507.3.2	Grounded conductor.	Section 250-32(b)(2)
E3507.4	Grounding electrode conductor	Section 250-24(c)
E3507.5	Main bonding jumper	Section 250-28
E3507.6	Common grounding electrode	Section 250-58
SECTION E3508	**GROUNDING ELECTRODE SYSTEM**	
E3508.1	Grounding electrode system	Section 250-50
E3508.1.1	Metal underground water pipe	Section 250-50(a)
E3508.1.2	Concrete-encased electrode	Section 250-50(c)
E3508.2	Made and other electrodes	Section 250-52
E3508.2.1	Rod and pipe electrodes	Section 250-52(c)(1) and (2)
E3508.2.2	Installation	Section 250-52(c)(2)
E3508.2.3	Aluminum electrodes	Section 250-52(e)
E3508.3	Resistance of made electrodes	Section 250-56
E3508.4	Metal underground gas piping system	Section 250-104(b)
SECTION E3509	**BONDING**	
E3509.1	General	Section 250-90
E3509.2	Bonding of service equipment	Section 250-92(a)
E3509.3	Bonding to other systems	Section 250-92(b)
E3509.4	Method of bonding at the service	Section 250-94
E3509.4.1	Grounded service conductor	Section 250-94(1)
E3509.4.2	Threaded connections	Section 250-94(2)
E3509.4.3	Threadless couplings and connectors	Section 250-94(3)
E3509.4.4	Other devices	Section 250-94(4)
E3509.5	Sizing bonding jumper on supply side of service and main bonding jumper	Section 250-102(c)
E3509.6	Metal water piping bonding	Section 250-104(a)
E3509.7	Metal gas piping bonding	Section 250-104(b)
E3509.8	Bonding other metal piping	Section 250-104(c)
SECTION E3510	**GROUNDING ELECTRODE CONDUCTORS**	
E3510.1	Installation	Section 250-64(a) and (b)
E3510.2	Enclosures for grounding electrode conductors	Section 250-64(e)
SECTION E3511	**GROUNDING ELECTRODE CONDUCTOR CONNECTION TO GROUNDING ELECTRODES**	
E3511.1	Methods of grounding connection to electrodes	Section 250-70
E3511.2	Accessibility	Section 250-68
E3511.3	Effective grounding path	Section 250-68
E3511.4	Protection of ground clamps and fittings	Section 250-10
E3511.5	Clean surfaces	Section 250-12
CHAPTER 36	**BRANCH CIRCUIT AND FEEDER REQUIREMENTS**	
SECTION E3601	**GENERAL**	
E3601.1	Scope	None
E3601.2	Branch-circuit and feeder ampacity	Sections 210-19(a) and 215-2
E3601.3	Multioutlet branch circuits	210-19(b)
E3601.4	Multiwire branch circuits	Section 210-4
SECTION E3602	**BRANCH CIRCUIT RATINGS**	
E3602.1	Branch-circuit voltage limitations	Sections 210-6(a) and (b)
E3602.2	Branch-circuit ampere rating	Section 210-3
E3602.3	Fifteen- and 20-ampere branch circuits	Section 210-23(a)
E3602.4	Thirty-ampere branch circuits	Section 210-23(b)
E3602.5	Branch circuits serving multiple loads or outlets	Section 210-23(a)
E3602.6	Branch circuits serving a single motor	Section 430-22(a)
E3602.7	Branch circuits serving motor-operated and combination loads	Section 220-4(a)

International Residential Code		National Electrical Code
E3602.8	Branch-circuit inductive lighting loads	Section 210-22(b)220-4(b)
E3602.9	Branch-circuit load for ranges and cooking appliances	Table 220-19, Note 4
E3602.9.1	Minimum branch circuit for ranges	Section 210-19(c)
E3602.10	Branch circuits serving heating loads	Sections 422-13 and 424-3(a) and (b)
E3602.11	Branch circuits for air-conditioning and heat pump equipment	Sections 440-35 and 440-52(a)(3)
E3602.12	Branch circuits serving room air conditioners	Section 440-62(a)
E3602.12.1	Where no other loads are supplied	Section 440-62(b)
E3602.12.2	Where lighting units or other appliances are also supplied	Section 440-62(c)
E3602.13	Branch-circuit requirement—summary	Section 210-24 and 210-25
Table E3602.13	Branch-circuit requirement—summary	Table 210-24
SECTION E3603	**REQUIRED BRANCH CIRCUITS**	
E3603.1	Branch circuits for heating	Section 422-12
E3603.2	Kitchen and dining area receptacles	Section 210-11(c)(1)
E3603.3	Laundry circuit	Sections 210-23(a), Exception and 210-11(c)(2)
E3603.4	Bathroom branch circuits	Section 210-11(c)(3)
E3603.5	Number of branch circuits	Section 210-11(a)
E3603.6	Branch-circuit load proportioning	Section 210-11(b)
SECTION E3604	**FEEDER REQUIREMENTS**	
E3604.1	Conductor size	Table 310-15(b)(6), Sections 215-2 and 220-10(a)
E3604.2	Minimum feeder conductor size	Section 215-2(b)
E3604.3	Feeder loads	Table 220-11
Table E3604.3(1)	Feeder load calculation	Table 220-11, Table 220-30(c), Sections 220-14, 220-15, 220-18 and 220-19
Table E3604.3(2)	Demand loads for electric ranges, wall-mounted ovens, counter-mounted cooking units and other cooking appliances over 13/4 kVA rating	Table 220-19
E3604.4	Feeder neutral load	Section 220-22
E3604.5	Lighting and convenience receptacle load	Section 220-3(a)
E3604.6	Ampacity and computed loads	Section 220-10(a)
E3604.7	Feeder and branch-circuit conductors	Section 240-21(a)
SECTION E3605	**CONDUCTOR SIZING AND OVERCURRENT PROTECTION**	
E3605.1	General	310-15(a)
Table E3605.1	Allowable ampacities	Table 310-16
E3605.2	Correction factors for ambient temperatures	Table 310-16
Table E3605.2	Ambient temperature correction factors	Table 310-16, Correction factors
E3605.3	Adjustment for conductor proximity	Section 310-15(b)(2)(a)
Table E3605.3	Conductor proximity adjustment factors	Table 310-15(b)(2)(a)
E3605.4	Temperature limitations	Section 110-14(c)
E3605.4.1	Conductors rated 60°C.	Section 110-14(c)(1)
E3605.4.2	Conductors rated 75°C.	Section 110-14(c)(2)
E3605.4.3	Separately installed pressure connectors	Section 110-14(c)(3)
E3605.4.4	Conductors of type NM cable	Section 33630(b) and 336-26
E3605.5	Overcurrent protection required	Table 310-15(b)(6), Sections 240-3 and 240-4
E3605.5.1	Cords	Section 240-4
E3605.5.2	Overcurrent devices of the next higher size	240-3(b)
E3605.5.3	Small conductors	240-3(d)
E3605.5.4	Air conditioning and heat pump equipment	240-3(g)
E3605.6	Fuses and fixed trip circuit breakers	Section 240-6
E3605.7	Overcurrent device location	Section 240-24(e)
E3605.8	Location of overcurrent devices in or on premises	Section 240-24
E3605.9	Ready access for occupants	Section 240-24(b)
E3605.10	Enclosures for overcurrent devices	Section 240-240-30

International Residential Code		National Electrical Code
SECTION E3606	**PANELBOARDS**	
E3606.1	Panelboard rating	Section 384-13
E3606.2	Panelboard circuit identification	Section 384-13
E3606.3	Panelboard overcurrent protection	Section 384-16(a)
CHAPTER 37	**WIRING METHODS**	
SECTION E3701	**GENERAL REQUIREMENTS**	
E3701.1	Scope	Section 300-1
E3701.2	Allowable wiring methods	Sections 110-8 and 300-3
Table E3701.2	Allowable wiring methods	None
E3701.3	Circuit conductors	Section 300-3(b)
E3701.4	Wiring method applications	Chapter 3 and Section 300-2
Table E3701.4	Allowable applications for wiring methods	Chapter 3 and Section 300-2
SECTION E3702	**ABOVEGROUND INSTALLATION REQUIREMENTS**	
E3702.1	Installation and support requirements	Chapter 3
Table E3702.1	General installation and support requirements for wiring methods	Chapter 3
E3702.2	Cables in accessible attics	Sections 333-12 and 336-6(d)
E3702.2.1	Across structural members	Section 333-12(a)
E3702.2.2	Cable installed parallel to framing members	Section 333-12(b)
E3702.3	Exposed cable	Sections 333-11 and 336-6
E3702.3.1	Surface installation	Section 336-6(a)
E3702.3.2	Protection from physical damage	Section 336-6(b)
E3702.4	In unfinished basements	Section 336-6(c)
E3702.5	Bends	Sections 333-8 and 336-16
SECTION E3703	**UNDERGROUND INSTALLATION REQUIREMENTS**	
E3703.1	Minimum cover requirements	Section 300-5(a)
Table E3703.1	Minimum cover requirements, burial in inches	Table 300-5
E3703.2	Grounding	Section 300-5(b)
E3703.3	Protection from damage	Section 300-5(d)
E3703.4	Splices and taps	Section 300-5(e)
E3703.5	Backfill	Section 300-5(f)
E3703.6	Raceway seals	Section 300-5(g)
E3703.7	Bushing	Section 300-5(h)
E3703.8	Single conductors	Section 300-5(i)
E3703.9	Ground movement	Section 300-5(j)
CHAPTER 38	**POWER AND LIGHTING DISTRIBUTION**	
SECTION E3801	**RECEPTACLE OUTLETS**	
E3801.1	General	Section 210-50 and 210-52
E3801.2	Convenience receptacle distribution	Section 210-52(a)
E3801.2.1	Spacing	Section 210-52(a)(1)
E3801.2.2	Wall space	Section 210-52(a)(2)
E3801.2.3	Floor receptacles	Section 210-52(a)(3)
E3801.3	Small appliance receptacles	Section 210-52(b)
E3801.3..1	Rooms	Section 210-52(b)(1)
E3801.3.2	No other outlets	Section 210-52(b)(2)
E3801.3.3	Limitations	Section 210-52(b)(3)
E3801.4	Countertop receptacles	Section 210-52(c)
E3801.4.1	Wall counter space	Section 210-52(c)(1)
E3801.4.2	Island counter spaces	Section 210-52(c)(2)
E3801.4.3	Peninsular counter space	Section 210-52(c)(3)
E3801.4.4	Separate spaces	Section 210-52(c)(4)
E3801.4.5	Receptacle outlet location	Section 210-52(c)(5)

International Residential Code		National Electrical Code
E3801.5	Appliance outlets	Section 210-50(c)
E3801.6	Bathroom and toilet room	Section 210-52(d)
E3801.7	Outdoor outlets	Section 210-52(e)
E3801.8	Laundry areas	Section 210-52(f)
E3801.9	Basements and garages	Section 210-52(g)
E3801.10	Hallways	Section 210-52(h)
E3801.11	HVAC outlet	Section 210-63
SECTION E3802	**GROUND-FAULT AND ARC-FAULT CIRCUIT-INTERRUPTER PROTECTION**	
E3802.1	Bathroom receptacles	Section 210-8(a)(1)
E3802.2	Garage receptacles	Section 210-8(a)(2)
E3802.3	Outdoor receptacles	Section 210-8(a)(3)
E3802.4	Crawl space receptacles	Section 210-8(a)(4)
E3802.5	Unfinished basement receptacles	Section 210-8(a)(5)
E3802.6	Kitchen receptacles	Section 210-8(a)(6)
E3802.7	Bar sink receptacles	Section 210-8(a)(7)
E3802.8	Exempt receptacles	Section 210-8(a)(2)
E3802.9	Bedroom receptacles	Section 210-12(b)
SECTION E3803	**LIGHTING OUTLETS**	
E3803.1	General	Section 210-70(a)
E3803.2	Habitable rooms	Section 210-70(a)(1)
E3803.3	Additional locations	Section 210-70(a)(2)
E3803.4	Storage or equipment spaces	Section 210-70(a)(3)
SECTION E3804	**GENERAL INSTALLATION REQUIREMENTS**	
E3804.1	Electrical continuity of metal raceways and enclosures	Section 300-10
E3804.2	Mechanical continuity—raceways and cables	Section 300-12
E3804.3	Securing and supporting	Section 300-11(a)
E3804.4	Raceways as means of support	Section 300-11(b)
E3804.5	Raceway installations	Section 300-18
E3804.6	Conduit and tubing fill	Section 300-17 and Chapter 9, Table 1
Tables E3804.6(1)-E3804.6(10)	Maximum number of conductors in conduit or tubing	Section 300-17 and Chapter 9, Table 1, Note 1
E3804.7	Air handling—stud cavity and joist spaces	Section 300-22(c), Exception
CHAPTER E3805	**BOXES, CONDUIT BODIES, AND FITTINGS**	
E3805.1	Box, conduit body, or fitting, —where required	Section 300-15
E3805.1.1	Equipment	Section 300-15(b)
E3805.1.2	Protection	Section 300-15(c)
E3805.1.3	Integral enclosure	Section 300-15(e)
E3805.1.4	Fitting	Section 300-15(f)
E3805.1.5	Buried conductors	Section 300-15(g)
E3805.1.6	Fixtures	Section 300-15(j)
E3805.1.7	Closed loop	Section 300-15(m)
E3805.3	Metal boxes	Section 370-4
E3805.4	Nonmetallic boxes	Sections 370-3
E3805.4.1	Nonmetallic-sheathed cable and nonmetallic boxes	Section 370-17(c)
E3805.4.2	Securing to box	Section 370-17(c)
E3805.4.3	Conductor rating	Section 370-17(c)
E3805.5	Minimum depth of outlet boxes	Section 370-24
E3805.6	Boxes enclosing flush-mounted devices	Section 370-19
E3805.7	Boxes at lighting fixture outlets	Section 370-27(a)
E3805.8	Floor boxes	Section 370-27(b)
E3805.9	Boxes at fan outlets	Section 370-27(c)
E3805.10	Conduit bodies, junction, pull and outlet boxes to be accessible	Section 370-29

International Residential Code		National Electrical Code
E3805.11	Damp or wet locations	Section 370-15
E3805.12	Number of conductors in outlet, device, and junction boxes, and conduit boxes	Section 370-16(a)
E3805.12.1	Box volume calculations	Section 370-16(a)
Table E3805.12.1	Maximum number of conductors in metal boxes	Table 370-16(a)
E3805.12.1.1	Standard boxes	Section 370-16(a)(1)
E3805.12.1.2	Other boxes	Section 370-16(a)(2)
E3805.12.2	Box fill calculations	Section 370-16(b)
E3805.12.2.1	Conductor fill	Section 370-16(b)(1)
Table E3805.12.2.1	Volume allowance required per conductor	Table 370-16(b)
E3805.12.2.2	Clamp fill	Section 370-16(b)(2)
E3805.12.2.3	Support fittings fill	Section 370-16(b)(3)
E3805.12.2.4	Device or equipment fill	Section 370-16(b)(4)
E3805.12.2.5	Equipment grounding conductor fill	Section 370-16(b)(5)
E3805.12.3	Conduit bodies	Section 370-16(c)
SECTION E3806	**INSTALLATION OF BOXES, CONDUIT BODIES AND FITTINGS**	
E3806.1	Conductors entering boxes, conduit bodies or fittings	Section 370-17
E3806.1.1	Insulated fittings	Section 300-4(f)
E3806.2	Openings	Section 370-17(a)
E3806.3	Metal boxes, conduit bodies and fittings	Section 370-17(b)
E3806.4	Unused openings	Section 370-18
E3806.5	In wall or ceiling	Section 370-20
E3806.6	Plaster, gypsum board and plasterboard	Section 370-21
E3806.7	Exposed surface extensions	Section 370-22
E3806.8	Supports	Section 370-23
E3806.8.1	Surface mounting	Section 370-23(a)
E3806.8.2	Structural mounting	Section 370-23(b)
E3806.8.2.1	Nails	Section 370-23(b)(1)
E3806.8.2.2	Braces	Section 370-23(b)(2)
E3806.8.3	Mounting in finished surfaces	Section 370-23(c)
E3806.8.4	Raceway supported enclosures without devices or fixtures	Section 370-23(e)
E3806.8.5	Raceway supported enclosures, with devices or fixtures	Section 370-23(f)
E3806.8.6	Enclosures in concrete or masonry	Section 370-23(f)
E3806.9	Covers and canopies	Section 370-25
E3806.10	Metal covers and plates	Section 370-25(a)
E3806.11	Exposed combustible finish	Section 370-25(b)
SECTION E3807	**CABINETS AND PANELBOARDS**	
E3807.1	Damp or wet locations	Section 373-2(a)
E3807.2	Position in wall	Section 373-3
E3807.3	Unused openings	Section 373-4
E3807.4	Conductors entering cabinets	Sections 300-3(f) and 373-5
E3807.5	Openings to be closed	Section 373-5(a)
E3807.6	Cables	Section 373-5(c)
SECTION E3808	**GROUNDING**	
E3808.1	Metal enclosures	Section 250-86
E3808.2	Equipment fastened in place or connected by permanent wiring methods (fixed)	Section 250-110
E3808.3	Specific equipment fastened in place or connected by permanent wiring methods	Section 250-112
E3808.4	Performance of fault current path	Section 250-2(d)
E3808.5	Earth as a grounding conductor	Section 250-2(d)
E3808.6	Load-side neutral	Section 250-142
E3808.7	Load-side equipment	Section 250-142
E3808.8	Types of equipment grounding conductors	Section 250-118
E3808.8.1	Flexible metal conduit	Section 250-118(6)

International Residential Code		National Electrical Code
E3808.8.2	Liquidtight flexible metal conduit	Section 250-118(7)
E3808.8.3	Nonmetallic sheathed cable (Type NM)	Section 336-30(b)
E3808.9	Equipment fastened in place or connected by permanent wiring methods	Section 250-119
E3808.10	Methods of equipment grounding	Section 250 Part G
E3808.11	Equipment grounding conductor installation	Section 250-120
E3808.12	Equipment grounding conductor size	Section 250-122
Table E3808.12	Equipment grounding conductor sizing	Table 250-122
E3808.12.1	Multiple circuits	Section 250-122(c)
E3808.13	Continuity and attachment of branch-circuit equipment grounding conductors to boxes	Section 250-148
E3808.14	Connecting receptacle grounding terminal to box	Section 250-146
E3808.15	Metal boxes	Section 250-148(a)
E3808.16	Nonmetallic boxes	Section 250-148(b)
E3808.17	Clean surfaces	Section 250-12
E3808.18	Bonding other enclosures	Section 250-96(a)
E3808.19	Size of equipment bonding jumper on load side of service	Section 250-102(d)
E3808.20	Installation—equipment bonding jumper	Section 250-102(e)
SECTION E3809	**FLEXIBLE CORDS**	
E3809.1	Where permitted	Section 400-7 and 400-8
E3809.2	Loading and protection	Section 400-5
Table E3809.2	Maximum ampere load and overcurrent-protection-device rating for flexible cords	Section 240-4 and Table 400-5(A)
E3809.3	Splices	Section 400-9
E3809.4	Attachment plugs	Section 400-7(b)
CHAPTER 39	**DEVICES AND LIGHTING FIXTURES**	
SECTION E3901	**SWITCHES**	
E3901.1	Rating and application of snap switches	Section 380-14(a)
E3901.2	CO/ALR snap switches	Section 380-14(c)
E3901.3	Indicating	Section 380-7
E3901.4	Time switches and similar devices	Section 380-5
E3901.5	Grounding of enclosures	Section 380-12
E3901.6	Access	Section 380-8
E3901.7	Wet locations	Section 380-4
E3901.8	Grounded conductors	Section 380-2(b)
E3901.9	Switch connections	Section 380-2(a)
E3901.10	Box mounted	Section 380-10(b)
E3901.11	Snap switch faceplates	Section 380-9(a)
E3901.11.1	Faceplate grounding	Section 380-9(b)
SECTION E3902	**RECEPTACLES**	
E3902.1	Rating and type	Section 410-56(a)
E3902.1.1	Single receptacle	Section 210-21(b)(1)
E3902.1.2	Two or more receptacles	Section 210-21(b)(3)
Table E3902.1.2	Receptacle ratings for various size multioutlet circuits	Section 210-21
E3902.2	Grounding type	Section 210-7(a)
E3902.3	CO/ALR receptacles	Section 410-56(b)
E3902.4	Faceplates	Section 410-56(d)
E3902.5	Position of receptacle faces	Section 410-56(e)
E3902.6	Receptacles mounted in boxes	Section 410-56(f)(1) and (2)
E3902.7	Receptacles mounted on covers	Section 410-56(f)(3)
E3902.8	Damp locations	Section 410-57(a)

International Residential Code		National Electrical Code
E3902.9	Wet locations	Section 410-57(b)
E3902.10	Bathtub and shower space	Section 410-57(c)
E3902.11	Flush mounting with faceplate	Section 410-57(e)
E3902.12	Outdoor installation	Section 410-57(f)
SECTION E3903	**FIXTURES**	
E3903.1	Energized parts	Section 410-3
E3903.2	Fixtures near combustible material	Section 410-5
E3903.3	Exposed conductive parts	Section 410-18(a)
E3903.4	Screw-shell type	Section 410-47
E3903.5	Recessed incandescent fixtures	Section 410-65(c)
E3903.6	Thermal protection	Section 410-73(e)
E3903.7	High-intensity discharge fixtures	Section 410-73(f)
E3903.8	Wet or damp locations	Section 410-4(a)
E3903.9	Lampholders in wet or damp locations	Section 410-49
E3903.10	Pendants	Section 410-4(d)
E3903.11	Fixtures in clothes closets	Section 410-8(a)
Figure E3903.11	Closet storage space	Figure 410-8
E3903.12	Fixture wiring—general	Section 410-22
E3903.12.1	Polarization of fixtures	Section 410-23
E3903.12.2	Fixtures as raceways	Section 410-31
SECTION E3904	**FIXTURE INSTALLATION**	
E3904.1	Outlet box covers	Section 410-12
E3904.2	Combustible material at outlet boxes	Section 410-13
E3904.3	Access	Section 410-16(b)
E3904.4	Supports	Section 410-15(a)
E3904.5	Means of support	Section 410-16(a)
E3904.6	Exposed ballasts	Section 410-76(a)
E3904.7	Combustible low-density cellulose fiberboard	Section 410-76(b)
E3904.8	Recessed fixture clearance	Section 410-66(a)
E3904.9	Recessed fixture installation	Section 410-66(b)
SECTION E3905	**TRACK LIGHTING**	
E3905.1	Fastening	Section 410-104
E3905.2	Grounding	Section 410-105(b)
CHAPTER 40	**APPLIANCE INSTALLATION**	
SECTION E4001	**GENERAL**	
E4001.1	Scope	Sections 422-1 and 424-1
E4001.2	Installation	Sections 110-3(b) and 422-17
E4001.3	Flexible cords	Section 422-16
Table E4001.3	Flexible cord length	Sections 422-16(b)(1) and (2)
E4001.4	Overcurrent protection	Section 422-11
E4001.4.1	Single motor-operated appliance	Section 422-11(e)
E4001.5	Disconnecting means	Sections 422-30 and 424-19
Table E4001.5	Disconnecting means	Sections 422-21, 422-25, 422-27, 424-19, and 440-14
E4001.6	Ceiling fans	Section 422-18
E4001.7	Snow-melting and deicing equipment protection	Section 210-8(a)(3) Exception and 426-28
CHAPTER 41	**SWIMMING POOLS**	
SECTION E4101	**GENERAL**	
E4101.1	Scope	Section 680-1
E4101.2	Definitions	Section 680-4

International Residential Code		National Electrical Code
SECTION E4102	**WIRING METHODS FOR POOLS, SPAS, HOT TUBS AND HYDROMASSAGE BATHTUBS**	
E4102.1	General	Sections 608-7, 680-20(b), 680-20(c), 680-25(b), 680-25(c), 680-25(d), 680-40 and 680-41
E4102.2	Flexible cords	Sections 680-6(b)(3), 680-7, 680-40 Exception No. 2, 680-41 Exception, 680-25(b)(3)
SECTION E4103	**EQUIPMENT LOCATION AND CLEARANCES**	
E4103.1	Receptacle outlets	Section 680-6(a)
E4103.1.1	Location	Section 680-6(a)(1)
E4103.1.2	Where required	Section 680-6(a)(2)
E4103.1.3	GFCI protection	Section 680-6(a)(3)
E4103.1.4	Indoor locations	Section 680-41(a)(1)
E4103.1.5	Indoor GFCI protection	Sections 680-41(a)(2), 680-70
E4103.2	Switching devices	Sections 680-6(c), 680-41(c), 680-71
E4103.3	Disconnecting means	Section 680-12
E4103.4	Lighting fixtures and ceiling fans	Section 680-6(b)
E4103.4.1	Outdoor location	Section 680-6(b)(1)
E4103.4.2	Indoor location	Section 680-6(b)(3)
E4103.4.3	Existing lighting outlets and fixtures	680-6(b)(2)
E4103.4.4	Indoor spas and hot tubs	Section 680-41(b)
E4103.4.5	GFCI protection	Sections 680-6(b)(4)
E4103.5	Overhead conductor clearances	Section 680-8
Table E4103.5	Overhead conductor clearances	Table 680-8
Figure E4103.5	Overhead conductor clearances	Figure 680-8
E4103.6	Underground wiring	Section 680-10
Table E4103.6	Minimum burial depths	Section 680-10
SECTION E4104	**BONDING**	
E4104.1	Bonded Parts	Sections 680-22(a), 680-40(b) and 680-41(d)(4)
E4104.2	Parts not required to be bonded	Sections 680-22(a) Exception No. 2, 680-22(a) Exception No. 4, 680-40 Exception No.1 and 680-41(d)(4) Exception
E4104.3	Methods of bonding	Sections 680-22(b), 680-40 Exception 3 and 680-41(e)
SECTION E4105	**GROUNDING**	
E4105.1	Equipment to be grounded	Section 680-24
E4105.2	Light fixtures and related equipment	Section 680-25(b)(1), 680-25(b)(2), 680-25(b)(3)
E4105.3	Nonmetallic conduit	Section 680-20(b)(1)
E4105.4	Flexible cords	Section 680-25(b)(3)
E4105.5	Motors	Section 680-25(c)
E4105.6	Panelboards	Section 680-25(d)
E4105.7	Cord-connected equipment	Section 680-25(e)
E4105.8	Other equipment	Section 680-25(f)
SECTION E4106	**EQUIPMENT INSTALLATION**	
E4106.1	Transformers	Section 680-5(a)
E4106.2	Ground-fault circuit-interrupters	Section 680-5(b)
E4106.3	Wiring on load side of ground-fault circuit-interrupters and transformers	Section 680-5(c)
E4106.4	Underwater lighting fixtures	Section 680-20(a)(1)
E4106.4.1	Maximum voltage	Section 680-20(a)(2)
E4106.4.2	Fixture location	Section 680(a)(3)
E4106.5	Wet-niche fixtures	Section 680-20(b)(1), 680-20(b)(3)
E4106.6	Dry-niche fixtures	Section 680-20(c)
E4106.7	No-niche fixtures	Section 680-20(d)
E4106.8	Junction boxes and enclosures for transformers or ground-fault circuit interrupters	Section 680-21
E4106.8.1	Junction boxes	Section 680-21(a)
E4106.8.2	Other enclosures	Section 680-21(b)
E4106.8.3	Protection of junction boxes and enclosures	Section 680-21(c)
E4106.8.4	Grounding terminals	Section 680-21(d)

International Residential Code		National Electrical Code
E4106.8.5	Strain relief	Section 680-21(e)
E4106.9	Underwater audio equipment	Section 680-23
E4106.9.1	Speakers	Section 680-23(a)
E4106.9.2	Wiring methods	Section 680-23(b)
E4106.9.3	Forming shell and metal screen	Section 680-23(c)
E4106.10	Electrically operated pool covers	Section 680-26
E4106.11	Electric pool water heaters	Section 680-9
E4106.12	Pool area heating	Section 680-27
E4106.12.1	Unit heaters	Section 680-27(a)
E4106.12.2	Permanently wired radiant heaters	Section 680-27(b)
E4106.12.3	Radiant heating cables prohibited	Section 680-27(c)
E4106.13	Double insulated pool pumps	Section 680-28
SECTION E4107	**STORABLE SWIMMING POOLS**	
E4107.1	Pumps	Section 680-30
E4107.2	Ground-fault circuit-interrupters required	Section 680-31
E4107.3	Lighting fixtures	Section 680-32
E4107.3.1	15 volts or less	680-32(a)
E4107.3.2	Not over 150 volts	680-32(b)
SECTION E4108	**SPAS AND HOT TUBS**	
E4108.1	Ground-fault circuit-interrupters	Section 680-42
E4108.2	Electric water heaters	Section 680-41(h)
E4108.3	Underwater audio equipment	Section 680-41(i)
SECTION E4109	**HYDROMASSAGE BATHTUBS**	
E4109.1	Ground-fault circuit-interrupters	Section 680-70
E4109.2	Other electric equipment	Section 680-71
E4109.3	Accessibility	Section 680-72
E4109.4.	Bonding	Section 680-73
CHAPTER 42	**CLASS 2 REMOTE CONTROL, SIGNALING, AND POWER-LIMITED CIRCUITS**	
SECTION E4201	**GENERAL**	
E4201.1	Scope	Section 725-1
E4201.2	Definitions	Article 100 and Section 725-2
SECTION E4202	**POWER SOURCES**	
E4202.1	Power sources for Class 2 circuits	Section 725-41(a)
E4202.2	Interconnection of power sources	Section 725-41(b)
SECTION E4203	**WIRING METHODS**	
E4203.1	Wiring methods on supply side of Class 2 power source	Section 725-51
E4203.2	Wiring methods and materials on load side of the Class 2 power source	Sections 725-52, 725-61, 725-71
E4203.2.1	Type CL2P cables	Section 725-71(a)
E4203.2.2	Type CL2 cables	Section 725-71(c)
E4203.2.3	Type CL2X cables	Sections 725-61(e), Exception 4 and 725-71(d)
E4203.2.4	Marking	Section 725-71(h)
SECTION E4204	**INSTALLATION REQUIREMENTS**	
E4204.1	Separation from other conductors	Section 725-54(a)(1)
E4204.2	Other applications	Section 725-54(a)(3)
E4204.3	Class 2 circuits with communication circuits	Section 725-54(b)(4)
E4204.4	Class 2 cables with other circuit cables	Section 725-54(b)(5)
E4204.5	Installation of conductors and cables	Sections 725-5, 725-7, 725-54(d)

INDEX

C

D

E

T